The
Evolution
of Medicine

ANDREW S. OLEARCHYK & RENATA M. OLEARCHYK

authorHOUSE

AuthorHouse™
1663 Liberty Drive
Bloomington, IN 47403
www.authorhouse.com
Phone: 833-262-8899

Published by AuthorHouse 01/12/2023

ISBN: 978-1-6655-7610-9 (sc)
ISBN: 978-1-6655-7622-2 (e)

Print information available on the last page.

Any people depicted in stock imagery provided by Getty Images are models, and such images are being used for illustrative purposes only. Certain stock imagery © Getty Images.

This book is printed on acid-free paper.

Because of the dynamic nature of the Internet, any web addresses or links contained in this book may have changed since publication and may no longer be valid. The views expressed in this work are solely those of the author and do not necessarily reflect the views of the publisher, and the publisher hereby disclaims any responsibility for them.

CONTENTS

ABOUT THE AUTHORS

Andrew S. Olearchyk (December 03, 1935, city of Peremyshl / Przemyśl, Syan river country, Ukraine / Poland –)[1], Medical Doctor (MD), graduated from the Medical University of Warsaw (1961), 1st degree specialist in anesthesia (1964) and general surgery (1965), Poland; Diplomate of the American Board of Surgery (ABS, 1976 – 86) and the American Board of Thoracic (and Cardiac) Surgery (ABTS, 1983 – 2023), Fellow of the American College of Surgeons (FACS, 1984 –) and Honorary Member of the Association of Cardiovascular Surgeons (ACVS) of Ukraine (1996 –). Authored 215 scientific works, including 204 papers, five monographies and six books.

Renata Maria Olearchyk, nee **Sharan (October 02, 1943, city of Lviv, Ukraine – September 14, 2019, city of Philadelphia, Pennsylvania, United States of America)**[2], Master of Arts (MA), graduated from the Department of Sociology of the University of Pennsylvania with the degree of Bachelor of Arts (BA, 1965), and from the Department of Sociology and Anthropology of the Brown University, city of Providence, State of Rhode Island, USA, with the degree of MA (1968) and Philosophiae Doctor (PhD) Candidate. She authored 13 scientific works, including seven papers, three monographies and three books. – Memory Eternal!

---

[1] City of Peremyshl / Przemyśl (founded 8th century), Syan river country, Ukraine (area 603,628 km^2) / since 1945 – Poland (area 312,698 km^2).

[2] City of Lviv (founded 5th century, 1231 – 35, 1240 – 47), Ukraine. — City of Philadelphia (settled 1682), Pennsylvania (PA, area 116,283 km^2), United States of America (USA, area 9,833,520 km^2). — City of Providence (settled 1636), State Rhode Island (R.I., area 3,144 km^2), USA.

ANNOTATION

Olearchyk A.S., Olearchyk R.M. The Evolution of Medicine. 2nd Ed. Book Publishing «AuthorHouse», Bloomington, IN 2023. – Pages (p.) 834.

The book entitled «The Evolution of Medicine» was composed using a novel approach of presenting in a chronological order the theoretical and clinical medicine from the prehistoric times to the 20th century and the beginning of the 21st century, based on the significant contribution of the known, lesser known, and unknown individuals. Dedicated for medical students and physicians.

FOREWORD

The work by Andrew and Renata Olearchyk «The Evolution of Medicine» is an extension of theirs previous writing entitled «Concise History of Medicine» (1991)[3], «Medicine» (1993)[4], «Basic Science and Clinical History of Medicine» (2016)[5], and «The Evolution of Medicine» (2020)[6].

The prominent personalities, physician-scientist, and others, the known, less known, and unknown, who have made a significant contribution to the development of theoretical or clinical medicine are identified by the first, middle and last name, birth date, birth and death dates, nationality or country of origin and specialty. Some of them were connected to two or more countries. When the national origin of a particular individual was not clear or debated, we added to his data the place of birth and death. In some of instances those data were not available.

The chronology of events was followed by the era (epoch) and traditional lay-out of branches of the theoretical and clinical medicine as clearly outlined in the «Contests», and within a particular epoch or branch of medicine by the person's birth date. Difficulties to properly place a person chronologically when the birth

---

[3] Olearchyk AS, Olearchyk RM. Concise History of Medicine. Journal of the Ukrainian Medical Association of North America (J Ukr Med Assoc North Am, or **JUMANA**) 1991;38,3(125): 77-159. [The work is permanently exposed at the display in the hall of the National Museum Medicine of Ukraine (NMMU, established 1982), city of Kyiv (founded 482), Kyiv Oblast (Obl.; area 28,131 km²), Ukraine (area 603,628 km²), by **Alexander (Olexandr) A. Grando [Oct. 10,1919, city of Mohyliv-Podilsky, Vinnytsia Oblast (Obl.), Ukraine – July 17, 2004, city of Kyiv, Ukraine]** — Ukrainian-physician scientist, specialist in social medicine, organization of the health care and history of medicine, director of the NMMU (1982 - 2004), the author of the book «The Journey in the Past of Medicine» Publ. Agency» «Triumph», Kyiv 1995), founder and editor of the Ukrainian Historical Medical Journal «Ahapit» (UHMJ «Ahapit») in Ukrainian, Russian and English languages (1995 - 2004). — City of Mohyliv-Podilsky (founded 1595), survived with vicinities organized by the Russian Communist Fascist regime of the Russian Soviet Federative Socialistic Republic (RSFSR, 1918 – 91), the Union of the Soviet Socialistic Republic (USSR, 1922 - 91), and the People's Comissariat for Internal Affairs / Narodnyy komissariat vnutrennikh del (NKVD, established 1917), Holodomor-Genocide 1921 – 23, 1932 – 33 and 1946 – 47, and political repressions of 1936 – 38 against the Ukrainian nation, Vinnytsia Obl. (area 26,472 km²), Ukraine.

[4] Olearchyk AS. Medicine. In, V. Kubijovyc, DH Struk (Ed), Encyclopedia of Ukraine. University of Toronto Press, Inc. Toronto & London, 1993. Vol. III. P. 363-6.

[5] Olearchyk AS, Olearchyk RM. Basic Science and Clinical History of Medicine. In, AS Olearchyk. Surgeon's Universe. Vol. I-IV. 4th Ed. Book Publ. «OutskirtsPress», Denver, KO 2016. Vol. IV. P. 356-535.

[6] Olearchyk AS, Olearchyk RM. The Evolution of Medicine. Book Publ. «AuthorHouse», Bloomingdon, IN 2020.

data was not available, but then we were guided by chronology of their description, discoveries, or innovation.

When the name of a particular individual had to be added further in the text for the second or more times within this same branch of medicine it was marked by his first and last name followed by direction «above», or with exemption – «below». If the name of a particular person had to be added further in the text for the second or more times in the different branch of medicine or different era, then it was marked by his first and last name, birth, and death date, that era and branch of medicine, without repeating its nationality and specialty.

We omitted the information of the cited individuals about family affairs, education, professional positions, membership, or prizes. It is because in this concise review we concentrated only on their descriptions, discoveries and innovations which benefit humanity. The other reason for that is to give the reader an opportunity to further enlarge his knowledge in chosen specialty while reading our work.

Medicine is the science and the art of the diagnosis, prevention and treatment of diseases.

PREHISTORIC MEDICINE

As defined by paleopathology, Prehistory extends from the *Paleozoic Era* 600 million (mln) years ago to the discovery of the calendar and the invention of writing circa (c.) 4,000 Before Christ (BC)[7].

The remains of animals from the Paleozoic Era (600 – 225 mln. years ago) and of humans from the Paleolithic Period (3 mln years ago – 10,000 BC) revealed bone injuries and diseases. A femur of Homo erectus (c. 1.5 mln years ago – 250,000 BC) showed a tumor or re-growth after trauma.

The prehistoric people have the experience with electric current observing lightning, creating fire by rubbing one stone against the other. The electrical current is intimately involved in the heart beating, muscle action and nerve transmission.

In the near-by village of Mauer[8], 10 km south-east from city of Heidelberg, Land Baden-Württemberg, Germany, one of the oldest known human's bone parts – mandible,

---

[7] Before Christ (BC), it is roughly the date of the birth of **Jesus Christ [between 7 BC - 1 Anno Domini (AD), Bethlehem, Palestine** – crucified **33 AD, Jerusalem, Israel/Palestine]** — Galilean teacher, prophet, creator, and main figure of the Christian creed (religion) and the doctrine, which should be based on love. / City of Bethlehem (in Hebrew – «house of bread», in Arabic – «house of meat»; founded the 14th century BC), Palestine [6,200 kilometers (km)[2]], Asia (area 44,529,000 km[2], the largest continent covering 8,7% of the Earth's total surface area, or 30% of its total land area). The surface of the planet Earth is 510,072,000 km[2], it formed approximately 4.54 billion (bln) years ago and life appeared on its surface within one bln years ago. / The city of Jerusalem (in Hebrew and in Arabic – «sacred»; founded in the 4th millennium BC), Israel (area 20,770 km[2]), Asia. / Galilee («the country of the Gaul's»; an elevated plain to the height of 600-700 m, located between the Sea of Galilee with the adjacent northern and southern segments of the Jordan River on the east and to the Mediterranean Sea and the Planes to the west), founded in the 11th century BC), by emigrants from Ukraine, the second largest country of Europe (area of 10,180,000 km[2]), who called themselves the Gaul's (European's name – the Celts; Palestine's name – the Philistines). / The Jordan River [the length of 257 km, drainage basin 18,000 km[2], average gradient 90 meters (m) / second (sec.)] originates in the north from of the Galilee Highlands (elevation 600 -700 m), flows to the south through the Sea of Galilee (freshwater lake at a height of -211,3 m below sea level, size 21 km x 15 km, surface area 170 km[2], maximum depth of about 43 m), and flows south to the Dead Sea (an endorheic lake at an altitude of - 450 below sea level, size 67 km x 10 km; the average depth of 120 m, the deepest 330 m; volume of water 147 km[3], average salinity 26%-27%, drainage area 40,650 km[2]). / The Mediterranean Sea (a surface of about 2,5 mln km[2], average depth of 1,500 m, maximum 5,267 m; an average salinity 3,9%), an offshoot of the Atlantic Ocean (surface area about 106,400 000 km[2], occupies 20% of the Earth's surface and about 29% of its water surface; the average depth of 3,339 m, the biggest 8,380 m; the volume of water with adjacent seas near 354,700 20,000 km[3], without them about 323,600 20,000 km[3]; average salinity 3,5%).

[8] Village of Mauer, borough of Stadtbezirke, city of Heidelberg (founded 5th century), county Baden-Württemberg (area 35,751.46 km[2]), Germany (area 357,588 km[2]).

named «Mauer1» of the oldest known specimen of Homo heidelbergensis, that is the «Haidelberg Man» who died 600,000 ± 40,000 years ago, was found in 1907 by **Otto Schoetensack (1850 – 1912)** — German physician and anthropologist.

In *Prehistoric Ukraine* the lower Paleolithic Age sites of the Acheulean (500,000 – 100,000 BC) and Mousterian (100,000 – 40,000 BC) Periods[9] were discovered (1975) near the village of Koroleve on the Tysa River)[10], Vynohradiv Region, Zakarpatska Obl. (area 12,777 km²).

Neanderthal Homo sapiens – intelligent humans (100,000 – 35,000 BC), suffered from an inflammation of a joint (arthritis). Remains of a Neanderthal adult male and one-year-old child from the middle Mousterian Period were also found in the Kyik-Kobe cave, Crimea[11], Ukraine.

Within the neighborhood of town Staryy Krym (appeared in 1st century AD) existed settlements from the Bronze Age (c. 3,200 – 600 BC), ruins of the late Scythe town on the Aharmysh River with remain of the old settlement (4th – 3rd, century BC – first centuries AD), Kirov Region (area 1,208 km²) in eastern Crimea.

In the city of Kyiv (Foot Note 3) – capital of Ukraine, on the Dnipro River[12], a man first appeared in during the late Paleolithic's early Magdalenian Periods (Kyrylivka settlement, 18,000 BC).

A Paleolithic Age drawing of a mammoth, found in the El Pindol cave in Spain (area 505,992 km²) and drawn in red ochre, is an anatomic illustration, showing a leaf-shaped dark area at the shoulder, and in all likelihood is intended to depict the heart.

Trepanning of the skull, also known as (a.k.a.), trepanation, trephination, trephining (in Greek – «trypanon», meaning borer), making a bur hole, or craniotomy is a prehistoric surgical procedure in which a hole is drilled, or scraped in the human skull in order to expose the dura mater – a thin most external membrane of three layers of the meninges covering the brain and spinal cord, in the belief it will cure epileptic seizures, migraine, and certain mental disorders.

Trepanation of the skull with a trace on large surface of the bone was performed on a 20 to 29-year-old man from the Bronze Age, near the village of Lhovs'ke (until

---

[9] Klein RG. Ace-Hunters of Ukraine. Univ Chicago Press, Chicago-London 1973.

[10] The Tysa River (length 966 km, drainage area 157,100 km²), flowing out of the Eastern Carpathian Mountains [length of 1,500 km, width 120-430 km, the highest Mt Gerlach (2,655 m) at an altitude of 2,020 m], flows from the left into the Danube River (length 2,850 km and drainage basin 817,000 km²) which originates in the west from the Black Forest of Europe at a height of 678 m, flows into the Black Sea (maximum length 1745 km, surface area 436,402 km², average depth 1,253 m, maximum – 2,212 m, water volume 544,000 km³), then into the Mediterranean Sea, and finally into the Atlantic Ocean.

[11] Peninsula Crimea (area 26,100 km²), Ukraine. — Von Engelhardt D. Yalta and Crimea in history and culture. UHMJ «Ahapit» 1994;1:46-9.

[12] The Dnipro River (length 2,285 km, drainage basin 504,000 km²) flows out in the north of the turf of the Vandal Hill (maximum height 347 m) at the elevation of 220 m, drains in the south to the Black Sea.

1945 – village of Chalebu-Ali), Kirov Region, Eastern Crimea in Ukraine. Near him were laid two flint stone tips, and two scrapers.

The ancient Ukrainian master-surgeon, with jeweler's precision carved with a stone tool quite large surface of the bone, leaving the thinnest, less than millimeter (mm) in thickness, bone plate without deadly penetration into the skull cavity, where are located large and multiple blood vessels. Despite of this, he probably died after surgery, since there was no visible growth of a new bone around the edges of trepanation opening.

The American Indians probably migrated to Northern America (area 24,709,000 km²) from Northern Asia across the Bering Strait (c. 38,000 BC – c. 18,000 BC), reached the southern tip of South America (area 14,840,000 km²; 6,000 BC). Their physicians, or «medicine men» – shame, witch doctors or sorcerers, religious healers, or ministers of worship, cared for the sick, introduced the leaves of the coca plant (Erythroxylon coca) – a strong stimulant containing alkaloid cocaine which they have been chewed, snored for recreation or recovery of one's health or strength and used for local (topical) anesthesia during operations on the skin and mucous membranes. They also introduced curare and quinidine, cured tropical «intermittent fever» with the tincture of the cinchona (Cinchona pubescent) tree bark which contains quinine, set up broken bones and on occasion formed a curing society at the time of the pre-Inca and the Inca civilization and empire (area 2,000,000 km² in 1527)[13].

Beside of Ukraine, in that time, trephinations were performed in North America by the Indians, and in France (area, metropolitan 551,697 km², total 674,843 km²)[14,15].

During the pre-Inca and the Inca civilization and empire, the Indians performed trephining of the skull under a local anesthesia, i.e., local analgesia (in Greek / Latin – meaning «without pain») with a mixture of the coca leaves and saliva, using stone instruments, with 50% long-term survival.

The practice of trepanations of the skull with survival on the banks of the Dnipro River in Mesolithic times dates to approximately 12,000 BC.

---

[13] The Inca civilization and empire existed from the early 13th century until 1,572 with the administrative, political and military center in Cusco (founded 1,100 at the elevation of 3,399 m), Cusco Region (area 71,986 km², elevation from 532 m to 4,801 m) in the vicinity of the Sacred (Urubamba) River Valley (length 72 km) and below the ancient Machu Picchu (in Incas language – «Old Peak»), built as an estate for **Pachacuti (1438 - 72)** — the Inca emperor, in the section of Andes (7,000 long, 200-700 km wide, an average height of about 4,000 m, the highest peak ⁻ Aconcagua with an elevation of 6,962 m) located in Peru (area 1,285,216 km²). Urubamba flows from the south into the Ucayali River (length 1,600.1 km), the right tributary of the Amazon River (length 6,400 km) with the mouth in the Atlantic Ocean.

[14] Muniz MA, McGee WJ. Primitive trephining in Peru. Annu Rep Ethonol 1894;16:1-72.

[15] Menshen F. The Human Skull. A Cultural History. Frederick A. Praeger, New York 1995. / **Volke Menshen (1881 - 1977)** — Swedish physician and pathologist.

The clear examples for trephination of the skull were identified from a Neolithic Age burial site near the town of Ensisheim (known from the Old Neolithic Age), the Haut-Rhin Department (area 3,525 km²) in the Grand Est Region (area 57,533 km²) of France, dating to 6,500 BC, 5,100 BC and 4,900 BC were well preserved skeletal remains of an approximately 50-year-old man whose cranium showed clear evidence of two bur holes. One burr hole had fully healed, and the other partially so, indicating the subject survived the operation.

In the Neolithic Age (c. 10,000 – c. 3000 BC) peoples shifted from food-gathering to food-producing, and one can assume that medical herbs were among the plants grown. The Neoliths are known to have suffered from tuberculosis of the spine. At approximately that time (c. 10,000 BC) the first known surgical instruments were believed to be developed[16,] and the health care providers in *Europe* and *Peru* continued to perform surgical procedures, including craniotomy which showed healing of the bone edges, indicating a recovery (c. 8000 BC).

---

[16] Ailawadi G, Nagjii AS, Jones DR. The legend behind cardiothoracic surgery instruments. Ann Thorac Surg. 2010;89:1693-1700.

ANCIENT (ARCHAIC) MEDICINE

It extends from the beginning of recorded time by calendar and writing (c. 4000 BC) to the fall of *Rome* (founded 753 BC) — the capital of the *Roman Empire* (area 6,500,000 km² in 117 BC) and of *Italy* (area 301,338 km² in 476).

A few northern wise men named **Abarys (3000 BC – 1000 BC)**[17] has spread in *Ukraine* the tradition of northern Paleolithic shamans (priestly sorcerers) and pagan priests.

The two most ancient data refers to the medical affairs of *Mesopotamia (4000 – 539 BC)*[18].

The first is the oldest Sumerian medical text engraved in southern Mesopotamia, a collection of cuneiform tablets, known as «Physicians Collection of Empiric Prescription» (c. 2700 BC)[19]. It contains, among others, a reference to tuberculosis, attacks of epilepsy, extract from the willow tree bark and other plants rich in salicylic acid [in Latin – «salix», meaning the willow tree; chemical formula C_8H_4 (OH)] used to treat pain, i.e., general analgesia (in Greek / Latin – «without pain»), fever, inflammation, and infection. They had the knowledge of soluble and insoluble urinary bladder stones.

The second refers to **Puabi**, a.k.a., **Shubad (c. 2600 BC)**[20] — Queen or a priestess and the first known woman-surgeon of the Sumerian coastal city-state Ur (known

---

[17] Ablitsov VH. «Galaxy Ukraine». The Ukrainian diaspora: prominent persons. Publ. «Kyt», Kyiv 2007.

[18] Mesopotamia (in Greek – «land between rivers») is the historic country situated between the Euphrates (length 2,800 km) and Tigris (1,800 km) Rivers. The first originates in the north from Taurus Mountains (in zoology – Bos taurus or the «taurine cattle», in astrology – «Taurus» means the constellation and sign of the Zodiak; extend along the Mediterranean coast, highest peak Mount Demirkazik 3,756 m), and the second also in the north from Pontic Mountains (extend along the Black Sea coast 1,100 km, average elevation close to 2,500 m, highest point Mount Kackar Dagi 3,937 m) in the Eastern Asia Minor or Anatolia (in Greek – «east» or «sunrise»; area 783,562 km²). The Euphrates flows southwards to join from the west the Tigris, both forming the Shatt al-Arab (in Arabic – «Stream of the Arabs», length 200 km, width 232-800 m), which empties into the Persian Gulf (maximum length 986 km, average depth 50 m, maximum – 90 m). The latter through the Strait of Hormuz (width 39 km) join the Arab Sea (maximum width 2,400 km, surface area 3,862 km², maximum depth 4,652 m) of the Indian Ocean (surface area 73,556,000 km², average depth 3,890 m, deepest point 8,047 m, water volume 292,131,000 km³, covering 20% of the water of the Earth surface).

[19] Sumerians (black-headed) — the ancient extinct people populated the Southern Lower Mesopotamia, or exactly the Land of Summer (4000-3000 BC).

[20] Wirzfeld DA. The history of women in surgery. Can J Surg 2009 Aug;52(4):317-20. — Ludi EK, Zalles MV, Antenzana LFV, et al. International female surgeon pioneers: paving the way to generations to come. J Am Coll Surg 2020 Aug;231(2):294-8.

from 3,800 BC.), near the mouth of the Euphrate River, now well inland from the Persian Gulf on the south bank of the Euphrates. Puabi (Shubad) was buried with bronze and flint surgical instruments, so she could perform surgery in the afterlife.

The god **Ea,** the Lord of Water (Babylonia, 2000 BC.) was the first great cosmic ancestor of physicians. His grandson **Nabu**, ruled over all science, including medicine, and to him temples were erected at the developed medical schools.

The healing god, **Ningishzida**, was depicted with a double-headed snake which since then became the emblem of the medical profession.

The code of Hammurabi (Babylonia, c. 1790 BC) contains the earliest known regulation of the practice of medicine. Among other causes which allowed to return of the purchased slave were epileptic attacks. It stated, *«If the doctor, in opening an abscess, shall kill the patient, his hand shall be cut off».*

Medical knowledge in *Egypt* (area 1001,450 км²; 4000 BC – 642 AD) was condensed in papyruses and the «Hermetic Collection».

The first images of the tracheostomy, id est (i.e.), incision and entrance into the trachea through the skin and fascia in the middle of the anterior neck, are carved on the two Egyptian stone slabs (stele) dating back to c. 3600 BC, and artificial teeth were created c. 3500 BC in Egypt.

Since the Early Kingdom (3000-2800 BC) Egyptians performed catheterization of the urinary bladder with the bronze, the gold or silver instruments to remove stones, and by the 3rd century BC, dilatation of the urethral strictures with an S-shaped catheter.

The Egyptians had experience with electricity dating back to 2750 BC through the electric fish that can generate electric field (electrogenic) or detect electric fields (electroreceptive), were aware of shock from the electric fish. They called this fish the «Thunder of the Nile»[21], and the «protector» of all other fish.

The god **Thoth** was a physician to the gods, a patron god of physicians and scribes. He later was replaced by **Imhotep (2665-35 BC – 2595 BC)**, a historic personage, who became the chief healing god of Egypt and was bestowed with a divine father, the god **Ptah**.

Egyptians linked the anatomical and physiological makeup of the body to a system of channels («metu») with the heart as its center. The speed of person's heart beats was related to his physical condition. The brain was recognized as the sensory and motor center.

Medical classification was based on symptoms and signs.

---

[21] The Nile River (6,650 km long) flows northerly in the northeastern Africa (area 30,221,582 km²) that enters the Mediterranean Sea.

Degeneration with arteriosclerosis and atherosclerosis/atheromatous (AS) and calcifications of the aorta, and coronary, carotid, iliac and femoral arteries were common in the Ancient Egypt[22].

The treatment consisted of a vast pharmacopoeia. Specially, selected mold, extract from the willow tree bark and other plant were used to treat pain, an infection and inflammation

Gigantism, a.k.a., giantism (in Greek – «gigos», meaning «giant», plural – «gigantes») and acromegaly (in Greek – «acron» + «megas» meaning «large») are rare diseases caused by a benign tumor, a.k.a., as pituitary gland (hypophysis) adenoma (95%) within the lateral wings of the anterior portion of the hypophysis of the brain which produces a chronic excess of growth hormone (GH) or human growth hormones (HGH) or somatotropin – a peptide that stimulate growth, cell reproduction and regeneration in humans and animals, and insulin-like growth factor-1 (IGF-1). While both gigantism and acromegaly have a similar etiology (cause of diseases) and pathology (in Greek – «pathos» meaning experience, suffering, «-logia» meaning to study of), gigantism occurs in childhood, and acromegaly occurs in adulthood.

Hyperfunction of GH and IGF-1 leads to the development of excessive enlargement due to thickening of the bones and soft tissues of the skull, face, jaw, long bones, hands, feet, and organs.

The condition is accompanied in some patients by diabetes mellitus (DM) and typically loss of vision laterally in one, or both visual fields (bitemporal hemianopsia) due to pressure or infiltration of the medial fibers by tumor at the cranial II (optic) nerve chiasm, or optic chiasma (in Greek – «chiasma» meaning crossing, decussation), located in the forebrain at the base of the skull before the Turkish seat (sella turcica).

Chronic excess of GH production induces the development of cardiomyopathy (CM) characterized by biventricular hypertrophy of the heart, myocardial necrosis, lymphocytic infiltrations, and interstitial fibrosis in about 3% of patients, congestive heart failure (CHF), hypertension (HTN), coronary artery disease (CAD), or ischemic heart disease (IHD), arrhythmias and cardiac valve disease.

Those patients had an average 10-year reduction of life expectancy compared to the general population, and at least a doubling of standardized mortality rates.

Treatment is surgical removal of not large pituitary tumor using endonasal transsphenoidal approach or stereotactic radiosurgery, and/or pharmacological with the somatostatin analogue or GH receptor antagonist, radiation therapy, and control of cardiovascular risk factors. With the adequate treatment life expectancy is typically normal.

---

[22] Allam AH, Thompson RC, Wann LS, et al. Atherosclerosis in ancient Egyptian mummies: the Horus study. J Am Cardiol Img. 2011;4:315-27.

The oldest case of gigantism, although in a mild form, was that of **Sanakht** or **Hor-Sanatht [?** – intered in the **large mastabaK2** (meaning «house for eternity» or «eternity house»), village of **Beit Khallaf**, 10 km west of the city of **Girga** in the **Sohag Governorate** (area 1,547 km^2) on the west bank of the Nile River in the Upper Egypt] — Egyptian king (pharaoh) who reigned 18 years, or less, beginning 2,650 BC [3rd Dynasty during Old Kingdom (2800-2250 BC)]. He was a very tall men who would have stood at around 187 cm (or even 198 cm) tall. His skull was very large and capacious, cranial index unusually broad and almost brachycephalic (in Greek – meaning «short» and «head»), that is the shape of his skull was shorter than typical. However, his overall cranial features were close to those of dynastic period of Egyptian skulls. The jaw had a trace of thickness. The long bones of the skeleton attested to a fierce growing up which is obvious sign of gigantism. The proportion of his long bones were tropically adapted like those of most other ancient Egyptians, especially those from prehistoric period.

Pharaoh Sanakht was a huge, much taller than other Egyptians of that time, who would have been closer to 160 cm on average. While the Egyptian royalty were generally taller on average 170 cm, the highest being **Ramesses** or **Ramsess II (1303 – 1213 BC)** — New Kingdom pharaoh of Egypt of the 19th Dynasty, who reigned between 1279 and 1213 BC, and whose height was 175 cm, because of better nutrition and stronger health than commoners. However, pharaoh Sanakh was much taller than all the others.

By c. 2500 BC Egyptian physicians had developed a systematic treatment of diseases. They introduced circumcision, pressure control of bleeding, adhesive tape, splinting and fixation of fractures, sewing up wounds, cauterization with a fire-drill, drainage of abscesses, surgical removal of an appendix (appendectomy), wiring of teeth, dental prostheses, contraception, and embalming.

The earliest reference to circumcision during transition from boyhood to adulthood date back to around 2,400 BC, including a flint-knife circumcision and the use by a surgeon of «*the ointment is to make it acceptable*», likely referring to some form of topical and antiseptic agents.

A statue of swollen limbs of **Nebnapetre Mentuhotep (Nebnapetre) II (2081 – 1938 BC)** — Pharaoh of the 11th Dynasty (reigned c. 2061 BC – 2010 BC) indicates that as early as c. 2000 BC the inhabitants of the Nile River Valley suffered from filariasis, a parasitic disease caused by infection with roundworms of the Filaricide type which spread by blood-feeding black flies and mosquitoes. The adult forms usually stay in the skin and release early larval form, known as microfilariae, which penetrate the lymphatic vessels and the lymphatic nodes (LN), subcutaneous (SC) tissue and serous cavities of the joints, thorax, and abdomen. But the most common is lymphatic filariasis which in its chronic stage may cause a huge swelling (elephantiasis) of the limps, usually the lower extremities and the scrotum, as evidenced by a statue of Mentuhotep II.

Medico-surgical **Edwin-Smith papyrus (c. 1700 BC)**[23], in its medical section describes vital organs (the brain, the heart and lungs), arterial and venous vessels, blood circulation and pulse; chest pain (angina pectoris, stenocardia) due to decreased blood flow (ischemia) to the heart muscle (myocardium), resulting from spasm or atherosclerotic (AS) narrowing of coronary arteries of the heart, the main symptom of coronary artery disease (CAD); atheromatous (AM) of the aorta; and peripheral arterial disease (PAD); mention tuberculosis of the lungs.

Surgical section describes 48 cases, mainly trauma to the human body (wounds, sprain and dislocation of joints, fracture of bones, penetrating wound of the cervical esophagus, stab wounds and blunt injuries to the chest with fractures ribs), tumors and its surgical treatment. It also refers to breast cancer with the conclusion that «there is no treatment». Some sections are dedicated to gynecology and cosmetics.

Medical **Eber's papyrus (c. 1550 BC)**[24] mention angina pectoris; contains detailed description of asthma, or bronchial asthma (in Greek- «aazein», meaning «sharp breath»), a paroxysmal, often allergic disorder of respiration characterized by bronchospasm, wheezing, and difficulty in expiration, treated by drinking of incense mixture of «kyphi», including six ingredients (cassia, cinnamon, mastic, mint, henna and minosa). In addition, it describe symptoms of slowly progressing chronic neurological disorder of the central nervous system (CNS) in the older population, resembling paralysis agitated, or shaking palsy, characterized by tremor at rest, shaking, muscular stiffness or rigidity, slowness, or limitation of movement (hypokinesis), difficulty in walking and postural instability, also inconsistency, fearfulness and anxiety, dementia, depression, behavioral, emotional, sensory, sleeping and thinking problems. It also contains description of epileptic attacks, and tracheostomy.

It mentions xerographic (in Greek – «xenos», meaning «foreign» or «strange»), or heterologous split-thickness skin graft (STSG), which include the epidermis and part of the dermis harvested from animal (frog) to cover the wound of a patient.

---

[23] **Edwin Smith (1822 - 1906)** — American Egyptologist, bought in 1862 from Egyptian merchant, ancient town of Luxor [known since the reign of **Amenhotep IV Ekhenaten (? - 1358 BC)** — Pharaoh-Reformer (1372 - 54 BC), Upper Egypt], Ancient Egyptian medical-surgical papyrus (c. 1700 BC) and Ancient Egyptian medical papyrus containing the herbal knowledge (1550 BC). He retained for himself the medical-surgical papyrus, named Edwin Smith papyrus (the 110-page scroll), but medical papyrus he sold to Georg Eberts (1837 - 98).

[24] **Georg M. Ebers (1837 - 98)** — German Egyptologist, and writer, who in the winter of 1873 - 74 bought from Edwin Smith (1822 - 1906) medical papyrus, named Eberts papyrus (c. 1550 BC).

The carved images on stele from the 18th Dynasty (1403-1365 BC) showed that the Egyptians were familiar with poliomyelitis, or infantile paralysis, as depicted by children walking with canes and adults with withered lower extremities.

The Pharaon **Merneptah** or **Mereptah**, the fourth ruler (1213 BC – 1203 BC of the 19th Dynasty died at young age with a bald head, obese abdomen and arteriosclerotic / atheromatous (AS) degeneration of the aorta.

The rectum (in Latin – intestinum rectum, meaning straight bowel) is the terminal portion of the large bowel, about 12 cm in length, which begins at the end of the sigmoid colon in the rectosigmoid junction at the level of the 3rd sacral vertebra or the sacral promontory, ends in the anorectal ring at the level of the puborectalis sling[25] or the dentate line. Consist of the pelvic and perineal divisions.

The anal canal (in Latin – ring), the lower end of the gastrointestinal (GI) tract situated between the rectum and anus, below the level of the pelvic diaphragm in the anal triangle of the perineum, in between the right and left ischio-anal fossa, is divided in three parts. The upper two third of the anal canal, located above the dentate (pectinate) line is lined by tunica mucosa of a simple columnar epithelium The lower one-third of the anal canal, located below the pectinate line is divided into two zones separated by the white line into the zona hemorrhagic lined by stratified non-keratinized epithelium, and the zona cutanea lined by stratified keratinized epithelium.

The blood supply to the rectum and the anus from the superior rectal artery, a branch of the inferior mesenteric artery (IMA), the middle rectal artery, a branch of the internal iliac artery (IIA), and the lower rectal (hemorrhoidal) artery, a branch of the lower pudental artery of the IIA. The venous return is formed in the subcutaneous, submucosal and subfascial plexuses, with flow away of the venous blood via the superior rectal vein into the portal vein system, the middle and the inferior rectal vein into the inferior vena cava (IVC) system. The most prominent venous plexuses of the submucosal layer are built in the form of cavernous bodies which are widely anastomosed with arteries and arterioles[26]. The cavernous bodies are located at the numerical 3,7 and 11 of the face of a clock in the supine patient. Therefore, in the wall of the rectum there are portocaval anastomoses which may form hemorrhoids (in Greek – «hemo» = blood, «reo» = flow), also called piles.

Internal hemorrhoids are located above the pectinate line, lack of pain and temperature receptors, but are prone to prolapse, while external hemorrhoids, located

---

[25] Os pubis or pubes, region pubica.

[26] Vayda RI, Hroysman SD. Rectum and pararectal space. Clinical anatomy. In, LYa Kovalchuk, VF Sayenko, HV Knyshov. Klinichna Khirurhiya. Vol. 1-2. Publ. «Ukrmedknyha», Ternopil 2000. Vol. 2. P. 299-300. — Korchynsky IYu. Hemorrhoids. In, LYa Kovalchuk, VF Sayenko, HV Knyshov. Klinichna Khirurhiya. Vol. 1-2. Publ. «Ukrmedknyha», Ternopil 2000. Vol. T. 2. P. 300-4.

below pectinate line are sensitive to pain and temperature. Internal hemorrhoids are classified into four grades depending on the degree of prolapse:

Grade   I: no prolapse, but visible prominent blood vessels;
Grade   II: prolapse upon bearing down, which reduces spontaneously;
Grade   III: prolapse upon bearing down, which require manual reduction;
Grade   IV: prolapse with inability to be manually reduced.

The first known mention of hemorrhoids which affect the sedentary people between 45 and 65 years of age, more common among the wealthy and constipated, is from a 1700 BC Egyptian papyrus. The Egyptian physicians treated hemorrhoids with acacia leaves, ground, filtered and cooked, then smear of a fine linen there with and placed on the anus.

Acute para-proctitis is a bacterial purulent inflammation of the cellular tissue surrounding the rectum, occurring usually in patients with diabetes mellitus (DM) and lowered resistance to infection because of depressed immune system[27]. The most common cause is the hematogenous or lymphogenic spreads of infection from the neighboring organs, surrounding cellular tissue, infected one, or all the 6-8 anal gland present in each of anal crypt, from trauma, rarely with formation of an abscess.

According to localization there are subcutaneous, submucosal, ischiorectal, retro-rectal, and pelvi-rectal acute para-proctitis. Acute ischiorectal para-proctitis is usually associated with trans-sphincteric or extra-sphincteric[28], while retro-rectal and pelvic-rectal with an extra-sphincteric route for rectal fistula.

Symptoms are acute pain in the rectal region, tenderness during defecation, fever, an infiltrate in the anal region or on the buttocks.

In the early stages acute para-proctitis should be treated conservatively, otherwise in the case of an abscess formation surgical intervention is necessary. Subcutaneous, submucosal, and trans-sphincteric ischiorectal abscesses should be incised and drained (I&D) over the fluctuation areas around the anus with excision of the internal opening of the anal canal, i.e., the anal crypt, while retro-rectal and pelvi-rectal abscesses should be I&D through the skin or mucous membrane from the side of the rectum. Undrained pararectal abscess may penetrate the rectal wall.

---

[27] Husak VK. Acute paraproctitis. In, LYa Kovalchuk, VF Sayenko, HV Knyshov. Klinichna Khirurhiya. Vol. 1-2. Publ. «Ukrmedknyha», Ternopil 2000. Vol. 2. P. 313-8.
[28] Sphincter (in Greek – meaning «squeeze») is a circumferential muscle surrounding and able to contract or close a bodily opening.

In general, anal/rectal fistulae, tubular purulent tracts are caused by an acute infection in the tissue surrounding rectum and anal canal[29].

Anal / rectal fistulae depending on relationship with the internal and external sphincter muscles are divided into:

(1) Extra-sphincteric fistulae which begin at the rectum outside the dentate line, or sigmoid colon, proceed downwards, through the levator ani muscle and open into the skin surrounding the anus.

(2) Supra-sphincteric fistulae begin between the internal and external sphincter muscles, extend above, and cross the puborectalis muscle, proceed downwards between the puborectalis and levator ani muscles, and open 2.5 cm or more away from the anus.

(3) Trans-sphincteric or «horseshoe» fistulae begin between the internal and external sphincter muscles or behind the anus, cross the external sphincter muscle, may take a «U» shape, and form multiple external opening a 2.5 cm or more away from the anus.

(4) Inter-sphincteric fistulae begin between the internal and external sphincter muscle, pass through the internal sphincter muscle, and open close to the anus.

(5) Submucosal fistulae pass superficially beneath the submucosa and do not cross either sphincter muscle.

Conservative treatment of chronic para-proctitis (anal / rectal fistulae) include sitz baths, warm compresses with 20% ethyl alcohol / ethanol (C_2H_5OH) and antibiotics. Surgical treatment is indicated for chronic and recurrent fistulae.

The surgical approach to anal / rectal fistulae was initiated in Egypt by using a seton (in Latin – «seta» meaning «bristle») – a drain made of gauze for infra-sphincteric and submucosal fistulae and a seton stitch for more complicated trans-sphincteric and supra-sphincteric fistulae.

The use of the seton stitch require passing a surgical-grade cord through the fistula tract, so that the cord creates a loop that joint up outside the fistula. The cord provides a path that allows the fistula to drain continuously while it is healing, rather than allowing the exterior of the wound to close over. Keeping the fistula tract open helps to prevent the trapping of puss or other infectious debris in the wound. The seton can be tied loosely or tightly, with different material, depending on the anatomical location and surgical requirement. It might be lied loosely as a palliative measure to avoid painful and septic exacerbations, or as temporary measure before

---

[29] Zakharash MP. Rectal fistulae (chronic paraproctitis). In, LYa Kovalchuk, VF Sayenko, HV Knyshov. Klinichna Khirurhiya. Vol. 1-2. Publ. «Ukrmedknyha», Ternopil 2000. Vol. 2. P. 318-22.

surgical excision. It may be tied with more tension and tightened periodically, cutting through tissue inside the loop while scaring behind the loop, finally «pulling out» the completely healed fistulous tract without operation.

The papyrus contains the first description of the physical examination, diagnosis, and treatment of mandible fractures.

While **Herodotus (c. 484 BC – c. 425 BC)** — Greek historian, «father of history», referred to Egyptians as the «healthiest of all men», **Pliny the Elder (23 AD – 79 AD)** — Roman writer, naturalist, and philosopher, called Egypt «the motherland of disease». The fact is that the Egyptians recognized and treated a great variety of illnesses.

In Ukraine, the ***Trypillian culture*** was created between 4000 BC and 2000 BC.

In ***Ukraine-Scythia*** (the 7th - 3rd century BC) herbal medicine and some surgical procedures were in common use. The sorcerer - physicians **Abarys the Hyperborean** (hyperborean = «the extremely northern»)[30] — likely descendant of the northern shamans and sorcerers of the 3000 - 1000 BC, **Anakharsys (c. 610 BC – 535 BC)**[31] and **Toxar/Toxarys (c. 650 BC – 575 BC)**[32], used medical herbs to increase blood coagulability, to treat ulcers and to induce general anesthesia / anesthesia (in Greek / Latin – meaning analgesia «without pain» and amnesia meaning «without memory»), i.e., poppy seeds, hen-bane, and mandragora. Bathhouses were widespread, and in common use.

Scythian physicians performed teeth pulling, debridement, suturing with a horsehair, and bandaging of the wounds, phlebotomy, reduction of dislocations (luxation) and fractures, amputation of extremities, trepanations, embalming, and mumification.

Medea (6th BC – 1st century BC)[33] — a mythical princess of the Kingdom of Colchida on the Eastern Black Sea shores (the 4th century BC – the 2nd century AD) and Scythian physician who discovered the dye for dyeing the gray hair.

Scythia of the Northern Black Sea shores was under the rule of **Mithridates VI (134 BC – 63 BC)**[34] — King of the Pontus Empire on the Southern Black Sea shores (291 BC – 63 BC) who left a legacy in the Scythian culture as the researcher of poisons (mithridatism), and anti-poisons (mithridatica, mithridatium) which he described in his treatise «Teriak».

---

[30] Ablitsov VH. «Galaxy Ukraine». The Ukrainian diaspora: prominent persons. Publ. «Kyt», Kyiv 2007.

[31] Ablitsov VH. «Galaxy Ukraine». The Ukrainian diaspora: prominent persons. Publ. «Kyt», Kyiv 2007.

[32] Ablitsov VH. «Galaxy Ukraine». The Ukrainian diaspora: prominent persons. Publ. «Kyt», Kyiv 2007.

[33] Ablitsov VH. «Galaxy Ukraine». The Ukrainian diaspora: prominent persons. Publ. «Kyt», Kyiv 2007.

[34] Ablitsov VH. «Galaxy Ukraine». The Ukrainian diaspora: prominent persons. Publ. «Kyt», Kyiv 2007. — Synyachenko OV. They glorified themselves and Ukraine: history of domestic therapy. Crimean Therapeutic Journal 2010;2(2):21-8.

The Scythians were also the inventors of trousers.

North Macedonia (area 25,000 km²) and **South Macedonia** (area 34,177 km²) is populated since 270,700 BC, its recorded history begins since 808 BC and the Argead dynasty since 750 – 700 BC.

During the reign of **Alexander the Great (356 BC – 323 BC)** — the second King of Macedonia (336 BC – 323 BC), tourniquets made from strips of bronze or leather were used to surround and apply pressure onto the proximal extremity to stanch (staunch) the bleeding of wounded soldiers.

Greece (area 131,957 km²) originated from the Aegean or Mycenean civilization (3000 – 1100 BC).

The principal god was **Apollo's** son, **Asclepios (Aesculapius) of Thessaly**. He was usually depicted with his sons, **Machaon** and **Podalirios**, the patron gods of surgeons and physicians, and with his daughters, **Hyegeia** and **Panacea**, goddesses of health and remedies. Asclepios became a god after his death.

The Greek experience with electric current was by noting that amber attached small object when rubbed with fur, and this phenomenon they named «electron» (in Greek – «amber»).

Homer (8ᵗʰ – 7ᵗʰ century BC)[35] — Greek poet, in his poem «The Iliad» was the first to use word asthma (Ancient medicine. *Egypt*) and depicted the wounds of the heart, inflicted by a spear during the battle with a subsequent sudden cardiac death (SCD) from hemorrhage (c. 450 BC). Particularly, **Serpedon** — son of Zeus — King of Gods and chief God of the heaven, thunder, and lightning and of Laodamia, King of Lycia in the south-west coastline of Asia Minor / Anatolia (area 756,000 km²), died of a spear impalement of the heart from the hand of his opponent Patroclus, during Troya War (c. 1,300 – 1,190 BC) near the north-west coastline of Anatolia.

The healing temples of Asclepios originated about the 6ᵗʰ century BC, either in Thessaly, Tricca or Epidauros (c. 360 BC). There were about 200 such temples. Offices of the physicians were situated close to them. Majority of illnesses treated in their temples were of psychic origin and the effectiveness of the cure was based on faith.

Acmean (5ᵗʰ century BC) — Greek medical scientist, a representative of the Cretan School, in his work «Concerning Nature» concluded that the brain was the organ of the mind and was responsible for thought, memory and sensation.

To prevent illnesses, **Pythagoras (c. 530 BC)** — Greek philosopher, recommended diet, exercise, music, and meditation.

---

[35] Homer. The Iliad. Translated by Alexander Pope. Publ. George Bell & Sons, London 1904. / **Alexander Pope (1688 - 1744)** — English poet who translated «The Iliad» (1715-20). / Homer. The Iliad. Introductory article by A. Biletsky. Translation from the Greek. Publ. «Dnipro», Kiev 1978. / **Andriy O. Biletsky (1911 - 95)** — Ukrainian linguist and political scientist.

The concept atom was introduced into science by **Anaxagoris (c. 500 BC – 428 BC)** — Greek philosopher, mathematician, and astronomer, **Leucippos (c. 500 BC – 440 BC)** — Greek philosopher, who assumed the interminable possibility of planets in the Universe, and **Democritus (c. 460 BC – c. 360 BC)** — Greek philosopher, the founder of the atomic hypothesis for explanation of the surrounding world, considered a possibility of the presence of unlimited quantity of unique worlds.

In Corinth (the 4th century BC), a communal latrine provided continuous running water beneath the seats for waste removal. It was the predecessor of the present-day bidet, a bathroom or toilet fixture consisted with a bowl, or receptacle usually with spigots, designed to be sat on for purpose to wash the human external genitalia, perineum, inner buttock, and anus, especially after passing each stool, i.e., after defecation.

The founder of the dogmatist school, **Diocle of Crystus (4th century BC)** — Greek physician, performed animal dissections, e.g., on mules.

Empedocles (c. 493 BC - 443 BC) — Greek philosopher, poet, physician, and political figure, stated that breathing occurs not only through the nose and the mouth, but also through respiratory pores in the skin.

The father of medicine, **Hippocrates (Hippocrates, II, 460 BC, Cos – 377 BC, Larisa, Thessaly)**[36] — Greek physician, was the last descendent of seventeen generations of physicians who practiced in Athens, and travelled to Egypt, Ukraine-Scythia, Asia Minor, and the Middle East.

Hippocrates' theories of medicine were summarized in «Corpus Hippocraticum» or «Hippocratic Collections», indicating he believed in facts, applied logic and reason in medicine and showed that diseases have only natural causes. Principles of his method were *«to observe all, to study the patient rather that the disease, to evaluate honestly and assist nature»*.

He considered phthisis (tuberculosis) of the lungs – consumption or wasting disease the most common illness of those times.

In his work «Of the Epidemics» (written 400 BC) Hippocrates mentioned infectious disease named «epidemic of the parotid gland», «Epidemic parotitis», or mumps, wherein he described swelling of the parotid glands and testicles.

Hippocrates was the first to use the term asthma (Ancient medicine. *Egypt*), referring to the medical condition caused by spasm of bronchi which according to him more likely occur in tailors, anglers, and metal workers.

---

[36] Hippocrates. «Corpus Hippocraticum». Last decades of the 5th century BC – the first half of the 4th century BC. — Chadwick J, Mann MN. The Medical Work of Hippocrates. Charles C. Tomas, Springlield, IL 1950. — Dachkevych J. Hippocrates. J Ukr Med Assoc North Am 1966 Apr;13,1-2(40-41): 20-2. — Lüderitz B. Historical perspectives of cardiac electrophysiology. Hellenic J Cardiol 2009;50:3-16.

He described the pericardium as a smooth mantle surrounding the heart and containing a urine-like fluid, showed that the aortic valve (AV) opened one way into the aorta.

Then he introduced the types of temperament (choleric, melancholic, phlegmatic, and sanguine), laid the foundations of etiology, clinical diagnosis, and prognosis.

Hippocrates noted a turtle-back nail («unquis Hippocratis», or «Hippocrates' nail»), a thickened soft tissues of the finger's distal phalanges without permanent bony changes («clubbed fingers», «drumstick fingers», or «digiti Hippocratici») in chronic diseases of the heart, lungs and liver; pulmonary edema; a sound of succussion (a splashing noise) from air-fluid collection inside the pleural cavity on a sudden motion of the torso, by shaking or during percussion of the chest («succussio Hippocratis»); melena (Hippocrates black disease).

He placed the origin of epilepsy in the brain, suggesting it is a congenital disease, thus could be treated («About the sacred disease»).

Depicted altered features of the face in advanced terminal conditions («facies Hippocratica»).

Particularly, he described a congenital heart disease (CHD) in children and younger adults[37] characterized by dyspnea, a latent or evident cyanosis, «clubbed fingers», edema of the lower extremities and bouts of weakness or faintness; and a disease complex consisting of arthritis, urethritis, and conjunctivitis.

Hippocrates established the diagnosis of pulmonary hypertrophic osteopathy by describing that «drumstick fingers» were a sign of a severe disease of the lungs.

In «Aphorism» (Section II, No. 41) he defined sudden cardiac death (SCD) as *«Those who are subject to frequent and severe fainting attacks without obvious causes die suddenly».*

Hippocrates treated his patients with proper diet, fresh air, changing climate, attention to habits and living conditions. His favorite diet was a barley gruel and his favorite medicine – honey. He used medical herbs, emetics, laxative, and enemas.

His patients who underwent surgical procedures or operations were anesthetized with opium and mandragora.

Hippocrates treated wounds by surgical debridement, that is removal of damaged tissue or foreign object, cauterization, irrigation, application of a tar, and coverage by dressing.

He burred holes in the skull (trepanation) to release the pressure caused by trauma or tumors.

Hippocrates performed incision and drainage (I&D) of abscesses with knife or scalpel to removed puss.

---

[37] Zinkovsky MF. Congenital Heart Disease. In, LYa Kovalchuk, VF Sayenko, HV Knyshov. Klinichna Khirurhiya. Vol. 1-2. Publ. «Ukrmedknyha», Ternopil 2000. Vol. 1. P. 184-227.

Tumors were catheterized using blazed steel rod.

He developed the principles of surgical branch called desmurgy or desmurgia (in Greek – desmurgia» meaning «to bind», or «tie together»), to dress and bind the wounds, the injured cranium of the head using «mitra Hippocratis», and orthopedics by reducing and fixing bone fractures on the «Hippocrates bench»; described the method for reduction of anterior dislocation of the shoulder joint by adduction of the arm with a longitudinal traction and gentle external rotation while the countertraction was performed by placing his own heel of the foot into the axilla (Hippocrates method); re-approximated and immobilized fractured bones using circumferential wires with an external bandaging.

Thoracic empyema or empyema thoracic is a collection of infected purulent material in the pleural cavity, generalized or localized (encapsulated) which spreads from infection of the lung, deep posterior space of the neck, thoracic spine and subphrenic abscess via diaphragmatic fenestrationi, from spontaneous or instrumental perforation of the esophagus, from penetrating chest trauma and after elective thoracic procedures.

Hippocrates distinguished between noninfected collection of pleural effusion (hydrothorax) and empyema thoracic based of clinical manifestations, such as fever, mild during the day and increasing at night, dry nonproductive cough, the hollow eyes, and the red spots on the cheeks, by auscultation hearing the «succussio Hippocratis» when patient was shake by the shoulders or the torso. As for hydrothorax he wrote: «*When applying the ear on the ribs, during a certain time you hear a noise like boiling vinegar, which suggests that the chest contains water and no pus*». He predicted that «*Patients with pleurisy who, from the beginning, have sputum of different colors or consistencies die on the third or the fifth day, or they become suppurative by the eleventh day*».

He treated surgically thoracic empyema by opening it at the site where pain or swelling were most evident using knife, cautery, or trephination of a rib. Noted that when «*the pus flows pale and white, the patient survives, but if it is mixed with blood, muddy and foul smelling, he will die*». After draining puss, he cleaned the lung surfaces, daily irrigated the empyema cavity with «warm oil or wine» and packed it with a strip of linen cloth to prevent air from entering the space. When the empyema had healed, the wound in the chest wall was closed with metal rods.

Of the two common carotid arteries (CCA), the right CCA arises from the brachiocephalic trunk (BCT), i.e., innominate artery (IA) that originates at the junction of the ascending aorta (AA) and the aortic arch, i.e., transverse aorta (TA), and the left CCA originates distally to the BCT from the TA. Both CCA's ascends the anterior portion of the neck forming the carotid bulb where they divide into the internal carotid artery (ICA) and external carotid artery (ECA). The ICA enters the base of the skull through the carotid foramen into the cranial cavity supplying the blood to the middle ear, brain, hypophysis, and orbit. The ECA supply the blood to the neck, face, and scalp.

It has been known since ancient times that applying pressure to the CCA, or the carotid bulb (in Greek – «carotid» means to stupefy, or place into a deep sleep) causes unconsciousness.

He operated on the anal and rectal disorders, treated hemorrhoids with a rubber band ligation or by transfixing their base with a needle and then tying them with very thick woolen thread until they drop off. Then he placed the patient on a course of treatment with Hellebore (Helleborus orientalis) – herbaceous or evergreen perennial flowering plant.

After his predecessors, Hippocrates with aim to treat surgically anal fistulae also used a seton drainage and a seton stitch[38].

The Hippocrates'motto was «*As to disease, make a habit of two things: to help, or at least not to harm*». He originated «The Hippocratic Oath».

During the 430 BC plague epidemic in the city of Athens (oldest human presence in the Cave of Schist dated between the 11th and 7th millennium BC), it was noted that the people who contracted plague but survived were often protected from getting this same infection again, apparently because survivors developed immunity against reinfection.

It was **Plato (c. 429 BC – 347 BC)** — Greek thinker, who envisioned an ideal state which would provide for the health of its citizens and would prevent poverty and overpopulation.

Among the first to separate arteries and veins was **Praxagroas of Cos (c. 340 BC)** — Greek physician, but he believed that both contained air.

During the first half of the 6th century BC, **Diokles of Carystos** distinguished pneumonia from pleurisy, intestinal cramps from obstruction and recognized fever as a symptom.

The founder of comparative anatomy, **Aristotle (382 BC – 322 BC)** — Greek scientist encyclopedist, philosopher and logician, described the punctum saliens (the first sign of the embryo), the early development of the heart and the great vessels, the beating of the embryo's heart, distinguished the cardiac valves, named the aorta, and traced the course of the ureter.

Representatives of the Alexandria School (331-30 BC) in the city of Alexandria (founded in 331 BC), Egypt, **Herophilus (300 BC – 250 BC)** and **Erasistratus (300**

---

[38] Zakharash MP. Rectal fistulae (chronic paraproctitis). In, LYa Kovalchuk, VF Sayenko, HV Knyshov. Klinichna Khirurhiya. Vol. 1-2. Publ. «Ukrmedknyha», Ternopil 2000. Vol. 2. P. 318-22.

BC – 240 BC) — Ancient Greek physicians, practiced systematic cadaver dissections (autopsy).

Archimedes (c. 287 – c. 212 BC) — Greek polymath (mathematician, physicist, engineer, inventor and astronomer), described a technique of amputation with the proximal use of a tourniquet and a temporary occlusion of the main proximal vessels to control bleeding.

The medicine of **China** (area 9,706,961) is based on the work of **Fu Tsi (c. 2900 BC)** — Chinese scientist, who originated the pa ku symbol, a basic ying-yang (female-male) dichotomy of the universe and formulated the eight trigraphs of all possible combinations of the two; of **Shen Nung (c. 2800 BC)** — Chinese physician, who compiled the «Pen tsao» – the first medicinal herbal; and of the **Yellow Emperor / Huang Ti** (reign, **c. 2698 BC – c. 2598 BC)** — Chinese physician, who authored the «Nei Ching» («Cannon of Medicine»), covering all phases of health and illnesses, its' prevention and treatment (including acupuncture). It stated that «*All blood is under the control of the heart*», and «*the blood current flows continuously in a circle and never stops*», recommended that «*A superior physician helps before the early budding of disease*».

All six sons of Huang Ti were born «cutting open a body», i.e., by Caesarean section (C-section), a surgical procedure performed through the lower abdominal horizontal incision in the midline along the «linea nigra» («black line») or suprapubic transverse incision below the hair line and above the Hills (dimples) of Venus (in Latin – mons pubis», or «mons veneris»)[39] and a transverse incision on the anterior lower pregnant uterus, to deliver alive baby, when the natural vaginal delivery would put the baby and mother at risk, e.g., obstructed labor, twin pregnancies, high blood pressure (BP) in the mother, breech presentation, abnormal location of the placenta and twisted umbilical cord, or on dying mother to save the child.

In league with Taoism which emphasized living in harmony with the nature, Chinese medicine focused on prevention of diseases.

Specific duties and organization of physicians were outlined in the «Institution of Chou» (c. 1050 BC – 255 BC).

The most important diagnostic sign was the pulse rate, as outlined by **Pien Ch'iao (410 BC – 310 BC)** — Chinese physician, in the «Secrets of the Pulse». The Chinese physicians determined the pulse rate of their patients in reference to their own.

Extensive occlusive arteriosclerosis (AS) of coronary arteries of the heart was found in a 50-year-old Chinese noble woman **Lady Dai**[40] who suffered from angina

---

[39] **Venus** – in Greek mythology she was the Goddes of love. — Hills of Venus «Mons pubis», seu «mons veneris» is a rounded mass of fatty tissue found over the pubic symphysis of the pubic bones.

[40] Cheng TO. Evidence of Type A personality in a Chinese lady who died of acute myocardial infarction 2,100 years ago. Tex Heart Inst. J 2002;29(2):154-5.

pectoris, with risk factors being the type A personality[41], a sedentary lifestyle, obesity, diabetes mellitus (DM) and hypertension (HTN). She was treated with herbal medicines containing cinnamon, magnolia bark, and peppercorns. The cause of death in 163 BC was an acute myocardial infarction (MI) due to occlusion of the left anterior descending (LAD) artery.

Infectious disease smallpox (variola major) was disseminated in China from the south-west of the country in the 1st century AD. It was endemic from that time until it was eradicated from the country in the early 1960s[42]. Inoculation against smallpox was introduced in China between 1000 and 1010 AD and variolization has been practiced since the Sung dynast (1023-55). The method consisting in griding up the scales and introducing them into the nostrils. Vaccination was started on a limited scale in 1803, but until 1951 disease continued much of illness and death.

Noteworthy accomplishments of Chinese clinicians include:

Tsang Kung (c. 200 BC) — Chinese physician, who described gastric cancer, aneurysm, and rheumatism;

Chang Chung-Ching (150 AD – 219 AD) — Chinese herbal physician, wrote a treatise on «Typhoid, and other Fevers»;

Hua To (110 AD - 207 AD) — Chinese surgeon, developed anesthesia, and furthered knowledge of anatomy;

Ge Hong or **Ko Hung** or **Zhichuan (283 AD – 343 AD)** — Chinese physician, Daoist, and alchemist, described beri-beri disease due to insufficiency of vitamin (avitaminosis) B_1 with multiply inflammations of nerves (polyneuritis) that causes weakness, lack of appetite, irritability, paralysis and very high probability of death, inflammation of the liver (hepatitis), plague, smallpox and variolization, wrote «The Handbook of Prescriptions for Emergency Treatment» (340) in which outlined preparation of a sweet worm wood plant (Artemisia annua) for the treatment of «intermittent fever» (malaria), by steeped extraction in a low-temperature (cold) water;

Shu Szu-Miao (581 AD – 682 AD) — Chinese physician, wrote a medical treatise called «A Thousand Golden Remedies» and headed a committee which produced the «Collection on Pathology».

Disease called goiter (in Latin – gutteria, struma) is a swelling of the neck or larynx resulting from diffuse hyperplasia (thyromegaly), unimodular, or multinodular,

---

[41] Type A and type B personality theory describes two common, contrasting personality types: the high-strung (type A) and the easy-going (type B). Type A patterns of behavior raise one's chances of developing CAD. — Friedman M, Rosenman R. Association of specific overt behavior pattern with blood and cardiovascular findings. JAMA 1959;169: 1286-96.

[42] Hui X, Yuty J. The eradication of small-pox in Shanghai, China October 1950 – July 1951. Bull WHO 1981;59(6):913-7.

inactive, or active (hyperfunction, hyperthyroidism), or toxic goiter of the thyroid gland (TG) due to in 90% by iodine salt deficiency.

Toxic goiter is characterized by triad of hyperthyroidism, goiter and bulging of eyeballs (exophthalmos), include cardiac arrhythmias, increased pulse rate, weight loss despite of increased appetite, intolerance to heat, profuse sweating, apprehension, weakness, tremor, diarrhea, vomiting, eyelid retraction, stare, elevated basal energy expenditure (BEE), and elevation of the blood serum protein-bound iodine level. Prevalent in fresh water and lake countries due to lack of iodine. May be related to malfunction of the immune system. Female to male ratio is 4:1, onset in third or fifth decade of life.

As outlined by **Zhen Quan (c. 541 AD – 643 AD)** — Chinese physician, in his book «Gu jin lu fang» confirmed that Chinese physicians of the Tang Dynasty (618 AD - 907 AD) successfully treated patients with goiter by using the iodine-rich thyroid gland of animals such as ship, and pigs – in raw pills or powdered form[43].

Among the drugs we have adopted from Chinese medicine are rhubarb, iron for anemia, castor oil, kaolin, aconite, camphor, Cannabis sativa, Chaulmoogra oil for leprosy, Ephedra vulgaris, ginseng and Rauwolfia serpentina for hypertension (HTN).

The Chinese also introduced hydrotherapy.

In in the southwestern region of **Pakistan** (area 881,913 km²) – **Balochistan** (area 347,190 km²) during the **Mehrgarh culture / period** (7,000 -3,300 BC) an ancient proto-dentistry was practiced by regularly shaping cavities of molar teeth with concentric ridges drilled into them.

The chief cities of the **Indus Valley civilization** (3,300-1,300 BC) were the ancient Moenjro-daro, called «mound of death» (built c. 2,600 BC – abandoned c.1900 BC, an archeological site since 1922) and the ancient Harappa (emerged c. 2,600 BC) on the Indus River[44], located in the region of today's Pakistan and northwestern **India** (area 3,287,263 km²). These cities had streets laid out in a rectangular pattern, well build and ventilated brick buildings, some with bathrooms, a drainage system which ran from the houses to the brick-linen sewers. Some houses had rooms built around a courtyard. The cities had public baths, sewers, and shutters for collection of trash. Their settlements were abandoned due to floods and drought.

Buddhism (the 6th century BC) in India forbade the taking of life, taught compassion and stressed the need for the development of charity hospitals.

---

[43] Temple R. The Genius of China: 3000 years of Science, Discovery, and Invention. Publ Simon & Shuster, New York 1986.

[44] The Indus River (length 3,180 km, drainage area 1,164,000 km²) originate in the Tibetan plateau, called the «Roof of the World» (stretches approximately 1,000 km north to south and 2,500 km east to west, area 2,500,000 km², altitude 5,000-1,500 m, highest peak Mount Alung Gangri 7,315 m) flows in a southerly direction to merge with the Arabian Sea.

Sushruta (c. 600 BC) — Indian surgeon, known as the main author of «The Compendium of Sasruta» («Sasruta-samhita»), described more than 120 surgical instruments, including the spatula (tongue depressor), the speculum, forceps, scissors, scalpels, saw, needles, syringes, catheters and trocar; reconstructed the injured ears and congenital cleft lip (harelip), removed thyroglossal cysts and sinuses (Sushruta operation), mentioned tracheostomy, ligated hemorrhoids with a rubber band or by transfixed their base with a needle and then tied them with very thick woolen thread until they drop off, emphasized the necessity of wound cleanliness. As the Egyptians physicians and Hippocrates with aim to treat surgically anal fistulae he also used a seton drainage and seton stitch.

The «wandering scholar physician» of Ancient Ukraine, Middle Asia and India, **Caryk (Caraka)**, a.k.a., or nom de guerre **Charaka** which in Ukrainian means «dear (little) tsar» **(c. 300 BC, Ukraine – ?)**[45] the chief co-author of a treatise «Ayurveda» («The Science of Renaissance») where he described the man nature, childbirth, methods to preserve and to restore the healthy human race, the first to propose the concept of hereditary, digestion, metabolism, and immunity against diseases. Mentioned epilepsy. Proposed to apply to the skin leeches for bloodletting to treat inflammation and pulmonary edema. Charaka's qualifications for a nurse included knowledge of drugs, cleverness, devotion, and purity (mind and body).

In «Ayurveda» there is mentioning of a very alkaline «kshara» extract made from ashes of pricky chaff-flower, or devil's horse whip (Achyranthes aspara) – a tropical plant, given to people to drink in a highly diluted form as a remedy for worms, indigestion, kidney stones, skin diseases and obesity. It is also administered topically in a less diluted form, to treat diseases of the anus and rectum like hemorrhoids, rectal prolapse, and to try to control infection after anal fistulectomy, excision of pilonidal sinus, after draining perirectal abscesses, to treat infected wounds in general.

Dermoid cyst, pilonidal (in Latin – pilus = hair, nidus = nest) or sacrococcygeal cyst, sinus, fistula, or abscess are tubules covered by multilayered (stratum) flat epithelium near or on the median natal cleft of the buttock, often containing hair or skin debris and opens to the outside with one or several openings[46]. They are itchy and often very painful, and typically occur between the ages 15 and 35.

The possible causes are congenital pilonidal dimple, ingrown hair, excessive horse riding, sitting and increases sweating and pressure on the coccygeal region. Occasionally, it affects the navel, armpit, or genital region.

---

[45] Ablitsov VH. «Galaxy Ukraine». The Ukrainian diaspora: prominent persons. Publ. «Kyt», Kyiv 2007.

[46] Korchynsky IYu. Pilonidal cysts. In, LYa Kovalchuk, VF Sayenko, HV Knyshov. Klinichna Khirurhiya. Vol. 1-2. Publ. «Ukrmedknyha», Ternopil 2000. Vol. 2. P. 322-7.

Treatment included hot compresses, application of depilatory cream and antibiotics. In more severe cases pilonidal cyst or sinuses should be, after impregnation with 1% solution of methylene blue, incised and drained (I&D); excised in the form of the oval flap of tissue together with pilonidal sinus tract, sewing the wound with the deaf stitch or sewing underneath the edges of the wound to the bottom; cut with the scalpel a slit into abscess or cyst and sutured the edges of the slit to the surrounding skin to form continuous surface from the interior to exterior surface for drainage and healing (marsupialization); or excised with a reconstructive flap technique such as a «cleft lift» or Z-plasty procedure.

In ancient medical books of India there are descriptions of an acute infectious disease named «sacred fire», or anthrax (in Greek – «anthracide», meaning «coal»), caused Bacillus anthracis, a rod-shaped Gram-positive[47] anaerobic bacterium about 1- 9 millimicron (µм) in size, which runs its course in humans most often as a diffuse affection of the skin with a purulent-necrotic inflammation of deep layers and subcutaneous tissue surrounded by redness and edema, neighboring (local) inflammation of the lymphatic nodes (LN), i.e., lymphadenitis, fatty glands, and hair follicles, on the background of fever and poisoning (intoxication) of an organism, less often with affection of the lung, bowel and generalization of disease.

The clinical picture is characterized, besides of fever, by chest pain, shortness of breath (SOB), nausea, vomiting and diarrhea.

Sometimes, anthrax is accompanying by more prominent local purulent-necrotic inflammation of its deep layers and subcutaneous tissue caused by superimposed infection by Staphylococcus aureus in the form of furuncle (in Latin – «minor thief») or carbuncle (in Latin – «carbonum» meaning «coal»).

Prevention from anthrax include maintenance of the proper personal hygiene, treatment – washing wound by oxidizing agent such as hydrogen peroxide (H_2O_2), frequent dressings using the gauze saturated with an extract from medical plants, e.g., from Plantain (ripple-grass), or its leaves alone, nowadays – by antibiotics and nonoclonal antibodies.

Circa 272 BC **Chanakaya** — Indian teacher, philosopher, economist, jurist, and royal advisor, performed C-section on the royal woman who was pregnant near the term of delivery of a baby, but dying from drug overdose, to save the child.

King **Asoka (273 BC – 232 BC)** built hospitals for humans and for animals and supplied them with healing herbs.

---

[47] Gram-stain, or staining is a method for distinguishing, and classifying bacterial species developed in 1884 by **Hans Ch. Gram (1853-1930)** — Danish bacteriologist, using safranin azonium compound of symmetrical (2,8-dimethyl-3,7-diamino-phenazine), or fuchsine, i.e., rosaniline hydrochloride ($C_{20} H_{19} N_3$ HCl) with addition of potassium iodine (KI) solution in water.

The laws of **Manu (200 BC – 200 AD)** contained an oath for graduating physicians, and rules for physicians and patients.

At medical examination, the patient, his sputum, vomitus, urine, and stool were checked. The pulse was an important diagnostic and prognostic sign.

Physicians recognized fever, cough, constipation, diarrhea, dropsy (edema), seizures, ascites, abscesses, tumors, skin diseases, consumption, or phthisis (pulmonary tuberculosis) with hemoptysis as being the «king of diseases», leprosy, and smallpox.

Gymnastic and breathing exercises were regarded as being important in the maintenance of good health.

Pharmacopoeia contained 500 to 700 herbs, including Rauwolfia serpentina. In use were emetics, purgatives, enemas, sneezing powders, inhalations, leeching, cupping, bloodletting, plague, and smallpox isolations and variolization.

Snakebites were treated by the proximal application of a tourniquet, then making a cut to connect the punctures, suctioning the wound and plastering it.

The «Indian method» of plastic and reconstructive surgery, that is closure of the body defect using the flap on one pedicle, cut from the skin, which is located along the body defect[48]. The flap on one pedicle can be used to cover the bodily defects of triangular, quadrangular, and rounded forms. When needed a rotation, transposition torsion is made, in which the flap designed to cover a defect is displaced in relation to the mean point of its axis with the angle from 30° to 180°. The Indian surgeons reconstructed the nose by a flap of skin taken from the forehead with its pedicle at the root of the nose or from the cheek («Indian rhinoplasty»). They also performed plastic reconstruction of the auricle (auricula) of the ear and of the lips.

Other surgeons extracted teeth, stitched wounds, ligated bleeding vessels, incised and drained (I&D) abscesses, tapped the abdomen for ascites, extracted foreign bodies, excised tumors, «couched» cataracts (an ancient method where the cataract is simply pushed down out of the line of sight), amputated extremities, repaired hernias, removed bladder stones, performed excision of anal fissure (anal fissurectomy), corrected abnormal fetus presentation, carried out caesarean section and removed dead fetuses through the vagina.

Apparently, in **Slovenia** (area 20,055 km²) in 4500 BC decaying tooth cavities were filled with bees' vax.

Originated in **Palestine / Israel** (Prehistoric medicine) «Good Samaritan Law» has among others, a legal application to physicians. It offered a legal protection to people who give reasonable assistance to those who are, or who they believe to be ill, in peril, or otherwise incapacitated. They take their name from a parable found in

---

[48] Bihuniak VV. Plastic surgery. In, LYa Kovalchuk, VF Sayenko, HV Knyshov. Klinichna kirurhia. Vol. 1-2. Publ. «Ukrmedknyha», Ternopil 2000. Vol. 2. P. 359-90.

«The Bible»[49], attributed to **Jesus Christ** (Prehistoric medicine), and referred as the «Parable of the Good Samaritan» (Luke 10-25-37) which records the aid given by a traveler from area known as Samaria[50] to another traveler from the same area, namely Jew who has conflicting religious and ethnic background and who has been beaten and robbed by bandits. Jews and Samaritans were generally enemies.

The Hebrews (1900's BC - 200's) according to the «Talmud» (written over many proceeded centuries – 4th century BC), pioneered public health by promoting personal and social hygiene, recognized hereditary bleeding at circumcision, reduced dislocations, and incised imperforate anuses. Rescue breathing on a child was carried out by Elijah («The Bible», 1 King. 17:17-22) 700 BC. During the Egyptian captivity (c. 1300 BC) the Hebrew midwives, utilized the expired air method to revive newborn[51].

The textile industries colored with cotton or silk clothing, along with the use of gauze (in Arabic «gazz», means «silk», in Persian – «kaz», meaning «raw silk»), originated in the seaport Gaza Сіти (inhabited from 3,000 years and founded 15 century BC), Palestine (Pre-historic Medicine)[52], on the account of refugees' work from the surrounding coastal plain villages. The most popular motifs of textures were scissors (muquass), combs (mushut) and triangles (hijabi, i.e., «head scarfs», and thobs – an ankle-length dress, usually with long sleeves, often arranged in clusters of fives, seventhees and threes, as the use of odd numbers is considered in Arab folklore to be effective against the evil eye.

Gauze is widely used in medicine as the light, loose and open woven cloth made of cotton or other fibers for gauze sponges, dressing lacerations and burns of the skin, subcutaneous tissues and mucles, deep wounds or surgical incisions and swabs, and silk as well as other fabrics for dress trimming. It is a rich weave design in which the weft yarns (threads) are arranged in pairs and crossed at the 90° angle before and after each warp thread keeping the weft attached firmly in the place.

The «Talmud» instructed that a boy must not be circumcised if he had two brothers who died due to complications arising from their circumcision, and **Moses'** ben **Maimon**, a.k.a., **Maimonides (1135/38 – 1204)** — Jewish philosopher, astronomer, and physician, said that this excludes paternal half-brothers.

The Etruscan civilization of Apennine Peninsula (spanning 1,000 km from north to south, area 134,337 km², the highest elevation Mount Corno Grande 2912 m) of Southern Europe, endured from the time of the earliest Etruscan inscription c. 700

---

[49] The Holy Bible. Edited by John D. O'Connell. Catholic Press, Inc, Chicago, Il 1954.

[50] Samaria is historical and biblical name used for central region of Palestine / Israel, bordered by Galilee to the north and Judea to the south, existed between 997 and 721/2 BC (Prehistoric medicine). Samaritans were settlers from the town of Gudua, Mesopotamia.

[51] Rosen Z, Davidson JT. Respiratory resuscitation in ancient Hebrew source. Anesth Analg 1972;51:502-5.

[52] Roquin A. Gauze: origin and the world. J Am Coll Surg 2021 Sep;233(3):494-5.

BC until the 1st century BC. Etruscans removed and treated wounded men from the battlefield, introduced golden teeth and bridges, and for public health built the sewer system.

The Romans (754 BC - 746 AD) made great strides in public health by advancing a system of water supply (aqueducts), sanitation, gymnasium, and public baths. Aqueducts carried 1,1 bln liters of fresh water to Rome each day.

The Roman physicians used tourniquets made of strips of bronze or leather to control bleeding from the wounded extremities, especially during its amputations.

Asclepiades (124/129 BC – 40 BC) — Greek physician who flourished in Rome where he established Greek medicine, introduced Democritus's atomic theory to medicine (Ancient medicine. *Greece*). According to this teaching it is the lack of harmony or the irregular movement of atoms which causes disease. Asclepiades's method of treatment were diet, friction, bathing, exercise, emetics, and bloodletting. He invented tracheostomy to help children in asphyxia from croup, resuscitated a dead man back to life and established humane treatment for the mentally ill. His principle was: «*Treat with hope, fast and pleasantly*».

The association of marshy land with diseases was recognized by **Marcus T. Varro (116 BC – 27 BC)** — Roman scholar and writer, who said that swampy land contains (harbors) «*certain minute animals invisible to the eye, breed there, and born of the air reach the inside of the body by the way of the mouth and cause disease*». He advised against building near swamps.

The first description of lymphatic filariasis with elephantiasis belongs to **Titus L. Carus (99 – 95 BC - 55 BC)** — Roman poet and philosopher, who saw it among the dwellers at the coast of the Nile River of of Egypt **[Nebhepetre Mentuhotep II** (before **2060 – 1995**? **BC.)]**.

The Romans created the position of the archiater, chief physician of their empire. The sick and the injured were treated in offices or homes of physicians or infirmaries.

As early as 40 BC the Romans recognized dementia praecox (imbecility, idiotism) as «being out of one's mind».

Beginning with the reign of **Augustus (27 BC – 14 AD)** — founder of the Roman Principate (30 BC – 284 AD) and the first Roman emperor (27 BC – 14 AD), wounded soldiers were treated in the military valetudinarian. Public physicians were appointed to attend the poor and modern-in-design hospitals were built as evidenced by one excavated near city of Düsseldorf (settled in the 7th and 8th centuries, first written mention dates to 1135), Land Nordrein-Westfalen (area 34,110.26 km²), Germany.

The Roman physicians counted the pulse rate using a float-regulated windlass.

The medical encyclopedist **Aulus C. Celsus (c. 25 BC – c. 50 AD)**[53] — Roman medical encyclopedist, known for his extant medical book «De Medicina octo libro» (Rome 1478), a primary source of history of medicine, pathology, diseases, anatomy, diet and pharmacy, surgery and orthopedics, where he described four cardinal signs of an inflammation («*rubor et tumor cum color et dolor*») and hemorrhagic shock as «*When much blood is lost, the pulse becomes feeble, the skin extremely pale, the body covered with a malodorous sweat, the extremities frigid, and death occurs speedily*». To treat pleural effusion (exudate) or empyema thoracis, he described the performance of segmental rib resection that included the use of a trocar and then metal cannula for drainage (15 AD)[54]. He also described psoriasis, a chronic skin disease characterized by dry red patches covered by scales which occurs especially on the scalp, ears, genitalia, and the skin over bony prominence.

Heron of Alexandria (10 AD – 70 AD) — Greek mathematician, and engineer, invented a syringe-like device to control delivery of air or liquid. He also noted that air partially filled a closed tube with one end in a container of water will expand or contract depending on the position of the water / air interface and move along the tube. This observation led some 500 years later to the construction of the thermometer.

Saint (St) **Luke (?– c. 84)**, the «beloved physician» became the new patron of physicians.

That jaundice is due to the common bile duct (CBD) obstruction was suggested by **Aretaeus of Cappadocia (30 AD – 90 AD)** — Greek physician who practiced in Rome and Alexandria. He also provided accurate description of diabetes mellitus (DM).

There was a House of Surgery (c. 62-79 AD) in in the city of Pompei (founded in the 7th - 6th century BC).

The forefather of obstetrics, **Sorranus Ephesus (98 AD – 138 AD)** — Greek physician from the city of Ephesus / Izmir (built in 10th century BC) in Asia Minor on the coast of the Aegean Sea (700 km x 400 km, surface area 214,000 km²), since 129 BC under reign of the Roman Republic, who practiced in Alexandria and in Rome, described the podalic version of the fetus in the transverse position, maneuvers to prevent laceration of the perineum at childbirth, caesarean section after a mother's death to save the child, ligation of the umbilical cord, eye irrigation in the newborn, neonatal care and embryotomy.

The anatomy of the eye and the course of the 2nd (optic) cranial nerve were outlined by **Rufus of Ephesus (c. 110 AD – 180 AD)** — Greek physician who practiced in Ephesus / Izmir in Asia Minor.

---

[53] Celsus AC. De medica octo libro, Liber I – History of medicine; Libra II – General pathology; Libra III – Specific diseases; Libra IV – Parts of the body; Libra V et VI Pharmacology; Libra VII – Surgery; Libra – VIII Orthopedics. Discovered in Vatican Library by Pope **Nicholaus V (1397 - 1455)** in 1426. Publ. in Rome 1478.

[54] Munnell ER. Thoracic drainage. Ann Thorac Surg 1997;63:1497-1502.

The first medical theories based on scientific experimentation were formulated by **Claudius Galen (c. 129 AD – 199 AD)** — Greek physician and surgeon, who discovered the aqueduct of the midbrain, a narrow canal which connects the 3rd and the 4th ventricles of the brain, and of the cisterna fossae lateralis cerebri – a space between the arachnoidian mater and the lateral fossa of the brain, which contain the cerebrospinal fluid (CSF); depicted the lamina of the encephalon, decussation (crossing) of the pyramidal (motor) fibers in the anterior median fissure of the lower medulla oblongata (conical stem, myelencephalon); documented existence of the cranial 5th (trigeminal) nerve, and the cranial 10th (vagus) nerve with the motor, sensory and mixed function, and its branches – the recurrent laryngeal nerves (RLN) of the 6th pharyngeal arch, all of seven described by him cranial nerves.

Described the congenital anomaly – cervical (C_8) rib.

Claudius Galen noted that bronchial asthma (Ancient medicine. *Egypt*) is caused by partial or complete stenosis (occlusion) of distal bronchi.

Also showed that the heart had the intrinsic ability to beat after being excised, indicating its independence from control by the brain or spinal cord; named all the vessels connected to the heart arteries, and those connected to the liver, veins; and considered that the interventricular septum (IVS) was riddled with invisible pores which allowed the blood to pass from the right ventricle (RV) to the left ventricle (LV); described the protective function of the pericardium and reported a pericardial effusion in animals; discovered three layers of arteries (intima, media and adventitia); showed that arteries contain blood and not air; considered to be the first to describe and define an aneurysmal lesion of an artery.

He was the first to describe congenital heart disease (CHD) – patent ductus arteriosus (PDA) formed from the left 6th aortic arch during embryonal development, which in the fetal life connects the main pulmonary artery (MPA, PA) to the aortic isthmus at the junction between the distal aortic arch and the proximal descending thoracic aorta (DTA), and of the ligamentum arteriosum – a post-fetal life remnant of the PDA. Failure of the PDA closure after approximately 4 days of the birth occurs in 4% of newborn.

The fetus (fetus) blood circulation begins at the end of the second month of its life when an arterial blood saturated with oxygen (O_2), i.e., oxygenated, with the partial pressure (concentration, tension) of O_2 (PO_2) of 25% to 28% mmHg is carried from placenta via the umbilical vein into the ductus venosus[55]. Of this blood, in human, 30% at 20 weeks and 18% at 36 weeks gets into the inferior vena cava (IVC) and into the right atrium (RA) from which it is deflected through patent foramen ovale (PFO)

---

[55] The ductus venosus, a.k.a., Arantius duct or ductus venosus Arantii, was discovered by **Julius Aranzius (1529/30 - 1589)** — Italian anatomist and surgeon, in 1564.

in the interatrial septum (IAS) into the left atrium (LA) and left ventricle (LV), and into the ascending aorta (AA) from which it is distributed through coronary arteries to the heart muscle (myocardium), then via common carotid arteries (CCA) to the head (and brain) and via subclavian arteries (SCA) to the upper body. Majority of this blood gets into the hepatic capillaries to supply the liver.

Some deoxygenated blood with PO_2 of about 22 mmHg from the superior vena cava (SVC) flow to the RA, right ventricle (RV), and then into the main pulmonary artery (MPA), but because of high pulmonary vascular resistance (PVR), only about 5% to 10% of this blood flows to the lungs, the majority is being shunted through the PDA down to the DTA for distribution to the rest of the body.

After birth ductus venosus closes with the clamping of the umbilical cord (in Latin – funiculus umbilicalis), its remnant is known as ligamentum venosum. The foramen ovale closes after the first few breaths as pressure within the LA rises above that in the RA. The PDA closes with the time range from just after first few breaths to 4 days in 90% to 95% of full-term newborn, and in 80% to 90% of premature newborn at 30-37 weeks gestational age, in response to the loss of placental source of the prostaglandins (PG), increased degradation of the PG in the lungs, and increased arterial PO_2 which stimulates constriction of sensitive muscle cells in the wall of the PDA. The PG-mediated mechanism is most active in the PDA of premature newborn, whereas PO_2 primarily promotes PDA closure in the term newborn and infant, most likely because of differences in cyclooxygenase (COX) isoforms present in the PDA tissue at various stages of gestation[56]. Thus, the PDA of a premature newborn usually respond to the treatment with indomethacin, whereas the PDA of the term newborn and infant do not.

Galen established the respiratory function of the diaphragm, intercostal and accessory muscles.

Made a classical clinical observation of breast cancer and paralysis agitans.

Leonidas of Alexandria (c. 2nd century AD – 3rd century AD) — Greek physician, is thought to be the first to discriminate between hard (fibrous) scirrhous tumor and cancer of the breast, drew attention to the nipple retraction as a sign of malignancy. He described a procedure in which excision of the breast tumor is alternated with a hot iron cautery to control hemorrhage, advised that «*if the scirrhous tumor begin at the edge of the breast and spread in more than half of it, we must try to amputate the breast without cautery*». But he suggested avoiding operation in case of tumor has

---

[56] Coceani F, Ackerley C, Seidlitz E, et al. Function of cyclo-oxygenase-1 and cyclo-oxygenase-2 in the ductus arteriosus from foetal lamb: differential development and change by oxygen and endotoxin. Br J Pharmacol 2001 Jan;132(1):241-51.

taken over the entire breast and adhered to the thorax. After excision of the breast poultices should be applied to the wound to promote healing.

During his reign, Emperor **Severus Alexander (222 AD – 238 AD)** passed a law requiring the training, certification, and control of physicians.

Roman armies were secured by physicians and the other medical personnel. Among them were sucking healers who treated deep wounds sustained in battle and possible «poisonous wound» by direct suction of the wound with success.

G. Galerius V.M.A. (c. 260 AD, Zaječar, Serbia – late Apr. or **early May 311 AD)**[57] — ambitious and skillful soldier, intolerant Roman emperor and persecutor of Christians (305 – 311 AD) fell ill in the spring of 311 AD from a painful gangrenous ulcer of the penis or the scrotum, accompanied by edema, causing an acute progressive necrotizing fasciitis, i.e., a necrotizing soft-tissue infection of the superficial and deep fascia of the genitourinary and perineal area, leading to necrosis of skin and subcutaneous tissues. It was advancing into the pelvis affecting genitourinary tract and the lower abdominal wall, causing bowel evisceration and necrosis. As a result of this, urine, bowel content and blood were seeping outside through open abdominal wound. He became pale, weak, and emaciated. The initial gangrenous ulcer and its subsequent fascial spread were resistant to remedies and repeated cauterizations. Finally, he went into shock and died. The clinical picture of Emperor Galerius' disease was described by **Eusebius of Caesaria (c. 263 – 339 AD)** — Greek historian, and **Lactantius (c. 240 AD – 320 AD)** — Roman historian, becoming known as necrotizing fasciitis of the genitourinary and perineal area.

St **Vitus (c. 290 AD – 303 AD)** from Isle of Sicily (area 25,711 km²) — one of the Fourteen Holy Helpers of the Rome Catholic Church (RCC) against various diseases, who credible cared for the people suffering from diseases, including neurological and psychiatric.

---

[57] Town of Zaječar (known as the site of fortified palace compound from the late 3rd and early 4th century), Serbia (area 77,474 km²). — The capital city of Sofia (first mentioned in 59 BC as Serdica), Bulgaria (area 110,994 km²). — Kousoulis AA, Economopoulos KP. The fatal disease of Emperor Galerius. J Am Coll Surg 2012 Dec;215(6):890-3.

MEDIEVAL MEDICINE

Medieval medicine encompasses the period between the fall of Rome in 476 and the fall of the city of *Constantinople* (founded in the 7th century BC by Greeks under the name of «Byzantion», 330) — the capital of *Byzantium Empire* (area 3,500,000 km² in 555), in 1453. Strongly influenced by Christianity, it resurrected beliefs in supernatural causes of diseases and cures. On the other hand, it brought an emphasis on compassion for the sick, charity, the institution of hospitals and organized medical care within monasteries.

The first hospital was founded c. 330, by Empress St **Helen (c. 246/50 – 330)**, the mother of the Byzantine Emperor **Constantine the Great (c. 272 – 337)**. In 369 St **Basil (330 – 79)** built the hospital in the city of Caesarea Mazaca / Kayseri (founded 3000 BC), Asia Minor, and in 394 another was built by the wealthy matron St **Fabiola (? – 399)**, in Rome, Italy.

The first legendary, non-scientific report of vascular anastomoses goes back to St **Cosmas** and St **Damiano (? Arabia** – martyred by tortures and beheaded by the sword **c. 278, city of Aegea, Syrian province of the Roman Empire)**[58] — twin brothers, physicians, early Christians, patrons of physicians and surgeons, who first tried to re-implant a limb at the level of the right middle thigh or below the left knee using ivory needles and a row flax stitches to suture vessels[59]. Apparently, result was poor since they were not reported it, and the technique never spread out. This operation was depicted by the **Master of Balbases**[60] — Spanish artist (active 1480 – 1500) in his painting «A Verger's Dream: Saints Cosmas and Damian Performing a Miraculous Cure by Transplantation of a Leg Assisted by Angels» (c. 1495. Oil on wood, 169 cm x 133 cm), and by the Masters **Stettener** and **Schnaiter** — German artists in their panting «Legendary Transplantation of Leg by Saints Cosmas and

---

[58] Arabia or Arabian Peninsula (area 3,237.500 km²), a subcontinent of Asia, the largest peninsula in the world. — City-port of Aegea (active since at least 2,000 BC, on the eastern coast of the Mediterranean Sea in Asia Minor), Roman province of Syria (area 185,300 km² from 64 BC to 135 AD, presently city of Yumurtalik, Turkey (area 783,365 km²).

[59] **Jacopo de Fazio**, a.k.a., **Jacobus da Voragine (c. 1230 – 1298)** — Italian chronicler and archbishop of the city-port Genoa (inhabited since the 5th or 4th millennium BC), Italy. — Da Voragine J. Legenda aurea (The Golden Legend). Assigned Latin Ed. c. 1469, Publ. Lyon 1473; Catalan Publ. 1494. — Callow AD. Historical overview of experimental and clinical development of vascular grafts. In, JC Stanley, WE Burkel, SM Lindenauer, et al (Eds). Biologic and Synthetic Vascular Prostheses. Grune & Stratton, New York 1982. P. 11-6.

[60] Los Balbases is a municipality in the province of Burgas (area 14,292 km²), Spain.

Damian, Assisted by Angels» (16th century), the city of Stuttgart (known since the 6th century BC as an important agricultural area, founded c. 950) in Germany.

Patriarch **Nestorius (c. 386 – c. 451)** and his followers (Nestorian Christians) who was expelled in 431 from city of Constantinople, established St Ephraim Hospital in the city of Edessa (founded in the 7th century BC), Mesopotamia, and in 489 Medical School (Academy) and Hospital in the city of Gundeshapur (founded 271), Persia (present-day Iran, area 1,648,195 km^2). These centers represented the fusion of ancient medicine with the model of efficient, new hospitals.

St Benedict Monastery Hospital in the city of Monte Carlo (founded 43 BC), Monaco (area 1,98 km^2), was erected in c. 529 on the site of an ancient temple of Apollo and controlled medical care for 500 years.

The oldest hospital in existence today is the Hôtel-Dieu in the city of Paris (inhabited from around 3rd century BC) — the capital (since 1163) of France, founded in 651 by St **Landry (? – c. 661)**, the bishop (650 – c. 661) of Paris.

Those were, unquestionably, the dark ages for medicine, as they were for other fields of knowledge. Ancient medicine was lost to be replaced by a simplified herbalism and folk medicine. A few rays of light did break through, however.

Alexander Tralles (525 – 605) — Ancient Greek physician, born on the town of Tralles / Aydin (founded c. 7th century BC), Asia Minor, introduced the use of finger pressure in edema and ascites, and the use of palpation for the detection of enlarged spleen.

The Arabists or Arabic physicians reintroduced the knowledge accumulated in ancient medicine.

Most Arabians consider the distinction between the natural small-pox and measles, morbillo (in Italian – «morbillo» meaning «small disease»), English measles, as being that small-pox came from the blood, while measles came from bile, and so the lesions of measles were small and less likely to affect the eyes.

However, **A.B. Muhammed ibn Zakariya al-Razi** or **Rhazes (c. 854 – c. 925)** — Persian physician, distinguishing between smallpox and measles, outlined in his «Treatise on the Smallpox and Measles» (in Latin – «Libro de variola vera et morbilli»), such differences between them: (1) regarding the predisposing factors, bodies (excretes) that are lean, bilious and dry eruptions are more characteristic for measles that to small-pox; (2) in the prodromal (in Greek – «prodromus», meaning «running before») stage inquietude (anxiety) and nausea are more frequent in measles, while pain in the back more peculiar to small-pox; (3) the measles eruptions comes out all at once, whereas the eruptions of small-pox appears gradually; (4) the eruption of measles is less elevated than that in small-pox; (5) measles is more to be dreaded than small-pox except in the eyes; (6) extract from barley is more suitable for measles than for small-pox.

Because of the occurrence and frequency of fatal enteritis as a complication of measles he warned physicians not to give the patient laxatives, especially towards the end of the disease, as diarrhea at this time may end fatally.

Rhazes made a detailed description of a sudden peripheral and the central, unilateral and bilateral paresis or paralysis (palsy) of the muscles of the face (prosoparesis or prosoplegia), idiopatic or caused by the involvement of the cranial 7th (facial) cranial nerve by viral infection (herpes zoster, varicella) or bacterial infection like systemic tick berylliosis (Borrelia burgdorferi sensu lato) or Lyme borreliosis (Lyme disease)[61], stroke, acoustic neuroma, brain tumor (ponto-cerebellar meningioma), sarcoidosis, and myastenia gravis. It is characterized by the facial drooping and curving (laming), inability to control movement in facial muscles, particularly closing the eyelids that leads to drying out of the eye.

The greatest medieval surgeon, **Abu al-Quasim al-Zahrawi (936 – 1013)** – Arab physician and surgeon, published a 30 chapter medical treatise «Kitab al-Tasrif» («The Method of Medicine», completed in 1000), in which he explained the hereditary nature of hemophilia; introduced the method for the reduction of shoulder dislocation by (1) external rotation, (2) adduction, and (3) internal rotation; designed a number of instruments for the treatment or extraction of teeth, inspection of interior of the ear and the urethra, removal of foreign bodies from the throat and the larynx; specialized in curing diseases by cautherization; discovered the dissolving nature of catgut sutures made from the sheep small intestines when his lute's strings, also made from a small intestines of the sheep, were eaten by monkey, and from then used catgut for internal stitching; surgically ligated the temporal arteries for migraine; described a supine position of the parturient women in which the lower extremities fall over the edge of the table) and an ectopic (extrauterine) pregnancy (963), used pincers for the vaginal delivery and removal of the dead fetus.

The most famous works of **Ibn-Sina** or **Avicenna (980 – 1037)** — Persian scientist-encyclopedist (polimath) and physician, is «The Book of Healing» (written 1014-20, published 1027) and «The Cannon of Medicine» (completed and published 1025). His legacy and innovations into medicine include the assumption of an existence of bacteria and viruses; pioneering the techniques of direct auscultation of the chest by the ear and percussion of the abdomen; paying attention that the food products may serve as a medicine; indicating that exercise and cleanness promote the well-being (health); determination that smallpox is an infectious disease and determination of the difference between black death (plague) and cholera; description of leprosy and dissociation of it from elephantiasis; differentiation of pneumonia from pleurisy;

---

[61] From the town of Old Lyme (set off 1665), New London County, State of Connecticut (CT; area14,374 km^2), USA, where Lyme borreliosis was first registered in 1975.

description of ulcer of the stomach; defining signs of meningitis; discovering and describing localization of the muscles of the eye; description of the cranial 7th (facial) nerve palsy; description of necrotizing infection of the genitals and perineum (1025).

Syyid I. al-Jurjani or **Zayn al-Din Gorgane** or **Ismaiz Gorgani (1040/42 – 1136)**[62] — Persian physician, who noted association of goiter with exophthalmos.

Ibn Zuhr or **Avenzoar (1091/94 – 1162)** — Arab Spanish physician and surgeon, the author of the book entitled «Practical Manual of Treatment and Diet», who accurately described the scabies mite which contributed to the scientific advancement of microbiology; made the earliest description of the bezoar stone (bezoar), an indigestible conglomerate of an organic origin, formed into tightly packed hairs or plant's fibers in the form of a ball (size, 5-7 cm), introduced intentionally into the digestive tract, could cause a small bowel obstruction, ulcers, or bleeding; accurately described inflammation of the pericardium (pericarditis), the esophageal and stomach cancer; introduced animal testing as an experimental trial method of surgical procedures before applying them to people; performed the first experimental tracheostomy on a goat.

In general, Arabic physicians underscored the importance of examining urine, understood tuberculosis, described mediastinitis, developed modern chemistry and pharmacy, performed bloodletting, cautery, and cataract couching.

The rise of the School of Salerno (904; Medical School, in the 11th-12th centuries), city of Salerno (founded 194 BC), Sicily (area 25,711 km^2) – the largest island in the Mediterranean Sea and one of the 20 regions of Italy; the University (univ.) of Bologna (1088), city of Bologna (inhabited since the 9th century BC), Italy; the Oxford Univ. (established 1096), city of Oxford (founded 900), England (area 130,279 km^2); the Montpellier School and Univ. in 1160, city of Montpellier (first mentioned in 985); and the Univ. of Paris (1215) and its Medical Faculty (1221), France; Univ. of Padua (1222), city of Padua (founded 1183) and Univ. of Florence (1321), city of Florence (founded 80 BC), Italy; Academy of Cracow (1364), city of Cracow (populated since the Stone Age settlement, founded 965), Poland (area 312,865 km^2); Univ. of Vienna – Universität Wien (established 1365), city of Vienna (continuous habitation since 500 BC, founded 1145), Austria (area 83,879 km^2); and Univ. of Prague (1397), city of Prague (founded c. 886), Czechia (area 78,866 km^2); kept alive rational medicine.

The first European usage of gauze (in Latin – «garza»: above - *Palestine / Israel*) took place in Bologna (1250), Italy.

In ***Ukraine-Rus' / Kyivan Rus'*** (area c. 2,100,000 km^2 in 1000, existed 880-1240) — predecessor of Ukraine, and its capital city of Kyiv (Foot Note 3), the first hospital

---

[62] Al-Jurjani Sl. Zakhireye Khwarazmshahi / Thesaurus of the Shah Kwarazm. Vol. 1-10. Persia 1110-1135 / 1110-1199.

was unveiled by **Olha (910 – 11.VII.969)**[63] — wife of **Ihor Yurykovych (878 – 945)** — Kyiv Prince (912 – 45), after his death – Kyiv Princess (945 – 64), where she care of the patients handed to women. She knew healing properties of many plants and possessed knowledge to use them for the treatment of diseases.

The profession of physician was officially acknowledged, and the establishment of hospitals was regulated by the statute of the Grand Prince **Volodymyr Sviatoslavych the Great (958 – 1015)**[64] from 992. He built hospitals, proposed the tenth part (tithe) of each person earning donate to the poor people, orphans, elderly and feeble, ordered to search and provide them with nutrition home.

The profession of physicians was again officially acknowledged, by the statute the Grand Prince **Yaroslav the Wise (c. 978 – 1054)**[65] from 1056. His daughter, **Anna Yaroslavna (1024, city of Kyiv, Ukraine – 05.IX.c. 1075, city of Etampes, department Essonne, France)**[66] — Princess of Kyiv and Queen of France (1051-66, 1074-75) upon marrying **Henri I (1008 – 60)** — King of France (1051 – 60), built in 1065 in city of Senlis (known since the early Roman Empire 6th century BC, and 987 when the fortified wall was built), department of Oise (area 5.860 km^2), the Royal Abbey of St Vincent of Senlis, which become her residence outside Paris. She possessed the wide knowledge of therapy, which she passed to the descendants, treated her permanent surrounding and poor people.

---

[63] Osadcha-Yanata NT. Ukrainian folks medicine. J Ukr Med Assoc North Am 1954 Dec; 1,2(1):19-26. — Osadcha-Yanata HT. Ukrainian folks medicine. J Ukr Med Assoc North Am 1955 May; 2,1(3):9-15. — Osadcha-Yanata. Ukrainian folks medicine. J Ukr Med Assoc North Am 1955 Dec.; 2,1(3):14-22. — Osadcha-Yanata HT. Ukrainian folks medicine. J Ukr Med Assoc North Am 1956 Dec;3,2(6):8-15. — Osadcha-Yanata. Ukrainian folks medicine. J Ukr Med Assoc North Am 1957 Sep;4,1(74):17-23. — Osadcha-Yanata. To the article by Dr L. Lutsev «Something about the Ukrainian herbarium». J Ukr Med Assoc North Am 1963 July;10,3(29):20-2. — Dzherman MSh. To the 100nd Anniversary since foundation of the National Academy of Science of Ukraine: accepted questions in discussion of the evolution of medical recognition of the Ukrainians (2nd announcement). Practical Physician 2018;7(4):53-70. — Ablitsov VH. «Galaxy Ukraine». The Ukrainian diaspora: prominent persons. Publ. «Kyt», Kyiv 2007. — Synyachenko OV. They glorified themselves and Ukraine: history of domestic therapy. Crimean Therapeutic Journal 2010;2(2):21-8.

[64] Mozharovska T. They healed both princes and beggars. UHMJ «Ahapit» 1996;4:19-24. — Khmilevsky Ya. On what were ill and from what have died some Ukrainian princes. J Ukr Med Assoc North Am 1962 Apr;9,2(25):21-8. — Ablitsov VH. «Galaxy Ukraine». The Ukrainian diaspora: prominent persons. Publ. «Kyt», Kyiv 2007. — Synyachenko OV. They glorified themselves and Ukraine: history of domestic therapy. Crimean Therapeutic Journal 2010;2(2):21-8.

[65] Rokhlin DH. Conclusions of anatomical and radiological studies of the Grand Prince Yaroslav the Wise's skeleton. J Ukr Med Assoc North Am 1960 Apr;7:14-24. — Ablitsov VH. «Galaxy Ukraine». The Ukrainian diaspora: prominent persons. Publ. «Kyt», Kyiv 2007.

[66] City of Etampes (existed in the early 7th century), department Essone (area 1,802 km^2), France. — Filipchak I. Anna Yaroslavna – Queen of France. Publ Vidrodzhennia, Drohobych 1995. / **Ivan Fylypchak (1871 – 1945)** – Ukrainian writer. — Dauxais J. Anne de Kiev. Reine de France. Presse de la Renaissance, Paris 2003. / Ablitsov VH. «Galaxy Ukraine». The Ukrainian diaspora: prominent persons. Publ. «Kyt», Kyiv 2007. — Synyachenko OV. They glorified themselves and Ukraine: history of domestic therapy. Crimean Therapeutic Journal 2010;2(2):21-8. — Delorme Ph. Anne de Kiev: Epouse de Henri Ier. Pygmalion, Paris 2015.

The daughter of Anna Yaroslavna and Henrich I, **Edigna** von **Puch [1053 – 61, city of Paris, France – 26.II.1109, town of Puch, District of Fürstenfeldbruck, State of Bavaria, Germany]** was against decision of her brother **Philip I (c. 1052 – 1108)**[67] — King of France (1060-1108), to force her into a matrimony, and in 1074 escaped from Paris, found a shelter in town of Puch, Bavaria, Germany, where she have taught reading and writing to the local inhabitants and their children, and treated sick people. For those kind of the deeds in 1600 she was proclaimed blessed by the Rome Catholic Church (RCC). Her sacred remains are situated at the RCC chapel-church in Pych where people up to this time are coming to her asking for help to give them strength to fight diseases.

Among the Ukrainian secular physicians were **Heorhiy Amertol (9th century)**[68], who bandaged wounds with a sheep's hair, amputated extremities and trephined the skull using pain killers and sleeping remedies[69].

Blessed **Antoniy (Anthony) Holy of the Cave (983 – 1073)**[70] — «the most wonderful physician», and **Teodoziy (Theodosius) of the Cave (1036 –)**[71] were monk-physicians who founded the Kyiv-Cave Monastery (KCM; 1051), the first school of higher education in Ukraine in which in 1070 was established the hospital for the feeble and sick, a shelter for elderly and homeless, and the printing house. The former personally supervised patients, treated them, gave them «to bite healing herbs».

Medicine was practised by folk-healers (lichtsi, volkuny, viduny, quaks and midwives), by monk-physicians and by secular physicians.

Among the Ukrainian monk physicians were **Ahapit (c. 1030 – 1095)**[72] — Ukrainian physician, considered to be the father of Ukrainian physicians who knew how to select the usual food plants and medical herbs for the healing of his patients;

[67] Town of Puch, District of Fürstenfeldbruck, State of Bavaria (area 70,550.19 km²), Germany. — Wolf JW. Beträge zur deutschen Mythology. Band I. 1852. S. 169-71. — Ablitsov VH. «Galaxy Ukraine». The Ukrainian diaspora: prominent persons. Publ. «Kyt», Kyiv 2007.
[68] Ablitsov VH. «Galaxy Ukraine». The Ukrainian diaspora: prominent persons. Publ. «Kyt», Kyiv 2007.
[69] Olearchyk A. History of anesthesiology in Ukraine. J Ukr Med Assoc North Am 1966;13,3-4 (42-3):19-21,24,32.
[70] Ablitsov VH. «Galaxy Ukraine». The Ukrainian diaspora: prominent persons. Publ. «Kyt», Kyiv 2007. — Synyachenko OV. They glorified themselves and Ukraine: history of domestic therapy. Crimean Therapeutic Journal 2010;2(2):21-8.
[71] Ablitsov VH. «Galaxy Ukraine». The Ukrainian diaspora: prominent persons. Publ. «Kyt», Kyiv 2007. — Synyachenko OV. They glorified themselves and Ukraine: history of domestic therapy. Crimean Therapeutic Journal 2010;2(2):21-8.
[72] Puchkivsky O. Olimpiy ta Ahapit – the first Ukrainian physicians. Zbir med sekcii Ukr nauk t-va, Kyiv 1922;5:2. — Novakovych R, Dudnyk S, Pashkovych S, Stolyanov V. St Ahapit of Pechersk as the first Doctor of Ukraine-Rus'. UHMJ «Ahapit» 1994;1:48-9. — Orlikovsky A. 900 Years of the Ukrainan medicne and pharmacy. J Ukr Med Assoc North Am 1995 Winter;42,1(135):42-4. — Aronov G. Will we come back to the sain doctor?. UHMJ «Ahapit» 1997-98;7-8:19-22. — Ablitsov VH. «Galaxy Ukraine». The Ukrainian diaspora: prominent persons. Publ. «Kyt», Kyiv 2007. — Synyachenko OV. They glorified themselves and Ukraine: history of domestic therapy. Crimean Therapeutic Journal 2010;2(2):21-8.

Damian (1071 – ?)[73] — Ukrainian physician, treated children by praying, with medical herbs, holy water and oil; **Kuzma (the 11ᵗʰ – 12ᵗʰ century)** Ukrainian physician; **Alipiy (Olimpiy) Pechersky (1050 – 1114)**[74] — Ukrainian physician, cured inflammations of the eyelids (blepharitis and stye) and the skin (pyoderma, furunculosis and carbuncle) with a paint that contained zinc oxide in a mixture of about 0,5% iron oxide (III) - FeO_3 (equivalent to calamine used as a lotion or compress) and an antimicrobial «green», a brilliant (diamant) green (tetraethylene-4,4-di-amino-three-phenyl-methan oxalate), a synthetic anilin dye of three-phenyl-methane series (in Latin – Viride nitens) and cured patients with leprosy; **Pymen**[75] — Ukrainian physician.

Princess **Anna (Hanna) Vsevolodivna (1046/67 – 1110/12)**[76], the grand-daughter of Yaroslav the Wise, since 1086 a nun St Andrew Monastery in Kyiv, established in 1090 a secular school for 300 girls where medicine was lectured as one of the subjects.

Among the Ukrainian secular physicians were **Yevrosyniya Polotsky (1001 – 73)**[77], Princess of Chernihiv city (settled since 2ⁿᵈ millennium BC, first chronicle's mention 907), Chernihiv Obl. (area 31,865 km²), Ukraine, who established the SHE at St Sophia Cathedral in Kyiv, where among others natural science was taught; **Danylo Zatonchyk**[78]— Ukrainian physician; **Maryna**[79]— Ukrainian physician; **Petro Syrianyn (Syrian)**[80] — Ukrainian physician, who was able to establish diagnoses based on the nature of the pulse («beating of the artery»); **Ivan Smera (980 – 1015)**[81] — Ukrainian physician; **Zoya Dobrodiya-Yevpraksiya (1108 – 72)**[82] — Ukrainian physician, wrote the medical treatise «Allima» («Ointments») in four parts (Constantinople 1130): the 1ˢᵗ part – describes personal hygiene, the 2ⁿᵈ part – a hygiene for married couples and childbearing, the 3ʳᵈ part – a hygiene of nourishment and diet, the 4ᵗʰ part – diseases of the skin and its treatment with ointments, toothache and its treatment,

---

[73] Ablitsov VH. «Galaxy Ukraine». The Ukrainian diaspora: prominent persons. Publ. «Kyt», Kyiv 2007.

[74] Puchkivs'kyi O. Olimpiy ta Ahapit – First Ukrainian physicians. Zbir med sekciï Ukr nauk t-va, u Kyivi 1922;5:2. — Ablitsov VH. «Galaxy Ukraine». The Ukrainian diaspora: prominent persons. Publ. «Kyt», Kyiv 2007. — Synyachenko OV. They glorified themselves and Ukraine: history of domestic therapy. Crimean Therapeutic Journal 2010;2(2):21-8.

[75] Ablitsov VH. «Galaxy Ukraine». The Ukrainian diaspora: prominent persons. Publ. «Kyt», Kyiv 2007.

[76] Ablitsov VH. «Galaxy Ukraine». The Ukrainian diaspora: prominent persons. Publ. «Kyt», Kyiv 2007. — Synyachenko OV. They glorified themselves and Ukraine: history of domestic therapy. Crimean Therapeutic Journal 2010;2(2):21-8.

[77] Ablitsov VH. «Galaxy Ukraine». The Ukrainian diaspora: prominent persons. Publ. «Kyt», Kyiv 2007.

[78] Ablitsov VH. «Galaxy Ukraine». The Ukrainian diaspora: prominent persons. Publ. «Kyt», Kyiv 2007.

[79] Ablitsov VH. «Galaxy Ukraine». The Ukrainian diaspora: prominent persons. Publ. «Kyt», Kyiv 2007.

[80] Ablitsov VH. «Galaxy Ukraine». The Ukrainian diaspora: prominent persons. Publ. «Kyt», Kyiv 2007.

[81] Ablitsov VH. «Galaxy Ukraine». The Ukrainian diaspora: prominent persons. Publ. «Kyt», Kyiv 2007.

[82] Ablitsov VH. «Galaxy Ukraine». The Ukrainian diaspora: prominent persons. Publ. «Kyt», Kyiv 2007. — Synyachenko OV. They glorified themselves and Ukraine: history of domestic therapy. Crimean Therapeutic Journal 2010;2(2):21-8.

and the 5th part – diseases of the heart and the stomach[83]; **Fevronia (? – 1228)**[84] — Ukrainian physician, who knew peculiarities of herbs and was skilled in the treatment of diseases, including leprosy.

Kyivan physicians were acquainted with scale, itching, calluses (callosities), dermatoses, eye disorders, scurvy, tuberculosis, pleurisy, bronchial asthma, angina pectoris, possible with congestive heart failure (CHF) and deep venous thrombosis (DVT), stroke, jaundice, diarrhea, intestinal worms, neuralgia, rheumatic disease (RD), arthritis, gout (podagra), brain contusion, meningocele, epilepsy, infectious diseases such as typhoid fever, malaria, plague and anthrax, as well as psychiatric diseases.

Treating consisted diet, hygiene, bathing and hydrotherapy, massage, medical herbs, sedatives, narcotics, hypnosis, isolation, maternal and child care were considered. Avicenna recommended a number of the «Rus' medicaments».

From antiquity in Ukraine the willow-tree was considered an unusual tree because of its healing peculiarities that is pain-relieving and anti-inflammatory effects. Its bark contains salicylic acid, the main ingredient of acetosalicylic acid (ASA) / aspirin.

The willow is harvested in April – June, dried in the open air. The broth (decoction) from the willow is prepared by taking a tablespoon (15 ml) of a crumbled (grinded) bark and poured into 500 ml of boiling water, it is used as a 15 ml dose three times a day. The infuse from a willow is prepared by taking one teaspoon (5 ml) of a grinded bark and poured into a glass (250 ml) of a boiling water, until cooled, it is used in a 5 ml dose three times a day. Externally after washing the willow's bark crumbled into powder and in dose of 5 ml the dust is poured into 500 ml of boiling water and remains there after boiling for 8 hours.

The decoction from the willow's leaves was used to cleanse the eyes and wash the hair to get rid of the scurf. The drink from the willow's bark has been a marvelous antipyretic remedy. Nervousness and headaches were alleviated by the broth or infusion from the willow branches. Abdominal pain was released by the tea or decoction with willow's leaves. Diarrhea and intestinal worms were also treated with the willow. The willow extract helped in woman's diseases. Edematous legs, likely due to CHF or DVT, were soaked in the cast-iron pot filled with the brewed branches of the osier – pain and swelling were gone. And the willow was helpful in neuralgia, rheumatism, podagra and baldness.

Externally, a fresh leaves of the willow were applied to callosities. The grinded to powder the willow's bark was used to stop nosebleed, to smear wounds to hold bleeding and to burns which quickly healed.

---

[83] Preserved at Lorenzo de' Medici (Laurentian) Library (opened 1571), Florence, Italy. / **Lorenzo de' Medici (1449 - 92)** — Italian stateman and de facto ruler of the Florence Republic (1469-92).

[84] Ablitsov VH. «Galaxy Ukraine». The Ukrainian diaspora: prominent persons. Publ. «Kyt», Kyiv 2007.

Surgeons (rizalnyky or rukodily) performed bandaging, tooth extractions, cautherization, removal of lymphatic gland tumors, amputation and trepanation of the skull (10th – 11th centuries).

The population of Kyiv prior to the Mongo-Tatarian invasion (1240)[85] was c. 100,000. The streets of the city were paved, water mains were installed and public baths were available.

Knowledge of natural science and medicine was accumulated during Kyivan-Rus' in Byzantine medical books «Shestodniv» by **Joan Ekzark**, a.k.a., **Ivan Bolharsky (9th – 10th century)**[86] — Bulgarian writer and translator, in «Physioloh», et al., and also in the local «Zilnyky» or «Travnyky» («Herbarium»), containing descriptions of healing herbs, baths, treatment of some diseases and bloodletting. Possibly as early as the 11th-13th centuries, Kyivan Rus' herbarium made the transition into medicinal, such as «Likarstvennyky», «Likuvalnyky» or «Lechebnyky», containing both, folk and scientific medicine. Only fragments of «Mefodiïvs'kyy likarstvennyk» have survived to the present.

Bishop **Yefrem II of Pereyaslav (XI century – 1105)**[87] — founded the physician affairs, established in 1089 a hospital in the town of Pereyaslav (founded 911; during made by Russian Communist Fascist of the USSR and NKVD Holodomor-Genocide 1932-33 against the Ukrainian Nation in town and region perished 26,878 person), Kyiv Obl., Ukraine, in 1091 ordered building of the public bath houses, and of churches free of charge hospitals for the poor and wanderers with a permanent division for seriously sick patients.

In the city of Peremyshl (Foot Note 1) – western bastion of Ukraine-Rus, founded with the erection of the fortified «Hight Castle» / «Zamchyshche» on the Mt Znesinnya (elevation 353 m) in the early 8th century (destroyed by the Poles in 1077), and of St Mykola Ukrainian Orthodox Church (UOC) on the Cathedral Square in 868 (destroyed by the Poles in 1460), must have a hospital and hospice to serve its residents and surrounding population. Nevertheless, a chronicle mentioned the hospital and hospice built in 1461, probably on premises of St John the Baptist UOC I and II (dated 1549).

Two hospitals were built in the city of Lviv (Foot Note 2), Lviv Obl. (area 21,833 km²), Ukraine, St Elisabeth Hospital (13th century) and St Stanislav Hospital (1404).

---

[85] The armed forces (AF) of Mongolia (formed 209 BC; area 1,564,116 km²) from Asia, supported by the Tatars, a Turkic ethnic group, native of Tatarstan (area 68,000 km²), Eastern Europe, invaded Kyivan Rus and their ruthless rule continued until 1380. The area of the Mongolian Empire was 24,000,000 km² in 1279.

[86] «Shestodniv» of John Ekzirha is stored in the library of the Rylsky Monastery (founded in the late 10th century), in the south-west extention of the Ryla Mountains (the highest, Mt Musola – 2925 m), at the elevation of 1150 m, 117 km to the South of Sofia, Bulgaria.

[87] Ablitsov VH. «Galaxy Ukraine». The Ukrainian diaspora: prominent persons. Publ. «Kyt», Kyiv 2007.

The whole network of therapeutic institutions existed at the monasteries on the principle of Christian charity and in the span of seven centuries have served the people.

The Crusaders (1096 – 1270), from their contact with the Muslims, brought to Europe their order of hospitals, elaborate pharmacology, and unfortunately contagious diseases.

In the 11th – 12th centuries with the rise of guilds, physicians became allied with apothecaries and surgeons with barbers. University trained physicians treated the elite of society. Confronted with a difficult situation, physicians formed a consilium. The most important diagnostic test was examination of urine, while treatment centered on diet and herbs. The surgeon's domain was bathing, bloodletting, tooth-pulling, care of the wound, ulcerated legs, dislocations and fractures, amputations, draining of abscesses, anal fissurectomy or fistulectomy, cataract couching, prescribing eyeglasses, and on occasion suturing, repair of hernia and removal of bladder stones. Folk-healers, barber surgeons, tooth-pullers, midwives and sorcerers treated the rest of the society.

An early forerunner of the encyclopedia was compendium by **Bartolomaeous Anglicus (1203 – 75)** — English scholar in Paris, entitled «De proprietatimus rerum» (written in 1240), published in 1275 in the capital city Magdeburg (founded 805), Saxony-Anhalt (area 20.451 km²), Germany. Its English translation entitled «On Properties of Thinks» (1398), for the first time mention infectious disease under the term of «anthrax» (Ancient medicine. *India*).

At the Univ. of Bologna, **Guglielmo Salicetti (1210 – 77)** — Italian surgeon, favored the use of a knife at surgery rather than cautery, first accomplished the use of intramaxillary fixation, and wrote a treatise on regional surgical anatomy. **Mondino de Luzzi (1270 – 1326)** — Italian anatomist, physician, and surgeon, performed the best medieval cadaver dissections resulting in another treatise «Anathomia corporis humani» (1316).

The first to correctly describe the pulmonary blood circulation and blood oxygenation in the lung was **Ibn al-Nafis (1213 – 88)**[88] — Arab physician, «the father of circulatory physiology», in his multivoluminous treatise «Al-Shamil fi-Tibb» stating that *«The blood from the right chamber* [RV] *of the heart must flow through vena arterialis* [i.e., the main pulmonary artery or pulmonary artery(MPA, PA)] *to the lungs, spread through its substances, be mingled there with air, pass through arteria venalis* [pulmonary vein (PV)] *to reach the left chamber* [left atrium (LA)] *of the heart and there form the vital spirit...».* He theorized a «premonition» of capillary (a microscopic

---

[88] Haddad SI, Khairallan AA. A forgotten chapter in the history of the circulation of the blood. Ann Surg 1936;104(1):104(1):1-8.

one-cell vessel paved by the squamous epithelium measuring 5-10 μm in diameter) circulation in his assertion that the PVs receives the blood from the MPA, and this being the reason for the existence of perceptible passages between the two.

The capillaries connect arterioles (small <50 μm) which are the most sensitive to metabolic mediators, intermediate (50-80 μm) where myogenic mechanism predominate, and large (80-150 μm) where flow induced dilatation most potently occurs)[89] and venules (diameter 8-100 μm) in almost all organs of the body. He properly described coronary arteries as a nourishing of the myocardium.

The Salerno physicians performed cataract couching, a nasal polypectomy and hemorrhoidectomy. The work that brought the most fame for this school, however, was a Latin poem by **Arnaldo de Villanova (1235 – 1312)** — Catalonian physician, on dietetics and hygiene called «Regimen Sanitaris Salernitarum» (1480).

In 1239 the Crimean Tatars, one of the three groups of ethnic Tatar, remnants of the Tatar-Mongol invasion of 1240 in Ukraine, catapulted corpses of their own soldiers who died of the plague into the besieged Theodosia (founded in the 6th century BC), Crimea, Ukraine, to infect its defenders. Because of the destruction of Kyivan Rus they were able to establish a Crimean Khaganate (existed 1441-1783) with an area on 400,000 km² (1500) in Crimea and Northern Black Sea's Coast, Ukraine.

A significant progress in learning the human anatomy and the techniques of surgical procedures was made by **Guido LaFrancchi** or **Lafranc of Milan (c 1250 – 1306)** — Italian French surgeon, who authored a treatise «Chirurgia Magna» (1296); **Henri de Mondeville (c. 1260 – 1316)** — French surgeon, the author of a treatise «Chirurgia» (published posthumously c. 1475); **Guy de Chauliac (c. 1300 – 1368)** — French physician and surgeon, who wrote in Latin a treatise «Chirurgia Magna» (1363) in which he described the use of bandages, intubation of the trachea, tracheostomy and the ways of surgical suturing; and **John Arderne (1307 – 92)** — English surgeon, who treated both the rich and the poor, including knights, his fees for the rich being as much as possible, but the poor men was remedied of charge[90].

John Arderne used an opium or poppy seed (Papaver sommiferum) which contain approximately 12% of the alkaloid morphine, as hypnotic (soporific) drug and as external anesthetic remedy in his patients during surgery, so they could sleep and feel no pain. Moreover, he used enema (clyster) made of hemlock (Conium maculatum) – a genus of flowering plants, opium, and henbane (Hyoscyamus nigger) for rectal anesthesia. He developed several methods of surgical procedures for knights who due too long amount of time sitting on a horse suffered from anal

---

[89] Jones CJ, Kuo L, Davis MJ, Chillian WM. Regulation of coronary blood flow: coordination of heterogenous control mechanism in vascular microdomains. Cardiovasc Res 1995;29:585-96.

[90] Pilcher JE. Guy de Chauliac and Henri de Monteville. Ann Surg 1895;21(1):84-102. — Agbo SP. Management of hemorrhoids. J Surg Technique and Case Report. 1 Jan 2011;3(3):68-75.

fistula, or fistula-in-ano, and from a pilonidal cyst, sinus, fistula, or abscess. Also, he created in 1376 an ointment for the treatment of arrow wounds.

«The black death» of the bubonic plague of 1347 killed 25 mln or 25% of Europe's population. In the city of Dubrovnik (founded in the 7th century), Croatia (area 56,594 km^2), the authorities imposed a period of 40 days of isolation, and this brought the term «quarantine» into general use.

On August 21, 1403, **Henry IV (1367 – 1413)** — King of England (1399-1413), fought the battle of Shrewsbury against the rebellion led by Sir **Henry Percy (1364 – 1403)** —English noblemen which former won.

During this battle the sixteen-year-old son of Henry IV, Prince Henry, later becoming **Henry V (1386 –**

1422) — King of England (1413 – 22), has been shoot by an arrow that penetrated his face below the left eye and beside the nose, which stuck in his head.

Initially, when few surgeons tried to remove the arrow, the shaft broke, leaving the bodkin point embedded in Prince Henry skull, some 12.5 cm to 15.0 cm deep, narrowly missing the brain stem and surrounding arteries. Then over a period of several days, **John Bradmore (? – 1314)**[91] — English crown surgeon and metalworker, while attending the wounded prince Henry poured into the wound honey with added roses as an antiseptic, progressively enlarged wound using «tents», i.e., shaped pads to keep it open and quickly crafted a tool to «vyse» (screw) into broken arrow shaft to extract the bodkin point. Two threated tongs held in the center of a threaded shaft, which could be inserted into the wound: the shape was not unlike a tapered threaded rod inside a splint cylinder. Once the end of the tongs located within skirt of the arrowhead, the threaded rod was turned to open the tongs within the bodkin socked locking it into place and it, along with the device, could be extracted. The wound was then filled with wine to cleanse it, healed within 20 days, leaving only the scar on the left side of his face.

Continuing the effort of his father, in August 1417 Henry V promoted the use of English as the official language of England.

---

[91] Bradmore J. Philomena gratia (meum librum vocatum Philomea gratia). Manuscript 1403-12. Published by his son-in-law, John Longe. Translated into English in 1446. — Lang SJ. John Bradmore and his book Philomena. Social Hist Med 1992;5:121-30. — Lang SJ. The «Philomena» of John Bradmore and its Middle English Derivative: a Perspective in Late Medevial England. PhD Thesis. Univ St Andrews, St Andrews, Scottland 1998.

THE RENAISSANCE

In the early Renaissance (the 15th-16th centuries), **Serefeddin Sabuncuoglu (1385 – 1468)**[92] — Turkish surgeon, depicted in detail the diagnosis and treatment of fractures of the sternum and the ribs, particularly the reduction of a fractured sternum, and of an pneumothorax (in Greek – «pneuma» meaning air, and «thoracos» – meaning the breast plate or chest) or collapsed lung secondary to a rib fracture, which he found by percussion and auscultation and treated by the method of simple aspiration using a cupping («mihcane») therapy to the skin of the chest wall, as color illustrated in his manuscriptal textbook of surgery entitled «Imperial Surgery» (Amasia, Anatolia 1465).

Marcin from Zhuravytsia (Zurawica), Martinus Rex de Peremislia (Foot Notes 1, between 87 & 88), **Ucraina (1422, village Zhuravytsia, Peremyshl County, Ucraina – 1460)**[93] — Ukrainian mathematician, astronomer and physician. After graduation in 1400 from the high school in city of Peremyshl (Medevial medicine. *Ukraine-Rus*), he studied in the Cracow Academy (1440-45), Poland, where he received Master of Free Arts Degree of Medicine at the University of Bologna (1448-49), Italy, where he defended dissertation for Medicine Doctor degree, and simultaneously lectured astronomy. After returning to Cracow, he practiced medicine, was nicknamed the «King of medicine», and now as the recent professor of his «Alma Mater» founded for her chair of astronomy (1450), later chair of astrology (after 1450).

Yuriy M. Drohobych, born as YuriyM. Donat-Kotemak (1450, city of Drohobych, Lviv Obl., Ukraine – 04.II.1494, city of Cracow, Poland)[94] — Ukrainian

---

[92] Kaya SO, Karatepe M, Tok T, et al. Were pneumothorax and its management known in 15th century Anatolia? Tex Heart Inst J 2009;31(2)152-3.

[93] Village Zhuravytsia (arised 14th century), Peremyshl County, Ukraine / Poland. — Zwiercan M. Marcin z Zurawicy zwany Król z Przemyśla, de Polonia (ok. 1422 – 1453). Polski Słownik Biograficzny. Zakład Narodowy imienia Ossolińskich, Wyd. Polskiej Akademii Nauk, Wrocław, Warszawa, Kraków, Gdańsk 1974;19,4(83):580-1. — Ablitsov VH. «Galaxy Ukraine». The Ukrainian diaspora: prominent persons. Publ. «Kyt», Kyiv 2007. — Synyachenko OV. They glorified themselves and Ukraine: history of domestic therapy. Crimean Therapeutic Journal 2010;2(2):21-8.

[94] City of Drohobych (founded 1081), Drohobych Agglomerate (area 45.5 км²), Lviv Obl., Ukraine. – City of Cracow, Poland). — Drohobich G. Iudicum prognosticon. Impressum Romae etc. Anno domini MCCCCLXXXIII. — Olearczyk A. Ukrainian – professor of three universities in the 15th century. Nasze Słowo (Warszawa), 11 May 1958; 3,19 (91):2. — Ablitsov VH. «Galaxy» Ukraine. The Ukrainian diaspora: prominent persons. Publ. «Kyt», Kyiv 2007. — Hanitkevych YaV. The contribution of the Ukrainian physicians to the world medicine. Ukr Med Chasopys 2009 VII/VIII;4(72):110-15. — Synyachenko OV. They glorified themselves and Ukraine: history of domestic therapy. Crimean Therapeutic Journal 2010;2(2):21-8.

philosopher, physician, and astronomer [Doctor of Philosophy (1476) and of Medicine (1482), University of Bologna; wrote «Iudicum prognosticon», Rome 1476] which deals with astronomy, geography and infectious disease (plague).

The voyages of **Christopher Columbus's (1451 – 1506)** — Italian merchant and navigator, to America (1492-1504) caused Genocide of the native American Indians and brought epidemics of syphilis to Europe.

Studying this new malady, **Girolamo Fracastoro of Verone (1476/8 or 1483 – 1553)** — Italian physician, in his «Syphilis sine Morbus Gallicus» (1530) and «Contagione» (1546), initiated a modern theory of infection by invisible germs.

It is important to note that the drawings of **Leonardo da Vinci (1452 – 1519)** — Italian polymath and anatomist, of the skeletal, muscular, cardiovascular, and nervous systems had a great impact on the development of anatomy and medicine. He portrayed, in a very accurate manner, the cardiac valves, muscle and coronary vasculature, showed that no air entered the heart from the lungs by way of pulmonary veins (PV's), stated that *«The heart…is a vessel made of thick muscle, kept alive and nourished by arteries and veins as the other muscles are…The heart is a muscle of preeminent power over other muscles»*.

Beside this, he diagnosed and sketched congenital heart disease (CHD) – dextrocardia.

To the risk of atheromasias (AM), atherosclerosis (AS) and the premature death he added xanthoma and lipoma that are seen on the portrait of «Mona Lisa» (1503-06), a woman 25-30-year-old, correspondingly, on the median end of her left upper eyelid and under the skin on the back of her left hand. She died prematurily when she was 37-year-old.

Increased interest in botanic and phytotherapy, i.e., healing of diseases by medical plants was expressed in the edited by **Peter Schöffer (?1452 – 1502)** — German printer, Latin «Herbarius», (Moguncia / Mainz, Germany 1484), the German «Herbarius» (Moguncia / Mainz 1485), the German «Gart der Gesundheit» (Augsburg, Germany 1486) and by **Jacob** von **Meydenbach** of the Latin «Ortus sanitalis de herbis et planti de animalibus» (Moguncia / Mainz, 1485).

Alessandro Achillini (1463 – 1512)[95] — Italian physician, described the submandibular salivary duct (5 cm long) opening into the mouth at the site of frenum lingue (1500) and two tympanal bones of the ear – malleus and incus, showed that the tarsus (middle part of the foot) consists of seven bones (1503), and re-discovered the fornix and infundibulum of the brain.

---

[95] Achillini A. De humani corporis anatomia. Venice 1516-24. — Achillini A. Anatomica anotationes. Impressae per Heronymus de Benedictis, Bologna 1520.

The founding fathers of botany, **Otto Brunfels (1488 – 1534)** — German botanist and theologian, authored «Herbarium vivae icones» (Teil 1-3 or 3Bde, 1530-36) and «Contrafayt Kräuterbuch» / «The Book of Different Herbs » (3Bde, 1532-1537), in which he included various woodcuts and illustrated the plants (herbs) in bright colored painting to help botanists across the world to identify the plants with similar medical properties; **Hieronymus Bock (1498 – 1534)** — German botanist and physician, published «Kneutterbuch» («Plant Book», 1539); and **Leonhart Fuchs (1501 – 66)** — German botanist and physician, wrote «Errata recentiorum medicorum» («Errors of recent doctors», The Hague 1530) and «De historia stirpium commentarii insignes» («Notable commentaries on the history of plants», Basel 1542). The latter was crowned with the name of two plants – «Fuchsia triphylla» (1703) and the purplish-red shrub's flower called «Fuchsia».

The allogeneic (in Greek – «different», «other») skin transplantation in which donor and recipient individuals are of the same species (human-to-human), but genetically and immunologically differs one from the other, was first described in 1503 in the manuscript of Branca, Sicily.

Stefan Falimyr / Stefan Falimierz (15th – 16th century)[96] — Ukrainian botanist and physician, made the first credible written evidence (testimony) about surface (outwards) overflow of the coal oil (naphtha, petroleum) in the Sub-Carpathian Ukraine. He prepared herbarium in Polish entitled «O ziołach i o mocy ich» (Drukarnia E. Englera, Kraków 1534), re-edited into Ukrainian (Printing by E. Engler, Cracow 1588) which contains descriptions and sketches of over 500 medicinal plants, their identifications, reservations, application in the prevention and the treatment of diseases. He asserted that vodka (in Ukrainian – «horilka») could serve «to increase fertility and awaken lust»[97]. He also refers to the naphtha industry in Halychyna (from the 13th century), advised about the preservation of naphtha and pointed to its healing virtues.

Jacobus Sylvius / Jaques Dubois (1478 – 1555) — French anatomist, who was the first to describe venous valves, pterygoid process, the sphenoid bone, the sphenoid sinus in adult, the bone clinoideo tear; re-described after **Clausius Galen (c. 129 – 199 BC**: Ancient medicine. *Romans*) the aqueduct of the midbrain (mesencephalon), a narrow canal which connects the 3rd and the 4th ventricles of

---

[96] Falimierz S. On Herbs and Their Potency, Printing by E. Engler, Cracow 1534. — Falymyr S. On Herbs and Their Potency, Printing by E. Engler, Cracow 1588. — Mykulych O. The Naphtha Industry in the Eastern Halychyna until the Middle of XIX Century. 2nd Ed, expanded. Publ. «Kolo», Drohobych 2004. — Ablitsov VH. «Galaxy Ukraine». The Ukrainian diaspora: prominent persons. Publ. «Kyt», Kyiv 2007.

[97] Vodka (means «water») or horilka (means «burns»), appeared in Ukraine between 950 and 1100, and is quite similar to horilka we drink today. Initially known as bread wine, it contained c. 40% of alcohol till the mid-18th century.

the brain (aqueduct of Sylvius), and of the cisterna fossae lateralis cerebri – a space between the arachnoidea mater and the lateral fossa of the brain (cisterna et fossae Sylvii).

Jean Fernal (1497 – 1588) —French physician, in his «Universa Medicina» (1554), divided medicine, for the first time, into physiology, pathology and therapeutics. He recognized that the aneurysm of the thoracic and abdominal aortic branches could be found in both thoracic and abdominal cavities where «a violent throbbing is frequently observable», credited with ascribing the arterial aneurysms to syphilis.

Metals were introduced into pharmacology by **Theophratus Paracelsus (1493 – 1541)** — Swiss German botanist and physician, who taught that reliance on one's own observation should take preference over that of ancient authorities. He promoted the development of iatrochemistry in the years between 1525 and 1660.

Bartholemeus Eustachius or **Bartolemeo Eustachio (1500** or **1514 – 1574)**[98] — Italian anatomist, discovered the tube connecting the middle ear cavity to the throat (the Eustachian tube), the valve of the inferior vena cava (IVC) or valvula venae cavae inferioris, named the Eustachian valve, and the suprarenal gland (glandulae renibus incumbents, 1563-64).

While still a medical student and later, **Michel de Nostradame** or **Michael Nostradamus (1505 – 66)** — French physician of Jewish origin, the author of «Les Propheties» (1555), showed the great courage and skill in caring for victims of plague in France (1525 and 1545). He may have cured plaque patients by encouraging drinking clean water (hydration), healthy diet, taking his «rose pill» made of rose hips and rose petals containing vitamin C, cypress sawdust, sweet calamus, cloves, removal of soiled bed lines, elimination of garbage, removal of infected corpses from populated area, and isolation of victims.

On the account of **Ambrose Paré (1509 – 90)** — French surgeon, the treatment of wounds by cauterization and with boiling water was replaced by the surgical debridement, a drainage and bandaging (1537). He experienced a fracture of both tibial and fibular bone four finger breaths above the ankle (Pare fracture, 1551), described a transcervical fracture of the femur (1552), reintroduced ligature on both sides of a bleeding vessel, mainly artery, for the control of hemorrhage (1557), first described two cases of post-traumatic diaphragmatic hernia (1579), and developed articulated artificial prostheses for extremities. Of particular interest is the fact that he sewed wounds through pieces of cloth, glued to the patient's skin, instead of stitching through the skin only. This practice may have initiated the future use of the pledged sutures in cardiovascular surgery, widely used today.

---

[98] Eustachii B. Tabulae anatomicae. Edited & published by G. Lancisi, Rome 1714.

Scarlet fever (in Latin – «scarlatium», meaning «bright-red») is an infection disease, commonly affecting children between five and 15 years of age, transmissible by close contact via respiratory droplets such as saliva or nasal discharge, characterized by fatigue, sore throat, headaches, fever, bright-red skin rash which by feeling remind the sand paper, initially localized at the flexion of the elbow joint, later spreads into the arms and legs. The tongue at the beginning is covered by white coating and after its shedding the bumpy bright-red papillae dominate the appearance. Lymph nodes (LN) became swollen. Complications include rheumatic fever, glomerulonephritis, and arthritis.

The description of scarlet appeared in 1553 by **Giovanni F. Ingrassia** or **Ioannis Ph. Ingrassiae (1510 – 80)**[99] — Italian anatomist and physician, in his book «Tumors Against Nature» where he refered it as «rossalia» and made a point to distinguish it from measles [Medieval Medicine – **Rhazes (c. 854 – c. 925)**].

One of the earliest discoverers of the blood circulation was **João Rodrigues de Castelo Branco / Amato Lusitano / Lusitanus Amatus (1511 – 68)** — Portuguese Jewish physician, who during postmortem dissection in 1547 p. found, demonstrated in front of the audience and published the existence of the venous valves inside the lumen of the azygos or unpaired vein (in Greek – «zyg» = pair, and «a» = no) .

The azygos vein is formed by the confluence of the ascending lumbar vein with the subcostal veins at the level of the 12th thoracic vertebral, ascent on the right side of the posterior mediastinum, arches over the right main stem bronchus (RMSB) posteriorly to join the superior vena cava (SVC). In 1%-2% of population the «arch of the azygos vein» is displaced laterally, thus creating a pleural septum separating part of the lung tissue from the rest of the right upper lobe (RUL) and forms the «azygos lobe» of the lung. The left-sided tributaries of the azygos vein include the hemiazygos and accessory veins, formed by the bronchial, pericardial, and posterior intercostal veins, which communicate with the vertebral venous plexus.

The father of anatomy **Andreas Vesalius (31 Dec. 1514 – 15, city of Brussels, Belgium** – as a result of a shipwreck on the Ionian Sea died **Oct. 15, 1564,** after being thrown out with the ship remains on the shore of **Zakynthos Island, Greece)**[100] — Flemish / Belgian anatomist and physician, in his classical anatomical treatise of the human body entitled «De humanis corporis fabrica libri sentem» (Basilae 1543) carefully described the total human body, including cervical (C_8) rib, pericardium and

---

[99] Ingrassiae Io Ph. De Tumoribus praeter naturam. Tomus primus. Napoli 1553.

[100] City of Brussels («settled on marshes», founded in 7th century, area 162 km²) – capital of Belgium (area 30, 689 km²) which in 862-1795 included Flanders (in Netherlander – Vlaanderen, area 13,624 km²), i.e., West (area 3,146 km²) and East (area 2,982 km²) parts. — The Island Zakynthos (area 405.55 km²), one of the Ionian «Seven Islands» (area 2,306.94 km²) on the Ionian Sea, an elongated bay with the deepest point (5,109 m) of the Mediterranean Sea, Greece.

coronary vessels coursing under and alongside the epicardial surface of the heart, as well as, applied by him intermittent positive pressure breathing or ventilation (IPPB, IPPV) and cardiac arrest due to ventricular fibrillation (VF) in a dog. He was the first to recognize and describe aneurysms of the thoracic and abdominal aorta.

The ovarian tubes were discovered by **Gabriello Falloppio (1523 – 62)** — Italian anatomist, and physician, and named after him the Falloppio tubes.

Local anesthesia in the form of refrigeration anesthesia or cryoanesthesia was described in 1595 by **Johannes Costaeus (de Costa)** or **Giovanni Costeo** or **Ioannis Costaei (1528 – 1603)**[101] — Italian botanist, chemist, and physician, as the local use on the skin of cold water, snow, or ice in relieving the pain of minor surgical procedures.

At the age 19, as first year medical student, **Guisio (Julius) C. Aranzi**, or **Arantius (1529/30 – 1589)**[102] — Italian anatomist and surgeon, described the elevator muscle of the upper eyelid (in Latin – musculus levator palpebrae superioris) and coracobrachialis (in Greek – «coracoid» meaning «raven» or «like raven's beak») muscle (1543). Subsequently, he described a rhomboid fossa, especially of the lower end of the ventricle of the brain and heart (Aranzi ventricle) – a curved elevation of the gray matter at the bottom of the temporal cornua (horn) of the lateral ventricles of the brain named the hippocampus feet (in Latin – pedes hippocampus), for which he coined the term «hippocampus», the choroid plexus (in Latin – plexus choroideus ventriculi lateralis, tertii et quarti) that produces the cerebrospinal fluid (CSF) and the 4th ventricle under the name of cisterna of the cerebellum

He noted small nodules in the free central border of the aortic valve (AV) – Arantius nodule.

After **Claudius Galen (c. 129 – 200**: Ancient medicine. *The Romans*), he re-described the arterial duct (in Latin – ductus arteriosus) and re-described the ductus venosus (Arantii) – a major blood channel during an embryonic life passing through the liver from the left umbilical vein to the inferior vena cava (IVC), which carries oxygenated blood from the placenta to a fetal circulation and closes after birth as pulmonary circulation is established and as the vessels in the umbilical cord collapse and became occluded.

He maintained the idea of impermeability of the dividing walls of the heart – interatrial and interventricular septage (IAS, IVS), corroborated with **Reonaldo Colombo (c. 1515 – 1559)** — Italian anatomist and surgeon, on the course of the

---

[101] Costaei I. De igneis medicinae praesidiis libri duo. Venecig Bologna 1595. / Davison MHA. The evolution of anaesthesia. Brit J Anaesth 1952;31:134-7.

[102] Arantius G.C. De humano foetu liber. Rome 1564.

blood flow from the right to the left side of the heart, described that blood of the mother and fetus remain separated during pregnancy.

Guisio Aranzi, and later **Jean (Geoges) L.N.F. Dagober** baron **Cuvier (1769 – 1832)** — French naturalist, zoologist, and paleontologist, described paired blood vessels of an embryo's liver (the common cardinal veins) which connects the umbilical vein with the right and the left (disappears) superior vena cava (SVC) – the venous canal or duct of Aranzi-Cuvier.

After **Claudius Galen (c. 129 – 200**; Ancient medicine. *Romans*), **Leonardo Botallo (1518/1530 – 1587/1600)**[103] — Italian anatomist and physician, re-described the patent ductus arteriosus (PDA) – the Botallo duct, described the foramen ovale cordis – an opening in the septum secundum of the fetus heart which secures connection between the atria (Botallo foramen) and re-described the ligamentum arteriosum (Botallo ligament).

After disintegration of the Roman Empire and decline of the world culture came a long break in the development of plastic and reconstructive surgery. Only with appearance of **Gaspare Taliacozzi (1545 – 99)**[104] — Italian plastic surgeon, who described reconstruction of the nose, auricula and lips with the use of the distant skin and the arm's biceps muscle flaps – the «Italian method» of plastic surgery and reconstructive surgery (1597), began the renaissance of the plastic and reconstructive surgery. In the «Italian method» of plastic and reconstructive surgery one uses flaps from the remote parts of the body, choosing for its formation such portion of the body, where it is possible to place a matching flap, which carry out closure of the defect at once.

Accurate drawings of lice, fleas and mites were made by **Thomas Muffet (1553 – 1604**) — French naturalist and physician, in 1589 who noted that mites cause scabies and recommended treatment with sulfur. He also discovered an infectious disease which infests silkworms.

The invention of the flush toilet by **John Harrington (1561 – 1612)** — English writer and inventor, facilitated the flushing away of human waste, but the flow from these indoor privies ran into cesspools, and ultimately into the waterway and wells.

The «father of modern physics», **Galileo Galilei (1564 – 1642)** — Italian polymath (mathematician, physicist, astronomer, engineer, and philosopher), after discovering

---

[103] Botallo L. De curandis vulneribus sclopettorum. Lyons 1560, 1564, 1566, 1575, 1583. — Botallo L. De foramine ovalis dissertatio. Lyons 1561. — Botallo L. Luis venerae curandaeratione. Paris 1563. — Botallo L. De catarrho commentarius. Paris 1565. — Botallo L. Commentariola duo, alter de medici, alter de aegroti munera. Lyons 1565. — Botallo L. De incidentae vanae, cuts scarificandae et hirudinum applicandarum. Lyons 1565.

[104] Taliacozzi G. De curtorum chirurgia per isitionem. Venezia 1597. — Bihuniak VV. Plastic surgery. In, LYa Kovalchuk, VF Sayenko, HV Knyshov. Klinichna Khirurhiya. Vol. 1-2. Publ. «Ukrmedknyha», Ternopil 2000. Vol. 2. P. 359-90.

in 1583, the isochronism of pendular objects, designed the measuring device for the pulse rate (pulsilogium, or pulsemeter) which consisted of a dial with a pointer and a pendulum attached. The string of the pendulum could be wound on a wheel behind the dial and when the pendulum swung in synchrony with the patient's pulse the pointer on the dial indicated the pulse rate. He invented an early thermometer, using the expansion and contraction of air in a bulb to move water in an attached tube (c. 1593) and wrote on the use of a compound microscope (1624-25).

The University of Köningsberg – Albertus Universität Köningsberg was established in 1544, the city of Köningsberg (founded 1255), East Prussia (area 36,993 km^2).

The Apothecary Order, introduced in 1581, became the highest medical board of the Tsardom of Russia[105] until 1716.

---

[105] Succeeding from Kyivan Rus, the Vladimir-Suzdal Principality created Vladimir-Suzdal Rus' (1168-1389), with the capital city of Vladimir (inhabitated by humans from 23,000 BC, founded 1108) and with the city of Suzdal (known from 1024), the area of approximately 29,000 km^2. / Grand Duchy of Moscow (1289-1547) with the capital city of Moscow (founded before 1147). / Tsardom of Russia (1547-1721). / Russian Empire (1721-1917) / RSFSR (1917-22) / USSR (1922-91) / Russia or Russian Federation (RF, 1991-present, 17,075,400 km^2).

THE 17ᵀᴴ CENTURY

This century of scientific revolution brought about a shift from speculation to experimentation[106]. Its medical achievements were marked by the discovery of the bursa (B) in birds that direct B cells humoral immunity (1621), of blood circulation (1628), by the first human transfusion (1667) and by the discovery of the microscope (1676). The first empirical society was Academia del Cimento in Florence, Italy, which published its work in 1667.

In his «Tabulae pictae» (1600) **Hieronymus Fabricius** or **Girolamo Fabrizio** or **Fabricus ab Aquapendente (May 20, 1537, Aquapendente, Province of Viterbo, Region Lazio, Italy – May 21, 1619, Padua**, Ibid)[107,108] — Italian anatomist and surgeon, described the lateral sulcus as dividing line between both the frontal lobe and the parietal lobe above from the temporal lobe below in both hemispheres of the brain which is larger in the left hemisphere in most people; described the membranous folds that he called «valves» in the interior of the veins, which prevents retrograde blood flow within the veins, thus facilitating antegrade flow of the blood towards the heart (1603); discovered the bursa, epithelial and lymphoid organs responsible for hematopoiesis in birds («De formatione ovi et pulli». Padua 1621, posthumous) which was named after him, the bursa of Fabricus or bursa cells (B-cells). The bursa of Fabricus is an equivalent in mammals to the bone marrow which produce B-cells lymphocytes that govern the human humoral immune system, as opposed to the thymus gland which produce thymic cells (T-cells) that govern the human cell-mediated immune response.

---

[106] Akhtemiichuk YuT, Pishak VP, Tsvihun AO. Anatomy of the 17ᵗʰ century in figures. Clinical Anatomy and Operative Surgery (Chernivtsi, Ukraine) 2003;2(1):1-4.

[107] City of Aquapendente (settled by Etruscans, the first document dates from the 9ᵗʰ century), Prov. Viterbo, Region Lazio (area 17,242 km²), Italy.

[108] Zanchin G, Decaro R. The nervous system in colours: the tabulae pictae of G.F. d'Aquapendente (ca.1533-1619). J Headche Pain 2006;7(5):360-6.

Guilelmus F. Hildanus, or **Wilhelm Fabry (1560 – 1634)**[109] — German surgeon, who explained diverticulum of the ileum, described three degrees of the skin burn[110]:

1st degree – superficial burn which involves only the epidermis;
2nd degree – intermediary burn which involves the papillary layer of the skin, or both papillary and reticular layers of the skin;
3rd degree – deep burn which penetrate through all layers of the skin, and the flesh, veins, the arteries and nerves forming eschar, and when this falls off a deep ulcer remains.

Then, he introduced the rational treatment of the skin burn, which included general measurers such as a good diet, purging, and blood-letting.

The local treatment consisted of the use of either (1) cold or moist juices, waters of house-leek, and lettuce, or (2) an external heat applied in the form of hot medicines which acted by scattering the sharp humors vital fluids «*in the matter that coals are quenched when scattered*», as mentioned in Ancient medicine (*Greece / Rome*) by **Hippocrates II (460 – 377 BC) / Claudius Galen (c 129 – 199 AD)**, by onion-salve mixed with a little salt applied to all burn except eyes.

In the 1st degree burns the aim was to prevent vesicle formation by moistening with water or saliva. Localized burns were anointed with onion-salve mixed with little salt. Generalized or a whole body's burns were treated, except eyes, with ointment containing Saponia liquida, Caepe cruda, Oleunde et tellisonorum, Amigdalar and Mucilage.

Burn of the face caused by a gun powder were treated by removing any grains sticking to the skin with a needle, powder in the eyes was washed out using a mother's milk or rose water. The skin was anointed with sweet almonds.

In the 2nd degree burns all vesicles was to be open with scissors and separated epidermis removed. The protective plaster with complicated plant formulas and the part of them wrapped in rollers moistened in water and vinegar was efficacious.

The treatment of the 3rd degree burns consisted in opening the vesicles and drying them with a sponge, or linen. The eschar was loosened gently each day, doubled cloths soaked in emulsion containing Medullae sem. cucurbita & Sem.

---

[109] Hildamus GF. Observationum et curationum chirurgicarum centuriae (XXV). Oppenheim, Basle - Geneva - Frankfurt - Lyons 1598 - 1641. — Jones E. The life and work of Guilhelmus Fabricius Hildanus (1560-1634). Part I. Med Hist 1960 Apr;4(2):112-34 / Part II Ibid. Med Hist 1960 July;4(3):196-209.

[110] **The nowadays (2020) four degree classification of burn of the skin**: 1st degree – superficial burn involve only the epidermis; 2th degree – intermediary burn involve the papillary layer of the skin, or deep burn involve both, the papillary and reticular layers of the skin; 3th degree – deep burn penetrate through all layers of the skin; 4th degree – very deep burn pass through all the skin, subcutaneous tissue, invade the muscles, and bone.

cidon. were applied hot. The joint burns and hardness of the scars were soothed and softened with the ointment containing the hen's and bear's grease, dark oil of thyme, lily oil, egg yolks, myrrh oil, and juice of earthworms. Afterwards a thin sliver of lead dipped in mercury was applied.

The prevention of contractures following burns of eyebrows, lips and fingers was enhanced by rolling the part back over a soft sponge, linen clots, lint or thin plates of lead were to be placed between those parts where conglutination might occur, e.g., lips and fingers.

Contractions of joints were prevented by use of emollients and physical therapy (PT). Treatment of established contractions was by making the part supple with plasters of mucilage, hysope (Hyssopus officinalis) or melilot, and the use of extension apparatus.

His wife, **Marie Colinet (c. 1560 – c. 1640)** — Swiss midwife-surgeon, introduced the use of heat for dilating and stimulating the uterus during labor; improved the technique Caesarean section delivery which she first successfully performed in 1603 and which has not been changed since the times of **G. Julius Caesar (100 BC – 44 BC)** — Roman state and political figure, military general and writer; used in 1624 a magnet to extract a piece of metal from patient's eye.

Obstetric forceps were invented in 1592 by the brothers, **Pierre Chamberlen, Sr (1560 – 1631)** — French barber-surgeon and man-midwife, and **Pierre (Peter) Chamberlen, Jr (1572 – 1626)** — French English surgeon, and obstetrician, but were originally restricted to deliveries of the royal children.

The thermometer was constructed **Santorio Santorio (1561 – 1631)** — Italian physiologist and physician.

Synkt Erazm from Lviv, Sixto Leopolien a6o **Sixtus Erasmus (c. 1570, city of Lviv, Ukraine – c. 1635)**[111] — Ukrainian physician, who researched mineral waters in the village of Shklo (founded 1360), Lviv Obl., Ukraine, a balneologic and muddy health resort, which have hydrogen sulphide calcium-sulphate water for bathing, and hydrocaronate-natrium-calcium water for drinking.

In 1541 **Daniel Sennertus / Daniel Sennert (Nov. 25, 1572, Wrocław, Poland – June 21, 1637, Wüttemberg, Germany)**[112] — German physician, described at autopsy an injury to the left hemidiaphragm with the hernia formation and protrusion of the

---

[111] Erasmus S. De thermis in pago Sklo. Zamonscia 1617. — Erasmus S. Commentarius medicum in L. Annaei Senecae opera. Leopolis 1627. — Ablitsov VH. «Galaxy Ukraine». The Ukrainian diaspora: prominent persons. Publ. «Kyt», Kyiv 2007.

[112] City of Vradislav, Breslau, Wroclaw [first mentioned 142-147, founded in 10th century by **Vradislav I (c. 888 - 921)** — Herzog (from 915 until death) of Bohemia – the most westward and the largest historically region of Czechia], Bohemia, Germany, Poland. — City of Wittenberg (settled 1180), Land Saxony-Anhalt (area 20,447.7 km²), Germany). — Sennertus D. Institutiones medicinae. Wüttemberg 1611 — Sennertus D. Practicae medicinae. Vol. 1-6. Wittenberg 1636. —Sennertus D. Cap. 42 Op Omn Lib 1650;5:306.

stomach after a penetrating trauma to the left chest, and a year later (1542) – a tear in the intima of the aorta, causing an intramural accumulation of the blood (hematoma) and an aortic dissection, forming a false lumen and an occlusion of the true lumen of the aorta and its arterial branches.

Aortic dissection occurs in 3.5 of 100,000-persons per year due to congenital defect of the connective tissue (mutations in the gene encoding fibrillin-1 on chromosome 15, mutation of the gene encoding type III collagen or by mutations in the genes encoding transforming growth factor-beta receptors 1 and 2) and degenerative disease of the media of the aortic wall and by hypertension. Without surgical correction, 40% of the patients die within 30 days from the onset and 95% within one year.

The first detailed description of systemic sclerodema (SSD) was made by **Abraham Zacuth** or **Zacutus Lusitanus** or **Zakuti Lisitani (1575 – 1642)** — Portuguese Jewish and Hollander physician. SSD is an autoimmune disorder of the connective tissue conditioned by diffuse disturbances in microcirculation, fibrosis of the skin and the internal organs (1634-37, posthumous 1649).

«The father of modern physiology», **William Harvey (1578 – 1657)** — English physician, accurately described the blood circulation in animals in his treatise «Anatomical Study on Motion of the Heart and Blood in Animals» (1628)[113], where he attested – «*It has been shown by reason and experiment that blood by the heat of the ventricles flows through the lungs and heart and is pumped into the whole body. There it passes through pores in the flesh into the veins through which it returns… finally coming to the vena cava and right ventricle. This occurs in such an amount, with such an outflow…that it cannot be supplied by the food consumed…It must therefore be concluded that the blood in the animal body moves around in a circle continuously, and that the action or function of the heart is to accomplish this by pumping. This is the only reason for the motion and beating of the heart*».

In this treatise, he was also the first to identify a potential role for the atrium as «*a receptacle and storehouse*», noted that the right atrium (RA) «*was the first to live, and the last to die*». Later, he reported on a case of hemopericardium (1649)[114].

The linea semilunaris was defined by by **Adriaan** van **der Spighel** or **Adriaan Spigelius (1578 – 1625)** — Flemish anatomist. It forms a longitudinal prominence of the lateral margin of the rectus abdominis muscle and extends from the pubic tubercle to the costal margin of the 9th costal cartilage and corresponds to the line of junction of the aponeurosis of the internal and the transverse muscles of the abdomen (spigelian line) This is the site of the rarest of all ventral hernias, called

---

[113] Harvei G. Exercitatio anatomica de motu cordis et sangvinis in animali. London 1628.
[114] Harvei G. Exercitationes anatomicae duae de curculatione sanguinis. London 1649.

the spigelian hernia, but which is known to have a high incidence of the intestinal incarceration.

The first report of a congenital lung defect from 1639 was by **Nicolaus Fontanus**[115] — Netherlander anatomist, in an infant with air-filled cysts in the lung. In this same year (1639) he provided the first description of amyloidosis in a young man with epistaxis, jaundice, and ascites in whom autopsy revealed liver abscess, and large spleen filled with white stones - the «sago spleen», with deposition in the tissues of an abnormal insoluble fibrous glycoproteins (amyloid).

The central tendon of the diaphragm (CTD; in Latin – centrum tendineum diaphragmatic), i.e., speculum helmontii, has been depicted by **Jan** van **Helmont (Johannes) B.** van **Helmont / Helmontii (1580 – 1644)**[116] — Belgian (Flemish) chemist, physiologistm and physician, the site of rare congenital diaphragmatic hernias (CDH).

The CTD, is a thin but strong aponeurosis situated slightly anterior to the vault formed by the muscle, resulted in longer posterior muscle fibers, shaped as a trefoil leaf, consist of three divisions or leaflets separated by slight indentations. The right leaflet is the larger, the middle leaflet extends toward the xiphoid process, and the left leaflet is the smallest.

He used magnetic curing of the wound (1621), introduced the word «gas» (in Greek – meaning «chaos»), perceiving that «gas Sylvester» for carbon dioxide (CO_2) was given off by burning charcoal and produced by fermenting must, introduced the measurement of the urine specific gravity.

His son, **Franciscus M.** van **Helmont (1614 – 98)** — Flemish alchemist and writer, published his father work.

Jean Riolan (1580 – 1657) — French anatomist and physician, described series of arterial mesenteric arches and arcades of the small intestine, between adjacent arteries of the jejunum and ileum, the ileal being shorter and more complex than the jejunal (Riolan's arches and arcades), and the arterial mesenteric arch of the large intestine between the proximal portions of the middle colic artery and the inferior mesenteric artery (IMA) – Riolan arch. The IMA supplies arterial blood flow to the section of digestive tract embryologically derived from the hind gut, i.e., the splenic flexure of the colon to the rectum.

Beside of this, Jean Riolan proposed a drainage of the pericardial sac by trephining the sternum (pericardiotomy) for a pericardial effusion compressing the heart (1649).

---

[115] Fontanus N. Repositionum et curationum medicalium. Typis Ioannis Ioanssonii, 1639.

[116] Van Helmont JB. Ortus medicinae. Publ. by Franciscus M. van Helmont. / Aput Ludovicum Elzevirium, Amsterodami 1648.

Congenital anomaly – situs inversus viscerum was described in 1643 by **Marco A. Severino (1580 – 1656)**[117] — Italian anatomist and surgeon. He also practiced refrigeration anesthesia using freezing mixtures of snow and ice during minor surgical procedures (1646).

The lymphatic system is a part of the blood circulatory system comprising of a network of lymphatic vessels that carry a pre-prandial slight opalescent fluid and a post-prandial milk-like fluid, called lymph (in Latin – «lympha» meaning water)[118].

Structurally, the lymph is close to blood's plasma, although does not contain red blood cells (RBC) / erythrocytes and platelets (thrombocytes), but contain about 5,000 macrocytes, white blood cells (WBC) / leukocytes) that participate in chemical snatching of foreign microorganisms. Also, lymph contains interstitial fluid from tissues, mainly fatty acids, and fats as chyle from digestive tract, WBC to and from lymph nodes to the bones, antigen-presented cells such as dendritic cells to the lymph nodes where cell immune response is generated.

Chylus (грецькою – «chilos» meaning juice) is a whitish milk-alkaline fluid of the lymphatic vessels of the small bowel created during digestion and absorbed by tiny channels in the lining of the bowel after eating fatty meals, and then transported by lymphatic vessels to the blood.

The lymph from lower extremities and the abdominal cavity is directed via the lumbar and intestinal trunks to cisterna / cisterna or receptacle (in Latin – «cisterna» or «receptaculum») chyli, a dilated sack at the upper abdomen behind the abdominal aorta (AA) on the anterior aspect of the 1st and 2nd lumbar vertebra (L_1 and L_2). From there, the collected lymph is moved through the thoracic duct (TD) upward via the aortic hiatus of the diaphragm at the level of the 12th thoracic vertebra (T_{12}) to the right chest between the descending thoracic aorta (DTA) medially, the azygos vein laterally and the esophagus posteriorly, until it crosses medially at the level of the 5th thoracic vertebra (T_5) into the left chest, where it continue to ascend on the left side of the esophagus, crosses aortic arch anteriorly and then moves posterior to the left subclavian artery (SCA), passes out of the thoracic inlet before arching and descending anterior to the medial border of the scalenus anterior muscle and finally joining the confluence of the left subclavian and internal jugular veins (SCV, IJV). The thoracic duct accepts the lymph from the lower extremities and abdominal cavity via the lumbar and intestinal trunks, the left hemithorax and lung, the left side of esophagus and trachea, the left upper extremity, the left side of neck and head. The right lymphatic duct, created from

[117] Severino MA. Synopseos chirurgiae libri sex. E. Weyerstroten, Amsterdam 1664.

[118] Yarema IV. Lymphatic system. In, LYa Kovalchuk, VF Sayenko, HV Knyshov. Klinichna Khirurhiya. Vol. 1-2. Publ. «Ukrmedknyha», Ternopil 2000. Vol.1. P. 372-87.

the right subclavian, jugular and broncho-cephalic trunks, collects the lymph from right side of the head and neck, the right upper extremity, hemithorax, heart and lung, the right half of esophagus and lower trachea, drains into the right venous angle.

Injuries of the TD may result in leakage of the chyle outside (external chylorrhea) or into the pleural cavity (chylothorax).

The first know discovery of the chylous (lacteal) vessels of the lymphatic system is that of **Gaspare Aselli (1581 - 1626)**[119] — Italian physician.

The definition of pruritus (itch, itching) of the skin was introduced by **Samuel Hafenreffer (26.IV.1587 – 26.IX.1660, town of Tübingen, State of Baden-Württemberg, Germany)**[120] — German physician, in 1660. Pruritus is the «unpleasant sensation» produced by different areas of the body's skin that provokes a person to scratch him selves the affected by itch areas of the skin. He is also credited as being the author of the first textbook in German speaking countries on diseases of the skin (dermatology, in Latin – morbus cutaneous, in German – die Hautskrankheit).

For the first time the congenital diaphragmatic hernia (CDH) in adult was observed in the beginning of the 17th century by **Lazare (Lazarus) Rivière (1589 – 1655)**[121] — French physician («Praxis medica», 1653), and in 1754 by **George MacAulay (1716 – 66)** — Scottish physician and man-midwife, in newborn who died from respiratory failure (RF) a little over one hour of birth (Macaulay G. An account of viscera herniation. Phil Trans Roy Phys 1754;6;25-35).

To Lazare Rivière belong the earliest account («Opera medica universe», 1646) of an aortic valve (AV) stenosis and an inflammation of the inner layer of the heart (endocardium) called endocarditis.

Military medicine attained a notable level of efficiency in Zaporizhian Sich (1552 – 1775) in Ukraine, under leadership of **Bohdan Z. Khmelnytsky (27.XII.1595 / 06.I.1596, village Subotiv, Cherkasy Region, Cherkasy Obl., Ukraine – 07.VII.**

---

[119] Aselli G. De lactibus sive lacteis venis. Milan 1627.

[120] Stadt Tübingen (first settled in the 12th millennium BC, the first official record appears in 1191), Staat (Land) Baden-Württemberg (area 37,751.46 km²), Germany. — Rost GA. Samuel Hafenreffer, author of the first textbook on dermatology in German speaking countries. Z Haut Geschlechtskrankheit 1933 Apr 1;14(7):227-30.

[121] Bonetus T. Suffocatione De, Observatio XLI. Suffocatio excitata a tenulum intestorum vulvus diaphramatis, in thoracem ingrestu. In B: Sepulchretum sive anatomia practica et cadareribus morto denatus. Geneva 1679.

(06.VIII).1657, town of Chyhyryn, Cherkasy Obl., Ukraine)[122] — Hetman of Ukraine during the Ukraine's Independence War (1648 – 54), where every regiment of his army had a physician, and each company (sotnia) had a tsyrulic (surgical assistant).

The old, feeble, sick and wounded Cossacks were placed or treated at the Trakhtemyrivsky Monastery's [mentioned by chronicle in 1096 and 1168, destroyed by the Polish szlachta armed forces (AF) in the 1660s], refuge (shelter), or military hospital (built 1576), near the settlement of Zarubyntsi (mentioned also in 1096 and 1168), near the village Trakhtemyriv (settled since 118,000 – 120,000 BC, destroyed by the Polish regular AF in 1637), and the village of Monastyrok (known since 1578), all submerged in 1972 under the Kaniv Reservoir (surface 675 km^2), Kaniv Region, Cherkasy Obl., Ukraine.

When the Polish AF destroyed the Trakhtemyrivsky Monastery, its refuge and military hospital, the old, feeble, sick and wounded Cossacks were placed, or treated in similar facilities at the Mezhyhirsky Monastery (built 988 – 1161, taken down in 1934 – 35 on the order of the Russian Communist-Fascist regime in city of Moscow (first documented in 1147) – capital of Muscovy (1480 – 1721), of RSFSR / USSR (1918/1924 – 1991), and Russia or Russian Federation / RF (1991 – present), village New Petrivtsi (founded 1746), Vyshhorodsky Region, Kyiv Obl., Ukraine.

Rene Descartes (1596 – 1650) — French philosopher, mathematician, and scientist, was an early exponent of iatro-physics and wrote the first textbook on physiology («De Homine», 1666).

Descriptions of the stomach, intestine and liver were made by **Francis Glisson (1597 – 1677)** — English anatomist, physiologist, and physician, including a layer of connective tissue surrounding the liver, enshrouding the hepatic artery, portal vein and bile ducts within the porta hepatis, i.e., transverse fissure of the liver (capsula fibrosa hepatis Glissoni et capsula fibrosa perivascularis hepatis Glissoni, or Glisson's capsule, 1654).

Johann G. Wirsung (1600 – 43) — German anatomist, described the main pancreatic duct (ductus pancreaticus Wirsungi) which together with the common bile duct (CBD) drain into, identified later by **Abraham Vater (1684 – 1751)** — German

---

[122] Village Subotiv (settled in the late Bronze Age, founded 1616, Illinska Church built in 1653, Cherkasy Region, Cherkasy Obl. (area 20,900 km^2), Ukraine. — Town of Chyhyryn (remains of settlements from the Bronze Age, founded in the first half of XVI century; as with all surrounding villages and towns, it was victim of organized by the Russian Communist Fascist regime of man-made famine – Holodomor-Genocide 1932-33, and Great Purges in 1936-38 against the Ukrainian nation), Cherkasy Obl., Ukraine). — Sichynsky V. Medicine in Ukraine during Cossack times 17th-18th century. J Ukr Med Assoc North Am 1955 May;2,1(3):16-9. —Dzherman M.Sh. To the 100nd Anniversary since foundation of the National Academy of Science of Ukraine: accepted questions in discussion of the evolution of medical recognition of the Ukrainians (2nd announcement). Practical Physician 2018;7(4):53-70. — Ablitsov VH. «Galaxy Ukraine». The Ukrainian diaspora: prominent persons. Publ. «Kyt», Kyiv 2007.

anatomist, the ampulla hepato-pancreatica or ampulla of Vateri, located at the descending (2nd) portion of the duodenum. The accessory pancreatic duct (ductus pancreaticus accessorius Santorini)[123], if functional, may open separately into the 2nd portion of duodenum (70%), or join the main pancreatic duct (30%).

Epifan Slavynetsky (c. 1600, city of Kyiv, Ukraine – 19.XI. 1675, city of Moscow, Muscovy)[124] — Ukrainian poet, philosopher and theologist, translated the human anatomy by **Andreas Vesalius (1514-64**; Renaissance) entitled «De humanis corporis fabrica libri sentem» (Basilae 1543) into the shortened Old Ukrainian language version entitled «Textbook of Human Anatomy, Translated from the Latin Book by Andreas Vesalius from Brussels», 1658).

Through a crude microscope **Othanasius Kircher (1602 – 1680)** — German polymath, saw in 1656 that the blood of people who died of the plague harbored animalcule.

Thousands of years have passed since electricity was appropriated by medicine. And, the first step towards that goal was made by **Otto** von **Gericke (1602 – 86)** — German physicist and engineer, who created in 1661 the first electrostatic generator.

Decarbohydrate sodium sulfate, or sulfate soda (Na_2SO_4) was discovered by **Johann R. Glauber (1604 – 70)** — Dutch German chemist and apothecary, in 1625 in Austrian spring water, named mirabilis salt (sal mirabilis) or Glauber's salt, and used as laxatives until the 1900's. Later, in 1659, he fused a mixture of mineral pyrolusite (manganese dioxide, MnO_2) and potassium carbonate (K_2CO_3) to obtain a material soluble in water, giving a green solution (potassium manganate, K_2MnO_4) which slowly shifted to violet and then finally red color, representing the first image of potassium permanganate ($KMnO_4$), a strong oxidizing remedy.

In 1643 **Evangelista Torricelli (1608 – 47)** — Italian mathematician and physicist, designed the manometer based on the vacuum, a space without the air or other gas (Torricellian vacuum) and defined the vena contracta as an area in a fluid stream issuing out of an orifice, exempli gratia (e.g.), normal or stenotic cardiac valve, where its contraction (diameter) is the smaller at the section slightly downstream of the orifice, where the jet is approximately horizontal and fluid velocity is maximal. The coefficient of contraction is the ratio between the jet area at the vena contracta to the orifice area, normally ca 0,64 for a sharp concentric orifice. The smaller the value, the more pronounced is the vena contracta.

---

[123] **Giovanni D. Santorini (1681 - 1737)** — Italian anatomist.
[124] City of Moscow (first mentioning in chronicle 1147), Muscovy. — Ablitsov VH. «Galaxy Ukraine». The Ukrainian diaspora: prominent persons. Publ. «Kyt», Kyiv 2007.

In 1657 **Job J. Meekeren (1611 – 66)**[125] — Dutch physician and surgeon, attended patient with familial evidence of excessive elasticity of the skin and joints (cutis et articuli hyperelastica), increased the body's sensitivity to the impact of external stimuli and disorders (diathesis), fragility of tissues with ecchymoses, bruises and contusions from minor trauma, subcutaneous hemorrhages, hardened spheroliths and pseudotumors. Such was description by Job von Meekeren of his findings – «*the skin was so elastic and stretchable that the subject, on request, could pull the skin over the right pectoral area, for example, all the way to the left ear, and on letting go, it would retract promptly to its original location without any telltale signs of having been subjected to this manoeuvre*».

He also showed great interest in hand surgery, especially in repair of the flexor tendon on corpses.

In c. 1658 Job van Meekeren published a case study of the incident involving a Russian nobleman named Butterlijin from city of Moscow, Muscovy, who had recently during a swordfight with a Tartar soldier obtained a skull wound. The cranial defect was patched by an anonymous inventive surgeon with a piece of the bone from a dog's skull and the wound healed well. This was the first reported bone grafting. However, the Russian Orthodox Church (ROC)[126] considered the merging of human and animal tissues unacceptable, treated to excommunicate the patient from the ROC, and ordered the removal of the graft. Because the wound had healed so well, physicians were unable to remove the transplant, and patient could not return to church.

Nathaniel Highmore (1613 – 85) — English anatomist and surgeon, described a paired paranasal maxillary sinus in the body of the maxillary bone of Highmori («Corporis humani disquisitio anatomica». Hague 1651).

Sixty-three year after description of the lateral fissure of the brain by **Hieronymus Fabricius (1537 – 1619**: above), **Franciscus Sylvius (1614 – 72)**[127] — Dutch anatomist, chemist and physiologist, re-described it in his «Disputationem Medicarum» (1663) as follow: «*Particularly noticeable is the deep tissue or hiatus which begins at the root of the eyes (oculorum radices). . . it runs posteriorly above the temples as far as the roots*

---

[125] Haeseker B. Mr. Job von Meekerer (1611-1666) and surgery of the hand. Plast Reconstr Surg 1988 Sep;82(3):539-46.

[126] Baptism of Ukraine-Rus 988, Kyiv Metropoly 997-1240,1458-1490, 1620-86; The Ukrainian Greek Catholic Church (UGCC) 1596 – 1946-49, 1989-present; Ukrainian Autocephalous Orthodox Church (UAOC) 1921-30, 1989-present; Ukrainian Orthodox Church - Moscow Patriarchate (UOC-MP) 1990 – present; Ukrainian Orthodox Church - Kyivan Patriarchate (UOC-KP) 1992-present; and Orthodox Church of Ukraine (OCU) 2018 – present. The ROC was formed in 1448 due to splitting break from the Kyiv metropolis, under the name of the Moscow Metropolis, from the 1920's – ROC.

[127] Collice M, Collice R, Riva A. Who discovered the Sylvian fissure? Neurosurgery 2008;63(4)623-8.

of the brain steam (medulla radices)... it divides the cerebrum into an upper, larger part and a lower, smaller part».

Description of the exocrine and endocrine glands was made by **Thomas Wharton (1614 – 73)**[128] — English anatomist and physician, who re-described, after **Alessandro Achillini (1463 – 1512**; Renaissance), the submandibular salivary (Wharthon's) duct (1654), named the thyroid gland (1656), and described the Wharton jelly of the umbilical cord.

Few years thereafter (1664) **Geoffey Websterson** discovered the three primary microscopic features of the thyroid gland (TG) – the follicles, follicular cells and parafollicular cells.

Discovery of the thoracic duct belongs to **Thomas Bartholin (1616 – 80)**[129] — Danish physician, although he missed cisterna chyli and thought it must be very rare (1552). We own him the only one thorough description of his friend **Marco Severino's (1580 – 1656**; above) of the refrigerated anesthesia practices for minor surgical procedures during his visit in the winter of 1643-44 in the city of Naples (established by the Greek settlers in the 2nd millennium BC, founded approximately in the 8th century BC), Italy.

Statistics were introduced into medicine by **John Graunt (1620 – 74)** — English demographer of Scottish origin.

The first blood transfusion in man was done by **Jean B. Denys (1643 – 1704)** — French physician, in 1667.

The incrimination of cerebral hemorrhage as a cause of apoplexy was made in 1658, by **J. Jacob Wepfer (1620 – 95)** — Swiss pathologist and pharmacologist. After he died, during the autopsy, his aorta was opened longitudinally, showing signs of diffuse softening as well as petrifaction, associated with an extensive arteriosclerosis (AS) with deposition of calcium (Ca^{++}) causing calcification – the Wepfer's aorta.

The circle of arteries at the base of the brain depicted in 1664 by **Thomas Willis (1621 – 75)**[130] — English physician, named after him the circle of Willis, a part of the cerebral circulation composed of the right and left anterior cerebral artery (ACA), anterior communicating artery, right and left internal carotid artery (ICA), right and left posterior cerebral artery (PCA) and the right and left posterior communicating artery. He also coined the term of diabetes mellitus (DM) and first wrote about myasthenia gravis (MG).

---

[128] Wharton T. Adenographia: sive glandularum totius descriptio. London 1656.

[129] Bartholino C. De lacteis thoracic in homine brutisque nuperrime observatis historia anatomica. M. Martzan, Copenhagen 1652.

[130] Willis T. Cerebri anatome: cui accessit nervorum descriptio et usus. Typis Ja. Flesher, Jo. Martyn & Ja. Allestry, Londini 1664.

Discovery of the receptacle chyli and the lacteal vessel in animal belongs to **Jean Pecquet (1622 - 74)** — French scientist. The cisterna chyli was named after him (reservoir of Pecquet, 1651).

According to **Blaise Pascal (1623 – 62)** — French mathematician and philosopher, the pressure exerted anywhere to the fluid, is transmitted equally in all directions (Pascal law, 1647).

The foundations of clinical medicine were made laid by **Thomas Sydenham (1624 – 89)** — English physician, who after Rhazes [Medieval Medicine – **Rhazes (c. 854 – c. 925)**] distinguished the natural smallpox from measles, and also from scarlet fever [Renaissance – **Giovanni Ingrassia (1510 – 80)**], as described in his book written in Latin – «Observationes medicae morborum acutarum historiam et curationem», 1676 (in English – «Medical Observations on the History and Cure of Acute Diseases», 1676)

In his description of measles in 1670 and 1674 is important on the account of a long prodromal period, catarrh – inflammation of mucous membranes of the oral cavity, throat, paranasal cavities and airways, the lack of an appearance of relief from eruption in contrast with what occurs in small-pox, the mortality of pneumonia. In his account of the epidemic of 1674, he described the sporadic cases of a «febris moribulosa» or measly fever, which to have been «forme cruste» of the disease, the eruptions being confined to the back of the neck and shoulders. The fever, on the other hand, was much more serious, as it lasted 14 days or longer.

He identified rheumatic disease (RD) form of «dancing» chorea or St Vitus «dancea» (Sydenham chorea, 1686).

Daniel Ludwig (1625 – 80) — German anatomist, described the angulus sterni, or manubrium-sternal junction, formed by the junction of the manubrium and the body (gladiolus) of the sternum with the 2nd rib which lies opposite the lower border the 4th thoracic (T_4) vertebral body (the angle of Ludwig). It is a useful reference point for counting ribs and at this level the trachea bifurcates into the right and the left main stem-bronchus (RMSB, LMSB) at the tracheal carina. At the level of the angle of Ludwig in the right hemithorax, the azygos vein crosses midline to join the superior vena cava (SVC), the thoracic duct crosses from the right to the left hemothorax, on it ascend to the left neck to join the venous system between the confluence of the of the left subclavian and the left internal jugular veins (SCV, IJV). At the level of the angle of Ludwig in the left hemithorax, the aorta and the pulmonary artery (PA) form the aortic pulmonary (AP) window.

As early as in 1678 **Oluf (Ole) Borch (1626 – 90)** — Danish scientist, physician, grammarian, and poet, extracted in 1678 an oxygen (O_2) out of saltpeter (potassium nitrate), and a year later (1679) reported on contusion of the heart with subsequent necrosis of the cardiac muscle (myocardium) and rupture of the heart chamber which were found at autopsy[131].

According to **Robert Boyle (1627 – 91)** — English Irish natural philosopher, chemist, and physicist, at the constant temperature the volume of an ideal gas changes inversely proportional to the pressure, and the pressure inversely proportional to the volume (Boyle law, 1662).

After **Ibn al Nafis (1213 – 88**: Medieval medicine**)** and **Park Ji-Sung** (in 1661**), Marcello Malpighi (1628 – 94)** — Italian biologist and physician, re-discovered capillaries.

Also, one of an early description of the lymphatic system was made by **Olof (Olaus) Rudbeck, Sr (1630 – 1702)** — Swedish scientist, in 1652 (published in 1653).

A pupil and research associate of **Thomas Willis (1621 – 75**: above**), Richard Lower (1631 – 91)**[132] — English physiologist and physician, depicted annulus fibrosus dexter et sinister cordis, et tuberculum intervenosum – more or less prominent ridge extending on the internal surface of the right atrium (RA) wall between the opening of the superior vena cava (SVC) and of the inferior vena cava (IVC); described the arrangement of muscle fibers in the ventricles of the heart; tried to explain how an exercise increases the heart rate (HR) and blood flow but he attributed it to the pumping action of the skeletal muscles, the movement of chyle; elucidated the appearance of edema after ligation of the veins. He related the color of blood to the affinity of venous blood for air, as he erroneously thought, the substance in the air was a laughing or nitrous gas / nitrous oxide (N_2O). Described cardiac tamponade and constructive pericarditis, re-described in 1674 by **John Mayow (1640 – 79)** — English chemist and physiologist.

The construction in 1676 by **Anton** van **Loeuwenhoek (1632 – 1723)**[133] — Dutch naturalist and microbiologist, a microscope with magnification of 270 diameters allowed him to discover on Oct. 08, 1776, bacteria in water, saliva, dental tar and in 1681 a protozoan Cercomonas intestinalis in his own diarrhea stool.

In 1665 **Robert Hooke (1635 – 1703)** — English naturalist, architect, and polymath, demonstrated during an experiment that a dog could survive an open thorax providing the air was pumped in and out of his lung by an artificial respiration, and noted a difference between venous and arterial blood. The same year, he examined under the microscope a cell (in Latin – «cella» or «small room») structure of cork tissue from the cork oak (Quercus suber) and coined the term «cell».

---

[131] Akenside M. An account of a blow upon the heart and its effects. Philos Transact 1764;53:353. — Osborn LR. Findings in 262 fatal accidents. Lancet Sep 4, 1943;6262(2):277-84.

[132] Lower R. Tractatus de corde item de motu & colore sanguinis et chyli in cum tranfitu. Londini 1669.

[133] Van Loeuwenhoek A. Anatomica seu interiora rerum. Leiden 1687.

The pioneer in the field of occupational diseases, **Bernardino Romazzini (1635 – 1714)**[134] — Italian physician, outlined the connection between bronchial asthma and organic dust.

Both, **Raymond de Vieussens (c. 1635 or 1641, commune LeVigan, department Occican, France – 16.VIII.1715, city of Monpellier, Department Herault, France)**[135] — French anatomist, provided an early description, and **FélixVicq d'Azyr (23. IV.1748, commune of Valognes, department Normandie, France** – died from tuberculosis **20.VI.1794, Paris, France)**[136] — French anatomist and physician, a more detailed one, of the brain centrum semiovale that is the white matter underneath of the grey one on the surface of the cerebrum (Vieussens-d'Azyr centrum, 1684, 1786); described cavum – a median fissure between both membranes, of septum pellucidum (5th ventricle of Sylvius-Vieussens); depicted nerve fibers passing on the surface and behind of the subclavian artery (SCA) creating a loop which connect the median and lower cervical plexus (the subclavian loop of Vieussens).

The work of Raymond de Vieussens also includes description of the pericardium, coronary vessels, muscle fibers of the heart, the limbus fossae ovalis and the intramural coronary circulation by injecting mercury and saffron dye into the coronary arteries and veins showing they passed directly into the cardiac chambers through small openings in the endocardium.

He detailed the anatomy of the epicardial coronary arteries and veins and in so doing discovered the adipose (conus) artery, the ramus coni arteriole of the right coronary artery (RCA). It arose from the aortic root, crossed over the conus arteriosus of the main pulmonary artery (MPA) to form the anastomotic ring of Vieussens.

Depicted the most constant Vieussens' valve of the great cardiac vein where it turns around the MPA, the venarum minimarum atrii dextri and venae ventriculi dextri arteriolas which collect the blood from the anterior wall of the right ventricle (RV) and drain into the right atrium (RA) (1705).

Raymond de Vieussens presented detailed anatomy of the lymphatic system and blood vessels of the heart (1706).

---

[134] Romazzini B. De Morbis Artificum (Disease of Workers), 1st Ed, Modena 1700; 2nd Ed. Padua 1713.

[135] Commune LeVigan, department Occican (area 5,217 km²), France. — City of Monpellier, department Herault (area 6,224 km²), France. — De Vieussens R. Traité nouveau de la structure et des causes du mouvement naturel du coer. Paris 1615. — De Vieussens R. Neurographiae universalis. Hoc est ómnium corpiris humani nervorum, simul et cerebri, medullaeque spinalis descriptio anatómica. Cum ipsorum actione et usu, physico discursu explicatis. Editio nova. Apud Joannem Certe, Lyon DCLXXXV (1685). — De Vieussens R. Nouvelles découvertes sur le coeur. Paris 1706.

[136] Commune of Valognes (fortified stronghold build 11th century), department of Normandie (area 29,906 km²), France. — d'Azyr F. Eloges, l'Académie, l'historie. Mémores sur l'Anatomie Humaine et Comparée. Traite d' Anatomie et Physiologie (avec des planches caloriées réprésentant au naturel les divers organes de l'Homme et des Animaux. Francois Didot l'ainé. Paris 1786). Systeme Anatomique des Qudrupédes. Vol. 1-7. Per Moreau de la Sarthe, Paris 1805.

Lastly, his treatise also included the first documented clinical presentations and autopsy results of patients who died from mitral valve (MV) stenosis and aortic valve (AV) insufficiency (1615).

Furthermore, Félix d'Azyr described locus coerulens – a pigmented elevation in the upper angle of the 4th ventricle of the brain which together with adrenal medulla secretes a hormone nor-adrenaline; a fiber system between the external granular layer and the external pyramidal layer of the cerebral cortex (the d'Azyr fibers); the compact, thick bundle of nerve fibers arising from the medial and lateral nuclei of the mammillary body, joined by fibers originating directly from the fornix of the brain, and then connecting the mamillary body to the dorsal and ventral tegmental nuclei, and the anterior thalamic nuclei (the mammillo-thalamic bundle or tract of d'Azyr); foramen cecum of the medullae oblongatae.

The tract of d'Azyr supports the spacious memory, and if infarcted by excessive use of alcohol (C_2H_6O) may cause alcoholic syndrome.

François Mauriceau (1637 – 1709) — French obstetrician, described an ectopic tubal pregnancy. Then he devised the classic head of a breech maneuver for the passing the head of the child at the breech presentation where its body lies on the physician's palm and the forearm, the index and the middle finger of the physician is situated over the upper jaw to convey flexion of the head, and with the other hand lying on shoulders of the child an extraction was fulfilled (Mauriceau maneuver, 1688). Together with **Justine Siegemund (1636 – 1705)** — German midwife, introduced the practice of puncturing the amniotic sac (amniocentesis) to arrest bleeding from placenta previa (1690, 1694).

The friend of **Thomas Sydenham (1624 – 89; above)**, **Richard Morton (1637 – 98)**[137] — English physician, was the first to recognize in 1689 that in phthisis (in Greek / in Latin meaning rot, decay), i.e., wasting or consumption disease, an ancient name for tuberculosis, tubercules are always present in involved organs, mainly in the lungs. Beside of this, he was first to describe wasting condition anorexia nervosa (in Greek – «a» for «without-», «non-»; «oresis» for «an urge is to the meal»), i.e., lack of appetite.

Heinrich Meibom (1638 – 1700) — German physician and scholar, is known for the discovery of the sebaceous (tarsal) gland in the eyelid between the eyelid cartilage and the eyelid conjunctiva (glandula tarsalis Meibomi), inflammation of which causes meibomian stye (blepharadenitis or chalazion).

---

[137] Morton R. Phthysiologia, seu, Exirtationes de phthisi tribus libris comprehensae: totumque opus varii historiis illustratum. Sam Smith & BengWalfotd, Londini 1689. — Morton R. Phthysiologia, or a Treatise of Consumptions. Sam Smith & Benf Walford, London 1694.

The agangionic large bowel (colon) or aganglionic megacolon is an inherited disease due to lack of ganglionic nervous cells with failure to of development of myenteric plexus of the rectosigmoid area of the colon was described in 1691 by **Frederik Ruysch / Frederici Rauschii (1638 – 1731)**[138] — Danish anatomist. It is characterized by constipation and vomiting.

Niels Steensen, a.k.a., **Nicolaus Stenonis** or **Nicolas Steno (1638 – 86)**[139] — Danish anatomist, geologist and ecclesiastic, discovered the accessory parotid gland duct (one of the pairs of salivary gland), the route that saliva passes through the buccal fat, buccopharyngeal fascia and buccinator muscle and opens into a vestibule of the mouth next to the maxillary 2nd molar tooth (ductus stenonianus or the Stensen duct, 1660); described the structure of muscles as consisting of longitudinal fibers and made an attempt to explain mechanical process of muscular contraction (1664), and established similarity between the mammalian ovary and the ovary of oviparous animals.

Subsequently, in 1671-72, Niels Steensen was first to describe a cyanotic newborn with congenital chest wall deformity – thoracic ectopia cordis and associated cyanotic congenital heart disease (CHD) – «blue baby» syndrome, consisting of four features: (1) ventricular septal defect (VSD), (2) a funnel-like stenosis of the right ventricle (RV) outflow tract and the main pulmonary artery (MPA), (3) hypertrophy of the RV, and (4) dextroposition of the aorta, which override the interventricular septum (IVS) receiving the venous as well as arterial blood.

There are three varieties of congenital chest wall deformity with the extra-thoracic location of the heart (ectopia cordis) – thoracic, cervical, or thoracoabdominal. In the thoracic ectopia cordis, the sternum is clefted, the heart shows a severe anterior rotation. In thoraco-abdominal ectopia cordis the sternum is usually clefted inferiorly, the heart is covered by an omphalocele-like membrane or thin often pigmented skin, the heart lacks the severe anterior rotation. Other associated conditions include defects in the pericardium, diaphragm, and abdominal wall. The major impediment to survival is the high incidence of intrinsic CHD.

Congenital esophageal atresia (EA) was first described in 1670, by **William Durston** — English physician, in newborn female twins conjoined by the thoracic wall and a blind upper esophagus in one of the twins, both died shortly thereafter.

---

[138] Ruyschii F. Observationum anatomico-chirurgicarum centuria. Aput Henrieum et Viduam Theodori Boom, Amsterdami 1691.

[139] Stensen N. Elementorum myologia specimen. Ex Typographia subsigno Stellae, Florentae 1667. — Stensen N. Embryo monstroaffinis Parisiis dissecturi. Acta medica et phylosophica Hafniensia 1671-72;1:202-3. — Willius F. An unusually early description of the so-called tetralogy of Fallot. Proc Staff Meeting Mayo Clin 1948;63:316-20. — Tubs RS, Gianaris N, Shoja MM, et al. «The heart is simple a muscle» and first description of the tetralogy of «Fallot». Early contributions to cardiac anatomy and pathology by bishop and anatomist Niels Stensen (1638-1686). International J Cardiology 2012 Feb 09;154(3):312-5.

The contribution of **Regnier de Graaf (1641 – 93)**[140] — Dutch anatomist and physician, to the anatomy and function of the woman reproductive organs include description of the lesser vestibular or periurethral glands; female ejaculation referring it to an erogenous (erotic) zone on the anterior wall of the vagina along the course of the urethra; the testicular ducts; the efferent ducts; the corpora lutea or yellow body – a temporary endocrine structure producing progesterone, estrogen and others, colored by carotenoids, developing during each menstrual cycle; function of the ovarian tubes of **Gabriello Falloppio (1523 – 62**: Renaissance**)**; and hydrosalpinx – a distally blocked ovarian tube filled with serous or clear fluid, linking its development to female infertility.

The lesser vestibular gland is located on the anterior wall of the vagina, along the lower end of the urethra, drain into the urethra and near the urethral opening. These glands are surrounded with tissue, which include part of the clitoris, reaches up inside the vagina and swells with the blood during sexual arousal. They are prone to infections.

The mature stage of the ovarian follicle is called the Graafian follicle in his honor. He also invented a practical syringe.

The founder of classical mechanics, Sir **Isaak Newton (1642 – 1726/27)**[141] — English mathematician, astronomer, and physicist, who in 1666 and 1670-72 developed the theory (Newton's) of color, and 1687 laid in physics the foundations of classical mechanics, the 1st branch of which evolved the theory of gravitation by watching the fall of the apple from the tree.

In January of 1666 Isaak Newton observed that although the white light ray entering a dispersive prism is circular, though it is exiting a prism in the position of minimum deviation as oblong decomposed the spectrum of colors, is to say, the prism retracts different colors by different angles. This led him to conclude that color is a property intrinsic to light. In other words, the color is the result of object interacting with already-colored light rather than objects generating the color themselves.

The classical Newtonian mechanics comprises the three sets of physical laws describing the motion of bodies under the influence of a system forces:

1st law: in an inertial reference frame, an object either remains at rest or continues to move at a constant velocity, unless acted upon by force; 2nd law: in an inertial reference frame, the vector of sum forces (F) on an object is equal to the mass (m) of that object multiplied by the acceleration (a) of the object: $F = ma$;

3rd law: when the body exerts force on the other (second) body, the second body simultaneously exerts a force equal in magnitude and opposite direction to the first body.

---

[140] De Graaf R. De mulierum generationi inservientibus. Leyden 1672.

[141] Newton Js. Philosophiae naturalis principia mathematica. Types Streater, Londoni 1687. — Romaniuk MO, Krochuk AS, Pashuk IP. Optics. Publ. Ivan Franko Lviv Nat Univ, Lviv 2012.

Newton's law of universal gravitation or the inverse-square law of universal gravitation states that a particle affects every other particle in the universe using a force (F) that is directly proportional to the product of their masses and inversely proportional to the square of the distance between them, according to the equation:

$$F = G \times m_1 \times m_2 / r^2, \text{ where}$$

R = force between the masses; G = the gravitation constant [6.674 x 10^{-11} N x (m/kg^2)]; M_1 = the 1st mass; M_2 = the 2nd mass; r^2 = the distance between the centers of masses.

Systeme international d'unites (SI)[142] for the force, as defined in 1687 by Isaak Newton and named Newton (N), equals to the amount of the net force required to accelerate in the vacuum a mass of 1 kilogram (kg) at the rate of 1 meter (m) per second (s., sec.) squared in the direction of the applied force, as outlined by the following equation:

$$1 \text{ N} = 1 \text{ kg} \times m/s^2$$

A severe traumatic complication of prolonged obstructed labor in which pressure from the impacted fetal head causes partial ischemic necrosis of the wall (septum) between the urinary bladder and vaginal vault, i.e., vesico-vaginal septum in women, causing vesico-vaginal fistula (VVF) that allows continuous involuntary flow of urine into the vagina, resulting in continuous and uncontrollable urinary frequency. Anthropological evidence suggest that inclination towards development of the VVF, have probably been occurring progressively over hundreds of thousands years by the development in anatomically modern women the mechanism of rotational child delivery.

The first successful surgical repair of VVF after child birth was performed by **Henry van Roonhuyen**[143] — Dutch surgeon, described in his treatise «Medico-Surgical Observations» (1676) that outlined six (6) steps in performing his technique: (1) use of lithotomy (elbow-knee) position; (2) expose of the VVF with a speculum; (3) sharp paring of the VVF edge prior to attempted closure; (4) careful approximation of denuded edges of the VVF; (5) dress of the wound with absorbed vaginal packing; (6) bed rest (BR) until the repair has healed.

---

[142] SI (established in 1960) is based on the m – kg – s. system.

[143] Van Roonhuyse H. Medico-Surgical Observations by Henry van Roohuyse practitioner of physick and chirurgery in Amsterdam, Englished out of Dutch by a careful hand. Moses Pitt, London 1676. — Wall LL. Henry van Roohuyen and the first repair of a vesico-vaginal fistula (c. 1676). Int Urogynecol J 2020 Feb 31(2):237-41.

Lorenzo Bellini (1643 – 1704)[144] — Italian anatomist, physiologist, and physician, discovered and researched the papillary ducts of the kidneys, named after him Bellini ducts (1662), which are the most distant portion of the collecting system. Papillary ducts receive a precursor to urine – a renal filtrate from several collecting ducts, reabsorb water and initiate electrolyte balance in the collecting tubules, and the rest empty into calyx.

On June 15, 1667, **Jean B. Denys (1643 – 1704)** — French physician performed the first blood transfusion in the human being by intravenous (IV) infusion about 375 milliliters (ml) of sheep blood into a 15-year-old boy, who had been bled with leeches 20 times, with survival.

As a physician, Sir **John Floyer (1649 – 1734)** — English physician and author, was best known for advocating cold bathing (1697), for presenting an early account of the pathological changes in the lung associated with emphysema (1698), for creating a special watch for pulse rate measurement and for introducing it into the medical practice (1707-10). It is how he described his pulse-watch – «*I have for many years try'd pulses by the minute in common watches and pendulum clocks, when I was among my patients; after some time, I met with the common sea minute-glass…and by that I made most of my experiments. But because this was not portable, I caused a pulse-watch to be made which ran 60 seconds, and I placed it in a box to be more easily carried, and by this I now feel pulses*».

Clopton Havers (1650 – 1702) — English anatomist and physician, in «Osteologia Nova» (1691), outlined the basic unit of compact bone (osteon), the haversion system which includes a haversian cannal and its concentrically arranged lamellae.

Augustus Q. Rivinus (1652 – 1723) — German botanist and physician, and **August F. Walther (1688 – 1746)** — German anatomist, botanist and physician, described the minor submandibular salivary ducts – small ducts of the sublingual gland that opens on the crest of the sublingual fold (ductus sublinguales minores or Rivinus or Walther ducts). The former also described the usually paired sublingual salivay gland – the smallest of the three major salivary glands which opens in the oral cavity with 10-30 lingual ducts (glandula sublingualis Rivini), the incisura (notch) tympanica – a defect in the upper part of the tympanic portion of the temporal bone filled by the upper portion of the tympanic membrane (incisura tympanica Rivini); he and **Henri J. Shrapnell (1761 – 1842)** — English anatomist, military sugeon and inventor, described the flaccid portion of the eardrum, situated between the anterior and posterior folds of the hammer, thinned and not in one-piece (pars flaccida membranae tympanicae, Rivinus or Shrapnell membrane). The latter also described

---

[144] Bellini L. De structura at usu renum. Florentia 1662. — Bellini L. Exercitatio anatomica de structura at usu renum. Florentia 1662.

the intravenous carotid plexus which is the portion of the internal carotid artery (ICA) plexus (carotid or Walther's plexus), the coccygeal (Walther) plexus and the posterior talofibular ligament – a strong fibrous horizontal streak, extending from the posteromedial surface of the lateral malleolus of the fibula into the posterior processus of the talus (ligamentum talofibulare posterius or Walther's ligamentum), and devised the instrument for dilatation of the female urethra (Walther dilator).

Discovered by **Johann C. Brunner (1653 – 1727)** — Swiss anatomist, tubule alveolar submucous glandules duodenalis (Brunneri, 1687) may hypertrophy (Brunner's gland hyperplasia), or cause a polyp-like tumor (Brunner gland adenoma).

Alexis Littre (1654 – 1725) — French anatomist and surgeon, described in 1700 the abdominal wall hernia sac containing a protruded diverticulum of the ileum or Meckel's diverticulum **[Wilhelm Fabry (1560 – 1634).** 19th century. Anatomy, histology, and embryology **– Johann Meckel (1781 – 1830)]**, was named the Littre's hernia. Its prevalence in premature and mature newborn and infants occurs in 10% to 30%, where beside of Meckel's diverticulum, may also contain ovaries and Falloppian tubes **[Gabrielle Falloppio (1523 – 62:** Renaissance)]**.

Francois Poupart (1661 – 1708) — French anatomist and surgeon, described the inguinal (Poupart) ligament (1695).

In 1697 **Thomas Gibson** — British obstetrician, was called by a midwife after delivery of a newborn baby girl who exhibited an excessive salivation, during breast feeding by her mother, could not swallow, was choking and vomiting milk, became blue (cyanotic) and had difficulty in breathing (dyspnea). The child was constantly very hungry, tried to suck the milk from the breast, however each time it ended with inability to swallow, choking, vomiting, cyanosis, and dyspnea. She became dehydrated, cachectic, and died. Autopsy revealed a congenital proximal esophageal atresia (EA) with distal tracheoesophageal fistula (TEF).

Despite these accomplishments, this era did not witness any significant advances in medicine.

However, in 1636 Harvard College / University was established, the city of Cambridge (settled 1630), Massachusetts (MA; area 27,336 km^2), USA, which became one of the world's most prestigious universities.

In 1661-1773 functioned the Lviv Academy with the statute of university, city of Lviv, Ukraine.

The only new improvements were the introduction of cinchoa (quinidine) from Peru to Europe in 1633 for treatment of malaria and tapping of the abdomen for ascites (1665).

The Vienna General Hospital (VGH; established 1686, 1693 and 1697), Austria, became a model for hospitals for Europe.

THE 18TH CENTURY

The 18th century of enlightenment brought startling advances in physics, in chemistry and in medicine[145].

In his work entitled «Exanthemologia…» of 1730, **Thomas Fuller (1654 – 1734)**[146] — English physician, distinguished a small-pox from measles of which the chief complain was pain in the back, not so violent in measles as is in small-pox; pain in the chest and shortness of breath (SOB) or dyspnea which are greater in measles than in small-pox; the constant presence of cough before measles; the greater frequency of sore throat and hoarseness before measles; the greater frequency of ocular and nasal catarrh in measles; the longer time for the eruption of small-pox to appear, and the relief of symptoms on appearance of the eruption in small-pox «if it be not a very bad sort».

Giovanni M. Lancisi (1654 – 1720)[147] — Italian anatomist and physician, described the median papillary muscle of the anterior and septal leaflets of the tricuspid valve (TV) in the right ventricle (RV) of the heart (Lancisi muscle, 1714). Moreover, he in 1717 named tropical «intermittent fever» into the term malaria (in Italian – mala aria, meaning «bad air»).

Caspar T. Bartholin, Jr (1655 – 1738)[148] — Danish anatomist, described the ductus sublingualis of the anterior sublingual salivary gland (Bartholin duct), which parallel **Thomas Wharton's (1614 – 73**: 17th century**)** duct draining the submandibular salivary gland, and opens together the latter in the sublingual bundle. Beside of this, he described the greater vestibular glands – a bean sized (1,5-2,0 cm long) bilateral oval structures, bordering the middle and posterior third of the labia majora at the base of the vagina, output ducts of which has a whitish and alkaline mucous secretion with a specific scent, providing vaginal lubrication during sexual intercourse (Bartolin's gland, 1637). The Bartholin's glands are prone to acute and

---

[145] Pishak VP, Akhtemiichuk YuT, Tsvihun AO. Anatomy of the 18th century in figures. Clin Anatomy Operative Surg (Chernivtsi) 2003;2(4):1-9.

[146] Fuller T. Exanthematologia or, an Attempt to Give Rational Account of Eruptive Fever, Especially of the Mesles and Small-Pox. Part I and II. Charles Rivington, London 1730.

[147] Klaassen Z, Chen J, Dixit V, et al. Giovanni Maria Lancisi (1524-1720): anatomist and papal physician. Clin Anat 2011 Oct;24(7):802-6.

[148] Bartholin C. De ovariis mulierum et generationis historia epistola anatomica. Amsterdam 1676.

chronic infections (adenitis), cyst and abscess formation, each of them bearing his name. Furthemore, he described in 1687 congenital cystic disease of the lung.

In 1717 **Marko Gerbec**, a.k.a. **Marcius Gerbezius (1658 – 1718)**[149] — Slovene physician and scientist, described patient's symptoms of an atrio-ventricular (AV) block or sick sinus syndrome (SSS), an occasional temporary standstill or extremely slowing of the pulse as a result of complete heart block (CHB), causing dizziness, fainting and sometimes convulsions.

The term orthopedics was proposed by **Nicolas Andry de Bois-Regard (1658 – 1742)** — French physician and writer, in 1741, when he published a book on prevention and correction of musculoskeletal deformities in children.

Rubella, a.k.a., three-day or German measles, first described in 1740 by **Friedrich Hoffman (1660 – 1742)** — German chemist and physician, it is often mild infection disease affecting usually children, the incubation period 15 – 24 days, characterized by fever, sore throat, fatigue, the rush, sometime itchy, usually starting on the face and spreads to the rest of the body, swollen lymph nodes, in adult – joint pain, bleeding testicular swelling and inflammation of nerves. Infection by rubella during early pregnancy may result in a mis-birth (miscarriage, abortion) or child born with congenital heart disease (CHD).

Francois P. du Petit (1664 – 1741)[150]— French anatomist, ophthalmologist and military surgeon, noted a striking correlation between the head injuries in soldiers and contralateral motor effect, which he documented in his treatise «Lettres d'u medicin des hospitaux du roi a un aute medecin de ses amis» (1710), and that let him to describe in detail decussation / cross over (in Latin – decussation pyramidon) of the two pyramidal pathways which contain the motor fibers that pass from the brain to the medulla oblongata and spinal cord. These are the corticobulbar and corticospinal fibers that make up 90% of the pyramidal tracts. Other 10% travels down to the anterior corticospinal tract.

He also provided the first account on afflictions by congenital anomalies; iatrogenic due to medical treatment; disease in the neck or chest such as superior sulcus tumor (SST) of the lung; thyrocervical venous dilatation; injuries of the sympathetic cervico-thoracic (stellate) ganglion, which receives the preganglionic sympathetic nerves from the lower cervical (C_7 and C_8) and the upper thoracic (T_1) ganglia (1727). The clinical picture of the latter affects the ipsilateral side of the face and include constriction of pupil (miosis), a dropping of the upper eyelid (ptosis), elevation of the

---

[149] Cibic B, Kenda MF, Fraz Z. Marco Grebec (Marcus Gerbezius) — Slovene physician who was first describe symptoms of complete atrioventricular block. Slovenska Kardiologija 2005;2:92-8.

[150] Du Petit FP. Memoire dans lequel il est demontre queles nerfs intercostaux fournissent des rameaux que portent des esprits dans les yeux. Hist Acad Roy Sci (Paris) 1727:1-9.

lower palpebral, narrowing of the eye fissure, decrease sweating of the face (facial anhidrosis), redness of the face, with or without inset eyeball (enophthalmos).

Antonio M. Valsalva (1666 – 1723) — Italian anatomist, described the right and left aortic (coronary) and posterior aortic (noncoronary) sinuses (RCS, LCS, NCS) of Valsalva; demonstrated that a forceful exhalation (expiration) against a closed glottis increases intrathoracic pressure and decreases venous return with (1) the initial blood pressure (BP) rise, (2) then reduced venous return and compensation, (3) rebound of the BP, and (4) return of cardiac output (CO) – Valsalva maneuver, useful in decreasing preload to the heart thus increasing CO and to arrest supraventricular tachycardia);

prove that a forceful expiration against the closed nostrils and mouth increases pressure in the hearing tubes and in the middle ear thus shifting the ear membrane outside (Valsalva test), used to detect patency of the hearing tubes).

His work was published posthumously (1740).

Herman Boerhaave (1668 – 1738) — Dutch botanist, chemist, and physician, introduced the use of the thermometer into daily medical practice, and described a post-emetic spontaneous perforation of the esophagus on the left side in the cardiac triangle limited anteriorly by the wall of the heart, posteriorly by the vertebral column and at the base by the left hemidiaphragm (Boerhaave syndrome, 1724).

Jacob B. Winslow (1669 – 1760)[151] — Danish born French anatomist, described omental (epiploic) foramen which is the passage between the greater and the lesser sac in the abdominal cavity. Its borders are: anterior – the free border of the lesser omentum, a.k.a., hepatoduodenal ligament with two layers within which are the common bile duct (CBD), hepatic artery and portal vein; posterior – the peritoneum covering the inferior vena cava (IVC); inferior – the peritoneum covering the commencement of the duodenum and the hepatic artery, the latter passing forward below the foramen before ascending between two layer of the lesser omentum; left lateral – the gastrosplenic and the splenorenal ligaments. Thus, the epiploic foramen is positioned between two great veins of the abdomen: the portal vein which is the most posterior structure in the hepatoduodenal ligament and the IVC lies under the posterior wall.

He had documented the existence of the foramen spinosum which is one of two foramina located on the base of the skull, in the sphenoid bone, just anterior to the spine and just lateral to the foramen ovale, through which pass the middle meningeal artery (MMA) and vein, and the meningeal branch of the mandibular nerve.

The posterior wall of the abdomen is bounded above by the 12th ribs, below by the crest of the iliac bones, and on the sides by the posterior axillary lines (PAL). There

---

[151] Winslow JB. Exposition anatomique de la structure du corpus humain. Un encombrement 1-3. G. Desprez et J. Desissart, Paris 1732.

are two potential places of less resistance (in Latin – locus minoris resistentiae) of tissues, where lumbar hernias may occur.

One of them is the inferior lumbar triangle, described by **Jean L. Petit (1674 – 1750)** — French surgeon, is composed of the iliac crest inferiorly, the margin of the latissimus muscle posteriorly and the external abdominal muscle anteriorly (Petit's triangle)[152]. The floor of the inferior lumbar triangle is the internal abdominal oblique muscle, where herniations occasionally occurs.

Jean Petit also performed craniotomy (1736); drained suppurative inflammation of the pleural cavity / empyema thoracis (1795); invented the screw tourniquet, a device for a temporary placing around and compressing the extremity proximally to a bleeding artery to prevent blood loss and death (1798).

To eradicate cancer of the breast, Jean Petit carried out a total mastectomy with removal of pectoralis major muscle and axillary lymph nodes (LN) – total mastectomy with axillary lymph node dissection (lymphadenectomy)[153].

He also described a thickening portion of the pelvic fascia between the uterine cervix and vagina, as it passes posteriorly in the retro-uterine fold for attachment to the anterior surface of the sacrum (Petit's fascia).

James Douglas (1675 – 1742) — Scottish anatomist and physicians, depicted the semilunar (Douglas) line which defined the anatomic transition below which all the aponeurotic layers of the abdominal muscles, except for the transversalis fascia, pass anteriorly to the rectus muscle. He also outlined the recto-uterine fold and pouch which bears his name (Douglas's cul-de-sac).

From an initial curiosity about plant fluids and hydraulics, Sir **Stephen Hales (1677 – 1761)**[154] — English botanist and physiologist, studied circulation, discovering the direct measurement of the arterial blood pressure (BP), measured the pulse rate, volume of the left ventricle (LV) of the heart, cardiac output (CO), velocity of the blood flow in the ascending aorta (AA) and systemic vascular resistance (SVR) in animals including the horse, ox, sheep and dog in 25 experiments, which he outlined in his treatise «Statical Essays: Containing Vegetable Statics». Vol. I (London 1727) and «Statical. Essays: Containing Haemastatics». Vol. II (London 1733).

---

[152] Vayda RI. Abdominal wall. Clinical anatomy and physiology. In, LYa Kovalchuk, VF Sayenko, HV Knyshov. Klinichna Khrurhiya. Vol. 1-2. Publ. «Ukrmedknyha», Ternopil 2000. Vol. 2. P. 9-11.

[153] Dryzhak VI. Cancer of the breast. In, LYa Kovalchuk, VF Sayenko, HV Knyshov. Klinichna Khirurhiya. Vol. 1-2. Publ. «Ukrmedknyha», Ternopil 2000. Vol. 2. P. 348-57.

[154] Hales S. Statical Essays: Containing Vegetable Statics; or, on Account of fomer Statistical Experiments on the Sap in Vegetables. Vol. I. 3rd Ed., with Amendments. Printed for W. Innys & R. Manby & T. Woodward & J. Peele, London M.DCC.XXX.VIII. — Hales S. Statical Essays: Containing Haemastatics; or, on Account of fomer Hydraulic and Hydrostatical Experiments on the Blood and Blood-Vessels of Animals. Vol. II. 2nd Ed., Corrected. Printed for W. Innys & R. Manby & T. Woodward, London M.DCC.XL

He inserted a cannula into the femoral artery of a horse after placing a temporary ligation. The cannula was connected to a glass tube of the same diameter and the ligature was untied. He then observed the height of the rise of the systolic BP and the fall of the diastolic BP of the blood above the level of the LV of the animal allowing him to estimate the arterial BP. Then he killed the animal, removed its heart, correctly described the roles of mitral valve (MV) and aortic valve (AV) during systole and diastole, and by injecting wax into the LV measured its volume.

The recorded LV volume was then multiplied by the pulse rate to determine CO/min. He showed that the peak level of the arterial BP correlated with the CO. He calculated that in man the velocity of the blood flow in the AA was 146 feet (511 m) per min. with CO about 4 liter (l)/min. Furthermore, he explained the pulsations of arteries in terms of their elasticity, showed that the diastolic arterial BP resulted from SVR, and attributed it to friction due to the passage of blood through small arteries. He did similar experiments on measurements of the venous pressure and showed that the ratio of between arterial and venous pressure was about 10 to one.

Well then, how he described discovery of the direct measurement of the arterial BP: «*In December* [1726] *I caused a mare to be tied down alive on her back…having laid open the left crural arterie about three inches from her belly, I inserted into it a brass pipe whose bore was one sixth of an inch in diameter…I fixed a glass tube of nearly this same diameter which was nine feet long: then untying the ligature of the artery, the blood rose in the tube eight feet three inches perpendicular above the level of the left ventricle of the heart…when it was at its full height it would rise and fall at and after each pulse 2, 3 or 4 inches*».

Chronic inflammation or induration of the tunica albuginea corporum cavernorum of the penis with growth of fibrous plaques and formation of chordea, described in 1743 by **Francois G. de la Peyronie (1678 – 1747)** — French surgeon, is a connective tissue disorder, causing pain, clubbing, loss of girth, shortening and erectile dysfunction of the penis, affecting up to 10% of men (in Latin – induratio penis plastica, or Peyronie disease).

The founder of modern dentistry, **Pierre Fauchard (1678 – 1761)** — French surgical dentist, used tin and lead to fill dental cavities («The Surgeon Dentist», 1728).

The first detailed description of artificial respiration by «mouth-to-mouth» breathing was made by **William Tossah**[155] — Scottish surgeon, in 1732.

The anterior wall of the abdomen contains places of less resistance (in Latin – locus minoris resistentiae) in the inguinal and femoral canals and around the umbilicus area where a congenital or acquired hernia may form with possibility of an incarceration causing necrosis of the bowel, acute intestinal obstruction, peritonitis, and death, if not operated in time.

---

[155] Trubutovich E. History of Medicine. Part 1. Critical Care Resuscitation 2005;7:250-7.

Claudius Amyand (1680 – 1740) — English surgeon, who performed in 1735 successful removal of the appendix vermiformis (appendectomy). Named after him is Amyand's hernia, containing the appendix vermiformis within the hernia sac and became incarcerated in less than 1% of all inguinal hernias.

Subphrenic abscess is a condition characterized by an accumulation of infected fluid (puss) in the subphrenic spaces between the diaphragm above and the transverse colon with its mesentery below, secondary to acute diseases in the abdominal cavity, such as perforated peptic ulcer of the stomach and duodenum, postoperative complications of the abdominal surgery, trauma, and associated with peritonitis.

On the right side the large space is divided by the liver into supra-hepatic and subhepatic compartments. The supra-hepatic compartment is divided by the falciform ligament into the right and left sectors. The subhepatic compartment is divided by the falciform ligament, the ligamentum teres hepatis and the descending duodenum into the combined subphrenic space anterior to the stomach and the lesser sac. The supra-hepatic compartment is further divided into anterior and posterior superior space, and the subhepatic compartment into anterior and posterior inferior space. In addition, the subphrenic space include the extraperitoneal space between the leaves of coronary ligament, known as the area bare of the liver that communicates with the retroperitoneal space and the perinephric areas and extends across the midline within the coronary ligament.

Because of existing communications between the lymphatic system of the abdominal and thoracic cavities and existing fenestration of the diaphragm, the infection may spread into the pleural cavity with the appearance of the pleural effusion, causing confusing thoraco-abdominal clinical complex (52.2 - 86.4%).

Subphrenic abscess presents with fever and chills, anorexia, cough, shoulder pain on the affected side, hiccough (hiccup), increased respiratory rate (RR) with shallow respiration, increased heart rate (HR) – tachycardia, diminished or absent breath sounds and dullness to percussion in the lower fields of the thorax, tenderness over the 8th - 11th ribs, sepsis, shock, and death if misdiagnosed and untreated.

Subphrenic abscess secondary to a perforation of the colon was recognized as a pathological entity by **P. Petit**[156] — French physician, at the autopsy in 1753.

---

[156] Petit P. Suite de l'essai sur les espanchemens. Mem Acad Roy chir 1753;2:92. — Olearczyk A, Konwolinka CW. Peripheral edema: an unusual sign of subphrenic abscess. Bull Geisinger Med Ctr 1971;23:217-8. — 36. Konvolinka CW, Olearczyk A. Subphrenic Abscess. Curr Probl. Surg. (Ed MM Rawitch). Year Book Med Publ, Chicago, 1972 Jan:1-52. — Olearchyk AS, Konvolinka CW. Subphrenic Abscesses. J Ukr Med Assoc North Am 1972;19,3 (66):3-56. — Olearchyk AS. Subphrenic abscesses; changing etiology and bacteriology J Ukr Med North Am 1973;20,1-2 (68-69):14-25. — Olearchyk AS. Subphrenic abscesses; cardiovascular signs and complications. J Ukr Med North Am 1973;20,3 (70):13-22. — Konvolinka CW, Olearchyk AS. Subphrenic abscess; cardiovascular complications and signs. Am Surg 1974;40 (4):216-20.

In his treatise «The Roots and Causes of Diseases» (1761), **Giovanni B. Morgagni (1682 – 1771)**[157] — Italian anatomist, pathologist and physician, proved that pathological changes of internal organs are responsible for the symptoms of diseases; described frontal hyperostosis (Morgagni's disease); discovered congenital diaphragmatic hernia (CDH) – a small opening (crevice) on both sites between the sternal and costal parts of the diaphragm, which serves as a passage of the superior epigastric blood vessels and a few lymphatic vessels, the site of the congenital retro-thoracic diaphragmatic hernia, through which abdominal viscera penetrate into pleural cavity (Morgagni hernia); credited **Stehenius** with the first observation that CDH was associated with pulmonary hypoplasia; depicted pulmonary agenesis, complete absence of the carina, main-stem bronchus, lung, and pulmonary vasculature; mentioned cyanotic congenital heart disease (CHD) – transposition of the great arteries (TGA); recognized the compressive effect of hemopericardium on the heart causing cardiac tamponade, and constrictive pericarditis (1756); observed in a patient who was merchant slowing of the pulse to 22 beats per minute with sudden episodes of unconsciousness (syncope) due to complete heart block (Morgagni attacks) – «When visiting by way of consultation I found with such a rarity of the pulse that within the 60th part of an hour the pulsations were only 22 – and this rareness which was perpetual – was perceived even more considerable, as often as even two (epileptic) attacks were at hand – so that physicians were never deceived from the increase of the rareness they foretold a paroxysm to be coming on».

Bilateral pulmonary agenesis is usually associated with cardiac abnormalities, and is incompatible with life, right-sided agenesis exhibit 50% of cardiac anomalies, left-sided agenesis, beside of cardiac malformation, may be a part of syndrome with ipsilateral facial, radial ray and renal agenesis.

Near the end of his life observed cases of an aortic dissection due to median degeneration (1771).

Clinical picture of hypovolemic shock described by **Aulus Celsus (c. 25 BC – c. 50**; Ancient medicine. *Romans*), was also seen by **Henri F. La Dran (1685 – 1770)** — French surgeon, following gunshot wounds in the battlefield, named by him in French as *«choc»* and depicted in his surgical treatise «Traité des opérations de chirurgie» (1749) as a collapse of vital functions, and as a «rude unhinging of the machinery of life».

---

[157] Morgagni JB. De sedibus, et causis morborum per anatomen indagatis. Tomus primus et secundus. Ex Topographia Remondigiana, Venetiis 1761. — Peacock BT. Report on cases of dissecting aneurysms. Trans Pathol Soc London 1863;14:87. — Leonard JC. Thomas Bevill Peacock on the early history of dissecting aneurysm. BMJ 1979;2;260-2.

The esophagogastric (EG) junction, a.k.a., gastroesophageal (GE) junction, an anatomic and physiologic barrier against gastroesophageal reflux disease (GERD) was described by **Jean C.A. Helvetius (1685 – 1755)**[158] — French physician, in 1719.

In 1724 **Daniel G. Fahrenheit (1686 – 1736)** — German chemist and physicist, invented the mercury-in-glass thermometer based on the Fahrenheit (°F) degree (in Latin – «gradus» is for steps) as a unit, in which the temperature (T) of freezing point of water is 32°F and the T of boiling point of water is 212°F at the standard atmospheric pressure at the sea level. The normal T of the human body measured orally is 98.2° ± 9 F°.

The earliest reports from 1757 on the left ventricular (LV) aneurysm[159] were made by **Domenico Galeati (1686 – 1775)** — Italian physician, and **John Hunter (1728 – 93)** — Scottish surgeon.

John Hunter's description was particularly precise: «*At the apex, it was forming itself into a kind of aneurysm, becoming very thin; that part was lined with a thrombus just the shape of the pouch in which it lay*».

Domenico Galeati also described glandulae duodenales.

John Hunter depicted the femoral adductor (subsartorial) canal – an aponeurotic tunnel in the middle of the antero-lateral third of the thigh extending from the apex of femoral triangle to the adductor hiatus (opening) that gives passage to the femoral vessels and saphenous nerve (canalis adductorius or Hunter's canal, c. 1761); described congenital heart disease (CHD) termed pulmonary stenosis (PS) with an intact interventricular septum (IVS) – PS -IVS (1784)[160]; depicted the gubernaculum testis – a ligamentum of the fetus, one end of which connects to the lower pole of the testicular epidymis and to the testicle, the other to the base of its pouch (Hunter's gubernaculum), present during the descend of the testicles into the scrotum and later atrophies; described the descent of testicles (1786), the ligamentum teres (Hunteri) and closed or ligated aneurysm (1794).

PS-IVS is characterized by variable size of the right ventricle (RV) that has no exit, fail to provide pulmonary blood flow and unable to decompress itself through the IVS. Only patent ductus arteriosus (PDA) permits blood flow to the lungs. But within

---

[158] Helvetius JCA. Observations anatomiques sur l'estomac de l'homme, avec des reflexion sur la systeme nouveau, qui regarde la tributation dans l'estomac, comme la cause de la digestion des aliments avec une planche. Mem acad roy sci 1719;336-49.

[159] Olearchyk AS, Lemole GM, Spagna PM. Left ventricular aneurysm: ten years experience in surgical treatment of 244 cases. Improved clinical status, hemodynamics, and long-term longevity. J Thorac Cardiovasc Surg 1984;88:544-53. — Olearchyk AS. Recurrent (residual?) left ventricular aneurysm: a report of 11 cases. J Thorac Cardiovasc Surg 1984;88:554-7.

[160] Freedom RM, Anderson RH, Perrin D. The significance of ventriculo-coronary arterial connection in the setting of pulmonary atresia with an intact ventricular septum. Cardiology Young 2005 Oct;15(5):447-68.

hours progressive lack of oxygen (O_2) – hypoxia causes closure of the PDA and death of a newborn from myocardial ischemia.

On the other hand, pulmonary stenosis (PS) with ventricular septum defect (PS-VSD)[161] is a cyanotic CHD characterized by a large VSD resulting from underdevelopment right ventricular outflow tract (RVOT), i.e., anterior deviation of the sub-pulmonary conal (infundibulum) septum; atresia of the pulmonary valve (PV); overriding of the ascending aorta (AA) or right aortic arch (45%); the major aortopulmonary collateral anastomoses (MAPCA's) that in some cases are the sole pulmonary blood flow supply, such as the aortopulmonary artery collaterals to the main pulmonary artery (MPA), the pulmonary artery (PA) lobe or without connection to the PA blood flow; atresia or stenosis of the tricuspid valve (TV); complete or corrected transposition of great arteries (TGA); left superior vena cava (SVC); anomalies of coronary sinus; dextrocardia, and asplenia or polysplemia syndrome. The frequency of PS-VSD ranges from 2,5%-3,4% of all CHD. The most common genetic defect is a 2q11 microdeletion.

Cardiac catheterization (CC) in patients with PS-VSD who have nonconfluent or hypoplastic PA's and the highest likelihood of having MAPCA's assist with surgical planning by understanding of the anatomy, size, degree of stenosis, and the presence of dual blood supply from native PA's or collateral arteries.

Medical treatment in neonates with PDA dependent pulmonary blood flow consist of prostaglandins (PG) infusion at 0.01 to 0.1 micrograms/kg/min to maintain ductal patency.

The aim of surgical repair is to create of biventricular circulation with unobstructed RV to pulmonary vein (PV) continuity by the unifocalization of MAPCA's and closure of intracardiac and extracardiac like shunts with a low RV pressure like unifocalization of collateral arteries by means of disconnection of MAPCA's from the descending thoracic aorta (DTA) and anastomosing them to a common confluence that receive blood supply from the RV or systemic Blalock-Taussig (BT) shunt[162] and provide it to PA's. In adolescent or adult patients with irreversible pulmonary hypertension and myocardial dysfunction heart and lung transplantation (HLT) should be considered.

Adam Ch. Thebesius (1686 – 1732) — German anatomist and physician, described foramina venarum minimarum cordis (Thebesian foramen), valvula sinus coronari – the Thebesian's valve of the coronary sinus (CS) and venae cordis minimae (Thebesian veins of the heart).

---

[161] Feld RH, Liao PK, Puga FJ. Clinical profile and natural history of pulmonary atresia and ventricular septal defect. Prog Pediatr Cardiol 1993;1(1):18-33.
[162] **Alfred Blalock (1899-1964)** — American cardiac surgeon. / **Helen B. Tausig (1898-1986)** — American pediatric cardiologist.

Rene J.C. de Garengeot (1688 – 1759)[163] — French surgeon, first described an incarcerated femoral hernia with the vermiform appendix within a femoral sac (De Gerengeot hernia, 1731).

John Huxham (1692 – 1768) — English surgeon, made the distinction between typhus and thyhoid fever.

In «Treatise on the Structure of the Heart, its Function and Diseases» (Jacques Vincent, Paris 1749), **Jean B. Senac (1693 – 1770)**[164] — French physician, was the first known to report on irregular pulse, outlined the belief that dilatation was the most common expression of cardiac pathology, described pericarditis, its frequent association with pleuritis and mediastinitis, emphasized the susceptibility of the heart to inflammation, remarked on atrophy and artherosclerosis / atheromatosis (AS).

In 1766 **Sauveur F. Morand (1697 – 1773)**[165] — French surgeon, made detailed description of the «pustulae maligna», i.e., anhrax in human [Ancient medicine. *India.* / Medieval medicine. – **Bartolamaeus Anglicus (1203 – 75)**]. Independently, he and **Albrecht** von **Haller (1708 – 77)**[166] — Swiss German botanist, anatomist, physiologist, and a poet, described the lower of two elevations on the medial wall of the posterior horn of the lateral ventricle of the brain, caused by the lateral depth and extension of the calcarine sulcus (Haller's unguis or Morand's spur). The later also described congenital anomaly of a foot having eight toes (Morand's foot).

William Smellie (1697 – 1763) — English gynecologist and obstetrician, described the mechanism of labor, designed obstetrical forceps, modified the classic maneuver of **François Mauriceau (1637 – 1709**: 17th century**)** for delivering the head of a breech (Mauriceau-Smellie's maneuver) and developed a method for measuring internal pelvic dimension (1749).

Frank Nicholls (1699 – 1778) — English anatomist and physician, proved that the inner and middle coats of an artery could be ruptured while the outer coat remained entire, thus explained the formation of chronic aneurysm[167]. In 1753 he was appointed physician to **George II (Oct. 30/Nov. 09, 1683 – Oct. 25, 1760)** — King of Great Britain (1727-60) who in his older age gained weight and became blind in one eye. In the early morning on Oct. 25, 1760, after light breakfast he went alone to his close stool, i.e., a toilet chair, or an early type of portable toilet, for bowel movement, and while straining on stool, he suddenly fell, and died at the age of 76. Autopsy was

---

[163] Akopian G, Alexander M. De Garengeot hernia: appendicitis with a femoral hernia. Am Surg 2006;71(6):526-7.

[164] Senac M. Traite de la structure du coeur, de son action et maladies. Jacques Vincent, Paris 1749. — Bowman IA. Jean-Baptiste Senac and his treatise on the heart. Tex Heart Inst J 1987 Mar;14(1):4-11.

[165] Morand SF. Opusculer de chirurgie. In-1-4°. Guillaume Desprez, Paris 1768-1772.

[166] Haller A. Bibliotheca anatomica qua scripta ad anatomen et physiologiam facienta a rerum initiis recensentur. Band 1-2. Orell, Gesser & Fuessli, Zürich 1774-77.

[167] Nichols F. Compedium Anatomico-oeconomicum. London 1638

performed by Frank Nicholls who detected and described in detail that the cause of death was dissecting aneurysm of the aortic arch (AA) with hemorrhage into the pericardial cavity[168].

In 1735 **Paul G. Werlhof (1699 – 1767)** — German physician and poet, made a detailed description of an idiopathic thrombocytopenic purpura (ITP), named Werlhof disease, an autoimmune disorder with normal bone marrow and low blood platelets counts with shortened lifespan due to the presence of antibodies or influence of the other factors causing its destruction (lysis). It is characterized by spontaneous ecchymoses by red or purple purpuric rash and petechiae in the skin and mucous membranes and bleeding tendency.

While investigating together, **Daniel Bernoulli (1700 – 84)** — Swiss mathematician and physicist and **Leonhard Euler (1707 – 83)** — Swiss, German and Russian mathematician and physicist, the relationship between the speed (velocity) of the blood flow (u) and its pressure (P), the latter punctured the wall of a pipe with a small open ended straw and noted that the height to which the fluid rose up the straw was related to the fluid's pressure in the pipe. Though painful, the method was instantly adopted for the arterial blood pressure (BP) measurement in the human. Not until 1896, was it modified into a less painful method of a continuous arterial BP monitoring by inserting a small plastic cannula into the radial, the brachial, or the femoral artery. In fluid dynamics Daniel Bernoulli stated that for an inviscid flow, an increase in the speed of the fluid occurs simultaneously with a decrease in pressure or a decrease in the fluid potential energy, according to the law of $\frac{1}{2} pu^2 + P =$ constant, where p = density of the fluid, or according to the equation – the pressure gradient (PG) = $4u^2$ (Bernoulli law and equation, Hydrodynamica, Basel 1738). In clinical situations when a bed of the blood stream narrows, i.e., stenotic valve or artery, the flow velocity increases in proportion to the degree of narrowing and PG across the stenotic site increases.

On Dec. 03,1732 **William A. Tossach (c. 1700 – 1771)**[169] — Scotish physician and surgeon, resuscitated a suffocated coalminer by the «mouth-to-mouth» breathing.

While performing experiments aiming at the definition of an international temperature (T), scale on scientific ground, **Anders Celsius (1701 – 44)** — Swedish mathematician, physicist and astronomer, proposed in 1742 the centigrade (in Latin – «centum» is for 100, «gradus» is for steps, i.e., «hundred steps») scale, based on the T for the freezing point of water as 0^0 and the T of boiling point of water as 100^0

---

[168] Nichols F. Phylosophical Transactions. London 1660.

[169] Tossach WA. A man in appearance recovered by distention of the lungs with air. Essays and Obs Soc Edinb 1744;V(part 2):605-8.

at the standard atmospheric pressure at the sea level, named the centigrade scale according to Celsius (C). The normal human body T measured orally is 36.8° ± 0.9°C.

Francois J. Hunauld (1701 – 42) — French anatomist, who in 1742 recognized the importance of the cervical rib in causing symptoms associated with thoracic outlet syndrome (TOS).

Joseph Lieutaud (1703 – 80)[170] — French anatomist, physiologist, and physician, delineated a smooth triangular area of the mucous membrane on the bottom of the urinary bladder between the opening of the two ureters and that of the urethra (trigonum vesicae, vesical triangle or Lieutaud triangle, 1760), first reported on a pleural tumor (1767).

Mathematical representation of the motion by **Leonhard Euler (1707 – 83**: above), and the mathematical representation of mechanics by **Joseph L. Lagrange (1736 – 1813)** — Italian mathematician and astronomer, found an application in modern radiology to better understand the cardiac ventricular mechanics, to be the 2nd and the 3rd branch, respectively, of the classical (Newtonian) mechanics **[Isaak Newton (1642 – 1727).** 17th century].

The 1st Eulerian's law of the motion states that the lineal momentum (in Latin – linear morphea) of a body is equal to the product of the mass of the body and the velocity of its center mass; and the 2nd Eulerian law of motion states that the rate of change of angular momentum about a point is equal to the sum of the external movements about the point.

A Lagrangian mechanics applies to systems whether, or not they conserve energy or momentum, and it provides conditions under which energy and/or momentum are conserved (1772-88). In classical mechanics, the Lagrangian (L) is defined as the kinetic energy (T) of the system minus its potential energy (V), that is L=T-V.

Johann J. von **Huber (1707 – 78)**[171] — Swiss botanist and physician, described an infant with an intra-lobal pulmonary sequestration with an arterial blood supply from the descending thoracic aorta (DTA) which account for 75% of all pulmonary sequestrations.

Sir **John Pringle (1707 – 82)** — British military physician, pleaded for decent ventilation for those confined in the ship's bulks and to military prisons. He proved that jail and hospital fevers were identical (exanthema typhus, 1750).

Nerve function and its relation to the brain was established by **Albrecht** van **Haller (1708 – 77)** — Swiss naturalist, anatomist, physiologist, and poet, in 1756.

---

[170] Lieutaud J. Elementa physiologuae. Fratrum Detournes, Amsterdami 1749. — Lieutaud J. Précis la médicine practique. Chez Vincent, Paris 1760.

[171] Huber J. Observations aliquot de arteria sigulari pulmoni concessa. Acta Helv 1777;9:85.

Benjamin Franklin (1709 – 90) — American polymath, designed bifocal (Franklin) glasses and discovered the therapeutic use of static electricity (franklinization).

William Heberden (1710 – 1801)[172] — English physician, depicted chicken-pox (1767), coronary artery disease (CAD) and its main symptom – angina pectoris (1768), night blindness (nyctalopia) that may exist from birth or be caused by injury or malnutrition, and the presence of the erythema annulare and authentic osteoarthritic nodules on the fingers, back of wrists, outside elbows and in front of knees in osteoarthritis as a result of an accumulation of collagenous fibers (Heberden nodules).

He was the forerunner in understanding of the clinical picture of CAD and the first to recognize and clearly described angina pectoris (1768. Chapter 70. Chest pain). But, he had no knowledge of the etiology and the pathology of this disease, therefore was unable to prevent it, or to treat it.

«The angina pectoris, as far as I have been able to investigate, belongs to the class of spasmodic, not of inflammatory complains. For, in the first place, the access and the recess of the fit is sudden. Secondly, there are long intervals of perfect health. Thirdly, wine, and spirituous liquors, and opium, afford considerable relief. Fourthly, it is increased by disturbances of the mind. Fifthly, it continues many years without any other injury to the health. Sixthly, in the beginning it is not brought on by riding on horseback, or in a carriage, as is usual in diseases arising from scirrhous, or inflammation. Seventhly, during the fit the pulse is not quickened. Lastly, its attacks are often after the first sleep, which is a circumstance common to many spasmodic disorders».

«On opening the body of one, who died suddenly of this disease, a very skillful anatomist could discover no fault in the heart, in the valves, in the arteries, or neighboring veins, excepting some small rudiments of ossification in the aorta».

«With respect to the treatment of this complains, I have little or nothing to advance not indeed is it to be expected we should have made much progress in the cure of a disease, which has hitherto hardly had a place or a name in medical books».

Although, he observed that cold bathing and bathing in the sea stops attacks of angina pectoris. That could increase the systemic vascular resistance (SVR), thus elevate the arterial blood pressure (BP) and augment blood supply to the myocardium through narrowed coronary arteries. On the other hand, an inhaled cold air is known to worsen attacks of angina pectoris by cooling the lungs which are close to the myocardium and that may cause coronary artery spasm on the background of CAD. In addition, the inhalation of cold air increases the pulmonary vascular resistance (PVR), hence increases the workload of the right ventricle (RV).

---

[172] Heberden W.Commentarii de morborum historia et curatione. Londoni 1768. — Heberden W. Commentaries on the history and Cause of Diseases. Printed for T. Payne, Pall-Mall, London 1800.

On the basis of the method of **Thomas Simpson (1710 – 61)** — English mathematician, for numerical integration and numerical approximation of definite integrals (1748), the biplane Simpson technique was developed for the echocardiographic estimation of the left ventricular (LV) volume and LV ejection fraction (LVEF) which require imaging of the LV from two orthogonal views that share a common long axis, for example, the apical four- and two-chamber views.

Johannes N. Lieberkühn (1711 – 56) — German physician, by using for preparation of histologic specimen with injections of wax-containing fluids into body cavities, created relatively durable shapes, which allowed him to observe a simple tubular hairy-like gland in the mucosa of the small intestine (crypts of Lieberkühn, 1745**).**

John E. Fothergill (1712 – 80)[173] — English physician, was the first to record atherosclerosis (AS), i.e., hardening of the wall of coronary arteries, supplying blood to the heart muscle (myocardium); described Streptococcal sore throat (1748)[174]; made description and named neuralgia of the cranial 5th (trigeminal) nerve, i.e., trigeminal neuralgia, or tic doloroux, a chronic burning pain disorder on one side of the face that last for second or a few minutes, more common in women than man, and believed to be caused by loss of the myelin around the trigeminal nerve (1773)[175].

Thomas Bond (1713 – 84)[176] — American physician and surgeon, who developed a splint for the immobilization of fractures of the lower arm, known as a «Bond splint».

Philipp Pfaff (1713 – 66) — German physician and dentist, made the plaster molds from impressions in the wax (1756) to produce dental prostheses.

Sir **Percival Pott (1714 – 88)**[177] — English surgeon, described the bimalleolar fracture of the ankle, a common fracture of one or both bones, the lower end of the fibula and the median malleolus of the tibia with rupture of the internal ligament of the ankle, caused by outward and backward displacement of the leg while the foot is fixed (Pott fracture, 1756, 1769); recognized scrotal cancer in chimney sweeps (1762); and tuberculous spondylitis deformans (Pott disease, 1779).

Robert Whytt (1714 – 66) — Scottish physician, stated that the body and soul hold equal influence over the movement of different parts of the body, and therefore may govern both voluntary action excited by one's will, and involuntary action that depends on the stimulus applied to the muscle or nerve of the muscle (1745);

---

[173] Fothergill JE. A Complete Collection of the Medical and Physiological Works of John Fothergill. John Walker, London 1781.

[174] Fothergill JE. An Account of Sore Throat Attended with Ulcers. C. Davis. London 1748.

[175] Fothergill JE. On a painful affection of the face. Medical Operations and Inquiries. Society of Physicians in London 1773;5:129-42.

[176] Norris GW. The Early History of Medicine in Philadelphia. Collins Printing House, Philadelphia 1886.

[177] Pott P. Some few general remarks on fractures and dislocations. Howes, London 1769.

explained that the pupillary light (Whytt) reflex is the contraction (narrowing) and the dilatation (re-sizing) of the pupil in different intensities of light (1751); in experiment on frogs concluded that spinal cord was a crucial component in facilitating response action stimuli (1761); wrote on dropsy of the brain or tubercular meningitis (published posthumously,1768).

To prevent scurvy Captain **James Lind (1716 – 94)** — Scottish military surgeon, recommended lemon juice which contain vitamin C.

The older brother of John Hunter **[Domenico Galeati (1686 – 1775) and John Hunter (1728 – 93)**, above], **William Hunter (1718 – 83)** — Scottish anatomist and physician, in his paper of 1743 on «On the structure and disease of articulating cartilages»[178] stated that on ulcerated cartilage is universally very troublesome illness, and cure is more difficult that carious bone, when destroyed, it is not recoverable. He also authored the treatise entitled «Anatomia humani gravidi» (1774).

The colleague of William Hunter (above), **Abraham Ludlow (18th century)**[179] — English anatomist and physician, was the first to recognize in 1764 anatomy and clinical picture of pharyngoesophageal diverticulum. Specifically, he described obstruction in swallowing from a «pocked» or «bag», apparently formed as a result of protrusion of the mucous layer of the pharynx and the esophagus through the opening in a triangular area in the wall of the pharynx between the oblique fibers of the constrictor pharyngeal inferior muscle and the transverse fibers of the cricopharyngeal muscle, or in the reverse triangle between the transverse fibers of the cricopharyngeal muscle and the esophageal inner circular layer of the muscularis propria without coverage by the external longitudinal muscle of the esophagus.

For the first time in 1758 or 1769 **Francois P. de La Salle (1719 – 88)** — French chemist, physician and pharmaceutics, received from three gallbladder stones (gallstones) a clear, white, thick and fat-like substance («fatty wax»), appearing to be cholesterol (in Greek – «chole» meaning bile, and «stereos» meaning solid).

Re-described in 1790 by **Robert Hamilton (1721 – 93)** — British physician, epidemic infectious disease of the parotid salivary gland, sometimes submandibular, and sublingual salivary glands, named mumps [Ancient medicine. *Greece –* **Hippocrates (460 – 377)**] is caused by mumps virus, transmitted via air or droplets, typically occurring from 16 to 18 days after exposure and lasting from seven to

---

[178] Hunter W. On the structure and diseases of articulating cartilages. Philosophical Transactions of the Royal Society of London 1743;42(470):514-52.

[179] Ludlow A. A case of obstructed deglutition, from a preternatural dilatation and bag formed in the pharynx. In: W. Johnson, T. Caldwell (eds). Medical Observations and Inquiries by a Society of Physicians in London. 2nd Ed. London Med Soc, London 1769;3:85-101. — Chitwood WRJ. Ludlow's esophageal diverticulum: a preternatural bag. Surgery 1979 Mar;85:549-53.

10 days, characterized by fever, general intoxication, tenderness and swelling of the parotid salivary gland. Complications include acute inflammation of the testes (testicles), epididymis, pancreas, durae of the brain, ovaries and / or breast, joints (acute orchitis, epididymitis, pancreatitis, meningitis, oophoritis and /or mastitis, arthritis, accordingly) and deafness.

The method of percussion of the patient's chest as a diagnostic technique was invented in 1754 by **Leopold J.E.** von **Auenbrugger (1722 – 1809)**[180] — Austrian physician, with the intention that a physician could relatively accurately and objectively determine an important sign of disease and the condition of the underlying organs based on auscultation of sounds. He published on his discovery i a little book entitled «A New Discovery that Enables the Physician from the Percussion of the Human Thorax to Detect the Diseases Hidden Within the Chest» (1761). His name is also associated with Auenbrugger sign, a bulging of the epigastric region below the xiphoid process resulting from a large pericardial effusion.

Petrus Camper (1722 – 89) — Dutch anatomist and physician, who depicted the superficial (fatty) fascia of the anterior abdominal wall (Camper fascia), invented the vectis (in Latin – veho) for a lever – a precursor of the obstetrical forceps.

Ivan A. Poletyka (1722, town of Romny, Sumy Obl., Ukraine – 24.IV.1789, town of Vasylkiv, Kyiv Obl., Ukraine[181] — Ukrainian physician, defented at the Kiel Universität (founded 1665) his doctoral dissertation «De morbis haeraeditariis» (1754), and thereupon was appointed to professorship at this university (1754-56). After return to Ukraine, he headed in 1763-83 the contiguous Vasylkiv Quarantine Service near Kyiv, distinguished himself in the management of plaque epidemic in 1770-71, created a hospital against a plaque in Kyiv, insisted upon an isolation of patients affected by a plaque.

The trigeminal (semilunar) ganglion of Gasser of the cranial 5th (trigeminal) nerve is the large flattened sensory ganglion lying close to the cavernous sinus along the median part of the middle cranial fossa in the trigeminal cavity of the dura mater of the brain, was discovered and described in his 1765 doctorate dissertation by

---

[180] Auenbrugger L. Inventum novum ex percussione thoracis humani interni pectoris morbos detegendi. Vienna 1761.

[181] Town of Romny (first mentioned 1096; unheard-of sufferings was inflicted to it by Russian Communist Fascist regime during organization of Holodomor-Genocide 1932-33 against the Ukrainian nation, when at least 2274 inhabitants have died), Sumy Obl. (area 23,834 km²), Ukraine. — Town of Vasylkiv (first mentioned 988; organized by Russian Communist Fascist regime of Holodomor-Genocide 1932-33, against the Ukrainian nation, assaulted the town, and took away life of thousands of the region residents), Kyiv Obl., Ukraine. — City of Kiel (settled by the Normand's / Vikings, founded 1233) – capital of the Schleswig-Holstein Land (area 15,763 km²), Germany. — Lapychak T. Doctor Ivan Poletyka. J Ukr Med Assoc North Am 1961 Jan;8,(20):43-7. — Ablitsov VH. «Galaxy Ukraine». The Ukrainian diaspora: prominent persons. Publ. «Kyt», Kyiv 2007.

Anton B.R. Hirsch (1744 – 78)[182] — Austrian anatomist and physician, who named this ganglion in the honor of his teacher, **Johann L Gasser (1723 – 69)** — Austrian anatomist.

Experimental fertilization was pioneered by **Lazzaro Spallanzani (1723 – 99)** — Italian physiologist.

Johann F. Meckel, Sr (1724 – 74) — German anatomist, discovered the cavum of the dura mater of the brain over the petrous portion of the temporal bone that covers the ganglion of the cranial 5th (trigeminal) nerve (Meckel space), and also the submandibular ganglion (Meckel ganglion, 1748).

The experimental work by **Lazzaro Spallanzani (1729 – 69)**[183] — Italian naturalist, an «Essay on microscopic observations regarding the generation system of Messrs. Needham and Buffon» (1765), was the first rebuttal of the theory of the spontaneous generation of microorganisms, in which he proved that they move through the air, could be destroyed by boiling water in an hour, if the material is hermetically sealed. In his important book entitled «Experiences to Serve to the History of the Generation of Animals and Plants» (1786), he was the first to show that fertilization requires both spermatozoa and an ovum, which he experimentally performed *in vitro* using frog, and an artificial insemination, using a dog.

In the «Treatise on the Management of Pregnant and Lying-in Women» (1773) **Charles White (1728 – 1813)** — English physician, appealed for surgical cleanliness to combat puerperal fever.

Named in 1767 after **Felice Fontana (1730 – 1805)** — Italian biologist, anatomist and physiologist, the trabecular shaped endothelium-lined spaces (of Fontana), between the processes ligamentum pectinates of the iris around the base of the cornea, near the ciliary body, through which the anterior chamber of the eye communicates with and convey the aqueous humors (lymph) to the sinus venosus sclerae or the Schlamm's channel [19th century. Anatomy, histology, and embryology. – **Friderich Schlemm (1895 – 1858)**]. The diameter of Fontana's spaces determines the rate of the lymph outflow.

Jean Descemet (1732 – 1810)[184] — French physician, who described the posterior elastic lamina, named Descemet membrane (DM) of the cornea with the thickness of 3 millimicrons *(μm)* in children to 8-10 μm in adult. It lies between the corneal proper substance, also called stroma which is composed by collagen alpha 2 (VIII)

---

[182] Hirsch ABR. Par quinti encephali disquisito anatomica. Diss. med. Vienna 1765.

[183] Spallanzani L. Saggio di osservazioni microscopiche concernenti il sistema della generazione de' signori di Needham, e Buffon. Modena 1765. — Spallanzani L. Experiencias Para Servir a La Historia de La Generation De Animales y Plantas . University of Padua 1786.

[184] Descemet J. Observation sur la choroide. De l'Imprimerie Royale, Paris 1868.

chain protein and the endothelial layer of the cornea, a single layer of squamous cells covering the cornea that faces the anterior chamber of the eye.

Mental hospitals were developed by **William H. Tuke (1732 – 1822)** — English psychiatrist.

The doctrine of epigenesis was advanced by **Caspar F. Wolff (1733 – 94)** — German embryologist and physiologist, the concept of development of the embryo, which provides number of consecutive growths of the eggs during some time internal factors and environmental conditions.

The interventricular foramen in the brain, discovered in 1783 by **Alexander Monro (1733 – 1814)** — Scottish anatomist and physician, and named after him foramen of Monro, are passages that connect the paired lateral ventricles with the 3rd ventricle in the middle of the brain. They allow the cerebrospinal fluid (CSF) produced in the lateral ventricles by the choroid plexus to reach in continuity the 3rd and the 4th ventricles.

In 1772 **Joseph Priestly (1733 – 1804)** — English chemist, discovered nitrous oxide / «laughing gas» (N_2O) by thermic disintegration at the temperature (T) 170^{0}C of a dry ammonium nitrate (NH_4NO_3).

The revival of interest in yatrophysics and iatrochemistry activated quackery, cults and charlatanism. The best example of this was the popularity gained by **Franz F.A. Messmer (1734 – 1815)** — German physician, who inadvertently used hypnosis for hysteria.

Domenico F.A. Cotugno (1736 – 1822) — Italian anatomist, physician, and surgeon, described the clinical picture of ischia's that is an affection of the nerve roots of the lumbosacral plexus of the spinal cord, mainly nervus ischiadicus (sciaticus), or dislocation of the intervertebral discs of the lumbar spine causing pain in the lumbo-sacral area and the posterior aspect of the lower extremities (Cotugno disease).

In 1773 **William Hawes (1736 – 1808)** — English physician, performed «mouth-to-mouth» breathing, bringing to life people who appeared to have drowned.

In 1774 **Benjamin Jesty (1736 – 1816)** — English farmer and experimentalize, who inoculated his own wife and two older sons with a discharge from the papule of the cow infected with cowpox, and in 1796 **Edward Jenner (1749 – 1823)** — English physician and scientist, who inoculated an 8-year-old son of his gardener with a discharge from the papule of the milkman's hand infected by cowpox, as a preventive measure (vaccination) against natural smallpox with success in each case.

Michael Underwood (1736 – 1820) — English physician and surgeon, described infantile paralysis in 1789.

James Watt (1736 – 1819) — Scottish engineer, who derived in the Systeme international (SI) of units the electric power in Watts (W) as one Joule (J)[185] per second (s). It is equivalent to the electric current force or strength (I) of one Ampere (A)[186] at the tension of one Volt (V)[187.]

Electrophysiology (EP) was commenced by **Luigi A. Galvani (1737 – 98)** — Italian physiologist, and **Alessandro G.A.A. Volta (1745 – 1827)** — Italian chemist and physicist.

The former, Luigi Galvani, described (1786) a phenomenon of electrical current from contact of the different metals as manifested by contraction of the muscles of posterior paws of the frog secured by the copper hooks when touched with the steel scalpel as a sign of the «animal electricity».

The latter, Alessandro Volta, basing of series experiments came into the conclusion (1800), that the reason for the muscle contraction is not the «animal electricity», but the presence of a different metals in the fluid creating the current of electric charges which are then transferred outside by a conductor.

Nevertheless, Alessandro Volta credited Luigi Galvani for discovery of the electric current phenomenon at the contact of different metals immersed into an electrolyte, during which the different potential between them creates the electric energy (nourishment), named the Galvani's element or the Volta's post.

Among several terms related to the experiment by Luigi Galvani and Alessandro Volta the most important are the galvanic cell – one of a series of cells generating

---

[185] **James P. Joule (1818 - 89)** — English physicist, who derived in the SI units one Joule (J) as a unit of energy (force), work, heat, electric current (I) which equals the work performed by a force of one Newton (N) during the move (transfer) of an object (body) through a distance of one meter (m) in the direction of this force, or by passing of the electric current of one Ampere (A) through a resistance (R) of one Ohm (Ω) for an one second (s), where $1J = 1N \times 1m$.

The first law of **Georg S. Ohm (1789 - 1854)** — German mathematician and physicist, states that the strength of the electric current (I) in A is directly proportional to the electromotive force acting between the extremities of any part of an electric circuit (tension) in Volts (V) and inversely proportional to the R of that part of the conductor measuref in the units of Ω, as depicted by the following equation: $I = V/R$ (1927). The Ohms law found an application in the cardiovascular hemodynamics for the estimation of an area in centimeter (cm)^2 of the cardiac valve orifice or the blood vessel lumen by dividing the cardiac output (CO) or flow (Q) in liter (l) / minute (min) by the diastolic filling period (DFP) or systolic ejection period (SEP) and multiplying it by the heart rate (HR) beats / min. and then dividing it by a constant time the pressure (P) gradient in millimeters of mercury (mmHg). The resistance (R) across the valve or the vascular bed can be calculated by relating the mean P across that cardiac valve or vascular bed to its cardiac output (CO) or gradient (Q).

[186] **André M. Amperé (1775 - 1836)** — French mathematician and physicist who measured the amount of electric charge passing a point in an electric circuit per unit time with 6.241×10^{18} electrons per sec., constituting the stroke volume (SI) unit of force of the electric current of one A (1820).

[187] Alessandro Volta (below) defined the SI unit for the electric potential or the electromotive force (I) of one A, as the difference in the tension of the electric current potential across a wire dissipates one W of power in one Volt (V).

electricity through chemical reaction, the galvanic battery – a series of cells, giving a combined effect of all units, and generating electricity through chemical reaction, and the galvanic current – direct electric current, usually from a battery.

Radical mastectomy is a surgical procedure in which the breast, underlying pectoralis major and minor muscles, and lymph nodes (LN) of the axilla are removed as a treatment of breast cancer. It was first performed by **Bernard Peyrilhe (1737 – 1804)** — French surgeon, in connection with his cancer research. He made the first systematic investigation of cancer, the cancer toxin, the transmission of cancer by injecting extracts of breast cancer into the animal, the nature of the disease, the manner of growth and the treatment.

An interrupted aortic arch (IAA) comprising 1,5% of all congenital heart disease (CHD), initially described in 1778 by **Raphael J. Steidele 1737 – 1823)**[188] — Austrian obstetrician, feature a luminal discontinuity distally to the left subclavian artery (SCA) between the aortic arch (AA) and descending thoracic artery (DTA), usually at the aortic isthmus (type A, 28%), in the distal AA between the left SCA and the left common carotid artery (CCA) (type B, 70%) and in the proximal AA between the left CCA and the brachiocephalic trunk (BCT) (type C, 2%), is almost always associated with ventricular septal defect (VSD) of the conoventricular type and often with the left ventricular outflow tract (LVOT) hypoplasia or obstruction.

The vascular rings are abnormal anatomic arrangement of the great arteries that leads to encirclement of the trachea and esophagus, such as double aortic arch; right aortic arch with left ligamentum arteriosum[189]; circumflex retroesophageal aorta; left aortic arch and anomalous (retroesophageal) subclavian artery (SCA) causing difficulty in swallowing (dysphagia) due to compression on the esophagus, difficulty in breathing (dyspnea), stridor (in Latin – creaking or grating noise) due to high-pitched breath sounds resulting from turbulent air flow in the larynx or in the tracheobronchial tree), occasionally hoarseness due to stretching or compression of the left recurrent laryngeal nerve at the ligamentum arteriosum, or combination of all; pulmonary artery (PA) sling; and innominate artery (IA) compression syndrome.

Of them, the left aortic arch with retroesophageal right SCA was discovered in 1761 by **David Bayford (c. 1739 – 1790)**[190] — English surgeon, first reported in 1784, and named dysphagia lusoria (in Latin – lucus naturae, meaning dysphagia by trick

---

[188] Steidele RJ. Sammlung verschiedener in der Chirurgisch-praktischen Lehrschule gemachten beobachtungen. Trattern, Wien 1778;2:114.

[189] Olearchyk AS. Right-sided double aortic arch in an adult. J Card Surg 2004;19(3):248-51.

[190] Bayford D. An account of a singular case of obstructed degludition, vol 2. Mem Med Soc London, 1st. Edn. Printed by Harry Fry for Charles Dilly in the Poultry 1789;271; Mem Med Soc London, 2st. Edn.1794. — Acherson N. Dysphagia lusoria. J Laryngology and Otology 1952;57;111. — Acherson N. David Bayford. His syndrome and sign of dysphagia lusoria. Ann R Surg Engl 1979 Jan;61(1):63-7.

of nature). The discovery made in his patient, a 33-year woman with long history of progressive dysphagia who eventually died from malnutrition. At autopsy, he found an aberrant right SCA compressing the esophagus.

Dyspnea can be a symptom of the PA or the left main stem bronchus (LMSB) stenosis by neoplasm, and hoarseness due to stretching or compression of neoplasm on the distal aortic arch aneurysm or of the left recurrent laryngeal nerve at the ligamentum arteriosum.

Independently of **Lazzaro Spallanzani (1729 – 69**: above), **Martinus M. Terechowsky, Martyn M. Terekhovsky,** or **Martyn M. Terekhovskii (1740, village of Velyka Pavlivka, Zinkivsky Region,** or **town of Hadiach, Poltava Obl., Ukraine – 02.VII.1796, city of St Petersburg, Russia)**[191] — Ukrainian microbiologist, in his doctoral thesis «De chao infusorio Linnaei», Universite de Strasbourg 1775) concluded that «liquoris animalcule» represent living organisms which on heating or cooling die and do not reappear. That immensely important discovery led to the use of boiling water for sterilization in medicine.

Opanas F. Shafonsky [13(24).XII.1740, village Sosnystsia, Sosnytsky Region, Chernihiv Obl., Ukraine – 27.III.(08.IV).1811, city of Chernihiv, Ukraine][192] — Ukrainian physician, wrote social and medical work «Chernihiv Viceregency Topographic Description with Short Geographic and Historic Specification of Little Russia, i.e., Ukraine)» (1781).

The original apothecaries grew herbs in their gardens, made from them pills, ointment, and infusions for sale. Therefore, they were not held in a high esteem, so they were succeeded by **William Withering (1741 – 99)**[193] — English chemist, geologist, botanist, and physician, who announced in 1785 that the plant foxglove

---

[191] **Carl Linnaeus (1707 - 78)** — Swedish botanists, zologist and physician, who laid the foundations for the modern scheme of binomial nomenclature (Systema naturae. Tomus I-XII. Leiden, Stockholm, Halle, Paris, Leipzig, Vienna 1735-70). — Village of Velyka Pavlivka (founded 1710; organized by the Russian Communist Fascist regime Holodomor-Genocide 1931-33 against the Ukrainian nations caused death of 95 innocent persons), Zinkivsky Region, and town of Hadiach (chronology 10th-13 century, first recorded in chronicle 1533; residents endured organized by Russian Communist Fascist Holodomor-Genocide 1932-33 against the Ukrainian Nation), Poltava Obl. (28,748 km²), Ukraine. — City of Saint (St) Petersburg (established 1703), Russia. — Terekhovsky MM. De virmibus infusoris - de Chao infus. Linnael. Doctoris Dissertatio, Universite de Strasbourg 1775. — Pluschch WM. Martyn Terekhovsky. J Ukr Med Assoc North Am 1970;17,3:50-5. — Ablitsov VH. «Galaxy Ukraine». The Ukrainian diaspora: prominent persons. Publ. «Kyt», Kyiv 2007. — Hanitkevych YaV. The contribution of the Ukrainian physicians to the world medicine. Ukr Med Chasopys 2009 VII/VIII;4(72):110-15. — Synyachenko OV. They glorified themselves and Ukraine: history of domestic therapy. Crimean Therapeutic Journal 2010;2(2):21-8.

[192] Village Sosnytsia (settled in Neolithic V-IV millennium BC, first chronicle's mention 992, 1234), Sosnytsia Region, Chernihiv Obl. (Foot Note 76), Ukraine . — City of Chernihiv (Foot Note 76), Ukraine. — Ablitsov VH. «Galaxy Ukraine». The Ukrainian diaspora: prominent persons. Publ. «Kyt», Kyiv 2007.

[193] Withering W. An Account of the Foxglove and Some of its Medical Uses; with Practical Remarks on the Dropsy, and Some Other Diseases. Publ. Swinney, Birmingham 1785.

(Digitalis purpurea) was effective in treating dropsy (edema, swelling) which is an abnormal accumulation of fluid in the interstitial under the skin and in the body cavities. He thought that it acted as a diuretic but learned of the use of digitalis in congestive heart failure (CHF) from an old woman who practiced as a folk herbalist, used the plant as part of a polyherbal formulation containing over 20 different ingredients to successfully treat this condition. Then he deduced that digitalis was the «active» ingredient in the formulation, and over the ensuing nine years carefully tried out different preparations of various parts of the plant (collected in different seasons), documented 156 cases where he had employed digitalis, gave detailed dose-response characteristics, described the effect and the best and safest way of using it. It was later identified as a cardiac glycoside.

August J. Richter (1742 – 1812)[194] — German surgeon, who described a hernia in which the hernia sac contains a portion of the anti-mesenteric circumference of the intestine protruding through a defect in the abdominal wall, most often in the fibrous ring of external abdominal muscle, causing strangulation, leading to ischemia and perforation without intestinal obstruction or any of its warning signs (parietal or Richter's hernia).

After the extraction by **Ole Borch (1626 – 90**: 17th century**)** of oxygen (O_2) in 1678, **Joseph Priestly (1733 – 1804)** — English naturalist and chemist, in 1771, and **Karl W. Sheele (1742 – 86)** — Swedish chemist-pharmacologist, independently in 1774, again isolated an O_2.

Francois Chopart (1743 – 95) — French surgeon, described the transverse tarsal joint created by the union of os calcaneus with os cuneiforme and os talus with os navicularis (Chopart joint); dislocation of the foot through the talo-navicular and calcaneo-cuboid joint with associated fractures (Chopart dislocation-fracture); and surgical separation of the forefood at the midtarsal joint (Chopart amputation).

In 1789 **Martin H. Klaproth (1743 – 1817)** — German chemist, discovered in the uranium ore of the earth crust a chemical uranium (U).

Antoine L. de Lavoisier (1743 – executed by the French revolutioniers with guillotine **1794)** — French biologist and chemist, developed the oxygen (O_2) theory of combustion, breathing and oxidation (1772-89).

Jean P. Marat (1743 – executed by French revolutionists using guillotine **1793)** — French physician, scientist, and political theorist, explained the mechanism of astigmatism and was one of the first to apply electrotherapy (1779).

---

[194] Richter AG. Abhandlung von den Brüchen. J.C. Dietrich, Göttengen 1778-79. Kapitel 24, Seiten 596-7. — Richter AG. Abhandlung von den Brüchen. Band I & II. Zweite Ausgabe. J.C. Dietrich, Göttengen 1785.

Nestor M. Ambodyk-Maksymovych (1744, village Vepryk, Hadiach Region, Poltava Obl., Ukraine – 1812, city of Petersburg, Russia)[195] — Ukrainian gynecologist and obstetrician, headed an Obstetrical School (1781) and organized the Clinical Obstetrical Institute (1797) in the city of St. Petersburg (founded 1703), Russia. He also authored «The Anatomo-Physiological Dictionary» (1783) and «The Medico-Pathological-Surgical Dictionary» (1785), introduced into the practice the gynecological bed and chairs, obstetrical forceps, and phantom (1784-86).

Bandage binding for fractured clavicles was designed by **Pierre J. Desault (1744 – 95)** — French surgeon, in which the elbow is reposed to the trunk and a soft pad is inserted into the axillary fossa (Desault bandage, 1779).

The creator and eminent organizer of the community (public) health service, **Andriy (Jędrzej, Andrzej, Andreas) Krupinski (11.XI.1744, village Bielane, gmina Kęty, powiat Oświęcim (Auschwitz), Voivodeship Malopolskie, Poland – 27.IV.1783, city of Lviv, Ukraine)**[196] — Ukrainian anatomist and physician of Polish origin, laid the base for the medical education, health care and sanitary security for the population of Halychyna (Galicia)[197].

---

[195] Village Vepryk (founded in first quarter of 17th century; suffered from conducted by Russian Communist Fascist regime of Holodomor-Genocide of 1932-33 and 1946-47 against the Ukrainian Nation), Hadiach Region, Poltava Obl., Ukraine. — Puchkivsky O. Three founders of Russian medicine (Nestor Ambodyk-Maksymovych, Petro Zahorsky, and Danylo Vellansky). Sci J Ukr Studies «Ukraine», AUASc, St Publ «Ukraine», Kyiv 1924. Book 4, P. 27-34. — Puchkivsky O. Three founders of medicine. J Ukr Med Assoc North Am 1961 July-Oct;8,(22-23):40-8. — Ablitsov VH. «Galaxy Ukraine». The Ukrainian diaspora: prominent persons. Publ. «Kyt», Kyiv 2007. — Synyachenko OV. They glorified themselves and Ukraine: history of domestic therapy. Crimean Therapeutic Journal 2010;2(2):21-8.

[196] Village Bielane (first documented in writing 1238), gmina Kęty, powiat Oświęcim, województwo Małopolskie (area 15,182.97 km²), Poland. — Krupinski A. Tractatus de febribus acutis generatim acceptis. Publ Antoni Piller, Leopoli 1774. — Krupinski A. Osteologia. Publ Antoni Piller, Lviv 1774 — Krupinski A. Splanchnology, or Science about Viscera. Publ Antoni Piller, Lviv 1775. — Krupinski A. The Blood Vessels of a Man. Publ Antoni Piller, Lviv 1776. — Krupinski A. Naevi in tendons. Publ Antoni Piller, Lviv 1777. — Krupinski J. Medicine and Surgery. Vol. 1-5. Publ Antoni Piller, Lviv 1774-77. — Krupinski A. The Knowledge about Floods in General, Especially Regarding to the Goats Mineral Water. Pochaïv 1782. — Szumowski W. Galicya pod względem medycznym za Jedrzeja Krupińskiego pierwszego protomedyka 1772-1783. Z portretem. Narładem Towarzystwa dla Popierania Nauki Polskiej, Lwów 1907 — Ablitsov VH. «Galaxy Ukraine». The Ukrainian diaspora: prominent persons. Publ. «Kyt», Kyiv 2007. — Synyachenko OV. They glorified themselves and Ukraine: history of domestic therapy. Crimean Therapeutic Journal 2010;2(2):21-8.

[197] As a part of constituent of the Austro-Magyar Empire in 1772 – 1918 the Ukrainian ethnographic space was separated under the name of Halychyna (Galicia), which consisted of Eastern Halychyna (include nowadays Lviv (LV) Obl., Ivano-Frankivsk (IF) Obl. (area 13,900 km²) and Ternopil (TE) Obl. (area 13,900 km²), a general area of 49,556 km², with the center in city of Lviv, Ukraine, and Western Halychyna (include nowadays Subcarpathian and Small Poland Voivodeships; general area 33,059 km²) with the center in city of Cracow, Poland. The name of Halychyna originates from town of Halych (first mentioned 896), Ivano-Frankivsk Obl., Ukraine — former capital of The Halych Kingdom (1084-1384).

After completion his studies at Medical Faculty of the Vienna University (1765 – 71) and successfully defending his thesis for the degree of Medical Doctor («theses medicinae pro honorabus doctoralibus obtinendis») entitled «Tractatus de febribus acutis generatim acceptis» (Viennae 1772), he was assigned by **Maria Theresa (1717 – 80)** — Empress of Austria, by the decree of 22.XII.1772 for the position of provincial physician (medicus provincialis; 1773 – 76), then chief physician (protomedicus; 1776 – 83) of Halychyna, and professor of medicine, with the aim to organize the medical education, and to improve the medical service of its population.

After arriving to Halychyna, he immediately realized and have seen that the medical aid is served to the people by unqualified and uncontrolled, mainly the foreign, barbers (blood-letters) and apothecaries, and the great problem was lack of competent and diplomate midwifes, and as a result of this, during delivery of child, often mothers were dying together with their newborn.

With the aid to improve the relationship between the health care providers he introduced into a life on 20.III.1773 the sanitary patent, postulated the creation of provincial pharmacies and implementation of prices on separate drugs.

This same day, 20.III.1773, on his importunity, already on 01.X.1773 was opened the Lviv Collegium medicum, which became the first in Western Ukraine middle-class medical facility to accelerate (hasten) teaching furnish with a diploma midwifes. He himself became a rector and professor-lecturer of anatomy, general pathology, therapy, and obstetrics.

During next few years he wrote and published number of textbooks with the aid to help physicians, barbers, midwifes and apothecaries to enhance the knowledge of medicine.

He stressed the need for teaching and training physicians among the local population to meet the health care demands of Halychyna. Wherefore, he required to create on basis of the Lviv Collegium medicum the Medical Faculty of the Lviv University (established 1781), to educate physicians, barbers, and apothecaries from graduates of the local gymnasiums (high schools).

Courageously, he rose against, despite of strong resistance from his colleagues, the use of German language in medical facilities.

He hoped to lead at the Medical Faculty of the Lviv University Department of Clinical Medicine and Pathology, but unfortunately his premature death did not let him to fulfil that dream.

While studying and fighting epidemics of the plague, **Danylo (Daniil, Danilo) S. Samoylovych (Samoilovich) - Shushkivsky, or Daniel Samoïlowitz [11(22). XII.1744, village Yanivka / Ivanivka, Chernihiv Region, Chernihiv Obl.,**

Ukraine – 20.II(04.III).1805), city of Mykolaïv, Ukraine][198] — Ukrainian physician and epidemiologist, was the first to prove its contagiousness (infectiousness) by bacilli-carrying people during direct contact with diseased person or from infected objects, described resistance to infection of those individuals who survived it, advocated disinfectants and inoculation against plague using attenuated vaccines (1781, 1803).

Jeans L. Baudelocque (1745 – 1810)[199] — French gynecologist and obstetrician, constructed a pelvimeter (Baudelocque), an antropometric callipers, to measure the external diameter (line) of the pelvis (conjugata externa Baudelocque, 1775-81).

Nykon K. Karpinsky (01.VII.1745, town of Lubny, Poltava Obl., Ukraine – 12.IX.1810, St Petersburg, Russia)[200] — Ukrainian Russian physician, who defended the doctoral dissertation on «De impedimentis in lythotomia occurentibus» (Strasbourg Univ., 1781), composed the book entitled «Pharmacopea Rossica». Pars I-II. Petropoli 1898), since 1805 general staff physician of the military department.

Phillipe Pinel (1745 – 1826) — French physician, pleaded for the humane treatment of the mentally ill.

Johann P. Frank (1745 – 1821) — German physician and hygienist, used statistics to establish the importance of public health and developed a modern plan for systematic health coverage, thus conceiving «a cradle-to-grave care» (1777-78).

By modifying the principle of **Daniel Bernoulli (1700 – 84**: above**), Giovanni B. Venturi (1746 – 1822)** — Italian physicist, established that the reduction in fluid pressure as it flows through a tube with constricted section (and the increasing velocity) creates a pull (jet) and drag (push) effect (Venturi effect, 1799), as depicted by the following equation:

---

[198] Village Yanivka / Ivanivka (founded 1635; suffered from organized by Russian Communist Fascist regime of Holodomor-Genocide 1932-33 against the Ukrainian nation), Chernihiv Region, Chernihiv Obl., Ukraine. — City of Mykolaïv (known since 5th century BC, founded 1789; during conducted by Russian Communist Fascist of Holodomor-Genocide 1932-33 against the Ukrainian Nation, died at least 6045 residents of the city, and peasants of the oblast died slowly from man-made hunger), Mykolaïv Obl. (area 24,598 km²), Ukraine. — Samoïlowitz D. Opuscules sur la peste. Qui, en 1771, ravagea Moscou; avec un discours aux éléves des hospitaux de l'empire de Russie. Chez Le Clerc, Libraire, quai des Augustins, a la Tolfon d'or, Paris 1787. Une page I-VII, 1-208. — Ablitsov VH. «Galaxy Ukraine». The Ukrainian diaspora: prominent persons. Publ. «Kyt», Kyiv 2007. — Synyachenko OV. They glorified themselves and Ukraine: history of domestic therapy. Crimean Therapeutic Journal 2010;2(2):21-8.

[199] Baudelocque J-L. Principe des accouchements. Paris 1775.

[200] Town of Lubny (founded 988; presence in town and surrounding villages of Russian Communist Fascist regime and NKVS agents were to organize Holodomor-Genocide 1921-23, 1932-33 and 1946-47, and political terror of 1937-38 against the Ukrainian Nation, during which thousands of people were killed), Poltava Obl., Ukraine. — Ablitsov VH. «Galaxy Ukraine». The Ukrainian diaspora: prominent persons. Publ. «Kyt», Kyiv 2007. — Synyachenko OV. They glorified themselves and Ukraine: history of domestic therapy. Crimean Therapeutic Journal 2010;2(2):21-8.

$$p_1 - p_2 = \frac{\rho}{2}(v_2^2 - v_1^2),$$

where p = density of fluid, v_1 = the (slower) fluid velocity where the pipe is wider, v_2 = the (faster) fluid velocity where the pipe is narrower. This assumes the flowing fluid (or other substance) is not significantly compressible - even though pressure varies, the density is assumed to remain approximately constant.

Antonio Scarpa (1747 – 1832) — Italian anatomist, described the deep (membranous, Scarpa) layer of the superficial abdominal fascia; and the subfascial femoral (Scarpa) triangle, bounded superiorly by the inguinal ligament, laterally by the medial border of the sartorius muscle and medially by adductor longus muscle, containing from outside to inside the femoral nerve, the common, deep and the proximal superficial femoral artery (CFA, DFA, SFA), femoral vein, covered by the femoral sheath, and the deep inguinal lymph nodes (LN).

The structural and functional unit of the kidney is nephron comprising of a renal corpuscle (glomerulus) and renal tubule (tubulus renalis). The kidneys are considered the most «intelligent» organ of the human body.

In his doctoral thesis «De structura renum» (Strasbourg 1782 et 1788) **Olexandr (Alexander) M. Schumlansky (1748, village of Mali Budyshcha, Zinkivsky Region, Poltava Obl., Ukraine – 1795, city of Moskva, Russia)**[201] — Ukrainian anatomist and surgeon, described a chalice-like capsula glomeruli or gromerular capsule of Shumlansky (corpusculum renale Schumlansky) encompassing the renal corpuscle, a tufts of calillaries formed by the afferent and efferent arterioles (arteriola afferent et efferent) in the cortex of the kidney (cortex renis). From the renal capsule extends the descending proximal convoluted tubule (tubulus contortus proximalis), then the descending proximal straight tubule of Schumlansky (tubulus rectus proximalis Schumlansky) which create the U-like loop (tubulus attentualis) of the medullary portion of the kidney (medulla renis), with further extension into the distal straight tubule of Schumlansky (tubulus rectus distalis Schumlansky) and the distal convoluted tubule (tubulus contortus distalis), which join the collecting tube of the kidney (tubulus renalis colligent) to take the extravascular fluid for the secretion of urine.

---

[201] Village Mali Budyshcha (known since 18th century; victims of organized by Russian Communist Fascist regime of Holodomor-Genocide 1932-33 against the Ukrainian Nations were 932 persons, and from political repressions 1937-39 of 68 countryman), Zinkivsky Region, Poltava Obl., Ukraine. — Schumlansky A. Dissertatio inauguralis anatomica. De Structura Renum quam pro licentia summos in medicina honores et priviledia doctoralia. Une 1-oe edition et une 2-oe edition. Universite de Strasbourg. Typis Lorenzii & Schulari, Direct. Nobilit. Typogr 1782 et 1788. — Ablitsov VH. «Galaxy Ukraine». The Ukrainian diaspora: prominent persons. Publ. «Kyt», Kyiv 2007. — Synyachenko OV. They glorified themselves and Ukraine: history of domestic therapy. Crimean Therapeutic Journal 2010;2(2):21-8.

Mykhaylo L. Hamaliya (1749, village Krypoderyntsi, Orzhytsky Region, Poltava Obl., Ukraine – 1830, city of Tula?, Russia)[202] — Ukrainian physician-epidemiologist, authored the monography «On the Account of Sibirka» (1792), i.e., anthrax in animal **[Sauveur Morand (1697 – 1773**: above**]**.

The symptoms, signs, and outcome of the aneurysm of the left ventricle (LV) of the heart and the аорта is explained by the law of LaPlace **[Pierre S. de LaPlace (1749 – 1827)** — French mathematician, physicist, and astronomer], according to which an average circular tension on their wall during systole is directly proportional to the intraluminal pressure and the curvature of the radius and reversely proportional to the wall thickness, as outlined by the following equations:

tension of the wall of the LV or aorta during systole (T) = intraluminal pressure (P) x radius (r) / 2 x wall thickness (h),

or

$$T = P \times r / 2 \times h$$

where

T = tension (wall stress), P = intraluminal pressure, r = radius and h = wall thickness,

or

$$\Delta P = \frac{T_1}{r_1} + \frac{T_2}{r_2}$$

where

delta P = pressure difference across the wall of the LV or aorta, T_1 i T_2 = tension of the LV or aortic wall in proportional directions, r_1 i r_2 = the largest and smallest principal curved radius of the LV or aortic wall.

The highest intraluminal pressure during diastole on LV or aortic wall, specifically in patients with hypertension, the thinnest wall and increased radius of the LV or aortic

---

[202] Village Krupoderyntsi (created 1663), Orzhytsky Region, Poltava Obl., Ukraine. — City of Tula (first mentioned in a chronicle during military campaign 1146), Russia. — Ablitsov VH. «Galaxy Ukraine». The Ukrainian diaspora: prominent persons. Publ. «Kyt», Kyiv 2007. — Synyachenko OV. They glorified themselves and Ukraine: history of domestic therapy. Crimean Therapeutic Journal 2010;2(2):21-8.

wall, the highest risk of a progressive dilatation, aneurysm formation, dissection, rupture of its wall and death.

That the anterior roots of the spinal nerves are motor roots, and the posterior roots are sensory was suspected in 1784 **Jiri (Georg) von Prochaska (1749 – 1820)**[203] — Czech anatomist, physiologist, and eye physician (ophthalmologist).

In 1772 **Daniel Rutherford (1749 – 1819)**[204] — Scottish chemist, botanist and physician, isolated nitrogen (N).

Pavlo M. Schumlansky (1754, village Mali Budyshcha, Zinkivsky Region, Poltava Obl., Ukraine – 1824, city of Kharkiv, Kharkivs'ka Obl., Ukraine)[205] — Ukrainian surgeon, defended Medical Doctoral Dissertation «On the account of the direct reason for local inflammation» (Universite de Strasbourg, 1789). The work was translated into German language and considered the better surgical work published in 1794 in city of Leipzig (appeared in 900), Saxony (area 18,449.99 km²), Germany.

Ivan L. Danylevsky (1751, city of Kyiv, Ukraine – 1807, ?)[206] — Ukrainian physician, a representative of the social hygiene, in his doctoral thesis «De magictratu medico - felicissimo» (Göttingen 1784)[207] motivated the need for state of care in the health of his nation, acknowledged the conditions, which according to him, would be favorable to lower morbidity and mortality of the population.

Physician-in-Chief of Russian Army under the leadership of **Alexander V. Suvorov 1729 – 1800)** — Russian general and generalissimo of the Russian empire (1746 – 1800), **Yukhym (Euphemins) T. Bilopolsky (1753 – ?)**[208] — Ukrainian physician, advocated

---

[203] Prochaska G. Disquisitio anatomico-physiologica organismi corpori humani eiusque processus vitalis. Anton de Haykul, Vienna 1812. — Schiller F. Jiri Prochaska: on 250th universary of his birth. J Hist Neuroscience 1999 Apr;8(1):70-2. — Chvatal A. Jiri Prochaska (1749-1820: Part 1: A significant Czech anatomist, physiologist, and neuroscientist of the eighteenth century. J Hist Neurosci 2014 Jun;23(4):1-10. — Chvatal A. Jiri Prochaska (1749-1820: Part 2: «De structure nervorum» – Studies on a structure of the nervous system. J Hist Neurosci 2014 Jul;24(1):1-25.

[204] Rutherford D. Dissertatio inauguralis de aerefixio, aut mephitico. University of Edinburg 1772.

[205] Village Mali Budyshcha (Foot Note 201), Zinkivsky Region, Poltavs'ka Obl., Ukraine. — City of Kharkiv (settled since the Bronze Age 2,000 BC, by Scythians 6th-3rt century BC, by Sarmatians 2nd-st century BC, by early Slav's 2nd-6th century, by fortified Slavic town 12th-17th century, established 1654; since 1930 greatly suffered from conducted by Russian Communist Fascist regime destruction of the Ukrainian culture, of Holodomor-Genocide 1932-33 against the Ukrainian Nation, political repressions, and as a result thousands of Ukrainian intellectuals were shot), Kharkivs'ka Obl. (area 31,415 km²), Ukraine. — Schumlansky PM. De proxima toricae inflamationis causa. Doctor medicinae dissertation, Universite de Strasbourg. Typis Lorenzii & Schulari, Direct. Nobilit. Typogr 1789. — Ablitsov VH. «Galaxy Ukraine». The Ukrainian diaspora: prominent persons. Publ. «Kyt», Kyiv 2007.

[206] Ablitsov VH. «Galaxy Ukraine». The Ukrainian diaspora: prominent persons. Publ. «Kyt», Kyiv 2007. — Synyachenko OV. They glorified themselves and Ukraine: history of domestic therapy. Crimean Therapeutic Journal 2010;2(2):21-8.

[207] University of Göttingen (established 1734), the city of Göttingen (first mentioned in 953), Germany (area 357,021 km²).

[208] Ablitsov VH. «Galaxy Ukraine». The Ukrainian diaspora: prominent persons. Publ. «Kyt», Kyiv 2007.

using prophylactic measures to strive removing causes of illnesses, and not be limited only to the treatment of patients («Principles of Medical Approaches», 1793).

Jean N. Corvisart (1755 – 1812) — French physician, was especially fond of **Leopold** von **Auenbrugger (1722 – 1809**: above**)** use of chest percussion, and he began to perfect and popularize this technique.

For the first time in a clear appearance, particles of cholesterine crystals from the gallstones **[Francois de La Salle (1719 – 88)**: above] were isolated in 1789 by **Antoine F. de Fourcroy (1755 – 1809)** — French physician and politician.

With the treatise entitled «Essay on a new Principle for Ascertaining the Curative Powers of Drugs» (1796), **Chrisitian F.S. Hahnemann (1755 – 1843)**[209] — German physician, created an alternative form of medicine called homeopathy by using small doses of drugs, which in larger doses would cause the manifestations (symptoms and signs) alike to a certain disease, and those smaller doses of drugs became curative. It appears that he got the idea of homeopathy from **Mithridates VI (134 – 63 BC**: Ancient medicine. *Ukraine-Scythia***)**.

The weighty contribution to medicine made **Caled H. Parry (1755 – 1822)**[210] — English physician, by description of diffuse toxic multinodular exophthalmic goiter (DTG) and thyrotoxic nodular goiter (Parry disease, 1786); inquire into ischemic heart disease (IHD) manifested by the syncope anginose as the heart demanding energy without the circulatory system able to deliver it (1799); correct conclusion that the pulse wave is generated by rhythmical contraction of the left ventricle (LV) of the heart and showing that the pressure on the carotid artery slows the heart rate (1716); and description of a rare condition characterized by progressive unilateral or bilateral facial atrophy due to scleroderma.

The renewed interest in sea bathing and hydrotherapy was sparked by **James Currie (1756 – 1805)** — Scottish physician and anthologist.

The satirical engraving by **Thomas Rowlandson (1756 – 1827)** — English artist and caricaturist, «Transplantation of teeth» (1787) in which the rich gentlemen and ladies paid for the healthy teeth to be pulled from poor children and transplanted in their gum by a dentist portrayed him as a quack. The teeth transplants were popular for some time in England (the 18th and in the beginning of the 19th centuries), although they rarely worked.

---

[209] Von Hahnemann DS. Versuch über ein neues Prinzip zur Auffindung der Heilkräfte der Arzneisubstanzen, nebst einigen Blicken auf die bisherigen. Hufelands Journal 1796;2:391-439.

[210] Parry CH. Inquire into the Symptoms and Causes of the Syncope Anginosa, Commonly Called Angina Pectoris: Illustrated by Dissections. Printed by R. Cruttwell, London 1799. — Parry CH. Enlargement of the Thyroid Gland in Connection with Enlargement or Palpitation of the Heart. Posthumous Collections from the Unpublished Medical Writing of C.H. Parry. London 1825. P. 111-29.

Yakiv T. Sandul-Sturza (1756, village of Kozatske, Bobrovsky Region, Chernihiv Obl., Ukraine – 1816)[211] — Ukrainian physician, defended dissertation for the degree of Doctor of Medical Science (DMS) «About Crimean's people's disease» (1792), and studied leper (leprosy) in Ukraine. Crimean's people's disease, a.k.a., Crimean hemorrhagic fever with signs of nettle-rash, ague (fever) and shivering or shaking.

The founder of dermatology **Robert Willan (1757 – 1812)**[212] — English physician, attempted to classify diseases of the skin to which he included impetigo, lupus vulgaris, psoriasis [Ancient medicine. *Romans* – **Aulus Celsius (c. 25 BC – c. 50 AD)**], systemic scleroderma (SSD) [17th century – **Zacutus Lisitanus (1577 – 1642)**], ichthyoses – heterogeneous family of mostly genetic disorders characterized by dry, thickened, or scaly skin, sycosis – a papulose inflammation of hair follicles of the beard, moustache and hairy part of the head, pemphigus, psoriasis diffuse – an occupational disease which affects hand and arms of bakers (1798), and exanthemas rash of childhood known as erythema infectious (1799).

Hryhoriy I. Bazylevych (1759, village of Boromla, Trostyanetsky Region, Sumy Obl., Ukraine – 26II.1802, city of Moscow, Russia)[213] — Ukrainian military physician, after study in the Kyiv-Mohyla Academy (1659 – 1817), the Kharkiv College [established 1722 in the town of Bilhorod / Belgorod (founded 1596), Bilhorod Obl. (area 27,100 km^2), Russia, moved in 1785 to the city of Kharkiv, Ukraine], since 1785 to the Petersburg Medical-Surgical School (created 1714, 1786), Russia, and then to the Strasbourg University (established 1621), city of Strasbourg (first mentioned under his ancient name »Angentoratum« in the year of 12 BC), France, where he defended dissertation for the scientific degree of doctor of medicine entitled «Dissertatio de systemate resorbente Argentorali» / «About the System of the Suction Vessels» (1793). Year later (1794), he was granted the position of the Chair Professor of Pathology and Therapy at Alma mater, introduced clinical sections at the military hospitals.

Franz K. Hesselbach (1759 – 1816)[214] — German anatomist and surgeon, described the femoral (Hesselbach) hernia (1798) – a hernia where a loop of

[211] Village Kozatske (founded 1587; greatly suffered from organized by Russian Communist Fascist regime of Holodomor-Genocide 1932-33 against Ukrainian Nation), Bobrovsky Region, Chernihiv Obl., Ukraine. — Ablitsov VH. «Galaxy Ukraine». The Ukrainian diaspora: prominent persons. Publ. «Kyt», Kyiv 2007.

[212] Botth CC. Robert Willam (1757-1812): Dermatologist of the Millennium. J R Soc Med 1999;96(2):313-8.

[213] Village Boromla (founded 1658; suffered from intervention by Russian Communist Fascist regime of the RSFSR in 1917-21, conducted by Russian Communist Fascist regime of Holodomor-Genocide 1932-33 against the Ukrainian nation), Trostyanetsky Region, Sumy Obl., Ukraine. — Ablitsov VH. «Galaxy Ukraine». The Ukrainian diaspora: prominent persons. Publ. «Kyt», Kyiv 2007. — Synyachenko OV. They glorified themselves and Ukraine: history of domestic therapy. Crimean Therapeutic Journal 2010;2(2):21-8.

[214] Hassebach FK. Anatomisch-Chirurgische Abhandlung über den Urspurg der Leistenbrüche. Baumgärtner, Würzburg 1806. — Hassebach FK. Neueste anatomish-pathologische Undersuchungen über den Ursprung und das Fortschreiten der Leisten- und Schenkelbrüche. Baumgärtner, Würzburg 1814.

intestines passes through cribriform (Hasselbach) fascia (1806) which is a part of the subcutaneous femoral fascia covering the femoral canal and the hiatus of the great saphenous vein (GSV); described ligamentum (Hesselbach) interfoveolare (1806) – the thickening of the transversalis fascia from the medial side of the deep inguinal ring, above connects with musculus transversus abdominis and below with ligamentum inguinale; described trigonum inguinale Hasselbachi (1806), bounded inferiorly the by ligamentum inguinale, medially by the exterior border of musculus rectus abdomini, also called linea semilunaris, and superolateral by the inferior epigastric vessels, the site of origin of an adult direct inguinal hernia.

He distinguished congenital / adult indirect inguinal hernia from direct adult inguinal hernia (1810). The congenital / adult indirect inguinal hernia protrudes through the inguinal canal and is the result of the failure of embryonal closure of the processus vaginalis after testicle passes through it, covered by the internal spermatic fascia, located laterally to the inferior epigastric vessel. The adult direct inguinal hernia enters through a weak point in the fascia of the abdominal wall (Hasselbach triangle), located medial to the inferior epigastric vessels.

Stepan S. Prokopovych-Andriyevsky (1760, village of Saltykova Divytsia, Kulikivsky Region, Chernihiv Obl., Ukraine – 19.XII.1818, city of Astrakhan, Russia)[215] — Ukrainian physician, in his work about anthrax **[Sauveur Morand (1697 – 1773). / Mykhailo Hamaliya (1749 – 1830)**: above] which he named «Siberian ulcer», that is «On Siberian Ulcers» (1786) where he described the pulmonary, gastrointestinal, and cutaneous form of anthrax in Western Siberia (Siberia - area 7,511,000 km^2), thus its other name is now «sibirka» in animals. He was the first to prove its contagion or contagious peculiarity (1789). In a self-experiment, described in his other work entitled «A Short Description of Sibirka» (1796), showed that human and animal anthrax were identical and proved the possibility of its transfer from animal to people.

Detailed description of the cyanotic congenital heart disease (CHD) – transposition of the great arteries (TGA) was made in 1793 by **Mathew Baillie (1761 – 1823)**[216] — Scottish pathologist and physician.

One of the early descriptions of the idiopathic cranial 7th (facial) nerve paralysis was made by **N. Anton Friedreich (1761 – 1838)** — German pathologist.

---

[215] Village of Saltykova Divytsia (settled by the Ukrainian Cossacks in 10th – 12th century; from conducted by Russian Communist Fascist regime of Holodomor-Genocide 1932-33 against the Ukrainian Nation suffered 21 farmers), Kulikivsky Region, Chernihiv Obl., Ukraine. — City of Astrakhan (settled by the Tatars in 13th century). — Lapychak T. Physicians Andriyevsky's. J Ukr Med Assoc North Am 1961 Jan;8,(20):43-7. — Ablitsov VH. «Galaxy Ukraine». The Ukrainian diaspora: prominent persons. Publ. «Kyt», Kyiv 2007. — Synyachenko OV. They glorified themselves and Ukraine: history of domestic therapy. Crimean Therapeutic Journal 2010;2(2):21-8.

[216] Baillie M. The Most Morbid Anatomy of Some of the Most Important Part of the Human Body. Printed by Barber & Smithwick, Albany 1797.

As early as in 1764 **Giovanni Aldini (1762 – 1834)** — Italian physician, showed that the intermittent electric stimulation of the precordium can resuscitate the patient.

Petro A. Zahorsky, or **Petr A. Zagorsky [09(20).VIII.1764, u.t.s., of Ponornytsia, Konopsky Region, Chernihiv Obl., Ukraine – 19.III(01.IV).1846, city of Moscow, Russia]**[217] — Ukrainian anatomist, described the anomalous aortic arch (1750-76), discovered the accessory nucleus of the cranial 6th (abducent) nerve to the musculus rectus oculi (1859-97), described the congenital heart disease (CHD) – atrial septal defect (ASD) of the fossa ovale type in an adult (1893), and also discovered the nucleus of the cranial 12th (hypoglossal) nerve (1801).

John Dalton (1766 – 1844) — English chemist, physicist, and meteorologist, known for his research into color blindness, from which suffered he and his brother, a.k.a., daltonism, defined the important characteristics of an x-linked genetic disorders in people with color blindness by studying himself and his own family (1794); introduced atomic theory into chemistry.

In 1798 **John A. Wilson**[218] reported on the CHD – a persistent (single) truncus arteriosus (PTA) arriving from the heart that gives the origin to the main pulmonary artery (MPA), ascending aorta (AA), coronary arteries, brachiocephalic trunk (BCT), and on the congenital chest wall deformity – thoracoabdominal ectopia cordis with associated somatic defect of the pericardium, diaphragm, and abdominal wall, as well as the intrinsic cardiac anomalies.

Thomas R. Malthus (1766 – 1834) — English cleric and scholar in the field of political economics and demography, developed hypothesis about the surpassed increase of the population in relations to the available consumption resources (Malthus law, 1798).

---

[217] Urban-type settlement (u.t.s.) of Ponornytsia (founded 1654; since 1929 Russian Communist Fascist regime began systematic terror against independent peasants, and during organized Holodomor-Genocide 1932-33 against the Ukrainian nation when died at least 99 victims), Konopsky Region, Chernihiv Obl., previously Novhorod-Siversky Vicregency (1781-96), Ukraine. — Zagorsky P. De arcus aortae abnormitate ei unius ramorum ejus ortu insolito. Mem de l'Acad 1750-76;2:318-20. — Zagorsky P. De supernumerario sive abducente accessorio oculi musculo in cadavere hominis observata. Mem de l'Acad 1859-97;7:396-99. — Zagorsky P. De foramen ovalis cordis in-adultus (Interatrial Septal Defect of the Oval Type in an Adult). Doctoral Dissertation, St Petersburg 1793. — Zagorsky P. The Concise Anatomy, or Textbook to became acquinted with the Structure of Human Body. Vol. 1-2. St Petersburg 1802. — Zagorsky P. Rare Parts of the Human Body. St Petersburg 1819. — Puchkivsky O. Three founders of Russian medicine (Nestor Ambodyk-Maksymovych, Petro Zahorsky, and Danylo Vellansky). Sci J Ukr Studies «Ukraine», AUASc, St Publ «Ukraine», Kyiv 1924. Book 4. P. 27-34. — Puchkivsky O. Three founders of medicine. J Ukr Med Assoc North Am 1961 July - Oct;8,(22-23):40-8. — Ablitsov VH. «Galaxy Ukraine». The Ukrainian diaspora: prominent persons. Publ. «Kyt», Kyiv 2007.

[218] Wilson JA. A description of a very unusual formation of the human heart. Philos Trans R Soc London 1798;2:346-56.

Nervism (reflex) theory (1800) was introduced into medicine by **Yefrem (Efrem) O. Mukhin (28.I(08.II).1766, village Zarozhne, Chuhuïvs'ky Region, Kharkiv Obl., Ukraine – 31.I.(12.II).1850, village Koltsovo, Tarus'ky Region, Kaluz'ka Obl., Russia)**[219] — Ukrainian anatomist and surgeon, who recommended resuscitation of drowning, strangled and chocked victims (1805), begun freezing of the cadavers for the axial examination of anatomy (1818) and also promoted physical therapy / physicotherapy (PT).

One of numerous prominent scientists of the German family that produced five generation of astronomers was **Christian A. Struve (28 or 22.I.1767, town of Görlitz, Germany – 06.XI.1807, Görlitz, Germany)**[220] — German physician and pediatrist, the author of «The Handbook of Children Diseases: Specially to be Used by the Parents and Educators» (1787); encouraged to perform cardio-pulmonary resuscitation (CPR; in German – the Herz-Lungen Wiederbelebung), i.e., try to revive people who suddenly lost consciousness, and vital signs (breathing, palpable pulse over the femoral and carotid arteries, and beating of the heart); maintained *«that it unnecessary to state one's position in the matter»*.

Khoma I. Borsuk-Moyseyev, genuine name was Khoma Moysa **(1768, Chernihiv Obl., Ukraine – 23.VI.1811, city of Moscow, Russia)**[221] — Ukrainian physician, graduate of the Kyiv-Mohyla Academy (1788), who in 1788-93 studied at the Medical Faculty of the Moscow University (opened 1855, composed of the philosophical, juridical and medical departments), where he successfully defended dissertation for the degree of medical doctor entitled «Dissertation medico-practicae de respiration»

---

[219] Village Zarozhne (founded 1647), Chuhuïvs'ky Region, Kharkiv Obl., Ukraine. — Village Koltsovo, Tarus'ky Region, Kaluz'ka Obl. (area 29,800 km²), Russsia. — Mukhin EO. De stimulus, corpus humanum vivum afficientibus. Gottingae 1804. — Mukhin EO. Description of Surgical Operations. Moscow 1808. — Ablitsov VH. «Galaxy Ukraine». The Ukrainian diaspora: prominent persons. Publ. «Kyt», Kyiv 2007. — Synyachenko OV. They glorified themselves and Ukraine: history of domestic therapy. Crimean Therapeutic Journal 2010;2(2):21-8.

[220] Town of Görlitz (settled before 1002), State of Saxony (area 18,415.66 km²), Germany. — Struve CA. Neues Handbuch der Kinderkrankheiten: besonders zum Gebrauch für Eltern und Erzieher. Verlag JF Korn dem Aeltern, Breslau 1787. — Struve CA. Versuch über die Kunst Scheintodte zu beleben und über die Rettung in schnellen Todesgefahren, ein tabellarisches Taschenbuch. Hannover 1797. — Struve CA. Triumph der Heilkunst. Verlag JF Korn dem Aeltern, Breslau 1800-1800. — Struve CA. Athenology: or the Art of Preserving Feeble Life; and of Supporting the Constitution under the Influence of Incurable Disease. Publ. Murray & Highlay, London 1801. — Tutzke D. Christian August Struve, 1767 – 1807; Leben und Werk eines Görlitzer Arztes im Dienste des Humanismus der Aufklärungszeit. Festschrifr zur 150. Wiederkehr seines Todestages am 6. November 1957. Verlag Rat der Stadt 1957. — Wunderlich P. Christian August Struve (1767-1807) and pediatrics. On the 28th anniversary of his birthday on 28 January,1967. Monatsschr Kinderheilkd 1967 Sep.;114(9):480-1.

[221] Ablitsov VH. «Galaxy Ukraine». The Ukrainian diaspora: prominent persons. Publ. «Kyt», Kyiv 2007. — Synyachenko OV.

Они glorified themselves and Ukraine: history of domestic therapy. Crimean Therapeutic Journal 2010;2(2):21-8.

(1793), devoted to physiology of breathing. Then, he in Alma mater lectured therapy, semantic and diet.

Phylip S. Physick (1768 – 1837)[222] — American physician and surgeon, who removed cataract; performed plastic surgery on harelip; increased the length of the splint to treat femoral bone fractures; used a seton to treat ununited bone fractures; pioneered the use of the stomach pump for gastric lavage, a.k.a., stomach pumping or gastric irrigation for cleaning out the content of the stomach on persons who ingested a poison, overdosed on a drug or became intoxicated by alcohol; devised a flexible catheter; invented a wire snare for removing tonsils and hemorrhoids; introduced absorbable catgut sutures to replace silk and a flax then used; designed novel forceps for removal of bladder stones (lithotomy) and successfully repaired an arterio-venous aneurysm (1893-1831).

M.F. Xavier Bichat (1771 – 1802) — French anatomist, histologist, physiologist, and pathologist, considered the tissue to be the prime element in pathology.

In 1797-1800 **Friedrich W. J. Schelling (1774 – 1854)** — German philosopher, re-vivid the ancient natural philosophy that has been undertaking to explain phenomena by natural causes without recourse to mythical being and extended it into his theory that there is an eternal and unchanging law of nature, processing from the absolute, from which laws governing natural phenomena and forces derive. The natural philosophy is the precursor of contemporary natural science.

The supporter of Friedrich Schelling, **Danylo M. Vellansky-Kavunnyk [11(22). XII.1774, town Borzna, Chernihiv Obl., Ukraine – 15(27).III.1847, city of Moscow, Russia]**[223] — Ukrainian physician, in 1805-37 lectured anatomy, physiology and philosophy at the Medico-Surgical Academy (founded 1798), city of St Peterburg, Russia, presented to the academy his doctoral dissertation about the interrelation between philosophy and medicine entitled «The prolusion (in Latin – «prolusio» meaning «exercise») for medicine as the principal science» (1805), for which he was granted the degree of «Doctor of Medicine and Surgery» without dispute.

Respected methods of treatment were still bloodletting, cupping and purging.

---

[222] Norris GW. The Early History of Medicine in Philadelphia. Collins Printing House, Philadelphia 1886.

[223] Town of Borzna (founded in 18th century; as a result of organized by Russian Communist Fascist of Holodomor-Genocide 1932-33 against the Ukrainian nation 12 inhabitants have died), Chernihiv Obl., Ukraine. — Vellansky-Kavunnyk DM. De reformatione theoriae medicinae et physicae auspicio philosophiae naturalibus inente. Doctoral dissertation. St Petersburg 1805. — Puchkivsky O. Three founders of Russian medicine (Nestor Ambodyk-Maksymovych, Petro Zahorsky, and Danylo Vellansky). Sci J Ukr Studies «Ukraine», AUASc, St Publ «Ukraine», Kyiv 1924. Book 4, P. 27-34. — Puchkivsky O. Three founders of medicine. J Ukr Med Assoc North Am 1961 July - Oct;8(22-23):40-8. — Ablitsov VH. «Galaxy Ukraine». The Ukrainian diaspora: prominent persons. Publ. «Kyt», Kyiv 2007. — Synyachenko OV. They glorified themselves and Ukraine: history of domestic therapy. Crimean Therapeutic Journal 2010;2(2):21-8.

In Russia, Ukraine and Belarus (area 207,959 km^2), in addition to Apothecary Order, Office of Principal Apothecary was established (1706) to treat the tsarist court and to distribute medicine to the military. In 1716 the Apothecary Order was subordinated to the Archimage and reorganized into the Medical Office (1721-63) and the Medical College. With the creation of provinces, the Order (Prykaz) of Social Provisions (Prykazna Medytsyna, 1775-1864) initiated organized medicine, including the organization of the medical school, hospitals infirmaries, pharmacies, and public health.

The Charité - Universitätsmedizin was established in 1709, the city of Berlin (earliest settlements took place in 1174, and 1192, first documented in the 13th century), Germany.

The Surgical Academy was founded in 1731 in Paris.

The first university in the USA with both undergraduate and graduate studies was University of Pennsylvania, city of Philadelphia (Foot Note 2), PA, was founded by **Benjamin Franklin (1709 – 90**: above**)** in 1740, with Medical College co-founded this same year by **John Morgan (1735 – 89)** — American physician. In the North America (area 24, 709,000 km^2) the first hospital founded was Pennsylvania Hospital by Benjamin Franklin and **Thomas Bond (1713 – 84**: above**)** in 1751, Philadelphia, PA. The University of Pennsylvania School of Medicine was established in 1765, also in Philadelphia, PA, through the pioneering efforts of **William Pepper, Jr (1843 – 98)** — American physician.

The application of Chamberlain obstetric forceps for general use (1727) reduced maternal mortality, the introduction of digitalis (1785) helped many who suffered from dropsy, and ligation of aneurysms (1794) preserved the limbs of thousands of people from unnecessary amputations. Then, the world was electrified by the discovery of smallpox vaccination (1774, 1796).

In 1755 the Kyiv Military Clinical Hospital was founded in the city of Kyiv, Ukraine.

In 1784 the Lviv Academy (17th century) was restored under the name of the Lviv University with the Medical School (Collegium medicum), the city of Lviv, Ukraine.

THE 19TH CENTURY

In the early years of the 19th century treatment included diet, rest, exercise, bath, massage, bloodletting, scarification, cupping, blistering, sweating, emetics, purges, laxatives, enemas and fumigations. Among the physiologically sound medicals were quinidine for malaria, digitalis for heart failure, colchicine for gout and opiates for pain. Arsenic compounds were used for diverse complains and antimony for parasitic infections.

In general, practitioners permitted illnesses to run their course without interference since little benefit was seen from therapies available.

It was the discovery of stethoscope (1816), anesthesia (1846), asepsis (1865), electrocardiography (ECG; 1869-72, 1894), and x-rays (1882, 1895) that launched the beginning of modern medicine. Surgery entered an unprecedented period of growth and expansion, and by the time of World War (WW) 1 (1914-18), the basic operative procedures, with the exception of thoracic and cardiovascular surgery (TCVS), had been developed.

Anatomy, histology and embryology

Ivan P. Kamensky [1771/73, village Nadezhda, Dykansky Region, Poltava Obl., Ukraine – 6(18). VIII.1819, city of Kyiv, Ukraine][224] — Ukrainian anatomist, physiologist, general and forensic physician, wrote a doctoral dissertation on cardiopulmonary resuscitation (CPR – CPR; in German – die Herz-Lungen Wiederbelebung) entitled «Squeezing the Heart», where he depicted cardiac massage, along with an artificial «mouth-to-mouth» respiration to revive suddenly dying people (Moscow Univ., 1802).

The suspition from 1784 by **Jiri** von **Prohaska (1749 – 1820**: 18th century**)** that the anterior roots of the spinal nerves are motor roots, and the posterior roots are sensory was proved in 1811 by Sir **Charles Bell (1774 – 1842)**[225] — Scottish anatomist,

---

[224] Village Nadezhda (founded before 1765), Dykansky Region, Poltava Obl., Ukraine. — Ablitsov VH. «Galaxy Ukraine». The Ukrainian diaspora: prominent persons. Publ. «Kyt», Kyiv 2007. — Moscow University (founded 1755), Moscow, Russia. — Synyachenko OV. They glorified themselves and Ukraine: history of domestic therapy. Crimean Therapeutic Journal 2010;2(2):21-8.

[225] Bell C. On the nerves; giving an account of some experiments on their structure and functions, which lead to new arrangement of the system. Philosoph Trans Royal Soc London 1821;111:398-424. — Dzul A. Bell's palsy. J Ukr Med Assoc North Am 1981 Summer;28,3(102):124-38.

neurologist, and surgeon, and confirmed in 1822 by **Francois Magendie (1783 – 1855)** — French physiologist, and it should be called the Prochaska-Bell-Magendie's law.

The paralysis of the cranial 7th (facial) nerve, previously described by **Razes (865 – 925**: Medieval medicine), **Avicenna (980 – 1037**: Medieval medicine) and by **Anton Friedreich (1761 – 1836**: 18th century), observed in 1821 by Sir Charles Bell who depicted the trajectory of facial nerve and the facial muscles, which show twitching, weakness, and paralysis during the outbreak of disease, named after him, the Bell's palsy. Its onset is sudden within 48 hours, usually unilateral, rarely (less than 1%) bilateral. Other symptoms are a change in taste, increased sensitivity to sound, pain around the ear and dropping of the eyelid. It occurs in 1-4 person between 15-60 years of life per 10.000 population in a year, approximately in 1,5% of individuals at any time in life.

In 1824 Charles Bell described amyotrophic lateral sclerosis (ALS), a progressive neurodegenerative disease of unknown etiology, with an onset in the most people at the age between 55 and 78-year-old, inherited in about 5%-10% from patient's parents, which affects the motor nerve cells in charge of movement of the voluntary muscles, gradually leading to shivering followed by paralysis of muscles of the extremities, mastication, speaking and breathing. The life expectancy of patient with ALS is usually 3-5 years.

Francois Magendie is also known due to discovery of a median aperture in the lower part of the roof of the 4th ventricle of the brain (apertura mediana ventriculi quarti cerebri or foramen of Magendie) which connects and drains the cerebrospinal fluid (CSF) into cisterna magna and subarachnoid space (space of Magendie).

Johann F. Meckel, Jr (1781 – 1832) — German anatomist, described the embryological origin of diverticulum of the ileum, a remnant of the non-closed vitelline (omphalo-mesenteric) duct of the human embryo (the 4th-9th week of fertilization age), present in 2% of the population, as a congenital pouch formation on the lower portion of ileum near caecum and the lower abdominal wall on the right side, named in 1809 the Meckel's diverticulum **[Wilhelm Fabry (1560 – 1634)**: 17th century], inflammation of which simulates a clinical picture of an acute appendicitis and require its removal (diverticulectomy). He also identified a cartilaginous bar from which the mandible is formed (Meckel cartilage).

Ductus epoophorum longitudinalis and its cysts were described by **Hermann Gardner (1785 – 1827)** and named after him.

Jan E. Purkyně / Johann E. Purkinje (1787 – 1869)[226] — Czech anatomist and physiologist, who traced the large branching neurons with many branching dendrites

[226] Kruta JE. Purkyně (1787–1869) Physiologist. A Short Account of His Contributions to the Progress of Physiology with a Bibliography of His Works. Publ Czech Acad Sci, Prague 1969.

in the middle layer of cortex cerebelli (Purkinje cells, 1837) and subendocardial branches, fibers or tissue, that conduct electrical impulses from the atrioventricular (AV) node to all parts of the ventricles of the heart (Purkinje fibers, 1839), which constitute the terminal ramification of the conducting system (network) of the heart (CSH). He first used the term «protoplasm».

Jules G. Cloquet (1790 – 1883) — French physician and surgeon, described the remnant (vestige) of the embryonic hyaloid artery, a small transparent canal running through the vitreous body from the cranial 2nd (optic) nerve disc to the lens, formed by an invagination of the hyaloid membrane which encloses the vitreous body (Cloquet's canal). He also described a hernia of the femoral canal (Cloquet's hernia), fibrous membrane bounding the annulus femoralis at the base of the femoral canal (Cloquet's septum) and small lymphatic nodes (LN) at the femoral canal (Cloquet's glands).

Karl E. Baer (1792 – 1876) — Estonian scientist, biologist, founder of modern embryology and explorer, who discovered the ovum (egg) of mammals and man, the blastula – a developmental stage with completion of fragmentation of the impregnated egg in multicellular chordates, and the notochord – a flexible dorsal cartilaginous rod-shaped body composed of cell derived from mesoderm in embryos of all chordates («De ovi mammalium et hominis genesi». Lipsio 1827). The Baer law, a conceptual predecessor of the recapitulation theory, states that those more general features that are common to all the members of a group of animals are developed in the embryo earlier than the more special features that distinguish the various members of the group.

Martin H. Rathke (1793 – 1860) — Prussian German co-founder of modern embryology and anatomist, described two cartilages at the anterior end of notochord (Rathke's column) and group of epithelial cells forming small colloid-filled cysts in the pars intermedia of the pituitary gland (Rathke cyst); identified two fetal folds of the mesoderm which unite at the median line to form **James Douglas (1675 – 1742**: 18th century**)** septum and to render the rectum a complete canal (Rathke folds); discovered a diverticulum from the embryonic buccopharyngeal membrane, from which the anterior lobe (adenohypophysis) of the pituitary gland is developed (Rathke pouch, 1837), a site of possible development of craniopharyngioma, a.k.a., Rathke's pouch tumor, causing compression of the lateral sections of the optic chiasm leading to bitemporal hemianopsia; and also described a trabeculae crani (Rathke trabeculae).

While studying the chick embryo, **Christian H. Pander (1894 – 1865)** — Latvian German biologist and co-founder of modern embryology, discovered the germ layers, i.e., three distinct regions of the embryo that give rise to the specific organs. Karl Baer (above) expanded Christian Panders concept to include all vertebrates.

The corneal nerves of the eye, and the canal with irregular spaces in the sclerocorneal junction of the eye (sinus venosus sclera), receiving the aqueos humor (lymph) from the anterior chamber of the eye, and then it flows into the bloodstream, were discovered by **Frederich Schlemm (1795 – 1858)**[227] — German anatomist. The later were named Schlemm's canals (1830).

The retropubic space was described by **Anders A. Retzius (1796 – 1860)** — Swedish anatomist, as an extraperitoneal space between the pubic symphysis, the urinary bladder, and the posterior layer of transversalis fascia of the anterior abdominal wall which contains loose connective tissue and fat, affording the access to the urinary bladder without opening of the peritoneal cavity (spatius retropubicum or Retzius space, 1849).

Phillipe F. Blandin (1798 – 1849) — French anatomist and surgeon, detailed the sublingual and submaxillary glands.

Alexander L. Bochdalek (1801 – 83) — Czech anatomist and pathologist, described ductus thyroglossalis; hiatus pleuroperitonealis (trigonum lumbocostale, or foramen of Bochdalek) and its congenital diaphragmatic hernia (CDH) – posterolateral diaphragmatic (Bochdalek) hernia (1848), occuring in 0,17% of adults, on the right side in 68%, on left side – in 18% and bilateral – in 14%; a fold within the lacrimal duct near the punctum lacrimale (Bochdalek valve); and plexus dentalis superior.

Pulmonary varix, dilated and deformed pulmonary vein (PV) / pulmonary artery (PA) or lymphatic vessels, attributed to the underlying cardiopulmonary conditions, specifically in PV varix, to pulmonary hypertension, inflammation, and mitral valve (MV) disease, may cause cerebral embolism, and from spontaneous rupture into the bronchus hemoptysis or hemorrhages. Diagnosis of PV and esophageal varix is confirmed by maneuver of **Johannes P. Müller (1801 – 58)**[228] — German comparative anatomist and physiologist, the reverse to maneuver of **Antonio Valsalva (1666 – 1723**: 18th century). Using the Müller's maneuver (maneuver), an inspiratory effort against a closed opening of the glottis after expiration create a negative intrathoracic pressure, expands pulmonary gas, cause engorgement of the intrathoracic vessels with blood, and using fluoroscopic / bronchoscopic examination gives an opportunity to differentiate vascular versus nonvascular structures, diagnose varicose vein of the lung and of the esophagus, look for collapsed sections of the trachea and bronchi, helps to determine cause of sleep apnea. He found ductus paramesonephric of Müller (1830), pioneered the use of the microscopic research in the investigation of pathological anatomy (1838).

---

[227] Schlemm F. Arteriarum capitis superficialum icon nova. Berlin 1830.

[228] Müller JP. Handbuch der Physiologie des Menschen für Vorlesungen. Band I-II. Verkag von Hölscher, Coblenz 1835-44.

Agenesis (aplasia) or hypoplasia of the Müllerian duct is a congenital condition in which genetically female with two x-chromosome are born with rudimentary or absence of the uterus and upper vagina, but with functional uterine (Fallopian) tubes, ovaries and eggs, normal external genitalia and height. While the main symptoms is the failure to menstruate, some have abnormally formed kidneys or a single kidney that may be located in unusual part of the abdominal cavity. Other signs include skeletal abnormalities, hearing loss, or congenital heart disease (CHD). It occurs in about 1 of 4,500 woman. The condition was described in 1828, by **August F.J.K. Mayer (1787 – 1865)** — German anatomist and physiologist, in 1838 by **Carl (Karel) F.** von **Rokitansky (1804 – 78)** — Czech pathologist and physician,

The transverse sinus of pericardium (sinus transversus pericardii) was identified by **Friederich W. Theile (1801 – 79)** — German anatomist, as a passage in the pericardial sac between the origins of the great vessels, posterior to the intrapericardial portions of the main pulmonary artery (MPA) and ascending aorta (AA), anterior to the superior vena cava (SVC) and superior to the atria and pulmonary veins (PV's), covered by a serous layer of pericardium.

The cell theory was developed by **Mathias Schleiden (1804 – 87)** — German botanist and one of the founders of the cell theory, and **Theodor Schwann (1810 – 82)** — German physiologist. The latter identified the cell as the basic structure of the tissues of plants, animal and humans. In addition, he revealed large nuclear cells, plasmalemma which twists spirally around the axons (Schwann cells).

Significant contribution to both, anatomy, and surgery, was made by **John Hilton (1805 – 780)**[229] — English anatomist and surgeon.

He noted that often a nerve will supply (innervate) both the muscle and skin overlying a particular joint, and in reverse, a nerve that supplies skin or a muscle will often innervate the underlying joint (Hilton's law, 1860-62).

The division of the lower one-third of the anal canal below the pectinate line is a landmark for the intermuscular border between internal and external anal sphincter muscles, dividing two zones separated by the white line into the zona hemorrhagic lined by stratified non-keratinized epithelium of the anal canal, and the zona cutanea lined by stratified keratinized epithelium of the anus and perianal skin (intersphincteric groove or the white line of Hilton, 1860-62).

As a surgeon, he opened carefully deeply seated abscesses with a probe and dressing forceps (Hilton's method, 1860-62).

Among described in 1782 and 1788 by **Olexandr Schumlansky (1748 – 95:** 18th century**)**, of the capsula glomeruli Schumlansky in the renal cortex and the

---

[229] Hilton J. On Rest and Pain. 2nd ed by W.H.A. Jacobson. London 1877. — Ewing MR. The white line of Hilton. Proc R Soc Med 1954 July;47(7):525-30. — Ancient medicine. *Egypt* – The anal canal.

descending proximal and the ascending distal straight tubules of Schumlansky, the capsula glomeruli was re-discovered in 1842 by Sir **William Bowman (1816 – 92)** — English anatomist, histologist, and surgeon, and should be named the glomeruli or gromerular capsule of Schumlansky-Bowmen (corpusculum renale Schumlansky-Bowmeni). The U-shaped renal loop (tubulus attenualis) of the renal medulla (medulla renis) was discovered in 1853 by **F.G. Jacob Henle (1809 – 58)** — German anatomist, pathologist, and surgeon, and was named the Henle loop (tubulus attenualis Henle).

Joseph Hyrtl (1810 – 94) — Austrian pathologist, saw a cervical nerve loop with the ascending portion formed by the cranial 12th (hypoglossal) nerve plus the fibers from the 2nd or 1st cervical nerve and the descending portion consisting of branches from the 2nd and 3rd cervical nerve (ansa nervi cervicalis or Hyrtl loop) in the carotid triangle, bordered posteriorly by the sternocleidomatoid (SCM) muscle, inferiorly by the superior belly of the omohyoideus muscle and superiorly by the stylohyoideus muscle and the posterior belly of the digastricus muscle. The 12th cranial nerve is the motor nerve of the tongue and ansa nervi cervicalis is a motor nerve to the subhyoid muscles. He also depicted the recessus epithympanicus (Hyrtl recessus) and a partial sphincter of the rectum (Hyrtl sphincter).

Benedict Stilling (1810 – 79) — Jewish German anatomist and surgeon, noticed a minute canal running through the vitreus body from the discus of the cranial 2nd (optic) nerve to the lens.

Karl L. Meckel (1812 – 76) — German anatomist and laryngologist, described the fulcrum ventriculi – a gap between the two prominences of the lateral wall of the vestibule of the larynx, formed by the scooping and wedgelike cartilage, and musculus ceratocricoideus – muscle fibers, an integral part of internal muscles of the larynx that originate from the cricoid cartilage and attach to the lower horn of the thyroid cartilage.

John Goodsir (1814 – 67) — Scottish anatomist, became a pioneer in the study of the cell and life organization.

Wenzel (Veneslaus) L. Gruber (1814 – 90) — Russian anatomist of Czech ancestry, described a lateral diverticulum in the suprasternal space; outlined the prepatellar and other bursas (1857); described the superior duodenal fold (recess) and along the ascending part of the duodenum extending downward behind the duodenojejunal angle – inferior duodenal fold (Gruber recess) which may be a site of an internal hernia; described internal mesogastric hernia with incarceration of the ileum and the omentum; also described the fissure petrosphenooccipitalis.

As the margin of the aperture of the thoracic inlet or sternocostovertebral circle slopes inferoanteriorly, the apex of each human lung is covered by a cupola made of fascia or membrana suprasternalis (suprasternal membrane), described by **Francis**

Sibson (1814 – 76) — English anatomist and physician, which represent a thickened connective tissue being a part of the endothoracic fascia attached to the internal surface of the 1st rib and to the transverse process of the 7th cervical vertebrae between the parietal pleura and the thoracic cage and covered by the cervical fascia (Sibson's fascia, 1848). The clinical significance of Sibson fascia is that its injury may cause herniation of cervical fascia; that the thoracic duct traverses it up to the level of cervical (C)$_7$ before turning around and emptying onto the major left duct, and the minor right duct traverses the thoracic inlet once; and that somatic dysfunction of Sibson fascia can decrease lymphatic drainage anywhere in the body. He also described musculus scalenus minimus, the smallest scalene muscle band between the anterior and median scalene muscles (Sibson muscle, 1869); a furrow (groove) at the lower border of the major pectoralis muscle (Sibson furror or groove, 1869); the vestibule of the aorta, the part of the left ventricle (LV) around the root of the aorta (Sibson vestibule, 1869); a notch of the internal curvature of the left superior precordial dullness in acute pericardial effusion (Sibson notch, 1869).

Tissues were classified by **Robert Remak (1815 – 65)** — Prussian German embryologist, physiologist and neurologist of Jewish Polish descent, according to embryological origin into three primary systems (germ layers) – ectoderm, mesoderm and endoderm.

Adolph Krukenberg (1816 – 74) — German anatomist, described the central (Krukenberg) vein of the liver (vena centralis hepatis).

Arthur H. Hassall (1817 – 94)[230] — British chemist, histologist, and physician, discovered the spherical or oval corpuscles (bodies) in the medullary portion of the thymus, made by concentric lines of the epithelial cells that contains keratohyaline and bundles of cytoplastic filaments (Hassall bodies, 1846), and independently of **Jacob Henle (1809 – 58**: above) – bodies of abnormal growth in lamina limitans posterior cornea (Hassal-Henle bodies, 1846).

The development of an embryo in the cell theory basis was explained by **R. Albert** von **Kolliker (1817 – 1905)** — Swiss anatomist, histologist and physiologist.

В 1856, **Mykola M Yakubovych** or **Nicolai (Nicolaj) M. Jacubowitsch (1817 – 79)**[231] — Ukrainian histologist and physiologist, in 1887 **Ludwig Edinger (1855 –**

---

[230] Hassall AH. The Microscopic Anatomy of the Human Body in Health and Disease. Longman, London 1846. — Olearchyk AS. Controversies in thoracic, cardiac, and vascular surgery. J Ukr Med Assoc North Am 1991; 38,2(124):88-103. — Olearchyk AS. Observations of rare thymic hyperplasia. Kinichna Khirurhiya (Kyiv) 1996; 10(651):50-1. — Olearchyk AS. Rare thymic tumors. In, AS Olearchyk. A Surgeon's Universe. Vol. I-IV. – 4th Ed. Publ. Outskirts Press, Inc. Denver, CO 2016. Vol. III. P. 299-301.

[231] Jacubowitsch N.M. Reserches comparatives sur la systéme nerveux. Compt. rend. Aout. Paris 1858. — Owsjannikow PV. Einige Worte über d. Mittheilungen d. Hrn. Dr. Jacubowitsch. Archiv für pathologische Anatomie und für klinische Medicine. Band XV. Seite 150. — Ablitsov VH. «Galaxy Ukraine». The Ukrainian diaspora: prominent persons. Publ. «Kyt», Kyiv 2007.

1918) — German anatomist and neurologist, and in 1889 **Karl F.O. Westphal (1833 – 90)** — German neurologist, discovered of a paired small-cell accessory nerve inside the nucleus of the 3rd (oculomotor) cranial nerve in the middle brain, supplying pre-ganglionic parasympathetics nerve fibers to the eye, which constricts the pupils and accommodates the lens (Yakubovych -Westphal-Edinger nucleus).

John Marshall (1818 – 91)[232] — English anatomist, described a small oblique vein of the left atrium (LA) – venae obliqua atrii sinistra draining the venous blood from the LA (oblique vein of Marshall, 1850) which together with the great cardiac vein, vein of the left obtuse margin, the posterior vein of the left ventricle (LV), middle cardiac vein and the right coronary vein drains into the coronary sinus (CS) of the right atrium (RA). He also described a triangular fold (Marshall, 1850) between the pulmonary artery (PA) and the pulmonary vein (PV), formed by duplication of the inner membrane of the pericardium over the remnant of the lower portion of the left superior vena cava (SVC) or duct of Arantius-Cuvier (plica venae cavae sinistrae) which extends as a fibrous band from the left supreme intercostal vein to Marshall oblique vein of the LA.

Karl W.L. Bruch (1819 – 94) — German anatomist, who described the innermost layer of the chorioid (lies between the retina and the sclera, thickness, 0.1-0.2 mm) of the eye, a.k.a. the laminae basalis seu vitrea lamina of a glassy microscopic appearance (thickness, 2-4 mm), named the Bruch's membrane (1844). It consists of five layers: (1) the basement membrane of the retinal pigment, (2) the inner collagenous lane, (3) a central band of elastic fibers, (4) the outer collagenous zone, and (5) the basement membrane of choriocapilaris.

Václaw (Wenzel) Treitz (09 Apr. 1819 – committed suicide by injecting himself with potassium cyanide **07 Aug. 1872)**[233] — Czech pathologist, discovered the suspensory muscle of duodenum a fibrous structure by which duodenojejunal junction is fixed to the posterior abdominal wall (musculus suspensorius duodeni, later named the ligamentum of Treitz, 1853), the sharp curve at the gastrojejunal junction (angle of Treitz), a fold of peritoneum that arches between the left side of the duodenal flexure and the medial border of the left kidney (plica paraduodenalis, Treitz's arc), fascia behind the head of the pancreas (Treitz fascia), a depression in the peritoneum extending posterior to the caecum subcaecal fossa, Treitz fossa), a duodenojejunal hernia, a.k.a., retroperitoneal hernia (Treitz hernia).

---

[232] Marshal J. On the development of the great anterior vein in man and mammalia: including on account of certain remnants of foetal structure found in the adult, a comparative view of these great veins in different mammalia, and an analysis of their occasional peculiarities in the human subject. Philos Trans R Soc London 1850;140;133-69.

[233] Treitz V. Ueber einen neunen Muskel am Duodenum des Menschen, über elastishe Sehnen und enige andere anatomishe Verhältnische. Prager Vierteljahrsschrift 1853.

Hubert von **Luschka (1820 – 75)** — German anatomist, discovered the right and left lateral foramen of the 4th ventricle of the brain at the end of the lateral recesses, by which they are connected and drains cerebro-spinal fluid (CSF) to the cerebro-pontine angle and to the subarachnoidal space (lateral foraminas of Luschka). He also described biliary ducts between the parenchyma of the liver, and the gallbladder (sinuses of Luschka). Apparently, not knowing about the description of the median papillary muscle by **Giovanni M. Lancini (1654 – 1720**: 18th century), described it again.

Franz von **Leydig (1821 – 1908)** — German zoologist and comparative anatomist, described epithelioid interstitial cells of the testicular hilum that create endocrine tissue and produce androgen, usually testosteron (Leydig's cells, 1852). From them develops the commonest non-germinal (Leydig) tumor.

The pioneer of neurology and oncology, **L. Pierre P. Broca (1824 – 80)** — French anatomist, physician, and anthropologist, established the speech function in the antero-inferior frontal lobe of the brain («Broca area», 1861).

In 1887 **Karl A. Möbius (1825 – 1908)** — German zoologist, who discovered cell organelles.

The axon reflex was defined (1877) by **Aleksander I. Babushkin (1827 – 91)** — Russian histologist.

Phillip V. Ovsyannikov, or **Phillip V. Ovsyannikow (1827 – 1906)**[234] — Russian histologist, who initiated artificial reproduction of starlet (in Latin – Acipenser ruthenus), discovered the vasomotor centre and determined its precise location in the medulla oblongata.

Henry Gray (1827 – 61) — English anatomist, wrote, **Henry V. Carter (1831 – 97)** — English anatomist and Doctor of Medicine, illustrated, the textbook «Anatomy: Descriptive and Surgical» (1st Ed. J.W. Parker & Son, London 1858), subsequently re-edited under the title «Anatomy: Descriptive and Applied» and the simplified title «Gray's Anatomy», the latest – «Gray's Anatomy» (40th Ed, Churchill Livingstone Elsevier 2008).

The enteric plexus of autonomic nerve fibers (sympathetic and parasympathetic) within the wall of the digestive tract is made up of the myenteric sympathetic and parasympathetic plexus of **Leopold Auerbach (1828 – 97)** — German anatomist and neuropathologist, that is situated between the longitudinal and circular layers of the muscularis propria, and it controls these muscles (plexus myentericus Auerbachi or Auerbach's plexus); of the submucous parasympathetic plexus of **Georg Meissner**

---

[234] Owsjannikov PhV. Reserches sur la structure intime du systéme nerveux des crustacés et principalement du Homard. Annales des Sciences 1862;15. — Owsjannikov PhV. Die tonishen und reflectorischen Centren des Gefässnerven. Berichte ueber die Verhandlugen der Königlich Sächsinischen Gesellschaft der Wissenschaften zu Leipzig 1871;23.

(1929 – 1905) — German anatomist and physiologist, that controls the muscularis mucosa and submucosal glands (Meissner plexus, 1862); and of the sub serosal plexus. The Auerbach plexus arises from parasympathetic cells in the trigonum of the 10th cranial (vagus) nerve (nucleus ala cirenea) in the medulla oblongata. The fibers are carried by both the anterior and posterior vagal nerves. A decrease in ganglion cell density in the Auerbach's plexus could cause achalasia, a condition where the lower esophageal sphincter fails to relax. Georg Meissner name is also associated with mechanoreceptors that are responsible for sensitivity to light touch (Meissner corpuscles, 1862).

Karl W. von **Kupffer (1829 – 1902)** — Latvian German histologist, described large stellar or gigantic pyramidal endothelial cells of the liver with a large oval nucleus and a small, clearly seen nucleolus (the stellar cells or cells of Kupffer, 1875) within the wall of the blood capillaries (hemocapillaries) of the sinusoid type, which connect branches of the portal vein with the central vein of segments of the liver. The hemocapillaries of the sinusoidal type, separate from the liver cells (hepatocyte), were described by **Joseph Disse (1852 – 1912)** — German anatomist and histologist, consist of a small perisinusoidal space filled with hepatic lymph (space of Disse, 1880).

Invention of microtome by **Wilhelm His, Sr (1831 – 1904)**[235] — Swiss anatomist, and embryologist, allowed by treating tissue with acid and salt to harden it and then slicing it very thinly to further research anatomy and function of tissue and cell under microscope.

He rejected all form of soft inheritance in his statement the «Until it has been refuted, I stand by the statement that characters cannot be inherited that were acquired during the lifetime of the individual» (1874).

His discovery in 1886 that each nerve fibers stems from a single nerve cell was essential to the development of the neuron theory, which state that neuron, or nerve cell, is the basic unit of the nervous system.

He described the bursa at the end of the archenteron (His bursa), ductus thyroglossalis, the perivascular (His) space between the adventitia of the blood vessels of the brain and spinal cord and the perivascular limiting membrane of the glia tissue, discovered the thalamus of the brain [17th century. – **Raymond de Viessens (1635/41 – 1715) / Felix d'Azyr (1748 – 94)**: 1893], four thickenings which run the entire length of the embryonic spinal cord (His's zones; 1880 – 87), and the acute angle and a valve between the esophagus and the cardia at the entrance to

---

[235] His WSr. Unsere Körperform und das physiologische Problem ihner Eustehnung. Verlag FCW Vogel, Leipzig 1874. — His WSr. Anatomische menschlicher Embryonen. Teil 1-7. Verlag RCW Vogel, Leipzig 1880-85. — His WSr. Zur Geschichte der den menschlischen Rückenmarkes und der Nervenwurzeln. Abhandlungen der Königlichen Säcchischen Gesellschaft deer Wissenschaften Mathematisch-Physische Classe. Leipzig 1886-87:13:477-514.

the stomach, created by the sling fibers and the inner circular muscle around the esophagogastric junction (His angle), preventing reflux of duodenal bile, enzymes and stomach acid from entering the esophagus, which can cause gastroesophageal reflux disease (GERD).

In 1872 **Petro I. Peremezhko / Petro I. Peremeschko [12 (24).VII.1833, village of Ryboten', Okhtyrsky Region, Sumy Obl., Ukraine – 27.XII.1893 / 08.I.1893, city of Kyiv, Ukraine)**[236] — Ukrainian histologist, and in 1877, **Walther Flemming (1843 – 1905)** — German biologist and a founder of cytogenetics, described mitosis (in Greek – «thread»), a direct division of cells, consisting of the initial interphase of the single nuclear cell, an early and late prophase, metaphase, anaphase, telophase, and finally interphase of the daughter (double nuclear) cells in the interphase with a single complex of chromosomes[237].

In 1872 Petro Peremezhko, in 1876 **Oskar Hertwig (1849 – 1922)** — German zoologist, and in 1877 Walther Flemming described meiosis, a direct division of cells during the maturation of sexual cells (meiosis I - leptotena, zygotena, pachytena, diplotena, diakinesis, metaphase I, anaphase I and telophase I and meiosis II - prophase I, metaphase II, anaphase II and telophase II) when each of daughter cells receive a half of the chromosomes (two haptoid daughter cells).

Walther Flemming description of cell division refers to a germinal (Flemming) center, the area in the center of a lymphatic node containing an aggregate of actively

---

[236] Village of Ryboten' (settled in Neolitic Era, founded at the end of 18th century), Kroveletsky Region, Chernihiv Obl., now Okhtyrsky Region, Sumy Obl., Ukraine. — Peremeschko P. Die Entwickelung der quergestreiften Muskelfaserm aus Muskekernen. Archiv für pathologishe Anatomie and Physiologie und für klinische Medicin (Springer Verlag) 1867-01-01;27(1):116-26. — Peremeschko P. Ueber den Bau des Hirnanhanges. Archiv für pathologishe Anatomie and Physiologie und für klinische Medicin (Springer Verlag) 1867-03-01;38(3):329-42. — Schleicher W. Die Knorpelzelltheilung. Ein Beitrag zur Lehre der Theilung von Gewebezellen. Archiv für mikroskopische Anatomie (Springer-Verlag) 1879-12-01;16(1):248-330. — Peremeschko P. Ueber die Theilung der thiereschen Zellen I. Archiv für mikroskopische Anatomie (Springer-Verlag) 1879-12-01;16(1):437-57. — Peremeschko P. Ueber die Theilung der thiereschen Zellen II. Archiv für mikroskopische Anatomie (Springer-Verlag) 1880-12-01;17(1):168-86. — Aronov HYu, Peleshchuk AP. Legends and past times of Kyiv medicine. Publ. «Stolittia», Kyiv 2001. — Ablitsov VH. «Galaxy Ukraine». The Ukrainian diaspora: prominent persons. Publ. «Kyt», Kyiv 2007. Olearchyk AS. A Surgeon's Universe. Vol. I-IV. 4th ed. Publ. «Outskirts Press», Denver, CO 2016. Vol. I. P. 679.

[237] In this same 1877 year, Walther Flemming, discovered in the cell nucleus, by tincturing it with the coloring dye aniline, a basophilic structure, which he called chromosome (in Greek – «colored body»), represented by a linear thread of deoxyribonucleic acid (DNA), which transmits the genetic information and interacts with ribonucleic acid (RNA) and histones (simple proteins that contain many groups, soluble in water and insoluble in ammonium salt). The inheritance of chromosome was discovered in 1890, by **F.L. August Weismann (1834 - 1914)** — German evolutionary biologist, who by 1892 developed the germ cell theory of heredity, which states that multicellular organisms consist of germ cells that contain and transmit heritable information, and somatic cells which carry out ordinary body function. Walther Flemming, **Eduard A. Strassburger (1844 - 1912)** — Polish-German botanist, and **Edouard J.L.M.** van **Beneden (1846 - 1910)** — Belgian embryologist, cytologist and marine biologist, elucidated chromosome distribution during cell division.

proliferating lymphocytes (antibody-forming B-cells). Investigating mitosis, Walther Flemming, did not see the splitting into identical halves, the daughter chromatid (one of the pairs of chromosomal strands) surmising for the first time that nuclei cells came from another predecessor and coined the phrase «omnis nucleus e nuclei» (1878-82).

Petro Peremezhko also described proprioceptors.

Volodymyr O. Betz [14(26.IV).1834, village Tatarivshchyna close to town of Oster, Chernihiv Region, Chernihiv Obl., Ukraine – 30.IX(12.X).1894, city of Kyiv, Ukraine][238] — Ukrainian anatomist and histologist, discovered large pyramidal ganglion cells or giant pyramids or giant pyramidal cells found in the fifth ganglionic layer of the cerebral cortex (Betz cells, 1874).

Heinrich F. Hoyer (1834 – 1907) — Polish anatomist and histologist, and **J.P. Sucquet (1840 – 70)** — French anatomist, described arteriovenous anastomoses or canals (Hoyer-Sucquet), which control blood flow in the glomus bodies under a nail of the digits and toes, from which a rare solid glomus tumor may arise.

James B. Pettigrew (1834 – 1908)[239] — Scottish naturalist, morphologist, and physician, developed a program of research into the laminar nature of the myocardium.

In the embryonic development of the rabbit's heart, during establishment of the lateral plate in the precardiac mesoderm of the cardiac crescent, **Victor Hensen (1835 – 1924)** — German anatomist and physiologist, discovered the thickening of cells in the blastomere on the cranial end of the primitive streak (primitive Hensen's node, 1876), which appear to control the genesis of left-right differences in gene expression, i.e., the rightward- or D-looping and a leftward flow of the fluid, despite the midline position. Beating cilia that project from the primitive node undergo a contra clockwise vortex-like movement that creates a leftward flow of fluid across the Hensen's node. A molecular cascade for left-right determination begins with asymmetric expression of the morphogen sonic hedgehog (Shh) on the left side of the primitive node.

---

[238] Village Tatarshchyna close to town of Oster (settled since Bronze Age 2 thousand years BC, first mentioned in chronicle 1098; during conduction by Russian Communist Fascist regime of Holodomor-Genocide 1932-33 against Ukrainian nation died at least 172 residents of town, region – up to 250 people, together during Holodomor-Genocide and political repressions in 1930's suffered in town up to 400 people), Chernihiv Region, Chernihiv Obl., Ukraine. — Betz W. Anatomischer Nachweis zweier Gehirncentra. Centralblatt für die medizinischen Wissenschaften 1874;12:578-80, 595-99. — Kukuyev LA. **Betz** Vladimir Alekseyevich (1834 - 94), prominent Ukrainian anatomist... In, SI Vavilov, BO Vvedensky. Bolsh Sov Encycl («BSE»). 2nd. Ed. Gosudar nauch izd «BSE», Moscow 1950, Vol. 5, P. 126. — Olearchyk A. Ukrainian Scientists. Volodymyr O. Betz (1834-1894). Nashe Słovo (Warszawa) 8 Sep. 1957;2,36(56):2. — Ablitsov VH. «Galaxy Ukraine». The Ukrainian diaspora: prominent persons. Publ. «Kyt», Kyiv 2007. — Olearchyk AS. A Surgeon's Universe. Vol. I-IV. 4th ed. Publ. «Outskirts Press», Denver, CO 2016. Vol. I. P. 679-80.

[239] Pettigrew J. On the arrangement of the muscular fibers in the ventricles of the vertebrate heart, with physiological remarks. Phil Trans 1864;154:445-50.

Louis A. Ranvier (1835 – 1922) — French anatomist, histologist, and physician, discovered a regularily spaced discontinuities of the myelin sheath that wraps around the axon of a neuron and very efficiently insulates it (nodes of Ranvier, 1878).

Nykanor A. Khrzhonshchevsky / Nykanor Chrzonsz(e)czewsky [26.VII.(07. VIII).1836, city of Perm, Permsky krai, Russia – 19.VIII.(01.IX).1906 city of Kyiv, Ukraine][240] — Ukrainian histologist of Polish ancestry, introduced in vivo intravenous (IV) injection of dye to monitor the function of the kidneys, liver, suprarenal glands, vascular and lymphatic system (1864).

Heinrich W.G. von **Waldeyer (1836 – 1921)[241]** — German anatomist, famous for naming the chromosome (1888) and consolidation of the neuron theory, its axons and dendrites as the building unit of the nervous system (1891). Described the lymphoid (Waldeyer) ring of the nosopharynx (1884) which include the tonsils – palate (1st and 2nd), pharyngeal (3rd), lingual (4th) and tubular (5the and 6th)[242]; described the right upper and lower paraduodenal (Waldeyer) recesses of the mesentero-peritoneal fossa as one space that is the source of paraduodenal herniation[243]; described the sweat (Waldeyer) glands near the margin of the eyelids; described dorso-lateral (postero-lateral) fasciculus – a longitudinal bundle of thin unmyelinated or poorly myelinated fibers capping the apex of the of the posterior horn of the spinal gray matter, composed of posterior root fibers and short association fibers that interconnect neighboring segments of the posterior horn; identified the sheath – a tubular space between the urinary bladder wall and the intramural portion of the ureter as it courses obliquely through and encircles the terminal urether (Waldeyer sheath). He delineated the presacral (Waldeyer) fascia which lines the anterior aspect of the sacrum, enclosing the sacral vessels and nerves. It continues anteriorly as pelvic parietal fascia, covering the pelvic cavity. The presacral fascia is limited postero-inferiorly as it fuses with the mesosacral fascia, lying above the levator ani muscle at the level of the anorectal junction. Identification and preservation of the presacral fascia is of the utmost importance in preventing complications and reducing local recurrence of rectal cancer surgery. Finally, he established that cancer growth originates from the epithelium of the ectoderm.

---

[240] City of Perm (inhabited since Stone Age, founded 1723), Permsky krai (area 160,600 km²), Russia. — Chrzonszczewsky N. Zur Anatomie der Niere. Virchows Archiv 1864;31(2):153-98.198. — Chrzonszczewsky N. Zur Lehre von dem Lungenepithel. Virchows Archiv 1866;35(1):165-8. — Chrzonszczewsky N. Zur Anatomie und Physiologie der Leber. Virchows Archiv 1866;35(1):153-64. — Chrzonszczewsky N. Ueber meine Methode der physiologischen Injection der Blut- und Lymphgefässe. Virchows Archiv 1898;153(1):110-29. — Chrzonszczewsky N. Zur Lehre von den vasomotorischen Nerven. Virchows Archiv 1899;157(2):373-6. — Ablitsov VH. «Galaxy Ukraine». The Ukrainian diaspora: prominent persons. Publ. «Kyt», Kyiv 2007.

[241] Van Waldeyer-Hanz H. Lebenserinnerungen. Bonn 1920.

[242] Van Waldeyer-Hanz H. Ueber den lymphatishen Apparat des Pharynx. Dtsch med Wschr, Berlin, 1884;10:13.

[243] Olearchyk AS, Cogbill CL. Right paraduodenal hernia: case report. Milit Med 1979; 144:192-4.

The law of **Julius Wolff (1836 – 1902)** — German surgeon of Jewish ancestry, states that a normal or abnormal bone develop the structure most suited to resist the forces acting upon it (1892).

In 1875 **Victor A. Cornil (1837 – 1908)** — French histologist, pathologist and politician, and **Louis Ranvier (1835 – 1922**: above**)** described the phenomenon of metachromasia that is a characteristic change in the color of staining carried by biological tissue and exhibited by certain aniline dyes when bound to the substances, called chromotropes, present in those tissues.

In the posterior abdominal wall, after description of the inferior (Petit's) lumbar triangle [18th century – **Jean Petit (1674 – 1750)**: 18th century], **Peter F. Lesgaft (1837 – 1909)**[244] — Russian physician of German origin, and **Joseph C. Grynfeltt (1840 – 1913)** — French surgeon, delineated a superior lumbar triangle / rhomb formed superiorly-laterally by the 12th rib, superiorly-anteriorly by the border of the serratus inferior posterior muscle, medially by the quadratus lumborum muscle, anteriorly by the internal abdominal oblique muscle (Grynfeltt-Lesgaft rhomb). The floor of the superior rhomb is the thoraco-lumbar fascia and aponeurosis of the transverse abdominal muscle, the roof is formed by the latissimus dorsi muscle and external oblique abdominal muscle. Occasionally, it is the site of the lumbar (Grynfeltt-Lesgaft) hernia or an abscess. The former laid the scientific foundations for physical education and fitness.

Francesco Todaro (1839 – 1918) — Italian anatomist, identified a tendinous subendocardial collagenous bundle in the wall of the right atrium (RA) that extends from the right fibrous trigone of the central fibrous body of the heart to the median brim of the inferior vena cava (IVC) valve, a.k.a. the Eustachian valve **[Bartolemo Eustachio (1520 – 74). Renaissance]**.

The founder of comparative embryology and experimental histology, **Olexandr (Alexander) O. Kovalevsky (07/19.XI.1840, village Värkava near town of Daugavpils, Preysky kray, Latgale Region, Latvia – 09.XI.1901, St Petersburg, Russia)**[245] — Ukrainian embryologist and histologist of Ukrainian Cossack ancestry, established the existence of a common pattern in the embryological development of all multicellular animals, discovered the neuroenteric (Kovalevsky) canal (1865).

Olexandr Kovalevsky studied at the Universität Heildelberg (established 1386), city of Heildelberg (Prehistoric medicine), and Universität Tübingen (established 1477),

---

[244] Lesgaft PF. The Relationship of Anatomy to Physical Education and Principal Task of Physical Education in School, Moscow 1888. — Vayda RI. Abdominal wall. Clinical anatomy and physiology. In, LYa Kovalchuk, VF Sayenko, HV Knyshov. Klinichna Khrurhiya. Vol. 1-2. Publ. «Ukrmedknyha», Ternopil 2000. Vol. 2. P. 9-11.

[245] Town Värkava near town of Daugavpils (former Dvinsk, founded 1275), Preysky kray (area 365,3 km²), Latgale Region (14,547 km²), Latvia (area 64,589 km²). — Kovalevsky A. Les Hedylidés, étude anatomique. Zapiski Imperatorskoi Adademii Nauk 1901;12:1-9. — Ablitsov VH. «Galaxy Ukraine». The Ukrainian diaspora: prominent persons. Publ. «Kyt», Kyiv 2007.

city of Tübingen (founded 1084), both in Land of Baden-Württemberg, Germany, and at the University of St Petersbug, Russia (1859-63), defended dissertation entitled «The History of Development of Brachiostoma / Lancelet – Amphioxus laceolatus» for the degree Master of Natural Science (1866), and dissertation on the theme of «About the Development of Phoronis» (1867) for degree Doctor of Natural Science (1867). Thereafter, he was professor of zoology at his «Alma Mater» (in 1865-67, 1891-93), at the University of Kazan (established 1804; in 1868-69), city-capital of Kazan (founded in late 13th century), Tatarstan, University of Kyiv (established 1833; in 1869-74), city of Kyiv, Ukraine, and the University of Odesa (established 1865; in 1874-90), city of Odesa (settled since the Upper Paleolithic Epoch 58-30 century BC., 6th-3th century BC, founded in 15th century, first written mention relates to 1415), Odesa Obl. (area 33.310 km²), Ukraine, finally became Director (1892-91) of Biological Station (established 1871; in 1892-1901), city of Sevastopol (founded 1783), Crimea, Ukraine.

Among the others, **Karl Toldt (1840 – 1920)** — German anatomist and anthropolist, described continuation of Treitz fascia behind the body of the pancreas (Toldt fasia), the anterior layer of fascia renalis (Toldt membrane) and lateral reflection of the posterior parietal peritoneum of the abdomen over the mesentery of the ascending and descending colon, where it attaches to the retroperitoneum (the white line of Toldt), that is where the incision starts during their surgical mobilization.

Louis Ch. Malassez (1842 – 1909) — French anatomist and histologist, designed the hemocytometer, a device used to quantitatively measure the blood cells.

Camilla Golgi (1843 – 1926) — Italian histologist and neurologist, developed the mixed dyeing of the nerve cells (Golgi method, 1873), discovered an organelle in most eukaryotic cells (Golgi apparatus or body, 1898) that processes and packages proteins, lipids, and carbohydrates before sending them to their destination; defined the pyramidal neurons of the cerebellum with the long axons (Golgi type I) and with the star-like short neurons (Golgi type II).

Gustav A. Schwalbe (1844 – 1916) — German anatomist, demonstrated that the major pathway to absorb cerebrospinal fluid (CSF) was lymphatic pathway (1869); described a circular ring or ridge consisting of collagenous fibers surrounding the outer margin of Descemet's membrane [**Jean Descemet (1732 – 1810)**. 18th century]. The subarachnoid or subdural spaces between the internal and external sheath of the cranial II (optic) nerve are referred as intervaginal spaces of the optic nerve (in Latin – spatial intervaginate nervi optici), or Schwalbe's spaces.

Found by **Friedrich S. Merkel (1845 – 1919)**[246] — German anatomist and histopathologist, are cells in the form of numerous membranous granules with a dense core, separate desmosomes and cytoplastic filaments in or at the epidermo-

---

[246] Merkel FS. Über die Endgungen der Sensiblen Nerven in der Haut der Wirbeltiere. Rostock 1880.

dermal junction, forming parallel bundles that conduct inside the stratum granulosum and spinosum, where they interdigitate with other keratocytes, and together with flat discoid-like nerve endings (tactile / touch menisci; in German – das Tastzellen), which perform the function of touch receptors (Merkel cells, 1875).

The anatomical structures discovered by Friedrich Merkel and **Louis Ranvier (1835 – 1922**: above) are melanocyte-like receptor cells in the basal layer of the epidermis and oral mucosa derived from neural crest that contain catecholamine granules and have synaptic contacts with somatosensory afferents, associated with the sense of light touch discrimination of shapes and textures (Merkel-Ranvier cells). They can turn malignant and form the skin or mucous membrane neuroendocrine tumor known as Merkel cell carcinoma.

Mikhail D. Lavdovskii (1846 – 1902) — Russian histologist, noted centrosome (Lavdovskii nucleoid) containing the genetic material (nucleic acid) of a virus, situated in the center of a vibrion (1887-88).

The name of **August** von **Froriep (1846 – 1917)** — German anatomist, is lent to «Froriep ganglion», a temporary group of nerve cells associated with the cranial 12th (hypoglossal) nerve of an embryo.

In 1881-82 **Adalbert (Albert) W. Adamkiewicz (1850 – 1921)**[247] — Polish anatomist, discovered the variable vascularity of the spinal cord. Arteria radicularis magna Adamkiewiczi (ARM) or the great radicular artery of Adamkewicz originate on the left side of the aorta (in about 80%) from the branches of intercostal arteries at the level of T7-12 in 85% of the people, from the branches of the intercostal or lumbar arteries at the level of T8-L4 – in 60%. It seldom is seen at the level of L3 (1.4%) and at the level of L4-5 (0.2%).

The variable blood supply to the spinal cord by the radicular arteries was explained in 1888-89 by **Henryk Kadyj (23.05.1851, city of Peremyshl, Ukraine / Poland –** died from an infected cut of the finger at autopsy **25.10.1912, city of Lviv, Ukraine)**[248] — Polish Ukrainian anatomist, in relation to phylogenesis by the «phenomenon of

---

[247] Adamkiewicz A. Die Blutgefässe des menschlichen Rückenmarkes. I Theil. Die Gefässe der Rückenmarkssubstantz. Sitz K Akad Wiss Wien, Math-naturwiss Klasse 1881;84 (3):469-501. — Adamkiewicz A. Die Blutgefässe des menschlichen Rückenmarkes. II Theil. Die Gefässe der Rückenmarkesoberflache. Sitz K Akad Wiss Wien, Math-naturwiss Klasse 1882;85 (3):101-30. — Olearchyk AS. Saddle embolism of the aorta with sudden paraplegia. Can J Surg 2004;47(6):472-3.

[248] City of Peremyshl (Foot Notes 1, between 87 & 88), Ukraine / Poland. — City of Lviv (Foot Notes 2, between 87 & 88), Ukraine. — Kadyi H. O naczyniach krwionośnych rdzenia pacierzowego ludzkiego. Drukarnia Uniw Jagiellonski, Kraków 1888. — Kadyi H. Über die Blutgefässe des menschlichen Rückenmarkes. Math-naturwiss Klasse Akad der Wissenschaften in Krakau. Band XV. Gubrynowicz und Schmidt, Lviv (Lemberg) 1889. Kadyi H. O naczyniach krwionośnych rdzenia pacierzowego ludzkiego. Drukarnia Uniw Jagiellonski, Kraków 1888. — Kadyi H. Über die Blutgefässe des menschlichen Rückenmarkes. Math-naturwiss Klasse Akad der Wissenschaften in Krakau. Band XV. Gubrynowicz und Schmidt, Lviv (Lemberg) 1889. — Ablitsov VH. «Galaxy Ukraine». The Ukrainian diaspora: prominent persons. Publ. «Kyt», Kyiv 2007.

regression and progression», according to which when one important radical artery is progressing, the other less important regresses and vice versa.

The membranes corpuscles in the mucosa of genital organs and in the skin around the nipple were discovered by **Alexander S. Dogiel (1852 – 1922)** — Russian histologist, called genital or Dogiel corpuscles.

In 1887-94 **Santiago R.-y-Cajal (1852 – 1934)** — Spanish histologist, physician, and creator of neuroscience, discovered the axial growth core thus providing definitive evidence for the neuron theory, and identified the interstitial cells mediating neurotransmission from nerves to the bowel smooth muscle cells, named the intestinal cells of Cajal (ICC). They are responsible for the slow motor activity (peristalsis) of the gastrointestinal (GI) tract by creating the bioelectrical wave potential leading to contraction of the smooth muscle. Thus, ICC are the pacemaker cells of the gut.

The ICC are cells of origin of the gastrointestinal stromal tumors (GIST), the common mesenchymal tumors of the GI tract representing 1-3% of GI malignancies. The GIST of connective tissue, i.e. sarcomas; unlike most of the GI tumors, are nonepithelial. About 70% of the GIST occurs in the stomach, 20% in the small intestine and less than 10% in the esophagus. Small GIST are generally benign, especially when cell division is slow (70-80%), but large GIST disseminate to the liver, omentum and peritoneal cavity.

Serafina (Seraphina) Schachowa (1854, city of Dnipro, Ukraine – practiced medicine 1878 - 1910 in **city of Kharkiv, Ukraine, since 1910 the details of her life are unknown)**[249] — Ukrainian histologist, who described in 1876 p. the extension from the renal pole of the Schumlansky-Bowmen capsula glomeruli **[Olexandr Schumlansky (1748-95)** and **William Bowmen (1816 – 92)**: above] of the descending proximal convoluted tubule of Schachowa (tubulus contortus proximalis Schachowa), then the descending proximal straight tubule of Schumlansky (tubulus rectus proximalis Schumlansky) which create the U-like loop of Henle (tubulus attentualis Henle) **[Jacob Henle (1809 – 58)**: above] of the renal medulla (medulla renis), with further extension into the distal straight tubule of Schumlansky (tubulus rectus distalis Schumlansky) and the distal convoluted tubule of Schachowa (tubulus contortus distalis Schachowa), which join the collecting tube of the kidney (tubulus renalis colligent) to take the extravascular fluid for the secretion of urine.

---

[249] City of Dnipro (settled since Paleolithic Age / c. 100,000 BC, established in 1645 as the New Kodak, founded 1778), Dnipropetrovska Obl (area 31,914 km²), Ukraine. — City of Kharkiv (Foot Note 205), Kharkivs'ka Obl., Ukraine. — Schachowa S. Untersuchungen über die Nieren. Inaugural Doktoral Dissertationa l'Universite de Berne 1876. — [«…elle decrivit le segment du nephron intercale entre le tube contourne proximal et l'anse de Henle, segment par fois nomine tube is de Schachowa, 1876»] — Dobson J. Anatomical Eponyms. 2nd Ed. E&S Livingstone, Ltd, Edinburgh and London 1962. — Creese MRS. Ladies in the Laboratory IV. Imperial Russian's Women in Science, 1800-1900. Rowmant & Littlefield, Lanham, MD 2015. — Ablitsov VH. «Galaxy Ukraine». The Ukrainian diaspora: prominent persons. Publ. «Kyt», Kyiv 2007.

Mykola (Nikolai) F. Kashchenko [25.IV(07.V).1855, village Moskovka, Vilniansky Region, Zaporizhia Obl., Ukraine – 29.III.1935, city of Kyiv, Ukraine][250] — Ukrainian zoologist and histologist, described in 1884 p. the great rounded cells with multiple vacuoles, similiar to phagocytes, in the stroma of the placental chorion in the first 3 months of pregnancy. Later its presence was supported by **Isfred J. Hofbauer (1878 – 1961)** — American gynecologist and obstetrician, and they should be named the Kashchenko-Hofbauer cells.

Mykola K. Kulchytsky / Nikolai K. Kultschitzky (Jan. 16, 1856, fortress-city of Kronstadt, Kotlin Island on the Gulf of Finland, near St Petersburg, Russia – died mysteriously **Jan. 30**, 24:00, **1925, Oxford, England**, after sustaining injuries the day before, Jan. 29, c. 8:00, 1925, in a bizarre lift-shaft accident at the University College London)[251] — Ukrainian anatomist and histologist, who discovered in 1882 the argentaffine (enterochromaffin, silver chromaffin) cells in the crypts of Lieberkühn (1745)[252] between the bases of villi of the gastrointestinal (GI) tract mucosa, especially in the ileum, sometimes in the bronchi of the lung (Kulchytsky cells), from which originates a benign or malignant carcinoid tumor and carcinoid syndrome, a component of neuroendocrine tumors.

The Kulchytsky cells secretes serotonin or 5-hydroxytryptamine (5-HT) – a monoamine neurotransmitter.

The neuroendocrine tumors of the bronchial epithelium are:

1. Typical carcinoid tumor (Kulchytsky tumors I), the microscopic picture of which shows a well-differentiated bundle of braided together cells and absent

---

[250] Village Moskovka (founded 1780; peasantry survived made by Russian Communist Fascist Holodomor-Genocide 1932-33 against the Ukrainian Nation), Vilniansky Region, Zaporizhia Obl. (area 27,180 km^2), Ukraine — Ablitsov VH. «Galaxy Ukraine». The Ukrainian diaspora: prominent persons. Publ. «Kyt», Kyiv 2007.

[251] The fortress-city of Kronstadt (founded 1704) on the Kotlin Island (area 15 km^2) in the Gulf of Finland (surface area 30,000 km^2, average depth 38 m, maximal 115 m) of the Baltic Sea (surface area 337, 000 km^2, average depth 459 m, maximal 459 m). — Kulchytsky AK. On the problem of the structure of the mucous membrane of the small intestines and the mechanism of absorption. Protocols of Meetings of the Medical Section of the Society of Experimental Science at Kharkiv Univ 1882. — Kulchytsky AK. Basics of the Practical Histology. Vol. I (Part 1-2). Publ. DN Poliekhnov, Kharkiv 1889-90. — Kultschitzky N. Ueber die Färbung der markhaltigen Nervenfasern in den Schnitten des Zentralnervensystems mit Hämatoxylin und mit Karmin. Anat Anz (Jena) 1890;5:519-24. — Kulchytsky AK. Technique of the Microscopic Investigation. 2nd Ed. Publ PA Breidigan, Kharkiv 1897. — University College London (established 1826), England. — The Kharkiv University (established 1805), city of Kharkiv (cultural artifacts date back to the Bronze Age, c. 3200 - 600 BC; founded in 1654), Ukraine. — Ablitsov VH. «Galaxy Ukraine». The Ukrainian diaspora: prominent persons. Publ. «Kyt», Kyiv 2007. — Olearchyk AS. A Surgeon's Universe. Vol. I-IV. 4th ed. Publ. «Outskirts Press», Denver, CO 2016. Vol. I. C. 690-1.

[252] **Johannes N. Lieberkühn (1711 – 56**: also mentioned in the 18th century) — German anatomist, the author of «De fabrica et actione vollorum intestinorum tenuium hominis. Amsterdam» 1745.

mitoses, usually grows in the central bronchi (80%), seldom invades lymphatic nodes, and metastasize to distant organs (10-15%).

2. Atypical carcinoid (Kulchytsky tumor II) microscopically shows frequent mitoses, develops in the peripheral bronchi (60 ± 10), often involves the lymphatic nodes and metastasize to the distant organs (approximately 60%).

3. Large-cells neuro-secreting carcinoma (Kulchytsky tumor III), is characterized by frequent mitoses, varied localization in the lungs, often (approximately 60%) recurs after surgical removal.

4. Small-cell (oat-cell) carcinoma shows multiple mitoses and central localization in the lungs, almost always invades lymphatic nodes, and metastasize to the distant organs (about 80%). It is treated by chemotherapy with the first line being somatotropin osteocyte, or combination of interferon-alpha or metaiodobenzylguanidine, the second line include fluor urea and streptozotocin; and the third line include cisplastin and etoposide, seldom surgical resection.

In the patients with carcinoid tumor, even with the involvement of the lymphatic nodes and with distant metastases, the carcinoid syndrome occurs rarely (2%).

Carcinoid tumors are rarely a part of the familial multiple endocrine neoplasia-1 (MEN-1), could elicit development of a single or multiple bronchial carcinoid tumorlets, a type of fragile, often microscopic benign neoplasm. Its multiple form is a diffuse idiopathic pulmonary neuroendocrine cell hyperplasia (DIPNECH). Supposedly, DIPNECH produce neuropeptides that can elicit interstitial lung disease (ILD) or pulmonary fibrosis.

Bulbous, encapsulated, thin accessory membranous corpuscles and nerve endings in the skin, subcutaneous tissue, and mucous membranes (mechanoreceptors) in humans, discovered by **Dymitriy A.** von **Timofeew (1859 – 1937)**[253] — Russian histologist, and **Angelo Ruffini (1864 – 1929)** — Italian histologist and embryologist, are the slowly-adapted receptors for the sensation of prolonged pressure (Timofeew-Ruffini corpuscles, 1894-96, 1902). Beside this, the former described the peculiar form of the autonomic nervous sensory membranous corpuscles and plexuses in the submucosa of pars intermedia and prostatic of the urethra, of the prostate and seminal vesicles (Timofeew corpuscle).

The parathyroid rest cells were named for **Karl Hürthle (1860 – 1945)** — German histologist, and physiologist. They represent the large granulomatous eosinophilic

---

[253] Timofeew D von. Zur Kentnis der Nervenendigungen in den mänlichen Geschlechtsorganen der Säuger. Anatomischer Anzeiger (Anat Anz) 1894;9:342-8. — Timofeew D von. Über eine besondere Art von eigenkapselten Nervenendigungen in den männlichen Geschlechtsorganen bei Säugetieren. Anat Anz 1896;11:44-9.

cells with the increased number of mitochondria (Hürthle, or Azkanazy cells)[254] cells inside the thyroid gland which may give rise to a Hürthle's cell tumor.

Noted by **Albert F.S. Kent (1863 – 1958)**[255] — English physiologist, in 1892, more fully researched by **Wilhelm His, Jr (1863 – 1934)**[256] — Swiss cardiologist, in 1893, described by both in 1893, a neuromuscular atrioventricular (AV) bundle of His or His bundle. It is a small band of atypical cardiac muscle cells specialized for electrical conduction that originate in the AV node, then runs under the endocardium of the right ventricle (RV) on the membranous portion of the interventricular septum (IVS). At the upper end of the muscular part of the IVS it divides into the right and left branches which descend into the IVS wall of the right and left ventricles, then lead to the Purkinje fibers **(Jan Purkinje (1787 – 1869**: above), which innervate and provide electrical conduction to the ventricles, causing the cardiac muscle of the ventricles to contract at a paced interval. The interruption of the bundle of His produces a second-degree, type II heart block.

In addition, Albert Kent described the congenital anomaly of the conductive system of the heart (CSH) that is the accessory anterior bundle (pathway) between the right atrium (RA) and RV outside of the CSH, located anteriorly and near the fibrous ring of the tricuspid valve (TV) – the Kent bundle or AV bundle (1894), and Wilhelm His, Jr – the congenital anomaly of the CSH that is the atriofascicular bypass tract (atrio-Hisian's fibers, 1894), and Volhynia / Volyn (trench, five-day) fever (1916)[257].

Remembered for in his work in comparative anatomy, histology and embryology, **Martin Heidenhain (1864 – 1949)** — Prussian German anatomist, created the iron hematoxylin stain for the cytological demonstration of cellular structures (nuclei, chromosomes, centrioles, fibrils, mitochondria, cilia etc.) – Heidenhain stain (1885-86). Introduced the word «telophase» for the last stage of mitosis.

Ruggiero Oddi (1864 – 1913) — Italian anatomist and physiologist, identified the sphincter of the common bile duct (CBD) and of the pancreatic duct (musculus sphincter ampullae hepato-pancreatica) at the site of passing through the median wall of the descending (2nd) part of the duodenum, encompassing the sphincter of the CBD and of the pancreatic duct (1887).

The membranous thin capsule (sheath) covering the kidneys, described c. 1898, by **Dmitrie C. Gerota (1867 – 1939)** — Romanian anatomist, radiologist, consist of a layer of connective tissue and perinephric (perirenal) fat (fascia renalis seu Gerota).

---

[254] **Max Askanazy (1860 - 1945)** — German pathologist.

[255] Kent AFS. Researche on the structure and function of the mammalian heart. J Physiol (London) 1893, 14: 233-254.

[256] His W, Jr. Die Thätigkeit des embryonalen Herzens und deren Bedeutung für die Lehre von der Herzbewegung beim Erwachsenen. Arbeiten aus der medizinischen Klinik zu Leipzig, Jena, DE 1893;1:14-50.

[257] Volhynia or Volyn' / Volyns'ka Obl. (area 20,144 km²) – the historic (since 1077) / contemporary name of the North-Western territory of Ukraine.

Physiology

According to the law of **Pierre S.** marquis **de Laplace (1749 – 1829**: also mentioned in the 18th century) — French mathematician, physicist, statistician, and astronomer, the average circumferential systolic left ventricular (LV), aortic or arterial wall tension or stress (T) is directly proportional to the intraluminal pressure (P) and to the internal radius (r) and inversely proportional to the wall thickness (h): $T = Pr/2h$ (1806).

Stephan Z. Stubylevych (Stefan Stubielewicz) (Nov. 02, 1762, Volyns'ka Obl., Ukraine – Apr. 17, 1814, Vilnius, Lithuania)[258] — Ukrainian Lithuanian physicist, described the action of the continuing electric current on the organism and proposed the use of electrophoresis in the treatment of some diseases (1819).

Taking **Thomas Young (1773 – 1829)** — English polymath, physiologist, and physician, work on the physiology of optics (1801), **Hermann L.F.** von **Helmholz (1821 – 94)** — German physician and physicist, developed the doctrine of color vision (Young-Helmholtz theory).

Between 1822-23 **William Beamont (1785 – 1853)** — American physiologist and military surgeon, by treating and observing a patient with an extensive wound of the stomach, which was permanently exposed through the abdominal wall, documented the presence of hydrochloric acid (HCl) in the gastric juice, showed that there was a relationship between the emotional state and gastric secretion and digestion, and described the motor activity of the stomach.

The prone and postural methods of artificial breathing («ready methods») was introduced by **Marshak Hall (1790 – 1857)** — English physiologist and physician, in 1856.

Independently, **August. F. Möbius (1790 – 1868)** — German mathematician and theoretical astronomer, and **Johann B. Listing (1808 – 82)** — German physiologist, discovered two non-orientable two dimensional surfaces with only one side and only one boundary when imbedded in three dimensional space, depending on the direction of the performed half-turn – clockwise and counterclockwise, as shown by the half-twisted strip of paper, which can be realized as a ruled surface (Möbius strip, 1858). Hence, the Möbius's strip shows chirality or handedness, which have a similar space peculiarity as a hand. But, whether a ventricular anatomic pattern of the heart is right-handed or left-handed is not decided by ventricular looping or twisting, as one can exemplify from the Möbius strip, because the former was present at the earlier stage of the ventricular loop formation, and this is illustrated by the superior-inferior location of ventricles.

---

[258] Volyns'ka Obl. (Foot Note 257), Ukraine. — Vilnius University (established 1579), city of Vilnius (known with primary name of Voruta since 1251, first written note 1323), capital of Lithuania. — Ablitsov VH. «Galaxy Ukraine. The Ukrainian diaspora: prominent persons. Publ. «Kyt», Kyiv 2007.

Velocity of the fluid volume flow through the capillary tube was outlined by **Jean L.M. Poiseuille (1799 – 1869)** — French physicist and physiologist, as the drop in pressure alongside the tube directly proportional to its radius and inversely proportional to the length of the tube and the viscosity of the fluid (Poiseuille equation-law, 1838).

Alfred W. Volkmann (1800 – 77) — German physiologist, described bone canals, other than those described by **Chlopton Havers (1650 – 1702)**: 17th century), by which blood flows through bones (Volkmann canals).

Visible changes in the frequency of the sound, light and radio waves caused by motion were discovered by von **Christiaan A. Doppler (1803 – 53)**[259] — Austrian mathematician and physicist, named Doppler effect (1842), which states that the observed frequency is increased when either the source of the sound, light or radio waves, are moving towards the observer, or the observer is moving towards the source of the sound, light or radiovalves, or the frequency is decreased if either one, or both are moving away from each other, as evident in the following equation:

$$ f = \left(\frac{v + v_r}{v + v_s} \right) f_0 $$

where f is observed frequency, f_0 – the emitted frequency, v – the velocity of waves in the medium, v_s – the velocity of the source relative to the medium, v_γ – the velocity of the receiver relative to the medium.

The Doppler effect or shift became the basis for the noninvasive ultrasonic (US) examination of the cardiovascular hemodynamics, the body organs, as well as the introduction of transthoracic and esophageal echocardiography (TTE, TEE), e.g., it is used to measure the valve area, the lumen of the blood vessels, and the size of the atrial septal defect (ASD) and the ventricular septal defect (VSD).

Manuel P.R. Garcia (1805 – 1906) — Spanish baritone singer, music educator and vocal pedagogue, is credited with the invention of laryngoscope for examination of the throat, vocal aperture and cord and larynx (1854), which was then introduced into medicine by **Ludwig Türck (1810 – 68)** — Austrian neurologist, and **Johann N. Czermak (1828 – 73)** — Czech physiologist, in 1860 and 1863, respectively.

Aleksii M. Filomafitskii [17(29).III.1807, village Malakhovo, Tutayevsky Region, Yaroslavska Obl., Russia – 22.I(03.II).1849, city of Moscow, Russia][260]

---

[259] Doppler Ch. Über das farbige Licht der Doppelsterne und einiger anderer Gestirne des Himmels. Gelesen bei der Königl. Böhm. Gesellschaft der Wissenschaften zu Prag, in der naturwissenschaftlichen Sectionssitzung vom 25. Mai 1842. In Commission bei Dorrosch & Andre, Prag 1842. V. Folge, Bd. 2, S. 465-482.

[260] Village Malakhovo, Tutayevsky Region, Yaroslavska Obl. (area 36,117 km^2), Russia). — Ablitsov VH. «Galaxy Ukraine. The Ukrainian diaspora: prominent persons. Publ. «Kyt», Kyiv 2007.

— Russian Ukrainian physiologist, introduced the theory of cyclical function of the nervous system (1836-40), and with **Vassilii A. Basov (1812 – 79)** — Russian surgeon, performed a gastric fistula in dogs (Basov fistula, 1842), and investigated the intravascular administration of ether, chloroform and benzine for anesthesia (1849).

With the intention to develop pulmonary function test (PFT), **John Hutchinson (1811 – 61)[261]** — English physician, designed the water spirometer, a device for measuring lung capacity, which consist of two cylinders – external and internal. The lighter external cylinder is filled up with water, and heavier internal cylinder dipped into the water with bottom up. In the created space above the water level a tube is positioned, the external end of which is connected to the rubber tube with the mouthpiece. Studied made the maximal (deepest) possible inspiration – the inspiratory reserve volume (IRV), squeeze nostrils of the nose and follows this with the deepest possible expiration slowly exhaled into the tube – the expiratory reserve volume (ERV). The internal cylinder ascended upwards the certain level of the scale, located on one side. The air from the cylinder let out through the opening in rubber tube.

Number of useful PFT measurement was performed by John Hutchinson, particularly the vital capacity (VC), a sum of the tidal volume (TV), the IRV and the ERV, or the amount of an air that residue in the lungs at the height of the maximal inspiration, in over 4000 patients.

The VC on the average in a healthy adults accounted 4.8 liters (L) per minute (min.), in women 3.1 L/min.; the IRV in men was 3.1 L/min., in women – 1.9 L/min.; the TV in men and women was equal 0.5 L/min.; and the ERV in men amounted 1.2 L/min., in women – 0.7 L/min.

The total lung capacity (TLC) consists of the VC plus residual volume (RV). The RV is a volume of the air remaining in the lungs after the most forcible expiration, which cannot be exhaled. Therefore, measurement of the RV volume should be performed by using indirect methods such as radiographic planimetry, body plethysmography, closed circuit dilution with helium or nitrogen washout. The average TLC amounts in men to 6.0 L/min., in women 4.2 L/min., the RV amounts in man 1.2 L/min., and in women 1.1 L/min.

Thomas Laylock (1812 – 76) — English neurophysiologist and physician, is known for a concept involving «reflex action of the brain», postulating that reflexes are intelligent, but unconscious reaction to stimuli.

---

[261] Hutchinson J. Lecture on vital statistics, embrancing an account of a new instrument for detecting the presence of disease in the system. Lancet 1844;1:567-70. — Hutchinson J. On the capacity of the lungs, and the respiratory function with a view to establish a precise and easy method of detecting disease by spirometer. Med Surg Trans 1846;29:137-252. — Bishop PJ. A bibliography of John Hutchinson. Med Hist 1977 Oct;21(4):384-96. —Spriggs EA. The history of spirometry. Br J Dis Chest 1978;72(C):165-80. — Speizer FE. John Hutchinson, 1811-1861: the first respiratory disease epidemiologist. Epidemiology 2011 May:22(9):e1-e9.

The founder of experimental physiology, **Claude Bernard (1813 – 78)** — French physiologist, developed a concept of hemostasis (in French – «milieu interieur»), when noted that all higher living organisms actively, constantly, coordinately, and purposefully counteract the factors of the external environment to infringe upon the conditions necessary for the security, support, or restoration of the stability of their internal environment vitality. Together with **Johann F. Horner (1831 – 86)** — Swiss ophthalmologist, re-described the paralysis of the cervical sympathetic nerve [Bernard-Horner syndrome: 18th century – **Francois Petit (1664 – 1741)**].

Julius R. von **Mayer (1814 – 78)** — German chemist, physicist and physician, and **Hermann** von **Helmholz** (above), independently, formulated the theory of conservation of energy (1842-47). Later, Hermann von Helmholz developed a theory of sound perception (Helmholz theory) and invented the ophthalmoscopy (1851).

Degenerative shivering of muscles supplied by severed nerve fibers was described in 1850 by **Augustus V. Waller (1816 – 70)**[262] — English histologist and neurophysiologist, named after him Wallerian degeneration. Usually, Wallerian degeneration push through antegrade (distally), rarely retrograde (proximally) to the site of nerve injury.

There are three stages of nerve injuries: (1) neuropraxia, the least severe form of nerve damage from trauma or disruption of blood supply in which the nerve axon remains intact, but surrounding nerve myelin layer is damaged, causing greater involvement of motor than sensory nerve function, recovery occurs on average in 6-8 weeks; (2) axonotmesis, severe form of nerve injury causing transection of the nerve axon with distal, occasionally also proximal Wallerian degeneration, associated motor, sensory and autonomic nerve dysfunction, where the proximal lesion may grow as fast as 2-3 mm per day, distal lesion as slow as 1.5 mm per day, but regeneration occurs over weeks to years; (3) neurotmesis is the most severe nerve lesion caused by transection, or internal disruption in which the nerve axon, surrounding myelin layer, and encapsulated connective tissue lose their continuity, without possibility for full recovery.

In this same year (1846) **Carl (Karl) F.W. Ludwig (1816 – 95)** — German physiologist and physician, found a ganglion connected with the cardiac plexus situated near the right atrium (RA) – Ludwig ganglion, and developed the first revolving drum recorder holding paper, calling it the «kymographion» that is «kymograph» or «valve-writer» (in Greek – «kyma» = wave, «grapheion» = stylus) to record pulmonary artery pressure (PAP) through direct cannulation.

---

[262] Waller A. Experiments on the section of the glossopharyngeal and hypoglossal nerves in the frog and observarions on the alterations produced thereby in the structure of their primitive fibers. Phylos Trans R Soc London 1850;140:423-9.

The «kymograph» was a difficult machine to set up and use. The smoked paper preserved minute fluctuations of the stylus in a very sensitive manner. The tracing recorded was in synchrony with the motion of a float on a mercury manometer which transmitted changes in arterial blood pressure (BP). A «Valve-writer» allowed measurement of intra-arterial BP without the intrusive effects on heartbeat and respiration. The main limitation of the method was the need to insert a catheter into a blood vessel. It became the model for the «sphygmograph» developed in 1854 by **Karl** von **Vierord (1818 – 84)** — German physician, for monitoring the pulse. The «kymographion» remained the basic laboratory tool, that by graphic representation allowed physiological research to progress more rapidly, was essential in recording many phenomena, revolutionized recording methods in the 19th century and was still useful tool for the physician well into the 20th century.

In 1849 Karl Ludwig and **M. Hoffa**[263], while investigating the influence of the cranial X (vagal) nerve on cardiac activity in an animal provoked and graphically documented ventricular fibrillation (VF) by applying an electric «faradic» current[264] to the heart.

Later (1867), M. Hoffa with **Iwan / Johann** von **Dogiel (1830 – 1916)** — Russian neurophysiologist and pharmacologist, invented a blood-stream meter (stromuhr) for the measurement of blood volume flow.

Karl von Vierord, also contributed to psychology of time perception by proposing that measurement of short durations tends to be overestimated, while long durations tend to be underestimated (Vierord's law, 1869).

Olexandr P. Walter or **Alexander P. Walther (29.XII.1817 / 09.1818, city of Tallinn, Estonia – 23.IX/14.X.1889, city of Warsaw, Poland**, buried in the Fair-tale Cemetery, city of Kyiv, Ukraine)[265] — Ukrainian anatomist and physiologist of German

---

[263] Hoffa M, Ludwig C. Einige neue Versuche uber Herzbewegung. Zeitschrift Rationelle Medizin 1850;9:107-44.

[264] **Michael Faraday (1791 - 1867)** — English physicist and chemist. / In chemistry, a «faradic» current is an electric current that corresponds to the reduction or oxidation of a chemical species; in electricity, a «faradic» current means an electric intermittent and nonsymmetrical alternating current like that obtained from the secondary winding of an induction coil.

[265] City of Tallinn (first appeared in 1145) – capital of Estonia (area 45,339 km²). — City of Warsaw (first fortified settlement in 9th-10th century, founded in 13th century) – capital of Poland. — The Fairy-tale Cemetery (mentioned 1831, opened 1834), city of Kyiv, Ukraine. — Anatomical Theatre of St Volodymyr Kyiv University Museum (built 1851-53), now The National Museum of Medicine (NMM) of Ukraine. — Walther AP. Beiträge zur Lehre von der thierischen Warme. Virchov Arch Path Anat 1962;25:414-28. — Walter A. On the influence of cold on living matters. Sovremennaya meditsina (Modern medicine) 1863;45(480:51-2. — Olearchyk AS. Endarterectomy and external prosthetic grafting of the ascending and transverse aorta under hypothermic circulatory arrest. Texas Heart Inst J 1989;16 (2):76-80. — Ablitsov VH. «Galaxy Ukraine». The Ukrainian diaspora: prominent persons. Publ. «Kyt», Kyiv 2007. — Hanitkevych YaV. The contribution of the Ukrainian physicians to the world medicine. Ukr Med Chasopys 2009 VII/VIII;4(72):110-15. — Olearchyk AS. A Surgeon's Universe. Vol. I-IV. 4th Ed. Publ. «Outskirts Press», Denver, CO 2016. Vol. I. P. 677.

origin, Adjunct (1843-46) and Professor (1843-74) in the Chair of Physiological Anatomy and Microscopy, first Director (1853-74) of the Anatomic Theatre, St Volodymyr Kyiv University, founder and editor of the journal «Sovremennaya meditsina» / «Modern Medicine» (1830-80), city of Kyiv, Ukraine, medical inspector of public hospitals (1874-89), city of Warsaw, Poland. He demonstrated the vasoconstrictive effect of the sympathetic nerves on blood vessels (1842), described the effect of cooling rabbits down to 20⁰C and concluded that the lowering of temperature increases the safety of an operation (1862) and thus introduced hypothermia.

The first scientist to work out the physiology of the spinal cord by demonstrating that the decussation of the nervous fibers carrying pain and temperature sensation occurs in the spinal cord itself, and the originator of the cellular (cell) or mesenchymal stem cell (MSC) therapy was **Charles É. Brown-Séquard (1817 – 94)** — Mauritian physiologist and neurologist of American French descend. He found that an injury to the half of the spinal cord causes ipsilateral paralysis and loss of deep sensation, and contralateral decrease or loss of the pain and temperature sensation (Brown-Séquard's syndrome, 1850-51). In regard to cell therapy, he advocated the hypodermic injection of an extract from the testicles of guinea pigs and dogs, as a means of prolonging life (1889).

The founder of electrophysiology (EP), **Emil H. DuBois-Reymond (1818 – 96)**[266] — German physiologist and physician, formulated the molecular theory of biopotentials and discovered the nerve action potential (AP). He stated that a variation in current density of «electric molecules», which we now known are natrium / sodium (Na), kalium / potassium (K) and other ions, rather than the absolute value of of current density at any given moment, act as a stimulus to the motor neuron and nerve and the muscle (DuBois-Reymond law, 1842). An AP is a short-lasting event in which the electrical membrane potential of the cell rapidly rises and falls, following a consistent trajectory.

Important studies involving the different phases of the cardiac cycle and intracardiac pressure were performed **Jean A. Chauevau (1827 – 1917)** — French veterinarian and **Etienne J. Marrey (1830 – 1904)** — French scientist, physiologist, and chronophotographer, by introducing air-filled ampules connected to a thin elastic tube into the heart chambers of a horse which they called «cardiograph» and directly recorded the variations in pressure using a «kymograph». The latter (Etienne Marey), was the first to document premature ventricular contractions (PVC, 1865), noted that an increase in the blood pressure (BP) lowers the heart rate (HR) – Marey law (1869), and studies the function of muscles by «myography» and movements of

---

[266] Coenen A, Zayachkivska O. Adolf Beck: A pioneer in encephalography in between Richard Caton and Hans Berger. Adv Cogn Psychol 2013;9(4):216-21.

the body. He wrote that «*Science meets with two obstacles, the deficiency of our senses to discover facts and the insufficiency of our language to describe them. The object of the graphic methods is to get around these two obstacles; to grasp fine details which would be otherwise unobserved; and to transcribe them clarity superior to that of our words*» (1885).

The work of **Adolf E. Fick (1829 – 1901)**[267] — German physiologist and physician, was influenced by **Karl Ludwig (1816 – 95**: above), on the two laws of diffusion which governs the diffusion of matter across a membrane (Fick laws of diffusion, 1855).

The Fick's 1st law relates the diffusive flux to the concentration, by postulating that the flux goes from regions of high concentration to regions of low concentration, with a magnitude that is proportional to the concentration gradient.

The Fick 2nd law predicts how diffusion causes the concentration to change with time. These two diffusion laws led to understanding of the relationship between blood and gas exchange in the lungs.

Then, the latter devised a technique for an indirect measuring of cardiac output (CO) and cardiac index (CI), called the Fick principle, as presented in the equation:

$$CO = HR \times SV \ (EDV - ESV)$$

where CO = cardiac output (L/min.); (HR = heart rate (normal 60-100 beats per min.); SV = stroke volume; EDV = end-diastolic volume; ESV = end-systolic volume.

Factors affecting CO are autonomic innervation, hormones, fitness, and age; factors affecting the SV are heart size, fitness gender, contractility, duration of contraction and after-load (resistance).

$$CI = VO_2 / C_a - C_v,$$

where CI = cardiac index [L/min. / body surface area (BSA) m^2]; VO$_2$ = the oxygen (O$_2$) consumption index (normal 110-150 ml/min/m^2, in man – 125 ml/min/m^2, in older women – 110 ml/min/m^2); Ca – Cv) = the difference between the O$_2$ content in the arterial (C$_a$, 99%) and in the venous (C$_v$, 75%) blood.

Normal CO = 4.0-6.0 L/min., and normal CI = 2.0-3.0 L/min. / BSA m^2 (1870).

He also invented the tonometer which is used to measure tension or pressure of the eye and inside the stomach.

---

[267] Fick A. Ueber die Messung das die Blutquantums in den Herzventrikela. Verhandlungen der Physikalisch-Medizinische Gesellschaft zu Würzburg. 1870 Juli 9;2:16-17.

In 1862 **Ivan M. Sechenov (1829 – 1905)** — Russian biologist, and physiologist, discovered the inhibitory reflex center in the medulla oblongata and the spital cord (Sechenov reflexes), stating that all the processes taking place in the brain, including thinking, were of reflex nature (Sechenov IM. «Reflexes of the Brain», 1863).

Ivan P. Shchelkov (21.II/05.III.1833, city of Kharkiv, Ukraine – 14/27.III. 1909, town of Sudak, Crimea, Ukraine)[268] — Ukrainian physiologist, established in the Faculty of Physiology (founded 1862) at the Kharkiv University the Physiology Research Laboratory (1863). Together with **Karl G.G. Sadler (1801 – 77)** — Russian physician of German origin, studied reciprocal action between blood circulation and muscle contraction, making discovery of so called «working hyperemia of skeletal muscles» (1869).

The most important accomplishment of **Rudolf P. Heidenhain (1834 – 97)** — Prussian German physiologist, was the measurement of heat production during muscle activity. He was the first to detect, by sensitive thermoelectric measurement, a minute increase in temperature (T) – $0.001 - 0.005^0C$ during every simple twitch of the muscle. The second important accomplishment was his discovery in 1867 that the chief cells of the gastric mucosal glands secrete enzyme pepsin, and the parietal cells of the gastric mucosal glands secretes hydrochloric acid (HCl). The third important accomplishment was his finding of an increased acid production in the working muscle and that stimulation of the motor nerve produced a special motor reactions, so called speedometer phenomenon (1883).

Carl E.K. Hering (1834 – 1918)[269] — German physiologist, authored the theory of color vision based on yellowish-blue, reddish-green and whitish-black colors (1872), presented the principles of unequal bilateral innervation of the muscles of two eyes so that one eye never moves independently of the other (Hering law, 1868).

Carl Hering with **Joseph Breuer (1842 – 1925)** — Austrian physician of Jewish ancestry, described the cranial 10th (vagal) nerve (Hering-Breuer, 1868) cranial reflex which controls respiration. This latter reflex means that the sensory endings in the lungs passing up to the cranial 10th nerve tend to limit both, the inspiration and expiration in ordinary breathing. Reflex inhibition of inspiration resulting from stimulation of baroreceptors (mechanoreceptors, pressoreceptors) by an excessive inflation of the lungs.

---

[268] City of Kharkiv (Foot Note 205), Ukraine. — Town of of Sudak / Sudaq (founded 212), Crimea, Ukraine. — Ablitsov VH. «Galaxy Ukraine». The Ukrainian diaspora: prominent persons. Publ. «Kyt», Kyiv 2007.

[269] Hering E. Beiträge zur Physiologie. Zur Lehre Ortsinne Metzhaut. Teil I-V. Engelmann Verlag, Leipzig 1861-64. — Hering E. Die Lehre vom binokularen Sehen. Engelmann Verlag, Leipzig 1868. — Hering E. Die Selbststeuerung der Athmungs durch Nerves vagus. Akademie Wissenschaften. Mathematisch-Natur Wissenschaftlishe Classe, Wien 1868;57,II:672-70.

In 1870 **Gustav T. Fritsch (1838 – 1927)** — German anatomist, physiologist, anthropologist and traveler, and **Edward Hitzig (1839 – 1907)** — German neurologist and neuropsychiatrist, showed clearly that the sensory and motor function could be localized in the cortex of the brain.

Defibrillation of the heart was first demonstrated in 1899 by **Jean L. Prévost (1838 – 1927)** — Swiss physiologist and neurologist, and **Frederic Batelli** — Swiss physiologist, as they discovered that small electrical shock could induce ventricular fibrillation (VF) in dogs, and larger charges would reverse the condition to the normal sinus rhythm (NSR).

Both, **Carl Ludwig (1816 – 95: above)** and **Henri P. Bowditch (1840 – 1911)**[270] — American physiologist and physician, formulated the «all-or-none law», the treppe / tripe («staircase») phenomenon, and the latter – the Bowditch law (1871–74). The «all-or-none law» states that that the heart muscle, under whatever stimulus, will contract to fullest extent or not at all. Stimulation of any single atrial or ventricular muscle fiber causes the action potential (AP) to travel over the entire atrial or ventricular mass or it does not travel at all. In other muscles and nerves, this principle is limited to an individual fiber, that is stimulation of a fiber causes an AP to travel over the entire fiber, or it does not travel at all. The tripe is the gradual increase to extend of a muscle contraction, following rapidly by repeated stimulation. The Bowditch law states that a nerve cannot be tired out by stimulation.

The adrenal (suprarenal) glands **(Bartholomeo Eustachii (1500 or 1514 – 1574:** Renaissance**)** has the external layer (cortex) which produces steroid hormones, i.e., mineralocorticoids, glucocorticoids and androgens, and the internal layer (medulla) which produces hormones adrenaline (epinephrine) and noradrenaline (norepinephrine).

George Oliver (1841 – 1915) — English physician, along with **Edward A. Sharpey-Schäffer (1850 – 1935)** — English physiologist, and endocrinologist, discovered in 1893-94 the existence of adrenaline / epinephrine by demonstrating that adrenal gland extract, obtained from calves, sheep's, guinea-pig's Cavia porcellus), cat's, dog, and man, when injected into blood stream raises blood pressure (BP). Also, the later coined the term «insulin» since considered it to be a single substance from pancreas responsible for diabetes mellitus (DM), proposed the method of external manual artificial respiration (Schäffers method, 1895).

The pioneer of electroencephalography (EEG), **Richard Caton (1842 – 1924)**[271] — English physiologist, and physician, using unipolar electrodes attached to cerebral

---

[270] Cannon WB. Biographical memoir Henry Pickrting Bowditch 1840-1911. Nat Acad Sc 1922;17:181-96.

[271] Caton R. The electrical currents of the brain. Brit J Med 1875 Aug 25;2(765):278. — Coenen A, Zayachkivska O. Adolf Beck: A pioneer in encephalography in between Richard Caton and Hans Berger. Adv Cogn Psychol 2013;9(4):216-21.

cortex and surfaces of the skull in living animals, such as the rabbit, monkey, and dog, measured the electrical current with a sensitive galvanometer, showed distinct variations in the electric current activities which increased during sleep and with the onset of death strengthened, after death became weaker and then disappeared (1875).

The first human electrocardiographs (ECG) were recorded in 1869-72 by **Alexander Muirhead (1848 – 1920)** — Scottish electrical engineer, and in 1872 by **Gabriel Lippmann (1854 – 1921)** — Luxembourg physicist, using a capillary (string, thread) electrometer.

In his work from 1880s **Magnus G. Blix (1849 – 1904)** — Swedish physiologist, discovered that electrical stimulation on different points on the surface of the skin caused distinct warm or cold sensation. Following construction of a temperature stimulator he showed that a decreased skin temperature created cool sensations from localized spot at separate skin locations, and that increased temperature induced warm sensations also from different cutaneous locations.

Ivan P. Pavlov (1849 – 1936) — Russian physiologist, proved that nerves control the heart function (1883) and digestion (1897), developed the concept of conditioned and unconditioned reflexes (1913-27).

In 1902 **Charles R. Richet (1850 – 1935)** — French physiologist, coined the term anaphylaxis (in Greek – «ana» meaning «against» and «phylaxis» meaning «protection») in humans or animals manifested by an acute, systemic (multi-organ) and severe allergic reaction with increased sensitivity to the repeated arrival of foreign substances, most commonly of protein origin acting as an antigen. Sometimes second small dose may lead to lethal reaction. The most severe type of anaphylaxis is an anaphylactic shock which usually causes death in a matter of minutes if not treated. Certainly, his research helped elucidate hay fever, asthma, and some other allergic reaction to trigger substances and explained some previously not understood cases of intoxication and subsequent sudden death.

Nicolai I. Vvedenskii or **Nicolai I. Wedensky (1852 – 1922)** — Russian neurophysiologist, described the optimal and the inhibitory muscle response on repetitive stimulation of the nerve (Vvedensky phenomenon, 1886).

In 1872 **J. W. Gordon**[272] — British physiologist, registered the whole body (rhythmic) vibrations which led to the development of ballistocardiograph (BCG) in the next century. The BCG measure of ballistic forces generated by the heart in which downward movement of the blood through the descending thoracic aorta (DTA) produces upward recoil, moving the body upwards with each heartbeat.

---

[272] Gordon JW. Certain molar movements of the human body produced by the circulation of the blood. J Anat Physiol 1877 Apr;11(pt3):533-6.

Vasyl Ya. Danylevsky / Vassily (Bazil) Ia. Danilewsky [13(25).I.1852, city of Kharkiv, Ukraine – 25.II.1939, city of Kharkiv, Ukraine][273] — Ukrainian physiologist, physician and parasitologist, who was the first to give comprehensive description of bioelectric nerve impulses from the brain of dogs (1876). He was the first to investigate systematically on blood parasites of vertebrate such as birds, reptiles, and amphibians in Ukraine, and in his paper of 1884 (in Russian) and 1885 (in German) titled «About blood parasites (Haematozoa)» he laid the foundation of modern parasitology in bird malaria and other protozoan infections. He was the first to observe in 1884 the species of Haemoproteus, parasitic protozoan in the blood of birds, named after him Haemoproteus danilewsky; was the first to discover in 1885 the protozoan Trypanosoma avium, the first known flagellate protozoan parasite in birds; by 1885 he recognized for the first time the existence of three separate genera of protozoan parasites in birds, now known as Plasmodium, Haemoproteus and Leukocytozoon; was the first to observe in 1888 a new genus Leukocytozoon from the owl blood, also named after him Leukocytozoon danilewsky; and was the first to describe the bird malaria and its symptoms of acute anemia, enlargement of the liver (hepatomegaly) and the spleen (splenomegaly), accumulation of pigments in the body, and gave the first clue to the similarity of malaria of birds to that of humans. He identified the bird malaria parasites as «pseudovacuoles».

Jacobius H. van't **Hoff, Jr (1852 – 1911)** — Dutch physicist and organic chemist, has described the method for determining the order of a reaction using graphics, applied the laws of thermodynamics to chemical equilibria, introduced the modern concept of affinity that is the electronic property by which dissimilar chemical species are capable of forming chemical compound, and proposed an equation (van't Hoff equation) for the temperature dependence of equilibrium constant (K_{eq}) of a chemical reaction to change in temperature (1884)[274]. He also showed a similarity between the behavior of dilute solution and gases (1886).

In 1895 **Napoleon O. Cybulski (Sep. 14, 1854,** manor **Kryvonosy,** now town of **Kryvonosy, Postava Region, Vitebsk Obl., Belorus – Apr. 26, 1919, city of Cracow, Poland)**[275] — Belorusan Polish physiologist, and his pupils **Władysław**

---

[273] Danylevsky V.Ya. Investigation into Physiology of the Brain. Thesis University of Kharkiv for doctoral degree, Kharkiv 1877. — Danilewski B. Zur Parasitologie des Blutes. Biol Zentralblatt 1885;5:529-37. — Danilewsky B. La Parasitologie Comparee du Sang. I. Nouvelles Recherches sur les Parasites du Sang des Oiseaux. Kharkoff 1889. — Coenen A, Zayachkivska O. Adolf Beck: A pioneer in encephalography in between Richard Caton and Hans Berger. Adv Cogn Psychol 2013;9(4):216-21. — Ablitsov VH. «Galaxy Ukraine». The Ukrainian diaspora: prominent persons. Publ. «Kyt», Kyiv 2007. — Hanitkevych YaV. The contribution of the Ukrainian physicians to the world medicine. Ukr Med Chasopys 2009 VII/VIII;4(72):110-15.

[274] Van't Hoff J. Etudes de dynamique chimique. Paris 1884.

[275] Coenen A, Zayachkivska O. Adolf Beck: A pioneer in encephalography in between Richard Caton and Hans Berger. Adv Cogn Psychol 2013;9(4):216-21.

Szymonowicz (Mar. 21.03.1869, city of Ternopil, Ukraine – Mar. 10.03. 1939, city of Lviv, Ukraine)[276] — Ukrainian Polish histologist, and embryologist, discovered an active extract and its derivatives secreted by the adrenal gland medullas and named it adrenaline.

In 1896-97 **R. Gottlieb**[277] combined intracardiac injection of the adrenal gland extract and external manual cardiac compression to resuscitate cardiac arrest due to asystole in rabbit and proposed to use this method in similar cases in man.

In 1896 **William H. Bates (1860 – 1931)**[278] — American ophthalmologist, discovered the astringent[279], hemostatic properties of adrenaline, and used it in eye surgery.

Finally, adrenaline was isolated in 1900 by **Jokichi Takamine (1854 – 1922)**[280] — Japanese American chemist.

Beside of being a hormone, adrenaline function as neurotransmitter and medication, which by binding alpha- and beta-receptors plays an important bodily role in the «flight-or-fight» response by increasing blood flow to muscles, cardiac output (CO), arterial blood pressure (BP) and the blood glucose level.

In 1891 **Christian H.L.P.E. Bohr (1855 – 1911)**[281] — Danish physiologist, characterized the anatomical dead space (DS) as the volume of air not taking part in gas exchange, because it either remains in the airways from the nose or mouth down to the trachea, right and left main stem bronchi (RMSB, LMSB), lobal bronchi and to the terminal bronchiole (up to the 16th generation) that have no contact with blood perfusion, and the physiological DS where because of the underlying lung disease, the air volume in addition to being naturally trapped in airways, cannot access a distal non-respiratory part of the bronchial airway and reach poorly perfused or not perfused by the blood alveoli, i.e., not all the air in each breath is available for

---

[276] City of Ternopil (founded 1540), Ternopil Obl. (Foot Note 197), Україна. — Szymonowicz L. Lehrbuch der Histologie und der microskopischen Anatomie: mit besonderer Berücksichting des menschlichen Körpers einschliesslich der microscopishen Technik. A. Verlag Stuber's Verlag (C. Kabitzch), Würzburg 1901. — Ablitsov VH. «Galaxy Ukraine». The Ukrainian diaspora: prominent persons. Publ. «Kyt», Kyiv 2007.

[277] Gottlieb R, Über die Wirkung der Nebennieren-Extrakte auf Herz und Blutdruck. Arch Exp Pathol Pharmacol 1896/97;38:99-112.

[278] Bates WH. The use of extract of suprarenal capsule in the eye. NY Med J 1896:647-50.

[279] Astringent is a substance that tends to shrink or constrict small blood vessel and body tissue thus decreasing local oozing or bleeding.

[280] Takamine J. Adrenalin, the active principle of the suprarenal gland, and its mode of preparation. Am J Pharmacy 1901;73:523-31. — Takamine J, et al. Adrenalin. Parke, Davis & Co, Detroit 1901. — Takamine J. The isolation of the active principle of suprarenal gland, Proceeding Physiol Soc Great Britain, Cambridge Univ Press Dec 14, 1901. PP. 29-30.

[281] Bohr C. Über Lungenatmung. Scand Arch Physiol 1891;2:236-68. — Bohr C, Hasselbach K, Krogh A. Über einen biologisher Beziehung wichtigen Einfluss, den die Kohlensäurespannung des Bluted auf dessen Sauerstoffbindung übt. Scand Arch Physiol 1904;16:401-12.

the exchange of oxygen (O_2) and carbon dioxide (CO_2). The DS is calculated by Bohr equation:

$$Vd/Vt = PaCO_2 - PeCO_2/PaCO_2, \text{ where}$$

Vd = DS, Vt = tidal volume, $PaCO_2$ = the partial pressure of CO_2 in the arterial blood, and $PeCO_2$ = the partial pressure of CO_2 in the expired air.

In the average healthy adult human, every Vt of approximately 450 to 500 ml of an air, minus the normal DS (Vt) of about 150 ml (33%) of an air, is exchanged during each breath without changes in the level of O_2 and CO_2.

From the studies on the effect of endolymph motion on the body, head, and eyes, also on the excitation-inhibition asymmetries in the vestibular system conducted on pigeons, **Ernst J.R. Ewald (1855 - 1921)** — German physiologist and **Rudolf Kobert (1854 – 1918)** — German pharmacologist and toxicologist, derived the three laws, named after the former (1892-95):

the 1st Ewald's law – «The axis of nystagmus parallels the anatomic axis of the semicircular canal that generated it»; the 2nd Ewald's law – «Ampullopedal endolymphatic flow produces stronger response than ampulofugal flow in the horizontal canal»; the 3rd Ewald's law – «Ampullofugal flow produces a stronger response then ampulopedal flow in the vertical canals (anterior and posterior semicircular canal)».

Augustus D. Waller (1856 – 1922)[282] — English physiologist, discovered the refractory period in the heart muscle in animal (1875), using a capillary electrogram made the first recoding of an electrocardiography (ECG) of the heart in animal (1876) and in a man (1887).

The electrolytic dissociation theory developed by **Svante A. Arrhenius (1859 – 1927)** — Swedish physicist and chemist, explained properties of electrolytes, founded on the presence in the solution of free ions, distinguished acids and bases as compounds that dissociate with the release of hydrogen (H^+) and hydroxide (OH^-) ions (1884-87).

According to the work of Svante Arrhenius, for many common chemical reactions at the room (tepid) temperature (T), the reaction rate doubles by increasing or decreasing for every 10°C increase or decrease in T (Q10)[283].

---

[282] Waller AD. A demonstration on man of electromotive changes accompanying the heart's beat. J Physiol 1887;8:229-34. — Waller AD. The electrical action of the human heart in 1887 and 1915. Br Med J 1915;i:35-7. — Besterman E, Creese R. Waller — pioneer of electrocardiography. Br Heart J 1979;42:61-4.

[283] Arrhenius SA. Uber der Dissociationwärme und den Einfluss der Temperatur auf den Associationsgrad der Electrolyte. Z. Phys Chem 1889;4:96-116.

He developed a formula of the T dependence for both forward and reverse reaction rates (Arrhenius equation, 1889):

$$K = Ae^{-Ea/(RT)},$$

where K = constant; T = absolute T (in Kelvin)[284]; A = pre-exponential factor, a constant for each chemical reaction thet defines the rate due to frequency of collisions in the correct orientation; Fa = the activation energy for the reaction [in Joules: 18th century. – **James Joule (1818 – 89)**]; R = the universe gas constant.

The introduction of the Q 10 T coefficient allowed to measure the rate of change of a biological (brain, nerves, muscle, heart, etc.) or chemical system as consequence of increasing or decreasing the T by 10°C. With increased T the rate of those reaction increases, with decreasing the T the rate decreases.

In humans, the decrease of a normal body T (36.6°C - 37°C) to a mild (28°C - 32°C), moderate (22°C – 25°C) or profound (14°C - 20°C) is protective, mainly for the brain and heart from an ischemic injury, during complex operations by slowing the rate of high energy adeno-triphosphate (ATP) consumption or degradation and decrease of myocardial oxygen (O_2) consumption (MVO_2). The normal MVO_2 of 100% [10-15 milliliters (ml) / kilo (kg) / minute (min.)] at room T, will decrease with drop of the body T by 10°C to 50%, and with the drop of the body T to 6°C it will decrease to only 6%.

On the other side hypothermia increase wound infection, blood loss and transfusion requirement. At least 50% of individuals whose body T dropped to 21°C will show prolongation the nerve conduction velocity, decline contraction of the muscle fiber, increase cardiac events including bradycardia, atrial fibrillation (AF), ventricular tachycardia or fibrillation (VT, VF) and myocardial infarction (MI). Most people exposed to the water at T of 2°C (28°F) will die within 15-50 min.

Beside of this, he developed the theory explaining the Ice Age, and in 1896 he became the first scientist who speculated that changes in the level of carbon dioxide (CO_2) in the atmosphere could seriously change the surface temperature of the Earth by causing the greenhouse effect.

Hartog J. Hamburger (1859 – 1924) — Dutch Netherlander physiologist, showed that the ionic interchange between the corpuscles and the plasma of blood denotes that a carbohydrate passes from the erythrocytes into the plasma and the chloride ion passes in the opposite direction (secondary buffering).

---

[284] **William T. Kelvin (1824 - 1907)** — Scottish physicist, who in 1848 found a unit of measure temperature (T) based upon an absolute scale using as its null point absolute zero, the T at which all thermal motion ceases in the classical description of thermodynamics, defined as the fraction 1/273.16 of the thermodynamic T of the triple point of water (exactly 0.01°C or 32.018°F), named the Kelvin unit.

Bronislav F. Veryho / Bronislav F. Werigo [14(26).II.1860, manor Uzhvald, Dynabursky county, Vitebsk Obl., Belarus, now **village Izvalta, Ilzvalta volost', Kraslavsky Krai, Latvia – 13.VI.1925, city of Perm, Russia]**[285] — Ukrainian physiologist, discovered the phenomenon of cathodic depression (1883); developed the theory of leap-like mechanism for the transmission of excitation through the nerve fibers (accommodation of the nerve, 1883); described displacement of the oxyhemoglobin (HbO_2) dissociation curve by a change in the temperature (T), concentration and partial pressure (P) of hydrogen (H^+) and carbon dioxide (CO_2), and the pH in the blood capillaries (Veryho's effect, 1892).

In 1894 **Willem Einthoven (1860 – 1927)** — Dutch physiologist and physician of Jewish descent after his father, developed the string galvanometer made of a single strand of silver-coated quartz thread suspended between two strong horseshoe magnets. Displacement (oscillations) of the thread resulting in the passage of an electric current, in proportion to the strength of the current, were recorded photographically, after magnification. This was achieved by aiming a beam of light through a hole in the poles of the magnet and reflecting it into a photographic plate. The apparatus occupied two rooms and needed five people to operate it. Other major problems with this early equipment were its sensitivity to vibration. The technique was modified by a telephone connection between the hospital and the laboratory 1.2 km away. Nevertheless, he made with it the first electrocardiographic (ECG) recordings the intraventricular conduction defect (IVCD). By 1901-03, after many innovations and improvement with the use of a stringed galvanometer, he made a complete ECG recording in man.

The instrument was perfected by the Cambridge Instrument Company (founded 1881) by 1926 who produced a more portable model, weighting 80 lbs (36.4 kg) and moved on a wheeled trolley.

He established a law, named Einthoven law stating that if the ECG is taken simultaneously with the three leads, at any given instance the potential in lead II is equal to the sum of the potentials in lead I and III ($e^1+e^3=e^2$).

---

[285] Farmyard Uzhvald, Dynabursky county, Vitebsk Obl., Belarus, now village Izvalta (first mentioned 1625), Ilzvalta volost' / district (in Latvian – Izvaltas pagasts, area 71,38 км²), Kraslavsky Krai (in Latvian – Kraslavas, area 1078,4 km²), Latvia (area 64,589 km²). — Verigo BF. Effecte der Nervenreizung durch intermittirende Kettennströme: ein Beitrag zur Theorie des Electrotunus und der Nervenerregung. Verlag August Hirschwad, Berlin 1891. — Verigo BF. Zur Frage über die Wirkung des Sauerstoff auf die Kohlensäureausscheidung in den Lungen. Archiv für die gesammte Physiologie des Menschen und der Thiere 1892;51:321-61. — Coenen A, Zayachkivska O. Adolf Beck: A pioneer in encephalography in between Richard Caton and Hans Berger. Adv Cogn Psychol 2013;9(4):216-21. — Ablitsov VH. «Galaxy Ukraine». The Ukrainian diaspora: prominent persons. Publ. «Kyt», Kyiv 2007.

John S. Haldane (1860 – 1936)[286] — Scottish physiologist, determined that the capacity of the blood for transporting of carbon dioxide (CO_2) depends on the arterial oxygen saturation ($Sa\%O_2$) of the hemoglobin (Hgb): the more deoxygenated the blood, the more CO_2 it can transport and absorb at a given CO_2 tension, while the capacity decreases with increasing oxygenizing (Haldane effect, 1895). John Haldane with **Claude D. Douglas (1882 – 1963)** — British physiologist, discovered an indirect measurement of the energy output of the body, based on the counting of the pulmonary ventilation and of the content of oxygen (O_2) and CO_2 in the exhaled air which is compared with oxygen content (O_2 content) in the ambient air (Douglas-Haldane method, 1911-12), and thus on this basis the maximum oxygen consumption (VO_2 max) by the body is calculated:

$$VO_2 \text{ maximum } (VO_2 \text{ max}) = Q \times (CaO_2 - CvO_2),$$

where Q = cardiac output (CO) or cardiac index (CI), CaO_2 = arterial O_2 content, CvO_2 = venous O_2 content, and $CaO_2 - CvO_2$, is an arteriovenous O_2 difference.

Normal VO_2 max = 3,5 – 4,0 ml/kg/ minute, or 250 ml/minute, or normal VO_2 max index = 110 – 150 ml/minute/m^2.

By recoding negative electrical potentials, using the DuBois-Raymon electrodes **[Emil H. DuBois-Raymond (1818 – 96**): above**]**, in several brains areas evoked due to peripheral sensory impulses, **Adolf A. Beck (01.01.1863, city of Cracow, Poland –** suicide by capsule of cyanide on **08.1942, city of Lviv, Ukraine)**[287] — Jewish Polish and Ukrainian physiologist, discovered continuous electrical oscillations in the electrical brain activity, noted that these oscillations ceased after sensory stimulation, and this was the first description of desynchronization in electrical brain potential (1890).

A protégé of **Karl Ludwig (1816 – 95**: above**), F.W.F. Otto Frank (1865 – 1944)**[288] — German physiologist and cardiologist, invented an isometric and isotonic contraction of the heart muscle, presented the synthesis of factors in atrial and ventricular contractions, coming to the conclusion that *«the maximum tensions of isometric contraction at first increase with augmentation of the initial length or initial tension but beyond a certain level of filling, the peak declines»*, that is *«in the physiological range, the force of contraction directly correlate with the initial length of the muscle fibers»*.

---

[286] Haldane J. The action of carbon oxide on man. J Physiol (Cambridge) 1895;18(5-6);430-62. — Haldane J, Smith JL. The absorption of oxygen by lungs. J Physiol (Cambridge) 1897;22:231-58.

[287] Beck A. Die Bestimmung der Localisation der Gehirn- und Rückenmarksfunctionen vermittelst der elektrischen Erscheinungen. Centralblatt Physiol 1890;4:473-6. — Coenen A, Zayachkivska O. Adolf Beck: A pioneer in encephalography in between Richard Caton and Hans Berger. Adv Cogn Psychol 2013;9(4):216-21.

[288] Frank O. Zur Dynamik des Herzmuskels. Z Biol 1895;32:370-437.

In 1898 **F.H. Pratt**[289] developed the concept of reverse blood perfusion (retroperfusion) of the heart muscle by the insertion of an arterial graft into coronary sinus (CS) to enhance the myocardial blood flow in isolated feline hearts which were contracting for 90 min.

Biochemistry

In 1802 **Bernard Courtois (1777 – 1838)** — French chemist, while researching opium found that it contains morphine, and in 1811 he isolated iodine.

Morphine is one of the main alkaloid opium, contained in sophomoric poppy (Papaver somniferum), from the morphine / isoquinoline group of alkaloids, with the chemical formula of $C_{17}H_{19}O_{20}$.

Iodine, named two years later (1813) by **Joseph-Louis Gay-Lussac (1778 – 1850)** — French chemist and physicist, and Sir **Humphry Davy (1778 – 1829)** — English physicist and chemist, from Greek word «ioedis», meaning «violet-colored» or resembling violet. It is a chemical element with symbol I and atomic number 53, the heaviest and the least electronegative of the stable halogens and heaviest of nonradioactive metals, which exist as lustrous, purple-black non-metallic solid at standard conditions that sublimes readily to form a violet gas / vapor. Iodine occurs in many oxidation states, including iodide (I^-), iodate (IO_3) and various perioxide anions.

In addition, Joseph-Louis Gay-Lussac formulated the law which states that if mass and volume of a gas are held constant then gas pressure increases linearly as temperature raises (Gay-Lussac law, 1802).

Moreover, Sir Humphry Davy discovered in 1808 magnesium (Mg), described phenomenon of the electric arch (1808-09) and constructed a safe mining lamp with a metal net (1815).

In 1807 **Ferdinand (Fedor) F. Reuss (1778 – 1852)** — German biochemist, discovered electroosmosis and electrophoresis.

The term bicarbonate (HCO_3) was coined by **William H. Wollaston (1766 – 1728)** — English chemist, in 1814. It is the most important physiological buffer in the blood regulated by the kidneys.

From 1820 **Frederich Accum (1769 – 1838)**[290] — German chemist, and since 1855 **Arthur Hassall (1817 – 94**: 19th century. Anatomy, histology, and embryology)[291]

---

[289] Pratt HR. The nutrition of the chest through the vessels of Thebesius and the coronary veins. Am J Physiol 1898;1:86-103.

[290] Accum F. A treatise on adulterations of food and culinary poisons. Longman, London 1820.

[291] Hassall AH. Food and its adulterations; comprising the reports of the analytical sanitary commission of «The Lancet» for the years 1851 to 1854. Longman, London 1855.

fought against food adulteration which terminated with the signing of a partial (1860) and full (1899) pact against food adulteration.

Friedrich Wöhler (1800 – 82) — German chemist, who synthesized urea (1829).

The pioneering studies on dialysis and the diffusion were made by **Thomas Graham (1805 – 69**) — Scottish chemist. By studying the colloids he was able to separate them from crystalloids using a so-called «dialyzer», the precursor of hemodialysis (HD) machine. His studies on the diffussion of gases resulted in «Graham law» (1833), which states that the rate of effusion of a gas is inversely proportional to the square root of its molar mass.

For the first time in 1933 **Anhelme Payen (1795 – 1871)**[292] — French chemist and biochemist, synthesized enzyme diastase (amylase) which splits starch into oligosaccharides. Later, in 1838 while trying to divide the wood, he found during mixing it up with nitric acid, a fibrous substance – carbohydrate cellulose ($C_6P_{10}O_5$) – the most abundant organic polymer on Earth.

Nicolai M. Zinin (1812 – 80) — Russian chemist, discovered a method for synthesizing aromatic amines by the restoration of nitrocompounds (Zinin reaction, 1843).

It was known for millennia (Ancient Medicine. *Mesopotamia*) and for centuries (Medieval medicine. *Ukraine*) that the willow's bark grinded to powder, swallowed by human, relieves pain, fever, inflammation and thins the blood. It is because the willow's bark contains salicylates from which **Charles F. Gerhardt (1816 – 56)** — French chemist, synthesized in 1853 acetylsalicylic acid (ASA) – aspirin, and its curative values were confirmed in 1887 by **Henrich Dreser (1860 – 1924)** — German chemist, and **Felix Hoffmann (1868 – 1946)** — German chemist.

Ernst H.F. Hoppe-Seyler (1825 – 95)[293] — German physiologist and biochemist, who described the optical absorption spectrum on the red blood cells (RBC) pigment, and its two distinctive bands, discovered in 1862 hemoglobin (Hgb), recognized the binding of oxygen (O_2) to the RBC's Hgb, creating the compound oxyhemoglobin ($HgbO_2$).

In 1859 p. **П. Е. Marcelen Bertlo (1827 – 1907)** — French chemist, establish that cholesterol belong to the alcoholic (spirits) class.

The first who introduced in 1857-61 the term of «chemical structure» and incorporated double bounds in structural formulas was **Alexander M. Butlerov (1828 – 86)** — Russian chemist, who stated clearly that the properties of a compound are determined by its molecular structure and reflect the way in which atoms are

---

[292] Payen A. Mamoire sur la composition du tissue propre des plantes du liguneus. Comptes rendus 1838;7:1052-6 + Dec 24;Suppl. / Comptes rendus 1839;8:169. / Comptes rendus 1839 Feb;9:149.

[293] Hoppe-Seyler F. Handbuch der physiologisch und pathologish-chemischen Analyse für Artzte und Studirende. Verlag August Hirschwald, Berlin 1858. — Hoppe-Seyler F. Physiologische Chemie. I-IV Theil. Verlag August Hirschwald, Berlin 1877-81.

bounded to one another in the molecules of the compound. Then, the development of the structural theory of organic chemistry progressed rapidly, and this theory has been of inestimable value in aiding organic chemists to implement their experimental results and to plan new experiments.

He also discovered hexamine (1859), formaldehyde, and of formose (a portmanteau of formaldehyde and aldose) reaction, a.k.a., Butlerov reaction (1861), which involves the formation of sugars from formaldehyde.

From the coca plant leaves (Prehistoric Medicine. *The American Indians*), **Friedrich G. C. Gaedcke (1828 – 90)** — German chemist, was the first to isolate in 1855 the cocaine alkaloid.

Cocaine hydrochloride causes blockage of dopamine transfer protein in the central nervous system (CNS), has a short half- life of 0.7 – 1.5 hour and is metabolized by cholinesterase enzymes in the liver and plasma, primarily excreted unchanged in the urine (about 1%).

Alexander P. Borodin (1833 – 87) — Russian chemist, introduced the urometer and a method of estimating urea in the blood and in the urine (1876).

The periodic law of elements was outlined by **Dimitri I. Mendeleev (1834 – 1907)** — Russian chemist, in his treatise «Elements of Chemistry» (1868).

The periodic acid (HIO_4) is a stain developed by **Hugo (Ugo) Schiff (1838 – 1915)**[294] — Italian naturalized chemist, German by nationality, to detect glycogen and other saccharides present in tissues of the agent causing diseases, is based on the reaction of periodic acid oxidizing the diol functional groups in glucose and other sugars, creating aldehydes that react with the Schiff reagent to give a purple-magenta color. This reagent is prepared by dilution of 0.25 gm of fructose in 1 liter of H_2O and discoloring by letting through it a sulfur dioxide. In the presence of aldehyde, the grey color is restored. The periodic-Schiff's stain and methenamine silver are the best for demonstrating living fungi in tissue.

The theory of strain was formulated by **J.F.W. Adolf Baeyer (1835 – 1917)** — German chemist, explains, stoichiometrically, the relative stability of cyclic and unbounded compounds (1885).

Olexandr Ya Danylevsky or **Alexander Ya. Danilewski [28.XI.(10.XII).1838, city of Kharkiv, Ukraine – 18.VI.1923, city of St Petersburg, Russia]**[295] — Ukrainian

---

[294] Schiff H. Mittheilungen aus dem Univesitäts-laboratorium in Pisa. Eine neue Reihe organisher. Ann Chem Pharm 1864;131:118-9.

[295] «Danilewski gelang 1862 als Erster die Trennung der Enzyme der Bauchspeicheldrüse und damit der Nachweis des Trypsins. Erschloss Frisches Pancreasgewebe mit Sand und Wasser in einen Mörser auf, entfernte die groben Bestandteile und sältigte die Lösung mit Magnesiumoxid, wodurch er die Pancreaslipase und andere Bestandteile abtrennte. Das Filtrat mit Trypsin und Elastase trennte er mit einen Alkohol-Ether-Kollodium Lösung».

biochemist, physiologist and pharmacologist, isolated amylase, and trypsin (1863)[296], observed the hydrolysis of proteins by pancreatic juice, demonstrated the synthesis of proteins from peptones in the presence of ferments (1886)[297], formulated the theory of protein structure (1888-91), and discovered antiferments.

The semi essential aminoacids glutamine (Gln) and phenylalanine (Phe) were discovered by **Ernst Schultze (1840 – 1912)**[298] — German chemist, including the first isolation with his assistant **Ernst Steiger** in 1886 of another semi-essential aminoacids from the seedlings of yellow lupin (Lupinis luteus) of arginine (Arg) that is used in the biosynthesis of proteins.

Mykola A. Bunge [03(15).XII.1842, city of Warsaw, Poland – 31.XII.1914 (13.I.1915), city of Kyiv, Ukraine][299] — Ukrainian chemist of German origin, studied the structure of nitro compounds, electrolysis of spirts, mercaptans and improved method of gas analysis (dissection and dissolution).

Known from antiquity, hyoscine (Hyoscyamusniger), a.k.a. scopolamine (Scapolia)[300], was first isolated by **Albert Landenburg (1842 – 1911)**[301] — German Jewish chemist, in 1880-81, from the flowering plant of the night shade (Solanaceae) family. It is used to treat motion sickness, e.g., sea sickness, normal or excessive secretion of saliva in certain neurological diseases, decrease salivation during and after surgery, and postoperative nausea and vomiting in pregnancy.

The first to isolate various phosphate-rich chemicals, named nucleic acids was **J. Friedrich Miescher (1844 – 95)**[302] — Swiss organic chemist, in 1869 from nuclei of white blood cells (WBC) / leukocytes of human wounds, thus paving the way for identification of deoxyribonucleic acid (DNA) as the carrier of inheritance (1871). He also demonstrated that carbon dioxide (CO_2) concentrations in blood regulate breathing.

After Sir **Thomas E. Thorpe (1845 – 1925)** — English chemist, was named a tubular instrument used to directly measure the flow rate of a gas, including oxygen (O_2) in medical devices (Thorpe's tube flowmeter).

---

[296] Danilewski O.Ya. On Specifically Acting Liquids on the Natural and Gastric Juices from the Pancreatic Gland. Doctoral dissertation. Kharkiv Univ., Kharkiv 1863. — Ablitsov VH. «Galaxy Ukraine». The Ukrainian diaspora: prominent persons. Publ. «Kyt», Kyiv 2007.

[297] Green JR. The Soluble Ferments and Fermentation. Cambridge Univ Press 2014.

[298] Schulze E, Steiger E. Ueber das Arginin. Zeitshr Physiol Chemie 1887;11:43-65.

[299] Ablitsov VH. «Galaxy Ukraine». The Ukrainian diaspora: prominent persons. Publ. «Kyt», Kyiv 2007.

[300] Bezuhlyy PO (Ed.). Pharmaceutical Chemistry. Handbook. «The New Book», Vinnytsia 2008. — Nekoval IV, Kazaniuk TV. Pharmacology: Handbook. 4th ed. Publ. VSB «Medytsyna», Kyiv 2011.

[301] Landenburg A. Die natürlich vorkommenden mydriatisch wirkenden Alkaloïde. Ann Chemie 1881;206(3):274-307.

[302] Miescher JF. Ueber die chemische Zusammensetzung der Eiterzellen. Medicinisch-chemische Untersuchungen 1871;4:441-60. — Miescher JF. Die histologischen und physiologischen Arbeiten. Band 1 und 2. Gesammelt u. hrsg. von seinen Freuden. Vogel, Leipzig 1897. — Dahm R. Friedrich Miescher und discovery of DNA. Developmental Biology 2005;278(2):274-88.

Marcellus V. von **Nencky (1847 – 1901)** — Polish biochemist and physician, established that urea is synthesized in the liver from ammonia and carbonic acid, with **Ivan Pavlov (1849 – 1936**: 19th century. Physiology**)**, in 1881-96 defined the prosthetic structure of hemoglobin (Hgb), separated hemin, proposed a test for indole (Nencky's test), demonstrated the similarity between Hgb and chlorophyll (1897).

In 1881 **Carl M.** von **Hell (1849 – 1926)** — German chemist, in 1887 **Jacob Volhard (1834 - 1910)** — German chemist, and **Mykola (Nikolay) D. Zelinsky [26.I.(06.II).1861, town of Tyraspol, Prydnistrovia-on-Dnister River, Ukraine / Moldavia – 31.VII. 1953, city of Moscow, Russia]**[303] — Ukrainian biochemist, discovered the halogenation reaction of carboxylic acids at the α-carbon (Hell-Volhard-Zelinsky halogenation, 1887).

H. Emil L. Fischer (1852 – 1919) and **Arthur Speier** — German organic chemists, discovered esterification (from German – Essig-Aether) by refluxing a carboxylic acid and alcohol in the presence of an acid catalyst (Fisher-Speier esterification, 1895)[304]; and Emil Fisher with **Joseph B.** von **Mering (1849 – 1908)** — German physiologist and internist, helped to lunch in 1902-05 the first barbiturates, a class of sedative drugs used for insomnia, epilepsy, anxiety, and general anesthesia; and developed a symbolic way of drawing a symmetric carbon atoms (the Fisher projection).

A better understanding of cellular chemistry became possible through the work of **L.K.M.L. Albrecht Kossel (1853 – 1927)** — Prussian German biochemist and genetics, who isolated and described in 1885-1901 the five organic compounds present in nucleic acid – adenine, cytosine, quanine, thymine and uracil, in 1896 discovered or worked upon quantitative separation of alpha- aminoacids – arginine, histidine and lysine. He also was the first to extract from tea leaves (Camellia sinensis) and chemically identified by around 1888 alkaloid theophylline (1,3-dimethylxantine), also present in cocoa (Theobroma cacao).

Later (1895), **Emil Fischer (1852 – 1919**: above**)** and **Lorenz Ach**[305] — German organic chemist, synthesized theophylline starting with a purine alkaloid – 1,3-dimethyluric acid. An alternative method to synthesize theophylline, was introduced by **Wilhelm Traube (1866 – 1942)** — Prussian German chemist of Jewish origin, by the Traube purine synthesis (1900).

Methyl xanthine is a drug that block the action of adenosine[306], relaxes bronchial smooth muscle, increases heart muscle contractility and efficiency by positive

---

[303] Town of Tyraspol (first mention 1790), Prydnistrovia (area 4,163 km^2), Ukraine / Moldovia or Moldova (area 33.843 km^2). — Ablitsov VH. «Galaxy Ukraine». The Ukrainian diaspora: prominent persons. Publ. «Kyt», Kyiv 2007.

[304] Fisher E, Speier A. Darstellung der Ester. Chemishe Berichte 1895;32-52-8.

[305] Fischer E, Ach L. Synthese des Caffeins. Berichte der Deutschen Chemichen Gesellschaft (Ber Dtsch Ges) 1895;28(3):3139.

[306] Adenosine is a neuromodulator that promote sleep and suppress arousal, and by vasodilation increases blood flow to various organs, used to treat supraventricular tachacardia (SVT).

inotropic effect, increases heart rate (HR), arterial blood pressure (BP) and renal blood flow promoting diuresis[307], has anti-inflammatory effect, and stimulate the respiratory center in a long stem-like structure (in Latin – myelencephalon, medulla oblongata, medulla, i.e., bulbus) located in the lower part of the brain steam. It used to treat chronic obstructive pulmonary disease (COPD), asthma and infant apnea (in Greek – apnoea), i.e., a temporary cessation of breathing movements brought about by depression (hindrance) of the respiratory medullary center.

Albrecht Kossel also pioneered the study on proteins and nucleic acid substances (1889), discovered histidine (1896).

Ivan Ya. Horbachevsky / Jan (Johann, Ivan) Horbaczewski [05.V.1854, village Zarubintsi, Zbarazh Region, Ternopil Obl., Ukraine – 25.V.1942, city of Praha, Czechia][308] — Ukrainian biochemist, synthesized uric acid from glycerol and urea (1882), proved that it is derivable from nucleic acids, i.e., the product of metabolism

---

[307] Minkowski O. Über Theocin (Theophyllin) als Diureticum. Ther Gegenwart 1902;43:490-3.

[308] Village Zarubintsi (founded 1882), Zbarazh Region, Ternopil Obl., Ukraine. — Horbaczewski I. Ueber den Nervus vestibuli. Wien 1875. — Horbaczewski J. Synthese der Harnsäure. Berichte der deutschen chemischen Gesellschaft (Ber Dtsch chem Ges) 1882 Juli-Dez;15(2):2678. — Horbaczewski J. Über künstiche Harnsä.ure und Methylharnsäure. Monatshefte für Chemie (Mh Chem) 1885;6:356-62. — Horbaczewski J. Über eine neue Synthese und über die Konstitution der Harnsäure. Wien Monatshefte (Wien Monatsh) 1887;8:201-7. — Horbaczewski J. Untersuchungen über die Entstehung der Harnsäure im Säugetierorganismus. Mh Chem 1889;10:624-41. — Horbaczewski J. Beträge zur Kenntnis der Bildung der Harnsäure und der xanthinbasen, sowie der Entstehung der Leukocytosen im Säugethierorganismus. Mh Chem 1891;12:221-75. — Studnička FJ, Čelakovský J. Horbaczewski Jan. Ottuv slovník naučný. J Otto, Praha 1897; 11:567-8. Horbaczewski J. Lekarska chemie. Vol. I. Anorganicka chemie. Praha 1904. — Horbaczewski J. Lekarska chemie. Vol. II. Organicka chemie. Praha 1905. — Horbaczewski J. Lekarska chemie. Vol. III. Physiologicka chemie. Knyha 1. Praha 1907. — Horbaczewski J. Lekarska chemie. Vol. IV. Physiologicka chemie. Knyha 2. Praha 1908. — Horbaczewsky I. Organic chemistry. Publ Ukrainian Free University, Praha 1924. — Kacl K. Professor Dr Jan Horbaczewski. Casopis lekaru ceskych. 1954;93:578-80. — Dobrovolskyy Y. To 80-anniversary of Professor Doctor Ivan Horbachevsky. J Ukr Med Assoc North Am ?1954;1,2(3):73. — Olearczyk A. Ukrainian Scientists. Ivan Horbachevs'kyi (103 years from the date of birth). Nasze Słowo (Warszawa) 14 Apr 1957;2,15 (35):2. — Shlachtychenko M. Professor, Doctor Ivan Horbachewski (Remembrance). J Ukr Med Assoc North Am 1958 Feb;5,(9):7-13. — Olearczyk A. I.I. Mechnykov (Metchniknoff), O.M. Bakh (Bach), I. Horbaczewski. Nasza Kultura (Warszawa) May 1959;5;(13):16. — Honcharenko M. Academician Ivan Horbaczewski (15.V.1854 – 24.V.1942). J Ukr Med Assoc North Am 1970 Jan; 17,1(56):37-42. — Bab'iuk A. Life and Work of Academician I. Horbaczewski. J Ukr Med Assoc North Am 1970 Jan; 17,1(56):42-44. — Teich M. Horbaczewski Jan. In, CC Gillispie. Dictionary of Scientific Biography. Charles Scribners Sons, New York 1972;6:506. — Schweitzer J. Prof. Dr Iwan Horbachewsky. J Ukr Med Assoc North Am 1989 winter;35,1(119):40-2. — Yurkevych OH. World Renown Ukrainian Scientist. In, O. Romanchuk (Ed.). Axioms' for Descendants. Ukrainian Names in the Worlds Science. Collection of Outlines. Publ. «Memorial», Lviv 1992. P. 339-47. — Hubsky Yul. Biologic Chemistry. Publ «Ukrmedknyha», Kyiv – Ternopil 2000. — Honsky Yal, Maksymchuk TP, Kalynsky MI. Biochemistry of Man. Publ «Ukrmedknyha», Ternopil 2002. — Honsky Yal. Ivan Horbachevsky in Remembrances and Correspondences (essay). Publ «Ukrmedknyha», Ternopil 2004. — Ablitsov VH. «Galaxy Ukraine». The Ukrainian diaspora: prominent persons. Publ. «Kyt», Kyiv 2007. — Delanghe JR, Speeckaert MM. Creatine determination according to Jaffe – what does it stand for? Clin Kidney J 2011April ;4,2:83-6.

in the cell's nuclei (1889) and increases in neoplasms (1891-92). Thus, he introduced a biochemical screening into oncology.

Oleksiy M. Bakh or **Aleksey N. Bakh [05(17).III.1857, town of Zolotonosha, Zolotonoshivsky Region, Cherkasy Obl., Ukraine – 13.V.1946, city of Moscow, Russia]**[309] — Ukrainian biochemist, explained chemistry of the process of assimilation of carbon dioxide (CO_2) by chlorophyll plants with creation of sugar, on account reaction with water elements; came to conclusion that peroxides play especially important role in the process of breathing, on the basis of which created peroxide theory; developed theory of slow oxygenation and cellular respiration (1897).

Fedir F. Selivanov or **Feodor F. Seliwanow** or **Theodor F.** von **Selivanoff (9/21.X.1859, town of Gorodishche, Penzen Obl., Russia – 21.II.1938,?)**[310] — Russian Ukrainian chemist, professor of chemistry at Odesa University (1895-1918), city of Odesa (Foot Note 245), Ukraine, who developed a reaction (test) for ketose (fructose) in the urine (Selivanov reaction or test, 1887). Ketose, a monosaccharide that contains a ketone group, during heating with the resorcin solution (0,05 g resorcinol in 30 ml H_2O and diluted hydrochloric acid (few drops of concentrated HCl with density of 1,19 g/ml) form a dark red (Burgundy red) color indicating the presence of fructose. Aldose, a monosaccharide that contain only one aldehyde group per molecule, under those circumstances interacts more slowly giving a light pink color.

In 1887 **Serhiy M. Reformatsky [21.III(02.IV).1860, village Borisoglebskoe, Ivanov Obl., Russia – 28.VII.1934, city of Moscow, Russia]**[311] — Russian Ukrainian biochemist, discovered the synthesis of β-hydroxy acids with the use of zinc (Reformatsky reaction), laid the basis for the synthesis of vitamin A and its derivatives.

Ivan A. Klimov (1865 – ?) introduced a test for blood in the urine adding an equal quantity of hydrogen peroxide (H_2O_2) and a little powdered aloin (Klimov test).

In 1893 **Alfred Werner (1866 – 1919)** — Swiss chemist, proposed the octahedral form of the intermediary metal compounds, and that led into the creation of ligands,

---

[309] Town Zolotonosha (on its territory and vicinities were found working tools made from stones of Bronze Age, grave of Zarubetsky Culture, settlement from the Chernihivsky Culture, former site of a town in ruins (horodyshche) of Ukraine-Rus, first written mention 1576), Zolotonoshivsky Region, Cherkasy Obl., Ukraine. — Ablitsov VH. «Galaxy Ukraine». The Ukrainian diaspora: prominent persons. Publ. «Kyt», Kyiv 2007. — Hanitkevych YaV. The contribution of the Ukrainian physicians to the world medicine. Ukr Med Chasopys 2009 VII/VIII;4(72):110-15.

[310] Town of Gorodishche (founded 1681), Penzen Obl., Russia. — Seliwanoff Th. Notiz über eine Fruchtzuckerreaction. Ber Dtsch chem Ges 1887;10:181-2. — Ablitsov VH. «Galaxy Ukraine». The Ukrainian diaspora: prominent persons. Publ. «Kyt», Kyiv 2007.

[311] Village Borisoglebskoe, Ivanovo Obl. (area 21,437 km²), Russia. — Hofmann K. A. Sergeius Reformatsky. Berichte der Chemischen Gesellschaft 1935;A(5):61. — Ablitsov VH. «Galaxy Ukraine». The Ukrainian diaspora: prominent persons. Publ. «Kyt», Kyiv 2007.

an equal part of ion or molecule, bounded into the central atom of the metal for the formation of equal compounds.

Microbiology and immunology

Although **Thomas Muffet (1553 – 1604**: Renaissance) discovered an infectious disease of domesticated silkworms (1589), nevertheless it was **Agostino Bassi (1773 – 1856)** — Italian botanist and entomologist, who linked a fungus Botrytis paradoxa (named Beauveria bassiana in his honor) with a white muscardine disease of a domesticated silkworm («Del mal del segno, calcinaccio o moscardico», 1835), and suggested that smallpox, typhus, plague and cholera were also caused by, as yet undiscovered, live organisms.

In 1819 **Jean G.A. Lugol (1786 – 1851)** — French physician, introduced antiseptic and disinfectant, also an antiviral agent containing iodine (5 g) and potassium iodide (10 g) in distilled water (Lugol solution or Lugol iodine).

Healthy sheeps infected with anthrax in experiments by **Pierre F. Rayer (1793 – 1867)** — French physician, and **Casimie J. Devaine (1812 – 82)** — French physician, with the blood of sheep dying from this same disease (1850). Bacillus anthracis and its spores were seen in the blood of the infected animals.

That yellow fever was transmitted by the insect Aedes-aegypti was discovered in 1853 by **Louis D. Beauperthuy (1807 – 71)** — Venezuelan microbiologist and physician, its viral cause was suspected in1881 by **Carlos J. Finley (1833 – 1915)** — Cuban epidemiologist, and proven in 1901, by **Walter Reed (1851 – 1902)** — American military physician.

Oliver W. Holmes, Sr (1809 – 94)[312] — American physician, essayist, poet, and polymath, in his assay «The contagiousness of puerperal fever» (1843) appealed for surgical cleanliness to combat the contagiousness of puerperal fever.

Statistically based research on asepsis conducted by **Ignaz F. Semmelweis (1818 – 65)** — Hungarian physician of German ancestry, in 1847-62 proved the contagious nature of postpartum infection in the obstetric wards of the 1st and 2nd Clinic at the VGH, Vienna, Austria. He advocated chlorine solution handwashing for all physicians, gynecologist and obstetricians and midwives to prevent childbirth fever and mortality.

Independently one from the other, **Otto H. Wuchener (1820 – 73)** — German Brazilian parasitologist and **Joseph Bancroft (1836 – 94)** — English Australian

---

[312] Holmes OW. The contagiousness of puerperal fever. New Engl Quart J Med Surg 1843;1:503-30. — Hebra A, Othersen Jr HB. Oliver Wendell Holmes: physician, writer, and poet, teacher, lecturer, and more. J Am Coll Surg 2017 Apr;224(4):751-754.

pharmacologist, parasitologist and surgeon, discovered in 1866 and 1876, respectively, human parasitic roundworms of the Filarioidea type, named Wucheneria bancrofti, microfilariae of which penetrate from the skin into the blood and lymphatic vessels, obstruct the lumen of the lymphatic vessels and flow of the lymph. Wucheneria bancrofti is the main of the three most common causes of lymphatic filariasis, which in chronic form leads to elephantiasis, mainly of the lower extremities. Filariasis affect over 120 million of people in tropical regions.

Louis Pasteur (1822 – 95) — French biologist, chemist, and microbiologist, showed that lactic fermentation depends on the presence of a microorganism (1857) and declared that those organisms were not spontaneously generated but came from a similar organism from the air (1864). He developed the germ theory of infectious diseases; explained the effectiveness of asepsis and antisepsis; proved that heat destroys microbes (pasteurization); introduced the principles of immunity using attenuated cultures of microorganisms which affords protection against diseases caused by a virulent microorganism – inoculation, on the beginning for sheep in 1881 against Bacillus anthracis, a causative factor of anthrax [18th century – **Mykhailo Hamaliya (1749 – 1830)** and **Stepan Prokopovych-Andriyevsky (1760 – 1818)**], that effectively helped in the prophylactics against this disease in animals. Then he developed inoculations against plague (1881), swine erysipelas (1882) and rabies (1885).

Later (1883), **Leon (Lev) Cienkowski / Tsenkovsky (01/13.X 1822, city of Warsaw, Poland – 02.IX/07.X 1887, city of Leipzig, Freistadt Sachsen, Deutschland)**[313] — Ukrainian biologist of Polish descent, also prepared vaccines against anthrax, which with some alterations is used till today.

Joseph M. Leidy (1823 – 91) — American anatomist, parasitologist, forensic innovator, and paleontologist, is renowned for determining in 1846 that roundworms (Trichinella spiralis), a parasite found in undercooked pork, bear, or dog meat, causes trichinosis, a parasitic disease in humans that commences about one week after ingesting. The enteral phase is manifested by abdominal pain, vomiting and diarrhea, and parenteral phase by periorbital edema, swelling around the eyes, splinter hemorrhage due to vasculitis, ataxia, respiratory paralysis, stroke due to cerebral sinus thrombosis. Trichinosis can be deadly, usually 4-6 weeks after ingestion from myocarditis, encephalitis, or pneumonia.

Also, in 1846 he became the first person to use microscope to solve a murder mystery by finding human red blood cells (erythrocytes) without nucleus (anucleate

---

[313] Stadt Leipzig (or Lypsk from the Ukrainian word «lypa», i.e., «linden», first documented 1015), Freistadt Sachsen; area 18,449.00 km²), Germany. — Tsenkovski L. Zur Morphologie der Bakterien. St Petersburg 1877. — Ablitsov VH. «Galaxy Ukraine». The Ukrainian diaspora: prominent persons. Publ. «Kyt», Kyiv 2007.

erythrocytes) in blood-stained clothes of the killer, instead of as claimed by the killer being the blood of slaughtered chick which erythrocytes retain their nuclei for hours when outside the body (nuclear erythrocytes).

Bacteria were classified by **Ferdinand J. Cohn (1828 – 98)** — German biologist, bacteriologist, and microbiologist.

Jointly, **T.A. Edwin Klebs (1834 – 1913)** — Prussian German Swiss microbiologist and pathologist, and **Fridrich A.Jo. Löffler (1852 – 1915)** — German microbiologist and physician, isolated bacillus diphtheria (Corynebacterium diphteriae or Klebs-Löffler's bacillus), causing diphtheria (1883-84); and the latter also isolated a virus (Aphtovirus) causing a «foot-and mouth» disease in the cloven-hooted animals, rarely in humans (Aphtae epizooticae, 1897).

During experiments on himself, **Hryhoriy M. Minkh [07(19).IX.1836, village Griazi, Saratov Obl., Russia – 11(23).XII.1896, city of Saratov, Russia]**[314] — Russian Ukrainian pathologist, and epidemiologist, and **Josyp Jo. Mochutkovsky [07. III.1845, Kherson Province / Obl., Ukraine – 23.V(05.VI).1903, city of Petersburg, Russia]**[315] — Ukrainian physician, proved that the blood of patients with recurrent fever (1874) and typhoid fever (1876), whose carriers were lice, was contagious. These observations were confirmed by **Charles J.H. Nicolle (1866 – 1936)** in 1909 — French bacteriologist.

Brigadier General **George M. Sternberg (1838 – 1915)**[316] — United States (US) Army physician-bacteriologist, wrote the first publications on chemical disinfection (1879), discovered the Streptococcus pneumoniae, or pneumococcus, a common cause of pneumonia (1880), speculated about phagocytosis (1881), documented the cause of malaria (1881), discovered lobal pneumonia (1881) and confirmed the roles of the bacilli of tuberculosis and typhoid fever (1886).

The causative agent of leprosy, Mycobacterium leprose, was discovered in 1872-73 by **Gerhard H.A. Hensen (1841 – 1912)** — Norwegian bacteriologist and dermatologist.

After observations by **Mykhailo Hamaliya (1749 – 1830)** and **Stepan Prokopovych-Andriyevsky (1760 – 1818**: 18th century), **Robert Koch (1843 – 1910)**

---

[314] Village Griazi, Saratov Obl. (area 100,200 km^2), Russia. — City of Saratov (founded 1590), Russia. — Aronov L. The deed of Doctor Minkh. Ukr Historic-Medical J «Ahapit» 2001;13(1):23-30. — Ablitsov VH. «Galaxy Ukraine». The Ukrainian diaspora: prominent persons. Publ. «Kyt», Kyiv 2007. — Synyachenko OV. They glorified themselves and Ukraine: history of domestic therapy. Crimean Therapeutic Journal 2010;2(2):21-8.

[315] Kherson Province (created 1802; area 71,436 km^2) / Kherson Obl. (created 1944; area 28,461 km^2), Ukraine. — Ablitsov VH. «Galaxy Ukraine». The Ukrainian diaspora: prominent persons. Publ. «Kyt», Kyiv 2007. — Hanitkevych YaV. The contribution of the Ukrainian physicians to the world medicine. Ukr Med Chasopys 2009 VII/VIII;4(72):110-15. — Synyachenko OV. They glorified themselves and Ukraine: history of domestic therapy. Crimean Therapeutic Journal 2010;2(2):21-8.

[316] Flaumenhaft E, Flaumenhaft C. Military Medicine 1993 July;158(7):449-57.

— German microbiologist and physician, established the sporulation and pathogenesis of Bacillus anthracis that caused the outbreak of anthrax disease in 1875, town of Wollstein (settled in 1285), Germany, since 1945 town of Wolsztyn, Greater Poland Voivodeship (area 35.580 km²), Poland.

Subsequently, he developed and refined techniques of culturing bacteria and the essential steps required to prove that organisms are the cause of disease (Koch postulates, 1878); discovered the causes of wound infection (1879), of tuberculosis (acid fast bacillus Mycobacterium tuberculosis, 1882), cholera (1883), Egyptian ophthalmia and sleeping fever; advanced methods of steam sterilization; introduced preventive measures in typhoid fever, plague, malaria, and other infectious diseases.

Tuberculosis is an infectious disease caused by mycobacteria (Mycobacterium tuberculosis) which typically enters the human body via lung, characterized by the development of the primary small interstitial granulomas[317], usually of the left upper lobe (Ghon-Assmann focus)[318]. In this stage untreated tuberculosis may spontaneously heal or proceed into caseous necrosis with formation of caverns and calcifications, spreading to the subsegmental, segmental, lobal, interlobar and hilar lymph nodes (Ghon complex). Subsequently, it may spread through the bronchi to other segments and lobes or via the blood and lymphatic vessels into the other organs. As a result of an erosion of the Ghon complex into the pulmonary vein (PV) mycobacteria may reach the left heart, systemic arterial circulation, spread into the liver, spleen, and other organs. Its entrance into the lymphatic nodes can lead to drainage into systemic vein, the left heart, then seeding or re-seeding in the lung. That will end up with disseminated (miliary) tuberculosis, mainly the lung, by the tiny size of the lesions (1-5 mm) with the appearance similar to millet seeds.

In 1880 **Charles L.A. Laveran (1845 – 1922)** — French physician, discovered that the cause of malaria is a protozoan – a simplest unicellular animal organism, after observing the parasites in a blood smear taken from a patient who had just died from malaria which he named Oscillaria malariae.

Thereafter, since 1880 the blood of malarian patients was studied by **Ettore Marchiafava (1847 – 1935)** — Italian pathologist, physician and neurologist and **Angelo Celli (1857 – 1914)** — Italian physician, hygienist and parasitologist, as well as by **Giovanni B. Grassi (1854 – 1925)** — Italian physician and zoologist, **Giuseppe Bastianelli (1862 – 1959)** — Italian physician and zoologist and **Amico Bignami (1862 – 1929)** — Italian microbiologist, pathologist, and physician. Subsequently, they showed that Oscillaria malariae indeed was the causative agent of malaria and using methylene

---

[317] The interstitium include the epithelial cells and capillary within the alveolar wall, the septal and the connective tissue that surround the vascular, bronchial, and lymphatic structures within the lung parenchyma.

[318] **Anton Ghon (1866 - 1936)** — Austrian physician. / **Herbert Assmann (1882 - 1950)** — German physician.

blue staining[319] identified malaria parasite as distinct blue colored pigment in the blood cells, and by splitting (fission) recognized several stages of the development of malaria parasite in the human blood, renamed it in 1885 to a new term – Plasmodium.

In 1882 **Illya I. Mechnykov / Ilya** or **Elie I. Metchnikoff (03/15.III.1845, village of Ivanivka, Dvorichansky Region, Kharkiv Obl., Ukraine – 02/15.VII.1916, city of Paris, France)**[320] — Ukrainian French biologist, microbiologist and immunologist of Ukrainian after his father and Jewish after his mother origin, discovered the occurrence of phagocytosis (In Greek – «phagein», meaning – «to eat» or «to devour»; «kytos», meaning «cell» or «hollow vessel»; «-osis», meaning «process») and phagocytes, that are responsible for an active capture and ingestion of microscopic harmful foreign objects, such as bacteria, fragments of cells, dead or dying cells, and solid particles by the unicellular organisms (bacteria, archaea and eucarya) or by some cells of the multicellular organisms (animals and humans), necessary for the cellular defense (immunity) to fight against infection. To assure an adequate immunity of the human body one liter (L) of human blood contain s about six billion of phagocytes.

In 1889, Illya Mechnykov discovered antilymphocyte globulin (ALG) that is used by the intravenous (IV) injection to treat post-transplant rejection.

Illya Mechnykov together with **Volodymyr K. Vysokovych / Wladimir Wyssokowitsch [2(14).III.1854, town of Hai(y)syn, Haisyns'kyy Region, Vinnyts'ka Obl., Ukraine – 13(26).V.1912, city of Kyiv, Ukraine]**[321] — Ukrainian pathologist, developed the concept of reticuloendothelial system (RES), or system of mononuclear phagocytes (1886-88). In his experimental studies on inflammation

---

[319] Honcharov AI, Kornilov MYu. The Reference Book for Chemistry. Publ. Higher Education, Kyiv 1974.

[320] Village of Ivanivka (founded 1850), Dvorichansky Region, Kharkiv Obl, Ukraine. — Metchnikoff E. L'Immunité dans les maladies infectieuses. G. Masson, Ed., Paris 1901. — Metchnikoff E. Leçons sur la pathologie comparée de l'inflammation. G. Masson, Ed., Paris 1892. — Metchnikoff E. Études sur la nature humaine. G. Masson, Ed., Paris 1903. — Metchnikoff E. Immunity in Infective Diseases. Univ Press, Cambridge 1905. — Metchnikoff E. the new hygiene: three lectures on the prevention of infection diseases. TW Keener Co, Chicago 1906. — Chykalenko Ye.Kh. Memoirs (1861-1907). Part 1-3. Publ Assoc «Dilo», Lviv 1925-26. — Metchnikoff EI. Academical Collection of Work. Vol. 1-26. Moskva 1950-64. — Tauber AI, Chernyak L. Metchnikoff, and the Origin of Immunology: From Metaphor to Theory. Oxford Univ Press, New York - Oxford 1991. — Butenko E. The ideas of Mechnykov and development of contemporary medicine and biology. UHMJ «Ahapit» 1995;3:34-8. — Markova O, Fayfura V. The Ukrainian period of life and activity of Illya Mechnykov. Medical Gazette. 1995,05;17(63). — Ablitsov VH. «Galaxy Ukraine». The Ukrainian diaspora: prominent persons. Publ. «Kyt», Kyiv 2007. — Mechnykov I. Robert Koch. UHMJ «Ahapit» 1997-1998;7,8:29-33. — Synyachenko OV. They glorified themselves and Ukraine: history of domestic therapy. Crimean Therapeutic Journal 2010;2(2):21-8.

[321] Town of Hai(y)syn (first mention 1545), Haisyns'kyy Region, Vinnyts'ka Obl., Ukraine. — Wyssokowitsch W. Ueber die Schicksale der in's Blut injicirten Microorganismen im Körper der Warmblüter. Ztschr f. Hyg (Leipzig) 1886;1:3-46. — Wyssokowitsch W. Beträge zur Lehre von der Endocarditis. Arch f. path Anat (Berlin) 1886;103:301-32. — Orth J. Ueber die Aetiologie der experimentellen mycotischen Endocarditis. Nachschrift zu der Mittheilung des Dr. Wyssokowitsch. Arch f. path Anat (Berlin) 1886;103:3332-343. — Vysokovych VK. Pathological Anatomy. Vol. 1-2. Kyiv 1915-18. — Vysokovych VK. The Selected Works. Moskva 1954. — Ablitsov VH. «Galaxy Ukraine». The Ukrainian diaspora: prominent persons. Publ. «Kyt», Kyiv 2007.

Volodymyr Vysokovych described the mechanism of bacteria capture by the endothelium of blood vessels, their extraction from the blood and destruction. These research results on experimental acute endocarditis were proven by his work entitled «To the etiology of acute endocarditis» (1886), included in the published work by **Johannes Orth (1847 – 1923)**[322] — German pathologist, while serving the apprenticeship in his laboratory (1884-86), and in his work entitled «The fate of microorganisms injected into the blood» (1886) included in the published work by **Karl G.F.W. Flüge (1847 – 1923)**[323] — German microbiologist, hygienist and epidemiologist, while also serving the apprenticeship in his laboratory (1884-86).

Illya Mechnykov and his other co-worker **Mykola F. Hamaliya / Nikolay F. Gamaleya (05(17).II.1859, city of Odesa, Ukraine – 29.IX.1949, city of Moscow, Russia)**[324] — Ukrainian microbiologist and physician-epidemiologist, organized the Bacteriological Station in Odesa, Ukraine (1886-1920). Mykola Hamaliya discovered viruses of the plague in the horned cattle (1886). By reporting in 1898 of the lysis (in Greek – «division») of Bacillus anthracis, he discovered a transmissible «ferment», the bacteria destroying antibodies, known as bacteriolyses. He also discovered a cholera-like vibrio in birds, which led to the development of anti-cholera vaccine.

The hypothesis of **Volodymyr Vysokovych** (above) from 1886 about the ability of the liver's star macrophage cells lining the walls of the sinusoids, discovered in 1874 by **Karl W.** von **Kupffer (1829 – 1902**: 19th century. Anatomy, histology, and embryology), called stellate or Kupffer cells, for phagocytosis, was confirmed in 1898 by **Tadeusz Browicz (15.IX.1847, city of Lviv, Ukraine – 19.III.1928, city of Cracow, Poland)**[325] — Ukrainian Polish pathologist, discoverer of bacillus causing abdominal typhus (1874), and in 1899 by the former (Karl von Kupffer), as well.

In 1887 **Anton Weichselbaum (1845 – 1920)** — Austrian bacteriologist and pathologist, was the first to isolate the causative agent of cerebrospinal meningitis, bacteria he named Diplococcus intracellular meningitis. Mortality from meningitis was high (over 90%).

---

[322] Orth J. Cursus der nornalen Histologie zur Einführung in den Gebrauch des Microscopes, sowie in das practische Studium der Gewebelehre. 2te Ausg. Berlin. 1881. — Orth J. Compendium der Patologish-Anatomischen Diagnostik nebst Anleitung zur Ausführung von Obdüctionen sowie von Pathologisch-Histologischen Undersuchungen. Verlag von August Hirschwald, Berlin 1888. — Orth J. Lehbuch der speciellen Pathologischen Anatomie. Verlag von August Hirschwald, Berlin 1902-06.

[323] Flüge K. Lehrbuch der hygienischen Untersuchungsmethoden. Verlag von Veit & Comp., Leipzig 1881. — Flüge K. Die Microorganismen mit besonderer Berücksichtigung der Aetiologe der Infectionkrankheiten. 2te völlig umgearbete Auflage der «Fermente und Microparaziten». Verlag F.C.W. Vogel, Leipzig 1886. — Flüge K. Grundriss der Hygiene. Verlag F.C.W. Vogel, Leipzig 1889.

[324] Gamaleya NF. Collection of Works. Vol. 1-6. Moscow 1951-61. — Ablitsov VH. «Galaxy Ukraine». The Ukrainian diaspora: prominent persons. Publ. «Kyt», Kyiv 2007. — Hanitkevych YaV. The contribution of the Ukrainian physicians to the world medicine. Ukr Med Chasopys 2009 VII/VIII;4(72):110-15.

[325] Ablitsov VH. «Galaxy Ukraine». The Ukrainian diaspora: prominent persons. Publ. «Kyt», Kyiv 2007.

Adolf Weil (1848 – 1916) — German physician, and **Nicolai P. Vasiliev (1852 – 91)** — Russian physician, described leptospirosis (Weil disease, 1886, 1888).

Charles R. Richet (1850 – 1935) — French physiologist, formulated the concept of «passive immunity» (1888), described allergy and anaphylaxis (1902).

Vasily (Basil) Danilewsky (1852 – 1939: 19th century. Physiology)[326] was the first to investigate systematically on blood parasites of vertebrate such as birds, reptiles, and amphibians in Ukraine, and in his paper of 1884 (in Russian) and 1885 (in German) titled «About blood parasites (Haematozoa)» laid the foundation of modern parasitology in bird malaria and other protozoan infections. He was the first to observe in 1884 the species of Haemoproteus, parasitic protozoan in the blood of birds, named after him Haemoproteus danilewsky; was the first to discover in 1885 the protozoan Trypanosoma avium, the first known flagellate protozoan parasite in birds; by 1885 he recognized for the first time the existence of three separate genera of protozoan parasites in birds, now known as Plasmodium, Haemoproteus and Leukocytozoon; was the first to observe in 1888 a new genus Leukocytozoon from the owl blood, named after him Leukocytozoon danilewsky; was the first to describe the bird malaria and its symptoms of acute anemia, enlargement of the liver (hepatomegaly) and the spleen (splenomegaly), accumulation of pigments in the body, gave the first clue to the similarity of malaria of birds to that of humans. He identified the bird malaria parasites as «pseudovacuoles».

Working with **Robert Koch (1843 – 1910**: above), **Julius R. Petri (1852 – 1921)**[327] — German microbiologist, developed in 1877 a small cylindrical glass dish which may be re-used after sterilization in an autoclave or one hour dry heating in a hot-air oven at temperature (T) 160°C, later also plastic dish, both lidded, used to culture (growth) cells, bacteria, fungi, small mosses, mold e.t.c., when filled with suitable nutrients, usually agar, blood, salt, carbohydrates, aminoacids, dyes, antibiotics or other ingredients, and closed with lids to prevent contamination of the study.

Basing on the previous work of **Nykyfor Khronzhevsky (1836 – 1906**: 19th century. Anatomy, histology, and embryology), **Cheslav I. Khentsynsky / Cheslav I. Chenzinsky / Czesław I. Chęciński [06(18).III.1851, city of Warsaw, Poland – 24.V(06.VI). 1916, city of Odesa, Ukraine, Ukraine]**[328] — Ukrainian pathologist of Polish descent, developed the method of double staining of the blood cells and

---

[326] Danilewski B. Zur Parasitologie des Blutes. Biol Zentralblatt 1885;5:529-37. — Danilewsky B. La Parasitologie Comparee du Sang. I. Nouvelles Recherches sur les Parasites du Sang des Oiseaux. Kharkoff 1889. — Ablitsov VH. «Galaxy Ukraine». The Ukrainian diaspora: prominent persons. Publ. «Kyt», Kyiv 2007.

[327] Petri RR. Eine kleine Modification des Koch'schen Platenverfahrens. Centralblatt für Bacteriologie und Parasitan Kunde 1887;1:279-80.

[328] Chenzinsky Ch.I. Zur Lehre über den Microorganismus des Malariafiebers. Centralblatt für Bakteriologie 1888;3(15):457-60. — Ablitsov VH. «Galaxy Ukraine». The Ukrainian diaspora: prominent persons. Publ. «Kyt», Kyiv 2007.

the blood parasites (Khentsynsky method, Dissertation on the degree Dr of medical Science, Odesa 1888-89).

The first effective treatment against diphtheria was antidiphtheric serum (antitoxin) obtained by **Emil A.** von **Behring (1854 – 1917)** — Prussian German physiologist, **Paul Erlich (1854 – 1915)** — German physician of Jewish descent, and **Shibasaburo Kitasato (1853 – 1931)** — Japanese bacteriologist, in 1890, by **Pierre P.E. Roux (1853 – 1933)** — French physician, bacteriologist and immunologist, and **Alexandre E. J. Yershin (1863 – 1943)** — Swiss French physician and bacteriologist, in 1891, and by **Yakiv Yu. Bardach (XII.1857, city of Odesa, Ukraine – 17.VI.1929, city of Odesa, Ukraine)**[329] — Ukrainian microbiologist, in 1893.

Independently of each other, Shibasaburo Kitasato and Alexandre Yershin discovered the bacillus causing the bubonic plaque (Yersinia pestis) in people and rodents during the epidemic of 1894 in the city of Hong-Kong[330], China.

Paul Erlich also discovered mast cells (1878) and proposed the theory of the lateral chain (Erlich) in the formation of antibodies (1896).

In 1879 **Albert L.S. Neisser (1855 – 1916)** — German dermatologist of Jewish descend, isolated the pathogen of gonorrhea, colloquially called the «clap», that was named in his honor Neisseria gonorrhoeae, and a year later (1880) re-discovered Mycobacterium leprose.

Serhiy M. Vynohradsky / Sergey N. Winogradsky (01.IX.1856, city of Kyiv, Ukraine – 25.II.1953, town of Brie-Comte-Robert, Department Seine-et-Marie, Region Ile de France, France)[331] — Ukrainian microbiologist, pioneered the concept of the «cycle of life», discovered the first known form of lithotrophy (1887), proved the ability of bacteria Clostridia pastereurian to bind nitrogen in the absence of O_2 (1895).

---

[329] Bardach YaYu. Investigation about Diphtheria. Moscow 1884. — Bardach YaYu. Public Lessions in Bacteriology. Odesa 1891.— Kuznetsov VO. Life and activity of Professor Yakiv Yuilievych Bardach (1857 – 1929). In, O.Ya. Pylypchak (Red.). History of Ukrainian Science within Boundary of Thousand Years. Kyiv 2008. Issue 34. P. 100-27. — Ablitsov VH. «Galaxy Ukraine». The Ukrainian diaspora: prominent persons. Publ. «Kyt», Kyiv 2007. — Hanitkevych YaV. The contribution of the Ukrainian physicians to the world medicine. Ukr Med Chasopys 2009 VII/VIII;4(72):110-15.

[330] City of Hong-Kong (meaning a «fragrant harbor»; total area of 1,104 km²; archeological studies support a human presence in its western island Chek Lap Kok with an area of 3.02 km², from 35,000 to 39,000 years ago; 214 BC), China.

[331] Town of Brie-Comte-Robert (founded c.10th century), Department Seine-et-Marie, Region Ile de France (area 12,012), France. — Winogradsky S. Über Schwefelbacterien. Bot Zeitung 1887;45:489-610. Waksman SA. Sergei Nikolaevitch Winogradsky: The study of a great bacteriologist. Soil Sc 1946; 62:197-226. — Ackert, L. From the Thermodynamics of Life to Ecological Microbiology: Sergei Vinogradskii and the Cycle of Life 1850-1950. PhD dissertation, the Johns Hopkins University 2004, Springer 2004. — Ackert L. The role of microbes in agriculture: Sergei Vinogradskii's discovery and investigation of chemosynthesis, 1880-1910. J Hist Biol 2006;39:373-406. — Ackert L. The «cycle of life» in ecology: Sergei Vinogradskii's soil microbiology, 1885-1940. J Hist Biol 2006; 40:109-145. — Ablitsov VH. «Galaxy Ukraine». The Ukrainian diaspora: prominent persons. Publ. «Kyt», Kyiv 2007.

Olexandr D. Pavlovsky (01.X.1857, village Chufarovo, Rostov Region, Yaroslav Obl., Russia – 08.X.1946, town of Soroky-on-Dnister River, Ukraine / Moldova)[332] — Ukrainian microbiologist and surgeon, introduced sterilization of the surgical instruments by boiling them in a 1% sodium solution (1890), used an antidiphtheric serum (1895).

Sir **Ronald Ross (1857 – 1932)** — Indian English physician, discovered in 1895-98 a malaria parasite in the stomach of mosquitoes and a carrier of an infection (vector) Anopheles in the blood of an infected man, inquired into life cycle of the malaria parasite in birds and showed its transmission by a bite of infected mosquito female which he called Plasmodium reticulum. He also revealed pigmented oocytes of malaria in the wall of the stomach (black spores of Ross).

An empirical staining process was devised in 1884 by **Hans Ch. J. Gram (1858 – 1938)** — Danish bacteriologist, in which microorganisms are stained with crystal violet, treated with 1:15 dilution with the Lugol solution [**Jean Lugol (1786 – 1851)**, above], decolorized with ethanol or ethanol-acetone and counterstained with contrasting dye (safranin O). Those that retain crystal violet are Gram-positive, those which do not are Gram-negative (Gram method or stain).

During an influenza (flu) pandemic in 1892 **Richard F. J. Pfeiffer (1858 – 1945)** — German physician and bacteriologist, and Sir **Almroth E. Wright (1861 – 1947)** — English bacteriologist and immunologist, independently discovered a Gram-negative, rod-shaped coccus-like bacterium Haemophilus influenzae (Pfeiffer bacillus or Bacillus influenzae) from the nose of flu-infected patient which usually colonizes the upper respiratory tract. It was mistakenly considered to be the cause of flu until 1933, when the viral etiology of this disease became apparent. Most strains of Haemophilus influenzae are opportunistic pathogens; that is, they usually live in their host without causing disease, but can cause a wide range of clinical problems, including inflammation of the lungs (pneumonia), only when other factors, such as a viral infection, reduced immune function or chronically inflamed tissues from allergies, create an opportunity.

Nevertheless, Haemophilus influenzae rod type B is one of the most common causes of generalized infections (bacteremia) in children, half of the develop purulent inflammation brain's meninges (meningitis), quite often (10-20%) pneumonia and rarely other focal infections.

Richard Pfeiffer discovered the phenomenon bacteriolysis, when he found that live cholera bacteria could be infected without ill effect into guinea pigs previously

---

[332] Village of Chufarovo (founded c. 1814), Rostov Region, Yaroslav Obl., Russia. — Town of Soroky-an-Dnister River (settled since the Age of Mesolithic and early Neolithic, Soroky's Fortress build 15th century, town's status since 1835), Україна / Moldova. — Ablitsov VH. «Galaxy Ukraine». The Ukrainian diaspora: prominent persons. Publ. «Кут», Kyiv 2007.

immunized against cholera, and that blood plasma from these animals added to live cholera bacteria caused them to be motionless and to lyse (Pfeiffer phenomenon, 1894). Then he isolated Micrococcus catarrhalis that is the cause of laryngitis, bronchitis, pneumonia, and occasionally meningitis (1896). Also invented a universal staining for histological preparations.

In 1885 **Theobald Smith (1859 – 1934)** — American epidemiologist and pathologist, isolated Bacterium suipestifer (Salmonella cholerae suis) in pigs dying from the plague. His findings were published by **Daniel E. Salmon (1850 – 1914)** — American veterinary surgeon, and the pathogen was named after him, Salmonella. It is «Gram-negative» bacteria, facultative (optional) anaerobe from a genus of Enterobacteriaceae, displaying a non-spore-forming bacilli, moving with the help of peritrichous projections (flagellum). Theobald Smith re-discovered, after **Charles Richet (1850 – 1935**: above), an anaphylaxis (1903).

Sir **Volodymyr (Waldemar) M.W. Haffkine (15.III.1860, city of Odesa, Ukraine – 26.X.1930, city of Lousanne, Swizerland)**[333] — Ukrainian bacteriologist of Jewich descent, developed vaccines for the preventive inoculation against cholera (1892) and the bubonic (lymphatic) plague (1897).

In 1890-91 **Dmytro (Dmitry) L. Romanovsky (1861, Pskov Obl., Russia** – died from heart attack **12.II.1921, town of Kyslovods'k, Stavropolsky krai, Ukraine / Russia)**[334] — Russian Ukrainian physician-therapeutant, microbiologist and hematologist, in 1902 – **Gustav Giemsa (1867 – 1948)**[335] — German chemist and bacteriologist, working with the double staining of **Cheslav Khentsynsky (1851 – 1916**: above), developed the eosin-methylene blue stain for coloring of cells and parasites in the blood (Romanovsky-Giemsa method).

---

[333] City of Lousanne (site of Celtic settlement since c. 450 BC outward, 2nd century, 280 and 400 AD), Swizerland. — Waksman SA. The brilliant and tragic life of W.M.W. Haffkine, bacteriologist. Rutgers Univ. Press 1964. — Lutzker E, Jochnowitz C. Waldemar Haffkine: pioneer of cholera vaccine. Am Soc Microbiology News 1987;53(7):366-9. — Ablitsov VH. «Galaxy Ukraine». The Ukrainian diaspora: prominent persons. Publ. «Kyt», Kyiv 2007. — Hanitkevych YaV. The contribution of the Ukrainian physicians to the world medicine. Ukr Med Chasopys 2009 VII/VIII;4(72):110-15. — Synyachenko OV. They glorified themselves and Ukraine: history of domestic therapy. Crimean Therapeutic Journal 2010;2(2):21-8.

[334] Pskov Obl. (area 44.211.2 km²), Russia. — Town of Kyslovods'k (discovery of the spring carbonic mineral water 1798, founded 1803), Stavropolsky krai (SK: area 66,500 km²), Ukraine (1918-22) / Russia. — Romanovsky DL. Question on parasitology and suffering from marsh fever. Dissertation for degree of DMS. St Petersburg 1891. — Romanowsky DL. Zur Frage der Parazitologie und Therapie der Malaria. St Petersburg Med Wochenschr 1891;16(34):297-302; 35:307-16. — Diachenko SS. Dmitrii Leonidovich Romanovsky (1861-1921). Likars'ka Sprava / Vrachebnoye delo (Kyiv) 1952;4:367-70. — Hoare CA. Romanovsky and his stain: 70th anniversary of its discovery. Trans Roy Soc Trop Med Hyg 1960;54:282-3. — Ablitsov VH. «Galaxy Ukraine». The Ukrainian diaspora: prominent persons. Publ. «Kyt», Kyiv 2007.

[335] Giemsa G. Über Romanowsky-Giemsa Färbung. Zbl. f. Bakteriol 1902;3(1):429.

Tetanus, a.k.a., lockjaw (in Latin – tetanus; in Greek – stretched)[336] is a bacterial infection affecting nervous system, caused by Clostridium tetani, one of over 150 species, a common soil rod shaped, up to 0.5 micrometers wide and 2.5 micrometers long Gram-positive anaerobic endospore, that cannot survive it the presence of oxygen (O_2). Disease is characterized by periodic generalized total clonic muscle spasm on the background of permanent strain (tension) of skeletal muscles and usually terminates in death of the patient.

Clinical descriptions of tetanus could be found in the **Eberts papyrus** (Ancient Medicine. *Egypt*) c. 2600 BC, in the work of **Hippocrates [(c. 460 BC – 377 BC).** (Ancient medicine. *Greece*]** who son likely died from lockjaw, gave name of this disease, and various description from the 5th century BC. But the first clear connection of tetanus to the soil was made in 1884 when **Arthur Nicolaier (1862** – suicide prior to departure into German concentration camp in **1942)** — Jewish German internist, isolated the strychnine-like toxin named Clostridium tetani from free-living anaerobic bacteria and showed that animals injected with garden sample known to contain bacteria would develop tetanus. He also was the first to use hexamethylenetetramine (Urotropin) to treat urinary tract infections (UTI). Later this same year (1884), first transmissibility of tetanus was demonstrated by **Antonio Carle** and **Gorgio Rattone** — Italian pathologists, in rabbits by injecting puss from a patient with fatal tetanus into their sciatic nerves. In 1889 Clostridium tetani was isolated from human victim by **Shibasaburo Kitasato (1853 – 1931**: above) who a year later (1890), together with **Emil** von **Behring (1854 – 1917**: above) showed that organism could produce disease when injected into animals, toxoid could be neutralized by specific antibodies, and obtained antitoxin against tetanus. Furthermore, in 1897 **Edmond Nocard (1856 – 1903)** — French microbiologist and veterinarian, showed that tetanus antitoxin induced passive immunity in humans, and could be used for prophylaxis and treatment of this disease.

Ivan H. Savchenko (02.III.1862, Romansky County, Poltava Obl., Ukraine – 13.IX.1932, city of Krasnodar, Krasnodarsky krai, Ukraine / Russia)[337] — Ukrainian physician-immunologist and microbiologist, and **Danylo K. Zabolotnyy** or **Daniil K. Zabolotnyy [16(28).XII.1866, village Chobotarka, now Zabolotne, Kryzhopilskyy Region, Vinnytsia Obl., Ukraine – 15.XII.1929, city of Kyiv,**

---

[336] Kartysheva AF. Specializing Epizootiology. Textbook. Publ. «Higher Education», Kyiv 2002. — Holubovska OA, Andreychyn MA, Shkurba AV. Infectious Diseases. Textbook. 2nd Ed. Publ. VSV «Medytsyna», Kyiv 2018.

[337] City of Krasnodar (founded 1793, as camp «New Sich» by Ukrainian Zaporizhia Cossack), Krasnodarsky krai (KK: area 75,500 km²; as outcome of organized by Russian Communist Fascist of Holodomor-Genocide 1932-33 in Krasnodar city and KK where majority of population at that time were Ukrainians, documented 62,000 victims of man-made hunger), Ukraine (1917-22) / Russia. — Ablitsov VH. «Galaxy Ukraine». The Ukrainian diaspora: prominent persons. Publ. «Kyt», Kyiv 2007.

Ukraine)[338] — Ukrainian microbiologist, discovered vibrio cholera and proved bacilli-carrying in cholera (1893). Danylo Zabolotnyy proved that bacteria causing the plague (Yershinia pestis) is transmitted to man by wild rodents (1911).

Working independently on tropical diseases in Uzbekistan and India, **Petro (Peter) F. Borovsky (Borovskii) [27.V(08.VI). 1863, town of Pohar, Bryansk Obl., Ukraine / Russia – 15.XII.1932, city of Tashkent, Uzbekistan]**[339] — Ukrainian surgeon, **Charles Domovan (1863 – 1951)** — Irish military physician, and Sir **William B. Leishman (1865 – 1926)** — Scottish pathologist, discovered Calymmatobacteria – a family of the most primitive encapsulated, facultative anaerobes, Gram-negative bacilli, subkingdom Protozoan, subline Trypanosomiasis, which appears in the amastigote stage as of intracellular cytoplasmatic inclusions of large mononuclear phagocytes (Borovsky-Leishman-Domovan bodies, 1892, 1901, 1903) in the vertebral host (people) and in the promastigote stage in the gastrointestinal (GI) tract of the nonvertebral host (some bloodsucking sand fly of the subfamily Phlebotomine). They cause an infection generally called leishmaniasis of the skin (Borovsky skin oriental sore, 1892), of the skin and mucosa membranes and of the viscera (kala-azar or visceral leishmaniosis, 1905). The second also identified Calymmatobacterium granulomatis as responsible for granuloma inguinale (Domovan granuloma, 1905). Petro Borovsky proposed a new modification for suturing of the urinary bladder.

Independently, **Simon Flexner (1863 – 1946)** — Czech American pathologist and physician of Jewish descent, and **Hugo Wintersteiner (1865 – 1946)** — Austrian ophthalmologist, discovered the characteristic rosettes formations in retinoblastoma, the most common rare, rapidly growing primary malignant tumor, developing from immature cells of the light detecting retina of the eye (Flexner-Wintersteinet rosettes, 1891, 1897).

[338] Village Chobotarka, since 1929 – Zabolotne (founded 1756), Kryzhopilskyy Region, Vinnytsia Obl., Ukraine. — Olearchyk A. Ukrainian Scientists. Danylo Kyrylovych Zabolotny. Nasze Słowo (Warszawa) 17.2.1957;2,7(27): 3. — Rozhin I. Danylo Zabolotny – Ukrainian scientist of international recognition (on the occasion of hundred year from his birth date). J Ukr Med Assoc North Am 1967 Apr;14,2(45):22-5,27. — Ablitsov VH. «Galaxy Ukraine». The Ukrainian diaspora: prominent persons. Publ. «Kyt», Kyiv 2007. — Hanitkevych YaV. The contribution of the Ukrainian physicians to the world medicine. Ukr Med Chasopys 2009 VII/VIII;4(72):110-15.

[339] Town of Pohar (Slav's settlements 8th-9th century, first written mention 1155), Pohar Region, Bryansk Obl. (area 34,857 km²), Ukraine / Russia. — City of Tashkent (founded 2nd century BC, Uzbekistan (area 448,924 rm²). — Borovskii PF. About Saratov ulcer. Milit-Medi Zh 1898;195:925. — Hoare CA. Early discoveries regarding the parasite of oriental sore. Trans Royal Soc Trop Med Hyg 1938 June 25;32(1):67-92. — Ablitsov VH. «Galaxy Ukraine». The Ukrainian diaspora: prominent persons. Publ. «Kyt», Kyiv.

Dimitry (Dmytro) Jo. Ivanovsky (28.10.1864, Village of Nizy, Gdov Uyezd, Novgorod Ob., Russia – 20.06.1920, city of Rostov-on-Don, Russia)[340] — Russian Ukrainian botanist, histologist, physiologist, and virologist, isolated a filtrable virus of the tobacco's mosaic disease (1892).

Viacheslav K. Stefansky (1867, city of Odesa, Ukraine – 07.VI.1949, city of Odesa, Ukraine)[341] — Ukrainian microbiologist, discovered Mycobacterium leprae murinum (Stefansky's bacillus, 1892), the causative factor of lepra in rats and the only model of it in the human which cause chronic granulomatous or neurotrophic affection of the skin, mucous membranes, nerves, bones, and internal organs (Hansen disease, 1873)[342]. He was the first to introduce in Ukraine in 1890-92 intubation of the trachea for children affected by diphtheria croup, in lieu of tracheostomy. He proved than bugs (bedbugs) are not carriers of spotted typhus, made contribution for the treatment of plague (1902).

In 1898 **Kiyoshi Shiga (1870 – 1957)** — Japanese physician and bacteriologist, isolated within the genus Enterobacteriaceae, Gram-negative, facultative aerobic bacilli-like bacteria, the four species named after him Shigella, based on the biochemical reaction. The first subgroup of the Shigella species was called Shigella dysenteries (subgroup A), the second subgroup was named in honor of **Simon Flexner (1863 – 1946**: above) Shigella flexneri (subgroup B), the third subgroup was discovered by **Mark F. Boyd (1889 – 1968)** — Scottish bacteriologist, Shigella boydii (subgroup C), and the fourth group was discovered by **Carl O Sonne (1882 – 1948)** — Danish bacteriologist and parasitologist, Shigella sonnei (subgroup D). Their normal environment for existence is intestinal tract of the human and the superior monkey. All species of Shigella are responsible for 90% cases of dysentery.

---

[340] Village of Nizy, Gdov Uyezd (first mentioned beginning 14th century), Novgorod Obl. (area 54,501 km²), Russia – City of Rostov-on-Don (founded 1749), Rostovs'ka Obl. (area 100,800 km²), Ukraine / Russia. — Iwanowski DJo. About diseases of the snuff. The tobacconist / snuff-taker's ash-tray, mosaic disease. Agriculture and Forestry 1892 February; 219(2):104-121. — Iwanowski D. Über die Mosaikkrankheit der Tabakspflanze. Zeitschrift für Pflanzenkrankheiten und Pflanzenschutz 1903;13:1-41. — Ablitsov VH. «Galaxy Ukraine». The Ukrainian diaspora: prominent persons. Publ. «Kyt», Kyiv 2007.

[341] Stefansky VK. Acute Infectious Disease. 1929. — Ablitsov VH. «Galaxy Ukraine». The Ukrainian diaspora: prominent persons. Publ. «Kyt», Kyiv 2007.

[342] **Gerhard H. A. Hansen (1841-1912)** — Norwegian bacteriologist and physician, who discovered in 1873 bacillus leprosy.

Pathology[343]

Friedrich Uhden or **Fedi(o)r C. (K.) Uden (1754, town of Stendal, Land of Sachsen-Anhalt, Germany – 02.IX.1823, St Petersburg, Russia)**[344]— Russian Ukrainian pathologist and physician of Prussian descent, graduated from the Medical-Surgical (established 1710), city of Berlin (first settlement dated 1174) and Medical Faculty of the University of Halle (established 1694), city of Halle (first mentioned 806), Sachsen-Anhalt, Germany, where received the degree of Medical Doctor (1776). Then, since July 13, 1786, through 1793 practiced medicine in the Chernihiv Viceregency (1781– 1796), city of Chernihiv (first mentioned 907), Ukraine, in 1793-94 lecturer of mathematics and physics at St Petersburg medical school, since 1800 Professor of pathology and therapy at St Petersburg Medical-Surgical Academy (established 1798). In 1812-22 described chronic ulcer of the stomach and duodenum.

Acute ulcers of the stomach described in 1830 **Jean B. Cruveilhier (1791 – 1874)** — French anatomist, pathologist, and physician.

Earlier (1816), Jean Cruveilhier attributed the left ventricular (LV) aneurysm to myocardial fibrosis.

Independently, **Pégot**[345], Jean Cruveilhier[346] and **Paul C. Baumgarten (1848 – 1928)**[347] — German microbiologist and pathologist, depicted liver cirrhosis with portal hypertension, congenital patency of the umbilical or paraumbilical veins without ascites (Pégot-Cruveilhier-Baumgarten's syndrome). In addition, the latter two later described during auscultation of the abdominal wall around the umbilicus a murmur over veins that connect the portal and caval system (Cruveilier-Baumgarted murmur), and Jean Cruveilhier provided the first pathological account and clinical picture of multiple sclerosis (MS).

---

[343] Zerbino DD, Bahriy MM, Bodnar YaYa, Dibrova VA. Pathomorphology and histology. Atlas. «Nova Knyha», Vinnytsia 2016.

[344] Town of Stendel (known as fortified town c. 1160), Land Saxhsen Anhalt, Germany. — Uden F. Primae lineae fundamentirum pathologiae therapiae. Petropoli 1809. — Uden F. The Academic Lectures about Chronic Diseases. Part 1-7. In Igersens's Typography, St Petersburg 1816-22. — Uden F. Pharmakopea. St Petersburg 1818. — Olearchyk AS. Ulcers of the Stomach and Duodenum in Children. J Ukr Med Assoc North Am 1974; 21,4 (75):3-100. — Ablitsov VH. «Galaxy Ukraine». The Ukrainian diaspora: prominent persons. Publ. «Kyt», Kyiv 2007. — Synyachenko OV. They glorified themselves and Ukraine: history of domestic therapy. Crimean Therapeutic Journal 2010;2(2):21-8.

[345] Pégot. Tumeur variqueuse avec anomalie du système veineux et persistance de la veine ombilicale, développement des veines souscutanées abdominals. Bul mém Soc anat Paris 1833; 44: 49-57. — Pégot. Anomalie veineuse. Ibid 1833, 8: 108.

[346] Cruveilhier J. Anatomie patologique du corps humain. Vol. 1. Bailliére, Paris 1829-35. — Cruveilhier J Traité d'anatomie patologique générale. Vol. 2. Bailliére, Paris 1852.

[347] Baumgarten P. Über vollständinges Offenbleiben der Vena umbilicalis; zugleich ein Beitrag zur Frage des Morbus Banti. Arb Geb Path Inst. Tubingen, Leipzig 1907. — Von Baumgarten P. Arbeiten 1908.

In 1832 **Thomas Hodgkin (1798 – 1866)** — British physician, described neoplastic growth of lymphatic glands (malignant lymphoma or Hodgkin's disease).

Pathophysiology of constrictive pericarditis due to tuberculosis was clarified in 1842 by **Norman Cheevers** (1818 – 86) — English physician and surgeon.

A classification of the pathological changes in organs, produced by different diseases, was developed by **Carl** von **Rokitansky (1804 – 78**: 19th century. Anatomy, histology, and embryology)[348].

After the report by **Johann Huber (1707 – 78**: 18th century**)** about an intra-lobal pulmonary sequestration, Carl Rokitansky described an extra-lobal pulmonary sequestration (Rokitansky lobe), accounting for 25% of all pulmonary sequestrations. Extra-lobal sequestrations are round, smooth, and soft masses with its own separate visceral covering situated just above the dome diaphragm, whereas intra-lobal sequestrations are cystic aberrations of the lung tissue within the visceral pleura. Both are derived from foregut tissue, in 90% situated at the base of the left lung, prone to pulmonary infection and hemoptysis. The arterial blood supply is from the descending thoracic (DTA, 74%), abdominal aorta (AA, 19%), intercostal arteries (3%) and from multiple arteries (20%). Drainage of the venous blood from the extra-lobal sequestrations is into the systemic venous system, usually to the azygos or hemiazygos veins (80%) or to pulmonary veins (PV, 20%), drainage from the intra-lobal sequestrations is most commonly into PV, occasionally to the systemic veins. Male to female ratio is 3:1.

He also described a traction diverticulum of the esophagus; a massive hepatic necrosis; syndrome of the superior mesenteric artery (SMA), a rare life-threatening disease caused by the compression of the 3th portion of duodenum due to a decrease of a normal angle between the abdominal aorta (AA) and SMA from 38°-56° to 6°-25° (1861); a pelvic spondylolisthesis; a coexistence of pulmonary artery (PA) stenosis with patent foramen ovale (PFO) and patent ductus arteriosus (PDA) – Rokitansky triad. The SMA supplies the arterial blood flow to the section of digestive tract embryologically derived from the mid gut, i.e., from the descending duodenum to the splenic flexure of the colon.

George Budd (1808 – 82) — English physician, and **Hans Chiari (1851 – 1916)** — Austrian pathologist, described a thrombosis of the hepatic (80-90%) or portal (10-20% veins causing cirrhosis of the liver, ascites and abdominal pain (Budd-Chiari's syndrome, 1845, 1898), **Julius Arnold (1835 – 1915)** — German pathologist, and Hans Chiari – a brain malformation in which the tonsilla cerebelli and the medulla oblongata protrudes through the foramen magnum at the bottom of the skull into

---

[348] Rokitansky C. Handbuch der allgemeinen pathologishen Anatomie. Bei Braumüller & Seidel, Wien 1846.
—Rokitansky C. Die Defecte der Scheidewände des Herzens. Bei W. Braumüller, Wien 1875.

the spinal canal (Arnold-Chiari's malformation (1894, 1891), and finally Hans Chiari – an embryologic strand like reticulated remnant connecting the Eustachio's valve **[Bartolemo Eustachio (1500 – 74**: Renaissance] of the inferior vena cava (IVC) with the interatrial septum (IAS) occurring in 2% of the patients (Chiari network, 1897).

Josyp V. Varvynsky (1811, town of Khorol, Poltava Obl., Ukraine – 1878, city of Moscow, Russia)[349] — Ukrainian physician and pathologist, defended DMS Dissertation entitled «De nervi vagi physiologia et pathologia», University of Tartu, city of Tartu 1838).

The basis for clinical-pathological directions in pathology were founded by **Yuliy (Ferdynand) I. Mazonn (22.04.1814, city of Riga, Latvia – 20.12.1885, city of Kyiv, Ukraine)**[350] — Ukrainian physician-therapeutant, and pathologist of German descent.

Sir **James Paget (1814 – 99)**[351] — English pathologist and surgeon, described in 1854, and **F. Ernst Krugenberg (1871 – 1946)** — German pathologist, described in 1896, what they thought was primary fibrosarcoma or carcinoma multicellulary / mucoidocellulare of the ovary or bilateral (80%), but subsequently proved to be secondary to a primary mucous adenocarcinoma of the gastrointestinal (GI) tract, mainly from the gastric pylorus which contains signed-ring cells filed with mucus, or from an invasive intralobular breast carcinoma that occurs in 1-20% of women, named Krukenberg tumor.

In 1875 **James Paget** and in 1884 **Kristelli L. von Schroetter (1837 – 1908)**[352] — Austrian internist and laryngologist, diagnosed thrombosis of the axillary-subclavian veins, secondary to thoracic outlet syndrome (TOS), producing external compression in the anteromedian compartment of the costoclavicular space, where the costoclavicular ligament congenitally inserts into the 1st rib farther laterally than normal, and further laterally from the hypertrophied scalenus anticus muscle, after excessive use or prolonged abduction of the arm, manifested by pain, swelling, bluish discoloration of the skin and distention of the superficial veins of the shoulder belt

---

[349] Town of Khorol (mentioned 1083; made by Russian Communist Fascist regime Holodomor-Genocide 1932-33 caused colossal human suffering, victims, and political repressions of 1930s touched residents of the town and surrounding villages), Poltava Obl., Ukraine. — Tartu (former Yur'iv) Univ. (opened 1632), city of Tartu (mentioned in 5th century BC, founded 1030), Estonia (Foot Note 265). — Ablitsov VH. «Galaxy Ukraine». The Ukrainian diaspora: prominent persons. Publ. «Kyt», Kyiv 2007.

[350] City of Riga (first mentioned 1198, founded 1201) – capital of Latvia (Foot Note 245). — Mazonn Fl. De primo gradu degenerationis renum in morbo Brightii. Doctoral dissertation, St Volodymyr University, Kyiv 1850. — Zhukovsky L. Contribution of Kyiv pathologist and therapeutist Yu.I. Mazonn to scientific development of method of thoracic percussion. UHMJ «Ahapit» 1995;2:57-60. — Ablitsov VH. «Galaxy Ukraine». The Ukrainian diaspora: prominent persons. Publ. «Kyt», Kyiv 2007.

[351] Paget J. Clinical Lectures and Essays. Longmans Green, London 1875.

[352] Von Schraetter L. Enkrankungen der Gefossi. In AK Nathnogel (Ed.). Handbuch der Pathologie und Therapie. Holder, Wein 1884.

and the upper extremity, a moderate tenderness over the axillary vein with palpable cordlike structure that corresponds to the course of that vein (effort thrombosis or syndrome, or Paget-Schroetter syndrome, 1875, 1884)[353].

James Paget also described intraductal carcinoma of the breast spreading into the epidermis around the nipple (Paget disease of the nipple, 1875), a chronic inflammation of bones (osteitis deformans) or Paget disease of the bone (1877), James Paget with **H. Radcliffe Crocker (1846 – 1909)** — English dermatologist, diagnosed an extramammary Padget disease of the epidermis and occasionally of the dermis with the predilection for apocrine glands bearing areas, mostly axilla, perineum, anus, scrotum, penis, and vulva (1889).

Radcliffe Crocker observed trichina infestations in human muscles.

Friedrich Krugenberg also described vertical, fusiform depositions of melanin pigmentation on the endothelium of the posterior surface of the cornea in the pupillary area in degenerative or chronic inflammatory diseases of the bulbus oculi, and in pigmented glaucoma (Krukenberg spindle, 1899).

In 1857 **Ludwig** von **Buhl (1816 – 80)**[354] — German pathologist, discovered a rare direct communication between the left ventricle (LV) and the right atrium (RA) through the membranous septal defect – Buhl's atrioventricular defect (AV) defect.

Beside of this, he described in the newborn an acute fatty infiltration of the parenchyma of the liver, kidney, or heart with sepsis, hemorrhages into the skin, mucous membranes, umbilicus and internal organs, cyanosis and jaundice (Buhl disease, 1861), and with **Franz Dittrich (1815 – 59)** — Austrian pathologist, while researching desquamative pneumonitis, disseminated miliary tuberculosis stated that «*in every case of acute miliary tuberculosis, there exist at least one old focus of causation in the body*» (Buhl-Dittrich law, 1872).

A genetic autosomal recessive progressive disease in which copper builds in the body affecting the liver and brain, first described in 1854 by **Friedrich T.** von **Frerichs (1819 – 85)** — German pathologist, re-described in 1912 by **Samuel A.K. Wilson (1878 – 1937)** — American English neurologist, named Wilson's disease. The disease typically begins between the ages of 5 and 35. The liver related symptoms are vomiting, weakness, yellowish skin, itchiness, ascites and swelling of the legs. The brain symptoms include tremor, muscle stiffness, difficulties in speaking, personality changes, anxiety, seeing and hearing imaginable things. Examination of the eyes, directly or using slit lamp commonly, shows a brown ring on the edge of cornea caused by copper depositions, first detected in 1902 by **Bernard Kayser**

---

[353] Aziz R, Straenley CJ, Whelan TJ. Effort-related axillosubclavian vein thrombosis. Am J. Surg. 1986;152:57.

[354] Buhl L, cited by Meyer H. Uber angeborene Enge oder Verschluss der Lungenarterienbahn. Virchow's Arch path Anat 1857;12:532.

(1869 – 1954) — German ophthalmologist, and **Bruno O. Fleishner (1874 – 1965)** — German physician, named Kayser-Fleschner ring. The level of the copper in the blood is decreased, in the urine increased. Treatment is by elimination of copper cookware and low copper diet, chelating (in Greek «chele» – claw of crabs or lobsters)[355] agents such as triamine and d-penicillinase, zinc supplements and liver transplantation.

The father of pathology and founder of social medicine, **Rudolph L.R. Virchow (1821 – 1902)**[356] — Prussian German pathologist, established that the cell arises only from preexisting cells, coining the phrase «*omnis cellua e cellula*», and regarded the body as a cell-state in which every cell is a citizen and disease is a civil war among the cells brought about by external forces.

The Virchov triad defined three factors conductive to deep venous thrombosis (DVT): (1) stasis (sluggish blood flow), (2) hypercoagulation and (3) injury to the intima of the vein. He explained the mechanism of pulmonary emboli (PE), created the term embolism, noted that the blood clots in the pulmonary artery (PA) comes from detached larger or smaller fragments (pieces) in the top of a softened venous thrombus which in the form of an emboli migrated here with the blood flow. The first to describe pulmonary lymphangiectasia as diffuse cystic lymphatic duct dilatations in infancy. A palpable left supraclavicular lymph node (Virchov node) usually implies metastatic carcinoma of the stomach.

While studying the histology of arteriosclerotic or atherosclerotic (AS) lesions, he recognized that the lesion was within or under the intimal lining and that primary deposits occurred by imbibition of certain blood elements through this lining. He described the next stage as a softening of the connective tissue matrix at the site of deposition followed by active proliferation of this tissue within the intima accompanied by a fatty metamorphosis of the connective tissue cells leading to localized thickening. He called this «endarteritis deformans» meaning that the atheroma was a product of inflammation within the intima and that the thickening evolved due to «reactive fibrosis».

Henrich von Bamberger (1922 – 88)[357] — Czech Austrian pathologist, described spasm of the legs producing tics and jumping or spinning motion associated with repetitive meaningless words (echolalia) and lack of ability to make decisions (abulla) observed in schizophrenia, and histeria due to irritation of motor cells of the spinal

---

[355] Chelates are claw like compounds that form during interaction between metal ions and some organic molecules.

[356] Virchow R. Handbuch der speciellen Pathologie und Therapie. Berlin 1854-62. — Virchow R. Die Cellular Pathologie in ihrer Begründung auf physiologische und pathologische Gewebelehre. Berlin 1858. — Virchow R. Vorlesungen über Pathologie. Berlin 1862-72.

[357] Von Bamberger H. Saltatonischer Reflekskramps, eine merkwürdige Form von Spinal-Irritation. Wien med Wschrift 1859;9:49-52, 65-67. — von Bamberger H.Ueber zei seltene Herz affektionen mit Bezugnahme auf die Theorie des ersten Herztons. Wien med. Wschrift 1872;22:1-4, 25-28.

cord (Bamberger's disease I, 1859), and provided an early description of idiopathic hematogenous albuminuria, uremic pericarditis and progressive polyserositis (Bamberger's disease II, 1872).

The first description in 1764 by **Abraham Ludlow** (18th century), and subsequent in 1867 by **Friedrich A.** von **Zenker (1825 – 98)**[358] — German pathologist and physician, made possible a detailed characterization of pharyngoesophageal diverticulum as a protrusion of the mucous layer of the pharynx and esophagus through the opening in a triangular area in the wall of the pharynx between the oblique fibers of the constrictor pharynges inferior muscle and the transverse fibers of the cricopharyngeal muscle, or in the reverse triangle between the transverse fibers of the cricopharyngeal muscle and the esophageal inner circular layer of the muscularis propria without coverage by the external longitudinal muscle of the esophagus. In 1877 pharyngoesophageal diverticulum was named the Zenker diverticulum.

He also described degeneration (necrosis) and hyaline dystrophy of the striated muscles due to an acute infectious disease (Zenker degeneration or necrosis) and of paralysis of the superficial, deep and accessory peroneal nerve (Zenker paralysis).

Isolated pulmonary artery (PA) aneurysm is a rare vascular anomaly due to dilatation of an artery. About 40-50% of such aneurysms are congenital and familial. Sporadic cases include penetrated chest trauma, chest tube placement or PA catheter related injury. Other cases are associated with infection and inflammatory destruction of vasa vasorum in the lung, occurring in syphilis, tuberculosis, fungus, giant-cell arteritis or widespread vasculitis, causing hemoptysis and may lead to rupture and severe hemorrhage. Dilatation of a branch of a PA artery adjacent or within tuberculous cavity in the lung was described in 1868-69 by **Fritz W. Rasmussen (1833/34/37 – 1877)** — Danish physician, named Rasmussen's aneurysm. It occurs in up to 5% of patients with such lesion.

Fredrich D. von **Recklinghausen (1833 – 1910)** — German pathologist, described small lymphatic spaces in the connective tissue (canals of Recklinghausen); osteitis fibrosa cystica from an active secondary activity of osteoclasts due to hyperfunction of the parathyroid glands, causing a soft inflammation of the bone, fibrous degeneration, formation of cysts and fibrous nodules (Recklinghausen disease); and neurofibromatosis (NF, 1882) – congenital autosomal dominant disorder, characterized by multiple abnormalities in the nervous plexuses of the subcutaneous (SC) tissue, muscles, bone and internal organs. The Recklinghausen's NF is divided into three types:

---

[358] Von Zenker FA, Ziemssen HW. Krankheiten des Oesophagus. In: HW von Ziemssen (Ed.). Handbuch des Speciellen Pathologie und Therapie. FC Vogel, Leipzig 1877;7 (suppl): 1-87.

NF1 type is characterized by non-malignant blots, nodules, or tumors in the form of «coffee with milky speckles», superficial pedunculated soft tissue growths, surrounded by areas of pigmentation along the peripheral nerves of the SC tissue, in axilla (armpit) and groin, associated with scoliosis. Fifteen percent of NF type 1 patients presents with a variety of neurofibroma – a plexiform neurofibroma, which is associated with multiple masses or diffuse enlargement along the course of the nerve trunk. Because they grow lengthways the nerves and spread out, plexiform neurofibromas are difficult to remove without complete excision of the nerve. In the thorax they can occur along either sympathetic or somatic nerve trunks in the posterior mediastinum or along the cranial X (vagal) or phrenic nerves in the middle mediastinum. Lesions of other nerves of the chest wall may enlarge enough to cause deformity or organ displacement. Patients with plexiform neurofibromas are 20 times more likely to have a malignant peripheral nerve sheath tumor (MPNST) and those with NF1 type have a 10% lifetime risk of developing MPNST which often metastasize, and cause death.

NF 2 type are nonmalignant tumors of the peripheral nerves, caused by the glial or Schwann cells **[Theodor Schwann (1810 – 82)**. 19th century. Anatomy, histology, and embryology] that myelinate[359] that is produce the insulating myelin sheath covering the axons of nerve cells. The myelin is a lipid which speeds the conduction of action potential (AP). Proliferating out of control Schwann cells are called schwannomas. Although schwannomas are benign tumors, but become detrimental when growing tumor compresses the nerves, especially the sensory nerve axon, causing chronic severe pain. To this type belong acoustic neuroma – schwannoma of the cranial VIII (vestibulocochlear) nerve which causes tinnitus and often leads to hearing loss on the affected side.

NF 3 type or is a disorder causing multiple schwannomas of the spinal cord and peripheral nerves with subsequent severe neurological complications.

It was **Julius F. Cohnheim (1839 – 84)** — German Jewish pathologist, who proposed the theory that inflamation results from the local action of noxious agents which allow blood to enter tissues (1873), and the theory that tumors develop from embryonic rests which do not participate in the formation of normal surrounding tissue (teratoma).

Theodor Langhans (1839 – 1915) — German pathologist, described polyhedral epithelial cells constituting cytotrophoblast (Langhans layer) and formed by fusion of epithelioid cells (macrophages), giant cells containing nuclei arranged in a

---

[359] Myelin was discovered in 1854 by **Rudolph Virchow [(1821 - 1902)**: above] using a light microscope, when he noted a sheath around the nerve fiber.

horseshoe-shaped pattern in the cell periphery, seen in granulomatous diseases (Langhans cells, 1868).

Moritz Roth (1839 – 1914)[360] — Swiss pathologist, described seen by fundoscopy using ophthalmoscopy or slit-lamp examination in the retina near the optic nerve disc hemorrhages without surrounding run out of the blood vessel, in the center with round or oval white or pale spots of the precipitated fibrin, including platelets, indicating vascular ischemic (embolic) or hemorrhagic stroke with the subsequent healing, inflammatory infiltrate, infection or neoplastic cells (Roth spots, 1872). The Roth are seen in inflammation of the internal layer of the heart (endocardium), i.e., native valve endocarditis (NVE), leukemia, diabetes mellitus (DM), pernicious (in Latin – «pernicious», meaning dangerous, deadly) anemia, and hypertension (HTN).

Friedrich R.G. Wegner (1843 – 1917) — German pathologist, noticed a narrow whitish line at the junction of the epiphysis and diaphysis of a long bone, related to congenital syphilitic epiphysitis.

Classical studies of starvation, anoxia and scurvy were made by **Victor V. Pashutin (1845 – 1901)** — Russian pathologist, and on hypoxia (1884) by his pupil **Peter M. Albitsky (1853 – 1922)** — Russian pathologist, who also explained the heat exchange in fever (1918).

The important contribution to the blood circulation, i.e., blood and its clots or thrombi, embolism and infarction, was made by **Friedrich W. Zahn (1845 – 1904)**[361] — German Swiss pathologist, by description of false (non-ischemic) infarction of the liver without necrosis, characterized by reddish-blue colored areas with stasis and hepato-cellular atrophy after occlusion of the intrahepatic venous branches of the portal vein (Zahn's infarct); and also depiction of transverse lines on the free surface of thrombi created by alternation of grayish-whitish layers of platelets (thrombocytes) mixed with the fibrin, and narrow darker zones of a reddish-blue clot made by the red blood cells (RBC) indicating its formation inside the cavities of the heart or aorta where the blood flow is fast and during atrial fibrillation (AF) its detachment with flow and distal lodgment as an emboli (pale lines of Zahn). Conversely, in diseased smaller arteries or veins where blood flow is slower, the transverse lines on the surface of the thrombi are formed locally by the layer of platelets mixed with fibrin, they are pale, less visible, while those made by the RBCs are red, readily distinctive (red lines of Zahn).

---

[360] Roth M. Ueber netzhautaffectionen bei wundfieber. Deutsche Z Chir 1872;1(5);471-84.

[361] Zahn FW. Untersuchungen über Thrombose. Bildung der Thrombosen. Archiv fur pathologische Anatomie und Physiologie und für klinische Medicin 1875;62:81-124. — Zanh FW. Über die Folgen des Ferschlussen der Lungenarterien und Pfortaderäste durch Embolie. Vershandlungen der Gesellschaft Deutscher Naturforscher und Aerzte 1898;2(2):9-11.

It is necessary to differentiate between the proximal origin of thrombi - emboli (pale lines Zahn) and the distal origin of thrombi (red lines of Zahn's lines), since it may have serious medical and legal implications. Both should be treated emergently by systemic anticoagulation with heparin. However, in addition the former should be treated as emergency surgical removal of embolus (embolectomy) and the later should be treated by intraarterial thrombolysis. The exception to those are embolism of the pulmonary artery (PA) which depending on the degree of the PA stenosis will require either intraarterial lysis of an emboli or emergency open surgical removal of emboli from PA (pulmonary embolectomy, and iliofemoral venous thrombosis which should be treated also surgically by removal of thrombus (iliofemoral thrombectomy)[362].

Discovered by **Paul Langerhans (25.VII.1847, city of Berlin, Germany –** died from kidney infection **20.VII.1888, capital city Furchal, Island of Madeira, Northern Atlantic Ocean, Portugal)**[363] — German biologist, physiologist and pathologist, new antigen-presenting mononuclear stellate dendritic cells of the external layer of the dermo-epithelial junction and muscous membranes, containing large bodies or granules in an external part of the epithelial layer, so called the stratum granulosum (layer of Langerhans), responsible for the bodily immunity of a body (cells of Langerhans, 1868)[364]. In the next year he discovered numerous, c. 1,000,000 bundles of islets in the pancreatic gland of a normal human, each of which contain alpha, beta and delta cells, producing hormones insulin and glucagon, although insulin is made only by beta cells (islets of Langerhans, 1869)[365].

For the first time in 1875 **Edwin Klebs (1834 – 1913**: 19th century. Microbiology) had seen bacteria in the airway of patients who died from pneumonia. Seven years later (1882) **Carl Friedländer (1847 – 87)** — German microbiologist and pathologist, identified those bacteria as the cause of pneumoniae, named Klebsiella pneumoniae.

---

[362] Olearchyk AS. Insertion of the inferior vena cava filter followed by iliofemoral venous thrombectomy for ischemic venous thrombosis. J Vasc Surg 1987;5:645-7. — Olearchyk AS. Complex cases in cardiac surgery. Angiology 1993 ;44,7:S 1-42. — Olearchyk AS. Saddle embolism of the aorta with sudden paraplegia. J Ukr Med Assoc North Am 2004;49,2(152): 52-6. — Olearchyk AS. Saddle embolism of the aorta with sudden paraplegia. Can J Surg 2004;47(5):472-3.

[363] City of Berlin (founded in 13th century), Deutschland. — City-capital Funchal (established in 1452-54), Island of Madeira [area 801 km^2, highest point Pico-Ruivo 1862 m, discovered 1418-19, populated 1425], one of three islands of the North Atlantic Ocean within Autonomous Region of Madeira, Portugal (area 92,212 km^2).

[364] Langenhans P. Über die Nerven der menschlichen Haupt. Archiv für pathologische Anatomie und Physiologie und for klinische Medicin 1868;40(2-3):325-37.

[365] Langerhans P. Beiträge zur microscopischen Anatomie und Bauchspeicheldrüse. Inaugural – dissertation. Gustav Lange Verlag, Berlin 1869. — Langerhans P. Über den feineren Bau der Bauchspeicheldrüse. Doctoral thesis. Gustav Lange Verlag, Berlin 1869. — The cells of Langerhans contain a stick-like cytoplasmic organellae with the middline thickness of phonocytic origin, discovered in 1961 by **Michael S.C. Birbeck (1925 – 2005)** — English scientist and electronic microscopist, named in his honor Bilbeck's granules or bodies.

However, a year later (1883) he re-identified them as somewhat different bacteria, named Friedländer bacteria. Finally, in 1884 **Friedrich Löffler (1852 – 1915**: 19th century. Microbiology) came to conclusion that indeed bacteria seen by Edwin Klebs are Klebsiella pneumoniae. Still some researchers argue that those bacteria were Streptococcus pneumoniae.

In addition to that, Carl Friedländer (above) was the first to describe thromboangiitis obliterans (1876); and to introduce ampoule (ampule) into medicine (1886). He apparently died from Friedländer pneumonia.

Following the description of thromboangiitis obliterans by Carl Friedländer, it was re-described in 1879 by **Felix** von **Winiwarter (1852 – 1931)** — Austrian physician and surgeon, and in 1908 by **Leo Buerger (1872 – 1943)** — Austrian American pathologist, surgeon, and urologist.

In fact, Felix von Winiwarter in his description of thromboangiitis obliterans in a 57-year-old man detected an unusual obliteration of the arteries and veins of the leg by new growth of tissue from the intima, thus proposing the name «endarteritis» obliterans.

Thromboangiitis obliterans is disease of small and medium arteries and veins characterized by proliferation of the intima in young or medium aged (average age 41-42 years) usually dark-complexioned man (68-77%) or brunette woman (23-32%), addicted to cigarette smoking, leading to a slowly progressing gangrene of the lower extremities, necessitating bilateral multiple segmental amputations (thromboangiitis obliterans or Friedländer-Winiwarter-Buerger disease). The addiction to nicotine is so strong that they rather accept the multiple piece-by-piece amputations of the extremities than abstain from smoking.

Paul A. Grawitz (1850 – 1932) — German pathologist, described clear-cell renal carcinoma hypernephroma or Grawitz tumor, 1883), the most common tumor of the kidney (c. 80%) and the most lethal tumor of the genitourinary tract.

Volodymyr V. Pidvysotsky, a.k.a., **W. Podwyssotzki (24.V/05.VI.1857, village Maksymivka, Ichniansky Region, Chernihiv Obl., Ukraine – 22.I.913, city of St Petersburg, Russia)**[366] — Ukrainian pathologist, published the textbook «The Basics of General and Experimental Pathology» (Vol. 1-2. Moscow 1891-94, 1905).

Mikhail N. Nikiforov (1858 – 1915) — Russian pathologist, established the histogenesis of chorioendothelioma. His greatest contributions were devising a

---

[366] Village Maksymivka (founded 1600), Ichniansky Region, Chernihiv Obl., Ukraine. — Podwyssotzki W. Beitrage zur Kenntnis des feinerer Bauchspeicheldrüse. Arch mikr Anat 1882;21. — Rozhin I, Rozhin V. Volodymyr Pidvysotsky (1857 – 1913). J Ukr Med Assoc North Am 1957 May;4,1(7):43-6. — Ablitsov VH. «Galaxy Ukraine». The Ukrainian diaspora: prominent persons. Publ. «Kyt», Kyiv 2007. — Hanitkevych YaV. The contribution of the Ukrainian physicians to the world medicine. Ukr Med Chasopys 2009 VII/VIII;4(72):110-15.

method for the observation of the growth of anaerobes under the microscope, a method of staining microbes in frozen section, and a method of fixing blood films by placing them for 5-15 minutes in absolute alcohol, pure ether, or equal parts of alcohol and ether (Nikifotov method, 1885).

In 1888 **Eugene P. Ménétrier (1859 – 1935)**[367] — French pathologist, while performing post-mortem examination on malnourished patient noted massive hypertrophic gastric folds up to 3 cm high and up to 2 cm wide. This disease, named hypoproteinemia hypertrophic gastropathy or Menetrie's syndrome, is rare, of unknown etiology, acquired, premalignant, characterize by excessive mucous secretion with resultant protein loss and little or no acid production. Between the folds deep cracks occurs from which massive hemorrhage can occur. Presenting symptoms are epigastric pain, anorexia, nausea, vomit, and weight loss. Protein losing gastropathy develop in 20% to 100% of patients, accompanied by the low level of blood serum albumin and by excessive production of transforming growth factor alpha (TGF-alpha). Diagnosis is confirmed by the upper endoscopy and multiple deep biopsies of the gastric mucosa. The initial treatment should be with a monoclonal antibody against epidermal growth factor receptor (EGFR) – cetuximab. In severe disease or substantial protein loss a total gastrectomy may be required.

In 1886 **Felix Fränkel (?1863 – ?1912)**[368] — German physician, described a case of bilateral completely latent adrenal (suprarenal) gland tumor and concurrent nephritis with changes in circulatory system and retinitis, which **Ludwig Pick (1868 – 1944)** — German pathologist, named in 1912, a phaeochromocytoma (in Greek – «phaios =dark, chroma = color, kytos = cell, oma = tumor»).

Mykola F. Melnykov-Razvedenkov, a.k.a., Nicolai F. Melnikoff-Razvedenkoff (24.12.1866, Cossack village U'st-Medvedyts'ka, now town Serafynovych, Volhohrad Obl., Росія – died after complicated biliary tracts operation **20.12.1937, city of Kharkiv, Ukraine)**[369] — Ukrainian pathologist, proposed the method of preparation and conservation of the anatomical preparations with the preservation of its natural coloring (1896, 1899).

---

[367] Kurko VS. Gastrointestinal heorrhages of non-ulcerative origin. In, LYa Kovalchuk, VF Sayenko, HV Knyshov. Klinichna Khirurhiya. Vol. 1-2. Publ «Ukrmedknyha», Ternopil 2000. Vol. 2. P. 84-93.

[368] Fränkel F. Ein von doppelseitigen völig latent ferlaufenen Nebennierentumor und gleichzeitigen Nephritis mit Veränderungenam Circulationsapparat und Retinitis. Arch Pathol Anat Physiol Klin Med 1886;103:244-63.

[369] Cossack village U'st-Medvedyts'ka-on-Don River (founded 1589), now town Serafynovych, Volhohrad Obl. (area 112,877 km²), Russia. — Melnykov-Razvedenkov RF. New method of preparation of anatomic preparations. Med obozr 1899;2:278-81. — Ablitsov VH. «Galaxy Ukraine». The Ukrainian diaspora: prominent persons. Publ. «Kyt», Kyiv 2007.

In 1890 **S. Ia. Berezovsky**[370] — Ukrainian histologist and physician, in 1898 – **Carl von Sternberg (1872 – 1935)** — Austrian pathologist, and in 1902 **Dorothy M. Reed Mendenhall (1874 – 1964)** — American pediatric cellular pathologist, described the giant binuclear histiocytic cells, the nuclei of which resembles a mirror image of one and the other (Berezovsky-Stenberg-Reed cells) indicating Thomas Hodgkin (1798-1866) disease.

In pathological research concerning development of tumors, **Max Willms (1867 – 1918)** — German pathologist and surgeon, proposed the concept that tumor cells originate in the embryonal period and described a malignant mixed tumor of the kidney, consistent with embryonal elements, may develop in the embryo, typically occurring in children up to 5 years of live, rarely in adults, named nephroblastoma or Wilms tumor (1899). It may be bilateral (5%), usually do not cross the abdominal midline and metastasizes to the lung. He is credited for developing manometer for measuring cerebrospinal fluid (CSF) pressure.

Genetics

The basis of modern genetics was outlined by **Gregor J. Mendel (1822 – 84)** — Czech Austrian genetics, in his law of inheritance of a single gene trait (1865).

Fritz J.F.T. Müller (1821 – 97) — German biologist, and **Ernst H.Ph.A. Haeckel (1834 – 1919)** — German biologist and physician, formulated the biogenic law which states that anthogeny (morphology, ontogenesis), i.e., the origin and development of an organism from the fertilized egg to the mature form recapitulates (repeats) phylogeny that is the evolutionary development of species (Müller-Haeckel recapitulation law, 1886).

Physics and radiology

Between 1802-03 **Vasili P. Petrov [July 08(19), 1761, town of Oboyan', Kursk Obl., Russia – July 22 (Aug. 03) 1834, St Petersburg, Russia]**[371] — Russian Ukrainian physicist and electrotechnician, who graduated (1785) from Kharkiv College (founded 1722, closed 1817), city of Kharkiv, Ukraine, after construction the Galvani's battery [18 century. –**Luigi Galvani (1737 – 98)**], with power about 1,000 volt, composed with 4,200 copper and zinc little disc (orbs) approximately 35 mm in diameter and

---

[370] Kirianov NA. Origin of the tumor cells in lymphogranulomatosis. Arkh Patol 1982;44(4):71-74. — Ablitsov VH. «Galaxy Ukraine». The Ukrainian diaspora: prominent persons. Publ. «Kyt», Kyiv 2007.

[371] Town of Oboyan (founded 1639), Kursk Obl. (area 29,800 km²), Росія. — Ablitsov VH. «Galaxy Ukraine». The Ukrainian diaspora: prominent persons. Publ. «Kyt», Kyiv 2007.

2,5 mm in thickness, mixed with papers saturated by solution containing ammonium chloride (nashatyr), and first used isolation with a sealing wax, with the help of which he discovered the phenomenon of electric arc or the arc spontaneous, unloading in the form of a cord or digit, caused by high temperature between two electrodes, separated by a short distance, and accompanying by glaring shining in the form of the arc. After this discovery he indicated the practical use of the electrical arc for lighting, floating, welding and restoration of metals from oxides (oxidation). Showed the dependence of electric current strength on the diametral area of electrical conductor. Proposed covering of the electric conductors with isolating layer.

The principles and methods of the electric arc welding of metals, discovered (1882) by **Mykola M. Benardos [26.VI (08.VII) 1842, village Benardosivka,** now **village Mostove, Mykolaïv Obl., Ukraine – 08(21).X 1905, town of Fasтiv, Kyiv Obl., Ukraine]**[372] — Ukrainian inventor and electrotechnician- scientist of Greek origin, consist of connecting edges of two parts of metal by melting them using the heat from the electric arc rising out between welded metal and electrode. He created carbonic electrodes of different form (Benardos method) and combined electrodes «carbon – metal»; the contrivance (instrument) for vertical welding; first to use an electric magnet for securing the welded produce in proper position; build several welding semi automatons; worked up methods for submerged welding and cutting of metals, welding in the stream of gas, the precise and sewn contact welding.

The electric arc welding is being investigated in in surgery for performing esophageal, gastric, and intestinal anastomoses.

In 1857 **J. Heinrich W. Geissler (1832 – 79)** — German glassblower and physicist, invented a tube, made of glass, and used for creation of a low-pressure gas-discharge, named the Geissler tube.

In 1875 Sir **William Crookes (1833 – 1919)**[373] — English chemist and physicist invented an early experimental electrical discharge tube, with partial vacuum in which the cathode rays' stream of electrons, known as the Crookes tube.

---

[372] Village Benardosivka, now village Mostove (during carried out by Russian Communist Fascist authority of Holodomor-Genocide 1932-33 against the Ukrainian Nation died at least 47 inhabitants settled 18,000-23,000 years BC, founded 1390; affected by two terrible interwar Holodomor-Genocide carried out by Russian Communist Fascist regime 1921-32 and 1932-33 against the Ukrainian Nation, Bratsky Region, Mykolaïv Obl. (Foot Note 198), Ukraine. — Town of Fastiv (inhabited between 18,000 -23,000 years BC, in Paleolithic Age, founded 1390; suffered as a larger part of Ukrainian territory from organized by Russian Communist Fascist regime of Holodomor-Genocide 1921-23 i 1932-33 against the Ukrainian Nation), Kyiv Obl., Ukraine. — Ablitsov VH. «Galaxy Ukraine». The Ukrainian diaspora: prominent persons. Publ. «Kyt», Kyiv 2007.

[373] Crookes W. On the illumination of lines of molecular pressure, and the trajectory of molecules. Phil Trans Dec 1878;170:135-64.

In 1882 **Ivan P. Pului (Puluj)**, or **Johann Puluj (02.XI.1845, u.t.v. Hrymaliv, Chortkiv Region, Ternopil Obl., Ukraine – 31.I.1918, city if Praha, Czech Republic)**[374] — Ukrainian physicist, designed a gas-filled cathode fluorescence tubular lamp («Pulyui tubular lamp») that produced the Puluj rays or x-rays. With its use he performed x-ray photography of the fractured shoulder in a 13-year-old boy, and the hand of his daughter with a nail under it (1881-83) and a skeleton of the fetus who died on the 7th month of pregnancy (1889).

In 1895 **Wilhelm C. Röntgen (1845 – 1923)**[375] — German physicist, taking advantage of Ivan Pulyui own «Pulyui tubular lamp», «re-discovered» x-rays and photographed his wife's left hand with her wedding ring.

---

[374] U.t.v. Hrymaliv (founded 1595), Chortkiv Region, Ternopil Obl., Ukraine. — Puluj J. Strahlende Elektrodenmaterie. Wiener Berichtel. 1880;81:854-923; II. 1881;83:402-20; III. 1881;83:693:708; IV. 1882(85);871-81. — Puluj J. Strahlende Elektrodenmaterie und der Sogenannte vierte Aggregatzustand'. Verlag Carl Gerold Sohn, Wien 1883. — Puluj J. Radiant Elektrode Matter and die soCalled Fourth State. Physical Memoirs, London 1889;I(2):233-331. — Puluj J. On Kathode Rays. Proceedings Physic Soc, London 1895;14(1):178. — Puluj J. Über die Entstehung der Röntgen'schen Strahlen und ihre photographische Wirkung. Wiener Berichte II Abl. 1896;105;228-38. — Puluj J. Nachtatrag zur Abhandlung «Über die Entstehung der Röntgen'schen Strahlen und ihre photographische Wirkung». Wiener Berichte 1896;105:243-5. — Puluj J. New and Transformed Stars. Topography NTSh, Lviv. — Studnicka FJ, Čelakovský J. Puluj Ivan. V, Studnicka FJ, Čelakovský J. Ottuv slovník naučný. J Otto, Praha 1903;20:83-4. — Puluj Ivan. Not Completely Lost Power. Publ. Rus'ka Bookstore, Winnipeg 1919. — Blokh O, Hayda R, Platsko R. Röntgen or Pului? About one «whitespeckle» in history of science. Science and Society 04 1989;18-25. — Blokh OH, Гайда RP, Platsko RM. The Fate of Scientist in the fate of Ukraine. In, O. Romanchuk (Ed..). Axioms' for Descendants. Ukrainian Names in the Worlds Science. Collection of Outlines. Publ. «Memorial», Lviv 1992. C. 183-207. —Dolchuk MZ. Ukrainian physicist I.P. Pului – presedecessor of W.K. Roentgen. J Ukr Med Assoc North Am 1994 Autumn;41,3(134):181-4. — Hnatyshak HYeA. Ivan Pului Пулюй – Great son of Ukrainian Nation. Ukrainian Radiological Journal (URJ) 1995;3:60-61. — Dolchuk MZ. Before W.K. Röntgen there was Ukrainian Professor I.P. Pului. URJ 1995;3:59-60. — Milko M, Topchii TV. Ivan Pului – Ukrainian scientist and experimentator. URJ 1995;3:58-59. — Pylypenko MI, Artamanova NO. To the discovery of x-rays. URJ 1995;3:61. — Pundij P, Horokhovsky A. Discovery of x-rays belong to Professor Ivan Pului. UHMJ «Ahapit» 1995;2:65-72. — Radysh Ya. Fascinating tube of Ivan Pului . Question regarding priority in discovery of x-rays. Vashe zdorrov'ia 29.07 - 04.08.1995;29(115). — Spiegel P. The first clinical x-ray made in America – 100 years. Am J Roentgenology 1995;164(1):241-3. — Baranetsky A. More on Puluj's pioneering work. Ukrainian Weekly Feb 18,1996;64(7):7. — Puluj I. Collected Works. Publ. «Rada», Kyiv 1996. — Mayba II, Gaida R, Kyle RA, Shampo MA. Ukrainian Physicist contributed to the discovery of x-rays. Mayo Clin Proc 1997 June 30;72(7):658. — Kalyniak D. Ivan Pului, the discoverer of x-ray. Ukrainian Weekly July 9,2000;68,28:6. — Ablitsov VH. «Galaxy Ukraine». The Ukrainian diaspora: prominent persons. Publ. «Kyt», Kyiv 2007. — Holovach Yu, Honchar Yu, Krasnytska M, and al. Physics and physicists in NTSh in Lviv. J Physicist Research 2018;22(4):4003-31. https://doi.org/10.30970/jps.22.4003sm — Holovach Yu, Honchar Yu, Krasnytska M, and al. Physics and physicists in NTSh in Lviv. In, O. Petryk, A. Trokhymchuk (Eds). Leopolis Scientifica. Lviv 2019.

[375] Röntgen W. Ueber eine neue Art von Strahlen. Vorläufige Mitteilung, in: Aus den Sitzungsberichten der Würzburger Physik.-medic. Gesellschaft Würzburg 1895:137-45. — Röntgen W. Eine neue Art von Strahlen. 2. Mitteilung, in: Aus den Sitzungsberichten der Würzburger Physik.-medic. Gesellschaft Würzburg 1896:11-7. — Röntgen W. Weitere Beobachtungen über die Eigenschaften der X-Strahlen, in: Mathematische und Naturwissenschaftliche Mitteilungen aus den Sitzungsberichten der Königlich Preußischen Akademie der Wissenschaften zu Berlin 1897:392-406.

Almost 100 years after discovery by **Martyn Klaproth (1743 – 1817**: 18th century**)** of uranium, **Antoinne H. Beclere (1852 – 1908)** — French physicist, discovered that uranium salts emitted natural radioactivity, x-rays in their penetrating power, and demonstrated that this radiation, unlike phosphorescence, did not depend on an external source of energy but seemed to arise spontaneously from uranium itself.

In 1880 **Paul J. Curie (1856 – 1941)** — French physicist, and his brother **Pierre Curie (1859 – 1906)** — French physicist, had developed the electrometer, a sensitive device for measuring electric charge or electric potential difference.

Using Curie's electrometer, **Marie Curie Skłodowska (1867** – died from myelodysplastic syndrome of the bone marrow due to expose to radiation **1934)** — Polish French physicist, discovered that uranium rays caused the air around a sample to conduct electricity. The findings showed that the activity of uranium compounds depend only on the quantity of uranium and that the radiation was not the outcome of some interaction of molecules but must come from the atom itself.

On July 1898 Pierre Curie and his wife Marie Curie-Skłodowska announced the existence of element named «polonium», in honor of her native Poland, and in December this same year the Curies announced the existence of a second element, which they named «radium» (in Latin meaning «ray») and they also coined the word «radioactivity».

Marie Curie-Skłodowska developed the theory of radioactivity and technologies for isolating radioactive isotopes. Under her direction, the world's first studies were conducted into the treatment of neoplasm, using radioactive isotopes.

The pioneer of radiation oncology, **Victor Despeignes (1866 – 1937)**[376] — French physician, was the first who treated a 52-year-old-man suffering from carcinoma of the stomach, the size of about the head of an 8-months-old-fetus, by two exposures each lasting 30 minutes daily beginning July 4, 1896 using a Crooke's tube (above) and six rechargeable electric battery elements. The patient experienced pain relief and tumor shrunk by about 50%, but he died twenty days later (July 24, 1896).

On Feb. 17, 1896 **Gaston H. Niewenglowski (1871 – ?)** — French physicist, discovered luminous rays emitted by substances that have been exposed to sunlight **71** (Niewenglowski rays) which later this same year **Ivan P. Tarkhanov (1846 – 1908)** — Russian Georgian physiologist, used to investigate its influence on the central nervous system (CNS).

In 1897-98 **Walter B. Cannon (1871 – 1945)** — American physiologist and neurologist, adapted the x-ray technique using a bismuth barium mixture to study the digestive tract.

---

[376] Despeignes V. Observation concermant un cas de cancer de l'stomac traite par les rayons Röntgen. Lyon Medical - Gazette Medicale et Journal de Medicine Reunis 28 Jul 1896; 428-30. / 9 Aug 1896;503-6. / / Dec 1896;503-6.

X-rays were applied by **Antoine Beclere (1856 – 1939)** — French physicist, to diagnose an aortic arch aneurysm (1897), pulmonary tuberculosis (1898) and for radiotherapy (1902).

Pharmacology

In 1805 morphine **[Bernard Courtois (1777 – 1838)**. Biochemistry. 19th century] was the first alkaloid extracted and isolated by **Friedrich W.A. Sertürner (1783 – 1814)** — German pharmacist, from opium – a dried latex obtained from seeds capsules of the opium poppy *Papaver somniferum*. He named the bitter white crystalline alkaloid «morphium» after **Morpheus** (in Greek meaning «form or «shape) — Greek god of sleep and dreams. He studied morphine first in stray dogs and then on himself.

In 1847 **Ascanio Sobrero (1812 – 88)** — Italian chemist, has been investigating nitrocellulose, a.k.a. cellulose nitrate, gun (flash) cotton, paper or string, a low flammable and explosive compound formed by nitrating cellulose through exposure to nitric acid or another powerful nitrating agent, used as a propellant or low-order explosive. During investigation he discovered glycerin trinitrate, a.k.a. nitroglycerine (NTG) which he initially called «pyroglycerine». It was still stronger than black powder, a chemical explosive invented in the 9th century in China, and certainly found application in the military industry.

Through the work of **Julius K. Trapp (1814 – 1908)**[377] — Prussian Russian pharmacist and pharmacologist, **Rudolph Buckheim (1820 – 79)** — German pharmacologist, and **Oswald Schmiedeberg (1836 – 1920)** — Latvian German pharmacologist, that pharmacology became a separate field. The former contributed the calculation that the product obtained by multiplying the last two digits of the number expressing its specific gravity by 2 (Trapp coefficient) closely represents the number of grains of solids in one liter (I) of urine (Trapp formula).

Inventors of the industrial naphtha-remake **Ivan (Johann, Jan) Zeh / (02.IX.1817, town of Lańcut, Subcarpathian Voivodeship, Poland – 25.I.1897, city of Boryslav, Lviv Obl., Ukraine)**[378] — Polish Ukrainian pharmaceutics and **Ihnatiy Lukasevych** or **Ignacy Łukasiewicz (1822-82)** — Polish Ukrainian pharmaceutics and chemist-technologist of Armenian origin, in the Pharmacy of **Petro Mirolash** — Ukrainian pharmacist – «Under the Gold Star» (opened 1826) in its chemistry pharmaceutics

---

[377] Trapp JK. The Handbook of Pharmacognosy. Vol. 1-2. St. Petersburg 1863. — Trapp JK. Pharmaceutical Operations and Recipes. St. Petersburg 1876, 1800. — Trapp JK. Pharmaceutical Chemistry. Vol. 1-2. St. Petersbug 1882-85.

[378] Town of Lańcut (existence of humans since 4000 BC, founded 1381), Subcarpathian Voivodeship, Poland. — City of Boryslav (existence of humans since 1000 BC, first written mention 1387; in middle of 19th century became important center of extraction and processing of naphta), Lviv Obl., Ukraine.

laboratory, Lviv, Ukraine, performed in 1852 distillation (rectification) and purification of the mountain (rocky) naphtha (in Old- Ukrainian – ropa; in Latin – petroleum) by chemical means, using concentrated sulfuric acid sodium solution, so it would be useful for technical purposes. Indeed on 30 of March 1853 at the «Under the Gold Star» Pharmacy in a tin-lamp made by tinsmith **Adam Bratkowski**, for the first time «lightened» naphtha-product for illumination of its display window. On 31 of July 1853 the Main city of Lviv Hospital began illumination all placings by naphtha-lamps, and at night same day with its help were illuminated the first emergency surgical operation.

Potassium permanganate ($KMnO_4$) was re-discovered by **Henry B. Condy (1826 – 1907)** — English Hungarian chemist and industrialist, in 1857, a.k.a., ozonized water or Condi fluid **[Johann Glauber (1604 – 70)**. 17th century], for internal or external use, to rinse the oral cavity and gargle the throat to treat sore throat, prevention, and treatment of scarlet fever, treat number of skin conditions such as washing and debridement of superficial wound, burns and ulcers, dermatitis, fungal infection of the foot, impetigo, pemphigus, tropical ulcers.

With the name of **Sydney Ringer (1835 – 1910)** — English physiologist, pharmacologist and clinician, is connected the isotonic, alkalizing (pH 6,5) Lactated Ringer Solution (LRS; content – Na^+ of 130 mEq, Cl^+ of 109 mEq, lactate of 28 mEq, K^+ of 4 mEq and Ca^{++} of 3 mEq / 1000 ml H_2O; 1882-85), used for an IV infusion to correct an acute decrease of an intravascular fluid volume depletion due to blood loss, trauma, surgery and burn injury.

Alkaloid ephedrine was first isolated in 1885 by **Nagal Nagayoshi (1844 – 1929)** — Japanese organic chemist, from Ephedra vulgaris, an everlasting green shrub in the family Ephedraceae, about 25 cm to 50 cm hight, that grows in southern Europe, beginning from Portugal (area 92,212 km^2), through the central and eastern Europe and towards the western and central Asia, including Kazakhstan (area 2,724.900 km^2). Ephedrine is sympathomimetic amine, acting through sympathetic nervous system by stimulation of the adrenergic receptors to increase the activity of norepinephrine (noradrenaline) at postsynaptic alpha and beta receptors. It has been administered by mouth (per orum), subcutaneously (SC), intramuscular (IM) and intravenously (IV) to treat nasal congestion, hives, bronchitis, bronchial asthma due it broncho-dilatator effect, narcolepsy, serum sickness, obesity by weight loss, and to prevent low blood pressure (BP) during spinal anesthesia.

Thirty-two years after the discovery of nitroglycerine (NTG) by Ascanio Sobrero (above), in 1879 **William Murrell (1853 – 1912)** — English physician, clinical pharmacologist and toxicologist, recognized the clinical benefits of the prepared by the Liebes' process NTG in the management of patients suffering of angina pectoris, hypertension and heart failure[379].

Internal medicine. Medicine

Doctor of Medicine **Josyp K. Kamenetsky (04/15.IV.1750, town Semenivka, Novhorod Siversky Region, Chernihiv Obl., Ukraine – 14/26.VI.1823, city of Moskva, Russia)**[380] — Ukrainian physician, who authored «A Short Counsel for the Treatment with Simple Resources» (Moscow 1803).

Systematic description of the clinical symptoms, complications, and treatment of the peptic ulcers of the stomach and the duodenum was given by **Fedir Uden (1754 – 1823**: 19th century. Pathology**)** in the treatise «The Academic Lectures about Chronic Diseases» (Part 1-7. St Petersburg 1816-22). Moreover, he specified the clinical picture consisting of: (1) ulcer of the stomach or duodenum with abdominal pain and meteorism and gastro-esophageal reflux disease (GERD) and (2) precordial heaviness and pain (angina pectoris) with radiation to the right arm, tachycardia, extrasystole, dyspnea and a low BP (syndromus gastro-cardialis or Uden syndrome, 1817).

In 1802 **Jean-Pierre Maunoir (1768 – 1861)**[381] — French Swiss physician and surgeon-oculist, described and named dissecting aneurysm of the thoracic aorta with the term dissecting aneurysm («dissequant anevrisme»). He also described ligature of arteries.

F.J. Victor Broussain (1772 – 1838) — French pathologist and physician, substituted leaches for bloodletting (1808-28).

Mytrofan M. Yellinsky (1772, c. Kurs'ke, Starooskolsky Region, Bilhorod Obl., Russia – 03.III.1830, St Petersburg, Russia)[382] — Ukrainian physician, described

---

[379] Murrell W. Nitro-glycerine as a remedy for angina pectoris. Lancet 1879;113(2894):225-7. — Murrell W. Nitro-glycerine as a remedy for angina pectoris. HR Lewis, London 1882.

[380] Town of Semenivka was founded in 1680 by **Semen I. Samoylovych c. 1660 - 1685)** — Colonel and appointed Hetman of the Starodub Company of 100 soldiers (Sotnia) / Regimen (1648 – 1781), southerly (50 km) from town of Starodub (found 1080) – center of one of 10 administrative regiments of the Ukrainian State, the Hetmanate Age (1649 – 1764), Bryansk Obl. (area 34,900 km²), Ukraine / Russia. In 1926 Semenivka was incorporated into Novhorod Siversky Region, Chernihiv Obl., Ukraine. — Timoshok A. Physician-in-ordinary from Semenivka (Details on J.K. Kamenetsky's Portrait). UMMJ «Ahapit» 2004;14-15:76-9. — Ablitsov VH. «Galaxy Ukraine». The Ukrainian diaspora: prominent persons. Publ. «Kyt», Kyiv 2007. — Synyachenko OV. They glorified themselves and Ukraine: history of domestic therapy. Crimean Therapeutic Journal 2010;2(2):21-8.

[381] Maunoir JP. Memoires physiologiques et pratiques sur l'anevrisme et la ligature des arteres. JJ Paschound, Geneva 1802.

[382] Village of Kurske (founded 17th century), Starooskolsky Region, Bilohorodska Obl. (area 27,134 km²), Russia. —Ablitsov VH. «Galaxy Ukraine». The Ukrainian diaspora: prominent persons. Publ. «Kyt», Kyiv 2007. — Synyachenko OV. They glorified themselves and Ukraine: history of domestic therapy. Crimean Therapeutic Journal 2010;2(2):21-8.

chronic constrictive calcified pericarditis («The Armored Heart») as well as wrote «About the ureteral stones» and «About an accessory hepatic duct».

The first reference to autism from 1797 belongs to **Jean M.G. Itard (1774 – 1838)** — French physician, with a particular reference to boy named **Victor (? – 1828)**, or so called «Wild Boy of Avalon» in France, who showed several signs of autism and assumed to have been living his entire childhood alone in the woods, and he treated him with a behavioral program designed to help form social attachment and to induce speech via imitation.

Although pneumothorax was first described by **Serafeddin Sabuncuoglu (1385 – 1468:** Renaissance), the term was coined by Jean Itard in 1803, when he called attention to five cases in which free air was found by percussion and auscultation in the thorax following trauma.

By 1803 **John C. Otto (1744 – 1844)**[383] — American physician, became aware that hemophilia is a hereditary x-linked disease, existing in certain families, transmitted by healthy females, affecting mostly males whom he called «bleeders».

John Cheyne (1777 – 1836) — Scottish physician and surgeon, in 1818 and **William Stokes (1804 – 78)** — Irish physician, in 1854, described the rhythmic waxing and waning of the breathing depth with regular periods of apnea, as observed in comatose patients with the central nervous system (CNS) affections (Cheyne-Stokes respiration).

Carditis is the inflammation of the heart subdivided into the inflammation of the pericardium (pericarditis) caused by viruses, bacteria such as tuberculosis, uremia, post myocardial infarction (MI) carcinoma, autoimmune disease, chest trauma, post-cardiotomy, or of unknown cause; the inflammation of the heart muscle (myocarditis) due to viral, bacterial infection or by immune disorders; the inflammation of the endocardium (endocarditis) which usually involve heart valves caused typically by infection (bacteria or fungi) or non-infective (thrombotic endocarditis); or the inflammation of the entire heart involving the pericardium, myocardium and endocardium, carditis (polycarditis), affecting mainly children suffering from rheumatic fever, and adults affected by chronic rheumatic disease (RD). Carditis is also used to describe inflammation of the distal esophagus and proximal stomach (cardia) because of long-standing gastro-esophageal reflux disease (GERD), first described by **Friedrich Uden (1754 – 1823:** above).

A student of **Joseph Frank (23 Dec. 1771, town of Rastatt, State Baden-Württemberg, Germany – 18 Dez 1842, city of Como, Lombardy, Italy)** — German

---

[383] Otto JC. An account of a haemorrhagic disposition in certain families. Medical Depository. Publ T & J Swords, New York 1803;5:1-4. — Otto JC. An account of a n haemorrhagic disposition in certain families. Am J Med 1951 Nov;11(5):557-8.

Lithuanian physician, professor of pathology (1804-23), Viliaus universitetas, Vilnius, Lithuania, founder of many medical and charitable institutions in Vilnius, **Khoma I. Volkovynsky / Tomasz A. Wołkowiński (1778, village of Khoriv, Lokachynsky Region, Volyns'ka Obl., Ukraine – 1841)**[384] — Ukrainian physician, graduate (1817) of University of Warsaw (1817), defended his doctoral disseration «Carditidis rheumaticae» (Viliaus universitetas, Vilnius 1817), where he first proved direct relation between rheumatic disease (RD) and carditis with enlarged heard (cardiomegaly), i.e., rheumatic heart disease (RHD).

Asbestos was discovered by **M. Hinrich C. Lichtenstein (1780 – 1857)**[385] — German physician, explorer, and zoologist, during his travels in 1803-06 in the vicinity (32 km^2 south-westerly) of the town Prieska (founded 1882) in the Northwest Cape Province (area 372,889 km^2) of South Africa (area 1,221.037 km^2). From that area the deposits of asbestos extend further north-east in the slopes of the hills range named the Asbestos Mountains, to the town of Kurukam (missionary post 1821, founded 1884), where the ranges are known as the Kuruman Hills (area of both ranges is approximately 4,968.9 km^2).

Of the two main types of asbestos, the first type is predominant (95%) chrysolite or white asbestos with short and serpentine fibers, and the second type, less common (5%), amphibole asbestos contains subtypes, including crocidolite (blue) and amosite (brown-green) with long, thin and straight fibers, extracted miners only in the mines (pits) of the South Africa (Cape asbestos) and the Western Australia (area 7,692.024 km^2), which prove a connection of those continents in the past.

The first identification of malignant mesothelioma as a discrete disease apparently occurred in 1870, and the mining in the Asbestos Mountains has begun in 1893. After extraction in pits asbestos was processed in the carding rooms (in Latin – «carduus» meaning «thistle» or «teasel») of the asbestos plants, and then widely used in various industries and construction works, where asbestos fibers were inhaled into the airways by the miners, industrial workers, builders, and everyday inhabitants.

---

[384] Town of Rastatt (known from before 1689), State of Baden-Württemberg, Germany. — City of Como (Roman foundation 196 BC), Lombardy (area 23.844 km^2), Italy. — Viliaus universitetas (founded 1579), city of Vilnius (settled 2000 BC, founded 13th century, first written mention 1323), Lithuania (area 65,301 km^2). — Village Khoriv (first mentioned 1545), Volodymyr Volyns'kyy Region, Volyns'ka Obl., Ukraine. — University of Warsaw (established 1816), city of Warsaw, Poland. — Volkovynsky Kh. Carditidis rheumaticae historia. Viliaus universitetas 1817). — Stemborowicz W. Doctor's degree thesis of Tomasz Adolf Wołkowiński — «Caditidis rheumaticea historia». Arch Hist Filoz Med 2001 Feb;64(1):3-8. — Ablitsov VH. «Galaxy Ukraine». The Ukrainian diaspora: prominent persons. Publ. «Kyt», Kyiv 2007. — Synyachenko OV. They glorified themselves and Ukraine: history of domestic therapy. Crimean Therapeutic Journal 2010;2(2):21-8.

[385] Lichtenstein H. Reisen in südlichen Africa: in den Jahren 1803, 1804, 1805, und 1806 / Travels in Southern Africa in the Years of 1803, 1804, 1805, and 1806. Translated from the original German by Anne Plumptre. H Colburn, London 1812-1815.

The first recorded case of asbestos-associated disease was seen in 1899 by **H. Montague Murray (1855 – 1907)** — English chest physician specializing in occupational diseases, in a man with marked shortness of breath who was employed for 12 years in the carding room of established asbestos factory. He died a year later and at autopsy fibers minerals were seen, together with asbestos bodies.

Founder of modern medicine, **René T.H. Laennec (1781 – 1826)**[386] — French physician, greatly contributed to the understanding of peritonitis and cirrhosis of the liver which he named (in Greek – «kirrhos», means «tawny»), referring to the tawny, yellow nodules seen in this disease; described three stages of alcoholic and polyarthritis (Laennec) liver cirrhosis (fatty infiltrations, fibrosis and formation of nodularity's); named a very malignant tumor of the skin melanoma (in Greek – «mela», «melan» for «black») and described metastases of melanoma to the lungs (1804); postulated the relationship between preexisting blebs and their unprovoking rupture, hence the name spontaneous pneumothorax (1819); used the term «aneurysme dissequant», or dissecting aneurysm, believing that this entity represent the early stage of a saccular aneurysm of the aorta (1819).

But his greatest contribution to medicine is the invention and construction of the wooden monoaural stethoscope for listening to the chest (auscultation), thus he was able to define rales, rhonchi, crepitations and egophony, and hear the heart sounds.

He defined bronchiectasis as a permanent dilatation of the bronchi, caused by a recurrent transmural infection and inflammation, characterized by cylindrical or tubular form due to tuberculosis, saccular or cystic secondary to obstruction or bacterial infection, most located in the left lower lobe (LLL), lingula or the right middle lobe (RML) of the lung (1819); and of pulmonary embolism (PE, 1819).

He truthfully depicted the clinical situation that led him to the invention and construction of the stethoscope: «*I was consulted in 1816 by a girl who presented the general symptoms of heart disease and in whom palpation and percussion gave little information on account of the patient's obesity. Her age and sex forbade an examination [by direct auscultation]. Then I remembered a well-known acoustic fact, that if the ear be applied to one end of a plank it is easy to hear a pin's scratching at the other end. I conceived the possibility of employing the property of matter in the present case…I was surprise, and pleased to hear the beating of the heart much more clearly than if I had applied my ear directly to the chest*».

In 1813 **John E. Hay**[387] — American physician, found that males affected by hemophilia could pass the trait of this disease into their unaffected by hemophilia daughters.

---

[386] Laennec R T H. De l'Auscultation Médiate ou Traité du Diagnostic des Maladies des Poumons et du Coeur. Brosson & Chaudé, Paris 1819.

[387] Hay JE. Account of remarkable haemorrhagic disposition, existing in many individuals of the same family. N Engl J Med Surg 1813 July;2(3):221-5.

Pierre Ch.A. Louis (14.IV.1787, former municipality Mareuil-sur-Ay, department Marne, region Grand Est, France – 22.VIII.1872, city of Paris, France)[388] — French Ukrainian pathologist and physician, who maintained in 1816-20 successful practice of medicine in Odesa, Ukraine; introduced the statistical «numerical method», that is the concept that knowledge about a disease, its history, clinical picture and treatment, could be derived from aggregated patient data; argued that all therapies be open to scientific evaluation, and on that basis proved that bloodletting was harmful to healing; advocated «méthode expectorante» in place of bloodletting (1835). He re-described, after **Daniel Ludwig (1625 – 80**: 17th century**)** the angulus sterni (the angle of Ludwig-Louis), formulated law that tuberculosis begins in the left upper lung and tuberculosis of the rest of the body is usually associated with the involvement of both lungs (Louis law).

The primary interest of **Richard Bright (1789 – 1858)**[389] — English physician, was a combination of (1) renal dysfunction due to acute or chronic glomerulonephritis, presenting by the presence in urine of the serum albumin (albuminuria), as well as red blood cells (erythrocytes) and casts, that can lead to end-stage kidney disease (ESRD), (2) heart disease with «edema (dropsy) and protein in urine, i.e., «coagulable urine», and (3) hypertension (HTN) due to increased systemic vascular resistance (SVR), known as Bright disease (1827-43).

He correlated clinical features with morphological changes found in autopsy. His description of an increased SVR is precise as for the time of his writing: «...*some less local causes, for the unusual efforts to which the heart has been impelled; and the two most ready solutions appear to be, either that the altered quality of the blood affords irregular and unwanted stimulus to the organ immediately; or, that it so affects the minute and capillary circulation, as to render greater action necessary to force the blood through the distant sub-divisions of the vascular system*».

---

[388] Former municipality Mareuil-sur-Ay (area 11.48 km²), former department Champagne-Ardenne, now Marne, region Grand Est (area 57,433 km²), France. — Ablitsov VH. «Galaxy Ukraine». The Ukrainian diaspora: prominent persons. Publ. «Kyt», Kyiv 2007.

[389] Bright R. Reports of Medical Cases, selected with a View of Illustrating the Symptoms and Cure of Diseases by a Reference to Morbid Anatomy. Vol. I. Longrams-Green, London 1827. — Bright R. Cases and observations, illustrative of renal disease accompanied with the secretion of albuminous urine. London Medical Gazette 1835-36;18:72-4. — Bright R. Cases and observations, illustrative of renal disease accompanied with the secretion of albuminous urine. Guy's Hospital Reports 1836;1:338-79. — Bright R. Tubular view of the morbid appearance in 100 cases connected with albuminous urine, with observations. Guy's Hospital Reports 1836;1:380-400. — Bright R. Observations on abdominal tumours and intumescenses; illustrated by cases of renal disease. Guy's Hospital Reports 1839;4:208-64. — Bright R. Cases and observations, illustrative of renal disease accompanied with the secretion of albuminous urine. Memoir the second. Guy's Hospital Reports 1840;5:101-61. — Bright R. Letter concerning the pathology of the kidney. Guy's Hospital Reports 1841-42;29:707-8. — Bright R. Accounting the observations made under the superintendence of Dr. Bright on patients whose urine was albuminous. Guy's Hospital Reports 1843;1:189-316.

Postinfarction ventricular septal defect (VSD) – anterior (50%), posterior (30%) and apical (20%) complicates 1-2% of cases of an acute myocardial infarction (MI) and accounts for 5% of deaths. Posterior postinfarction VSD was first diagnosed at autopsy in 1845, by **Peter M. Latham (1789 – 1875)**[390] — English physician, who also depicted the circle with the diameter of 5 cm, an absolute cardiac dullness located between the left nipple and the lower border of the sternum (circle of Latham).

Petro P. Pelekhin or **Pelekh (1789, village Makiïvka, Smiliansky county, Cherkasy Obl., Ukraine – 22.IX.1871, city of Kyiv, Ukraine)**[391] — Ukrainian physician, defended dissertation for the degree of DMS «About Neurosis» (Edinburgh University (1892), Scotland in which expressed thought about absence of the permanent connection between neurosis and the type of human body constitution.

In 1818 **James Blundell (1790 – 1878)** — English physiologist, physician, gynecologist, and obstetrician, performed the first successful blood transfusion on a woman with post-partum hemorrhage, the donor being her husband who gave four ounces (556 ml) of blood directly from an incision of the vein (venesection) of his arm directly to the venesection of his wife arm using syringe. Between 1818 and 1830 he conducted 10 direct blood transfusion of which five were successful.

Prokhor O. Charuki(o)vsky (1790, village Polohy, Poltavs'kyy Region, Poltavs'ka Obl., Ukraine – 11.VI.1842, city of St Petersburg, Russia)[392] — Ukrainian physician-therapeutant, after graduation in 1802-10 from Chernihiv Ecclesiastical Seminary (active 1776-1917), in 1812-16 from Petersburg Medico-Surgical Academy (established 1798), remained in Alma Mater as Assistant-Professor of mathematics and therapy until set for further professional experience abroad, where he mainly was occupied with physiology, pathology, and therapy (1818-22). After returning to St Petersburgh in 1922, he was assigned Adjunct-Professor of therapy and directed into military-land hospital, where he defended dissertation «De haemoptysi»

---

[390] Latham PM, editor. Lectures on subjects connected with clinical medicine, comprising diseases of the heart. Longmans, Brown, Green & Longmans, London 1845-6; Vol. 2:168-76.

[391] Village Makiïvka (created beginning17th century), Cherkasy Obl., Ukraine. — Pelekhin P. De neurosibus in genere. Diss. d.m., Edinburgii 1829. — University of Edinburgh (established 1583), city of Edinburgh [archeological remains of the camp from the Middle Stone (Mesolithic) Age c. 6,500 BC, first written reference dated 320 BC], Scotland (area 77,933 km²). — Ablitsov VH. «Galaxy Ukraine». The Ukrainian diaspora: prominent persons. Publ. «Kyt», Kyiv 2007. — Synyachenko OV. They glorified themselves and Ukraine: history of domestic therapy. Crimean Therapeutic Journal 2010;2(2):21-8.

[392] Village Polohy (founded 1859), Poltavs'kyy Region, Poltavs'ka Obl., Ukraine. — Charukovsky PA. General Pathologic Semiotics, or the Doctrine of General Signs of Diseases. Typography of Imperial Educational House, St Petersburg 1825 & 1841. — Charukovsky PA. About Stethoscope and Signs, with Help of which they are Detected. St Petersburgh 1828. — Charukovsky PA. Treatise for System of Practical Medicine. St Petersburg 1833. — Ablitsov VH. «Galaxy Ukraine». The Ukrainian diaspora: prominent persons. Publ. «Kyt», Kyiv 2007. — Synyachenko OV. They glorified themselves and Ukraine: history of domestic therapy. Crimean Therapeutic Journal 2010;2(2):21-8.

and was honored with the degree of DMS (St Petersburg 1823), from 1838 was appointed Professor of clinical therapy. Promoted into internal medicine percussion and auscultation as methods of objective examination of the patient's body in practice and writing («Stethoscope and signs, which with its help are uncovered». St Petersburg 1828-33).

The complete heart block was first described by **Marko Gerbec (1658 – 1718:** 18th century**)** in 1717, then by **Giovanni Morgagni (1682 – 1771**: 18th century**)** in 1761, and that was confirmed later to be correct by **Robert Adams (1791 – 1875)** — Irish surgeon, and **William Stokes (1804 – 78;** above**)**, and should be called Gerbec-Morgagni-Adams-Stokes syndrome, or sick sinus syndrome (SSS).

The association of pernicious anemia with adrenal insufficiency was noted by **Thomas Addison (1793-1860)** — English physician, named Addison's disease.

Johann L Schönlein (1793 – 1860) — German naturalist and physician, and **Edward H. Henoch (1820 – 1910)** — German physician, depicted nonthrombocytopenic purpura.

The topographical method of percussion was practiced by **Pierre A. Piorre (1794 – 1849)**[393] — French physician, who was the first to describe in 1828 relative and absolute cardiac dullness and to estimate the size of the heart, based on his percussive method. That included introduction of the pleximeter into percussion, initially made of wood, later of ivory, which was placed on the chest with one hand and struck with the finger of the other hand. He advocated the use of percussion rather than auscultation, suggesting that the former was more accurate in determining underlying morbid anatomic change. However, he was not narrow minded in stating that *«To employ one mode of exploration to the exclusion of others would be proof of poor judgment»*.

J.F. Wilhelm Malcz (1795 – 1852) — Polish physician of German origin, described a sign symptomatic of louse-born typhus abdominal fever, when patient is unable to move forward his tongue when asked to do so (1847).

Jean B. Bouillard (1796 – 1881)[394] — French physician, introduced the term endocarditis but did not distinguish between the infective and rheumatic basis for the lesions he observed.

Endocarditis is an exudative and proliferative inflammatory change of the endocardium, usually characterized by the presence of vegetations on its surface or therein and often affects the heart valves, though sometimes injure the inner lining of the cardiac cavity, or other areas of endocardium. It may arise as a primary

---

[393] Piorre PA. De la percussion mediate et des signes obtenus a la l'aide de ce nouveau moyen d'exploration, dans les maladies des organes thoraciques et abdominaux. JS Chaude, Libraire, et al., Paris 1828.

[394] Bouillaud J. Traité clinique des maladies du coer. Paris 1835.

disorder or as a complication of another disease or in connection with it. Against the background of acute feverish rheumatic disease (RD) with joint involvement, can flare up rheumatic endocarditis with spread to the myocardium, but usually the heart valve became involved, and then this condition is called rheumatic valvulitis (Bouillard syndrome). He distinguished within it three stages: (1) «sanquinary congestion, softening, ulceration and suppuration»; (2) «a period of organization of secreted products or a portion of fibrinous concretions»; and (3) «cartilaginous, osseous or calcareous induration of the endocardium in general and the valves in particularly, with or without narrowing of the orifices of the heart».

Independently, in 1835 **Robert J. Graves (1796 – 1853)**[395] — Irish surgeon, and in 1840 **Karl A.** von **Basedow (1799 – 1854)**[396] — German physician, described of diffuse toxic enlarged thyroid gland (TG) – goitre or goiter (in Latin – «gutturia», meaning throat; DTG), named the Graves-Basedov disease in honor of both physicians (Ancient Medicine. – *China*).

The DTG is an autoimmune disease of the TG and is the most common cause of its increased bodily function (metabolism), i.e., hyperthyroidism. It is caused by an antibody, called thyroid-stimulating immunoglobulin (TSI), similar in effect to thyroid stimulating hormone (TSH), a glycoprotein hormone secreted by thyrotrope cells in the anterior pituitary gland, causing the TG to secrete excess of hormones named triiodothyronine (T_3) and thyroxine (T_4), partially composed by iodine ($_{53}I$), its growth and vascularization. T_3 and T_4 stimulate the basic metabolism rate (BMR). A deficiency in $_{53}I^-$ leads to decreased secretion of T_3 and T_4, enlarges the thyroid gland and will cause simple or multinodular goiter.

The characteristic symptoms of DTG are irritability, nervousness, weakness, intolerance to heat, increased appetite with weight loss, vomiting, diarrhea, mild elevation of the body temperature, dryness of the skin, profuse sweating, tremor of the hands, stare, bulging (protrusion) of the eyeballs (exophthalmos), abnormal widening of the palpebral fissure, unilateral lagging [19th century. Diseases of the eye. – **Friedrich** von **Gröefe (1828 – 70)**] or retraction [19th century. Diseases of the eye. – **John Dalrymple (1803 – 52)**] of the upper eyelid on downwards rotation of the eye, the «Merseburger triad»[397] of tachycardia, goiter and exophthalmos, cardiac arrhythmia, elevated BMR and thyrotoxic coma.

---

[395] Graves RJ. Newly observed affection of the thyroid gland in females. London Med Surg J 1835;7(2):516-7.
[396] Von Basedow KA. Exophtalmus durch Hypertrophic des Zellgewebes in der Augenhöhle. Wochenschrift für die gesampte Heilkund. 1840;6:197-204, 220-8.
[397] The «Merseburger triad» was described in 1840 by von Karl Basedow to distinguish the town of Merseburg (founded 850), District of Saalekreis, State of Saxony-Anhalt, Germany, from where apparently the patients with DTG were referred to him for diagnosis and treatment.

On the other side both lids lag, retraction may occur in hyperthyroidism from other causes.

A severe form of hypothyroidism may cause myxedema, characterized by deposition of mucopolysaccharides in the dermis (skin), resulting in swelling of the affected area, like pre-tibial non-pitting edema and exophthalmos, which are a hallmark of this disease.

Younger brother of **Prokhir Charuki(o)vsky (1790 – 1842**: above), **Yakym O. Charuki(o)vsky (1798, village Polohy, Poltavs'kyy Region, Poltavs'ka Obl., Україна** – died from aneurysm of the heart **1848)**[398] — Ukrainian physician surgeon, published the treatise «The Military-Campaign Medicine» (Part 1-5. St Petersburg 1836-37), contributing to development of military medicine.

Prosper Meniere (1799 – 1862) — French physician, described nonsuppurative disease of the ear's labyrinths, manifested by hearing loss, tinnitus, and vertigo (Meniere disease).

The theory of multiple human racial creations from variety of species (polygenism) originated in antiquity from the culture and mythology of the African tribes Mbuti and Bambuti, Pygmies of Congo (area 2,345, km^2) which was populated as early as 90,000 years ago. Both tribes live in the tropical rainforest Ituri covering 70,000 km^2 in the north/northwestern part of Congo.

In contrast, the theory of single human race creation from one of species (monogenism) is found in the «Bible».

Samuel G. Morton (1799 – 1851)[399] — American physician, natural scientist, writer and the originator of «American School» of ethnology, supported a theory of multiple human racial creation. However, he also claimed the «Bible» supported polygenism, and within working in a biblical framework his theory held that each race had been created separately and each was given specific irrevocable characteristics. By measuring the volume of the posthumous brains of the three major human races, he estimated that the Caucasians (Whites) had the biggest brains (averaging 1,426 cm^3), American Indians were in the middle (averaging 1,344 cm^3), and African American (Black or Negroes) had the smallest brains (averaging 1,278 cm^3).

For comparison, on the postmortem examination, the brain of **Albert Einstein (1879 – 1955)** — Jewish, German American theoretical physicist, who developed

---

[398] Ablitsov VH. «Galaxy Ukraine». The Ukrainian diaspora: prominent persons. Publ. «Kyt», Kyiv 2007. — Synyachenko OV. They glorified themselves and Ukraine: history of domestic therapy. Crimean Therapeutic Journal 2010;2(2):21-8.

[399] Morton SG. Crania Americana; or a Comparative View of the Skulls of Various Aboriginal Nations of North and South America. J Dobson, Philadelphia, and Sipkin, Marshal & Co, London 1839. — Morton Morton SG. Crania Americana. An Inquiry into the Distinctive Characteristics of the Aboriginal Race of America and Crania Aegyptiana. Vol. 1-3. Publ: Gale, Sabin Americana, Philadelphia 1839-49. — Morton SG. An Illustrated System of Human Anatomy. Publ: Grigg, Elliot, Philadelphia 1849.

the theory of relativity, weighted 1,230 grams (g), towards the low end of a normal man of his age.

Contribution to medicine of **Armand Trousseau (1801 – 67)**[400] — French internist, include description of a sign of spontaneous recurrent or appearing in different locations over time (migratory) superficial venous thrombotic inflammation of the chest wall and arm associated with lung, stomach or pancreatic cancer (superficial migratory thrombophlebitis), and nonbacterial arterial embolic or thrombotic endocarditis, or both (Trousseau sign of malignancy, or Trousseau's syndrome, 1860's).

The other sign of Armand Trousseau is of nerve hyperexcitability of latent spastic contractions (tetany) of muscles observed in patients with low blood plasma calcium (Ca^{++}) level (hypocalcemia) in hypoparathyroidismus, pseudohypoparathyroidismus, hypovitaminosis D or respiratory acidosis. It is elecided when a blood pressure (BP) cuff is placed around the arm and inflated to an arterial BP above the systolic and held in place for 3 min. In the absence of blood flow in the brachial artery, the patient's hypocalcemia and subsequent neurovascular irritability will induce spasm of the hand and the forearm, the wrist and metacarophalangeal joints will flex, the proximal and distal interphalangeal joints (PIJ, DIJ) will extend, and fingers will adduct (Trusseau sign of latent tetany, 1861). This sign may be positive before other manifestations of hypocalcemia such as hyperreflexia and tetany occur, such as tetany of unilateral facial muscles provoked by pressure on related nerves, usually the VII (facial) cranial nerve (Trousseau phenomenon).

Fedir I. Inozemtsev [12(24).II.1802, village Bielkino, Borovs'kyy Region, Kaluz'ka Obl., Russia – 06(18).VIII.1869 city of Moskva, Russia][401] — Ukrainian Russian physician, probably of Persian origin, graduated from Medical Faculty of the Kharkiv University, Kharkiv, Ukraine, where received in 1828 Diploma of Physician; prepared drops for the treatment of gastrointestinal (GI) dysfunction in cholera and other illnesses (guttae Inozemcovi, 1835) which consist of rhubarb infusion (liquor) 3-6 gm, valerian (Valeriana officinalis) 1-8 gm, oil 0,9 gm and wormwood (Artimisia santhonica) seed 3 gm DS: one time dose on water 0,25-0,5 gm (15-20 guttae) for abdominal pain or diarrhea; on February 07, 1847 performed general inhalation anesthesia using ether.

Aortic regurgitation (AR) causes primarily hemodynamic changes of the left ventricle (LV) of the heart by dilatation and eccentric hypertrophy of the myocardium,

---

[400] Trusseau A. Clinique medicale de L'Hotel-Die de Paris. Vol. 1-2. B.J. Bailliere, Paris 1861.

[401] Village Belkino founded 2nd half of 16th century), Borovsky Region, Kalush Obl. (area 29,800 km²), Russia. — Inozemtsev I Fl. Treatment of cholera by complex rhubarb infusion. Vladimir 1835, Moscow 1853. — Ablitsov VH. «Galaxy Ukraine». The Ukrainian diaspora: prominent persons. Publ. «Kyt», Kyiv 2007. — Synyachenko OV. They glorified themselves and Ukraine: history of domestic therapy. Crimean Therapeutic Journal 2010;2(2):21-8.

elevation of diastolic volume, filling pressure and wall stress. The main symptoms are congestive, that is shortness of breath (dyspnea or dyspnea) with a mild exertion, anxiety or at rest, dyspnea that occurs lying flat, causing the person to sleep with a head of the bed elevated or sitting in a chair (orthopnea or orthopnea) and paroxysmal nocturnal dyspnea (PND), signs of widened pulse pressure in peripheral arteries with low diastolic pressure.

Dominic J. Corrigan (1802 – 80) — Irish physician, diagnosed AR by a jerky pulse with full expansion, followed by a sudden collapse, who called it a water-hammer pulse (Corrigan pulse, 1832).

In AR and in an aortic aneurysm occasionally one can note a nodding or bobbing of the head which is synchronous with the heart beating as result of amplification of the pulse, named after **Alfred Ch.A. Musset (1810 – 57)** — French dramatist, the poet and novelist, who suffered from this disease (Musset sign).

In AR as result of dilatation of the LV the cardiac apex is displacement towards the left.

Described by **Austin Flint (1812 – 88)**[402] — American physician, in AR is a presystolic or mid-systolic murmur at the apex of the heart, causing fluttering of the anterior leaflet of the mitral valve (MV) when blood flow simultaneously retrogradely from the aorta and anterogradely from the left atrium (LA) – Austin Flint murmur (1862).

Noted in AR by **Paul L. Duroziez (1826 – 97)** — French physician, is a double murmur over the common femoral artery (CFA) and the other large peripheral artery (Duroiziez murmur, 1861).

Henrich I. Quincke (1842 – 1922) — German internist, the neurologist and surgeon, observed in AR an intermittent whitening and reddening of the nail's bed and the skin around the nail upon the finger compression due to a pulsation of capillary arterial and venous plexuses (capillary pulsation or Quincke pulse, 1862).

Reported by **Friedrich** von **Müller (1858 – 1941)** — German internist, in AR was the pulsation of palatine uvula and the soft palate during systole as result of the wide difference in systolic and diastolic arterial blood pressure (ABP), thus wavering stroke volume (SV) – Müller sign.

Also, Paul Duroiziez was the first to write about a case of congenital stenosis of the MV stenosis (1877), Henrich Quincke was the first to recognize a rapid angioedema of the dermis (skin), the subcutaneous (SC) tissue, mucosa and submucosa (Quincke edema, 1882).

In regard to murmurs Austin Flint wrote: «*The murmurs, in themselves, give no information respecting the amount of obstruction from contracted orifices, or of*

---

[402] Flint A. On cardiac murmurs. Am J Med Sc 1862; 44:29-54.

regurgitation from valvular insufficiency...The truth is, the evils and danger arising from valvular lesions, for the most part, are not dependent directly on these lesions, but on the enlargement of the heart resulting from the lesions...serious consequences of valvular lesions do not follow until the heart becomes weakened either by dilatation or by degenerative changes».

A master of chest percussion, **Joseph Scoda (1805 – 81)** — Czech physician, described an increased percussion resonance at the upper, and a flatness at the lower chest (Skoda resonance), and tympanic sound above a large effusion or consolidation (Skoda sign).

For the treatment of cardiac and kidneys diseases, **Philipp J. Karell (1806 – 86)** — Estonian physician, advocated an initial diet of skim milk to be gradually amplified by eggs, dry toast, meat, rice, and vegetables (Karell diet) and bed rest (BR).

Thoracentesis (in Greek – thorax + centesis, punction, punctures) or pleural tap is an invasive medical procedure implying introduction of a hollow needle or canula into the pleural cavity through the 8th-10th intercostal space (ICS) in the midaxillary line (MAL) to remove fluid or air with the diagnostic or therapeutic aim. The first known thoracocentesis was performed in 1852 by **Henry I. Bowditch (1808 – 92)**[403] — American physician, to treat pleural effusion.

Devised in 1850-52 by **Morrill Wydman (1812 – 1903)** — American physician and social reformer, a trocar and cannula, allowed to perform aspiration of an acute or chronic pleural effusion (thoracentesis) in the 6th-7th intercostal space (ISC), 15 cm from the posterior middle line (PML), usually in the triangle of auscultation of the lungs relatively free of muscles of the back, situated laterally by the median border of the scapula, superiorly and medially by the inferior portion of the trapezius muscle and inferiorly by the latissimus dorsi muscle.

Pulmonary insufficiency was detailed by **Anton Wintrich (1812 – 82)** — German therapeutant, in 1854.

The anomalous plasma and urine protein which consist of light monoclonal chains of the immunoglobin and secreted in some dyscrasias (diseases, disorders) of plasmatic cells of the blood, such as multiple myeloma and macroglobulinemia, was discovered by **H. Bence Jones (1814 – 73)** — English chemist and physician, named Bence Jones's protein (1847). It is characterized by peculiar dilutional abilities: during warming to 50-60°C it precipitates and again dilutes at the 90-100°C; during cooling it again precipitates into a sediment, and then again becomes soluble.

---

[403] Bowditch HI. On pleuritic effusion and the necessity of paracentesis for their removal. Am J Med Sc 1852;23:834. — Jarko S. Henry I. Bowditch on pleuritic effusion and thoracocentesis (1852). Am J Cardiol 1965 Jun;15:832-6.

Antonin J. Desormeaux (1815 – 94) — French physician and inventor, improved early endoscopes, specifically cystoscope, and was the first to succeed in 1853 to operate it in a living patient.

Carl R.W. Wunderlich (1815 – 77) — German physician and psychiatrist, is known for his measurement and establishing the mean normal human body temperature (T) at the level of 37^0C, later corrected to more normal level at about 36.8^0C / 98.2^0F (Fahrenheit)[404]. He also popularized the use of the thermometer and demonstrated a typical T variation in typhoid fever (Wunderlich curve).

Sir **William W. Gull (1816 – 90)** — English physician, recognized tabes dorsalis (1856), transient hemoglobinuria (1866), AS degeneration of kidneys (1872) and atrophy of the thyroid gland with myxedema (Gull disease, 1873).

The taxonomist of medical documentation, **Ludwig Traube (1818 – 76)** — German physician of Jewish origin, initiated the temperature, pulse and respiratory rate curve into clinical praxis (1852), described a loud double-sound tone, like pistol-shot sound at auscultation over the common femoral artery (CFA) at the heart rate (HR) 100 beats per minute or more, due to reverse blood flow in the aorta in an aortic valve (AV) insufficiency or mitral valve (MV) stenosis – Traube sign, and depicted a crescendo-shaped space 12 mm wide, just above the left costal margin in the antero-inferior area of the chest, due to gas in the stomach, producing a vesiculo-tympanic sound (Traube semilunar space, 1868).

Augustin M. Morvan (1819 – 97) — French physician, is best known for description in 1870 of myxedema [**Robert Graves (1796 – 1853**: above**)** and **Karl von Basedov (1799 – 1854**: above**)**] and for treating in 1890 the first recorded case of eponymous Morvan syndrome[405], a rare neurological condition marked by acute insomnia.

The Morvan syndrome is a familial or acquired condition, caused by the thalamo-limbic circuits dysfunction, male to female ratio is 9:1, with a mean onset age of 57 years. The pathology indicates an association with myasthenia gravis (MG), small-cell lung carcinoma and thymoma. It is characterized by severe insomnia, extreme sleep-stage dissociation with loss of the typical rapid and non-rapid eye movement (REM, NREM) sleep, an involuntary eyelid muscle contraction, typically involving the lower eyelid or less often the upper eyelid / persistent quivering of other muscles (myokymia)

---

[404] **Daniel G. Fahrenheit (May 24, 1686, Gdansk, Poland – Sep. 16, 1736, the Hague, Netherlands)** — Prussian German chemist, physicist, engineer and glass blower, who first proposed to refine the mercury, invented the spirits-in-glass (1709) and the mercury-in-glass (1714) thermometers and developed a temperature (T) scale, later named for him (F), in which the T of thawing ice was 32^0F, and distance to the T of the boiling water – 180^0F (1724). / Fahrenheit DG. Experimenta circa gradum caloris liquorum nonnulorum ebullientium instituta. Philosoph Trans R Soc 1724-1725;33:1-3.

[405] Vale Tc, Pedroso JL, Dutra LA, et al. Morvan syndrome as paraneoplastic disorder of thymoma with anti-CASPR2 antibodies. Lancet Apr 1-7, 2017;389:1367-8.

or involuntary rippling, twitching, and stiffness of muscles (neuromyotomia), abnormally increased sweeting (hyperhidrosis), and encephalopathy causing confusion, hallucinations, and fluctuating cognition. Electromyography (EMG) shows spontaneous muscle fiber activity with fasciculations, multiplex, myokymia, and neuromyotonic discharges. Electroencephalography (EEG) is either normal or shows diffuse involvement. Magnetic resonance imaging (MRI) is usually normal. Treatment of choice is usually immunotherapy (corticosteroids, immunoglobulins, plasmapheresis), azathioprine or cyclophosphamide for therapy of paraneoplastic autoimmune encephalopathies, and low doses of carbamazepine or phenytoin to relief neuromyotonic.

In fluid mechanics, the dimensionless number that gives a measure of ratio of internal forces to viscous forces and consequently quantities the relative importance of these two types of forces for a given flow conditions was introduced by Sir **George G. Stokes (1819 – 1903)** — Irish English mathematician and physician, in 1851. This number was then popularized in 1883, by **Osborn Reynolds (1842 – 1912)** — English Irish innovator and named after him the Reynolds number (Re or N_R). It found an application in the physiology of murmurs and thrills in the heart and vessels in an acoustic phenomenon produced by turbulent blood flow. The chance that bloods flow will be turbulent in a situation is dependent upon its N_R, as expressed in the following equation: $N_R = p \times d \times v / n$, where «p» is the density of fluid, «d» – the diameter of orifice or vessel, «v» – the velocity of flow, and «n» – the viscosity of fluid. The increased turbulency is more likely when the N_R is higher. When murmur occurs outside the heart, for example in an aneurysm of the abdominal aorta (AAA) or a narrowing of the carotid arteries they are referred to as bruit. Despite the name being different they represent this same type of phenomenon.

Etiology of murmurs is related to decreased blood viscosity («n») – anemia, decreased diameter of valve, vessel, or orifice («d») – valvular stenosis, coarctation of the aorta (COA) and ventricular septal defect (VSD), increased velocity of blood through normal structures («v»), and regurgitation across an incompetent valve – tricuspid regurgitation (TR), pulmonary regurgitation (PR), mitral regurgitation (MR) and aortic regurgitation (AR).

Clinically important features of murmurs are timing, location and radiation, shape, pitch, intensity, quality and response to specific simple physiologic maneuvers.

The most important characteristic of murmurs is timing relatively to cardiac cycle – systole, diastole, or both. The commonest are systolic murmurs occurring in 95% hospitalized patients. They are caused by hyperdynamic state of anemia («flow murmurs»), aortic / pulmonary stenosis (AS, PS), TR, mitral valve (MV), VSD, outflow tract obstruction of he left ventricle (LV), for example, in idiopathic hypertrophic subaortic stenosis (IHSS) / hypertrophic obstructive cardiomyopathy (HOCM) / hypertrophic cardiomyopathy (HCM). Diastolic murmurs include AR (most common in American and Canadian males) / PR, mitral stenosis (MS) / tricuspid stenosis (TS), and continuous murmur, meaning that there are components of murmurs present in both, systole, and diastole – usually caused by patent ductus arteriosus (PDA) in infancy, occasionally in adults in combination of AS and AR, when there is systolic and separate diastolic murmur in this same region.

The next important characteristic is the location where murmur is most easily heard (loudest) and its radiation:

aortic arch	–	the right 2nd intercostal space (ICS) to near the sternum, the left 4th ICS near the sternum, or the left 5th ICS in the middle axillary line (MAL), that is at the apex of the heart;
pulmonary artery (PA)	–	the left 2nd ICS near the sternum;
tricuspid valve (TV)	–	the left 4th ICS near the sternum;
mitral valve (MV)	–	the left 5th ICS in the MAL at the apex of the heart.

Radiation describes other location(s) where the murmur is audible, despite not being directly over the heart. Murmurs usually radiated in the direction of the turbulent blood flow, for example, murmur of AS may radiate to carotid arteries, murmur from TR to the anterior right thorax, murmur from MR to the left axilla.

The three basic shapes of the murmur describe how murmur's intensity changes from the onset to completion which is generally determined by the pattern of the pressure gradient driving the turbulent blood flow with the loudest segment occurring during the greatest gradient, since this will be the point of the highest velocity. The first shape is crescendo-decrescendo murmur occurs during systole, the second is decrescendo murmur occurs during diastole, and the third murmur is uniform («holosystolic») murmur heard usually during systole.

While severe MS typically sounds as «holosystolic» diastolic murmur with presystolic accentuation, the slope of mild MS may be decrescendo as the pressures in the left atrium (LA) and the LV may equalize by the time the atrial kick occurs.

Also, in MS, there may be an extra sound in early diastole and S$_3$ which is an opening snap caused by a stenotic and stiff MV once pressure in the LV drops below that in the left atrium (LA).

Pitch of the murmur is most directly related to high pressure gradients creating a high pitch (VSD); large volume of blood flow across low pressure gradient creating a low pitch (MS); high pressure gradient and a large volume of blood flow creating a simultaneously high and low pitches – a «harsh» murmur (AS).

The intensity of murmur describes how high the volume is, it depends on a variety of physiologic properties, such as velocity of blood flow, the origin, and acoustic properties of the intervening tissue, hearing and experience of the examiner, the stethoscope used, and the presence of an ambient noise.

The quality of a murmur is the most subjective and non-specific characteristic, and it is difficult to describe a hummer of the murmur. The murmur of MR is described as «blowing», «musical», of MS as «rumbling», of AS as «harsh», of AR as «blowing», of PR/MS with pulmonary hypertension (HTN) as «musical», «vibratory» (Graham-Steell[406] murmur), and of PDA as «machinery».

The response of murmurs intensity by specific simple physiologic maneuvers, such as clenching fists which by increase of an afterload distinguishes MR (increase intensity) from AS (decreased or unchanged intensity), squatting or quick assuming a supine position increases venous return and by this stroke volume (SV) and LV-end diastolic volume, distinguishes AS (increase intensity) from IHSS / HOCM / HCM (decrease intensity).

Established by **John Tyndall (1820 – 93)** — Irish English physicist, in 1862-64, capnography is a noninvasive method using infrared radiation for direct measurement and monitoring of the inhaled and exhaled concentration or partial pressure of carbon dioxide (CO_2), and indirect monitoring of the CO_2 partial pressure in the arterial blood as measured in millimeters of mercury (mmHg), plotted on the plotting paper against time. The body cells in healthy individuals under normal circumstances produce a small amount, about 200 mL of CO_2 per minute[407] and the difference between arterial blood and expired CO_2 partial pressure is small. In disease, under physical activity or stressful environmental condition, with increased metabolic rate, the production of CO_2 increases.

Fedir F. Mering (26.02.1822, town of Dohna, Saxony, Germany – 19/31.10.1887, city of Kyiv, Ukraine, buried on the cemetery «Askold's Grave» in Kyiv)[408] — Ukrainian physician and cardiologist of German descent, in «Lectures about Hygiene» (1863-1865) and «The course of Clinical Lectures» (1864) educated his students to treat not only disease but the patient, spoke against the serfdom, difficult workers and soldiers living conditions, accentuated the importance of society to care for people's health». He said that «The society will enjoy prosperity only when all its members are

---

[406] **Graham Steell (1851 - 1942)** — English cardiologist.

[407] Guyton AC. Texbook of Medical Physiology. 5th Ed. WB Saunders, Philadelphia 1976. — **Arthur C. Guyton (1919 - 2003)** — American physiologist.

[408] Town of Dohna (first written mention 960), close to city of Dresden (settled since c. 7500 BC), Saxony (area 18,415.66 km²), Germany. — Cemetery «Askold's Grave» (opened 983-1935), Kyiv, Ukraine. — Ablitsov VH. «Galaxy Ukraine». The Ukrainian diaspora: prominent persons. Publ. «Kyt», Kyiv 2007. — Synyachenko OV. They glorified themselves and Ukraine: history of domestic therapy. Crimean Therapeutic Journal 2010;2(2):21-8.

healthy, that is when people are well developed physically and morally», remarked, that only «in terms of sufficient payment for work the number of married couple and births will increase, and morbidity of newborn will decrease, and that will be a great benefit for the society». Established clinical diagnosis of myocardial infarction (MI) due to thrombosis in the coronary arteries in alive patient (Kyiv, Ukraine May 1883).

C. Ph. Adolph K. Kussmaul (1822 – 1902) — German physician, established in chronic constrictive pericarditis and in cardiac tamponade: the «pulsus paradoxus» – (1) a drop during inspiration of the systolic blood pressure (BP)>20 mmHg [without a corresponding decrease in the diastolic BP] due to decreased oxygenated blood return to the left atrium (LA) and therefore decrease in the stroke volume (SV) of the left ventricle (LV), cardiac output (CO) and cardiac index (CI); (2) the absence of normal inspiratory collapse of the external jugular veins (EJV) and IJV due to decrease of the venous pressure in the RA and the SVC as measured by the central venous pressure (CVP), but an opposite, a pathological distention of the EJV and IJV due to an elevation of the venous pressure in the right atrium (RA) and the superior vena cava (SVC) as measured by the CVP, as result of the RV failure, or the homogenous scarring of the pericardium with decreased filling of the cardiac cavities by the blood (Kussmaul sign, 1868, 1873).

Independently, Adolf Kussmaul and **Alphonse M.J. Kien (1840 – 1923)** — French physician, described paroxysms of severe dyspnea, initially with a rapid and shallow breathing (respiration), gradually becoming deep, the labored and finally gasping, a form of hyperventilation to reduce carbon dioxide (CO_2) cranial in the blood, associated with severe metabolic acidosis, particularly diabetic ketoacidosis in renal failure (RF) and other diseases (Kussmaul-Kien respiration, 1874, 1878).

Adolf Kussmaul with **Rudolph R. Maier (1824 – 88)** — German pathologist, diagnosed periarteritis nodosa (Kussmaul-Maier disease, 1866).

B 1856 p. **Vilem D. Lambl (05.XII.1824,** village Letiny, District Plzen-South, Kreis Plzen, Czechia – died 12.II.1895, city of Warsaw, Poland)[409] — Czech Ukrainian anatomist, pathologist and physician, described age-related small whisker-like (capillary) projection that may develop within the line of heart valves closure, usually on the its anterior ring, especially on the aortic valve (AV), named the Lambl excrescences, which should be differentiated from papillary fibroelastoma (PFE). Although in 1681 p. **Antoni van Leeveuwenhoek (1632 – 1723:** 17th century) observed under his

---

[409] C. Village Letiny, District Plzen-South (area 996.55 км²), Kreis Plzen (area 990.04 км²), Czechy (area 78,866 км²). — Lamb D Mikroscopische untersuchungen der Darmexcrete. Praeger Vierteljahresschrift fur practische Heilkunde. 1859;61:1-58. — Lambl VD. Papillare exkreszenzes an der Semilunar-Klappe der Aorta. Wien Med Wochenschr 1856;6:244-7. —Lambl VD. Pathologic Anatomy. Vol. 1-4. Kharkiv 1861-64. — Ablitsov VH. «Galaxy Ukraine». The Ukrainian diaspora: prominent persons. Publ. «Kyt», Kyiv 2007. — Hanitkevych YaV. The contribution of the Ukrainian physicians to the world medicine. Ukr Med Chasopys 2009 VII/VIII;4(72):110-15. — Synyachenko OV. They glorified themselves and Ukraine: history of domestic therapy. Crimean Therapeutic Journal 2010;2(2):21-8.

microscope protozoan – trophozoite, unicellular parasite organism from the class of Jute, named by him Cercomonas intestinalis, which causes gastrointestinal (GI) inflammation (gastroenterocolitis) in his own diarrheal stool. Nevertheless, it was Vilem Lambl who in 1859 isolated Lamblia intestinalis, an organism within protozoan (trophozoite) parasites, in patient with GI infection, sometime asymptomatic, but often manifested by diarrhea, dyspepsia, and imperfect absorption of nutrients by the small intestine (malabsoption), named lambliasis. It was later re-discovered (1879), named in 1882-83 Gardia intestinalis in honor of **Alfred Giard (1846 – 1908)** — French biologist.

Among major discoveries of Sir **Samuel Wilks (1824 – 1911)** — British physician, are description of the formation of nodes along the hair shaft (trichorrhexis nodosa, 1852); recognition of ulcerative colitis in a 42-year old woman complaining of fever and diarrhea of several months duration, in which autopsy demonstrated a transmural ulcerative inflammation of the colon and terminal ileum (1856); a report of on primary amyloidosis in a 52-year-old man with dropsy and albuminuria in whom autopsy revealed a lardaceous spleen and viscera (1956); description symptoms of alcoholic paraplegia (1868); and a report on myasthenia gravis (in Greek myos «muscle» and asthenia «weakness», and the Latin gravis «serious»; MG; 1877).

MG is an autoimmune disease caused by anti-acetylcholine receptors (anti-AChR) complement-fixing antibodies to block or destroy nicotinic acetylcholine nerve receptor at the post-synaptic membrane of the striated muscles, thus preventing nerve impulses from triggering muscle contractions.

Rarely an inherited defect in the neuromuscular junction causes congenital MG. However, the most common tumor of the mediastinum, usually of the anterior mediastinum is tumor of the thymic gland, completely encapsulated or invasive thymoma (95%), which is associated in 30% to 50% with clinical symptoms of MG, whereas only 5% to 15% of clinically symptomatic MG are found to have thymoma.

About 50% of patients with MG older than 40 years presents with either local symptoms due to invasive thymomas such as painless weakness of the ocular muscles (ptosis, nystagmus, diplopia) and facial expression (poor mimicking); dysphagia, chocking, change in timbre of a voice, dysarthria, hoarseness or stammering; painless weakness of the muscles of mastication, neck, torso, extremities and breathing; systemic symptoms due to associated non-thymic malignancies, diseases of the thyroid gland or the central nervous system (CNS) or systemic lupus erythematosus (SLE); and myasthenic crisis in which takes place an acute worsening of the patient condition as a result of respiratory failure and worsening of vital functions.

Treatment of MG should be directed towards reducing the risk of myasthenia crisis, which include anticholinesterase medications (physostigmine), anabolic steroids, glucocorticosteroids, intravenous immunoglobulins (IVIG), plasmapheresis, respiratory support, and after stabilization of the patient's condition removal of the thymic gland (thymectomy).

Pierce Ch. E. Potain (1825 – 1901) — French cardiologist, designed a pleural suction apparatus.

Further progress in percussion of the chest were made by **Friedrich** von **Korányi (1828 – 1913)** — Hungarian physician, with the introduction of the auscultative percussion which is produced by tapping with the index finger of one hand over the proximal joint of the second hand index finger, laid on the surface of the body, and defining of a triangular dullness at a paravertebral field posteriorly at the base of the chest cavity on the opposite site of pleural effusion in children and youths (Korányi sign).

In 1862 **John L.H. Down (1828 – 96)**[410] — English physician, fully portrayed chromosome disorders due to an aberration of trisomy 21 and nondisjunction, characterized by the presence of congenital heart diseases (CHD) termed a complete atrioventricular (AV) canal (7,3% of all CHD, 40% at the trisomy 21 aberration), a mongoloid appearance and mental retardation known as Down syndrome.

In 1889, **Gregori A. Zakharin (1829 – 96)** — Russian physician, and in 1893-96, Sir **Henry Head (1861 – 1940)** — English neurologist, determined areas of cutaneous sensitivities (hyperalgesia) associated with diseases of the viscera (Zakharin-Head zones).

Interstitial lung disease (ILD) consists of a group of more than 200 primary or secondary clinical entities that manifest itself with chronic, progressive, and diffuse inflammation of the pulmonary interstitial. It is divided into the granulomatous, and the alveolitic patterns. Granulomas are abnormal collections of inflammatory cells, consisting of activated immune cells, typically macrophages, encircling particles that are generally not recognized by the antigen-specific or adoptive immune system.

The granulomatous pattern ILD include tuberculosis representing granulomatous disease with caseation and sarcoidosis, a granulomatous disease without caseation. Sarcoidosis, affecting most commonly the lungs or the lymph nodes as a collection of granulomas forming multiple nodules were recognized by **Ernest H. Besnier (1831 – 1909)** — French dermatologist, in 1889 and **Caesar P.M. Boeck (1845 – 1917)** — Norwegian dermatologist, in 1899 (Besnier-Beck sarcoidosis or disease).

Granulomatous polyangiitis or Wegener granulomatosis was described by **Peter McBridge (1854 - 1945)** — Scottish otolaryngologist, in 1897, by **Heintz R.E. Klinger (1907 – ?)** — German pathologist, in 1931, and in details by **Friedrich Wegener**

---

[410] Down JHL. Observation on an ethnic classification of idiots. Clinical Lecture Reports, London Hospital 1866;3:259-62.

(**1907 – 90**) — German pathologist, in n 1936-39), is characterized by necrotizing granulomatous involvement of the pulmonary parenchyma and both pulmonary and renal vasculature, causes hemoptysis and renal failure (RF). **Lotte Strauss (1915 – 85)** — German American pathologist, and **Jacob Churg (1910 – 2005)** — Belarusan American pathologist, published on an allergic granulomatosis which is a rare systemic disorder, characterized by the white blood cells (WBC) infiltration, specifically eosinophilia, vasculitis, and pulmonary parenchymal granuloma formation (Churg-Strauss syndrome, 1951).

Wilhelm Löffler (1887 – 1927) — Swiss physician, described a disease in which eosinophils, accumulate in the pulmonary parenchyma, characterized by interstitial edema in response to parasitic infection – Ascaris lumbricoides, Strongyloides stercoralis, and hookworms Ancylostoma duodenalis and Necator americanus (Löffler eosinophilic pneumonia or Löffler syndrome, 1932), and eosinophilia with fibroblastic infiltration of the endocardium, causing restrictive cardiomyopathy (CM) with congestrive heart failure (CHF), constant tachycardia, hepatosplenomegaly, serous pleural effusion, edema of the hands and legs (Löffler parietal fibroplastic endocarditis, 1936)[411].

The underlying pathology for histiocytosis X or Langerhans histiocytosis is parabronchial accumulation of stellate dendritic cells. Children variant of this a genetic autosomal-recessive disorder caused by histiocytic infiltration was described by **Erich Letterer (1895 – 1982)** — German physician, and **Sture A. Siwe (1897 – 1966)** — Swedish pediatrician, which is manifested by lymphadenopathy, hepatosplenomegaly and seborrhea-like lesions on the skin (Lettere-Siwe disease, 1924 and 1933), and by **Alfred Hand Jr (1868 – 1949)** — American pediatrician, **Artur Schüller (1874 – 1958)** — Austrian neurologist, and **Henry A. Christian (1876 – 1951)** — American internist, which present itself by a triad of exophthalmos, lytic bone lesions (usually of the skull) and diabetes insipidus from pituitary stalk infiltration (Hand-Schüller-Christian disease, 1893, 1915-15 and 1919).

Alveolar pattern of ILD comprises the immunologic injury primarily directed towards the alveolar epithelium, causing inflammation of the alveolar wall and air space (alveolitis). Among other diseases. Here belong the syndrome of **Ernest W. Goodpasture (1886 – 1960)** — American pathologist and physician, characterized by the presence of pulmonary hemorrhage and glomerulonephritis, and idiopathic pulmonary fibrosis of **Louis V. Hammans (1877 – 1946)** — American internist, and of **Arnold R. Rich (1893 – 1968)** — American pathologist, hereditary/immunologic disorders, categorized as between ILD and acute respiratory distress syndrome

---

[411] Löffler W. Endocarditis parietalis fibroplastica mit Bluteosinophile. Ein eigenartiges Krankheitsbild. Schweiz med Wschrift, Basel 1936;66:817-20.

(ARDS), manifested by progressive dyspnea, nonproductive cough fever, rales, clubbing of fingers, weight loss and the chest x-ray (CXR) evidence of a preferential involvement of the lower lobes by lineal opacities in a diffuse reticulonodular or reticular pattern with honeycombing, cysts, and ground-grass opacities (Hammans-Rich syndrome, 1935, 1944).

The ILD is treated with glucocorticoids 0.5-1.0 mg/kg/day for 4-12 weeks, cyclophosphamide, azathioprine, methotrexate, cyclosporine, tumor-necrosis factor (infliximab) and lung transplantation.

Sergei P. Botkin (1832 – 89) — Russian therapist, differentiated between hypertrophy and dilatation of the left ventricle (LV) of the heart, made a clinical description of atherosclerosis (AS) and detailed virus hepatitis A (1867-75 and 1887-88).

The obstruction of the arteries at their origin from the aortic arch can cause circulatory disturbance in the head or arms was recognized in 1875 by **William Broadbent (1835 – 1907)** — English cardiologist and neurologist.

In his doctoral dissertation of 1862 **A.G. Maurice Raynauld (1834 – 81)**[412] — French physician, studied a peripheral vascular disease (PVD) of the primary form (Raynaud's disease) and the secondary form (Raynaud's phenomenon).

The primary form is an idiopathic disorder of unknown etiology, partially hereditary, affects approximately 4% of the population, with the onset typically at 15-30 years of life, more frequently in young women and older people, precipitated by exposure to cold or emotional stress. It causes the sympathetic spasm of small arteries or arterioles, or both with subsequently reduction of blood flow to the tissues. Characterized by sudden, bilateral, symmetrical, and sequential attacks of numbness or burning, painful changes in color of the skin from paleness to cyanosis and finally to bright redness, the most commonly of the digits of the hand and in 40% toes of the foot, rarely of the nose, ears, nipple and lips, induced by the cold weather or nervousness, lasting from 15 min. to several hours, and improves by the warmth. The radial, ulnar and pedal pulses are somewhat decreased or palpable on both sides. Occasionally, an attack may progress to the formation of ulcers or gangrene in the involved area.

The secondary form is caused by excessive smoking of tobacco, trauma, in particularly to the fingers, complications of PVD, systemic connective tissue disorder, such as scleroderma or lupus erythematosus, pulmonary hypertension, myxedema or birth control pills. In 7.5% of the younger individuals a cervical rib is manifested by the thoracic outflow syndrome (TOS).

Symptoms are unilateral, precipitated by hyperabduction of the involved arm, turning of the head and carrying of heavy object, include coldness, weakness,

---

412 Raynoud A.G.M. De la l'asphyxie locale et la gangrene symetrique des extremites. Rignoux, Paris 1862.

fatigability and usually diffuse pain of the arm and hand, and blanching followed by cyanosis and rubor of fingers. During the attacks the radial pulse on the affected side could be diminished or absent.

The primary treatment is avoiding cold, discontinuation of nicotine or other stimulants, calcium channel blockers and iloprost, a synthetic analogue to prostacyclin (PGI$_2$). Depending on the clinical status and the nerve conduction velocity studies, patients with the TOS should be initially treated with physical therapy (PT), but patients suffering from symptomatic TOS should undergo a segmental removal of the 1st rib with / without cervical sympathectomy.

In 1873 **G.E. Playfair**[413] — English physician, attended a 7-year-old rachitic child with right-sided bulging and fluctuating pleural empyema that he initially aspirated five times in the 6th intercostal space (ICS) by thoracentesis, but each time the pleural fluid re-accumulated. Then he opened the empyema cavity and inserted a flexible caoutchouc tube with the distal end connected to a glass tube that went through the cork cap connected to the siphon (syphon) system that is into a bottle with a sealing level of the water on the floor. The tube drained 25-50 ml of puss daily without any entry of air into the abscess cavity for one month. The patient's clinical condition improved, and the tubes were removed, allowing the re-entry of air, necessitating a counter-opening with placement of a drainage tube which was removed a week later with full recovery.

Thus he was the first physician to perform a closed water-seal or under-water drainage in the treatment of thoracic empyema.

In 1875 **Gotthard** von **Bülau (1835 – 1901)**[414] — German internist, facing a 35-year-old carpenter admitted with bronchopneumonia with subsequent pleural empyema, punctured the chest wall with an external trocar, introduced a catheter into the empyema space and connected it to an underwater siphon apparatus for a drainage (Bülau drainage, published in 1891). Since then, Bülau's drainage was used to treat not only empyema thoracis, but also pleural effusion and pneumothorax.

Name of **Francisek Chvostek (1835 – 84)**[415] — Czech Austrian military physician, is connected to a sign of existing nerve excitability (tetany) seen in hypocalcemia, referring to an abnormal reaction to the stimulation of the cranial VII (facial) nerve which by its tapping at the angle of the jaw, i.e., masseter muscle, one of the muscles of mastication, the facial muscles on the same side of the face will contract momentarily, typically a twitch of the nose or lips (Chvostek sign, 1871).

---

[413] Playfair GE. Case of empyema treated by aspiration and subsequently by drainage: recovery. Br Med J 1875;1:45.

[414] Bülau G. Für die Heber-Drainage bei Behandlung des Empyems. Z klin Med 1891;18:31–45. /. Van Schil PE. Thoracic drainage and the contribution of Gotthard Bülau. Ann Thorac Surg 1997;64:1876.

[415] Chvostek F. Betrag zur Tetanie. Wien Med Press 1876;17:1201-3, 1225-7, 1253-8, 1313-6.

In 1866 **Wilhelm Ebstein (1836 – 1912)**[416] — German physician of Jewish descent, reported a case of cyanotic congenital heart disease (CHD) in a 17- year-old man with dyspnea, pulsatile jugular veins, cardiomegaly, and cardiac murmurs. It is caused by the failure of delamination of the septal, posterior, and partially anterior leaflets of the tricuspid valve (TV) from the inner layer of the inlet portion of the right ventricle (RV) in the fetus before the 18 weeks pregnancy, and atrialization of the RV (Ebstein anomaly).

The arterial blood pressure (BP) meter, a.k.a., sphyngomanometer was invented in 1881 by **Samuel S.K. S.K. Ritter** von **Basch (1837 – 1905)**[417] — Czech Austrian physician of Jewish descent. It is a device used to measure the arterial BP.

Antin (Anton, Antoniy) Kobyliansky (29.01.1837, village Pererisl, Nadvirniansky Region, Ivano-Frankivsk Obl., Ukraine – 08.02.1916, city of Lviv, Ukraine)[418] — Ukrainian physician, designed physiological apparatus-camera which improve breathing by absorbing and transforming smoke.

In 1872 **Moritz Kohn Kaposi (1837 – 1902)** — Hungarian dermatologist of Jewish origin, reported six patiens, mostly elderly men of Jewish and Italian origin, with bluish red multicentric cutaneous nodules on the lower extremities, characterized by the proliferation of the blood vessels, dermal hemorrhages and hemosiderin depositions (Kaposi's sarcoma).

In 1873 **Josyp (Osyp) O.** von **Rustyts'kyy / Rustitzky**, a.k.a., **Osip J. Rusticki (31.03.1839, village Horodnia, Chernihivs'kyy Region, Chernihivs'ka Obl., Ukraine – 13.04.1912, city of Kyiv, Ukraine)**[419] — Ukrainian surgeon, in 1889 – **Otto Kahler (1849 – 93)** — Austrian surgeon, described multiple myeloma disease (plasma-cell myeloma, or plasmocytoma), or Rustitzky-Kahler disease. Josyp von Rustyts'kyy / Rustitzky introduced into medicine the term «myeloma multiplex».

---

[416] Ebstein W. Ueber einen sehr seltenen Fall von Insufficienz der Valvula tricuspidalis, bedingt durch eine angeborene hochgradige Missbildung derselben. Arch Anat Physiol 1866:238-54.

[417] Booth J. A short history of blood pressure measurement. Proc Royal Soc Med 1977;70(11):793-9.

[418] Village Pererisl (founded 1424), Nadvirniansky Region, Ivano-Frankivsk Obl., Ukraine. — Ablitsov VH. «Galaxy Ukraine». The Ukrainian diaspora: prominent persons. Publ. «Kyt», Kyiv 2007.

[419] Town of Horodnia (known from time of Ukraine-Rus, declined in 14th century, restored in 1552, town status since 1957; suffered during organized by Russian Communist Fascist regime of Holodomor-Genocide 1932-33 against Ukrainian peasants who revolted, but stifled by military force, at least 8 victims died from hunger, then 175 insurgents were repressed), Chernihivs'kyy Region, Chernihivs'ka Obl., Ukraine. — Rustitzky von J. Multiples Myelom. Deutsche Zeitschrift für Chirurgie 1873;3(1-2):162-72. — Rustitzky J von. Epithelial carcinom der Dura mater mit hyaliner Degeneration. Archiv für pathologische Anatomie und Physiologie und für klinische Medizin 1874;59(2):191-201. — Rustitzky J von. Untersuchungen über Knochenresorption und Riesenzellen. Archiv für pathologische Anatomie und Physiologie und für klinische Medizin 1874;59(2):202-27. — Rustitzky J von. Ein Fall von Abscessus retrosternalis mit Resection des Manubrium und der oberen Hälfste des Corpus sterni. Deutsche Zeitschrift für Chirurgie 1887;26(2-6):594-7. — Kyle RA. Multiple myeloma: an odyssey of discovery. Brit J Haematology 2000 Dec;111(4):1035-44. — Ablitsov VH. «Galaxy Ukraine». The Ukrainian diaspora: prominent persons. Publ. «Kyt», Kyiv 2007. — Synyachenko OV. They glorified themselves and Ukraine: history of domestic therapy. Crimean Therapeutic Journal 2010;2(2):21-8.

In 1875 **Karl Ferdinand (Fedir) O. Lesh / Lösch [11(23).IV.1840, city of St Petersburg, Russia – 16(29).I.1903, city of Kyiv, Ukraine]**[420] — Ukrainian physician and bacteriologist, was the first to isolate causative factor of intestinal amebiasis – Entamoeba histolytica.

Cirrhotic jaundice was described by **Victor Ch. Hanot (1844 – 96)** — French physician, who named it «cirrhosis biliare primitive» (1875).

In 1896 **Henri J.L.M. Rendu (1844 – 1902)**[421] — French physician, in 1901 Sir **William Osler (1849 – 1919)**[422] — Canadian physician, and in 1907 **Frederick P. Weber (1863 – 1962)**[423] — English dermatologist, reported on hereditary hemorrhagic telangiectasia (HHT), William Osler and **Louis H. Vaques (1860 – 1939)** — French physician, on polycythemia.

HHT is a genetic autosomal dominant vascular anomaly, characterized by permanent dilatation of capillaries, arterioles and venules of the skin, mucous membranes, gastrointestinal tract, lungs, liver and brain. It is characterized by telangiectasia, fragility of capillaries and hereditary hemorrhagic diathesis (Osler triad). There is an association between the pulmonary arterio-venous (AV) malformation, a direct connection between the branches of pulmonary artery and vein (PA, PV) and HHT in that 33% of the patients with pulmonary AV malformation suffer from HHT.

William Osler understood that atherosclerosis (AS) of coronary arteries causes angina pectoris, and AS of the lower extremities – intermittent claudication. He called angina pectoris a «*claudication of the heart*».

---

[420] Lösch FA. Massenhafte Entwickelung von Amöben im Dickdarm. Virchow's Arch path Anat 1875;65:196-211. — Ablitsov VH. «Galaxy Ukraine». The Ukrainian diaspora: prominent persons. Publ. «Kyt», Kyiv 2007. — Synyachenko OV. They glorified themselves and Ukraine: history of domestic therapy. Crimean Therapeutic Journal 2010;2(2):21-8.

[421] Rendu H. Epistaxis répétés chez un sujet porteur de petits angiomes cutanés et muqueux. Lancette française: gazette des hôpitaux civils et militaires (Paris) 1896; 69:1322-23. — Rendu H. Epistaxis répétées chez un sujet porteur de petits angiomes cutanés et muqueux. Bulletin de la Société médicale des hôpitaux de Paris 1896;13(1):731. — Rendu H. Il est l'auteur de Leçons de clinique médicale. En deux volumes. Doin, Paris, 1890. — Kurko VS. Gastrointestinal bleeding of non-ulcer origin. In, LYa Kovalchuk, VF Sayenko, HV Knyshov. Klinichna Khirurhiya. Vol. 1-2. Publ. «Ukrmedknyha», Ternopil 2000. Vol. 2. P. 84-93.

[422] Osler W. Gulstonial lectures on malignant endocarditis. Lecture I & II. Lancet 1885;1(415-18):459-64. — Osler W. The Principles and Practice of Medicine. Publ. Appleton & Co, New York 1892. — Osler W. On a family form of recurring epistaxis, associated with multiple teleangiectases of the skin and mucous membranes. Johns Hopkins Hosp 1901;12:333-7. — Kurko VS. Gastrointestinal bleeding of non-ulcer origin. In, LYa Kovalchuk, VF Sayenko, HV Knyshov. Klinichna Khirurhiya. Vol. 1-2. Publ. «Ukrmedknyha», Ternopil 2000. Vol. 2. P. 84-93.

[423] Weber FP. Multiple hereditary developmental angiomata (telangiectases) of the skin and mucous membranes associated with recurring haemorrhages. Lancet 1907; 2:160-62. — Kurko VS. Gastrointestinal bleeding of non-ulcer origin. In, LYa Kovalchuk, VF Sayenko, HV Knyshov. Klinichna Khirurhiya. Vol. 1-2. Publ. «Ukrmedknyha», Ternopil 2000. Vol. 2. P. 84-93.

In 1885 William Osler observed an association between perioperative bacteremia and native valve endocarditis (NVE) of the heart. Common physical findings of NVE are fever, malaise, petechiae or ecchymoses, splinter hemorrhages; the Osler nodes, a small slightly elevated swollen painful pea size areas of bluish, red, or orange coloring, some with a pale center, on the hands thenar and the palm surfaces or the sole and the cushion of toes of the foot (1908-09); Roth's spots [**Moritz Roth (1839 – 1914).** 19th century. Pathology]; the sign or spots of **Moritz Litten (1845 – 1907)**[424] — Prussian German physician, a white-centered retinal hemorrhage, also seen in hemorrhagic stroke, malignant anemia and leukemia; lesions of **Theodore C. Janeway (1872 – 1917)** — American physician, a small erythematous or hemorrhagic painless areas located on the palm or sole, caused by septic embolic deposition of bacteria that form micro-abscesses in the derma of the skin; clubbing of the digits; a new heart murmur or change to an existing one; enlarged spleen (splenomegaly).

Moritz Litten was one of the first who described in 1880 a paradoxical embolism through the patent foramen ovale (PFO).

Victor A. Subbotin [01(13).III.1844, town of Pryluky, Prylutsky Region, Chernihiv Obl., Ukraine – 17(29).IX.1898,? city of Kyiv, Ukraine][425] — Ukrainian physician-hygienist, who studied and graduated from the Medical Department of St Volodymyr University in Kyiv (1861-67). After successfully defending the dissertation entitled «Substances for Physiology of Fatty Tissues» (University Herald, Kyiv 1869;4:1-70) for the Doctor Medical Sciences degree, he became the first extra-ordinary (1873) and ordinary (1880-93) professor of hygiene in his alma mater. Took part

---

[424] Litten M. Ueber akute maligne endokarditis und die dabei vorkommenden retinal veranderungen. Charite Ann 1878;3:135-72. — Litten M. Ueber einige vom allgemein-klinische Standpunkt aus interessante Augenveränderungen. Berl klin Wschr 1881;18:23-7.

[425] Town of Pryluky (first chronicle mention 1085; during organized by Russian Communist Fascist regime, Holodomor-Genocide 1932-33 against the Ukrainian nation, this entirely Ukrainian town was surrounded by the food provisions siege, as result of which the nutritive situation became extremely bad, 41 people suffered from hunger swelling, more than 51 residents suffered from hunger cachexia being entirely unable walk to work, refectories were unable to nourish sufficiently own workers. Selective research of the peasant 20 villages in the Prylutsky Region revealed 797 hungry families – 1050 persons and 1800 children needing emergency help, tenth of thousand terrorized by hunger peasants were trickling into the town, but even factories refectory could not decrease those massive killings), Prylutsky Region, Chernihiv Obl., Ukraine. — Subbotin V. Einiges über die Wirksamkeit des über-marganzsaueren Kalis auf Albumin Körper. Chemischer Centralblatt 1865;38. — Subbotin V. Zur Frage über die Anwesenheit der Peptone im Blut und Chylysserum. Zeitschrift der rationale Medicin 1886;30. — Subbotin V. Beträige zur Physiologie des Fettgwebes. Zeitschrift für Biologie 1870;6:73-94. — Subbotin V. Mittheilung über Ein den Einfluss der Nahrung auf Kämoglobingehalt des Blutes. Zeitschrift für Biologie 1870;7. — Subbotin V. Ueber die physiologische Bedeutung des Alkohols für den tierishen organismus. Zeitschrift für Biologie 1870;7. — Subbotin V. Textbook of Hygiene. 1880-93. — Nikberg I. Viktor Andriyovych Subbotin the first Ukrainian professor of hygiene. UHMJ «Ahapit» 1996;4:31-6. — Ablitsov VH. «Galaxy Ukraine». The Ukrainian diaspora: prominent persons. Publ. «Kyt», Kyiv 2007. — Synyachenko OV. They glorified themselves and Ukraine: history of domestic therapy. Crimean Therapeutic Journal 2010;2(2):21-8.

in liquidation of cholera and typhus in Kyiv, then dedicated his scientific work to physiology, hygiene, epidemiology, organization of health care.

Pioneer in the branch of an internal medicine – gastroenterology was **Carl A. Ewald (30.X.1845, city of Berlin, Germany – 20.IX.1915, city of Berlin, Germany)**[426] — German gastroenterologist, who initiated investigation of gastric secretion by intubation of the stomach using thin rubber tube, designed a flexible rubber gastric tube of large caliber for emptying the contest of the stomach and its irrigation (Ewald tube, 1885).

In 1888 **Fedir I. Pasternacky [25(13).XII.1845, town of Dukova, Minsk Obl., Belarus – 07(20).VIII.1902, village Piatevshchyna, Minsk Obl., Belarus]**[427] — Byelorusian Ukrainian physician-internist, found that tenderness or pain during percussion at the costovertebral angle (in Latin – succusio renalis, in English – costovertebral angle tenderness), and in addition the presence of blood in urine (hematuria) is suggestive of a nonspecific bacterial renal parenchymal disease, usually of the interstitial tissue, renal pelvis and calices (pyelonephritis), hemorrhagic fever, inflammation of the ureter (urethritis), perirenal abscess, and also kidney stones disease (urolithiasis) – Pasternacky sign (1888, 1907).

Eugene Baumann (1846 – 96)[428] — German chemist, investigated the interrelation between consumption of iodine and function of the thyroid gland (TG; 1895).

In detailed studies, **Frederick H.H. Akbar Mahomed (1849 – 89)** — British physician, separated chronic nephritis with essential (secondary) hypertension (HTN); described the constitutional basis and natural history of essential HTN and pointed out that this disease could progress into nephrosclerosis and renal failure (RF).

Victor H. Hutinel (1849 – 1933) — French pediatrician, and **Charles Sabourin (1849 – 1920)** — French pathologist and pulmonologist, described hypertrophic fatty liver cirrhosis of alcoholic or tuberculous origin with rapid course, death usually occurring in 2-3 months (Hutinel-Sabourin alcoholic-tuberculous cirrhosis).

Both, **Joseph** von **Mering (1849 – 1908**: 19th century. Biochemistry)[429], and **Oskar Minkowski (1858 – 1931)** — Lithuanian therapeutant of Jewish descent, in study in 1889 in dogs by removing their pancreatic gland, induced diabetes mellitus (DM),

---

[426] Ewald C. Handbuch der allgemeine und speciellen Artzneiverordnungslehre: aus Grunlage des Arzneibuchs fur das Deutsche Reich.Hirschwald. 1-14. Aufl. Berlin 1831-1911.

[427] Town of Dukova (first mentioned 1554), Minsk Obl. (area 39,854 km^2), Belarus. — village Piatevshchyna, Minsk Obl., Belarus. — Pasternacky Fl. Pyelitis. St Petersburg 1907. — Synyachenko OV. They glorified themselves and Ukraine: history of domestic therapy. Crimean Therapeutic Journal 2010;2(2):21-8.

[428] Baumann E. Ueber das normale Vorkommen von Jod im Thicrkörper. Hoppe-Seylers Z Physic Chem 1895-96; 21:319-330.

[429] Von Mering J, Minkowski O. Diabetes mellitus nach Pancreasextripation. Centralblatt für klinische Medicizin, Leipzig 1889;10(23):393-4.

thus established that pancreas not only secretes digestive juices, but also a hormone insulin which control blood sugar level.

Vasyl P. Obrazcov (01/03.I.1849, town of Hriazovtsi, Hriazovetsky Region, Volohodska Obl., Russia – 27.XII.1920, city of Kyiv, Ukraine)[430] — Ukrainian physician, introduced the deep gliding palpation of the abdominal cavity (1887) and the direct percussion with one finger of the chest or abdominal cavity (1910).

In an acute appendicitis he described the psoas muscle (Obraztsov's) sign, i.e., right lower abdominal quadrant pain caused with either passive extension of the right hip or by the active flexion of the patient's right hip while supine. The elicited pain is due to inflammation of the peritoneum overlying the iliopsoas muscles and inflammation of the psoas muscles itself. Straightening out the leg causes pain because it stretches these muscles, while flexing the hip activates the iliopsoas muscle causing pain.

Arthur V. Meigs (1850 – 1912) — American physician, described a cystic degeneration of the heart, spleen and liver (1893), and after **Ibn al Nafis (1213 – 88**: Medieval medicine), **Park Ji-Sung (1661 – ?:** 17th century) and **Marcello Malpighi (1628 – 94**: 17th century) – of capillaries of the myocardium (Meigs's capillaries, 1907).

Serhiy (Serge, Sergei) Podolynskyj (Podolynsky, Podolinski) [July 19 (31), 1850, village Yaroslavka, Shpolansky Region, Cherkasy Obl., Ukraine – June 30(12), 1891, city of Kyiv, Ukraine][431] — Ukrainian scientist, physicist, physician, economist, sociologist and writer, who in 1867-71 studied at the Division of Natural

---

[430] Town of Hriazovtsi (founded 1530), Hriazovetsky Region, Volohodska Obl. (area 147,700 km²), Russia. — Zhukovsky L. From the history of development of the life-time myocardial infarction diagnosis (on the 75th anniversary of the death of V.P. Obraztsov). UHMJ «Ahapit» 1995;3:39-45. — Kontsevyy OO. Acute appendicitis. In, LYa Kovalchuk, VF Sayenko, HV Knyshov. Klinichna Khirurhiya. Vol. 1-2. Publ. «Ukrmedknyha», Ternopil 2000. Vol. 2. P. 125-363. — Ablitsov VH. «Galaxy Ukraine». The Ukrainian diaspora: prominent persons. Publ. «Kyt», Kyiv 2007. — Synyachenko OV. They glorified themselves and Ukraine: history of domestic therapy. Crimean Therapeutic Journal 2010;2(2):21-8. — Wolfson AB, Cloutier RL, Hendey GW, et al. Harwood-Nuss' Clinical Practice of Emergency Medicine. 6th Ed. Wolters Kluwer Health / Lippincott Williams & Wilkins, Philadelphia 2015. P. 5810.

[431] Village Yaroslavka (founded 1810), Shpolansky Region, Cherkasy Obl., Ukraine. — Podolynsky Serge aus Kiev. Beiträge pancreatischen Eiweissfermentes (Inaugural – Dissertation). Der medicinischen Facultät hiessiger Universität zur Erlangung der Universität Breslau. Buchdruckereil Lindner, Breslau 1876. — Serhiy Podolynskyj. The Life and Health of the People of Ukraine. Geneva 1878. — Podolynsky SA. The Work of Man and their Attitude Towards Distribution of Energy. Slovo, St. Petersburg 1880; Nr 4-5:135-211. — Serge Podolynsky. Menschliche Arbeit und Einheit der Kraft. Die neue Zeit. Revue des geistigen und öffentlichen Lebens. Stuttgard. 1883; Band 1, Teil 1(Heft 9):413-24. / Teil 2(Heft 10):449-57. — Serbyn NP. Serhiy Andriiovych Podolynskyj (1850-91). Biography. Ukrainian Historian 1986;03-04;136-44. — Zlupko SM. Serhiy Podolynskyj – scientist, thinker, revolutionier. Lviv 1990. — Podolynskyj SA. Selected Works. Kyiv 2002. — Chesnokov WS. Sergey Andreevych Podolynskyj, 1850 – 1891. Nauka, Moscow 2001. / 2nd Ed., supplemented. Nauka, Moscow 2006. — Ablitsov VH. «Galaxy Ukraine». The Ukrainian diaspora: prominent persons. Publ. «Kyt», Kyiv 2007. — Synyachenko OV. They glorified themselves and Ukraine: history of domestic therapy. Crimean Therapeutic Journal 2010;2(2):21-8.

Sciences of the Physics-Mathematic Faculty if St Volodymyr Kyiv University where he defended dissertation entitled «The Synthesis of Sugar» for the Candidate of Natural Sciences (1871), in 1872-76 he studied medicine at the University of Zürich (founded 1833 from existing colleges which go back to 1525), city of Zürick (permanently settled 4,400 BC, founded by Romans in 15 BC), Switzerland, in 1876 at the Department of Physiology of the Medical Faculty of the Universität Breslau (established 1702), under Rudolf Heidenhain [19th century. Physiology. – **Rudolf Heidenhain (1834-97)**], defended Dissertation named «The Splitting of Proteins by the Pancreatic Gland for the degree of Doctor of Medicine, and then in the St Volodymyr Kyiv University was granted Diploma of Physician.

From December 1876 until December 1877, he worked as physician in the Kyiv Shelter for the Workers Children and lectured at course for nurses.

Thereafter, he resettled near the town of Montpellier (first written mention dated November 26, 986), Department Herault (area 6,101 km^2) on the Coast of the Mediterranean Sea, France. From then, he concentrated all his attention towards scientific work, making fundamental research for his important writing, among them on «The Life and Health of the People in Ukraine» (1878), «The Work of Humans and its Relation to the Distribution of Energy» (1880, 1883).

Serhiy Podolynsky initiated the physical economy designating the balance between amassed endogenic energy and coming exogenous energy and mechanism of its accumulation through the phenomenon of nature and the work of humans; established the equivalent of physical and mental work; elaborated on cosmogenic (space) energetic energy; researched ecologic-economic problems; analyzed the idea of physical economy for the cultivation of steadily development.

Paul G. Unna (1850 – 1929) — German dermatologist, and **Arthur Pappenheim (1870 – 1916)** —German physician, and hematologist, developed staining of plasma cells and nucleoproteins (Unna's alkaline methylene blue stain,1894; Unna-Pappenheim stain, 1908) and a stain for differentiating basophilic granules of erythrocytes and nuclear fragments (Pappenheim stain). In addition, Paul Unna designed a dressing, containing a paste made from gelatin, zinc oxide and glycerin for varicose ulcers.

In 1888 **Étienne L.A. Fallot (1850 – 1911)**[432] — French physician, re-described previously discovered by **Nel Stenson (1638 – 86**: 17th century**)** the four-featured cyanotic congenital heart disease (CHD), named after the formed tetralogy of Fallot (TOF).

---

[432] Fallot BA. Contribution l'anatomie pathologique de la maladie bleue (cyanose cardiaque). Marseilles med 1888:25-77,138,207,270,341,403.

A highly pitched early diastolic murmur («musical», «vibratory»), heard best at the left sternal edge in the 2nd or 3rd intercostal space (ICS) with downward extension along the chest in the patient in full inspiration, characteristic for pulmonary regurgitation (PR) / mitral stenosis (MS) with pulmonary hypertension (HTN) and the cor pulmonale as a result of chronic obstructive pulmonary disease (COPD), was revealed in 1888, by **Graham Steell (1851 – 1942)** — British physician and cardiologist, and named after him Graham Steell murmur. Besides this, he described arterial murmur caused by external compression or intraluminal narrowing (Steell stenose murmur) and a functional low frequency, vibrating or whizzling cardiac murmur in children during the middle-diastole, usually with maximal intensity in the lower border of the sternum (Steell functional murmur).

Henrich O. Shapiro (1852, Grodno Obl., Belorus – 1901, city of St Petersburg, Russia)[433] — Ukrainian Russian physician of Jewish descent, noted bradycardia in course of myocarditis (Shapiro's sign, 1880).

The most common cause of the lysosomal storage diseases is glucosylceramide lipidosis, from the deficiency of glycocerebrosidase (glucosylceramidase) with accumulation of glycocerebrosides in the cells which was described by **Phillippe Ch.E. Gaucher (1854 – 1918)** — French dermatologist, as large spherical cells with eccentrically located nuclei and thin waving fibrillae of keratin in the protoplasm, oriented longitudinally parallel to the long axis of the cell, giving the gray or dark-blue cytoplasm a wrinkle appearance, in the cells of liver, spleen, lymph nodes, alveolar capillaries and bone marrow (Gaucher disease, 1882). It is divided into the non-neuropathic «adult» type with hypersplenism, thrombocytopenia, anemia, jaundice and bone damage; acute neuropathic type of «infants» with hepatosplenomegaly and the central nervous system (CNS) damage that leads to death within the first year of life; and an acute neuropathic «juvenile» type, variable, like to the first two types.

George A. Gibson (1854 – 1913) — Scottish physician, discovered a long loud continuous murmur, occupying most of the systole and diastole, usually localized in the 2nd left intercostal space near the sternum, indicating the presence of a small patent ductus arteriosus (PDA, 1898).

Anatole M.E. Chauffard (1855 – 1933) — French therapeutant, and **Oskar Minkowski (1858 – 1931**: above) reported on hereditary spherocytosis.

Abbott Alkaloidal Company (Co.) / Laboratories were founded in 1888 by **Wallace C. Abbott (1857 – 1921)** — American physician, to formulate the known drugs. Today,

---

[433] Grodno Obl. (area 25,000 km^2), Belorus. — Shapiro HA. Spreading of Cardiac and Vascular Diseases. St Petersburg 1899. — Zetkin M, und Schaldach H. (Hrsg.). Wörterbuch der Medizin. 1. Auflage. Verlag Volk und Gesundheit, Berlin 1956. — Ablitsov VH. «Galaxy Ukraine». The Ukrainian diaspora: prominent persons. Publ. «Kyt», Kyiv 2007. — Synyachenko OV. They glorified themselves and Ukraine: history of domestic therapy. Crimean Therapeutic Journal 2010;2(2):21-8.

it sells medical devices, diagnostics, branded generic medications and nutritional products.

In 1899 **John A. MacWilliam (1857 – 1937)**[434] — Scottish physiologist and pioneer in cardiac electrophysiology, reported his experiment in which a transthoracic application of controlled electrical impulses to the human heart in cardiac arrest (asystole) caused a ventricular contraction and a heart rhythm of 60-70 beats per minute. Therefore, he laid foundations in understanding and treatment of life-threatening heart conditions with an artificial pacemaker (PM).

Co-pioneer of gastroenterology **Isidor I. Boas (28.III.1858, town of Exin, Preussen** – suicide by swallowing overdose of Veronal **15.III.1938, city of Wien, Austria)**[435] — German gastroenterologist of Jewish origin, who was the first to describe bacillus / bacteria Lactobacillus acidophilus in the stomach juce in individuum with gastric cancer, established a painful point to the left of the 12th thoracic vertebra in persons with ulcer of the stomach; Carl Ewald (above) together with Isidor Boas developed a standard «test meal» for gastric investigation which involves of giving to the patient and followed by analysis of gastric juice at scheduled intervals (Ewald-Boas test meal, 1885); first explained clinical significance of the blood in stool, that is presence of gastrointestinal (GI) disorder which require thorough evaluation (1901).

Francis Voelcker (1861 – 1946) — English physician, in 1894 **Ernest G.G. Little (1867 – 1950)** — English dermatologist, in 1901, **Rupert Waterhouse (1873 – 1958)** — English physician, in 1911, and **Carl Friederichsen (1886 – 1979)** — Danish pediatrician, in 1918, described a lightening form of cerebrospinal bacterial septicemia, due usually to invasion by Neisseria meningitidis, causing massive hemorrhages into the adrenal glands with subsequent acute insufficiency, known under eponym of Waterhouse-Friederichsen syndrome. It is characterized by fever, widespread purpura, petechial bleeding into the skin, mucosa and serosa with necrosis, cyanosis, hypotension, coagulopathy, shock, coma, multiple organ failure and death.

Giovanni Paladino (1863 – 1917)[436] — Italian cardiologist, and **Wilhelm His, Jr. (1863 – 1935**; 19th century. Anatomy, histology and embryology) described the congenital anomaly of the conduction system of the heart (CSH) that is the accessory septal pathway between the right atrium (RA) and right ventricle (RV) outside of the CSH (Palatino-Hisian pathway, 1876, 1894).

---

[434] MacWilliam JA. Electrical stimulation of the heart in man. Br Med J 1899;1(1468):348-50.

[435] Town of Exin (appeared in 11th century), Province of Posen, Prussia, today town of Kcynia, Nakło County, Kuyavian-Pomeranian Voivodeship (area 17,969 km^2), Poland.

[436] Eulchori E, Rossi L. The Paladino-His atrioventricular conduction bundle. Pathologica 1997;89(5):572-4. / Sonnimo RE, Mawk JR. Giovanni Paladino: true father of the accessory myocardial conduction pathways. Chest 1988;93(1):199-200.

The sphygmomanometer [**Samuel Ritter** von **Bash (1837 – 1905)**, above] was modified by easily used version in 1896 by **Scipione Riva-Rocci (1863 – 1935)**[437] — Italian pathologist, physician, and pediatrician, to measure the graphical and palpable systolic blood pressure (BP). It consisted of a pneumatic cuff placed on the middle third of the upper arm to slowly inflate and deflate it to compress and decompress the brachial artery. The pneumatic cuff is connected via tube to the mercury manometer and inkwell scale, to register graphically or by palpation at the end of compression or the beginning of decompression the brachial artery systolic BP, and to register graphically at the end of decompression the brachial artery diastolic BP.

Although the clinical picture of congenital heart disease (CHD) was mentioned by **Hippocrates (c. 460 – 370 BC**: Ancient medicine. – *Greece*), it was **Victor Eisenmenger (1864 – 1932)**[438] — Austrian physician, who described its symptoms, signs, and pathology – ventricular septal defect (VSD) with a severe hypertrophy of the right ventricle (RV) and pulmonary insufficiency (Eisenmenger complex, 1897).

Myxedema [**Robert Graves (1796 – 1853**: above), **Karl** von **Basedov (1799 – 1854**: above) and **Augustin Morvan (1819 – 97**: above)] was first treated in 1891 by **George R. Murray (1865 – 1939)**[439] — English physician, of a 46-year-old woman with the disease by hypodermic injections of an extract from a sheep thyroid gland (TG). The patient improved significantly within few weeks and lived another 28 years while taking the sheep TG extract.

Cardiovocal syndrome, described in 1897 by **Norbert Ortner (1865 – 1935)** — Austrian internist, better known as Ortner syndrome, is a left vocal cord palsy due to compression of the left recurrent laryngeal nerve between the aortic arch and dilated left pulmonary vein (PV), enlargement of left atrium (LA) from mitral stenosis (MS) and aneurysms or dissections of the thoracic aorta (c. 5%).

In 1893 **Hans N. Kohn (1866 – 1935)**[440] — German physician, discovered the alveolar (intraalveolar) pores (Kohn pores, 1893) between adjacent alveoli, or interalveolar connections, which function as a means of collateral ventilation, i.e., if the lung is partially deflated, ventilation can still occur to some extent through those pores, and also allow the passage of fluid, bacteria or blood. In congenital abnormality of the airway – bronchial atresia where a lobal or segmental bronchus

---

[437] Riva-Rocci S. Un nuovo sfingomanometro. Gazzeta med di Torino 1896;47:1001-17.

[438] Eisenmenger V. Die angeborenen Defekte der Kammerscheidewände des Herzens. Zeitschr klin Med 1897;32(Suppl):1-28.

[439] Murray GR. Note on the treatment of myxoedema by hypodermic injections of an extract of the thyroid gland of a ship. Br Med J 1891 Oct 10;2(1606):796-7.

[440] Kohn HN. Zur Histologie der indurierenden fibrinösen Pneumonie. Münchener med Wschr 1893 40: 42-5. — Ramsay BY, Byron FX. Mucocele, congenital bronchiectasis, and bronchogenic cyst. J Thorac Surg 1953;26(1):21-30.

ends blindly in the lung tissue, alveoli distal to the bronchial atresia expands as a result of air entering through the pores of Kohn and becomes emphysematous.

Bertram W. Sippe (1866 – 1924) — American physician, treated peptic ulcer initially with milk and cream which over a period of a month was advanced to regular diet (Sippy diet).

In 1896 **Friedel J. Pick (1867 – 1925)** — Czech Austrian physician, reported three patients with constrictive pericarditis and hepatic cirrhosis without edema or jaundice (Pick disease).

Anesthesia

In 1800 **F.W.H. Alexander F.** von **Humboldt (1769 – 1859)**[441] — Prussian botanical geographer, naturalist, and explorer, first seen and described preparation of curare (d-tubocurarine) – a powerful muscle relaxant and paralyzer of respiration, by the Indians living alongside the Orinoco River[442] in South America, from a plant Strychnos toxifera and Chondrodendron tomentosum. The Indians used a soaked (wetted) in curare tip with the bow arrows or spears to asphyxate hunted animals.

Breathing and cardiac complications from pneumothorax during procedures on organs of the chest cavity led to the development of positive and negative pressure ventilators.

The earliest non-invasive ventilator was created in 1832 by **John Dalziel**[443] — Scottish physician, in an airtight metal box that a patient sat in while a manual bellows generated intermittent negative pressure breathing.

The term anesthesia (in Greek – «without sensation») was coined in 1846 by **Oliver W. Holmes, Sr. (1809 – 94)** — American physician, poet, writer, and polymath, in 1846.

In 1842 **Crawford W. Long (1815 – 78)** — American pharmacologist, anesthesiologist, and surgeon, performed three minor surgical procedures using the sulfuric ether anesthesia.

A hollow needle was invented in 1844 by **Francis Rynd (1801 – 61)**[444] — Irish physician.

In Dec. 1844 **Horace Wells (1815 – 48)** — American dentist and anesthesiologist, with the help of **Gardney Q. Colton (1814 – 98)** — American anesthesiologist, and

---

[441] Du Humboldt A, Bonpland A. Personal Narrative of Travels of the Equinocial Regions of the New Continent during Years 1799-1804. Vol. 1-2. Publ. Longman et al, London 1914.

[442] The Orinoco River (length 2,140 m, draining area 881,000 km²) originates in the west from the mountain ridge Parima (elevaion 1,047 m), South America, drains by wide delta into the Atlantic Ocean.

[443] Dalziel J. On sleep and apparatus for promoting artificial respiration. Br Assoc Adv Sc 1838;1:127.

[444] Rynd F. Neuralgia: Introduction of fluid to the nerve. Dublin Med Press 1845; 13:167-8.

John M. Riggs (1811 – 85) — American dentist, used nitrous oxide (N_2O) anesthesia during teeth extraction. The latter described periodontal disease, characterized by a purulent inflammation of the dental periosteum, which produce progressive necrosis of the alveoli and looseness of the teeth, which may fall of the sockets (pyorrhea of a tooth socked, or gingivitis expulsive, or Rigg's disease).

William T.G. Morton (1819 – 68) — American dentist, arranged the first public demonstration of ether anesthesia during the removal of a vascular tumor in the neck (Oct. 16, 1846). A year later (1847) **James Y. Simpson (1811 – 70)** — Scottish anesthesiologist, gynecologist, and obstetrician, who introduced chloroform as an anesthetic in obstetrics and surgery.

Mykola I. Pyrohov / Nikolai I. Pirogov [13(25).XI.1810, city of Moscow, Russia – 3.XI(5.XII).1881, city of Vinnytsia, Ukraine][445] — Ukrainian surgeon, introduced general anesthesia by the administration of an ether through a tube into a dog's stomach, small bowel, and rectum, and into its arteries and veins (1847).

An important event was the invention in 1853 of the first practical syringe for the hypodermic infusions by **Charles G. Pravaz (1791 – 1853)**[446] — French physician. A hollow needle was fastened to the end of the cylinder, made of a hard India rubber (caoutchouc) piston with leather and asbestos and its metal rod and a scale.

The first description of the intercostal space thoracotomy (ICST) / thoracotomy, open cardiac massage and aortic compression was provided by **Moritz Schiff (1828 – 96)**[447] — Jewish German physiologist, in 1874.

To put patients into sleep **Johann F.A.** von **Esmarch (19.I.1828, town of Tönning, State of Schleswig-Holstein, Germany – 23.II.1908, city of Kiel,**

---

[445] City of Vinnytsia (inhabited since Neolithic Age, presence of burials from the Bronze Age, existence of early Slavic settlements from Cherniakhiv and Trypillian Cultures, founded 1355 or 1363), Vinnytsia Obl. (Foot Note 3), Ukraine. The incorrigible loses endured in 1937-41 following political repressions by Russian Communist Fascist regime, when NKVS carried out massive killing of peaceful population during the «Vinnytsia tragedy» by shooting 9,432 people, among them 169 women, aged 25-45-year-old, while the general number of repressed in Vinnytsia and Vinnytsia Obl. in 1932-33 and 1937-41 reached 20,000 people. — Pirogov NI. Collection of Works. Vol. 1-8. Publ. «Medgiz», Moscow 1953-62. — Dragan A. Vinnytsia. A Forgotten Holocaust. Jersey City 1986. — Bihyniak VV. Plastic Surgery. In, LYa Kovalchuk, VF Sayenko, HV Knyshov. Klinichna khirurhiya. Vol. 1-2. Publ. «Ukrmedknyha», Ternopil 2000. Vol. 2. P. 359-90. — Yermilov V. Unknown N.I. Pyrogov: political and social ideas. UHMJ «Ahapit» 2004;14-15:30-5. — Ablitsov VH. «Galaxy Ukraine». The Ukrainian diaspora: prominent persons. Publ. «Kyt», Kyiv 2007. — Hanitkevych YaV. The contribution of the Ukrainian physicians to the world medicine. Ukr Med Chasopys 2009 VII/VIII;4(72):110-15.

[446] Pravaz CG. Sur un nouveau moyen d'opérer la coagulation du sang dans les artères, applicable à la guérison des anéurismes. Comp Rend Acad Sci (Paris) 1853; 36: 88-91.

[447] Schiff M. Beiträge zur Physiologe. Band 1-4. Benda, Libraire-Editor, Lusanne 1894-98. — Vallejo-Manzur F, Varon J, Fromm R Jr, et al. Moritz Schiff and the history of open-chest cardiac massage. Resuscitation 2005;53(1):3-5.

Schleswig-Holstein, Germany)[448] — German surgeon, designed the silk-meshy mask for dropping ether (Esmarch mask) into it. He also re-designed a narrow hard rubber tourniquet which, if temporarily applied, and gently fastened around the limb, controls bleeding, or assures surgical hemostasis during the operation (Esmarch tourniquet, 1869 / Ancient medicine. *The Romans*), one of the first described cardiopulmonary resuscitation (CPR; in German – die Herz-Lungen-Wiederbelebung, 1882, 1912).

Endotracheal intubation was performed in 1880 by **William Macewen (1848 – 1924)** — Scottish surgeon, using the endotracheal tube (ET). He delineated the supra-meatal triangle in the temporal bone between the posterior wall of the external acoustic meatus and the posterior root of zygomatic process (Mac Ewan's triangle), through which an instrument may be pushed into the mastoid atrium in the adult.

In 1886 **George E. Fell (1849 – 1918)** — American surgeon and inventor, who devised apparatus for forced artificial respiration through an opening cut in the anterior wall of the trachea, and in 1889 designed the first wooden electric chair for execution of criminals.

Soon after 1887, **Joseph O'Dwyer (1841 – 98)** — English pediatrician, added to it his own method of intubating the larynx (the Fell-O'Dwyer's apparatus). Air from a foot bellows was passed into the lungs via tubes whose calibrated end pieces were introduced into the larynx. He developed this apparatus to prevent from suffocation or alleviate choking in diphtheria patients due to growth of a false membrane inside the lumen of the larynx which occurs in severe causes of disease.

In 1884 **Olexandr Lukashevych (1851 – ?)**[449] — Ukrainian surgeon, and in 1890 **Maksymilian Oberst (1849 – 1925)**[450] — German surgeon, introduced a local conductive anesthesia of the fingers by injecting cocaine hydrochloride solution **[Friedrich Gaedcke (1828 – 90). 19th century. Biochemistry]** on both sides of extensors distally to a tourniquet placed at the base of the finger. By 1886 Olexandr Lukashevych reported the use of cocaine hydrochloride solution for local conductive

---

[448] Town of Tönning, (destroyed in the Burchardi flood in 1634, rebuilt), State of Schleswig-Holstein (area 15,763.18 km²), Germany. — City of Kiel (founded 1233), Schleswig-Holstein, Germany. — Von Esmarch F. Die erste Hülfe bei plötzlichen Unglücksfällen. Ein Leitfaden für Samariter-Schulen in fünf Vorträgen. Leipzig 1882; später unter dem Titel «Die erste Hülfe bei plötzlichen Unglücksfällen. Ein Leitfaden für Samariter-Schulen in sechs Vorträgen». ebenda 1912. S. 69-82.

[449] Lukashevych AI. About the subcutaneous injection of cocaine. Medical Review 1886;10. — Pernice L. Ueber Cocainanästhesie. Dtsch med Woschr 1890; 16: 287-9. — Buess H. Über die Anwendung der Koka und des Kokains in der Medizin. Ciba Z 1944; 8: 3362-5. — Ablitsov VH. «Galaxy Ukraine». The Ukrainian diaspora: prominent persons. Publ. «Kyt», Kyiv 2007.

[450] Oberst M. Die Amputationen unter dem Einflusse der antiseptischen Behandlung. Max Miemeyer, Halle 1882.

anesthesia in 150 cases on himself and volunteers, as well in 36 operations under local anesthesia with satisfactory results (city of Kyiv, Ukraine).

The first spinal analgesia was administered in 1885 by **James L. Corning (1855 – 1923)**[451] — American neurologist, who experimenting with cocaine hydrochloride solution on the spinal nerves of a dog when he accidentally pierced dura mater. A year later (1886) to prevent pain during urological procedures he administered the first spinal anesthesia in man by injecting 30 minims of 3% solution hydrochloride cocaine in the space between the 11th and 12th spinous process of the dorsal vertebrae.

Local (topical) anesthesia with cocaine hydrochloride for procedures on the eye was reported by **Karl Koller (1857 – 1944)** — Austrian medical student of Jewish origin, in 1884, subsequently American ophthalmologist.

In 1889, **J. Heinrich Draeger (1847 – 1917)** — German engineer, inventor and industrialist, and **Carl A. Gerling (1853 – 90)** — German engineer, inventor, and industrialist, founded the Draeger Manufacture in the city of Lübeck (settled after the last Ice Age, the 10th century, 1143), Germany, for the rescue production of artificial respiration apparatuses for miners and firefighters.

The next planned spinal anesthesia for surgery in man was administered on August 16, 1898, by **August K.G. Bier (1861 – 1949)** — German surgeon, when he injected 3 ml of 0.5% cocaine hydrochloride solution into 34-year-old laborer for segmental resection of infected left ankle[452]. Then in 1908-09 performed the regional and intravenous (IV) cocaine anesthesia for brief operations upon the hand, wrist, foot, ankle, and leg, but gave it up due to toxicity of cocaine. He also pioneered the field of sport medicine (1919).

1893 **Victor Eisenmenger (1864 – 1932**: 19th century. Medicine**)** designed a cuffed endotracheal tube (ET) with the aim of complete elimination of the space around the trachea, the insertion into the trachea through the oral cavity or via one of the nostrils of the nose for the administration of anesthesia, maintenance of ventilation, suctioning of the tracheobronchial secretions, prevention of aspiration of the gastric contennt or foreign bodies into the tracheobronchial tree.

The positive pressure ventilation was developed by **Edovard Quenu (1852 – 1933)** — French surgeon, in 1896. Quenu apparatus was made of a diver helmet which had an airtight seal around the neck of the patient, and compressed air was pumped in over a sponge soaked with chloroform.

---

[451] Gorelick PB, Zych D. James Leonard Corning and the early history of spinal puncture. Neurology 1987;37(4):672-4.

[452] Bier A. Versuche über cocainsung des Rückenmarkes. Deutsch Zeitschrift für Chirurgie 1899;51:361.

The first arterial blood pressure (ABP) monitoring during anaesthesia using the sphyngomanometer was recorded in 1897 by Sir **Leonard E. Hill (1866 – 1952)**[453] — English physiologist, and **Harold L. Barnard (1868 – 1908)** — English surgeon.

Sir Leonard Hill believed that «*the arterial pressure can be taken in man as rapidly, simply, and accurately as the temperature can be taken with the clinical thermometer*». Subsequently, he noted a rise of systolic ABP with exciting. That observation, apparently led him to discover that the peripheral popliteal systolic ABP ≥30 mmHg than the systolic ABP of the brachial artery, precisely indicate aortic insufficiency / aortic regurgitation (AI, AR), named the Hill's test. In AI / AR the leaking of the aortic valve (AV) of the heart causes the portion of the blood to flow in the reverse direction during the left ventricular (LV) diastole, from the ascending aorta AA) to the LV, making the heart muscle to work harder than normal and create an aortic diastolic murmur in the 2nd intercostal space (ICS) at the parasternal line.

When he was introduced to a diving facility where he have seen people suffering from caisson (decompensation) disease which occurs in the human body by the quick transition from the environment with an increased pressure of the air and accumulation of nitrogen (N) in the blood, to the environment with lower pressure, where N is violently getting out of the blood and do not have time to diffuse through the lungs outside, causing muscle pain, itching of the skin, nose bleeding, paralysis of the legs, injury to the lungs and the heart. He found that decompressed-air illness could be overcome by the slow uniform decompression which would allow time for the excess of N to escape from the body through the lungs without bubble formation. By this method Sir Leonard Hill and **Major Greenwood (1880 – 1949)**[454] — English epidemiologist and medical statistician, succeed fully decompressed after exposure to 6 atmosphere the environment by allowing 20 minutes of decompensation per one atmosphere (1913).

Sir Leonard Hill rightly believed that exercise and physical therapy (PT) is the best way to prevent the development of many diseases or treat them and extend the lifespan of the people. Thereby, he walked every day 3-4 miles along the cliffs in the bricking east wind and was mentally alert and physically vigorous until his 86th year he died without warning from a cerebral thrombosis.

---

[453] Hill L, Barnard H, Sequera JH. The effect of venous pressure on the pulse. J Physiol 1897 Mar 17;21(2-3):147-59. — Hill L, Barnard H. A simple and accurate form of sphygmomanometer or arterial pressure gauge contrived for clinical use. Brit Med J Oct 2, 1897;2:904. — Hill AB, Hill B. The life of Sir Leonard Erskine Hill FRC (1866-1952). Proc R Soc Med 1968 Mar;61(3):307-16. — Nuqui NH. Who was the first to monitor blood pressure during anesthesia? Eur J Anaesthesiology 1998;15:255-59.

[454] Farewell V, Johnson T. Major Greenwood (1880-1949): a biographical and bibliographical study. Stat Med 2016 Feb 28;35(5):645-70.

Surgery

In the spring of 1801 **Francisco Romero (ca. 1670 – after 1819)**[455] — Catalonian surgeon, was the first to perform, in a 35-year-old farmer suffering from pericardial effusion, through an incision next to the curvature of the left 6th rib at the cartilage level an incision into the pericardium with a small, curved scissors and drained 2,268 gm of brick-colored pericardial fluid (pericardiotomy), without complications, going back to work in 4 months. During 3 year of follow-up the only complain was an incisional pain.

Initial surgical interventions in the congenital chest wall deformity – thoracoabdominal ectopia cordis should consist of primary excision of the omphalocele with skin closure to avoid infection and mediastinitis, or by local application of astringent, thus allowing secondary epithelialization. The first intervention with the intact skin coverage despite the abnormal location and coverage of the heart was achieved by **Michel-Jean Cullerier (1758 – 1827)**[456] — French surgeon, in 1806 with the long-term viability of the patient.

John Abernathy (1764 – 1831) — English surgeon, described fascia iliaca and also sarcoma of the trunk (Abernathy's sarcoma).

The main surgeon (1797-1815) of the French Army and Emperors Guard in the service of **Napoleon Bonaparte (1759 – 1821)** — French general and emperor, **Dominique J. Larrey (1766 – 1842)**[457] — French military surgeon, named the right and left trigonum sternocostale, a space between the sternal and ribs attachment of the diaphragm, below which the internal thoracic vessels became the superior epigastric vessels (Larrey triangle); pioneered the technique of pericardial window through the site between the base of the xiphoid process and the united ends in the 7th and 8th left costal cartilages to release pericardial tamponade (Larrey point); advocated early amputation in gunshot fractures of extremities, mainly disarticulation of the shoulder and hip joints.

During the Napoleonic Wars (1803-15), Dominic Larrey was the first to introduce into medicine gauze [Ancient Medicine. *Palestine / Israel. //* Medieval medicine. *Bologna (1250), Italy*], instituted «flying ambulances» and attended the wounded of both sides of the battle, thus, he was harbinger of the principles of the Red Cross (RC).

He also described «trench foot», scurvy, contagious eye infection, feeding of a malnourished patient through a stomach tube; treated soldiers with cardiac wounds

---

[455] Aris A. Francisco Romero, the first heart surgeon. Ann Thorac Surg 1997;64(3):870-1.

[456] Cullerier M. Observations sue un deplacemnt remarquable du coeur; par M. Deshamps, medicin a Laval. J general med chir pharmacie 1806;26:275-9.

[457] Larrey DJ. Mémoires de chirurgie militaire, et campagnes. V. 1-4. J Smith, Paris 1812-17. — Larrey DJ. Mémoires de chirurgie militaire. Bull Sci Med 1819;6:284.

by pericardial drainage and an open pneumothorax by closure of the wound with an occlusive dressing (1829).

Captain of the 2nd Grenadier regiment of the French Army and assistant to **Dominique Larrey (1766 – 1842**: above), **Dominique (Demian) P. de la Flise (18.XII.1787, city of Nancy,** department Meurthe-et-Moselle, region Grand-Est, **France – 1861, town of Nizhyn, Chernihiv Obl., Ukraine)**[458] — French military medic-surgeon, in November 1812 near city of Smolensk (founded 863), Russia, was wounded, became prisoner of war (POW) of the Russian Army, and afterwards, settled and married in Ukraine, became Ukrainian physician and ethnographer. In 1831-38 he headed the Economical Hospital of die Countess Über-hoffmeisterin von **Olexandra V. Branytsky**, born **Engelhard (1754 – 1838)**, town of Bila Tserkva (first chronical mention in 862 and 1115, founded 1589; because of creation by Russian Communist Fascist regime of Holodomor - Genocide 1932-33 against the Ukrainian Nation, 108 innocent residents were annihilated by artificial hunger), where he provided duties of medical service in Bilatserkivs'kyy Community / Region (area 6,514.8 km^2), Ksaverivs'kyy Region (presently incorporated into Bilatserkivs'kyy Region) and Rokytnyans'kyy Region (area 661,1 km^2), Kyiv Obl., Ukraine. During cholera epidemics in only Bila Tserkva, he in over 300 patients saved life of 247 persons, while in course of seventh years inoculated (vaccinated) against smallpox over 10,000 peasantry children. In the Albums of «Medico-topographic Description…Kyiv District» (I-IX: 1848-57; pages 1245) laid out his studies about the influence of the surrounding environment on the health of peoples.

Yefrem / Efrem Mukhin (1766 – 1850: 18th century) resuscitated drowned, suffocated, and choked peoples (1805), treated those with trauma and broken bones (1806), introduced freezing of cadavers for an axial study of the anatomy («frozen axial sections», 1818).

Sir **Astley P. Cooper (1768 – 1841)** — English surgeon, described the cremasteric fascia covering the spermatic cord, formed of delicate connective tissue and muscular fibres derived from the internal oblique muscle of the superficial inguinal ring (Cooper fascia), the pectinate ligament on the crest on the pubic bone (pubic ligament of Cooper), inguinal hernia with a direct and oblique sac (pantaloon hernia) and supensory ovarian fascia of Cooper, developed inguinal hernia repair using the

---

[458] City of Nancy (earliest human settlement in the area date to 800 BC, a small, fortified town built around 1050, burned 1218, rebuilt in stone by 1477), department Meurthe-et-Moselle (area 5,26 km^2), region Grand-Est (area 57,433 km^2), France. — Town of Nizhyn (vicinities / town inhabited since the Epoch of Bronze and early of Iron, and Ancient Slav times, first chronical mention 1078; on account of creation by Russian Communist Fascist regime of Holodomor-Genocide 1932-33 against the Ukrainian Nation died at least 487 residents), Chernihiv Obl., Ukraine.

pubic ligament of Cooper (1804-07), and also ligated an abdominal aortic aneurysm (AAA, 1817).

Ivan A. Bush (1771 – 1843) — Russian surgeon of German origin, described the detachment of the triangular fragment from the base of the nail's phalanx of the finger together with the extensor tendon (Bush fracture).

In 1805-06 **Phillip Bozzini (1773 – 1809)** — German physician of Italian ancestry, designed the first endoscope – the «light conductor » (In German – «einen Lichtleiter») for investigation of the urethra and the urinary (the urethro-cystoscopy) and the other hollow organs.

In the field of anatomy **Abraham Colles (1773 – 1843)** — English anatomist and surgeon, is known for description of the deep (membranous) layer of the superficial fascia of the perineum (Colles fascia or fascia diaphragmatis urogenitalis inferior). It is in front continuous with the dartos fascia of the penis and fascia of **Antonio Scarpa (1752 – 1832**: 18th century**)** upon the anterior wall of the abdomen; on either side is firmly attached to the margins of the rami of the penis and ischium lateral to the crus penis and as far back as the tuberosity of the ischium; posteriorly, it curves around the superficial transverse perineal muscle to join the lower margin of the inferior fascia of the urogenital diaphragm; in the middle line, it is connected with the superficial fascia and the median septum of bulbospongious muscle. He also made description of the ligamentum inguinale reflexum. In orthopedics he is known for his description of a fracture of the lower end of the radius at the wrist in which a distal fragment is displaced posteriorly (Colles fracture).

An anal or rectal fissure is a break or tear in the skin and mucosal membrane in the anal canal and the rectum, or both, in the lineal or triangular form, caused by an inflammation of an anal glands inside the anal crypts, stretching of the anal mucosa and spasm of the internal sphincter muscle, resulting in non-healing ulcer infected by fecal bacteria[459] It may be noticeable by a bright red anal bleeding on toilet paper, toilet or under garments. An acute anal fissure cause pain after defecation, chronic fissure are less painful. They usually extend from the anal opening in the middle posterior line and may pass down into the underlying internal sphincter muscle, because of weakly supported tissue and poor blood supply in that area. Anal fissure may occur from childbirth in women, poor toileting in children and adult, sexually transmitted disease such as syphilis, herpes, chlamydia, and human papilloma virus. The lateral fissure is usually caused by occult abscess, leukemic infiltrates, tuberculosis, carcinoma, acquired immune deficiency syndrome (AIDS) or in inflammatory bowel disease.

---

[459] Korchynsky IYu. Fissures of the rectum. In, LYa Kovalchuk, VF Sayenko, HV Knyshov. Klinichna Khirurhiya. Vol. 1-2. Publ. «Ukrmedknyha», Ternopil 2000. Vol. 2. P. 304-9.

In infantile fissure could be prevented by frequent diaper change and hydration, in adult by eating food rich in dietary fibers, hydration and stool softeners to prevent constipation, proper hygiene after defecation using soft toilet paper, sanitary wipes, cleaning with water, a bidet or an «anal douche» (introduced in 1710 in Italy), or sitz bath.

Local treatment with nitroglycerine and a calcium channel blocking ointments may relax sphincters. If an anal fissure doesn't heal after 1-3 month of prevention and conservative treatment, then surgical intervention is indicated. That include manual anal dilatation or stretching up to four the finger breath, a left lateral partial internal sphincterotomy (LIS), and total dissection of an anal fissure in the longitudinal direction in the form of an oval flap of the mucous membrane and perianal skin, a partial LIS and closure of defect in the transverse direction by sewing the mucous membrane of the rectum to the perianal skin.

Manual anal dilatation or stretching was described in 1838 by **Joseph C.A. Recamier (1774 – 1852)**[460] — French gynecologist, who previously (1829) coined the term «metastasis».

Guillane Dupuytren (1777 – 1835) — French surgeon, and **Jacques Lisfranc (1790 – 1847)** — French surgeon and gynecologist, performed an amputation of the arm at the shoulder joint.

Dupuytren contractures denotes a shortening, thickening and fibrosis of the palmaris or plantaris fascia which produce flexion deformity of fingers and toes.

Jaques Lisfranc described tuberculum of the 1th rib for the insertion of musculi scaleni anteriorii (Lisfranci); described the tarso-metatarsal joint – a fibrous band extending from infero-external surface of the middle cuneiform bone to the internal surface at the base of the second metatarsal bone (Lisfranc joint); performed a surgical separation of the foot at the tarso-metatarsal joint (Lisfranc amputation).

The most common benign chest wall tumors are osteochondroma (nearly 50%), chondroma (15%) and desmoid tumor. The most common malignant chest wall neoplasms are malignant fibrous histiocytoma (fibrosarcoma), chondrosarcoma and rhabdomyosarcoma.

After consulting a patient who was physician with fungating mass protruding from the left chest, **Anthelme B. Richerand (1779 – 1840)**[461] — French surgeon, on March 31, 1818, resected the 6th and 7th left ribs, but acute respiratory distress occurred unexpectedly as the chest cavity was entered. The patient was saved by covering the aperture in the chest wall with a linen cloth plastered with cerate. Patient survived and returned home 27 days postoperatively. A primary rib malignancy was suspected.

---

[460] Recamier J.C.A. Extension, massage et percussion cadencee das traitment des contractures muscullaires. Revue Medicale 1838;1:74-90. — Recamier J.C.A. Reserches sue de Traitment du Cancer. Paris 1829.

[461] Richerand AB, Deshamps JFP, Percy P.F. Histoire d'ne resection des cotes et de la plevre. Academie royale des science (France), Paris 1818.

He is also credited with removal of the rectum, incising the urinary bladder for removal of a stone (lithotomy) in women and amputation of cervix uteri.

In 1828 **Jean N. Marjolis (1780 – 1850)** — French pathologist and surgeon, described a secondary aggressive ulcerated lesion due to well differentiated squamous cell carcinoma in the area previously, usually 10-20 years later, traumatize, chronically inflamed or scarred skin, commonly present in the chronic wounds, including burn injuries, varicose veins, venous and bed sore or decubitus (decumbent) or pressure ulcers, ulcers from chronic osteomyelitis, and with introduction of radiotherapy by the end of the 19th century – in post radiotherapy scars, named Marjolis ulcer. It occurs in 40% on the lower extremities, spreads locally, marginal changes are usually painless, ulcer spreads locally, and is associated with poor prognosis. Confirmation requires both, central and marginal biopsy of the lesion. The treatment of choice is wide excision of the lesion with a one cm margin all around the lesion.

Reactive arthritis, known to **Hippocrates (c. 460 BC. – 375 BC**: Ancient medicine – *Greece*), was precisely described in 1818, by Sir **Benjamin C. Brodie (1783 – 1862)** — English physiologist and surgeon, and in 1916, among the soldiers of WW I by **Hans C.J. Reiter (1881 – 1969)** — German physician, as the complex of arthritis, conjunctivitis, and urethritis, named Reiter syndrome or reactive arthritis.

Benjamin Brodie diagnosed an oval abscess of the metaphysis of the long tubular bones caused by Staphylococcus aureus or albus (Brodie abscess), chronic degenerative synovitis of the knee joint (Brodie disease), and hysterical fracture of the vertebra; described the oval shaped destruction of the bone, filled with puss or connecting tissue, affecting mainly in the area of metaphysis of long tubular bones caused by Staphylococcus albus or aureus (Brodie abscess), a lingering inflammation of the joint synovia[462], where the affected area becomes of soft spongy density (Brodie knee), a chronic synovitis, especially the knees, with degeneration of the affected surfaces (the 1st Brodie disease); diagnosed a hysterical pseudo-fracture of the vertebra (the 2nd Brodie's disease); noted a black spot on the head of the penis as a result of gangrene caused by the transudation[463] of urine into the spongious body (Brodie sign).

In 1818 **Valentine Mott (1785 – 1865)**[464] — American vascular surgeon, ligated in a patient the proximal innominate artery (IA) or brachiocephalic trunk (BCT) for an

---

[462] Synovium (in Greek and Latin: «syn», means «egg») is a secretion of the synovial membrane of the joints, cavities, and ligamentous sheaths, consists of a clear content, alkaline viscous liquid, similiar to the albumen of the eggs.

[463] Transudate means a passing of liquid ingredients (serum) and other body fluids through the blood vessel membrane or being forced out of the blood by force due to the action of hydrodynamic forces, e.g., inflammation, tumors. Unlike the exudate, the property of transudate is high fluidity and low content of protein, cells, or solids of cellular origin.

[464] Rutkow IM. Valentine Mott (1785-1865), the father of American vascular surgery: a historical perspective. Surgery 1979 Apr;85(4):441-50.

aneurysm of the right subclavian artery (SCA). The patient died 26 days thereafter, probably from erosion of an artery by the ligature. Subsequently, he performed a successful ligation of aneurysms of the carotid arteries and an aneurysm of the common iliac artery (CIA, 1827).

Description of a cylindrical aneurysmal dilatation in the proximal part of the aorta without saccular bulging, often accompanied by dilatation or hypertrophy of the heart, belongs to **Joseph Hodgson (1788 – 1869)**[465] — English physician, and was called Hodgson disease.

Further relative and pupil of **Petro Zahorsky (1764 – 1846**: 18th century), **Illia V. Buyalsky [26.VII(06.VIII).1789, village of Vorob'ïvka, Novhorod-Siverskyy Region, Chernihiv Obl., Ukraine – 08(20).XII.1866, city of Petersburg, Russia]**[466] — Ukrainian anatomist, surgeon, and forensic (judicial) medicine physician, authored «Anatomo-Surgical Tables» (1826, 1835 i 1852), with **Efrem Mukhin (1766 – 1850**: 18th century)** improved the method of «frozen axial sections» of cadavers to study anatomy (1836), resected aneurysms and the upper jaw because of a peripheral ossified fibroma (epulis) of the gum of (1843).

In 1832 **John MacFarlane**[467] — Scottish surgeon, described two patients with desmoid tumors, a benign appearing disease but locally invasive in the aponeurosis of the abdominal muscles. The incidence is approximately 2-3 case per million of the population per year. Twenty percent of patients with dermoid tumor involvement of the chest wall, but less than half complain of pain. Optimal treatment is a complete resection with generous margins. Recurrent tumor is identified in over 20% of patients when surgical margins are negative; recurrent persistent disease is expected when margins are positive.

Wilhelm F. von **Ludwig (1790 – 1865)** — German physician, depicted a severe form of cellulitis of the submaxillary space with involvement of the sublingual and submental spaces, resulting from an infection of the mandibular molar teeth or penetrating injuries of the floor of the mouth (Ludwig angina, 1836).

An important contribution to the treatment of urinary bladder stones was made by **Jean Civiale (1792 – 1867)** — French surgeon and urologist, who in 1832

[465] Hodgson J. Treatise on the Diseases of Arteries and Veins, Containing the Pathology and Treatment of Aneurisms and Wounded Arteries. London 1815.

[466] Village Vorob'ïvka (founded 1650), Novhorod-Siversky Region, Chernihiv Obl., Ukraine. — Bohoslavskyy YeV. Authority of the European Scale (Illia Vasylevych Buyalsky). Chernihiv 2004. — Olkhovskyy VO, Dunayev OV, Biriukova LI, Kobetskoi SP. Founder of judicial medicine-scientist and physician I.V. Buyalsky. Ukr med almanac 2011;14 (2): 137-39. — Ablitsov VH. «Galaxy Ukraine». The Ukrainian diaspora: prominent persons. Publ. «Kyt», Kyiv 2007. — Hanitkevych YaV. The contribution of the Ukrainian physicians to the world medicine. Ukr Med Chasopys 2009 VII/VIII;4(72):110-15.

[467] MacFarlane J. Clinical report of surgical practice of Glasgow Infirmary, Glasgow, Scotland. D. Robertson 1932:63-6.

invented a surgical instrument, the lithotrite and performed transurethral lithotripsy by crushing stones inside the urinary bladder without having to open the abdomen (lithotomy). He inserted his instrument through the urethra and bored holes in the stone, crushed it with the same instrument and aspirated the resulting fragments or let them flow normally with urine. The relative mortality rate for lithotripsy was 2.2%, while for lithotomy it was 18.8%.

Robert Liston (1794 – 1847) — Scottish surgeon, excised the upper jaw (Liston operation).

After the first accurate diagnosis of leukemia (1827) by **Alfred A.L.M. Velpeau (1795 – 1867)** — French anatomist and surgeon, a bandage to support and to provide immobilization of the upper arm, elbow, and shoulder to the chest («Velpeau bandage») was devised by him.

After conducting extensive studies (1833-36), in 1836 brothers **Wilhelm E. Weber (1794 – 1891)**[468] — German physicist, and **Eduard F. Weber (1796 – 1871)** — German anatomist and physiologist, described the exact anatomical location of the cruciate ligaments of the knee joint, and discovered that the anterior cruciate ligament (ACL) was made-up of two distinct fiber bundles. They also found that cutting or tear of the ACL would do abnormal knee movements, allowing the tibia to move forwards on the femur, a sign known as anterior draw.

Modern plastic surgery was pioneered by **Johann F. Dieffenbach (1795 – 1847)** — German surgeon, **Charles E. Sedilott (1804 – 83)** — French surgeon, **Karl Thiersch (1822 – 95)** — German surgeon, **John F. Wolfe (1824 – 1904)** — Scottish ophthalmologist, **Juliy K.** von **Shymanovsky** / von **Szymanowski (27.I.(08.II). 1829, city of Riga, Latvia** – died from metastases after removal of malignant testicular tumor **13(25).IV.1868, city of Kyiv, Ukraine)**[469] — Ukrainian surgeon, **L.X.E.Leopold Ollier (1830 – 1900)** — French surgeon, **Jacob A. Estlander (1831 – 81)** — Finnish surgeon, **Jaques L. Reverdin (1842 – 1929)** — Swiss surgeon, **Olexandr S. Yatsenko (28.IX.1843, city of Simferopol, Crimea, Ukraine – 06.X.1897, city of Kyiv,**

---

[468] Weber WE, Weber EF. Mechanik der menschlishen Gehwerkzeuge. Eine anatomisch-physiologische Untersuchung. Verlag Diederich, Göttingen 1836.

[469] City of Riga, Latvia (Foot Note between 245 & 350). — Tartu (previously Dorpat) Univ., city of Tartu, Estonia (Foot Note between 265 & 349). — Szymanowski J. Additamenta ad ossium resectionem. Dissertationem pro medicinae doctoris grade. Dorpati Liv: typ. viduæ J.C. Schünmanni et C. Mattieseni, 1856. — Szymanowski J. Adnotationes ad rhinoplasticen, typ. viduæ J. C. Schünmanni et C. Mattieseni Dorpati Liv. 1857. — Von Szymanowski J. Handbuch der operativen Chirurgie. Deutsche Ausg. von dem Verfasser und C. W. F. Uhde. Erster Teil. Druck und Verlag von F. Vieweg u. Sohn, Braunschweig 1870. — Rogers BO. Julius von Szymanowski (1829 – 1868). His life and contribution to plastic surgery. Plast Reconstr Surg 1979 Oct;64(4):468-78. — Der Gypsverband, mit desonderer Berücsichtigung der Militar-Chirurgie von Media Corporation. Tachenbuch, 18 Februar 2001. — Bihuniak VV. Plastic Surgery. In, L.Y.Я Kovalchuk, VF Sayenko, HV Knyshov. Klinichna Khirurhia. Vol. 1-2. Publ. «Ukrmedknyha», Ternopil 2000. Vol. 2. P. 359-90. — Ablitsov VH. «Galaxy Ukraine». The Ukrainian diaspora: prominent persons. Publ. «Kyt», Kyiv 2007.

Ukraine)[470] — Ukrainian surgeon, and **Fedor V. Krause (1857 – 1937)**[471] — German surgeon.

Johann Dieffenbach introduced a closure of the triangular defect by displacing the quadrangular flap towards the one side of the angle.

Leopold Ollier and Karl Thiersch introduced an epidermal (a thin split, or partial thickness (Ollier-Thiersch, 1844, 1872) skin graft.

Jacob Estlander performed a turning of the triangular flap from the side of the lower lip to cover the defect of the lateral part of the upper lip (Estlander operation).

Jaques Reverdin designed an epidermal free graft (a «pinch graft») which is a procedure for removing tiny pieces of skin from a healthy area of the body and seeding them in a location that needs to be covered (Reverdin's graft) and performed the first skin allograft (1869).

Olexandr Yatsenko developed the methods of free skin grafting, which he successfully presented during defence of dissertation for degree of Doctor of Medical Sciences «An Inquire in transfer or acceptance of the separate skin patches on the granulated surfaces» (Medical Surgical Academy, St Petersburg, 1871).

John Wolfe, Olexandr Yatsenko and Fedir Krause introduced blepharoplasty with a full-thickness large flap free skin graft without subcutaneous tissue after blepharoplasty (Krause-Yatsenko-Wolfe, 1871).

Olexandr Yatsenko and **Victor** von **Hacker (1852 – 1933)** — Austrian surgeon, performed bougienage of the stenosed esophagus through gastrostomy (Yatsenko-Hacker bouginage).

Charles Sedilott and Juliy Shymanovsky designed a flap for the reconstruction of the cleft lip (Sedilott-Shymanovsky operation). Juliy Shymanovsky introduced plastic repair of the cleft palate and transplanted the cornea.

Rocco Gritti (1828 – 1920) — Italian surgeon, Juliy Shymanovsky and Sir **William Stokes (1839 – 1900)** — Irish surgeon, performed an amputation of the femur through the knee using a patella as an osteoplastic flap over the end of the femur (Gritti- Shymanovsky-Stokes amputation, 1857-63).

Leonard Ollier described solitary end-chondroma (true chondroma) – a benign growth of the cartilage in the metaphysis of the bone and multiple congenital enchondromas (chondromatosis), osteochondroma or osseocartilaginous exostosis, named Ollier disease. Osteochondroma, or osseocartilaginous exostosis, accounting

---

[470] City of Simferopol (remains of people's residences from Neolithic Era, capital city Scythian Neopolis 3th century BC – 3th century AD, founded 1784), Taurica / Taurida / Tauris / presently Crimea, Ukraine. — Bihuniak VV. Plastic Surgery. In, LYa Kovalchuk, VF Sayenko, HV Knyshov. Klinichna Khirurhiya. Vol. 1-2. Publ. «Ukrmedknyha», Ternopil 2000. Vol. 2. P. 359-90. — Ablitsov VH. «Galaxy Ukraine». The Ukrainian diaspora: prominent persons. Publ. «Kyt», Kyiv 2007.

[471] Krause F. Zur Freilegung der hinteren Falsenbeinflache und des Kleinhirns. Betr Klin Chir 1903;37:728-67.

for up to 50% of benign bone tumors, occurs most frequent in patients under 20 years of age and are usually painless. Multiple tumors and positive familial history are apparent in many patients. Secondary chondrosarcoma develops in up to 1% of solitary benign cartilage tumors, they grow slowly and may reach significant size before becoming obvious on clinical examination. Rapid growth and pain should lead to complete surgical excision.

Identified by **Angello Maffucci (1847 – 1903)** — Italian physician, and **Alfred Kast (1856 – 1903)** — German physician, multiple congenital enchodromas associated with multiple cutaneous and visceral hemangiomas (Maffucci-Kasta syndrome) are at much higher risk of chondrosarcoma (5-15%) than the normal population and often present in the 3rd or 4th decades of life.

Mykola (Nikolai) I. Yellinsky (1796 – 1834)[472] — Ukrainian surgeon, who depicted the signs of inflammatory condition, that is a local response to cellular injury marked by dilatation of small blood vessels (capillary), infiltration by white blood cells (WBC), i.e., leukocytes, redness, heat, and pain, that initiate the elimination of noxious agents and of damaged tissue («On the account of the precise signs of inflammatory processes», 1826), further outlined tasks of desmurgia [Ancient Medicine. *Greece*. –

Hippocrates (460 – 322 BC)], which include the dressing and bandaging of the wound, the use of gypsum (plaster), braces, splints, or orthosis (in Greek – «orthos» meaning to straighten, or to align) to immobilize dislocation or fractured bone, presented the attainments of traumatology of those approaches («The Review of Desmurgia», 1832).

Frederic Salmon (1796 – 1868) — English surgeon, expanded surgical procedure for hemorrhoids (Ancient medicine. *Egypt*) into a combination of excision and ligation, where the peri-anal skin is incised, the hemorrhoidal plexus and the muscles are dissected, and the hemorrhoid is ligated at the base.

James Syme (1799 – 1870) — Scottish surgeon, amputated a foot at the ankle joint at the tibia-calcaneal joint, and removed both malleoli (Syme amputation, 1831).

The first known pericardiocenthesis (in Greek: peri + centesis = around + punction) was performed in 1827 by **Thomas Jawett (1801 – 32)**[473] — English surgeon and botanist, with removal of pericardial effusion. Pericardiocentes is an interventional procedure in which a long hollow needle is inserted from the left costoxiphoid

---

[472] Yellinsky MI. De inflammationis causa proxima. Addresses. Kharkiv University 1826. P. 1-47. — Yellinsky MI. The Method of Treatment Cholera. Typography of Kharkiv University 1831. —Yellinsky MI. The Review Desmurgia. Vol. 1-2. Typography of Kharkiv University 1832. — Ablitsov VH. «Galaxy Ukraine». The Ukrainian diaspora: prominent persons. Publ. «Kyt», Kyiv 2007.

[473] Jawett T. Cause of dropsy of the pericardium, in which the operation of tapping was performed. Medico-Chirurgical Review 1927;623-6. — Jarco S. Thomas Jawett on pericardiocentesis (1927). Am J Cardiol 1973; Feb;31(2):273-6.

angle in the direction of the left scapula tapping the diaphragmatic wall of the pericardium to enter pericardial cavity, is used to remove the fluid with the aim of differential diagnosis of idiopathic, infectious, malignant or immune effusion, and for the removal of an excessive accumulation of the blood between layers of pericardium that lead to pressure on the heart (cardiac tamponade), cardiogenic shock and death.

Pavlo (Pavel) A. Naranovych (1801, village Chapliyivka, Shostkyns'kyy Region, Sumy Obl., Ukraine – 14(26).I.1874, city of Kharkiv, Ukraine][474] — Ukrainian surgeon, graduated from St Petersburg Medical-Surgical (1826), from 1937 – professor of anatomy, since 1843 – surgery at the Kharkiv University. The work of Pavlo Naranovych was dedicated to study the motion system of humans, an inquiry into practical surgery, especially removal of the urinary bladder stones, and proposed number of surgical instruments. Authored about 30 fundamental scientific works related to devoted to anatomy, surgery, military-sanitary matters. Among them best known are: «Chronic inflammation of the heart» [Military-Medical Journal (MMJ) 1823], «Concerning the radical treatment of hernias» (MMJ 1823), «An action of vomiting on stunning of the heart » (MMJ 1840), «Anatomical considerations of an aneurysm of the ascending aorta bursting into the right ventricle of the heart» (MMJ 1842), «Sarcoma on the head, cured by operation» (MMJ 1843) and «Ligation of the brachial artery above the elbow joint» (MMJ 1843), «Diseases of the bone» (1845 – 1846), «Anesthesia with the sulphuryl ether vapor during surgical operations» (MMJ 1847), «Treatment of cold abscesses» (Protocols of the Society of Rus' / Ukrainian physicians 1856 – 1857), «New method of surgical operations without bleeding» (Protocols of the Society of Rus' / Ukrainian physicians 1856 – 1857). Honorary member of his «Alma Mater».

Antoine Lambert (1802 – 51) — French surgeon, introduced the inverting suture patterns (interrupted or continuous) into gastrointestinal surgery.

The first recorded extirpation (removal) of the kidney (nephrectomy) was performed in 1861 by Brigadier General (Brig. Gen) **Erastus B. Wolcott (1804 – 80)**[475] — American military physician and surgeon, in a 58-year-old man suffering from a large retroperitoneal abdominal tumor of six years duration. From the surgical point of view the operation was successful, however the patient died on the fifteenth postoperative day from exhaustion caused by wound infection.

---

[474] Village of Chapliyivka (settled since Neolithic, Bronze and Scythe times, written documents dated to 1605-10), Shostkyns'kyy Region, Sumy Obl., Ukraine. — Ablitsov VH. «Galaxy Ukraine». The Ukrainian diaspora: prominent persons. Publ. «Kyt», Kyiv 2007. — Synyachenko OV. They glorified themselves and Ukraine: history of domestic therapy. Crimean Therapeutic Journal 2010;2(2):21-8.

[475] Tinker MB. The first nephrectomy and the first cholecystostomy, with a sketch of the lives of Doctors Erastus B. Wolcott and John S. Bobbs. John Hopkins Hosp Bull 1901;12:247. — Tinker MB. A sketch of the lives of a German and American Master of surgery. John Hopkins Hosp Bull 1906;17:53. — Best S. Fri 14 Erastus B. Wolcott: a pioneer in renal surgery and Wisconsin medicine. J Urology Apr 2015;193(4S).

The parathyroid gland (PTG) are two pair of endocrine glands (weight 25-40 mg) positioned behind the outer wings of the thyroid gland. They derive from the 3rd and 4th brachial arches, with the superior PTG arising from the 4th brachial arch, and the inferior PTG from the 3rd brachial arch[476].

About 5% of the PTG descend into the ectopic location in the mediastinum. Of them, the inferior PTG descend into the anterior mediastinum, usually inside the thymus, or the middle mediastinum around the great vessels and airways (80%), while the superior PTG descend into the posterior mediastinum around the esophagus (20%).

They were discovered in 1852 in Indian Rhinoceros by Sir **Richard Owen (1804 – 92)** — English biologist, in a human in 1880 by **Ivar V. Sanström (1852 – 89)** — Swedish medical student.

The parathyroid hormone (PTH) regulates the blood calcium and phosphorus level which is essential for the normal bone metabolism. The most common pathology of PTG is their hyperplasia or benign epithelial tumor (adenoma), rarely malignancy – carcinoma. Excess of the PTH causes primary hyperparathyroidism by increase secretion of phosphorus with urine, which leads loss of calcium, bone reabsorption of calcium with subsequent increase of the blood calcium level (hypercalcemia), some increase secretion of calcium with urine and deposition of calcium in tissues. Hypercalcemia is responsible for dehydration, osteoporosis with bone tenderness, pain and fractures, deposition of kidney stones and subsequent failure, arteriosclerosis / atherosclerosis (AS), hypertension, coronary artery disease (CAD), atrial fibrillation (AF), stroke, gastroesophageal reflux disease (GERD), increase risk of cancer, inhibits function of the central nervous system (CNS), and decreases life expectancy by 5-6 years.

Charles M.E. Chassaignac (1805 – 79) — French surgeon, recognized the tuberculum caroticum, the enlarged anterior tubercle on the transverse process of the C$_6$ vertebra, located laterally and slightly above to the posterior tubercle, against which the common carotid artery (CCA) may be compressed by the finger (Chassaignac tubercle).

Auguste Nelaton (1807 – 73) — English anatomist and surgeon, designed a bullet-like catheter with a porcelain head for the location of bullets (Nelaton's bullet catheter) and a soft rubber urethral catheter (Nelaton ureteral catheter).

In the reverse **Abraham Colles (1773 – 1843**: above**)** fracture, i.e., fracture of **Robert W. Smith (1807 – 73)** — Irish surgeon and pathologist, the lower end of the radius is displaced anteriorly (Smith fracture, 1849)[477].

---

[476] Olearchyk AS. Non-recurrent laryngeal nerve. J Ukr Med Assoc North Am 1973;20,4 (71):45-7.

[477] Smith RW. Treatise on Fractures in the Vicinity of Joints and on Certain Forms of Accidental and Congenital Dislocations. Hodges & Smith, Dublin 1847.

The contribution of Robert Smith to pathology includes description of neuromatosis / neurofibromatosis – congenital disease characterized by the development of multiple oval soft subcutaneous nodules of neurogenic origin sizing from the millet grain to 7 cm in the diameter, localized thoroughly along the nerve trunks, not displacing themselves alongside, but only in the transverse direction[478].

Sir **William Ferguson (1808 – 77)** — Scottish surgeon, who repaired cleft palate in child (1852) and resected the upper jaw.

Cholecystostomy or cholecystotomy is a surgical procedure consisting of creation of an artificial round opening into the hollow organ, i.e., stoma in the gallbladder fundus and sewing it to the surrounding skin to facilitate placement of a tube for drainage of the stagnant bile, puss and retained gallstones. Those procedures are performed in lieu of removal of the gallbladder (cholecystectomy) in high risk patients.

Cholecystostomy along with extraction of the gallstones, was first performed on June 15, 1867, by **John S. Bobbs (1809 – 70)**[479] — American surgeon of German descent, in a 31-year-old seamstress who complained of a excruciated abdominal pain due to growing large abdominal mass that prevented her from working. Under chloroform general anesthesia, he performed a lower abdominal incision, found a mass measuring 12.5 cm x 5.0 cm, created a rounded opening in the gallbladder fundus which was filled with stagnant pussy bile and multiple gallstones, drained, and extracted them, realizing it was an empyema of the gallbladder. Then he sewn the stoma in the gallbladder fundus to the skin for drainage. In his correct judgement it was too risky to proceed with cholecystectomy. She fully recovered in six weeks, married 12 years later, died in 1919 at the age of 77, surviving nearly 46 years more.

John Bobbs also resected the superior maxillary bone, together with the eye on the affected side, for extensive carcinoma. Despite that the operation lasted several hours, there was little blood loss and good recovery.

Tito Vanzetti (29 Apr. 1809, city of Venice, Italy – 06 Jan. 1888, city of Padua, Italy)[480]— Italian Ukrainian surgeon, noted that in sciatica of the pelvis is always in horizontal position despite of scoliosis, but in other lesions with scoliosis, the pelvis is in inclined position (Vanzetti sign). He edited «Annales Clinicae Chirurgicae Ceasarea Universitatis Charcoviensis» (1846) in the city of Kharkiv (cultural artefacts from the Bronze Age c. 3200 – 600 BC, founded 1654), Ukraine.

---

[478] Smith RW. Treatise on Pathology, Diagnosis and Treatment of Neuroma. Hodges & Smith, Dublin 1849.

[479] Tinker MB. The first nephrectomy and the first cholecystostomy, with a sketch of the lives of Doctors Erastus B. Wolcott and John S. Bobbs. John Hopkins Hosp Bull, 1901;12:247. — Tinker MB. A sketch of the lives of a German and American Master of surgery. John Hopkins Hosp Bull 1906;17:53.

[480] City of Venice (inhabited since 10th century BC, Italy. — City of Padua (founded 1185), Italy. — Ablitsov VH. «Galaxy Ukraine». The Ukrainian diaspora: prominent persons. Publ. «Kyt», Kyiv 2007.

Bernhard R.K. von **Langenbeck (1810 – 87)** — German surgeon, removed the larynx with part of the tongue, epiglottis and esophagus (1875); developed the method of amputation in which the flaps are formed from outside inside; described muco-periostial flap and the femoral triangle (Landenbeck); introduced lower abdominal paramedian incision with the possibility of its extension above or below in a procedure for a complicated acute appendicitis, cholecystitis and salpingoophoritis or twisted ovarian cyst.

He is considered the «father of the surgical residency», since under his tutelage at the Charité – Universitätsmedizin (XVIII century), Berlin, Germany, he conceived and developed a system where new medical graduates would live at the hospital as the «house staff» gradually assuming a greater experience, confidence, and role in the day-to-day care of surgical patients.

Mykola Pyrohov (1810 – 81: 19th century. Anesthesiology**)** performed a mastectomy with removal of the pectoralis major muscle for breast cancer (1847).

In his «Topographic Anatomy of the Human Body» (Vol. 1-4. 1851-54) re-described the method of sequential «frozen axial sections» of deseased for the study of anatomy, as outlined by **Efrem Mukhin (1766 – 1850**: 18th century**)** and **Illia Buyalsky (1789 – 1866**: above**)**; outlined a venous (Pyrohov) angle formed by the junction of the internal jugular vein (IJV) and the subclavian vein (SCV); researched the angles of the teeth and their cavities (Pyrohov point); proposed an osteoplastic amputation of the leg at the ankle in which a part of the calcaneous bone is left in the lower end of the stump, that is osteoplastic amputation of the foot (Pyrohov's amputation, 1852).

During the Crimean War (1853-56) Mykola Pyrohov made a classic description of hemorrhagic and traumatic shock, proposed a conservative (non-amputative) management of gunshot fractures of extremities; and first began in 1855 to use the gypsum (plaster) bandages, which accelerated healing of fractures and prevented in many soldiers and officers crooking of extremities.

In final year of his life, introduced an inguinal extraperitoneal approach to the external iliac artery (EIA) – Pyrohov's incision (1881).

Volodymyr O. Karavayev [08(20).VII.1811, town of Viatka / Kirov, Kirov Obl., Russia – 03(15).III.1892, city of Kyiv, Ukraine][481] — Ukrainian surgeon, contributed to eye and to plastic surgery. He also performed pericardiocenthesis for hemorrhagic

---

[481] Town of Viatka / Kirov (founded 1181, 1374, since 1934 – Kirov), Kirov Obl. (area 120,800 km^2), Russia. — Karavayev VA BA. Couse of Operative Surgery. Kyiv 1858. — Karavayev VA. Operative Surgery. Kyiv 1886. — Verkhradsky SA. V.O. Karavayev. Kyiv 1976. — Ebert PA, Najafi H. The Pericardium. In, DC Sabiston, Jr., FC Spencer (Eds). Surgery of the Chest. 5th Ed. Vol. I-II. WB Saunders Co, Philadelphia - Tokyo 1990. P. 1230-49. — Kulchytsky K. On the history of Kyiv School of Operative Surgery and Topographic Anatomy. UHMJ «Ahapit» 1995;3:11-6. — Ablitsov VH. «Galaxy Ukraine». The Ukrainian diaspora: prominent persons. Publ. «Kyt», Kyiv 2007.

effusion accompanying an outbreak of scurvy (1839-40). Seven of his 30 patients survived.

A superior sulcus tumor (SST) of the lung was first described by **Edward S. Hare (1812 – 38)**[482] — English surgeon, named Hare syndrome.

In 1850, 1853, 1855 and 1865 **Antonin J. Desormeaux (1815 – 94)** — French physician, modified the endoscope of **Phillip Bozzini (1773 – 1809**: above**)** to investigate vagina, urethra, the rectum and larynx.

Alphonse F.M. Guerin (1816 – 95) — French surgeon, described a horizontal fracture of the alveolar process of the maxilla immediately above the teeth and palate (Guerin fracture, 1866); discovered the cotton-wood bandage (dressing) for prevention of wound infections (1870). After **Rainer de Graaf (1641 – 93**: 17th century**)** re-described the of the lesser vestibular (periurethral) glands women.

Richard H. Coote (1817 – 72) — English physician, did in 1862 the first cervical rib resection with the aim to relieve the patient of thoracic outlet syndrome (TOS).

In South-Eastern Asia (area 5 million km²) and in Northern Australia-Asia, from the Isle of Taiwan (area 36,008 km²) towards the South to the Malay Peninsula and to the East towards the Solomon Islands (area 28,400 km²) grows a tropical caoutchouc (india-rubber, latex) genus of the tree called gutta-percha, and particularly its species – the evergreen tree Palaquium gutta (height 5-30 m, trunk diameter up to 1 m) that secretes a sap from which a inelastic latex – polyterpene or the polymer of isoprene (trans-1,4 polyisoprene) is processed. After its discovery (1842), gutta-percha was introduced in 1847 by **Edwin Truman (1818 – 1905)** — English dentist and inventor, as filling material for dental cavities and as the base for making the teeth impressions (dentures). But as dentures gutta-percha was unsatisfactory because of the distortion after its removal from the mouth and the shrinkage when cooled. To improve its quality **Charles Stent (1807 – 85)** — English dentist and inventor, added in 1856 into gutta-percha a stearin to perfect its plasticity and stability, the talc (talcum)to increase its volume and for red coloring, thus receiving quite a satisfactory compound for dentures.

The first planned nephrectomy was carried out on August 02, 1869 by **Gustav Simon (1824 – 76)**[483] — German general military physician and surgeon, on the 46-year woman with a urinary fistula after surgical removal of her uterus and ovaries (hysterectomy with oophorectomy or ovariectomy).

---

[482] Hare ES. Tumor involving certain nerves. London Med Gaz 1838;1:16-18.

[483] Simon G. Extirpation einer Niere am Menschen. Deutsche Klinik 1870;22:137-8. — Simon G. Extirpation einer Niere bei Steinkrankheit. Arch Klin Chir 1874;16:48-57. — Simon G. Chirurgie der Nieren, Bd I-II. Gedrukt bei F. Enke zu Erlangen 1871 und 1876. — Moll F, Rathert P. The surgery and his intention. Gustav Simon (1824-1876), his planned nephrectomy and further contribution to urology. World J Urology. 1999;17(3):162-7.

Wilhelm J. S. (F.) Grube [30.V(11.VI). 1827, village Vecmulza («Neugut»), Viekumnieki Municipality, Latvia – 28.V. (11.VI.) 1898, city of Kharkiv, Ukraine][484] — Ukrainian surgeon of Latvian origin, created the Bacteriological Station to study antiseptic treatment of purulent wounds at the Kharkiv University (1859), carried out an operation under an inhalation general anesthesia using nitrous oxide (N_2O; 1871).

In 1865-67 Sir **Joseph Lister (1827 – 1912)**[485] — English surgeon, and a pioneer of aseptic surgery, introduced the principles of asepsis that is a prevention of infection by mechanical removal of microorganisms with cleaning and antisepsis by removal of microorganisms using the chemical means (carbolic acid) or hot air, from the surgeon's hands and clothes, an operative field, linen, dressings, surgical instruments, infected wounds, and complex bone fractures.

In a 52-year-old woman with a primary partially calcified sarcoma of the right side of the body of the sternum with the size of a turkey egg, **J.S. Holden**[486] — English surgeon, performed on August 06, 1878, a partial removal of the sternum), i.e., partial sternectomy by extirpation down to the posterior compact layer of the bone.

The founder of abdominal surgery **Ch.H. Theodor Bilroth (1829 – 94)**[487] — Prussian Austrian surgeon, described adeno-cystic carcinoma («cylindroma», 1859) – malignant tracheal and proximal bronchial tumor that occurs with about the same incidence as squamous cell carcinoma (40%).

Because of malignant tumors, he removed (resected) esophagus (1872), whole larynx (total laryngectomy) on December 31, 1873, and the tongue (1874).

However, the greatest achievement of Theodor Bilroth was resection of the distal portion of the stomach (distal gastrectomy, or pylorectomy) for cancer and the «end-to-end» anastomosis to the duodenum (gastroduonenostomy), i.e., «Bilroth-1» operation (1881), or distal gastrectomy, closure of the line of resection, and the «end-to-side» anastomosis of the anterior wall of the remaining body of the stomach to the proximal portion of the jejunum («Bilroth -2» operation (1881).

At those times he opposed any operations on the heart.

J. Emile Pean (1830 – 98)[488] — French surgeon, introduced straight and curved clamps for hemostasis (Pean forceps), performed the first distal resection of the stomach with gastroduodenostomy for carcinoma of the stomach (1879, unsuccessful)

---

[484] Village Vecmulza («Neugut»), Viekumnieki Municipality (area 884,4 km²), Latvia. — Grube W. Ankylosis mandibulae vera. Langenbeck's Archiv 1859;4. — Ablitsov VH. «Galaxy Ukraine». The Ukrainian diaspora: prominent persons. Publ. «Kyt», Kyiv 2007.

[485] Lister J. On the antiseptic principle in the practice of surgery. The Lancet 1867 21 Sep;90(2299)353-6.

[486] Holden JS. Sarcoma of the sternum. Br Med J 1878 Sep 7;2(923):358.

[487] Schwartz SI. «Paroperative Pantheon». J Am Coll Surg Aug 2011;213(2)319-32.

[488] Morgenstern L. The clamp, the stomach, and the spleen: Jules-Emilie Pean (1830-1898). J Am Coll Surg 2012;215(4): 587-90.

and the first successful toral splenectomy (1867), and implanted in 1893 to the patient, a waiter, an artificial shoulder, which was removed 2 years later because of infection.

Richard von **Volkmann (1830 – 89)** — German surgeon and poet, performed the first excision of carcinoma of the rectum (1878); accomplished the first successful drainage of a subphrenic abscess by resecting a part of the 7th rib in the middle axillary line (MAL) and several days later established drainage transpleurally through the diaphragm after it had become adherent to the chest wall (1879); described an ischemic contracture of the hand (palm, fingers, sometimes wrist) after injury to the elbow and contracture of the foot – Volkmann contractures (1875).

Thoracic surgery evolved rapidly when thoracoplasty was introduced in 1879 by **Jacob Estlander (1831 – 81**: above) and in 1890, by **Max Schede (1844 – 1902)** — German surgeon, for the treatment of chronic pleural empyema, as well as introduction of decortication of the lung. It include an extrapleural resection of the segment of posterior one or more ribs, in one or two-three stages, to collapse the chest wall over diseased pleural space or lung. The latter implies removal of the constraining fibrous peal surrounding the lung in attempt to re-expand it.

A primary hydrocele is usually congenital affection, characterized by accumulation of peritoneal fluid inside the most internal membrane, called tunica vaginalis or sac, which cover the front and sides of the male testicles. It occurs because of defective absorption of tissue fluid or irritation of the membrane leading to overproduction of fluid. In addition to filling tunica vaginalis, the fluid may also fill a portion of the spermatic cord duct (epididymis) in the scrotum. A secondary hydrocele is caused by trauma, inflammation, or neoplasm of the testicle. Hydrocele is found in children (80% or more) and in adult males of over 40 years of age (1%).

Hydrocele usually appears as a soft swelling inside the membrane surrounding the testicle, not painful and does not damage the testicle. Inflammation is not usually present, although if the hydrocele occurs in conjunction with inflammation of the epididymis (epididymitis) the testicles can be swollen and painful.

Initial needle aspiration of hydrocele may be useful for examination of aspirate for malignant cells or occasionally for cure, but recurrence rate is high. However, the proper treatment is hydrocelectomy that is surgical removal of the sac for a large, thick-walled or multilocular hydrocele, and eversion / plication of the sac for a small hydrocele.

Hydrocelectomy with excision of the sac for a large, thick-walled or multiloculated hydrocele was pioneered by **Ernst** von **Bergmann (1836 – 1907)** — Latvian German surgeon. It consists of incision into the hydrocele sac after complete mobilization of the hydrocele and partial resection of the hydrocele sac, leaving a margin of 1-2 cm, which is oversewn for hemostasis (Bergmann technique).

Beside of this, Ernst von Bergman was the first physician to introduce heat sterilization of surgical instruments, use of steam sterilized dressing material, adapted the «white coat», and performed the first successful esophageal diverticulum resection.

Adam Politzer (1835 – 1920) — Hungarian Austrian otologist, who invented the head mirror.

Mykola (Nikolai) V. Sklifosovsky [25.III(06.IV).1836, farmstead Karantyn near **town of Dubasari, Prydnistrovia, Ukraine / Moldova – 30.XI(13.XII).1904, village Yakivtsi,** since 1929 incorporated into **city of Poltava, Ukraine]**[489] — Ukrainian anatomist and surgeon, and **Ivan I. Nasilov (1843 – 1902**) — Russian surgeon, introduced the compression lock for fixation of fractures complicated by delayed union or non-union (Sklifosofsky's lock, 1875) and proposed metalosynthesis (1893). The former resected a prolapsed rectum followed by mucomuscular anastomosis (Sklifosovsky operation), and the latter resected an esophagus through the extrapleural approach (1888).

The first operations on the stapes were performed by **Johannes** von **Kessel (1839 – 1907)**[490] — German otologist, removal in animals (1876), two years later – mobilization (fenestration) of the stapes in man for deafness (1878). Ten years later (1888), **E. Boucheron**[491] reported on 60 fenestrations of the stapes.

Walter Whitehead (1840 – 1913) — English surgeon, who performed an excision of advanced hemorrhoids (Ancient medicine. *Egypt*) by two circular incisions, one above and below the involved veins, allowing normal mucosa to be pulled down and sutured to the skin (Whitehead hemorrhoidectomy, 1882)[492], and excision of the tongue with scissors for carcinoma (Whitehead glossectomy, 1891).

Louis H. Farabeuf (1841 – 1910)[493] — French surgeon, depicted the triangle (Farabeuf) in the upper neck bound by the internal jugular vein (IJV), the cranial 7th (facial) nerve, and the cranial 12th (hypoglossal) nerve; performed amputation of the

---

[489] Farmstead Karantyn near town of Dubasari (founded 1408), Prydnistrovia, Ukraine / Moldova. — Village Yakivtsi [preserved are nine mounds of the Scythian Period 5^n–3rd century BC, founded at the end of 17th century; when in 04(17).X.1919 fighters of Red Army of Communist Fascist RSFSR broke into the estate of deceased Mykola Sklifosovsky, his widowed and ill wife Sophia was hewed (killed) to death with shovels, and the daughter Tamara who attended ill mother was raped and hanged head down on the oak tree], in 1929 incorporated into city of Poltava, Ukraine. — Ablitsov VH. «Galaxy Ukraine». The Ukrainian diaspora: prominent persons. Publ. «Kyt», Kyiv 2007.

[490] Kessel J. Ueber das Mobilisieren des Steigbugels durch Ausschneiden des Trommelfelles, Hammers und Ambosses bei Undurchgangigkeit der Tuba. Arch Ohrenh 1878;13:69-73. — Heermann H. Johannes Kessel and the history of endaural surgery. Arch Otolaryngol 1969;90(5):136-42.

[491] Boucheron E. La mobilisation de l'etrier et son procede operatoire. Union Med 1888;46:412-6

[492] Whitehead W. The surgical treatment of haemorrhoid. Brit Med J Feb 4,1882;1:148. — Whitehead W. Three hundred consecutive cases of haemorrhoids cured by excision. Brit Med J 1887;1:449.

[493] Farabeuf LH. Precise de manuele operatoire. Masson, Paris 1889.

lower extremity with the formation of a large external flap (Farabeuf); designed a hand- and finger-held retractor with flat blades, on each end, for holding back tissues in deep cavities (Farabeuf, 1889).

Emil T. Kocher (1841 – 1917)[494] — Swiss physician and surgeon, performed partial and total excision of the thyroid gland (TG) – thyroidectomy (1872, 1876); popularized the manoeuvre (maneuver) to reduce anterior shoulder dislocation by **Abu al-Zahravi (936 – 1013**: Medieval medicine**)** – Zahravi-Kocher maneuver; described the method to expose structures in the retroperitoneum behind the duodenum and the pancreas to control hemorrhage from the inferior vena cava (IVC) or to facilitate removal of pancreatic tumor, by incising at the right edge of the duodenum, the duodenum and the head of the pancreas are reflected to the opposite direction, i.e., to the left (Kocherization of the duodenum or Kocher's maneuver).

Acute appendicitis begins with an attack-like pain in the epigastric area below the xiphoid process or in the umbilical area, and in 1-3 hours translocate to the right iliac area, becomes constant and irradiates nowhere (Kocher symptom, 1892).

Femoral hernias protrude into the thigh just below the inguinal ligament[495]. They are divided into (1) the medial vasculo-lacunar femoral hernia which is the most common; (2) femoral hernia which protrude through middle portion of vascular gap or through the vascular sheath; and (3) the lateral femoral vasculo-lacunar hernia outside the femoral vessels. The incidence of femoral hernias is the second after inquinal hernias, and 10-20 times more often in women.

The repair of femoral hernia according to Emil Kocher is performed using an infra-inguinal vertical incision, excision of the hernia sac, and approximating the inguinal ligament to the pectinate fascia using U-like non-absorbable sutures.

The first work about antiseptics in surgery after **Sir Joseph Lister (1827 – 1912**: above), was published by **Pavlo P. Pelekhin [1842, city of St Petersburg, Russia – 27.IX(10.X).1917, city of St Petersburg, Russia]**[496] — Ukrainian surgeon, who by using strict antiseptic precocious was able to decrease operative mortality (OM) from 50-94% to 7% («Success of the new ideas in surgery during the treatment of wound, complicated fractures and suppurative collections gathering» (1868).

Serhii P. Kolomnin (25.IX.1842, city of St Petersburg, Russia – shoot himself, blaming oneself for death operated by himself patient **11.XI.1886, city**

[494] Kocher Th. Chirurgische Operationslehre. Vierte Vielfach Umgearbeitete Aufgabe. Verlag von Gustav Fischer, Jena 1902.

[495] Shkrobot VV. Hernias of the abdominal wall. In, LYa Kovalchuk, VF Sayenko, HV Knyshov. Klinichna Khirurhiya. Vol. 1-2. Publ. «Ukrmedknyha», Ternopil 2000. Vol. 2. P. 11-23.

[496] Pelekhin P. Success of the new ideas in surgery during the treatment of wound, complicated fractures and supporative collections. Med vesti 1868;34:325, 1868;35:332. — Ablitsov VH. «Galaxy Ukraine». The Ukrainian diaspora: prominent persons. Publ. «Kyt», Kyiv 2007.

of St Petersburg, Russia)[497] — Russian Ukrainian surgeon, successfully ligated the common carotid artery (CCA (1874), carried out an intraarterial transfusion of defibrinated blood during the military campaign (1877-78).

Wincenz Czerny (1842 – 1916) — Czech surgeon, resected and reconstructed the cervical esophagus for carcinoma (1877), designed the abdominoperineal (AP) resection for carcinoma of the lower part of the sigmoid colon and the rectum with the removal of the pelvic part of colon, its mesentery and adjacent lymph nodes, a wide excision of the rectum and anus with the creation of permanent colostomy (1883). This AP resection was improved and popularized by **William E. Miles (1869 – 1947)** — American surgeon, in 1908, and named the Miles operation.

The law of **Ludwig G. Courvoisier (1843 – 1918)** — French surgeon, states that when the common bile duct (CBD) is obstructed by a stone, dilatation of the gallbladder is rare, while if the duct is obstructed by hilar cholangiocarcinoma (Klatskin tumor)[498], gallbladder carcinoma, carcinoma of papilla duodeni major or of the head of the pancreas, ascariasis or oriental cholangiohepatitis, dilatation is common (Courvoisier law, 1890).

William Alexander (1844 – 1919) — Scottish surgeon, performed the first dorsal sympathectomy by removing the superior sympathetic ganglia and ligating the vertebral arteries (VA) with short-lasting improvement (1889), and dorsal sympathectomy for the treatment of glaucoma (1898).

The repair of a direct inguinal hernia according to the method of **Eduardo Bassini (1844 – 1924)**[499]— Italian surgeon, from 1884, is performed by excision and ligation of the hernia sac, approximation using knotted sutures through the edges of the transverse and internal oblique abdominal muscle and into the inguinal ligament under the spermatic cord with subsequent tying, thus closing the hernia defect. The spermatic cord placed on the newly formatted posterior wall of the inguinal canal is covered by sewing over it the aponeurosis of the external oblique abdominal muscle.

Fredrich Trendelenburg (1844 – 1924) — German surgeon, introduced the «head-down position» of the patient in which the body laid supine or flat on the back with head down 10°-30° and the feet high, and the reverse «feet-down position» (Trendelenburg positions) instituted by him in 1881, described by his student **Willy Meyer (1854 – 1932)** — German American surgeon in 1885 and by Friedrich Trendelemburg himself in 1890; designed the tracheal canula to prevent an aspiration

---

[497] Ablitsov VH. «Galaxy Ukraine». The Ukrainian diaspora: prominent persons. Publ. «Kyt», Kyiv 2007.

[498] **Gerald Klatskin (1910 - 86)** — American physician who described hilar cholangiocarcinona at the confluence of the right and left hepatic duct (Klatskin tumor, 1965).

[499] Shkrobot VV. Hernias of the abdominal wall. In, LYa Kovalchuk, VF Sayenko, HV Knyshov. Klinichna Khirurhiya. Vol. 1-2. Publ. «Ukrmedknyha», Ternopil 2000. Vol. 2. P. 11-23.

of the blood while operating on the larynx; described a test for the determination of sufficiency of the deep vein and of the great saphenous vein (GSV) valves in the patients with varicose veins of the lower extremities (Trendelenburg test, 1891), and designed the surgical treatment of the later by ligation of the proximal GSV and its branches – Trendelenburg operation-I (1891).

Charles McBurney (1845 – 1913) — American surgeon, described the point of greatest tenderness in appendicitis (McBurney point, 1889) and the right lower (quadrant) oblique abdominal incision for appendectomy (McBurney incision, 1894).

On July 15, 1882, **Carl J. A. Langenbuch (1846 – 1901)** — German surgeon, performed the first open removal of the gallbladder (cholecystectomy) in a 43-year-old man.

Diffuse symmetrical lipoma of the neck depicted by **Otto W. Madelung (1846 – 1926)** — German surgeon, was named the Madelung's neck, as was a deviation of the wrist to the side of the radial bone, secondary to overgrowth of a distal portion of the ulnar bone or a shortening of the radial bone (carpus curus or Madelung deformity, 1878).

Oscar T. Bloch (1847 – 1926)[500] — Danish surgeon, developed a staged colectomy for rectal carcinoma: (1) an exteriorization of the affected bowel to be resected in the form of a loop on the surface of the abdomen, (2) resection of the exteriorized loop with the creation of the double (proximal and distal) fecal fistula, (3) a closure of the fecal fistula by the anastomosis of both ends of the bowel in the «end-to-end» fashion (Bloch operation).

As a supplement to gastroenterostomy (gastrojejunostomy), **Ch. Heinrich Braun (1847 – 1911)** — German surgeon, proposed an anastomosis between the afferent and efferent loop of the jejunum distally to the gastrojejunostomy (Braun anastomosis), to prevent a closed loop for the gastric and duodenal juices.

Rudolf U. Krönlein (1847 – 1910) — Swiss surgeon, is known for performing the first documented appendectomy in a 17-year-old boy who died two days after surgery (1884), repairing of inquino-properitoneal double-sac (Krönlein) hernia and resection of the external wall of the orbital cavity for removal of tumor without removal of the eyeball (lateral orbitotomy of the eye or Krönlein operation).

Nestor D. Monastyrsky / Nestor D. Monastyrski (1847, suburb Rosha of city Chernivtsi, Ukraine – died May 24, 1888, two days after operation for carcinoma of the kidney at the Surgical Division of the Institute for post-graduation studies for

---

[500] Block HM. Experimentelles zur Lungenresektion. Dsche med Wschrift 1881;7(47):634-6.

physician in **St Petersburg, Russia)**[501] — Ukrainian surgeon, performed on March 13, 1882, in a patient with gastric outlet obstruction caused by peptic ulcer, an anastomosis between the gallbladder fundus and the jejunum (cholecystojejunostomy). Immediately after the surgery the patient was doing well, however, unfortunately he died in the evening the same day. In the subsequent two years, he performed three more gastrojejunostomies.

The increase in heart rate (HR) in response to a digital compression of arteriovenous (AV) fistula was noted by **Carl Nicoladoni (1847 – 1902)** — Austrian surgeon, and this phenomenon was named after him Nicoladoni sign (1875).

Ovxentiy T. Bohayevsky [13(25).XII.1848, village of Ustyvytsia, Myrhorods'kyy Region, Poltavs'vka Obl., Ukraine – 05(19).XII.1930, city of Kremenchuk, Poltava Obl., Ukraine)][502] — Ukrainian surgeon, was one of the first to carry out in patient with a gastric carcinoma a successful gastrectomy (1888).

A young brilliant young man **H.V. Block (? – suicide by self-shooting 1883, Gdansk, Poland)**[503] — German surgeon, after successful laboratory experiments with pulmonary resection in animals which he presented at the Congress of the German Surgeons in Berlin, Germany, in 1882, decided to operate on humans suffering from lung disease.

Ironically, this was the time (1882) when **Ivan Puliui (1845 – 1918**: 19th century. Radiology) discovered x-rays, which were not yet available for the use in medical practice, so localization of the lung lesions was made by percussion and auscultation only. In addition, controlled lung ventilation was not known, so by opening the chest cavity, the mediastinum would shift toward the opposite hemithorax and obstruct the venous blood return to the right side of the heart, the patient's respiration would become fast and labored, the heart rate would become rapid, the pulse faint, cyanosis, shock and death would quickly ensue.

In the summer of 1883, he chose a young female, his relative, with a clinical diagnosis of bilateral pulmonary tuberculosis, made an intercostal space (ICS) opening

---

[501] City of Chernivtsi (settled since Neolithic Era 10,200 – 2,000 BC, founded in 12th century, first chronicle's recording 1407), Chernivets'ka Obl. (area 8,097 km²), Ukraine. –– Monastyrski ND, Tilling G. Zur Frage von der chirurgischen Behandlung des Ductus choledochus. Zentralbl Chir 1888;15:778-9. — Pishak VP, Akhtemiychuk YuT. Nestor Dmytrovych Monastyrsky (to the 155th anniversary of the famous Bokovinian). Bukovinian Medical Herald 2002;6(1-2):125-6 — Gachabayov M, Kubachev K. Professor Monastyrski N.D. (1847-1888): one of the forgotten pioneers of biliary surgery. Clinical Medicine & Research. Apr 2017;15(1-2):33-6. — Ablitsov VH. «Galaxy Ukraine». The Ukrainian diaspora: prominent persons. Publ. «Kyt», Kyiv 2007.

[502] Village Ustyvytsia (founded 1625), Myrhorods'kyy Region, Poltavs'ka Obl., Ukraine. — City of Kremenchuk (founded 1571; during conduction by Russian Communist Fascist regime of Holodomor-Genocide 1932-33 against of the Ukrainian Nation perished 497 persons, in 1930 – 1940s during repressions suffered over 2000 people), Poltava Obl. Ukraine. — Ablitsov VH. «Galaxy Ukraine». The Ukrainian diaspora: prominent persons. Publ. «Kyt», Kyiv 2007.

[503] Block HM. Experimentelles zur Lungenresektion. Dtsch med Wschr 1881;7(47):634-6.

of the chest cavity (thoracotomy), removed part of the lung, but the patient died as result of mediastinal shifting and shock. The removed part of the lung was healthy. A few days later he died from a self-inflicted gunshot wound to the head. Such was the end of the first planned operation to remove a part of the lung for tuberculosis.

The multiple stage electrocautery to remove the lung (pneumonectomy) for tuberculous empyema was performed by Sir **William MacEwen (1848 – 1924)** — Scottish surgeon, in 1895. The patient survived 45 years after this operation.

Urology was advanced by **Maximilian C. F. Nitze (1848 – 1906)** — German urologist, who invented the cystoscope with an electrical source of light (1877-79).

Maxym S. Subbotin (1848 – 1913)[504] — Russian Ukrainian surgeon, designed an aspiration apparatus with a negative pressure suction for drainage of wounds, abscesses and body cavities (1887-88), the idea of which was later adopted into the suctioning devices of **August Bier (1861 – 1949)** — German surgeon and **Georg C. Perthes (1869 - 1927)** — German orthopedic surgeon and pioneer of diagnostic radiography (1903); performed a thoracoplasty for pleural empyema (1888); developed an artificial urinary bladder using the anterior portion of the rectum (Subbotin operation, 1911).

While experimenting with the porto-caval shunt, **Nicolai V. Eck (1849 – 1908)** — Russian surgeon, performed the first vascular (venous) anastomosis in the «end-to-side» fashion (Eck fistula, 1877).

In 1882 **Anton R. von Frisch (1849 – 1917)**[505] — Austrian urologist, identified Klebsiella rhinoscleromatis as a causative factor of rhinoscleroma, a chronic granulomatous inflammation of the nose, upper lips, mouth, pharynx, and upper airways.

A penetrating knife wound of the right ventricle (RV) was sutured through the left 4th intercostal space thoracotomy (ICST) in 1896 by **Ludwig Rehn (1849 – 1930)**[506] — German surgeon, with the success.

In the repair of an indirect inguinal hernia according to the method of **Charles Girard (1850 – 1916)**[507] — Swiss surgeon, the rim of the internal oblique and transverse abdominal muscles are approximated and sutured to the inguinal ligament over the spermatic cord. The aponeurosis of the external oblique abdominal muscle is approximated with the second row of sutures.

---

[504] Subbotin MS. General Surgical Pathology and Therapy. Lectures. 1st Ed. Kharkiv 1887. — Subbotin MS. New Operative Method for Empyema of the Pleural Cavity. Kharkiv 1888. — Subbotin MS. Handbook of General Surgery. Vol. 1–3. St Petersburg 1902-06. — Subbotin MS. Clefts of the Urinary Bladder пузыря. St Petersburg 1911. — Ablitsov VH. «Galaxy Ukraine». The Ukrainian diaspora: prominent persons. Publ. «Kyt», Kyiv 2007.

[505] Frish A. Rhinosklerom. Arch Klin Chir Berlin 1889:355-418.

[506] Rehn L. Über penetrierende Herzwunden und Herznacht. Arch für klin Chir 1897;55:315-29.

[507] Shkrobot VV. Hernias of the abdominal wall. In, LYa Kovalchuk, VF Sayenko, HV Knyshov. Klinichna Khirurhiya. Vol. 1-2. Publ. «Ukrmedknyha», Ternopil 2000. Vol. 2. P. 11-23.

Ivan (Johannes, Jan) A. von **Mikulicz-Radecki** or **Ivan** von **Mykulych-Radetsky (May 16, 1850, city of Chernivtsi, Ukraine – Jun 04, 1905, city of Vratislav, Bohemia; Breslau, Germany; Wroclaw, Poland)**[508] — Ukrainian surgeon, who calculated a normal angle of declination (130⁰) between the long planes of the epiphysis and the diaphysis of the femur (Mikulicz angle); performed osteoplastic resection (tarsectomy) of the foot in diseases of the talus and calcaneus independently of **Alexander D. Vladimirov (1837 – 1903)** — Russian surgeon, named Vladimirov-Mikulicz operation (1872-73); after intubation by **Adolph Kussmaul (1822 – 1906**: 19th century. Internal medicine) a sword swallower stomach via esophagus with a 13 mm hollow metal tube (1868), developed electrically lightened esophagogastroscope for a detailed examination of the oral cavity, esophagus and stomach (1880-81); popularized aseptic and antiseptic methods by using iodoform gauze (1881), surgical gauze mask (1896) and surgical gloves (1897); introduced the drain formed by pushing a single layer of gauze into a wound or cavity, then packing with several layers of gauze as the original layer is pushed farther and farther into the defect, while gauzes saturated by the blood, discharge and puss are removed and replaced by fresh ones (Mikulicz drain, 1881); introduced a folded gauze pad for packing off viscera in abdominal operations and used as a sponge to soak the blood and discharge (Mikulicz pad, 1884); carried out independently of **Oleksandr Yatsenko**

[508] City of Chernivtsi (Foot Note 501), Chernivets'ka Obl., Ukraine. / City of Vratislav / Breslau / Wrocław [first mentioned in 142-147 AD, founded in the 10th century by **Vratislaus I (c. 888 - 921)** — Duke (from 915 until his death) of Bohemia – the westernmost and largest historical region of Czechia], Bohemia, Germany, Poland. — Mikulicz-Radecki J. Über das Rhinosklerom (Hebra). Arch für klin Chir, Berlin 1877;20:485-534. — Mikulicz-Radecki J. Über die Anwendung der Antisepsis bei Laparatomien mit besonderen Rücksicht die Drainage der Peritoneahöhle. Arch für klin Chir 1881;2:111-50. — Mikulicz-Radecki J. Weitere Erfarungen über die Verwendung des iodoform in der Chirurgie. Berl klin Wschr 1881;18:721-5, 741-4. — Mikulicz-Radecki J. Über Gastroscopie und Oesophagoscopie. Wien Med Presse 1881;22:1405-8. — Mikulicz-Radecki J. Leczenie nowotworów i kamicy pęcherza moczowego. Przegl Lek 1885;24(13):185-6. — Mikulicz-Radecki J. Ein Fall von Resection des carcinomatösen Oesophagus mit plastischen Ersatz des excidierten Stückes. Prager med Wochenschr 1886;11:93-4. — Mikulicz-Radecki J Zur Befestigung der Wanderniere. Ber. über D. Verhandl. d. Geselsch. für Geburtsch. u. Gyn. zu Berlin. Z. Gynäkol 1890;19:357-62. — Mikulicz-Radecki J. Über eine eigenartige symmetrische Erkrankung der Thränenrische und Mundspeicheldräsen. Beträge zur Chirurgie, Stuttgart 1892:610-30. — Mikulicz-Radecki J. Das Operieren in sterilisierten Zwirhandschuhen und mit Mundbinde; ein Betrag zur Sicherung des aseptischen Verlaufs von Operationswunden. Zentralbl Chir 1897;24:713. — Mikulicz-Radecki J. Beträge zur Technik der Operation der Magencarcinomans. Arch für klin Chir 1898;57:524. — Mikulicz-Radecki J. Die Kranheiten des Mundes. Jena 1898. — Mikulicz-Radecki J. Zur Operation der angeborenen Blasenspalte. Zentralbl Chir, Leipzig 1899;26:641-3. — Mikulicz-Radecki J. Demonstration eines Falles von Uterenstein der mit Erfolg durch operation entfernt wurde. Allg med Zentral-Ztg 1902;71:1231-2. — Mikulicz-Radecki J. Über perineale Prostatektomien. Zentralbl Chir 1904;31:1367-9. — Mikulicz-Radecki J. Chirurgische Erfahrungen über die Sauerbruch'sche Kammer bei Unter- and Überdruck. In: Verhandlungen der Deutschen Gesellschaft für Chirurgie. 1904;33(1):34-41. — Pishak VP, Kardash VE, Bilous VI, Bilous VV. Mikulych-Radets'kyi Ivan Andriiovych (Johan Mikulicz-Radecki) 1850-1905. Clin Anatomy Operative Surg (Chernivtsi) 2003;2(4):1-3. — Ablitsov VH. «Galaxy Ukraine». The Ukrainian diaspora: prominent persons. Publ. «Kyt», Kyiv 2007.

(1843 – 97: above) — Ukrainian surgeon, a retrograde (via gastrostomy) dilatation of the esophageal cardiospasm in achalasia and the other diseases; introduced the thoracoabdominal incision for the resection of carcinoma of the cardia and the distal portion of the esophagus (1896 i 1899); was the first to suture a perforated gastric ulcer (1885); modified Billroth-II partial gastrectomy by performing a retrocolic «side-to-side» gastrojejunostomy; transected the sternocleidomastoid muscle in congenital torticollis (Mikulicz operation); removed independently of **Alexy V. Martynov (1868 – 1934)** — Russian surgeon, thyroid goiter by a wedge thyroid lobectomy (1886); first to restore the cervical esophagus after resecting its for cancer (1886); drained surgically empyemas of the maxillary and paranasal sinuses (Mikulicz operation, 1887); together with **Walther H. Heinecke (1824 – 1901)** — German surgeon, enlarged the outlet of the stomach for congenital pyloric stenosis by a longitudinal incision and transverse closure (Heinecke-Mikulicz pyloroplasty, 1888); independently of **Carl Nicoladoni (1847 – 1902: above)** — Austrian surgeon, repaired a prolapsed rectum and colon (Mikulicz-Nicoladoni operation, 1888); performed a plastic surgical procedure involving the use of the portion of intestine to enlarge the urinary bladder (enterocystoplasty, 1889); described a large, round or oval vacuolated phagocyte with small pyknic nucleus in nodules of rhinoscleroma (Mikulicz cells); described with **Werner F. F. Kümmel (1866 – 1930)** — German ear, nose and throat (ENT) physician, an exfoliative cheilitis; described a recurrent periadenitis mucosa necrotica of the oral cavity (Mikulicz aphta); described primary benign, chronic and symmetrical dacroadenitis with lymphoepithelial infiltration and hypertrophy, painless swelling of lacrimal and salivary glands and decreased or absent lacrimation (Mikulicz disease); described a secondary Mikulicz's disease as a complication of another disease (Mikulicz syndrome, 1892); independently of **Oscar Bloch (1847 – 1926)** — Danish surgeon, performed a staged resection of carcinoma of the descending and sigmoid colon using the Mikulicz's intestinal clamp (Bloch-Mikulicz operation, 1903).

In 1882 **Anton R.** von **Frisch (1849 – 1917)**[509] — Austrian urologist, identified Klebsiella rhinoscleromatis as causative factor of rhinosclerosis, chronic granulomatous inflammation of the nose, the upper lips, the mouth, the throat and the upper respiratory airways.

Ludwik Rydygier, born **Riedigier (21.VIII.1850, village Schöntal, Kingdom of Prussia; since 1945 village of Dusocin, Kuyavian-Pomeranian Voivodeship, Poland – 25.VI.1920, city of Lviv, Ukraine)**[510] — Polish surgeon of Prussian

---

[509] Frish A. Rhinosklerom. Arch Klin Chir Berlin 1889:355-418.

[510] Village of Schöntal / Dossoczyn (mentioned 1285), Kingdom of Prussia (area 348,779 km² - 1910); since 1945 village of Dusocin, County Grudziądz, Kuyavian-Pomeranian Voivodoship (17,969 km²), Poland.

nationality, performed Billroth-1 operation (resection of the stomach's pylorus) in patient suffering from carcinoma (1880), local resection of peptic gastric ulcer (1881), and Bilroth-2 operation (partial distal resection of the stomach with gastro-jejunal anastomosis) for peptic ulcer of the stomach (1900).

Still at the end of the 1880s. **Dmitriy S. Shchetkin (1851 – 1923)**[511] — Russian physician-obstetrician, utilized for a differential diagnosis of an acute abdomen a progressive digital pressure to the wall of right lower abdominal quadrant with subsequent quick release, and when rebound tenderness was evident, then it indicated an acute appendicitis. However, the official notification he made only in 1908 at the meeting of the Penza Medical Society, town of Penza (founded 1663), Russia, without any written evidence.

William S. Halsted (1852 – 1922) — American surgeon, introduced an operation for an inguinal hernia with transposition of the spermatic cord above the external oblique aponeurosis, a radical mastectomy for carcinoma of the breast (1882) and cloth gloves in operations (1890).

In 1886 **Hermann Kümmell (1852 – 1937)** — German surgeon, was the first to make an incision into the common bile duct (CBD), i.e. choledochotomy with the aim explore the CBD and remove of gallstones; described a delayed collapse of a traumatized (after a compression fracture) vertebral body, from osteoporosis due to decreased blood supply (traumatic spondylitis) that involves spinal pain, intercostal neuralgia and motor disturbances of the legs (Kümmell disease, 1891, 1895, 1928).

Apolinariy H. Podriz (Podrez) [18(39).XI.1852, town of Kupiansk, Kharkiv Obl., Ukraine – 09(22).XI.1900, city of Kharkiv, Ukraine][512] — Ukrainian surgeon-urologist (specialist in the surgical treatment diseases of the kidneys, ureter, urinary bladder, urethra, and reproductive system), proposed the method of reconstruction / restoration of the ureter (1884), removed the spleen (splenectomy, 1887) and the prostate (prostatectomy, 1887), published the handbook «Surgical Diseases of the Urinary-Genital System (Vol. I. 1887, 1896; Vol. II. 1897), introduced modification of gastrointestinal anastomosis (gastroenterostomy) that bears his name (1895). In a 6-year-old-girl removed a gunshot (revolver) bullet lodged in the wall of the right

---

[511] Kontsevyy OO. Acute appendicitis. In, LYa Kovalchuk, VF Sayenko, HV Knyshov. Clinical Surgery. Vol. 1-2. Publ. «Ukrmedknyha», Ternopil 2000. Vol. 2. P. 125-363. — Leshchyshyn IM, Okhotska OI, Kryvorchuk IT, et al. Surgical treatment of ovarian apoplexy and acute appendicitis. Khirurhiya Ukrainy 2012;4:76-82.

[512] Town of Kupiansk (first settlements during Neolithic Age 5th-4thth thousand years BC., Bronze Age 2nd thousand years BC 2nd, 8th, and 11th century AD, founded 1655), Kharkiv Obl., Україна. — Groznyi AD. Professor Apollinariï Grigor'evich Podrez (1852-1900). Klin khir 1975;(12):78-9. — Telichkin IA. Apollinariï Grigor'evich Podrez, professor of surgery in the Kharkov University (on the 150th anniversary of the birthday). Lik Sprava 2002;(7):135-8. — Zaitsev EI, Lazarev SM, Ozerov VF. Apollinariï Grigor'evich Podrez (1852-1900). Vest Khir Grekov 2003;162(5):9-11. Ablitsov VH. «Galaxy Ukraine». The Ukrainian diaspora: prominent persons. Publ. «Kyt», Kyiv 2007.

ventricle (RV) of the heart, followed by closure of the wound using silk sutures, with success (city of Kharkiv 1897). *«Surgery of the heart,* said Apolinariy Podriz**, is only being born now**, *but to her belongs the great future».*

The older brother of **Felix** von **Winiwarter (1852 – 1931**: 19th century. Pathology**), Alexander** von **Winiwarter (1842 – 1917)** — Austrian Belgian surgeon, introduced by the end of the 19th century a specialized massage and compression procedure to treat lymphedema that causes swollen upper and lower extremities due to fluid retention in the lymphatic system. In 1932 **Emil Vodder (1894 – 1986)** — Danish physiotherapist, together with his wife **Estrid Vodder (1897 – 1996)** — Danish naturopath, refined the Winiwaters technique to treat lymphedema by manual lymphatic drainage.

The first documented total wrist arthroplasty, meaning re-forming of joint was performed in 1890 by **Themistocles Gluck (Nov. 30, 1853, town of Iasi, Moldavia – Apr. 25, 1943, Berlin, Germany)**[513] — German Moldovian physician and surgeon. Re-forming of joint include replacement, remodeling, or realignment of the joint by cutting to shorten or lengthen or change alignment of a bone (osteotomy). Also, the earliest attempts at hip replacement commenced in this same year (1890) by Themistocles Gluck who invented endoprosthesis from ivory, and used it to replace the femoral head, attaching it with nickel-plated screw, Plaster of Paris, and glue.

Karel Maydl (1853 – 1903) — Czech surgeon, introduced a method of correcting a congenital malformation of the urinary bladder where it is located partially outside the body (extrophy) by its implantation with the surrounding connective tissue to the sigmoid colon; devised a colostomy for an inoperable colonic carcinoma in which the colon is drawn out through a small incision in the abdominal wall and maintained in position by placing a glass-rod beneath it on the mesenteric side until it becomes secured by adhesions (Mandl operation or a glass-rod colostomy); described an incarceration of an intraperitoneal hernia (Mandl hernia).

A degenerative disease of one of the lower or upper lumbar vertebrae and of the hip joint, where growth and loss of growth leads to some degree of its collapse and to the deformity of lumbar vertebrae, the ball of the femur and the surface of the hip socket was first described by Karel Maydl in 1897, then by **Arthur T. Legg (1874 – 1939)** — American orthopedic surgeon, in 1908-09, by **Georg Perthes (1869 – 1927**; above**)** and **Jacques Calve (1875 – 1954)** — French orthopedic surgeon, in 1910, is characterized by idiopathic aseptic avascular osteonecrosis of lumbar vertebrae and of capital femoral epiphisis of the femoral head leading to an interruption of the

---

[513] Town of Iasi / Jassy / Iassi (human settlement since the 6th - 7th century, founded in 1408, capital of Moldovian Principality in 1564-1862, and Moldavia in 1881 – 1916), from 1945 – de facto capital of the Moldavian Region (area 46,146,173 km^2), Romania (area 238,397 km^2).

blood supply of the head of the femur close to the hip joint (Legg-Calve-Perthes syndrome). The disease is typically found in children 6-10- year-old, and it can lead to osteoarthritis in adults.

In orthopedics, **Adolf Lorenz (1854 – 1946)** — Austrian orthopedic surgeon, and **Albert Hoffa (1859 – 1907**) — German surgeon, orthopedist and physiotherapeutic, remembered for introducing a repair and pinning the head femur opposite the rudimentary acetabulum with formation of the true acetabulum for congenital hip dislocation in children (Lorentz-Hoffa operation, ?,1890), and the latter also for developing a system of massage therapy (Hoffa system), and description of a chronic knee pain primarily beneath the patella known as «Hoffa fat pad disease».

John B. Deaver (1855 – 1931) — American surgeon, who designed surgical blade; hemostat; an incision in the right lower quadrant of the abdomen with medial displacement of the rectus abdominal muscle; operative scissor; marvelous retractor; skin incision and T-tube, all named after him.

The largest of the thoracic wall muscles, the latissimus dorsi muscle, supplied along the undersuface by the thoracodorsal artery, a branch of the subscapular artery from the axillary artery, was introduced by **Inginio Tansini (1855 – 1943)** — Italian plastic surgeon, in 1896, as the myocutaneous flap for the coverage of chest wall defects. This muscle can reach both, the anterior and the posterior thorax, for the coverage of the chest wall and the intrathoracic organs. This includes reconstruction of the mastectomy defect[514].

Herbert M.N. Milton (1856 – 1921)[515] — English surgeon, introduced a median longitudinal sternotomy incision (MLSI) or median sternotomy incision using for its closure «six strong silver wires», the principal incision for operations on the heart and on the great vessel (1897).

Important contribution into plastic, orthopedic, general and vascular surgery was made by Ivan F. **Sabaneyev / Ivan Ssabanejew (18(30).10.1856, city of Moscow, Russia – 1937, city of Saloniki /Thessaloniki, Greece)**[516] — Ukrainian surgeon, who after graduation from Medical Division of St Volodymyr Kyiv University, Kyiv, Ukraine,

---

[514] Tansini I. Sopra il mio nuovo processo di amputazione della mammella. Gazzetta Med Ital 1906;57:141-2.

[515] Milton H. Mediastinal surgery. Lancet 1897;1:872-5. — Olearchyk AS. Replacement of lacerated superior vena cava with synthetic graft. J Med Soc New J 1985;82(11):881-4. — Van Wingerden JJ. Sternotomy and intrathoracic omentum: two procedures, two innovators, and the river that runs throught it - a brief history. Ann Thorac Surg 2015 Jan;99(2):738-43.

[516] City of Saloniki / Thessaloniki (founded 315 BC) – capital of Greek (South) Macedonia (area 34,177 km²). — Sabaneev IK. An inquiry of suturing blood vessels. Russian Surg Archiv 1895;11:625-39. — Ismail-Zade IA, Davies B, Earnshaw JJ. Ivan F. Sananeev (1856-1937). The surgeon who first described thromboembolectomy. Eur J Vasc Endovasc Surg 1997;13(3):261-2. — Vasiliev KK. The surgeon Doctor of Medicine I.F. Sabaneev (1856-1937). Pages of biography. Visnyk Sumy State University, Ukraine 2005;75(3):10-15. — Vasiliev KK. Ivan Fedorovich Sabaneev (1856-1937). Vest khir Grekov 2006;165(2). — Ablitsov VH. «Galaxy Ukraine». The Ukrainian diaspora: prominent persons. Publ. «Kyt», Kyiv 2007.

completed the surgical apprenticeship in Military-Medical Academy, St Petersburg (1886-87), then practiced surgery in city of Odesa, Ukraine, where simultaneously carried out duty of Privat-Dozent at the Odesa University (1887-1908).

In 1887-87, Ivan Sabaneev performed the osteoplastic restoration of crooked nasal septum; in 1890 – an osteoplastic intercondyloid amputation of the femur (Sabanyev operation); in 1887-90 – an osteoplastic intercondyloid husking (exarticulation) of the tibia from the knee joint – Gritti-Szymanowski-Stokes operation **[Rocco Gritti (1828 – 1920), Julius** von **Szymanowski (1829 – 68),** and **Sir William Stokes (1829 – 1900):** above]; in 1890 – for the release of esophageal stenosis devised, independently of **Rudolf Frank (1862 – 1913)** — Austrian surgeon, gastrostomy with conus like drawing (pulling) up the stomach into an incision in the left rectus abdominal muscle (RAM) with sewing it into the skin (Sabaneev-Frank (gastrostomy).

When in 1895, in a 34-year-old man, an excision of enlarged inguinal lymphatic glands (inguinal lymphadenectomy) was complicated by tearing wound of the common femoral vein (CFV) and bleeding, he stopped it by with the proximal and distal control, by surrounding the CFV with surgical band / ribbon and occlusion, thus stopping bleeding. Then, under dry field, he successfully closed the wound of the CFV.

Also in 1895, while attending a 28-year-old woman suffering from rheumatic heart disease (RHD) complicated by acute thromboembolism to the right common femoral artery (CFA) with acute ischemia of the extremity, he surgically removed embolus and thrombi (thromboembolectomy) from the occluded CFA, supplemented by the distal isolated perfusion thrombolysis of the superficial femoral artery (SFA).

With the aim to save ischemic limb, he performed under the proximal and distal control of the CFA a 3-4 cm long vertical incision (arteriotomy) of the CFA, and through it removed the proximal part of the thromboembolism. The distal part of it, he removed via a separate vertical incision of the right superficial femoral artery (SFA), extracting manually and with an infusion of a normal saline (0,9% NaCl) solution, the distal portion of thromboembolism. Thus, he introduced distant (peripheral) isolated perfusion for dissolving thromboembolism (thrombolysis), e.i., isolated perfusion thrombolysis of peripheral arteries.

The father of neurosurgery, Sir **Victor A.H. Horsley (1857 – 1916)** [517] — English neurosurgeon, as neuroscientist performed in 1884-86 studies on motor response to the faradic electrical stimulation[518] of the cerebral cortex, internal capsule, and spinal cord, which translated into his later pioneering work in neurosurgery for epilepsy.

---

[517] Horsley V, Sturge D, Newsholme A. Alcohol and Human Body. An Introduction to the Study of the Subject, and a Contribution to National Health. Macmillan, London 1908.

[518] **Michael Faraday (1791 - 1867)** — English physicist, who in 1831 built the Faraday disc, the first electric generator.

While studying the thyroid gland (TG) of monkeys, established that myxedema and cretinism, which are caused by a decreased level of TG hormone (hypothyroidism), could be treated with TG extract.

As neurosurgeon he performed in 1887 a laminectomy, a neurosurgical procedure that removes a part of lamina of the posterior vertebral arch for an access to remove a spinal cord tumor, introduced hemostatic bone wax in surgery to control bleeding from the bone marrow, skin flaps, ligation of of the carotid artery to treat cerebral aneurysms, the transcranial approach to the pituitary gland (hypophysis) and the intradural division of the cranial 5th (trigeminal) nerve for surgical treatment of trigeminal neuralgia.

In 1908 Victor Horsley developed together with **Robert H. Clarke** — English physiologist, the Horsley-Clarke apparatus for a stereotactic (in Greek «stereos» meaning solid, and «taxis» meaning arrangement, order) surgery or stereotaxy for animal experimentation, whereby a set of precice numerical coordinated are used to locate each brain structure. The Horsley-Clarke apparatus was based on a Cartesian three coordinate (x, y and z) system in orthogonal frame of reference. It specifies each point of uniquely in a plane by a pair of numerical coordinates, which are the signed distance from the point to two fixed perpendicular directed lines, measuring in the same unit of length. The Cartesian system was developed by **Rene Descartes** or **Renatus Cartesius (1596 – 1650)** — French mathematician.

As a neurosurgeon Victor Horsley, having observed that many injuries of patients admitted to the hospital were due to alcohol intoxication, he became a temperance reformer, and in 1907, along with **Mary D. Sturge (1865 – 1925)** — English physician, and Sir **Arthur Newsholme (1857 – 1943)** — British epidemiologist and public health expert, published book covering injurious physiologic effect of alcohol on the human being and its widespread tragic consequences to the society.

Removal of the coccyx and part of the sacrum for access to carcinoma of the rectum was proposed by **Paul Kraske (1857 – 1922)** — German surgeon, was named after him Kraske operation.

In 1884 **John B. Murphy (1857 – 1916)**[519] — American physician, abdominal surgeon and inventor, described tenderness and pain during delicate punching with fisted hand at the costovertebral angle overlying the kidneys, useful for diagnosing of perinephric abscesses (Murphy's punch sign, 1884). [19th century. Internal Medicine. – **Fedir Pasternacky (1845 – 1902)**].

Subsequently, he designed a mechanical device for suture-less anastomosis between hollow organs, consisting of two hollow cylinders, each one sutured to an

---

[519] Murphy JB. Cholecysto-intestinal, gastro-intestinal anastomosis, and approximation without sutures. Med Rec, New York 1892;42:665-6. — Murphy JB. Five diagnosis methods of John B. Murphy. Surgical Clinics of J.B. Murphy. 1912;1:459-66.

open end of the intestine and fitted together by a quick connector (Murphy's button, 1892). After firm healing of intestines, cylinders are passed in stools. It was used by him for the treatment of acute cholecystitis by anastomosing of the gallbladder to the duodenum, and later also for anastomosing of the other segments of the intestine. The Murphy's button is the forerunner of the modern mechanical «end to-end» anastomoses.

Then, he was the first to successfully re-anastomose a femoral artery severed by a gunshot (1896), and the first to perform the biliary tract endoscopy (1912).

Finally, he described a sign in which the examinant's fingers are gently placed below the right costal margin and the edge of the liver at the site of the middle clavicular line (MCL), the appropriate location of the gallbladder. When during inspiration an expansion of the lungs displaces the diaphragm and the gallbladder downwards pushing it into a contact with the examinant's fingers, elicit pain in cases a diseased gallbladder, usually in acute cholecystitis (Murphy's sign, 1880s). A similar test, named in his honor, is the sonographic Murphy's sign, where a sonographer presses the sonographic probe over the gallbladder elicit pain if diseased.

In 1890, **Olexandr Pavlovsky (1857 – 1944**: 19th century. Microbiology and immunology] introduced cleaning (sterilization) of surgical instruments from microorganisms in a boiling 1% sodium chloride solution.

The not constant lymphatic nodes (LN) between the major and minor pectoralis muscle, discovered by **Joseph Rotter (1857 – 1924)** — German surgeon, often harbor metastases from carcinoma of the breast (Rotter lymphatic nodes, 1896).

In 1892, **Cesar Roux (1857 – 1934)**[520] — Swiss surgeon, introduced an «end-to-side» esophago-jejunal anastomosis, in the form of the letter «Y», named the «Roux-en-Y» anastomosis. It has an application after partial or total gastrectomy as gastro-jejunal or esophago-jejunal anastomosis post resection of tumors of the common bile duct (CHD), pancreas or liver, and as palliative approach in obstruction by neoplasm and following an injury of the bile or pancreatic ducts by creating anastomosis between the gallbladder, CBD, bile or pancreatic ducts and the jejunum. Additional indication for performing the «Roux-en-Y» anastomosis is morbid obesity, by doing gastric bypass, i.e., gastro-jejunal anastomosis (since 1991).

Kyrylo (Kirill) M. Sapezhko, or **Sapieżko (30. III.1857 – 1928, city of Kyshyniv, Moldova)**[521] — Ukrainian surgeon, graduate of Medical Division of St Volodymyr

---

[520] Roux C. De la gastroenterostomie. Rev chir 1893;13:402-3.

[521] City of Kyshyniv (first mentioned in reading and writing 1436) – capital of Moldova. — Shkrobot BB. Hernias of the abdominal wall. In, LYa Kovalchuk, VF Sayenko, HV Knyshov. Klinichna Khirurhiya. Vol. 1-2. Publ. «Ukrmedknyha», Ternopil 2000. Vol. 2. P. 11-23. — Marynzha L. K.M. Sapezhko, surgeon and patriot. «Zdorov'ia Ukrainy» 2010;4(233):73. —— Ablitsov VH. «Galaxy Ukraine». The Ukrainian diaspora: prominent persons. Publ. «Kyt», Kyiv 2007.

University, city of Kyiv, Ukraine (1884), where he successfully defended dissertation for the degree of Doctor of Medical Sciences (1894). Subsequently, he organized, chaired Facultative Surgery, became well-deserving professor the Odesa University, city of Odesa, Ukraine (1903-19).

He was the first to use the mucous membrane of the oral cavity to cover defect of the ureter – oral mucosal graft urethroplasty (1884)[522], to correct entropion by tarsocheiloplasty (van Millingen-Sapezhko operation, 1887)[523], reconstruct of eyelids and conjunctival sac.

During the repair of umbilical hernia according to Kyrylo Sapiezhko (Sapiezhko's operation, 1900)[524], the skin over the hernia protrusion is dissected in the vertical direction, then the hernia sac is carefully separated from the subcutaneous tissue and the aponeurosis of the rectus abdominis muscle (RAM) in all directions for 10-15 cm. The umbilical ring at the white line is cut up and down, the excess of the hernia sac is removed, and the peritoneum closed. Then, the edge of dissected aponeurosis of the ipsilateral RAM is sutured to the posterior sheath of the contralateral RAM. The remaining free rim of the aponeurosis of the contralateral RAM is sutured to the anterior sheath of the ipsilateral RAM. As a result, the sheath of both RAM is laid one upon other on the white line as a flap of the coat.

After the October 1917 coup d' état, he emigrated to Turkey, from there to Moldova.

Theodor M. Tuffier (1857 – 1929)[525] — French thoracic surgeon, who did under general anesthesia with a positive pressure breathing a successful intercostal space thoracotomy (ISCT) with partial lung resection (1891); introduced intratracheal intubation with an inflatable balloon (1896); performed in a young woman with disabling symptoms the first closed («blind») digital commissurotomy of the stenosed aortic valve (AV) using his finger as a dilator and invaginated the wall of the aorta into the orifice of the AV, thus widening the cusp's commissures and relieving stenosis with long-term survival (1912).

Werner H. Zoege-Manteufel (1857 – 1926) — Estonian surgeon, introduced the use of rubber surgical gloves (1897) and **Paul L. Friedrich (1864 – 1916)**[526] — German surgeon, of latex surgical gloves (1898). The latter demonstrated the importance of

---

[522] Korneyev I, Ilyin D, Schultheiss D, Chapple C. The first oral mucosal graft urethroplasty was carried out in the 19th century: the pioneering experience of Kirill Sapezhko (1857-1928). Eur Urol 2012 Oct;62(4):624-7.

[523] Van Millingen E. The tarsocheiloplastic operation for the cure of trichiasis. Ophthal Rev. 1887;6:309-14.

[524] Shkrobot BB. Hernias of the abdominal wall. In, LYa Kovalchuk, VF Sayenko, HV Knyshov. Klinichna Khirurhiya. Vol. 1-2. Publ. «Ukrmedknyha», Ternopil 2000. Vol. 2. P. 11-23.

[525] Lichtenstein SV. Closed heart surgery: back to the future. J Thorac Cardiovasc Surg 2006;131:941-3.

[526] Friedrich PL. Die aseptische versorgung frischer wunden. Arch Klin Chir 1898;57:288-310. — Schmitt W. P.L. Friedlich and the wound infection. Zentralbl Chir 1978;103(2):65-9.

surgical removal of all infected tissues from the wound (debridement) within 6-8 hours, and then after debridement such wounds can be treated as a sterile one.

Friedrich P. Reichel (1858 – 1934) — German surgeon, and **Jenő (Eugene) A. Pólya (1876 – 1944)** — Hungarian surgeon, proposed a modification of the Bilroth II operation that involves a subtotal (2/3) gastrectomy with blind closure of duodenal stump, and gastrojejunal retrocolic, antiperistaltic, posterior and full circumference anastomosis (Reichel-Polya operation, 1897, 1911-13). Its modification is **Franz** von **Hofmeister (1867 – 1929)** — German surgeon, and **Hans Finsterer (1877 – 1955)** — Austrian surgeon, partial gastrectomy with blind closure of the duodenal stump, closure of a portion of the lesser curvature and retro-colic «end-to-side» anastomosis of the remainder, i.e., only the greater curvature, to the jejunum (Hofmeister-Finsterer's operation, 1918).

Pericardial resection for constrictive pericarditis was proposed in 1895 by **Edmond Weill (1858 – 1924)** — French pediatrician, and in 1898, by **Edmond Delorme (1847 – 1929)** — a French surgeon. The former described absence of expansions in the sub-clavicular area in infants with pneumonitis (Weill sign). The latter is known because of his procedure for rectal prolapse which consist of excision of the mucosa overlying the exposed rectal prolapse from the anorectal junction to the apex of the prolapse, longitudinal plication of the muscular wall of the rectum, reduction of the prolapse and retropexy in the sacral space (Delorme operation, 1900)[527].

Attending a patient with a knife wound alongside of the left 2nd rib cartilage, **Daniel H. Williams (1858 – 1931)** — American surgeon, closed a severed left internal thoracic artery (ITA) and sutured a small oozing laceration on the epicardium of the right ventricle (RV) with survival (1893), and **Axel H. Cappelen (1858 – 1919)** — Norwegian surgeon, via the left thoracotomy ligated a transected coronary artery and closed a ventricular wound but the patient died on the 3rd postoperative day from mediastinitis (1895).

Mykola M. Volkovych [09(21).XII.1858, village of Horodnia, Chernihivs'kyy Region, Chernihivs'ka Obl., Ukraine – 11.XII.1928, city of Kyiv, Ukraine][528] — Ukrainian surgeon, developed the method of an extracapsular resection of the knee joint (1896), of an open reconstruction of the skull for a herniated brain (1903),

---

[527] Delorme R. Sur le traitement des prolapsus du rectum la totaux par l'excision de la muquese rectale ou recta-colique. Bull Soc Chir Paris 1900;26:498-9.

[528] Village of Horodnia (Foot Note 419), Chernihivs'kyy Region, Chernihivs'ka Obl., Ukraine. — Rozhin I. Professor Mykola Volkovych (on occasion of 100-anniversary of his birthday). J Ukr Med Assoc North Am 1960 Jan;7,(16):22-4. — Iskiv B. The luminary of the Ukrainian medical science (to the 140-anniversary of the birthday of Mykola Markyanovych Volkovych). J Ukr Med Assoc North Am 2002 Spring;47,2(148):44-8. — Ablitsov VH. «Galaxy Ukraine». The Ukrainian diaspora: prominent persons. Publ. «Kyt», Kyiv 2007. — Hanitkevych YaV. The contribution of the Ukrainian physicians to the world medicine. Ukr Med Chasopys 2009 VII/VIII;4(72):110-15.

of closure of a recto-vaginal fistula (1904) and a splint for immobilization of the fractured shoulder, hip and tibia (1906). He re-described the symptom of **Emil Kocher (1841 – 1917**: above**)** of acute appendicitis which begins with an attack-like pain in the epigastric region below the xiphoid process or in the umbilical region, and in 1-3 hours translocated to the right iliac region, becomes constant and irradiates nowhere (Kocher-Volkovych symptom, 1892, 1926).

Contribution of **Eugene L. Doyen (1859 – 1916)** — French surgeon, into surgery include the design of the operating table, of the retractor for the maintenance of operative access by the separation of wound edges and retraction organs and tissues located below, of the rib dissector for the separation of the intercostal muscles during intercostal space thoracotomy (ICST), of the clamp with curved arms to clamp blood vessels for temporary control of bleeding during operations on the GI tract, of the device for aspirating the blood from the operative field.

Mykola P. Trinkler (07/19.IX 1859, city of St Petersburg, Russia – 10.VIII 1925, town Yevpatoria, Crimea, Ukraine)[529] — Ukrainian surgeon, was one of the first introduced the prophylactic approach of mechanical and chemical cleanliness with the aim to prevent entry of microorganisms into the wound and the operative field (asepsis); proposed the method to drain chambers of the brain; treated malignant tumors of the brain using radiotherapy in in 1884 – 1926 at the Kharkiv Medical University, city of Kharkiv, Ukraine.

An invention of a device for the adjustment of cystoscope during catheterization of the ureter is credited to **Jaquin Albarrán y Dominguez (1860 – 1912)** — Cuban urologist, known as the Albarrán level (1893), who also depicted the median part of the prostatic gland lobe below the little tongue (uvula) of the urinary bladder (Albarrán's gland) and small branching tubules of the cervical portion of the prostate (Albarrán tubules). He described in 1905 and **John K. Ormond (1886 – 1978)** — American physician, re-described in 1948 an inflammatory retroperitoneal fibrosis, presented as lower back pain, rheumatic fever (RF), hypertension, deep venous thrombosis (DVT) and other obstructive symptoms (Albarrán-Ormond syndrome).

Mathie Jaboulay (1860 – 1913) — French surgeon, introduced a mobilization of the duodenum (kocherization) according to **Emil Kocher (1841 – 1917**: above**)** with «side-to-side» gastroduodenostomy for severely scarred pylorus and proximal duodenum (Jaboulay gastroduodenostomy, 1892); along with **Karl Wilkenmann (1863 – 1925)** — German urologist, to treat large, thickened-wall or multiloculated testicular hydrocele partially excised tunica vaginalis with edges oversewn together

---

[529] Town of Yevpatoria (founded 497), Crimea Ukraine. — Ablitsov VH. «Galaxy Ukraine». The Ukrainian diaspora: prominent persons. Publ. «Kyt», Kyiv 2007. — Hanitkevych YaV. The contribution of the Ukrainian physicians to the world medicine. Ukr Med Chasopys 2009 VII/VIII;4(72):110-15.

behind the spermatic cord (Jabolay-Wilkenmann operation, 1892); carried out an amputation of the lower extremity within the sacral-iliac joint for sarcoma (hemipelvectomy, hindquarter, inter-ilio-abdominal or Jaboulay amputation, 1894); and attempted kidney transplantation in human, anastomosing the renal vessels of a sheep and a pig kidney to the brachial vessel of two female patients, respectively, who were dying of renal failure (RF), none of them survived (1906).

Rigid bronchoscopy is an endoscopic procedure of visualization the inside into the airway for diagnostic and therapeutic purpose using instrument called rigid bronchoscope inserted through the mouth, allowing visual examination of the patient's larynx, trachea, right and left main stem branches (RMSB, LMSB) for abnormalities such as foreign bodies, inflammation, bleeding, and tumors. It allows to define endobronchial anatomy, take discharge for culture and sensitivity to antibiotics and cytology, perform pulmonary toilet or endoluminal debridement, treat of atelectasis and lung abscess, take biopsy of the proximal lymph nodes (LN) for staging of primary lung carcinoma before operation and of an endobronchial tumor, aids in performing endotracheal intubation, extubating, placement of percutaneous tracheostomy (PCT).

The first rigid bronchoscopy using a bronchial tube was performed in 1897 by **Gustav Killian (1860 – 1921)**[530] — German laryngologist and founder of bronchoscopy, to remove a foreign body (pork bone) from the (RMSB).

He also excised the anterior wall of the frontal sinus with diseased tissue to form for drainage a permanent communication with the nasal cavity (Killian's operation); depicted a triangular area of weakness of the posterior wall of the pharynx between the oblique fibers of the constrictor pharyngeals inferior muscle and the transverse fibers of the cricopharyngeal muscle (Killian area, dehiscence or triangle), where a juxtrasphincteric (pulsion) pharyngoesophageal Zenker **[Friedrich** von **A. Zenker (1825 – 98)**: 19th century. Pathology]**[531] diverticulum most commonly occurs.

Inferiorly, at the proximal 1-2 cm portion of the posterior cervical esophagus, there is a reverse triangle between the transverse fibers of the cricopharyngeal muscle and the esophageal inner circular layer of the muscularis propria without coverage by the external longitudinal muscle of the esophagus which create a mirror-image triangular area of weakness, named after **Eduard Laimer** — German anatomist, the Laimer triangle. Through the Laimer triangle may arise a juxtrasphincteric (pulsion) pharyngoesophageal diverticuli.

---

[530] Killian G. The mouth of the oesophagus. Laryngoscope 1907;17:421-8. — Killian G. La boudre de l'oesophage. Ann Mal Orelle Larynx 1908;34:1. — Killian H. Gustav Killian, sein Leben, sein Werk, zugleich ein Beitrag zur Geschichte der Bronchiologe und Laryngologie. Dustri-Verlag, Remscheid-Lennep 1958.

[531] Kovalchuk LYa. Esophageal diverticuli. In, LYa Kovalchuk, VF Sayenko, HV Knyshov. Klinichna Khirurhiya. Vol. 1-2. Publ. «Ukrmedknyha», Ternopil 2000. Vol. 2. P. 132-7.

The name of **Jean F. Callot (1861 – 1944)** — French surgeon, is associated with an anatomic space bordering the common bile duct (CBD) medially, the cystic duct inferiorly and to the cystic artery or the inferior border of the right lobe of the liver superiorly (the cystohepatic, hepatobiliary Callot triangle, 1890)[532]. It usually contains the gallbladder lymph node (LN), called the Lund LN **[Fred B. Lund (1865 – 1950) — American surgeon]** or the Callot LN, which increases in size during cholecystitis or cholangitis and is removed along with the gallbladder in cholecystectomy.

John M.T. Finney (1863 – 1942) — American surgeon, reconstructed the pyloric channel by means of a longitudinal incision through the pylorus and adjacent wall of the stomach and duodenum by creating an inverted U-shaped anastomosis between the stomach and duodenum (Finney pyloroplasty).

Leonardo Gigli (1863 – 1908) — Italian surgeon and obstetrician, designed a flexible wire saw with teeth for cutting bone on both sides – pubiotomy (transection of the pubic bone laterally from the pelvic symphysis) to widen the pelvis in order to assist in childbirth, median sternotomy or four small rounded trepanation holes in the skull for the wider exposure of the brain (Gigli saw, 1893-97).

The open jejunostomy is the surgical procedure, initially performed by **Friderich O. Witzel (1963 – 1925)** — German surgeon, by inserting distal tip of a rubber or silicone catheter through the skin at the front of the abdomen into the jejunum 30 cm distal to the ligamentum of Treitz **[Václaw Treitz (1819 – 72).** 19th century Anatomy, histology, and embryology] on the antimesenteric border with the catheter tunneled in a seromuscular grove and bringing its proximal tip to the skin surface, for administration of medication and feedings in patients who cannot swallow.

In 1889-90, **Olexandr O. Jasinovsky** or **Alexander A. Jassinowsky (1864, city of Odesa, Ukraine – 1913, city of Odesa, Ukraine)**[533] — Ukrainian surgeon, was the first to perform the scientifically based arterial anastomoses. He made 26 longitudinal and transverse arteriotomies on the carotid arteries of dogs, horses, and calves, and then sutured them using fine curved needles, fine silk, placed his sutures 1 mm apart through the adventitia and media only, avoiding the penetration of the intima for fear of injuring the epithelial cells that might provoke the clot formation, and stressed

---

[532] Calot JF. De la cholecystectomie. These de doctorat, Paris 12 decembre 1890.

[533] Jassinowsky A. Die Arteriennaht: Eine Experimentell-Chirurgische Studie. Inaugural-Dissertation zur Erlangung des Grades eines Doctors der Medicin verfasst und mit Bewilligung einer Hochverordneten medicinischen Facultät der Kaiserlichen Universität zu Dorpat zur ö 1889. Druck von C. Mattiesen, Dorpat, Estonia 1889. — Jassinowsky A. Ein Beitrag zur Lehre von der Gefässnaht. Langenbecks Arch klin Chir 1891;42:816-41. — Barker WF. A century's worth of arterial sutures. Ann Vasc Surg 1988;2(1):85-91. — Friedmann P. Presidential address: Decay and revival of vascular surgery. J Vas Surg 1993 June;17(6):985-93. — Jassinowsky A. Classic Reprint Series. Die Arteriennaht. Eine Chirurgische Studie. Publ. Forgotten Books, London Feb. 23, 2018. — Ablitsov VH. «Galaxy Ukraine». The Ukrainian diaspora: prominent persons. Publ. «Kyt», Kyiv 2007.

the necessity of aseptic. During the follow-up period of up to 100 days there was no complication – infection, thrombosis or twisting of the artery in 22 animals.

The first successful total gastrectomy was performed in 1897, by **Carl B. Schlatter (1864 – 1934)** — Swiss surgeon. Then, he and **Robert B. Osgood (1873 – 1956)** — American orthopedic surgeon, reported concurrently on one of the most common causes of knee pain in active adolescents due to sprain of the knee joint resulting in osteochondritis of the tibial tuberosity (apophysitis tibialis adolescentium or Osgood-Schlatter condition, 1903).

During the repair of umbilical hernia according to the method of **Charles H. Mayo (1865 – 1939)**[534] — American surgeon, a semicircular transverse skin incision is used. After removal of the hernia sac and closure of the peritoneum with continuous catgut sutures, a defect of the anterior abdominal wall is closed using the 1st and 2nd row of П-like sutures in the transverse direction.

The painful inflammation of a tendon (tenosynovitis) due to a narrow tendon sheath (tunnel) of the long adductor muscle and a short extender muscle of the thumb described in 1895, by **Fritz de Quervain (1868 – 1940)** — Swiss surgeon, called «the washer woman's sprain» or de Quervain tenosynovitis (1895).

The surgical repair of congenital chest wall deformity – thoracoabdominal ectopia cordis in 1896 by **C. Arndt**[535] — German surgeon, resulted in death during an attempt to return the heart to the thoracic cavity.

In 1898 **Daniel W. Samways**[536] — English physician, proposed notching of the stenotic mitral valve (MV) orifice affected by rheumatic heart disease (RHD), which could provide symptomatic relief.

In 1898 **Frederick W. Parham**[537] — American surgeon, reported resection of a bone chest wall tumor involving three ribs in continuity with a pulmonary tumor. Visible pneumothorax was controlled by soft tissue coverage.

[534] Shkrobot BB. Hernias of the abdominal wall. In, LYa Kovalchuk, VF Sayenko, HV Knyshov. Klinichna Khirurhiya. Vol. 1-2. Publ. «Ukrmedknyha», Ternopil 2000. Vol. 2. P. 11-23.

[535] Arndt C. Nabelschnurbruch mit Hertzhernie: Operation durch laparatomie mit todlichen Ausgang. Centralbl Gynekol 1896;20:632-3.

[536] Samways DM. Mitral stenosis: a statistical inquire. Br Med J 1898 Feb 5 (1936):364-5. — Samways DM. Cardiac peristalsis: its nature and effect. Lancet 1898 Apr;2:927. — Samways DM. Mitral stenosis. Br Med J 1898 Jul 23;2(1960):275. — Samways DM. The surgical treatment of mitral stenosis. Br Med J 1925 Oct;2(3383);818.

[537] Parham FW. Thoracic resection for tumor growing from the bony chest wall. Trans South Surg Gynecol Assoc (New Orleans) 1898;2:223-263.

Gynecology and obstetrics

In 1809 **Ephraim McDowell (1771 – 1830)** — American surgeon and gynecologist, after consulting a young woman, removed in his own house, without the benefit of anesthetic or antisepsis, through a median laparotomy, an ovarian tumor (weight 10,2 km) by ligating of the **Gabriele Fallopio (1523 – 62**: Renaissance**)** tube near the uterus and cut off the tumor (ovariotomy). The operation lasted 25 min., included removal of blood from the peritoneal cavity and bathing the intestines with warm water, without complications and with the survival of 35 years.

Franz K Naegele (1778 – 1851) — German obstetrician, introduced the standard rule for calculating the expected date of delivery (EDD) when assuming a gestational age is 280 days (40 weeks) of childbirth, by adding a year, subtracting three months, and adding seven days to the origin of the gestational age, i.e., the first day of the last menstrual period (LMP), or more simply, from the first day of the LMP deduct three months and add seven days (Naegele's rule, 1825).

Following the first classic repair of post-partum of vesico-vaginal fistula (VVF) in c. 1676 by **Henry van Roonhuyen (17th century)**, successive repairs of post-partum VVF were performed in 1838 by **John P. Mattauer (1787 – 1875)**[538] — American surgeon and gynecologist, in 1839 by **George J.W. Hayward (1791 – 1863)**[539] — American urologist and gynecologist, in 1845-49 and 1852 by **J. Marion Sims (1813 – 83)**[540] — American gynecologist, all of which basically followed Henry van Roonhuyen technique of repair, except that the latter in addition to the «elbow – knee» position, used the semi-prone position, designed the Sim's vaginal mirror speculum, and closed VVF with silver-wire sutures (1849).

While studying the dimples of Venus (Ancient medicine. *China*), **Gustaw A. Michaelis (1798** – when realized that theory of the prophylaxis of puerperal sepsis to be true and desperate that had caused death of so many women in childbirth committed suicide in **1848)** — German obstetrician, depicted a rhombus shaped contour on the posterior surface of the human pelvis, created by concavities of the superior posterior hip crest and lines between the gluteal muscles and the furrow at the distal end of the vertebral column (the rhombus of Michaelis).

---

[538] Mettauer P. Vesiculo-vaginal fistula. Boston Med Surg J 1840;22:154-5.
[539] Hayward DM. Care of vesico-vaginal fistula. Am J Med Sci 1839;24:282-8. — Wall LL. Dr George Hayward (1791-1863): A forgotten pioneer of reconstructive pelvic surgery. Int Urogynecal J Pelvic Floor Dysfunct 2005 Sep-Oct;16(5):330-3.
[540] Sims JM. On the treatment of vesico-vaginal fistula. Am J Med Sci 1952;23:59-82. — Anonymous. Marion Sims and his silver sutures. NEJM 1945 Nov 22;233:631-3.

Charles Clay (1801 – 93) — English general surgeon, gynecologist and obstetrician, who successfully remove an ovarian tumor (weight 23,8 kg) in 10 minutes (ovariectomy, 1842) and leiomyomata's uterus (1845).

In 1853 **Olexandr P. Matveyev (20.IV(02.V).1816, Province of Orlov, Russia – 23.V(04.VI).1882, city of Kyiv, Ukraine)**[541] — Ukrainian gynecologist and obstetrician, and in 1854 **Karl S.F. Crede (1819 – 92)** — German gynecologist and obstetrician, introduced a prophylaxis for neonatal blenorrhea by routine subconjuctival application of a 2% silver nitrate (Lapis lunaris) solution (Matveyev-Crede method).

The Crede's maneuvers were designed for expressing the retained placenta after delivery by manually forcing the uterus down to the pelvis and for expressing urine by manually forcing the the urinary bladder down to the pelvis (1881).

A protracted postpartum excessive secretion of the milk independently of the breast feeding (galactorrhea) and a lack of menstruation (amenorrhea) were described by **Johann B. Chiari (1817 – 54)** — German gynecologist and obstetrician, and **Richard J.E. Frommel (1854 – 1912)** — German gynecologist, and obstetrician, and named in their honor the Budd-Frommel syndrome.

John Brixton Hicks (1823 – 97) — English obstetrician, described uterine contractions during pregnancy (Braxon Hicks contractions, 1872) and the podalic (Braxon Hicks) version of a nonviable fetus.

Ivan P. Lazarevych or **Lazarevich (17(29).III.1829, town of Mohyliv-Podilsky, Vinnytsia Obl., Ukraine – 26.II(10.III).1902, city of Kharkiv, Ukraine)**[542] — Ukrainian gynecologist and obstetrician, revealed a connection between the status of the pregnant uterus and the function of the fetal heart; designed the straight obstetrical forceps (1865), the obstetrical forceps without pelvic curvature with the parallel spoons (1887), blepharotomy, a hook for pulling out a fetus, the uterine probe.

E.L. Alfred Hegar (1830 – 1914) — German gynecologist, designed dilators (Hegar) – a set of the bougie of different sizes for dilatation of the uterine ostium.

After **Rainer de Graaf (1641 – 93**: 17th century) and **Alphonse Guerin (1816 – 95)**: 19th century. Surgery), **Alexander J. Ch. Skene (1837 – 1900)**[543] — Scottish

---

[541] Ablitsov VH. «Galaxy Ukraine». The Ukrainian diaspora: prominent persons. Publ. «Kyt», Kyiv 2007.

[542] Town of Mohyliv-Podilsky (Foot Note 3), Vinnytsia Obl., Ukraine. — Lazarevich IP. De pelvis feminae metiendae retionalibus. Dissertatio. Kioviae 1857. — Lazarevich IP. Examination of the pregnant women abdomen. Kharkiv 1865. — Lazarevich IP (Ivan Pavlovich) 1829-1902. On the Obstetric Forceps. Publ. JW Kolcman, London 1881. — Lazarevich IP. The Course of Obstetrics. Vol. 1-2. St Petersburg 1892. — Ablitsov VH. «Galaxy Ukraine». The Ukrainian diaspora: prominent persons. Publ. «Kyt», Kyiv 2007. — Hanitkevych YaV. The contribution of the Ukrainian physicians to the world medicine. Ukr Med Chasopys 2009 VII/VIII;4(72):110-15.

[543] Skene A. The anatomy and pathology of two important glands of the female urethra. Am J Obst Dis Women Child 1880;13:265-70.

gynecologist, re-described in 1880 the lesser vestibular (periurethral) glands, which thereafter were named after him Skene's glands. Its inflammation is called skenitis.

Alexander Skene performed the first successful laparo-elytrotomy in pregnant women with a stalled delivery, by the right iliac extraperitoneal incision with exposure of the upper vagina, longitudinal vaginal incision, and delivery of a child in that way, saving the life of both mother and child[544] and craniotomy using the Sims's vaginal mirror speculum (above).

The important contribution of Alexander Skene was also diagnosing, description and naming in 1887 of interstitial inflammation of the urinary bladder (interstitial cystitis) or bladder pain syndrome (BPS), a type of chronic pain of the urinary bladder, manifested by feeling extreme urgency and frequency to urinate, especially at night, and often pain during sex, associated with depression and lower quality of life, irritable bowel syndrome, fibromyalgia, urinary tract infection (UTI), sexually transmitted infections, bladder ulcer or cancer and prostatitis. Except in UTI, urine cultures are usually negative. In persistent cases visual survey of the urinary bladder (cystoscopy) is necessary for proper diagnosis and treatment. Patients with BTS may improve by modifying their diet to remove food and beverages that trigger symptoms, such as regular and caffeinated coffee, tea including green tea, soda, artificial sugar and fruit juices, mainly before going into sleep. Medications include:

(1) AZO bladder control dietary supplement. 54 capsules. Ingredients: go-less proprietary blend [pumpkin (Cucurbita pepo) seed extract, maltodextrin, soy germ (Glycine max) extract], dicalcium phosphate, gelatin, microcrystalline cellulose, magnesium stearate, silicon dioxide, titanium dioxide (color), caramel (color). Contains soy. Directions: for the first two weeks: take 1 capsule morning, noon and night. After two weeks, begin taking 1 capsule twice a day. For best results, daily use for a minimum of 30 days is important.

(2) Azo Cranberry. Urinary tract health. Help flush urinary tract cleanness. Vitamin C 120 mg. Cranberry (Vaccinum macrocarpon) fruit equivalent 25,000 mg plus vitamin C 2 soft gels. Whole powder 500 mg. Other ingredients: soybean, oil, rice bran oil, gelatin, glycerin soy lecithin, yellow bees wax, water and color (carob, carmine and annato). Contains soy. Directions: take 2 soft gels daily with water.

(3) Pentosane polysulfide (Elmiron) / peridium per orum (p.o.) 300 mg, 600 mg or 900 mg daily with or without antibiotics for 1-2 days.

---

[544] Jewett Ch. Two cases of laparo-elytrotomy, with remarks. Gynecological Transactions 1885;10:1-11.

The law of **Christian G. Leopold (1846 – 1911)** — German gynecologist and obstetrician, states that when a placenta is attached to the posterior wall of the uterus, the oviducts assume a direction converting upon the anterior wall; but when the attachment is on the anterior wall during recumbency, the tubes turn backward, parallel to the axis of the body.

Radical removal of the uterus (hysterectomy) for carcinoma of the uterine cervix was successfully performed by **Karel J. Pawlik (1849 – 1914)** — Czech Austrian gynecologist and obstetrician, in 1889, and by **Ernst Werthheim (1864 – 1920)** — Austrian gynecologist, in 1898 (Pawlik- Werthheim operation) which include removal of the uterus, tubes, parametrium (tissues surrounding the upper vagina) and pelvic lymph nodes. Karel Pawlik is also known for description of the triangle (Pawlik) on the anterior vaginal wall corresponding to the urinary bladder triangle of **Joseph Lieutaud (1703 – 80**: 18th century**).**

Otto E. Kustner (1850 – 1931) — German obstetrician, proposition states that if an ovarian tumor is left-sided, the torsion of its pedicle takes place towards the right; if right-sided, then towards the left.

In 1892 **Albert S.G. Döderlain (1860 – 1941)** — German obstetrician, discovered Gram-negative bacillus (Lactobacillus acidiphilus) in the vagina (Döderlain bacillus).

Charles B. Penrose (1862 – 1925) — American gynecologist, who introduced a thin rubber tube (Penrose) drain and a triple-lumen (sump-Penrose) drain for evacuation of the discharge from wound and abscesses.

Children's diseases (pediatrics)

Stepan Kh. Khotovytsky (1794 or 1796, town of Krasyliv, Khmelnytsky Obl., Ukraine – 30.III/11.IV.1885, city of St Petersburg, Russia)[545] — Ukrainian pediatrician, founded and chaired the first Department of Pediatrics in Russia (1836) and published the textbook «Pediatrica» (1847). According to him *«Pediatrica – is the science about the different peculiarites in the structure, secretions and diseases of the child's organism and about the resulting peculiarities of preserving health and the treatment of diseases in children».*

Since 1825 **Pierre F. Bretonnean (1778 – 1862)** — French physician, employed successful tracheostomies, including one in 1832 as a last resort to treat a case of diphtheria.

---

[545] Town of Krasyliv (first mentioned 1444; suffered from conducted by Russian Communist Fascist regime Holodomor-Genocide 1932-33 against the Ukrainian Nation), Khmelnytsky Obl. (area 20,629 km², Ukraine. — Voskresenskaya N.S. Kh. Khotovytsky – medical scientist: pediatrist and hygienist. UHMJ «Ahapit» 1997-1998;7-8:65-9. — Ablitsov VH. «Galaxy Ukraine». The Ukrainian diaspora: prominent persons. Publ. «Kyt», Kyiv 2007.

The congenital heart disease (CHD) – aortopulmonary (AP) window, described by **John Elliotson (1791 – 1868)**[546] — English physician, is characterized by a side-to side communication between the adjacent portion of the ascending aorta (AA) and a main pulmonary artery (MPA).

Poliomyelitis or infantile paralysis, caused by poliovirus / ribonucleic acid (RNA) enterovirus of the picornavirus family, was more thoroughly investigated, and treated from 1840 by **Jacob** von **Heine (1800 – 79)** — German orthopedist, who treated scoliosis, clubfeet, paralysis of the arms and legs and used hydrotherapy and gymnastics, since 1890 by **Karl O. Medin (1847 – 1927)** — Swedish pediatrician, who described its epidemic character. It was named in their honor the Heine-Medin's disease.

In 1839 **Robert F. Froriep (1804 – 61)**[547] — German anatomist and physician, in 1841, Sir **Alfred Poland (1822 – 72)** — English surgeon, described congenital absence or hypoplasia of the major and minor pectoralis muscles, subcutaneous (SC) fat and axillary hair, the manubrium of the sternum, the 2nd - 5th ribs and costal cartilages, anomalies of the breast, i.e., absence or underdevelopment of the mammary gland (amastia) and missing of the nipple and areola of the mammary gland (atelia mammaria) with ipsilateral depression of the chest wall, hypoplasia of fingers (brachydactyly), most commonly fusion of the three central fingers (syndactyly), a wrist in which several fingers are fused and have a common nail («mitten hand») and «claw hand» (Poland syndrome). Associated disorders could include deformity described by **Otto G. K. Sprengel (1852 – 1915)** — German surgeon, in which the scapula is small, elevated and winged (1891), or syndrome described by **Paul J. Möbius (1853 – 1907)** — German neurologist, a congenital bilateral or unilateral paralysis of the cranial 7th (facial) nerve with facial immobility and of the cranial 6th (abducent) nerve with inability to move eyes from side to side (1888), and leukemia.

Cerebral palsy was defined by **William J. Little (1810 – 94)**[548] — English surgeon, as a brain injury caused by asphyxia (deprivation of O_2) during pathological birth, which result in spastic rigidity and sometimes paralytic contraction usually of the lower extremities (spastic diplegia or cerebral palsy, 1853, 1861).

Cor triatriatum sinistrum, one of the rarest congenital heart diseases (CHD), comprising 0.1% of them, in the classic form described in 1868 by **W.S. Church**[549]

---

[546] Eliotson J. Case of malformation of pulmonary artery and aorta. Lancet 1830;1:247-8.

[547] Müller-Dietz HE. Sieben unver öffentlische Briefe des Naturforschers Karl Ernst bon Baaer an L.F. Froriep und dessen Sohn aus den Jahren 1823 bis 1831. NTM Int Hist Ethiccs Natural Sci, Technol Med 1993 Dec;1(1):167-9.

[548] Little WJ. On the nature and treatment of the deformities of the human frame. Longman, Brown, Green and Longmans, London 1853.

[549] Church WS. Congenital malformation of the heart: abnormal septum in the left auricle. Trans Pathol Soc London 1868;19:188-90.

— English pathologist, is created by a separation of the pulmonary vein (PV) from the left atrium (LA) by a fibromuscular membrane with a communication between them by an orifice in the center of that membrane.

Progeria (in Greek – «pro» meaning «before», or «premature», and «geras» meaning «old age») was independently described in 1886 by Sir **Jonathan Hutchinson (1828 – 1941)**[550] — English pathologist, ophthalmologist, dermatologist, and surgeon, and in 1897 by **Hastings Gilford (1828 – 1941)** — English surgeon, is an extremely rare autosome dominant congenital disorder of premature old age in an early childhood.

The main cause for the development of progeria or Hutchinson-Gilford syndrome appears to be a mutation of the gene coded by a protein A-type laminas and disruption of immune system. Lamins are major component of a protein scaffold on the inner edge (lamina) of the nucleus which help organize and process nuclear ribonucleic acid (RNA) and deoxyribonucleic acid (DNA). A defective A-type laminas increase DNA damage and chromosome aberrations, making the cell's nucleus unstable, followed by replication of genes leading to premature aging.

The Hutchinson-Gilford syndrome is marked in the first few months of life by failure to thrive and localized scleroderma, a.k.a., morphea – localized hardening of skin and subcutaneous tissue with a bluish, purple or brown discoloration of the skin.

In around 18-36 months of life of a child there appears a small face with a shallow recessed jaw and a pinched nose, limited growth and small stature, gray hair or baldness, absence of facial and pubic hair, then alopecia (full body hair loss), later – wrinkled skin, atrophy of muscles, loss of bone mass and of eyesight. Intelligence tends to be average or above normal. Death, most commonly, in at least 90% of patients, occurs from complications of atherosclerosis such as aortic stenosis (AS), coronary artery disease (CAD), myocardial infarction (MI) or stroke, and kidney failure, between the mid-teens to early twenties years of life.

There is no known cure. But possibly vitamin D may allow reverse of the process by the blockage of interferon activation and decrease of gene replication stress and thus rejuvenate the cells.

Jonathan Hutchinton was the first to describe his triad of signs for congenital syphilis (Hutchinton's triad), i.e., notched incisor teeth, labyrinthine deafness and interstitial keratitis, before discovery of its causative factor Spirochete pallida in 1905 **[Fritz Chaudinn (1871 – 1906)** and **Erick Hoffman (1868 – 1959)**. 20th century. Microbiology and immunology].

---

[550] Hutchington J. Case of congenital absence of hair, with atrophic condition of the skin and its appendages, in a boy whose mother had been almost wholly bald from apolecia areata from the age of six. Lancet 15 May 1886;127(3272):923.

In 1888 **Harald Hirschsprung (1830 – 1916)**[551] — Danish pediatrician, re-described congenital aganglionic large bowel (1886) or congenital megacolon, named Hirschprung disease [17th century. **Frederik Ruysch (1638 – 1731)]**, and hypertrophic pyloric stenosis, a narrowing of the pyloric canal from muscular hypertrophy and edema of the mucosa in infants which is manifested by nausea, vomiting, epigastric pain, loss of appetite and weight, dehydration and hypochloremic alkalosis, visible peristalsis and presence of the palpable pyloric mass.

In 1874 **Mykola P. Tolochynov (1838, city of Starodub, Chernihiv Province, Ukraine – 1908, city of Kharkiv, Ukraine)**[552] — Ukrainian gynecologist and obstetrician who described in newborns the clinical picture of congestive heart disease (CHD) termed ventricular septal defect (VSD) or Tolochynov disease, and in 1879 **Henri L. Roger (1809 – 91)** — French pediatrician, heard in a child a loud long systolic murmur at the 3rd intercostal space (ICS) on the left side near the sternum, characteristic for a small muscular VSD (malasie de Roger).

VSD is a hole in the interventricular septum (IVS) between the left and the right ventricle (LV, RV) of the heart. Anatomical classification of VSD identify conoventricular (membranous or peri-membranous) VSD's centered inside or around membranous IVS, comprising 80% of all defects; conal (subarterial or subcristal) VSD, located in the conal (intrabulbar or outlet) septum are entirely surrounded by cardiac muscle or limited upstream by aortic or pulmonary annuli of the ascending aorta (AA) or main pulmonary artery (MPA), comprising 8% of all defects: inlet or atrioventricular (AV) canal VSD, located immediately underneath the septal leaflets of the tricuspid valve (TV), comprising 6% of all defects; and cardiac muscular (single or multiple) VSD, entirely surrounded by myocardium, comprising 10% of all defects[553].

VSD causes a shunt of the blood between the LV and RV, direction of which depend on the size of the defect, the pulmonary and systemic vascular resistance (PVR, SVR), and the ratio of pulmonary-to-systemic blood flow (Qp/Qs). Large VSD's are called nonrestrictive since they offer little or no resistance to blood flow. On the other hand, smaller VSD's offer a resistance to blood flow across the defect and are therefore called restrictive VSD's with Qp/Qs ratio rarely exceeding 1.5. The Qp/Qs ratio of moderate-sized VSD usually ranges between 2.5 to 3.0. Within the first year

---

[551] Hirschsprung H. Stuhlträgheit Neugeborener in Folge von Dilatation und Hypertrophic des Colons. Jahrbuch für Kinderheilkunde und physische Erziehung (Berlin) 1888;27:1-7.

[552] Town of Starodub (founded 1080), Province of Chernihiv, Ukraine / since 1919 – Bryansk Obl. (area 34,900 km²), Russia. — Ablitsov VH. «Galaxy Ukraine». The Ukrainian diaspora: prominent persons. Publ. «Kyt», Kyiv 2007.

[553] Soto B, Becker AE, Moulaert AJ, et al. Classification of ventricular septal defects. Br Heart J 1980 Mar;43(3):332-7.

of life approximately 30% of infants with VSD develop intractable congestive heart failure (CHF) or failure to thrive which will require surgical closure of the defect.

Infantile scurvy due to lack of ascorbic acid (vitamin C) in food was described by Sir **Thomas Barlow (1845 – 1945)** — English pediatrician, in 1883, and since called Barlow disease.

In 1885 **Nils F. Filatov (1847 – 1902)** — Russian pediatrician, described infectious mononucleosis (Filatov's disease), in 1889 he and in 1900 **Clement Dukes (1845 – 1925)** — English physician, the «fourth disease» (scarlatinella, Filatov-Dukes disease).

In 1895 Nils Filatov, and a year later (1896) **Henry Koplik (1858 – 1927)**[554] — American pediatrician, formulated the prodromal signs of measles [Medieval medicine – **Rzazes (c. 854 – c. 925)** / 17th century – **Thomas Sydenheim (1624 – 89)** / 17th century – **Richard Morton (1634 – 98)** / 18th century – **Thomas Fuller (1654 – 1734)**], which are small, irregular, bright red spots on the buccal and lingual mucosa with a minute bluish-white speck in the center of each (Filatov-Koplik spots).

Vitamins were discovered by pediatrician **Nicolai I. Lunin (1854 – 1937)** — Russian pediatrician, in 1880 and rediscovered (1912) by **Kazimierz Funk (1884 – 1967)** — Polish American biochemist.

The pseudo genital dysgenesis or multiple pterygium syndrome was noted, researched and described in 1883 by **C. Oskar A. de Kobylinski [26.08.1856, village-parish) Trikata, Beverina Minicipality, Latvia** – died from heart attack **02.04.1926, city of Rome, Italy**; temporarily buried (interred) in the Nom-Catholic Cemetery, 2nd (Russian) zone, Testaccio district of Rome, Italy, with subsequent transfer into the family tomb at Evele parish, Cemetery of the Evangelical Lutheran Church of Latvia, the town-manor Kempeni, Valkas municipality, Latvia][555] — Latvian physician of Polish Ukrainian descent, while a student of medicine at University of Dorpat (existed 1802-1919), now Tartu (founded 1632, closed 1710, reopened 1919), Estonia, a 22-year-old patient, who suffered from a congenital deformity of the neck – a wing-like extension of the neck with lung infection. Oskar de Kobylinski received the medical

---

[554] Koplik H. The diagnosis of the invasion of measles from a study of the exanthema as it appears on the buccal mucous membrane. Arch Pediatr 1896;13:918-22.

[555] Village-parish Trikata [arised while building in 1283-84 the castle by **Wilken** von **Endorp (?-1287)** — Grand Master of the Chivalrous Crusaders (Teutonic) Order], Beverina Municipality or Region (area 300.68 km²), Latvia. — Evele parish of the Evangelical Lutheran Church of Latvia (church originated 1517). — The town-manor of Kempeni (begins with the formation of Kempeni parish, which united with Evele parish in 1892), Valkas Municipality (area 908.83 km²), Latvia. — The Non-Catholic Cemetery for foreingers in the Testaccio Quorter (established 1716, first interment 1731), Italy. — Kobylinski O. Ueber eine flughautähnliche Ausbreitung am Halse. Arch Anthropol 1883;14:342-8. — Opitz JM, Pellister PD. Brief historical note: The concept of the «gonadal dyskinesis». Am J Med Genetics 1979;4(4):333-43. — Ablitsov VH. «Galaxy Ukraine». The Ukrainian diaspora: prominent persons. Publ. «Kyt», Kyiv 2007. — Marcinkowski F. Oskar Kobylinski (1856-1923) and the first description of Noonan syndrome in the medical literature. J Med Biogr 2020 Nov;28(40:202-207.

diploma, but did not present a doctoral dissertation, therefore was not presented with doctoral title.

This syndrome is caused by a single gene defect, is autosomal dominant and is inherited, it affects both sexes, is characterized by mental challenge (retardation), short stature without genital dysgenesis, hypogonadism, ptosis, ocular and neck pterygium, pulmonary infection, most commonly right-sided congenital heart disease (CHD) – atrial septal defect (ASD), stenosis of the pulmonary artery (PA) valve or PA branches without dysplasia, less commonly left-sided CHD – left ventricle (LV) hypertrophy, obstructive or non-obstructive cardiomyopathy (CM), cryptorchism's (77%) and deformities of the skeleton.

A year later (1884), he returned to Kempeni to renew medical practice developed previously by his father, **Antoni Kobylinski [12(31).1823, city of Sandomierz, Poland – 16.I.1880]**[556] — Latvian physician, who also studied at University of Dorpat (1845-50), published few medical scientific papers.

Theodor Escherich (1857 – 1911) — German Austrian pediatrician, who published a monograph entitled «Die Darmbakterien des Säuglings und ihre Beziehungen zur Physiologie der Verdauung» (Stuttgard 1886) in which he described discovered by him Gram-negative facultatively anaerobic bacillus-like bacteria called «commune bacterium coli» and its relationship to the physiology of digestion in infants. The «commune bacterium coli» was renamed Escherichia coli in 1919.

Ovksentiy V. Korchak-Chepurkivsky (28.II.1857, town of Konstiantynohrad / Krasnohrad, Kharkiv Obl., Ukraine – 27.XI.1947, city of Kyiv, Ukraine)[557] — Ukrainian pediatrician and hygienist, formulated the regularity of the periodical epidemics of diphtheria and the concept of antagonism between epidemics of diphtheria and other children's infections (1898); established the specialty of public (social) hygiene and organization of health welfare (1906-17).

Antoine B.J. Marfan (1858 – 1942)[558] — French pediatrician, who described in 1896 congenital disorder of the connective tissue acquired as a dominant autosomal trait, characterized by the mutation of fibrillin (FBN)-1 gene, located on the long arm of the chromosome 15, coded for the glycoprotein FBN-1 – the main component of an elastic fibrillar cellular matrix of different organs – Marfan syndrome (MFS).

---

[556] City of Sandomierz (inhabitated since Neolithic times, first mentioned in writing 1115-18), Swiętokrzyskie Voivodeship or Holy Cross Province (area 11,672 km²), Poland.

[557] Town of Konstiantynohrad, since 1922 Krasnohrad (founded 1732-33; during made by Russian Communist Fascist regime of Holodomor Genocide 1932-33 against the Ukrainian nation died at least 1062 residents), Kharkiv Obl., Ukraine. — Ablitsov VH. «Galaxy Ukraine». The Ukrainian diaspora: prominent persons. Publ. «Kyt», Kyiv 2007.

[558] Marfan A. Un cas de deformation congenitale des quartre membres plus prononcee aux extremities, caracterisee par l'allongement des os avec uncertain degre d'amincissement. Bull mem Soc med hosp Paris. 1896;13(3):220-6. — Olearchyk AS. Aortic valve replacement in Marfan's syndrome. Vasc Surg 1992;26:756-7.

The main criteria to establish the diagnosis of MFS are cardiovascular – usually a bicuspid aortic valve (BAV), dilatation of the aortic root, sinotubular junction and ascending aorta (AA), that is annuloaortic ectasia with/ without aortic valve (AV) regurgitation, aneurysm, perforation or dissection of the AA, large thoracic or abdominal aorta (mega-aorta); ocular (lens dislocation, retinal detachment); four skeletal (severe pectus excavatum or carinatum, pes planus, sign of a long finger or a the wrist[559], scoliosis or spondylolisthesis, femoral head protrusion into abnormally deep sockets, decreased abduction of the elbow) and the central nervous system (CNS) – dilatation of the dura matter of the lumbosacral spine.

The lesser criteria of MFS include cardiovascular – mitral valve (MV) regurgitation with prolapse due to dilatation of its ring with or without calcifications in people less than 40-year-old, leaflet thickening and tissue excess without myxoid degeneration, dilatation of the pulmonary artery (PA), dilatation / dissection of the descending thoracic aorta (DTA), two ocular (flat cornea, shortsightedness, protruding eyeball), two main or one main and two minor skeletal (moderate pectus excavatum or carinatum, high narrow arched palate, typical facies, subluxation of joints), pulmonary (spontaneous pneumothorax, apical bullae), cutaneous (unexplained stretch marks), recurrent hernias.

Patients with the apparent degenerative aortic root aneurysm without other sign of MFS are classified as an erroneous form (forma frusta) of MFS.

The most important predictive sign for perforation or dissection of the AA is its transverse diameter 5,9 cm or more, or documented enlargement rate of the small aneurysms (<4 cm) to 0,08 cm/year, large (8 cm) to 0,16 cm/year.

The Egyptian Pharaoh **Amenhotep Akhenaten IV (? – 1336/1334 BC)**, who married his stepsister Egyptian Queen **Nefertiti (c. 1370 – c. 1330 BC)**, died from MFS (Ancient medicine – *Egypt*).

Sir **George F. Still (1868 – 1941)** — English pediatrician, identified a rare chronic inflammatory systemic disorder which may cause joint or connective tissue damage and visceral lesions thorough the body in children (Still disease or syndrome, 1897).

The field of neonatology (neo – «new», and natal – «pertaining to birth or origin»), a subspecialty of pediatrics that consist of medical care of newborn infant, especially premature or ill, was conceived by **Joseph B. DeLee (1869 – 1942)** — American Jewish obstetrician, who established in 1898 the first premature newborn incubation station.

---

[559] The thumb (Steinberg's) sign is elicited by asking the patient to fully flex the thumb and then cover the fingers over it. A positive sign for MFS is where the distal portion of thumb is sticking beyond the ulnar border of the hand, caused by an excessive thumb length or mobility, or both. — The wrist (Walker's) sign is elicited by instructing the patient to curl the thumb around the wrist of the other hand. A positive sign for MFS is where the little finger and the thumb overlap, caused by a thin wrist or long fingers, or both.

Diseases of the skin (dermatology)

Edward Nettleship (1845 – 1913) — English dermatologist and eye surgeon, who identified of ocular albinism (disorder in synthesis of the pigment melanin), retinitis pigmentosa and hereditary nigh blindness (1869); described urticaria pigmentosa, i.e., pigmented macules or nodules (Nettleship's syndrome 1, 1869) and x-linked inheritance in males born of normal or mottled skin with pigmented naevi (nevi) or freckles, nystagmus (in Greek – drowsiness) or eye trembling, photophobia or extreme sensitivity to light, moderate to severe reduction of vision, nodding and tilting head (50%), strabismus or abnormal alignment of the eyes (60%) and common sex anomalies (Nettleship's syndrome 2, 1908-09).

The development of dermatology initiated by **Robert Willan (1775 – 1812**: 18th century**)** was extended by his disciple **Thomas Bateman (1778 – 1821)**[560] — British physician, who provided names and descriptions of skin diseases, such as lichen urticatus, apolecia areata, erythema multiforme and epithelioma contagiosum or moluscum contagiosum – an infectious disease of the skin in any part of the body, sometimes mucous membranes in children, characterized by hyperplasia and degeneration of the skin epithelium and mucous membranes, measuring from a pinhead to a pea-sized pearly white umbilicated nodules that exudate waxy material (Bateman syndrome). Histologically the lesions are marked by large, cytoplastic, viral inclusions (molluscum bodies) in the epithelial cells.

The histological school of dermatology was founded by **Ferdinand R.** von **Hebra (1816 – 80)** — Austrian dermatologist, who described an acute eruption of macules, papules or subdermal vesicles presenting a multiform appearance (erythema multiforme, familial nonhemolytic anemia, or Herba disease), and a severe form of chronic dermatitis with secondary infection (prurigo agria, prurigo mitis, prurigo ferox, or Hebra prurigo).

The fatal gangrene of the urogenital and perineal area that affected the Roman soldier and **Emperor Galerius (c. 260 – 311**: Ancient medicine. *Romans***)**, noted by **Avicenna (980 – 1037**: Medieval medicine**)**, was re-described in 1883 by **Jean A. Fournier (1832 – 1915)**[561] — French dermatologist, and since called Fournier gangrene or disease of the superficial (Colles) perineal fascia, genitalia and perineum. It commonly occurs older men, but it can occur in women and children. Gangrene is precipitated by trauma or surgery to genitourinary or perineal area with secondary mixed infection by aerobic and anaerobic bacteria, aggravated

---

[560] Willan R, Bateman T. Delineation of Cutaneous Diseases: Exhibiting the Characteristic Appearances of the Principal Genera and Species. Printed for Longman, Hurst, Orme & Brown, London 1817.
[561] Fourier JA. Gangrene fourdroyante de la verge. Sem Med 1883.

by diabetes mellitus (DM). It is a surgical emergency requiring intravenous (IV) infusion of antibiotics, surgical debridement of necrotic tissues and hyperbaric oxygen therapy. Fortunately, recently mortality decreased from 20%-88% to 10% or less.

Veniamin M. Tarnovsky [09(21).VII.1837, village of Nizhnii Daimen, Zolotukhinskii Raion, Kursk Obl., Russia – 05.V.1906, city of Paris, France][562] — Russian dermatologist and venerologist, considered syphilis to be a systemic disease and proved the possibility repeated re-infections after successful treatment with iodine and mercury based preparations (1867-1900); explained thoroughly the idea about mechanism for the development of congenital malignant syphilis; described an accelerated growth of hairs on the forehead to the eyebrows in the late congenital syphilis (Tarnovsky sign, 1901-02).

Independently, in 1898 **William Anderson (1842 – 1900)**[563] — English surgeon, and **Johannes Fabry (1860 – 1930)**[564] — German dermatologist, described, in a 39-year-old man, and in a 13-year-old boy, respectively, a rare genetic x-linked recessive metabolic lysosomal storage disease. It involves dysfunctional metabolism of sphingolipids, a class of lipid containing a backbone of sphenoid base, a set of aliphatic amino-alcohol, that include sphingosine, discovered in the brain extract in the 1870s, and were named after **Sphinx** — a mythical creature with head of the men and the body of a lion, because of their enigmatic nature. Thus, this disease was named sphyngolipidosis or Anderson-Fabry disease.

Typically, sphingolipidosis is manifested in early childhood by generalized burning or burning of the extremities; fatigue, anorexia, diarrhea, inability to gain weight; dizziness or vertigo (sensation rotation or movement of one's self or one's surroundings in any plane); tinnitus (ringing in the ears). Then, appears skin lesions such as angiokeratomas (tiny, painless papules) or teleangiectasia (abnormally dilated red, blue or purple superficial capillaries, arterioles od venules below the skin's surface); anhidrosis (lack of sweating), or hyperhidrosis (excessive sweating). Usually, in the third decade of life disease causes clouding of the cornea (cornea verticilata) or vortex keratopathy (corneal deposits at the level of the basal epithelium forming a faint golden-brown pattern); hypertension, restrictive cardiomyopathy (CM) and strokes; involves the central nervous system (CNS), and

---

[562] Village Nizhnii Daimen, Zolotukhinskii Region, Kursk Obl. (area 29,800 km^2), Russia. — Tarnovsky VM. Diagnosis of Venereal Diseases in Woman and Children. 2nd ed. St Petersburg 1867. — Tarnovsky VM. Course of Venereal Diseases. St Petersburg 1870. — Tarnovsky VM. Cure of Syphilis. St. Petersburg 1900.

[563] Anderson W. A case of «angio-keratoma». Brit J Dematol, Oxford 1898;10:113-7.

[564] Fabry J. Ein Beitrag zur Kenntnis der Purpura haemorrhagica nodularis (Purpura papuloso haemorrhagica Hebrae). Archiv Dermatol Syphilis, Berlin 1898;43:187-200.

finally leads to kidney insufficiency and failure. The life expectancy of patients is greatly reduced.

Ferdinand J. Darrier (1856 – 1938) — French pathologist and dermatologist, discovered a peculiar figurate erythema (Darrier disease, 1889).

Independently of **James C. White (1833 – 1916)** — American dermatologist, discovered a follicular keratosis (Darrier-White syndrome).

Discovered acanthosis nigrans – an abnormal but benign thickening of the prickle-cell layer of the skin as in psoriasis [Ancient medicine. *Romans* – **Aulus Celsius (c. 25 BC – c. 50 AD)**].

Together with **Marcel Ferrand (1878 – 1940)** — French physician, Ferdinand Darrier discovered dermatofibrosarcoma (Darrier-Ferrand disease, 1889, 1924).

Ferdinand Darrier discovered erythema annularis.

With **Gustave Roussy (1874 – 1948)** — French neuropathologist, discovered subcutaneous sarcoidosis (Darrier-Roussy sarcoid).

Ferdinand Darrier observed a sign characterized by stroking of the skin that results in erythema, edema in mastocytosis – a chronic disease affecting the skin, internal organs, and bones[565], and in uriticaria pigmentosa.

Petro V. Nikolsky (01.IV.1858, town of Uman', Lypetska Obl., Russia – 17.III.1940, city of Rostov-on-Don, Rostov Obl., Russia)[566] — Ukrainian dermatologist, described an early separation of the epidermis from the basal layer of the skin in pemphigus vulgaris and some other bullous diseases of the skin (Nikolsky sign, 1896).

Researching the light therapy (phototherapy), **Niels R. Finsen (1860 – 1904)**[567] — Faroese Danish physician of Islandic origin, discovered a curing effect of the concentrated ultraviolet rays, particularly on lupus vulgaris, i.e., tuberculosis of the skin (1893, 1896 and 1903).

An excellent general purpose culture medium for growing fungi is one developed by **Raymond J.A. Sabouraud (1864 – 1938)** — French microbiologist and mycologist, consisting of dextran agar which contain glucose, peptone, hydrolyzed

---

[565] Nettleship E, Tay W. Report of Medical and Surgical Practice in the Hospitals of Great Britain. Brit Med J 1869;2(455):323-4.

[566] Town of Usman (founded 1645), Lypetska Obl. (area 24,100 km²), Russia. — City of Rostow-on-Don (founded 1749), Rostov Obl. (area 100,967 km²), Russia. — Nikolsky PV. Materials for studies of pemphigus foliaceus. Doctoral thesis, Kyiv 1896. — Torsuyev NA. P.V. Nikolskii (1858-1940). «Medgiz», Moskva 1953. — Grando SA, Grando AA, Glukhenky BT, et al. History and clinical significance of mechanical symptoms in blistering dermatoses: a reappraisal. J Am Acad Dermatol Jan 2003;48(1):86-92. — Ablitsov VH. «Galaxy Ukraine». The Ukrainian diaspora: prominent persons. Publ. «Kyt», Kyiv 2007.

[567] The Faroe / the Faeroes islands (1,399 km²) are an archipelago between the Norwegian Sea and North Atlantic Ocean. — Finsen NR. Die Bekampfung Des Lupus Vulgaris. 1903.

pancreatic casein, hydrolysated pepsin of animal tissues and may contain antibiotics (Sabouraud's culture medium, 1892).

Diseases of the eyes (oculistics, ophthalmology)

Important work in microanatomy of the eye was carried out by **William E. Horner (1793 – 1853)** — American anatomist, **Jacob Henle (1809 – 58**: 19th century. Anatomy, histology and embryology), **William Bowman (1816 – 92**: ibid), **Arthur Hassal (1817 – 94**: ibid), **Heinrich Müller (1820 – 64)** — German anatomist and physiologist, **Friedrich W.E.A.** von **Gröefe / Graefe (1828 – 70)** — Prussian ophthalmologist, **Robert Blessig (1830 – 78)** — German physician, **Olexandr V. Ivanov / Alexander Iwanoff (1836, Crimea, Ukraine** – died from pulmonary tuberculosis **15/27.X.1880, city of Mentona,** department Alpes-Maritims**, France)**[568] — Ukrainian ophthalmologist, Director of the Eye Department Kyiv Military Hospital (1867-76) and Extraordinary Professor's Chair of Ophtalmology of the Kyiv University (1875-76), city of Kyiv, Ukraine, and **Mykhailo Borysykevych / Michal Borysiekiewicz**

---

[568] Kyiv Military Clinical Hospital (founded 1755), city of Kyiv, Ukraine. — City of Mentona (settled since the Paleolithic Era, first major settlement during 11th century), department Alpes-Maritims (area 4.299 km²), France. — Iwanoff A. Zur Anatomie des Glaskorpers. Klin Monatsbl. Zebender'a 1864. — Iwanoff A. Ueber die verschidenen Entzundungsformen der Retina. Klin Monatsbl. Zebender'a 1864. / Iwanoff AW. Perivasculitis Retinae. Ibid. 1865. — Ivanov AB. Materials for pathological anatomy of the membrane of the retina. Med vest (St Petersburg) 1865;5:281-83. — Iwanoff A. Zur pathologischen Anatomie der Retina. Arch. f. Ophth (Berlin) 1865;11(1):136-55. — Iwanoff A. Zur normalen und pathologischen Anatomie des Glaskörpers. Arch. f. Ophth (Berlin) 1865;11(1):155-70. — Iwanoff A. Zur Ablösung des Chorioidea. Archiv f. Ophth (Berlin) 1865;11(1):191-9. — Iwanoff A. Zur Pathologie der Retina. Archiv f. Ophth (Berlin) 1866;12. — Iwanoff A. Beiträge zur pathologischen Anatomie der Hornhaut und Linsenepithels. Klinische Beobachtungen aus d. Augenheilanstalt zur Wiesbaden. Darmstadt 1866. — Ivanov AB. Materials for the normal and pathological anatomy of lenses of the eye. Doctoral dissertation, St Petersburg 1867. — Iwanoff A, Rollett A. Bemerkungen zur Anatomie der Irisanheftung und des Annulus ciliaris. Arch f. Ophth (Berlin) 1869;15(1):17-74. — Iwanoff A. Das Oedem der Netzhaut. Arch f. Ophth (Berlin) 1869;15(2):88-107. — Iwanoff A. Beiträge zur Anatomie des Ciliarmuskels. Archiv f. Ophth (Berlin) 1869;15 (3):284-98. — Iwanoff A. Ueber Conjuctivitis und Keratitis phtyctaenularis. Klin Monatsbl f. Augenh 1869;7:462-70. — Iwanoff A. Glaskorper und Tunica vasculosa. Stricker's Handbuch der Lebre von den Geweben 1872. — Iwanoff A, Arnold J. Microscopische Anatomie des Uvealtractus und der Linse. Handbuch der gesammten Augenheikunde. Hrsg. Von Arlt, J Arnold [et al.]. Redigirt von A Graefe und Th Saemisch. W Engelman, Leipzig 1874-80. Pp. 265-320. — Wecker L de, Landolt E. Traite complet d'ophtalmologie. Anatomie microscopique par les professeurs J. Arnold, A. Iwanoff et W. Waldeyer. Vve. A. Delahaye et Cie, Paris 1880-89. — Ablitsov VH. «Galaxy Ukraine». The Ukrainian diaspora: prominent persons. Publ. «Kyt», Kyiv 2007.

**(13.I.1848, village of Bilobozhnytsia, Chortkiv Region, Ternopil Obl., Ukraine –
18.IX1892, city of Graz, Austria)**[569] — Ukrainian ophtalmologist.

William Horner described pars lacrimalis musculi orbicularis oculi (Horner muscle),
Jacob Henle – microscopic crypts in the conjunctiva (Henle crypts), a membrane
in the innermost layer of the choroid, so called vitreous lamina (Henle lamina),
with Arthur Hassal – transparent excrescences in the periphery of the descement
membrane of the eye (Hassal-Henle body, 1846). William Bowman identified smell
glands in the olphactory mucosa of the nose (Bowman glands) and the anterior
limiting membrane of the cornea (Bowman membrane).

Heinrich Müller discovered rhodopsin (visual purple) of the external portion of
the retinal rods (1851), described the thin elongated glial cells penetrating all layers
of the retina and creating its most important supporting element (Müller glia, 1856).

Friedrich von Gröefe — the founder of German ophthalmology, introduced
iridectomy – surgical removal of portion of the eye's iris for the treatment of closed-
angle glaucoma and melanoma; noted a combination of retinitis pigmentosa
and perceptive deafness (1858); diagnosed paralysis of the cranial III (oculomotor)
nerve that starts with the external muscles, before the internal muscles become
involved (Gröefe syndrome, 1860); noted failure of the upper lid to follow a
downward movement of the eyeball when the patient changes sight (vision) from
looking up to looking down (a dynamic lid lagging) in exophthalmic toxic goiter
and hyperthyroidism (Gröefe sign, 1864)[570]; described chronic progressive external
ophthalmoplegia, papilledema and a swelling of the optic nerve disc in patients with
brain tumor (1868); designed tissue forceps (of Gröefe); and a specialized knife (of
Gröefe) for cataract surgery.

Robert Blessig and Olexandr Ivanov described the cystic spaces on the periphery
of the retina close to the serrated line without significant impact on sight (Blessig-
Ivanov cysts, 1864-67).

---

[569] Village Bilobozhnytsia (first settlement belong to the Trypilian Culture, first written mention 1453),
Chortkiv Region, Ternopil Obl. (area 13,823 km²), Ukraine. — City of Gratz / Graz (earliest settlement
dated to the Cooper Age, founded 12th century), Австрія. — Borysiekiewicz M. Über Pemphigus conjunct.
Zehender's Mtsschr. 1879. — Borysiekiewicz M. Beiträge zur Extraction des grauen Staar. Ib 1880. —
Borysiekiewicz M. Ophthalmoscop. Beobb. an 171 Geisteskranken in der Klinik von Meynert. W. M. Bl.
1882. — Borysiekiewicz M. Über die Anwendung des Cocains in der ocluist. Praxis. W. M. W. 1887. —
Borysiekiewicz M. Untersuchungen über den feineren Bau der Netzhaut. Toeplitz & Deuticke, Leipzig und
Wien 1887. — Borysiekiewicz M. Weitere Untersuchungen über den feineren Bau der Netzhaut. F. Deuticke,
Leipzig und Wien 1894. — Borysiekiewicz M. Beiträge zum feineren Bau der Netzhaut des Chamaeleon
vulg. Praxis. W. M. W. 1899. — Pagel J: Biographisches Lexikon hervorragender Ärzte des neunzehnten
Jahrhunderts. Urban & Schwartzenberg, Berlin - Wien 1901, Sp. 214-215. — Ablitsov VH. «Galaxy Ukraine».
The Ukrainian diaspora: prominent persons. Publ. «Kyt», Kyiv 2007.

[570] Von Graefe A.F.W.E. Über Basedow'sche Krankheit. Deutsche Klinik 1864;16:158-9.

Mykhailo Borysykevych considered the ends of the elongated Müller glia of the retina to be the light sensitive formation which gathers and directs the light into retinal rods and cones (1894).

In honor of **John Dalrymple (1803 – 52)**[571] — English ophthalmologist, one of the eye signs of diffuse toxic goiter [19th century. Internal medicine. – **Robert Graves (1796 – 1853) Karl** von **Basedow (1799 – 1854)**] was named Dalrymple's sign, i.e., widened the eyelid (palpebral) opening or retraction or eyelid spasm causing abnormal wideness of the palpebral fissure (1849). He also described inflammation of the ciliary body and cornea (Dalrymple's disease).

Carl Zeiss (1816 – 88) — German optician, **Ernst K. Abbe (1840 – 1905)** — German physicist and optical scientist, and **Otto Schott (1851 – 1935)** — German chemist and glass technologist, laid the foundation of modern optics, research microscopes, astronomical telescopes, planetariums and other optical systems.

The work of **Adolf E. Fick (1829 – 1901)** — German physiologist and physician, on the tonometer influenced his nephew **Adolf G.E. Fick (1852 – 1937)** — German ophthalmologist, to make in 1888 fully fit contact glasses from blown glass.

Johann N. von **Nussbaum (1829 – 90)** — German surgeon, attempted to replace a cataract defect with artificial lenses (keratoprosthesis, 1853).

The combined recessive inheritability of blindness and deafness was demonstrated by **Richard Liebereich (1830 – 1917)** — German physiologist and ophthalmologist of Jewish ancestry, since it occurred particularly in the siblings of blood-related marriages or in families with individual of different generations. He supplied the first proof for the coupled transmission of blindness and deafness, since no isolated cases of either could be found in the family trees.

He also constructed ophthalmoscope (Liebereich, c. 1860).

The initial description by **Francois du Petit (1664 – 1741**: 18th century) of injury to the sympathetic cervico-thoracic (stellate) ganglion and its clinical picture, was re-described by **Claude Bernard (1813 – 73**: 19th century. Physiology) and **Johann F. Horner (1831 – 86**: 19th century. Physiology)[572], and named the Horner syndrome (1869). It usually occurs in patients with the superior sulcus tumor (SST) or during a dorsal sympathectomy in 0.5-2.0% of the patients.

Johann Horner discovered that the usual color night blindness is transmitted from man to man through a normal woman (Horner law, 1876).

Jacob H. Knapp (1832 – 1911) — German ophthalmologist, noticed tiny breaks in the elastin tissue on the back of the eye in the Bruch's membrane of the retina [19

---

[571] Dalrymple J. The Anatomy of the Human Eye, being an account of the History, Progress, and Present State of Knowledge of Organ of Vision in Man. Vol. 1-8. Churchill, London 1834. — Dalrymple J. The Pathology of the Human Eye. Churchill, London 1851-52.

[572] Horner JF. Uber eine Form von Ptosis. Klin Monatsbl Augenheilk 1869;7:193-8.

century. Anatomy, histology and embryology. – **Karl Bruch (1819 – 94)]**, called by him an «angioid streaks» or «angioid striae», that may calcify and crack causing retinal hemorrhages, named the Knapp's streaks (1889). The likely explanation for «angioid streaks»'may be degeneration of Bruch's membrane along with iron deposition in the elastic fibers from hemolysis (in Greek – «haima» meaning blood, «lysis» is for decomposition), secondary mineralization and impaired nutrition due to stasis and vessel occlusion.

Theodor K.G. von **Leber (1840 – 1917)**[573] — German ophthalmologist, described congenital amaurosis, a rare inherited eye disease that appears as a blindness at birth or in the first few months of life, named Leber's congenital amaurosis (LCA, 1869), and Leber's hereditary optic neuropathy (1871).

A small venous plexus in the eye located between Schlemm's canal [19th century. Anatomy, histology and embryology. – **Frederich Schlemm (1895 – 1858)]** and Fontana's spaces [18th century. – **Felice Fontana (1730 – 1805)]** was named the Leber's plexus his honor.

In 1886 **Arthur** von **Hippel (1841 – 1916)** — German ophthalmologist, grafted a full-thickness cornea from rabbit into the lamellar bed of the cornea of young women, suffering from near-sightedness with a significant improvement of the vision, thus becoming a pioneer of corneal transplantation.

Jacob Stilling (1842 – 1915) — German ophthalmologist of Jewish descent, **Siegmund Türk (1867 – ?)** — German Swiss ophthalmologist, and **Alexander Duane (1858 – 1926)**[574] — American ophthalmologist, described a congenital syndrome of ocular and systemic anomalies with fibrosis of the external rectus muscle of the eye and limb abnormalities which is associated with congestive heart failure (CHD) – atrial septal defect (ASD) – Stilling-Türk-Duane syndrome (1887, 1896, 1905).

In 1881 **Warren Tay (1843 – 1927)** — British ophthalmologist, in 1887 **Bernard Sachs (1858 – 1941)** — American neurologist of Jewish descent, identified a genetic autosome recessive disorder that results from destruction of nerve cells in the brain and spinal cord by deficient enzyme called beta-hexosaminidase which led to accumulation of the toxins GM2 gangliosides within the cells (Tay-Sachs disease).

The Tay-Sachs disease affects Ashkenazi Jews, French Canadians in southeastern Province of Quebec (area 1,542,056 km^2), Canada (area 9,984,670 km^2), and Cajuns – descendants of Acadia (in Greek – «refuge», established 1604) who were in the 2nd half of the 17th century expelled from eastern Quebec to the southern State of Louisiana (area 13,382 km^2), since 1803 USA.

---

[573] Leber T. Über Retinitis pigmentosa und ungeborene Amaurose. Archiv für Ophthalmologe. 1869;15(3):11-25.

[574] Okihiro MM, et al. Duane syndrome and congenital upper limb anomalies. Arch Neurol 1977;34:174-9.

The most common children type of the Tay-Sachs disease affects 3- to 6-year-old infants who are constantly startled, lose the ability to turn over, sit or crawl, then follows seizures, blindness, hearing loss, inability to walk and swallow, atrophy and paralysis. During ophthalmoscopic examination one can see cherry-red spot in the retina of the eyes. Death usually occurs by the age of 4-year of life.

The less common juvenile type begins with the onset of cognitive and motor skills deterioration, speech dysfunction (dysarthria), difficulty in swallowing (dysphagia) and gait abnormality (ataxia), then seizures, spasticity, and death between 5 and 15-year of life.

A rare, late, adult type is characterized by prolonged course, begins in 30s or 40s with unsteadiness, dysphagia, neuro-psychic dysfunction, and an early attachment to the wheelchair.

The only available treatment is supportive.

Named after **Hubert Sattler (1844 – 1928)** — Austrian German ophthalmologist, is one of the five layers of medium-diameter blood vessel of choroid, and a layer of the eye situated between the Bruch [19th century. Anatomy, histology and embryology. – **Karl Bruch (1819 – 94)**] membrane below and suprachoroidal above, respectively.

In 1899 **Robert W. Doyne (1857 – 1916)** — English ophthalmologist, discovered colloid bodies lying in the Bruch's membrane [19 century. Anatomy, histology and embryology. – **Karl Bruch (1819 – 94)**], that appears to merge, forming mosaic pattern resembling a honeycomb, a rare hereditary form of macular degeneration, that result in progressive and irreversible loss of sight, known as Doyne's honeycomb retinal choroiditis or Doyne's honey-comb retinal dystrophy.

In 1881 **Moritz Litten (1845 – 1907**: 19th century. Internal medicine), and in 1900, **Albert Terson (1867 – 1935)** — French ophthalmologist, described hemmorrhage from the retina and the vitreous body of the eye from a nontraumatic subarachnoid hemorrhage (SAH) due to rupture of the intracranial arterial aneurysm (Terson syndrome). This syndrome became a model for investigating of the brain - heart interrelationship. 90% of the patients with a nontraumatic SAH had cardiac abnormalities, among them, the release of cardiac biomarkers, electrocardiographic (ECG) changes and left ventricular (LV) dysfunction.

Neurology

Following description of «paralysis agitans» in the **Ebers papyrus** (Ancient medicine. *Egypt*), **James Parkinson (1755 – 1824)**[575] — English surgeon, re-described this slowly progressing chronic neurological disease of the central nervous system

---

[575] Parkinson J. An Essay on the Shaking Palsy. Printed by Whittingham & Rowland, London 1817.

(CNS) in older people, subsequently named in his honor Parkinson disease (PD) and «parkinsonism».

PD is an idiopathic (spontaneous) degenerative disorder of unknown cause which is not caused by genetic disturbances or environmental factors, like certain pesticides, prior head injuries, or other diseases. It results from the destruction and death of nerve cells (neurons) producing the neurotransmitter dopamine[576] in the substantia nigra of the midbrain, and in other divisions of the CNS which leads to an active influence of the basal ganglia on the brain cortex that mainly affects the extrapyramidal motor system. «Parkinsonism» is a general concept for number of diseases or conditions with the chief symptoms of PD, among them the most important is just PD.

According to their origin PD and «parkinsonism» is divided into the following four subtypes: (1) primary or idiopathic, (2) secondary or acquired, (3) hereditary «parkinsonism», or (4) PD plus syndromes of multiple system degeneration[577].

The average life expectancy following diagnosis of PD or «parkinsonism» is 7-14 years.

That seizures in epilepsy involve one or several parts of the body on one side had been described in 1827 by **Louis F. Bravais (1801 – 43)**[578] — French physician. Contractions begin from a focus such as tip of the toe, corner of mouth, and they gradually spread to other muscles. If it remains hemilateral, consciousness is preserved. Sometimes it proceeded to a generalized fit with loss of consciousness.

The first citation of juvenile myoclonic epilepsy (JME) was made in 1859 when **Theodore J.D. Herpin (1799 – 1865)**[579] — French neurologist, described a 13-year-old boy suffering from myoclonic jerks which progressed to tonic-clonic seizures three months later.

Guillame B.A. Duchene (1806 – 75) — French neurologist, depicted spinal pseudo-hypertrophic muscular atrophy, and with **Wilhelm H. Erb (1840 – 1921)** — German neurologist, the upper arm brachial paralysis. The latter also described limb-girdle muscular dystrophy (Erb's dystrophy, 1884) and peroneal muscular atrophy. Erb dystrophy is a slowly progressing juvenile form of the upper extremity and scapulo-humeral muscular dystrophy affecting each sex equally with successive involvement

---

[576] Dopamine (3-hydroxytyramine) was synthetized in 1910 by **George Barger (1878 - 1939)** — British chemist, by removing a carboxyl group from the L-3,4-dioxyphenylalanine (L-DOPA). The L-DOPA is the precursor to the neurotransmitters dopamine, norepinephrine (noradrenaline) and epinephrine (adrenaline), known as catecholamines.

[577] Jankovic J. Parkinson's disease: clinical features and diagnosis. J Neurol Neurosurg and Psychiatry 2007 Apr;79(4):368-76. and

[578] Bravais LF. Recherches sur les symptomes et le traitement de l'epilepsie hemiplegique. These de Paris, no 118, 1827.

[579] Herpin Th. Des Acces Incomplets D'Epilepsie. JB Bailliere et Fils, Paris 1867.

of the lower extremities and pelvic girdle, associated with cardiomyopathy and disorder of the conduction system of the heart (CSH), i.e., the 2nd - 3rd degree AV block.

In cardiology, the aortic valve (AV) sound can be heard best at the 2nd intercostal space (ICS) near the right border of the sternum, and the 2nd AV sound may be heard at the Erb's point at the 3rd ICS, just left to the sternum.

In neurology, the nerve point of Erb is a site at the upper supraclavicular space (fossa), 2-3 cm above the clavicle, where the upper trunk of the brachial plexus (C_5 and C_6) is located.

In the head and neck surgery the point of Erb is located on the posterior border of the sternocleidomastoid (SCM) muscle, approximately at the junction of the middle and lower third of the anterior border of this muscle. From behind the SCM muscle at this point emerges the four superficial branches of the cervical plexus – the greater auricular, lesser occipital, transverse cervical and supraclavicular nerves, into the posterior (lateral) triangle of the neck. This triangle is bounded at the apex by the union of the SCM muscle and the trapezius muscle, anteriorly by the posterior border of the SCM muscle, posteriorly by the anterior border of the trapezius muscle, the base by the middle third of the clavicle, and the roof by investing layer of the deep cervical fascia. One cm above the Erb's point, the cranial XI (accessory) nerve can be found which travels through the lateral triangle of the neck to enter the anterior border of the trapezius muscle at the junction of the middle and lower third of this muscle. It innervates the SCM muscle and trapezius muscle.

The name of **Ludwig Türck (1810 – 68)** [19th century. Physiology – **Manuel Garcia (1805 – 1906)**] is lent to the uncrossed fibers forming a small bundle of the pyramidal tract, called the anterior corticospinal tract, or Türck bundle or tract; c. 1864).

Contribution to specialty of **Bernard A.** von **Gudden (1824 – 86)** — German neuroanatomist and psychiatrist, include developing of the specialized microtome for microscopic section of the brain for pathological studies; description of commissural fiber above and behind optic chiasm (commissurae supraopticae, or Gudden commisures), of a median, unpaired, ovoid cell group at the base of the midbrain tegmentum between the cerebral peduncles (interpeduncular nucleus, or Gudden ganglion), and two small round cells group in the caudal part of the midbrain associated with the mamillary body by the way of the mamillary peduncle and mamillary tegmental bundle (tegmental nuclei, or Gudden tegmental nuclei), interconnected with the mamillo-thalamic tract of d'Azur [17th century. – **Raymond de Viessens (1635/41 – 1715) / Felix d'Azyr (1748 – 94)**].

He advocated humane treatment and no-restrain of the mentally ill people, communal social interaction among patients, and well trained medical (and paramedical) providers.

The founder of modern neurology **Jean M Charcot (1825 – 93)** — French neurologist, described diabetic neuropathic arthropathy («Charcot joint», 1865).

Jean Charcot together with **Charles J. Bouchard (1837 – 1915)** — French pathologist, described small aneurysms on cerebral perforated vessels that could be the cause of intracranial hemorrhages (Charcot-Bouchard aneurysm, 1866).

In 1868 Jean Charcot described multiple sclerosis and its three signs that is nystagmus, intention tremor and telegraphic speech (Charcot triad).

After initial description of amyotrophic lateral sclerosis (ALS) in 1824 by Sir **Charles Bell (1774 – 1842**: 19th century. Anatomy, physiology, and embryology), in 1869 Jean Charcot made connection between the symptoms and clinical picture of this neurological disease, named it ALS. It is a nonhereditary (95%), rarely hereditary autosomal dominant (5%), motor neuron paralysis of unknown etiology, with degeneration of neurons in the anterior horn of the spinal cord and the cortical neurons that provide their afferent input. ALS presents itself by rapidly progressive weakness, muscle atrophy and fasciculations with subsequent paralysis, sometimes spasticity, dysarthria, dysphagia, respiratory compromise, pneumonia, and death occurring usually 2-3 years from the onset. The only available current treatment is ribuzole to reduce further damage to motor neurons by decreasing the release of glutamate via activation of glutamate transporters. It extends the need for respiratory support and survival by a few months.

Jean Charcot with **Pierre Marie (1853 – 1940)** — French neurologist, and **Howard Y. Tooth (1856 – 1925)** — British neurologist, described incurable hereditary motor and sensory neuropathy characterized by loss of muscle tissue and touch sensation, predominantly feet and legs, but also in the hands and arms in the advanced stage (Charcot-Marie-Tooth disease, 1886).

Pierre Marie recognized in 1886 that disorder of the anterior part of the pituitary gland (hypophysis) causes acromegaly which is complicated in some patients by cardiomegaly and heart failure (Ancient medicine. *Egypt*), described pulmonary hypertrophic osteoarthropathy, cleidocranial dysostosis – a congenital disorder in which typically the clavicular bones are poorly developed, or absent, rhizomeric spondylosis with term «rhizomeric» related to the hip or shoulder joint and «micromeric» means disproportionately short or small extremities, and the speech disorder in which patients develop speech pattern that is perceived as a foreign accent (foreign accent syndrome, 1907).

An inherited disease of unknown etiology, causing progressive damage to the nervous system, resulting from symptoms ranging from gait disturbances to speech problem was described by **Nicolaus Friedreich (1825 – 88)** — German pathologist and neurologist, can lead to heart disease and diabetes mellitus (Friedreich ataxia or paramyoclonus multiplex, 1860).

Ascending paralysis or an acute inflammatory demyelinating polyneuropathy was diagnosed by **Jean B.O. Landry (1826 – 65)** — French physician, in 1859, by **Georges Ch. Guillain (1876 – 1961)** — French neurologist, **Jean A. Barre (1880 – 1967)** — French neurologist, and **Andre Strohl (1887 – 1977)** — French physiologist, in 1916 (Laudry paralysis or Guillain-Barre- Strohl syndrome). It is a life-threatening disease, affecting the peripheral nervous system, characterized by an ascending weakness and paralysis beginning in the feet and hands, migrating towards the trunk with the possibility of affecting the breathing muscles.

John H. Jackson (1835 – 1911) — English neurologist, found that the nerve functions are the last to develop and the first to be destroyed (Jackson law). Described epilepsy as the spread of abnormal electrical activity from one area of the cerebral cortex to adjacent areas (cortical Jacksonian's epilepsy), the «Jacksoniam march» of symptoms in focal motor seizures and the so called «dreamy state» or «déjà vu» of psychomotor seizures of temporal lobe origin, and local lesions on the contralateral side of the brain, usually in the area of the motor cortex (Jacksonian seizures), all as a part of the Jacksonian epilepsy (1863, 1881).

Aleksiy Ya. Kozhevnikov (1836 – 1902) — Russian neurologist, described cortical epilepsy (epilepsies partialis continua, Kohzevnikov epilepsy, 1894).

Disharmonic development of the gray and white matter of the brain was described by **Jan L. Mierzejewski (1839 – 1908**) — Polish neurologist and psychiatrist, in which the gray matter prevailed (Mierzejewski effect).

From 5% to 15% of patients with myasthenia gravis (MG), diagnosed in 1878 by **Wilhelm Erb (1840 – 1921**; above**),** analyzed in 1893 by **Samuel V. Goldflam (1852 – 1932)** — Jewish Polish neurologist, an autoimmune disease (of Erb-Goldflam) caused by acetylcholine receptor antibodies, are found to harbor an enlarged thymus (thymoma), the most common tumor of the anterior mediastinum, while 30% to 50% of thymomas are associated with myasthenia gravis MG).

It is interesting that Samuel Goldflam re-described in 1900 tenderness or pain in the costovertebral angle during percussion in kidney's disease [19th century. Surgery. – **John Murphy (1857 – 1916)**. / Internal Medicine. – **Fedir Pasternacky (1845 – 1902)]**, which in Poland is called the Goldflam' sign.

In 1882 **Woldemar (Vladimir) M. Kernig (1840 – 1917)** — Russian neurologist, established that when the leg is bent at the hip and knee at 90° angle and subsequent extension in the knee is painful leading to resistance, or painless inability to extend the knee beyond 135° angle, constitutes a positive (Kernig) sign in meningitis or in subarachnoid hemorrhage (SAH). Patients also may show opisthotonos, a spasm of the whole body that leads to legs, head and torso being bent backwards like a bow.

Congenital or acquired diaphragmatic eventration is an abnormal elevation of the hemidiaphragm, usually from damage to the phrenic nerve, as a result of traumatic delivery, thoracic surgery, tumors, inflammation, or central nervous system (CNS) conditions, among them, an autosomal dominant spinal muscular atrophy in newborn or infant, described by **Guido Wardnig (1844 – 1919)** — Austrian neurologist, and **Johann Hoffmann (1857 – 1919)** — German neurologist, and named Warding-Hoffmann disease (1891, 1894 / 1891, 1893, 1897). In the latter disease, respiratory distress resulting from diaphragmatic eventration may be the initial manifestation of this disorder.

Sir **William R. Gowers (1845 – 1915)** — British neurologist and artist, described tractus spinocerebellaris (Gowers column) and vasovagal attacks.

In 1872 **George Huntington (1850 – 1916)**[580] — American physician, described an inherited autosomal dominant disorder after whom it is named Huntington's disease (HD) or chorea, like «paralysis agitans» [**James Parkinson (1755 – 1824**: above]. Apparently, the expansion of cytosine-adenine-guanidine triplets repeats in the gene coding for the Huntington protein results in an abnormal protein which gradually damage cells of the striatum (in Latin – corpus striatum) – subcortical basal ganglia of the anterior brain and the component of the extrapyramidal and reward systems. Symptoms usually begin between 30 and 50 years, although about 8% of cases start before the age of 20. HD is characterized by subtle problems with mood or mental abilities, a general lack of coordination, unsteady gait, abnormally increased muscular activity, or function (hyperkinesis), jerky body movements – bowing, twisting and grimacing, initially called the «choreatic dancing», speech, chewing and swallowing difficulties, dementia and death. With the appearance of the first symptoms the length of the life last approximately 15-20 years. The only available treatment is to reduce the hyperkinetic movements by the drug tetrabenazine, physical therapy (PT) and full-time supportive care in the later stages of the disease. At the present time there is no cure.

The sign of weakness of the convergence of the eyeball was depicted by **Paul Möbius (1853 – 1907)** — German neurologist, when one eye converges and the other diverges while looking at the tip of one's nose, from failure of median rectus muscles (Möbius sign, 1879).

Constantin von **Monakow (1853 – 1930)** — Russian Swiss neurologist, described a syndrome of hemiplegia, hemianesthesia and hemianopsia on the side opposite

---

[580] Huntington G. On chorea. Medical and Surgical Reporter of Philadelphia 13 Apr 1872;26(15):317-21.

the lesion of an occluded anterior choroidal artery, tractus rubrospinal (Monakow tract), and proposed the diaschisis (Monakow) theory which says that there is a loss of function and electrical activity caused by a cerebral lesion in areas remote from the lesion but neurologically connected to it.

Hematomyelia of the central part of the spinal cord was reported by **Lazar S. Minor (1855 – 1942)** — Lithuanian neurologist of Jewish origin, named Minor disease.

The painful subcutaneous lipomas in association with obesity, hypercholesteremia and asthenia was brought to light by **Francis X. Dercum (1856 – 1931)** — American physician and neurologist, called adiposa dolorosa or Dercum disease (1892).

Joseph J.F.F. Babinsky (1857 – 1932) — French neurologist of Polish origin, introduced a reflex of a dorsiflexion of the big toe when one stimulates the sole of the foot in organic lesions of the pyramidal tract (Babinsky reflex, 1896).

In 1885 **Georges G. de la Tourette (1857 – 1904)** — French physician, pediatrician and neurologist, published on account of 9 patients with an inherited or environmental neurodevelopmental disorder in childhood, which affects about 1% school-aged children and adolescent, named by him «maladie des tics», later re-named in his honor into the Tourette disease or syndrome. It is characterized by transient or chronic multiple physical motor and neural, one or several vocal tics a day or in a year, which wax and wane, can be suppressed temporarily, and are preceded by a premonitory urge. It is often accompanied by hyperactivity, obsessive ideas, inattention and ìmpulsiveness. The severity and frequency of tics is often reduced, may even stop altogether at adolescent or adult age. It does not adversely affect intelligence or life expectancy. The tics are believed to result from dysfunction in the cortical regions, thalamus, basal ganglia and frontal subcortex. The treatment of mild tics should be with psychobehavioral therapy and reassurance, otherwise pharmacological, initially with clonidine (Catapress) per orum or with patches, if no improvement – alpha-2 adrenergic agonists, antipsychotic, topiramate and botulinum toxin.

In 1889 **Eugene** von **Bamberger (1858 – 1921)**[581] — Austrian internist, in 1890 – **Pierre Marie (1853 – 1940**; above)[582], described hypertrophic pulmonary osteoarthropathy (Bamberger-Marie- disease).

---

[581] Von Bamberger E. Veränderungen der Röhrenknochen bei Bronciectasie. Wien klin Wochenschr 1889;2:226.

[582] Marie P. De l'osteo-arthropathie hypertrophiante pneumonique. Rev Med Paris 1890;10:1-36.

Olexandr Ye. Shcherbak (30.VIII.1863, town of Nizhyn, Nizhyns'kyy Region, Chernihivs'ka Obl., Ukraine – 23.IV.1934, city of Sevastopol, Crimea, Ukraine)[583] — Ukrainian neurologist, psychiatrist and physiotherapist, studied the changes of the phosphoric acid (H_2PO_2) and nitrogen (N) level in the humans during an intellectual work, organized the Institute of the physical methods of treatment in the city of Sevastopol (established as a Greek colony in the 6th century BC), Crimea, Ukraine.

Petro O. Preobrazhensky (1864, city of Moscow, Russia – 11.09.1913, city of Moscow, Russia)[584] — Russian Ukrainian neurologist, described the syndrome of discirculation in the supply area by the anterior spinal artery characterized by paresis or paralysis with dissociated dysfunction of sensation and function of the pelvic organs (Preobryazhensky syndrome, 1904).

Edward Flatau (1868 – 1932) — Jewish Polish neurologist, discovered that *«the greatest the length of the fibers in the spinal cord the closer they are situated to the periphery»* (Fratau law, 1897). He also described the 5th (trigeminalis), the 7th (facialis) and the 8th (vestibulocochlealis) cranial nerves and outlined their nuclei.

---

[583] Town of Nizhyn (Foot Note 458), Nizhyns'kyy Region, Chernihivs'ka Obl., Ukraine. — City of Sevastopol (established as Greece colony in 4th century, founded 1783), Crimea, Ukraine. — Shcherbak AE. The Villain Men (inborn criminal – morally – confused – epileptic) according to Lombroso. Typography PI Smith, St Petersburg 1889. — Shcherbak AE. Materiali for studying dependency upon phosphate exchange on increased or decreased function of the brain. Typography of Society «The Common Benefit», St Petersburg 1890. — Shcherbak A. Contribution a l'étude de l'influence de l'activité cérébrale sur l'échange d'acide phosphorique et d'azote. Arch de médéc Experim 1893. — Shcherbak AE. The Importance of Anatomy and Pathology of Nervous System on Physiological Psychology. Lecture, reading for yearly meeting of Russian Medical Society in Warsaw 22 February 1898. Typography of Warsaw Institute for Deaf, and Mute, and Blind, Warsaw 1898. — Shcherbak AE. Clinical Lectures about Neurologic and Psychiatric Disorders. Sevastopol 1905. — Shcherbak A. Signe de Babinski et dissociation des réflexes profonds et cutanés, provoqués expérimentalement chez l'homme; valeur séméiologique du réflexe dorsal du pied. Revue Neurologique 1908;16:408. — Shcherbak A. Mouvements pendulaires bi- et monoculaires; fermeture volontaire des yeux; centres corticaux oculo-moteur chez l'homme. Progrès méd 1908;24: 303-8. — Shcherbak A. Cas d'acathisie (akathisia paraesthética), guérie par l'autosuggestion; paresthésies vibratoires. Progrès méd 1908;24:265-8. — Shcherbak AYe. Basic Works in Physiotherapy. Publ. Sechenov Inst, Sevastopol 1935 - 1936. — Ablitsov VH. «Galaxy Ukraine». The Ukrainian diaspora: prominent persons. Publ. «Kyt», Kyiv 2007.

[584] Preobrazhensky PA. Zur Pathologie der Gehirns. Neurologisches Centralblatt 1893;6:759-60. — Preobrazhensky PA. An Inquiry about Formation Cavities in the Spinal Cord in Gliomatous Syringomyelia. Moscow 1900. — Preobrazhensky PA. Casuistry of hereditary syringomyelia. Med obozr 1900;53(2):239-47. — Preobrazhensky PA. Zur Frage der Bedeutung der Syphilis in der Aetiologie der Tabes dorsalis; über einen Fall von Tabes dorsalis im Kindesalter. Monatschr f. Kindern 1905;4:133-45. — Preobrazhensky PA. About acute encephalitis. Med obozr 1910;74(21)1059-74. — Eldaroff N. [Obituary] J neuropath psychiat Korsakov 1913;13:614-7. — Sarichev I. [Necrology] Med obozr 1913;29:4-6. — Ablitsov VH. «Galaxy Ukraine». The Ukrainian diaspora: prominent persons. Publ. «Kyt», Kyiv 2007.

Psychiatry and psychology

The psychical school headed by **Johann C.R. Heinroth (1773 – 1843)** — German physician, asserted that all insanity is the product of moral or psychological weakness and rejected any motion of a physical pathological causes. The somatic school, part of which were **Carl W.M. Jacobi (1775 – 1858)** — German psychiatrist, **Christian F. Nasse (1778 – 1851)** — German psychiatrist, and **Carl F. Flemming (1799 – 1880)** — German psychiatrist, held that insanity is a symptom of biological diseases located outside the brain, particularly diseases of the abdominal and thoracic viscera, akin to the delirium or delirium tremens caused by many acute biological illnesses. The latter, Carl Flemming introduced in 1844 the term dysthymia mutable to describe a disorder that is an alteration of dysthymia atra (black depression) and dysthymia candida (low-level mania)[585].

The physiological school represented by **Carl Wunderlich (1815 – 77**: 19th century. Medicine), **Wilhelm Griesenger (1817 – 68)** — German neurologist and psychiatrist, and **Wilhelm Roser (1817 – 88)** — German surgeon and ophthalmologist, insisted on there being a brain lesion underlying every case of insanity, even if in some instances that lesion is the product of pre-existing, extracerebral biological illness.

The evolution of **Franz Messner's (1734 – 1815**: 18th century**)** ideas and practices led **James Braid (1795 – 1860)** — Scottish surgeon, hypnotist, and hydrotherapeutics, to develop in 1841-42 hypnosis and hypnotherapy.

The term eugenics (in Greek – «well born», «suitable»), invented by Sir **Frances Galton (1822 – 1911)** — English biologist, psychologist, sociologist, statistician and polymath, designated the science aimed in the improvement genetic quality of human population, popularized the collocation of the words «nature and nurture» in his book entitled «Heredity Genius» (1869), being the first social scientist to study genius and greatness, where he showed that the number of eminent relatives dropped off when going from the first degree to the second degree relatives, and from the second degree to the third degree, as evidence of the inheritance of abilities. Nevertheless, he recognized that cultural circumstances influence the capacity of a civilization's citizens, their reproductive success.

He envisaged the best form of civilization in respect to the improvement of the race, would be one in which society was not costly; where incomes were chiefly derived from professional sources, and not much through the inheritance; where every lad had a chance of showing his abilities, and, if highly gifted, was enabled to achieve a first-class education and entrance into professional life, by the liberal help

---

[585] Dysrhythmia (neurotic depression, dysthymia disorder or chronic depression) is a mood disorder of the same cognitive and physical problems as in depression, with less severe but longer-lasting symptoms.

of the exhibitions and scholarships; where marriage was held in as high honor; where the weak could find a welcome and refuge in celibate monasteries or sisters hands, and where a better sort of immigrants or refugees from other lands were invited, welcomed, and their descendants naturalized.

Karl L. Kahlbaum (1828 – 99)[586] — Prussian German psychiatrist, who described catatonia – a state of psychogenic motor immobility and behavioral abnormality manifested by stupor and its symptom of continuous and purposeless repetition of words and sentences that are either meaningless or have no significance in a certain rhythm (Kahlbaum syndrome, 1874).

The development of psychiatry was greatly influenced by the work of **Theodor H. Meynert (1833 – 92**) — Austrian neurologist, who discovered numerous conveying pathways in the central nervous system (CNS) and outlined cytoarchitectonics of the cerebral cortex.

Leopold von **Zacher-Masoch (27 Jan. 1836, city of Lviv, Ukraine – 09 Sep. 1895, municipality of Lindheim** near **municipality of Allenstadt, district Wetterauskreis, state of Hessen, Germany)**[587] — Ukrainian writer, neurologist, and psychiatrist, described in 1870 psychiatric deviation in which a patient (masochist) experiences a pleasure, including sexual, form the bodily or psychological pain incurred from another person (sadist), named masochism.

Probable under the influence of creative power by Count **Donatien A.F. Marquis** de **Sade (1740 – 1814)** — French nobleman, masochist / sadist, and romantic writer, who wrote novels entitled «120 Days of Sodom» (1791), «Justine» (1791), «Philosophy of Bedroom» (1795), «Juliette» (1799), Leopold von Zacher-Masoch authored thoughtful selection of novels «The Testament of Cain», in one of which «Venus in Furs» (in German — «Venus in Pelz» (1870) described different forms of masochism and sadism.

The founder of sexual psychopathology **Richard F.** von **Krafft-Ebing (1840 – 1902)** — German neuropsychologist, authored the book «Psychopathia Sexualis» (1886), introduced the terms for psychiatric disorders called «masochism» and «sadism».

[586] Kahlbaum K. Die Katatomie oder das Spannungsirresein. Eine klinishe Form psychischer Kreankheit. First druck. A. Hirschwald, Berlin 1874.

[587] Municipality of Lindheim near municipality of Allenstadt (first documented mention 767), district Wetterauskreis (area 1,101.71 km²), land / state of Hesse (area 21,100 km²), Germany. — Sacher-Masoch L. Venus im Petz. In: Das Vermächtniss Kains – Erster Theil. Die Liebe. Verlag Cotta, Stuttgard 1870. Seite 121-368. —Von Sacher-Masoch L. Venus in Furs. Penquin Classics, London 1901. — Nieuwenhuijs V, von Sacher-Masoch L, Seyferth M. Film «Venus in Furs». Die Niederlande 1995. — Von Sacher-Masoch L. Venus in Furs. Publ. Literary Agency «Pyramide», Lviv 2008. — Ablitsov VH. «Galaxy Ukraine». The Ukrainian diaspora: prominent persons. Publ. «Kyt», Kyiv 2007.

Pityfully, the introduction of the term «masochism» which was erroneous and inaccurate, led to oblivion of Leopold von Sacher-Masoch as the writer, the style of which remains the style of **Ivan S. Turgeniev (1818 – 83)** — Russian writer.

Carrying experiments on animals behavior, **Douglas A. Spalding (1841 – 77)** — English biologist, discovered the phenomenon of attachment or imprinting which outlines the dynamics of long-term and short-term relationship between humans, specifically how human beings respond within relationship when hurt, separated from loved ones, or perceived a treat. Almost all infants became attached if provided any caregiver, and seeks proximity when they are alarmed, with the expectation receiving protection and emotional support.

The father of child psychology and psychopathology, **Ivan O. Sikorsky (26.V.1842, village of Antoniv, Bilotserkivs'kyy Region, Kyivs'ka Obl., Ukraine – 14.II.1912, city of Kyiv, Ukraine)**[588] — Ukrainian psychiatrist and psychologist, indicated the need for a tight interrelationship between intellectual and physical development in children; considered that mental tiredness is a manifestation of changes in the psychomotor function and that correct upbringing is a synthesis of the harmonious development of the mind, feeling and will, that secures the moral integrity of the personality and its dignity; proposed a three-staged classification of memory (objective, analytical and assimilated) and identified the corresponding rules for intellectual work, especially educational; emphasized that the development of the will of the person should be one of the most important tasks in the education of humans; created the first Institute of Childhood Psychology (Kyiv 1912); applied the experimental methods to investigate children's psychopathology.

In 1885 **Victor Kh. Kandynsky (1849 – 89)** — Russian psychiatrist, and in 1927 **Gaetan G. De Clerambault (1873 – 1934)** — French psychiatrist, described a syndrome of psychical automatism (Kandynsky-Clerambault syndrome) in which his own thoughts appear to the patient to be of the other's (the stranger's), as if arising against his will and are read from the distance by the others.

Pavlo I. Kovalevsky (1849, u.t.v., of Petropavlivka, Petropavlivs'kyy Region, Dnipropetrovs'ka Obl., Ukraine – 17.X.1931, city of Liege, Wallonia, Belgium)[589] — Ukrainian psychiatrist, proposed the concept of dysfunction of the blood flow and the tissue metabolism in the brain as a cause of psychiatric disorders (1885)

---

[588] Village of Antoniv (founded 1471), Bilotserkivs'kyy Region, Kyivs'ka Obl., Ukraine. — Ablitsov VH. «Galaxy Ukraine». The Ukrainian diaspora: prominent persons. Publ. «Kyt», Kyiv 2007.

[589] U.t.v. Petropavlivka (founded 1775), Petropavlivs'kyy Region, Dnipropetrovs'ka Obl., Ukraine – 17.X.1931, city of Liege (inhabited since 798,000 BC, first written evidence 558), Wallonia (area 16,901 km², Belgium (area 32,547 km²). — Kovalevsky PI. The Basis for the Mechanism of Spiritual Activity. Kharkiv 1885. — Kovalevsky PI. Psychiatry. Vol. 1-2. Kharkiv 1890-92. — Kovalevsky PI. The Imagination and Persecution, Ligancy and Greatness. The Judicial Medicine. St Petersburg 1902. —Ablitsov VH. «Galaxy Ukraine». The Ukrainian diaspora: prominent persons. Publ. «Kyt», Kyiv 2007.

Karl Wernicke (1848 – 1905) — German anatomist, neurologist and psychiatrist, and **Sergei S. Korsakov / Korsakoff (1854 – 1900)** — Russian neuropsychiatrist, originated the Wernicke-Korsakov syndrome (due to thiamine deficiency, usually reversible) and the Korsakov syndrome or psychosis (1887), a severe antegrade or retrograde amnesia associated with alcoholic [17th century. – **Raymond de Viessens (1635/41 – 1715) / Felix d'Azyr (1748 – 94)**] or nonalcoholic polyneuritis (also due to thiamine deficiency, usually irreversible).

The founder of modern psychiatry, psychiatric genetics and psychopharmacology, and the author of the classic «Textbook of Psychiatry» (1893), **Emil W.M.G. Kraepelin (1856 – 1926)** — German psychiatrist, considered the main primary source of psychiatric diseases to be biological and genetic dysfunction. He classified psychoses into (1) manic depressive psychoses that include chronic recurrent depression and bipolar disorders, and (2) dementia praecox or presenile.

Mykola V. Krajinsky (01.05.1869, city of Kyiv, Ukraine – 19.07.1951, city of Kharkiv, Ukraine, Україна)[590] — Ukrainian psychiatrist, considered that the cause of epileptic attacks lies in accumulation of toxic substances in the body, mainly of carbonic acid and ammonia, during which they eliminate itself (1893-95).

Health Welfare

The 19th century will also be remembered for the development of organized medicine and public health, the beginning of the nursing profession, the founding of the Red Cross (RC) and the Mayo Clinic.

In the Russian Empire during 1803-1917 medical education was managed by the Ministry of Education, health care by the Office of Internal Affairs (OIA) and military medicine by the Department of Land and Navy Forces. Provincial (zemska) medicine served rural areas (1864-1917).

In 1805 the University of Kharkiv was established, in 1835 the University of Kyiv with Medical School (1840), and in 1865 the University of Odesa, Ukraine.

The first attempt to introduce postgraduate training of physician in Ukraine was made by the Kharkiv Medical Society (created 1860) in 1861.

The term and practice of the public health was introduced in 1875 by **John Simon (1816 – 1904)** — English pathologist, surgeon, and the pioneer of the public health.

The USA physicians organized the American Medical Association (AMA, 1847) to help rise medical standards. The International Medical Sanitary Conference in Paris

---

[590] Krainsky NV. The Investigation of Metabolism in Epileptics. Kharkiv 1893. — Krainsky NV. The Investigation into Pathology of Epilepsy. Dissertation for Degree of Medicine Doctor, Moscow 1895. — Krainsky NV. The Ground, Herald and Raging as Appearance of Russian National Life. Provincial typography, Novgorod 1900.

(1851) helped to coordinate programs on health and diseases. The first National Health Insurance plan was established in Germany (1883).

It was **Frorence Nightingale (1820 – 1910)** — English social reformer and statistician, and the founder of modern nursing who established nursing as a profession during the Crimean War (1853-56) in Ukraine, and organized (1860) the School of Nursing at St Thomas Hospital (founded c. 1010), the city of London (first major Roman settlement in 43, recorded in c.121 as Londinum) — the capital of England and the United Kingdom of Great Britain and Northern Ireland (UK, area 243,610 km^2). A training school for visiting nurses was first organized by **Lillian D. Wald (1867 – 1940)** — American nurse, humanitarian and author, in 1893.

The pitiable conditions of wounded soldiers in the battle of Solferino (June 24, 1859), Italy, aroused **Jean H. Durant (1828 – 1910)** — Swiss businessman and social activist, the founder of the International RC, which was approved in the city of Geneva (fortified town in 121 BC), Switzerland (area 41,285 km^2), by the Geneva Convention (1864). The impoverished Jean Durant won the first Peace Price (1901) of **Alfred Nobel (1833 – 96)** — Swedish chemist, engineer, businessmen and philanthropist, which he donated to the charity.

It was **Otto** von **Bismarck (1815 – 98)**[591] — Minister President of Prussia (1862-90) and Chancellor of the German Empire (1871-90), who created the first welfare state in the modern world, establishing the Sickness Insurance Law (1883), the Accident Insurance Law (1884) and the Old Age Disability Law (1889).

The Sickness Insurance Law of 1883 provided the health services on a local basis, with the cost divided between employers who contributed one third, and the employede (laborers) contributed two third. The minimum payments for medical treatment and sick pay for up to 13 weeks were legally fixed. The individual local health bureaus were administered by committee elected by the members of each bureau, and this move had established majority of representation for the workers on the account of their large financial contribution. This also allowed them to achieve their first small foothold in public administration.

The Accidental Insurance Law of 1884 was a program directed by the Organization of Employers in Occupational Organizations which established the central, federal, and state insurance offices to administer the program whose benefits replaced the sickness insurance program as of the 14th week. It paid medical treatment and a pension of up to two third of earned wages if the worker were fully disabled. This program was expanded in 1886 to include agricultural workers.

The Old Age and Disability Insurance Law of 1889 included program which equally financed employers and all workers (industrial, agrarian, artisans and

---

[591] Steinberg J. Bismarck: A Life. Oxford Univ Press, New York 2011.

servants), was designed to provide a pension annuity for workers who reached the age of 70. The principle that the national government should contribute a portion of the underwriting cost with the other two portions prorated accordingly, was accepted without question. The disability insurance was intended to be used by those permanently disabled, and directly supervised by the state or province.

In 1889 **William W. Mayo (1819 – 1911)** — British American chemist and physician, **William J. Mayo (1861 – 1939)** — American physician and surgeon, and **Charles H. Mayo (1865 – 1939)** — American physician, established the Mayo Clinic in the city of Rochester (founded in 1854), MN, USA.

THE 20TH CENTURY AND THE BEGINNING
OF THE 21ST CENTURY

Advances in science and engineering created a medical revolution in this century. The highlights in the basic sciences are the discovery of blood groups (1900), immunity (1906-20), genes and chromosomes (1926), antibiotics (1925), acquires immunological tolerance (1949) and the structure of nucleic acids (1953). Clinical diagnosis was enhanced by the introduction of improved electrocardiography (ECG, 1901-03), arteriography (1923), artificial cardiac pacemakers (1926), cardiac catheterization (CC, 1929), nuclear magnetic resonance / magnetic resonance / magnetic resonance imaging (NMR, MR, MRI, 1944-46), selective coronary angiography (SCA, 1958) and computer tomography (CT, 1972). Startling achievements in clinical medicine began with the development of a safe brain (1905-18) and chest (1931) surgery, the heart-lung machine (HLM) for cardiopulmonary bypass (CPB) in 1928-42, 1953, kidney transplantation (1933, 1954), designing of a total artificial heart and biventricular system (TAH, BVS, 1937-55), the surgical treatment of congenital heart disease (CHD) beginning with the ligation of patent ductus arteriosus (PDA, 1938), development of digital subtraction angiography (DSA, 1939, 1971-77), treatment of an acute and chronic renal failure with hemodialysis (HD, 1943), an experimental heart transplantation (HT, 1945), vascular procedures (1947-51), introduction of medical ultrasonography (SG) and echocardiography (1953), beginning of space exploration and space medicine (1957-1961), repair and replacement of cardiac valves (1959-60), coronary artery bypass (CAB) using the internal thoracic artery (ITA) graft and veins graft (VG, 1960-68, 1975-77), liver transplantation (1963), cardiac (heart) transplantation (CT, HT) in man (1967), landing the man on the Moon (1969), replacement of the heart with a total artificial heart (TAH) in man (1969, 1982), percutaneous (PC) double catheter-balloon embolization of the intracranial aneurysms using detachable micro-balloon (1970-73), percutaneous (PC) transluminal coronary angioplasty (PTCA, 1977), robotic (rob.) surgery (1985) and intraluminal self-expanding and self-fixing synthetic tubular grafting (ISS-STG) of the descending thoracic aorta (DTA) aneurysms (1987).

Anatomy, histology and embryology

The greatest anatomist and histologist of the 20th century were **August Rauber (1841 – 1917)** — German anatomist and embryologist, **Frederich W. Kopsch (1868 – 1955)** — German anatomist, histologist and embryologist, **Vladimir N. Tonkov (1872 – 1954)** — Russian anatomist, **Adam Bochenek (1875** – suicide **1913)** — Polish physician, anatomist, histologist and anthropologist, **Volodymyr P. Vorobiov [15(27).VI.1876, city of Odesa, Ukraine – 31.VIII.1937, city of Kharkiv, Ukraine]**[592] — Ukrainian physician, anatomist and histologist, **Eduard Pernkopf (1888 – 1955)** — Austrian anatomist, **Michał Reicher (1888 – 1973)** — Polish anatomist and anthropologist of Jewish descent, and **Rafail D. Synelnykov (1896, city of Berdiansk, Zaporizhia Obl., Ukraine – 21.02.1981, city of Kharkiv, Ukraine)**[593] — Ukrainian physician and anatomist of Jewish descent, who developed macro-microscopic approach to anatomy.

August Rauber and later Frederich Kopsch published the classic «Lehrbuch der Anatomie des Mensches» (Bänd 1-2. 5th Ed. 1897-98), August Rauber and Frederich Kopsch «Lehrbuch der Anatomie des Mensches» (Bänd 1-6. Georg Thieme Verlag 1910-20).

Vladimir Tonkov composed «Handbook of the Normal Human Anatomy» (Vol. 1-2. 1st Ed. 1918), «Handbook of the Normal Human Anatomy» (Vol. 1-2. 5th Ed. «Medgiz», Moscow 1953).

Adam Bochenek worked out «Anatomia człowieka» / «The Human Anatomy» (Vol. 1 of the planned four. 1st Ed. State Academy of Knowledge, Crakow 1909. After his death, since 1913 Michał Reicher became co-author and continuator of his work – Adam Bochenek and Michał Reicher. «Anatomia człowieka» (Vol. 1-7. PZWL, Warsaw 1953-65) and «Anatomia człowieka» (Vol. 1-5. PZWL, Warsaw 2014).

In addition, Volodymyr Vorobiov described the innervation of the stomach (1913) and subepicardial (Vorobiov) nervous plexus of the heart (1922), published the classic «Human Anatomy» (Vol. 1. «Derzhmedvydav», Kharkiv 1934), with co-authors «A

---

[592] VorobiovVA. Human Anatomy. «Derzhmedvydav» of Ukraine, Kharkiv 1934. — Vorobiov VA. (& assoc.). Short Handbook of Human Anatomy. Vol. 1-2. « Derzhmedvydav» of Ukraine, Kyiv 1936-38. — VorobiovVA., Symelnykov RD. Atlas of Human Anatomy. Vol. 1-5. «Medgiz», Moscow-St Petersburg 1938-42. — VorobiovVA., Symelnykov RD. Atlas of Human Anatomy. Vol. 1-5. «Derzhmedvydav»of Ukraine, Kharkiv 1946-48. — VorobiovVA. Selected Works. St Petersburg 1958. — Ablitsov VH. «Galaxy Ukraine». The Ukrainian diaspora: prominent persons. Publ. «Kyt», Kyiv 2007. — Hanitkevych YaV. The contribution of the Ukrainian physicians to the world medicine. Ukr Med Chasopys 2009 VII/VIII;4(72):110-15.

[593] City of Berdiansk (settled since 3rd millennium BC, the first known settlers were Ukrainian Zaporizhian Cossacks who built fishing piers on the coast of the Azov Sea in the middle of 16th century, 1673, founded 1825; during organized by Russian Communist Fascist regime of Holodomor-Genocide 1932-33 died from man-made famine at least 1770 residents of the city), Zaporizhia Obl., Ukraine. — Symelnykov RD «Atlas Human Anatomy» (Vol. 1-2. «Medgiz», Moscow 1952- 58). — Ablitsov VH. «Galaxy Ukraine». The Ukrainian diaspora: prominent persons. Publ. «Kyt», Kyiv 2007.

Short Texbook of Human Anatomy» (Vol. 1-2. «Derzhmedvydav», Kyiv 1936-38). After Volodymyr Vorobiov's death, his pupil Rafail Synelnykov edited his work – «Atlas of Human Anatomy» (Vol. 1-5. «Medgiz», Moscow-St Petersburg 1932-42) and «Atlas of Human Anatomy» (Vol. 1-5. «Derzhmedvydav», Kharkiv 1946-48). On his own Rafail Synelnykov worked out «Atlas of Human Anatomy» (Vol. 1-2. «Medgiz», Moscow 1952-58).

Edward Pernkopf created «Topographische Anatomie des Menshen» (Bänd 1-4. Urban und Schwartzen Verlag / Urban und Fisher Verlag, Wien 1937 - Bänd 1, 1941 - Bänd 2, 1952 - Bänd 3, und 1960- Bänd 4).

Both outer lobes of the thyroid gland (TG) or glandula thyroidea are covered by a thin fibrous capsule, which has an inner and outer layer. The outer layer is continuous with the pretracheal fascia, attaching bilaterally the TG to the posterior / lateral suspensory ligament, identified by Sir **James Berry (1860 – 1946)** — Canadian English surgeon, named in his honor the Berry ligament, to the cricoid and thyroid cartilage at the level of the 2nd and 3rd tracheal ring. The Berry ligament causes the TG to move up and down with swallowing.

The recurrent laryngeal nerves (RLN)[594], first documented by **Claudius Galenus (c.129 – 99 AD**: Ancient medicine. *Romans*), branches of the cranial X (vagal) nerves pass behind the posterior middle part of the outer lobes of the TG or deep to the Berry ligament and enter the larynx underneath the inferior pharyngeal constrictor muscle, just posterior to the paired cricothyroid joint. The RLN supplies all the intristic muscles of the larynx, except the cricothyroid muscle, which is innervated by the external branch of the superior laryngeal nerve. In addition, the RLN supply sensation to the larynx below the vocal cords, gives cardiac branches to the deep cardiac plexus and branches to the trachea, esophagus, and the inferior pharyngeal constrictor muscles. The posterior cricoarytenoid muscles, the only muscles that can abduct (open) the vocal cords to allow respiration by inspiration and expiration of an air, are innervated by the RLN.

In 1907 Sir **Arthur Keith (1866 – 1955)** — Scottish anatomist and anthropologist, and **Martin W. Flack (1882 – 1931)** — British physiologist, discovered in the vertebrates a natural pacemaker of the heart, sino-atrial (SA) or Keith-Flack node, situated in the terminal groove at the junction of the the right atrium (RA) appendage and the superior vena cava (SVC).

In 1902-05 **Ivar Broman (1868 – 1946)**[595] — Swedish anatomist and embryologist, performed studies which linked congenital diaphragmatic hernia (CDH) to deviation of the embryologic development of pleuroperitoneal membrane.

---

594 Olearchyk AS. The non-recurrent laryngeal nerves. J Ukr Med Assoc North America 1973;20,4 (71):45-7.
595 Broman I. Über die Entwicklung des Zwerchfells beim Menschen. Verh Anat Ges 1902;16:9-17. — Broman
      I. Über die Entwicklung und Bedeoutung der Mesenterien und der Körperhöhlen bei den Wirbeltieren.
      Eugeb Anat Etwr 1905;15:332-409.

Olexandr H. Cherniakhivsky [13.XI(01.XII).1869, village Mazepyntsi, Bilotserkivsky Region, Kyiv Obl., Ukraine – 22.XII.1939, city of Kyiv, Ukraine][596] — Ukrainian histologist, investigated the presence of multinuclear neurons of the central nervous system (CNS, 1911); proved the existence of «wandering» nervous fibers and branches from the restraining neurons in the early stage of embryogenesis. He independently of **Henrich E. Hering (1866 – 1948)**[597] — German physiologist, who discovered afferent nerve branches from the carotid sinus by the way of the cranial IX (glossopharyngeal) nerve – ramus sinus carotici glossopharyngei, to the brain (Hering nerves, 1927), studied chemo- and baroreceptors in the intima of the carotid sinus, explaining its physiological role in regulation of the arterial blood pressure (BP); found the ingrowth of nervous fiber from the surrounding healthy tissues into malignant tumors. The Hering nerves are pressoreceptor nerves responding to changes in the arterial BP that reflexively control heart rate (HR). An increase in the arterial BP diminishes HR.

Daughter of Olexandr Cherniakhivsky, **Veronika O. Cherniakhivsky [25.IV.1900, city of Kyiv, Ukraine, Україна** – twice arrested (1929 and 1938), accused in spying without evidence, raped by Russian Communist Fascist inquirings / investigators of People's Commissariat for Internal Affairs, as consequence of which she became demented, sentenced to be shot and executed this same day **22.IX.1938, city of Kyiv, Ukraine]**[598] — Ukrainian poetess, together with father, from the People's Commissariat of Health, where worked as a reviewer-translator, as owned Ukrainian, Russian, Ancient Greek, Latin, French, German and English languages, rode in a business trip to Germany.

In 1908 **Alexander A. Maximow (1874 – 1928)**[599] — Russian German American histologist, proposed the term «stem cells» and originated hypothesis of the existence

---

[596] Village Mazepyntsi (founded 1500; during conducted by Russian Communist Fascist regime of Holodomor-Genocide 1932-33 against the Ukrainian Nation, perished 424 inhabitants), Bilatserkivsky Region, Kyiv Obl., Ukraine. — Vilensky Y, Konstantynovsky H, Chaikovsky Y. Ukrainian neurohistologist Olexandr Cherniakhivsky. UHMJ «Ahapit» 1997-1998;7-8:29-33. — Ablitsov VH. «Galaxy Ukraine». The Ukrainian diaspora: prominent persons. Publ. «Kyt», Kyiv 2007.

[597] Hering H. Der Karotisdruckversuch. München Medicinische Wochenschrift 1923;70:1287-90. — Hering E. Die Aenderung des Herzschlagzahl durch Aerderung des arteriellen Blutdruckes erfolgt aus reflekteorischen Wege; gleichzeiting eine Mitteilung über die Funktion des Sinus caroticus, beziehungsweise der Sinusnerven. Pflügers Archiv für gasampte physiologic des Menschen und der Tiere 1924;206:721-3. — Hering E. Die Karotissinusreflex auf Herz und Gefässe. Dresden-Leipzig 1927.

[598] Ablitsov VH. «Galaxy Ukraine». The Ukrainian diaspora: prominent persons. Publ. «Kyt», Kyiv 2007.

[599] Maximow A. Der Lymphozyt als gemainsame Stammzelle der verschiedenen Blutelemente in der embryonalen Entwicklung und in postfetalen Leben der Säugetiere. Folia Hematologica 8.1909:125-34. / **Maximov AA**, Bloom W. Texbook of Histology. WB Saunders Co., Philadelphia, PA 1930. **William Bloom (1899 - 1972)** — American histologist of Jewish Lithuanian ancestry. / Singer R. William Bloon (1899-1972). A Biographical Memoir. Nat Acad Sc USA, Washington, DC 1993:1-36. / Konstantinov IE. In search of Alexander A. Maximow; the man behind the unitarian theory of hematopoiesis. Perspectives in Biology and Medicine 2000;43(2):269-76.

of hematopoietic stem cells (HSC) – a general precursor (predecessor, progenitor) of all cells of the blood, and in 1924 identified a singular type of predecessor cell within the mesenchyme that develops into different mesenchymal stem cells (MSC: proved in 1961). Thus, he originated the unitarian theory of hematopoiesis that all blood cells develop from a common progenitor cell – MSC, a basic theory of the modern concept of blood cell's origin and differentiation. The HSC, as well as MSC are multipotent precursors, giving the origin of cell's lines, the final elements of which create the formed element of the blood, and the number of specialized cells for the tissue cells of an immune system.

André Latarje / Latarjet (1877 – 1947) — French anatomist and surgeon, described the anterior branch (nerve) of the left (anterior) trunk of the cranial 10th (vagal) nerve and the posterior branch of the right (posterior) trunk of the vagal nerve running along the lesser curvature of the stomach, innervating the fundus, the body, and the pylorus (nerves of Latarjet, 1928). In in highly selective vagotomy for the surgical treatment of peptic ulcer, only those branches of the vagal nerve trunk should be transected which innervate the lower esophagus and the stomach, leaving intact the nerves of Latarjet, especially the posterior nerve of Latarjet so that the function of the gastric emptying remains intact.

The earliest embryonal period includes stages of zygote (in Greek, meaning - «joined»), an eukaryotic cell formed by fertilization events between two gametes, which undergoes many cleavages to develop into a ball of cells called a morula. A blastula (in Greek – meaning «blastos», «embryon» or «sprout») is a hollow sphere of one layer cells, which is formed as a result of division of the zygote during the second week after a fertilization, referred to as blastomeres or blastoderm, surrounding an inner fluid-filled cavity called blastocele. The upper disc of blastoderm (epiblast) gives the origin of the ectoderm (external layer). The gastrula appears at the third weeks after fertilization when from the embryonal ectoderm develops the mesoderm (middle layer) and the endoderm (internal layer).

Approximately 20 days from fertilization begins the development of the heart, precisely endocardial and myocardial cells from the components of cardiac precursors of the anterior right and left lateral plates of the abdominal (ventral) or visceral mesoderm on a signal from the bone morphogenic protein (BMP) and fibroblast growth factor (FGF)[600].

The main embryonal development stages of the heart appear in the form of the epiblast, heart mesoderm, cardiac crescent, lineal heart tube, looped (S-like) heart tube and chambered heart.

---

[600] Moe GK, Preston JB, Burlington HJ. Physiologic evidence for a dual AV transmission system. Circ Res 1956;4:357. — Mayer A. Rhythmical pulsation in Scyphomeduse. Carnegie Institution of Washington, 1906.

The components of epiblast include extraembryonic ectoderm (EE), primitive streak (PS), FGF and brachyury or a gen of the short tail (transcription of genes required for mesoderm formation and cellular differentiation).

The mutation of brachyury was first discovered in 1927 by **Nadiya Dobrovols'ka-Zavads'ka / Nadine O. Dobrovolskaïa-Zavadskaïa [Sep. 12(25), 1878, city of Kyiv, Ukraine – Oct. 21, 1954, city of Milano** (founded by Gauls c. 600 BC), Italy][601] — Ukrainian French physician, vascular surgeon, scientist, radiobiologist, genetics, and biologist, as a change which influence the length of the tail and the sacral column in the heterozygote animals and is fatal in the homozygote animals at about 10 day of the embryonal development as a result of anomaly in the formation of the mesoderm, differentiation of the notochord[602] and the absence of structures behind the bud (conception) of the anterior extremity.

The components of the heart mesoderm (HM) are mesoderm, foregut endoderm (FG), BMP, inhibitor Wnt (proteins secreted by the neural plate), beta-catenin and homeobox protein Nkx2.5.

The crescent heart develops from the lower (caudal) cells of the precardial mesoderm, which create the right and left atria (RA, LA), from the upper rostral cell of the precordial mesoderm, which form the right and left ventricle (RV, LV).

Although the vertebrate heart develops primarily from the precardiac mesoderm (the cardiac crescent), the «anterior heart field» cells, located towards the crescent heart center, promote the development of the anterior surface of the primordial heart muscle (myocardium), right ventricular outflow tract (RVOT), located between the supraventricular crest and pulmonary valve (PV), comprised by the right conus arteriosus (infundibulum), interventricular septum (IVS) and RV free wall, and the left ventricular outflow tract (LVOT) or left conus arteriosus, which is nearly indistinguishable from the rest of the ventricle. The neural crest (cell streak of the spinal lateral neural tube) procreates the ganglia of the cranial and cerebrospinal nerves, the primordial main pulmonary artery (MPA) and aortic root of the ascending aorta (AA), dorsal aorta, vascular and hematopoietic system.

---

[601] Dobrovolskaya NA. Zur Technic der Nöchte an Gefässen kleiner Kalibers. Z Chir 1912;119:31-54. — Dobrovolskaïa-Zavadskaïa N. Sur la mortification spontanee de la chez la souris nonveau-nee et sur l'existence d'um caratere (facteur) hereditaire non viable. Crit Rev Soc Biol 1927;97:114-6. — Dobrovolskaya-Zavadskaya N. The problem of species in view of the origin of some new forms in mice. Biol Rev 1929;4:327-51. — Korzh V, Grunwald D. Nadine Dobrovolskaïa-Zavadskaïa and the dawn of developmental genetics. BioEssays 2001;23 (4):365-71. — Ablitsov VH. «Galaxy Ukraine». The Ukrainian diaspora: prominent persons. Publ. «Kyt», Kyiv 2007.

[602] The notochord is a flexible cartilaginous rod along the anteroposterior («head to tail»), closer to the dorsal than ventral axis of the body in chordate.

The existence of the anterior heart field was suspected by **Szabolcs Virah** — Hungarian scientist and **Cyril E. Challice**[603] — Canadian scientist, when they observed the transformation of epithelial cells to myocardial cells at the arterial pole of the developing mouse heart. **Maria V. de la Cruz (1910s – 1999)**[604] — Cuban cardiac embryologist, used marking studies in living chick embryos to demonstrate that the distal outflow tract was a late addition to the developing heart.

On approximately 4 week of the embryonal development with convergence and fusion of the primordial two-sided heart the lineal heart tube is formed, which compose of an aortic sac (AS), the future conotruncus (C), the RV, the LV, the nodal gene – a secretory protein in the human, located on the chromosome 10q22.1, belonging to the transforming growth factor (TGF)-beta family, involved in cell differentiation in the early embryogenesis, plying a key role in signal from the node in the PS to the lateral plate mesoderm, the lefty – the left-right determination protein factors related to the TGF-beta family, which are secreted and play a role in the left-right asymmetry determination of organs during development, the notch – signaling pathway present in most of multicellular organisms, the pitx – a transcription protein factor than control transfer of genetic information from deoxyribonucleic acid (DNA) to ribonucleic acid (RNA). In the horizontal plain of the lineal heart tube there appears a narrowing between the atrium and the ventricle – a site of the future atrioventricular canal. On roughly the 27- day of the embryonal development on the spinal and ventral sides of the precardial canal grows out an endocardial or atrioventricular shafts, i.e., endocardial folds, from the roof of atria and from the apex of the ventricle, in the vertical plain grows off interatrial septum (IAS) and IVS. For the first time, the lineal heart tube reveals contractions of a beating myocardium. Likely, during this time (21-23 days after conception), together with cardiac contractions begins the development of the conduction network (system) of the heart (CNH, CSH).

The looped (S-like) heart tube comprises of the platelets derived growth factor (PDGF), the TGF-beta, the YB1 antibody, the iroquois homeobox genes, the pax genes that belongs to the paired box family transcription factors.

In turn between the 31th - 35th day of the embryonal development period on a signal originating from the mesodermal heart of the primitive Hensen's node **[Victor Hensen (1835 – 1924). 19th century. Anatomy, histology, and embryology]** takes place a three-dimensional right-sided turn (D-looping) of the lineal heart tube on account of which appears the tubular (S-like) heart loop with a three-dimension

---

[603] Viragh S, Challice CE. Origin and differentiation of cardiac muscle cells in the mouse. J Ultrastruct Res 1973 Jan;42(1):1-24.

[604] Angelini P. Maria Victoria de la Cruz. Texas Heart Inst 2000 Feb 27(1):1-2. — De la Cruz MV, Gomez CS, Arteaga MM, et al. Experimental study of the development of the truncus and the conus in the chick embryo. J Anat 1977;123:661-86.

structure and asymmetry in the anteroposterior, left to right and vertebral ventral axis.

The formation of the IAS and IVS, tricuspid and mitral valves (TV, MV), infundibular septum and aortic root are accomplished by contribution of the RV folds. With the help of the fibrous body of the heart, TV and MV are firmly and hermetically attached to atria and ventricles, the RV is separated from the cardiac infundibulum and the MPA by the supraventricular crista, and the LV by the fibrous aortic ring from the aortic root.

The terminal stage in the development of the heart is the chambered heart comprised of the superior and inferior vena cava (SVC, IVC), RA, TV, RV, RVOT, pulmonary valve (PV), LA, MV, LV, LVOT, aortic valve (AV), ascending aorta (AA), coronary arteries and veins, epicardium, myocardium and endocardium, IAS, IVS, conal septum and aortopulmonary (aortic, spiral) septum and CSH.

In 1927 **Albert Kuntz (1879 – 1957)**[605] — German histologist, identified inconstant sympathetic branches from the 2nd intercostal nerve to the 1st intercostal nerve, hence the somatic branchial plexus that bypassed the sympathetic stellate (cervico-thoracic) ganglion (intrathoracic nerve of Kuntz). The incidence of Kuntz's nerves may be in the range of 40% - 80% of the population.

After **Johannes Müller (1801 – 58**: 19th century. Anatomy, histology, and embryology), **August Mayer (1787-1865**: 19th century. Anatomy, histology, and embryology) and **Karel** von **Rokitansky (1804 – 78**: 19th century. Anatomy, histology, and embryology), agenesis (aplasia) or hypoplasia of the Müllerian duct was described again in 1910 by **Hermann Küster (1879 – 1964)** — German gynecologist, and in 1961 by **Georges A. Hauser (1921 – 2009)** — Swiss physician, and since then named Müllerian agenesis, uterovaginal aplasia/hypoplasia or Mayer-Rokitansky-Küster-Hauser (MRKH) syndrome.

In 1907 **Walter K. Koch (1880 – 1962)** — German surgeon, defined a triangular shaped area in the right atrium (RA) of the heart (triangle of Koch) formed by the ring of the tricuspid valve (TV), **Adam Tebesian (1686 – 1732**: 18th century) valve of the coronary sinus (CS) and tendon of **Francesco Todaro (1839 – 1918**: 19th century. Anatomy, histology, and embryology), the location of the atrioventricular node (AVN).

Herbert M. Evans (1882 – 1971) — American anatomist and embryologist, discovered human growth hormone (HGH) from the pituitary gland (1909), i diazo dye which has been principal method of determining blood volume in animals and

---

[605] Kuntz A. Distribution of the sympathetic rami to the brachial plexus: its relation to sympathectomy affecting the upper extremity. Ann Surg 1927;15:871-7.

humans (Evans blue, 1915), with **Katarina S. Bishop (1889 – 1979)** — American histologist and physician, discovered vitamin E (1923).

In 1927 **M.A. Kugel**[606] — described the atrial artery (arteria anastomotica auricularis magna) that arose from the proximal circumflex artery (CFA) or from its branches, coursed through the lower part of the interatial septum (IAS), and in the 66% of cases anastomosed directly or via its branches with the distal right coronary artery (RCA) or with branches from the anterior portion of the CFA and RCA and the posterior portion of the CFA (8%).

Oscar V. Batson (1894 – 1979) — American anatomist and otolaryngologist, described in 1940 four interconnected networks of the vertebral venous system surrounding the vertebral column (vertebral venous plexus or Batson's plexus): (1) anterior external vertebral venous plexus (plexus venosus vertebralis externus anterior) – the small system around the vertebral bodies; (2) posterior external vertebral venous plexus (plexus venosus vertebralis externus posterior) – the extensive system around the transverse and spinous vertebral processes; (3) anterior internal vertebral venous plexus (plexus venosus vertebralis internus anterior) – the system running the length of the vertebral canal anterior to the dura mater; (4) posterior internal venous plexus (plexus venosus vertebralis internus posterior) – the system running the length of the of the vertebral canal posterior to the dura mater.

The latter two, anterior and posterior internal vertebral venous plexus constitute epidural venous system, which connect the internal vertebral venous plexus with the deep pelvic veins (which drain the base of the urinary bladder and prostate) and the thoracic veins (which drain the breast). Due to its location and lack valves, Batson's plexus is regarded as a conduit for metastases from carcinoma of the rectum and prostate to the vertebral column or brain, as well as for infection.

The investigations of **Fernando de Castro (1896 – 1967)** — Spanish neuro-histologist, led to description in details of the innervation of the aorto-carotid region, circumscribing the presence of baroreceptors to the carotid sinus (in Latin – sinus caroticus) of the bifurcation of the common carotid artery (CCA), but that of chemoreceptors to a small nodule, named the carotid body (in Latin – glomus caroticum) adjacent to the carotid sinus.

The structural and functional organization of cells was outlined by **Albert Claude (1899 – 1983)** — Belgian cell biologist, **George E. Palade (1912 – 2008)** — Romanian American cell biologist, **Alex B. Novikoff (28.02.1913, u.t.v. of Semenivka, Poltava**

[606] Kugel MA. Anatomical studies on the coronary arteries. 1. Arteria anastomotica auricularis magna. Am Heart J 1927;3:260-70. — Nerantzis CM, Marianou SK, Koulouris SN, et al. Kugel's artery. An anatomical and angiographic study using a new technique. Tex Heart Inst J 2004;31(3):267-70.

Obl., Ukraine – 09.01.1987, city of New York, N.-Y., USA)[607] — Ukrainian American cell biologist of Jewish ancestry, and **Christian R.M.J.V. de Duve (1917 – 2013)** — Belgian cell biologist.

Albert Claude in 1930 developed the cell fractionation, a process used to separate cellular components while preserving individual functions of each component, by which he discovered eight years later (1938) a virus causing Rous sarcoma **[Francis Roux (1879 – 1970)** — American virologist and pathologist], components of cell organelles (usually free-floating part of the eukaryotic cell which has a specific function), such as mitochondria, хлоропласт chloroplast, the Golgi apparatus **[Camillo Golgi (1842 – 1926)**. 19th century. – Anatomy, histology and embryology], and endoplasmic reticulum, a type of organelle in eukaryotic cells that forms an interconnected network of flattened membrane-enclosed sacs or tube-like structures, named cisternae. That membranes are continuous with the outer nuclear membrane. He was the first to employ electron microscope in the field of biology and published the first detailed structural and functional description properties of the cell (1955).

The most notable discovery of George Pallade in the beginning of the 1955s was that of the ribosome – a non-membranous organelle in the endoplasmic reticulum of a cell, made of ribonucleic acid (RNA). It serves as a site of biological protein synthesis. Ribosomes link amino acids together in the order specified by messenger ribonucleic acid (mRNA) molecule. The small ribosomal subunit reads the RNA, and the large subunit join amino acids to form polypeptide chain. Each subunit is composed of one or more ribosomal RNA (rRNA) molecules and a variety of ribosomal proteins.

They also transport proteins and participate in protein sorting pathway.

Christian de Duve made serendipitous discovery of two cell organelles, with the help of Alex Novikoff discovery of the lysosome (1955) and by himself discovery of the peroxisome (1967).

Alex Novicoff successfully produced the first hard evidence of the organelle using an electron microscope (1955), then by using staining method for acid phosphatase with Christian de Duve, confirmed the location of the hydrolytic enzyme lysosome (1965). The peroxisome plays an important metabolic role, including the beta-oxidation of very long-chain fatty acids by a pathway in mitochondria.

Christian de Duve also coined the word lysosome, peroxisome, autophagy, endocytosis and exocytosis.

---

[607] U.t.v. of Semenivka (founded 1753; Russian Communist Fascist NKVS in 1933-46 shoot 1 person, imprisoned 6 persons), Kremenchuts'kyy Region, Poltavs'ka Obl., Ukraine. — City of New York [general area 1,213.37 km², elevation 10 m.; settled since 10,000 BC by Algonquian Indians, first recorded visit in 1524 by **Giovanni da Verrazzano (1485 -1528)** — Italian explorer; settlement by Hollanders began in 1609-24], State of New York (N.Y., area 141,300 km²), USA. — Ablitsov VH. «Galaxy Ukraine». The Ukrainian diaspora: prominent persons. Publ. «Kyt», Kyiv 2007.

Borys (Boris) I. Balinsky (10.IX.1905, city of Kyiv, Ukraine – 01.IX.1997, city of Johannesburg, South African Republic)[608] — Ukrainian South-African histologist, embryologist, and developmental biologist, introduced into electron microscopical studies of ultrastructure of the organism in an early developmental stage and predicted the importance of molecular biology in connecting the key mechanisms of embryogenesis.

Contribution to medicine of **Bolesław R. Jałowy (11.IX.1906, city Peremyshl, Ukraine / Poland** – assassinated **01.X.1943, city of Lviv, Ukraine)**[609] — Polish histologist, embryologist and dermatologist, Doctor of Medical Sciences (D.M.S., 1931), Professor and Head Division of Histology and Embryology (1937-43), Dean of the Department of Medicine (1939-41), Danylo Halytsky Lviv National Medical University (LNMU), and member of the National Democracy (in Polish – Narodowa Demokracja, or «Endecja») of Poland, was description of silver staining technique for the silver-absorving reticulum fibers of the skin (1936).

---

[608] City of Johannesburg (founded 1881), South-African Republic (area 1,221.037 km^2). —Ablitsov VH. «Galaxy Ukraine». The Ukrainian diaspora: prominent persons. Publ. «Kyt», Kyiv 2007.

[609] City of Peremyshl (Foot Notes 1, between 87 & 88), **Ukraine-Rus'**, a.k.a., **Kyivan Rus'**, Ukraine / Poland. — Danylo Halytsky LNMU [established near the end of 17th century. –1661-73: 18th century. – **Andriy Krupynsky (1744-83**: 1773), and the end of 18th century – 1784]. — National Democracy (in Polish – Narodowa Demokracja, or «Endecja») of Poland (founded 1886, dissolved 1947) – Polish political party which in 1928-39 forcefully polonized and oppressed minority of Germans in Poland, polonized and oppressed majority of the Ukrainians, Belarusans, and Lithuanians who lived in their own ethnic homeland's territories occupied by the 2nd Polish Republic (1918-39), and also organized anti-semitic actions. — Jałowy B. Changes in blood clotting after ovariectomy. Acta Biol Exp 1933;8(5):45-8. — Jałowy B. Über die heterogene Regeneration der Nervennendigungen in den Tasthaaren des Meerschweinchens (Covia cobaya). Zeitshrift für Zellforschung and Mikroskopische Anatomie 1934 Jan;21:149-68. — Jałowy B, Harasimowicz A. Das Mikroskipische Bild der Klinisch unveränderten Haut bei Pemphigus und dessen Zasammenhang mit dem Nikolsky-Phönomen. Archiv für Dermatologie und Syphilis 1935 Jan;1971:179-86. — Harasimowicz A, Jałowy B. A leukocyte Gouin reaction in the blood of syphilis infected patients. Pol Gaz Lek 1935;14(18):123-9. — Jałowy B. The silver-absorving fibers of skin. Przeg Derm 1936;31(4):397-413. — Jałowy B. Morphological structure of the nerve ending and their functional properties (with a description of the case of nodular eczema). Pol Gaz Lek 1936;15 (28-29):539-42. — Hołobut WS, Jałowy B. Die Re- und Regeneration des Peripheren motorischen Nervensysten auf einer Morphologischfunktionellen Grundlage. Z Zellforsch Mikrosk Anat 1936;25:541-64. — Hołobut WS, Jałowy B. Degeneration and regeneration of peripheral motor neuron on the morphological and functional basis. Pol Gaz Lek 1936;15(44):849-56. — Lenartowicz J, Jałowy B. Essals de production d'argyria artificielle chez lez animaux. Ann Dermatol Syphil (Paris);1938;9(6):483-94. — Jałowy B. Über die heterogene Regeneration der Nervenendigungen in den Tasthaaren des Meerschweinchens (Covia cobaya). Bulletin de L'Academie Polonaise des Sciences et des Letteres – Classe des Sciences Mathematiques at Naturelles 1938;2(4-6):203-19. — Jałowy B. Ein Beitrag zur Innervation Speicheldrusen. Z Zellforsch Mikrosk Anat 1938;28(1):114-9. — Jałowy B. The cysts of hair follicles in the form of a so-called Sebaceous cystadenoma (multiple steatocystomas) and milla. Przeg Derm 1938;331(1):21-9. — Wincewicz A. Pro Memoria. Professor Bolesław Jałowy (1906 – 1943): Mortui viventes obligant – the living are obligated to the dead. Romanian Journal of Morphology and Embryology 2016 April – June;56:29 Suppl:899-901.

New rector of Danylo Halytsky LNMU, **Olexandr Makarchenko (1903 – 79**: 20ᵗʰ century. Pathology**)** — Ukrainian pathologist, who arrived in Lviv from the mainland Ukraine, became impressed with academic achievements of Bolesław Jałowy, remained friendly to him, and offered him the position of Dean of the Medical Department in new reorganized institution which he accepted.

Unfortunately, during the German invasion and occupation of Ukraine (1941-44), Bolesław Jałowy while walking on the Lviv street with Jan Chmiel — student of medicine, on October 01, 1943, was shot by a member of the Organization of Ukrainian Nationalists (OUN, established 1929), possible by the son of **Marian Panchyshyn (1882 – 1943**: 20ᵗʰ century. Internal medicine**)** — prominent Ukrainian physician.

The assassination of Bolesław Jałowy was revenge for previous, September 11, 1943, assassination on the Lviv street of **Andrij M. Lastoweckyj (31.VIII.1902, city of Ivano-Frankivsk, Ukraine** – shoot **11.IX.1943, city of Lviv, Ukraine)**[610] — Ukrainian physicist, Philosophies Doctor (Ph.D., 1927), assistant (1929-1939), Professor (1939-43), Chief of Department of Experimental Physics (1939-43), and Dean (1942-43) of Danylo Halytsky LNMU. The crime was made on groundless accusation or unsubstantiated suspicions of not admitting Poles for medical studies in Lviv, on the order of Directorate of Diversion (in Polish – **Kie**rownictwo **Dyw**ersji, i.e., «kedyw»), by a member of the Home Army (in Polish – Armia Krajowa, active February 14, 1942 – January 19, 1945).

A clear and conceptual understanding of human anatomy for medical students and physicians was created by **Frank H. Netter (1906 – 91)**[611] — American surgeon and medical illustrator, in his «Atlas of Human Anatomy» (1-7 Eds), published between 1989 and 2018.

Initially in 1966 **Anthony G.E. Pearse (1916 – 2003)**[612] — English histochemist, described for the calcitonin-secreting parafollicular C-cells of dog thyroid gland (TG) using 5-hydroxytryptophan (5-HTP) as a precursor, that pepside secreting

---

[610] City of Ivano-Frankivsk (established by inclusion of two neighboring fortresses from 1435 and 1449 into new one fortress in 1663), Ivano-Frankivsk Obl. (Foot Note 197), Ukraine. — Gail OW. Fizyka na wesoło. Przełożył dr Andrzej Lastowiecki Nakładem Mathesis Polskiej. Warszawa 1936. — Lastovetsky A. The Dead Nature. Handbook for the 5ᵗʰ grade the basic schools. State Publishing of the School Books in Lviv, 1938. — Otto W. Gail (1896-1956) — German science journalist and writer, who authored «Wir plaudern uns durch die Physik» (1931). — Ablitsov VH. «Galaxy Ukraine». The Ukrainian diaspora: prominent persons. Publ. «Kyt», Kyiv 2007.

[611] Netter FH. Atlas of Human Anatomy. 1ˢᵗ Ed., Ciba-Geigy, Submit, NJ 1989. — Netter FM, Friedlaender GE. Frank H. Netter, MD, and a brief history of medical illustration. Clinical Orthopaedics and Related Research. 2014 Mar;472(3):812-9.

[612] Pearse AG. 5-hydrotryptophan uptake by dog thyroid «C» cells, and its possible significance in polypeptide hormone production. Nature 1966;211:598-600. — Pearse AG. The cytochemistry and ultrastructure of polypeptide hormone-producing cells of the APUD-series and the embryologic, physiologic, and pathologic implications of the concept. J Histochem Cytochem 1969;17(5):303-13.

endocrine cells share ability to take up precursors of biologically active amines, to produce active amine through subsequent intracellular decarboxylation and then store the amine product in secretory vesicles, hence amine precursor uptake, and decarboxylation (APUD) cells. This kind of uptake was also found in other peptide hormone-secreting cells, for example as corticotropic hormones of the anterior pituitary gland, the gastrointestinal (GI) tract, the beta-cells of the pancreas, and in more than 40 endocrine cell types, defined as a diffuse neuroendocrine system, the third branch of the nervous system acting within the second, the autonomic nervous system in the control of all internal organs and with a developmental lineage from neuroectoderm

To put out in 1864 by **James Pettigrew (1834 – 1908**: 19th century. Anatomy, histology and embryology) concept of the laminal nature of the myocardium was supported by **Francisco Torrent-Guasp (Oct. 07, 1931, Gandia, Province of Valencia, Spain – 2005, Madrid, Spain)** — Spanish cardiologist, who discovered the ventricular myocardial band (VMB), conceiving the biventricular myocardium as a continuous muscular band, extending from the pulmonary artery (PA) to the aorta with a basal and apical loop, a double helix derived from the spiral fold and an apex in a figure-8 configuration (VMB of Torrent-Guasp or helical heart, 1972).

In his research in 1970-80 **Günter Blober (1936 – 2018)** — German American biologist, discovered that proteins have intrinsic signals governing transport and localization.

In 1971 **Alexey M. Olovnikov (1936 –)** — Russian biologist was the first to recognize the problem of shortening the telomere (in Greek – nouns «telos» meaning «end» and «meros» meaning «part») – a structure at the end of chromosomes that protects it; the first to predict the existence of telomerase, also called terminal transferase – ribonucleoprotein, the enzyme that replenishes the telomere by adding a species-dependent telomere repeat sequence to the 3'end of telomeres; and was the first to suggest telomere hypothesis of aging and relationship of telomeres to cancer.

In 1976 **André J.M.G. Moulaer**t[613] — Belgian Netherlandish pediatric cardiologist, described a muscle in the left ventricular (LV) outflow tract between the left main coronary artery (LMCA) left semilunar cusp of the aortic valve (AV) and the anterior leaflets of the mitral valve (MV), named the anterolateral muscle bundle of the LV, or Moulaert muscle, which is present in about 40% of normal hearts. It is suggested that this bundle is a muscular remnant of the left extremity of the bulboatrioventricular

---

[613] Moulaert AJ. Ventricular Septal Defects and Anomalies of the Aortic Arch. Thesis. University of Leide 1974. — Moulaert AJ, Oppenheimer-Dekker A. Anterolateral muscle bundle of the left ventricle, bulboventricular flange and subaortic stenosis. Am J Card Jan 1976;37(11):78-81.

flange, and it can be very prominent, causing obstruction of both the LV inflow and outflow tracts.

Laland H. Hartwell (1939 –) — American biologist, discovered protein molecules that control the division (duplication) of cells (1967-70). Sir **Richard T. Hunt (1943 –)** — British biochemist and molecular physiologist, using the sea urchin (Arbacia punetulata) discovered the cyclin molecule – a family of protein that control the progression of cells through cell cycle by activating cyclin dependent kinase (cdk) enzyme (1982). Sir **Paul M. Nurse (1949 –)** — English geneticist, identified the gene cdk1 or cell division cycle protein homolog (cdk)2 – an essential factor in cell cycle regulation, in fission yeast that control the progression of the cell cycle from growth (G)1 phase – the cell divisions that takes place in eukaryotic cell division, to synthesis (S) phase – the part of the cycle in which deoxyribonucleic acid (DNA) is replicated, occurring between G1 and G2 phase, and the transition from G2 phase to mitosis (M) (1973-76); identified the homologous gene in human, which codes for a cdk (1987). Together Laland Hartwell and Richard Hunt discovered two checkpoint proteins, cyclic and cdk that check and control the transition from one stage to another. If the cell does not divide correctly, other proteins will attempt to repair it, and if unsuccessful, they will destroy the cell. But, if the cell divides incorrectly and survive, it can cause cancer or other serious disease (2001).

Sidney Altman (May 07, 1939, city of Montreal, Province Quebec, Canada – Apr. 05, 2022, borough Rekleigh, Bergen County, New Jersey, USA)[614] — Canadian American molecular biologist of Jewish Ukrainian descent, and **Thomas R. Cech (1947 –)** — American chemist of Czech origin, discovered in 1989-2007 catalytic properties of ribonucleic acid (RNA).

The discovery and a significant contribution to the development of cryoelectronic microscopy (cryo-EM) were made by **Joachim Frank (1940 –)** — German American biophysicist, **Jacques Dubochet (1942 –)** — Swiss biophysicist and biochemist, and **Richard M. Henderson (1945 –)** — Scottish molecular biologist and biophysicist.

Joachim Frank is regarded as the founder of a single-particle cryo-EM[615].

The problem with observing a group of individual molecules with electron microscopy (EM) is that intense electron beam destroy specimen. To solve this problem Joachim Frank and colleagues devised between 1975 and 1985 the method

---

[614] City of Montreal (first occupied by Indians 2000 years BC, founded 1642), Province of Quebec (area 1,542,056 km²), Canada (area 1,984,670 km²). — Borough Rekleigh incorporated 1923), Bergen County, New Jersey (N.J., area 22,591.38 км²), USA. — Ablitsov VH. «Galaxy Ukraine». The Ukrainian diaspora: prominent persons. Publ. «Kyt», Kyiv 2007.

[615] Frank J. Untersuchunngen von electromicroskopischen Aufnahmen hoher Auflösung mit Bildidfferenz- und Reconstruktions verfahren. Hochschulschrift (148 gez. Bl. mit Abb.:4). Dissertation des Doctor der Philosophie, Technische Universität Münchner, München 1970.

to observe individual molecules, that were only faintly visible with the help of the EM, by using poor quality images with a weaker electron beam. Therefore, they developed an imaging method in which the EM's fuzzy two-dimensional (2-D) images are analyzed and merged to reveal a sharp three-dimensional (3-D) structure. Subsequently, they successfully applied this approach to image enzyme glutamine synthetase (1978).

In the early 1980s Joachim Frank and **Marin G.** van **Heel (1949 –)** — Dutch biophysicist and structural biologist, devised statistical method and a software to determine a particle's 3-D structure from 2-D images, where the image of a particle is represented by vector[616].

In 1981 Joachim Frank with colleagues used the averaging technique to obtain the high-quality EM images of ribosome. In the remaining 1980s they switched to cryo-EM which uses frozen specimens and thus allows the ribosomes to maintain their shape.

In 1970-76 the group headed by Jacques Dubochet developed the method of cryo-EM for determination the structure of biomolecules with a high resolution in solution. They added liquid water to the EM that evaporated in its vacuum making the biomolecule collapse.

In 1980-81 they succeeded in vitrifying water by cooling water so rapidly that it solidified in its liquid form around a biological sample, allowing the molecules to retain their natural shape even in a vacuum[617].

In a key work from 1984, they got, by means of the cryo-EM and ultrafast freezing into the layer of amorphous ice, an image of adenovirus[618].

Using the EM Richard Henderson together with **P. Nigel T. Unwin (1942 –)** — New Zealander and English neuroscientist, established low resolution structural model for bacteriorhodopsin, showing the albumen's protein to consist of seven transmembrane alpha-helices (1975)[619].

In 1990 Richard Henderson and colleagues succeeded in using the EM to generate a 3-DM image of protein at atomic resolution[620].

---

[616] Frank J, van Heel M. Intelligent averaging of single molecule using computer alignment and correspondence analysis: I The basic method. Elect Micros 1980;2:690-3. — van Heel M, Frank J. Use of multivariate statistics in analyzing the images of biological macromolecules. Ultramicroscopy 1981;6:187-94.

[617] Dubachet J, McDowall A. Vitrification of pure water for electron microscope. J Microbiology Dec 1981;124(3):3-4. Dubachet J, Lepault J, Freeman R, et al. Electron microscopy of frozen water, and aqueous solution. J Microscopy Dec 1982;128(3):219-237.

[618] Adrian M, Dubochet J, Lapault J, et al. Cryo-electron microscope of viruses. Nature 1984;308(7568):32-6.

[619] Henderson RM, Unwin PNT. Three-dimensional model of purple membrane obtained by electron microscope. Nature 1975;257:28-32.

[620] Henderson RM, Baldwin TA, Ceska F, et al. Model for the structure of bacteriorhodopsin based on high-resolution electron cryomicroscopy. J Mol Biol 1990;213:899-929.

By 2013, and now 3-D structures of biomolecule can routinely be achieved.

Sir **Martin J. Evans (1941 –)** — British biologist, who with **Mathew Kaufman (1942 – 2013)** — British anatomist, and embryologist, were the first to cultivate mice embryonic stem cells (SC) and cultivate them in laboratory (1981).

Liudmila O. Stechenko (05.XI.1943, town of Tarashcha, Bilotserkivsky Region, Kyiv Obl., Ukraine –)[621]— Ukrainian scientist-medic in the field of histology and embryology, specialist the field of electron microscopy (since 1968), in dissertation on the candidate of medical science entitled «Ultrastructure of endothelium of the blood capillaries of the myocardium in congenital and acquired diseases of the heart» (National Kyiv Medical University, 1974) first described particular morphologic formations during decompensation of the myocardium in cardiac diseases. Has six (6) patents for discoveries and two (2) rationalization propositions.

In 1984 **Elizabeth H. Blackburn (1948 –)** — Australian American molecular biologist, **Jack W. Szostak (1952 –)** — Canadian American biologist of English Polish descent and **Carol W. Greider (1961 –)** — American molecular biologist, discovered telomerase.

While working in the field of fluorescence microscopy and photoactivated microscopy **William E. Moerner (1953 –)** — American physical chemist, **R. Eric Betzig (1960 –)** — American physicist, and **Stefan W. Hell (1962 –)** — Romanian German physicist, developed super-resolution fluorescence microscope. In this method fluorescence in individual molecules is steered by light, and an image of very high resolve is achieved by combining images in which different molecules are activated, making possible to track processes occurring inside living cells and to study the division in human embryos in three-dimension (3-D) formation (c. 2006).

Physiology

In 1902 **Charles R. Richet (1850 – 1935)** — French physiologist, coined the term anaphylaxis (in Greek – «ana» meaning «against» and «phylaxis» meaning «protection») in humans or animals manifested by an acute, systemic (multi-organ)

---

[621] Town of Tarashcha [inhabited since Enolite Age, founded 1709; greatly suffered from sudden attacks and bestiality of the 1st cavalry army of Communist Fascist RSFSR under command of Semion M. Budionnyy (1883-1973) between 01-15 October 1920], Bilotserkivsky Region, Kyiv Obl., Ukraine. — Stechenko LO. Radiation, heart, and laser. Kyiv 1996. — Stechenko LO. Radiosensitivity and membranes of lymphocytes. Kyiv 2001. — Stechenko LO. The Structure of the Immune System after Action of Small Doses of Ionizing Radiation. Kyiv 2008. — Stechenko LO. Heart at Hypothyroidism. Kyiv 2008. — Ablitsov VH. «Galaxy Ukraine». The Ukrainian diaspora: prominent persons. Publ. «Kyt», Kyiv 2007. — Petrenko VA, Melnyk HO. Professor Stechenko Liudmila Oleksandrivna (to the birthday jubilee). Klinichna Anatomiia ta Operatyvna Chirurhiia 2008;7(4):112-3. — Synyachenko OV. They glorified themselves and Ukraine: history of domestic therapy. Crimean Therapeutic Journal 2010;2(2):21-8.

and severe allergic reaction with increased sensitivity to the repeated arrival of foreign substances, most commonly of protein origin acting as an antigen. Sometimes, a second small dose may lead to lethal reaction. The most severe type of anaphylaxis is an anaphylactic shock which usually causes death in a matter of minutes if not treated. Certainly, his research helped elucidate hay fever, asthma, and some other allergic reaction to trigger substances and explained some previously not understood cases of intoxication and subsequent sudden death.

The data of **Bronislav (Veriho) Werigo (1860 – 1925)** [19th century. Physiology] of 1892 about the displacement of the oxyhemoglobin (Hb / Hgb O_2) dissociation curve by a change in the temperature (T), concentration and partial pressure (P) of hydrogen (H^+), carbon dioxide (CO_2), 2,3-diphosphoglyceric acid (2,3-DPG) of the red blood cells (RBC's) or erythrocytes and the pH in the blood capillaries, were enhanced by **Christian Bohr (1955 – 1911)** [19th century. Physiology] in 1904, **John S. Haldane (1860 – 1936)** [19th century. Physiology] in 1905, and Sir **Joseph Barcroft (1872 – 1947)**[622] — British physiologist, in 1910.

Increase in the T, pH, pCO_2, quantity of 2,3-DPG of RBCs and decrease of the pH in the blood capillaries decreases the Hb affinity to oxygen (O_2), resulting in the HbO_2 dissociation curve being displaced to the right, which improve the shift of O_2 from the capillaries to the tissues (Veryho-Bohr effect, 1892, 1904)[623].

Decrease in the T, pH, pCO_2, quantity of 2,3-DPG of RBCs and increase of the pH in the blood capillaries increases the Hb affinity to O_2, resulting in the HbO_2 dissociation curve being displaced to the left, which impair the shift of O_2 from the capillaries to the tissues (Haldane effect, 1905).

The term synapse (in Greek – «synapsis», meaning «conjunction») was introduced in 1897 by **Charles S. Sherington (1857 – 1952)** — English histologist and neurophysiologist, for a structure which permits a neuron or nerve cells to pass a chemical or electrical signal to another neuron. He has made an important contribution into understanding of functional integration within the brain, showing that every posterior nerve root supplies a special region of the skin and that when a muscle receives a nerve impulse to contract its antagonist receives an impulse to relax, thus demonstrated the action of reflex unit within the central nervous system (CNS) – reciprocal innervation or Sherrington's law (1904)[624].

The vitamin B_1 (thiamine) was discovered in 1906 by **Christian G. Eijkman (1858 – 1930)** — Dutch physiologist and physician, is used for the treatment of an acute and chronic neuritis and polyneuritis and degenerative diseases of the nervous system.

---

[622] Bancroft J. The Respiratory Function of the Blood. Cambridge Univ Press, London 1928.

[623] Bronislav Werigo assumed that CO_2 and O_2 compete for binding on Hgb which cannot be derived from the existence of a CO_2 - Hgb complex as experimentally shown by Christian Bohr.

[624] Sherrington CS. The Integrative Action of the Central Nervous System. 1st Ed. Oxford Univ Press 1906.

In 1902 **William M. Bayliss (1860 – 1924**) — English physiologist, and **Ernest H. Starling (1866 – 1927)** — English physiologist, discovered a hormone secretin which stimulates the pancreatic secretion and intestinal peristalsis. To Ernest Staring belongs the formulation of the cardinal hypotheses and laws related to fluid transfer and cardiac hemodynamics. According to this hypothesis, the direction and rate of fluid transfer between blood plasma in the capillary and fluid in the tissue spaces depends on the hydrostatic pressure on each side of the capillary wall, on the osmotic pressure of protein in plasma and in tissue fluid, and on the properties of the capillary wall as a filtering membrane. In 1912-18 he re-stated the previous (1895) findings of **Otto Frank (1865 – 1944**: 19th century. Physiology**)** that the stroke volume (SV) ml/ per each beat, cardiac output (CO) L/min. or cardiac index (CI) L/min./ body surface area (BSA) in m^2 is directly proportional to the diastolic filling that is the left ventricular end-diastolic volume or pressure (LVEDV, LVEDP), and the law of heart according to which the energy which is set free at each contraction of the heart is a single function of the length of the fibers composing its muscular walls (Frank-Starling law)[625].

Sir **Frederick G. Hopkins (1861 – 1947)** — English biochemist, discovered tryptophan (1903); proved that vitamins A and D, help to attain normal growth in animal (1912)[626]; discovered glutathione (1921) – an antioxidant among plants, animals, some bacteria and archaea (in Greek – «archaria» meaning «ancient», i.e., unicellular prokaryotes), an exogenous antioxidant produced by the cells, participating in the neutralization of free radicals and reactive oxygen compounds, as well as maintaining exogenous antioxidants such as vitamin C and E in their reduced (active) form, regulate the nitric oxide (NO) cycle which is critical for life, enhances the function of the immune system.

In his work on optics and dioptric, **Alvar Gullstrand (1862 – 1930)** — Swedish ophthalmologist and optician, elucidated the formation of optical imaging in the eye and incorporated it in the general laws governing optical image formation (1899-1919).

The hormonal function of the antrum of the stomach was discovered by **John S. Edkins (1863 – 1940)**[627] — English physiologist, in 1905. The hormone is a tissue factor (gastrin) which stimulates hydrochloric acid (HCl) secretion.

---

[625] Paterson S, Starling E. On the mechanical factors which determine the outputs from ventricles. J Physiol (London) 1914;48:357-79. — Paterson S, Piper H, Starling E. The regulation of the heartbeat. J Physiol (London) 1914;48:465-513.

[626] Hopkins FG. Feeding experiments illustrating the importance of accessory factors in normal dietaries. J Physiol 1912;44(5-6):425-60.

[627] Edkins JS.The chemical mechanism of gastric secretion. J Physiol (London) 13 Mar 1906;34 (1-2): 133-44. — Modlin IM, Kidd M, Marks IN, et al. The pivotal role of John S. Edkins in the discovery of gastrin. World J Surg 1997;21(2):226-34.

The hearts of sacrificed animals were resuscitated by **Aleksei (Oleksiy) A. Kulabko (12.03.1866, city of Omsk, Russia – 06.08.1930, city of Moscow, Russia)**[628] — Ukrainian Russian physiologist, by topical cooling (freezing) while brains were resuscitated by an artificial circulation **Frank S. Locke (1871 – 1941)** — English physiologist, solution, containing sodium chloride (NaCl), calcium chloride (CaCl$_2$), potassium chloride (KCl), magnesium chloride (MgCl$_2$), sodium bicarbonate (NaHCO$_3$), d-glucose and water. Aleksei Kulabko was able to revive a child's heart 20 hours after death from pneumonia (1902).

Expanding the clinical research of **Willem Einthoven (1860 – 1927**: 19th century. Physiology), **Olexandr F. Samoilov [27.III.(07.IV).1867, city of Odesa, Ukraine – 22.VII.1930, city of Kazan, Tatarstan, Russia]**[629] — Ukrainian physiologist, designed electrocardiography (ECG) with one lead and applied it for the diagnosis of mitral stenosis (MS) and ventricular hypertrophy of the heart, and applied the capillary galvanometer to investigate electrical phenomenon during excitation of the skeletal muscles (1908).

He determined the temperature coefficient for the transmission of a nerve impulse from a nerve to a muscle and demonstrated that this is a chemical process (1925).

In 1900-01 **Karl Lansteiner (1868 – 1943)** — Austrian biologist, physician, and immunologist of Jewish ancestry, discovered the human ABO group system, based on identification of the presence of agglutinins in the blood. The fourth AB group was discovered in 1907, by **Jan Jansky (1873 – 1921)** — Czech serologist, neurologist and psychiatrist, who proposed a classification of the blood group using Roman numerals while **Emil** von **Dungern (1867 – 1961)** — German internist and **Ludwig Hirszfeld (1884 – 1954)** — Polish microbiologist and serologist of Jewish ancestry, based their classification on the Latin alphabet as follow: I (O), II (A), III (B) i IV (AB); both discovered the heritability of the ABO blood group (1910-11). The Rhesus (Rh) factor was discovered by Karl Lansteiner and **Alexander S. Wiener (1907 – 76)** — American forensic physician, serologist, and immunogenetics, in 1937. Safe blood transfusions were made possible through the work of **Reuben Ottenberg (1882 – 1959)** — American physician and hematologist, who in 1907 found, that patient antibodies against donor cells could be harmful but not vice

---

[628] City of Omsk (founded 1716), Russia. — Kuliabko AA. Experience in revival of the heart. Izv Pos AN 1902;16(5):175. Ablitsov VH. «Galaxy Ukraine». The Ukrainian diaspora: prominent persons. Publ. «Kyt», Kyiv 2007.

[629] Samojloff A. Beiträge zur Elektrophysiologie des Herzens. Arch f. Anat. u. Physiol. 1906. — Samojloff A. Elektrokardiogramme. Verlag von Gustav Fisher, Jena 1909. — Samojloff A. Weitere Beiträge zur Elektrophysiologie des Herzens. Pflügers Arch für die gesamte Physiologie des Menschen und der Thiere. 1910 Dez;105(8-10):417-68.— Samoilov AF. Selected Papers and Speeches. Moscow – St Petersburg 1946. — Samoilov AF. Selected Works. Moscow 1967. —Ablitsov VH. «Galaxy Ukraine». The Ukrainian diaspora: prominent persons. Publ. «Kyt», Kyiv 2007.

versa, since 1908 began testing the blood of the donor and the recipient before transfusion (hemocompatibility test) and had coined the term «universal donor» for persons with the blood group O (I).

Olexandr V. Leontovych (20.X(01.IX).1869, city of Kyiv, Ukraine – 15.XII.1943, city of Moscow, Russia)[630] — Ukrainian physiologist, described cerebrospinal and vegetative innervation of the skin, blood vessels and other organs of warm- and cold-blooded animals (in Latin – plexus nervosum autonomous peripheric, 1900), which is capable of regenerating itself (1906-37); proposed the theory of the neuron as a source of alternative electric current (1928-39); developed the method of in vivo staining and fixation of the nervous structures using methylene blue (1939).

The method of the cells culture outside the body by mounting a fragment of tissue in the clotting lymph fluid on a cover slip of a hanging drop (concave) slide containing an artificial tissue culture which **Ross G. Harrison (1870 – 1952)** — American biologist and anatomist, devised in 1907 to determine how the nerve fibers regenerate after injury, became essential in virology, and thoracic and cardiovascular surgery (TCVS) research. By successfully culturing the frog's neuroblasts in the lymph node (LN) medium he took the first step toward research on precursors and stem cells.

In 1906 **Georg (George) L. Zülzer** or **Zuelcer (1870 – 1949)** — German physician, used pancreatic extract on diabetic dogs, and a patient in diabetic coma with temporary improvement. Then in 1916 and 1921 **Nicolae Paulescu (1869 – 1931)** — Romanian physiologist, used pancreatic extract to normalize blood sugar in a diabetic dog, and in 1919 **Israel S. Kleiner (1885 – 1966)** — American biochemist of Jewish descent, demonstrated the effect of extract from the pancreas on animals causing hypoglycemia. Finally, 1921-22 **John J.R. Macleod (1876 – 1935)** — Scottish physiologist and biochemist, together with Sir **Frederick G. Banting (Nov. 14, 1891,** settlement **Alliston, Ontario, Canada** – died of wounds and exposure to the fire next day after plane crush **Feb. 21, 1941,** town of **Musgrave Harbor, Island of Newfoundland (NF,** area 108,860 km²), **Canada)**[631] — Canadian physician and surgeon, **James B. Collip (1892 – 1965)** — Canadian biochemist, and **Charles H. Best (1899 – 1978)** — Canadian physiologist, isolated insulin from the bovine/dog pancreas that saved the life of a 14-year-old diabetic patient.

[630] Ablitsov VH. «Galaxy Ukraine». The Ukrainian diaspora: prominent persons. Publ. «Kyt», Kyiv 2007.
[631] Village Alliston (beginning of settlement 1821), Simkoe County (area 4,859.64 km²), Province of Ontario (ON, area 1,076,395 km²), Canada. — Town of Musgrave Harbour (settled 1834), Island of Newfoundland (NF, area 108,860 km²), Canada.

Walter B. Cannon (1871 – 1945) — American physiologist who coined the term «fight or fight response» (1915), expanded **Claude Bernard's (1813 – 78**: 19th century. Physiology) concept of hemostasis (1932).

Vasyl Y. Chahovets [18(30).IV.1873, manor Patychykha near village Zaruddia, Romensky Region, Sumy Obl., Ukraine – 19.V.1941, city of Kyiv, Ukraine][632] — Ukrainian physiologist, applied the **Svante Arrhenius (1859-1927**: 19th century. Physiology) theory of electric dissociation into the liquid environment (medium) of living organism, creating the ionic (1896) and compensatory (1903) theory of the electric conduction and stimulation in living tissues, described the electrogastrography (1935).

Beside the law regulating the blood pressure (BP) and cardiac output (CO) of **Otto Frank (1865 – 1944**: 19th century. Physiology) and **Ernest Starling (1866 – 1927**: above), the blood volume is regulated by the low pressure stretch reflex[633], and the arterial (systemic) BP is regulated by high pressure baroreceptor reflex[634].

Frances A. Bainbridge (1874 – 1921) — English physiologist, noted that an increase in circulating blood volume stimulates stretch receptors in the atria, signaling the medulla ablongata to decrease the tonus of the parasympathetic nervous system via cardiac branches of the cranial 10th (vagal) nerve, thus causing by preference the sympathetic system to increase the heart rate (HR) – atrial or Bainbridge stretch reflex (1915).

In contrast, **Ivan De Burgh Daly (1883 – 1974)** — English physiologist, published on increase in the systemic blood pressure (BP) due to stimulation of the high pressure baroreceptors in the carotid body (sinus) innervated by the cranial 9th (glossopharyngeal) nerve, and the aortic arch innervated by the vagal nerve, signaling the brain stem to increase the tonus of the parasympathetic nervous system, causing the HR to decrease (the baroreceptor or De Burgh Daly reflex, 1927).

The resultant change in the HR is the sum of the competing reflexes help to maintain the systemic BP at the nearly constant level.

Together, **Joseph Erlarger (1874 – 1965)** — American physiologist, with **Herbert S. Gasser (1888 – 1963)** — American physiologist, identified in 1927 several varieties of nerve fibers, developed the apparatus for measurement of electrical charges that passes through the nerve fiber, establishing that each nerve fiber has their own

---

[632] Manor Patychykha near village Zaruddia (known from beginning of 19th century; suffered from conducted by Russian Communist Fascist regime of Holodomor-Genocide 1923-33 and 1946-47 against the Ukrainian Nation), Romensky Region, Sumy Obl., Ukraine. — Ablitsov VH. «Galaxy Ukraine». The Ukrainian diaspora: prominent persons. Publ. «Kyt», Kyiv 2007.

[633] Bainbridge F. The influence of venous filling upon the rate of the heart. J Physiol 1915;50:65-84.

[634] De Burgh Daly I, Verney EB. The localization of receptors involved in the reflex regulation of the heart rate. J Physiol 1927;62:330-40.

potential for excitability, and that the velocity the action potentials (AP) was directly proportional to the diameter of nerve fiber.

S. August S. Krogh (1874 – 1949) — Danish physiologist, discovered the mechanism of regulation of capillaries in skeletal muscle by the adaptability of blood perfusion in muscle and other organs according to demands through opening and closing arterioles and capillaries (1922)[635], estimated the distance between capillaries which nutrients diffuse to the surrounding cells, based on cellular consumption of the nutrients (Krogh's length), and proposed the physiology concept for a new era that «For such a large number of problems there will some animals of choice or a few such animals on which it can be most conveniently studied» (Krogh's principle, 1929)[636].

The existence of the second pancreatic hormone, beside of insulin, was suspected in 1923 when **Charles P. Kimball** — American student in biochemistry, later physiologist, and **John R. Murlin (1874 – 1960)**[637] — American physiologist, were studying methods of insulin extraction from the pancreas and found that acetone precipitated a fraction soluble in 95% alcohol, a reaction which separated the unknown substance from insulin. Injection of this free from insulin fraction into dogs and rabbits caused a rapid rise in blood glucose level. This inferred that the preparation contained a second pancreatic hormone, named by them the **gluc**ose **agon**ist, hence glucagon. Glucagon is produced by the pancreatic islets by cells, differing from the insulin-produced beta-cells, presumably by the alpha-cells, and by the gastric mucosa and other parts of the digestive tract. It is the major hormone influencing the breakdown of glycogen (glycogenolysis).

In 1904 Sir **Henry H. Dale (1875 – 1968)** — English physiologist and pharmacologist, discovered oxytocin (in Greek – «quick birth»), a polypeptide hormone of nine amino-acids, produced by the posterior lobe of the pituitary gland, that stimulates contraction of the smooth muscles of the uterus and milk ejection from the female breast during feeding of newborn and infant. It also evolves feeling of contentment, romantic attachment, calmness, security around the mate, human bound and trust, decreases anxiety, fear, and stress. Commercial form of oxytocin, obtained from beef and hog pituitary glands or especially by synthesis is used in obstetrics to induce labor and to control postnatal hemorrhage.

Then, in 1914 Henry Dale discovered acetylcholine (ACh), an organic polyatomic cation that act as a chemical nerve transmitter (neurotransmitter)

---

[635] Krogh A. The Anatomy and Physiology of Capillaries. Yale Univ. Press, New Haven 1922.
[636] Krogh A. The progress of physiology. Am J Physiol 1929;90(2(:243-51.
[637] Kimball CP, Murlin JR. Aqueous extracts of pancreas. J Biol Chem 1923;58(1):337-48.

in the central and peripheral nervous system (CNS, PNS) at the neuro-muscular signaling synapses.

Subsequently, in 1921 **Otto Loewi (1873 – 1961)** — German American pharmacologist of Jewish ancestry, showed experimentally in the frog heart that ACh chemically transmits impulses from the cranial X (vagus) nerve to the cardiac muscle, causing slowing of the heart rate (HR). Thus, he proved, «...*that nerves do not influence the heart directly but liberate from the terminals specific chemical substances... characteristic of the stimulation of the nerves*».

Boris P. Babkin (1877 – 1950) — Russian physiologist, noted that pressure exerted by the examiner's thumbs on the palms of both hands of an infant resulted in the opening of the infant's mouth, except in cases of lethargy or coma (Babkin's or palmar cephalic reflex).

Jean G. Bachmann (1877 – 1959) — American physiologist, described the anterior internodal band of the sinoatrial (SA) node providing a specialized interatrial conduction to the left atrium (LA) – Bachmann bundle (1916).

In 1904 **Thomas R. Elliott (1877 – 1961)**[638] — British physiologist and physician predicted a chemical nerve transmitter (neurotransmission) and suggested that adrenaline **[George Oliver (1845 – 1915) / Edward Sharpey (1850 – 1935)**. 19th century. Physiology] act as a sympathetic neurotransmitter.

Year later (1905) **Carl J. Wiggers (1883 – 1963)** — American physiologist, illustrated the importance of using purified or synthetic form of adrenaline in bioactivity studies, and demonstrated vasoconstrictor effect of purified adrenaline on cerebral blood flow[639]. He also developed a standard diagram, named the Wiggers diagram, that is used in teaching cardiac physiology, where the «x»-axis is used to place time, while the «y»-axis contains all the following on a single grid – arterial, aortic, cardiac ventricular and atrial blood pressures (BP); ventricular volume; electrocardiography (EKG); arterial blood flow (optional); and heart sounds (optional)[640].

In 1905 **J. Winter**[641] supported by animal studies, advocated, and resuscitated the heart by direct intracardiac injection of adrenaline.

Continuing earliest studies of **Keith Lucas (1879** – killed in the middle-air plane collision crash in **1916)** — English physiologist, on the nerve action in the animal, **Edgar D. Adrian (1889 – 1977)** — English electrophysiologist, who previously proved the accuracy of the «all or none» law of nerve excitation (1913-22), using

---

[638] Elliott TR. On the action of adrenalin. Proceedings Physiol Soc. May 21, 1904:1.

[639] Wiggers C. On the action of adrenalin on the cerebral vessels. Am J Physiol 1905;14:452-63.

[640] Mitchell JR, Wang Jiun-Jr. Expanding application to teach cardiovascular physiology. Adv Physiol Educ 2014-06-01;38(2):70-5.

[641] Winter J. Wien klin Wschr 1905;18:525.

designed by the former, a capillary electrometer and cathode-ray tube to amplify the signal produced by the nervous system, accidentally discovered in 1928 the electrical discharges of a single fiber under physical stimulus, thus proved the presence of an electric current in the nervous system. Furthermore, it also showed that excitation of the skin under constant stimulus was initially strong but gradually decreased over the time, whereas the sensory impulses passing along the nerves from the point of a contact were constant in the strength but reduced in frequency over the time. Subsequently, he received such signals in the brain and special distribution of the sensory areas of the cerebral cortex which led to the creation of a sensory map, called the «homuncule», «homuncle» or «homunculus» (in Latin meaning «little man» or «dwarf») in the somatosensory system.

The electroencephalogram (EEG) and evoke potential in dog's brain was recorded by **Volodymyr V. Pravdych-Nemynsky** or **Wolodymyr W. Prawdicz-Neminski (02.07.1879, city of Kyiv, Ukraine – 17.05.1952, city of Moscow, Russia)**[642] — Ukrainian physiologist, using a stringed galvanometer (1913-25), and EEG by **Hans Berger (1873** – suicide due to depression from skin disease **1941)**[643] — German psychiatrist, in a man at which time the alpha (Berger) rhythm was recorded (1924).

Usually, EEG record of normal conscious person at rest taken from occipital region shows homogeneous rhythm waves with frequency 8-13 seconds (average, 10 second), named alpha-rhythm (alpha-waves, alpha-activity, or Berger rhythm).

---

[642] Neminsky VV. Ein Versuch der Registrierung der elektrischen Gehirnerscheinungen. Zbl Physiol 1913;27:951-60. —Neminsky VV. Sur la conaissance du rythme d'innervation. Journ. de méd. d'Ekaterinoslaw 1923;13-4. — Práwdicz-Neminski WW. Zur Kenntnis der elektrischen und der Innervationsvorgänge in den funktionellen Elementen und Geweben des tierischen Organismus. Pflug Arch ges Physiol 1925;207(1):671-90. — Práwdicz-Neminski WW. Zur Kenntnis der elektrischen und der Innervationsvorgänge in den funktionellen Elementen und Geweben des tierischen Organismus. Elektrocerebrogramm der Säugetiere. Pflüg Arch ges Physiol 1925;209(1): 362-82. — Práwdicz-Neminski WW. Zur Kenntnis der elektrischen und der Innervationsvorgänge in den funktionellen Elementen und Geweben des tierischen Organismus. Pflug Arch ges Physiol 1925;210(1):223-37. — Práwdicz-Neminski WW. Anschauliche Methode der fraktionierten Blutgerinnungsbestimmung. Z. Ges. Exp. Med 1927;54:820-25. — Pravdich-Neminsky VV. Structural modifications in the nerve during exposure to direct current. Doklady AN SSSR 1951;78(3):397-99. — Pravdich-Neminsky VV. Tonoelectrocerebrogram. Tonoelectrocerebrogram. Doklady AN SSSR 1951;79(6):1061-64. — Pravdych-Nemynskyy VV. Electrocerebrography, Electromyography and Importance of Ammonia Ions in Vital Processes of the Organism. Moscow 1958. — Coenen A, Zayachkivska O. Adolf Beck: A pioneer in encephalography in between Richard Caton and Hans Berger. Adv Cogn Psychol 2013;9(4):216-21. — Ablitsov VH. «Galaxy Ukraine». The Ukrainian diaspora: prominent persons. Publ. «Kyt», Kyiv 2007. — Hanitkevych YaV. The contribution of the Ukrainian physicians to the world medicine. Ukr Med Chasopys 2009 VII/VIII;4(72):110-15.

[643] Berger H. Ueber das electroenkephalogramm des Menschen. Arch Psychiatr Nervenkr 1929;87:527-70. — Coenen A, Zayachkivska O. Adolf Beck: A pioneer in encephalography in between Richard Caton and Hans Berger. Adv Cogn Psychol 2013;9(4):216-21.

The basal metabolic rate (BMR), synonymous with basal energy expenditure (BEE) can be based on urinary nitrogen (N) excretion or calculated from the equation derived in 1918-19, by **J. Arthur Harris (1880 – 1930)** — American botanist and biometrician, and **Francis G. Benedict (1870 – 1957)** — American chemist, physiologist, and nutritionist. The Harris-Benedict equation for males: BEE = 66,5 + (13,7 x **W**eight in kg) + (5,0 x **H**eight in cm) – (6,7 x **A**ge in years) kilocalorie (kcal)/day; for females: = 65,1 + (9,6 x **W** in kg) + (1,8 x **H** in cm) – (4,7 x **A** in years) kcal/day.

Patients without postoperative complications have a normal requirement for the BEE, mild peritonitis and long bone fractures increases BEE by approximately 25%, severe infections or multiorgan failure – by 50%, and most severe cases (e.g., more than 50% body burn) – by 100% or more.

Basal protein requirement is 0,8-1,0 gm/kg/day, which may double with severe injury or infection. The optimal N in gm /caloric (cal.) ratio is 1:120-150 for adult. Normal N to protein ratio is 1:6,25 gm. Standard enteral formulas have 1 cal./ml (cc^3) water, and approximately 50 gm of protein per liter.

The mapping of the brain in cats was carried out by **Walter P. Hess (1881 – 1973)** — Swiss physiologist, to localize the areas involved in the control of internal organs. As a result of these experiments, he discovered the functional organization of the interbrain (diencephalon), which is portion of the forebrain (pro-encephalon), as the coordinator of activities of the internal organs (1924). The diencephalon consists of the thalamus (in Latin – «thalamus dorsalis»; in Greek –«room» or «marriage bed»), the epithalamus, the hypothalamus and the subthalamus. According to his theory the sleep center is in the hypothalamus.

Claude G. Douglas (1882 – 1963) — English physiologist, designed the measurement of dead respiratory space to tidal volume ratio (V_D / V_T), and a direct measurement of the oxygen (O_2) consumption index, range 110-150 mL / min. / m^2 (an average 125 mL / min. / m^2), necessary for the measurement of carbon monoxide (CO). In the latter, the O_2 content of expired air into the bag (Douglas) is compared with the O_2 content in the ambient air.

In 1923 **Leon A. Orbeli (1882 – 1958)** — Armenian physiologist, and **Alexander H. Ginetsynski (1895 – 1962),** documented that when the response of a nerve-muscle preparation is diminished because of fatigue, stimulation of the sympathetic nerve increases height of contractions.

Otto H. Wartburg (1883 – 1970)[644] — German physiologist and physician of Jewish origin, performed research on oxygen (O_2) consumption in sea urchin eggs after fertilization, and proved that upon fertilization the respiratory rate (RR) increases

---

[644] Wartburg OH. On the Origin of Cancer Cells. Science 1956;123(3191):309-14.

sixfold, and that iron is essential for the development of the larvae (1911). While investigating the metabolism of tumors and respiration of cells, particularly cancer cells, discovered the nature and mode of action of the respiratory enzyme (1931).

While studying cellular oxidation Otto Wartburg discovered iron containing respiratory ferment cytochrome (1932-33) and nicotinamide-adenine-dinucleotide phosphate (NADP) – Wartburg enzyme (1936-37).

Otto Warburg with **Dean Burk (1904 – 88)** — American biochemist, and **Murno Gladst** in the study of photosynthesis discover the I-quantum reaction that splits carbon dioxide (CO_2), activated by respiration.

He believed that cancer should be interpreted as a mitochondrial disease, because cancer cells growths caused by tumor cells generating energy, e.g., adenosine triphosphate (ATP) mainly by anaerobic breakdown of glucose, known as fermentation, or anaerobic respiration. Contrary, the healthy cells generate energy mainly from oxidative breakdown of pyruvate which is the end product of glycolysis, and is oxidized within the mitochondria.

Therefore, he assumed that «*Cancer, above all other diseases, has countless secondary causes. But, even, for cancer, there is only one prime cause. Summarized in a few words, the prime cause of cancer is the replacement of the respiration of O_2 in normal body by fermentation of sugar*».

He was convinced that cancer resulted from pollution of the environment, and this induced him to advocate organic food, particularly tea and butter.

Heorhii V. Folbort [23.I.(04.II)1885, city of St Petersburg, Russia – 17.IV.1960, city of Kyiv, Ukraine][645] — Russian Ukrainian physiologist of German descent, described empiric (in Greek – experience) generalizations about negative (restraining) conditional reflexes, explaining physiological exhaustion and restoration of function of digestion and organism after long-lasting exhaustion (Folbort rule, 1935).

Using galvanometer **Danylo S. Vorontsov [12(24).XII.1886, farmstead Pohozhne near town Poboysk, now town Slavhorod, Mohyliv Obl., Belarus – 21.XII.1965, city of Kyiv, Україна]**[646] — Ukrainian electrophysiologist, studied the

---

[645] Ablitsov VH. «Galaxy Ukraine». The Ukrainian diaspora: prominent persons. Publ. «Kyt», Kyiv 2007. — Synyachenko OV. They glorified themselves and Ukraine: history of domestic therapy. Crimean Therapeutic Journal 2010;2(2):21-8.

[646] Farmstead Pohozhne near town of Poboysk, now town of Slavhorod (mention in 1136), Mohyliv Obl. (area 29,138 km²), Belarus. — Vorontsov DS. Inhibiting influence of vagal nerve on the heart. Dissertation for the title of DMS. St Petersburg 1912. — Vorontsov DS. Veranderunger der T-Zacke in V.E.G. in Abhangigkeit vonder Lageder ableitenden Elektrodenam Herzen (Volsufige Mitteilung). Zbl Pzysiol 1914;28(2):305-07. — Vorontsov DS. Formveranderungen des V.E.G. in Abhangigkeit von der Lage der ableitenden Electrode am Herzen. Pfugers Arch 1915;160(3):581-92. — Vorontsov DC, Yemchenko AI. Physiology of Animals and the Human: Handbook. Publ. «Radianska shkola», Kyiv 1952. — Ablitsov VH. «Galaxy Ukraine». The Ukrainian diaspora: prominent persons. Publ. «Kyt», Kyiv 2007.

electric current action potential (AP) of the cardiac muscle, function of the cranial X (vagal) nerve (1913-16); proved influence of chlorides of different cations, sugars and narcotics o stability (refractivity) of nerve fibers under influence of continuous electric current (Odesa 1917, published 1924). Concluded that excitability can be reversed by action on anode by continuous electric current, studied the nature of slow electric oscillations in the central nervous system (CNS, Kyiv 1936-65).

In the field of musculo-skeletal thermodynamics, **Archibald V. Hill (1886 – 1977)**[647] — English physiologist, showed that running a dash relies on energy stories which afterwards is replenished by increased oxygen (O_2) consumption, elucidated the processes whereby mechanical work is produced in muscles, and introduced concept of maximal O_2 uptake and O_2 dept (1922); proved that the muscle contraction and conduction of a nerve impulse is associated with the increase of heat production (1926).

In 1915-17 **Carey P. McCord (1886 – 1979)** and **Floyd P Allen**[648] — American physiologists, established an influence of the pinocytic substance in the pineal gland of the intermediary part of the pituitary gland or hypophysis at the base of the brain, upon grow and differential process in tadpoles. Out of this work have come evidence of pineal gland influence upon pigmentation, upon the phases of colloidal state, and upon the vegetative nervous system. Specifically, they discovered that feeding extract of the pineal gland to cows lightened tadpole skin by contracting the dark epidermal melanophores[649].

Later, in 1958 **Aron B. Lerner (1920 – 2007)** — American physician-dermatologist, led the team of researchers who isolated and named hormone from bovine pineal gland extract as melatonin (N-acetyl-5-methoxytryptamine; in Latin – «melanos» meaning «black») which may be useful in treating skin diseases.

---

[647] Hill AV, Long CNH, Lupton H. Muscular exercise, lactic acid, and the supply and utilization of oxygen. Proc R Soc B: Biological Sciences 1924 Sept 1;96(679):438-75.

[648] McCord CP, Allen FD. Evidence associating pineal gland function with alteration in pigmentation. J Exp Zool 1917 Jan;23:267.

[649] Melanophores are pigment-containing and light-reflecting cells or group of cells of many animals including amphibians, fish, reptiles, crustaceaus and cephalopodas (in Greek – «cephalopoda» meaning «head-feet»), while mammals and birds have a class of cells called melanocytes for coloration.

Valentina V. Radzimowska (01.X.1886, village of Matiashivka, Lubensky Region, Poltava Obl., Ukraine – 22.XII.1953, town of Champaing, IL, USA)[650] — Ukrainian physiologist, constructed the precise electrode for the pH measurement of tissue cultures (1923).

During his student years in 1916, **Jay F. McLean (1890 – 1957)** — American physician, discovered a blood anticoagulant heparin in mast cells while researching dog liver (in Greek – «hepar»). Its was named heparin and introduced into clinical use by **William H. Howell (1860 – 1945)** — American physiologist, two years later (1918) established the trophic influence of the nervous system on the penetration of the vessels and the tissue (1928), de-sensibilization and anti-inflammatory influence of the extract of the hypohysis (1935).

The investigations performed in 1920 and 1935 on two dog model by **Corneille J.F. Hymans (1892 – 1968)** — Belgian physiologist, revealed that: (1) the stretch baroreceptors in the carotid sinus of the common carotid artery (CCA) bifurcation and

---

[650] Village of Matiashivka (founded before 1866), Lubensky Region, Poltava Obl., Ukraine. — Town of Campaign (founded 1855), State of Illinois (IL; area 149,968 km^2), USA. — Radzimovska V. About importance of the environment reaction for tissue cells, cultured outside the organism. Likarska sprava / Vrach delo 1922. — Krontovsky AA, Radzimovska, VV. On the influence of changes of concentration of the H$^+$ resp. OH$^-$ ions on the life of the tissue cells of vertebrates. The Journal of Physiology 1922;56 (5): 275-82. — Radzimowska W. Eine Ansatzelektrode zur pH Bestimmung in Festen Nährboden. Bioch Ztscht 1923;142(1-2). — Radzimovska VV. Discussion of lecture by VG Shtefko. About influence of starving on physical development of growing up generation of Russia. Zhurnal for studying early children ages 1923;2(1/2):20. — Radzimovska V. About influence of hydrogen ions in the life of organism. Vrach delo 1924;8-9. — Radzimowska WW, Jazimirska MC. Über die Bestimmung der Wasserstoffionen-Konzentration in Einzelnen Bakterienkolonien. Klin Wochenschrift 1925;4 (2):72–73. — Radzimovska V. About possibility of dependency of tissue breathing from an active reaction of environment. Ukr med visti 1928;7-8. — Radzimovska V. Organization of Physiologist Society Branch in Kyiv. Bulletin of Kyiv Section of Workers-Scientist 1929:3:7-8. — Radzimovska V. Lactic acid in tuberculosis patients. Probl tuberk 1936; 4. — Radzimowska W, Iwanow W. Die Alkalireserven des Blutes und die periodische Tätigkeit des Verdauungsapparates. Ztscht ges exp Med 1927;55:103-6. — Radzimovska V, Balinska YeB, Chrnysheva ZYu. Study of experimentally induced alkalosis in animals and observation upon an alkalotic direction of metabolism in men. Fiziol J USSR CCCP 1937;22(6):863-70. — Radzimovska VV, Vydro ED, Odryna SI, Rybinsky SV. In book. Allergy, Kyiv 1938. C. 134. — Radzimovska VV, Vorobiev NA, et al. Perception upon changes composition of calcium, phosphorus, and phosphates in the blood of patients with bone-joint tuberculosis during course of heliotherapy. Heliotherapy in bone-joint tuberculosis. 1939;100. — Radzimovska VV, Nichkevych OI. Normality in course of tissue reaction in patients with bone-joint tuberculosis during heliotherapy. Trudy Ukr inst traumatology and orthopedics. Kyiv 1939;26. — Radzimovska V. Arterial blood oxygen in pulmonary tuberculosis. Klin med 1939. — Radzimovska V. Physiology of Humans and Domestic Animals. Handbook. Vol. 1-2. Ukrainian Technical-Agricultural Institute (UTAI), Munich 1948-49. — Osinchuk R. Professor dr med. Valentyna Radzimowska. J Ukr Med Assoc North Am 1954;1,1:48-50. — Rozhin I. Valentyna Radzymovska: short outline of life, scientific work, and community activity. Ukrainian Free Academy of Science (UFAS), Winnipeg, MB 1968;15:48. — Hanitkevych YaV. Valentyna Ivanovna Radzymovska. Фізіол ж 1994;40(1):118-21. — Ablitsov VH. «Galaxy Ukraine». The Ukrainian diaspora: prominent persons. Publ. «Kyt», Kyiv 2007. — Hanitkevych YaV. The contribution of the Ukrainian physicians to the world medicine. Ukr Med Chasopys 2009 VII/VIII;4(72):110-15. — Dontsov D. In Memoty of V.V. Radzimovska. In, Dontsov D. Selected Works (Ed. O. Bahan). Vol. 1-10. VF «Vidrodzhennia», Drohobych – Lviv 2010. Vol. 10.

the aortic arch (AA) triggered by changes in the arterial blood pressure (BP) and the heart rate (HR) with each arterial pulse causing repolarization and action potentials (AP) which transmit the information via the cranial IX (glossopharyngeal) and cranial X (vagus) nerves, respectively, to the neurons within the solitary sensory nucleus in the brain stem, and by the axons reflex effect via the autonomic nervous system and hormone secretion targets the heart and the blood vessels for appropriate counter reaction; (2) the chemoreceptors of the carotid sinus of the CCA bifurcation and AA bodies detect the arterial blood oxygen (O_2) partial pressure (PO_2), to a lesser degree carbon dioxide (CO_2) partial pressure (Pco_2)and decrease pH, transmit the information to the respiratory center in the medulla oblongata, which sent the nervous impulses to the external respiratory muscles and to the diaphragm though the intercostal and the phrenic nerves.

That experiment demonstrated that arterial BP and O_2 content of the arterial blood are measured by the carotid sinus of the common carotid artery CCA bifurcation and the AA baroreceptors and chemoreceptors and transmitted to the brain via nerves and not by the blood itself, and that the peripheral chemoreceptors regulate the respiration.

In 1933 by **Joseph T. Wearn (1893 – 1984)**[651] — American physiologist and physician, described the trans-endocardial myocardial sinusoid allowing blood flow directly from the left ventricle (LV) into the myocardium in more primitive vertebrate heart, i.e., reptilians who do not have coronary arteries.

Isaaк Starr (1895 – 1989) — American physician, designed the ballistocardiograph for noninvasive measurement of the ballistic forces of the heart (1936-39).

In 1924 **Josef M. Balo (1896 – 1979)**[652] — Hungarian physician and neuropathologist, described that human after the pituitary gland (hypophysis) necrosis are more sensitive to insulin and the intensity of diabetes mellitus (DM) is diminished. Then, in 1930-36 **Bernardo A. Houssay (1887 – 1971)**[653] — Argentine physiologist and physician, and **Alfredo B. Biasotti (1903 – 91)** — Argentine physiologist, discovered that an animal after removal of the pancreas gland (pancreatectomy), and after removal of the pituitary gland or hypophysis (hypophysectomy), i.e., that has been hypophysectomies (Houssay's animal) are more sensitive to insulin after hypophysectomy, and that after this operation the intensity of DM in those after

---

[651] Wearn JT, Mettier SR, Klumpp TG, et al. The nature of the vascular communications between the coronary arteries and the chambers of the heart Am Heart J 1933;9:143-64.

[652] Balo J. Über Necrosen des Hypophysenvorderlappens und ihre Folgen. Beträge zur pathologishen Anatomie und zur allgemeinen Pathologe. Stuttgart 1924;72:599.

[653] Haussay BA, Biasotti A. La diabetes pancreatica de los porros hipofisoprivos. Revista de los Societad Argentina de Biologia. Buenos Aires 1930;6:251-96. — Haussay BA, Biasotti A. Pancreas diabetes und Hypophyse beim Hund. [Pflügers] Archiv für die gesampte Physiologie des Menschen der Tiere. Berlin 1931;227:664. — Haussay BA. The hypophysis and metabolism. NEJM 1936;214:961-71.

pancreatectomy is diminished (Houssay's phenomenon), and demonstrated that a hormone secreted by the anterior lobe (part) of the pituitary gland prevents the metabolism of sugar and that injection of the pituitary extract induces symptoms of DM.

In 1932 **Royal M. Calder**[654] — American physician, reported on the treatment of pituitary gland cachexia with the extract of its anterior part.

Georg von **Bekesy (1899 – 1972)** — Hungarian biophysicist, described a physical mechanism of stimulation within the cochlea of the ear (1960).

In 1940-52 (1943), both, **Emile Gagman (1900 – 79)** — French engineer and **Jaques-Yves Costeau (1910 - 97)** — French naval diver, scientist, explorer and innovator invented the diving regulator (Aqua-Lung), a.k.a. the demand valve, and with the help of Major of the United States Navy, (USN) **Christian J. Lambertsen (1910 – 2011)** — American scientist specializing in environmental and diving medicine, lead to the development of the self-contained underwater breathing apparatus (SCUBA, 1941) used for the first SCUBA equipment. The later, Christian Lambertsen, was principally responsive for development of the USN frogmen's rebreathing in the early 1940's for the underwater warfare and used the SCUBA equipment to study human physiology in high-pressure environments.

Important research concerning the physiology and the chemistry of vision by the cranial II (optic) nerve were done by **Ragnar A. Granit (1900 – 91)** — Finish Swedish neurophysiologist, **Haldon K. Hartline (1903 – 83)** — American physiologist, and **George Wald (1906 – 97)** — American neurobiologist of Jewish Ukrainian ancestry, during the years of 1934-67.

From 1920 to around 1947 Ragnat Granit main research was in the field of vision using psychophysics and electrophysiology (EP). He next took up muscular afferent nerves, in particular the muscle spindles and their motor control, passing to the spinal cord, studying their projection, and separated tonic and phasic motor neurons, establishing algebraic summation of excitation and inhibition upon those cells, finally also making use of the intracellular approach for the investigation of those and several other problems of motor control

Haldan Hartline investigating the electrical responses of the retinas of certain arthropods, vertebrates, and mollusks, concentrating on the horseshoe crab (Limulus polyphemus), using minute electrodes obtained the first record of the electrical impulses send by a single cranial 2nd (optic) nerve fiber when the receptors connected to it were stimulated by light. He found that the photoreceptor cells in the eye are interconnected in such a way that when one is stimulated, others nearby

---

[654] Calder RM. Pituitary cachexia (Simmond's disease) treated with anterior pituitary extract. JAMA 1932 Jan 23;98(4):314-5.

are depressed, thus enhancing the contrast in light pattern and sharpening the perception of shapes. Certainly, he built detailed understanding of the function of individual photoreceptors and nerve fibers inside the retina, showed how simple retinal mechanism constitute vital steps in the investigation of visual information.

George Wald discovered in 1932-33 that vitamin A is a component of the retina, showed that when the pigment is exposed to light, it yields a protein opsin and a compound containing vitamin A. This suggest that vitamin A is essential in retinal function.

In the 1950s he proposed the chemical method to extract pigment from the retina, and with the help of a spectrophotometer measured the light absorbance of the pigments. Since the absorbance of the light by retina pigments corresponds to the wavelength that best activate photoreceptor cells, this experiment showed the wavelength that the eye could best detect. However, since rod cell made up most of the retina, it was, specifically measuring the absorbance of rhodopsin, the main photopigment in rods. The technique of micro-spectro-photometry allowed to measure absorbance directly from cells, rather from an extract of the pigment. This allowed to determine the absorbance of pigments of the cone cells.

To study synapses in the peripheral nervous system (PNS), Sir **John C. Eccles (1903 – 97)** — Australian neurophysiologist, used the stretch reflex as a model which consist only of two neurons, a sensory neuron that is the muscle spindle fibers and the motor neuron (the early1950s). The later synapses on the motor neuron in the spinal cord. When an electric current was passed on to sensory neuron in the quadriceps femoral muscle, this motor innervating muscle produced a small excitatory postsynaptic potential (EPSP). When this same electric current was passed through the hamstring muscle, the opposing muscle to the quadriceps femoris muscle, an inhibitory postsynaptic potential (IPSP) appeared in the quadriceps femoral muscle motor neuron. Although a single EPSP was not enough to fire an action potential (AP) in the motor neuron, the sum of several EPSPs from multiple sensory neurons synapsing onto the motor neuron could cause the motor nerve to fire, thus contracting the quadriceps femoral muscle. Conversely, IPSPs could subtract from this sum of ESPSs, preventing the motor neuron firing.

Ulf S. von **Euler (1905 – 83)** — Swedish physiologist and pharmacologist, who with Sir **John H. Gaddum (1900 – 65)** — English pharmacologist, discovered in 1931 substance P (undecapeptide) – a peptide composed of 11 amino acid residues and member of the tachykinin neuropeptide family, acting as a neurotransmitter and as neuromodulator. Then, on his own he discovered endogenous active substance, prostaglandin (1935), vesiglandin (1935), piperidine (1942) and catecholamine adrenaline (noradrenaline) – a chemical intermediary (neurotransmitter) in the sympathetic nervous system (1946). Then, again with **Göran Liljestrand (1886 – 1968)** — Swedish pharmacologist, described a physiological arterial shunt in response to the decrease in local oxygenation of the lungs, a primary mechanism underlying ventilation/perfusion matching, named hypoxic pulmonary vasoconstriction (HPV) or Euler-Liljestrand mechanism (1946)[655].

With the help of **William W.H.F. Addison (1880 – 1951)** — American anatomist, **Julius H. Comroe, Jr (1911 – 84)**[656] — American physiologist and physician, and **Carl F. Schmidt (1993 – 1988)** — American pharmacologist, demonstrated 1937-38's experiments carried out on cats and dogs, that the tiny carotid body at the common carotid artery (CCA) bifurcation [20th century. Anatomy, histology, and embryology – **Fernando de Castro (1896 – 1967)** was sensitive to a fall in arterial blood oxygen (O_2) partial pressure (PO_2) and to rise in arterial blood carbon dioxide (CO_2) – an emergency system to protect against a fall in the arterial blood PO_2 as contrasted with O_2 concentration and had little effect on respiration.

Again, in collaboration with William Addison, Julius Comroe carried out detailed anatomic studies on the aortic arch (AA) chemoreceptors which was shown to be from aorta in dogs and coronary arteries in cats. Then he showed that the AA chemoreceptors caused an increase in ventilation of the lung when the arterial blood PO_2 fell (1938-39).

While investigating the sensitivity of the respiratory center near the dorsal surface of the medulla oblongata near the brainstem he was led to the general conclusion that this area is the major determinant of normal ventilation of the lungs, and that the function of the peripheral chemoreceptors was to protect the brain against anoxia (1943).

In practical clinical terms Julius Comroe introduced pulmonary function test (PFT) that enabled proper evaluation of patients for anesthesia and surgery, and the treatment of lung diseases.

Sir **Bernard Katz (1911 – 2003)** — German English physician and biophysicist of Jewish ancestry, discovered with **Paul Fatt (1924 – 2014)** — British neuroscientist, that neurotransmitter release at synapses is «quantal» (in Latin – «all or none»), meaning that at any synapse, the amount of neurotransmitter released is never less that a certain amount, and if more is always integral number times this amount. This circumstance arises, because prior to their release into the synaptic gap, transmitter molecules reside in like-sized subcellular packages known as synaptic vesicles, released in a similar way to any other vesicle during exocytosis (in Greek – «endon» - «inside»,

---

[655] Von Euler US, Liljestrand G. Observations on the pulmonary arterial blood pressure in the cat. Acta Physiol Scand 1946;12:301-20.

[656] Ketty SS, Foster RE. Julius H. Comroe, Jr. 1911-1984. A Biographical Memoir. Nat Acad Sc & Press, Washington, DC 2001;79:1-19.

«kytos» - «volume», «vessel») – it is an energy-dependent process for releasing macromolecule, as a result of which an intracellular releasing vesicles are fused by the cell plasmatic membrane, and their content is released from the cell.

Working on monoamine oxidase (MAO) inhibitors in 1957 **Julius Axelrod (1912 – 2004)** — American biochemist of Jewish descent, showed that catecholamine neurotransmitter – adrenaline (epinephrine) and noradrenaline (norepinephrine) do not stop working after they were released its content into synapse. Instead, they are recaptured («re-uptake») by the presynaptic nerve ending and recycled for later transmission. Perhaps, they are held in tissues in nonactive form and released by the nervous system when needed. His research laid the groundwork for later selective serotonin (5-hydroxytryptamine = 5-HT, i.e., serotonin) re-uptake inhibitors (SSRI), such as Prozac which block the re-uptake of another neurotransmitter. Prozac is an antidepressant drug used to treat a major depressive, obsessive-compulsive, panic and premenopausal dysphoric disorders, bulimia, and premature ejaculation.

In 1959 Julius Axelrod reported that melatonin **[Carey McCord (1886 – 1979)** and **Floyd Allen** (above), and **Aaron Lerner (1920 – 2007)**: above] is a neurotransmitter for serotonin that has undergone a chemical conversion and is stored along with epinephrine in nerve terminals for later release.

Subsequently, **H. J. Lynch**[657] — American biochemist, demonstrated that melatonin regulate sleep and wakefulness, exhibits circadian rhythm in human pineal gland, reaching its maximum after the midnight (1975), **Burkhard Poeggeler (1961 –)**[658] — German American biologist and psychologist, discovered that melatonin is a powerful free-radical scavenger and wide-spectrum antioxidant, including being the most effective lipophilic antioxidant (1993), and **Richard J. Wurtman (1936 –)**[659] — American biochemist, was granted the first patent for the use of melatonin as a low dose sleep aid, 1-3 mg p.o. h.s. for a short-term treatment of insomnia (3- month) in people 55 years old or older (1995).

Melatonin is also produced in plants, where its function as a first line of defense against oxidative stress.

Respiratory minute volume (minute ventilation, or minute volume) is the volume of gas inhaled (inhaled minute volume) or exhaled (exhaled minute volume) from a person's lung per minute volume (V_E). It shows relationship of V_E to the partial pressure carbon dioxide ($PaCO_2$) level in the blood gases, since $PaCO_2$ level generally vary inversely with V_E.

[657] Lynch HJ, Wurtman RJ, Moskowicz MA, et al. Daily rhythm in human urinary melatonin. Science 1975 Jan;187(4172):169-71.

[658] Poeggeler B, Reiter RJ, Tan DX, et al. Melatonin, hydroxyl radical-mediated oxidative damage, and aging: a hypothesis. J Pineal Res 1993;14(4):151-68.

[659] Wurman RJ. Methods of inducing sleep using melatonin. US patent 54496883 issued 12 Sep 1995 to MIT.

In practice V_E is measured as a flow rate (given that it represents a volume change over time), in an equation, where V_E = 0,5 L, i.e., tidal volume (V_T) x 12 breaths/min. = 6 L/min., by $V_E = V_T$ x f or RR, where f or RR = respiratory rate, by Wright spirometer [devised by **Basil Wright (1912 – 2001)** — English bioengineer and physician, who also invented the peak flow meter and syringe driver], or other device capable of cumulatively measuring gas flow, such as mechanical ventilators.

V_E comprises the sum of alveolar ventilation (V_A) and death space (V_D), as expressed in the following equation: $V_E = V_A + V_D$.

The concept of cardiac and musculo-skeletal physiology founded by Adolf Fick [19th century. Physiology – **Adolf Fick (1829 – 1906)**], Archibald Hill [20th century. Physiology – **Archibald Hill (1886 – 1977)**] and Otto Meyerhof [20th century. Biochemistry – **Otto Meyerhof (1884 – 1951)**] was further developed in 1944-45 by **Henry L. Taylor (1912 – 83)**[660] — American physiologist, in 1952 by **Per-Olof Astrand (1922 – 2015)** — Swedish physiologist (1922-2015) and in 1964 by **Bengt Saltin (1931 – 2014)** — Swedish exercise physiologist.

In the «Minnesota Starvation Experiment» (1944-45) designed to determine the physiological and psychological effect of severe and prolonged dietary restriction and the effectiveness of dietary rehabilitation strategies on 36 men volunteers, produced significant increase in depression, hysteria, hypochondriasis, decreased basal metabolic rate (BMR) and edema of extremities.

Then, Per-Olof Astrand and Bengt Saltin introduced the estimation or measurement of the maximal oxygen (O_2) aerobic capacity for consumption or uptake or a peak O_2 uptake, utilized in a minute (min.) during exercise (MVO$_2$) and the total amount of O_2 metabolized per min. per kilogram (kg) of body weight at the peak of exercise (VO$_2$ max).

Earl H. Wood (1912 – 2009) — American cardiopulmonary physiologist, helped to develop a gravitation suit (G-suit), or anti-g suit, a flight suit worn by aviators and astronauts who are subject to high level acceleration force, designed to prevent a black-out and g-induced loss of consciousness (g-LOC) caused by the blood pooling in the lower part of the body when under acceleration, thus depriving the brain of blood (1946); studied and refined techniques for measuring blood flow within and to the heart (1950s - 1960s); and received a patent for a blood oximeter used to measure oxygen saturation (SaO$_2$) in the blood stream (1955).

In 1959 **Roger W. Sperry (1913 – 94)** — American neurobiologist and neuropsychologist, showed that dopamine was not just only a precursor for norepinephrine, but also a neurotransmitter in the brain, in 1974 he together with **David H. Hubel (1926 – 2013)** — Canadian neurophysiologist and **Torsten N. Wisel**

---

660 Keys A, Brozek J, Henschel A, et al. The Biology of Starvation. Vol. 1-2. Univ Minnesota 1950.

(1924 –) — Swedish neurophysiologist, completed an important research on the organization of the cerebral hemispheres and the visual system, particularly on the divided and lateralizing function of the brain.

Both, Sir **Allan L. Hodgkin (1914 – 98)** — English physiologist and biophysicist, and Sir **Andrew F. Huxley (1917 – 2012)** — English physiologist, discovered the basis for the propagation of nerve impulses, called an action potential (AP), which they recorded in 1939. Then, in 1952-54 Andrew Huxley, **Emmeline J. Henson (1919 – 73)** — English biophysicist and zoologist, **Rolf Niedergerke (1921 – 2011)** — German physiologist and physician and **Hugh E. Huxley (1921 – 2013)** — English molecular biologist, discovered the mechanism of muscle contraction, called the «sliding filament» theory, based on muscle proteins that pass each other to generate movement. According to this theory, the thin actin filament of muscle fibers slide passes the thick myosin filament during muscle contraction, while two groups of filaments remain in relatively constant length.

Maurice M. Rapport (1919 – 2011) — American biochemist, **Arda A. Green (1899 – 1958)** — American biochemist and **Irvine H. Page (1901 – 91)** — American physiologist, isolated 5-hydroxytryptamine (5-HT), that is serotonin (1948).

General direction of scientific activity of **Volodymyr V. Frolkis (27.I.1924, city of Zhytomyr, Zhytomyr Obl., Ukraine – 02.X.1999, city of Kyiv, Ukraine)**[661] — Ukrainian physiologist, and gerontologist (science of aging), were creation of the adaptive and regulatory theory of aging, to study pathology of aging, search remedies to prolong life, and establishment the role of changes in regulation of the genome in the mechanism of developing an aging pathology (1972-80).

Platon H. Kostyuk (20.VIII.1924, city of Kyiv, Ukraine – 10.V.2010, city of Kyiv, Ukraine)[662] — Ukrainian physiologist, introduced the microelectrode technique for

---

[661] City of Zhytomyr (founded 884; during organized by Russian Communist Fascist of Holodomor-Genocide 1932-33 against the Ukrainian Nation, many residents suffered from starvation death, massive diseases and psychoses from long-lasting starvation, surrounding villages and farmsteads were completely or partially destroyed), Zhytomyr Obl. [area 29,832 km²; between 03.X.1937 and XI.1938 Russian Communist Fascist NKVS in Zhytomyr Obl., convicted 20,168 persons, of which 16,676 (82,7%) were shoot, other convicted served punishment in correctional-laborious concentrating camps], Україна. — Ablitsov VH. «Galaxy Ukraine». The Ukrainian diaspora: prominent persons. Publ. «Kyt», Kyiv 2007. — Hanitkevych YaV. The contribution of the Ukrainian physicians to the world medicine. Ukr Med Chasopys 2009 VII/VIII;4(72):110-15.

[662] Kostyuk PG, Krishtal OA, Shakhovalov YuA. Separation of sodium and calcium currents in the somatic membrane of mollusc neurones. J Physiol London 1977;Sep 1;270:545-68. — Fedulova SA, Kostyuk PG, Vesolovsky NS. Two types of calcium channels in the somatic membrane of new-born rat dorsal root ganglion neurons. J Physiol London 1985;359:431-46. — Kostyuk P. Pronchuk N, Savchenko A, Verkhratsky A. Calcium currents in aged rat dorsal root ganglion neurons. J Physiol 1993 Feb 1;461(1):467-83. — Tsien RW, Barret CF. A Brief History of Calcium Channel Discovery. In, Gerald W. Zamponi. Voltage-Gated Calcium Channels. Eurekah.com and Kluwer Academic / Plenum Publishers, Stanford 2005. P. 27-47. — Ablitsov VH. «Galaxy Ukraine». The Ukrainian diaspora: prominent persons. Publ. «Kyt», Kyiv 2007.

investigating the function of the nervous cells in Ukraine (1959-60); with co-workers first in the world proved the presence of separated calcium (Ca⁺⁺) channels in the cell membrane of the nerve fibers (1977); with co-workers separated Ca⁺⁺ strings into two types – the intra-high threshold and the intra-low threshold Ca⁺⁺ cannels (1985); discovered the selective signaling of Ca⁺⁺ conduction in so called Ca⁺⁺ channels within the soma of nervous cells (1986).

In 1949 **E.J. Christopher Polge (1926 – 2006)**[663] — English biologist, developed technique of cryopreservation, a process where organelles, cells, tissues, organs, or other biological materials susceptible to damage by chemical substances are preserved by cooling to very low temperature.

Major research contribution to the ventilation-perfusion relationship in the lung was made by **John B. West (1928 –)**[664] — Australian respiratory physiologist, who measured pulmonary inequality of ventilation-perfusion using short-lived radioactive gases.

In 1977 **Thomas Hökfell (1940 –)** — Swedish neuroanatomist and neuropharmacologist discovered that the central and peripheral nervous system (CNS, PNS) neurotransmission at the signal synapses take place not only by acetocholine (ACh) and epinephrine, but also by neuropeptide like somatostatin and amino acid, referred as «coexistence principle».

Robert J. Lefkowitz (1943 –) — American biochemist, internist and cardiologist of Jewish ancestry, who discovered and characterized the two families of proteins which regulate them – the G protein-coupled receptors (GPCR) kinases and β-arrestins, cloned the gene for the β-adrenergic receptor, and for a total of 8 adrenergic receptors - adrenaline and noradrenaline (1982). This led to the discovery that all GPCRs (which include the β-adrenergic receptor) have a similar molecular structure, defined by an amino acid sequence which weaves its way back and forth across the plasma membrane seven times. Today we know that about 1,000 receptors in the human body belong to this same family, and that these receptors use the same basic mechanisms so that pharmaceutical researchers now understood how to effectively target the largest receptor family in the human body. Working in the same field, **Brian K. Kobilka (1955 –)** — American physiologist, determined the molecular structure of the β_2-adrenergic receptor that are important targets for pharmaceutical therapeutics (2007).

---

[663] Polge C. Low-temperature storage of mammalian spermatozoa. Proc Roy Soc London. Series B, Biological Sci. 1957;147(929):498-509.

[664] West JB. Respiratory Physiology: The Essentials. 1st ed. Blackwell Scientific, Oxford 1974. — West JB. Respiratory Physiology: The Essentials. 9th ed. Walters Kluwers Health / Lippicott & Wilkins, Philadelphia 2011.

Erwin Neher (1944 –) — German biophysicist and cell physiologist and **Bert Sakmann (1942 –)** — German cell physiologist, were the first to record the current of single ion channel through biological membranes on the living cells, using the lipid bilayer method with the help of developed by them the patch-clamp technique, that is adopted laboratory technique in electrophysiology (EP) allowing the study of single or multiple ion channels in cells (1978).

The discovery in 1993 of the mechanism of autophagy (in Greek – «auto» means «self», and «phagy» means «eating», literally «self-eating»)[665] by **Yoshinori Ohsumi (1945 –)** — Japanese cell biologist, occurring when budding yeast, lacking a degradation enzyme contained in the vacuole, were cultured in a nitrogen-free starvation medium. Autophagy in the cells is an essential metabolic function for maintaining life during and without starvation, concerns all eukaryotes – any organism whose cells have a nucleus and other organelles enclosed within membranes, ranging from yeast to mammals. This process involves 15 essential cellular genes which with the help of lysosomes digest or break down old organelles and portions of ectoplasm with the subsequent orderly recycling. Those genes control the vital processes in animal and human cells, play crucial role in embryo development, cell differentiation and in the immune system.

In 1977-80 **Richard Axel (1946 –)** — American molecular biologist and neuroscientist, along with **Saul J. Silverstein** — American microbiologist and **Michael H. Wigler (1947 –)** — American molecular biologist, discovered the technique of contra location which allows foreign deoxyribonucleic acid (DNA) to be inserted into a host cell to produce protein. That means they developed methods for engineering animal cells which are the basis in mammalian genetics, and the means for producing the therapeutic protein for the treatment heart disease, tumor, and strokes.

In the years of 1991-96 **Linda B. Buck (1947 –)** — American biologist and **Richard Axel (1946 – : above)** cloned the cranial nerve I (olfactory) receptors, showing that they belongs to the family of G-protein coupled receptors and estimated by analyzing rat DNA existence of about 1,000 different genes for olfactory receptors in the mammalian genome[666]. Each olfactory receptor neuron expresses one kind of olfactory receptor protein and the input from all neurons expresses by the same receptor is collected in a single dedicated glomerulus of the olfactory bulb.

David J. Julius (1955 –) — American physiologist, biochemist, and neuroscientist of Jewish descent, cloned and characterized the transient receptor potential (TRP)

---

[665] Phagocytosis and phagocytes [19th century. Microbiology – **Illya Mechnykov (1845 - 1916)**] versus autophagy

[666] Genome is the complete set of genes, hereditary factors contained in the haptoid set of chromosomes. The human genome has an estimated 30,000 to 40,000 genes.

V1 – cation channels derivatives which is receptor that detects capsaicin, the chemical in chili peppers making them hot (termoreceptor), TRPM8 cold and mentol receptor (CMR)1 and TRPA1 that detect menthol and cooler temperature, and TRP1 that detect mustard oil (allyl isothiocyanate), made contribution to the study of detection of the touch (notiception) by discovering toxins that modulate these channels, described unique adaptations of the channels in diverse species and solved using the cryo-elecron microscope (cryo-EM) structures at numerous channels (1989-97).

Independently, **Ardem Patapoutian (1967 –)** — American molecular biochemist and neuroscientist of Armenian descent, discovered in the skin the gene PIEZO1 (in Greek means squeeze or press) and gene PIEZO2 channels that sense touch, arterial blood pressure (BP), respiratory rate (RR), urinary bladder control та TRPM8 that sense temperature, pressure, and menthol.

The afferent and efferent nervous pathways that convey an electrical signal from the skin are tiny somatosensory afferent nerve endings and Merkel, Merkel-Ranvier, or tactile cells [19th century. Anatomy, histology, and embryology. – **Louis Ranvier (1835 – 1922)** and **Friedrich Merkel (1845 – 1919)**] through the posterior (dorsal) and anterior (ventral) horn of the spinal cord to the brain.

In summary, David Julius and Ardem Patapourtian discoveries showed how we mammals (humans and animals) perceive and adopt to the surrounding world, made it possible to understand how heat, cold and mechanical forces trigger muscle impulse, gave us the knowledge to further advance the treatment range of diseases, including treating of chronic pain.

The extensive work **Oleksiy (Akeksiy) Verkhratsky (30.VII. 1961, city of Ivano-Frankivsk, Ukraine –)**[667] — Ukrainian neurophysiologist, is concerned with glial physiology, cellular mechanism of brain ageing (aging), glial pathophysiology, and physiology of neurons.

As of the glial physiology, he discovered functional expression of low- and high-threshold Ca^{++} channels in oligodendroglia precursors, the earliest finding underlying the concept of electrical excitability of NG2-glia (1990)[668].

As concerning the cellular mechanism of brain ageing, he conducted the first recording of Ca^{++} currents in ageing sensory neurons (1993), pioneering cytosolic Ca^{++} recording in ageing neurons *in situ*, which gave direct experimental support

---

[667] City of Ivano-Frankivsk (IF) Obl., Ukraine. — Verkhratsky A, Orkland RK, Kettenman H. Glial calcium; homeostasis and signaling function. Physiological Reviews 1998;78(8):99-141. —Verkhratsky A, Petersen OH (Eds). Calcium Measurement Method. Publ Humana Press / Springer, New York 2009. —Verkhratsky A, Butt A. Gial Physiology and Pathophysiology. Publ. Wiley-Blackwell, Hoboken, NJ. 2013. — Ablitsov VH. «Galaxy Ukraine». The Ukrainian diaspora: prominent persons. Publ. «Kyt», Kyiv 2007.

[668] Verkhratsky A, Orkland RK, Kettenman H. Glial calcium; homeostasis and signaling function. Physiological Reviews 1998;78(8):99-141.

for multiple aspects of a Ca^{++} theory of ageing; and he was the first to perform an in-depth analysis of astrocytic structure and function in the aging brain.

Concerning the glial pathophysiology, he developed a new concept of astroglia atrophy associated with loss of function and glial paralysis as key elements neuropathology; astroglia atrophy contributes to pathophysiology of several diseases from neuropsychiatric disorders to neurodegeneration (2008-2017).

Referring to physiology of neurons, he contributed to the identification of energy reading (ER) Ca^{++} release mechanism in sensory neurons and was the first to perform real-time measurement of intra-ER Ca^{++} dynamics in neurons to demonstrate the graded nature of Ca^{++}- induced Ca^{++} release (1993-2005). Together with **Denis Burdakov** — Russian Ukrainian neurophysiologist, he also found and characterized the link between physiological glucose changes and excitability of hypothalamic neurons[669].

Biochemistry

The concept of the p[H] [in Latin – pondus hydrogenii, which means quantity of hydrogen (H^+)] was introduced in 1909 by **Soren P.L. Sorensen (1868 – 1939)**[670] — Danish chemist and revised in 1924 to the modern pH to accommodate definitions and measurements in terms of electrochemical cells.

The pH is a H^+ indicator and a numeric scale used to specify the acidity or basicity of an aqueous solution, being the negative of the base 10 logarithm (log) of the molar concentration, measured in units of moles per liter (L), of H^+ ions:

$$pH = \log [H^+]$$

It is the negative of the log to base 10 of the acidity of the H^+ ion. Pure water is neutral, being neither an acid nor a base. Solutions with a pH less than 7 are acidic and solutions with a pH greater than 7 are basic. From the meaning of pH one can calculate pOH:

$$pOH = 14\text{-}pH$$

[669] Burdakov D, Gerasimenko O, Verkhratsky A. Physiological changes in glucose differentially modulate the excitability of hypothalamic melanin-concentrating hormone (MCH) and orexin neurons *in situ*. J Neuroscience 2005 Mar 2;25(9):2429-33.

[670] Sorensen SPL. Über die messing und die Bedeuting der Wasserstoffionenkonzentration bei enzymatischen Prozessen. Biochem Zeitschr 1909;21:131-304.

The successful management of the cardiopulmonary bypass (CPB) and hypothermic circulatory arrest (HCA) in cardiac surgery require diligent observation of the arterial blood gases (ABG) and the blood acid-base balance (ABB), correction of an oxygen (O_2) saturation of the arterial and venous blood O_2 saturation (SaO_2, SvO_2) by adding an O_2, the alpha-stat (alpha-statim) strategy, i.e., without active correction of pH with the hypothermic blood temperature (T), or the pH-stat (pH-statim) strategy, i.e., active correction of pH with the hypothermic blood T, by adding carbon dioxide (CO_2) for neutrality and and to keep pH unchanged)[671].

In general, the safest way is adding O_2 to the CPB circuit to assure the value of arterial PO_2 in hyperoxic limits (400-600 торr)[672], specially during the CPB and HCA[673] to decrease the possibility of injury to the brain by micro emboli of nitrogen (N_2).

Therefore, the partial pressure of components of air gases stands up for CO_2 0,3 torr, for O_2 158 torr, for N_2 596 torr. CO_2 is easily solvable. O_2 is bound to hemoglobin (Hb) – poorly dilutable (coefficient of solubility is 0,0031 ml/dl at T 37^{0}C), N_2 – relatively non-solvable (coefficient solubility is 0,0017 ml/dl at T 37^{0}C)[674]. Thus, the most difficult is to remove N_2 from the blood plasma, and it most commonly creates emboli and neurological damage. Hypothermia increases the O_2 solubility coefficient on account of the coefficient of solubility of N_2 (up to 0,0021 ml/dl at T 20^{0}C). During hyperoxia CPB O_2 displaces N_2 from the blood plasma, better oxygenates the brain and decreases quantity of cerebral micro emboli.

Conversely, the hyperoxia CPB through elevation of the level of O_2 radicals in the blood plasma may slow down re-perfusion of the myocardium in children with congenital heart disease (CHD) of cyanotic type. But still, the ability of the hyperoxia CPB to improve delivery of O_2 to the brain and decrease the incidence of micro emboli exceeds its shortcoming.

---

[671] Patel RL, Turtle MR, Chambers DJ, et al. Alpha-stat acid-base regulating during cardiopulmonary bypass improves neuropsychological outcome in patients undergoing coronary artery bypass grafting. J Thorac Cardiovasc Surg 1996;111:1267-79. — Mahmood F, Zhao X, Matyal R. Adult Cardiac Anesthesia. In, FW Sellke, PJ del Nido, SJ Swanson (Eds). Sabiston & Spencer Surgery of the Chest. Vol. I-II. 8th Ed. ElsevierSaunders, Philadelphia 2010. P. 919-31. — Rubens FD. Cardiopulmonary Bypass: Technique and Pathophysiology. Ibid. P. 957-75. — Del Nido PJ, McGowan Jr FX. Surgical Approaches and Cardiopulmonary Bypass in Pediatric Cardiac Surgery. Ibid. P. 1709-33. — Emani SM. Patent Ductus Arteriosus, Coarctation of the Aorta, and Vascular Rings. Ibid. P. 1781-95.

[672] 1 torr is the unit of pressure based on the absolute scale which was defined as 1/760 of the standard atmospheric, that is 0,9 mmHg. Named in honor of **Evangelista Torricelli (1608 - 47**: 17th century) — Italian mathematician and physicist.

[673] **Volodymyr Walther (1817 - 89**: 19th century. Physiology) — Ukrainian anatomist and physiologist, who proved in 1862 that cooling of the animal's body decreases the risk of the operation.

[674] Tovar EA, Del Campo C, Borsari A, et al. Postoperative management of cerebral air embolism. Gas physiology for surgeons. Ann Thorac Surg 1995;60:1138-42.

The state of ABG and the blood ABB is defined by shifting of the pH (normal 0,35-0,35) and relates to the dynamics of equation developed by **Lawrence J. Henderson (1878 – 1942)** — American chemist and physiologist, and **Karl A. Hasselbalch (1874 – 1962)** — Danish biochemist and physician (Henderson-Hasselbalch equation (1908, 1917)[675], where $pH = pK + [HCO_3^-] / (0.03 \times PCO_2)$. The normal pH means neutrality of electrochemical status of water (H_2O), where $[H^+] = [OH^-]$. The Henderson-Hasselbalch equation describes the origin of pH, calculate acidity of buffer solutions (bicarbonates, phosphates and aminoacids), finds the equilibrium of pH in the acid-base reactions and calculates the isoelectric point of proteins.

Within normal limits of the body T (36,6-37°C) pH is established by the primary acceptors of protons (buffers) of the blood, that is bicarbonate salt (bicarbonates; HCO_3) and phosphates. However, during decreasing the body T the most important intracellular buffer is the alpha-imidazole moiety of aminoacid histidine (the imidazole moiety of histidine).

With increase of the body T H_2O became more ionized, solubility of CO_2 in the blood decreases, the gaseous phase is increased by 4,5% / 1°C increase in the body T, the blood acidity increases, pH decreases by 0,015 unit / 1°C increase in the body T, the quantity of 2,3-diphosphoglyceric (2,3-DPG) acid in erythrocytes increases, as a result of which Hb / Hgb-affinity to O_2 decreases, the dissociation curve of the oxyhemoglobin (HbO_2) is shifted to the right, O_2 is released to the tissues.

Conversely, with decrease of the body T H_2O became less ionized, solubility of CO_2 in the blood increases, the gaseous phase decreases by 4,5% / 1°C decrease in the body T, the blood alkalinity increases, pH increases by 0,015 unit / 1°C decrease in the body T, the quantity of 2,3-DPG acid in erythrocytes decreases, causing increase Hb-affinity to O_2, the dissociation curve of the HbO_2 is shifted to the left, O_2 is not released to the tissues.

During deep hypothermia the blood pH stands at about 7,7-7,8.

The management of ABG and blood ABB with the alpha-stat strategy during hypothermia consist of preservation of the electro-chemical neutrality of the blood and buffer ability of the alpha-imidazole moiety of histidine. Usually, the sample of cooled patient's blood is wormed to the T 37°C shows pH of about 7.4 and CO_2 40 torr.

The management of ABG and blood ABB with pH-stat strategy during hypothermia allows to keep the arterial CO_2 at the physiological level, and that provide autoregulation of the blood supply to the brain in line with the blood flow

---

[675] Henderson LJ. Concerning the relationship between the strength of acids and their capacity to preserve neutrality. Am J Physiol 1 May 1908;21(4):173-9. / Hasselbalch KA. Die Berechnung der Wasserstoffzahl des Blutes aus der freien und gebundenen Kohlensäure desselben, und die Sauerstoffbindung des Blutes als Funktion der Waterstoffzahl. Biochemische Zeitschrift 1917;78:112-44.

within the arterial blood pressure (BP) from 40 to 140 torr without splitting of the cerebral metabolism.

However, alpha-stat management during hypothermia with a low blood flow and arterial BP may not uniformly cool the brain and supply O_2 to the tissues at the borderline level.

Generally, it is safer to use alpha-stat in adult patients undergoing cardiac operations under hypothermic CPB with/without HCA.

The aim of ABG and blood ABB during pH-stat is to rearrange an «alkalotic» level of pH and CO_2 of a patient during hypothermia by adding CO_2 to the oxygenator of the heart-lung machine into «acidic» (low pH and high CO_2) owing to disruption of the electrochemical neutrality of the blood. Usually, a blood sample of cooled patients is rewarmed to the T 37°C showing pH approximately of 7,1-7,4 (acidity) and CO_2 approximately of 60-70 torr (hypercarbia). Due to shifting of the HbO_2 dissociation curve to the left an «alkalotic» pH decreases during deep hypothermia, the brain blood flow increases, cooling of the brain became uniformed and the O_2 supply to the brain increases if the arterial BP and blood flow are maintained at the responsible level. Beside of this, hypercarbia decreases metabolism of the brain, energy expenditure, the glycolytic flux and production of lactic acid. Both, hypercarbia and acidity depress the function of N-methyl-D-aspartate (NMDA) receptors, output of glutamate and calcium flux of neurons.

The indications for pH-stat are newborn and infants during correction of complex CHD of cyanotic type under CPB with HCA, such as pulmonary stenosis with intact interventricular septum (PS/IIVS), tetralogy of Fallot (TOF) with stenosis or atresia pulmonary artery (PA), PS/IIVS associated with major aorto-pulmonary secondary arteries / anastomoses (MAPSA) and a single ventricle, due to better blood flow to the brain.

The pathway of glucose metabolism, discovered in 1935-40, by **Gustav G. Embden (1874 – 1933)** — German physiological chemist, **Otto F. Mayerhof (1884 – 1951)** — German physician and biochemist of Jewish parents and **Jakub K. Parnas** or **Yakiv O. Parnas (Jan. 16, 1884, village Mokriany, Drohobych Region, Lviv Obl., Ukraine** – died from heart attack during first the interrogation in headquarters of the Russian Communist Fascists NKVS **Jan. 29, 1949, city of Moscow, Russia)**[676]— Ukrainian biochemist born to Jewish parents, implies a series of enzymatic reactions in the anaerobic conversion of glucose to lactic acid in muscles (glycolysis), resulting in the release of an energy in the form of adenosine

---

[676] Village of Mokriany (founded 1515), Drohobych Region, Lviv Obl., Ukraine. — Mirsky M. Ukrainian medical scientists – victims of Stalin's repressions. UHMJ «Ahapit» 1995;2:45-54.— Ablitsov VH. «Galaxy Ukraine». The Ukrainian diaspora: prominent persons. Publ. «Kyt», Kyiv 2007.

triphosphate (ATP) – Embden-Mayerhof-Parnas pathway). In addition, based on early findings of **Louis Pasteur (1822 – 95**: 19th century. Microbiology and immunology**)** Otto Mayerhof delineated the metabolism of lactic acid in the Pasteur – Mayerhof reaction.

Adolf Windaus (1876 – 1959) — German chemist, while studying the steroid hormones and their relation to vitamins, became involved in the discovery of the transformation of cholesterol **[Francois de la Salle (1719 – 88)**. 18th century] into vitamin D_3 (cholecalciferol). Subsequently, vitamin D_3 was introduced into the medical practice in 1927 as Vigantol.

В 1916-18 pp. **Henry D. Dakin (1880 – 1952)** — English chemist, developed an antiseptic solution consisting of sodium hypochloride (NaOCl) known as blech or clorox (0.4 to 0.5%) and boric acid (4%) to treat the wound by intermittent irrigation (Dakin solution).

Donald D. van **Slyke (1883 – 1971)**[677] — Dutch American biochemist, who introduced a non-Standard International (SI) unit of measuring buffering activity, the Van Slyke chemical determination test of aminoacids containing a primary amine group.

Theodor Svedberg (1884 – 1971) — Swedish chemist, constructed the ultracentrifuge to study the migration of the protein molecule (1922) and did important work on the dispersed system. The Svedberg unit equals 10^{-13} second used for expressing the sedimentation coefficients of micromolecules.

The principal directions of scientific activities of **Olexandr (Alexander) V. Palladin [29.VIII(10.IX).1885, city of Moscow, Russia – 06.XII.1972, city of Kyiv, Ukraine]**[678]

---

[677] Van Slyke DD, Hiller A. An identified base among the hydrolic procucts of gelatin. Proc Natl Acad Sci USA July 1921;7(7):185-6.

[678] Palladin AV. Investigations on Creation and Excretion of Creatinine in Animals. Kharkov 1916. — Palladin AV. The Textbook of Physiological Chemistry for Students and Physicians. «Gosizdat» of Ukraine, Kharkov 1924. — Palladin OV. The Basics of Nutrition: Physiological Outlines. «Dezhvydav» of Ukraine», Kharkiv 1926. — Palladin OV. Biochemistry: Textbook for Students and Physicians. «Dezhvydav» of Ukraine, Kyiv 1927. — Palladin OV. Biochemistry: Textbook for Students and Physicians. 2nd Ed. «Dezhvydav» of Ukraine, Kyiv 1930. — Palladin OV. Biochemistry: Textbook for Students and Physicians. 3rd Ed. «Dezhvydav» of Ukraine, Kyiv 1932. — Palladin OV. Biochemistry: Textbook for Students and Physicians. 4th Ed. «Dezhmedvydav» of Ukraine, Kyiv-Kherson 1935. — Palladin OV. Biochemistry: Textbook for Students and Physicians. 5th Ed. «Dezhmedvydav» of Ukraine, Kyiv 1938. — Palladin AV. Biochemia. PZWL, Warszawa 1953. — Olearchyk A. Ukrainian scientists. Olexandr Volodymyrovych Palladin — prominent biochenist. Nasze Slovo (Warszawa) 16.12.1956; 1(18):3. — Olearchyk A. Interview with O. V. Palladin — Ukrainian scientist world-renown. Nasza Kultura (Warszawa) 12.1958;8:1-2. — — Palladin AV, Belik Ya.V, Polyakova NM. Protein Metabolism of the Brain and its Metabolism. «Naukova Dumka», Kyiv 1972. — Palladin AV. The Seleted Works. «Naukova dumka», Kyiv 1975. — Palladin AV, Belik YaV, Polyakova NM. Protein Metabolism of the Brain. Consultants Bureau, New York – London 1977. — Palladin AV, Belik YaV, Polyakova NM. Protein Metabolism of the Brain. Springer-Verlag, New York 2013. — Ablitsov VH. «Galaxy Ukraine». The Ukrainian diaspora: prominent persons. Publ. «Kyt», Kyiv 2007.

— Ukrainian biochemist, were concerned with the creatinine metabolism during work, rest and training, disturbances in the metabolism (exchange of substances) and vitamin deficit in certain diseases, biochemistry of the nervous system and the synthesis of the vitamin K analogue.

Olexandr Palladin proved in 1916 p., that arginine - Arg [19th century. Biochemistry – **Ernst Schultze (1840 – 1912)]** is one of the precursors of creatinine present in muscles of humans and warm-blood animals, take part in the energy exchange of substances and renewal muscles and other tissues, participate in the synthesis of hormones, especially of the growth, it being secreted into the blood and excreted into the urine. Therefore, the concentration of creatinine in the blood (normal 0.60 – 1.30 mg/dL) is an important indicator of the renal function. The level of creatinine in the blood increases during, renal or cardiac insufficiency, in diabetes mellitus (DM), infections, sever pneumonia and hyperfunction the thyroid gland (TG). By studying of the peculiarities of creatinine during work, rest and physical he created the base for physical culture, especially for athletes.

By studying the interrelation between disturbances in metabolism and deficit of vitamins during creation of experimental scurvy and disseminated inflammation of nerves (polyneuritis), he created the scientific base for treating those diseases with vitamins (1919-38).

The first with his students since 1922 he has begun systematic studies of the biochemistry of the nervous system, established the natural biochemical topography of the nervous tissue, its philo- and ontogenesis, linked up peculiarities of the substance exchange under the influence of diverse factors of the environment, varied functional, extraordinary and pathological state of the organism.

He educed watery analogue if vitamin K – «Vikasol», or «Vicasolum» (menadione), an effective ingredient to control bleeding or oozing (1943).

In 1913 **Edward C. Kendall (1886 – 1972)** — American chemist, purified the thyroid factor using a bioassay that measured the urinary nitrogen in dogs, describing the isolation and crystallization of the thyroxine (T_4) from 6500 pounds of hog thyroid gland (TG). Unfortunately, he incorrectly concluded that the structure was triiodo-hexahydroxy-indole propionic acid[679].

Nevertheless, in 1926 Sir **Charles R. Harington (1897 – 1972)** — English biochemist and **George Barger (1878 – 1939)** — English biochemist, who for the first time correctly described hormones of the TG tetraiodothyronine (T_3), and thyroxine (T_4), and produced synthetic T_4. Excess of T_4 cause simple goiter but is used to tread TG insufficiency (hypothyroidism).

---

[679] Kendall EC. Isolation of the iodine compound which occurs in the thyroid. J Biol Chem 1919;39:125-47. — Kendall EC, Osterberg AE. The chemical identification of thyroxin. J Biol Chem 1919:40:265-34.

In 1917 Sir **Robert L. Robinson (1886 – 1975)** — British organic chemist, synthesized tropinone, a precursor of cocaine, which was a big advancement in alkaloid chemistry, and showed that tandem reactions in one-pot synthesis can form bicyclic molecules.

Working on the adrenal glands, a.k.a., supra-adrenal glands, endocrine glands that produce in the cortical layer steroids and aldosterone, in the medullary layer catecholamines (adrenalin and noradrenalin), **Tadeusz Reichstein (1897 – 1996)** — Polish Swiss biochemist of Jewish ancestry, isolated in 1937-38 from zona fasciculate of the adrenal cortex the glucocorticoid hormone cortisol, but failed to recognize its biological significance. It was **Edward Kendall (1886 – 1972**: above) and **Philip S. Hench (1896 – 1965)** — American physician, who again in 1948-48 isolated cortisol, described its chemical structure and biological function. A year later (1949), the first commercial production of cortisol was commenced as hydrocortisone, for the treatment of arthritis and rheumatoid arthritis.

Radoslav (Leopold) Ruzicka (1887 – 1976)[680] — Croatian Swiss organic chemist, who demonstrated in 1916-17 that civetone, a macrocytic ketone and the main constituent of civet oil, is the principal odoriferous compound found in the musk deer (Moschus moschiferus) and civet cat (Viverra civetta). He also devised a method for synthesizing macrocytes, an organic reaction in which a dicarbon acid is converted to a cyclic ketone at high temperature and suitable catalyst, such as thorium oxide (the Ruzicka large-ring synthesis, or Ruzicka reaction, or Ruzicka cyclization, 1926-27). In 1930-34 synthesized the male hormones – androsterone and testosterone, and one year later (1935) defined a molecular structure of testosterone.

James B. Sumner (1887 – 1955) — American chemist, isolate and crystalized an enzyme urease and showed it to be a protein.

In 1925 **Yaroslav Heyrovsky (1890 – 1967)** — Czech chemist and inventor, discovered polarography.

Independently, within two weeks each other in the spring of 1932, **Albert Szent-Györgyi (1893 – 1986)** — Hungarian American biochemist, and **Charles G. King (1896 – 1988**) — American biochemist, isolated L-ascorbic acid (vitamin C) and described its catalytic virtues in the suprarenal glands, as antiscorbutic factor that

---

[680] Ruzicka L. Zur Kenntnis des Kohlenstoffringes I. Über die Konstitution des Zibetons. Helv Chim Acta 1926;9(1):230-48. — Ruzicka L, Stoll H, Schinz H. Zur Kenntnis des Kohlenstoffringes II. Synthese der carbocyclischen Ketone vom Zehner-bis zum Achtzehnerring. Helv Chim Acta 1926;9(1):249-641. — Ruzicka L. Zur Kennnis des Kohlenstoffringes VII. Über die Konstitution des Muscous. Helv Chim Acta 1926;9(1):715-29. — Ruzicka L, Schinz H, Seidel CF. Zur Kenntnis des Kohlenstoffringes IX. Über den Abbau von Zibeton, Zibetol und Zibetan. Helv Chim Acta 1927;10(1):695-706. — Ruzicka L. The isoprene rule and the biogenesis of terpenic compounds. Cellular and Molecular Life Science 1953;9(10):357-67.

prevents scurvy. Next year (1933), Sir **W. Norman Haworth (1883 – 1950)** — English biochemist, having properly deduced the correct structure and optical-isomeric nature of vitamin C reported its synthesis.

When Albert Szent-Györgyi begun work in 1938 on the biophysics of muscle movements, he found that muscle contain actin, which combined with the protein myosin and the energy source adenosine-triphosphate (ATP) contract muscle fibers.

In the late 1931 **Harold C. Urey (1893 – 1981)**[681] — American physical chemist, was the first to detect spectroscopically deuterium [hydrogen-2, H-2 or D, a.k.a., heavy oxygen (O_2)] which is one of two stable isotopes of hydrogen (H), the other being protium (hydrogen-1).

In 1922 **Alexander I. Oparin (1894 – 1980)**[682] — Russian biochemist, developed a theory of the origin of life on Earth by the abiogenesis[683] evolution of carbonic acids approximately 4.28 billion years ago.

In 1952-59 **Stanley L. Miller (1930 – 2007)**[684] — American chemist with assistance of Harold Urey **p**erformed the first classic experiment to investigate whether the chemical self-organization would have been possible on the early development of Earth. Thus they showed that from a mixture of several simple non-organic components, such as the water (H_2O) vapor, methane (CH_4), ammonia (NH_3) and hydrogen (H) that would be present in a reduced atmosphere of the pre-icteric Earth, subjected to the input of only heat to provide reflux and electrical energy (sparks to stimulate lighting), originally produced, identified by a paper chromatography in the period of one week, five familiar organic compounds, such as aminoacids: clearly delineated – glycine, alpha-alanine and beta-alanine, less clearly delineated – aspartic acid and alpha-aminobutyric acid, later (Miller-Urey experiment, 1953-59). After Stanley Miller death re-examination of sealed vials preserved from the original Miller-Urey experiment revealed well over 20 different aminoacids, and more that 20 that naturally occurs in life.

Vitamin K was discovered by **C.P. Henrik Dam (1895 – 1976)** — Danish physiologist and biochemist, in 1935, while feeding a cholesterol-free diet to chickens and using chloroform to remove all fat from chick chow. The chicks fed

---

[681] UreyH, Brickwedde F, Murphy G. A hydrogen isotope of mass 2. Physical Rev 1932;39(1):164-5.

[682] Oparin AI. The Origin of Life. Publ Moscow rabochii, Moskwa 1924.

[683] Theory of abionenesis (in Greek – «a» for non-integral part, «bios» for life, «genesis» for birth, beginning) or the origin of life explains that the life have had arisen from the transformation non-organic (non-living) material (molecule) into organic (living) molecule by the entry of the electrical or nuclear energy approximately 4,28 billion years ago, after formation of the oceans approximately 4.41 billion years ago, and after formation of the Earth approximately 4.54 billion years ago.

[684] Miller SL. Production of aminoacids under possible primitive Earth conditions. Science 1953;117(3046):528-9. — Miller SL, Urey HC. Organic compound synthesis on the primitive Earth. Science 1959;130(3370):245-51.

only with fat-depleted chow developed hemorrhages and started bleeding from tag sites. Induced defect could not be restored by adding purified cholesterol to the diet. It appeared that with the cholesterol a second compound had been extracted from the foot, called coagulation factor (in German – «Koagulations-Vitamin»). Thus, the new vitamin received the letter «K». The chemical structure of vitamin K was defined by **Edward A. Doisy (1893 – 1986)** — American biochemist, in 1939.

Adenosine triphosphate (ATP) is an organic compound and hydrotropic produced in mitochondria, present in all living cells to provide energy for physiological processes, such as muscle contraction, nerve impulse propagation, condensate dissolution and chemical synthesis.

ATP was discovered and isolated from muscle and liver extract in 1929 by both **Cyrus H. Fiske** and **Yellapragada Subbarow (12.I.1895, city of Bhimavaram, Andhara Pradesh, India** – died from cardiac arrest **08.VIII.1948, New York City, N.Y., USA)** — Indian American biochemist, and independently by **Karl Lohman (1898 – 1978)**[685] — German biochemist. Yellapragada Subbarow also synthesized methotrexate for the treatment of cancer and immune diseases.

Then, in 1935 **Vladimir A. Engelhardt (03.XII.1894, city of Moscow, Russia – 10.VII.1984, city of Moscow, Russia)** — Russian biochemist, noted that muscle contractions require ATP.

Two years later (1937) **Herman M. Kalckar (1908 – 91)**[686] — Danish biochemist, established that ATP synthase is linked with cell respiration and had been considered to be the final product of catabolic reactions.

Both, **Carl F. Cori (1896 – 1984)** — Czech American biochemist and pharmacologist and **Gerti T. Rodnitz-Cori (1896 – 1957)** — Czech American biochemist born into a Jewish family, discovered the catalytic conversion of glycogen to lactic acid (Cori glucose-lactic acid cycle) and proved that an enzyme deficiency could be inborn and responsible for metabolic disorders. By showing how the enzymes convert animal starch in blood sugar made it possible to understand diabetes mellitus (DM).

The work of Sir **Cyril N. Hinshelwood (1897 – 1967)**[687] — British chemist, on chemical changes in the bacterial cell were of a great importance in the later research on antibiotics and the other therapeutic agents.

---

[685] Lohman K. Über die Pyrophosphatfraction im Muskel. Naturwissenschaften 1929;17:624-51.

[686] Kalckar H. Phosphorylation in kidney tissue. Enzymologia 1937;2:47-52.

[687] Hinshelwood CN. The Chemical Kinetics of the Bacterial Cell. Clarendon Press, Oxford 1946. — Dean ACR, Norman Hinsheklwood CN Growth, Function and Regulation in Bacterial Cells, Clarendon Press, Oxford 1966.

Paul H. Mueller (1899 – 1965) — Swiss chemist, discovered the insecticidal qualities of a chlorinated hydrocarbon dichlorodiphenyltrichloroethane (DDT) in 1939 which was synthesize as early as 1874. It was subsequently used against insects but is now banned in the USA, except for a few specialized purposes, because of the ecological harm.

Sir **Hans A. Krebs (1900 – 81)** — Jewish German English biochemist, discovered the tricarboxylic acid (TCA) cycle, a.k.a., citric acid cycle (CAC) in the mitochondria of cells – a final general pathway of oxygenation during which the carbon chains of carbohydrates, fatty acids and aminoacids are metabolized with the creation of carbon dioxide (CO_2) energized molecules, usually entering the cycle as the acetyl-coenzyme A, and creates adenosine triphosphate (ATP) by means of supplying electrons into the electron-transporting chain (Krebs cycle, 1937)[688].

Albert Szent-Györgyi (1893 – 1986: above**)** also discovered the catalysis of fumaric acid, a chemical compound found in fumitory (Fumaria officinalis, specifically Boletus fomentarius), lichen and Iceland moss, an intermediary in the TCA Krebs cycle used to produce energy in the form of ATP from food.

In 1941 **Fritz A. Lipmann (1899 – 1986)**[689] — Jewish Prussian American biochemist, proposed hypothesis about the central role of adenosine triphosphate (ATP) as a source of delivery and exchange of the energy for the support metabolic reactions of the living cell, that is chemical compounds created by this phosphate group supplies the energy for utilization by cells of the organism. However, in five years (1945) owing to his discovery and synthesis of the catalyst (catalyzer) coenzyme A (CoA) of acetylation, he confirmed the correctness of his hypothesis. The CoA molecule consist of the remains of adenylic, and pantothenic acids bound together by pyrophosphate, takes part in the transfer of acrylic groups. He showed how ATP with the support of CoA which is takes part in the synthesis and oxidation of fatty acids and the oxidation of pyruvate, helps to convert the energy of the phosphate compounds into other necessary for the organism forms of the chemical energy and determined the influence of the CoA on the intermediate metabolic stage of the TCA Krebs cycle (above). He coined phrase «energy-rich phosphate bonds».

In 1923-24 **Karl P.G. Link (1901 – 78)**[690] — American biochemist, discovered the oral anticoagulant agent coumadin (warfarin) – a related compound which he

---

[688] Wainwright M. William Arthur Johnson – postdoctorate contribution into the Krebs cycle. Trends Biochem Sc. 1993;Feb;18(2):61-2.

[689] Lipmann F. Metabolic Generation and Utilization of Phosphate Bound Energy. In, FF Nord, CH Werkman. Advances in Enzymatology and Related Areas of Molecular Biology 1941 Jan 1;1:99-162.

[690] Link KP. The discovery of dicumarol and his sequels. Circulation Jan 1958;19(1):97-107.

synthetized and tested, and which turned to be dicumarol [3,3' methylenebis-(4 hydroxycoumadin)].

One of the founders of quantum chemistry and molecular biology was **Linus C. Pauling (1901 – 94)** — American chemist, biochemist, and engineer, who contributed to the theory of chemical bounds by introducing the concept of orbital hybridization and the accurate scale of electronegativities of the elements (1932). He showed the importance of the alpha helix and beta sheet in the secondary structure of a protein. His approach combined methods and results from radiology, crystallography, molecular model building and quantum chemistry.

Also, Linus Pauling discover that amino acid – ethylene diamine tetraacetic acid (EDTA), was not only a powerful way to eliminate heavy metals and poisons from the body, but also may flush away obstructing atherosclerotic plaques that adheres to the inner wall (intima) of arteries, thus preventing a stroke or myocardial infarction (MI).

In 1937, **Ciryl N. Hugh Long (1901 – 70)** — English American biochemist, with **Abraham White (1908 – 80)**[691] — American biochemist od Jewish ancestry, isolated bovine prolactin, the first protein of pituitary gland hormones to be obtained in pure crystalline form.

The electrophoresis of proteins was described in 1930 by **Arne V.K. Tiselius (1902 – 71)** — Swedish biochemist, who also developed the absorption analysis (1948) and chromatography (1958).

Adolf F.J. Butenandt (1903 – 95)[692] — German biochemist, extracted from ovaries sex hormone estrone and from urine the other primary sex hormones (1929), then sex hormones progesterone (1934) and testosterone (1935). Part of the organic chemistry are higher terpenes and steroids.

In 1933-35 **A. Hugo T. Theorell (1903 – 82)** — Swedish biochemist, discovered the oxidation cellular enzymes called the «yellow ferment» and succeeded in splitting it reversible into coenzyme part, which was found to be flavin mononucleotide and colorless protein part. Subsequently, he carried out research on various oxidative cellular enzymes, contributing to knowledge of cytochrome, peroxidases flavoproteins, «pyridine» proteins and alcohol dehydrogenases. He also has contributed to progress on antidiuretic hormone (ADH), which breaks down alcohol

[691] White A. Principles of Biochemistry, Mc Graw Hill Co, New York 1973. — Smith EL. Abraham White (1908-1980). Nat Acad Sc USA, Washington, DC 1985:505-36.

[692] Butenandt A. Über «Progynon» ein krystallisiertes weibliches Sexualhormone. Die Naturwissen Schaften 1929 Nov;17(45):879. — Butenandt A. Über die chemische Undersuchung der Sexualhormone. Zeitschaft für Angewandte Chemie. 1931;44(46):905-16.

in the kidneys, and to the theory of the toxic effect of sodium fluorine on cofactors crucial to human cellular enzymes.

David L. Ferdman [07(25).I.1903, town of Terespil, Pidlashsia, Ukraine / Lublin Voivodeship, Poland – 11.I.1970, city of Kyiv, Ukraine][693] — Ukrainian biochemist, proved the existence of glycogen and glutamine in the muscles and the other tissues (1929), the role of the latter in the metabolism (1942), proposed to treat muscular dystrophies with adenosine triphoshate (ATP, 1953).

Beside of physics, **George A. Gamov (04.03.1904, city of Odesa, Ukraine – 19.08.1968, town of Boulder, CO, USA)**[694] — Ukrainian and American physicist, made the contribution to biochemistry by proposing the theory of origin for the genetic code set by the order of returnable nucleotide triplets, the basic component of deoxyribonucleic acid (DNA, 1954). He affirmed that sets of three bases is necessary to encode 20 standard aminoacids used by living cells to build proteins, which would allow to create a maximum of $4^3 = 64$ aminoacids.

In 1935 **Wendel M. Stanley (1904 – 71)** — American biochemist, isolated the tobacco mosaic virus in crystalline form and proved it to be a protein molecule. Likewise, in 1939, **John H. Northrop (1891 – 1987)** — American biochemist, isolated and crystallized an enzyme and isolated virus protein in pure form.

Severo Ochoa (1905 – 93) — Spanish biochemist, and **Arthur Kornberg (1918 – 2007)** — American biochemist, disclosed in 1955 the mechanism of biosynthesis of deoxyribonucleic acid (DNA): ribonucleic acid (RNA).

In 1939 **Carl E. Bachman** — American gynecologist and obstetrician, and **Samuel Gurin (1905 – 70)** — American biochemist, identified the gonadotropin-releasing hormone (GnRH) in pregnancy which became basis for the every-day tests to determine the pregnancy.

---

[693] Town of Terespil (in Polish – Terespol; appeared in 2nd half of 13 century), Pidlashshia (area 2,753.63 km²), Ukrainian National Republic (UNR, 1918-19) / Lublin Voivodeship (area 25,127 km²), Poland. — Ablitsov VH. «Galaxy Ukraine». The Ukrainian diaspora: prominent persons. Publ. «Kyt», Kyiv 2007.

[694] Town of Boulder (settled 1858), State of Colorado (CO, area 269,837 km²), USA. — — Gamow G. The Origin of Elements and the Separation of Galaxies. Physical Review 1948;74:505. — Gamow G. The evolution of the Universe. Nature 1948;162:680. — Gamow G. The Great Physicists from Gallileo to Einstein. Dover Publ., Inc., New York 1961. — Gamow G. Thirty Years That Shook Physics. The Story of Quantum Theory. Dover Publ., Inc., New York 1966. — Gamow G. My World Line; An Informal Autobiography. The Viking Press, New York 1970. — Frenkel V, Chernin A. Return of «nonreturner». Sputnik (Moscow) 1991;4:62-7. — Ablitsov VH. «Galaxy Ukraine». The Ukrainian diaspora: prominent persons. Publ. «Kyt», Kyiv 2007.

Erwin Chargaff (11.08.1905, city of Chernivtsi, Ukraine – 20.06.2002, New City, N.Y., USA)[695] — Ukrainian and American biochemist, who was the first to synthetize deoxyribonucleic acid (DNA) in the pure form (1947); by using the paper chromatography and ultraviolet spectroscopy showed that the number of adenine (A) units were equal to the number of thymine (T), and the number of units of guanine (G) was equal to the number of cytosine (C) and thus established the Chargaff Parity Rule 1 and the Chargaff Parity Rule 2 (1950-51, 1979; fig. 1).

The Chargaff Parity Rule 1 holds that a double-stranded DNA molecule globally %A = %T and %G = %C, and that constitutes the basis of Watson-Crick[696],[697] pairs in the DNA double helix model (1953).

The Chargaff Parity Rule 2 holds that globally both %A ~ %T and %G ~ %C are valid for each of two DNA strands and describes only a global feature of the base composition in a single DNA strand.

Fig. 1.
The Chargaff Parity Rule 1and 2: AT base pair / GC base pair.

---

[695] Chargaff E, Zamenhof S, Green Ch. Composition of human desoxypentose nucleic acid. Nature 1950 May;165 (4202): 756-**7**. — Chargaff E, Lipshitz R, Green C, Hodes ME. The composition of the deoxyribonucleic acid of salmon sperm. J Biol Chem 1951;192:223-230. — Chargaff, E. Heraclitean Fire: Sketches from a Life Before Nature. Rockefeller University Press. New York 1978. — Chargaff, E. Unbegreifliches Geheimnis. Wissenschaft als Kampf für und gegen die Natur. Klett-Cotta, Stuttgard 1980. — Chargaff, E. Bemerkungen. Klett-Cotta, Stuttgard 1981. — Chargaff, E. Warnungstafeln. Die Vergangenheit spricht zur Gegenwart. Klett-Cotta, Stuttgard 1982. — Chargaff, E. Kritik der Zukunft. Essay. Klett-Cotta, Stuttgard 1983. — Chargaff, E. Zeugenschaft. Essays über Sprache und Wissenschaft. Klett-Cotta, Stuttgard 1985. — Chargaff E. Engineering a molecular nightmare. Nature 1987;327(6119): 199–200. — Chargaff E. Abscheu vor der Weltgeschichte. Fragmente vom Menschen. Klett-Cotta, Stuttgard 1988. — Chargaff E. Alphabetische Anschläge. Klett-Cotta, Stuttgart 1989. — Chargaff E. Vorläufiges Ende. Ein Dreigespräch. Klett-Cotta, Stuttgart 1990. — Chargaff E. Vermächtnis. Essays. Klett-Cotta, Stuttgart 1992. — Chargaff E. Über das Lebendige. Ausgewählte Essays. Klett-Cotta, Stuttgart 1993. — Chargaff E. Armes Amerika – Arme Welt. Klett-Cotta, Stuttgart 1994. — Chargaff E. Ein zweites Leben. Autobiographisches und andere Texte. Klett-Cotta, Stuttgart 1995. — Chargaff E. Die Aussicht vom dreizehnten Stock. Klett-Cotta, Stuttgart 1998. — Chargaff E. Brevier der Ahnungen. Eine Auswahl aus dem Werk. Klett-Cotta, Stuttgart 2002. — Chargaff E. Stimmen im Labyrinth. Über die Natur und ihre Erforschung. Klett-Cotta, Stuttgart 2003. — Ablitsov VH. «Galaxy Ukraine». The Ukrainian diaspora: prominent persons. Publ. «Kyt», Kyiv 2007. — Pylypchuk OYa, Fando RO, Deforzh HV (Scientific Editor). Ervin Chargaff (1905-2002): monography. Publ. «Talkom», Kyiv 2017.

[696] **James D. Watson (1928-)** — American molecular biologist.

[697] **Francis H.C. Crick (1916-2004)** — English molecular biologist.

Research performed in 1939 by **Volodymyr O. Belitser (30.09.1906, city of Ryazan, Russia – 04.03.1988, city of Kyiv, Ukraine)**[698] — Ukrainian biochemist, concerning the use of adenosine three-phosphate (ATP) for cellular respiration showed that during «oxidative phosphorylation» (fraze created by him), three moleculei of ATP are synthetized, which is universal source of energy in living systems. At the same time, he demonstrated that «oxidative phosphorylation », is a process wherein breathing of cells could be supported by adding phosphates into proteins. He also, with **K.I. Kotkova,** proposed the protein blood substitute (PB)-8 (1955).

The growth factor is a naturally occurring protein or steroid hormone capable of stimulating cellular growth, the proliferation and cellular differentiation. Initially, **Rita Levi-Montalcini (1909 – 2012)** —- Italian neurobiologist and neurologist of Jewish descent, isolated the nerve growth factors (NGF) on nerve fibers in chicken embryos, then working together with **Stanley Cohen (1922 –)** — American biochemist, isolated the NGF from observations of certain cancerous tissues that cause extremely rapid growth of nerve cells (1952). The latter, then went on to discover epidermal growth factor. Research on cellular growth factors has proven fundamental to understanding the development of cancer and designing anti-cancer drugs.

The important work on the chemical structure of ferments was done by **William H. Stein (1911 – 80**) — American biochemist, **Stanford Moore (1913 – 82)** — American biochemist, and **Christian B. Afinsen (1916 – 95)** — American biochemist, in 1972.

Both, **Feodor F.K. Lynen (1911 – 79)**[699]— German biochemist and **Konrad E. Bloch (1912 – 2000)** — German American biochemist of Jewish ancestry, working separately, but cooperating with one another, between 1943-61 explained the metabolism of cholesterol and fatty acids in the human body.

After WW II they determined the sequence of thirty-six steps by which animal cells produce cholesterol.

In 1951 Feodor Lynen described the first step in the chain of reactions that resulted in the production of cholesterol with the discovery, that acetyl-coezym A (CoA) which is formed when acetate radicals reacts with CoA, was needed to begin chemical reaction. Furthermore, he discovered that as with cholesterol, acetyl-CoA was also necessary first step in the biosynthesis of fatty acids. The catabolism of fatty acids in

---

[698] City of Ryazan (founded 1095), Russia. — Belitser VA, Tsibakova ET. The mechanism of phosphorylation associated with respiration. Biokhimiia 1939;4:516-35. — Kalckar HM. 50 years of biological research – from oxidative phosphorylation to energy requiring transport regulation. Annual Review of Biochemistry 1991 July;60:1-37.— Ablitsov VH. «Galaxy Ukraine». The Ukrainian diaspora: prominent persons. Publ. «Kyt», Kyiv 2007.

[699] Lynen F. The biochemical basis of the biosynthesis of cholesterol and fatty acids. Wiener Klin Wschr 1966 July;78(27):489-97.

foods produce energy when fatty acids are burned up to form carbon dioxide (CO_2) and water.

Their work showed that the body first makes squalene (in Latin – word «squalenus» meaning «dirty», «ugly», «rough» «insatiable» genus of shacks), a natural 30-carbon organic compound obtained from shark liver oil and plants such as vegetable oil, from acetate over many steps and then converts the squalene oil into cholesterol. Also, acetyl-CoA is turned to mevalonic acid, a precursor in the mevalonate pathway that produce terpenes and steroids. Both showed that mevalonic acid is converted into chemically active isoprene, the precursor to squalene oil.

Additionally, Konrad Bloch discovered that bile and sex hormones were made from cholestrerol which led to the discovery that all steroids are made from cholesterol and showed that biotin (vitamin B_7) was needed to produce of fat.

Those discoveries are important because understanding metabolism of cholesterol and fatty acids, reveal its cause of cardiovascular diseases, specifically coronary artery disease (CAD), myocardial infarction (MI), degenerative diseases of the heart valve and stroke.

Herbert Ch. Brown (22.05.1912, city of London, England –died from acute MI **19.XII.2004, town Lafayette, IN, USA)**[700] — English biochemist of Jewish Ukrainian descent, discovered organoborane (1956). They are valuable precursors of ketones.

Horst Jatzewitz (1912 – 2002) and **Konrad Sandoff (1939 –)** — German biochemists, made the first biochemical description of GM1- gangliosides (Jatzewitz-Dandhoff's disease, 1963) and GM2 - gangliosides (Dandhoff's disease, 1968), a lysosomes genetic lipid storage disorder caused by the inherited deficiency of beta hexosaminidase A and B in children with mental retardation (amaurotic idiocy).

Britton Chance (1913 – 2010) — American biochemists and biophysicist, who invented the stop-flow device to measure the existence of the enzyme-substrate complex in enzyme reaction (1942) and introduced the numerical stimulation of biomechanical reaction and metabolic pathways (1952-60). Subsequently, he improved metabolic control phenomena in living tissues as studied by noninvasive techniques such as phosphorus nuclear magnetic resonance (NMR), optical spectroscopy and fluorometry, including the use of infrared light to characterize the properties of various tissues and breast tumors.

Studies of enzymes and proteins by nuclear magnetic resonance (NMR) originated by **Milfred Cohn (1913 – 2009)** — American biochemist born of Jewish parents, which concentrated on the way enzymes work to speed biological reactions in human body, provided the inside into the underlying thermodynamics of enzyme reaction.

---

[700] Town of Lafayette (founded 1825), State of Indiana (IN, area 94,321 rm^2), USA). — Ablitsov VH. «Galaxy Ukraine». The Ukrainian diaspora: prominent persons. Publ. «Kyt», Kyiv 2007.

Thus in 1958 using NMR she noted the first three peaks of adenosine triphosphate (ATP) and could distinguish for the first time the three phosphorus atoms of ATP with spectroscopic method. Then using radioactive oxygen (O_2) she discovered how the phosphorylation and water are part of the electron transport system of the metabolic pathway of oxidative phosphorylation, the ubiquitous (in Latin – «ubique» for anywhere, everywhere) process used by an aerobic organism to generate energy in the form of ATP from nutrients. She also explained the involvement of the divalent (in Latin – «valeo» for value) metal ions in the enzymatic reactions of adenosine diphosphate (ATP) and ATP by studying NMR spectra of the phosphorus nuclei and the structural change in the presence of various divalent ions.

The work of **Earl W. Sutherland, Jr (1915 – 74)** — American biochemist and pharmacologist, in 1956-70 led to the discovery that hormones, specifically adrenaline (epinephrine) and glucagon, act via a cyclic adenosine nonophosphate (cATP).

Initially, he helped to identify the importance of enzyme liver phosphatase (LP) in glycogenolysis and found that of the tree enzymes involved in the glycogenolysis, LP was rate-limiting, meaning the progression of glycogen metabolism is dependent on this enzyme.

Subsequently, he found that the enzymatic activity of LP depends on the addition or removal of a phosphatase group, i.e., phosphorylation, and that more phosphatase was taken up in the liver slides when epinephrine and glucagon were added, suggesting that these hormones were promoting phosphorylation of LP, activating an unknown enzyme.

Further researching disclosed that this response was characterized by particulate fraction produced by an unknown heat stable factor in the presence of epinephrine and glucagon. Unexpectantly, LP activation increased with the addition of 5-AMP.

At the end, the research showed that this unknown heat stable factor stimulated the secondary formation of LP in a fraction of the homogenate where the hormones were not present, finally discovering cAMP, the precursor of which was 5-AMP.

The interest of **Sune K. Bergström (1916 – 2004)** — Swedish biochemist and **Bergt I. Samuelsson (1934 –)** — Swedish biochemist, in the transformation of arachidonic acid led to the the discovery and identification of prostaglandins and related substances in the body in 1957 by the former, and he establishment if its chemical structure 1962 by the latter. Prostaglandins and related substances such as endoperoxides, thromboxane's and leukotriene have implication in thrombosis, the inflammation and allergy.

In 1953 the molecular structure of deoxyribonucleic acid (DNA) was discovered by **Francis H.C. Crick (1916 – 2004)** — British molecular biologist, physicist and neuroscientist, **Maurice H.F. Wilkinson (1916 – 2004)** — New Zealander-British molecular biologist and **James D. Watson (1928 –)** — American molecular biologist. Francis Crick and James Watson contrived a double (Watson-Crick) helix, each chain of which contains two complete chain information about the other chain, provides a structural formation of the mechanism by which the genetic information in DNA represents itself. However, Francis Crick, erroneously had claimed only that sequence information cannot flow out of protein into DNA or RNA, but he was commonly interpreted as saying that information flows exclusively from DNA to RNA to protein.

The research of **Robert F. Furchgott (1916 – 2009)** — American biochemist of Jewish faith, and **Louis J. Ignarro (1941 –)** — American pharmacologist, has led to the discovery in 1978-81 and 1983, respectively, of nitrous oxide (NO) – a substance secreted by the endothelial cells that relaxes blood vessels, called endothelium-derived relaxing factor (EDRF), a transient cellular signal in mammalian system and important compound in cardiovascular physiology. Also, both demonstrated in 1984 that NO is indirectly involved in dilatation of penile blood vessels. That led Louis Ignarro to introduction in 1986 a drug phosphodiesterase inhibitor type 5 – sildenafil (Viagra) to treat erectile dysfunction of the penis (impotency) and pulmonary hypertension. Independently, **Ferid Murad (1936 –)** — Albanian American physician and pharmacologist demonstrated that nitroglycerin (NTG) and related drugs worked by releasing NO into the blood vessels with subsequent relaxation of smooth muscle by elevating guanosine monophosphate (GMP, 1997).

Furthermore, the NO derived agents are not only effective in dilation of the smooth muscles of the blood vessels useful for the management of pulmonary hypertension and impotency in male, but they were found to mobilize the immune systems in humans against injury, infections and neoplasms. Specifically, in combination with doxorubicin (brand name – Adriamycin) became a chemotherapy medication effective in the treatment of carcinoma of the prostate (in Greek – prostates», meaning «protector» or «one who stands before») and other malignancies.

Reinhold Benesch (1917 – 86) — American biochemist and **Ruth E. Benesch (1925 – 2000)** — American biochemist, discovered in 1967 the enzyme 2,3-diphosphoglycerate (2,3-DPG), which is present in the human red blood cells (RBC). It binds with greater affinity to deoxygenated hemoglobin (HB, Hgb), e.g., when RBC is near respiratory tissue, releasing oxygen (O_2) into the tissue thus shifting O_2-Hb dissociation curve to the right along with decrease of pH and increase of temperature (T) – Bohr effect, than it does to oxygenated Hb, e.g., in lungs, where it keeps binding O_2 to Hb thus shifting the O_2-Hb dissociation curve to the left along with increase of pH and decrease T – Haldane effect.

Robert B. Woodward (1917 – 79) — American organic chemist, and **Roald Hoffmann (18.07.1937, town Zolochiv, Lviv Obl., Ukraine –)**[701] — Ukrainian American biochemist of Jewish descent, developed the theory regarding the course of chemical reaction (Woodward-Hoffmann's rules, 1969).

Isozymes, a.k.a., isoenzymes are proteins, first described in 1957 by **Robert. L. Hunter** and **Clement L. Markert (1917 – 99)**[702] — American biologists, who defined and developed concept of isoenzymes-based electrophoresis and staining enzymes, as different variant of the same enzyme having identical function and present in the same individual, or a different form of enzymes that catalyze the same reaction. Their creation of different forms is brought about by differences in the structure of genes, which are coded those isozymes.

Beginning in 1947 Sir **John Kendrew (1917 – 97)** — English biochemist and crystallographer, began study the molecular structure of biological systems, specifically on the heme-containing proteins using the horse heart, then the whale meat, but he encountered difficulties with the phase problem in analysis of the distraction pattern. In 1953 he was joined by **Max F. Perutz (1914 – 2002)** — Austrian British molecular biologist of Jewish descent, who solved this problem by multiple isomorphous replacement, comparison of patterns from several crystals, so an electron density at 6 A[703] / 0,9 nanometers (nm) resolution in 1957 was improved at 2 A / 0,2 nm resolution (full) in 1959. This allowed to determine the first atomic structures of proteins using an x-ray crystallography, the structure of oxy- and deoxy- hemoglobin at high resolution, and the protein of the myoglobin molecule (quarter the size of Hb molecule) which stores oxygen (O_2) in the muscle cells. By 1970 Max Perutz was able to demonstrate the work of Hb, how it switches between its deoxygenated and its oxygenated states, in turn triggering the uptake of O_2 in the lungs, then release to the muscles and other organs.

Although the protein nature of antibody molecule had already been established, but their chemical structure was puzzling. This had ignited **Rodney R. Porter (1917 – died in a road accident in 1985)** — English biochemist, to work on uncovering the basic chemical structure of antibody molecule.

Rodney Porter had worked with **Frederic Sanger (1918 – 2013)** — English biochemist, who determined the structure of protein's aminoacids, especially that

---

[701] Town of Zolochiv (occupied since 1180, mentioned 1326, founded 1443), Lviv Obl., Ukraine. — Ablitsov VH. «Galaxy Ukraine». The Ukrainian diaspora: prominent persons. Publ. «Kyt», Kyiv 2007. — Olearchyk AS. A Surgeon's Universe. Vol. I-IV. 4th Ed. Publ. «Outskirts Press», Denver, CO 2016. Vol. I. P. 721-2. — Stakhiv EZ. Roald Hoffmann: Ukraine's Nobel laureate. The Ukrainian Weekly 2017 Sep. 17;85(38): 7-8.

[702] Hunter RL, Markert CL. Histochemical demonstration of enzymes separated by zone electrophoresis in starch gels. Science June 28, 1957;125(3261):1294-5.

[703] **Andreas J. Angström (1814 - 74)** — Swedish physicist, in whose name is defined an **A** or **angst** – a unit of the light walve-length equals to ten millionth of a millimeter.

of insulin and developed the method to decipher (decode) the primary structure of deoxynucleic acid (DNA), on determining the N-terminal aminoacids in protein sequences. The former using the proteolytic enzyme papain (Carica papaya) found that rabbit IgG was cleaved into three similarly sized pieces: two of these, the Fab (antigen-binding) fragments, retained their original antibody specificity but the third portion, Fc (fragment crystallizable), crystallized.

In 1953 **Edmund C. Kornfeld (1919 – 2012)** — American organic chemist, isolated antibiotic Vancomycin, a branched tricyclic glycosylated non-ribosomal peptid produced by Actinobateria species Amycolatopsis (formerly – Nocardia) orientalis. Vancomycin is used IV to treat complicated skin infections, septicemia, endocarditis, bone and joint infections and meningitis caused by methicillin-resistant Staphylococcus aureus (MRSA), and p.o. as a therapy for severe Gram-positive rod Clostridioides (Clostridium) difficile pseudomembranous colitis.

The peptide chain within IgG molecules were linked by disulfide bridges as shown by **Gerald M. Edelman (1929 – 2014)** — American immunologist of Jewish family. The bridges are both inter- and intra-chain. By adopting techniques used to reduce SS bounds so that the inter-chain links were preferable reduced. Rodney Porter found there were four chains in each antibody molecule, two identical larger chains, the heavy chains, and the two identical smaller ones, the light chains. Therefore, in 1962, he proposed the 4-chain model for IgG molecules, with two light and two heavy chains.

Frederic Sanger isolated and identified the complete aminoacid sequence of the two-polypeptide chain of bovine insulin, A and B (1951 and 1952, respectively), and created the method to identify the order of aminoacids in more complicated protein molecules (1975).

Both, Frederick Sanger and **Walter Gilbert (1932 –)** — American biochemist, physician and molecular biology pioneer, contributed to concerning of base sequences in nucleic acids.

Working together in 1964-66 on the glycogen phosphorylase, **Edwin G. Krebs (1918 – 2009)** — American biochemist and **Edmond H. Fisher (1920 –)** — Swiss American biochemist, defined a series of reactions leading to the activation or inactivation of this enzyme as triggered by hormones and calcium (Ca^{++}), and in the process discovered reversible protein phosphorylation, which works by a protein kinase moving a phosphate group from adenosine triphosphate (ATP) to a protein. Thus, the shape and the function of the protein is altered enabling it to participate in some biological process. When the protein has completed its role a protein phosphatase removes the phosphate, and the protein reverts to its original state. This cycle represents the biological control of various cellular metabolic processes.

In 1959 **Rosalyn S. Yallow (19.07.1921, New York City, N.Y., USA – 30.05.2011, borough of Bronx, New York City, N.Y., USA)**[704] — American medical physicist of Ukrainian Jewish origin, developed of radioimmunoassay (RIA; from Anglo-French – «assaier», «assai» or «essai», meaning «trial») or radioisotope tracing technique that allow measuring of tiny quantities of various biological substances in human blood as a magnitude of other aqueous fluids.

The RIA consists of the creation of two reagent: (1) a radioactive molecule, the product of covalently a radioactive isotope atom with a molecule of the target substance, e.g., peptide hormone; (2) an antibody which specifically chemically attaches itself to the target substance where two are in contact. The initial radioactivity of a mixture of two reagents is then measured, and then added to a measured quantity of fluid, e.g., blood containing of unknown, or very small concentration of target substance. Since the antibodies preferentially attach to non-radioactive molecules, the proportion of radioactive target antibody links is reduced by the amount proportional to the concentration of the target substance in the fluid. When the final radioactivity of the isolated target antibody is measured, the concentration of the target substance, i.e., the amount of insulin in the blood can be calculated.

Initially the RIA was used for the blood volume calculation, then since 1960 for sensitive determination of the insulin level in the blood serum patients suffering from diabetes mellitus (DM), finding the level of the biological compounds in the blood, testing the blood for peptide neurohormones, kinetics of the iodine metabolism in patients with thyroid gland diseases, determination of vitamins and enzymes in the blood, which could not be established at that time because its low concentration in the blood.

Independently and simultaneously, **Roger C.W. Guillemin (1924 –)** — French American biologist and neurologist and **Andrew V. Schally (1926 –)** — Canadian American endocrinologist of Jewish Lithuanian Polish origin, clarified the structures of thyrotropin (thyroid) – releasing hormone (TRH) – a topic tripeptide neurohormone produced by hypothalamus that stimulates TRH and prolactin from the anterior pituitary gland (hypophysis), used clinically for the treatment of spinocerebellar degeneration and disturbances of consciousness in human (1969), and gonadotropin-releasing hormone (GnRH) – a topic peptide neurohormone responsible for the release of follicular-stimulating hormone (FSH) and luteinizing hormone (LH) – a tropic peptide neurohormone from the anterior pituitary gland synthetized and released from the hypothalamus **[Carl Bachman** and **Samuel Gurin (1905 – 97):**

---

[704] Borough of Bronx (area 150 km², created 1898), city of New York (Foot Note 607), N.Y., USA. — Ablitsov VH. «Galaxy Ukraine». The Ukrainian diaspora: prominent persons. Publ. «Kyt», Kyiv 2007.

above], used for the treatment of precocious puberty, breast and prostate carcinoma and endometriosis (1977).

The research of **Robert W. Holley (1922 – 93)** — American biochemist, **Har G. Khorana (1922 – 2011)** — Indian American biochemist, **Marshall W. Nirenberg (1927 – 2010)** — American biochemist and geneticist, **J. Heindrich Matthaei (1929 –)** — German biochemist and **Philip Leder (1934 –)** — American geneticist, deciphered the genetic code and its function in protein synthesis. The genetic code is the set of rules by which genetic material – ribonucleic acid (RNA), or messenger / transfer ribonucleic acid (mRNA / tRNA) sequences is translated into proteins by living cells.

In 1961 Heinrich Matthaei personally deciphered the genetic code (codon) made of four bases (adenine, cytosine, guanine, and thymine).

Robert Holley's research of RNA concentrated first on isolating tRNA, later determining the sequence and structure of tRNA, the molecule that incorporates the amino-acid alanine into proteins. The tRNA structure was determined by using two ribonucleases to split the tRNA molecules into pieces. Each enzyme split the molecule at location points for specific nucleotides. The entire structure of the tRNA molecule was determined in 1964 by «puzzling out» the structure of the pieces split into two different enzymes, then comparing the pieces from both enzymes split. It was a key discovery in explaining the synthesis of proteins from mRNA and the third ever performed determination of RNA nucleotide sequence.

Using the Robert Holley method, the structures of the remaining tRNA were determined, and a few years later it also was modified to track the sequence of nucleotides in various bacterial, plants and human viruses.

Har Khorana had established that the mother of all genetic codes – the biological language common to all living organism, is spelled out in three nucleotides codes for a specific aminoacid. Beside of this, he was the first to create oligonucleotides (1979).

Marshall Nirenberg is known for breaking the genetic code and describing how it operate in protein synthesis. He with Philip Leder performed experiment which elucidated the triplet nature of the genetic code and allowed the remaining ambiguous codons in the genetic code to be deciphered (the Nirenberg and Leder experiment, 1964).

Pioneering on gene splicing of recombinant deoxyribonucleic acid (rDNA) was performed by **Paul Berg (1926 –)**[705] — American biochemist of Jewish origin, who was the first to create a molecule containing DNA from two different species by inserting DNA from another species into a molecule (1974). The gene-splicing technique was a fundamental step in the development of modern genetic engineering. It sparked

---

[705] Berg P. Potential Biohazards of Recombinant DNR molecules. Letters, Science 26 Jul 1974;185(4148):303.

an industry to manufacture new drugs. The ability to recombine pieces of DNA and transfer them into cells is the basis of the important new medical approach to treating disease by a technique named gene therapy.

Together, **Irwin A. Rose (16.07.1926, borough of Brooklyn, New York City, NY, USA – 02.06.2015, town of Deerfield, Franklin County, MA, USA)**[706] — American biochemist of Jewish Ukrainian descent, **Abraham Hershko (1937 –)** — Israeli biochemist, and **Aaron Ciechanover (1947 –)** — Israeli biochemist, working on lysosome-independent intracellular proteolysis using as a model reticulocytes – predecessors of red blood cells (RBC), devoid of lysosomes, led to discovery in 1980 ubiquitin-dependent proteolysis[707], i.e., an importance of ubiquitin in cellular system of protein degradation in proteosomes. Thus, ubiquitin-proteosome system, that is ubiquitin-intermediate degradation of proteins by proteosome have a critical role in the maintenance of cells homeostasis, influences the development and progress of many diseases, in particular cancer, muscular and neurological illnesses, immune and inflammatory response of an organism.

While working on the green fluorescent protein (GFP) in the bioluminescent jellyfish (Aequorea victoria), **Osamu Simomura (1928 – 2018)** — Japanese American organic chemist, discovered in 1961 the protein, a key tool to trace processes in the individual cells. This protein does not need any additives to glow. Instead, it must be radiated with ultraviolet or blue light before forming bright green. Because it can attach itself to individual proteins, GFP is used as a «tag» allowing to tract sorts of cellular processes under a microscope. It can illuminate growing malignant tumor, pinpoint of Alzheimer's **[Alois Alzheimer (1864 – 1915). 20th century. Psychiatry and psychology]** disease in the brain and a spot of light in the growth of bacteria.

Since 1970 **Jaroslav O. Babjuk (06.VI.1929, city of Kolomyia, IF Obl., Ukraine – 2004, city of Praha, Chechy)**[708] — Ukrainian Czech biochemist and pharmacologist, performed bioanalytical research and conrolled dispertion of drugs.

---

[706] Borough of Brooklyn (settled 1634), New York City (Foot Note 607), N.Y., USA. — Town of Deerfield (settled 1673), Franklin County, State of Massachusetts (MA, area 27,337 km^2), USA. — Ablitsov VH. «Galaxy Ukraine». The Ukrainian diaspora: prominent persons. Publ. «Kyt», Kyiv 2007.

[707] Ubiquitin (in English – «present everywhere»).

[708] City of Kolomyia (first mention dated 1240), IF Obl., Ukraine. — Babjuk J, Knobloch E, Palat K. Ve farmacii; 6. Avicenum, Praha 1984. — Babjuk J, Perlik RF, Sidlo Z. Bioanalityka leku. Vyd. 1. Nakladatelske udaje: Avicenum, Praha 1990. — Ablitsov VH. «Galaxy Ukraine». The Ukrainian diaspora: prominent persons. Publ. «Kyt», Kyiv 2007.

The experiments of **Jurij Rozhin (19.III.1931, city of Kharkiv, Ukraine –)**[709] — Ukrainian American biochemist and cancer researcher, showed that the suppression of prostaglandin biosynthesis and restraining ornithine decarboxylation (ODC) by used alone indomethacin, and difluoromethylornithine (DFMO) has been shown to inhibit cancer formation is several systems, including colon in rats and humans.

Sviatoslaw Trofimenko (15.12.1931, city of Lviv, Ukraine – 26.02.2007, city of Wilmington, DE, USA)[710] — Ukrainian chemist, discovered heterocycles – pyrazolyborate and polypyrazolyborate ligands (1966), created the concept of «cyclometalation» (1973) and «scorpionates» (1999).

Sidney Altman (Foot Note 614)[711] and **Thomas R. Cech (1947 –)** — American chemist of Czech origin, discovered in 1989-2007 catalytic properties of ribonucleic acid (RNA).

In 1985, **George P. Smith (1941 –)**[712] — American biochemist, developed a technique, named display of phagocyte or phage display, where a specific protein sequence is artificially inserted into the coat protein gene of a bacteriophage – a

---

[709] Rozhin I. To the history of medicine in Ukraine. J Ukr Med Assoc North Am 1956;3,2(6):19. — Bollet AJ, Rozhin J, Simpson W. The hexosamine content of the serum globulin in normal and pathological sera. J Clin Invest 1957 Jan 1;36(1):51-8. — Rozhin I, Rozhin V. Volodymyr Pidvysoc'kyi. J Ukr Med Assoc North Am 1957;4,1(17):43. — Rozhin I. Professor D-r Olexandra Smyrnova-Zamkova. J Ukr Med Assoc North Am 1959;3,1(5):32. — Rozhin I. Dozent Doctor Andriy Zhuravel. Memory – Feature. J Ukr Med Assoc North Am 1960;7(17):25-30; 1960;7(18):37-40. — Rozhin I. Materials for the history of veterynary-medical science. J Ukr Med Assoc North Am 1963;10,3(30):2-7. — Rozhin J, Sonderstrom RL, Brooks S. Specificity studies on bovine adrenal estrogen sulfotransferase. J Biol Chem 1974 Apr10;249(7):2079-87. — Rozhin J, Wilson PS, Bull AW. Ornithine decarboxylase activation in the rats and human colon. Cancer 1984;44:3226-30. — Sloane BF, Rozhin J, Johnson K et al. Cathepsin B: association with plasma membrane in metastatic tumors. Proc Natl Acad Sci USA 1986;83:2483-7. — Rozhin J, Wilson PS, Bull A, Nigro NA. Ornithine decarboxylase activation in the rats and human colon. J Ukr Med Assoc North Am 1987 Autumn;34,2(116):67-77. — Sloane BF, Rozhin J, Krepela E et al. The malignant phenotype and cysteine proteinases. Biomed Biochim Acta 1991;50:549-54. — Sloane BF, Rozhin J, Gomez AP et al. Effects of 12-hydroxyeicosatetraenoic acid on release of cathepsin B and cysteine proteinase inhibitors from malignant melanoma cells. In, K.V. Honn et al (eds). Eicosanoids and Other Bioactive Lipids in Cancer and Radiation Injury. Kluver Acad Publ 1991. — Sloane BF, Rozhin J, Moin K et al. Cysteine endopeptidase and their inhibitors in malignant progression of rat embryo fibroblast. Biol Chem Hoppe-Seyler 1992 Jul;373(2):589-94. — Rozhin J, Sameni M, Ziegler G et al. Pericellular pH affects distribution and secretion of cathepsin B in malignant cells. Cancer Research 1994 Dec 15;54:6514-25. — Ablitsov VH. «Galaxy Ukraine». The Ukrainian diaspora: prominent persons. Publ. «Kyt», Kyiv 2007.

[710] City of Wilmington (settled 1638), State of Delaware (DE, area 6,452 km^2), USA. — Trofimenko S. Boron-Pyrazole Chemistry. J Am Chem Soc 1966;88(8):1842-44. — Trofimenko S. Scorpionates. The Coordination Chemistry of Polypyrazolyborate Ligands. Imp Coll Press, London 1999. — Ablitsov VH. «Galaxy Ukraine». The Ukrainian diaspora: prominent persons. Publ. «Kyt», Kyiv 2007.

[711] Ablitsov VH. «Galaxy Ukraine». The Ukrainian diaspora: prominent persons. Publ. «Kyt», Kyiv 2007.

[712] Smith GP. Filamentous fusion phage: novel expression vectors that display cloned antigens on the virion surface. Science 14 Juni 1985;228(4705):315-17. — Smith GP, Petrenko VA. Phage display. Chem Rev 1997 Apr 1;97(2):391-410.

virus (virion) that infect bacteria, causing the protein to be expressed on the outside of the bacteriophage, can be employed to evolve new proteins and could be used for directed evolution of antibodies.

The prions («proteinoids» and «infectious»), discovered and named in 1982 by **Stanley B. Prusiner (1942 –)** — American biochemist and neurologist, are a class of infectious self-reproducing pathogens primarily or solely composed of protein in a misfolded form, responsible for the transmissible spongiform encephalopathies in humans and animals, specifically in bovines («mad cow disease»).

Serhiy V. Komisarenko (09.VII.1943, city of Ufa, Bashkortostan, Росія –)[713] — Ukrainian biochemist, who researched the biochemistry of molecular immunology(1976-90), proved the decrease of quantity and activity of innate cytotoxic T-lymphocytes, necessary for the innate immune system, under the influence of low dose of radiation (1994), examined the structure and biological activity of biopolymers (2004).

Randy W. Schekman (1948 –) — American cytologist, **James E. Rothman (1950 –)** — American biochemist, and **Thomas Ch. Südhof (1955 –)** — German American biochemist, discovered the regulation mechanism of vesicles[714] trafficking by protein at the heart of neurotransmission, hormone secretion, cholesterol homeostasis and metabolic regulation, an important transport system of cells.

Randy Schekman in 1980-96 isolated selenocysteine (sec)[715] mutants that accumulate secretory pathway intermediates, cloned the corresponding genes, and established biochemical reactions that reproduces specific secretory pathway events, thus transforming the secretion pathway, previously descriptive and morphologic, into molecular and mechanistic one. Established by him the cell-free reactions led to isolation of the sec61[716] translocation of a complex coatomer protein (COP) II[717]

---

[713] City of Ufa (early history dates to Paleolithic / Stone Age c. 3.3 million years ago, founded 1574), Bashkortostan (area 142,947 km²), Russia. — Ablitsov VH. «Galaxy Ukraine». The Ukrainian diaspora: prominent persons. Publ. «Kyt», Kyiv 2007.

[714] Vehicles are tiny sac-like structures that transport hormones, growth factors and other molecules within the cells, known how to reach their correct destination and where and when to release their content. This cellular trafficking underlies the propagation of the cell itself in division, communication between the nerve cells in the brain, secretion of hormones and nutrient uptake. Defects in this process leads to metabolic diseases like diabetes mellitus (DM) and infectious diseases such as botulisms.

[715] The conservation of the sec proteins and the trafficking mechanisms are important in neurotransmission, hormone secretion, cholesterol homeostasis and metabolic regulation.

[716] The sec61 is an endoplasmic reticulum (ER) membrane protein translocator, allowing translocation of polypeptide into ER lumen.

[717] COPII is a coatomer (a protein complex that coats membrane-bound transport vesicles) – a type of vesicle coat protein that transfer antegrade proteins from the rough ER to the Goldi apparatus **[Camillo Goldi (1843 - 1926).** 19th century. Anatomy, histology, and embryology], an organelle found in the most eukaryotic cells, a part of the endomembrane system in cytoplasm which packages proteins into membrane-bound vesicles inside the cells before the vesicles are sent to their destination.

that coat membrane-bound transport vesicles and the first purified inter-organelle transport vesicles.

James Rothman detailed the function of vesicles and the cellular trafficking (1978-2008).

Thomas Südhof described the role of low-density lipoprotein (LDL) receptor in cholesterol metabolism (1983), elucidated the principle of receptor-mediated endocytosis (1985), discovered of steroid regulatory elements and LDL receptor function that led to the subsequent development of statin derived cholesterol lowering medications such as atorvastatin (lipitor), and discovered the machinery mediated neurotransmitter release from the presynaptic neuron, synapse formation and specification (1986).

Evolution, meaning the adaptation of species to different environment, has created an enormous diversity of life.

The work of Sir **Gregory P. Winter (1951 –)** — English biochemist, is dedicated to the therapeutic use of the monoclonal antibodies, and the invention of technique for humanity (1986). He also used phage display for evolution of antibody with the aim to produce new pharmaceuticals, such as adalimumab being used for the treatment of rheumatoid arthritis, psoriasis, and inflammatory diseases of the bowel[718].Since then, he produced antibodies able to neutralize toxins, counteract autoimmune disease and cure metastatic carcinoma.

Both, George Smith, and Gregory Winter found proteins on the surface of bacteriophage, where it is possible to bring a cod of the other protein, for example a functional part of antibody. As a result, they succeeded to implant a cod of other protein, not one, but whole population the very similar genes, i.e., coded in all antibodies of a human being.

In 1993 **Frances N. Arnold (1956 –)** — American biochemist and engineer, has used the same approach for the genetic change and selection to develop protein that solve chemical problems of the mankind, by conducting the first directed molecular evolution of enzymes, which are proteins, catalyzing (speeding up) the chemical reaction and active in water-organic mixtures, or capable to catalyze non-natural reactions, e.g., proteins synthesizing silica organic compound in bacteria[719].

---

[718] Winter GP, Griffiths AD, Hawkins RE, et al. Making antibodies by phage display technology. Ann Rev Immunol Apr 1994;12:433-55.

[719] Arnold FH. Engineering proteins for non-natural environments. FASEB J 1993;7(9):744-9. — Arnold FH. Engineering proteins for unusual environments. Current Opinion in Biotechnology. Aug 1993;4(4):450-55. — Chen K, Arnold FH. Turning the activity of an enzyme for unusual environments: sequential random mutagenesis of subtilisin E for catalysis in dimethylformaline. Proc Natl Acad Sci USA 1993;90:5618-22. 1993.

Directed molecular evolution of enzymes consist in that a sequence of gene correcting synthesis of protein, by casual selection bring in mutation, whereupon takes away protein with necessary properties, e.g., increase activity.

Subsequently, she discovered the basic principles of the biological design and incarnated successful findings into the medical molecular diagnostics, pharmaceutical and fuel industry. She has refined the methods that are now used to develop new catalysts. Enzymes that she developed allows manufacturing more environmentally friendly chemical substances, such as pharmaceuticals, and the production of renewable fuel for greener transport.

The extend of scientific research of **Mykhailo M. Korda (13.II.1965, village Veryn, Mykhaylivsky Region, Lviv Obl., Ukraine –)**[720] — Ukrainian biochemist, include biochemistry free oxygen radicals (H_2O_2, O_2^-); toxicology of the liver; formation mechanism for reactive oxygen (O_2) and azote (N) forms in hepatocytes; role of monoxide azote (N) NO [20th century. Biochemistry (above) – **Robert Furchgott (1916 – 2009) / Luis Ignarro ((1941 –) / Ferid Murad (1939 –)]**, per-oxynitride and super-oxide in diseases of cardio-vascular system; epithelial dysfunction in obesity, hypertension and atherosclerosis (AS); free-radical mechanism for the development of AS; nanotoxicology, the use of nanomaterials in medicine and pharmacology and berylliosis. He holds the position of Rector of I.Ya. Horbachevsky Ternopil National Medical University (TNMU, founded 1957) since 2015, city of Ternopil, Ukraine.

Microbiology and immunology

Ivan Dudchenko (19.IX.1857, town of Novomyrhorod, Kirovohrad Obl., Ukraine – killed **05.IV.1917, town of Chita, Zbaikalsky kray, Russia)**[721] — Ukrainian physician-microbiologist and epidemiologist, after graduation from the Medical Faculty of St Volodymyr Kyiv University (1885), maintained medical practice in Ukraine – city of Odesa and town Zlatopil (founded seemingly 1545, connected in

---

[720] Village Veryn (founded 1451), Mykhaykivsky Region, Lviv Obl., Ukraine. — Korda MM, et al. Isolated hepatocytes: the use in experimental and clinical hepatology. J Ukr Acad Med Sc 1997;1():0-0. — Korda MM, et al. Hepato-protection property of crezacine under circumstances of toxic action of galactosamine. Liky. 1998;1():0-0. — Korda MM. Lipid peroxidation is a prerequisite for galactosamine-induced damage. Current Topics in Biophysics. 2000;24(2):0-0. — Korda MM, et al. Effect of beta-blockers on endothelial function during biological aging: a nanotechnological approach. J Cardiovascular Pharmacology. 2008; № 51(2). — Ablitsov VH. «Galaxy Ukraine». The Ukrainian diaspora: prominent persons. Publ. «Kyt», Kyiv 2007.

[721] Town of Novomyrhorod (founded 1740s; inhabitants were assaulted during sponsored by Russian Communist Fascist regime Holodomor- Genocide 1932-33 against the Ukrainian Nation), Kirovohrad Obl. (area 24,588 km²), Ukraine. — Town of Chita (founded 1653), Zabaikalsky Kray (area 431,500 rm²), Russia. — Kazimirov VV. To follow at the heels of forgotten life. Doctor I.S. Dudchenko. Chita 1957. — Ablitsov VH. «Galaxy Ukraine». The Ukrainian diaspora: prominent persons. Publ. «Kyt», Kyiv 2007.

1959 to town Novomyrhorod, Kirov Obl.), in the Middle Asia, Ural Mountains (781,100 км²), Siberia (4,360,000 km²), Far East (6,942,900 km²) and in town of Nerchynsk (founded 1653), Zabaikalsky Krai (1896-98).

In 1908 was directed by the Special Anti-plaque Commission to bordering with Mongolia (area 1,566,000 km²) regions of Zabaikalsky Krai to study plaque, in 1912 became the first Director of the Chita Antiplaque Laboratory. During this position he described series of outbreaks of plaque, subjected them to the epidemiologic analysis, studied morphology and peculiarities of its causative factor, proved the pathological picture of plaque pneumonia, characteristics of leucocyte reaction, put a question about liquidation of fleas as an intermediary in transmission of this disease, one of the first expressed hypothesis about mutual connection of plaque in people with so called «disease of Mongolian marmot / prairie dog (Marmota sibirica)», which in 1911 was proved by **Danylo Zabolotny (1866 – 1929**: 19th century – Microbiology and immunology). One of the first suspected bacterial origin of diarrhea in children, studied flaming in Latin («ignis fatuus») of leprosy (leper, leprous) in Zabaikalsky Krai.

Initially, **Leon A.Ch. Calmette (1863 – 1933)** — French physician, bacteriologist, and immunologist, developed the first antivenin for snake venom (the Calmette serum, 1895). Then together with **Camille Gu**érin **(1872 – 1961)** — French veterinarian, bacteriologist, and immunologist, were the first to discover an attenuated form of Mycobacterium tuberculosis – the bacillus Calmette-Quérin (BCG, 1921) used for an inoculation against tuberculosis.

In 1903 **Simon Flexner (1863 – 1946**: 19th century. Microbiology and immunology**)** developed serum treatment for meningococcal meningitis, and in 1910 he and **Georg Jochman (1874 – 1915)** — German physician, published on serum diagnosis and therapy for epidemic meningitis.

In 1910 Simon Flexner with his assistant **Paul A. Lewis (1879 – 1929)**, identified the poliovirus, a causative factor of an acute infectious cerebrospinal disorder, causing paresis (weakness) or paralysis (absence) of voluntary movements (poliomyelitis), determined its entry into the body through the nose, then attacking the cranial 1st (olfactory) nerve, and demonstrated that monkeys can be infected by administering poliomyelitis virus in the nasopharynx.

Ipolit O. Deminsky [16.IV.1864, town Novomyrhorod, Kirovohrad Obl., Ukraine – 09(22).X.1912, village Rakhinka, Serednioakhtubynsky Region, Volgograd Obl., Russia][722] — Ukrainian physician-epidemiologist, who in August 1922 was able to isolate plague from a death marmot, thereby proving that

---

[722] Town Novomyrhorod (Foot Note 721), Kirovohrad Obl., Ukraine. — Village Rakhinka (founded before 1859), Serednioakhtubynsky Region, Volgograd Obl. (area 112,877 km²), Russia. — Ablitsov VH. «Galaxy Ukraine». The Ukrainian diaspora: prominent persons. Publ. «Kyt», Kyiv 2007.

marmots were natural plague hosts. Died after contracting plague during laboratory experiments.

Pavlo (Pavel) N. Laschtschenko (16/28.06.1865, city of Kharkiv, Ukraine – 27.04.1927, city of Tomsk, Russia)[723] — Ukrainian physician-hygienist and bacteriologist, who after graduation from the Medical Department of Kharkiv University (1886), where he defended dissertation for the degree of Medical Doctor (1893), became Privat-Dozent (1895-1904) his of alma mater, and then Extraordinary (1904) and Ordinary (1906) Professor of hygiene at the Tomsk University (founded 1878). He was the first to prove the possibility of transmission of infection via the air drops (1898), described bactericidal action of the hen-egg white, indicating that it contains proteolytic enzyme (ferment), which condition this action, although enzyme itself he had not isolated (1909). Actually, that enzyme was isolated in 1922, and termed «lysozyme».

In 1905 **Fritz R. Schaudinn (1871 – 1906)** — German Lithuanian zoologist, and **Erick Hoffmann (1868 – 1959)** — German dermatologist, proved the cause of syphilis to be Spirochete pallida (Treponema pallidum).

The following year (1906) **August P.** von **Wassermann (1866 – 1925)** — German bacteriologist and hygienist with Jewish origins, and **Albert Neisser (1855 – 1916**: 19th century. Microbiology and immunology**)** introduced the diagnostic serological test for syphilis (Wasserman's test or reaction).

Next year (1907) **Paul Erlich [(1854 – 1915)**. 19th century. Microbiology and immunology] and **Alfred Bertheim (1879 – 1914)** — German chemist, opened a new era an in antimicrobial chemotherapy with the discovery of arsenic-derived agent to treat infections caused by trypanosome in mice and spirochete in rabbits.

Three years later (1910) **Paul Erlich** (above) and **Sahachiro Hata (1873 – 1938)** — Japanese bacteriologist announced the discovery of an arsenical preparation 606 (salvarsan), now known as arsphenamine, against Treponema pallidum and effective in the treatment of syphilis.

Among the bacterial causes for the cranial VII (facial) nerve (Bell's) palsy, described by **Rhazes (865 – 927**: Medieval medicine), **Avicenna (980 – 1037**: Medieval medicine),

---

[723] City of Tomsk (founded 1604), Tomsk Obl. (318,000 km²), Russia. — Laschtschenko P. Über Producte aus sogen Waldwolle. Archiv für Hygiene 1898;33:193-205. — Laschtschenko P. Über Luft infection durch beim Husten, Niesen and Sprechen Verspritze Tropfchen. Zeitschrift für Hygiene und Infektion Krankheiten 1899 Dez;30:128-38. — Laschtschenko P. Über Extraction von Alexinen aus Kaninchenleukozyten mit dem Blutserum anderen Tiere. Archiv für Hygiene 1899. — Laschtschenko P. Zeitschrift für Hygiene und Infektion Krankheiten 1909;64,1(12):1432-1831. — Laschtschenko P. Über die keimtötente und entwicklungshemmende Wirkung von Hühnereiweiss. Zeitschrift für Hygiene und Infektion Krankheiten 1909;64:419-27. — Fleming A. One remarkable bacteriolytic element found in tissue and secretions. Proc Roy Soc London 1922 May;93(653):306-17. — Ablitsov VH. «Galaxy Ukraine». The Ukrainian diaspora: prominent persons. Publ. «Kyt», Kyiv 2007.

Anton Freidreich (1761 – 1836: 18th century) and Sir **Charles Bell (1777 – 1842**: 20th century. Anatomy, histology, and embryology), stands out Borrelia burgodorferi, a genus of the Spirochete phylum, initiating borreliosis or Lyme disease, characterized by relapsing fever and the classic «bull-eye» rash, called erythema migrant. Borrelia burgdorferi was named after **Amedee Borrel (1867 – 1939)** — French biologist, who discovered the former eponym of the bacteria, and after **Wilhelm «Willy» Burgdorfer (1925 – 2014)** — American scientist, who discovered in 1981 the latter eponym of the bacteria.

Of founders of the experiments related to methanogenesis, **Vasyl L. Omeliansky (26.II/10.III.1867, city of Poltava, Ukraine – 21.IV.1867, town of Gagra, Abkhazia)**[724] — Ukrainian microbiologist, synthesized cultures of anaerobic and sporous bacteria, which caused fermentation of tissues with creation of organic acids and hydrogen (H), investigating methanogenetic fermentation of cellulose [19th century. Biochemistry. – **Anselme Payen (1795 – 1871)**] and ethanol discovered H as product of cellulose fermentation (c. 1909).

In 1896-97 **Johannes A.G. Fibiger (1867 – 1928)** — Danish physician, conducted the first controlled clinical trial on diphtheria serum in which random allocation was used and emphasized as a pivotal methodological principle. He thought Spirotera carcinoma from the class Nematoda which he had discovered, caused tumors of the stomach in rats (1907). It was a false conclusion since the hypertrophy of the epithelium of the stomach was due to lack of vitamin A (retinol and carotenes). Although a similar parasite Spiraea loupes can cause esophageal osteosarcoma in dogs.

Mykhailo (Michel) V. Weinberg (30.I.1868, city of Odesa, Ukraine – 21.IV.1940, Commune Bullion, Department Yvelines, Region L'lie-de-France, France)[725] — Ukrainian French microbiologist, developed a complement fixation serological test (Weinberg) test for diagnosis of hydatid disease, an infectious disease, affecting mostly the liver and the lung characterized by formation of cyst filled with fluid and echinococcus formed larvae (hydatid cysts), growing in peoples, big horned domestic

---

[724] City of Poltava (founded 899), Ukraine. — Town of Gagra (colonized by the Greeks 8th-7th century BC), Abkhasia (area 8,660 km²) — Omelianski VL. Fermentation methanique de l'alcool ethylique. Annales de l'Institut Pasteur 1916;30:56-60. — Ablitsov VH. «Galaxy Ukraine». The Ukrainian diaspora: prominent persons. Publ. «Kyt», Kyiv 2007.

[725] Commune Bullion (settled since 11 BC. – Gaoo-Roman Times, Department Yvelines (area 2,284.4 km²), Region L'lie-de-France (area 12,011 km²), France. — Weinberg M, Gilberg A. Traite du sang. 1923. L'ouvrege est reedite en 1932. — Weinberg M, Seguin P. La gangrene gazeuse 1918. — Weinberg M, Ginsbourg B. Donnees recentes sur les microbes anaerobies et leur role en pathologie. 1927. — Ablitsov VH. «Galaxy Ukraine». The Ukrainian diaspora: prominent persons. Publ. «Kyt», Kyiv 2007. — Bereznytsky YaS, Zakharash MP, Mishalov VH, Shidlovsky VO. General Surgery: Handbook for Students of the Higher Educational Institutes. Publ. «Nova knyha». Vinnytsia 2018.

cattle, lambs, pigs, horses, and other mammals. Introduced nutritional culture (Weinberg) media for growth of anaerobes, described anaerobic spore forming bacteria genus Clostridium perfringers (40-50% cases), Clositridium oedematicum (15-50% cases), Clostridium septicum (10-30% cases), Clostridium hystolyticum (2-6% cases), and Clostridium novyi, which cause gas gangrene, and developed against them vaccines (1918).

In 1908 **Karl Landsteiner (1868 – 1943**: 20th century. Physiology) and **Erwin Popper (1879 – 1955)** — Austrian physician, isolated infantile paralysis virus – poliovirus from the genus of enteroviruses.

In 1936 **Marko P. Neshchadymenko (28.04.1869, village Serdehivka, Shpolansky Region, Cherkasy Obl., Ukraine – 01.10.1942, city of Kyiv, Ukraine)**[726] — Ukrainian microbiologist, proved the feasibility of the simultaneous immunization by two, or more antigens – smallpox vaccine, diphtheria antitoxin, cholera vaccine and dysentery anatoxin.

Olexandr (Aleksandre) M. Bezredka [27.III.(08.IV).1870, city of Odesa, Ukraine – 28.II.1940, city of Paris, France][727] — Ukrainian French microbiologist, discovered antiviruses (Besredka antiviruses, 1903), introduced the filtration and penetration by heated broth environments for bacteria to study local resistance (1925), described the complement deviation test in tuberculosis (Bezredka reaction),

---

[726] Village Serdehivka (founded before 1788; victims of organized by Russian Communist Fascist regime Holodomor-Genocide 1932-33 against the Ukrainian nation were over 350 person), Shpolansky Region, Cherkasy Obl., Ukraine. — Neshchadymenko MV. One time and once immunization by smallpox vaccine and diphtheria anatoxin. Annals of Mechnykov Institute 1936;3(1):31-8. — Rozhin I. Marko Neshchadymenko and his role in the development of microbiology in Ukraine. J Ukr Med Assoc North Am 1965 Oct.;12,4(39):28-34. — Hanitkevych YaV. Marko Neshchadymenko — founder of the Ukrainian school of microbiologists and epidemiologists: To the 130th years birthday anniversary. Lviv Medical Chasopys 1999;5(2):103-7. — Ablitsov VH. «Galaxy Ukraine». The Ukrainian diaspora: prominent persons. Publ. «Kyt», Kyiv 2007. — Hanitkevych YaV. The contribution of the Ukrainian physicians to the world medicine. Ukr med chasopys 2009 VII/VIII;4(72):110-15.

[727] Besredka A. — Evaluation de la conception stereochimique. Thesis for doctoral science. Odesa 1892. — Besredka A. Absces sous-phreniques. Dissertation for Doctor of Medicine degree, Paris 1897. — Besredka A, Steinhardt E. De l'anaphylaxie et de anti-anaphylaxie vis-à-vis de serum de chevez. Annales de L'Institut Pasteur, Paris 1907;21:117-27, 384-91. — Besredka A. Antianaphylaxie. Jahrebericht über die Ergebnisse der Immunitätsforschung. Stuttgard 1912;1(8):90-102. — Besredka A, Jupille F. Le bouillon a L'oneuf. Annales de L'Institut Pasteur 1913;27:10009-17. — Besredka A, Manoukhine J. De La reaction de fixation chez les tuberculeux. Annales de L'Institut Pasteur 1914;569-75. — Besredka A. Les endotoxines bacteriennes. Paris 1914. — Besredka A. Anaphylaxie et antianaphylaxie. Paris 1918. — Besredka A. Immunisation locale; pansements specifiques. Mason & Cie, Paris 1925. — Besredka A. De la vaccination par voie buccale contre la dysenterie, la fievre typhoid et la cholera. Revue d'hygiene et medicine preventive 1927;49:445-63. — Besredka A. etudes sur l'immunite dans les maladies infectieuses. Paris 1928. — Besredka A. Antivirustherapie. Masson & Cie, Paris 1930. — Besredka A. Les Immunites Locales. Paris 1937. — Besredka A. Le chic anaphylactique et la principe de la desensibilisation. Paris 1930. — Ablitsov VH. «Galaxy Ukraine». The Ukrainian diaspora: prominent persons. Publ. «Kyt», Kyiv 2007. — Hanitkevych YaV. The contribution of the Ukrainian physicians to the world medicine. Ukr Med Chasopys 2009 VII/VIII;4(72):110-15.

introduced the term «*anaphylactic shock*» and explained the role of the nervous system in its development, introduced the method of desensitization during inoculation (vaccination) with tetanus anatoxin (toxoid) and with immunoglobulin (Bezredka's method, 1930).

Whooping cough (in Latin – pertussis) is a highly contagious bacterial infectious disease of the respiratory tract, caused by Bordella pertussis with the ability to suppress the immune system, spreads by an air-born droplets during close contact with other patient (90-100%), sometime from, the incubation period is 7-12 days. It is characterized by a severe paroxysmal hacking cough followed by a high-pitched intake of breath that sounds like «whoop» which is so hard that may cause a patient to vomit, break ribs and became very tired. The disease usually last whereby 100 days. First-year infants are likely to develop apnea (in Greek – absence of breathing), i.e., temporary cessation of breathing motions), pneumonia, seizure, encephalopathy (in Greek – «encephalos» meaning brain, «pathos» meaning disease, suffering) and death (0.5%). The treatment with antibiotic such as erythromycin, azithromycin, clarithromycin, or trimethoprim / sulfamethoxazole should be started within three weeks of the initial symptoms and in children less than one year-old.

Pertussis was discovered in 1906 by **Jules J.B.V. Bordet (1870 – 1961)** — Belgian microbiologist and immunologist, and **Octave Gengau (1875 – 1957)** — Belgian bacteriologist. Later, in 1912 Octave Gengau was the first to develop whooping cough vaccine. It was through the work of Jules Bordet between 1906 and 1920 that immunity was discovered.

Howard T. Ricketts (1871 – 1910) — American pathologist, showed in 1906, that spotty typhus – spotty fever of the Rocky Mountains is caused by a chainlike Gram-negative bacterium from Rickettsia ranks which is transmitted by tick's bite and feeds on the human and monkey's erythrocytes and cytoplasm. A year later (1907) **Alberto E.B. Barton Thompson (1871 – 1950)** — Peruvian physician, discovered the bacteria of the Rickettsia ranks from the Rickettsia genus – Bartonella quintana, transmitted by the human louse, caused Volyn / Volhynia (trench, fifth-day) fever – an acute infectious disease affecting soldiers alongside the trench military operations during WW I (1914-18) in the land of Volyn (area c. 70.000 km², the name first appeared in the chronicle from 1077) of north-western Ukraine[728] The clinical picture of Volyn fever described in 1916 p. **Wilhelm His, Jr (1863 – 1934**: 19th century. Anatomy, histology, and embryology).

Independently, in 1915 **Frederick Twort (1877 – 1950)** — English microbiologist, and in 1917 **Felix d'Herelle (1873 – 1949)** — French microbiologist, discovered

---

[728] Olearchyk AS. A Surgeon's Universe. Vol. I-IV. – 4th Ed. Publ. Outskirts Press, Inc., Denver, CO 2016. Vol. I. P. 347-8.

viruses that infect and destroy bacteria, called bacteriophages or phages[729]. The later, experimented with the possibility of phage therapy.

Southern and Central American Trypanosomiasis, caused by the tropical flagella protozoan Trypanosoma crura from bites by the form of conenose beetle Triatoma, the genus of bug Panstrongulus, and the South American insect Rodnius and carriers of the causative factor – domestic and wild animals (cats, rodents, the genus of Dasypodidae, bats, foxes et al.), was discovered by **Carlos J.R. das Chagas (1879 – 1934)** — Brazilian bacteriologist and sanitary physician, and named Chagas disease (1909). Its acute form prevail in children, is manifested at the area of inoculation by the erythematous nodule (chagoma), fever, unilateral face edema, palpebral swelling (Romaña sign)[730], local lymphadenopathy, hepatomegaly, meningeal irritation and death or full recovery or transmission into a subacute or chronic form. Subacute form last several weeks to years, is characterized by moderate fever, asthenia and general lymphadenopathy, chronic form usually by myocarditis, cardiomyopathy (CM), achalasia of the esophagus, megacolon, and meningoencephalitis, it may turn into an acute form.

The demonstration of bacterial transformation was reported in 1928 by **Frederick Griffith (1879 – 1947)**[731] — English bacteriologist, whereby a bacterium distinctly changes its form and function, i.e., that Staphylococcus pneumoniae, implicated in many cases of pneumonia, could transform himself from one strain into a different strain. This was proved in his experiment when one strain of Streptococcus pneumoiae's deoxyribonucleic acid (DNA) containing the genes that form the smooth protective polydaccharide capsule, although destroyed by heat, but its DNA survives, and if the remains of this strain are taken by the other strain, the surviving DNA gene protects him from the host's immune system and could kill the host (Griffith's experiment).

The subsequent studies conducted in 1944 by **Oswald T. Avery (1877 – 1955)** — American physician, **Colin M. MacLeod (1909 – 72)**[732] — Canadian American geneticist and **Maclyn McCarthy (1911 – 2005)** — American geneticist, discovered that the transformation of the physical characteristics of bacteria, demonstrated

---

[729] Sifferlin A. The war on super bugs. How a 100-year-old therapy is saving lives. Times Dec 25, 2017 / Jan 1, 2018;100(27-28):54-9.

[730] **Cecilio F. Romaña (1899 - 1997)** — Argentinian physician, who described an early symptom of Chagas disease, i.e., unilateral periorbital edema due to severe inflammation of the eye or conjunctiva or deep structures of the eye (ophthalmia) with palpebral edema, conjunctivitis, and enlargement of local lymphatic nodes (Romaña's sign, 1935).

[731] Griffith F. The significance of Pneumococcal types. J Hygiene (Cambridge Univ Press) 1928 Jan; 27(2):113-59.

[732] Avery OT, MacLeod CM, McCarthy M. Studies on the chemical nature of the substance including transformation of Pneumococcal types – induction of transformation by a deoxyribonucleic acid fraction isolated from Pneumococcus type III. J Experimental Med 1944 Feb 1;79(1):137-58.

by the Griffith's experiment of 1928, is carried out by the DNA – the basic genetic component of chromosome, which then led to its identification as the molecule responsible for heredity.

In 1916 **Henrique** da **Rocha Lima (1879 – 1956)** — Brazilian pathologist and physician-infectionist, discovered the pathogen of epidemic typhus, transmitted from flying squirrels (tribe of Pteromyini) by the louse stool, from which he and his colleague **Stanislaw (Stanislaus) J.M.** von **Prowazek (1875 – 1915)** — Czech zoologist and parasitologist, got sick. The latter died. The pathogen was named by the surviving Henrique da Rocha Lima in the memory of the dead companion – Rickettsia prowazekii.

Tomaso Casoni (1880 – 1933) — Italian physician, described in 1912 the intradermal test for echinococcus.

While following footsteps of **Pavel Laschtschenko (1865 – 1927**: above), Sir **Alexander Fleming (1881 – 1955)**[733] — Scottish botanist, biologist and pharmacologist, isolated in 1922 from nasal discharge of the patient affected by cold a substance (enzyme, ferment), that could destroy some bacteria, such as Micrococcus lysodeikticus, for which he coined in 1924 the term «lysozyme». The ferment «lysozyme» is a glycoside hydrolase, that catalyzes hydrolysis of 1,4-Beta-linkages between acetyl-D-glucosamine residue in peptidoglycan, proteolytic antimicrobial enzyme, part of innate (immune) system, major component of bacterial cell wall, abundantly present in the chicken egg-white albumin, human tissue, secretion, nasal mucus, tears, saliva, and milk. Later, in 1929, he reported on antibacterial properties of a mould, which contain antibiotic penicillin (Penicillium notatum / Penicillium chrysogenum), effective in gains Staphylococcus aureus and other Gram-negative pathogens causing scarlet fever, pneumonia, meningitis and diphteria.

Rudolf S. Weigl (02.09.1883, town of Prerov, Olomouc Region, Czechy — town of Zakopane, Lesser Poland Voivodeship, Poland)[734] — Ukrainian Polish microbiologist of Austrian origin, discovered the vaccine against exanthemata (epidemic) typhus (1928).

---

[733] Fleming A. One remarkable bacteriolytic element found in tissue and secretions. Proc Roy Soc (London) 1922 May;93(653):306-17. — Fleming A, Allison VD. On the antibacterial power of egg white. Lancet 1924;206(5261):1303-7 — Fleming A. On the antibacterial action cultures of Penicillin, with special reference to their use in isolation of B. influence. Brit J Exper Pathology 1929;1093):226-36. — Osadcha-Yanata NT. Antibiotics of plant origin. J Ukr Med Assoc North Am 1967 Apr;14,2(45):9-16.

[734] Town of Prerov (settlements dates to prehistoric times, and the oldest written reference to 1141), Olomouc Region (area 5,265.57 km²), Czechia. — Town of Zakopane (earliest documents describing a glade, an open area within a forest, named Zakopisko, to 1676), Lesser Poland Voivodeship (area 15,108 km²), Poland. — Ablitsov VH. «Galaxy Ukraine». The Ukrainian diaspora: prominent persons. Publ. «Kyt», Kyiv 2007. — Hanitkevych YaV. The contribution of the Ukrainian physicians to the world medicine. Ukr Med Chasopys 2009 VII/VIII;4(72):110-15.

Victor H. Drobot'ko (23.XI.1885, village of Dihtiary, Srebniansky Region, Chernihiv Obl. Ukraine – 23.XI.1966, city of Kyiv, Ukraine)[735] — Ukrainian microbiologist, developed the synthetic nutritional environment (Drobot'ko) allowing studies of physiology and metabolism of intestinal (endobacteria); formulated the bacterial dissociation theory, suggesting the changes of bacterial morphology due to mutation and selection in the colonial morphology of the bacteria in the suboptimal condition of laboratory culture; discovered in 1949 the causative factor (Stachybotrys chartarum) of fungous disease in horses and people named Stachybotryotoxicosis.

For the first non-toxic drug treatment for fungal infection in a human was Myostatin (Nystatin), developed by **Elizabeth L. Hazen (1885 – 1975)** — American microbiologist and **Rachel F. Brown (1898 – 1980)** — American chemist, in 1948-50, which up to present day is the therapy of choice for this infection.

Infectious disease called diphtheria [in Greek – «leather hide», which describes the distinctive fibrous / diphtheria inflammatory coating that appears in the mouth, throat, and respiratory tract of its victims, «kiss of death», «the strangling angel of children» or strangulating disease; **Emil** von **Behring (1854 – 1917)**. 19th century. Microbiology and immunology], caused by a stick-like diphtheria bacteria (Corynobacterium diphtheria), and usually transmitted by a close air contact with an infective individual. The overall mortality rates from diphtheria range between 5% to 10%, and in children under 5 years of age, and adults over 50 years of age, can be as high as 20%. In 1923, **Gaston Ramon (1886 – 1963)** — French biologist and veterinarian, discovered that when diphtheria toxins were exposed to minute quantities of formalin and heated, the toxoid became non-toxic, yet could stimulate active immunity like vaccines. Subsequently, he was able on a small scale to test diphtheria toxoid and demonstrate its antigenic value. Then, in 1924-27 Gaston Ramon has created the method of inactivating tetanus toxoid (ПТ), or tetanus antitoxin against Clostridium tetani [**Arthur Nicolaier (1861 – 1942)**. 19th century. Microbiology and immunology] with formaldehyde, thus developing an active immunization against neurotoxin of Clostridium tetani. This led to development in 1937 by **P. Descombey** of TT vaccine. The use of TT vaccines resulted in a 95% decrease in rate of tetanus morbidity.

Lev V. Hromashevsky [01(13).X.1887, city of Mykolaïv, Ukraine – 01.V.1980, city of Kyiv, Ukraine][736] — Ukrainian microbiologist and epidemiologist, set up

---

[735] Village of Dihtiary (in vicinity was revealed the Old Rus town in ruins, remains of settlements from 9th-13th century; during organized by Russian Communist Fascist regime of Holodomor-Genocide 1932-1932 against the Ukrainian Nation died at least 79 residents), Srebniansky Region, Chernihiv Obl., Ukraine. — Ablitsov VH. «Galaxy Ukraine». The Ukrainian diaspora: prominent persons. Publ. «Kyt», Kyiv 2007.

[736] City of Mykolaiv (Foot Note 198), Ukraine. — Ablitsov VH. «Galaxy Ukraine». The Ukrainian diaspora: prominent persons. Publ. «Kyt», Kyiv 2007.

foundations for transmission mechanism of infection and classification of infectious diseases (1930s).

In 1945 **Serhiy N. Ruchki(o)vsky [07(19).I.1888, town of Tarashcha, Bila Tserkva Region, Kyiv Obl., Ukraine – 05.IV.1967, city of Kyiv, Ukraine]**[737] — Ukrainian microbiologist and epidemiologist, established natural locality of Volyn fever [19th century. Anatomy, histology and embryology. – **Wilhelm His, Jr (1863 – 1934)]**, later named Paroxysmal rickettsiosis [**Howard Ricketts (1871 – 1910)**: above], proved that immunological agent for Paroxysmal rickettsiosis differs from other rickettsia, and named it Rickettsia quintanae.

The term «antibiotic» was introduced in 1942 by **Selman A. Waksman (22.07.1888, village Nova Pryluka, Lypovetsky Region, Vinnytsia Obl. Ukraine – 16.08.1973, C**ensus-designated place **Woods Hole, town of Falmouth, Barnstable County, MA, USA)**[738] — Ukrainian American microbiologist of Jewish origin. Two years later (1944) he obtained from the genus species Streptomyces griseus strain, a rod shaped, spore formed Gram positive bacteria, commonly found in soil or deep-sea sediment, an agent named streptomycin, proved to be the first effective drug against tuberculosis.

Discovered in 1950 by **Walter H. Burkholder (1891 – 1983)** — American plant pathologist, Burkholderia cepacia complex, is a group of catalase-producing, lactose-nonfermenting, Gram negative bacilli-like bacteria composed of at least 17 different species, including Burkholderia cepacia, being an important human pathogen which most often causes septic pneumonia in immunocompromised individuals with underlying lung disease, such as cystic fibrosis, mucoviscidosis or chronic granulomatous disease. It is characterized by significant sputum production and multidrug resistance, and especially if pan-resistant, contraindicates lung transplantation because of high early mortality.

Intrauterine infection by rubella [18th century – **Friedrich Hoffmann (1660 – 1742)]** virus of the woman during early pregnancy may result in the development of

---

[737] Town of Tarashcha (Foot Note 621), Bila Tserkva Region, Kyiv Obl., Ukraine. — Ruchkovsky SN. Materials for epidemiologyof the so called Volyn fever. Vest Acad Med Sc USSR 1948;2. — Ablitsov VH. «Galaxy Ukraine». The Ukrainian diaspora: prominent persons. Publ. «Kyt», Kyiv 2007.

[738] Village Nova Pryluka (founded 1146), Lypovetsky Region, Vinnytsia Obl., Ukraine. — Census designated place Woods Hole (10.3 km²) in town of Falmouth (settled 1660), Barnstable County, MA, USA. — Waksman SA. Life with the Microbes. Simon & Shuster, New York 1954. — Farb P. Lekarstwa w łyżce ziemi. Ameryka, Warszawa 1963;52:48. — Osadcha-Yanata NT. Antibiotics of plan origin. J Ukr Med Assoc North Am 1967 Apr;14,2(45):9-16. — Kogan V. The world celebrity, unknown in his homeland. UHMJ «Ahapit» 1997-1998 (7-8):65-9. — Grando O. The Distinguished Names in history of the Ukrainian Medicine. Publ. Agency «Triumph», Kyiv 1997. — Olearchyk AS. Left-sided extra pleural thoracoplasty for pulmonary tuberculosis. J Ukr Med Assoc North Am 2001;46,1(146):20-5. — Ablitsov VH. «Galaxy Ukraine». The Ukrainian diaspora: prominent persons. Publ. «Kyt», Kyiv 2007. — Hanitkevych YaV. The contribution of the Ukrainian physicians to the world medicine. Ukr Med Chasopys 2009 VII/VIII;4(72):110-15.

congenital rubella syndrome CRS) in the newborn, described in 1941 by Sir **Borman M. Gregg (1892 – 1960)** — Australian ophthalmologist, comprises heart, brain, eye and auditory defects.

In 1937 **Lev A. Zilbert (1894 – 1966)** — Russian microbiologist of Jewish descent, described viral disease – Far Eastern tick encephalitis or Russian spring-summer encephalitis.

Gerhard J.P. Domagk (1895 – 1964) — German bacteriologist and pathologist, discovered in 1932 sulfonamidochrysoidine (KI-730), the sulfonamide, under the brand name prontosil (1935) – the precursor of sulfonamides, effective against bacterial infections caused by Streptococcus infections. Its clinical effectiveness was first proven when in 1935 he treated his own daughter with it, saving her from amputation of an arm.

John F. Enders (1897 – 1985) — American biochemical scientist, **Thomas H. Weller (1915 – 2008)** — American virologist and **Frederick C. Robbins (1916 – 2003)** — American virologist, showed that tissue cultures could be used for the reproduction of viruses (including poliomyelitis virus) in the laboratory, thus making it possible to produce vaccines (1948-52).

Ten years (1939) after the report of **Alexander Fleming (1881 – 1955**: above), **Howard W. Florey (1898 – 1968)** — Australian pathologist and pharmacologist, Sir **Ernst B. Chain (1906 – 79)** — German and British biochemist of Jewish ancestry, proved that penicillin had a great therapeutic potential and promoted production beginning from 1941. Howard Florey carried out the first ever trials in 1941 of penicillin on the first patients who started to recover but subsequently died because at that time there was not enough of drug to continue treatment. Ernst Chain along with **Edward P. Abraham (1913 – 99)** — English biochemist, was involved in theorizing the beta-lactan structure of penicillin. Later (1954), Ernst Chain went on to discover the chemical composition, method to isolate and concentrate, and mechanism of therapeutic action of penicillin.

Serhiy S. Di(y)achenko (06.X.1898, village Nova Basan', Bobrovytsky Region, Chernihiv Obl., Ukraine – 21.I.1992, city of Kyiv, Ukraine)[739] — Ukrainian microbiologist, established in the bacterial structure of abdominal (intestinal) typhus or Salmonella typhi[740] antigen Vi (virulency), or bacterial capsular (superficial) antigen

---

[739] Village Nova Basan' (Ancient Slavs' settlements of Cherniakhivsky Culture 2nd - 5th century AD; first mention 1400; in consequence of organized by Russian Communist Fascist regime of Holodomor-Genocide against the Ukrainian nation were revealed cases of cannibalism due to hunger psychosis), Bobrovytsky Region, Chernihiv Obl., Ukraine. — Ablitsov VH. «Galaxy Ukraine». The Ukrainian diaspora: prominent persons. Publ. «Kyt», Kyiv 2007.

[740] Discovered in 1885 by **Theobald Smith (1859 - 1934)** — American microbiologist and pathologist, named by him Salmonella enterica (choleraesuis) in honor of **Daniel E. Salmon (1850 - 1914)** — American veterinary surgeon.

K, outside cellular wall, previously thought to be setting direct boundaries of bacterial toxicity. Proved that Vi-antigen is present in the early stages of infection (1944), and that the body immunity is decreased by taking hormonal drugs.

Pierre Grabar (10.IX.1898, city of Kyiv, Ukraine – 28.I.1986, city of Paris, France)[741] — Ukrainian French biochemist and immunologist, developed the method of immunoelectrophoretic (1952-53), and created the theory of the «transferable function of antibodies» (1953).

During the Middle Asian expedition 1938 under the leadership of **Maria C. Jazimirska-Krontowska (1898, village of Havrylivka, Novovoronovsky Region, Kherson Obl., Ukraine – 1961, city of Moscow?, Russia)**[742]— Ukrainian microbiologist, was discovered a tick, causative factor recurrent typhus, carrier which was tick Demacentor nuttalli, and natural source of an infection – population of rodents. She also revealed a natural habitation of a tick rickettsiosis (Rikettsia sibirica (1940-45) in Kyrgyzstan (area 199,951 km^2).

Max Theiler (1899 – 1972) — South-African American physician virologist, who proposed in 1937 an attenuated virus strain vaccine against a virus causing yellow fever. It ended its epidemics of yellow fever in South America in 1940-47. He also developed in this same year (1937) the virus, referred to as Theiler's encephalomyelitis virus, as a standard model for studying multiple sclerosis (MS).

Also, ten years (1939) after the report of **Alexander Fleming (1881 – 1955**: above), **Rene J. Dubos (1901 – 82)** — French American microbiologist, pathologist, and environmental humanist, isolated antibiotic tyrothricin from the soil's Bacillus brevis, effective against many forms of gram-positive bacteria.

In 1953-65 **Andre M. Lwoff (1902 – 94)** — French microbiologist, who partnered in experiments with his wife **Marguerite Lwoff**, nee **Bourdaleix (1905 – 75)** — French microbiologist and virologist known for her studies on metabolism, did

---

[741] Grabar P, Burtin P. Imuno-electrophoretic Analysis: Applications to Human Biological Fluids. Elsevier Publ. Co, Amsterdam – New York 1964. — Courtois JE. Eulogy for Pierre Grabar (1898-1986). Bull Acad Natl Med 1986;170(5):635-9. — Ablitsov VH. «Galaxy Ukraine». The Ukrainian diaspora: prominent persons. Publ. «Кут», Kyiv 2007.

[742] Village Havrylivka (founded 1780), Novovoronovsky Region, Kherson Obl. (Foot Note 715), Ukraine. — Radzimowska WW, Jazimirska MC. Über die Bestimmung der Wassenstoffionen-konzentration in einzeinen Baktrienkolonien. Klin Wochenscht 1925;4(2)72-3. — Krontowski AA, Jazimirska-Krontowska MC. Stoffwechselstudien an Gewebskulturen. II. Über Zucherverbrauch durch Gewebskulturen eines mittels Passagen nach Carrel in vitro gezichlichtened reinen Fibroblastenstammes. Arch Exp Zellforsdch Besonders Gewebezwecht 1926;5:114-25. — Krontowska-Kazimirska MK. Influence du potassium et du calcium sur la croissance et la metabolissme des tissues in vitra. C.R. Soc Biol Paris 1930;103:1182-82. — Tovarnickij VI, Krontovskaya MK, Ceburkina N. Chemical composition of Rikettia prowazeki. Nature Dec 1946;158(4025):912. — Ablitsov VH. «Galaxy Ukraine». The Ukrainian diaspora: prominent persons. Publ. «Кут», Kyiv 2007.

research on flagellates, bacteriophages, microbiota, and poliovirus, discovered the mechanism that some viruses, which he called proviruses, use to infect bacteria.

Jonas E Salk (1904 – 95) — American medical researcher and virologist of Jewish ancestry, obtained his noninfectious vaccine for poliomyelitis (1954), and **Albert B. Sabin (1906 – 93)** — Polish medical researcher of Jewish ancestry, live attenuated oral poliovirus.

Max L.D. Delbrück (1906 – 81) — German American biophysicist, **Alfred D. Hershey (1908 – 97)** — American microbiologist and geneticist, and **Salvador E. Luria (1912 – 91)** — American microbiologist of Italian Jewish origin, formulated the method for testing genetic theory of bacteriophages, i.e., viruses that infect bacteria, so called oncoviruses which are the cause of some forms of human cancers and observed that when two different strains of bacteriophage have infected the same bacteria, the two viruses may exchange genetic information (1940). They also described the genetic structure, the mechanism and substratum of inheritance of bacteriophages (1953). The mechanism of carcinogenesis mediated by oncoviruses closely resemble the process by which normal cells degenerate into cancer cells. This discovery allowed better understanding and treatment of cancer. Salvador Luria with Max Delbrück demonstrated statistically that inheritance in bacteria must follow Darwinian[743] rather than Lamarckian[744] theories and that mutant bacteria occurring randomly can still bestow viral resistance without the virus being present (the Lurie-Delbrück experiment, 1943). The idea that natural reaction affects bacteria has profound consequences, as it explains how bacteria develop antibiotic resistance. Alfred Hershey with **Martha C. Chase (1927 – 2003)** — American geneticist, provided additional evidence that deoxyribonucleic acid (DNA), not protein, is the genetic molecule of life (the Hershey-Chase experiment, 1952). Finally, Salvador Luria found that bacteriocins impair the function of cell membranes, by forming holes within it, allowing ions to flow through and destroy the electromagnetic gradient of cells (1963).

In 1931 **Ernst A.F. Ruska (1906 – 88)** — German physicist constructed the prototype of electron microscope which in the improved version from 1933 make it possible to examine the structure of viruses.

Herald R. Cox (1907 – 86) — American bacteriologist, isolated the named after him genus Coxiella of the bacterial family Coxiella, encountered in the form of short bacilli in vacuole inside the cell of the microorganism, common among different mammals and ticks, causing an infectious disease, Q fever, of domesticated

---

[743] **Charles D. Darwin (1809 - 82)** — English biologist, author of the theory of evolution, according to which the higher organisms have evolved from lower by natural selection (Darwinism). — Darwin CD. On the Origin of Species. Publ. House by John Murray, London 1859.

[744] **Jean B.P.A.M.** de **Lamarck (1744 - 1829)** — French naturalist, author of the theory, according to which the acquired qualities can be transmitted (Lamarck's theory).

animals and people (1938). Independently, **Hilary Koprowski (1916 – 2013)** — Polish virologist and immunologist of Jewish ancestry, and Herald Cox, had developed from the live attenuated poliovirus (rendered nonvirulent) a successful oral vaccine against poliomyelitis (respectively, 1950, 1961).

A yeast-like fungus of the genus Pseudocysts, the causative factor of Pneumocystis pneumonia, interstitial plasma-cell pneumonia, in rats and other animals, was discovered in 1952, by **Otto Jirovec (1907 – 72)** — Czech parasitologist and protozoologist, in immunosuppressed humans, specifically in neonates with plasma cell pneumonia, and named Pneumocystis jiroveci. It causes infectious granulomas, comprising approximately 80% of all benign nodules in the lung.

In 1956 **Denis P. Burkitt (1911 – 93)** — Irish surgeon in Uganda[745], revealed in Central Africa a cancer of the lymphatic system, in particular B-cell lymphocytes, a form of non-divided cell lymphoma (Burkitt lymphoma or tumor, African lymphoma). Burkitt lymphoma is divided into the endemic, sporadic, and human immunodeficiency virus (HIV) associated variants. It is most present as lymphoma, nosepharyngeal carcinoma, a large osteolytic lesion of the jaw, abdominal mass or lung cancer in patients infected with HIV.

The infection of normal cells with certain types of viruses, called oncological viruses (oncoviruses), led **Renato Dulbecco (1914 – 2012)** — Italian American virologist, to incorporate virus derived genes into the host-cell genome. That event caused the transformation, i.e., the acquisition of a tumor phenotype of those cells.

Being taught by Renato Dulbecco and learning his methods beginning since 1958, both **Howard M. Temin (1934 – 94)** —American virologist and geneticist of Jewish parent and **David Baltimore (1938 –)** — American biologist of Jewish ancestry, then working independently and simultaneously, with ribonucleic acid (RNA)-dependent deoxyribonucleic acid (DNA) polymerase with Rous sarcoma virus **[Francis Rous (1879 – 1970)**. 20th century. Pathology] and the Rauscher murine leukemia virus[746], arrived at the discovery in 1969, 1970, respectively, that the transfer of viral genes to the cells is mediated by an enzyme called reverse transcriptase (RNA-dependent DNA polymerase) reduplicate the viral genome made by RNA into DNA, later incorporated in the host genome.

Thus, distinct classes of viruses – retroviruses, were discovered that use an RNA template to catalyze the synthesis of DNA.

The two teams independently isolated the rubella [18th century – **Friedrich Hoffmann (1660 – 1742) / Norman Gregg (1892 – 1666)**: above] virus and developed vaccines against it.

---

[745] Republic of Uganda (area 241,038 km²), East Africa.
[746] **Frank J. Rauscher, Jr (1932 - 93)** — American microbiologist, who discovered of animal cancer virus.

The first team headed by **Thomas Weller (1915 – 2008**: above)[747] and **Franklin A. Neva (1922 – 2011)** — American physician and virologist, isolated in 1962 rubella virus and developed vaccine against rubella. Prior to this, Franklin Neva discovered Boston exanthem disease which first occurred in 1951 as epidemic in Boston, MA, caused by echo 16 virus. It is a cutaneous condition affecting children more often than adults, clinical picture shows some similarity to rubella and Human herpesvirus 6.

The other team headed by **Harry M. Meyer, Jr (1928 – 2001)** — American pediatric virologist, and **Paul D. Parkman (1932 –)** — infection diseases specialist, and with the assistance of **Edward L. Buescher (1925 – 89)** — American virologist and physician, and **Malcolm S. Artenstein (1930 – 76)** — specialist in infection diseases and immunologist, isolated in 1967-69 rubella virus and developed vaccine against rubella.

The fact that tumor viruses act on the genetic material of the cell through reverse transcription is a groundbreaking discovery in molecular biology, that corrected erroneous statements made by **Francis Crick (1916 – 2004**: 20th century. Biochemistry), since the further research by **Renato Dulbecco** (above) using a stem cell (SC) model suggested that a single malignant cell with SC properties is sufficient to induce carcinoma in mice and generate distinct population of tumor-initiating cells also with cancer SC properties. His identification and characterization of the origin mammary gland cancer SC in solid tumors led him conclude of carcinoma being a disease of an acquired mutation, and that in addition to genomic mutations, epigenic modifications of a cell may contribute to the development or progression of cancer (2007-11).

Some form of human cancers are caused by oncoviruses, and the mechanism of carcinogenesis mediated by them resemble the process by which normal cells degenerate into cancer cells. Renato Dulbecco discoveries allowed further advancement in understanding, prevention, and treatment of neoplastic diseases.

Reversed transcriptase is the important enzyme in several wide-spread diseases among them an acquired immune deficiency syndrome (AIDS) – serotype 1 (more common and more virulent) and serotype 2 (less common and less virulent), and hepatitis B, and component of such diagnostic techniques in molecular biology as the reverse transcription polymerase chain reaction (PCR). Ultimately, causes of AIDS, led to development if the first group of drugs successful against the virus – the reverse transcriptase inhibitors, inclusion into the prevention and treatment of an active antiretroviral medication zidovudine or azidothymidine synthesized in 1964 and others.

---

[747] Weller TH, Neva FA. Propagation in tissue culture of cytoplasmic agents from patients with rubella-like illness. Proc Soc Exp Biol Med 1961;111(1):215-25.

In the 1940s **Örjan Ouchterlony (1914 – 2004)** — Swedish bacteriologist and immunologist, worked out the double immunodiffusion test (Ouchterlony).

Herman L. Gardner (1918 – 2005)[748] — American microbiologist, who discovered in 1955 a genus of Gram—variable facultative small bacteria (1.0-1.5 μm in diameter), non-spore-forming, nonmotile coccobacilli, named after him Gardella vaginalis. It is a predominant cause of asymptomatic bacterial vaginosis, but include irritation, discharge and a «fish-like odor».

In 1974-80 **Edward B. Lewis (1918 – 2004)** — American geneticist, **Christiane Nüsslein-Volhard (1942 –)** —German development biologist and **Eric F. Wieschaus (1947 –)** — American development biologist, identified the genes involved in the embryologic development of the fruit fly Drosophila melanogaster. Subsequently, in 1985 Christiane Nüsslein-Volhard and Eric Wieschaus identified mutants in the Toll (in German – the adjective «toll», means «amazing» or «great») gene in Drosophila melanogaster, which was cloned in 1988 by **Kathryn V. Anderson (1952 –)** — American development biologist and geneticist. Then, in 1997 **Ruslan M. Medzhitov** — Uzbek American cell biologist and immunologist and **Charles A. Janeway (1943 – 2003)** — American immunologist, identified Toll-like receptors (TLR) in mammals, known as lipoglycans or endotoxins, a class of proteins that play role in the innate (non-specific, or not-born) immune system.

Measles [Medieval medicine – **Rzazes (c. 854 – c. 925)** / 17ᵗʰ century – **Thomas Sydenheim (1624 – 89)** / 17ᵗʰ century – **Richard Morton (1634 – 98)** / 18ᵗʰ century – **Thomas Fuller (1654 – 1734)**] is a highly contagious infection disease caused by a single -stranded, negative-sense enveloped, non-segmented ribonucleic acid (RNA) virus of the genus Morbillivirus within the family of Paramyxoviruses. The prodromal period after exposure to an infected person last usually 10-14 days. Initial symptoms include fever, often greater than 40°C or 104°F, cough, catarrh (cold, runny nose), inflamed conjunctiva of the eyes last usually 7-10 days. Two or three days after first symptoms small whitish gray Koplik's spots [19ᵗʰ century. Children's diseases – **Henry Koplik (1858 – 1927)**] may form on the inside of the cheek on the buccal mucous membrane of the mouth. Thereafter, three to five days a reddish flat maculopapular rush begins on the face and then spreads on the rest of the body. Usual common complications include diarrhea, middle ear infection and pneumonia, in more severe cases – meningitis.

Influenza (flu) is an infection disease caused in humans by Hemophilus influenza virus A-C [19ᵗʰ century. Microbiology and immunology – **Richard Pfeiffer (1858 – 1945)**], transmitted from person who sneezes or coughs. Symptoms usually develop

---

748 Gardner HL, Duke CD. Haemophilus vaginalis. A newly defined specific infection previously classified as «Non-specific vaginitis». Am J Obstet Gynecol 1955;69:962-76.

two days after exposure, and are followed by fever, runny nose, sore throat, headache, muscle pain, coughing and tiredness lasting usually about one week. However, cough may last two weeks or more. Children may suffer from vomiting, diarrhrea and abdominal pain. Complications include primary viral followed by bacterial pneumonia, maxillary (paranasal), frontal, ethmoid or sphenoidal sinus infection, worsening asthma, congestive heart failure (CHF) and death. Prevention is by the frequent hand washing, wearing a surgical mask and vaccination.

Maurice R. Hilleman (1919 – 2005) — American microbiologist specialized in vaccinology, who developed the first successful vaccines against measles (licensed in 1963, improved in 1968). He also developed vaccines against mumps [Ancient medicine. *Greece* – **Hippocrates (460 – 77 BC)**. 18th century – **Robert Hamilton (1721 – 93)**] (1963-67), hepatitis A (1996), hepatitis B (1981), chickenpox (1995), meningitis B (2007), Streptococcus pneumoniae type 3 (1987) and Hemophilus influenza bacteria (1957, 1974).

Also, Maurice Hilleman discovered the genetic changes occurring when the influenza virus mutates, known as a «shift and drift» that helped him to recognize the 1957's epidemic outbreak of influenza in Honk Kong, China, that could become pandemic, if he and his colleague did not developed vaccine against influenza.

Burkitt lymphoma **[Denis Burkitt (1911 – 93)**: above] is associated with isolated in 1964, by Sir **Michael A. Epstein (1921 –)** — English physician, and **Yvonne M. Barr (1932 –)** — English virologist, a human herpes virus (Epstein-Barr's herpes virus, or simple Epstein-Barr's virus) of the genus Lymphocryptoviral, with a doubling time <30 days, which cause infectious mononucleosis. Epstein-Barr herpes virus infections are associated with post-transplant lymphoproliferative disease (PTLD) which are either polyclonal or monoclonal proliferations of B lymphocytes in a context of post-transplantation immunosuppression, and they may have many of the features of immune system malignancy. The incidence of PTLD in children is high (13-26%) versus adults (10%), after transplantation of intestines (20%), heart (2-10%), lungs (4-8%) and kidneys (1%).

In 1957 **Alick Isaacs (1921 – 67)** — Scottish virologist of Jewish Lithuanian origin, and **Jean Lindenmann (1924 – 2015)** — Swiss virologist and immunologist, discovered interferon from the glycoprotein family that exerts virus-nonspecific but host-specific antiviral activity.

In about 1965 **Daniel C. Gaidusek (1923 – 2008)** — Slovak American physician and medical researcher and **Baruch S. Blumberg (1925 – 2011)** — American physician, geneticist and anthropologist of Jewish ancestry, concluded that the infectious disease kuru («shaking») was caused by a prion (protein infectious agent), transmitted by the ritualistic consumption of the brains of deceased by relative cannibals, members of the Fore tribe of New Guinea (786,000 km²), resulting in spread of an incurable spongiform encephalopathy, and thus discovered new mechanisms of prevalence and spreading of infectious diseases. In addition, Baruch Blumberg identified in 1976 hepatitis B virus.

Passion of **Ivan Kochan (Aug. 20, 1923, village of Tudorovychi, Sokal Region, Lviv Obl., Ukraine – Jan. 05, 2017, town of Hooper, NE, USA)**[749] — Ukrainian microbiologist and immunologist, was to study nutritional immunology and iron

[749] Village Tudorkovychi (dated since 1645), Sokal Region, Lviv Oblast, Ukraine. — Town of Hooper (appeared in 1871), NE, USA. — Kochan I, Raffel S. A property of immune sera inhibitory for the growth of the tubercle bacillus. J Immunol 1960;84:374-83. — Kochan I, Rose SR. Fate of tubercle bacilli in lipopolysacharide and BCG-treated mice. J Bacteriol 1962;84(2):291-4. — Kochan I. Tuberculostatic activity of normal human sera. J Immunol 1963;90:711-9. — Kochan I, Smith I. Antibacterial activity of tuberculostatic factor on intracellular bacilli. J Immunol 1965;94:220-7. — Kochan I, Bendel WL Jr. Passive transfer of tuberculin hypersensitivity in guinea pig. J Allergy 1966;37(5):284-95. — Kochan I, Christopher JA, Kupchyk L. Study of the cellular factor of delayed hypersensitivity. J Allergy 1966;38(5):280-9. — Kochan I, Khan A. Respiration-enhancing effect of tuberculin on splenic cells of hypertensive guinea pigs. J Bacteriol 1969;97(1):1-5. — Kochan I. Mechanism of tuberculostasis in mammalian serum. I. Role of transferrin in human serum tuberculostasis. J Infect Dis. 1969;119;(1):11-8. — Kochan I, Golden CA, Bukovic JA. Mechanism of tuberculostasis in mammalian serum II. Induction of serum tuberculostasis in guinea pigs. J Bacteriol. 1969;100(1):64-70. — Kochan, Pellis NR.; Golden CA. Mechanism of tuberculostasis in mammalian serum III. Neutralization of serum tuberculostasis by mycobactin. Infect Immun. 1971;3(4):553-8. — Kochan I, Cahall DL, Golden CA. Employment of tuberculostasis in serum-agar medium for the study of production and activity of mycobactin. Infect Immun. 1971;4(2):130-7. — Kochan I, Pellis NR, Pfohl DG. Effects of normal and activated cell fractions on the growth of tubercle bacilli. Infect Immun. 1972;6(2):142-8. — Kochan I. The role of iron in bacterial infections, with special consideration of host-tubercle bacillus interaction. Curr Top Microbiol Immunol. 1973;60:1-30. — Kochan I, Golden CA. Antimycobacterial effect of lysates prepared from immunologically activated macrophages. Infect Immun. 1973;8(3):388-94. — Kochan I. Role of macrophages in clinical immunity. J Ukr Med Ass North Am 1975 Apr;21,1(77):16-31. — Kochan I, Golden CA. Immunological nature of antimycobacterial phenomenon in macrophages. Infect Immun. 1974;9(2):249–54. — Golden CA, Kochan I, Spriggs DR. Antimycobacterial effect of lysates prepared from immunologically activated macrophages. Infect Immun. 1973;8(3):388-94. — Kochan I, Berendt M. Fatty acid-induced tuberculocidal activity in sera of guinea pigs treated with bacillus Calmette-Guérin and lipopolysaccharide. J Infect Dis. 1974;129(6):696-704. — Kochan I. Role of iron in the regulation of antibacterial resistance. J Ukr Med Ass North Am 1975 Apr;22,2(77):16-31. — Kochan I, Kvach JT, Wiles TI. Virulence-associated acquisition of iron in mammalian serum by Escherichia coli. J Infect Dis. 1977;135(4):623–32. — Kochan I, Wasynczuk J, McCabe MA. Effects of injected iron and siderophores on infections in normal and immune mice. Infect Immun 1978;22(2):560-7. — Kvach JT, Wiles TI, Mellencamp MW, Kochan I. Use of transferrin-iron enterobactin complexes as the source of iron by serum-exposed bacteria. Infect Immun. 1977;18(2):439-45. — Hemsworth GR, Kochan I. Secretion of antimycobacterial fatty acids by normal and activated macrophages. Infect Immun. 1978;19(1):170-7. — Kochan I. Wagner SK, Wasynczuk J. Effect of iron on antibacterial immunity in vaccinated mice. Infec Immun 1978;22(2):560-7. — Mellencamp MW, McCabe MA, Kochan I. The growth-promoting effect of bacterial iron for serum-exposed bacteria. Immunology. 1981;43(3):483-91. — Kochan I, Wagner SK, Wasynczuk J. Effect of iron on antibacterial immunity in vaccinated mice. Infect Immun. 1984;43(2):543-8. — Kochan I. Immunology. Handbook of Immunology, Serology, Immunocheemistry, Immunobiology, Immunogenetics. 1st Ed. Kyiv 1994. — Kochan I. Role of iron in immunity. Bulletin of the Shevchenko Scientific Society 2007;23(39):19-23. — Town of Grass-Valley (established 1851), Nevada District, CA, USA. — Ablitsov VH. «Galaxy Ukraine». The Ukrainian diaspora: prominent persons. Publ. «Kyt», Kyiv 2007.

metabolism in bacteria responsible for tuberculosis. Those studies lead him to introduce by 1973 the concept of nutritional immunity, referred to attempt by the host to withhold iron from parasites.

With the help of phylogenetic taxonomy of 16S ribonucleic acid (RNA) – a part of ribosome, **Carl R. Woese (1928 – 2012)**[750] — American microbiologist, discovered the group of single-cell prokaryotic (in Greek – «pro-» meaning «before», and «kayon» meaning «nucleus») organisms known as archaea (in Greek – «archaios», meaning – «ancient») which constitute the system of a third domain of life. This technique became standard in microbiology.

The son of **George Gamov (1904 – 68**: 20th century. Biochemistry), **Rustam I. Gamow (04.11.1935, neighborhood Georgetown, city of Washington, DC, USA –)**[751] — American microbiologist and inventor of Ukrainian origin, developed the bag enabling mountain climbers to avoid high altitude sickness by rising the surrounding pressure (Gamow bag, 1990) and the shallow under water breathing apparatus (SUBA) – a pressurized snorkel system permitting swimmers to breath easily as deep as 30,48 m under water (1990-93).

Currently worldwide live approximately 71 million of people with acute or chronic infectious inflammation of the liver (hepatitis).

Together, **Harvey J. Alter (12.IX.1935, New York City, N.Y, USA –)**[752] — Jewish American virologist, **Michael Houghton (1949, city of London, England –)**[753] — English Canadian virologist, and **Charles M. Rice (25.VIII.1952, capital city of Sacramento, CA, USA –)**[754] —American virologist, cooperated in the discovery of blood-born human hepatitis virus C, thus helped to prevent and treat millions of people affected by this infectious disease.

In 1964 Harvey Alter with **Baruch Blumberg (1925 – 2011**; above), took part in the discovery of Australian antigen which played a key role in isolation of hepatitis B virus.

---

[750] Woese C, Fox GE. Phylogenetic structure of the prokaryotic domains: the primary kingdoms. Proc NAS USA 1977;74(11):5088-90.

[751] Neighborhood Georgetown (founded 1751), city of Washington (founded 1791), Federal District of Columbia (DC, area 158,1 km²), USA. — Ablitsov VH. «Galaxy Ukraine». The Ukrainian diaspora: prominent persons. Publ. «Kyt», Kyiv 2007.

[752] Kuo G, Choo QL, **Alter HJ**, et al…**Houghton M**. An assay for circulating antibodies to a major etiologic virus of human non-A, non-B hepatitis. Science 1989;244:362-4.

[753] **Alter HJ**, Purcell RH, …**Houghton M**, et al. Detection of antibody to hepatitis C virus in prospectively followed transfusion recipients with acute and chronic non-A, non-B hepatitis. NEJM 1989 Nov 30;321:1494-1500.

[754] Capital city Sacramento (founded 1849), State of California (CA; area 423.970 km²), USA –). — Kolykhalov AA, Agapov EF, et al…**Rice CM**. Transmission of hepatitis C by intrahepatic inoculation with transcribed RNA. Science 1997;277:570-4.

In the middle of 1970s Harvey Alter with co-workers were able accidentally prove, that majority of hepatitis cases after blood transfusion are not connected with known genera hepatitis virus A and B.

With the help of experiments on the chimpanzee Harvey Alter's team was able to prove in 1988 the existence in samples of the blood from people the new virus, genetic sequence of which was isolated by Michael Houghton, and in April 1989 named hepatitis C virus which belong to the genus Hepaciviral, member of the family Flaviviridae.

Finally, Charles Rice, provided the evidence that hepatitis C virus alone could replicate and cause disease.

In 1975 **Harald** zur **Hausen (1936 –)** — German virologist, announced hypothesis that human papilloma virus (HPV) plays an important role in the cause of genital warts identified as HPV6 (1979), and of uterine cervical cancer identified as HPV16 and HPV18 (1983-84). Subsequently, in 2006 HPV vaccine was introduced to prevent this disease. More, in 2008 he advised to eat pork meat, rather than beef, to avoid oncological diseases, since a little studied virus of cattle's may evoke malignant tumors of large bowel.

It is known that viruses infect host cells and reproduce inside them. The killer thymus (T)-cells destroy those infected by virus cells so that viruses cannot reproduce itself. Both, **Peter C. Doherty (1940 –)** — Australian veterinary surgeon and medical researcher and **Rolf M. Zinkernagel (1944 –)** — Swiss experimental immunologist, discovered that in order for killer T-cells recognize infected by virus cells, they had to recognize two molecules on the surface of the cell – the virus antigen and molecule of the major histocompatibility complex (MHC), the one previously identified with rejection of incompatible tissue during organ transplantation (1975). They also discovered that the MHC was responsible for the body defense against meningitis viruses.

In 1996 **Jules A. Hoffmann (1941 –)** — Luxembourger French biologist and **Bruno Lemaitre** — French biologist, discovered the function of the Drosophila melanogaster Toll gene in the innate immunity. Toll-like receptors (TLR) identify constituent of other organisms like bacteria and fungi, trigger an immune response, explaining the way bacterial remnants can trigger septic shock.

In 1973 **Ralph M. Steinman (14.01.1943, city of Montreal, Province of Quebec, Canada – 30.09.2011, borough of Manhattan, New York City, N.Y., USA)**[755] — Canadian cell biologist and immunologist of Jewish Ukrainian ancestry, discovered and termed the dendritic (tree-like, accessory) cells, an antigen-presenting cells of the

---

[755] Borough of Manhattan (settled 1624), New York City (Foot Note 607), N.Y., USA. — Ablitsov VH. «Galaxy Ukraine». The Ukrainian diaspora: prominent persons. Publ. «Kyt», Kyiv 2007.

mammalian immune system which process antigen material and present it on the cell surface of the thymus (T) cell of the immune system, so they act as messengers between the innate and adaptive immune systems.

Since 1981 the incidence of Kaposi's sarcoma [19th century. Internal medicine – **Moritz Kohn Kaposi (1837 – 1902)]**, endemic in Central Africa and Europe, linked to an acquired immune deficiency syndrome (AIDS) began to rapidly spread throughout the world.

In 1983 **Francoise Barre-Sinoussi (1947** –) — French virologist, together with **Luc Montagnier (1932** –) — French virologist, discovered human immune deficiency virus (HIV) as a cause of an AIDS.

Subsequently, it was established that Kaposi's sarcoma is really associated with AIDS.

A genetic predisposition to AIDS caused by the HIV with the human leukocyte (lymphocyte) antigen (HLA) - Dr5 alleles has been identified in Blacks, Italians, and Jews.

As of June 1988, there were an estimated 96,433 reported cases of AIDS in the USA while the actual cumulative number worldwide was 200,000. It was estimated that by 1992 there was 200,000 cases of AIDS in the USA.

In July of 1976 in the cotton factory of the town Nzara in South Sudan (area 619,745 km^2), a landlocked country in north eastern Africa (area 30,370.000 km^2), occurred an epidemic outbreak of Sudan virus disease (SVD), and in August of 1976 in the village of Yambuku, situated 31 km north of the Ebola River, called by the ingenious people the Legbala River («white river», 250 km in length) which is the headstream of the Mongala River, a northern tributary of the Congo River (length 4,700 km, basin 4,014.500 km^2, the world's deepest river in excess of 220 m, draining into the Atlantic Ocean), northern Province of Mongala (area 58,141 km^2), Congo, then (1971-79) called Zaire (area 2,345,409 km^2) of the central Africa, occurred an epidemic outbreak of Ebola virus disease (EVD). Between July and October 1976 in the region of Nzara 151 people died, and from August through October of this same year in the region of Yambuku 97 people died, including 18 health care providers.

SVD and EVD, a.k.a., Ebola hemorrhagic fever, or simple Ebola is a viral hemorrhagic fever disease of human and other primates, caused by the species Sudan virus (SUDV) and species Ebola virus (EBOV), both of 5 genera of Zaire ebolavirus, family Filoviridae, order Mononegaviruses. Both viruses carry negative-sense ribonucleic acid (RNA) genome in virions that are cylindrical or tubular, contain viral envelope, matrix and nucleocapsid component. They cause a sharp increase of the vascular permeability and massive hemorrhages.

The EBOV was discovered and named by Baron **Peter R. Piot (1949 –)** — Belgian microbiologist and his team from a sample of blood taken from a sick nun in Zaire.

The natural source of both viruses are bats, particularly fruit bats, and is primarily transmitted between humans and from animal to humans through body fluids.

The SVD and EVD is characterized by high temperature of about 39^0C, sore throat, headaches, prostration with «heavy articulations», muscular, retrosternal, or abdominal pain, then vomiting usually with blood (hematemesis), diarrhea with blood, rush, dehydration, finally liver and kidney failure and death. To confirm diagnosis the blood samples are taken for viral RNA, viral antibodies or for the virus itself.

Both viruses show a rapid evolution with a death initially after a mean of 3 days, from hypotension due to body fluid loss. In the initial 1976's outbreaks in South Sudan and Zaire, a total of 280 (90%) of 318 people have died. Since that time, as of January 2015, more than 8000 people have died, a mortality rate of 83-90%. Later data indicate the onset of recovery or death typically in 6 to 14-16 days from the first symptoms.

The prevention is by isolation and protection using the recombinant vesicular stomatitis virus-Zaire - EBOV (rVSV - ZEBOV) vaccine which is effective in 70-100% (December 2016).

The treatment supportive with rehydration.

The molecular and genetic studies of inflammation and innate immunity was pioneered in by **Bruce A. Beutler (1957 –)** — American immunologist and geneticist, who isolated in 1985 a mouse tumor necrosis factor (TNF), demonstrated the inflammatory potential of this cytokine, proving its important role in the endotoxin induced shock (1987).

Subsequently, in 1991 he invented recombinant molecules expressly designated to neutralize TNF, fusing the binding portion of TNF receptor proteins to the heavy chain of an immunoglobulin molecule to force receptor dimerization. These molecules were later used to as drug Etanercept in the treatment of an autoimmune granulomatous enteritis, psoriasis, and other granulomatous inflammations.

Furthermore, Bruce Beutler used TNF production as a phenotypic endpoint to identify he lipopolysaccharides (LPS) receptor. That hinged on the positional cloning of the mammalian LPS locus (known since 1968), a key genetic determinant of all biological responses to LPS. Thus, he discovered in 1998 the key sensors of microbial infection in mammals, demonstrating that one of the mammals Toll-like receptors (TLR), TLR4, acts as membrane spanning component of which 10 are known to exist in humans, the mammalian LPS receptor complex. The TLRs function in perception of microbes, each detecting signature molecules heralding infection, mediate severe illness, including shock and systemic inflammation in course of an infection, are important to the pathogenesis of sterile inflammatory and autoimmune disease such as systemic lupus erythematosus (SLE).

Pathology

The implications of arteriosclerosis / atherosclerosis (AS) in cardiovascular disease, including stroke, was not understood until 1905, when **Hans Chiari (1851 – 1916:** 19th century. Pathology**)** showed that an ulcerating plaque in the carotid artery could be the source of embolic stroke in the brain.

In 1902 by **Theodor H. Boveri (1862 – 1915)**[756] — German biologist, cytologist, embryologist, and geneticist, state that malignant tumors arise from aberrant cell division (mitosis) and uncontrolled growth caused by radiation, physical or chemical insults or microscopic pathogens.

K.A. Ludwig Aschoff (1866 – 1942)[757] — German pathologist and physician, described a microscopical granulomatous structures consisting of fibroid change, with foci of T-lymphocytes, occasionally plasma cells and activated macrophages, slowly replaced by a fibrous scar in the interstitial tissue of the heart in rheumatic fever (RF), mainly rheumatic endocarditis (Aschoff's bodies, or nodules, 1904). Beside of this, he with **Sunao Tawara (1873 – 1952)** — Japanese pathologist, discovered the atrioventricular (Aschoff-Tawara) node (AVN, 1906-07).

Description of an exceptionally malignant, metastatic, primitive small rounded-cell tumor of the bone was made by **James Ewing (1866 – 1943)** — American pathologist, that is seen in the diaphysis of long bones, ribs and flat bones of children and adolescents (Ewing sarcoma or tumor, 1920). It is manifested by pain, edema, leukocytosis, and fever.

In 1903-05 **Antoni Leśniowski (1867 – 1940)** — Polish surgeon, and in 1932 **Burrill B. Crohn (1884 – 1983)** — American physician-gastroenterologist of Jewish origin, described autoimmune chronic granulomatous enteritis with thickening and scarring of the gastrointestinal (GI) tract wall of unknown etiology (regional ileitis or Lesniowski-Crohn's disease).

Volodymyr (Wolodymyr, Wladimir) K. Lindenmann or **Włodzimierz Lindenman (30.VII.1868, village Petrivske-Rozumovske,** now incorporated into **city of Moscow, Russia** – died from ischemic stroke of the brain **18.IV.1933,**

---

[756] Boveri T. Zur Frage der Entstehung maligner Tumoren. Verlag Gustav Fisher, Jena 1914.
[757] Aschoff KAL. Zur Myocarditisfrage. Verbandlungender der deutschen pathologischen Gesellschaft (Stuttgard). 1904;8:46-53.

city of Warsaw, Poland)[758] — Ukrainian pathologist of German descent, recreated experimentally inflammation in capillaries of the renal glomeruli using nephrotoxic serum (1901).

---

[758] Village Petrivske-Rozumovske, in 1917 incorporate into city of Moscow, Russia. — Lindenmann W. Ueber die Löszienkeitsverhältnisse des Paracaseins im Künstlichen Magensafte. Archiv für pathologische Anatomie und Physiologie und für klinische Medicin 1897;149(1):51-65. — Lindenmann W. Ueber pathologische Fettbildung. Beiträge zur pathologischen Anatomie und allgemeinen Pathologie 1898;25:392-430 + 1 pl. — Lindenmann W. Ueber das Pulegon. Arch. fur exper. Pathologie 1898. — Lindenmann W. Ueber die Secretionserscheinungen der Giftdrüse der Kreuzotter. Archiv für mikroscopishe Anatomie 1898;53(1):313-21. — Lindenmann W. Zur Toxikologie der organischen Phosphorverbindungen. Archiv fur experimentelle Pathologie und Pharmakologie 1898;41(2-3):191-217. — Lindenmann W. Ueber die Veränderungen des Gesammtstoffwechsels bei Vergiftung mit Pulegon. Zeitschrift für Biologie 1899;21:1-17. — Lindenmann W. Ueber die Wirkungen des Oleum Pulegii. Naunyn-Schmiedeberg Archiv 1899;42(5/6): 356-74. — Lindenmann W. Ueber die Wirkung des Phosphors und des Pulegons auf die Cephalopoden. Beiträge zur pathologischen Anatomie und allgemeinen Pathologie 1900;27:484-490. — Lindenmann W. Sur l'action de quelques poisons renaux. Annales de l'Instit. Pasteur, Paris 1900. — Lindenmann W. Cytolyses as the Cause of Toxic Nephritis. Moskva 1901. — Lindenman W. Ethiology of Malaria. Typography of St Volodymyr University Printing & Publishing Office of NT Korchak-Novytsky, Kyiv 1903. — Lindenmann W. Ueber die Resorption in der Niere. Beiträge zur pathologischen Anatomie und allgemeinen Pathologie.1904. — Lindenman W. Mechanism of Urinary Secretion. Typography of St Volodymyr University Printing & Publishing Office of NT Korchak-Novytsky, Kyiv 1908. — Lindenmann W. Beiträge zur Theorie der Harnabsonderung. Naunyn-Schmiedeberg Archiv 1908;59 (2/3):196-208. — Lindenmann W. Handbook of General Pathology. Vol. 1: Introduction into Subject. General Nosology. General Etiology. Typography of St Volodymyr University Printing &Publishing Office of NT Korchak-Novytsky, Kyiv 1910. — Lindenmann W. Handbook of General Pathology. T. 2: General Pathogenesis. Typography of St Volodymyr University Printing & Publishing Office of NT Korchak-Novytsky, Kyiv 1911. — Lindenmann W. Short Course of Medical Zoology.Typography of St Volodymyr University Printing & Publishing Office of NT Korchak-Novytsky, Kyiv 1912. — Lindemann W. Zur Lehre von den Funktionen der Niere. Ergebnisse der Physiologie, biologischen Chemie und experimentellen Pharmakologie 1914;14 (1):618-56. — Lindenmann W. Zur Lehre von den Funktionen der Niere. Monatsschrift Kinderheilkunde 1914;14 (1):618-656. — Lindenman W. Walka chemiczna w przyrodzie. Warszawa, 1924. — Lindenman W. Toksykologja chemicznych środków bojowych. Wojskowy Instytut Naukowo-Wydawniczy, Warszawa 1925. — Lindenman W. Podstawy ratownictwa zatrutych gazami. Warszawa 1926. — Lindenman W. Walka chemiczna w przyrodzie. Wyd. 2. Tow. Obrony Przeciwgaz., Warszawa 1926. — Lindenman W. Toksyczne własności siarczków organicznych. Lekarz Wojskowy 1928;11(3/4):259-267. — Lindenman W. Iperyt. Wojskowy Instytut Naukowo-Wydawniczy, Warszawa 1929. — Lindenman W. Toksykologiczna klasyfikacja chemicznych środków bojowych. Główna Księgarnia Wojskowa, Warszawa 1927. — Lindenman W. Hormony płciowe. Biologja Lekarska 1930:9 (8):293-337. — Lindenman W. Saponiny. Biologja Lekarska 1930;9:101-59. — Lindenman W. Jod jako pierwiastek życiowy. Biologja Lekarska 1931;10:53-95. — Lindenman W. Synteza i rozkład hemoglobiny w ustroju. Biologja Lekarska 1931:10:245-312. — Lindenman W. Krzywa oddechowa. Lekarz Wojskowy 1931:18(1/2, 3/4, 5/6): 12-22, 98-106, 184-95. — Lindenman W. O działaniu oligodynamicznem w sensie Naegeliego. Biologja Lekarska 1932;11(6):273-287. — Lindenman W. O lipoidach. Biologja Lekarska 1932:11:79-147. — Lindenman W. Hormon trzustkowy. Biologja Lekarska 1934;13(2):1-34, 49-80, 99-122. — Shilinis T. V.K. Lindenman – one of the creators of general biological trends in pathology. UHIMJ «Ahapit» 1995;3:53-8. — Ablitsov VH. «Galaxy Ukraine». The Ukrainian diaspora: prominent persons. Publ. «Kyt», Kyiv 2007. — Hanitkevych YaV. The contribution of the Ukrainian physicians to the world medicine. Ukr Med Chasopys 2009 VII/VIII;4(72):110-15.

Lev O. Tarasevych [02(14).II.1868, town of Tyraspol, Prydnistrovia, Ukraine / Moldova – 12.VI.1927, city of Dresden, Germany][759] — Ukrainian microbiologist and pathologist, organized in 1918 Institute for the Standardization and Control of Biological Medical Preparations in Moscow, Russia.

Francis C. Wood (1869 – 1951)[760] — American pathologist and cancer researcher, who opened the laboratory for pathology (1910), and was one of the initiators of radiotherapy and radium therapy for the treatment of malignant tumors (c. 1921).

Syndrome of postpartum pituitary necrosis was described in 1913 by **Leon K. Glinsky (1870 – 1918)** — Polish pathologist, in 1914 by **Morris Simmonds (1855 – 1925)** — German pathologist, and in 1937 by **Harold L. Sheehan (1900 – 88)** — English pathologist, a.k.a., Glinsky-Simmonds-Sheehan syndrome.

In 1920 **Ernrst Krukenberg (1871 – 1946**: 19th century. Pathology)[761] described intramural hematoma (IMH) as «dissection without intimal tear» of the aorta which probably originated from rupture of vasa vasorum within the outer third of the aortic media, resulting in the circumferential accumulation of blood with no apparent intimal defect. Ruptured intima of IMH causes the development of dissecting aneurysm of the thoracic aorta.

Aleksei I. Abrikosov (1875 – 1955)[762] — Russian pathologist of Jewish descent, described granular cell myoblastoma of the tongue (Abricosov tumor, 1922).

Research in into causes of arteriosclerosis / atherosclerosis (AS) began in 1907, when **Alexander I. Ignatovski (1875 – 1955)**[763] — Russian physician, began to feed rabbits with animal proteins, that is a diet of full-fat milk, eggs, and meat, noting that they soon developed pronounced AS of the aorta.

In 1907 **George H. Whipple (1878 – 1976)** — American pathologist, physician, and biomedical researcher, described Lipodystrophies intestinalis or intestinal

---

[759] City of Dresden (settled in the Neolithic era by Lineal Pottery culture tribes c. 7500 BC, in 1206, late 12th century and 1350), Germany. — Ablitsov VH. «Galaxy Ukraine». The Ukrainian diaspora: prominent persons. Publ. «Kyt», Kyiv 2007. — Hanitkevych YaV. The contribution of the Ukrainian physicians to the world medicine. Ukr Med Chasopys 2009 VII/VIII;4(72):110-15.

[760] Wood F. Laboratory Guide to Clinical Pathology, Stettiner Bros 1901.

[761] Krukenberg E. Beiträge zur Frage des aneurysma dissecans. Betr Pathol Anat Allg Pathol 1920;67:329-51.

[762] Anichkov AN, Abrikosov AI. Pathological Anatomy of the Heart and Vessels. Publ «Meditsina», St Petersburg 1940. — Abrikosov AI, Strukov AI. Pathological Anatomy. Part 1-2. Kyiv 1955-56. — Strukov AI, Sierov VV. Pathological Anatomy. Part. 1-2. 4th Ed. Publ. «Fact». Kharkiv 2004.

[763] Ignatowski A. Changes in parenchymatous organs and in the aorta of rabbits under the influence of animal protein. Izv Imp Voyenno-Med Akad (St Petersburg) 1908;18:231-44. — Ignatowski A. Iufluence de la nourriture animale sur l'organisme des lapins. Arch med exp anat pathol 1908;20:1-20. — Ignatowski A. Wirkung des thierischen nahrung auf den kaninchenoorganisms. Ber Milit Med Akad 1908;16:154-76. — Ingatowski A. Ueber die wirkung des thierischen eiweisses auf die aorta und die parenchymatosen organe der kaninchen. Virchows Arch Pathol Anat Physiol Klin Med 1909;198;248-70. / Konstantinov IE, Jankovic GM. Alexander I. Ignatowski. A pioneer in the study of atherosclerosis. Tex Heart Inst J 2013;40(3):46-9.

lipodystrophy, characterized by abnormal lipid deposits in the small intestine and correctly pointed bacteria as its causative factor, named in his honor the Whipple's disease.

Subsequently, in 1923-30 George Whipple together with his close associate **Frieda S. Robscheit-Robbins (1888 – 1973)** — German born American pathologist, and with **George R. Minot (1885 – 1950)** — American medical researcher and **William P. Murphy (1892 – 1987)** — American physician, introduced the liver extract to treat macrocytic anemia, specifically pernicious anemia which at that time was uncurable terminal disease. Then, **William B. Castle (1897 – 1990)** — American physiologist and physician, discovered gastric intrinsic factor – a glycoprotein produced by the parietal cells of the stomach which is necessary for the absorption of an extrinsic (vitamin B_{12} – cobalamin) factor in diet in the ileum of the small intestine. So, the absence of gastric intrinsic factor causes pernicious anemia.

Initially, the research into physiology and pathology of the liver and etiology of pernicious anemia was commenced by George Whipple and Frieda Robscheit-Robbin establishing an animal model for anemia. Dogs were bled exhibiting symptoms alike to severe anemia, and after that fed a diet of liver, spleen, lung, intestines etc. The recovery was quicker in those dogs fed with liver. Their initial conclusion was that malfunctioning liver and lack of iron were responsible for anemia. Having completed the study, they began to treat patients suffering from pernicious anemia using the liver extract with improvement. Later, it was found that active ingredient was not iron but a water-soluble liver extract containing vitamin B_{12}.

In 1910 **Francis P. Rous (1879 – 1970**: 19th century. Anatomy, histology and embryology)[764] established that malignant tumor (sarcoma) growing on a domestic chicken could be transferred to another fowl simply by exposing the healthy bird to a cell-free filtrate (retrovirus, or Rous sarcoma). Thus, he proved that viruses could induce growth malignant tumors.

Olexandra I. Smyrnova-Zamkova [05(17).V.1880, town of Pereyaslav, Kyiv Obl., Ukraine – 22.IX.1962, city of Kyiv, Ukraine][765] — Ukrainian pathologist, founded the concept of basic argyrophilic substance, that is, an internal environment of organs and tissues, functionally connected by the nervous system (1955).

The pioneering work on the theory of stress – a nonspecific reaction of the organism on a strong exceeding the norm external action or irritant, and suitable

---

[764] Rous P. A transmissible avian neoplasm (sarcoma of the common fowl). J Exp Med 1910 Sep 1;15(5):696-705.

[765] Town of Pereyaslav (Foot Note 87), Kyiv Obl., Ukraine. — Smyrnova-Zamkova AI. The Principal Argynophylic Substance and its Functional Importance. Kyiv 1955. — Rozhin I. Professor Dr Oleksandra Smyrnova-Zamkova. J Ukr Med Assoc North Am 1956 May;3,1(5):32-4. — Ablitsov VH. «Galaxy Ukraine». The Ukrainian diaspora: prominent persons. Publ. «Kyt», Kyiv 2007. — Hanitkevych YaV. The contribution of the Ukrainian physicians to the world medicine. Ukr Med Chasopys 2009 VII/VIII;4(72):110-15.

reaction of the nervous system, was carried out by **Olexandr O. Bohomolets [12(24).V.1881, city of Kyiv, Ukraine – 19.VII.1946, city of Kyiv, Ukraine]**[766] — Ukrainian pathologist. He advanced the concept of the reticuloendothelial system (RES), demonstrating trophic and defensive function of the connective tissue and on this basis synthesized the antireticular cytotoxic serum (ACS) or Bohomolets serum (1924), established the importance of the adrenal cortex in regulating defensive reactions against infections and chemical insults (1933-36).

Pavlo O. Kucherenko [15(27).X.1882, city of Rostov-on-Don River, Russia – 28.V.1936, city of Kyiv, Ukraine][767] — Ukrainian pathologist, combined clinical and pathological data on the status of the incretory (internally secreted) system in malignant tumors (1936).

The first demonstration in 1913 of cholesterol role in the development of atherosclerotic (AS) changes in the arterial wall and of cardiovascular diseases belongs to **Nicolai N. Anichkov**, a.k.a., **Nicolai N. Anitschkow (1885 – 1964)**[768] — Russian pathologist. Beside of this, he described a cardiac myocyte found in the

[766] Bohomolets OO (ed.). Basics of Pathological Physiology. Vol. 1-3. Kyiv 1933-36. — Bohomolets OO. Pathological Physiology. Vol. I. First Ukrainian edition. DVOU. Medical Publishing, Kharkiv – Kyiv 1934. — Bogomolets AA. Selected Works. Vol. 1-3. Kyiv 1956-58. — Rozhin I, Rozhin W. Academician Olexandr Bohomolets. J Ukr Med Assoc North Am 1958 Sep;5(11);5-12. — Rozhin I, Rozhin V. Scientific work of OO Bohomolets (General characteristic). J Ukr Med Assoc North Am 1958 Dec – 1959 Jan;5(12-13);11-6. — Rozhin I, Rozhin V. Scientific work of OO Bohomolets (General characteristic). J Ukr Med Assoc North Am 1958 Dec – 1959 Jan;5(12-13);11-6. — Gazhenko AI, Gurkalova IP, Zukow W, et al. Pathology. Medical Student's Library. Radom Medical University in Radom and Odesa State University in Odesa. Radom 2009. P. 42-82. — Ablitsov VH. «Galaxy Ukraine». The Ukrainian diaspora: prominent persons. Publ. «Kyt», Kyiv 2007. — Hanitkevych YaV. The contribution of the Ukrainian physicians to the world medicine. Ukr Med Chasopys 2009 VII/VIII;4(72):110-15.

[767] City of Rostov-on-Don Rivery (traces of human presence 2,1-1,97 million years BC, established after construction of ford-post 1713 and custom service 1749), Russia. — Kycherenko PO. Foundations of Pathological Morphology. Publ. «Derzhmedvydav» of Ukraine, Kyiv 1929. — Kucherenko PO. Pathological Anatomy. Special Part. Publ. «Derzhmedvydav» of Ukraine, Kyiv 1936. — Kucherenko PO. About Obesity and Paradoxically Good Condition of Alimentation in Malignant Tumors. Publ. «Derzhmedvydav» of Ukraine, Kyiv 1937. — Kucherenko PO. Incretory Glands and Malignant Newgrowths. Publ. «Derzhmevydav» of Ukraine, Київ 1937. — Rozhin I. Professor, Doctor of Medicine Pavlo Kucherenko. Bio-bibliographical outline-memoir. J Ukr Med Assoc North Am 1957 Oct;4,2(8);7-14. — Ablitsov VH. «Galaxy Ukraine». The Ukrainian diaspora: prominent persons. Publ. «Kyt», Kyiv 2007. — Hanitkevych YaV. The contribution of the Ukrainian physicians to the world medicine. Ukr Med Chasopys 2009 VII/VIII;4(72):110-15. — Hanitkevych YaV. The contribution of the Ukrainian physicians to the world medicine. Ukr med chasopys 2009 VII/VIII;4(72):110-15.

[768] Anitschkow N., Chalatow S. Über experimentelle Cholesterinsteatose und ihre Bedeutung für die Entstehung einiger pathologischer Prozesse. Zentralblatt für allgemeine Pathologie und pathologische Anatomie (Zentrb allg Pathol Anat), Jena. 1913;24:1-9. — Anichkov AN, Abrikosov AI. Pathological Anatomy of the Heart and Vessels. Publ «Meditsina», St Petersburg 1940. — Friedman M, Friedland GW. Medicines's 10 Greatest Discoveries. Yale Univ Press, New Haven, CT 1998. — Konstantinov IE, Mejevoi N, Anichkov NM. Nikolai N. Anichkov and his theory of atherosclerosis. Tex Heart Inst J 2006;33(4):417-23.

Aschoff bodies **[Ludwig Aschoff (1866 – 1942)**: above]**, which have a serrated bar of chromatin in their nucleus, representing a large mononuclear cell in connective tissue in the wall of the heart (Anichkov myocyte, 1912).

An innovative numerical microscopic grading system for carcinoma of the lip, based on cellular differentiation on a scale 1 to 4, introduced **Albert P. Broders (1885 – 1954)**[769] — American pathologist, in 1940. A couple of years later, he extended his grading system to cancers of other anatomical sites, and he reached a conclusion that tumors with the microscopical cellular morphology of carcinoma should be called carcinoma whether, or not the cancer cells invaded beyond the basement membrane. He named noninvasive carcinoma as carcinoma in situ. In the 1930s he with his associates performed retrograde grading of soft tissue and bone sarcomas as well as breast carcinoma, and were able to determine, by grading, the occurrence or non-occurrence of axillary lymph nodal (LN) metastases.

Oleksiy A. Krontovsky (12/24.III.1885, city of Penza, Russia – died from colitis **15.VIII.1933, city of Berdyansk, Zaporizhia Obl., Ukraine)**[770] — Ukrainian pathologist, studied the metabolism of normal and neoplastic tissue cultures using developed sophisticated physic-chemical, and microbiochemical methods (1916-17); cultivated in the tissue culture Rickettsia prowazekii (1922) that allowed him to develop the vaccine against typhus (1941); worked out methods to grow tissues outside the organism and proved that the metabolism of healthy tissues deprived the regulatory influence of an organism assumes the metabolic features approaching

[769] Broders AC. Squamous cell cancer of the lip: a study of five hundred and thirty-seven cases. JAMA 1920;74:656-64. — Broders AC. Carcinoma in situ contrasted with penetrating epithelium. JAMA 1932;99:1670-4. — Broders AC. The microscopic grading of cancer. Surg Clin North Am 1941;21:947-62.

[770] City of Penza (founded 1663), Russia. — City of Berdyansk (Foot Note 593), Zaporizhia Obl., Ukraine. — Krontowski AA, Radzimovska VV. On the influence of changes of concentration of the H+ resp. OH- ions on the life of tissue cells of vertebrates. J Physiology 1922;56(5):275-82. — Krontovski AA. Über die Kultivierung der Gewebe ausserhalb des Organismus bei Ahwendung der Kombinierten Medien. Virchovs Archiv für Pathologische und Physiologishe und für Klinische Medizin 1923;24(1):488=501. — Krontowski AA, Jazimirska-Krontowska MC. Stoffwechsel-Studien an Gewebskulturen. II Über Zuckerverbrauch durch Gewebskulturen eines mittels Passagen nach Carrel in vitro geziichteren reinen Fibroblastenstammes. Arch Exp Zehlforsen. Besonders Gewebezwecht 1926;5:114-24. — Krontowski AA. Explanation und deren Ergebnisse fürder normale und pathologische Physiologe – Ergebnisse der Physiologe 1928 Jan;26(4):370-500. — Krontowski AA, Jazimirska-Krontowska MK, Ssawitska HP. Beträge zur Wirkung der Monojod- und Monobromessigsäure auf Tumoren. I. Mitteilung Versuche mit Explantation und Transplantation. Zeitshrift für Krebsforschung 1930;37:457-91. — Krontowski AA, Magath MA, Smailowskaja EJ. Beträge zur Wirkung der Monojod- und Monobromessigsäure auf Tumoren. II. Mitteilung Versuche an Impftumoren in vivo. Z Krebsforschung 1933;38:495-500. — Krontowski AA, Jazimirska-Krontowski MC, Savitska HP, et al. Application de la methode des cultures de tissus a l'etude du typhus exanthematique. V. Nouvelles experiences de culture du virus typhys exanematique per de nouveausx procedes. Ann Inst Pasteur 1934;53,6:654-63. — Ablitsov VH. «Galaxy Ukraine». The Ukrainian diaspora: prominent persons. Publ. «Kyt», Kyiv 2007. — Hanitkevych YaV. The contribution of the Ukrainian physicians to the world medicine. Ukr Med Chasopys 2009 VII/VIII;4(72):110-15.

the corresponding malignant tumors (1927); introduced the method to study the metabolism of the malignant neoplasms in conditions of the «isolated tumor» (1928).

Neurilemoma is the commonest well-encapsulated tumor of the intercostal nerves in the posterior mediastinum, usually benign, occasionally associated with neurofibromatosis of **Fredrich D.** von **Recklinghausen (1833 – 1910**: 19th century. Pathology**)**. It should be differentiated from neurilemoma, or schwannoma [19th century. Anatomy, histology, and embryology. – **Theodor Schwann (1810 – 82)**] which involves intracranial ganglions of the cranial I-XII nerves. Morphologically neurilemoma is differentiated by the cells described in 1920 by **Nils R.E. Antoni (1887 – 1968)**[771] — Swedish neurologist, as the Antoni cells type A (highly cellular with nuclear palisading pattern, fibrillary, polarized and elongated in the tissue) characteristic for neurilemoma or the Antoni cells type B (loosely organized with reticular, microcystic pattern, possible degenerated fields of Antoni cells type A), usually seen in schwannoma.

By studying the depth of invasion and lymphatic spread of carcinoma of the rectum **Culthbert Duke (1890 – 1977)**[772] — English pathologist, classified in 1935 tumors into three groups: (A) carcinoma confined to the rectum; (B) carcinoma has progressed by direct spread to the perirectal tissues; and (C) carcinoma metastasized to perirectal lymph nodes (LN).

Danylo O. Alpern (15.XII.1894, city of Kharkiv, Ukraine – 24.VII. 1968, city of Kharkiv, Ukraine)[773] — Ukrainian pathophysiologist, established trophic influence of nervous system on permeability of vessels and tissues (1928); revealed the role of nucleon exchange products in development inflammation; obtained data regarding significance an hydro-carbonate degradation in pathogenesis of allergic reactions; studied de-sensibility and anti-inflammatory action of some extracts from hypophysis 1935.

In 1938 **Fredrich Feyrter (1895 – 1973)** — Austrian pathologist, formulated his doctrine of the peripheral endocrine cell system. It was advanced in 1966-68 by **Antony G.E. Pearse (1916 – 2003**: 20th century. Anatomy, histology, and embryology**)** who developed the concept of amine precursor uptake decarboxylation (APUD) system.

In 1947 **Yellapragada Subbarow (1895 – 1948)** — Indian biochemist, and **Sidney Farber (1903 – 73)** — Polish-American pediatric pathologist of Jewish ancestry,

---

[771] Olearchyk. AS. Neurilemoma. In, AS Olearchyk. A Surgeon's Universe. Vol. I-IV. - 4th Ed. Publ. Outskirts Press, Inc. Denver, CO 2016. Vol. II. P. 670-3.

[772] Dukes C, Bussey HJ, Lymphatic spread in cancer of th rectum. Br J Surg 1935;24:751-96.

[773] Alpern DYe. Pathological Physiology. Publ. «Derzhmedvydav» of Ukraine, Kharkiv 1934.— Kharkiv DYe. Pathological Anatomy.Part Special. Publ. «Derzhmedvydav» of Ukraine, Kharkiv 1934. — Alpern DYe. Pathological Physiology. 3rd Ed redone. Publ. «Derzhmedvydav» of Ukraine, Kyiv 1949. — Ablitsov VH. «Galaxy Ukraine». The Ukrainian diaspora: prominent persons. Publ. «Kyt», Kyiv 2007. — Hanitkevych YaV. The contribution of the Ukrainian physicians to the world medicine. Ukr Med Chasopys 2009 VII/VIII;4(72):110-15.

pioneer of modern chemotherapy, reported in 1947 that folic acid derivatives compete with folic acid itself and thus inhibit leukemia in children.

Mykola M. Syrotynin, a.k.a., **Nikolai N. Sironinin [14(28).XI.1896, city of Saratov, Russia – 04.IV.1977, city of Kyiv, Ukraine]**[774] — Ukrainian pathologist, proved that with an increase of evolutionary development of an organism increases its sensitivity towards decreased oxygen (O_2) level in the blood (hypoxia) and oxygen starvation. Thereby, gradual acclimatization increases resistance to O_2 starvation, therefore has an important practical use in astronautics.

Important contribution to medicine and veterinary was made by **Ivan F. Rozhin (18.IX.1897, village of Kumaniv, Horodotsky Region, Khmelnytsky Obl., Ukraine – 10.VII.1972, town of Hamtrack, Wayne County, MI, USA)**[775] — Ukrainian biologist and pathologist, who authored «Metabolism in animals affected by malignant tumors» (1930), «Role of bacteriophage in biology of Bacillus anthracis» (Dissertation for D.M.S., AS of Ukraine, Kyiv 1937) / [18th century. – **Mykhaiklo Hamaliya (1749 – 1830)** & **Stepan Prokopovych-Andriyevsky (1760 – 1818)** / 19th century. Mikrobiology and Immunology – **Pasteur (1822 – 95)]**, «Role of bacteriophage in variability of virus and vaccine against Bacillus anthracis» (1937), «Basics of Veterinary» (Moscow 1939), «Materials for studying resistance of organism towards carcinomas» (1953), «Grassing in Ukraine» (1969-71).

---

[774] City of Saratov (founded 1590), Russia. — Syrotinin MM. Life on Heights and Altitude Disease. Publ. Academy of Sciences of Ukrainian RSR, Kyiv 1939. — Ablitsov VH. «Galaxy Ukraine». The Ukrainian diaspora: prominent persons. Publ. «Kyt», Kyiv 2007. — Hanitkevych YaV. The contribution of the Ukrainian physicians to the world medicine. Ukr Med Chasopys 2009 VII/VIII;4(72):110-15.

[775] Village Kumaniv (known since 1469), Horodotsky Region, Khmelnytsky Obl., Ukraine. — Town of Hamtrack (organized 1894), Wayne County, State of Michigan (MI, area 250,493 km²), USA). — Rozhin I. Professor Dr Oleksandra Smyrnova-Zamkova. J Ukr Med Assoc North Am 1956 May;3,1(5):32-4. — Rozhin I. History of medicine in Ukraine. J Ukr Med Assoc North Am 1956 May;3,2(6):19-24. — Rozhin I, Rozhin W. Volodymyr Pidvysotsky (1957-1913). J Ukr Med Assoc North Am 1957 May;4,1(7):43-6. — Rozhin I. Professor, Doctor of Medicine Pavlo Kucherenko. Bio-bibliographical outline-memoir. J Ukr Med Assoc North Am 1957 Oct;4,2(8);7-14. — Rozhin I, Rozhin W. Academician Oleksandr Bohomolets. Bio-bibliographic outline-memoir. J Ukr Med Assoc North Am 1958 Sep; 5(11):5-12. — Rozhin I, Rozhin W. Scientific works of OO Bohomolets (General characteristic). J Ukr Med Assoc North Am 1958 Dec - 1959 Jan;5(12-13):11-6. — Rozhin I. Professor Mykola Volkovych (on occasion of 100-anniversary of his birth). J Ukr Med Assoc North Am 1960 Jan;7,(16):22-4. — Rozhin I. Dozent Doctor Andriy Zhuravel. Memoir – outline. J Ukr Med Assoc North Am 1960 Apr;7(17):25-9. / 1960 July:7(18): 37-40. — Rozhin I, Rozhin W. Physiological system of connective tissue. ткании. J Ukr Med Assoc North Am 1962 Jan;9,1(24):2-7. — Rozhin I. Materials to the history of Ukrainian veterinary-medical science. J Ukr Med Assoc North Am 1963 Jul;10,3(30):2-7. — Rozhin I. Professor Dr Ivan Bazylevych. J Ukr Med Assoc North Am 1964 Apr;11,2(3):40-8. — Rozhin I. Mykola Stachowsky. J Ukr Med Assoc North Am 1965 Jan;12,1(36):32-44. — Rozhin I. Marko Neshchadymenko and his part in the development of microbiology in Ukraine. J Ukr Med Assoc North Am 1965 Oct;12,4(39):28-34. — Rozhin I. Danylo Zabolotny – Ukrainian scientist of international recognition (on occasion of hundred year from his birth date). J Ukr Med Assoc North Am 1967 Apr;14,2(45):22-5,27. — Ablitsov VH. «Galaxy Ukraine». The Ukrainian diaspora: prominent persons. Publ. «Kyt», Kyiv 2007.

Julian Walawski, previous **Julian R. E. Jordan-Walawski (01.VIII.1898, town of Bolekhiv, Ivano-Frankivsk Obl., Ukraine – 29.I.1975, city of Warsaw, Poland)**[776] — Ukrainian Polish pathologist, discovered hormone enterogastrone in the small bowel, a hormone suppressing gastric secretion and mobility (1928), and he proposed the etiology of gastric ulcer based on function of enterogastrone. In the current understanding enterogastrone is an any hormone secreted by the mucosa of the duodenum in the lower gastrointestinal tract in response to dietary lipids that inhibit the forward motion of the contents of chyme.

Concept of interaction between a tumor and a host was developed in 1937 by **Rostyslav Ye. Kavetsky (01.I.1899, city of Samara, Russia, Росія – 12.X.1978, city of Kyiv, Ukraine)**[777] — Ukrainian pathologist. Proved regression of carcinogenesis and ability tumorous cells towards de-differentiation. In 1968 applied a light amplification by stimulated emission of radiation (laser) to treat oncological conditions.

In alcoholic patients, those with eating disorders and hiatus hernia, **G. Kenneth Mallory (1900 – 86)**[778] — Hungarian American physician, and **Soma Weiss (1898 – 1942)** — American pathologist, observed after severe belching, coughing, or vomiting, a hematemesis or melena, from a fissure-like tear in the mucosa of the stomach below the esophagogastric junction (Mallory-Weiss syndrome, 1929)

The congenital absence or malformation of kidneys in newborn, usually males, was described by **Edith L. Potter (1901 – 93)** — American physician and pioneer in the perinatal[779] period of pathology, probably due to deficiency of amnionic fluid (oligohydramnios) and fetal anuria, characterized by other congenital abnormalities, including facies with wide-set eyes, flattened palpebral fissures and nasal bridge, prominent epicanthus, mandibular micrognathia and large low-set ears deficient in

---

[776] Town of Bolekhiv (first mentioned 1371), Ivano-Frankivsk Obl., Ukraine. — Walawski J. La sécrétine intestinale, excitant de la sécrétion des glandes stomacales. Préparation et effect. CR Soc De Biol 1928;8:48. — Walawski J. Fizjologia Patologiczna: Częś Szczegółowa. PZWL, Warszawa 1956. — Walawski J. Patofizjologia Patologiczna. Ogólna i Szczegółowa. PZWK, Warszawa 1960. — Walawski J, Kaleta Z. Miażdżyca. PZWL, Warszawa 1966. — Walawski J. Patofizjologia ogólna i narządowa. PZWL, Warszawa 1969. — Ablitsov VH. «Galaxy Ukraine». The Ukrainian diaspora: prominent persons. Publ. «Kyt», Kyiv 2007.

[777] City of Samara (mention about settlement 1361, founded 1586), Russia. — Kavetsky RYe. Role of an Active Mesenchyme in Disposition of Organism Towards Malignant Neoplasms. Kyiv 1937. — Kavetsky RE. Tumorous Process and Nervous System. Kyiv 1958. — Kavetsky RE. Tumor and Organism. Kyiv 1961. — Kavetsky RE. Lasers in Biology and Medicine. Kyiv 1969. — Kavetsky RE . Biophysics of Carcinoma. Kyiv 1976. — Kavetsky RE. Interrelation of an Organism and Tumor. Kyiv 1977. — Ablitsov VH. «Galaxy Ukraine». The Ukrainian diaspora: prominent persons. Publ. «Kyt», Kyiv 2007.

[778] Mallory GK, Weiss S. Hemorrhages from lacerations of the cardiac orifice of the stomach due to vomiting. Am J Med Sc 1929;178:506-15. — Marchuk V. Mallory-Weiss syndrome with motion sickness. J Ukr Med Assoc North Am 1981 Spring;28,2(101):59-65. — Kurko VS. Gastrointestinal bleeding of non-ulcerous origin. In, LYa Kovalchuk, VF Sayenko, HV Knyshov. Klinichna Khirurhiya. Vol. 1-2. Publ. «Ukrmedknyha», Ternopil 2000. Vol. 2. C. 84-93.

[779] Perinatal period of the child development is the duration of the time, which begins from the 28th week of the intrauterine development and terminates during the first week of life.

cartilage (Potter facies), skeleto-muscular anomalies such as clubbing of the hands or feet with contractures and pulmonary hypoplasia (Potter syndrome, 1934, 1946). The affected children are usually stillborn or die shortly after birth. She also classified a group of four renal cystic diseases: infantile polycystic kidneys, multicystic dysplastic kidneys, adult type polycystic kidneys and renal cysts in infants with posterior urethral valves (1964, 1971).

In 1928 **Otto Gsell (1902 – 90)**[780] — Swiss pathologist, described the clinical picture, in 1929, **Jacob Erdheim (May 24, 1874, Boryslav, Lviv Obl., Ukraine – May 18, 1937, Vienna, Austria)**[781] — Ukrainian Austrian pathologist of Jewish origin, worked over the pathological anatomy and coined the term «medionecrosis aortae idiopathica cystica» (Erdheim), a congenital familial autosomal dominant degeneration of the media of the thoracic aorta due to breakdown of collagen, elastin and smooth muscle with acid mucopolysaccharides accumulation within the elastic media, causing an annuloaortic ectasia, aneurysm and dissection (Gsell-Erdheim syndrome).

Described in 1931 by **William Chester**[782] — American pathologist, a disease characterized by the abnormal multiplication of a specific type of white blood cells (WBC) called histiocytes, or tissue macrophages – useful cells of the connective tissue as defenders from pathogens, which however could be displaced into other tissues where they are not present, injuring those tissues, causing a histiocytic neoplasm (Erdheim-Chester disease).

Olexandr F. Makarchenko [(22.X.1903, city of Mariupol, Donetsk Obl., Ukraine – 05.VII.1979, city of Kyiv, Ukraine): 20th century. Anatomy, histology and embryology. – **Bolesław Jałowy (1906 – 43)]**[783] — Ukrainian pathophysiologist,

---

[780] Gsell O. Wandnecrosen der Aorta als selbständige Erkrankung und ihre Bezierung zur Spontanruptur. [Virchow's] Arch path Anat Physiol Klin Med 1928;270:1-36.

[781] Town of Boryslav (area has been inhabited since Bronze Age; the remnants of pagan shrine from the 1st millennium BC; between the 9th and 13th century the site of a fortress named Tustan, which was a part of belt of similar strongholds defending Ukraine from the west and south; first mentioned in a document from 1387), Lviv Obl., Ukraine. — Erdheim J. Medionecrosis aorthae idiopathica (cystica). [Virchow's] Arch path Anat Physiol Klin Med 1929;273:454-79. — Ablitsov VH. «Galaxy Ukraine». The Ukrainian diaspora: prominent persons. Publ. «Kyt», Kyiv 2007.

[782] Chester W. Über Lipidgranulomatose. (Virchow's) Archiv für pathologische Anatomie und Physiologie, and für die klinische Medizin 1931;279:561-602.

[783] City of Mariupol (founded in 1500 as the Ukrainian Zaporizhian Cossacks guard encampment of **Domakha**, granted city rights 1778; destroyed during Russian-Ukrainian War by Russian Nazi regime of the RF and «army»-horde of Russia Feb. 25 – May 17, 2022), Donetsk Obl., Ukraine. — Makarchenko OF. The Influence of the Cerebral Cortex on Blood Chemistry. Dis. Candidate Med. Sc. Kyiv 1941.— Makarchenko OF. The influence of Manganese Intoxication on the Nervous System Function. Dis. Doc. Med. Sc. Kyiv 1954.— Makarchenko OF. Influenza and the Nervous System. Monography. Kyiv 1963.— Makarchenko OF. The role of the Hypothalamic Neuro-hormonal Systems in Physiology and Pathology. Monography. Kyiv 1978.— Makarchenko OF. The Hypothalamic-cortical Influences: Neurologic and Neuro-chemical.Monography. Kyiv 1980. — Ablitsov VH. «Galaxy Ukraine». The Ukrainian diaspora: prominent persons. Publ. «Kyt», Kyiv 2007.

who studied the influence of the cerebral cortex on blood biochemistry (1941), the bioelectric action of on the cerebral cortex during infections and intoxications by manganese (1954), the reciprocal action between cerebral cortex and subcortical structures in normal and pathological states (1963-79).

The giant or angio-follicular lymph node hyperplasia was discovered by **Benjamin Castleman (1906 – 82)** — American physician and pathologist and named after him Castleman disease (1954-56). It involves hyperproliferation of non-malignant growth of certain B cells that often produce cytokines at single or multiple lymph nodes in the mediastinum and retroperitoneum.

The research pioneered in 1944 by **Dale R. Colman (1906 – 93)** — American pathologist, indicated that cancer spreads by observing their lack of adhesiveness to each other, one of the earliest biological properties of those cells to be measured.

Mustard gas (sulfur mustard), discovered during WW 1, was noted during WW 2 (1939-45) to arrest malignant growth for a short time. In 1942 **Alfred Gilman, Sr (1908 – 84)** — American pharmacologist and **Louis S. Goodman (1906 – 2000)** — American pharmacologist, initiated the clinical investigation of this compound (nitrogen mustard).

Leon L. Dmochowski (01.VII.1909, city of Ternopil, Ukraine – died from heart attack **26.VIII.1981, city of Mexico, Mexico)**[784] — Ukrainian American virologist and experimental oncologist, proved that viral particles are present in mammary cancer in mice (1953) and in woman (1968). The confirmation of his studies occurred in 1982 with the discovery of cancer genes.

---

[784] City of Mexico (founded 1325), Mexico (area 1,972,500 km²). — Dmochowski L, Grey KE. Electronic microscopy of tumors with sclerosing and suspicious virus etiology. J UkrMed Asssoc North Am 1957 May;4,4(7)1-9. — Dmochowski L. Tumors and viruses in the past, present and in the future. J Ukr Med Assoc North Am 1958 Dec – 1959 Mar; 6(12-13):1-11. — Dmochowski L. The search for human viruses. Ukr Med Assoc North Am 1963 Apr;10,2(29):2-7. — Dmochowski L. Studies using electron microscopy of leukemia in man and animals. J Ukr Med Assoc North Am 1965 Jan;12,1(36):2-8. — Dmochowski L. Viral studies in human leukemia and lymphoma. J Ukr Med Assoc North Am 1969 Jul-Oct;16,3-4-(54-55):3-24. — Dmochowski L, Seman G, Gallager HS. Viral etiology of breast cancer. J Ukr Med Assoc North Am 1970 July;17,3(58):3-13. — Dmochowski L, Maruyama K. Immunological studies on leukemia and solid tumors of animals and man. J Ukr Med Assoc North Am 1971 Oct;18,4(63):3-21. — Dmochowski L. Implications of animal model system in studies of the relationship of viruses to human leukemia and solid tumors. J Ukr Med Assoc North Am 1972 Oct;19,4(67):3-34. — Dmochowski L. Viruses and cancer. J Ukr Med Assoc North Am 1974 Jul;21,3(74):3-24. — Dmochowski L. Unifying concept of leukemia and related neoplasm in animal and man. 1975 Apr;22,2(77):3-15. — Dmochowski L, Bowen J. Relevant results of experimental oncology for detection of cancer. J Ukr Med Assoc North Am 1979 Jan;26,1(92):3-12. — Worobec R. Leon-Lubomyr Dmochowski, MD, PhD (1909-1881). A biographical sketch of his life and work. J Ukr Med Assoc North Am 2005 Winter;50,2(154):62-5. — Ablitsov VH. «Galaxy Ukraine». The Ukrainian diaspora: prominent persons. Publ. «Kyt», Kyiv 2007.

In 1957 **Charles Heidelberger (1920 – 83)** — American cancer researcher who synthesized the antitumor agent, 5-fluoropyrimidine or 5-fluorouracil (5-FU).

Martin Rodbell (1925 – 98) — American biochemist and molecular endocrinologist of Jewish descend, and **Alfred G. Gilman (1941 – 2015)** — American pharmacologist and biochemist, the son of **Alfred Gillman, Sr (1908 – 84)**, discovered protein-G (1969-70).

Foundations of the contemporary oncological hematology, i.e., study of structure and function of the stem hematopoietic cells and its role in leukemia / leucosis[785] laid **Zoya A. Butenko (01.IX.1928, village Bahata Cherneshchyna, Sakhnoshchynsky Region, Kharkiv Obl., Ukraine – 21.IV.2001, city of Kyiv, Ukraine)**[786] — Ukrainian scientific oncologist, who first connected the human ribonucleic acid (RNA) to malignant transformation of cells, appearance of cancer diseases of the blood, specifically in development of leucosis (1969).

Oleksiy O. Moibenko (07.X.1931, city Roston-on-Don River, Russia – 08.V.2015, city of Kyiv, Ukraine)[787] — Ukrainian pathologist and cardiologist, proved essential differences in regulation of vascular tone in the arterial aortic system versus the pulmonary artery (PA) system; described adrenergic reaction in pulmonary vessels and changes of pulmonary circulation in hypertension of humoral and neurogenic origin (1964); investigated role of the receptors field in the heart during reflex regulation of blood circulation (1979). Together with associated he developed a method for an immune local injury to the heart causing immunogenic myocardial infarction (MI); designated main mechanism for injury to the cellular membrane with disruption of ion-transporting processes and degradation of phosphorus-lipide bi-layer with activization lipo-oxygenase and cyclo-oxygenase routes for metabolism of arachidonic acid and enforced creation of leukotrienes and prostanoids; assigned role for leukotrienes in constriction of coronary arteries and dilatation of peripheral vessels, especially venous, which is the main reason for immunogenic shock. Thus, he

---

[785] Leukemia / leukosis is a malignant disease of the blood with proliferation and formation by white blood cells (WBC) of a leucocytes tissue, characterized by paralysis, blindness, creation of tumors in an internal organs and calcification of bone.

[786] Village Bahata Cherneshchyna (founded 1756; during organization by Russian Communist Fascist of Holodomor-Genocide 1932-33 against the Ukrainian Nation perished at least 600 innocent children and adult), Sakhnoshchynsky Region, Kharkiv Obl., Ukraine. — Butenko ZA. Role of ribonucleic acid in development of leucosis. Disseration for the degree of D.M.S., Kyiv 1969. — Ablitsov VH. «Galaxy Ukraine». The Ukrainian diaspora: prominent persons. Publ. «Kyt», Kyiv 2007.

[787] Moibenko OO, Бутенко ГМ, Povzhytkov MM. Cytotoxic Injury of the Heart and Cardiogenic Shock. Kyiv 1975. — Moibenko AA. Cardiogenic Reflexes and their Role in Regulation of the Blood Circulation. Publ. «Nauyova dumka», Kyiv 1979. Moibenko OO, Cahach VF. Immunogenic Disturbances of the Cardio-vascular System Function. Kyiv 1992. — Ablitsov VH. «Galaxy Ukraine». The Ukrainian diaspora: prominent persons. Publ. «Kyt», Kyiv 2007.

first proved the possibility to correct this form of shock by blocking lipo-oxygenase, and cyclo-oxygenase pathways (1975, 1992).

First in Ukraine, together with **Mykola Syrotinin (1896 – 1977**: above), **Vadym Ya. Berezovsky (29.VIII.1932, city of Kyiv, Ukraine – 14.XI.2020, city of Kyiv, Ukraine)**[788] — Ukrainian pathophysiologist, utilized the method for measuring oxygen (O_2) tension in human alive tissue and in laboratory animals at while conducting those researches. During those experiments they developed and introduced technology for calibration of the indicator electrode, method for stabilization of the active superficial electrode in biological environment, means for the prevention of catalytic release of O_2 while working in acid environment. Designed and introduced into industrial production oxytensiometer – device for measurement systolic and diastolic O_2 tension in the myocardium.

Jointly with **Yu. Horchakov**, they designed and created an instrument surfameter for detection activity of superficially active substances (SAS) or surfactants in the lung at norm and pathology (1982, 1983).

Researching the influence upon secretion of endorphins[789] erythropoietins, and cathecholamines, general activation of energized metabolism, established two different types of O_2 deprivation or loss, i.e., (1) sanogenic hypoxia, favorable for human life and longevity (1984, 1988), i (2) pathogenic hypoxia, capable to be harmful to the health. Furthermore, he created generators of artificial mountain's air with the aim to secure individualized doses of the normo-baric sapogenic hypoxia (1990, 1998). Mutually with associates proved the possibility to restrain the development of allergic conditions and osteopathy of inactivity during preadaptation in boundaries of sapogenic hypoxia. They worked a method of prophylactic (prevented) increase resistance of an organism was used in preparation of astronauts before departure into the space.

Two pathological entities (conditions) are associated with the name of **J. Aldon Carney (1934 –)** — Irish-American pathologist. One is the coexistence, mainly in young women, of gastric epithelioid leiomyosarcoma, pulmonary chondroma

---

[788] A surface-active substance (SAS) or surfactants of the lung are lipoproteins securing superficial tension within alveoli close to zero which prevents their sticking together during exhalation, dilute carbon dioxide (CO_2) and oxygen (O_2), by this alleviate transition of those gases through the wall of the alveoli and capillary, create draught of the lung and have antibacterial function. — Berezovsky VYa. Superficially active substances of the Lung. 1982. — Berezovsky VYa. Surfactants in the norm and pathology 1983. — Berezovsky VYa. Explanatory Dictionary in Physiology and Pathology of Breathing. 1984. — Berezovsky VYa. Physiological mechanisms of sapogenic effects from the mountains air climate. 1988. — Berezovsky VYa. Biophysical charactristic of human tissue. 1990. — Berezovsky VYa. Introduction into opotherapy. 1998. — Ablitsov VH. «Galaxy Ukraine». The Ukrainian diaspora: prominent persons. Publ. «Kyt», Kyiv 2007.

[789] Endorphins are polypeptic chemical compounds, similiar to opiates, produced by neurons of the brain (discovered in the middle of 1970s).

and extraadrenal periganglioma (Carney triad, 1983). The other is that the most common primary cardiac tumor is a benign myxoma, made of a primitive connective tissue, usually sporadic and located in the left atrium (LA) in 75-86% of patients. But, if myxoma occurs as a part of the autosomal dominant trait (7-20%), then it is characteristically associated with spotty pigmentations of the skin and endocrine hyperactivity (Carney complex, 1986).

Primitive neuroectodermal malignant small-cell soft tissue tumor (sarcoma) of the thoraco-pulmonary, retroperitoneal and pelvic region and extremities (PNET of the chest wall) was described by **Frederic B. Askin (1939 –)**[790] — American pathologist, appearing in the pleural cavity as a single lesion or as multiple pleural nodules in children and young adults under 25 year of age (Askin tumor, 1976). PNET is a member of **James Ewing (1866 – 1943**: 20th century. Pathology) tumor of the bone (ETB) or Eving sarcoma of the bone family, along with extraosseous Ewing tumor (EOET), because of the chromosomal aberrations t(11;22)(q 24; q;12) is peculiar to both tumors. However, PNET originate from soft tissue, whereas ETB is a tumor of the bone. The treatment include adjuvant chemotherapy, surgical resection, and adjuvant chemoradiotherapy.

In 1985 **Mirella Marino** — Italian pathologist and oncologist, and **Hans K. Muller-Hermelink (1947 –)**[791] — German pathologist, proposed a histologic classification of thymoma [19th century. Internal medicine and subspecialties – Sir **Samuel Wilks (1824 – 1911)**] with its division into cortical, medullary, and mixed types. The cortical type contains medium to large epithelial cells of characteristic appearance, and usually abundant lymphocytes. The medullary type contains small to medium cells with different features and fewer lymphocytes. The former tends to be of higher clinical stage, showing the least favorable prognosis, whereas the latter tend to be of lower invasiveness, having the best prognosis.

Genetics and immunology

In 1901-02 **Clarence E. McClung (1870 – 1946)**[792] — American zoologist and cytologist of Irish Scottish descent, isolated and established the chromosome theory of inheritance when he pointed out that the accessory (x) chromosome was possible the nuclear element responsible for determining the sex of humans. In other words,

---

[790] Askin FB, Rosai J, Sibley RK, et al. Malignant small cell tumour of the thoracopulmonary region in childhood: a di stinctive clinicopathologic entity of uncertain histogenesis. Cancer 1979;43:2438-51.

[791] Marino M, Muller-Hermelink HK. Thymoma and thymic carcinoma: relation of thymoma epithelial cells to the cortical and medullary differentiation of thymus. Virchows Arch A Pathol Anat Histopathol 1985;407(2):119-49.

[792] McClung CE. The accessory chromosome – sex determinant. The Biological Bulletin 1902 May 1;3(1-2):43-84.

it is the sex chromosome, either alone or pair of chromosomes that decides whether an individual is a male or female. In humans the sex is determined by one pair of the total of 23 pairs of chromosomes, the other 22 pair of chromosomes are non-sex chromosomes, called autosomes.

In 1911 **Edward B. Wilson (1856 – 1939)** — American cell biologist, was the first to identify the gene responsible for a human disorder, color blindness, in a human chromosome.

In this same year (1911), **Thomas H. Morgan (1866 – 1945)** — American evolutionary biologist, geneticist, embryologist, and science author, observing the crossover in the fruit fly Drosophila melanogaster meiosis, proved that genes are arranged in 46 chromosomes in a clear fixed lineal order, and that certain characteristics are transmitted through genes on chromosomes. He detailed his study in his work «The Theory of the Gene» (1926).

Hryhohiy (Gregory) A. Lewitzky (07/19.XI.1878, village Bilky, Skvyrsky Region, Kyiv Obl., Ukraine – repressed by Russian Communist Fascists regime, died in prison from hunger **20.V.1942, town of Zlotoust, Chelabinsk Obl., Russia)**[793] — Ukrainian botanist, geneticist, cytologist, and karyotinst, described the presence of mitochondria [(in Greek – «mitos» meaning thread, «khondrion» meaning granule) which is bi-membranous organelle, seen in majority of cells eulariots (in Greek – «eu» meaning good or fullness, «karyot» meaning nucleus)] in the cells of the plants; worked up several methods of staining of the chromosome and mitochondria; showed prominent changes in rebuilding of the mitochondrial structure following exposure to x-rays (1931); estimated that chromosome has double shoulder structure; pustulated that nuclear fragments which are not constricted, i.e., centromere, are not inherited, and therefore could not be considered being chromosomes; showed, that evolutionally close species retain the close load of chromosomes. Together with **Lev**

---

[793] Village of Bilky (known since 14th century on account of confirmation mentioned in archives document from 1546 p.; during conduct by Russian Communist Fascist regime of Holodomor - Genocide 1932-33 against the Ukrainian Nation villagers were dying en masse), Skvyrsky Region, Kyiv Obl., Ukraine. — Town of Platoust (founded 1754), Chelabinska Obl. (areaa 88,529 km²), Russia. — Levytskyy GA. The Material Basis for Heredity. Gosizdat Ukrainy, Kyiv 1924. — Lewitzky GA. Enfolge der genetischen Zytologie und ihre Anwendung bei Kulturpflanzen in der practischen Botanik, Entstehungs -und Züchtungs-Lehre. St Petersburg 1929. — Lewitzky GA. Die «Kariotype» Systematics. Bull Appl Bot Plant Breed 1931;27:19-174 oder 187-240. — Lewitzky GA, Araratian AG. Transformation of the morphology of chromosomes under the influence of x-rays. Bull Appl Bot Plant Breed 1931;27:289-303. — Lewitzky GA. Experimentally induced alterations of the morphology of chromosimes. The American Naturalist (University of Chicago Press) 1931 Nov-Dec;65(701):564-7. — Mirsky M. Ukrainian medical scientists – victims of Stalin's repressions. UHMJ «Ahapit» 1995;2:45-54. — Ablitsov VH. «Galaxy Ukraine». The Ukrainian diaspora: prominent persons. Publ. «Kyt», Kyiv 2007. — Hanitkevych YaV. The contribution of the Ukrainian physicians to the world medicine. Ukr Med Chasopys 2009 VII/VIII;4(72):110-15. — Velychko MV, Stefanyk VI. Hryhoriy Andriiovych Levytsky — Ukrainian cytogeneticist and kario-systematicist (1878 – 1943). Cytology and Genetics 2011 3-9 Feb;2:71-6.

M. Delome (11/23.V.1891, city of Petersburg, Russia – 01.XII.1968, city of Kyiv, Ukraine) — Ukrainian geneticist, and cytologist, introduced the term «karyotin» – the reticular usually stainable material of cell nucleus with a load of chromosome, specifical for the organism of every species (1931).

One of the main individuals who created modern synthetic theory of evolution was **Ivan I. Schmalhausen [11(23).IV.1884, city of Kyiv, Ukraine – 07.X.1963, city of Moscow, Russia]**[794] — Ukrainian Russian evolutionary biologist, zoologist and morphologist, founder of the stabilizing selection theory. In 1930 he established Institute of Zoology of All-Ukrainian Academy of Sciences (All-UAS), now the National Academy of Sciences (NAS) of Ukraine, named I.I. Schmalhausen.

Sir **Ronald A. Fisher (1890 – 1962)** — English statistician and biologist, who used mathematics to combine Mendelian genetics [19th century. Genetics – **Gregor Mendel (1822 – 84)**] and natural selection which helped to create the new Darwinist synthesis of evolution[795] known as the modern evolutionary synthesis.

In 1927 **Hermann J. Muller (1890 – 1967)** — American geneticist of Jewish ancestry, discovered that the x-rays / radiation causes physiological and genetic mutation in genes (mutagenesis).

First described in 1958 by **Werner Catel (1894 – 1981)**[796] and **J. Schmidt** — German pediatricians, re-described in 1964 by **Michael Lesh (1939 –)**[797]— American cardiologist and **William L. Nyhan (1926 –)** — American pediatrician, a rare familiar genetic x-linked recessive disease due to defective gene, affecting almost exclusively males.

Initially, presented itself in two brothers at the of two and four months of life by irritability. Usually, this disease began during infancy. Females almost never suffer from this illness but may be carriers.

The underlying cause disease is genetic disorder of purine metabolism whereby a partial hypoxanthine guanine phosphoribosulftransferase 1 (HGPRT 1) deficiency resulting in overproduction of purine and uric acid with its rise in the blood.

---

[794] Schmalhausen II. Factors of Evolution: Theory of Stabilizing Selection. Publ. Blackstone, Philadelphia 1949. — I.I. Schmalhausen Institute of Zoology of the National Academy of Science of Ukraine 75-years. Publ. «ArtEK», Kyiv 2005. — Ablitsov VH. «Galaxy Ukraine». The Ukrainian diaspora: prominent persons. Publ. «Kyt», Kyiv 2007.

[795] **Charlles Darwin (1809 - 82)** — English naturalist who developed a theory of biological evolution stating that all species of organism arise and develop through the natural selection of small, inherited variations that increase the individual's ability to compete, survive and reproduce (Darwinian theory), as outlined in his treatise – Darwin C. On the Origin of Species by Means of Natural Selection, or the Preservation of Favored Races on the Struggle for Life. John Murray, London 1859.

[796] Catel W, Schmidt J. Über familiar gichtische Diathese in Verbindung mit zerebralen und Symptomen bei einen Kleinkind. Dtsch med Wschr (Stuttgard);1958:84:2145.

[797] Lesh M, Nyhan WL. A familial disorder of uric acid metabolism and central nervous function. Am J Med (New York) 1964;536-70.

Further clinical picture shows progressive mental retardation, impaired growth, tendency to self-mutilation such as lips, thumb, and foot biting, face scratching and head banging. Other signs include dislocation of the eyes, central nervous system (CNS) dysfunction – epileptic seizure, extrapyramidal signs with lack of coordination, aggressive behavior, exaggerated tendon reflexes and Babinski sign [19th century. Neurology – **Joseph Babinski (1857 – 1931)**], and renal failure RF).

It was named the Lesh-Nyhan disease or syndrome (1964).

Hypogonadotropic hypogonadism with anosmia (loss of scent) first described in 1944 by **Franz Jo. Kolmann (1897 – 1965)** — German American geneticist of Jewish heritage, is a genetic disorder that prevent person from starting or fully complete puberty, also termed Kolmann syndrome. The additional symptoms are a total lack of sense of smell or reduced sense of smell.

The great contribution of Sir **Frank M. Burnet (1899 – 1985)** — Australian virologist, is the advancement of hypothesis that during embryologic life and immediately after birth, cell gradually acquire the ability to distinguish between their own tissue substances and unwanted cells substances or foreign material, thus predicted acquired immune tolerance (1944) and developed a conceptual immunologic theory of clonal selection (1949).

Theodosius H. Dobzhansky (Jan. 25, 1900, town of Nemyriv, Vinnytsia Obl., Ukraine – Dec. 15, 1975, town of San Jacinto, CA, USA)[798] — Ukrainian American geneticist and evolutionary biologist, by combining the theory of evolution with genetics (1937) established an evolutionary genetics as an independent discipline; found in the process of natural selection an extensive and rapid variability of genes affected by environmental conditions; proved that among heterozygotes «good mixers» genes «get ahead» and become more widespread in the population; defined the process by which species split into two or more species (speciation) which led to the origin of man – «descent of man» (1962).

Researching beget in a species of fruit fly *Drosophila pseudoobscura*, **Michael O. Vetukhiv (25.07(07.08).1902, city of Kharkiv, Ukraine – 11.06,1959, New**

---

[798] Town of Nemyriv (not far in boundary-line were found remains of the settlement of the Trypillia Culture and Scythes former towns in ruins from 7th – 6th century BC; founded 1390, 1506; during conducted by Russian Communist Fascist regime of Holodomor-Genocide 1932-33 against the Ukrainian Nation died at least 487 inhabitants, and as the outcome of terrible political repressions of 1937-38 vanished number of people), Vinnytsia Obl., Ukraine. — Town of San Jacinto (founded 1870), California (CA, area 163,696 km²), USA. — Dobzhansky Th. Genetics and the Origin of Species. Columbia Univ. Press, New York 1937.— Dobzhansky Th. Evolution, Genetics and Man. Wiley & Sons, New York 1955. — Donzhansky Th. Making Evolving. Yale Univ. Press, New Haven 1962 — Dobzhansky Th. Heredity and Nature of Man. Harecourt, Brace & World, Inc, New York 1966. — Dobzhansky Th. The Biology of Ultimate Concern. New American Library, New York 1967. — Donzhansky Th. Genetics of the Evolutionary Process. Columbia Univ. Press, New York 1970. — Dobzhansky Th. Genetic Diversity and the Equality. Basic Books, New York 1973. — Ablitsov VH. «Galaxy Ukraine». The Ukrainian diaspora: prominent persons. Publ. «Kyt», Kyiv 2007.

York City, N.Y., USA)[799]— Ukrainian American biologist and geneticist, Lecturer of genetics and Rector (1942-43) of the Kharkiv National University (KhNU) named for V.H. Karazin (founded 1804), co-founder and President of the Ukrainian Free Academy of Sciences (UFAS) and Arts in the U.S.A. (1950-59) and geneticist-researcher (1951-59) at the Columbia University (established 1754), New York City, N.Y., USA., established rule for substantial decrease of disjoining ability of hybrids in the second generation («Vetukhiv weakening», 1953) / established the substantial loss of breeding potential among second-generation hybrid plants («Vetukhiv breakdown», 1953).

Barbara McClintock (1902 – 92) — American geneticist, discovered the «jumping genes» that can move from one spot to another on chromosomes and thus produce changes that can alter the cells' function and can be passed on to offspring (1953).

In their work on the bread mold Neospora crassa, **George W. Beadle (1903 – 89)** — American geneticist, and **Edward L. Tatum (1909 – 75)** — American geneticist, showed that genes control a cell's enzyme production and chemistry and transmit hereditary characteristics (1941-52).

The major histocompatibility complex and genetic regulation of the body's immune system was explained by **George D. Snell (1903 – 96)** — American mouse geneticist and native transplant immunologist, **Jean B.G.J. Dausset (1916 – 2009)** — French immunologist, and **Baruj Denacerraf (1920 – 2011)** —Venezuelan American immunologist. Additionally, Jean Dausset with discovery of the human leucocyte (lymphocyte) antigen (HLA) in 1967 made a breakthrough in the modern cellular therapy rejection («graft versus host»). Baruj Denacerraf showed heredity of the immune reactions in guinea pigs and theorized that the genetic variations in humans

---

[799] Vetukhiv M. Milk and Milk Products; How to Utilize and Prepare of Dairy Production for Sale and Utilization in the Household. Kharkiv 1926. — Vetukhiv M. Methods for Improving Animals Based on Genetic Methodic. Kharkiv 1928. — Vetukhiv M. Increase Resistance of the Large Horned Cattle to Tuberculosis. Kharkiv 1942. —Vetukhiv, M. Viability of hybrids between local populations of *Drosophila pseudoobscura*. Proc Natl Acad Sci USA, Washington 1953 Jan15; 39(1):30-34. — Vetukhiv, M. Integration of the genotype in local populations of three species of *Drosophila*. Evolution 1954;8:241-51. — Vetukhiv, M. Integration of the genotype in local populations of three species of *Drosophila*. Evolution 1954; 8, 241-51. — Vetukhiv, M. Heterosis and hybrid breakdown in hybrids between local populations of certain species of *Drosophila*. Caryologia 1954;6(suppl):760-61. — Wallace B, Vetukhiv M. Adaptive integration of the gene pools of *Drosophila* populations. Cold Spr Harb Symp quant Biol 1955;20, 303-11. — Vetukhiv, M. Fecundity of hybrids between geographic populations *of Drosophila pseudoobscura*. Evolution 1956;10:139-46. — Vetukhiv M. Heterosis in hybrids between local populations of *Drosophila pseudoobscura*. Cytologia 1956;Suppl Vol:394-5. — Vetukhiv, M. The longevity of hybrids between local populations of *Drosophila pseudoobscura*. Acta genet 1956;6-252-4. — Vetukhiv, M. Longevity of hybrids between geographic populations of *Drosophila pseudoobscura*. Evolution 1957;11:348-60. — Vetukhiv M, Beardmore, JA. Effect of environment upon the manifestation of heterosis and homeostasis in *Drosophila pseudoobscura*. Genetics 1959;44:759-68. — Ehrman L Genetic divergence in M. Vetukhiv's experimental populations of *Drosophila pseudoobscura* 1. Rudiments of sexual isolation. Publ Cambrige Univ Press Apr 2009. https://doi.org/10.1017/500016672300001099.

and other vertebrates is an evolutionary adaptation to ensure that some members of a species survive infections from mutated viruses.

Together in 1939-59 **Serhiy M. Hershenson [29.I(11.II).1906, city of Moscow, Russia – 07.IV.1998, city of Kyiv, Ukraine)**[800] — Ukrainian biochemist and geneticist, with **Mykola D. Tarnavsky (06.VIII.1906, village Muksha Kytayhorodska, Kamyanets-Podilsky Region, Khmelnytska Obl., Ukraine –** died from acute MI **13.VII.1953, city of Bila Tserkva, Kyiv Obl., Ukraine)**[801] — Ukrainian biologist and geneticist, proved mutagenic action of thymic deoxyribonucleic acid (DNA) on Drosophilae, established the possibility of reverse transmission of genetic exogenic peculiarity (information, familiarity) from ribonucleic acid (RNA) into exogenic DNA. Besides of chemical mutagenesis he revealed phenomenon of «jumping genes» and reverse transcription. Serhiy Hershenson gathered from nucleic acids alive virus, that is he discovered synthesis of DNA virus on the matrix of infected polyhedroses RNA.

The most important scientific positions of Mykola Tarnavsky formed in his experiments on Drosophila by disclosing the influence of exogenic DNA on crossing-over, discovery of mutagenic ability of DNA as able to cause directed mutations and experimental proof reciprocal non homologous chromosomes during conjugation. His views on genome as singular wholeness and genes as plastic structures, work of which depends on the status of metabolism and trial to explain the influence of exogenically introduced DNA and aminoacids on crossing-over frequency, as realization of mutual functionally active groups of those compounds with genes, means the announcement of a new epoch in molecular biology.

In 1939 at Institute of Zoology, Academy of Sciences of Ukraine, **Panteleymon Sit'ko (27.VI.1906, village of Ksaveriivka, Vasylkivsky Region, Kyiv Obl., Ukraine –** died from acute MI **25.IX.1985, city of Kyiv, Ukraine)**[802] — Ukrainian

---

[800] Ablitsov VH. «Galaxy Ukraine». The Ukrainian diaspora: prominent persons. Publ. «Kyt», Kyiv 2007.

[801] Village Muksha Kytayhorods'ka (founded 1926), Kamyanets-Podils'kyy Region, Khmelnyts'ka Obl., Ukraine. — City of Bila Tserkva (first mentioned 1032 p.; during realization by Russian Communist Fascist o regime of Holodomor-Genocide 1932-33 against the Ukrainian Nation streets were sown with bodies of people dead from man-made hunger), Kyiv Obl., Ukraine. — Tarnavsky MD. Role of biological factors in the process of inheritance. Institute of Zoology. Academy of Sciences of Ukraine, Kyiv 1947. — Ablitsov VH. «Galaxy Ukraine». The Ukrainian diaspora: prominent persons. Publ. «Kyt», Kyiv 2007. — Holda DM, Potopalsky AI, Katsan VA. The letters to the Eternity of the Ukrainian Geneticist Nicolai Tarnavsky (To the centenary of his birth and 70th anniversary of the publication of the first article on influence of DNA on the genetical processes). Fizyka zhyvoho / Physics of the Alive 2008;16(2):191-7.

[802] Village Ksaverivka (founded 1780s), Vasylivsky Region, Kyiv Obl., Ukraine. — Sit'ko PO. Dependence of Mutability from Genotype. Dissertation for degree of Candidate of Biological Sciences. Institute of Zoology All-Ukrainian Academy of Sciences, Kyiv 1932. — Sit'ko PO. Principles and Methods of Selection and Genus in the he Silk Industry. Dissertation for degree of Doctor of Biological Sciences. Institute of Zoology, Academy of Sciences of Ukraine, Kyiv 1960. — Ablitsov VH. «Galaxy Ukraine». The Ukrainian diaspora: prominent persons. Publ. «Kyt», Kyiv 2007.

zoologist and geneticist, with the help of radiological setting, performed series of experiments with radiation and obtained mutations of Drosophila melanogaster, in which was proven dependency of mutability from the nature of genome and revealed variability mutability in lines, obtained from one and this same population of Drosophila. Then in 1960 he brought two new highly productive breed of silkworm and proposed modern method of «synthetic selection», assuring stable form of hybrids, which doesn't degrade during further multiplications.

In 1948 **Murray L. Barr (1908 – 95)**[803] — Canadian anatomist, and geneticist, discovered with **Ewart G. Bertram (1923 –)** — Canadian graduate student and anatomist, an intracellular sex chromatin body, the inactive X chromosome in female somatic cell, rendered inactive in a process of the chromosome X-inactivation, in those species in which sex is determined by the presence of the Y chromosome, including humans, or the W chromosome rather than diploidy of the X chromosome (the body of sex chromatin or Barr body). This discovery initiated a new approach in research and diagnosis of genetic disorders, specifically a greater understanding and ability to manage certain conditions associated with mental retardation.

Eldon J. Gardner (1909 – 89) — American geneticist, described a familial autosomal dominant multiple adenomatous polyposis of the gastrointestinal (GI) tract, mostly colon, with predisposition to development of cancer, together with desmoid tumors (15%), osteomas of the scull, thyroid carcinoma, epidermoid or sebaceous cysts, lipomas, fibromas, dental abnormalities, and periampulary carcinoma of the duodenum (Gardner syndrome, 1951). The incidence of Gardner syndrome is 1:14,025 with an equal sex distribution.

In 1961 **Jacques L. Monod (1910 – 76)** — French biologist and **Francois Jacob (1920 – 2013)** — French biologist of Jewish ancestry, established that the mechanism of protein synthesis and its genetic regulation in viruses is controlled by enzyme expression in cells as result of transcription of deoxyribonucleic acid (DNA) sequences, that originated from DNA to its product the ribonucleic acid (RNA) which in turn is decoded to protein.

Bacterial and other cells could respond to external conditions by regulating level of their key metabolic enzymes, and/or the activity of those enzymes. If a bacterium find itself in a broth containing lactose, rather than single sugar glucose, it must adapt itself to the need to (1) import lactose, (2) cleave lactose to its constituents glucose and galactose, and (3) convert galactose to glucose. They showed that in case of the lactose, the common bacterium Escherichia coli, there are specific proteins that are

---

[803] Barr ML, Bertram EG. A morphological distinction between neurons of the male and female, and the behavior of the nucleolar satellite during accelerated mucoprotein synthesis. Nature 1949 Apr 30;163(4138):676-7.

devoted to repressing (the lactose repressor) the transition of DNA to its product RNA which in turn is decoded to protein. This repressor protein is made in all cells, binding directly to DNA at the genes it controls, and physically preventing the transcription apparatus from gaining access to the DNA. In presence of lactose, this repressor binds lactose, making it no longer able to bind DNA, and the transcription repression is lifted. In this way, a robust feedback loop is constructed that allows the set of lactose-digesting proteins products to be made only when they are needed. This repressor model is extended to all genes in all organisms in their initial exuberate. The regulation of gene activity has developed into a sub-discipline of molecular biology.

Robert W. Briggs (1911 – 83) — American scientist, in 1952, together with **Thomas J. King (1921 – 2000)** — American biologist, cloned (in Greek - «klon» means «twig») a frog by nuclear transfer of embryonic cells. The same technique, using somatic cells, was later used to create «Dolly the Sheep». Their experiment was the first successful nuclear transplantation performed in metazoans. In the extension of their work, Sir **John B. Gurdon (1933 –)** — English developmental biologist, cloned a frog using intact nuclei from the somatic cells of a Xenopus tadpole (1958), showed that deoxyribonucleic acid (DNA) from specialized cells of ordinary frog skin or intestinal cells, could be used to generate new tadpoles – potent stem cells (SC, 1962). Then, in 2006, **Shinya Yamanaka (1962 –)** — Japanese SC researcher, showed that the adult mouse and adult human cells (fibroblasts) could be returned into immature (primitive) pluripotent stem cells. Discovery how to reprogram ordinary cells to behave like embryonic SC, offer a way to skirt around ethical problems by avoiding destroying human embryos, harness the reprograming to create replacement tissues for treating diseases such as paralysis agitans, cystic fibrosis and diabetes mellitus (DM) and for studying the roots of different diseases in the laboratory cells.

The three theories of the immune system developed by **Niels K. Jerne (1911 – 94)** — Danish immunologist, were: (1) that instead of the body producing antibodies in response to a foreign antigen, the immune system already has the specific antibodies against a foreign antigen; (2) the immune system learns to be tolerant to the individual's own self which takes place in the thymus (T); (3) the T-cells and the bursa (B)-cells communicate with each other through the network of the active sides antibodies attached to both specific antigens (idiotypes)[804] and to the other antibodies that bind to the same side, and that antibodies are in equilibrium until

---

[804] The term idiotype was given to individual antibodies by **Jaques H.L. Oudin (1908 - 85)** — French immunologist.

foreign antigens perturb the normal equilibrium, stimulating immune reaction (the network immune theory, 1984)[805].

The work of **Volodymyr Hordynsky (Mar. 18, 1915, village Pidberezhzhia, Bolekhiv Region, Ivano-Frankivsk Obl., Ukraine – 01.10.1994, town of Livingston, NY, Essex County, NY, USA)**[806] — Ukrainian American immunologist and oncologist, is dedicated to the embryonal basis for diagnosis of cancer, the role of immune globulins in modern medicine and radioimmune assay of carcino-embryonal antigen.

John Maynard-Smith (1920 – 2004)[807] — English geneticist, mathematical and biological evolutionist, who was instrumental in the application of gene theory to evolution with **George R. Price (1922 – 75)** — American geneticist and theorized on the evolution of sex and signaling (1973).

The hypothesis of acquired immune tolerance was proved by Sir **Peter B. Medawar (1915 – 87)** — British biologist, discovering the acquired immune tolerance (1953). **Rupert E. Billingham (1921 – 2002)** — British American immunologist, and **Leslie B. Brent (1925 –)** — British immunologist, showed it was possible to coax the body of an animal to accept foreign tissue transplanted from another. Skin taken from a white mouse could be accepted as «self» by the immune system of a brown mouse, if transplanted early in life, a condition defined as «immunological tolerance». They also discovered later that in some cases a transplanted organ could reject its new body (in the process known as «graft-versus-host») that posed additional complications for bone marrow transplants.

Beatrice Mintz (1921 –) — American embryologist of Jewish origin who pioneered of genetic engineering techniques and was among the first scientist to generate both chimeric and transgenic mammals.

In experiments by **M. Laurence Morse (1921 – 2003)** — American microbiologist, **Esther M. Zimmer-Lederberg (1922 – 2006)** — American microbiologist and a pioneer in bacterial genetics who discovered bacterial virus (bacteriophage) Lambda (1951), and **Joshua Lederberg (1925 – 2008)** — American molecular biologist and

---

[805] The Niels Jerne's network immune theory is a resurrection of the natural selection side-chain theory of immunology lied out by **Paul Erlich (1854 - 1917**: 19th century. Microbiology and immunology. / Above).

[806] Village Pidberezhzhia (documented 1515), Bolekhiv Region, Ivano-Frankivsk Obl., Ukraine. — Town of Livingston (established 1727), Essex County, N.J., USA) — Hordynsky BZ? Hypertonia. J Ukr Med Assoc North Am 1954 Dec.;2:14-8. — Hordynsky V. Immunoglobulins in the modern medicine. J Ukr Med Assoc North Am 1971 Jul; 18,3(62):20-5. — Hordynsky V. Embryonal basis for diagnosis of cancer. J Ukr Med Assoc North Am 1973 Jan-Apr; 20,1-2(68-69):61-4. — Hordynsky V. Radioimmune assay of carcino-embryonal antigen. J Ukr Med Assoc North Am 1977 Jan;24,1(85);33-5. — Ablitsov VH. «Galaxy Ukraine». The Ukrainian diaspora: prominent persons. Publ. «Kyt», Kyiv 2007.

[807] Maynard-Smith J. The Evolution of Sex. Cambridge Univ Press, Cambridge 1978, — Maynard-Smith J. The Evolution and Theory of Games. Cambridge Univ Press, Cambridge 1982.

geneticist of Jewish descent, on the Lambda-mediated transduction (gene transfer from one bacterial strain to another by a bacteriophage as a vector) of bacterial determinants for galactose fermentation were encountered defective lysogenic strains among their transductions . Thus, they confirmed the genetic recombination and organization of bacteria.

In 1955 **Oliver Smithies (1925 – 2017)**[808] — English American geneticist and physical biochemist, introduced starch as a matrix (medium) for gelatinous electrophoresis, which is now one of the basic techniques used in biochemistry and molecular biology. Subsequently, he and his wife, **Nobuyo N. Maeda (early 1950s –)** — Japanese American geneticist, turned to gene targeting in mice, by replacing single mouse genes using homologues recombination with cloned copies modified in vitro (1987). Their research is the basis for investigation the role of genes in such diseases as cancer, cystic fibrosis, diabetes mellitus and hypertension.

In 1963 **Sidney Brenner (1927 – 2019)** — American biologist of Jewish faith, proposed the usage of free-living (non-parasite) 1-mm long ground (soil) rounded worm (Caenorhabditis elegans or C. elegans) as model organism for research primary neural development of in animals. Afterwards mutually Sir **John E. Sulston (1942 – 2018)** — South-African biologist, **H. Robert Horvitz (08.05.1947, city of Chicago, IL, USA –)**[809] — American biologist of Jewish Ukrainian descent, after establishing C. elegans as their research model, then they identified one with functional defect for incoordination («inc»), that is unfavorable. This favored identification of new proteins recruitment, such as protein «inc», among them «inc»119 – gen specifically distinguish in in photoreceptors of the retina. Finally, in 2002 all three noted in C. elegans hereditary organ for development and programmed cell death.

Thus, they discovered the genetic regulation of development organisms and mechanism of apoptosis, i.e., genetically steered destruction of cells through fragmentation of deoxyribonucleic acid (RNA) of its nucleus. This discovery is important for medical research and better understanding of pathogenesis of many diseases.

Independently, in 1980-1981 Oliver Smithies, **Mario R. Capecchi (1937 –)**[810] — Italian American molecular geneticist, Sir **Martin J. Evans (1941 –)** — English biologist, who introduced the technique of homologous recombination of

---

[808] Smithies O. Zone electrophoresis in starch gel: group variations in the serum proteins of normal human adults. Biochem J Dec 1955;64(4):629-41.

[809] City of Chicago (founded 1770-80), IL (Foot Note 650), USA. — Ablitsov VH. «Galaxy Ukraine». The Ukrainian diaspora: prominent persons. Publ. «Kyt», Kyiv 2007.

[810] Capecchi MR. High efficiency transformation by direct microinjection of DNA into cultured mammalian cells. Cell Nov 01,1980; 22(2):479-88. — Thomas KA, Capecchi MR. Site-directed mutagenesis by gene targeting in mouse embryo- derived stem cells. Cell Nov 06,1987; 5(3):503-12.

transgenic deoxyribonucleic acid (DNA) with genomic DNA (1975) and **Gail R. Martin (1944 –)**[811] — American molecular biologist, Martin Evans and **Matthew H. Kaufmann (1942 – 2013)**[812] — English Scottish anatomist and embryologist born in Jewish family, discovered and cultivated embryonic stem cells (ESC), were able to switch off genes in mouse ESC and by in vivo fertilization changed (engineered) genetically modified mouse (Mus musculus), creating the so called knockout mice, or knock-out mice.

Specifically, Mario Capecchi created the knock-out mice in which an existing gene was reactivated by replacing it or disrupting it with an artificial piece of DNA, an important animal model for studying the role of genes which have been sequenced.

Their experiments have been useful in studying and modeling different kind of neoplasms, heart disease, obesity, diabetes mellitus (DM), arthritis, substance abuse, anxiety, Parkinson's disease (PD) [19th century. Neurology – **James Parkinson (1755 – 1824)**] and aging.

Together, **Cesar Milstein (08 Oct. 1927, city of Bahia Blanka, Province of Buenos Aires, Argentine – 24 Mar. 2002, city of Cambridge, England)**[813] — Argentinian British biologist, son of Jewish Ukrainian emigrants and **Georges J.E. Köhler (1946 – 95)** — German biologist, invented and developed in 1975 the hybridoma technique[814] for production of individual (monoclonal) antibodies. The hybridoma technology starts by injecting an animal, usually mouse, with an antigen that provokes immune reaction. A type of white blood cells (WBC), the bursa (B)-cell that makes antibodies able to bind the antigen are then harvested from this mouse. These isolated B-cells are in turn fused with immortal B cancer cells, a myeloma, to produce a hybrid cell line called hybridoma which has both abilities – the antibody-made of the B-cells and the exaggerated longevity and reproductivity of myeloma. They can be grown in the tissue culture. Monoclonal antibodies are used for the diagnosis, prevention, and treatment of diseases.

It was not until 1956, when **John H. Edwards (1928 – 2007)** — British medical geneticist, suggested that prenatal diagnosis of genetic disease could be done by

---

[811] Martin GR, Evans MJ. Differentiation of clonal lines of teratocarcinoma cells: formation of embryonal bodies in vitro. Proc Natl Acad Sci USA 1975;72(4):1441-5. — Martin GR. Isolation of pluripotent cell line from early mouse embryos cultured in medium conditioned by teratocarcinoma stem cells. Proc Natl Acad Sci USA Dec 1, 1981;78(12):7634-8.

[812] Evans MJ, Kaufman MH. Establishment in culture of pluripotential cells from mouse embryos. Nature 1981 July 09;292(5819):154-6.

[813] City of Bahia-Blanka (founded 1828), Province of Buenos Aires (area 307,511 km²), Argentine (area 2,780,400 km²) – 24 March 2002, city of Cambridge (remains of a 3,500-year-old-farmstead, founded 410), England). — Ablitsov VH. «Galaxy Ukraine». The Ukrainian diaspora: prominent persons. Publ. «Kyt», Kyiv 2007.

[814] The term hybridoma was coined by **Leonard A. Herzenberg (1931 - 2013)** — American immunologist and geneticist.

studying the amnionic fluid. That led to the development of genetic «fingerprinting», deoxyribonucleic acid (DNA) mapping and sequencing allowing to pinpoint many potential genetic disorders. He described a genetic disorder caused by the presence of all or part of an extra 18th chromosome, almost always caused from nondisjunction during meiosis (Edwards syndrome, 1960).

In the simple process of cleaning slides bearing of chronic myeloid leukemia (CML) cells, **Peter C. Nowell (1928 – 2016)**[815] — American oncologist and pathologist, helped uncover the first clear sign of a genetic cause of cancer (1959-60). While studying samples of CML, he mistakenly washed his slides with tap water instead of laboratory solution, and when studied them under microscope noted that the cell's chromosomes expand. His associate, **David A. Hungerford (1927 – 93)** — American oncologist, while analyzing in 1973 the white blood cells (WBC) of patients with CML consistently noted that the chromosome 22 was notably short, a tiny flaw of chromosomes from the blood cells of patient with a type of leukemia under the microscope of the cells in the act of dividing. They named it for city of its discovery – the Philadelphia chromosome.

In parallel, also in 1973, **Janet D. Rowley (1925 – 2013)** — American human geneticist, was the first to identify a chromosomal translocation, in which portion of two chromosomes exchange places, as a cause of CML and other cancers.

Further progress in linking genetics to cancer was made by **Alfred G. Knudson (1922 – 2016)** — American physician and geneticist, in his «two-hit hypothesis» (1971), explaining the incidence of hereditary retinoblastoma, a rare cancer that rapidly developers from the immature cells of an eye's retina, a light-detecting tissue in the fundus of the eye. According to this hypothesis humans inherit two copies of every gene, one from each parent, except for genes on the x and y chromosomes in males. Some people inherit one mutated version and one normal version of the retinoblastoma gene, which produces the retinoblastoma protein involved in controlling cell cycle progression («the first-hit»). Over time, a mutation may arise in the normal version in one cell, thus producing «the second-hit», which leaves the cell unable to control the process of cell division in orderly manner, leading to cancer.

Beginning since 1962 **Daniel Nathans (1928 – 99)** — American microbiologist of Jewish ancestry, **Werner Arber (1929 –)** — Swiss microbiologist and geneticist, **Hamilton O Smith (1931 –)** — American microbiologist and **Daisy Roulland-Dussoix (1936 – 2014)** — Swiss molecular microbiologist, made discoveries on restriction and modification enzymes, resulted in cleavage of deoxyribonucleic acid (DNA) by enzymes at sites characterized by specific sequences unless these are

---

[815] Nowell PC. Discovery of the Philadelphia chromosome: a personal perspective. J Clin Invest 2007 Aug 1;117(8):2033-5.

protected by prior enzymatic modification to the DNA bases[816]. This system protects bacterial cells from viral infection.

Work of **James M. Wilson (1928-2019)** — American clinical geneticist, led to creation «in vivo» adeno-associated viruses from monkey Rhesus macaques / Macaca mulatta as vector for gene therapy in human (1994)[817].

In the late 1960's, techniques were developed to diagnose diseases prenatally, by culturing cells obtained from the amniotic fluid, detecting abnormal fetuses and inducing selective abortion. It may be possible in the future to introduce genetic material into defective cells to induce self-cure.

In the field of heart transplantation (HT) pathology, **Margaret Billingham (1930 – 2009)**[818] — American pathologist, developed the grading system for diagnosis and reporting of an acute rejection of the cardiac transplant by endomyocardial biopsy (EMB) – I-III. The type II-III is characterized by the presence of nodular endomyocardial lymph histiocytic infiltrates in the form of a mysterious «guilty» lesions which may indicate vascular rejection (1988-90). They are present in 4%-26% of human EMB.

Both, **J. Michael Bishop (1936 –)** — American microbiologist and immunobiologist, and **Harold E. Varmus (1939 –)** — American scientist of Jewish parents, are known for the work on retroviral oncogenesis, since 1980s, when they discovered the first human oncogene sarcoma (Src), that is proto-oncogene of the thyrosine-protein kinase (Src), also known as proto-oncogene (c-Src) or c-Sr – a non-receptor thyrosine kinase that in humans is encoded by Src gene. Those findings led to understanding the way malignant tumors are formed from changes to the normal genes of cell, induced by viruses, radiation or certain chemicals.

In 1967 **William N. Kelley (1936 –)** along with J**arvis «Jay» E. Seegmiller (1920 – 2006)**[819] — American physicians and biochemical geneticists, identified disease clinically close to, but without CNS effects such as those in the Lesh-Nyham disease or syndrome **[Werner Catel (1894 – 1981)**: above], with a partial hypoxantine guanine phosphoribosulftransferase 1 (HGPRT 1) deficiency, who typically develop

---

[816]   Arber W, Dussoix D. Host specificity of DNA produced by Escherichia coli. I. Host controlled modification of bacteriophage Lambda. J Mol Biol 1962;5:18-36. — Dussoix D, Arber W. Host specificity of DNA produced by Escherichia coli. II. Controll over acceptance of DNA from infecting phage Lambda. J Mol Biol 1962;5:37-49. — Dussoix D, Arber W. Host specificity of DNA produced by Escherichia coli. IV. Host specificity of infections DNA from bacteriophage Lambda. J Mol Biol 1965;11:238-46

[817]   Gao GP, Alvira MR, Wang L, et al. Novel adeno-associated viruses for Rhesus monkeys as vector for human gene therapy. Proc Natl Acad Sci USA 2002 Sep 3;99(18):11854-9.

[818]   Billingham ME. The postsurgical heart. Amer J Cardiovasc Pathol 1988;1(3):319-34.

[819]   Seegmiller JE, Rosenbloom FM, Kelley WN. Enzyme defect associated with sex-linked human neurological disorder and excessive purine synthesis. Science 1967 Mar 31;155(770):1682-4. — Kelley WN, Rosenbloom FM, Seegmiller JE. A specific enzyme defect in gout associated with overproduction of uric acid. Proc Nat Acad Sci USA 1967;57:1735-9.

gouty arthritis in the second or third decade of life and have a high incidence of uric acid nephrolithiasis (Kelley-Seegmiller syndrome).

Subsequently in 1987-97 William Kelley alongside with **Thomas D. Patella** and **Myron M. Levine** — American specialist in infectious diseases and vaccines, established the first proof of principal «in vivo» virus deoxyribonucleic acid (DNA) vector mediated gene therapy to obtain an expression of the intraneural cells in culture and in intact animal, documented by the United States (U.S.) patent No. 5.672.344, entitled «Viral-mediated gene transfer system».

The work of **James M. Wilson** — clinical geneticist, led to the creation of «in vivo» adeno-associated viruses from the Rhesus macaques (Macaca mulatta) monkeys as vectors for gene therapy in human (1994)[820].

The genetic mechanism of the adaptive immune system that produces antibody diversion was discovered by **Susumu Tanegawa (1939 –)** — Japanese geneticist and immunologist, in experiments beginning 1976, when he showed that genetic material under 19,000 genes in the human body, rearranges itself to form millions of antibodies. When he compared deoxyribonucleic acid (DNA) of the bursa (B)-cells, a type of white blood cells (WBC) in embryonic and adult mice, he found that genes in the mature B-cells of the adult mice are moved around, recombined, and deleted to form diversity of the variable region of antibodies. He also discovered the first cellular enhancer element – a transcriptional one associated with antibody gene complex.

In 1971 **Michael S. Brown (1941 –)** — American geneticist, succeeded in solubilizing and partially purifying 3-hydroxy-3-methylglutaryl coenzyme A reductase that catalyzes the rate-controlling enzyme in cholesterol biosynthesis. Year later (1972), he and **Joseph L. Goldstein (1940 –)** — American biochemist, had developed the hypothesis that abnormalities in the regulation of this enzyme were the cause of familial hypercholesteremia – a genetic disease with elevated level of cholesterol in the bloodstream, arteries and tissues[821]. They also found that human cells have low-density lipoprotein (LDL) receptors that extract cholesterol from the bloodstream, and lack of sufficient LDL receptors causes familial hypercholesterolemia which predisposes for cholesterol-related disorders, such as generalized arteriosclerosis / atherosclerosis (AS), among them to coronary artery disease (CAD). Subsequently, their findings led to the development of the cholesterol-lowering compounds in the blood, a.k.a., statins.

They also uncovered an important aspect of cell biology, i.e., receptor-mediated endocytosis by which cells absorbs metabolites, hormones, proteins and sometimes viruses by the inward budding of the plasma membrane (invagination).

---

[820] Gao GP, Alvira MR, Wang L, et al. Novel adeno-associated viruses for Rhesus monkeys as vector for human gene therapy. Proc Natl Acad Sci USA 2002 Sep 3;99(18):11854-9.

[821] Brown MS, Goldstein JL. Expression of the familial hypercholesteremia gene in heterozygotes: mechanism for a dormant disorder in man. Science 1974:185(4145):61-3.

Tasuku Honjo (1942 –) — Japanese immunologist, who has established the basic conceptual framework of class switch recombination, presented a model explaining antibody gene rearrangement in class switch and verified its validity by elucidation deoxynucleic acid (DNA) structure (1980-82); succeed in cDNA cloning of interleukins (IL)-4 and IL-5 cytokines involved in class switching, and IL-2 receptor alpha chain (1986); first identified of the programmed cell's protein's death (PD-1) as an inducible gene on activated thymus (T) - lymphocytes, and this discovery significantly contributed to the establishment of cancer immunotherapy principle by PD-1 blockage (1992); discovered of activation-induced cytidine deaminase, which is essential for class switch recombination and somatic hypermutation (2000).

Rudolf Jaenisch (1942 –) — German biochemist, geneticist, physician and pioneer of transgene science who created in 1974 the first genetically varied animal by the insertion of deoxyribonucleic acid (DNA) of the virus into the early stage embryo of the mouse and showed that inserted genes were present in each cell. Despite this, the mouse has not passed over the transcend into its offspring. Only in 1981 when the refined DNA was injected into a single-cell embryo of the mouse, then the transmission of the genetic material was detected in the next generations.

Robert A. Weinberg (1942 –)[822] — American biologist, is best known for discoveries of the first human oncogene Ras and the first tumor suppressor gene RD.

Both, **Richard J. Roberts (1943 –)** — English biochemist and nuclear biologist, and **Phillip A. Sharp (1944 –)** — American geneticist and molecular biologist, discovered introns that is a deoxyribonucleic acid (DNA) sequence within a gene and corresponding sequence of ribonucleic acid (RNA) transcripts in eukaryotic DNA. The realization that individual genes in eukaryotes are not contiguous strings, could exist as separate, disconnected segments within longer strands of DNA first arouse in the study of adenovirus[823], one of viruses causing common cold. Subsequently, they discovered the alternative splicing of genes containing introns, and that splicing of messenger RNA to delete those introns can yield different proteins from this same DNA sequence in higher organism, including humans (1993).

In collaboration with **Michael M. Rosbash (1944 –)** — American geneticist and biochronologist of Jewish descent, and with **Paul Hardin (1960 –)** — American biochronologist, he discovered that the period proteins played role in suppressing its own transmission, and were able to develop a negative transcription-translation feedback loop (TTFL) model that serves as a central mechanism of the circadian clock in the fruit fly (1990).

---

[822] Mukherjee S. The Emperor of All Maladies: A Biology of Cancer. Simon & Shuster 2010.

[823] Chow LT, Gelinas RE, Broker TR, et al. An amazing sequence arrangement at the 5' ends of adenovirus 2 messenger RNA. Cell 1977;12(1):1-8.

The leader of the research group, сир Sir **Ian Wilmut (1944 –)** — English Scottish embryologist and geneticist, that first in 1996 cloned a mammalian from an adult somatic cells a Finish lamb named Dolly.

In 1970s **Jeffrey C. Hall (1945 –)** — American geneticist and biochronologist, focusing his research on Drosophila megalogaster, a.k.a., fruit fly examined the neurological component of fruit fly courtship and behavioral rhythm, identified the nervous system that contributed to the regulation of males courtship song which was produced rhythmically with the normal period of about one minute. He uncovered essential mechanisms of biological clocks and shed light on the foundation for sexual differentiation in the nervous system.

Suspecting the period mutation of abnormal sleep-wake cycles, generated in the late 1960s by **Ronald J. Konopka (1947 – 2015)** — American geneticist and biochronologist, may also alter courtship song cycles, he tested the effect of mutations in the period of a courtship song, and found that period mutations affected the courtship song in the same way they changed the circadian rhythms. This finding aroused Jeffrey Hall interest in the per locus for being a central component in the generation of multiple rhythms in fruits flies and eventually led to the isolation the period gene and successful cloning of fruitless fly as a master regulator gene for courtship (mid-1990s).

In 1997 **Jeffrey Hall, Michael Rosbash** et al. discovered genes that are a part of the TTFL model are expressed in the cells throughout the body of each being.

In 2003 **Jeffrey Hall** found that the pigment dispersing factor (PDF) protein, localized to small ventral lateral neurons (sLNus) in the fruit fly brain, helps to control the circadian rhythms and in turn locomotor activity, of these genes in cells.

In 1977 **James P. Allison (1948 –)** — American immunologist, and his colleague **Gerald N. Callahan** — American writer, microbiologist, pathologist and immunologist, found evidence that the immune system was prevented from attacking cancer cells due to antigen's association with additional proteins. Then, in the early 1990s, the former showed that the cytotoxic T-lymphocyte-associated protein-4 (CTLA-4) – a protein receptor expressed in regulatory thymus cells (T-cells) that function as an immune checkpoint and downregulates immune responses, acts as inhibitory molecule to restrict T-cell responses. Subsequently, James Allison was first to show that antibody blockage of a T-cell inhibitory molecule, now identified as CTLA-4, could enhance antitumor immune responses and tumor rejection. Blocking CTLA-4 to unleash anti-tumor immune responses and elicit clinical benefit laid the foundation for the clinical development of other drugs that target T-cell inhibitory pathway, named «immune checkpoint therapies», such as ipilimumad for the treatment of metastatic melanoma, carcinoma of the lung, kidney, and some other malignant tumors.

In the span of 1980s – 2001 **Michael W. Young (1949 –)** — American biochronologist, identified key genes associated with regulation of the internal biological clock responsible for circadian rhythm, elucidated the function of the periodic gene necessary for the fruit fly to exhibit normal sleep cycles, attributed with the discovery of the timeless and double-time genes, which makes proteins that are necessary for circadian rhythm.

In the other words, those researchers were able to glance into small ventral lateral neurons of the fruit fly brain controlling the circadian rhythms and extracted from the fruit fly cell the gene coding a protein, which during the night accumulates in cells, and during the day disintegrates. In this way the normal work of the biological clock is regulated.

In 1999 **H. Lee Sweeney (c. 1950 –)** — American physiologist and biochemist, developed a gene-therapy blocking age-related loss of muscle size and strength in mice. Such therapy could reverse the feebleness associated with old age or counter the muscle-wasting effects of muscular dystrophies (in Greek – «dystrophe», from «dys» - prefix, meaning disturbance, «trophe» - meaning nutrition) and similar diseases.

Sphere of the scientific activity of **Aleksei M. Dugan (18.05.1951, town Khanty-Mansiys'k, Tiumen Obl., Russia –)**[824] — Ukrainian geneticist, Dean (since 1999), Faculty of Biotechnology, and Biotechnique (FBT), National Technical University of Ukraine (NTUU) – «Kyiv Technical Institute» («KTI», founded 1898) named for Ihor Sikorsky, encompasses microbiological synthesis of useful for humanity products and experimental mutagenesis. He is studying actual inquiry of ecological and genetic outcome of chemical contamination of an air and water of our planet, experimental mutagenesis, evaluate ecological and genetic risk associated with contamination of the environment by people activity products. The FBT NTUU-«KTU» is important base for preparation of specialists in the sphere of biotechnology and include acquiring useful resources using alive organisms, creation of alternative source of energy, protection of surroundings with the help of live animals, and design and construction of new generation bioreactors for different biotechnologic processes.

The work of **Jeffrey A. Drebin (1957 –)** — American general surgeon and scientist, has been focused on targeted cancer therapy and the synthesis of the anti-cancer drugs transhuman (Herceptin) and epratuzumab (Perjeta).

---

[824] Town of Khanty-Mansiysk (inhabited since 2nd millennium BC, established 1930), Khanta-Mansia Okrug (area 534,800 км²), Russia. — Dugan AM. Products of water chloration as inductors of gene mutation. Cytology and Genetics 1996;30(5);76-81. — Dugan AM. Chemical contamination of atmospheric air as potential inductors of gene mutations. Dovkillia i zdorov'ia 1998;2. — Dugan OM. Potential mutagenic action of ingredients of dye for hair in alternative test-systems. Problemy ecol. ta med. genetics i klin. imunolohii. Zb. nauk. pr. Kyiv – Luhansk – Kharkiv 2006. Vyp. 1(64). — Ablitsov VH. «Galaxy Ukraine». The Ukrainian diaspora: prominent persons. Publ. «Kyt», Kyiv 2007.

James A. Thomson (1958 –) — American biologist, who is best known for deriving the first human embryonic stem cells (ESC) line (1998), and contemporaneously with **Shinya Yamanaka (1962 – : above)**, reported a method for converting human skin cells into cells that very closely resembling human embryonic ESC – induced pluripotent stem (iPS) cell lines derived from human somatic cells (2007).

In 1998 **Andrew Z. Fire (1959 –)** — American biologist, pathologist, and geneticist, and **Craig C. Mello (1960 –)** — American biologist of molecular medicine, discovered a genetic interference by double-stranded ribonucleic acid (RNA) causing switching genes on or off in Caenorhabditis elegans (a blend of the Greek – «caeno», meaning «recent», and «rhanditis», meaning «rod-like», and Latin – «elegans», meaning «elegant») – a free-living transparent nematode.

Working with a rodent model **Kim M. Olthoff (1960 –)** — American general and transplant surgeon, developed in 1998 the two-stage approach for control of rejection during liver transplantation: 1) first a gene-therapy applied to a donor of the liver to alter its immunity before transplantation surgery; (2) then proceeding with the liver transplantation without the need for immunosuppressive drugs because the immune system of the recipient became permanently tolerant of the new organ.

The research of **Craig B. Tompson (1963 –)** — American cell biologist, has helped to advance the understanding and deployment of immunotherapy to treat people suffering from cancer, by explanation of the process the genes control the programmed cell death and apoptosis (in Greek – «atoptwois» meaning falling off), i.e., a form of programmed cell death occurring most often in multicellular organisms (1993).

Investigating the structure of an immune system, specifically lymphatic nodes (LN), described by **Gaspare Aselli (1561 – 1626 / 17th century)**, **Imogen Moran**[825] — Australian immunologist, using in motion high resolution 3-D microscopy in living animals, discovered in 2018 a new «micro-organ» or structure within the immune system, that is the subcapsular sinus (SCS)[826] of the LN. Inside the SCS are positioned the macrophages that trap and present pathogenic antigen to the bursae (B) cells, thus preventing the systemic spread of lymph-born pathogen and are capable of activating an innate lymphoid effector cells[827] and adaptive memory cells, including the follicular memory thymic (T) cells and memory B_{mems} cells, that are either

---

[825] Moran I, Crootveld AK, Nguyen A, et al. Subcapsular sinus macrophages: the seat of inane and adaptive memory cells in murine lymph nodes. Trends in Immunology 2019 Jan 01;40(1):35-48.

[826] Subcortical sinus (SCS) is an endothelium lined space directly below the capsule of the lymph node (LN) where an afferent lymph drain into the LN.

[827] Innate lymphoid immune cells are derived from common lymphatic progenitor and belongs to the lymphoid lineage, but without antigenic specific B cell or T cell receptor.

prepositioned or rapidly recruited to the subcapsular niche[828] following infection and inflammation. Furthermore, B_{mems} cells are rapidly reactivated to differentiate into plasma cells in the subcapsular proliferative foci (SCF)[829]. Thus, understanding how SCS macrophages coordinate both innate and adaptive memory responses in subcapsular niche, can provide new approaches to enforce immunity against microorganisms and tumors.

The SCS «remembers» past infections and vaccinations and where immune cells gather to mount a rapid response against an infection the body dealt with before.

They also revealed the existence of thin, flattened structures extending over the surface of the LN in mice, which appears only when need to fight infection against which the animal has previously been exposed.

The SCS within the immune system is also an important step towards understanding and making better vaccines.

Physics and radiology

Atomic theory of **Max K.E.L. Planck (1858 – 1947)** — German physicist, **Ernst Rutherford (1871 – 1933)** — English physicist, **Albert Einstein (1879 – 1955)** — German American physicist of Jewish descent, **Niels H.D. Bohr (1885 – 1962)** — Danish physicist of Jewish faith, and **George Gamow (1904 – 68:** 20th century. Biochemistry**)** — Ukrainian American physicist, had a profound influence on the development of radiology. Max Planck proposed a quantum theory of atomic energy (1900). Ernst Rutherford discovered the product of such energy breakdown, radium - emanation (1900), explained radioactivity as a spontaneous decomposition of the atom (1903), and suggested a planetary composition of the atom with a positive nucleus in the center and negative electrons at the periphery. Albert Einstein theory of relativity expressed as an equation $E = mc^2$ (energy equals mass times the velocity of light squared), became the foundation for the development of atomic energy (1905, 1915). Niels Bohr created the theory of the structure of atoms and of the radiation emanating from them (1913). George Gamow (1904-68) believed that changes in atomic nuclei explain the origin of the solar system and the universe (1948).

Ionization and avalanche ionization was established by Sir **John Townsend (1868 – 1957)** — Irish mathematical physicist.

---

[828] Subcapsular niche is a region in the B cell follicle created by SCS macrophages, lymphatic endothelial and marginal reticular cells to provide a barrier function to prevent pathogen entry, to recruit and activate innate effector and adaptive memory cells to generate adaptive immunity.

[829] Subcapsular proliferative foci (SPF) is an extensive structure within the subcapsular region covering the cortical surface of the B cell follicle where B_{mems} are reactivated and differentiate into plasma cells.

In 1904 **George Perthes (1869 – 1927**: 19th century. Surgery x 2**)** noted disturbances in the cell division due to radiation.

Guido Holzknecht (1872 – 1931) — Austrian radiologist, described the median of the three clear pulmonary fields on the chest x-ray in the oblique projection; when x-rays are directed from the left and behind towards the right anteriorly (retrocardiac, prevertebral or Holzknecht space).

Robert Kienböck (1871 – 1953) — Austrian radiologist, who in 1910 described disorder which consisted of breakdown (malacia) of the lunate bone in the wrist (Kienböck's disease).

Henry K. Pancoast (1875 – 1939) — American radiologist, described the radiological picture of the SST of the lung (Pancoast tumor, 1932)[830].

Serhiy P. Hryhoriev / Sergei P. Grigoriev (27.VIII.1878, city of Kharkiv, Ukraine – 28.X.1920 city of Khrkiv, Ukraine)[831] — Ukrainian radiologist, introduced radiological examination of the appendix (1911), kidneys and ureters (1912).

George C. de Hevesy (1885 – 1966) — Hungarian radiochemist, introduced with **Frederick Paneth (1887-1957)** — Austrian-British chemist of Jewish descent, the method of radio isotopic indicators in 1913, and radio isotopic diagnosis in 1931.

The triad of **Charles F.M. Saint (1886 – 1960)** — South African radiologist, is simultaneous display during gastrointestinal (GI) tract examination using barium mixture of a diaphragmatic (hiatal) hernia, diverticulosis of the large bowel (colon) and cholelithiasis.

B 1921 **Andre Bocage (1892 – 1953)** — French tomographer, devised tomography to refine radiological diagnosis.

Arteriograms and venograms were obtained in man by **Janos Berberich (1877 – 1950)** and **S. Hirsh** in 1923, using 20% strontium bromine solution injected into the vessels of the upper extremity.

The pioneers of television (TV), **Vladimir K. Zworykin (1889 – 1969)** — Russian American physicist, built the first successful TV camera and picture tubes (1923), and **Borys P. Hrabovsky (26.V.1901, town of Tobolsk, Tiumen Obl., Russia – 13.I.1966, city of Bishkek, Kyrgyzstan)**[832] — Ukrainian physicist, son of **Pavlo A. Hrabovsky (1864 – 1902)** — Ukrainian poet and publicist, who patented, realized in practice by demonstrating the electronic system, with help of which fulfilled the

---

[830] Morris JH, Harken DE. The superior pulmonary sulcus «Tumor of Pancoast» in relation to Hare's syndrome. Ann Surg 1940 July;112(1):1-21.

[831] Grigotiev SP. The Newest Point of View in Radiology. Kharkiv 1920. — Ablitsov VH. «Galaxy Ukraine». The Ukrainian diaspora: prominent persons. Publ. «Kyt», Kyiv 2007.

[832] Town of Tobolsk (founded 1587), Tiumen Obl. (area 1,464,200 km²), Russia. — City of Bishkek (founded 1825), Kyrgyzstan (area 198,500 km²). — Ablitsov VH. «Galaxy Ukraine». The Ukrainian diaspora: prominent persons. Publ. «Kyt», Kyiv 2007.

first in the world distant public TV transmission, without electrical conduit (wiring), of moving image on 26 June and 04 August 1928, city of Tashkent (populated since 5th-3rd century BC) street, Uzbekistan (area 448,978 km²). Their discovery paved the way for electronics and TV monitoring in medicine.

In 1925 **Reynaldo dos Santos (1880 – 1969)** — Portuguese internist and radiologist, performed percutaneous (PC) aortogram, renal and lower extremities arteriograms, and in 1935 venocavography.

Christian I. Baastrup (1885 – 1950) — Danish physician-radiologist, described the condition of the neighboring posterior spinous processes of the lumbar vertebral column touching (kissing) each other on its upper and lower edge in elderly people, that result in the formation of osteophytes, i.e., anomalous bone growth causing pain (Baastrup sign and syndrome, 1933).

Fedir P. Bohatyrchuk (14.XI.1892, city of Kyiv, Ukraine – 04.IX.1984, city of Ottawa, ON, Canada)[833] — Ukrainian physicist and chess-player, elaborated on the microradiological method of studying bone.

The percutaneous (PC) retrograde passage of a catheter from the common femoral artery (CFA) for abdominal aortography was reported (1941) by **Pedro L. Farinas (1892 – 1951)**[834] — Cuban radiologist.

In 1926-54 **Vadym Ye. Dyachenko [02.VI.(18.VII., or 30. XII.1896, city of Nyzhniy Novhorod, Russia – 02.VI.1954, city of Kyiv, Ukraine]**[835] — Ukrainian mathematical physicist, became associate, since 1928 ordinary professor of numerical mathematics[836] at T.H. Shevchenko Kyiv National university (KNU), where in 1946-51 developed physical-mathematic methods for computed mathematic and construction of instruments, created laboratory for electro-designing and computed mathematics. Under his leadership in the KNU were developed the first electro-integrals / electro-generators for solving differential equations of the elliptical type.

One of the founders of radiation chemistry was **Moïse M. Haissinsky [23.X.(04. XI).1898, town of Tarashcha, Kyiv Obl., Ukraine – 10.II.1976, city of Paris,**

---

[833] City of Ottawa (founded 1826), Province of ON, Canada. — Ablitsov VH. «Galaxy Ukraine». The Ukrainian diaspora: prominent persons. Publ. «Kyt», Kyiv 2007.

[834] Farinas PL A new technique for the arteriographic examination of the abdominal aorta and its branches. *Am J Roentgenol* 1941; 46: 641-45.

[835] City of Nyzhniy Novhorod (founded 1221), Russia. — Ablitsov VH. «Galaxy Ukraine». The Ukrainian diaspora: prominent persons. Publ. «Kyt», Kyiv 2007.

[836] Numerical mathematics is a division of mathematics connected with the circle of questions concerning the performance the closest calculations, in narrow understanding, a theory of numerical method for solving typical mathematical problems.

France][837] — Ukrainian French of Jewish descent who authored the treatise «Radiation Chemistry and its Use» (Masson et Cie, Paris 1957).

Dimitri E. Olshevsky (1900 – ?) — American physician, radiologist, constructed radiological tube to use only the stronger x-rays which pass through the target and armoring the rest of the tube (Olshevsky tube, 1931, 1937).

The use of an opaque organic iodine for intravenous (IV) urography was reported by **Moses Swick (1900 – 85)** in 1929.

A pioneer of cardiac catheterization (CC) **Werner T.O. Forssman (1904 – 79)** — German physician, opened a vein of his own arm and inserted a ureteral catheter until it reached the right atrium (RA) of the heart and confirmed this by x-ray of his chest (1929). He repeated this experiment on himself 17 times, including passage of a catheter into the pulmonary artery (PA). Further development of CC techniques was made by **Andre F. Cournand (1895 – 1988)** — French physiologist and physician and **Richard W. Dickinson, Jr (1895 – 1973)** — American physiologist and physician.

Artificial isotopes, discovered by **Frederich Joiliet (1900 – 58)** — French physicist and **Irene Joiliet- Curie (1897 – 1956)** — French chemist of Polish ancestry, in 1934, became available for diagnosis and treatment.

The percutaneous (PC) median intercostal approach for the needle puncture injection of sodium iodide solution into the ascending aorta (AA) and the left ventricle (LV) for thoracic aortography in patients was developed by **I. Nuvoli**[838] in 1935.

In 1936 **Burckhard (Burkhard) F. Kommerell (1901 – 90)** — German internist and radiologist, performing esophagography observed the left-sided aortic arch and aberrant origin of the right-sided subclavian artery (SCA) from the aortic diverticulum at the junction between the distal aortic arch and the proximal descending thoracic aorta (DTA) – Komerell's diverticulum, a remnant of the primitive right branchial aorta as a result of regression in the embryonal period of the 4th branchial arch between the right common carotid artery (CCA) and the right SCA[839]. The Komerell's diverticulum is the most common congenital vascular aberration of the aortic arch occurring in 1% of individuals.

---

[837] Town of Tarashcha (Foot Note 621), Kyiv Obl., Ukraine. — Haissinsky M. La chimie nucleaire et ses applications. Masson et Cie, Paris 1957. — HaissinskyM. Chemia jądrowa i jej zastosowanie. PWN, Warszawa 1959. — Ablitsov VH. «Galaxy Ukraine». The Ukrainian diaspora: prominent persons. Publ. «Kyt», Kyiv 2007.

[838] Nuvoli I. Arteriografia dell'aorta toracica mediante puntura del'aorta ascendente e del ventriculo sinistro. Policlinico 1935;43 : 227-37.

[839] Kommerell B. Verlagerung des Oesophagus durch eine abnorm ferlaufende Arteria subclavia dextra (Arteria lusoria). Forschritte auf dem Gebiet der Röntgenstrahlen und Nuclearmedicin (Fortsch Geb Röntgenstrahlen) 1936;54:590-5. — Van Son JAM, Konstantinov IE, Burchard F. Burckhard Kommerell and Kommerell's Diverticulum. Tex Heart Inst 2002;29(2):109-12. — Olearchyk AS. Right-sided double aortic arch in an adult. J Card Surg 2004;19(3):248-51.

Radiological ring of **Richard Schatzky (1901 – 99)** — German American radiologist with Jewish background, is the narrowing of the distal part of the esophagus, usually in the site of transition of mucosa of the esophagus into mucosa of the stomach that is at the junction of squamous-columnar epithelium (esophageal or Schatzky ring, 1953).

In 1936 **Lev V. Shubnykov (29.IX.1901, city of St Petersburg, Russia** – shoot by Russian Communist Fascist NKVS **18.X(10.XI).1937,?]**[840] — Ukrainian physicist, with **Borys H. Lazariev (06.VIII.1906, village of Myropillia, Kracnopilsky Region, Sumy Obl., Ukraine – 20.III.2001, city of Kharkiv, Ukraine)**[841] — Ukrainian physicist, discovered nuclear para-magnetism (NPM), in 1944 **Yevhen K. Zavoysky (05(28).IX.1907, town of Mohyliv-Podilsky, Vinnytsia Obl., Ukraine – 09.X.1976, city of Moscow, Russia)**[842] — Ukrainian physicist, discovered the electron-tumble magnetic (paramagnetic) resonance (MR) on the salt particles of the iron group, that lead to the development in 1946 by **Felix C. Bloch (1905 – 83)** — Swiss American physicist of Jewish descent, and **Edward M. Purcell (1912 – 97)** — American physicist, of nuclear magnetic resonance (NMR) using protons. On account of those researchers it became possible to implement in the diagnostic radiology NMR / magnetic resonance imaging (MRI), and magnetic resonance angiography (MRA) for investigation of the heart, and blood vessels.

Then, in 1971 **Paul C. Lauterbur (1929 – 2007)** — American chemist, applied magnetic field gradients in three dimensions and a back-projection technique to create the NMR images, published the first images of the tubes in water in 1973, followed by the picture living eatable mollusks, in 1974 the imagery of the thoracic cavity in a mouse. However, it was not until the 1978's when Paul Lauterbur and Sir **Peter Mansfield (1933 – 2017)**[843] — English physicist, developed MNR / MRI technology that could be used to provide images of the body. Paul Lauterbur constructed the prototype of the MRI machine that could distinguish between heavy water and ordinary human's body water. Peter Mansfield invented «slide selection» for the MRI machine and made understandable the radio signals from MRI for mathematical analysis, thus making interpretation of the signals into a useful image; discovered the way the fast imaging could developed the MRI protocol called

---

[840] Shubnikov LV. Selected Works. Remembrance. Publ. Naukova Dumka, Kyiv 1990. — Mirsky M. Ukrainian medical scientists – victims of Stalin's repressions. UHMJ «Ahapit» 1995;2:45-54. — Ablitsov VH. «Galaxy Ukraine». The Ukrainian diaspora: prominent persons. Publ. «Kyt», Kyiv 2007.

[841] Village Myropillia (founded c. 1650; during conducted by Russian Communist Fascists of Holodomor-Genocide 1932-33 against the Ukrainian Nation perished at least 45 inhabitants), Krasnopilsky Region, Sumy Obl., Ukraine. — Ablitsov VH. «Galaxy Ukraine». The Ukrainian diaspora: prominent persons. Publ. «Kyt», Kyiv 2007.

[842] Town of Mohyliv-Podilsky (Foot Note 3), Vinnytsia Obl., Ukraine. — Ablitsov VH. «Galaxy Ukraine». The Ukrainian diaspora: prominent persons. Publ. «Kyt», Kyiv 2007.

[843] Mansfield P, Grannell PK. «Distraction» and microscopy in solids and liquids by NMR. Review B 1975;12(9):3618-34.

echo-plantar imaging (EPI). The latter allows T2 weighted images (T2WI), one of the basic pulse sequences in MRI, where the sequence weighting highlights difference in the T2 relaxation of tissues, to be collected many times faster than previously possible. The contribution of those scientist has made the functional MRI possible.

In 1937-38 **Augustin W. Castellanos (1902 – 2000)** — Cuban pediatrician, radiologist and cardiologist, pioneer of angiocardiography and **R. Pereira** described the angiographic appearance of congenital heart disease (CHD) and in 1939 they created the countercurrent and retrograde brachial thoracic aortography.

Serhiy O. Lebiedev (02.XI.1902, city of Nizhnyy Novhorod, Russia – 03.VII.1974, city of Moscow, Russia)[844] — Russian Ukrainian cybernetics, invented the computer flow block, and under his direction the small electronic digital computers were created, beginning 1948 in the Laboratory for Design and Regulation at the Electrotechnical Institute of the National Academy of Sciences (NAS) of Ukraine, which was launched on December 25, 1951.

Digital subtraction angiography (DSA) was made possible, initially through the work of **G.P. Robb** and **I. Steinberg**[845] — American radiologists, in 1939, which was further developed by **B.G. Ziedses** des **Plantes (1902 – 93)** — Netherlanders radiologist, in 1961, and **Charles A. Mistretta (1941 –)** — American radiologist, in 1971-77.

In 1941 **Vadym Ye. Lashkariov (07.X.1903, city of Kyiv, Ukraine – 01.XII.1974, city of Kyiv, Ukraine)**[846] — Ukrainian physicist, by discovering the *p-n* transition, disclosed the mechanism of the electronic-porous diffusion and created the first semiconductive light emitting diode (LED).

The team of scientist consisting of **William B. Shockley, Jr (1910 – 89)** — American physicist and inventor, **Walter H. Brattain (1902 – 87)** — American physicist, and **John Bardeen (1908 – 91)** — American physicist and electronic engineer, created in 1947, the first bipolar point contact transistor[847]. The LED was made by **Nick Holonyak, Jr (03.XI.1928, town of Zeigler, Franklin County, IL, USA–)**[848] — American physicist of Ukrainian descent, practically useful as visible one in 1962 for the wide use in industry, including monitoring in medicine. Their inventions lead to the development of the modern cardiac pacemakers.

---

[844] City of Nyzhniy Novhorod (Foot Note 835), Russia. — Ablitsov VH. «Galaxy Ukraine». The Ukrainian diaspora: prominent persons. Publ. «Kyt», Kyiv 2007.

[845] Robb GP, Steinberg I. Visualization of the heart, the pulmonary circulation and the great blood vessels in man. Am J Roentgenol 1939;41:1-17.

[846] Ablitsov VH. «Galaxy Ukraine». The Ukrainian diaspora: prominent persons. Publ. «Kyt», Kyiv 2007.

[847] Transistor is a semiconductive device used to amplify and switch electric signals and electric power.

[848] Town of Zeigler (incorporated 1914), Franklin County, IL, USA. — The son of emigrant Mykola Holonyak, Sr, from Ukrainian village in the Ukrainian Carpathian Mountains, baptized by the Ukrainian Catholic Church (UCC) of IL, when he was given first name «Nick». — Ablitsov VH. «Galaxy Ukraine». The Ukrainian diaspora: prominent persons. Publ. «Kyt», Kyiv 2007.

In 1940 **Louis H. Gray (1905 – 65)** — English physicist, the founder of radiobiology who derived units of absorbed dose of energy, named by the standard International units (SIU) in his name as gray (Gy, 1975), typically associated with ionizing radiation, such as x-ray or gamma particles or with other nuclear particles, defined as the absorption of 1 joule (J) of such energy by 1 kg of matter, usually human tissue, where the rad is equivalent to 0,01 Gy.

The ultrasound (US) studies were introduced in 1941 by **Karl T. Dussik (1908 – 68)** — Austrian physician, neurologist, and psychiatrist of Czech origin, to diagnose brain tumors.

After World War (WW) II fission from splitting the atom of uranium and isotopes from bombardment in nuclear reactors began to replace x-ray and radium therapy. For radiation, the orthovoltage (140-400 kilovolts), super-voltage (c. 500 kilovolts) and megavoltage (greater than one megavolt) therapies became available.

According to **Benjamin Felson (1913 – 88)** — American radiologist and **Max J. Palayew (?– 2015)**[849] — Canadian radiologist of Jewish descend, in type I right-sided aortic arches with an aberrant right subclavian artery (SCA) originating from the Kommerell diverticulum at the distal aortic arch or proximal descending thoracic aorta (DTA), a vascular ring encircles and compresses the esophagus and trachea, more so if aneurysmal. Type II right-sided aortic arches is a mirror image of an aberrant right SCA with a Kommerell diverticulum, an aberrant left SCA comes off the proximal right-sided DTA.

Independently, in 1951-54 **Charles H. Townes (1915 – 2015)** — American physicist of German origin, and **Alexander M. Prokhorov (1916 – 2002)** — Australian Russian physicist, with **Nicolai G. Basov (1922 – 2001)** — Russian physicist, presented the principles for using the energy to produce microwave amplification simulated emission of radiation (maser); and in 1957-58 **Charles Townes** with **Arthur L. Schawlow (1921 – 99)** — American physicist of Jewish Latvian origin, laid the foundation for the creation of the light amplification by stimulated emission of radiation (laser). Lasers are being used today for the removal of endobronchial tumors, eye surgery and thoracic and cardio-vascular surgery (TCVS) / [20th century. Pathology. – **Rostyslav Kavetsky (1899 – 1978)**].

Harold E. Johns (1915 – 98) — Canadian medical physicist, together with **Sylvia O. Fedoruk [May 05, 1927, town of Canora, SK, Canada – Sept. 26, 2012, city of Saskatoon, SK, Canada]**[850] — Ukrainian Canadian medical and oncologist, were the

---

[849] Felson B, Palayew MJ. The two types of right aortic arch. Radiology 1963;81:745-59.

[850] Town of Canora (first ranchers arrived in 1884; the first Ukrainian block settlement was established in 1897 when 150 families arrived to Canora District from Western Ukraine), Province of Saskatchewan (SK, area 651,900 km²), Canada. — City of Saskatoon (established 1883), SK, Canada. — Ablitsov VH. «Galaxy Ukraine». The Ukrainian diaspora: prominent persons. Publ. «Kyt», Kyiv 2007.

first to design and construct the external bean radioactive cobalt-60 unit as a source of a gamma ray for medical nuclear imaging (scanning) and for radiation therapy of patients with cancer (October 27 and November 1951). Also, she was instrumental in the development of the dosimeter for measuring of radiation and a radioactive iodine device.

Founded in 2011 the Canadian Centre for Nuclear Innovation (CCNI) was renamed in 2012 to her honor into Sylvia Fedoruk Centre at the University of Saskatchewan (founded 1907), Saskatoon, SK, Canada.

Harold H. Hopkins (1918 – 94)[851] — British physicist, who in 1950-54 had designed a «fibrescope» consisting of a bundle of flexible glass fibers able to coherently transmit an image.

In 1958 **F. Mason Sones (1918 – 85)** — American cardiologist, developed selective coronary angiography (SCA by an accidental insertion of the catheter into the right coronary artery (RCA).

Stanley Baum (1929 –)[852] — American radiologist, pioneered the specialty of interventional radiology by radiographic arterial demonstration and non-operative treatment of gastrointestinal bleeding (1963-69).

The development of a percutaneous (PC) technique of catheter placement over a previously inserted guidewire by **Sven I. Seldinger (1921 – 98)**[853] — Swedish radiologist, in 1952 initiated a new era in angiography.

In 1960 **L. Contorni**[854] published the first radiological image of blood flowing in a retrograde direction in the vertebral artery (VA) associated with a proximal ipsilateral selective coronary angiography (SCA) occlusion, a condition subsequently named the subclavian steal syndrome (SSS).

---

[851] Hopkins HH. Walve Theory of Aberrations. Oxford Univ Press, London 1950. — Hopkins HH, Kapany NS. A flexible fiberscope, using static scanning. Nature 02 Jan 1954;173:39-41.

[852] Nusbaum M, Baum S. Radiographic demonstration of unknown sites of gastrointestinal bleeding. Surg Forum 1963;14:374-5. — Baum S, Nusbaum M, Blakemore WS, et al. The preoperative radiographic demonstration of intra-abdominal bleeding from undetermined sites by percutaneous selective celiac and superior mesenteric arteriography. Surgery 1965 Nov;58(5):797-805. — Baum S, Greenstein RH, Nusbaum M. Diagnosis of ruptured, noncalcified splenic artery aneurysm by selective celiac arteriography. Arch Surg 1965 Dec;91(6):1026-8. — Baum S, Nusbaum M, Clearfield HR, et al. Angiography in the diagnosis of gastrointestinal bleeding. Arch Int Med 1967 Jan;119(1):16-24. — Baum S, Nusbaum M, Kuroda K, et al. Control of hypertension by selective mesenteric arterial drug infusion. Arch Surg 1968 Dec;97(6):1005-13. — Baum, Nusbaum M, Kuroda K, et al. Direct serial magnification arteriography as an adjunct in the diagnosis of surgical lesions in the alimentary tract. Am J Surg 1969 Feb;117(2):570-6.

[853] Seldinger SI Catheter replacement of the needle in percutaneous arteriography. *Acta Radiologica* 1953; 39: 368-76.

[854] Contorni L. II circolo collaterale vertebro-vertebrale nella obliterazione dell'arteria succlavia alla sue origine. Minerva Chir 1960; 15:268:71.

On Jan. 16, 1964, **Charles T. Dotter (1921 – 85)**[855] — American radiologist, and **Melvin P. Judkins (1922 – 85)** — American radiologist, carried out a percutaneous (PC) transfemoral dilatation of an atherosclerotic (AS) stenosis of the left superficial femoral artery (SFA) in a 82 year-old woman using a tapered radiopaque 3,2 outside diameter (OD) Teflon dilating catheter. Angiography showed that that stenosis was no longer present. By Apr. 21, 1964, they recanalized trans-luminally AS stenosis in 11 patients, with success in six.

After co-discovering of the photorefractive effects in the piezoelectric crystal and with the beginning his work on manipulation of microparticles with low light and optical trapping in the late 1960s, **Arthur Ashkin (02.09.1922, borough of Brooklyn, New York City, N.Y., USA – 21.09.2020, borough of Rumson, Monmouth District, N.J., USA)**[856] — American physicist of Ukrainian Jewish background, invented in 1987 a single-beam gradient force trap, a.k.a., the «laser tweezers» or «optical tweezers» that grasp particles, atom viruses and other living cells with their laser beam finger, and with this he was able to use radiation pressure of light to move physical object. The laser beam of a special form separate cells and even atom, without harming them, with the aim to carry out surgery on tissues, even not necessarily opening the cell itself, that is possible to move something inside directly over the capsule. The «optical tweezers» found application in biology and surgery of the eye.

The leading role in development of the theory of functional electronic density have played **Walter Kohn (09.03.1923, city of Vienna, Austria – 19.04.2016, town of Santa Barbara, CA, USA)**[857] — American theoretical physicist and chemist of Jewish Ukrainian descent. This theory made it possible to calculate quantum mechanical electronic structure by equations involving the electronic density, rather than the many-body wave function, and it revolutionized quantum chemistry.

Developing further the method of sequential «frozen axial sections» of the dead for the study of anatomy, as outlined by **Yefrem Mukhin (1766 – 1850**: 18th century. 19th century. Surgery**), Illia Buyalsky (1789 – 1866**: 19th century. Surgery**)** and **Mykola**

---

[855] Dotter CT, Judkins MP. Transluminal treatment of arteriosclerotic obstruction. Description of a new technic and preliminary report of its application. Circulation 1964;30:654-60.

[856] Borough of Rumson (settled by native Americans since prehistoric times, bought by English settlers 1665), Monmouth District, N.J., USA. — Ashkin A. A measurement of positron-electron scattering and electron-electron scattering. PhD thesis, Cornell University, New York, N.Y. 1953. — Ashkin A. Acceleration and Trapping of Particles by Radiation Pressure. Phys Rev Letters 1970;24(4)156-9. — Ashkin A, Dziedzic JM, Bjorkholm JE, Chu S. Observation of a single-beam gradient force optical trap foe dielectric particle. Optics Letters 11 May 1986:288-90. — Ashkin A. Optical Trapping and Manipulation of Neutral Particles Using Lasers: A Reprint Volume with Commentaries. Illustrated Ed. World Sci Publ Co, Singapore 2006. — Ablitsov VH. «Galaxy Ukraine». The Ukrainian diaspora: prominent persons. Publ. «Kyt», Kyiv 2007.

[857] Town of Santa Barbara (human habitation by Native Americans begins at least 11.000 years BC, Spaniard's discovery of the area took place 1602 and 1769, founded 1782), State of California (CA, area 423,970 km^2), USA. — Ablitsov VH. «Galaxy Ukraine». The Ukrainian diaspora: prominent persons. Publ. «Kyt», Kyiv 2007.

Pyrohov (1810 – 84: 19th century. Surgery), based on the mathematical work of **Allen M. Cormak (1924 – 98)** — South African American physicist, led Sir **Godfrey H. Hounsfield (1919 – 2004)** — English electrical engineer, to construct the first computerized computer tomography (CT) scanner in 1972. From the name of the latter comes the Hounsfield scale, a quantitative measure of radiodensity used in evaluating CT scans. The scale is defined in Hounsfield units (HU), running from air at − 1000 HU, through water at 0 HU, through benign tumors of the lung <15 HU and malignant (carcinoma) pulmonary tumors >20 ОГ and up to a dense cortical bone at +1000 HU and more.

Clinical signs of tracheal and bronchial injury include subcutaneous (SC) emphysema, hemoptysis, pneumothorax, and air leak on the chest tube insertion, plain chest x-ray (CXR) may reveal pneumothorax, pneumomediastinum, atelectases, in an underinflated lung or lobe, and also the fallen lung sign, if the lung has collapsed away outward and downward (rather than inward and upward) from the hilum, described by **David A. Kumpe (1942 –)**[858] — American radiologist, and named the sign of Kumpe (1970).

The trans-jugular intrahepatic portocaval shunt (TIPS) was first described in 1969 by **Josef Rösch (1925 – 2016)** — Czech American researcher and pioneer intravascular and interventional radiology, for the treatment of portal hypertension in patients with liver cirrhosis and other conditions to bypass the liver prior to its transplantation, however, it could not be used in patients with a complete portal vein occlusion[859].

For the TIPS, through a percutaneous (PC) access from the internal jugular vein (IJV), the right hepatic vein is «connected» to the right portal vein branch via a trans-parenchymal tract. This tract is held open by a self-expandable metallic non-coated/non-covered endoprosthesis (brained construction with Elgiloy Wallstent™) placed by a delivery system.

Among patients with a complete portal vein occlusion, the PC trans-splenic approach also allows to bypass the liver. A splenic sheath and lasso snare are used to transverse the splenic pulp, then the procedure is continued under guidance of

---

[858] Oh KS, Fleischner FG, Wyman SM. Characteristic pulmonary finding in traumatic complete transection of a mainstem bronchus. Radiology 1969; 92:371-2. — Kumpe DA, Oh KS, Wyman SM. Characteristic pulmonary finding in unilateral complete bronchial transection. Am J Roentgenol 1970; 110:704-6.

[859] Chu HH, Kim HC, Jae HJ, et al. Percutaneous trans-splenic access to the portal vein for management of vascular complications in patients with chronic liver disease. Cardiovasc Interv Radiol 12012 Dec; 35(6):1388-95. — Habib A, Desai K, Hickey R, et al. Portal vein recanalization – trans-jugular intrahepatic portosystemic shunt using the trans-splenic approach to achiev transplant candidacy in patients with chronic portal vein thrombosis. J Vasc Interv Radiol 2015. 1-8. — Singh H, Neutze J (eds). Radiology Fundamentals: Introduction to Imaging and Technology. 5th ed, Verlag Springer, New York 2015. — Salem R, Vouche M, Baker T, et al. Pretransplant portal vein recanalization – trans-jugular intrahepatic portosystemic shunt in patients with complete obliterative portal vein thrombosis. Transplantation 2015;99(11):2347-55.

the x-ray, achieving «end-to-end anastomosis», which allows for balloon angioplasty in the blocked portal vein and facilitates completion of the TIPS.

The intravascular titanium stents were first implanted in dogs in October 1983 by **Iosip Kh. Rabkin (1926 –)**[860], and **Dimitri Io. Rabkin (1963 –)** — Russian radiologists. Then, the former on March 27, 1984, dilated in a 56-year-old man, stenosed by atheromatous (AM) the left external iliac artery (EIA) with success.

Constantin Cope (1927 – 2016)[861] — American interventional radiologist (IR), who introduced the simultaneous two-stage IR approach for the treatment of the thoracic duct (TD) injury with persistent and uncontrolled leaking fistulas with pleural chylous effusion or chylous ascites, or both after neck, esophageal, pancreatic and aortic surgery: (1) Conventional catheterization of the peripheral lymphatic vessels (lymphangiography) by inserting a needle and catheter through a lymphatic vessel on the dorsum of the foot following by infusion o contrast flush of normal saline solution (NSS) for more rapid and better visualization of the cisterna chyle [17th century. – **Gaspare Anselli (1581 – 1626)**]; (2) the percutaneous transabdominal puncture with a 21-gauge needle and catheterization with 3Fr catheter of the cisterna chyle or lymphatic duct (PTPCLD) under the guide of computer tomography (CT) or magnetic resonance imaging (MRI) with the TD embolization using sclerosants or micro coil. Even if a needle passes through a hollow viscus on the way of cisterna chyle the puncture site healed without adverse effect.

In 1951-70 **David E. Kuhl (1929 – 2017)**, **Luke Chapman** and **Roy Edwards** — American radiologist specializing in nuclear medicine, conceived, and set up the development of the positron emission tomography (PET) and the single-photon emission computed tomography (SPECT) for the imaging of the distribution of radioactive isotopes in the body, using constructed by them the emission tomographic devices – scanner, i.e., the Mark II (1964), Mark III (1970) and Mark IV (1976). The Mark II was the first device for the computed axial tomography (CAT), performed the first quantitative three-dimensional (3-D) mapping of the brain function (1964-65) and became a forerunner of the commercial computer tomographic (CT) scanners.

In the middle of 1960s David Kuhl succeeded in the axial transverse tomographic imaging of humans, and in the early of 1970s he measured regional cerebral blood flow volume using the reconstructed tomographic brain images of radioactive

---

[860] Rabkin IKh, Zaimovsky VA, Khmelevskaya IYu, et al. The experimental validation of the first clinical experience endovascular prosthetics with x-ray. Vestn Roentgenol Radiol 1984;4:59-64. — Rabkin Ikh. Personal communication, Feb. 14-15, 1898, Scottsdale, AZ.

[861] Cope K. Diagnosis and treatment of postoperative chyle leakage via percutaneous transabdominal catheterization of the cisterna chyle: a preliminary study. J Vasc Interv Radiol 1998 Mar;9(5):727-34. / Kaiser LR. Non-surgical approach to the thoracic duct: a technique whose time has come. J Thorac Dis 2017 Mar;9(3):428-9. / **Larry R. Kaiser (1937 -)** — American thoracic surgeon.

isotopes obtained by his SPECT scanner, and this was the first time the physiologic function of a living body has been measured by 3-D mapping.

At about this same time, **Louis Sokoloff (14.10.1921, city of Philadelphia, PA, USA – 30.07.2015, city of Washingnon, DC, USA)**[862] — American neuroscientist of Jewish Ukrainian origin, discovered that radiographic imaging using the x-ray emitting radionuclide 140-deoxyglucose was effective for evaluating of the regional metabolic function in animal brain.

Aiming to employ Louis Sokoloff's technique clinically, David Kuhl group conducted joint research with **Seymour S. Kety (1915 – 2000)** — American neuroscientist, who discovered a method for measuring blood flow in the brain, Louis Sokoloff, and **Alfred P. Wolf (1923 – 98)** — American nuclear organic chemist, reaching the conclusion that synthetized in 1976 18-fluorodeoxyglucose (18-FDG) was the most appropriate positron emitting tracer for human use.

Thereafter, 18-FDG was used with the Mark IV SPECT to successfully image the metabolism of human brain for the first time, and then used as a radiotracer for PET scanning of the brain, heart, and tumors.

These tomographic techniques David Kuhn invented were in 1975 further advanced into nowadays PET by **Michael M. Ter-Pogossian (1925 – 96)** — Armenian American nuclear physicist, **Michael E. Phelp (1939 –)** — American biophysicist, and **Edward J. Hoffman (1943 – 2004)** — American chemist.

Building on the work of David Kuhl, **Abass Alavi (1938 –)** — Iranian American radiologist and **Martin Reivich (1933 –)** — American neurologist, administered in August 1976 the first doses of radiotracer 18-fluorodeoxyglucose ([18]FDG) to healthy volunteer and acquired through PET images of the human brain.

Discoveries of David Kuhl and clinical translations of Abass Alavi and Martin Reivich helped lead to the routine clinical use of the PET and [18]FDG-PET scans in neurology, cardiology, and oncology.

Thus, Louis Sokoloff on account of the initial discovery and creation of technique to detect and treat major brain disorders, made him a pioneer in functional imaging of the brain.

The first to demonstrate during angiography intramyocardial (IM) sinusoids in atresia of the pulmonary artery (PA) valve with an intact interventricular septum (IVS) and hypoplasia of the right ventricle (RV) was **Ronald M. Lauer (1930 – 2007)**[863] — American pediatric cardiologist, in 1964.

---

[862] City of Philadelphia (Foot Note 2), PA, USA. — City of Washington (Foot Note 751), DC, USA. — Ablitsov VH. «Galaxy Ukraine». The Ukrainian diaspora: prominent persons. Publ. «Kyt», Kyiv 2007.

[863] Lauer RM, Fink HP, Petry EL, et al. Angiographic demonstration of intramyocardial sinusoids in pulmonary-valve atresia with intact ventricular septum and hypoplastic right ventricle. N Engl J Med 1964 July 9;271:68-72.

Notable contribution to radiology made **Rostyslav P. Matiushko (25.08.1934, village Sofiivka, Nizhynsky Region, Chernihiv Obl., Ukraine – 30.06.2008, city of Kyiv, Ukraine)**[864] — Ukrainian physician-radiologist, by studying methods of radionuclide diagnostics using small and above lethal doses of ionizing radiation upon an organism of animals and humans, wherefore he was decorated with the State Prize of Ukraine in the Field of Sciences and Technique (1986).

Larissa T. Bilaniuk (15.07.1941, village Potiк, Kozivsky Region, Ternopil Obl., Ukraine –)[865] — Ukrainian American neuroradiologist, who one of the first reported on diagnostic utilization in internal medicine, neuroradiology, head and neck surgery,

---

[864] Village Sofiivka (initially mentioned as manor in 1875-77, created in 1920s; due to conduction by Russian Communist Fascist regime of Holodomor-Genocide 1932-22 against the Ukrainian Nation, 25 residents died from man-made hunger), Nizhynsky Region, Chernihiv Obl., Ukraine. — Matiushko RP, et al. Radionucleid Diagnostics (Evaluation of the Treatment Effectiveness of Some Disease). Kyiv 1991. — Matiushko RP, et al. Learning of Possibilities Utilizing Metal-Silicate Compounds for Deactivation. Chornobyl and Health of Population. Kyiv 1993. Part 2. — Matiushko RP, et al. Radio-Pharmaceutics Preparations: Instructive Handbook. Kyiv 1997. — Matiushko RP. Basics of Radiation Meditsine: Instructive Handbook. Odesa 2002, 2003. — Ablitsov VH. «Galaxy Ukraine». The Ukrainian diaspora: prominent persons. Publ. «Kyt», Kyiv 2007.

[865] Village Potik (first written document 1626, when the village was destroyed during an attack by Tatars, as of 01.01.1939 in the village lived 1210 residens), Koziv Region, Tepnopil Obl., Ukraine — Bilaniuk LT, Zimmerman RA. The role of computed tomography in intracranial infections. J Ukr Med Assoc North Am 1975 Apr;24,2(85):12-9. — Bilaniuk LT, Patel S, Zimmerman RA. Computed tomography in systemic Lupus erythematosus. Radiology 1977 Jul;124(1):119-21. — Bilaniuk LT, Zimmerman RA, Brown I, et al. Computed tomography in meningitis. Neuroradiology 1978;16:13-4. — Bilaniuk LT. Computed tomography of the postoperative acoustic neuroma. Adv Otorhinolaryngol 1978;24:34-8. — Bilaniuk LT, Zimmerman RA, Littman P, et al. Computed tomography of brain stem gliomas in children. Radiology 1980 Jan;134(1):89-95. — Bilaniuk LT, Zimmerman RA. Role of computed tomography in diagnosis of head. J Ukr Med Assoc North Am 1980 Winter;27,1(96):3-10. — Bilaniuk LT, Zimmerman RA. Computer-assisted tomography: sinus lesions with orbital involvement. Head Neck Surg 1980 Mar-Apr;2 (4):293-301). — Bilaniuk LT, Zimmerman RA, et al. Computed tomography in pediatric infratectorial brain tumors. J Neurorad 1981;8(3):229-42.— Bilaniuk LT, Zimmerman RA. Computed tomography in evaluation of the paranasal sinuses. Radiol Clin North Am 1982 Mar;10(1):51-66. — Bilaniuk LT, Bilaniuk OM. Diagnostic nuclear magnetic resonance. Principles and prospects. J Ukr Med Assoc North Am 1983 Winter;30,1(106):3-11. — Bilaniuk LT. Magnetic resonance imaging of the brain – the present and the future. Psychophacol Bull 1984 Summer;20(3):371-5. — Bilaniuk LT. Cerebral magnetic resonance: comparison of high and low field strength imaging. Radiology 1984 Nov;153(2):409-14. — Bilaniuk LT, Zimmerman RA, Werhii FL, et al. Magnetic resonance imaging of pituitary lesions using 1. to 1.5 T field strength. Radiology 1984 Nov;153(2):415-8. — Bilaniuk LT, Moshan T, Cara J, et al. Pituitary enlargement mimicking pituitary tumors. J Neurosurg 1985 Jul;63(1):39-42. — Bilaniuk LT, Schenk JF, Zimmerman RA, et al. Ocular and orbital MR imaging. Radiology 1885 Sep;156(3):669-74. — Bilaniuk LT, Atlas SW, Zimmerman RA. Magnetic resonance imaging of the orbit. Radiol Clin North Am 1987 May;25(3):509-28. — Bilaniuk LT. Adult infratectorial tumors. Semin Rentgenol 1990 Apr;25(2):155-73. — Bilaniuk LT, Farber M. Imaging of developmental anomalies of the eye and the orbit. AJNR Am Neuroradiol 1992 Mar – Apr;13(2):793-803. — Bilaniuk LT, Rapoport RJ. Magnetic resonance of the orbit. Top Magn Reson Inaging 1994 Summer;6(3):167-81. — Bilaniuk LT, Malloy PT, Zimmerman RA, et al. Neurofiromatosis type 1; brain stem tumours. Neuroradiology 1997 Sep;39(9):642-53. — Bilaniuk LT. Orbital vascular lesions. Role of imaging. Radiol Clin North Am 1999 Jan;37(1):169-83. — Bilaniuk LT. Magnetic resonance imagings of the fetal brain. Semin Roentgenol 1999 Jan;34(1):48-61. — Bilaniuk LT. Vascular lesions of the orbit in children. Neuroimaging Clin North Am 2005 Feb;15(1):107-20.

otorhinolaryngology (ORL) of the computed tomography (CT), and in neuroradiology, ophthalmology, head and neck surgery of the magnetic resonance imaging (MRI).

Clinical signs of tracheal and bronchial trauma include subcutaneous (SC) emphysema, hemoptysis, pneumothorax, and air leak on the chest tube insertion. Plain chest x-ray (CXR) may reveal pneumothorax, air in the mediastinum (pneumomediastinum), underinflated lung and absent or poor aeration of the lung, lobe or segment (atelectasis), fallen lung sign, if the lung has collapsed away outwards and downward (rather than inward and upward) from the hilum, described by **David A Kumpe (1942 –)**[866] — American radiologist, named Kumpe sign (1970).

Gerard A. Mourou (1944 –) — French physicist and pioneer in the field of electrical engineering and lasers, and **Donna T. Strickland (1959 –)**[867] — Canadian physicist and pioneer in the field of pulsed lasers, invented in 1985 the chirped pulses amplification (CPA) technique which allowed amplifying an ultrashort laser pulses to very high energy (intensity, power), presently petawatts(PW)[868] with the laser pulse being stretched out temporarily and spectrally prior to amplification and lasting in a trillionth of a second or less. Thus, their work pawed the way for the shortest, most intense laser beams ever created. The ultra-brief, ultra-sharp beams can be used to make extremely precise cuts, so their technique is now used in laser installations and enable physicians to perform corrective laser eye surgery.

Direct percutaneous (PC) jejunostomy **[Fredrich Witzel (1863 – 1925).** 19th century. Surgery]** with the use of endoscope, guidewire and a 10F pigtail catheter was first performed in 1987 by **Robin R. Gray**[869] — Canadian invasive radiologist, for feeding purposes.

Recently, the modified magnetic resonance imaging (MRI) technique has been used for the non-invasive intramural two dimensional (2D) or three dimensional (3D) assessment of the regional wall motion of the heart, aneurysms and intracranial arteries by localized perturbations of the tissue magnetization, e.g., with spatial

---

[866] Oh KS, Fleischner FG, Wyman SM. Characteristic pulmonary finding in traumatic complete transection of a mainstem bronchus. Radiology 1969; 92:371-2. — Kumpe DA, Oh KS, Wyman SM. Characteristic pulmonary finding in unilateral complete bronchial transection. Am J Roentgenol 1970; 110:704-6.

[867] Strickland D, Monrou G. Compression of amplified chirped optical pulses. Optics Communications 1985;56(3):219-21. — Strickland DT, Development of an ultrabright laser and an application to multi-photon ionization. PhD, University of Rochester, NY 1988.

[868] Peta (in Greek – «penta», meaning «five»), since it is an equivalent to, i.e., denotes the fifth power of 1000^5, which gives extreme of 10^{15} (one quadrillion). — The PW is equal to one quadrillion (10^{15}) watts (W) **[James Watts (1736 - 1919**; 18th century)]** and can be produced by current generation of lasers for time scales of the orders of piesoseconds (10^{12}s). Achieved by CPA power output is 1.25 x 10^{15} W), or a laser pulse that last approximately наблизно one attosecond (in Danish –«atten», meaning «eighteen», and is equal to quintillionth of a second), i.e., one billionth of one billionth of a second.

[869] Gray RR, Ho CS, Ye A, et al. Direct percutaneous jejunostomy. AJR (Am J Roentgenol) 1987 Nov;149(5):931-2. — Van Overhagen H, Schippers J. Percutaneous jejunostomy. Semin Intervent Radiol 2004 Sep;21(3)199-204.

modulation and magnetization (SPAMM) tagging to produce MR-visible marks in the heart or arterial wall and phase-shift to study the intramural motion by receiving signal that can be produced by the motion along the modified magnetic field gradient used in the MRI[870]. It was quantitatively characterized by material-point correspondence data using 3D cine phase-contrast MRI to map serial displacement of each point of the myocardial wall as a set of vectors over time, linking corresponding locations of each material point in a Lagrangian representation of the motion[871], and the velocity data provided by phase-shift MRI to follow the serial evolution of velocity vectors at the special location in the imaging as in Eulerian representation of the motion[872].

In the 1990s **John A. Detre (1959 –)** — American neurologist and **John S. Leigh (1939 – 2008)** — American radiologist, developed an application for magnetic resonance imaging (MRI) that actually assesses the blood perfusion of oxygen (O_2) and the nutrients delivery to the tissues.

Pharmacology

Olexandr I. Cherkes (20.IV(02.V).1894, city of Kharkiv, Ukraine – 25.IX.1974, city of Kyiv, Ukkraine)[873] — Ukrainian pharmacologist, directed attention towards the nutritional function of cardiac glycosides (1949); introduced hypotensive drugs – benzohexonium and pirylen, and the antidote against arsenic – unitol.

After finding ephedrine **[Nagal Nagoyoshi (1844 – 1929)**. 19th century. Pharmacology] in China, it was re-discovered and re-introduced in 1924 into the USA by **Ko Kuei Chen (1898 – 1988)**[874]— Chinese and American pharmacologist, and **Carl F.Schmidt (1893 – 1988)** — American pharmacologist.

The first drug effective for the treatment of malaria or «intermitted fever» of people in tropical countries in South Asia, Africa and South America was discovered in 1932 by **Hans J. Andersag (1902 – 55)** — Italian German scientist, chloroquine (Resorchin), but with the time a resistance to chloroquine has developed and there was a need for a new effective drug.

---

[870] Greve JM, Lea AS, Tang BT, et al. Allometric scaling of wall shear stress from mice to humans: quantification using cine phase-contrast MRI and computational fluid dynamics. Am J Physiol Heaat Circ Physiol 2006;291(4):H1700-8. — Isoda H, Hirano M, Takeda H, et al. Visualization of hemodynamics in a silicon aneurysm model using time-resolved, 3D, phase contrast MRI. AJNM (Am J Neuradiol) 2006;27(5):1119-22. — Yamashita S, Isoda H, Hirano M, et al. Visualization of hemodynamics in intracranial arteries using time-resolved three-dimentional phase-contrast MRI. J Magn Reson Imaging 2007;25(3):473-8.

[871] **Joseph L. Lagrange (1736 - 1813)** — Italian mathematician and astronomer.

[872] **Leonhard Euler (1707 - 83)** — Swiss mathematician, physicist, and astronomer.

[873] Ablitsov VH. «Galaxy Ukraine». The Ukrainian diaspora: prominent persons. Publ. «Kyt», Kyiv 2007.

[874] Chen KK, Schmidt CF. The action of ephedrine, the active principle of the Chinese drug Ma Huang. J Pharmacol Exp Ther 1924;24:339-57.

Beginning from 1944 **George H. Hitchings (1905 – 98)** — American physician, in collaboration with **Gertrude B. Elion (1918 – 99)** — American biochemist and pharmacologist of Jewish, Lithuanian and Polish descent, developed a compound 2,6-draminopurine to treat leukemia; folic acid antagonist p-chlorophenoxy-2,4-diaminopyridine; pyrimethamine (daraprin) for malaria; 6-mercaptopurine (imuran) – the first immunosuppressive drug to treat post-transplantation rejection, the former (Imuran) and thioguanine to treat leukemias; allopurinol (zyloprin) for gout; azathioprine – immunosuppressive drug to treat post-transplantation rejection, rheumatoid arthritis, granulomatous enteritis and ulcerative colitis per orum and via intravenous (IV) infusion; trimethoprim / sulfamethoxazole (co-trimoxazole) to treat middle ear infections, meningitis, bacterial infections of the urinary and respiratory tracts, methicillin resistant Staphylococcus aureus (MRSA) skin infections, traveler's diarrhea, cholera, with sulfamethoxazole or dapsone for the treatment of Pseumocystis jirovecii pneumonia in patients with human immunodeficiency virus (HIV) / acquired immune deficiency syndrome (AIDS); antiviral drugs acyclovir (zovirax) for herpes zoster and zidovudine for AIDS; and chemotherapy drug nelarabine to treat patients with T-cell acute lymphoblastic leukemia.

In 1937 **Daniel Bovet (1907 – 92)** — Swiss Italian pharmacologist, researching histamine discovered the antihistamine agent. Histamine is an organic nitrogenous compound found in mast cells, smooth muscles, and endothelium, involved in local immune response, regulation of the physiological gut function and acting as a neurotransmitter for the uterus. The antihistamine agent binds histamine H1-receptors and is incorporated in allergy medications to treat allergic rhinitis and other allergies. Later, in 1946 he also described curare-like muscle relaxant (flaxedil) used by anesthesiologist in combination with general anesthetics during surgical procedures which require muscle relaxation.

It was **Bernard B. Brodie (1907 – 89)** — British American pharmacologist, who established the concept that blood drug level must guide therapeutic dosages and established the basis for chemotherapy. Together with **Julius Axelrod (1912 – 2004**: 20th century. Physiology**)** he discovered that analgesic (pain killers) acetanilide and phenacetin, both metabolize to acetaminophen (paracenatol) which unlike its precursor does not cause methemoglobinemia in humans. This medication is used to treat mild to moderate pain and fever (discovered in 1877). He also was the first scientist to determine how the neurohormones, like serotonin and norepinephrine affect the functioning of the brain, thereby leading to an understanding of how anti-psychic drugs could be used effectively in the treatment of emotional and mental disorders.

Classification of antiarrhythmic / antidysrhythmic agents in use to suppress cardiac arrhythmia, i.e., atrial fibrillation or flutter (AF), ventricular tachycardia or fibrillation (VT, VF), was proposed in 1970 by **M. Vaughan Williams (1918 – 2016)** — British cardiac pharmacologist, and **Bramah N. Sigh (1938 – 2014)** — Fijian cardiac pharmacologist[875], named the Vaughan Williams classification:

Class I agents cause sodium (Na$^+$) and potassium (K$^+$) or only Na$^+$ channels blockage: Ia agents cause fast N$^+$ and K$^+$ channel blockage, i.e., quinidine, ajmaline, procanamide and disopyramide; Ib agents cause slow Na$^+$ and K$^+$ channel blockage, i.e., lidocaine (lignocaine, or xylocaine), phenytoin, mexiletine and tocanide; Ic agents cause fast Na$^+$ channel blockage, i.e., encainide, flecaidine, propafenone and moricizine.

Class II agents are anti-sympathetic nervous system beta-blockers, i.e., carvedilol, propranolol (also shows some class I action), esmolol, timolol, metaprolol, atenolol, bisoprolol and nebivolol.

Class III agents affects K$^+$ influx, i.e., amiodarone (also show class I-IV action), sotatol (K$^+$ channel blocker), ibutilide, dofetilide, dronedarone and E-4031.

Class IV agents affect calcium (Ca^{++}) channel and the atrioventricular (AV) node, i.e., verapamil and diltiazem.

Class V agents, i.e., adenosine, digoxin, and magnesium sulfate (MgSO$_4$) work by other or unknown mechanism (direct nodal inhibition).

Sir **James W. Black (1924 – 2010)** — Scottish pharmacologist, developed in 1958-64 the beta adrenergic receptors antagonist (blocker) propranolol (metoprolol) drug for the treatment of angina pectoris and hypertension[876], and in 1966-75 in collaboration with **C. Robin Gamellin (1934 –)** — English chemist, the first class drug histamine (H)2-receptors antagonist (blocker) cimetidine (tagamet) to block the action of histamine on parietal cells in the stomach, decreasing production of hydrochloric acid (HCl) by these cells, to treat dyspepsia, peptic ulcer disease, gastroesophageal reflux disease (GERD), laryngo-esophageal reflex, columnar epithelium lined lower esophagus, gastrinomas, ulcerogenic non-beta islet cell tumors of the pancreas, and to prevent stress gastritis.

The pain-relieving anti-inflammatory and antiplatelet effects of aceto-salicylic acid (ASA, aspirin) – a non-steroid anti-inflammatory drug (NSAID), first synthetized by **Charles Gerhardt (1816 – 56**: 19th century. Biochemistry), was elucidated in 1971 by Sir **John R. Vane (1927 – 2004)** — English pharmacologist of Jewish faith, as being dependent on the irreversible acetylation of platelet cyclooxygenase, hence blocking the synthesis of prostaglandins and thromboxane A$_2$ in the spleen (1971).

---

[875] Republic of Fiji (area 18,274 km^2) is the archipelago of 332 islands in the south-western Pacific Ocean.

[876] Black JW, Crowthen AC, Shanks RG, et al. A new adrenergic beta-receptor antagonist. Lancet 1964;283(7342):1080-1.

By the inhibition of platelet's aggregation ASA prolongs the bleeding time which last about 10 days. He also introduced an angiotensin-converting-enzyme (ACE) inhibitor which found its use in the prevention and treatment of cardiovascular diseases.

The discovery and development of various drugs originally from plants and microorganisms were made by **Tu Youyou (1930 –)** — Chinese pharmaceutical chemist, **Satoshi Omura (1935 –)** — Japanese biochemist and **William C. Campbell (1930 –)** — Irish American biologist, biochemist and parasitologist.

In 1967 Tu Youyou started a new discovery project to treat a form of malaria resistant to the chloroquine during Vietnam War (1955-75). After screening 2,000 traditional Chinese recipes and making 380 extracts tested on mice, she found two years later (1967) that a hallmark for treatment of malaria was a sweet worm wood plant (Artemisia annua). She followed the recommendation of **Ge Hong (283 – 343**: Ancient medicine. *Chinese*) and by stepping that plant in a low-temperature (cold) water for the extraction, not damaged by heat, a new effective semi-synthetic drug was obtained against Plasmodium falciparum malaria – artemisinin, thus making a significant breakthrough in the tropical medicine of the 20th century, that brought a health improvement to the inhabitants of the tropical countries.

Since 1975 Satoshi Omura during invention a bacterial strain of Streptomyces avermitilis discovered produced by it an anti-parasitical compound avermectin. Acquiring that strain of bacteria, William Campbell developed the derived drug invemectin (mectizan) used against river blindness (first clinical trial performed in 1981) which prevents blindness in over 300,000 people each year, against lymphatic filariasis [19th century. Microbiology – **Otto H. Wuchener (1820 – 73)** and **Joseph Bancroft (1836 – 94)**] and against other parasitic infections.

In 1971 **Akira Endo (1933 –)** — Japanese biochemist, working on a fungal secondary metabolites excreted to the extracellular medium (extrolites), e.g., the oyster musher (Pleurotos ostreatus) extracted the statins group of drugs – mevastatin from which later its derivative pravastatin was synthetized, and also levastatin (Mevacor), a lipid-lowering agents working by inhibiting 3-hydroxy-3-methyl-glutaryl-coenzyme A reductase (HMG-CoA reductase), a key enzyme in the metabolic pathway the body uses to produce dyslipidemia, an abnormal amount of lipids such as triglycerides, cholesterol and/or fat phospholipids in the blood, for the treatment of cardiovascular diseases.

Scholar and follower of **Olexandr Cherkes (1894 – 1974**: above), **Ivan S. Chekman (04.X.1936, village Chan'kiv, Dunayevsky Region, Khmelnytska Obl.,**

Ukraine – 26.X.2019)[877] — Ukrainian pharmacologist, was the first to establish the leading role of complex creation of medical means with the bio-membranous, as a trigger mechanism for the primary pharmacological effect, which allows to decide the basic question of general and clinical pharmacology of cardiovascular neurotropic and metabolic preparations.

He established that the ether oils of plants have virtues to create complexes with xenobiotics, to decrease its toxicity and to remove them from an organism. He developed the composition of ether oils for the use in the clinical practice. He obtained the original preparation derived from the plants – carbolize, which remove radionuclides, nitrates, and salts of the heavy metal from an organism.

Investigations into the quantum pharmacology have established that the specific activity of drugs is conditioned by the evidence and peculiarity in location of the reactive centers of its molecule responsible for curative properties of medicaments. The suspension of high dispersion flint (quartz) decreases toxicity of xenobiotics, as well as other medical ingredients. The oxides of nano-cooper, and nano-silver reveal more explicit antimicrobial activities, than the usual compounds of given metals. Basing on bio-acidic metal-silicate, compounds were developed, original preparations of which having no analogy in the world, which could be used for disinfection, deactivation, and identification of different chemical substances.

Later (1985) **Bruce D. Roth (1954 –)** — American organic and medicinal chemist, synthetized other statins group drug – atorvastatin (Lipitor), and most recently (2013) one more statin – rosuvastatin (Crestor) was developed.

However, it was reported in 2014 that the use statins may cause, among many other side-effects, memory loss, forgetfulness, and confusion.

---

[877] Village Chan'kiv (remains of Tripillian Culture, appears in historical documents of 16th century), Dunayevetsky Region, Khmelnytsky Obl., Ukraine. — Chekman IS. Biochemical Pharmacodynamics. Publ. «Zdorov'ia», Kyiv 1991. — Chekman IS. Handbook of Clinical Laboratory Diagnostics. Kyiv 1992. — Chekman IS et al. Pharmacology. Publ. «Nova knyha», Vinnytsia 2006. — Chekman IS et al. Magnesium Containing Preparations: Pharmacologic Properties, Usage. Zaporizhia – Kyiv 2007. — Chekman IS et al. Metabolic Trophic Preparations. Zaporizhia 2007. — Chekman IS et al. Pharmacology: Handbook. Department of Health of Ukraine and of OO Bohomolets NMU, Vinnytsia 2009.— Chekman IS. Nanopharmacology. PVP «Zadruha», Kyiv 2011. — Chekman IS et al. Pharmacology. Handbook for Students of the Medical Faculties. Publ. «Nova knyha», Vinnytsia 2011. — Chekman IS. Quantum Pharmacology. Publ. «Naukova dumka». Kyiv 2012. — Chekman IS et al. Nanoscience, Nanobiology, Nanopharmacy. Kyiv 2012. — Chekman IS. English-Ukrainian Nanoscience Reference Guide. Kyiv 2013. — Chekman IS et al. Pharmacology. Handbook for Students of the Stomatological Faculties of High Medical Educational Institutions. Publ. «Nova knyha», Vinnytsia 2014. — Chekman IS et al. General Pharmacology: Handbook. Zaporizhia - Kyiv 2016. — Ablitsov VH. «Galaxy Ukraine». The Ukrainian diaspora: prominent persons. Publ. «Kyt», Kyiv 2007.

Space of scientific, practical and educational medical activity of **Ludmila M. Maloshtan (25.I.1957, city of Kharkiv, Ukraine –)**[878] — Ukrainian chemist, biologist, physiologist, pharmaceutics, and geneticist, is indeed very wide, about what attests authorship or co-authorship in over 35 an author certificates and patents, published 302 scientific works, including at least 6 handbooks. She researches medical remedies, among them phytoncides[879], which act upon healing onto cardio-vascular system and gastro-intestinal (GI) tract.

Internal medicine. Medicine

In 1900 **Ivan F. Zelenev [19(31).V.1860, Tversk Obl., Russia – 30.IV.1918, city of Moscow, Russia]**[880] — Russian Ukrainian physician-dermatologist and cardiologist, established that syphilis, in its primary and secondary periods, involves the heart.

The New York Heart Association (NYHA) Functional Classification, developed in 1902 (published 1928) consist of four classes[881]:

Class I.   –   Cardiac disease. Asymptomatic with normal (ordinary) physical activity, i.e., no fatique, dyspnea (shortness of breath), palpitation, or angina pectoris while walking or climbing stairs.

Class II.   –   Cardiac disease. Comfortable at rest; symptomatic with ordinary physical activity.

Class III. –   Cardiac disease. Comfortable at rest; symptomatic with less than ordinary physical activity which outcomes in symptoms, e.g., walking short distances (20-200 m.).

---

[878] Maloshtan LM. Pharmacological Aspects Pathologically Confirmed Chemotherapy of Diabetes Mellitus Using Phito-preparations Based on Extracts from Plant Bean. Dissertation for the degree of Doctor Biological Sciences, National University of Pharmaceutics (NUPh) of Ukraine. Kharkiv 1994. — Maloshtan LN, Riadnykh EK, Zhegunova GP, et al. Physiology with Basis of Anatomy. Publ. NUPh, «Golden Pages», Kharkiv 2002. — Maloshtan LM, Riadnykh OK, Zhehunova HP, et al. Physiology with Basis of Human Anatomy. Publ. NUPh «Gold Pages», Kharkiv 2003. — Maloshtan LN., Volkovoy VA. Human Anatomy. Textbook. Publ. BURUN & K, Kharkiv 2009. — Maloshtan LN, Filiptsova OV. Biology and Genetics Principles. Publ. NUPh «Golden Pages», Kharkiv 2011. — Maloshtan LN, Filiptsova OV, Shakina LA. Biology and Genetics Principles. Publ. Smith, Kharkiv 2014. — Maloshtan LN, Filiptsova OV, Shakina LA. Biology and Genetics Principles. Publ. NUPh «Golden Pages», Kharkiv 2016. — Ablitsov VH. «Galaxy Ukraine». The Ukrainian diaspora: prominent persons. Publ. «Kyt», Kyiv 2007.

[879] Phytoncides (in Greek – «phyton» plant, in Latin – «aedo» to kill) are biologically active plants substances which kill, or depress growth and development of microorganisms (bacteria, viruses, microscopic fungi).

[880] Tversk Obl. (area 84,201 km²), Russia. — Ablitsov VH. «Galaxy Ukraine». The Ukrainian diaspora: prominent persons. Publ. «Kyt», Kyiv 2007.

[881] City of New York (Foot Note 620), N.Y., USA. — Sytar LL. Aquired heart diseases. In, LYa Kovalchuk, VF Sayenko, HV Knyshov. Klinichna Khirurhiya. Vol. 1-2. Publ. «Ukrmedknyha», Ternopil 2000. Vol. 1. P. 228-42.

Class IV. – Cardiac disease. Symptomatic at rest; unable to carry any physical activity without discomfort.

In 1902 **Alfons F.J.K. Zivert** or **Ziwert (1872, city Kyiv, Ukraine – 1922, city of Kyiv, Ukraine)**[882] — Ukrainian physician, in 1933 **Manes Kartagener (1897, city of Peremyshl, Ukraine / Poland – 1975, city of Zurich, Zwitzerland)**[883] — Ukrainian physician of Jewish origin, diagnosed congenital disease transmitted as an autosomal recessive trait, characterized by the complex of bronchiectasis, complete or partial with dextrocardia (situs viscum inversus) and polyposis of the mucous mebrane of the nose, i.e., primary ciliary dyskinesis, or immobile cilia syndrome (Zivert-Kartegener syndrome or triad).

Also, in 1912 Alfons Zivert stated that the active mechanic energy needed by cardiac myocardium is used not only for contraction during systole, but also for relaxation during active diastole.

The first case of primary plasma-cell leukemia, a malignant leukemia, or leucosis pPVL) was described in 1906 by **Władysław A. Gluziński (1856 – 1935)**[884] — Polish internist, in a 47-year-old man professional railway ticket controller, who presented with bone pain, and during palpation swelling over his hip was detected, as well as

---

[882] Zivert AK. Case of congenital bronchiectasis in the patient with a reverse localization of viscera. Russ vrach 1902;38(1:1361- 2. — Zivert AK. Über einen Fall von Bronchiectasie bei einem Patienten mit Situs inversus viscerum. Berliner klinische Wochenschrift 1904;41:139-41. — Siewert AK. Über ein die Verfahren der manometrischen Registrirung der Zusammenziehungen des isolirten Säugethierherzens. Archiv für gesamte Physiologe des Menschen und der Thiere (Berlin) 1904;102(7):364. — Siewert A, Zebrowski E. Über den komparativen Einfluss des weissen und dunzen Fleisches auf die Ausssscheidung von Harnsäure und von anderen stickstoffhaltigen Substanzen im Harn. Zeinschrift für klinische Medizin 1912;75:331-58. — Siewert AK. Über active Diastole. Zeintschrift für die gesamte experimentelle Medicin 1922;28(1):324-46. — McManus C. Eponymous but anonymous: who was Dr Siewert? Lancet 2004;363(9409):662. — Vasyliev KK. Zivert Alphons-Ferdynand-Julius Karlovych. In, MH. Zhelezniak. Encyclopedia of Modern Ukraine. Vol. 10. National Acad Sci Ukraine and Shevchenko Sci Soc, Kyiv 2010. P. 583. — Ablitsov VH. «Galaxy Ukraine». The Ukrainian diaspora: prominent persons. Publ. «Kyt», Kyiv 2007.

[883] City of Peremyshl (Foot Notes 1, between 87 & 88) / Przemysl, Ukraine / Poland. — City of Zürich, or Zurich (human presence in the area 4,400 BC, founded by the Romans in 15 BC; the earliest name was Turicum, meaning a tombstone, present name is attested since the 6th century; area 87,88 km^2), Switzerland. — Kartagener M. Zur Pathogenese der Bronchiectasien: Bronchiectasien bei Situs viscerum invertus. Beträge zum Klinik der Tuberkulose 1933;83:489-501. — Kartagener M. Zur Frage der Bronchiectasien: Bronchiectasien bei Situs viscerum invertus. Beträge zum Klinik der Tuberkulose 1933;84:73. — Kartagener M, Horlachen A. Zur Pathogenese der Bronchiectasien: Situs viscerum invertus und polyposis nasi in einem Falle familiärer Bronchiektasen. Beträge zum Klinik der Tuberkulose 1935;87:489. — Ablitsov VH. «Galaxy Ukraine». The Ukrainian diaspora: prominent persons. Publ. «Kyt», Kyiv 2007.

[884] Gluzinski W, Reichenstein M. Myeloma und leukemia lymphatica plasmocellularis. Wien klin Wschr 1906;12:336-9. — Van de Donk NWCJ, Lokhorst HM, Anderson KC, et al. How I treat plasma cell leukemia. Blood 2012 Sep 20;120(12). Doi.org/10.1182/blood-2012-05-408682.

rib fractures, anemia and enlarged spleen (splenomegaly). pPV is the most aggressive form of plasma-cell dysplasia[885].

The clinical premortem diagnosis of an acute postinfarction ventricular septal defect (VSD) was made by **F. Brunn**[886].

Simon S. Leopold and **Charles S. Leopold (1896 – 1916)** — American engineers, designed and installed in 1923 room equipped with an air conditioner in the Hospital Univ. of Pennsylvania (founded 1874), Philadelphia, PA, USA, to investigate causes of and treatment for bronchial asthma (Ancient medicine. *Egypt*) and other allergic conditions.

The term asbestosis coined in 1924 **Thomas Oliver (1853 – 1942)**[887] — English physician specializing in occupational medicine. An association between asbestos exposure and mesothelioma development documented / confirmed in 1924 **William E. Cooke**[888] — English pathologist, in 1930 and 1935 **Kenneth M. Lynch**[889] — American pathologist and **W. Atmar Smith** — American physician, in 1960 **J. Christopher Wagner (1923 – 2000)**[890] — South African pathologist, **Christopher A. Sleggs** — South African pulmonologist, and **Paul Marchand** — South African thoracic surgeon, in asbestos miners of South Africa, and in 1964 **Irwing J. Selikoff (1915 – 92)**[891] — American physician and researcher, and in 1964 among the miners of the USA.

The oncogenic potential of asbestos fibers based on their physical properties of dimension, shape and length has been explained by **Merle F. Stanton**[892] and **C. Wrench**, in the Stanton hypothesis (1972), and **J. Nicholson** (2001)[893].

The blue, thin (1000 tinner than human hair), long and straight fibers of amphibole asbestos – crocidolite (blue green) and amosite (brown) have greater malignancy potential because the longer, and bio-persistent fibers penetrate deeper into the bronchial and alveolar mucosa or pleura and excite mesothelial cell proliferation into interstitial lung diseases (ILD), carcinoma of the lung or malignant pleural

---

[885] Dysplasia denotes anomalous changes in the nulear cell, cytoplasm, or abnormal growth the cells itself.

[886] Brunn F. Diagnostik der erworbeden ruptur der Kammerscheidewand des Herzens. Wien Arch Inn Med 1923;6:533.

[887] Oliver T. Some dusty occupations and their effect upon the lungs. J Roy Sanitary Inst 1925;46:226-30.

[888] Cooke WE. Fibrosis of the lungs due to inhalation of asbestos dust. Br Med J 1924 Jul 26;2(3317):140-2.

[889] Lynch KM, Smith WA. Asbesthosis bodies in sputum and lung. JAMA 1930 Aug30;95(4):650-61. — Lynch K, Smith W. Pulmonary asbestosis III: carcinoma of lung in asbesto-silicosis. Am J Cancer1935;24:56-64.

[890] Wagner JC, Sleggs CA, Marchand P. Diffuse pleural mesothelioma and asbestos exposure in the Northwestern Cape Province. Br J Ind Med 1960 Oct;17(4):260-71.

[891] Selikoff IJ, Churg J, Hammond EC. Asbestos exposure and neoplasia. JAMA 1964 Apr 6;188:22-6.

[892] Stanton MF, Wrench C. Mechanisms of mesothelioma induction with asbestos and fibrous glass. J Natl Cancer Inst 1972;48:707-821. — Stanton MF, Layard MW. Carcinogenesity of natural and man-made fibers. Adv Clin Oncol 1978;1:181-7.

[893] Nicholson WJ. The carcinogenicity of of chrysolite asbestos; a review. Ind Health 2001;39:57-64.

mesothelioma (MPM) with invasion into of the pericardium and peritoneum. The white, short, and serpentine chrysolite fibers are broken down and are cleaned from the body by macrophages too quickly to provoke malignancy. The pathologic / histologic classification of mesothelioma reveals the epithelial cell pattern (61,5%) with the least malignant potential, the sarcoid cell pattern (16,4%) with the most malignant potential, and the biphasic (mixed) epithelial cell pattern (22%) with an intermediary malignant potential.

The next step after confirmation of the histologic pattern of MPM by the needle, open or thoracoscopic biopsy, should be the staging of this disease according to the International Mesothelioma Interest Group (IMIG, 1994) and to determine the best therapy for each patient.

Stages Ia correlate with tumor $(T)_{1a}$, in which neoplasm involve an ipsilateral parietal pleura with or without diaphragmatic involvement, and T1b in which tumors involve the visceral pleura without lymph nodal (LN) involvement, i.e., NO. The treatment of choice should be pleurectomy and decortication (P/D).

Stage II includes tumor T_2, in which disease invades the lung parenchyma, N_0, and if only one lobe is involved by tumor the treatment should be P/D; but if the pleura and entire lung are invaded by tumor, preoperative (neoadjuvant) multimodality therapy, i.e., low dose radiotherapy (20 Gy), as not to damage contralateral healthy lung, and chemotherapy (Pemetrexed), should preceed surgical intervention, i.e., extra-pleural pneumonectomy (EPP).

Stage III defined as any tumor T_3, if neoplasm locally non-aggressively advance into endothoracic fascia, mediastinal fat, local chest wall, or pericardium, or any N_1 – ipsilateral mediastinal LN involvement, it should be preceded by neoadjuvant multimodality therapy, P/D and local resection; but tumors locally aggressively invading the mediastinum, diaphragm, or chest wall with N_1 mediastinal LN involvement should be treated by aggressive local resection, including chest wall, pericardium and diaphragm with appropriate reconstitution; those with any N_2 – contralateral mediastinal (N_2) or extra-thoracic (N_3) LN involvement, should be treated by multimodality therapy.

Stage IV involve any tumor T_4, N_3 and metastatic $(M)_1$ disease which should be treated by multimodality therapy.

Other adjunctive therapies may include intraoperative heated chemotherapy, antiangiogenic therapy, photodynamic therapy (PFT), immunotherapy and gene therapy.

The first diagnosis of pulmonary infarction in the alive patient was made in 1902 by **Teofil H. Yanovsky / Theophilus G. Janowski [12(24).VI.1860, village Myn'kivtsi, Dunayevs'kyy Region, Khmelnyts'ka Obl., Ukraine – 08.VII.1928, city of Kyiv,**

Ukraine][894] — Ukrainian internist. His other contributions to the internal medicine include description of an increased bronchophony in the early stage of pneumonia over the lung (bronchophony or Yanovsky sign), friction of the peritoneum in pericholecystitis or perforation of the gallbladder (Yanovsky friction of peritoneum) and percussion of the chest over the finger or hand of an examiner (Yankovsky percussion), publishing of a classical monography on pathology and diagnostics of kidney's diseases (1927).

---

[894] Village Myn'kivtsi (settled since Bronze Age 4,000-3,000 years BC, presence is traces to Scythian and Cherniakhivsky Culture, first written mention 1393, founded 1407; during conducted by Russian Communist Fascist of Holodomor-Genocide 1932-33 against the Ukrainian Nation 63 people died from artificial hunger), Dunayevsky Region, Khmelnytsky Obl., Україна. — Janowski ThG. Ueber den Bakteriengehalt des Schnees. Centralblatt für Bacteriologie und Parasitenkunde 1888;4: 547-552. — Janowski ThG. Zur diagnostischen Verwerthung der Untersuchung des Blutes bezüglich des Vorkommens von Typhusbacillen. Centralblatt für Bacteriologie und Parasitenkunde 1889;5:657-663. — Janowski ThG. Zur Biologie der Typhusbacillen. I. Die Wirkung des Sonnenlichts. Centralblatt für Bacteriologie und Parasitenkunde 1890; 8:167- 172,193-199,230-234,262-266; ~ II. Zur Wirkung hoher und niedriger Temperatur. Ibid., 417-423, 449-457. — Janowski ThG. Ein Fall von chronischen Dickarms. Wratsch 1891;10:279. — Yanovsky FH. About the technique auscultation. Vrach 1893;28:288-9. — Janowski ThG. Zur Technik der Auscultation. Dtsch med Wschr 1894;20(32) :651-2. — Janowski ThG. Terpinol bei Haemoptoe. Klinisch-therapeutische Wochenschrift 1900;7(8):230-232. — Janowski ThG, Wyssokowicz WK. Ein Fall von Dermatomyositis. Deutsches Archiv für klinische Medizin 1901;71:493-512. — Janowski ThG, Wyssokowicz WK. Zur Symptomatologie und Pathogenese der käsigen Pneumonie. Zeitschrift für Tuberkulose und Heilstättenwesen 1902;4:33-41. — Janowski ThG. Ueber die Anwendung der Nebennierenpräparate bei Erkrankungen des Oeso-phagus. Archiv für Verdauungskrankheiten 1904;10:508-519. — Janowski ThG. Les nouvelles voies dans la diagnostic des maladies internes. Reveue de Medicine 1907; 9. — Janowski ThG. Über nervöse Diarrhöe. Medizinische Klinik 1911;7:1377-81. — Janowski ThG. Zur Perkussion der Lungenspitzen bei Tuberkulose. Medizinische Klinik 1912;8:1786-89. — Yanovsky FH. Methods of functional diagnosis of kidney diseases. Russ vrach врач. 1913;6:169-77. — Yanovsky FH. About contemporary situation of diet therapy in nephritis. Russ vrach 1913;34. — Janowski ThG. Zum gegenwärtigen Stande der Diätetik bei Nephritis. Medizinische Klinik 1913;9,2(35):1404-07. — Janowski ThG. Ueber die funktionelle Diagnostik der Nierenkrankheiten. Medizinische Klinik 1914;10,1(6):234-37; Nr. 7, 272-76. — Yanovsky FH. Tuberculosis of the Lung. (pathology, clinic and treatment). «Gosizdat», Moskva-St Petersburg 1923. —Janowski ThG. Zur Diagnostik des Flecktyphus. «Konjunktival-Syndrom». Wiener medizinische Klinik 1926;76(12):367-69. — Yanovsky FH. Ways for Scientific Studies in Clinics. «Dezhmedvydav» Ukrainy, Kharkiv 1927. — Yanovsky FH. Diagnostics of Kidney Diseases in Connection with Pathology. «Derzhmedvydav» Ukrainy, Kyiv 1927. — Janowski ThG. Die Klinische Bedeutung des Geruchs. Klinische Wochenschrift 1929;8:172-74. — Janowski ThG. Die Bahnen wissenschaftlicher Untersuchung in der Klinik. Wiener medizinische Klinik 1930;80(7):233-36. — Yanovsky FH. Tuberculosis of the Lung. 3rd Ed. «Medgiz» Moscov – St Petersburg 1931. — Bazylevych I. Teofil Yanovsky (1860-1928). J Ukr Med Assoc North Am 1955 Dec;2,2(4):23-8. — Shypulin VP, Nazarenko D. Academician F.H. Yanovsky: borders of personality. Hastroenterolohiya 2014;2(52):94-101. — Tarchenko IP, Добriansky DV, Humeniuk HL, et al. F.H. Yanovsky – physician, scientist, personality. Asthma ta alerhiya. 2018;4:46-53. — Ablitsov VH. «Galaxy Ukraine». The Ukrainian diaspora: prominent persons. Publ. «Kyt», Kyiv 2007. — Hanitkevych YaV. The contribution of the Ukrainian physicians to the world medicine. Ukr Med Chasopys 2009 VII/VIII;4(72):110-15. — Synyachenko OV. They glorified themselves and Ukraine: history of domestic therapy. Crimean Therapeutic Journal 2010;2(2):21-8.

In 1899-1906 **Karen F. Wenckebach (1864 – 1940)**[895] — Dutch physiologist and cardiologist, and in 1930 – **Alexander / Olexandr Samoilov (1867 – 1930**: 20th century. Physiology**)** described II (partial) degree atrio-ventricular (AV) alternating heart block with a progressive lengthening of the conduction time (P-R interval) in the cardiac muscle until a ventricular response occurred with irregular pulses (Wenkenbach-Samoilov phenomenon or periodicity).

The common causes of dysfunction the cardiac muscle leading to cardiac (heart) failure (CF, HF) are myocardial such as ischemia of the myocardium or myocardial infarction (MI), viral myocarditis, idiopathic, hypertrophic or peripartum cardiomyopathy (CM), hypertension (HTN), toxins (alcohol, cocaine, chemotherapeutic agents), infiltrative disease (amyloidosis, hemochromatosis), infectious (systemic tick or Lyme borreliosis[896]; Trypanosoma cruzi or American trypanosomiasis or Chagas disease)[897], thyroid dysfunction, metabolic abnormalities (thiamine or selenium deficiency), cardiac valves – aortic stenosis or regurgitation (AS, AR) or both, mitral stenosis or regurgitation (MS, MR) or both, arrhythmic, tachycardia mediated CM, and pericardial – constructive pericarditis.

Those affections depress myocardial contractility (systolic dysfunction) or impair ventricular diastolic filling (diastolic dysfunction). That initiate a rise in the left ventricular end-diastolic volume (preload) and pressure (LVEDV, LVEDP), transmitted to the left atrium (LA) and subsequently to the pulmonary venous and capillary vessels. Due increased LVEDV and LVEDP, the atria secrete an atrial nitrogenic peptide and the ventricle of the brain secrete a brain nitrogenic peptide (BNP) which act through a receptor-mediated guanosine monophosphate (GMP) in vascular smooth muscles cells and are degraded by neutral endopeptidase. The ANP and BNP have a potent balanced vasodilating actions resulting in decreasing vascular preload and afterload, and in kidneys cause vasodilatation of the afferent arteriole (diameter 15-70 μm), thereby increasing glomerular filtration rate (GFR), and inhibit sodium (Na^{++}) reabsorption in the renal collection ducts and aldosterone secretion by the adrenal glands with resulting natriuresis and diuresis.

As disease progresses the increased intravascular pressure results in venous stasis, transudation of fluid into the pulmonary interstitial, where it interferes with gas exchange, causing hypoxia, i.e., diminished oxygen (O_2) availability and dyspnea with reduction in the left ventricular (LV) stroke volume (SV) and cardiac output

---

[895] Wenckebach KF. On the analysis of irregular pulses. Z Klin Med 1899;37:475-88. — Wenckebach KF. Arhythmia of the Heart. A Physiological and Clinical Study. William Green and Sons, Edinburgh & London 1904. — Wenckebach KF. Beiträge zur Kenntnis der menschlichen Herztätigkeit. Arch Anat Physiol 1906;297-354.

[896] **Rhazes (854 - 927)** — Persian physician (Medieval medicine. *Egypt*).

[897] **Carlos Chagas (1879 - 1934)** — Brazilian physican (20th century. Microbiology and immunology).

/ index (CO, CI), with inadequate delivery of blood to the arterial vessels, blood hypoperfusion, tissue anoxia due to lack of O_2 and acidosis.

The compensatory heart response is withdrawal of parasympathetic nerve system influence, and direct sympathetic nerve system stimulation of the heart by beta-adrenergically induced release of catecholamines (epinephrine and norepinephrine) from the adrenal glands, resulting in tachycardia and an increase in myocardial contractility to increase CO.

Induced by catecholamines peripheral arterial vasoconstriction redirect the available CO from less vital skin, skeletal muscle and kidneys to the most vital heart and brain.

The beta-adrenergic stimulation of the juxtaglomerular apparatus in the kidneys releases renin and activates angiotensin I and II system, and in concert with a direct alpha-adrenergic stimulation of the vasculature maintain arterial blood pressure (BP).

The reduced renal blood flow from depressed CO and redistribution of blood volume result in activation of renal baroreceptors which further stimulate release and augmentation of catecholamine, thus contributing to continuous vasoconstriction.

To increase blood volume and to maintain tissue perfusion angiotensin II directly promotes reabsorption of Na^{++} in the proximal nephron and indirectly by angiotensin II-induced release of aldosterone from the adrenal cortex, promotes absorption of Na^{++} from distal nephron. In addition, angiotensin II and norepinephrine stimulate hypothalamic release of arginine vasopressin from the posterior part of the hypophysis resulting in further vasoconstriction and free water reabsorption, expansion of intravascular volume and augmentation of venous return, thereby increasing LVEDV (preload), and by the Frank-Starling curve (law, mechanism, or relationship), of the CO **[Otto Frank (1865 – 1944)**. 19th century. Physiology. / **Ernest Starling (1966 – 1927)**. 20th century. Physiology].

The CO is regulated by intrinsic cardiac and extrinsic regulatory mechanism. However, the Frank-Sterling relationship is the principal intrinsic cardiac regulatory mechanism between the SV and the end-diastolic volume (V_{ED}) and end-systolic volume (V_{ES}), as in the following equation:

$$SV = V_{ED} - V_{ES}$$
$$then$$
$$CO = HR \times SV, where$$
$$HR = heart\ rate.$$

The molecular determinants of the Frank-Sterling relationship depend on the magnitude of the force generated by the myocytes at their initial length at the time of initial contraction. A quadratic polynomial function has been used to approximate this relationship.

In chronic HF the hemodynamic overload causes the left ventricular (LV) remodeling, i.e., changes in shape and size, depending on the condition of the heart affected by a particular disease. In HTN and AS it results in LV hypertrophy. If LV is insufficient to normalize the wall stress from AR and MR, or both, a rise in diastolic wall stress induces dilatation and leads according to the Laplace law **[Pierre de Laplace (1749 – 1827)**. 19th century. Physiology], to the development of interstitial fibrosis and change of the LV from the normal ellipsoid shape to an abnormal spherical one.

In conservative treatment of HF the best hemodynamic effect could be obtained by in increasing the LV SV without significant increase of the LV filling volume with combination of inotropic, vasodilation and diuretic agents which can shift the Frank-Starling curve to up and to the left, thus improving the pump function of the heart. A shift of the Frank-Starling curve down and to the right indicate that pump function of the heart is worsening.

Of the non-Frank-Starling regulation of the systemic BP and CO are volume and pressure receptors, which provide two important HR regulatory mechanism, the atrial or Bainbridge stretch reflex **[Frances A. Bainbridge (1874 – 1921)**. 20th century. Physiology. 1915], and the baroreceptor or De Burgh Daly reflex **[Ivan De Burgh Daly (1883 – 1974)**. 20th century. Physiology. 1927].

An increase in circulating blood volume stimulates receptors in the atria, increasing the sympathetic nervous system tone and cause the HR to accelerate.

An increase in the systemic BP stimulates baroreceptors in the carotid sinus and aortic arch, increase the parasympathetic nervous system tone, and cause the HR to decrease.

The resultant change in the HR is the sum of those two competing reflexes.

Perhaps, the last significant contribution to the percussion of the chest was laid out by **Edward M. Brockbank (1866 – 1959)** — Australian physician and surgeon, in his work on «The Treatment and Diagnosis of Heart Disease» (1930) in which he presented the method of differentiating between dilatation and hypertrophy of the heart. However, since the middle of the 20th century, the percussion methods were in many aspects replaced by transthoracic echocardiography (TTE).

Alexander I. Jarotsky (1866 – 1944) — Russian physician, who recommended that gastroduodenal ulcers be treated with a diet of egg whites, fresh butter with bread and milk or noodles (1910).

An effect of coronary artery occlusion on myocardial contraction was noted by **F. Robert Tennant (1866 – 1957)**[898] — English scientist and **Carl Wiggers (1883 – 1963**: 20th century. Physiology**)**. Particularly, they showed dyskinesis, i.e., paradoxical

---

[898] Tennant R., Wiggers C.J. The effect of coronary occlusion on myocardial contraction. Am J Physiol 1935;112:351-61.

motion – protruding outside during systole and dragging inside during diastole of the portion of the left ventricle (LV) affected by an acutely ischemic myocardium.

The first designation in sanitary practice by **Konstiantyn E. Dobrovolsky [11(23).V.1867, city of Kropyvnytsky, Kirovhradska Obl., Ukraine – 13.XII.1946, city of Kyiv, Ukraine]**[899] — Ukrainian physician, to use the titer of Bacillus coli as an indicator for contamination of the water.

Independently, while working in Mashfield Clinic (founded 1916), the city of Mashfield (settled 1818), Wisconsin (WI, USA), **Karl W. Doege (1867 – 1932)**[900] — German American physician and **Roy P. Potter (1879 – 1968)**[901] — American radiologist, reported in 1930 of a solitary or multiple non-islet cell fibrous tumor of the sub-mesothelial layer of the pleura, the commonest benign pleural tumor, associated with hypertrophic pulmonary osteoarthropathy (22% of cases) and severe hypoglycemia (3-4% of cases), attributed to tumor-derived (insulin-like) growth factor II (paraneoplastic or Doege-Potter syndrome).

Both, **Paul Carnot (1869 – 1958)** — French physician, with his graduate student **Clotilde C. Deflandre (1871 – 1946)** — French scientist, contributed in 1895 to the modern organ transplantation by examining the ability of skin graft from black guinea pigs to persist in white guinea pigs, i.e. the white skin that surrounds black graft blackens centrifugally, while white skin grafted to a black area in due course blackens[902], and discovered in 1905 erythropoietin, a.k.a., hematopoietic (EPO) which is a glycoprotein cytokine secreted by the kidney in response to anemia and decreased oxygen (O_2) concentration in the blood (hypoxia), by observing an increase of immature red blood cells (RBC), called reticulocytes from the bone marrow, in normal rabbits occurred following the injection of blood plasma taken from anemic rabbits after earlier bloodletting[903].

Patient-related causes of bleeding include congenital or acquired bleeding disorders (coagulopathies) due to an abnormality of the platelets, anemia with hemoglobin (Hgb) less than 6 g/dl, thrombocytopenia (platelets count below 50,000),

---

[899] City of Kropyvbytsky (settled since the time of middle Neolithic Age, and by the Ukrainian Cossack of Zaporizhia Sich since 16th century, founded 1754), Kirovhradska Obl., Ukraine. — Ablitsov VH. «Galaxy Ukraine». The Ukrainian diaspora: prominent persons. Publ. «Kyt», Kyiv 2007. — Synyachenko OV. They glorified themselves and Ukraine: history of domestic therapy. Crimean Therapeutic Journal 2010;2(2):21-8.

[900] Doege KW. Fibro-sarcoma of mediastinum. Ann Surg 1930;92(5):955-60.

[901] Potter RP. Intrathoracic tumours. Case report. Radiology 1930;14:16-1.

[902] Carnot P, Deflandre Cl. Persistance de la pigmentation dans les greffes epidermiques. Compt Rend Acad Sci 1895;48:430-3.

[903] Carnot P, Deflandre Cl. Sur l'activite hemopoïtique du serum au cours de la regeneration des differens organes au cours de la regeneration du sang. Compt Rend Acad Sci 1906;143:384-6. — Carnot P, Deflandre Cl. Sur l'activite hemopoietique du serum au cours de la regeneration du sang. Compt Rend Acad Sci 1906;143:422-5.

complex cardiac procedures, e.g., combined cardiac valve procedure and coronary artery bypass (CAB) grafting, repair of aortic dissection under deep hypothermic cardiac arrest (HCA), repeated cardiac procedures, Jehovah's witnesses, accentuated response to antiplatelets drugs like acetyl salicylic acid (ASA), clopidogrel, platelets glycoprotein (GP) Ib/IIIa – tirofiban (Aggrastat), eptifibatide (Integrin) and abciximab (ReoPro), hepatic and renal failure.

The most common hereditary autosomal dominant coagulopathy, seen in 70%-90% of patients, was detected (1924) and described (1926) by **Erik A.** von **Willebrand (1870 – 1949)** — Finnish internist, disease, called after him von Willebrand disease (vWD). It affects 1% of the population, causes the quantative or qualitative defect or absence of a multimeric protein, namely von Willebrand Factor (vWF) which is required for platelets adhesion by binding platelets GPIb as well as GPIIb/IIIa receptors and acting as a carrier protein for antihemolytic factor (AHF) VIII to the injured vascular endothelium. Beside humans it affects dogs, noticeable Dobermann Pinchers.

Type I vWD occurs in 60-80% of all patients, is caused by decreased quantity and quality of vWF, typically asymptomatic or with mild, occasionally prolong nosebleed from the anterior nasal cavity (epistaxis), bleeding from gums, easy ecchymoses, bruising, heavy periods (menorrhagia) and blood loss after trauma, operation, or childbirth.

Type II vWD has four subgroups, occurs in 20%-30% of patients, it is a qualitative disorder and the bleeding tendency can vary between individuals.

The type III vWD is the most severe form with no detectable vWF antigen, low factor VIII, characterized by severe mucosal and joint bleeding, as in cases of mild hemophilia, spontaneous or induced internal hemorrhage. The platelet-type pseudo-vWD is an autosomal dominant trait, caused by gain of function mutations of the vWF GPIb receptor on platelets, large vWF multimers are lost and the ristocetin level is decreased.

The treatment of vWD depends on its type, but in general include the prevention of hemorrhage following major surgery by raising factor VIII level with its concentrate to greater than 50 IU/dl intra- and post-operatively for 10 days, 1-deamino-8-deamino arginine vasopressin (DDAVP) which is however contraindicated in vWD type IIb, cryoprecipitate (factor I, VIII, XIII) and if necessary fresh frozen plasma (FFP) which contains all coagulation factors except for low level of factor V (80%) and factor VIII (60%).

The phenomenon in which the highest frequency component of an aortic stenosis (AS) murmur radiates to the apex of the heart, mimicking mitral regurgitation (MR), was described in 1925, by **Louis Gallavardin (1870 – 1957)** — French cardiologist, and named after him the Gallavardin phenomenon.

In 1910 **Antin Kh. Kakovsky [13(23).01.1871, town of Fastiv, Kyiv Obl., Ukraine – c. 1944]**[904] — Ukrainian internist, in 1925 **Thomas Addis, Jr (1881 – 1949)** — Scottish physician, proposed the method of quantitative estimation of erythrocytes, leucocytes, epithelial and cylindrical cells and the volume of proteins in the total volume of urine using the counting camera of the hemocytometer for a period of 12 or 24 hours for the diagnosis and the effective treatment renal diseases (Kakovsky-Addis count). In addition, Antin Kakovsky designed the examining metal teeth mirror and aseptic throat mirror.

Dmytro D. Pletniov [25.XI(07.XII).1871, village Moskovsky Bobryk, Lebedynsky Region, Sumy Obl., Ukraine – shoot by Russian Communist Fascist NKVS **11.IX.1941, Medvedivsky Lis (MedvedivskyForest) close to town of Orel, Orlovska Obl., Russia]**[905] — Ukrainian theraudant and cardiologist, who studied excision of the stellate ganglion for alleviation of angina pectoris / stenocardia (beginning of 1920s).

In 1909 **Emanuel Libman (1872 – 1946)**[906] — American physician and **H.I. Celler** — American physician, diagnosed subacute bacterial endocarditis, and in 1924, Emanuel Libman and **Benjamin Sacks (1873 – 1939)** — American physician, reported a final stage of chronic disseminated systemic lupus erythematic (SLE), an atypical verrucous nonbacterial endocarditis of the cardiac valves, chordae tendinea and walls with infiltration by fibroblastic and inflammatory cells, characterized by the presence of systolic and diastolic apical murmurs, causing thromboembolisms and rupture of chordae tendinea (Libman-Sacks disease or endocarditis).

In 1905 **John Hay (1873 – 1959)**[907] — English cardiologist, observed bradycardia and cardiac arrhythmia with a constant typically normal time interval between the onset of atrial contraction and the onset of ventricular contraction between 120 ms to 200 ms in duration, without benefit of an echocardiography (ECG), apparently by a careful evaluation of the heart sounds, the heart rate (HR) and pulse.

---

[904] Town of Fastiv (Foot Note 372), Kyiv Obl., Ukraine. — Kakovsky AF. Method of counting organized elements of urine. Russ vrach 1910;9(41):1444. — Ablitsov VH. «Galaxy Ukraine». The Ukrainian diaspora: prominent persons. Publ. «Kyt», Kyiv 2007. — Hanitkevych YaV. The contribution of the Ukrainian physicians to the world medicine. Ukr Med Chasopys 2009 VII/VIII;4(72):110-15. Synyachenko OV. They glorified themselves and Ukraine: history of domestic therapy. Crimean Therapeutic Journal 2010;2(2):21-8.

[905] C. Moskovsky Bobryk (settled since late Paleolithic Age, first mentioned in document 1645), Lebedynsky Region, Sumy Obl., Ukraibe. — Town of Orel (founded 1566), Orlov Obl., (area 24,700 km²), Russia. — Mirsky M. Ukrainian medical scientists – victims of Stalin's repressions. UHMJ «Ahapit» 1995;2:45-54. — Ablitsov VH. «Galaxy Ukraine». The Ukrainian diaspora: prominent persons. Publ. «Kyt», Kyiv 2007.

[906] Libman E, Celler HL. Etiology of subacute infective endocarditis. Am J Med Sc 1910;140:516-27. — Libman E, Sacks B. A hitherto to undescribed form of valvular and mural form endocarditis. Arch Intern Med 1924;33;701-37.

[907] Hay J. Bradycardia and cardiac arrhythmia produced by depression of certain of the functions of the heart. The Lancet. 1906;(1):139-43.

Oleksii K. Shenk [5(17).X.1873, city of St Petersburg, Russia – 24.VII. 1943, city of Moscow, Russia][908] — Russian Ukrainian physician, founded the Actinometry Station in Crimea, Ukraine (1917-24), constructed the fully equip installation needed for the corrective treatment of vertebral column curvatures.

Independently, in 1912 **Henry S. Plummer (1874 – 1936)**[909] — American physician and endocrinologist, and in1919 **Porter P. Vinson (1890 – 1959)**[910] — American surgeon, described a genetic and — nutritional disease due mainly to iron-deficiency, occurring more commonly in woman, characterized by difficulties in swallowing (dysphasia), inflammation of the lips (cheilitis), one or both angles of the mouth (cheilosis) and of the tongue (glossitis), esophageal web or webs, splenomegaly, hypochromic microcytic anemia, and increased risk for the development of esophageal squamous-cell carcinoma with the peak incidence at age over 50-year of life, named Plummer-Vinson syndrome.

Henry Plummer is also known for noticing the separation of the nail from the nailbed which occurs in thyrotoxicosis, a condition due to excessive function of the thyroid gland (TG) and psoriatic arthritis.

William G. MakCallum (1874 – 1944) — Canadian American pathologist and physician, discovered in 1905 functional difference between the thyroid gland (TG) and parathyroid glands (PTG), that muscle seizure (tetany) was due to inattentive removal of the PTG during resection of the thyroid gland (thyroidectomy), and that intravenous (IV) injection of calcium salt will correct the condition. This observation laid the understanding of the calcium role in muscle contraction.

Leonid F. Dmyterko (23.09.1875, city of Odesa, Ukraine – 04.12.1957, city of Odesa, Ukraine)[911] — Ukrainian physician, described the velvety (soft, sounded, deafened) 1st sound (tone) at the apex of the heart and at the cardiological auscultation point of Erb **[Wilhelm Erb (1840 – 1921)**. 19th century. Neurology], not accompanying by other cardiac tones, sign of rheumatic inflammation of the internal layer of the heart (endocarditis) with involvement of the left atrioventricular (mitral) valve (MV) – Dmytrenko sound.

---

[908] Ablitsov VH. «Galaxy Ukraine». The Ukrainian diaspora: prominent persons. Publ. «Kyt», Kyiv 2007.

[909] Plummer HS. Diffuse dilatation of the esophagus without anatomic stenosis (cardio-spasm). A report of ninety cases. JAMA 1912;58:2013-5.

[910] Vinson PP. A case of cardio-spasm with dilatation and angulation of the esophagus. Med Clin North Am 1919;3:623-7.

[911] Dmytrenko LP. Contemporary state of the specific treatment for granular inflammation of the lung. Ukrainian Medical Archives 1928;2(1-2). — Dmytrenko LP. About angina pectoris. Likars'ka Sprava / Vrach delo 1935;11. — Ablitsov VH. «Galaxy Ukraine». The Ukrainian diaspora: prominent persons. Publ. «Kyt», Kyiv 2007. — Synyachenko OV. They glorified themselves and Ukraine: history of domestic therapy. Crimean Therapeutic Journal 2010;2(2):21-8.

In 1906 **Mykola D. Strazhesko [17(29).XII.1876, city of Odesa, Ukraine –
27.VI.1952, city of Kyiv, Ukraine]**[912] — Ukrainians therapeutant and cardiologist,
described «cannon sound» due to a total atrio-ventricular block with the evidence
of congestive heart failure (CHF). Strazhesko «cannon sound» is a harshly reinforced
1st cardiac sound resulting from a total atrioventricular dissociation, which appears
during confluence in time of the atrial and ventricular systole of the heart.

The second time, after **Fedir Mering (1822 – 87**: 19th century. Internal medicine),
Vasyl Obrazcov (1849 – 1920: 19th century. Internal medicine) and **Mykola
Strazhesko**, made a detailed description of coronary artery thrombosis and
established a clinical diagnosis in an alive patient of an acute myocardial infarction
(MI, 1909-10). The third time, **James B Herrick (1861 – 1940)** — American physician,
made a clinical diagnosis of an acute MI due to thrombosis of a stenosed left main
coronary artery (LMCA) which was confirmed by post-mortem examination (1912)**.**

Risking his own life, **Mykola Strazhesko**, investigating reasons for swelling of
the body peoples suffered from arranged by Russian Communist Fascist regime
Holodomor-Genocide 1932-33 in Ukraine, which he treated, proved the possibility
for their recovery by normal feeding and delicately stretched, that they should be
nourished rather in the community shelters, then in hospitals («Question about
pathogenesis of swellings», Likars'ka Sprava / Vrachebnoe delo, Kyiv 1933), and also
supported position of Streptococcal etiology of rheumatic disease (RD, 1934).

The presence of air in the mediastinum (mediastinal emphysema) following
thoracic trauma or spontaneous was clinically diagnosed by **Louis V. Hamman (1877 –
1946)** — American physician, hearing a crunching, rasping sound synchronous with
the heartbeat over the precordium produced by the heart beating against air-filled
spaces (Homman syndrome and murmur, 1939).

Following **Wilhelm Malcz (1795 – 1852**: 19th century. Internal medicine),
independently, **Yakiv L. Okyniewskii (24.XI.1877, village of Trykhatky, Bilhorod-
Dnistrovsky Region, Odesa Obl., Ukraine – 23.IX.1940, city of СПб, Росія)**[913]
— Ukrainian physician-epidemiologist, specialist in military medicine, working in
Military Hospital (1908-12), Odesa, Ukraine, and **Władysław Sterling (1877 –** shoot by
German fascist gestapo **1943)** — Polish neurologist of Jewish descent, re-described

---

[912] Strazhesko ND. Selected Works. Vol. 1-2. Kyiv 1955-56. — Zhukovsky L. From the history of development of
the life-time myocardial infarction diagnosis (on the 75th anniversary of the death of V.P. Obraztsov). UHMJ
«Ahapit» 1995;3:39-45. — Bobrov V. Academician Mykola Strazhesko (1876 – 1952). To the 120-unniversary
from his birth. UHMJ «Ahapit» 1996-1997;5-6):31-6. — Ablitsov VH. «Galaxy Ukraine». The Ukrainian
diaspora: prominent persons. Publ. «Kyt», Kyiv 2007. — Hanitkevych YaV. The contribution of the Ukrainian
physicians to the world medicine. Ukr Med Chasopys 2009 VII/VIII;4(72):110-15. — Synyachenko OV. They
glorified themselves and Ukraine: history of domestic therapy. Crimean Therapeutic Journal 2010;2(2):21-8.

[913] Village of Trykhatky (appeared 1831), Bilhorod-Dnistrovsky Region, Odesa Obl., Ukraine. — Ablitsov VH.
«Galaxy Ukraine». The Ukrainian diaspora: prominent persons. Publ. «Kyt», Kyiv 2007.

a sign symptomatic of louse-born typhus abdominal fever, when patient is unable to move forward his tongue when asked to do so, named Malcz-Sterling-Okuniewski sign.

Poliyen H. Mezernytsky [28.04(10.05).1878, city of Kyiv, Ukraine – 08.10.1943, city of Moscow, Russia][914] — Ukrainian Russian physician-physiotherapeutic, introduced the method for measuring dose of sun radiation in calories (1927-28).

The name of **Werner Schultz (1878 – 1947)** — German physician and hematologist, connected with the triad (Schultz) of jaundice, gangrene of the stomach and leukopenia (agranulocytosis).

A mild-diastolic murmur heard at the apex of the heart during acute rheumatic disease (RD) was described by **Carey F. Coombs (1879 – 1932)** — English cardiologist, named after him Carey Coombs murmur.

In 1927 **Arthur C. Alport (1880 – 1959)** — English physician, identified a rare congenital autosomal dominant or x-connected disease, which involves the connective tissues of the ears, kidneys, and eyes, characterized by neuro-sensory loss of hearing, pyelonephritis, glomerulonephritis, sometimes changes to eye structure, proteinuria, and hematuria, in the presence of renal failure (RF) requiring hemodialysis (HD) and bilateral renal transplantation (Alport syndrome).

The extract of the parathyroid hormone (PTH) from the parathyroid gland (PTG) was isolated in 1923 by **Adolph M. Hanson (1880 – 1959)** — American physician and military surgeon, and in 1925 extracted from the PTG of cattle by **James B. Collip (1892 – 1965)** — Canadian biochemist and endocrinologist.

In 1921 **Abraham L. Levin (1880 – 1940)**[915] — Jewish, Polish American physician, invented a nasal-gastric tube for aspiration of gastric content (Levin Tube). Then, in 1934 **Thomas G. Miller (1886 – 1981)** — American internist, and **William O. Abbott (1902 – 43)** — American gastroenterologist, devised of the two-channel 3 meters (m) long intestinal tube with inflatable balloon at the end, inserted through a nostril, passed through the stomach into the small intestine, used for sampling of gastrointestinal (GI) contest, therapeutic aspiration, to relief intestinal distention, decompression and stenting of the small intestine, and also for the initial treatment of small bowel obstruction (Miller-Abbott tube).

Hakaru Hashimoto (1881 – 1934) — Japanese medical scientist, described chronic lymphocytic inflammation of the thyroid gland (TG) – lymphocytic thyroiditis, an autoimmune disease where one's T-cells attack the cells of the thyroid gland and destroy it (Hashimoto's disease, 1912).

---

[914] Ablitsov VH. «Galaxy Ukraine». The Ukrainian diaspora: prominent persons. Publ. «Kyt», Kyiv 2007.
[915] Levin AL A new gastroduodenal catheter. JAMA 1921;76:1007

Since 1906 Sir **Thomas Lewis (1881 – 1945)**[916] — English cardiologist, corresponded with **Willem Einthoven (1860 - 1927**: 19th century. Physiology) concerning the latter's invention of electrocardiogram (ECG).

In 1908 Sir **Thomas Lewis** and **Arthur S. MacNalty (1880 – 1969)** — Scottish physician, pioneered the clinical use of the ECG, were the first to diagnose irregular cardiac rhythm due to heart block in a man.

In 1913 Sir Thomas Lewis published the textbook on «Clinical Echocardiography». At the start of the WW 1, he directed a study on the condition known as «soldier's heart», and having established it was not a cardiological problem, renamed it as the effort syndrome, wrote the monograph on «The Soldier, Heart and the Effort Syndrome» (1918), devised remedial exercises and allowed many soldiers suffering from this condition to return to duty.

He himself suffered myocardial infarction (MI) at the age of 45 and gave up his 70-cigarette-a-day habit, being one of the first in 1926, to realize that smoking damages the blood vessels. Unfortunately, he died from coronary artery disease (CAD), a result of longstanding effect of cigarette smoking.

In 1914 **Albert Husting (1882 – 1967)**[917] — Belgian physician, performed the first successful non-direct blood transfusion by adding sodium citrate create ($Na_3C_6H_5O_7$) with glucose as anticoagulant to the blood, since $Na_3C_6H_5O_7$ create complexes with calcium (Ca^{++}) ions, thereby prevent blood from clotting.

Volodymyr V. Udovenko (1881, city of Samarkand, Uzbekizstan – shoot by Russian Communist Fascist NKVS 08 or 09.12.1937, **forest massif Sandarmokh, Medvezhyyehirs'k Region, Respublic of Karelia, Russia)**[918] — Ukrainian hygienist, Minister of Health of the Ukrainian National Republic (UNR, 1917-21), who worked up the method of scientific investigation of ventilatory properties of building materials, directed hygiene of the water and water supply, hygienic purification and disinfection of the water, hygiene of dwellings, especially in villages.

---

[916] Lewis T, MacNalty AS. A note on the simultaneous occurrence of sinus and ventricular rhythm in man. J Physiol 1908 Dec 15;37(5-6):445-58.

[917] Van Hee R. The development of blood transfusion: the role of Albert Husting and the influence of World War I. Acta chir Belgica 2015;115(3):247-55.

[918] City of Samarkand (founded 742 BC), Uzbekizstan (area 447,400 km²). — forest massif Sandarmokh, Medvezhyyehirs'k Region, Rebublic of Karelia (area 172,400 km²), Russia, where on the area of 10 hectares in 1930s Russian Communist Fascist shoot 9,500 persons of 58 different nationalities. — Mirsky M. Ukrainian medical scientists – victims of Stalin's repressions. UHMJ «Ahapit» 1995;2:45-54. — Ablitsov VH. «Galaxy Ukraine». The Ukrainian diaspora: prominent persons. Publ. «Kyt», Kyiv 2007.

«The people's physician» **Marian I. Panchyshyn (06.09.1882, city of Lviv, Ukraine** – died from acute MI **09.10.1943, city of Lviv, Ukraine)**[919] — Ukrainian physician-internist, reported on the treatment of concealed tuberculosis of the genitourinary system using quartz lamp rays (1920); described anomalies of the common and external carotid artery (1925); described clinical picture and radiological images of syphilis of the stomach (1925); by combining the clinical, laboratory and radiological data of ulcer / carcinoma of the stomach and duodenum choose the best approach for the treatment (1931).

Demetrius Chilaiditi (1883 – 1975)[920] — Greek Austrian radiologist, revealed the hepato-diaphragmatic interposition of the right bend of the colon (Chilaiditi sign, 1910).

In 1932 **Antoine M.B. Lacassagne (1884 – 1971)** — French physician, discovered that carcinoma of the breast can be induced by female sex hormone (estrogen) injection in male mice, confirmed later (1935) by injection of estrogen compounds into male and female mice. Subsequently, in 1936 the question was rise whether the breast cancer patients would benefit from treatment with hormonal antagonists.

In 1925 **Mykola (Nikolai) A. Shereshevsky (09.XI.1885, city of Moscow, Russia – 1961, city of Moscow, Russia)**[921] — Ukrainian Russian physician

---

[919] Panchyshyn M. Treatment of concealed genitourinary tuberculosis with the quarz lamp rays. Likarskyy Visnyk / Medical Journal (Lviv) 1920;1:1-4. — Panchyshyn M. Contribution to the anomaly of arteria carotis communis et arteria carotis externa.Collection of Mathematic, Natural and Medical Section of Shevchenko Scientific Society (Lviv) 1925;23-24-37-41. — Panchyshyn M, Preszniewski Ye. On the account of syphilis of the stomach. Likars'kyy Visnyk (Lviv) 1925;1:1-3. — Panchyshyn M. On the account of ulcer and carcinoma of the stomach and duodenum. Likarskyy Visnyk (Lviv) 1931;2-3:8-18. Olesnytsky B, Chovhan I. To the memory of Professor Dr Marian Panchyshyn. Man, physician, citizen. / Parfamovych S. Professor, educator, and mentor of growing-up physicians, «by goodness the poured heart» (supplement to the silhouette of Professor Dr Marian Panchyshyn). J Ukr Med Assoc North Am 1964 Jan;11,1(32):37-55. — Pundy P. П. Arrangement of the memory of Professor Dr Marian Panchyshyn. J Ukr Med Assoc North America 1979 Spring;26,2(93):123-4. — Malovanyy AV, Nadraha MS, Vasilieva SV, Kril OM. «The people's physician» Marian Panchyshyn (To 135th Anniversary of his birthday): Bibliographic Index. Lviv National Medical University, Scientific Library. Lviv 2017. — Ablitsov VH. «Galaxy Ukraine». The Ukrainian diaspora: prominent persons. Publ. «Kyt», Kyiv 2007.

[920] Chilaiditi D. Zur Fräge der Leberptose und Ptose in algemeinen in Anschluss und drei Fälle von temporärer, parttieller Leberverlangerung. Fortschritte auf dem Gebiete Räntgenstrahlen 1910-11;16:173-208. — Konvolinka CW, Olearchyk A. Subphrenic Abscess. Curr Probl Surg (Ed MM Rawitch). Years Book Med Publ, Chicago, Jan. 1972:1-52. — Olearchyk AS, Konvolinka KV. Subphrenic abscesses. J Ukr Med Assoc North Am 1972; 19,3 (66): 3-56.

[921] Mykola Shereshevsky parents were Ukrainian Jews by birth from Vinnytsia Obl., Ukraine. He was groundlessly arrested by Russian Communist Fascist NKVS on 02 February 1953 in Moscow, Russia, and accused for «foreign spying and killing patients». Freed after death of Joseph V. Stalin, born Joseb B. dze Jugashvili (06.XII.1878 – 05.III.1953) — Russian Communist Fascist military and political criminal dictator of Georgian descent (1922-53). — Shereshevsky NA. On the problem of association of monstrosity with endocrinopathy. Vestnik endocrinology 1925;1(4):296. — Shereshevsky NA. Clinical Endocrinology. Moscow 1957. — Magalini SI, Magalini SC. Dictionary of Medical Syndromes. 4th Ed. Lippincott–Raven Publ., Philadelphia – New York 1990. — Mirsky M. Ukrainian medical scientists – victims of Stalin's repressions. UHMJ «Ahapit» 1995;2:45-54. — Ablitsov VH. «Galaxy Ukraine». The Ukrainian diaspora: prominent persons. Publ. «Kyt», Kyiv 2007.

endocrinologist of Jewish origin, and in 1938 **Henry H. Turner (1892 – 1970)**[922] — American endocrinologist, described a congenital syndrome of true gonadal dysgenesis, or syndrome of genital dysgenesis in women, associated by short stature, hypogonadism and amenorrhea, ptosis, pterygium, usually left-sided congenital heart disease (CHD) – coarctation of the aorta (COA, 70%), lymph stasis, and skeletal deformities (Shereshevsky-Turner syndrome).

Marc M. Huberhrits [7(19).I.1886, city of Tartu, Estonia – 06.V.1951, city of Kyiv, Ukraine][923] — Ukrainian therapeutant of Jewish descent, who with **Ivan P. Pavlov (1849-1935**: 19th century. Physiology**)**, carried out experimental observations, which laid at the base for their work on «Reflex of Freedom» (1915), made noticeable contribution in studies of cardio-vascular, digestive, and endocrine systems.

In 1921 **Johannes L.A. Peutz (1886 – 1953)** — Netherlandish physician, gave full description, in 1949, **Harold J. Jeghers (1904 – 90)** — American physician, detailed account of the familial (40%-50%), inherited, autosomal dominant trait condition, which begins in adolescence, consisting of benign polyps of the gastrointestinal (GI) tract, mainly of the small intestine (0.5-7.0 cm in diameter), and mucocutaneous melanin hyperpigmentation macules on the nose and oral mucosa, lips and of an outer ear face, often causing severe and recurrent abdominal pain, pallor of the skin due to anemia from the lower GI tract bleeding, small bowel obstruction and sometimes degeneration of polyps into malignancy (Peutz-Jeghers syndrome, 1921, 1949).

On January 01, 1916, **Oswald H. Robertson (1886 – 1966)** — English medical researcher, carried out the first blood transfusion utilizing the blood that has been storied and refrigerated in the refrigerator. During his service in the rank of medical officer of the United States (U.S.) Army Medical Corps during World War I in France he established in 1917 the first blood bank in the «blood depots».

That goiter denoted iodine deficiency was documented by **David Marine (1888 – 1976)** — American pathologist, in 1910 who also proposed in 1917-22 taking iodine to prevent this condition and that resulted in the iodization of the table salt.

Thrombasthenia is a rare inherited bleeding (hemorrhagic) disorder (coagulopathy) due to platelets abnormality with autosomal recessive or autosomal dominant trait or acquired disease, seen in disseminated malignant tumors and uremia, described in 1918 by **Eduard Glanzmann (1887 – 1959)**[924] — Swiss pediatrician, where the

---

[922] Turner HH. A syndrome of infantilism, congenital webbed neck and cubitus valgue. Endocrinology 1938;23:66-74. — Magalini SI, Magalini SC. Dictionary of Medical Syndromes. 4th Ed. Lippincott–Raven Publ., Philadelphia – New York 1990.

[923] City of Tartu (founded 1261), Estonia. — Ablitsov VH. «Galaxy Ukraine». The Ukrainian diaspora: prominent persons. Publ. «Kyt», Kyiv 2007.

[924] Glanzmann WE. Hereditäre hämorrhagische Trombasthenie. Ein Beitrag zur Pathologie der Blutplättchen. Jahrbuch der Kinderheilkunde 1918;88(1-42):113-41.

thrombocyte membrane lacks of glycoproteins (GP) IIb and III type which are receptors for fibrinogen, so that platelets do not aggregate during blood coagulation and after addition of adenosine diphosphatase (ADP). That is characterized by easy bruising and epistaxis, prolonged and excessive bleeding after trauma or surgery (Glanzmann thrombasthenia). Prophylactics include avoidance of antiplatelet drugs, treatment of prolonged or severe bleeding with platelets transfusion.

In 1924, **Woldemar Mobitz (1889 – 1951)**[925] – Russian and German physician, with the aid of an electrocardiography (ECG) classified the two types of II degree atrioventricular (AV) block:

Type 1 Mobitz-Wenckebach AV block where the P-R interval increases gradually until there is a breakdown of AV conduction, which is identical to the II degree AV block described **Karel Wenckenbach (1864 – 1940**: above); and Type 2 Mobitz-Hay AV block where all conducted beats shows a constant, typically normal P-R interval, and conduction to the ventricles at regular interval, which is the type described by **John Hay (1873 – 1959**: above). The type 2 Mobitz-Hay AV block is almost always caused by a disease of the distal / terminal subendocardial ventricular part (His-Purkinje)[926] of the conduction system of the heart (CSH).

There is a limited measurable maximum velocity with a pulsed wave Doppler ultrasound (US) unit, as determined by **Harry Nyquist (1889 – 1976)** — Swiss American electronic engineer, beyond which the signal data and peak velocity cannot be accurately measured (the Nyquist limit).

Volodymyr Ya. Pidhayetsky [24.VII.1889, city of Kam'ianets-Podilsky, Kam'ianets-Podilsky Region, Khmelnytsky Obl., Ukraine – shoot by Russian Communist Fascist NKVS **09.XI.1937, forest masiff Sandarmokh, Medvezh'iehirs'kyy Region, Republik of Karelia, Russia]**[927] — Ukrainian hygienist, organizer of the Institute of Physical Education at the All-Ukrainian Academy of Sciences (All-UAS, 1921) and faculty of Hygiene and Work at the Kyiv University (1923), authored handbook «Hygiene of Work» (Kyiv 1929).

The electrocardiographic (ECG's) QRS complex is a combination of the graphical deflection due to depolarization of the right ventricle (RV) and left ventricle (LV)

---

[925] Mobitz W. Über die unvollständige Störung der Erregungsüberleitung zwischen Vorhof und Kammer des menschlichen Herzens. Z Gesamte Exp Med 1924; 41: 180-237.

[926] **[Jan Purkinje (1787 - 1869)** and **Wilhelm His, Jr (1863 - 1934)**. 19th century. Anatomy, histology, and embryology].

[927] City of Kam'ianets-Podilsky [first mentioned 1062; during organized by Russian Communist Fascist regime of Holodomor-Genocide 1932-33 against the Ukrainian Nation in the city of Kam'ianets-Podilsky died at least 775 residents, and in the Kam'ianets-Podilsky Region (area 4,521.2 km²)] – perished 2,533 people, Khmelnytsky Obl., Ukraine. — Mirsky M. Ukrainian medical scientists – victims of Stalin's repressions. UHMJ «Ahapit» 1995;2:45-54. — Ablitsov VH. «Galaxy Ukraine». The Ukrainian diaspora: prominent persons. Publ. «Kyt», Kyiv 2007.

of the heart which normally in adult last 60-110 millisecond (msec.), in children and during physical activity it may be shorter. In 1949 **Willam Dressler (1890 – 1969)** — Polish American cardiologist, published on a wide QRS complex lasting 120 msec. or more and irregular tachycardia (100 beats/min. or above), the site of its origin may be ventricular or supraventricular, while the incidence of ventricular tachycardia (VT) being about 32% (Dressler beat). Then, in 1955, he noted the post-myocardial-infarction (MI) triad of fever, pleuritic pain, and pericardial effusion (Dressler syndrome)[928].

The histochemical stain and reaction (test) for detection of cholesterol, and cholesterol ester was developed by **Arthur R.H. Schultz (1890 – ?)** — German physician, in which frozen reaction of formalin-fixed tissues are oxidized in iron alum, hydrogen peroxide (H_2O_2), or sodium iodate ($NaIO_3$), then treated with sulfuric acid (H_2SO_4) to give a blue green to red color in a positive reaction. The presence of glycerol inhibit reaction.

In 1921 **Frank N. Wilson (1890 – 1952)** — American cardiologist, and then in 1930 **Louis Wolff (1898 - 1972)** — American cardiologist, Sir **John Parkinson (1885 – 1976)** — English cardiologist, and **Paul D. White (1886 – 1973)** — American cardiologist, described the syndrome of paroxysmal palpitations, in which the electrocardiography (ECG) displays a tachycardia or atrial fibrillation (AF) with pre-excitation, short P-R interval and normal or wide QRS complex with delta waves, due to an accessory anterior pathway between the right atrium (RA) and the right ventricle (RV) of **Albert Kent (1863 – 1958**: 19th century. Anatomy, histology, and embryology**)** or atrioventricular re-entrant tachycardia (AVRT), named pre-excitation of ventricles (PEV) or Wolff-Parkinson-White (WPW) syndrome.

Classic electrocardiographic (ECG) signs of pulmonary embolism (PE) are a deep (large) S wave in lead I, a large Q wave in lead III and an inverted T wave in lead III ($S_1Q_3T_3$)[929].

Following the steps of **Richard Bright (1789 – 1858**: 19th century. Internal medicine), **Harry Goldblatt (1891 – 1977)**[930] — American experimental pathologist of Jewish Lithuanian ancestry, by reducing the blood flow to the kidney during the application of the clamp to the renal artery (Goldblatt clamp) and creating the renal ischemia (Golblatt kidney or phenomenon), with subsequent development of hypertension (HTN) – Goldblatt HTN, he attributed the rise in blood pressure to excessive secretion of the enzyme renin and considered ischemia to be the result of occlusion of the arteriolar vessels, secondary to nephrosclerosis (1934-43). He

---

[928] Dressler W. The post-myocardial-infarction syndrome: a report of fourty-four cases. AMA Intern Med 1959 Jan;104(1):28-42.

[929] McGinn S, White PD. Acute cor pulmonale resulting from pulmonary embolism JAMA 1935;104(17):1473-80.

[930] Goldblatt H. The Renal Origin of Hypertension. Springfield, Ill 1948.

attempted to treat hypertension by using anti-renal therapy, initially via passive immunization. He had mixed results, and the method fell out of favor. However, in 1990, the role for the immune system in the etiology and/or prevention of hypertension was reconsidered.

The intensity of the heart murmur is graded according to the scale developed by **Samuel A. Levine (1891 – 1966)** — American cardiologist of Jewish-Polish origin: grade I - barely audible; grade II – soft but easily heard; grade III - loud; grade IV – associated with palpable thrill; grade V - audible with the stethoscope only partly touching the chest; grade VI - audible without the stethoscope (the Levine scale, 1933).

Andriy A. Zhuravel (04.IV.1892, village Sofiys'k, Ulchs'kiy Region, Khabarovsky krai, Russia – shoot by Russian Communist Fascist regime **20.X.1938, city of Kharkiv, Ukraine)**[931] — Ukrainian physiatrist, showed that the primitive sanitary conditions in Ukraine's prisons caused dissemination of tuberculosis with transmission of the disease to the healthy prisoners, and to the population outside the prisons, and that the regime used this to eliminate patriotic stratum of the Ukrainian people (1930). He was the only known physician who openly urged the regime to implement emergency means to stop the man-made Holodomor-Genocide in Ukraine (1932-33), for which he was executed.

Principal scientific works of **Vadym V. Ivanov [18(30).IX.1892, city of Mariupol, Donetsk Obl., Ukraine – 15.I.1962, city of Kyiv, Ukraine]**[932] — Ukrainian therapeutant, were dedicated to physiology and pathology of digestive tract, clinical picture of and diagnostic of carcinoma of the lung and stomach, proposition for classification of gastric cancer (1932).

Peter G. Sergeev (1893 – 1973) and **Evgenii M. Tareev (1895 – 1986)** — Russian nephrologist, described in 1939 serum hepatitis.

In 1933 **Armand J. Quick (1894 – 1978)**[933] — American physician, introduced hippuric acid test for quantitative evaluation of liver function, and in 1941 a

---

[931] Village Sofiiys'k (founded 1858) on the right bank of the proper Amur River (length 2,824 km, drainage basin 1,855,000 km²), Ulchs'kiy Region, Khabarovsky krai (areaa 788,600 km²), Russia. — Rozhin I. Dozent Doctor Andriy Zhuravel. Memory – Feature. J Ukr Med Assoc North Am 1960;7(17):25-30. / 1960:7(18):37-40. — Mirsky M. Ukrainian medical scientists – victims of Stalin's repressions. UHMJ «Ahapit» 1995;2:45-54. — Ablitsov VH. «Galaxy Ukraine». The Ukrainian diaspora: prominent persons. Publ. «Kyt», Kyiv 2007.

[932] City of Mariupol (Foot Note 783), Donetsk Obl., Ukraine. — Ablitsov VH. «Galaxy Ukraine». The Ukrainian diaspora: prominent persons. Publ. «Kyt», Kyiv 2007.

[933] **Quick's test I:** Quick AJ. The prothrombin in hemophilia and in obstructive jaundice. J Biol Chem (Baltimore) 1935;109: 73-4. — Quick AJ. Margaret Stanley-Brown M, Bancroft FW. A study of the coagulation defect in hemophilia and in jaundice. Am J Med Sci (Thorofare, NJ) 1935; 190:501-11. — **Quick's test II:** Quick AJ. The synthesis of hippuric acid: a new test of liver function. Am J Med Sci (Thorofare, NJ) 1933, 185: 630-635. — **Quick's test III:** Quick AJ, Ottenstein HN, Weltchek H. Synthesis of hippuric acid in man following intravenous injection of sodium benzoate. Proc Soc Exp Biol Med (New York) 1938;38:77-8.

prothrombin time (PT) which is widely used to monitor anticoagulation with sodium warfarin (coumadin). His hippuric acid test was modified in 1941 by **Antin Ya. Pytel (23.I.1902, village Levashy, Lidskyy Region, Grodnenska Obl., Belarus – 26.X.1982, city of Moscow, Russia)**[934] — Ukrainian urologist, to evaluate the antitoxic function of the liver by measuring the acidity of excrete hippuric acid in the urine after the administration of natrium benzoate (Quick-Pytel test).

Augustus R. Felty (1895 – 1964) — American physician, identified atypical rheumatoid arthritis with fever, splenomegaly, leukopenia, and in some patients, anemia, and thrombocytosis (Felty syndrome, 1922-25).

In 1948-49 **Phillip S. Hench (1896 – 1965)** — American physician, reported on a successful treatment of rheumatoid arthritis with the adrenocorticotropic hormone (ACTH) and cortisone.

Work of **Haik Kh. Shakhbazian [09(21).I.1896, village Panik, Shyraks'ka Obl., Armenia – 10.IX.1982, city of Kyiv, Ukraine]**[935] — Ukrainian scientist-hygienist of Armenian descent, were dedicated to problems of the microclimate improvement in manufacturing factories and the influence on workers organism production factors of small intensity.

Risking his own life, **Mykola Strazhesko (1876 – 1952**: above**)**, while examining the cause of swelling of the peoples body suffering during Holodomor-Genocide in Ukraine (1932-33), proved they are caused by starvation, could be cured by normal feeding, and gently hinted they should be rather feed in the public refuges (shelters), than in hospitals («About the question of pathogenesis of swellings», «Likars'ka sprava» Kyiv 1933).

Mykola Strazhesko together with **Volodymyr Kh. Vasylenko [25.V(07.VI).1897, city of Kyiv, Ukraine – 19.XII.1985, city of Moscow, Russia]**[936] — Ukrainian therapeutist, reported the case of metabolic alkalosis in congestive heart failure (CHF) and proposed three-staged classification of CHF, based on clinical and metabolic disturbances of the patient (Strazhesko-Vasylenko Classification of CHF, 1935):

---

[934] Pytel AYa. Functional test of the liver in clinical surgical diseases. Klin med 1945;23(4-5):42. — Ablitsov VH. «Galaxy Ukraine». The Ukrainian diaspora: prominent persons. Publ. «Kyt», Kyiv 2007.

[935] Village Panik, Shirak Province (area 2,680 km^2), Armenia (area 29,743 km^2). — Ablitsov VH. «Galaxy Ukraine». The Ukrainian diaspora: prominent persons. Publ. «Kyt», Kyiv 2007.

[936] Vasylenko VKh. Method of clinical diagnosis. Medychnyy zhurnal 1934;4(2). — Vasylenko VKh. Materials on Metabolism in Chronic Insufficiency of Blood Circulation. Dissertation for Degree of DMS. Kyiv 1941. — Vasylenko VKh. Azote (nitrogen) Metabolism in Chronic Insufficiency of Blood Circulation. Kyiv 1941. — Vasylenko VKh. Acquired Diseases of the Heart. Publ. «Zdorov'ia», Kyiv 1972. — Ablitsov VH. «Galaxy Ukraine». The Ukrainian diaspora: prominent persons. Publ. «Kyt», Kyiv 2007. — Hanitkevych YaV. The contribution of the Ukrainian physicians to the world medicine. Ukr Med Chasopys 2009 VII/VIII;4(72):110-15. — Synyachenko OV. They glorified themselves and Ukraine: history of domestic therapy. Crimean Therapeutic Journal 2010;2(2):21-8.

I degree CHF –clinical symptoms absent in rest, present during physical loading, when appears palpitation, paleness, acrocyanosis, anxiety and weakness.

II-A degree CHF – hemodynamic disturbances in the form of congestion of the right- or left-sided cardiac circulation, compensated at rest: cyanosis, increased respiratory and pulse rate, paste-like subcutaneous and fatty base, swelling of the extremities, increased size of the liver, wet respiratory rales.

II-B degree CHF – visible i hemodynamic disturbances in the form of congestion in the both sides of the cardiac circulation at rest: persistent dyspnea and tachycardia, anorexia, anxiety, apathy, sleepless, sweatiness, cardiomegaly, deadened or deafened sounds of the heart, disturbances in conductive system of the heart (CSH) and its rhythm, increased cardiac silhouette, congestive changes in the lung, increased peripheral edema, hepatomegaly, icteric sclera and skin, decreased urinary output, proteinuria, disturbances in the water-electrolyte and acid-base balance and deadened cardiac sounds.

III degree CHF – visible disturbances of the cardiac hemodynamics, CSH and its rhythm with venous congestion in both sides of cardiac blood circulation, perfusion, and metabolic dysfunction with the presence of ascites, hydrothorax, and generalized swelling (anasarca).

Ivan Mahaim (1897 – 1965) — Belgian French pathologist, and cardiologist, described the congenital anomaly of the conduction system of the heart (CSH) that is the accessory compact atrioventricular node (AVN) - ventricular pathway (Mahaim's fibers, 1937) and the bundle-ventricular pathway (Mahaim fibers, 1937).

Congenital heart disease (CHD) termed double outlet right ventricle (DORV) is characterized by both great vessels, that is the ascending aorta (AA) and the main pulmonary artery (MPA), both with subaortic and sub-pulmonary coni with neither the aortic valve (AV) not the pulmonary valve (PV) are in fibrous continuation with either mitral or tricuspid valve (MV, TV), arise from the right ventricle (RV) and are associated with a ventricular septal defect (VSD).

DORV is classified into:

(1) DORV with subaortic VSD, presented as a child with large VSD (50%) and those with right ventricular outflow tract (RVOT) obstruction presented as tetralogy of Fallot (TOF)[937];

(2) DORV with physiological transposition of the great arteries (TGA) and sub-pulmonary VSD, i.e., VSD straddled by a large MPA, located immediately below both AA and MPA, described in 1949 by **Helen B. Taussig (1898 – 1986)**

---

[937] **Neils Steensen (1638 - 86**: 17th century) / **Etienne Fallot (1850 - 1911**: 19th century. Internal medicine).

—American pediatric cardiologist and **Richard J. Bing (1909 – 2010)** — German American cardiologist of Jewish origin, named Tausig-Bing syndrome (30% of children)[938]. Because of location of VSD where an oxygenated, i.e., saturated with oxygen (O_2) left ventricular (LV) blood preferably streams through VSD into the MPA, while desaturated, i.e., desaturated of oxygen (O_2) RV blood streams into the AA, thereby result in a TGA type of physiology and clinical presentation;

(3) DORV with a double-committed VSD;

(4) DORV with non-committed VSD located below the conal septum.

Absence of subaortic (sub-semilunar) canal myocardium can be associated with double-outlet left ventricle (DOLV) in which there is aortic-mitral and pulmonary-mitral fibrous continuity.

A diastolic murmur heard in stenosis of the left anterior descending (LAD) artery was described by **William Dock (1898 – 1990)** and **Samuel A. Zoneraich (1921 – 2000)**[939] — American cardiologists, and named after the former, Dock murmur.

Direction of scientific search of **Fedir Ya. Prymak [11(23).IV.1899, village Sofiivka, Nizhynsky Region, Chernihiv Obl., Ukraine – 09.IX.1981, city of Kyiv, Ukraine]**[940] — Ukrainian therapeutant, was to study the role of changes in blood content of proteins and vascular wall in pathogenesis of edema; changes of skin capillaries in rheumatic disease (RD), capillaries and vascular bed in anaphylaxis and allergic reactions; histophysiology of lymphatic system of the heart in norm and pathology; diagnostics, clinics and treatment of tuberculosis, hypertension, early sepsis, hypoxic state.

---

[938] Taussig HB, Bing RJ. Complete transposition of the aorta and levoposition of the pulmonary artery: clinical, physiological, and pathological findings. Am Heart J 1949;37:551-9.

[939] Dock W, Zoneraich S. A diastolic murmur arising in a stenosed coronary artery, Am J Med 1967 Apr;42(4):617-9. — Zoneraich S. Evaluation of century-old physical signs, S_3 and S_4 by modern technology. JACC 1992 Feb;19(2):458-9.

[940] Village Sofiivka (Foot Note 864), Nizhynsky Region, Chernihiv Obl., Ukraine – 09.IX.1981, city of Kyiv, Ukraine. — Prymak FYa. Dietetic Treatment of Kidney Diseases. Diplomate Work. Kyiv 1925. — Prymak FYa. About some biochemical changes in organism of different pathological processes. Zhurnal All-Ukrainian Academy of Sciences 1932;1. — Prymak FYa. Terminal Edema of Muscles. Dissertation for degree of DMSc, monography. Kyiv 1936. — Prymak FYa. Main clinical picture of Ukrainian endemic goitre. Medical Zhurnal of the Academy of Sciences of the Ukrainian RSR 1936;1. — Prymak FYa. Changes of blood capillaries of the skin during allergic state. In book, AYa Prymak. Alerhiya. Kyiv 1937. — Prymak FYa. Significance of lymphatic system in clinics of cardiac diseases. Therapevtychnyy Archiv 1938;6. — Prymak FYa. Cardiovascular Failure and Hypoxidosis (hypoxia) of the Internal Pathology. Kyiv 1963. — Ablitsov VH. «Galaxy Ukraine». The Ukrainian diaspora: prominent persons. Publ. «Kyt», Kyiv 2007.

The son of Olexandr Jasynovsky **[Olexandr Jasynovsky (1864 –** after **1899).** 19th century. Surgery]**, Mykhailo O. Yasynovsky (1899 – 1972)**[941] — Ukrainian therapeutant, colonel of the medical service and the main physician of the Black Sea Fleet (1943-46), work of which was dedicated to study diseases of the salivary gland membrane (1931), gastrointestinal (GI) tract, gallbladder and determination of the peculiar development, treatment and anticipation of congestive heart failure (CHF) in rheumatic fever (RF) and rheumatic heart disease (RHD), elaborated approaches to treatment an acute RF, and also prophylaxis of its recurrences (1972). Moreover, he proposed the functional diagnostic method for the use in the evaluation of a health resort treatment, made significant contribution into the field-military and sea-military therapy.

Unfortunately, the Ukrainian and Russian medical literature remains silent about his father, renowned vascular surgeon.

Francis D.W. Lukens (1899 – 1978)[942] — American chemist, physiologist, and physician, together with **Cyril Haugh Long [(1901 – 70).** 20th century. Biochemistry], first reproduced the Houssay phenomenon **[Joseph Balo (1895 – 1979)** and **Bernardo Houssay (1887 – 1971).** 20th century. Physiology] which showed that pituitary gland ablation had influenced the clinical course of diabetes mellitus (DM). Then they showed that removal of the adrenal gland (adrenalectomy) had a similar effect, in that prolonged survival of de-pancreatized (after removal of the pancreas) animal and reduced the degree of hyperglycemia and hyperketonemia. Likewise, they demonstrated that this effect was due to removal of the adrenal gland cortex since it was not produced by removal of the adrenal gland medulla (de-medullation) or de-nervation. These observations proved that both, the pituitary gland, and the adrenal gland cortex had modulating effect of insulin regulation of glucose and lipid metabolism.

Furthermore, Francis Lukens with **William Ch. Stadie (1886 – 1959)** — American physician, established the relationship between ketone body production and rate of ketone body oxidation in de-pancreatized cats and in man, thus helped to establish the view that increased ketone body production is the prominent factor in the development of hyperketonemia in patients with diabetic ketoacidosis.

In cooperation with **Samuel Gurin (1905 – 97)** and **Roscoe O. Brady (1923 – 2016)** — American biochemists, Francis Lukens studied the impairment of hepatic fatty acid synthesis resulting from insulin deficiency.

---

[941] Jasynovsky MO. To the physiology, pathology, and clinics of the salivary membranes. Doctoral dissertation. Publ. «Naukova dumka», Kharkiv 1931. — Jasynovsky MO. Antirheumatic remedies. Publ. «Zdorov'ya», Kyiv 1972. — Ablitsov VH. «Galaxy Ukraine». The Ukrainian diaspora: prominent persons. Publ. «Kyt», Kyiv 2007.
[942] Wood FC. Francis D.W. Lukens. Trans Am Climatol Assoc 1981:lx-lxi.

In addition, Francis Lukens developed standardized method for the experiments on animals with alloxan induced DM; produced persistent hyperglycemia in cat by repeated glucose injection and observed that it produced pathological alternation in the beta-cell structure and function of the pancreas; noted that insulin modified the effect of growth hormone (GH) on nitrogen balance.

Inappropriate sinus tachycardia (IST) is a rare non-paroxysmal cardiac tachyarrhythmia, within the category of supraventricular tachycardia (SVT), described in 1939 by **M.M. Cadvelle,** and **H. Boucher**[943] — French physicians, in a young adult man with a resting heart rate (HR) in the range 160 beats per minute lasting more than 2 years. An abnormally elevated HR at rest is exaggerated with physical activity, typically in young women. It a debilitating condition due to enhanced autonomy of the sinoatrial node (SAN) or hypersensitivity to beta-adrenergic stimulation. The SAN impulse may have a multifocal origin, and early activation sites can shift within the SAN complex in response to automatic influences.

Symptoms associated with IST include frequent sustained palpitations, dyspnea (shortness of breath), presyncope (feeling about to faint), physical fatigue, dizziness, exercise intolerance, occasional paresthesia and cramping, and symptoms associated with autonomic nervous system disturbances, including gastrointestinal problems.

The initial therapy is with beta-blockers, calcium channel blockers, or by pacemaker or funny electric current (ivabradine). Catheter radiofrequency (RF) ablation of the SAN has had a high percentage of recurrences at long-term follow-up.

Surgical treatment with RF ablation of the SAN is performed through a median sternotomy incision under cardiopulmonary bypass (CPB) or mini-invasive cardiac surgery (MICS) via a 6 cm mini-thoracotomy sub-mammary incision without CPB[944]. The right atrium (RA) is separated from the adjoined junction with the superior vena cava (SVC), after which the SAN and surrounding RA tissue are isolated by positioning the bipolar RF device-clamp in oblique fashion from the right to left, following by an application of three sequential electric charges. The RF isolation is confirmed with significant blunting response to intravenous (IV) administration of isoproterenol. Bipolar pacing wire above and below the RF ablation line confirms that the SAN and surrounding RA wall are isolated from the remainder of the heart. Surgical RF ablation of the SAN may necessitate of implantation of a permanent artificial pacemaker.

---

[943] Cadvelle MM, Boucher H. Tachycardia sinusale permanente a houte frequence sans troubles fonetionnels. Bull mem soc hosp Paris 1939;54:1849-52.

[944] Melby SJ, Kreisel D, Lindsay BD, et al. Surgical treatment for inappropriate sinus tachycardia. Heart Rhythm 2005;2(5):S108.

In 1950 **Vinzento Monaldi (1899 – 1969)** — Italian physiologist and physician suggested draining the thoracic cavity with more superior approach at the 2nd or 3rd intercostal space (ICS) in the midclavicular line (MCL).

The risk of radiation-induced malignancy, including radiation-induced sarcoma (RIS) of the breast, is extremely low for the general population. Nonetheless, the use of radiotherapy to treat benign lesions in patients with factors known to predispose to malignant transformation, such as fibrous dysplasia (Albright-McCune syndrome, 1937)[945], which consist of an autonomous endocrine hyperfunction, that is precocious puberty, polycystic fibrous dysplasia and unilateral pigmented birth skin marks (café-an-lait spots), should be considered carefully.

William M. Dameshek (May 22, 1900, city of Voronezh. Russia – died next day after repair of unsuspected dissected thoracic aortic aneurysm **Oct. 06, 1969, Boston, MA, USA)**[946] — American hematologist, used chemotherapy with nitrogen mustard in various malignant diseases of the blood (1949); developed the concept about myeloproliferative diseases, that is the medullary or extramedullary proliferation of the ingredient parts of the bone marrow, such as erythroblasts, granulocytes, megacaryocytes and fibroblasts (1951); described chronic lymphocytic anemia. **Theodore H. Spaet (1920 – 92)** — American hematologist, with the former noted the coexistence of the previous infected postoperative skin incision with scar formation and later repeated severe and prolonged infections of the skin, mouth, throat, ears, paranasal sinuses and lungs in the adolescent and adult (Spaet-Dameshek syndrome, 1952).

At the early stage of WW II, beginning in early July 1940, when the Battle for Britain has begun, **John Scudder (1900 – 76)** — British physician, developed the plasma for Britain program, needed to supply the colloid as a blood substitute, which is plasma, for the replacement of the blood volume in injured soldiers and civilians.

Eugene (Gene) M. Landis (1901 – 87) — American physician and vascular physiology researcher, was the first who in 1926 accurately measured the capillary blood pressure (BP).

Francis C. Wood (1901 – 90) — American physician, who contributed to the collaborative effort to develop the chest leads for electrocardiography (ECG).

In practice, the concept of barotherapy (hyperbaric oxygenation) was reintroduced in 1956-60 by **Ite Boerema (1902 – 80)** — Dutch cardiac surgeon, and **Willem H. Brummelkamp (1928 – 2010)** — Dutch general surgeon, when they reported on the use of high atmospheric pressure as an aid to cardiac surgery

---

[945] **Fuller Albright (1900 - 60)** — American endocrinologist. / **Domovan J. McCune (1902 - 76)** — Americaan pediatrician.

[946] City of Voronezh (founded 1580), Russia. — Dameshek W. Chronic lymphocytic leukemia – an accumulative disease of immunologically incompetent lymphocytes. Blood 1967;29:566-84.

Treating tuberculosis patients, **Jaros Veres (1903 – 79)**[947] — Hungarian internist, designed in 1932 a spring-loaded needle to create pneumothorax on the side of an infected lung to collapse it and allow lesion to heal (Veres needle).

The work of **Wasyl L. Pluschtch (Dec. 28, 1902 / Jan. 10, 1903, city of Warsaw, Poland – Nov. 16, 1976, city of München, Germany)**[948] — Ukrainian physiatrist, was dedicated to study and treatment tuberculosis, about social hygiene, history of Ukrainian medicine and health care, and organized by Russian Communist Fascist USSR regime of Holodomor-Genocide 1932-33 against the Ukrainian Nation.

Ian J. Wood (1903 – 86) — Australian physician and medical scientist, invented the gastric suction tube (1949), **Margo Shiner (1923 – 98)** – an intestinal biopsy tube (1958).

---

[947] Veres J. Neues instrument zur ausfuhrung von brust- oder bauchunktionen und pneumothorax behandlung. Deutsche Med Wochenschrift 1938;64:1640-1.

[948] City of Warsaw (Foot Note 299), Poland. — City of München (first mentioned in 1158), Germany. — Pluschtch WL. Artificial Pneumothorax. Kyiv 1928. — Pluschtch WL. Late Outcomes after Artificial Pneumothorax. Kyiv 1934. — Pluschtch WL. Handbook on the Sanatorium Treatment of Tuberculosis. Vol. 1-2. Moskva 1938. — Pluschtch WL. Clinics of Disseminated Haematogenous Tuberculosis of the Lungs. Kyiv 1939. — Pluschtch WL. Clinics of Disseminated Haematogenous Tuberculosis. Kyiv 1941. — Pluschtch WL. Medical education and science in Ukraine. J Ukr Med Assoc North Am. 1954 May;1,1(1):4-9. — Pluschtch WL. Health Welfare in Ukraine. München 1956. — Plushtch WL. Ukrainian physician scientist of the 18th century. J Ukr Med Assoc North Am 1958 May;5,(10):7-12. — Plushtch WL. Medicine at the time of Hetman Ivan Stepanovych Mazepa (1639-1709). J Ukr Med Assoc North Am 1959 Sep;6,(16):12-7. — Plushtch WL. Treatment of pulmonary tuberculosis. J Ukr Med Assoc North Am. 1961 Jan;8,(20):1-7. — PlushtchWL. Medicine in the time of Taras Hryhorovych Shevchenko. J Ukr Med Assoc North Am 1962 Apr;9,2(25):6-20. — Pluschtch WL. 3rd Conference of phthisists of the Ukrainian S.S.R. J Ukr Med Assoc North Am 1962 Apr;9,2(25):29-34. — Pluschtch WL. Principles and methods of contemporary treatment of tuberculosis. J Ukr Med Assoc North Am 1963 Jan;10,1(28):2-8. — Pluschtch WL. Contemporary Ukrainian poet-physician Vitaliy Korotych. J Ukr Med Assoc North Am 1963 Oct;10,4(31):37-46. — Pluschtch WL. Contemporary medical care and medical education in Ukraine. J Ukr Med Assoc North Am 1964 Jan;11,1(32):5-10. — Pluschtch WL. New anti-tuberculous drugs. J Ukr Med Assoc North Am 1964 Oct;11,3-4(34-35):2-7. — Pluschtch WL. Tuberculosis in the aged (Part I). J Ukr Med Assoc North Am 1966 Oct;13,3-4(42-43):3-13. — Pluschtch WL. Tuberculosis in the aged (Part II). J Ukr Med Assoc North Am 1967 Jan;14,1(44):3-13. — Pluschtch WL. Establishment by the Ukrainians of medical education in Russia (Part I). J Ukr Med Assoc North Am 1967 July;14,3(46):14-7. — Pluschtch WL. Establishment by the Ukrainians of medical education in Russia (Part II). J Ukr Med Assoc North Am 1967 Oct;14,4(47):23-7. — Pluschtch WL. Present day health care in Ukraine. J Ukr Med Assoc North Am 1969 Jan-Apr;16,1-2(52-53):38-49. — Pluschtch WL. MartynTerekhovsky. J Ukr Med Assoc North Am 1970 July;17,3(58):50-5. — Pluschtch WL. Determination of activity of pulmonary tuberculosis. J Ukr Med Assoc North Am 1970 Oct.; 17,4(59):3-12. — Pluschtch WL. Outline of the History of Ukrainian Medicine and Education. Book I (since the beginning of the Ukrainian Statehood to the 16th century). Publ. Ukrainian Free Academy of Sciences (UFAS) in Germany, München 1970. — Pluschtch WL. History of Ukrainian medicine. J Ukr Med Assoc North Am 1972 Oct;19,4(67):53-61. — Pluschtch WL. My experience in the treatment of tuberculosis. J Ukr Med Assoc North Am 1973 Jan-Apr; 20,1-2(68-69):26-43. — Pluschtch WL. Mord am Ukrainischen Volk. München 1973. — Pluschtch WL. Genocide of the Ukrainian People. München 1973. — Pluschtch WL. Collection of Materials on History of Ukrainian Medicine. Vol. I. Publ Ukr Med Assoc North Am, New York - München 1975. — Ablitsov VH. «Galaxy Ukraine». The Ukrainian diaspora: prominent persons. Publ. «Kyt», Kyiv 2007.

During WW II in the late 1940s **Charles R. Drew (1904 – 50)** — African American physician, surgeon and medical researcher, set up the first blood banks and convinced physicians to use blood plasma for the battlefield and other emergency transfusions.

In 1964 **Jean Lenegre (1904 – 72)**[949] — French cardiologist, and **Maurice Lev (1908 – 94)**[950] — American pathologist, described an acquired complete heart block (CHB) due to idiopathic fibrosis and calcification of the electrical conduction system of the heart (CSH), commonly seen in the elderly as senile degeneration of the CSH (Lenegre-Lev disease or syndrome).

Taro Takemi (1904 – 83) — Japanese physician, constructed in 1937 the first portable echocardiograph (ECG) and invented in 1939 the vector cardiogram.

Pulse oximetry (%SpO$_2$), developed by **Karl Matthes (1905 – 62)** — German physician, in 1935, and by **Glen A. Milligan (1906 – 47)** — American physiologist, in 1940-43, is a noninvasive method for monitoring a person's oxygen pulse saturation (%SpO$_2$) of the arterial blood by mounting a sensor pulse with plethysmogram on the palmar surface of the distal finger or toe, on the forehead or behind the ear; or by the invasive route, i.e., percutaneous (PC) insertion of a small bore catheter into the lumen of the patient's radial artery or common femoral artery (CFA).

Capnography [**John Tyndall (1820 – 93**: 19th century. Internal medicine] is useful in evaluation of pulmonary diseases, during anesthesia and in intensive care units (ICU). In the presence of most lung disease (bronchitis, emphysema, asthma) and some form of cyanotic congenital heart disease (CHD) the difference between the content of arterial and expired carbon dioxide (CO$_2$) increases and can exceed 1kPa units[951]. It has been found to be more useful than clinical judgment alone in the early detection of adverse respiratory events, such as hypoventilation, esophageal intubation, and ventilator circuit disconnection, thus avoiding or preventing patient injury, i.e., during surgical procedures under sedation provides more useful information, e.g., on the frequency and regulation of ventilation then pulse oximetry. Capnography provides a rapid and reliable method to detect life-threatening conditions, like malposition of tracheal intubation and tracheostomy tubes, unsuspected respiratory or circulatory failure and defective breathing circuits.

Thus, pulse oximetry together with capnography prevents potentially irreversible patient injury of 93% avoidable anesthesia mishaps.

---

[949] Lenegre J. Etiology and pathology of bilateral bundle branch block in relation to complete heart block. Prog Cardiovasc Dis 1964 Mar;6:409-44.

[950] Lev M. Anatomic basis for atrioventricular block. Am J Med 1964 Nov;37:742-8.

[951] The kPa (Pascal) unit – the SI derived unit of pressure used to quantify internal pressure, is defined as one newton per square metre. — **Blaise Pascal (1623 - 62**: 17th century). — **Isaak Newton (1642 - 1726/27**: 17th century).

Dmytro I. Panchenko (24.II .1906, u.t.v. Novhorodka, Novhorodkivsky Region, Kirovhradska Obl., Ukraine – 22.IX.1995 city of Kyiv, Ukraine)[952] — Ukrainian neurologist, described the combined involvement by the arteriosclerosis (AS) of the carotid, coronary and femoral-popliteal arteries (1941); introduced barotherapy inside the hermetic camera with the artificial environment for the treatment of primary hypertension (HTN), neuropsychiatric disturbances, expiratory bronchial asthma, endarteritis obliterans, severe infections, air embolism gas gangrene and carbon dioxide (CO_2) poisoning – biotron Panchenko (1952-64).

Macroglobulinemia of **Jan G. Waldenström (1906 – 96)** — Swedish physician-oncologist, or «hyper viscosity syndrome» (1944) is disease of plasmatic cell, like leukemia, with cells of lymphocytic, plasmocytic or intermediary morphology, that synthesize the M-component IgM, causing diffuse infiltration of the bone marrow, spleen, liver, lymphatic nodes, and central nervous system (CNS). Circulating macroglobulin are responsible in older people for symptoms of supra sluggishness syndrome (weakness, tiredness, disorders of blood coagulation and disarranged vision). The blood and urine contain usually the **Bence Jones (1814 – 73**: 19th century. Internal medicine**)** proteins.

A rare autosomal recessive bleeding (hemorrhagic) thrombolytic disorder (coagulopathy), called hemorrhage thrombolytic dystrophy, described in 1948 by **Jean A. Bernard (1907 – 2006)** — French physician and hematologist, and **Jean P. Soulier (1915 – 2003)** — French physician and hematologist, is characterized by the presence of gigantic platelets with membranes deficient in the glycoprotein (GP)Ib, a possible receptor for the plasma von Willebrand factor (vWF) which prevents platelets from fixation of a factor necessary for their binding with the subepithelial layers of the blood vessels, causing purpura, moderate to long-lasting cutaneal-serous

---

[952] U.t.v. Novhorodka (founded 1768; as a result of defeat of the liberation struggle from Russian tyranny in the beginning of February 1918 Russian Communist Fascist regime was established and began black streak in history of villages and its residents – repressions 1918-21, Holodomor-Genocide 1921-22, destruction of church and provincial education, liquidation of wealthy farmers and collectivization of 1929-31, then Holodomor-Genocide 1932-33 during which died at least 131 residents of village, repressions 1937-38, World War II (1941-44), repressions 1944-52, Holodomor-Genocide 1946-47), Novhorodkivsky Region, Kirovhradska Obl., Ukraine. — Panchenko DI. About combined affection of vessels of the brain, lower extremities, and the heart. Klin med 1940;7-8. — Pantschenko DI. Von der kombinierten Affection der Hirn- und Herzgefässe und der Gefässe der unteren Extremitäten. Z Ges Neurol Psychiat 1941;171:395. — Babych I. Ukrainian biotron. Ukraïna 1960;17(272):2. — Panchenko DI. Treatment of patients with hypertensive disease in the biotrone. Publ. «Derzhmedvydav», Kyiv 1962. — Panchenko DI, Isakov YuA, Lukashevych LP. Medical-biological significance of biotrone and its physical-technical resources. Publ. «Derzhmedvydav», Kyiv 1964. — Ablitsov VH. «Galaxy Ukraine». The Ukrainian diaspora: prominent persons. Publ. «Kyt», Kyiv 2007. — Synyachenko OV. They glorified themselves and Ukraine: history of domestic therapy. Crimean Therapeutic Journal 2010;2(2):21-8. — Zozula IS. Scientist, pedagogue, physician: towards 105-years from the birth date of Dmytro Ivanovych Panchenko. Ukr med chasopys 2011;1(81):1.

and visceral hemorrhages, that may require platelets transfusion (Bernard-Soulier syndrome).

Under the leadership of **Vasyl P. Komisarenko [01(14).I.1907, village Cherniakhiv, Chaharlyts'kyy, Kyïvs'ka Obl., Ukraïna – 07.IV.1993, city of Kyïv, Ukraïna]**[953] — Ukrainian pathologist and endocrinologist, the studies were conducted on the adrenal cortex hormones and their role in physiological and pathological processes of the organism (1956); the drug «splenin» was synthesized (1961); investigation were performed on the adrenal cortex inhibitors (1972) and molecular mechanisms of action of steroid hormones (1986).

In 1936 **Hans H. B. Selye (1907 – 82)**[954] — Austrian Slovakian Canadian endocrinologist, derived a concept of a general adaptation and stress theory from the **Olexandr Bohomolets (1881 – 1946**: 20th century. Pathology**)** theory of stress.

Paul H. Wood (1907 – 62) — West Indies, Australian, New Zealander and British cardiologist, proposed the indexing of the measured systemic vascular resistance (SVR) and pulmonary vascular resistance (PVR) in $mmHg/L/min./m^2$ into units, that is Wood units (WU) which found an application in pediatric cardiology. Maximal normal PVR is 3-4 WU. As a general rule, advanced (irreversible) pulmonary hypertension, characterized by PVR of 8-12 WU/m^2 with pulmonary to systemic arterial (Qp:Qs) blood shunt less than 1.2:1, despite aggressive vasodilator treatment, is contraindicated for the ventricular septal defect (VSD) closure. To safely close VSD, the PVR must fall below 7 WU. Shunt fraction during cardiac catheterization (CC) = aortic oxygen saturation (O_2%) – right atrium oxygen saturation (RA O_2%) / pulmonary veins oxygen saturation (PV O_2%) – pulmonary artery oxygen saturation (PA O_2%), or in the operating room (OR) = [1– superior vena cava oxygen saturation (SVC O_2%)] / 1 – PA O_2%). PVR = [pulmonary artery pressure (PAP) (mean) – pulmonary capillary wedge pressure (PCWP)] / cardiac output (CO) in WU. In patients with elevated PVR, CC should be carried out under controlled conditions with general anesthesia to manipulate level of carbon dioxide (CO_2). This is particularly important for patients with whom other sources of pulmonary hypertension may be addictive, as in the Down syndrome

---

[953] Village Cherniakhiv (Early-Slav Mound of the origin and name for the «Cherniakhivsky Culture» 2nd-4th century, written data from 1781 and 1794), Chaharlytsky Region, Kyiv Obl., Ukraine. — Komisarenko VP. Hormones from the cortex of the suprarenal glands and their role in physiological and pathological processes of an organism (Kyiv 1956). — Komisarenko VP. Splenin (Kyiv 1961). — Komisarenko VP. Inhibitors of the suprarenal gland function (1972) — Komisarenko VP, Demchenko VN. Quantitative designation of the 8 more important fractions of 17-ketosteroids with the help of two-dimensional thin-layer chromatography in urine of a man. Lab delo 1974;2:73-4. — Komisarenko VP. Molecular mechanism for action of steroid hormones (1986). — Ablitsov VH. «Galaxy Ukraine». The Ukrainian diaspora: prominent persons. Publ. «Kyt», Kyiv 2007. — Synyachenko OV. They glorified themselves and Ukraine: history of domestic therapy. Crimean Therapeutic Journal 2010;2(2):21-8.

[954] Selye H. The Stress of Life. McGraw-Hill, New York 1956.

[John Down (1828 – 96). 19th century. Internal medicine] patient with upper airway obstruction. Positive response and vasoreactivity is defined as a decrease in the systolic pulmonary artery pressure (PAP) of 20% or more with a fall to near normal levels (<40 mmHg), no change or increase in cardiac index (CI), no change or increase in PVR/SVR index, and normal RA pressure and CI.

Described by **Myron Prinzmetal (1908 – 87)** — American cardiologist, a variant cardiac chest pain or angina inverse, is a syndrome of angina pectoris at rest that occurs in cycles, caused by vasospasm, i.e., a narrowing of the coronary arteries due to contraction of the smooth muscle tissue in the vessel walls rather than by arteriosclerosis (AS), affecting more often younger women (Prinzmetal angina, 1959). Physical activity is well tolerated and do not cause angina. The attack is associated with the ST segment elevation, rather than depression, on the electrocardiography (ECG).

The intracardiac shunts, left-to-right, right to left or bidirectional, occur when the blood flow in the heart enters another chamber without traversing a cardiac valve. The diagnostic selective coronary angiography (SCA) and right heart catheterization (RHC) are necessary for detection, localization, and estimation of the shunt size.

In the left-to-right shunts the oxygenated arterial blood traverse the atrial septal defect (ASD), patent forament ovale (PFO), ventricular septal defect (VSD) and patent ductus arteriosus (PDA), causing a step-up in an average (mixed) oxygen (O_2) content (saturation) of the blood – SO_2% (normal 95-100%) at the different level in the recipient chamber which contain the less oxygenated venous blood during the RHC. To estimate a shunt it is necessary to obtain by the oximetry the most mixed venous blood sample. For the ASD and PFO this can be obtained in the right atrium (RA) as a combination of the superior and inferior vena cava (SVC, IVC) venous blood sample, for the VSD in the right ventricle (RV), and for PDA in the pulmonary artery (PA).

In 1947 by **Lewis Dexter (1909 – 95)**[955] — American cardiovascular physiologist and clinician, has determined the maximal acceptable increase in the SO_2% of the venous blood from the SVC to the PA to be 8%. Any value higher than that require to qualify the shunt by calculation of the pulmonary blood flow (Q_P) and the systemic blood flow (Q_S) which are necessary for estimation of the left-to-right shunt (Q_P/Q_S) ratio according to the following equations:

---

[955] Dexter L, Haynes FW, Burwell CS, et al. Studies of congenital heart disease. I. Technique of venous catheterization as diagnostic procedure. J Clin Invest 1947 May 26(3):547-53. — Dexter L, Haynes FW, Burwell CS, et al. Studies of congenital heart disease. II. The pressure and oxygen content of blood in the right auricle, right ventricle, and pulmonary artery in control patients, with observations on the oxygen saturation and source of pulmonary «capillary» blood. J Clin Invest 1947 May;26(3):554-60.

$$Qp = O_2 \text{ consumption (mL/minute)} / [(P_VO_2) - (P_AO_2)],$$

where P_VO_2 – pulmonary vein O_2 concentration, P_AO_2 – pulmonary artery O_2 concentration

$$Qs = O_2 \text{ consumption (mL/minute)} / [(S_AO_2) - (M_VO_2)],$$

where S_AO_2 – systemic artery O_2 concentration, M_VO_2 – the mixed venous O_2 concentration (in the RA).

Therefore, for the left-to-right shunt the equation is:

$$Q_P/Q_S = (S_AO_2) - (M_VO_2) / (P_VO_2) - (P_AO_2)$$

In childhood, **Volodymyr O. Nehovsky / Wolodymyr O. Nehowskyj, or Vladimir A. Negovskiy (06/19.III.1909, u.t.v. of Kozelets, Chernihiv Obl., Ukraine – 02.VIII.2003, city of Moscow, Russia)**[956] — Ukrainian Russian physician and pathologist, creator science of revival (reanimation) of an organism, i.e., respiration, heart and brain, or the cardio-pulmonary resuscitation (CPR; in German – die Herz-Lungen-Wiederbelebung), became sick from bone tuberculosis, underwent long treatment with bed rest (BR) in one of the hospital in Kyiv, Ukraine, where he became enchanted by medicine, decided to be physician, and his patients will not be dying.

After recovery, and graduation from a high school in Kyiv, he studied in 1928-33 at the Medical Faculty of the 2nd Moscow Medical Institute (founded 1906), and at his

---

[956] U.t.v. of Kozelets first mentioned 1098; during organization by Russian Communist Fascist of Holodomor-Genocide 1932-33 against the Ukrainian Nation died at least 145 peoples), Chernihiv Obl., Ukraine. — Negovsky VA. Resuscitation after lethal blood loss. Bull Exp Biol Med 1938;16:35-6. — Negovsky VA. Therapy of Agonal States and Clinical Death in an Army Region. Medgiz, Moscow 1945. — Negovsky VA. Some physio-pathologic regularities in the process of dying and resuscitation. Circulation 1961;23(3):452-7. — Negovsky VA. Resuscitation and Artificial Hypothermia. Consultant Bureau, New York 1961. — Grenda J, Olearczyk A. Postępowanie przy resuscytacji. Wiad Lek 1965;28, 21a:19-24. — Grenda J, Olearchyk A. Ergebnisse der Resuszitation. Sym Anaesth Int'l. Abst. Praha, Soc Anaesth, Sec Med Bohoslov Nom JE Purkyne Insign 17-20.VIII.1965; (4-30):414-5. — Grenda J, Olearchyk AS. Results in resuscitaion. J Ukr Med Assoc North Am 1975;22,1(76):26-8. —Negovsky VA. Reanimatology today: some scientific and physiologic consideration. Crit Care Med 1982;10 (editorial):130-3. — Negovsky VA, Gurvitch AM, Zolotokrilina ES. Post-resuscitation Disease. Elsevier, Amsterdam 1983. — Negovsky VA. Reanimatology: the science of resuscitation. Prehospital Disaster Medicine 1985;1:1-3. — Negovsky VA. Post-resuscitation disease. Crit Care Med 1988;16:942-6. — Negovsky VA. Reanimatology as a neurological science Minerva Anesthesiol 1994;60:478-82. — Negovsky VA. Neurologic stage of reanimatology. Resuscitation. 1995;29:169-76. — Safar P. Vladimir A. Negovsky the father of «reanimatology». Resuscitation. 2001 June 01:49(3):223-9. — Kovalchuk LYa. Anesthesiology, Reanimation and Intensive Therapy of Emergency Conditions. Publ «Ukrmedknyha», Ternopil 2003. — Ablitsov VH. «Galaxy Ukraine». The Ukrainian diaspora: prominent persons. Publ. «Kyt», Kyiv 2007. — Hanitkevych YaV. The contribution of the Ukrainian physicians to the world medicine. Ukr Med Chasopys 2009 VII/VIII;4(72):110-15.

«Alma-Mater» in the beginning worked as physician in a general practice, and then became scientist co-worker in physio-pathology laboratory of the Central Institute of Hematology and Blood Transfusion in Moscow (founded 1926), took part in all CPR experiment.

In his dissertations for a CMSc (1942) and DM Sc (1943) uttered general principles of pathology in terminal stages and principles of the complex method for the CPR, i.e., artificial ventilation of the lungs, indirect or direct cardiac massage, defibrillation of the cardiac ventricles in ventricular fibrillation (VF), intraarterial or intravenous (IV) infusion of the whole blood mixed with glucose, adrenaline, natrium bicarbonate ($NaHC_3$) for correction of acidosis, hydrogen peroxide (H_2O_2) as additional source of oxygen (O_2) for the body.

During World War II (1941-45) he organized brigade and as a part was driven in the ambulance to the active army and was the first to achieve the sterling revival by CPR of approximately 50 wounded soldiers who were in the terminal stage or clinical death.

On his account of in 1952-58 were created the mobile CPR brigades, equipped with ventilators, blood transfusion sets and defibrillators.

Beginning since 1960s he published concerning post-cardiac arrest or post-CPR syndrome, the abnormal physiological state which occurs when whole-body ischemia is followed by whole-body reperfusion. This is a systemic inflammatory state resembling other form of vasodilatory response seen in shock. The degree of organ dysfunction depends on the sensitivity of body organs to ischemia, the most sensitive being brain, and the duration of ischemic time.

Since 1958 **John Lukoudis (1910 – 80)** — Greek physician, who suffered from peptic ulcer disease and suspecting it is caused by a bacteria, treated himself with antibiotics, and when he noted improvement, he similarly treated his patient who suffered from this disease. This was before the Gram-negative bacteria Helicobacter pylori was re-discovered in 1982 by **J. Robin Warren (1937 –)** — Australian pathologist and **Barry J. Marshall (1951 –)**[957] — Australian physician, as a causative factor for peptic ulcer disease, which is now treated with per oral doses of the two of four of the following antibiotics: clarithromycin, amoxicilin, tetracycline or metronidazole.

Scientific direction of **Oleksandr S. Mamolat (12.IX.1910, village Kalnibolota, Holovanivsky Region, Kirovohradska Obl., Ukraine – 24.X.1991, city of Kyiv,**

---

[957] Marshall BJ, Warren JR. Unidentified curved bacilli on gastric epithelium in active chronic gastritis. Lancet 1883 June;321(8326):1273-5. — Marshall BJ, Warren JR. Unidentified curved bacilli in the stomach of patients with gastritis and peptic ulcerations. Lancet 1984 June; 323(8390):1311-5. — Lytvyak E, Jessop P, Troshyn NO. Helibacter pylori as cause of gastric ulcer: current thinking. J Ukr Med Assoc North Am 2013 Dec 27;53,1(157):18-34.

Ukraine)[958] — Ukrainian physician-physiatrist, Director (1936-79) of F.H. Yanovsky National Institute of Physiatry and Pulmonology (NIPhP, founded 1922), National Academy of Medical Sciences (NAMSc) of Ukraine, concentrated on epidemiology and statistics of tuberculosis, organization of prophylactics and treatment of tuberculosis in villages.

The first M-mode of medical ultrasonography (USG) and echocardiography – transthoracic echocardiography (TTE) was developed in 1953 by **Inge G. Edler (1911 – 2011)**[959] — Swedish cardiologist and **Carl H. Hertz (1920 – 90)** — German physicist, marking the beginning of a new diagnostic noninvasive technique. The former used it for the preoperative study of the mitral valve (MV) stenosis and MV regurgitation, influenced its use in neurology and obstetrics. Further progress in this field lead to the development of 2-dimentional (2-D), contrast, and transesophageal TEE. The later was introduced adult in 1976, in children in 1989. Advances in transducer technology and image processing in 2008 by **Lissa Sugeng (1966 –)** — American cardiologist, allowed real-time three-dimensional (3-D) TEE[960].

The use of fetal echocardiography dates to the late 1960s when continuous wave Doppler [**Christiaan Doppler (1803 – 53)**. 19th century. Physiology] was used to record fetal heart rate (HR) and to the late 1970s when **Charles S. Kleinman**[961] — American pediatric cardiologist, had some success in detecting congenital heart disease (CHD) by cardioechography in the M-mode. However, only on in the mid-1980s when high-resolution 2-D TTE became available that accurate delineation of CHD became clinically routine[962].

The acquired von Villebrand's disease (vWD; above) can occur in patients with autoantibodies where the function of vWF is not inhibited, but the vWF-antibody complex is rapidly cleared from the circulation, like that described by **Edward C.**

---

[958] Village Kalnybolota (beginning from the arrangement of the Horse hussar regiment 1752-64, continuing settlement since 1772), Holovanivsky Region, Kirovohradska Obl., Ukraine. — Mamolat OS. Elaboration and Scientific Generalization of Methodology and Organization of Treatment Patients with Destructive Tuberculosis. Dissertation for degree DMSc., Kyiv 1965. — Mamolat OS (Ed.). Tuberculosis. 1971. — Mamolat OS (Ed.). Treatment of Pulmonary Tuberculosis 1973. — Zhukovsky L. Professor O.S. Mamolat – scientist, clinicist, leader. UHMJ «Ahapit» 1996-1997;5-6:40-5. — Feshchenko YuI, Melnyk VM, Prykhod'ko AM. Professor Olexandr Mamolat – life, given to medicine (To the 100-Anniversary from the birthday of O.S. Mamolat). Ukr Pulmonol J 2010;2:12-13. — Ablitsov VH. «Galaxy Ukraine». The Ukrainian diaspora: prominent persons. Publ. «Kyt», Kyiv 2007.

[959] Gnoj J. Echocardiography. J Ukr Med Assoc North Am 1978 July; 25,3(90):97-135. — Sigh S, Goyal A. The origin of echocardiography. A tribute to Inge Edler. Tex Heart Inst J 2007;34(4):431-8.

[960] Sugeng L, Sherman SK, Salgo IS, et al. 3-dimensional transesophageal echocardiography initial experience using fully-sample matrix allay probe. J Am Coll Cardiol 2008;52(6):446-9.

[961] Kleinman Hobbins JC, Jaffe CC, et al. Echocardiography studies of the human fetus: prenatal diagnosis of congenital heart disease and cardiac dysrhythmia. Pediatrics 1980;65(6):1059-65.

[962] Maulic D, Manda N, Maulic D, et al. A brief history of fetal echocardiography and its impact on the managemet of congenital heart disease. Echocardiography 2017;34:1760-7.

Heyde (1911 – 2004)[963] — American Canadian internist, in the coexistence of the aortic valve (AV) stenosis and gastrointestinal (GI) angiodysplasia with hemorrhage, most commonly of the colon (Heyde syndrome, 1958) in patients with the IIA vWD, and also in Wilms's tumor, hypothyroidism and mesenchymal dysplasia.

The first realization that blood became saturated with oxygen (O_2) that is oxygenated as it passes through the cellophane chambers in his artificial kidney occurred to **Willem J. Kolff (1911 – 2009)**[964] — Dutch physician-researcher, in 1943-44 during its construction for hemodialysis (HD) to treat patients suffering from end-stage kidney disease (ESRD). Thereafter, he treated with it his first 15 patient with ESRD, all died. But, in 1945 he was able to save a 67-year-old woman's life suffering from ESRD he treated with HD using his artificial kidney.

Part of interstitial lung disease (ILD) / **[19th century.** Internal medicine. – **Ernest Besnier (1831 – 1909)** and **Caesar Boeck (1845 – 1917)]** – idiopathic interstitial pneumonias (IIP) is characterized by infiltration of immune cells into the pulmonary interstitium, inflammatory interstitial changes which if not alleviated will lead to fibrosis of the pulmonary parenchyma and death from pulmonary failure.

The histologic classification of IIP proposed by **Averill A. Liebow (March 21, 1911, town of Stryy, Lviv Obl., Ukraine – May 1978, city of La Jolla, CA, USA)**[965] — Ukrainian American pulmonary pathologist, include:

The usual idiopathic pneumonia (UIP) / idiopathic pulmonary fibrosis (IPF) is characterized by nonproductive cough, progressive dyspnea, and weight loss. Chest x-ray (CXR) shows lineal diffuse reticulonodular opacities, usually in the lower lobes (80%), computer tomography (CT) of the chest demonstrated a reticular pattern with honeycombing, cysts, and ground-glass opacities, and enlargement of mediastinal lymph nodes (LN). Bronchoalveolar lavage (BAL) large number of neutrophils, eosinophils, or lymphocytes. Transbronchial biopsy yields an inadequate piece of tissue to make a microscopic reliable. The diagnosis is established by an open, or video-assisted thoracic surgery (VATS) by exclusion of other causes of disease. The

---

[963] Heyde EC. Gastrointestinal bleeding in aortic stenosis. NEJM 1958;259(4):196. — Olearchyk AS. Heyde's syndrome. J Thorac Cardiovasc Surg 1992;103:823-4. — Olearchyk AS. Lactobacillus endocarditis necessitating combined mitral and aortic valve replacement - a case report. Vasc Surg 1993;27:219-25. — Olearchyk AS. Lactobacillus endocarditis: long-term survival after mitral and aortic valve replacement. Cardio-Vascular Surg (Kyiv) 1999;7:222.

[964] Kolff WJ, Berk HT, ter Welle M, et al. The artificial kidney: a dialyzer with a great area. 1944. J Am Soc Nephrol 1997;8(12):1959-65.

[965] Town of Stryy (first mentioned 1431), Lviv Obl., Ukraine. — M. La Jolla (settled by the Indian tribe Kumeyaay and named by them the «Land of Holes»), CA, USA). — Liebow A. Definition and classification of interstitial pneumonia in human pathology. Prog Respir Res 1975;8:1-33. / Smith GJ. Averill Abraham Liebow: contribution to pulmonary pathology. Yale J Biol Med. 1981 Mar-Apr;54(2)139-46. — Ablitsov VH. «Galaxy Ukraine». The Ukrainian diaspora: prominent persons. Publ. «Kyt», Kyiv 2007.

mainstay of treatment are steroids, and other immunosuppressive drugs, or both, including methotrexate, penicillamine, colchicine and cyclosporine.

Desquamative interstitial pneumonia (DIP).

Nonspecific interstitial pneumonia (NIP).

Acute interstitial pneumonia (AIP) is undistinguishable from an acute lung injury and acute respiratory distress syndrome (ARDS) with exception of a pre-existent trauma or sepsis.

Respiratory bronchiolitis – associated ILD (RB – ILD).

Cryptogenic organizing pneumonia (COP), or idiopathic bronchiolitis obliterans organizing pneumonia (BOOP), i.e., alveolitis pulmonary disease, characterized by a chronic alveolar inflammation, the production of granulation tissue in the bronchioles and alveoli, and the accumulation macrophages within the alveoli. The chief complain is a nonproductive cough and weight loss. CXR and CT shows bilateral diffuse opacities and air spaces with ground-glass opacities. The diagnosis is established by lung biopsy. The primary disease responds to steroid therapy, relapses occur in a secondary of BOOP from infection, drugs or connective tissue diseases, and are treated also with steroids and immunosuppressive medications.

Lymphocytic interstitial pneumonia (LIP) is a lymphoproliferative disorder with the infiltration of lymphocytes and plasma cells into the lung parenchyma, with minimal to no alveolar injury. It occurs predominantly in children, women between the ages of 40 and 80 years and in immunosuppressed patients.

Idiopatic hypertrophic subaortic stenosis (IHSS) / hypertrophic obstructive cardiomyopathy (HOCM) / hypertrophic cardiomyopathy (HCM) is a congenital, familial (60-80%) disease of the heart muscle affecting about 0,2% to 0,5% of population, usually young adults, in which a portion of the myocardium is thickened without any obvious causes, e.g., absence of systemic hypertension or aortic valve (AV) stenosis. In its obstructive form, IHSS / HOCM / HCM, a basal, diffuse or apical thickening of the interventricular septum (IVS) leads the left ventricular outflow tract (LVOT) obstruction that is significant when the basal gradient accross the AV is at least 30 mmHg. The latter is aggravated by the prolapse of the anterior leaflet of the mitral valve (MV), contributed by the insertion into it of the anomalous papillary muscle, so called systolic anterior motion (SAM), failure of the posterior (MV) leaflets to coapt with the anterior one, resulting in MV regurgitation. IHSS / HOCM / HCM is the leading cause of the left ventricle (LV) diastolic dysfunction and the premature sudden cardiac death (SCD). The description of IHSS / HOCM / HCM is credited to **Robert D. Teare (1911 – 79)** — English pathologist, who in 1958 linked it to a «tumor of the heart», asymetrical myocardial hypertrophy, cardiac cell (myocyte) disarray, familial association and premature SCD.

By interviewing about 700 lung-cancer patients Sir **Richard Doll (1913 – 2005)** — English oncological epidemiologist, established in 1950 a link between cigarette smoking and lung cancer.

The electrocardiograph (ECG)-monitor is a portable device for constant registration with two channels during 24 hrs of the frequency and duration of disturbances in the cardiac rhythm and for the evaluation of function of the cardiac pacemaker. The modification of the ECG-monitor was developed in 1947-62 by **Norman J. Holter (1914 – 83)** — American biophysicist. The first device (1947) weighted 38 kg, the second (1962) was small and generally accepted until the present time.

David A. Karnofsky (1914 – 69) — American clinical oncologist of the Lithuanian Jewish descent, who developed the functional performance scale with the appropriation of meanings in the diapason from 0 (a nonfunctioning organism or dead one) to 100 (a person with fully normal functions (Karnofsky scale, 1949).

In 1958 **Leslie Zieve (1915 – 2000)**[966] — American physician, described patients with an acute metabolic condition that can occur during withdrawal from prolonged alcohol abuse causing liver disease, characterized by abdominal pain, hepatosplenomegaly, jaundice, fatty infiltration of the liver, hemolytic anemia and hyperlipoproteinemia (Zieve syndrome). The underlying cause is liver delipidization. This is distinct from alcoholic hepatitis which, however, may be present simultaneously or develop later.

The programmed electrical stimulation, i.e., an initiation and termination by critically timed premature beats of the heart was performed in 1970 by **Dirk Durrer (1918 – 84)** and **Henrich (Hein) J.J. Wellens (1935 –)** — Dutch cardiologists and cardiac electrophysiologists голландські кардіологи і серцеві електрофізіологи, on patients with the pre-excitation of ventricles (PEV) or the Wolff-Parkinson-White (WPW) syndrome (above), in 1971 by Hein Wellens in patients with atrial flutter, intraventricular nodal tachycardia and accessory atrioventricular (AV) connection, and in 1972 in patients with ventricular tachycardia (VT). Later, in 1982 Hein Wellens described electrocardiographic (ECG) display of critical stenosis of the proximal left anterior descending artery (LADA) characterized initially biphasic T wave inversion, later becoming symmetrical, often deep (more than 2 mm) and T wave inversion in the anterior precordial lead (Wellens' syndrome).

With the advent of electron microscopy and development by **Gilbert N. Ling (1919 –)** — Chinese American cell physiologist and biochemist, **Ralph W. Gerard (1900–74)**—American neuro-scientist and behavioral scientist, of fine microelectrodes for intracellular recording of electromotive forces (Ling-Gerard electrode, 1942), only

---

[966] Zieve L. Jaundice, hyperlipemia and hemolytic anemia: a heretofore unrecognized syndrome associated with alcoholic fatty liver and cirrhosis. Ann Int Med 1958;48 (3): 471-6.

in 1958, **Brian F. Hoffman (1925 – 2013)** — American cardiac electrophysiologist, **Paul F. Cranfield (1935 –)**[967] — American cardiac electrophysiologist and **Antonio P. de Calvahlo (1935 – 2021)** — Brazilian physiologist, were able to provide the anatomical work at the cellular level, presented physiological evidence for intra-atrial conduction from transmembrane electrical potentials of a bundle between the sinus atrial (SA) node and the atrioventricular (AV) node, traversing the superior vena cava (SVC) edge of the crista terminalis. That led to present understanding of the conduction system of the heart (CSH).

Lubov T. Mala (13.I.1919, village Kopani, Orkhivsky Region, Zaporizhia Obl., Ukraine – 2003, city of Kharkiv, Ukraine)[968] — Ukrainian cardiologist, main work of which was dedicated to study prostaglandin, leukotrienes, interleukins, action of beta-blockers in congestive heart failure (CHF, 1980), the use of nitrogen oxide (NO) to teat CHF.

In 1955 the patient lift was designed by **R.R. Straton** — American automobile industrialist, and named the Hoyer lift, is a mobile (portable) or suspended from the ceiling device, powered by electricity or hydraulically, allowing the patients with limited mobility to be moved from the bed to the armchair or wheelchair in the horizontal with the head elevated or vertical positions on the Comfort Glide Repositioning Sheet (Medline Industries, Inc., Northfield, IL) in the hospital, long-term acute care hospital (LTACH), nursing homes or homes.

Bernard Lown (1921 – 2021)[969] — Lithuanian American cardiologist of Jewish descent, **William F. Ganong, Jr (1924 – 2007)** — American physiologist, and **Samuel Levine (1891 – 1966**: 20th century. Internal medicine), described a pre-excitation of the ventricle (PEV) or Lown-Ganong-Levine (LGL, 1952) syndrome, caused by an accessory electrical pathway between the left atrium (LA) and the left ventricle (LV) type A, discovered by **Thomas James (1924 – 2010)**[970] — American cardiologist (1961, 1962). That accessory electrical pathway between the LA and the LV type A represent congenital anomaly of the conduction system of the heart (CSH) that is the accessory inferoatrial-Hisonian pathway with bypass to the lower nodal bundle of

[967] Hoffman BF, Cranfield PF. Electrophysiology. McGraw-Hill, New York 1960.

[968] Village Kopani (founded 1851; during conduction by Russian Communists Fascists regime of Holodomor-Genocide 1932-33 against the Ukrainian Nation died at least 162 peasants), Orkhivsky Region, Zaporizhia Obl., Ukraine). — Ablitsov VH. «Galaxy Ukraine». The Ukrainian diaspora: prominent persons. Publ. «Kyt», Kyiv 2007. — Synyachenko OV. They glorified themselves and Ukraine: history of domestic therapy. Crimean Therapeutic Journal 2010;2(2):21-8.

[969] Lown B, Ganong WF, Levine SA. The syndrome of short PR interval, normal QRS complex and paroxysmal rapid heart action. Circulation 1952 May;5(5):693-706. — Lown B, Levine SD. Atrial Arrhythmia, Digitalis and Potassium. Landerger Medical Books, Inc., New York 1958.

[970] James TN. Morphology of the human atrioventricular node, with remarks pertinent to its electrophysiology. Am Heart J 1961;62:756-71. — James TN. The connecting pathways between the sinus node and the A-V node and between the right and the left atrium in the human heart. Am Heart J 1963;66:498-508.

His (NB-H), portion of the atrioventricular node (AVN), named the James fibers that can cause of the LGL syndrome.

Beside of this, the LGL syndrome grouped with the Wolff-Parkinson-White (WPW) syndrome (above), may also be caused by an accessory electrical pathway between the right atrium (RA) and the right ventricle (RV) type B, discovered by **Albert Kent (1863 – 1958**: 19th century. Anatomy, histology, and embryology**).** The LGL syndrome is characterized by palpitation, the electrocardiography (ECG) evidence of tachycardia without atrial fibrillation (AF), short P-R interval and normal QRS. It is rare, affecting 0.1%-0.3% of the general population, causing sudden cardiac death (SCD) in less than 0.6.

In the 1950s Bernard Lown recognized that a major cause of mortality in patients treated for congestive heart failure (CHF) with the long-acting drug digitalis glycoside (digitoxin) to increase cardiac contraction by a positive inotropic effect and as an antiarrhythmic agent to control the rapid irregular heart rate, was digitalis toxicity due to low level of the blood serum potassium (K), called hypokalemia. He corrected the situation by replacing the long-acting digoxin with a short-acting digitalis glycoside (digoxin, lanoxin) and maintaining the level of serum K at the upper limit of normal (3.5-5.5 mEq / L).

Then, in 1961, he developed direct electric current (DC) defibrillator and cardiac implantable cardioverter–defibrillator (ICD) to reverse a deadly rapid disorganized heart rhythm, specifically ventricular fibrillation (VF) and non-perfusing ventricular tachycardia (VT) to a normal sinus rhythm (NSR), widely used in cardiopulmonary resuscitation (CPR). At about that time, he introduced the drug lidocaine (discovered in 1946) for the treatment of ventricular arrhythmias by blocking sodium (Na) channels and thus decreasing the rate of contraction of the heart.

Finally, in the early 1961s Bernard Lown working with Samuel Levine, realized that the high mortality of patients after heart attack / myocardial infarction (MI), then 35%, most likely resulted from vigorous bed rest (BR) for 6 week or more, and major complication was unrecognized pulmonary embolism (PE), accounted for significant part of mortality, diagnosed usually at autopsy. Therefore, he introduced for those patients, an early ambulation ordering the out of bed (OOB) in the chair regime and this reduced the mortality by two-third (1967).

In 1959 **E. Watkins, Jr** and **R.D. Sullivan** introduced regional chemotherapy infusion for malignant tumors.

Colonoscopy was performed by **R. Turell** in 1963.

In 1960 **Wayne E. Quinton (1921 – 2015)** — American developer of medical devices and **Belding H. Scribner (1921 – 2003)** — American physician and pioneer in kidney dialysis, invented arteriovenous (AV) shunt allowing repeated hemodialysis

(HD) in patients with end-stage kidney disease (ESKD), consistent with an external U-form synthetic cannula and Teflon tips for an anastomosis between the radial artery and the cephalic vein (CV).

The first successful mesenchymal stem cell (MSC) bone marrow transplantation was performed in January 1958 by **Georges Mathé (1922 – 2010)** — French oncologist, immunologist, and founder of bone marrow transplantation, in six accidentally radiated nuclear researchers. Eleven months later (November 1958) he performed the first successful allogenic bone marrow transplant[971] ever performed on unrelated human being. Then, in 1959 and 1963 he successfully transplanted bone marrow in identical twins to treat leukemia in one of them.

Further progress in bone marrow transplantation was made by **E. Donnall Thomas (1920 – 2012)** — American physician who carried out the first bone marrow transplant using a matched sibling donor for one with leukemia in 1969, and the first matched transplant from unrelated donor in 1977. Presently, bone marrow transplants can cure 70-80% of the healthiest children and teenagers with leukemia.

H. Jeremy C. Swan (1922 – 2005) — American cardiologist and **William Ganz (1919 – 2009)** — American cardiologist, designed a multifunctional flexible flow-directed balloon pulmonary artery (PA) catheter for an invasive measuring and monitoring of cardiac hemodynamics (Swan-Ganz catheter, 1970).

According to the concept of **Konstantin P. Buteyko (27.I.1923, village of Ivanytsia, Ichniansky Region, Chernihiv Obl., Ukraine – 02.V. 2003, city of Moscow, Russia)**[972] — Ukrainian Russian physician, carbon dioxide (CO_2) is not only the end-product of the metabolism, a poison, which the organism gets rid-off during expiration, but also a part of the organic compounds necessary for the existence of humans and animals. The development of civilization causes a deficit of CO_2 in the pulmonary alveoli from a deep breathing (hyperventilation) with subsequent unfolding of numerous chronic diseases of the respiratory system. Therefore, to alleviate the CO_2 deficit, he introduced in 1968 the method of a voluntary elimination of deep breathing (VEDB) or Buteyko breathing) which is used, along with the traditionally accepted treatment of bronchial asthma and of other chronic diseases of the lung.

---

[971] Allotransplantation or heterotransplantation pertains to the donor of an organ of this same (identical) species, who differs genetically and immunologically from the recipient.

[972] Village Ivanytsia (mound from Bronze Age 2,000 years BC and Scythe Period 5th – 3rd century AD, founded 1600, carried out by Russian Communist Fascist regime Holodomor-Genocide 1932-33 against the Ukrainian Nation took life of approximately 250 inhabitants), Ichniansky Region, Chernihiv Obl., Ukraine. — Buteyko K. The Buteyko Method. An Experience of Use in Medicine. Publ. «Patriot», Moscow 1990. — Ablitsov VH. «Galaxy Ukraine». The Ukrainian diaspora: prominent persons. Publ. «Kyt», Kyiv 2007.

Olexandr I. Hrytsiuk (10.X.1923, city of Kyiv, Ukraine – 14.III.1990, city of Kyiv, Ukraine)[973] — Ukrainian cardiologist, first to prove meaning of the speed of fibrinolysis in pathogenesis of thrombus formation (not only its intensity) and described preactivated plasminogen of rapid action – physiological factor of the fibrinolytic system; proved the presence in rheumatic disease (RD) phased changes in blood coagulation, its dependency upon prevalence of immediate or slow type of allergy and importance in the development of thrombotic and hemorrhagic complications (1966); described the transition of the small-focal (subendocardial) myocardial infarction (MI) into the large-focal transmural MI (1973); assigned the blood coagulation homeostasis as relatively dynamic balance between procoagulants and anticoagulants, fibrinolysis agents in plasma, formed elements of the blood and vascular wall, prostacyclin-thromboxane balance; showed that between thrombolytic, erythrocytic and plasmocytic pro-factors and factors, and those which inhibited them, exist exchange and dynamic equilibrium, as between blood plasma and vessel wall (1979).

In 1963 **John B. Barlow (1924 – 2008)**[974] — South African cardiologist, depicted a progressive congenital heart disease (CHD) – viral valve (MV) prolapses syndrome (floppy-valve syndrome, Barlow syndrome), in which the MV ring is dilated, sometimes calcified, the cusps are myxomatous, proliferative, degenerative, thicken, with the excessive amount of tissue and billowing, chordae tendineae cordis are thickened, prolongated or ruptured, papillary muscles are elongated. During the systole MV cusps are prolapsing into the left atrium (LA) causing regurgitation of the blood. The clinical picture is usually asymptomatic, but with time occurs anxiety, retrosternal discomfort, and palpitation, during an auscultation an accentuated mid-systolic click with/without a late systolic murmur is heard. The electocardiography (ECG) shows inferior myocardial ischemia. In rare circumstances a sudden cardiac death (SCD) may occur. If conservative treatment is unsuccessful, a repair or replacement of the MV is indicated.

---

[973] Hrytsiuk OJ. Principal Pathogenic Factors for Thrombus Formation in Rheumatic Injuries to the Cardiovascular System. DMSc Dissertation. Kyiv 1966. — Hrytsiuk AI. Thromboses and Emboli in Rheumatism. Publ. «Zdoro'ia», Kyiv 1973. — Hrytsiuk AI, et al. Myocardial Infarction. Publ. «Zdoro'ia», Kyiv 1979. — Hrytsiuk AI. Texbook of Cardiology. Publ. «Zdoro'ia», Kyiv 1984. — Hrytsiuk AI. Inflammatory Diseases of the Heart. Publ. «Zdoro'ia», Kyiv 1986. — Hrytsiuk AI. Linical Angiology. Publ. «Zdoro'ia», Kyiv 1988. — Hrytsiuk AI, et al. Drug Resources in Clinical Cardiology and Rheumatology. Publ. «Zdoro'ia», Kyiv 1992. — Hrytsiuk AI, Amosova KM. Practical Hemostasology. Publ. «Zdoro'ia», Kyiv 1994. — Ablitsov VH. «Galaxy Ukraine». The Ukrainian diaspora: prominent persons. Publ. «Kyt», Kyiv 2007. — Synyachenko OV. They glorified themselves and Ukraine: history of domestic therapy. Crimean Therapeutic Journal 2010;2(2):21-8.

[974] Barlow JB, Polock WA. The significance of the late systolic murmur and mid-late systolic clicks. Md State MJ 1963 Feb;12:76-7. — Barlow JB, Bosman CK. Aneurysmal protrusion of the posterior leaflet of the mitral valve. An auscultatory-electrocardiographic syndrome. Am Heart J 1966;71 (2): 166-78.

The concept of the «mitral valve (MV) apparatus» was introduced by **Joseph K. Perloff (1924 – 2014)**[975] — American cardiologist. It is composed of the left atrium (LA), the inter-trigonal space between the left fibrous trigone and left coronary sinus (LCS) and the right fibrous trigone and noncoronary sinus (NCS) with the base at the anterior MV fibrous annulus (ring) with mitral/aortic continuity and its anterolateral and posteromedial commissures, anterior (aortic) and posterior (mural) mitral leaflets, primary (marginal) and secondary chordae tendinea and the anterolateral and posteromedial papillary muscles which arise from the area between the apical and middle thirds of the left ventricle (LV) wall.

Congenital heart disease (CHD) described in 1963 by **John D. Shone (July 04, 1924, North Wales, UK – July 12, 2002, Bobcaygeon, ON, Canada)**[976] — English American Canadian pediatric cardiologist, is a hypoplastic left heart syndrome (HLHS) where both left ventricle (LV) inflow and outflow are obstructed and consist of (1) a supravalvular mitral membrane, (2) parachute mitral valve, (3) membranous or muscular subaortic stenosis and (4) coarctation of the aorta (CA), so called Shone syndrome (complex or anomaly).

The initial treatment by **Myron W. Wheat, Jr (1924 – 2012)**[977] — American heart surgeon, for type I(A) aortic dissection and destination therapy (DT) for an uncomplicated type 3(B) aortic dissection is to reduce the shear force of the aorta, or the rate of rise of the aortic pressure dP/dt with beta-blocker propranolol intravenously (IV) 1 mg q 5/min till the heart rate (HR) is between 50-60 per min., then IV or po q 4 hrs; or short acting beta-blocker estmolol IV bolus 500 mcg/kg, followed by IV infusion of 50 mcg/kg/min. to keep the mean arterial blood pressure (BP) at 60-75 mmHg; or alpha- and beta-blocker labetolol (0,25 mg/kg/min/) 20 mg IV bolus q 2 min., followed by 0,50 mg/kg/min. (40 mg) bolus q 10 min. until HR, mean arterial BP and dP/dt are controlled, then IV infusion of 0,0125 mg/kg/min/ (2 mg/min.); or sodium nitroprusside with the usual starting IV infusion dose of 0.1-0,25 mcg/kg/min, titrate up by 0,25 mcg/kg/min. q 5-10 min. to the maximum dose of 10 mg/kg/min. to keep HR at 60-75 min. and mean arterial BP at 60-75 mmHg, but since it increases dP/dt, a beta-blocker must be added; an alternative to sodium nitroprusside in patients with renal insufficiency is fenoldopan at IV dose of 0,1 mcg/kg/min.; or calcium channel blocker, e.g., diltiazen 120 mg to 540 mg qd; or angiotensin-covering enzyme

[975] Perloff JK, Roberts WC. The mitral apparatus. Functional anatomy of mitral regurgitation. Circulation 1972 Aug;46(2):227-32.

[976] Wales (Welsh, Cymru: area 20,779 km²), United Kingdom (UK, area 242,495 km²). — Town of Bobcaygeon (known since 1615), ON, Canada. — Shone JD, Sellers RD, Anderson RC, et al. The developmental complex of «parachute mitral valve», supra-annular ring of left atrium, subaortic stenosis, and coarctation of aorta. Am J Cardiol 1963 June 1;11(6):714-25.

[977] Wheat MWJr, Palmer RF, Bartley TD, et al: Treatment of dissecting aneurysms of the aorta without surgery. J Thorac Cardiovasc Surg 1965; 50:364-73.

(ACE) inhibitor, e.g., catopril from 6,25 mg to 50 mg tid, enalapril from 2,5 mg to 10-20 mg bid, lisinopril from 2,5 mg to 40 mg q daily; or angiotensin receptor blocker, e.g., losartan from 50 mg to 100 mg q daily, concominant with a beta-blocker.

Directions of scientific researches of **Iryna I. Datsenko (Jan. 30, 1925, city of Rivne, Ukraine – Mar. 23, 2006, city of Lviv, Ukraine)**[978] — Ukrainian physician-hygienist and physician-ecologist, were hygiene of the surroundings (vicinity) and the atmospheric air, particularly, study of the biologic effect and hygienic importance of the factors of external environment with a low intensity in the conditions of the inhabited places; worked out new methods to determine chemical substances in the air and set forth a conception of development of chronic intoxications (poisoning) by carbon monoxide (CO); studied the state of pollution by pollutants in the city of Lviv, Ukraine, air environment and their influence on an organism; established the borderline (maximally) permissible concentration of numerous chemical substances in an environment.

During the 1957-1958s **Basil I. Hirschowitz (1925 – 2013)** — South-African American gastroenterologist, invented improved optical fiber which allowed the creation of useful flexible fiberoptic esophagogatroduodenoscope[979].

Invented in 1960 by **E. J. Moran Campbell (1925 – 2004)**[980] — Canadian physician, the Venturi mask with valve with (VM) / **[Giovanni Venturi (1746 – 1822): 18th century]**, a.k.a., an air-entrainment masks that deliver a known oxygen (O_2) concentration to patients on controlled O_2 therapy, as a replacement for intermittent O_2 treatment. However, the VM is not always able to guarantee the total flow with O_2 percentage above 35% in patient with high inspiratory flow demands.

---

[978] City of Rivne (within its boundary is situated Ancient late Paleolithic adobe 35,000 years BC, first mentioned by the chronicle 1283), Rivnens'ka Obl. (area 20,047 km^2), Ukraine. — Datsenko II. Pollution by Carbon Monoxide of an Air in a Cabin of Automobiles and Streets in Lviv and its Influence of an Organism of Workers of the Automobile Transport (Dissertation for Candidate of of Medical Science). Lviv 1955. — Datsenko II. Carbon Monoxide in the External Environment and the Possibility of Chronic Intoxications (Dissertation for Doctor of Medical Science), Lviv 1966. — Datsenko II, Martyniuk VZ. Intoxication by Carbon Monoxide and Ways of its Weakening: Monography. Publ. «Zdorov'ya», Kyiv 1971. — Datsenko II. Hygiene and Ecology of the Man. Educational Textbook. Publ. «Afisha», Lviv 2000. — Datsenko II, et al. General Hygiene. Textbook for Practical Studies. Publ. «Svit», Lviv 2001. — Datsenko II, Hrabovych RD. Prophylactic Medicine. General Hygiene with Basics of Ecology. 2nd Ed. Publ « Zdorov'ya », Kyiv 2004. — Datsenko II, Shehedyn MB, Shashkov YuI. Hygiene for Children and Adolescents. Textbook. Publ. «Medytsyna», Kyiv 2006. — Ablitsov VH. «Galaxy Ukraine». The Ukrainian diaspora: prominent persons. Publ. «Kyt», Kyiv 2007.

[979] Linder TE, Simmen D, Stool SE. The history of endoscopy. Arch Otolaryngol Head Neck 1997;123:1161-3.

[980] Campbell EJM. A method of controlled oxygen administration which reduces carbon-dioxide retention. Lancet 2 July1960;7140:12-14. — Campbell EJM. Apparatus for oxygen administration. BJM 1963 Nov 16;2(5367):1269-70.

In 1951 **Richard Gorlin (1926 – Oct. 16, 1997, New York City, NY, USA)**[981] — American physiologist and cardiologist, proposed the equation for calculating the area of the orifice of a cardiac valve which is based on flow of the blood across the valve and the mean pressures in the chambers on either side of the valve (Gorlin formula). It was simplified by the Hakki formula[982] for the estimation of the aortic valve (AV) area (AVA) in cm^2, where the AVA equals the cardiac output (CO) (L/mim.) divided by the square root of the peak-to-peak systolic gradient across the AV (mmHg).

First Coronary Care Unit (CCU) was opened in 1961 by **Desmond G. Julian (1926 – 2019)**[983] — Scottish cardiologist, in the Royal Infirmary (established 1729), city of Edinburgh (founded before the 7th century BC), Scotland, who attempted to deal with complications of coronary artery disease (CAD), such as cardiac arrhythmia, myocardial infarction (MI), cardiogenic shock and sudden cardiac death (SCD). He recommended all staff, including nurses, be trained in cardiopulmonary resuscitation (CPR) to treat patients with suspected heart attack as rapidly as possible.

Following **Desmond Julian** (above), **Pavlo Ye. Lukomsky [11(23).VII.1899, Suvorov Staff, Grodno Obl., Belarus – 08.IV.1974, city of Moscow, Russia]**[984] — Ukrainian Russian cardiologist who in 1964 opened the CCU on the premise of hospital therapy of the 2nd Moscow State Medical Institute, now Russian National Research Medical University of M.I. Pirogov (founded 1906), Moscow, Russia.

The central male-type or apple-shaped obesity as a result of an excessive deposition of fat in the body is a complex disease, a combination of visceral obesity, the surplus of weight with body mass index (BMI) ≥ 30 kilograms (kg)/meter (m)^2, diabetes mellitus (DM), insulin resistance or impaired glucose tolerance test, dyslipidemia, high blood pressure (BP), i.e., hypertension and endothelial dysfunction, linked to cardiovascular diseases, was defined in 1993 by **Gerald M «Jerry» Reaven (1928 – 2018)**[985] — American physician-endocrinologist, and initially titled syndrome X, later named metabolic syndrome (MetS). The MetS affects approximately 40% of the U.S.A. youth

---

[981] Gorlin R, Gorlin SG. Hydraulic formula for calculating of the area of the stenotic mitral valve, other cardiac valves, and central circulatory shunts. Am Heart J 1951 Jan;41(1):1-29.

[982] **Abdul H. Hakki** — Kurdish American cardiothoracic surgeon. / Hakki A, Iskandrian A, Bemis C, et al. A simplified valve formula for the calculation of stenotic cardiac valve areas. Circulation 1981 May;63 (5): 1050-5.

[983] Julian DG. Treatment of cardiac arrest in acute myocardial ischemia and infarction. Lancet 1961 Oct 4;2(7207):840-4. — Julian DG. History of coronary care units. Brit Heart J 1987;57:497-502. — Silverman ME. Desmond Gareth Julian: pioneer in coronary care. Clin Card 2001 Oct;24(10):695-6. — Julian DG, Campbell CJ, Mclenachan JM. Cardiology. 8th Ed. ElsevierSaunders 2005.

[984] Ablitsov VH. «Galaxy Ukraine». The Ukrainian diaspora: prominent persons. Publ. «Kyt», Kyiv 2007.

[985] Reaven GM. Role of insulin resistance in human disease (syndrome X): an expanded definition. Ann Rev Med 1993;44:121-31.

and adult population[986]. It is associated with increased incidence of cardiovascular, pulmonary (including sleep apnea), renal and infections comorbidities. There is also increased length of stay in the hospital, increased rate of postoperative complications, morbidity, and mortality, especially in severely injured trauma patients, increase of children born with congenital anomalies and decreased life span. The best way to prevent obesity is to decrease eating a high caloric food with considerable amount of fat and sugar, preferable vegetarian diet, and exercise.

Eugene Braunwald (Aug. 15, 1929, Vienna, Austria –) — American cardiologist of Jewish parent, found that wall tension and contraction of the left ventricle (LV) of the heart were determined of myocardial oxygen consumption (MVO_2), and this is the cornerstone in the treatment of myocardial ischemia with afterload reduction and reduction of heart rate (HR) with beta-blockers[987]. He developed the concept of thrombosis superimposed on atherosclerosis (AS) as the pathological basis of acute myocardial infarction (MI, 1984).

Julian Gnoj (14.X.1928, village of Olesha, Ivano-Frankivsk Region, Ivano-Frankivsk Obl., Ukraine –)[988] — Ukrainian American internist and invasive cardiologist, made significant contribution to determination of pulmonary blood volume (1970), treatment of an acute myocardial infarction / MI (1975), prevention complication of percutaneous (PC) selective coronary angiography / SCA / (1975), echocardiography (1978) and percutaneous transluminal coronary angioplasty / PTCA (1983).

---

[986] Hales CM, Carroll MD, Fryar CD, et al. Prevalence of obesity among adults and youth: United States, 2015-2016. NCHS Data Brief 2017;[288]:1-8.

[987] Sarnoff SL, Braunwald E, Welch GH, et al. Hemodynamic determinant of oxygen consumption of the heart with special reference to tension-time index. Am J Physiol 1958;192(1):148-56.

[988] Village of Olesha (settled from late Paleolithic Age, first chronical mention 1205, founded 1441), Trumatsky (Ivano-Frankivsk) Region, Ivano-Frankivsk Obl., Ukraine. — Gnoj J. New method of treatment cardiac arrhythmias. J Ukr Med Assoc North Am 1965 Oct;12,4(39):7-10. — Lewis MJ, Gnoj J, Fisher V, et al Determinants of pulmonary blood volume. J Clin Invest 1970 Jan;49,(1):170-82. — Gnoj J. Etiology and prophylaxis of coronary artery disease. J Ukr Med Assoc North Am 1971Jan. – Apr.;18,1-2(60-61):28-37. — Gnoj J. Effect of alcohol on cardiac function. J Ukr Med Assoc North Am 1972 Jan-Apr;19,1-2(62-63):13-7. — Weinstein J, Gnoj J, Ayres JT, et al. Temporary transvenous pacing via the percutaneous femoral vein approach. A prospective study of 100 cases. Am Heart J 1973 May;85(5):695-705. — Gnoj J. Management of acute myocardial infarction in coronary care units. J Ukr Med Assoc North Am 1975 Jan; 22,1(76):6-11. — Shah A, Gnoj J, Fisher VJ, et al. Complications of selective coronary arteriography by the Judkin's technique and their prevention. Am Heart J 1975 Sep;90(3):353-9. — Gnoj J. Echocardiography. J Ukr Med Assoc North Am 1978 July; 25,3(90):97-135. — Gnoj J. Percutaneous transluminal coronary angioplasty. J Ukr Med Assoc North Am 1983 Spring; 30,2(107):59-75. — Gnoj J. Ю. Directional coronary atherectomy. J Ukr Med Assoc North Am 1995 Spring;42,2(136):67-88. — Gnoj J. Thrombolytic therapy in acute myocardial infarction. J Ukr Med Assoc North Am 1991 Spring;38,2(124):67-87. — Baker HA III, Olearchyk AS. Emergency revascularization of coronary arteries in myocardial inferction. J Ukr Med Assoc North Am 1993 Spring;40,2(130):87-9. — Ablitsov VH. «Galaxy Ukraine». The Ukrainian diaspora: prominent persons. Publ. «Kyt», Kyiv 2007.

In 1975 **Yevhen I. Chazov (10.VI.1929, city of Nyzhniy Novhorod, Росія –)**[989] — Russian Ukrainian cardiologist, graduated in 1953 from O.O. Bohomolets National Medical University (NMU), city of Kyiv, Ukraine, introduced intracoronary thrombolysis for an acute myocardial infarction (MI).

In 1961 **Edwin C. Brockenbrough (1930 –)** — American cardiac surgeon, Eugene Braunwald (above) and **Andrew G. Morrow (1957 – 2017)** — American cardiac surgeon, noted in idiopathic hypertrophic cardiomyopathy (IHSS) / hypertrophic obstructive cardiomyopathy (HOCM) / hypertrophic cardiomyopathy (HCM) a paradoxical decrease in the arterial pulse pressure and associated increase in the left ventricular (LV) systolic pressure in the beat following paroxysmal ventricular contraction (PVC), giving rise to the sign called Brockenbrought-Braunwald-Morrow[990]. After a PVC, there is a compensatory pause that causes an increase in diastolic filling and therefore an increase in diastolic volume. The normal physiological response to increased stretch according to **Frank Starling (1866 – 1927**: 20th century. Physiology) law is to increase stroke volume (SV) by an increase in contractility, causing the arterial pulse pressure to rise. In patients with idiopathic hypertrophic cardiomyopathy (IHSS) / hypertrophic obstructive cardiomyopathy (HOCM) / hypertrophic cardiomyopathy (HCM), the increase in contractility after a PVC worsens the left ventricular outflow tract obstruction (LVOT) obstruction, causing a decrease in the arterial pulse pressure and the Brockenbrought-Braunwald-Morrow sign.

To further develop laid by Per-Olof Ostrand [20th century. Physiology – **Per-Olof Ostrand (1922 – 2015)**] and by Bengt Saltin [20th century. Physiology – **Bengt Saltin (1931 – 2014)**], the base of the physical exercise physiology, **Kenneth H. Cooper (1931 –)**[991] — American physician and former Colonel of the United States (US) Air Forces, have underlined the benefit of doing aerobic exercise to maintain and improve health and longevity of the people, particularly athletes (sportsmen's).

He introduced the aerobics exercise test for athletes by estimating or measuring of the maximal oxygen (O_2) aerobic capacity for consumption or uptake or a peak O_2 uptake, metabolized in minute (min.) per kilogram (kg) of body weight at the peak of exercise (VO_2 max) in which the distance in meters (m) is covered by running as fast as possible in 12 min. (d12), on an open field track or on a graded treadmill, incorporated in the following equation (Cooper test or formula, 1968):

---

[989] Chazov EL, Mateeva LS, Mazaev AL et al. Intracoronary administration of fibrinolysin in acute myocardial infarction. Ter Arkh 1976;48:8-19. — Olearchyk AS. Personal communication. All-Russian Institute of Cardiology (VRIC, founded 1967) Russian Academy of Medical Sciences (PAMSc., founded 1944, 1992), Cherepkovska Street 3, 15a, city of Moscow, Russia, 02.VIII.1990.

[990] Brockenbrough EC, Braunwald E, Morrow AG. A hemodynamic technique or the detection of hypertrophic subaortic stenosis. Circulation 1961;23:189-94.

[991] Cooper KH. The New Aerobics. Bantam Books, New York 1968.

$$VO_2 \text{ max (mL /kg / min.} = d12 - 509,4 / 44.73,$$

where mL = milliliter; d12 = distance in m, covered in 12 min.

Results of the Cooper test are being interpreted for juniors and athletes as the d12, according to the age (13 - 50+) and gender, placed against the very good, good, average, bad and very bad result, e.g., the d12 for males is very good (2400+ to 3000+ m), good (2000-3000 m), average (1300-2400 m), bad (1300-2499 m) and very bad (1300- to 2300- m); the d12 for females is very good (2200+ to 2700+ m), good (1700-2700 m), average (1400-2199 m), bad (1100-1799 m) and very bad (1100- to 1700- m).

Of the pulmonary function test (PFT) an exercise test, i.e., maximum oxygen consumption with an exercise VO_2 max with a loading >15 ml/kg/min. is considered satisfactory, VO_2 max <10 ml/kg/min. – unsatisfactory [20th century. Surgery – **James Hardy (1918 – 2003)**].

The low VO_2 max indicate the possibility of high risk in the development of cardiac or vascular diseases, or both, various types of malignant tumors with associated high morbidity and mortality rates.

In 1967 **Michael D. Klein (1935 –)**, **Michael V. Herman (1937 –)** and **Richard Gorlin (1926 – 97**: above)[992] — American cardiologists, were the first to state that when approximately 20%-25% of the left ventricular (LV) wall is inactivated by any pathologic process, including acute myocardial infarction (MI), the degree that the myofiber must shorten to maintain stroke volume (SV) exceed physiologic limit, and according to the law of **Otto Frank (1865 – 1944**: 19th century. Physiology**)** and **Ernerst Starling (1866 – 1927**: 20th century. Physiology**)**, cardiac enlargement must ensue to maintain adequate left ventricular ejection fraction (LVEF) and cardiac output (CO).

That lead to the development of the concept of LV remodeling, relation of MI to the origin of post-infarction LV aneurysm. The later produces asynergy, either dyskinesia (i.e., paradoxical bulging or expansion during systole) or akinesia (i.e., partial, or complete loss or suppression of muscle movement during systole, due to myocardial fibrosis, calcification within the scar, thickening of the myocardium, mural thrombi, endocardial thickening causing rigidity of the aneurysmal wall thus preventing its expansion). The remote to an acute MI's areas may show hypokinesis or even akinesis from critical narrowing of the coronary artery causing hibernation,

---

[992] Klein MD, Herman MV, Gorlin R, et al. A hemodynamic study of the left ventricular aneurysm. Circulation. 1967Apr;35(4):614-30. — Olearchyk AS, Lemole GM, Spagna PM. Left ventricular aneurysm: ten years experience in surgical treatment of 244 cases. Improved clinical status, hemodynamics, and long-term longevity. J Thorac Cardiovasc Surg 1984;88:544-53. — Olearchyk AS. Recurrent (residual?) left ventricular aneurysm: a report of 11 cases. J Thorac Cardiovasc Surg 1984;88:554-7.

or stunning of the myocardium or be dysfunctional without stenosis of the coronary artery because of a high local tension that reduces myofiber's shortening.

Richard Gorlin offered the following definition of post-MI LV aneurysm – «*An aneurysm is identified by a left ventricular angiogram as any akinetic or dyskinetic segment of myocardium. An akinetic segment is defined as a segment that appears to have no motion during systole, whereas a dyskinetic segments appears to bulge paradoxically during systole. Intraoperatively, an aneurysm is identified as a circumscribed area of scar, which is thin, often adherent to the pericardium and which may or may not bulge paradoxically during systole. The aneurysmal segment is easily outlined by looking for the area that puckers and collapses when the left ventricle is vented*».

The changes in the LV size and geometry, i.e., shape, volume and cardiac mass altered by cardiac disease, primarily degenerative, valvular, and ischemic in origin affect the LV, coronary arteries, and valve or all, leading to progression of cardiac (heart) failure (CF, HF)[993].

Myocardial changes (myocyte loss, necrosis, and apoptosis) and extracellular matrix changes (degeneration, replacement by fibrosis) affects the LV chamber geometry causing its dilatation, increased sphericity and wall thinning and mitral regurgitation (MR).

Alteration in the LV geometry are evaluated by measuring the sphericity index (SI) and conicity index (CI). The global SI is a constant ratio of a short to long axis of the LV (normal, 42 ± 3 mm / 80 ± 9 mm = 0.52). The local CI is a constant ratio of a short axis of the LV apex to the short axis of the global LV (normal, 24 ± 9 mm / 42 ± 3 mm, = 0.57)[994].

Among patients after an anterior post-MI ischemic cardiomyopathy (CM), where primarily the cardiac apex is involved with regional changes in the anterior and inferoseptal LV walls and in the interventricular septum (IVS), without MR, the response of the LV to those regional changes is its global proportional elongation and widening to maintain constant ratio of a short to long axis, i.e., a constant SI. However, in an anterior post-MI ischemic CM the SI fails to detect shape abnormalities. Therefore, the CI has been used to evaluate the conical shape of the cardiac apex and it changes after an anterior post-MI ischemic CM without MR.

In patients after an anterior post-MI ischemic CM with secondary MR, the SI is abnormal, and the LV is more spherical. In addition, the posterior papillary muscles are displaced laterally, towards the cardiac apex and away from each other, causing «tethering» of the posterior mitral valve (MV) leaflet and MV leaflet restriction, i.e., delayed closure («loitering»).

---

[993] Mann DL, Bristow MR. Mechanisms, and models in heart failure. The biomechanical model and beyond. Circulation 2005;111:2837-49.

[994] Di Donato M, Castelvechcio S, Kukulski T, et al. Surgical ventricular restoration left ventricular shape influence on cardiac function, clinical status, and survival. Ann Thorac Surg 2009;87:455-62.

Among patients after an inferior post-MI ischemic CM, the short axis is widened more than the long axis is elongated, leading to an increase in the SI and more frequent occurrence of severe MR.

Nonischemic dilated CM shows a more spherical shape of the LV than ischemic CM; the short axis is enlarged, MR is more frequent, wall motion abnormality (WMA) is diffuse and severely hypokinetic with no regional differences in WMA. But, the LV contractility, SV and LVEF are markedly reduced. Endomyocardial biopsy (EMB) shows nonhomogeneous disease with scaring and fibrosis ranging from 4% to 60% between the free wall of the LV and IVS.

In 1969-70 **Melvin D. Flamm (1934 –)**[995] — American military and civilian cardiologist, reported that the M_VO_2 value of the average SO_2% for an ASD is derived number from the Flamm equation (formula), where $M_VO_2 = [3 \, (SVC) + 1 \, (IVC)] / 4$.

The M_VO_2 value of the average SO_2% is derived from the RA for VSD, from the RV for PDA.

A Q_P/Q_S ratio greater than 2.0 is considered high, indicating possible surgical or percutaneous (PC) closure, values between 1.5 and 2.0 are intermediate, and surgical or PC closure may be performed for low surgical risk patients, or PC closure if symptoms or complications are present such as embolic stroke for ASD or PFO. A Q_P/QS ratio of less than 1.0 suggest a right-to-left shunt.

A significant right-to-left shunt is usually detected early since those patient exhibit cyanosis, arterial hypoxemia or the oximetric evidence of a Q_P/Q_S ratio of less than 1.0. A Q_P/Q_S ratio of less than 0.7 is critical, and less than 0.3 is incompatible with life.

In the bidirectional shunts there is evidence of both left-to-right and right-to-left shunting, and a formula comparing the effective blood flow (Q_{eff}) is used:

$$Q_{eff} = O_2 \text{ consumption (mL/minute)} / [(P_VO_2) - (M_VO_2)].$$

Then the left-to-right and right-to-left shunts are $Q_P - Q_{eff}$ and $Q_{eff} - Q_s$, respectively.

In the interrupted aortic arch (IAA) the incidence of the left ventricular outflow (LVOT) hypoplasia or obstruction is aggravated by displacement of the misaligned conal septum in almost always coexisting ventricular septal defect (VSD) into the LVOT, bicuspid aortic valve (BAV) or described by **André J. Moulaert (**20th century. Anatomy, Histology and Embryology)[996], prominent bundle of muscle that extends from the anterolateral papillary muscle into the LVOT (muscle of Moulaert).

---

[995] Flamm MD, Cohn KE, Hancock EW. Measurement of systemic cardiac output at rest and exercise in patients with atrial septal defect. Am J Cardiol 1969;23:256-65. — Flamm MD, Cohn KE, Hancock EW. Ventricular function in atrial septal defect. Am J Med 1970 Mar;48(3):286-94.

[996] Moulaert AJ: Ventricular septal defects and anomalies of the aortic arch. Thesis. Univ of Leiden 1974. — Moulaert AJ, Oppenheimer-Dekker A. Anterolater muscle of the left ventricle, bulboventricular flange and subaortic stenosis. Am J Cardiol 1976 Jan;37(8):78-81.

In 1975 **Andreas R. Grüntzig (June 25,1939, city of Dresden, Germany** – died in his own airplane crash on **Oct. 27, 1985,** near **city of Forsyth, GA, USA)**[997] — German Swiss American radiologist and cardiologist, developed a double-lumen fitted with a polyvinyl chloride balloon catheter.

Subsequently, he performed intraoperative coronary artery balloon angioplasty (beginning September 1977) in the operating room (OR) during elective coronary artery bypass (CAB) grafting surgery at the Hospital of the University of Zürich, city of Zurick, Switzerland.

Then, on September 16, 1977, he performed at this same hospital on an awake patient suffering from angina pectoris due to coronary artery disease (CAD) caused by a short (3 mm) high grade (80%) luminal stenosis in non-branching proximal section of the left anterior descending (LAD) coronary artery, under local anesthesia, the first successful percutaneous transluminal coronary artery angioplasty (PTCA).

Joseph L. Goldstein (1940 –) — American biochemist and **Michael S. Brown (1941 –)** — American geneticist, discovered the regulation of cholesterol metabolism and the treatment of diseases caused by its abnormal level in the blood (1973-76).

Identified in 1969 by **Friderick P. Li (1940 – 2015)** — Chinese American physician and **Joseph F. Fraumeni, Jr (1933 –)** —American physician and cancer researcher, while reviewing the medical records and death certificates of 648 children with rhabdomyosarcoma, an association of sarcoma, breast leukemia and adrenal gland – adrenocortical carcinoma (SBLA), named SBLA or Li-Fraumeni syndrome[998], a rare hereditary disorder that predisposes carriers to cancer development, linked to germline mutation of the p53 suppression gene. It encodes transcription factor (p53) that normally regulates the cell cycle and prevent genomic mutations. The SBLA syndrome comprise about 80% of all cancers with this syndrome, is characterized by early onset of cancer, a wide variety cancer types and developments of multiply cancers throughout once's life. Once a person is suspected for the SBLA syndrome, the genetic counselling and testing are indicated to confirm genetic mutation. An identified person should undergo an early and regular screening for cancer because of likely development of another primary malignancy in the future at the risk rate of 57% within 30 years of diagnosis.

---

[997] City of Dresden (area has been settled in the Neolithic era by Lineal / Hereditary Pottery culture c. 7,500 BC, and by Slavic tribes around late 12th century), Germany. — City of Forsyth (established 1828), GA, USA). — Grüntzig AR. Die perkutane transluminale Rekanalisation chronische Arterienverschlüsse mit einer neuen Dilatationtechnik. G. Witzstrock, Baden Baden, Köln, New York 1977. — Grüntzig AR, Senning A, Siebenthaler WE. Nonoperative dilatation of coronary artery stenosis: percutaneous transluminal coronary angioplasty. N Engl J Med 1979 Jul 12;301(2):61-8. — Gnoj J. Percutaneous transluminal coronary angioplasty. J Ukr Med Assoc North Am 1983 Spring; 30,2(107):59-75.

[998] Li FP, Fraumeni JF. Soft tissue sarcomas, breast cancer, and other neoplasms. A familial syndrome? Ann Inter Med 1969 Oct;71(4):747-52.

Mitral stenosis (MS) due to rheumatic heart disease (RHD) is the most prevalent valvular heart disease in developing countries, while mitral regurgitation (MR) from degenerative heart disease is the most prevalent valvular disease in developed countries.

In general, the symptomatic severe mitral and aortic valve (MV, AV) disease or asymptomatic with the left ventricular (LV) dysfunction treated surgically by the conventional or a mini-invasive open-heart surgery are well established with superior long-term results.

However, the percutaneous (PC) transcatheter MV and AV repairs or replacement are becoming an alternative to treat severe disease in patients who are at high surgical risk due to advances age, associated heart disease and / or comorbidity.

A short QT interval syndrome was discovered by **Preben Bjerregaard (c. 1942 –)** — Danish-American cardiologist, in 1999 as being a genetic autosomal dominant disorder of the conduction system of the heart (CSH) consisting of a constellation of signs and symptoms, structurally normal heart, the electrocardiography (ECG) showing a short QT interval ($\leq$300 ms) that does not significantly changes with heart rate, tall and peaked T valves.

Priority direction in the scientific activity of **Olha M. Kovalova (18.III.1942, city of Sarapul, Udmurtia, Russia –)**[999] — Ukrainian physician-scientist, is research of

---

[999] City of Sarapul (founded 1780), Udmurtia (area 42,061 km^2), Russia. — Kovalova OM. Lifetime differences in maintaining glucose-amino-glucans and lipids of the blood in coronary atherosclerosis Dissertation for the degree of CMSci., Kharkiv 1975. — Kovalova OM. Clinical-epidemiological and genetic analysis of the structure inclination towards arterial hypertension. Dissertation for degree of DMSci disturbances. KhNMU, Kharkiv 1989. — Kovalova OH, Ashcheulova TV. Tumor necrosis factor-α. Clinical investigation activity in arterial hypertension. «Original», Kharkiv 2003. — Kovalova OH, Ambrosova TN. Pathogenesis and treatment of thrombocyte disturbances in cardiology. «Tornado», Kharkiv 2003. — Bilovol OM, Kovalova OM, Popova SS, et al. Obesity in practice of cardiologist and endocrinologist «Ukrmedknyha», Ternopil 2009. — Kovalova OH, Ashcheulova TV, Ambrosova TH. Morphological and functional changes of the heart in obesity. «Hove slovo», Kharkiv 2009. — Kovalova ON, Demydenko AV. Mechanismof formation and methods of correction endothelial dysfunction in arterial hypertension. «Nove slovo», Kharkiv 2010. — Kovalova OM, Safarhalyna-Kornilova NA. Propaedeutics of Internal Medicine. VSV «Medytsyna», Kyiv 2010. — Kovalova OM et al. Cardio-vascular risks, stratification, pathogenesis, prognosis. «Rarity of Ukraine», Kharkiv 2011. — Kovalova ON, Safarhalina-Kornilova NA. Propaedeutics of Internal Medicine. VSV «Medytsyna», Kyiv 2013. — Kovalova OM et al. Biomarkers of cardiovascular risk in arterial hypertension. «Planeta-print», Kharkiv 2014. — Kovalova OM, Safarhalyna-Kornilova NA, Harasymchuk NM. Deonthology In Medicine. VSV «Medytsyna», Kyiv 2015. — Kovalova OM, Lisovyy VM, Shevchenko SI, et al. Patient care (practical course). 3rd Ed. VSV «Medytsyna», Київ 2015. — Kovalova ON, Safarhalina-Kornilova NA HA, Herasymchuk NN. Deontology in Medicine. VSV «Medytsyna», Kyiv 2017. — Kovalova OM, Lisovyy VM, Ambrosova TM, et al. Propedeutics to internal medicine. Part 1: Diagnostics. Vol. I. 3th Ed. «Nova Knyha», Vinnytsia 2017. — Kovalova ON, Ashcheulova TV. Propedeutics to internal medicine. Part 2: Syndromes and diseases. Vol. I. 3rd Ed. «Nova Knyha», Vinnytsia 2017. — Kovalova ON, Lesovoy VN, Shevchenko SI. Patient care (practical course). 2nd Ed. Publ. AUS «Medytsyna», Kyiv 2018. — Kovalova OM, Mykytenko DO. Medical and ethical aspects of genetic testing and consultations. KhNMU, Kyiv 2020. — Ablitsov VH. «Galaxy Ukraine». The Ukrainian diaspora: prominent persons. Publ. «Kyt», Kyiv 2007.

hemodynamic and cell-receptive, neurohumoral disturbances in patients suffering from hypertension with comorbidities. She became Director (1991) and Professor (1992), Department of Propaedeutics to Internal Medicine, Kharkiv National Medical University (KhNMU), which with her initiative was renamed in 2012 into Department of Propaedeutics to Internal Medicine No. 1, Foundations of Bioethics and Biosafety.

Pleural effusion is an excess fluid in the pleural cavity, between the visceral and parietal pleura, which could impair breathing by limiting expansion of the lungs during inspiration. It could be transudate passing across membranes or squeezed out by the effect of hemodynamical forces in congestive heart failure (CHF), liver cirrhosis and nephrotic syndrome or exudate characterized by the high fluidity and low content of protein, cells, or the solid matter of cellular origin as in bacterial pneumonia, viral infections, neoplasms, and pulmonary embolism (PE). According to **Richard W. Light (1942 –)**[1000] — American pulmonologist, a pleural effusion is likely exudative if at least one of the following exist:

(1) the ratio of pleural fluid to serum total protein is greater than 0.5;
(2) the ratio of pleural fluid to serum lactate dehydrogenase (LDH) is greater than 0,6; and
(3) the value of LDH in pleural fluid is greater than 200 U/L or greater than 2/3 of the laboratory's upper limit of normal for serum (Light criteria, 1972).

In 1982 **Kanji Inoue** — Japanese cardiac surgeon, first developed the idea that mitral stenosis (MS) could be inflated using a balloon made of strong yet pliant natural rubber. Two years later (1984), he described the first percutaneous (PC) single balloon valvulotomy of the stenotic mitral valve (MV), and since then the PC balloon mitral valvuloplasty (PBMV), a minimally invasive therapeutic procedure became the treatment of choice in selected patients suffering from MS[1001].

The PBMV can be performed using the antegrade (transvenous) route through the right femoral vein just below the inguinal ligament or via internal jugular vein (IJV) in the neck, or retrograde (trans-arterial) route. The latter route has been largely abandoned because its complexity and a risk of potential arterial damage.

Under local anesthesia, a ballon catheter is passed from the right femoral vein, up to the inferior vena cava (IVC) and into the right atrium (RA), from where the interatrial septum (IAS) is punctured, and the catheter is passed further into the left

---

[1000] Light R, Macgregor M, Luchsinger P, Ball W. Pleural effusions: the diagnostic separation of transudates and exudates. Ann Intern Med 1972;77 (4): 507-13. — Light RW (Ed.), Lee YCG. Pleural Diseases. 3rd ed. CRC Press, LLC, Boca Baton, FL 2016.
[1001] Inoue K, Owaki T, Nakamura T, et al. Clinical application of transvenous mitral commissurotomy by a new balloon catheter. J Thorac Cardiovasc Surg 1984 Mar;87(3):394-402.

atrium (LA). Here, the balloon is subdivided in three segments and inflated in three stages: (1) the distal portion of the ballon inside the LV is inflated and pulled against the cusps of the narrowed MV orifice; (2) the proximal segment of the ballon is dilated in order to fix the center segment at the valve orifice; (3) the central portion of the ballon is inflated to dilate the valve orifice for no longer than 30 second, since the full inflation obstruct the blood flow from the LA to the LV, causing pulmonary edema with subsequent cardiac arrest.

The PBMV is indicated in symptomatic patient who are in the New York Heart Association (NYHA) Class III-IV with isolated moderate – severe MS with mitral valve area (MVA ≤ 1.5 cm^2 and favorable valve morphology, and advised in asymptomatic patients with moderate – severe MS and pulmonary artery (PA) hypertension with PA systolic pressure > 50 mmHg or with exercise PA systolic pressure > 60 mmHg, with pulmonary capillary wedge pressure (PCWP) ≥ 25 mmHg, or with new onset of atrial fibrillation (AF).

The PBMV is contraindicated in patients with LA thrombus, moderate MR and poor valve morphology that include leaflet mobility, valve thickness, subvalvular thickness and valvular calcifications as assessed by the Wilkins echocardiographic (echoCG) score[1002]:

Score	Leaflets mobility	Valve thickness	Subvalvular thickness	Valvular calcifications
1	Highly mobile with little restriction	Normal (4-5 mm)	Minimal chordal thickening	A single area of calcification
2	Decreased in midportion & base of leaflets	Mid-leaflet / marginal thickening	Chordal thickening 1/3 of its length	Confined to leaflet margin
3	Forward movement of valve leaflets in diastole	Total leaflet thickening (5-8 mm)	Chordal thickening 2/3 up chordal length	Up to mid-leaflet
4	No or minimal forward movement of leaflets in diastole	Severe leaflet thickening (> 8 mm)	Complete chordal thickening up to papillary muscle	Throughout

A total score of > 8 is predictive of low success with this procedure.

[1002] **Gerard T. Wilkins** — New Zealander interventional cardiologist. // Wilkins GT, Weyman AE, Abascal VM, at al. Percutaneous balloon dilatation of the mitral valve: an analysis of echocardiographic variables related to outcome and the mechanism of dilatation. Br Heart J 1988;60(4):299-308. // Soliman OII, Anwar AM, Metawel AK, et al. New scores for the assessment of mitral valve stenosis using real-time three dimensional echocardiography. Curr Cardiovasc Imaging Rep 2011 Oct 4(5):370-7.

A feared complication of the PBMV is acute severe MR with the 1% incidence of periprocedural mortality. After the PBMV 70%-75% of the patients are free from restenosis 10 years following the procedure, and about 40% 15 years post- PBMV.

Then, **Alain Cribier (1945 –)**[1003] — French interventional cardiologist, performed the percutaneous transluminal balloon aortic valvuloplasty (PTBAV) with the aim of widening of stenotic AV in 1986, the PBMV in 1995, and the transluminal aortic valve implantation / replacement (TAVI / TAVR) through the common femoral artery (CFA) and transcatheter via a small incision below the xiphoid process or a mini-left 5th-6th intercostal space thoracotomy (ICST) in the left midclavicular line (MCL), through the apex of the left ventricle (LV) of the heart for severe native AV disease in 2002.

Following discovery that the PTBAV for severe AV stenosis was not effective in 80% patients after one year, it was replaced by the TAVI / TAVR.

Furthermore, **John G. Webb**[1004] — Canadian interventional cardiologist, performed the transfemoral and transapical TAVI in 2005, the transcatheter mitral valve-in-valve replacement in 2009, and PC transcatheter MV replacement (TMVR)[1005] in 2014.

Those procedures are done under general anesthesia and guided by guided by transesophageal echocardiography (TEE), fluoroscopy and angiography.

Most common access routes for the TAVI / TAVR are transfemoral and transapical. The other approaches for performing the transcatheter TAVI / TAVR in patients who are not eligible for the transfemoral and transapical approaches, are the transcatheter approach via the subclavian artery (SCA) below the collar bone, a direct transaortic approach through a minimally invasive surgical incision into the aorta, and the transcaval approach from the femoral vein below the groin, into the IVC and through a temporary hole made by a guided wire to cross into the adjacent abdominal aorta (AA) at the level of the umbilicus. Once the guided wire is across in the AA, a catheter is passed over the guided wire which is used to place the transcatheter heart valve through the femoral vein and IVC into the AA and from there to the aortic root for replacement of the patient's native diseased AV. Afterwards, the created hole in

---

[1003] Cribier A, Elchaninoff H, Bash A, et al. Percutaneous transcatheter implantation of an aortic valve prosthesis for calcific aortic stenosis. Circulation 2002 Nov 25;106:3006-8.

[1004] Webb JG, Candavimol M, Thompson Ch. Percutaneous aortic valve implantation retrograde from the femoral artery. Circulation 2006 Feb;113:842-500.

[1005] A 86-year-old man experiencing symptomatic MR grade IV+, who declined surgical MV repair, mitral valve replacement (MVR) or MV clip placement. Finally accepted the PC TMVR which was performed in Denmark on June 12, 2012, by the use of an antegrade interatrial septum (IAS) route with uneventful initial recovery. The hemodynamics was stable in the first 24 hours. But despite functioning trans-MV prosthesis, the patient died three days after the procedure from multiorgan failure.

the AA is closed with a self-collapsing nitinol device used to close the atrial and ventricular septal defect (ASD, VSD) of the heart.

Transcatheter TAVI / TAVR is indicated to treat a severe symptomatic native AV disease in an older patient over 75-year of life, and those with concomitant severe systolic congestive heart failure (CHD), coronary artery disease (CAD), as well as comorbidities such as cerebrovascular disease, peripheral arterial disease (PAD), chronic or end-stage kidney disease (ESKD) and chronic obstructive pulmonary disease (COPD).

Contemporary technologies of non-surgical treatment of MV diseases are by a PC mitral valve repair or PC TMVR. The former is based on the pathology of MV and include: (1) leaflets plication (edge-to-edge repair), coaptation or ablation; (2) indirect annuloplasty through the coronary sinus (CS) or direct annuloplasty (through the PC or hybrid approach); (3) PC chordae tendineae implantation; or PC LV remodeling.

The PC TMVR through is performed through a small subxiphoid incision, exposing the apex of the LV of the heart, where a U-shaped purse string suture is placed around the entry site, and a J-tipped guide wire is then advanced through the LV and across the native diseased MV orifice. The catheter, loaded with the MV bioprosthesis is then advanced over the guide wire positioned in the LA. When central positioning is confirmed using TEE and angiography, the prosthetic valve is deployed and anchored behind the anterior and posterior leaflets of the native diseased MV. Then the guide with catheter and delivery system are removed. The purse string suture is tied to secure hemostasis.

In 1978, **Stanley N. Schwartz (1946 –)**[1006] — American physician specializing in infectious diseases, found that pathogenic microorganisms, especially gram-negative bacteria can migrate in only three days after endotracheal (ET) intubation from the mouth, along the ET tube to the lungs, causing aspiration pneumonia (AP).

[1006] Schwartz SN, Dowling JN, Bencovic C. Sources of gram-negative bacilli colonizing the tracheae of intubated patients. J Infectiouss Diseases. Aug 1978;138(2):227-31.

Principal directions of scientific endeavor of **Volodymyr M. Kovalenko (02. II.1949, c.t.v. Voronizh, Shostkynvsky Region, Sumy Obl., Ukraine –)**[1007] — Ukrainian physician therapeutant and cardiologist, is concerning elaboration of theoretical and clinical evaluation of cardiac function and diagnosis of heart failure, systematization of non-coronary diseases of the heart, to relate and incarnate into practice of health care of Ukraine new organization and methodical form of analysis, management and perspectives to develop therapeutic, cardiologic and rheumatologic services.

Francis E. Marchlinsky (1951 –) — American cardiologist and cardiac electrophysiologist, documented in 2000 the effectiveness of a technique to cure atrial fibrillation (AF) by targeting and isolating the triggers or «hot spots» on the pulmonary veins (PV) entering the left atrium (LA) of the heart, thus preventing natural electrical circuitry and eliminating the need for medication or pacemaker (PM).

First noted in 1989, recognized as a distinct clinical entity in 1992 by **Pedro Brugada (1952 –)**[1008] — Spanish Belgian cardiologist and **Joseph Brugada** — Spanish-French cardiologist, and named the Brugada syndrome, is a genetic disease characterized by the abnormal electrocardiography (ECG) pattern (right bundle branch block / right bundle branch block (RBBB) and persistent ST-segment elevation in the right precordial leads) with type 1 (a coved-like ST elevation with at least 2 mm or 0.2 mV J-point elevation a gradually descending ST segment followed by a negative T-wave), type 2 (a saddle back pattern with at least 2 mm J-point elevation and at least 1 mm ST elevation with a positive or biphasic T-wave, this pattern can occasionally be seen in healthy subjects), and type 3 (either a coved-like ST elevation as the in

---

[1007] Ctv Voronizh (settled since late Paleolithic Age 13 thousand years BC, Age of Bronze, Scythes, and early Slavs 7th-8th century; suffered from organized by Russian Communist Fascist Holodomor-Genocide 1932-33 against the Ukrainian Nation), Shostkynvsky Region, Sumy Obl., Ukraine. — Kovalenko VM, et al. Rheumatic Disease in Ukraine: contemporary status, problems and giving medical help and the ways for improvement. Publ. «Vipol», Kyiv 2002. — Kovalenko VM, et al. Nomenclature, Classification, Criteria of Diagnostic and Programs to Treat Rheumatic Diseases. Publ. «Katran Grup», Kyiv 2002. — Kovalenko VM, Kornatsky VM. Academician Strazhesko and Contemporaneity. Kyiv 2006. — Kovalenko VM, et al. Handbook of Cardiology. Publ. «Morion», Kyiv 2008. — Kovalenko VM (Ed.). Demeanor of Cardiology. Publ. «Morion», Kyiv 2009. — Kovalenko VM, Bortkevych OP. Osteoporosis. Practical Adjustment. 3rd Ed. Publ. «Morion», 2010. — Kovalenko VM, Shuba NM, Kazimirko VK. National Texbook of Rheumatology. Publ «Morion», Kyiv 2013. — Ablitsov VH. «Galaxy Ukraine». The Ukrainian diaspora: prominent persons. Publ. «Kyt», Kyiv 2007. — Hanitkevych YaV. The contribution of the Ukrainian physicians to the world medicine. Ukr Med Chasopys 2009 VII/VIII;4(72):110-15. — Synyachenko OV. They glorified themselves and Ukraine: history of domestic therapy. Crimean Therapeutic Journal 2010;2(2):21-8.

[1008] Brugada P, Brugada J. Right bundle branch block, persistent ST segment elevation and sudden cardiac death: a distinctive clinical and electrocardiographic syndrome. J Am Coll Cardiol 1992;20:1391-6. — Brugada J, Brugada P, Brugada R. The syndrome of right bundle branch block ST segment elevation in V1 to V3 and sudden death – the Brugada syndrome. Europace 1999;1 (3): 156-66.

type I pattern, or a saddle back like pattern as in the type 2 with less than 2 mm J-point elevation and less than 1 mm ST elevation). This syndrome carry an increased risk of an unexpected sudden cardiac death (SCD) from ventricular fibrillation (VF), especially ECG type III in young men without known underlying cardiac disease. The treatment of choice is implantable cardioverter-defibrillator (ICD).

In 1995 **Patricia A. Ford (1955 –)** — American physician, hematologist and oncologist, carried out the first bloodless stem cell (SC) transplantation.

Visible evidence of the brain-heart interrelationship which was noted as early as in 1881 by **Moritz Litten (1845 – 1907**: 19th century. Internal medicine and subspecialties. / Ophthalmology), and in 1900, by **Albert Terson (1867 – 1935**: 19th century. Ophthalmology), was described in 1990-91 by **Hirashi Sadoh (1961 –)**[1009] — Japanese cardiologist, as a takotsubo (in Japanese – «tako tsubo»), a «bait for octopus» syndrome, a «broken-heart» syndrome or the takotsubo cardiomyopathy (CM), a sudden temporary weakening of the heart muscle or stress-related CM, challenged by a serious recent physical or emotional burden, especially after loss or death of a beloved person[1010].

The prevalence of takotsubo CM in the general population is estimated to be 1,2% to 2,2%, up to 40% of the cases of stress-induced takotsubo CM, most frequently seen in menopausal and post-menopausal women.

It is possible that because of such an event the brain directs the sympathetic nervous system to emergently secrete high doses of catecholamines (epinephrine / nor-epinephrine), that by narrowing small arteries of the myocardium leads to a local dysfunction of the intima and microcirculation, insufficient supply of oxygen (O_2) and ischemia, acute heart failure, similar to that after myocardial infarction (MI).

The clinical picture of takotsubo CM is characterized chest pain, can resemble EKG changes that of MI with elevation of cardiac troponin I levels in the blood; the transthoracic echocardiography (TTE), transesophageal echocardiography (TEE) and cardiac catheterization - selective coronary angiography (CC-SCA) evidence of transient focal wall-motion abnormalities (hypokinesis, akinesis, or dyskinesis) of the left ventricle (LV) that involve the apical and middle segments of the LV, in the absence of obstructive coronary artery disease (CAD). Characteristically, the

---

[1009] Sadoh, H, Tadeishi H, Uchida T, et al. Takotsubo type cardiomyopathy due to multivessel spasm. In, K. Kanama, K. Haze, M. Hon, et al. (Eds). Clinical Aspect of Myocardial Injury from Ischemia to Heart Failure. Kagakuhyouronsya, Tokyo 1900. P. 56-64. — Dote K, Sato H, Tateishi H, et al. Myocardial stunning due to simultaneous multivessel coronary spasms; a review of 5 cases. J Cardiol 1991;21(2):203-14.

[1010] Yamanaka NN, Nakayama O, et al. «Tako-Tsubo» transient ventricular dysfunction: a case report. Jpn Circ J 2000;64(9):715-9. — Gianni M, Dentali F, Grandi AM, et al. Apical balloning syndrome or takotsubo cardiomyopathy: a systemic review. Euro Heart J 2006;27(13):1523-9.

later study shows normal or spasmatic coronary arteries, during diastole an apical paradoxical bulging (narrow neck/wide base appearance), resembling a «bait for octopus».

The variant of takotsubo CM, that is an atypical (inverted or reverse) takotsubo CM was described in 2010[1011]. It presents itself at a younger age and is always triggered by stress. The TTE, TEE or CC-SCA shows hypokinesis, akinesis, or dyskinesis of the middle and basal LV segments, sparing the cardiac apex.

The presence of pheochromocytoma or myocarditis must be excluded.

This frightening situation can usually last approximately 2 months and can lead to death.

Treatment consists of psychological support, hydration, acetylsalicylic acid (ASA), negative inotropes such as beta blockers or calcium channel blockers, sometimes it is necessary to use an intraarterial balloon assist (IABA) device.

Cardiac resynchronization therapy (CRT), or biventricular, i.e., the right and left ventricular (RV, LV) pacing, as an adjunctive treatment for advanced congestive heart failure (CHF) / heart failure (HF), where the RV and the LV do not contract simultaneously due to mechanical ventricular desynchrony in association with drug resistance, was introduced in 1994 by **Serge Cazeau**[1012] — French cardiologist. His first patient was a 54-year-old man in severe CHF as evidenced by the NYHA Clinical Class IV who received CRT, i.e, four chamber pacing system, has had a significant increase of cardiac output (CO) and decrease in pulmonary capillary wedge pressure (PCWP). Six months later, the patient's clinical status improved markedly to the NYHA Class II, with weight loss of 17 kg and disappear of peripheral edema.

In severe CHR / HF ventricular desynchrony occurs in 25-50% of patients.

The indication for the CRS in CHF / HF is the NYHA Clinical Class > II due to systolic dysfunction with the left ventricular ejection fraction (LVEF) < 35%, intraventricular conduction delay (IVCD), generally the left bundle branch block (LBBB) with the ventricular complex QRS duration of more than 150 msec (wide complex QRS).

[1011] Ramaray R, Movahed MR. Reverse or inverted takotsudo cardiomyopathy (reverse left ventricular apical ballooning syndrome) presents at a younger age compared with the mid or apical variant and is always associateted with triggering stress. Congest Heart Fail 2010;16(6):284-6.

[1012] Caseau S, Ritter P, Bakdach S, et al. Four chamber pacing in dilated cardiomyopathy. Pacing Clin Electrophysiol 1994 Nov;17(12)1974-9. — Caseau S, Leclercq C, Laverge T, et al. Effects of multisite biventricular pacing in patients with heart failure and intraventricular conduction delay. N Engl J Med 2001;344(12):873-80. — Kravchuk BB, Malyarchuk RG, Paratsyi OZ, Sychyk MM. Resynchronization therapy effectivization by radiofrequency ablation of ventricular arrhythmia. Heart and Vessels 2014;4:67-72.

For patients at risk of arrhythmia, the CRT can be combined with an implantation of cardioverter defibrillator (ICD), which provide effective protection against life-treating arrhythmia.

However, there are no convincing evidence that the CRT benefits patients in CHF with IVCD < 120 msec (narrow complex QRS), even if echocardiography (echoCG) shows evidence of mechanical ventricular desynchrony.

According to **Kenneth Shay (c. 1958 –)**[1013] — American dentist, geriatrician and gerontologist, one cubic millimeter (mm^2) of the morning human teeth could harbors up to 10 billion microorganisms. He also found that, it takes only 48 hours of hospitalization for bacteria to change from gram-positive to gram-negative pathogens.

Fortunately, in healthy individual chemicals and bacteria facilitated by saliva maintains homeostasis in the oral cavity. However, in diseased, immune-suppressed or stressed patients with comorbidities, especially elderly and aging, treated in hospitals or those stayed in the long-term acute care hospitals (LTACH), pulmonary rehabilitation facilities and nursing homes, a heightened number of pathogenic microorganism, especially in those intubated or with tracheostomy, may colonize on them, causing serious diseases, such as the syndrome of oropharyngeal dysphagia (OD)[1014], manifest by inability to swallowing and intermittent tracheal-bronchial aspiration and recurrent aspiration, i.e., when a solid and/or liquid secretions from the oral cavity (saliva) or gastric content (including hydrochloric acid – HCl), or one and other (descending and ascending aspiration), and also food, causing recurrent AP.

The OD syndrome is a geriatric syndrome due to high prevalence and its relationship with AP, respiratory tract infections, undernourishment and malnutrition, functional disability and frailty, institutionalization, increased readmissions, comorbidity, poor outcomes, high morbidity, and mortality.

The etiology includes many concomitant factors with neurogenic and neurogenerative diseases, muscular weakness, and sarcopenia.

The pathophysiology includes mechanical deficits in the swallow response, mainly delayed or absent laryngeal vestibule closure time and weak tongue thrust, reduced pharyngeal sensitivity and sensory / motor central nerve system (CNS)

[1013] Shay K, Yoshikawa TT. Overview of VA health care for older veterans: lessons learned and policy implementations. Generations 2010;34(2):20-28.

[1014] Legemann JA. Evaluation and treatment of swallowing disorders. 2nd Ed. Tex. Pro-Ed, Austin, TX 1998. — Cichero J, Lam P, Steele P, at al. development of international terminology and definitions for texture-modified foot and thickened fluid used in dysphagia management; the IDDSI framework. Dysphagia Apr 2017;32(2):293-314. — Bock JM, Varadarajan V, Brawley MC, Blumin JH. Evaluation of the natural history of patients who aspirate. Laryngoscope 2017 Dec;127(suppl 8):S1-S10.

impairments, with failure to recognize «foreign bodies», like thickened fluids, clear or soft meals.

The treatment of OD syndrome in the elderly is mainly oriented to compensate swallowing impairments through adaptation of fluid viscosity and solid food textures to avoid aspiration and choking, improving nutritional status, vigorous oral hygiene by a suction brushing over foam swab and rinsing of the teeths, gums, mucosal membrane, and tongue to avoid infections.

The 2006 study by **Arnelia G. Ross** and **Janet Crumpler**[1015] — American intensive care unit (ICU) registered nurses (RN), showed that a nurse training program on the benefit of tooth-brushing over foam swabs of the throat led to a 50% reduction in ventilator-acquired AP rates among older patients.

The research conducted in 2014-2018 on frail elderly patient with the OD syndrome by **Omar Ortega**[1016] — Spanish gastroenterologist, revealed the highest bacterial concentration on the teeth, tongue and saliva.

reflux disease (GERD), uncontrollable aspiration of the tracheobronchial system from above by the saliva and below by the gastric contest, it may be necessary to perform tracheostomy for ventilatory support, suctioning from the oral cavity and the tracheobronchial tree, per endoscopic gastrostomy (PEG) for feeding or drainage, insert percutaneously (PC) 14F jejunostomy tube (14F J-tube) in adults by interventional radiologist, (IR) or via mini-invasive laparotomy by gastrointestinal (GI) surgeon. In recurrent or resistant OD syndrome a total laryngectomy **[Theodor Bilroth (1827 – 94). 19**th **century. Surgery]** should be taken into consideration. The undergoing studies and developments may bring methods to correct the OD syndrome with pharmacological or physical stimuli.

The group led by Sir **Peter J. Ratckliff (1954 –)**[1017] — English physician-scientist and nephrologist, discovered that m-ribonucleic acid (mRNA) from kidneys was part of erythropoietin (EPO) **[Paul Carnot (1869 – 1958) and Clotilda Deflandre**

[1015] Ross A, Crumpler J. The impact of an evidence-based practice education program on the role of oral care in the prevention of ventilator-associated pneumonia. Intensive Crit Care Nurs 2006;23:132-6.

[1016] Ortega O, Parra C, Zarcero S, et al. Oral health in older patients with oropharyngeal dysphagia. Age and Ageing Jan 2014;43(1):132-7. — Ortega O, Martin A, Clave P. Diagnosis, and management of oropharyngeal dysphagia among older patients. State of the Art. J Am Med Dir Assoc 1 Jul 2017;18(7);576-82. — Martin A, Ortega O, Clave P. Orophyngeal dysphagia, a new geriatric syndrome. Rev Esp Geriatr Gerontol 2018 Jan-Feb;53(1):3-5.

[1017] Maywell GL, Wiesener MS, Chang GW, et al. Tumor suppressor protein VHL targets hypoxia-inducible factors for oxygen-dependent proteolysis. Nature 1999;271-5. — Jaakkola P, Mole DR, Tian YM, et al. Targeting of HIE-alpha to the von Hippel-Lindau ubiquitilation complex by O_2–regulated propyl hydroxylation. Science 2001;468-72.

(1871 – 1962). Internal medicine. 20th century] production system capable of detecting low cell's oxygen (O_2) level was also present in the brain, spleen, and testicles (testes). The group was able to modify other cells, using the identified mRNA to give these cells O_2-sensing capacity.

In 1989 the Peter Ratcliff group joined **William G. Koelin, Jr (1957 –)**[1018] — American oncologist, and **Gregg L. Semenza (1956 –)**[1019] — American pediatrician and oncologist, to help uncover detailed molecular chain of events that cells use to sense O_2.

In the meantime, William Koelin found in 1993 in subsets with the von Hippel-Lindau (VHL) disease **[Eugene** von **Hippel (1867 – 1939)** and **Arvid Lindau (1892 – 1958)**. Ophthalmology. 20th century]**, genes expressed the formation of a protein critical in the EPO production, but the mutation suppressed it.

While evaluating gene expression in transgenic animals to determine how this affected the production of EPO, Gregg Semenza identified in 1995 the gene sequences that expressed hypoxia inducible factor (HIF) proteins, consisted of two parts – HIF 1-alpha that deteriorated at the presence a nominal O_2 levels, and HIF 1-beta, a stable factor under most conditions. HIF 1-alpha is a basic-helix – loop – helix – a Per-Apnt-Sim (PAS)[1020] heterodimer regulated by cellular O_2 tension. It was essential to the EPO production process, as test subjects modified to be deficient in HIF 1-alpha and were found to have malformed blood vessels and decreased EPO level, and these HIF proteins were found in multiple test animals. He further found that HIF 1-alpha overproduction could lead to cancer in other subjects.

A specific step identified was the binding of proteins expressed by the VHL tumor suppressor gene VHL to HIF, a transcription factor which trans-activates the EPO gene. Peter Ratcliff found that the VHL protein can bind hydrolyzed residues of HIF when O_2 is present at acceptable levels; the VHL protein then ubiquitinate[1021] the HIF protein, which ultimately leads to the HIF protein's destruction. When O_2 level

---

[1018] Mircea I, Kondo K, Yang H, et al. HIF a targeted for VHL-mediated destruction by proline hydroxylation: implications for O_2 sensing. Science 2001;292:464-8.

[1019] Semenza GL, Nejfeld MK, Chi SM, and al. Hypoxia inducible nuclear factors binds to enhancer element located 3' to human erythropoietin gene. Proc Natl Acad Sc USA 1991;88:568-4.—Wang GL, Semenza GL. Purification and characterization of hypoxia-induced facror 1. J Biol Chem 1995 Jan 20:270(3)1230-7.

[1020] A Per - Apnt – Sim (PAS) domain found in all kingdoms of life, act as molecular sensor, whereby small molecules and other proteins associate via binding.

[1021] Goldstein G, Scheid M, Hammerlink KM, et al. Isolation of a polypeptide that has lymphocyte-differentiating properties and is probably represented universally in living cells. Proc Natl Acad Sci USA 1975 Jan;72(1):11-5. // Ubiquitin (in Latin – «ubique» meaning «everywhere», from «ubi», meaning «where»), or ubiquitination is a small regulatory protein, occurring ubiquitously, found in most tissues of eukaryotic organisms, those whose cells have a nucleus enclosed within mebranes. —Ubiquitin discovered in 1975 **Gideon Goldstein (1938 -)** — American immunologist.

fall, O_2-requing HIF hydrolase enzymes no longer act and VHL does not bind HIF, allowing HIF remains and activate EPO gene. This process takes minutes to complete permitting the body to react quickly to hypoxia. The molecular pathway of EPO production is also switched on in many malignant tumors, allowing them to create new blood vessels to sustain their growth.

The combined work of Peter Ratckliff, William Koelin, Jr and Gregg Semenza led to development of drugs that can block the VHL proteins from binding with HIF to help treat patients with anemia and renal failure (RF). The discovery of HIF is implicated in understanding and treating low-oxygen conditions of the body, such as coronary artery disease (CAD), myocardial protection against ischemia - reperfusion injury, therapy of inflammation, malignant tumors, and neurological diseases.

The first endoscopic cholecystostomy was performed in 2007 by **Todd H. Baron (1960 –)**[1022] and **Mark Topazian** — American gastroenterologists, using ultrasonic (US) guidance to puncture the stomach wall and place a plastic biliary catheter for drainage.

Boris N. Mankovsky (10.01.1964, city of Kyiv, Ukraine –)[1023] — Ukrainian physician-endocrinologist, described disturbances of the immunologic phenotype of lymphocytes, interrelation of pro- and anty-inflammatory cytokinin's during formation of diabetes mellitus (DM) - 1th type; proposed new scheme for treatment patients suffering from DM complicated by neuropathy, especially by stimulation of the cranial X (vagal) nerve in patients with severe degree of diabetic paresis of the stomach (gastroparesis).

[1022] Baron TH, Topazian MD, et al. Endoscopic trans-duodenal drainage of the gallbladder. Implication for endoluminal treatment of gallbladder disease. Gastrointest Endosc 2007 Apr;65(4):735-7.

[1023] Mankovsky BN, Patrick JT, Metzger DE, Saver JL. The size of subcortical ischemic infarction in patients with and without diabetes mellitus. Clinical Neurology and Neurosurgery 1996;98(2):137-41. — Mankovsky BN, Piolot R, Mankovsky OL, Ziegler D. Impairment of cerebral autoregulation in diabetic patients with cardiovascular autonomic neuropathy and orthostatic hypotension. Diabetic Medicine. 2003 Mar;20(2):119-26. Mankovsky BN, Ziegler D. Stroke in patients with diabetes mellitus. Diabetes / Metabolism Research and Reviews 2004Jul;20(4):268-87. — Mandrik O, Severens JL, Doroshenko O, at al. Impact of hypoglycemia on daily life of type 2 diabetes patients in Ukraine. J Multidisciplinary Healthcare 2013 Jul;6:249-57. — Ablitsov VH. «Galaxy Ukraine». The Ukrainian diaspora: prominent persons. Publ. «Kyt», Kyiv 2007.

Colonel of the United States (US) Army **Andrew P. Cap (1971 –), MD, Phylosophy Doctor (PhD)**[1024] — Ukrainian American military hematologist, oncologist and specialist in bone marrow transplantation, Director of Research at the US Army Institute of Surgical Research, who directs 340 staff and over \$45 million in research funds, made an important contribution for correction of abnormalities in the blood coagulation and volume in soldiers sustaining combat injuries.

[1024] Cap AP. Crash-3: a win for patients with traumatic head injury. Lancet 2019;394(10210)1687-8. — Cop AP, Reddoch Cardenas KM. Can't get platelets to your bleeding patients? Just chill…the solution is in your refrigerator. Transfusion Clinique et Bioloque 10.1016/j tracli 2018.06.008(2018). — Yazer MH, Cap, Spinella PC. Raising the standartds on whole blood. J Trauma and Acute Care Surgery 2018;84(65):S14-S17. — Cap AP, Pidcoke HF, Spinella P, at al. Damage control resuscitation. Milit Med 2018;183(Suppl_2):36-43. — Cop AP, Beckett A, Benou A, et al. Whole blood transfusion. Milit Med 2018;183(Suppl_2):44-51. — Cap AP. Targeting hemorrhage: alternative storage of platelets for hemostativc transfusion. Blood 2017;130(Suppl 1):Sci-32–Sci-32. — Cop AP, Pidcoke HF, Keil SD, et al. Treatment of blood with a pathogen reduction technology using ultraviolet and riboflavin inactivates Ebola virus in vivo. Transfusion 2017;56:S6-S15. — Cop AP, Spinella. Justy chill – it's worth it! Transfusion 2017;57(12):2817(12):1817-20. — Cop AP. Platelets storage: a license to chill! Transfusion 2016;56(1):13-6. — Cap AP, Getz TM, Spinella, et al. platelet transfusion. Trauma Induced Coagulopathy 2016:347-6. — Cap AP, Hunt BJ. The pathogenesis of traumatic coagulopathy. Anesthesia 2015;70:96 – e34. — Fisher AD, Miles EA, Cap AP, et al. Tactical damage control resuscitation. Milit Med 2015;180(8):869-75. — Kragh JF, Kotwal RS, Cap JP, et al. Performance of functional tourniquets in normal human volunteers. Prehospital Emergency Care 2015;19(3):391-8. — Strandenes G, Berseus O, Cap AP, et al. Low filter group O whole blood transfusion in emergency situations. Shock 2014;41:70-5. — Nair PM, Pidcoke HF, Cap AP, et al. Effect of cold storage on shear-induced platelets aggregation and clot strength. J Trauma and Acute Care Surgery 2014;77(3);S88-S93. — Strandenes G, DiPasquqle M, Cap AP, et al. Emergency whole blood use in the field: a simplified protocol for collection and transfusion. Shock 2014;41:76-86. — Sailliot A, Martinand AP, Cap AP. The evolving role lyophilized plasma in remote damage control resuscitation in the French Armed Forces Health Service. Transfusion 2013;65S-71S. — Gerhard RT, Strandenes G, Cap AP, et al. Remote damage control resuscitation and the Solstrand Conference defining the need, the language, and a way forward. Transfusion 2013;53:9S-16S. — Cap AP, Spinella PC, Borgman MA, et al. Timing and location of blood product transfusions and outcomes in massively transfused combat causalities. J Trauma and Acute Care Surgery 2012;73(2):S89-S94. — Spinella PC, Cap AP. Whole blood: back to the future. Current Opinions in Hematology 2014;11(6):406-15. — Cap AP, Baer BG, Orman JA, et a Transexamic acid for trauma patients: a critical review of the literature. J Trauma and Acute Care Surgery 2011;71(1):S9-S14. — Cap AP, Spinella PC. Severity of head injury is associated with increases risk of coagulopathy in combat causalities. J Trauma and Acute Care Surgery 2011;71(1):S78-S81.— Perkins JG, Cap AP, Spinella PC, et al. Comparition of platelets transfusion as fresh whole blood versus apheresis platelets for massively transfused combat trauma patients (CME). Transfusion 2011;51(2):242-52. — Cap AP, Gurney JM, Meledeo MA. Hemostatic Resuscitation. Damage Control Resuscitation 2010:117-:44. — Perkins JG, Cap AP, Weiss BA, et al. Massive transfusion and nonsurgical hemostatic agens. Crit Care Med 2008;32(7):S325-S339. — Leopold JA, Cap AP, Scribner AW, et al. Glucose-6-phosphatase dehydrogenese deficiency promotes endothelial oxidant stress and decreases endothelial nitric oxide bioavailability. FASEB J 2001;36(7):1771-3. — Ablitsov VH. «Galaxy Ukraine». The Ukrainian diaspora: prominent persons. Publ. «Kyt», Kyiv 2007.

Anesthesiology

In 1902 **Emil Fisher (1852 – 1919**: 19th century. Biochemistry**)** and **Joseph** von **Mering (1849 – 1908**: 19th century. Biochemistry / Internal medicine**)** synthetized barbiturate (veronal) for intravenous (IV) anesthesia.

Alfred Einhorn or **Alfred Finkle (1856 – 1917)** — Jewish German chemist, synthesized in 1905 procaine (Novocain), a safe, effective, and widely used local anesthetic, although weaker than cocaine hydrochloride and some patients were highly allergic to it.

The morphine-scopolamine **[Albert Landenburg (1842 – 1911)**. 19th century. Biochemistry**]** anesthesia, known as «twilight sleep» (in German – Dämmerschlaf) was introduced in 1899-1900 via intravenous (IV) injection by **Schneiderlin**[1025] — German surgeon, to relieve the pain caused by surgery, either by itself or as adjuvant to ether administration, and in 1902-03 by **Richard A.L.A.** von **Steinbüchel (1865 – ?)**[1026] — German gynecologist and obstetrician, into obstetrics to relieve the pain of childbirth. It is characterized by amnesia with insensitivity to pain without loss of consciousness. The latter used the morphine-scopolamine anesthesia in 20 obstetrics' patients with favourable results.

In 1903 **Ernst F. Sauerbruch (1875 – 1951)**[1027] — German thoracic surgeon, built a large hermetic low-pressure (approximately -7 cm H_2O) Sauerbruch's cabinet (chamber) – the operative room (OR) where the patient was positioned on the operative table with his head outside the cabinet, separated by a tight soft collar placed around his neck, while the rest of the body and the operative team were inside the chamber. The low-pressure cabinet allowed the lung expansion during inspiration and recoil during expiration, as in normal breathing, allowing to perform a intercostal space thoracotomy an (ICST) or a midline longitudinal sternotomy incision (MLSI), operate inside the chest cavity or mediastinum, without causing collapse of the lung and circulatory instability.

[1025] Ries E. Scopolamine- morphine anesthesia, California St J Med 1906 Mar;4(3):109-110. — Keys TE. The History of Surgical Anesthesia. Wood Library, Museum of Anesthesia, Park Ridge, Ill . 1996.

[1026] Steinbüchel R. Vorläufige Mittheilung über die Anwendung Scopolamin-Morphium Injektionen in der Geburtshülfe. Zentralblatt fur Gynekologie 1902;26:1304. — Steinbüchel R. Schmerz verminderung und Narkose in der Geburtshülfe. 1903. — Steinbüchel R. Die Scopolamin-Morphium-Halbnarkose in der Geburtshülfe, 1 Beitrage 2 und Gynekologie Festschrift Z. Rudolf Chrobak. 1903;294.

[1027] Sauerbruch F. Über die Auschaltung der schandlichen Wirkung des Pneumothorax bei intrathorakelen Operationen. Zentralbl Chir 1904;31:146-9. — Sauerbruch F. Master Surgeon. The Cornwall Press, Inc., Cornwall, NY 1953.

In contrast to the Sauerbruch chamber, **Rudolf Brauer (1865 – 1951)**[1028] — German surgeon, introduced in 1904 a small hermetic high-pressure glass camera in which the head of the patient packed thickly around the neck was inserted, and ventilation during the MLSI or ICST was assured by periodically inflating the lungs with the pressurized air for an inspiration.

The next step for the safe opening of the chest was a tracheal intubation and the introduction of anesthetic gases into the lungs during inspiration under the pressure.

And so in 1904 **Chevalier Jackson (1865 – 1958)** — American otorhinolaryngologist (ORL), devised an open-ventilating rigid bronchoscope with a channel for suctioning and with distal illumination for evaluation of the upper and lower airways, described in 1909 the currently used tracheostomy, and in 1913 described the use of a U-shaped laryngoscope to facilitate the placement of an endotracheal tube, which then served as a conduit for the administration of inhaled oxygen and anesthetics.

Samuel J. Meltzer (1851 – 1920) — American physiologist and his son-in-law **John Auer (1875 – 1948)** — American physiologist and pharmacologist, developed in 1909 the method of endotracheal ventilation of the lung using the intermittent positive pressure breathing (IPPB).

Due primarily to the introduction of the endotracheal intubation, the method of IPPB became generally accepted in medicine.

The first automated respirator was constructed by the Firm «Draeger» in 1911. Around 1916 **Knut H. Giertz (1876 – 1950)** — Swedish inventor, developed the breathing apparatus called «spiro-pulsator».

In 1926 **Mark C. Lidwill (1878 – 1969)**[1029] — English Australian anesthesiologist, along with **Edward H. Booth** — Australian physicist, created the first rudimentary portable artificial cardiac pacemaker which was *«plugged into a lighting point»* and in which *«One pole was applied to a skin pad soaked in strong salt solution»* while the other pole *«consisted of a needle insulated except at its point, and was plunged into the appropriate cardiac chamber».* *«The pacemaker rate was variable from about 80 to 120 pulses per minute, and likewise the voltage variable from 1,5 to 120 volts».* It was used in 1928 to revive a stillborn infant.

The persons establishing anesthesiology as a specialty were **Francis H. McMechan (1879 – 1939)** — American anesthesiologist, who founded the American Association of Anesthetists (1912), **Elmer I. McKesson (1881 – 1935)** — American inventor, **Ralph M. Waters (1883 – 1979)** — American anesthesiologist and **Arthur**

[1028] Brauer L. Die Ausschaltung der Pneumthoraxfolgen met Hilfe des Überdruckverfahrens. Mitteilungen aus den Grenzebieten der Medizin und Chirurgie 1904;398-486.

[1029] Lidwell MC. Cardiac Disease in Relation to Anesthesia. In, Abbot GH, Archdall, M. Transactions of the Third Session; Australian Med Congress (British Med Assoc), Sydney Sep. 2-7, 1929. Alfred J. Kent, I.S.Q., Government Printer, Sydney 1930. P. 160.

E. Guedel (1883 – 1956) — American anesthesiologist. Elmer McKesson invented and developed gas-oxygen (O_2) machines, suction apparatus, metabolism measuring devices, intermittent and demand gas flow valves, as well as the oxygen tent. Ralph Waters established the first residence training in anesthesia at the University of Wisconsin (1927), introduced cuffed endotracheal tubes, laryngoscopy blades, oropharyngeal airways, carbon dioxide (CO_2) absorption canisters and precision-controlled, liquid anesthetic vaporizer. He showed that respiration could be controlled either by squeezing the anesthetic bag by hand or by using a ventilator (1931). Arthur Guedel designed oropharyngeal (Guedel) airways (1933) and described the four stages of a general inhalational anesthesia (1937).

Phillip Drinker (1894 – 1972)[1030] — American industrial hygienist and **Louis A. Shaw, Jr (1886 – 1940)** — American physiologist, constructed a Drinker respirator known as «iron lung» (1928), a form of noninvasive negative-pressure ventilator which was used widely during epidemics of poliomyelitis in the 1940s and 1950s. In a while it was improved by **John H. Emerson (1906 – 97)** — American inventor of biomedical devices who introduced his own (Emerson respirator, 1931).

Combined use of ether for a general anesthesia and d-tubocurarine, a non-polarizing long-acting muscle relaxant, were first applied in 1942 by **Harold Griffith (1894 – 1985)** — Canadian physician-anesthetist and **Enid J. McLeod (1909 – 2001)** — Canadian physician-anesthetist for an appendectomy. It proved to be safe.

Emery A. Rovenstine (1895 – 1960) — American anesthesiologist, organized the first academic department of anesthesiology in 1935 at the Bellevue Hospital (founded 1736), city of New York, NY, USA, helped to develop the use of gas cyclopropane in general anesthesia and initiated therapeutic nerve blocking for the pain relief[1031].

The modern cardio-pulmonary resuscitation (CPR; in German – die Herz-Lungen -Wiederbelebung)[1032] is an emergency medical procedure directed towards the reviving vital functions of a person, bringing him back to life from clinical death, during the first 3-5 minutes from cessation of breathing, loss of the carotid, or femoral

---

[1030] Drinker P, Shaw LA. An apparatus for the prolonged administration of artificial respiration: I. Design for adult and children. J Clin Invest 1929;7:229-47.

[1031] Rovenstine EA, Hershey SG. Therapeutic and diagnostic nerve blocking: a plan for organization. Aneasthesiology 1944;5(6):574-82.

[1032] Jude JR, Kouwenhoven WB, Knickerbocker GG. Cardiac arrest. Report of application of external cardiac massage on 118 patients. JAMA 1961;178;1063-70. — Grenda J, Olearchyk A. Ergebnisse der Resuszitation. Sym Anaesth Int'l. Abst. Praha, Soc Anaesth, Sec Med Bohoslov Nom JE Purkyne Insign 17-20.VIII.1965; (4-30):414-5. — Grenda J, Olearczyk A. Postępowanie przy resuscytacji. Wiad Lek 1965;28, 21a:19-24. — Grenda J, Musiał MM, Olearczyk A. Przypadek resuscytacji po utonięciu. Pol Tyg Lek 1966; 21,51:1975-6. — Grenda J, Olearchyк AS. Outcomes of resuscitation. J Ukr Med Assoc North Am 1975;22,1(76):26-8. — Beaudouin D. W.B. Kouwenhoven: Reviving the body electric. JHU Engineering. 2002 Fall:27-32. — Grenda J, Olearczyk A. Ergebnisse der Resuszitation, In, Olearchyk AS A Surgeon's Universe. Vol. 1-4. 4th Ed. Outskirts Press, Denver, Co 2016. Vol. 2, P. 216-8.

artery pulses, disappearance of the cardiac sound, or murmur (asystole), or the evidence by the echocardiography (ECG) of ventricular fibrillation (VF), and loss of consciousness, to prevent irreversible ischemia of the brain and assure intact brain function. The procedure consists of simultaneous ventilation of the lungs, closed chest (external), or open chest (internal) cardiac massage (squeezing), injection of norepinephrine in asystole and other essential therapy, and the direct current (DC) electrical external, or internal cardiac defibrillation in VF.

The first known efforts to revive suddenly dying child, mentioned in the «Bible Old Testament» (Ancient medicine. *The Jews*) was attempted by the artificial «mouth-to-mouth» breathing.

In 1732 **William Tossach (c. 1700 – 71)**. 18th century) revived dying coal miner giving him the «mouth-to-mouth» breathing, and since 1743 **William Hawes (1736 – 1808**. 18th century) performed the «mouth-to-mouth» breathing, bringing to life people who appeared to have drowned.

In 1802 **Ivan Kamensky (1773 – 1819**: 19th century. Anatomy, histology and embryology) described the CPR by cardiac squeezing, along with an artificial «mouth-to-mouth» respiration, to revive suddenly dying people.

The effective approach to CPR was developed by **William B. Kouwenhoven (1896 – 1975)** — American electrical engineer, **James R. Jude (1928 – 2015)** — American thoracic surgeon, and by **Gay G. Knickerbocker (1932 –)** — American engineer and the CPR researcher, who refined the closed-chest cardiac massage, invented an open-chest DC cardiac defibrillator (1947), and closed-chest cardiac defibrillator (prototype in 1957, portable in 1961: 120-200 Joules).

At the same 1961 **Bernard Lown (1961 – : 20th century. Internal medicine), also used closed-chest DC cardiac defibrillator to revive patients from a sudden cardiac death (SCD) caused by VF.

The Ambu bag or resuscitator for the CPR was designed and made in 1953-56 by **Holger Hesse** — German engineer, Dr rer. nat.[1033], and his partner **Haning Ruber** — Danish anesthesiologist, is a bag-valve-mask device, the first manual (non-electrical), self-inflating and with positive pressure ventilation device, providing oxygen (O_2) of an ambient, or room air (21%) to assist the patient's breathing in emergency situations (Ambu International, Copenhagen, Denmark). Moreover, the Ambu bag can be connected to a separate bag reservoir filled with pure O_2 from compressed O_2 source to increase its concentration to nearly 100%.

The «mouth-to-mouth» breathing for the CPR was re-introduced in 1956-57 by **James O. Elam (1918 – 95)** — American physician and respiratory researcher, and

---

[1033] In Latin – Doctor rerum naturarium (Dr rer. nat.); in English – Doctor of Natural Science.

Peter Safar (1924 – 2003) — Austrian physician of Czech descent, for an in hospital and out hospital resuscitation CPR.

On July 1959 **James Jude (1928 – 2015**: above**)** confirmed that manual pressure applied to the exterior of the patient chest could restore cardiac output (CO) in the event of cardiac arrest.

Finally, in 1965 and 1968 **J. Frank Pantrige (1916 – 2004)** — North Ireland physician and cardiologist, developed the portable automatic external defibrillator in 1965 (heavy) and by 1968 (light). The later was approved for the use by the staff of an emergency medical care (EMC) ambulances.

Thanks to **Clarence Crafoord (1899 – 1985)** — Swedish cardiovascular surgeon and **Louis A. Shaw, Jr (1886 – 1940)** — American physiologist, the clinical use of the mechanical ventilation for thoracic operations, the «spiropulsator», became a reality from 1938.

The earliest approach to an isolated ventilation with one lung was developed in the 1930s, utilizing the «bronchial blockers» fashioned from bundles of gaze or balloon-tipped catheters placed through a rigid bronchoscopy.

One lung isolation requires a specialized endotracheal tube (ET) or combination of a standard single ET and an airway-blocking device to physically isolate ventilation to the right or left lung. That means, one lung is isolated and collapsed while the other lung is ventilated, during partial or total pneumonectomy, resection of thoracic aortic aneurysm, massive hemoptysis from ruptured pulmonary artery (PA), arterio-venous (AV) malformation, endobronchial carcinoma, pulmonary embolism (PE), necrotizing pneumonia or lung abscess, broncho-pleural fistula, lung laceration, those with markedly different compliance of the lung or airway resistance, such as following single lung transplantation, during bronchoalveolar lavage, and recently for the video-assisted thoracic surgery (VATS) and minimally invasive cardiac surgery (MICS).

The important event in development of a one lung isolation technique was introduction in 1942 by **Frank J. Murphy (1900 – 72)**[1034] — American anesthesiologist, of the distal opening in the wall of an endotracheal tube (ET), between the leading edge of the bevel and the inflatable cuff, which allows air flow in the event of the tube opening lying against the tracheal wall or being obstructed by tumor, mucus plug or in other ways (the «Murphy eye»).

In contemporary modification, the «Murphy eye» is a disposable polyvinyl double lumen (DL) right-sided ET which must be placed under the fiberoptic bronchoscopy (FOB) guidance into the right upper lobe (RUL) bronchus orifice. It is used for right-sided

---

[1034] Murphy FJ. Two improved intratracheal catheters. Anesth Analg 1941;20:102-5. — Forestner JE. Frank J. Murphy, M.D., C.M., 1900-1972: his life, career, and the Murphy eye. Anesthesiology 2010 Nov;113(5):1019-25.

procedures when only bronchial lumen is ventilated or for a sleeve resection of the left main stem bronchus (LMSB). Although there are advantages in the use of the «Murphy eye», there are complications associated with its side vent. Most of them have dealt with fiberoptic bronchoscopy (FOB) or suction catheters being passed down the ET and becoming lodged in the «Murphy eye». Then, ET and catheter had to be removed together, exposing the patients to the hazards of reintubation, or the punched- out sliver from the «Murphy eye» with the onset of an intermittent positive pressure breathing (IPPB) could be expelled into the patient trachea.

Invention by **David S. Scheridan (1908 – 2004)** — American inventor, of «disposable» plastic catheter and endotracheal tube (ET) in the 1940s significantly decreased the rate of blood-born and post-anesthesia pulmonary infection.

A double-lumen endobronchial tube for intubation of the right main stem bronchus (RMSB) with a small carinal hook for its fixation to the tracheal bifurcation and separate bronchial and tracheal cuffs with an inflatable balloon, was designed by **Eric Carlens (1908 – 90)** — Swedish otorhinolaryngologist (ORL), for an independent ventilation of the right and left lung (Carlens tube, 1949). He also described in 1959 cervical suprasternal mediastinoscopy which employ a rigid, lighted endoscope placed in the avascular, pre-tracheal space to assess the superior mediastinal lymph node (LN) involvement by metastatic carcinoma[1035]. It is the most accurate pre-thoracotomy method for staging the mediastinum for bronchogenic carcinoma with sensitivity of 81% and specificity of 100%.

Further progress was achieved with the introduction of the double lumen endotracheal tube (DLET) which is the most common device used to allow separate ventilation of the lungs[1036]. It consists of two long, cuffed ETs fused together as to permit one lumen and cuff to reside in a pulmonary bronchus and the other to remain more proximal in the trachea. With each lumen independently connected to a ventilator or both lumens united via a bridge connector attached to a single ventilator, gas flow into each lung can be controlled.

The crucial feature of all ET cuffs design is to allow sealing and isolation of the RMSB or the left main stem bronchus (LMSB) without occluding any of the upper lobe bronchi. In actual use, the right upper lobe (RUL) is often partially or completely occluded and poorly ventilated. Some suggest using only left-side DLET unless LMSB intubation is contraindicated.

One design of the right-sided DLET with carinal hook utilizes an elongated, S-shaped endobronchial cuff attached at an acute angle to allow the addition of a

---

[1035] Carlens E. Mediastinoscopy: a method for inspection and tissue biopsy in the superior mediastinum. Dis Chest 1959 Oct;36(4):343-52.

[1036] Jaeger JM, Durbin CG Jr. Special purpose endotracheal tubes. Respir Care 1999;June;44(6):661-85.

large «Murphy eye» to oppose the orifice of the RUL bronchus. The other design of the right-sided DLET with carinal hook incorporates two small, round endobronchial cuffs that straddle a through- and through slit-like opening in the distal wall of the ET. In the left sided DLET with carinal hook ET cuffs are similar and consist of a spheroid or elliptical cuff approximately 1 cm proximal to the end of the tube.

The other devices to separate the lungs under the fiberoptic bronchoscopy (FOB) guidance is the «bronchial blocker» in the form of the single-use an univent tube that consist of a large, single-lumen silicone ET with a channel fused to its entire length. This channel contains a long, thin, cuffed hollow rod that can be advanced into the RMSB or LMSB, then inflated to block the airway. The «bronchial blocker» can be connected to suction (to evacuate air, blood, or secretion from the occluded lung), connected to a high-frequency ventilation, or it could be capped.

Robert D. Dripps (1911 – 73) — American anesthesiologist, who with his team solved the problem of «spinal headaches» after spinal anesthesia (1942)[1037].

In 1943 **Nills Löfgren (1913 – 67)** — Swedish chemist, discovered lidocaine **[Miles Williams (1918 – 2016)** and **Bramah Sigh (1938 – 2014).** 20th century. Pharmacology]**, a local anesthetic allowing a prolonged numbing of the skin and mucous membranes and to perform nerve blocks, starting within 4 min., lasting for 30 min. to 3 hours. Lidocaine is also a cardiac depressor used to treat ventricular arrhythmia.

Lidocaine alters signal conduction in neurons by blocking the fast voltage-gated natrium (Na^+) channels in the neuronal cell membrane, responsive for action potential (AP) propagation. This prevents the generation of impulses in the terminal sensory nerves and propagation of impulses through the nervous filaments.

James E. Eckenhoff (1915 – 96), George L. Hoffman, Sr, Robert Dripps, and **Lonnie W. Funderburg (1921 – 2009)**[1038] — American anesthesiologist, elucidated the action of N-allyl normorphine (presently effective Naxolone, and the mixed opioid agonist-antagonist) in reversing the respiratory depression induced by opioids, precisely an opioid and oxygen chemoreceptors (1952), and also in the prevention of neonatal depression in children born by mothers injected by a «twilight sleep», an analgetic-amnesic[1039] technique administered for painless childbirth (1953).

Ten years later (1962-63) James Eckenhoff investigated the physiologic implication of deliberate hypotension used to diminish blood flow during anesthesia with drugs blocking the autonomic nervous system and positioning the body, depicted overall

---

[1037] Dripps RD, Vandam LD. Hazard of lumbal puncture. JAMA 1951;147(12):1118-21.

[1038] Eckenhoff JE, Hoffman GL, Sr, Dripps RD, N-allyl normorphine: an antagonist to the opiates. Anesthesiology 1951 May;13;242-51.

[1039] Analgesia (in Greek / Latin – without pain) and amnesia (in Greek / Latin – without memory).

oxygen delivery, cerebral blood flow, pulmonary gas exchange and related changes, defined its benefit and limitations.

Involved in the 1952 poliomyelitis outbreak with respiratory paralysis in Denmark [Denmark proper area of 42,924 km^2), the Danish Realm, comprising Faroe Islands (area c.1,400 km^2) and Greenland (area 2,166,086 km^2) – total 2,210,549 km^2], **Bjorn A. Ibsen (1915 – 2007)** — Danish anesthesiologist, replaced largely unsuccessful type of negative pressure ventilators by successful type of intermittent positive pressure breathing / ventilation (IPPB, IPPV) by using the cuffed intubation of the trachea or a cuffed tracheostomy tube with a bag and the Waters' «to-and from» system **[Ralph Waters (1883 – 1979): above]** with absorption of carbon dioxide (CO_2) by the absorber connected to one of those two tubes. Mortality declined from 90% to around 25%.

In 1953 Bjorn Ibsen set up the world's first medical / surgical intensive care units (MICU, SICU) at Kommunehospitalet in the capital city of Copenhagen (established in the 10th century as a Viking fishing village), Denmark, to treat children affected by paralytic poliomyelitis and provided one of the first accounts of the management of tetanus with muscle relaxants and controlled ventilation.

Suxamethonium chloride or succinylcholine, a depolarizing muscle relaxant of a short duration was introduced in 1951 by **K.H. Ginzel**, **H. Klupp** and **Gerhard Werner (1921 – 2012)** — German physician, during electroshock therapy to prevent bone fractures and to facilitate tracheal intubation.

Halothane (fluothane) was synthesized in 1951, and first used for a general inhalation anesthesia in 1956. But, combined anesthesia with halothane and d-tubocurarine was considered unsafe because of possible depressive action on the myocardium and blockage of the nervous ganglions with subsequent bradycardia falling BP.

In 1955 **Viking O. Björk (1918 – 2007)** — Swedish cardiac surgeon and **Arne V. Engström (1920 – 1996)** — Swedish biomedical industrialist, first used the Engstom volume ventilator for poor risk patients after thoracic surgery.

By 1961 the routine and a rapid sequence induction and intubation of the trachea for the general anesthesia was developed.

The routine induction and intubation of the trachea involves preoxygenating the lung with a tightly fitting mask, an intravenous (IV) administration of sleep-inducing drugs (barbiturates) and a rapid-acting depolarizing muscle relaxation (suxamethonium chloride or succinylcholine chloride) while patient's lungs are manually ventilated by periodical squeezing the respiratory bag, endotracheal intubation and inflation of the cuff, then maintenance of the general anesthesia with mixture of oxygen (O_2) and fluid or gaseous anesthetic, such as ether, nitrous oxide (N_2O), cyclopropane or fluothane, continuation of muscle relaxation with nondepolarizing drugs of a long duration (d-tubocurarine or flaxedil) and a hand supported ventilation of the lungs by periodical squeezing the respiratory bag or by the ventilator.

A rapid sequence induction and intubation of the trachea for the general anesthesia, commonly employed in patients who are assumed to have a meal within few hours prior to emergency operations and other situations, involves preoxygenating the lung with a tightly fitting mask, the remaining sequence is as in the routine induction and intubation, with exception that the patient after preoxygenation is not oxygenated until intubated and the cuff has been inflated, for the fear of overinflating of the stomach with кисень O_2 and aspirating gastric content into the tracheobronchial tree. To it **Brian A. Sellick (1918 – 96)** — added a cricoid pressure to close the esophagus and possibly avoid regurgitation of stomach content during endotracheal intubation (Sellick maneuver, 1961).

Septicemia or sepsis (in Greek – decay) is a life-threatening pathological condition caused by dysfunction of the organism due to infection by microorganism or microorganisms, associated with depressed immune responses of the patient. The clinical picture is depicted as the systemic inflammatory response syndrome (SIRS). The prevention should consist of a maximal hospital cleanliness and aseptic precautious by the health care providers, especially during noninvasive and invasive surgical procedures. The treatment should include surgical removal of a source of infection, the use of wide spectrum antibiotics, the maintenance of fluid and electrolyte balance, adequate nutrition and passive immunotherapy, in septic shock corticosteroids may have to be used.

The SIRS is noted in 17,4% patients, who underwent an intensive therapy, moreover in 62,2% of cases sepsis is a hospital acquired infection. Mortality from SIRS is 35% - 65%.

Between 1982 – 89 **Anatoly S. Trishchynsky (20.V.1923, village Rusanivka, Lypovodolynsky, Sumy Obl., Ukraine – 31.III.2009, city of Kyiv, Ukraine)**[1040] — Ukrainian anesthesiologist, with associates treated 217 patients with different forms of septicemia / sepsis with specific antibiotics, combined with immune-modulator Leucoinferon (human leucocyte interferon), a group of signaling proteins, made and released by the host cells in response to several viruses, to enhance an antivirus defenses [20th century. Microbiology and immunology. –**Alick Isaack (1921 – 67) / Jean**

[1040] Village Rusanivka (first inhabitants appeared in the 4th century BC, founded in the first half of the 18th century), Lypovodolynsky Region, Symy Oblast', Ukraine. — Olearchyk A. History of anesthesia in Ukraine. J Ukr Med Assoc North Am 1966 Oct;13,3-4 (42-43):19-21,24,32. — Trishchynsky AI, Spivaк MYa, Kishko YaH et al. Therapeutic values of interferon in treatment of septicemia. J Ukr Med Assoc Norh Am 1991 Winter; 38,1(123):13-9. — Ablitsov VH. «Galaxy Ukraine». The Ukrainian diaspora: prominent persons. Publ. «Kyt», Kyiv 2007.

Lindenmann (1924 – 2015)]. Clinical improvement was noticeable in two to two and half times sooner than with other form of therapy. Body temperature and laboratory findings returned to normal on the 19.9 day after initiation of treatment with leukinferon.

Patriarch of anesthesiology in Ukraine **Leonard P. Chepkyy (02.03.1925, town of Haisyn, Vinnytsia Obl., Ukraine – 14.09.2019, city of Kyiv, Ukraine)**[1041] — Ukrainian anesthesiologist and the first professor of anesthesiology in Ukraine (1964), published monographies about curative hypothermia, ambulatory anesthesia, geriatric anesthesia, reanimation, anesthesia, and intensive postoperative therapy in obese patients.

The high dose opioid anesthesia was developed in 1969-71 by **Edward Lowenstein (1934 –)**[1042] — American anesthesiologist, based on the observation that patients on mechanical ventilation after cardiac surgery tolerated large dosages of morphine (0.3 to 3 mg/kg) for sedation and anesthesia with minimal hemodynamic effect. Such dosages were then tried immediately before operation as an adjunct to anesthesia for cardiac surgery, with significant success. However, opioids by themselves do not ensure amnesia and cause fluid retention in the body. Therefore, in late 1990s high dosages opioid anesthesia was gradually replaced by a balanced anesthesia that involves using modest doses of opioids in conjunction with inhalational anesthetics and other intravenous (IV) adjuncts that have a faster onset to facilitate early extubation and discharge from the intensive care unit (ICU).

As to combined use for the general inhalation anesthesia of halothane and for muscle relaxation of d-tubocurarine (above), then its accurate application in 1966 by **Andrzej Olearczyk / Andrew S. Olearchyk** (Foot Notes 1, between 87 & 88, 1032)[1043] — Ukrainian Polish anesthesiologist and surgeon, with decreased by about 25% doses, and careful monitoring for general surgery and urological operations demonstrated a complete safety, without any complications and operative mortality (OM).

John B. Glen (1940 –) — Scottish veterinarian and researcher, discovered in 1977 Propofol (Deprivan), a short acting narcotic drug that causes decrease the level of consciousness and lack of memory for events, used IV for induction and maintenance of general anesthesia, sedation for mechanical ventilation in adults and procedural sedation.

[1041] Town of Haisyn (Foot Note 321), Vinnytsia Obl., Ukraine. — Chepkyy LP. Reanimatolohiya ta Intensyvna Therapiya v Chasi Kolapsu i Shoky. V-vo. «Zdorov'ia», Kyiv 1994. — Chepkyy LP, Novyts'ka-Usenko LV, Tkachenko RO. Anesteziolohiya, Reanimatolohiya ta Intensyvna Terapiya. V-vo «Vyshcha Shlola», Kyiv 2003. — Ablitsov VH. «Galaxy Ukraine». The Ukrainian diaspora: prominent persons. Publ. «Kyt», Kyiv 2007.

[1042] Lowenstein E., Hallowell P, Levine FH, et al. Cardiovascular response to large doses of intravenous morphine in man. NEJM 1969;281:389-93. — Lowenstein E. Morphine «anesthesia»: a perspective, Anesthesiology 1971;35:563-5. — Lowenstein E. The birth of apioid anesthesia. Anesthesiology Apr 2004;100:1013-5.

[1043] Olearczyk A, Grenda J, Gawryś A. Własne spostrzeżenia przy stosowaniu halotanu i kurary. Pol Tyg Lek 1967;22,5:183-4. — Ablitsov VH. «Galaxy Ukraine». The Ukrainian diaspora: prominent persons. Publ. «Kyt», Kyiv 2007.

In 1985 **Seshhagiri S. Mallampati (1941 –)**[1044] — Indian American anesthesiologist, proposed the classification / score, named after him the Mallampati classification to predict the ease of endotracheal (ET) intubation. The patient in a sitting position is asked to open mouth and protrude the tongue as much as possible. The anatomy of the oral cavity is visualized whether the base of uvula, fauces, faucial pillars, i.e., the arches in front of and behind the palatine tonsils and soft palate (velum) are visible, with or without phonation.

In the modified Mallampati classification:

Class 1   –   soft palate, uvula, fauces, pillars and palatine tonsils clearly visible;
Class 2   –   hard and soft palate, uvula, fauces and upper half of the palatine tonsils visible;
Class 3   –   soft and hard palate, base of uvula visible;
Class 4   –   only hard palate visible.

During complex operations for the correction of congenital heart disease (CHD) in children and adult hypothermia reduces metabolic rate of the body, including the brain, thus protecting vital organs, allows cardiopulmonary bypass (CPB) flow rates to be reduced, thereby reducing the blood volume returning to the heart and improves surgical exposure. The accepted values for the low-flow rates in hypothermic CPB of about 50 mL/kg/min in adults and children. According to **Hwee L. Pua** — Singaporean[1045] Canadian anesthesiologist and **Bruno Bissonnette** — Canadian anesthesiologist, this range of low-flow CPB in combination with hypothermia is relatively safe in terms of cerebral protection compare with deep hypothermic circulatory arrest (DHCA). However, there is no agreement on what a safe degree of flow rates reduction might be for a given temperature in neonates, infants, and children in general.

Frank H. Kern[1046] — American pediatric anesthesiologist, have suggested that the critical pump flow rate in terms of the crucial juncture at which cerebral metabolism becomes flow dependent between 30 and 35 mL/kg/min at moderate hypothermia

---

[1044] Mallampati SR, Gatt SP, Gugino LD, et al. A clinical sign to predict difficult tracheal intubation: a prospective study. Can Anaesthesist's Soc J 1985 Jul;32(4):429-31.

[1045] Sovereign city-state and island country Singapore (area 719,9 km²) in Southeast Asia. — Pua HL, Bissonnette B. Cerebral physiology in pediatric cardiopulmonary bypass. Can J Anaesth 1998;45:960-78.

[1046] Greeley WJ, Underleider RM, Kern FH, et al. Effects of cardiopulmomary bypass on cerebral blood flow in neonates, infants, and children. Circulation 1989;80:1209-15. — Greeley WJ, Kern FH, Underleider RM, et al. The effect of cardiopulmomary bypass and total circulatory arrest on cerebral metabolism in neonates, infants, and children. J Thorac Cardiovasc Surg 1991;101:783-94. — Kern FH, Underleider RM, Reves JG, et al. Effect of altering pump flow rate on cerebral blood flow and metabolism in infants and children. Ann Thorac Surg 1993;56:1366-72.

(26^0 to 29^0C) and between 5 and 30 mL/kg/min during deep hypothermia (18^0 to 22^0C). It suggested that cerebral autoregulation is markedly diminished or absent at temperature below 20^0C, and hence cerebral blood flow (CBF) becomes pressure-dependent at very low temperature during CPB in neonates, infant and children[1047].

Frederick A. Burrows (1952 –) — American anesthesiologist and **Bruno Bissonnette** (above) found that significant percentage of neonates and infants undergoing low-flow CPB (< 22% of normal pump flow) have no detectable CBF as measured by transcranial Doppler **[Christiaan Doppler (1803 – 53)**. 19th century. Physiology]** and require higher perfusion pressures to establish CBF[1048]. Similar results had been detected at flow rates typical during profound (nosopharyngeal temperature, 14^0 to 20^0C) hypothermic CPB, where some infants lost CBF at flow rates as high as 25% to 35% of normal[1049]. So that, the development of critical closing and opening arterial blood pressure (BP) during CPB in the cerebral (and other) vascular beds may contribute to a no-reflow phenomenon and uneven brain cooling during low flow CPB[1050]. Such notion suggest that blood flow may be more dependent on arterial BP than pump flow rate in these circumstances, and that a minimum mean arterial BP is necessary to maintain adequate flow to the brain and other organs.

Vitaliy D. Maksymenko (09.01.1952, city of Kyiv, Ukraine –)[1051] — Ukrainian anesthesiologist, who developed and introduced into the clinical practice protocols of complete hyperthermic perfusion of the body for the treatment of septic shock states (1999) and the original method of myocardial protection using artificial blood – perfluorocarbon (2001).

The evidence-based principle shown that in a correct management of mechanically ventilated patients, especially those with risk factor for hospital acquired (nosocomial) pneumonia and acute respiratory distress syndrome (ARDS), those in the perioperative period, and suffering from chronic obstructive pulmonary

[1047] Pua HL, Bissonnette B. Cerebral physiology in paediatric cardiopulmonary bypass. Can J Anaesth 1998;45:960-78.

[1048] Burrows FA, Bissonnette B. Cerebral blood flow velocity patterns during cardiac surgery utilizing profound hypothermia with low-flow cardiopulmonary bypass or circulatory arrest in neonates and infants. Can J Anaesth 1993;40:298-307.

[1049] Taylor RH, Burrows FA, Bissonnette B. Cerebral pressure-flow velocity relationship during hypothermic cardiopulmonary bypass in neonates and infants. Anesth Analg 1992;74:636-42.

[1050] Pua HL, Bissonnette B. Cerebral physiology in paediatric cardiopulmonary bypass. Can J Anaesth 1998;45:960-78.

[1051] Maksymenko VB, et al. Anesthesiological security and intensive therapy in children cardiac surgery. Kyiv 2002. Maksymenko VB. Cardioanesthesia, artificial blood circulation, protection of the myocardium. Kyiv 2007. — Maksymenko VB, Yarmola TM. Nanomaterials and their use in medical products. Educational handbook. Kyiv 2013. — Uvarova IV, Maksymenko VB, Yarmola TM. Biocompatible materials for medical products. Educational handbook. Kyiv 2013. — Ablitsov VH. «Galaxy Ukraine». The Ukrainian diaspora: prominent persons. Publ. «Kyt», Kyiv 2007.

disease (COPD), to allow an accelerated weaning from mechanical respiration and a safe extubation it is necessary to: (1) keep lower plateau pressure (<30 cm H_2O), that is the pressure applied to small airways and alveoli during mechanical intermittent positive pressure ventilation (IPPV), measured during an inspiratory pause on normal breathing (0,5-1 sec.) or on the ventilator; (2) maintain a relatively low tital volume (VT) of 6-8 cm/kg; (3) avoid creation of an intrinsic positive end-expiratory pressure (PEEP), i.e. auto-PEEP or «breath stacking», especially in COPD, because emphysema delays exhalation and with volume ventilators setup there may be insufficient time for exhalation and when a new breath begins after only partial exhalation, when there is still significant airway pressure above the set level of PEEP, the peak airway pressure rises, but this can be remedied by decreasing tidal volume (V_T) or by increasing the relative time allowed for exhalation by decreasing the inspiratory (I) / expiratory (E) ratio, decreasing the respiratory rate (RR), and/or decreasing the inspiratory time; (4) perform a monitored spontaneous breathing trial (SBT) using a T-tube or flow-triggered ventilation (pressure support) and protocol-driven weaning. These guidance's are associated with more rapid and successful weaning, less barotrauma and lowered mortality.

Surgery

In 1900 **Walther Petersen (1867 – 1922)**[1052] — German surgeon, described an intraabdominal (Petersen) space hernia that occurs after partial or subtotal gastrectomy and antero- or retro-colic gastrojejunal anastomosis for the treatment of chronic peptic disease or after non-resection Roux-en-Y gastric bypass with antero-or-retrocolic gastrojejunostomy[1053] for morbid obesity in the space (Petersen) posterior to a gastroenterostomy.

First successful repairs of congenital diaphragmatic hernia (CDH) were performed in 1901 by **O. Aue**[1054] — German pediatric surgeon, in a 9-year-old boy, and in 1902 by **L. Heidenhein**[1055] — German pediatric surgeon, also in a 9-year-old boy.

In 1901 **Kristian Igelsrud (1867 – 1940)** — Norwegian-American surgeon, successfully applied a direct cardiac massage to a patient in cardiac arrest.

Sir **T. Lauder Brunton (1844 – 1916)** — Scottish physician, used clinically amyl nitrite to treat coronary artery disease (CAD), specifically angina pectoris,

---

[1052] Petersen W. Über Darmverschlingung nach der Gastroenterostomie. Arch Klin Chir 1900;62:94-114.

[1053] **Cesar Roux (1857 – 1934**: 19th century. Surgery).

[1054] Aue O. Über angeborene Zwerchfellhernien Dtsch Z Chir 1920;160:14-35.

[1055] Heidenhein L. Gesichte eines Falles von chronisher Incarceration des Magens in einer angeborenen Zwerchfellhernie welcher durcher Laparatomie geheilt wurde, mit anschiessender Bermerkungen uber die Moglichkeit, das Kardiacarcinom der Speiserihre zu reseciren. Deutsch Z Chir 1905;76;394-407.

reasoning that it dilates coronary arteries, and proposed to perform closed mitral commissurotomy (MC) for treatment of rheumatic mitral stenosis (MS), based on his experimental work in cadavers[1056].

In 1902 **Thomas H. Noble**[1057] — American surgeon, described endo-anal advancement flap for dealing with recтovaginal fistula following childbirth.

In 1902 **S. Spangaro**[1058] — Italian surgeon, found abnormal histological changes in an undescended, outside the scrotum, testicles – cryptorchidism (in Greek «cryptos» meaning «to hide», «orchis» meaning «testicle»), i.e., a hidden testicle after maturity may lead to infertility, development of germ-cell tumor, and psychological consequences in adolescents. Therefore, he recommended the necessity for surgical correction of cryptorchidism by placing an undescended testicle in its natural position inside the scrotum (orchiopexy). In addition, he described in 1906, the 4th-5th intercostal space thoracotomy (ICST) incision, designed an atraumatic clamp to control of the right ventricular (RV) angle to expose and repair of cardiac wounds.

First pulmonary embolectomy (PE) performed in 1907 by **Fredrich Trendelenburg (1844 – 1924**: 19th century. Surgery)[1059] in three patients with an acute pulmonary embolism (PE) via the left intercostal space thoracotomy (ICST) with a temporary occlusion of the superior vena cava (SVC) and inferior vena cava (IVC) orifices to stop the blood inflow into the heart (the method of an «inflow occlusion») – Trendelenburg operation-II, two died in the operating room (OR), one 37 hours after the operation due to hemorrhage from an unrecognized transection of the left internal thoracic artery (ITA).

The address «Surgical Aspect of Subphrenic Abscess» was delivered by **Harold Bernard (1868 – 1908)**[1060] **[19th century. Anesthesiology. – Lenard Hill (1866 – 1952)]** before the Surgical Section of the Royal Society of Medicine on January 14, 1907 in London, England, in which subphrenic abscess has been described rather from the classical position of the men upright, diagnosis by means of x-ray examination and

---

[1056] Brunton L. Preliminary note on the possibility of treating mitral stenosis by surgical methods. Lancet 1902;159(4093):352-5.

[1057] Noble GH. A new operation for complete laceration of the perineum designed for the purpose of eliminating danger of infection from the rectum. Trans Am Gynecol Soc 1902;27:357. / JAMA 1902;39(6):302-4.

[1058] Spangaro S. Üeber die histologischen Veränderunger des Hodens, Nebenhodens und Samenleiters von der Geburt an bis zum Greisenalter. Anat Hefte 1902; Abt 1 (Heft 60), 18:593). — Spangaro S. Sulla técnica da seguire negli interventi chirurgici per ferite del cuore e su di un nuovo processo di toracotomia. Clinica Chir Milan 1906; 14: 224.

[1059] Trendelenburg F. Über die operative Behandlung der embolie der Lungarterie. Arch Chir 1908;86:686-700.

[1060] Bernard HL. An address on subphrenic abscess. Br Med J 1908;1:371-77, 429-36. — Konvolinka CW, Olearchyk A. Subphrenic Abscess. Curr Probl Surg (Ed MM Rawitch). Year Book Med Publ, Chicago, Jan. 1972:1-52. — Olearchyk AS, Konvolinka CV. Subphrenic abscesses. J Ukr Med Assoc North Am 1972; 19,3 (66): 3-56.

the patient position at the operation are alike the horizontal one, and it is this position that the surgeon is called upon to approach and drain subphrenic suppurations.

Since first reported by **A. Abrashanoff**[1061] in 1911, transposing the local pedicled muscle flaps has been used for the treatment of persistent residual infected pleural cavity space closed or in the form of open window thoracostomies, reinforcement for the bronchial stump after closure of an associated bronchopleural fistula (BPF), or closure of a small (<2 mm) BPF.

In 1912 **Sigmunt Exner (1846 – 1926)** —Austrian physiologist, who explained the compound eye functions (1891) and **E. Schwatzman** suggested a truncal vagotomy for the surgical treatment of an ulcer disease of the stomach and duodenum.

In 1912 **Arthur W. Elting**[1062] — American surgeon, proposed for the surgical treatment of complex fistula-in-ano closure of the internal opening, when possible, by an endorectal advancement flap, cutterage of the fistulous tract and wide external drainage.

Originally, in 1913 **G.A. Heuer**[1063] successfully treated a patient with tuberculous empyema thoracis by repeated injection of bismuth paste into the cavity. By 1920 he reported 24 patients with chronic empyema thoracis, some of tuberculous origin, treated by space sterilizing techniques using antimicrobial solutions. In number of the patients, he tried, in addition to space sterilization operative maneuvers that involved the parietal pleura.

Pericardiectomy for constructive pericarditis was performed in 1913, by **Ludwig Rehn (1849 – 1930**: 19th century. Surgery**)** and in 1925 by **Ernst Sauerbruch (1875 – 1951**: 20th century. Anesthesiology). The left ventricle (LV) should be freed first to avoid the right ventricular (RV) distention. The anterior pericardium is freed from about 1,5 cm anterior to the right phrenic nerve to about 1,5 cm anterior to the left phrenic nerve laterally, and from the anterior diaphragm inferiorly to the base of the great vessels superiorly. Then the diaphragmatic pericardium between the inferior heart and the diaphragm should be excised to prevent the inferior vena cava (IVC) obstruction, as well as the atrial pericardium and pericardium posterior to the phrenic nerves bilaterally to decrease atrial constriction.

Beside of this, Ernest Sauerbruch[1064] first performed in an 18-year-old man suffering from congenital pectus excavatum who developed dyspnea and palpitation with limited exercise, a surgical repair of this disease. Three years after the operation his

---

[1061] Abrashanoff A. Plastische Methode zur Schlieszung vor Fistelgangen, welche von inneren Organen kommen. Zentralbl Chir 1911;38:186-91.

[1062] Elting AW. The treatment of fistula in ano: cutterage of the fistulous tract with special reference to the Whitehead operation. Ann Surg 1902 Nov;56(5):726-52.

[1063] Heuer GH. Observations on the treatment of chronic empyema. Ann Surg 1920;72:80-6.

[1064] Sauerbruch F. Die Chrurgie der Brustorgane. Spinger, Berlin 1920.

stamina increased, he could work 12 to 14 hours a day without tiring and without dyspnea and palpitation.

One of the first CPR was described in 1909 by **Ernst Engleman (24.IV.1852, city of Stuttgart, State of Baden-Wüttemberg, Germany – 1927)**[1065] — German surgeon, Head physician in the Department of Surgery (1885-1917) at the Göppingen Hospital, town of Göppingen, Baden-Wüttemberg, Germany.

Ottokar Chiari (1853 – 1916) — Austrian rhinolaryngologist, is credited with introduction in 1912 of the trans-ethmoid trans-sphenoidal removal of the pituitary gland tumors (hypophysectomy).

In about 1928 **John B. Deaver (1855 – 1931)**[1066] — American surgeon, developed the stainless steel deep tissue retractor having a curved retracting blade and flat handle in which the width of the handle is equal to the width of the retracting blade (Deaver retractor). It is used to hold the edges of an abdominal or chest incisions open during operations inside the abdominal and chest cavities on such organs as liver, gallbladder, esophagus, stomach, duodenum, lung etc.

History of laparoscopy began in 1901 when **Dimitri O. Ott (1855 – 1929)** — Russian gynecologist, examined the peritoneal cavity of a pregnant woman by using a head mirror and a speculum introduced into a culdoscopic incision, i.e., through the incision in the posterior fornix of the vagina.

Cesar Roux (1857 – 1934: 19th century. Surgery**)** is credited with performing the first successful removal of pheochromocytoma (1926).

Pheochromocytoma is a neuroendocrine hereditary tumor of the medulla of the suprarenal glands, originating in the chromaffin (a portmanteau of «chromium» and «affinity»; in Greek – «phaios» means «dusky», and «khroma» means «color») cells (10-15%), bilateral (10%), or of an extra-adrenal chromaffin tissue that failed to involute after birth in the ganglia of the sympathetic nervous system (15%), occurring in children (10%) and young or middle age adults (2-8 in 1 million) with malignant potential in 11%-30%. It secretes high amount of catecholamines, usually noradrenaline (norepinephrine) and adrenaline (epinephrine). The symptoms are headaches, abnormalities in skin sensation, anxiety, palpitation, excessive sweating and flank pain, signs – pallor, increased heart rate (HR) and blood pressure (BP), weight loss, subsequently amyloid deposition in tissues and malignant hypertension. The

---

[1065] City of Stuttgart (first settlements was a massive Roman Castra statina built c. 90, founded 950), Baden-Wüttenberg, Germany. — Town of Göppingen (found in 3rd or 4th century), Baden-Wüttemberg, Germany. — Englehorn E. Die Samariterbuch. Ein Leitfaden für die erste Hilfe bei Unglücksfällen und die Krankenpflege im Hause, insbesondere auch zum Gebrauche für Damenkurse. Verlag Ferdinand Enke, Stuttgart 1909. S. 71-82 und 97.

[1066] Newhook TE, Yeo CJ, Maxwell IV PJ. John Blair Deaver, and his marvelous retractor. Am Surg 2012 Feb;78:155-6.

diagnosis can be established by measuring catecholamines and metanephrines in the blood serum or through a 24-hour urine collection, computer tomography (CT) and magnetic resonance imaging (MRI) of the head, neck, chest and abdomen, positron emission tomography (PET) / CT with F-18-fluorodopamine. The treatment of choice is surgical adrenalectomy, either by open laparotomy or else laparoscopy, which in congenital pheochromocytoma includes planning for a potential metachronous bilateral presentation and the possibility of lifelong steroid dependency if bilateral adrenelectomy is needed. In hereditary bilateral pheochromocytoma, a cortical-sparing adrenelectomy in which the adrenal vein is preserved avoids long-term corticosteroid dependency with minimal risk of acute adrenal insufficiency and recurrence (7%)[1067].

Theodore Tuffier (1857 – 1929: 19th century. Surgery**)** performed in 1914 a successful closed digital commissurotomy of the aortic valve (AV).

In 1909 **Eugene Doyen (1859 – 1916**: 19th century. Surgery**)** designed the bipolar electrocoagulation device, and one year later (1910), **Vincenz Czerny (1842 – 1916**: 19th century. Surgery**)** introduced the electrocoagulation for tissue dissection.

Isolated compression of the left iliac vein by the right common iliac artery (CIA) against the 5th lumbar vertebra, resulting in endothelial damage and intraluminal spur formation that predispose the patient in theirs 2nd to 4th decade of life to left-lower extremity deep venous thrombosis (DVT) was first described in 1908, by **James P. McMurrick (1859 – 1939)**[1068] — Canadian zoologist, anatomist and embryologist, defined anatomically in 1956, by **R. May** — Austrian pathologist and **B. Thurner**[1069] — Austrian pathologist, and clinically in 1965, by **Frank B. Cockett (1916 –2014)** — Australian surgeon and **L. Thomas**, and named iliac vein compression (May-Thurner) syndrome. It causes the left iliofemoral venous thrombi in 2% to 5% of patients who are evaluated for lower-extremity venous disorders. The diagnosis is suspected by left lower extremity bluish discoloration, swelling and tenderness, confirmed by magnetic resonance venography. Open surgical thrombectomy, catheter-directed thrombolysis and balloon angioplasty have been effective in the acute phase of the disease, and IV stent placement can resolve the manifestations of chronic venous

---

[1067] Grupps EG, Rich TA, Ng C, et al. Long-term outcomes of surgical treatment for hereditary pheochromocytoma. J Am Coll Surg 2013; 216(2):280-9.

[1068] McMurrich JP. The occurrence of congenital adhesions in the common iliac veins and their relations to thrombosis of the femoral and iliac veins. Am J Med Sc 1908;135:342.

[1069] May R, Thurner J. Ein Gefässporn in der vena iliaca communis sinistra als wahrscheinliche ursache der überwiegend linksseitigen Beckenvenenthrombose. Zeitschrift für Kreislaufforschung, Darmstadt 1956;45: 912. — May R, Thurner J. The cause of the predominantly sinistral occurrence of thrombosis of the pelvic veins. Angiology Oct.1957;8(5):419-27. — Fernando RR, Koranne RP, Schneider D, et al. May-Turner syndrome in a 68-year-old woman after remote abdominal surgery. Tex Heart Ist J 2013;40(1):82-7.

compression. Anticoagulation, antiplatelet and antithrombotic therapy alone is ineffective.

Henri A. Hartmann (1860 – 1952)[1070] — French surgeon, designed an emergency resection of affected rectosigmoid by diverticulitis, necrosis, or tumor with oversewing of the proximal end or the rectum and bringing the distal end of the colon for colostomy while the patency of the bowel could be reestablished later (Hartmann two-stage colostomy or operation, 1921).

Rudolph Matas (1860 – 1957) — American surgeon and **Nicolai S. Korotkov (1874 – 1920)** — Russian surgeon, developed early operation for aneurysm (1902). The Korotkov-Matas (Matas-1) operation consist of a ligation of the artery proximally and distally to an aneurysm, opening of aneurysm, and ligation from the inside of the retrogradely bleeding branches. The Matas-2 or endoaneurysmorrhaphy (EAR), consist of opening the aneurysmal sac and intrasaccular suturing or closing of the mouth of the vessel entering inside an aneurysm. But Nicolai Korotkov has achieved permanent fame by discovery of the auscultative measuring of systolic and diastolic arterial blood pressure (BP) using sphingomanometer and stethoscope (Korotkov sounds, 1905).

First of its kind, the Fracture Clinic was established in 1917 at the Massachusetts General Hospital (MGH, founded 1811), the city of Boston (inhabitable since 5000 BC, settled 1630), MA, USA, by **Charles L. Scudder (1860 – 1949)**[1071] — American orthopedic surgeon, who stressed that a bone fracture should not be treated as an isolated injury, but that whenever bony was broken soft tissue damage followed.

In 1905 **Alessandro Codivilla (1861 – 1912)** — Italian general and orthopedic surgeon who discovered the method of distraction osteosynthesis.

Emerich Ullman (1861 – 1937) — Austrian surgeon, transplanted a dog's kidney from its normal position to the collar (nuchal) vessels of the neck with success. But transplantation of a kidney from one dog to another was short-lived. He also performed unsuccessfully the first xerotransplantation of the kidney by transplanting the pig's kidney to the hand of a woman with uremia (1902).

Among the patients with calculous cholecystitis with cholelithiasis, common bile duct (CBD) stone and jaundice, after cholecystectomy, intraoperative cholangiography and CBD exploration it is necessary to drain it, with designed by **Hans O. Kehr (1862 – 1916)** — German surgeon, the T-shaped (Kehr) tube.

[1070] Hartmann H. Nouveau procede d'ablation des cancers de la partie terminale du colon pelvien. Rapport du XXXe Congress Francais de chirurgie, Strasbourg 1921:411-4. — Hartmann H. Chirurgie du rectum. Masson, Paris 1931.

[1071] Meredith JW. If Charles L Scudder could see us now. J Am Coll Surg 2017;224(5): 761-70.

In 1907 **Niels T. Rovsing (1862 – 1927)**[1072] — Danish surgeon, published the test for acute appendicitis in which the pain in the right iliac region increases when one presses with the left hand the anterior wall in the left iliac region in the projection of the descending colon, and with the right hand presses in the boost-like manner on the anterior wall above the left hand, thus pushing gas through the colon towards the caecum (Rovsing sign).

To treat hypertrophic pyloric stenosis **John M.T. Finney (1863 – 1942)** — American surgeon, introduced a longitudinal incision through the pylorus and adjacent walls of the stomach and the duodenum to establish an inverted U-shaped anastomosis between them (Finney operation or pyloroplasty, 1902).

In 1917 **Hermann Krukenberg (1863 – 1935)** — German surgeon, developed an operation involving separation of the ulna and radius bones in order to convey a below-elbow amputation stump into a «sensory forceps» that receives its strength from the pronator teres muscle (musculus pronator teres).`

James H. Pringle (1863 – 1941) — Scottish surgeon, is known for the development of the hepatic portal inflow occlusion (Pringle maneuver) by placing a hemostat to temporarily clamp the hepatoduodenal ligament, thus interrupting the blood flow through the hepatic artery and portal vein, to control bleeding from the liver, the inferior vena cava (IVC) / superior vena cava (SVC) injury or both during surgical repair with/without atriocaval shunt (1908)[1073].

Subsequently, the Pringle maneuver found application during resection of the liver and IVC for hepatic malignancy[1074]. Initially, if the portion of the IVC involved with hepatic tumor was below the hepatic veins, the parenchyma of the liver was divided to expose the retrohepatic IVC using the continuous hepatic portal inflow occlusion.

Then, during the hepatic parenchymal transection with/without the hepatic portal inflow occlusion, the central venous pressure (CVP) was kept $\leq$ 5 cm H_2O to minimize blood loss. When the IVC was exposed, the hepatic portal inflow occlusion was released if used, the patient blood volume loaded. Clamps were placed above and below the area of tumor involvement to allow the portion of the liver and involved IVC to be removed and the IVC reconstructed, while allowing for continued perfusion of the liver with portal venous and hepatic arterial blood flow, thus minimizing the safe warm hepatic ischemic time to 60-90 min.

[1072] Rovsing NT. Indirektes Hervorrufen des typishen Schmerzes au Mc Burney's Punkt. Ein Beträge zur diagnostic der Appendicitis und Typhlitis. Zentralbl f Chirurgie Leipzig. 1907:34:1257-59. — Kontsevyy OO. Acute appendicitis. In, LYa Kovalchuk, VF Sayenko, HV Knyshov. Klinichna khirurhiya. Vol. 1-2. Publ. «Ukrmedknyha», Ternopil 2000. Vol. 2. P. 125-363.

[1073] Pringle JH, Notes on the arrest of hepatic hemorrhage due to trauma. Ann Surg 1908;48(4):541-9.

[1074] Hemming AW, Mekeel KL, Zendejas I, et al. Resection of the liver and inferior vena cava for hepatic malignancy. J Am Coll Surg 21013;217:115-25.

Should blood loss be excessive during the hepatic transection, the intermittent hepatic portal inflow occlusion is used with 15 min. of ischemia time followed by 5 min. of reperfusion.

The hepatic parenchyma was divided using a cavipulse ultrasonic surgical aspirator (CUSA) device, or later a waterjet device.

If clamps are placed above the hepatic veins on the IVC, then it is necessary to use a complete normotermic hepatic vein exclusion (HVE) with the blood volume loading without/with the veno-venous bypass, the hypothermia without/with ex vivo resection on the back table, or the ante situm approach with the hepatic portal inflow occlusion but the portal structures intact and the IVC divided above and below the tumor.

Later, in 1913 he presented two cases of the great saphenous vein (GSV) grafting for the maintenance of a direct arterial circulation[1075].

Experimental, pathological and clinical studies performed by **William G. Spiller (1863 – 1940)**[1076] — American neurologist, and **Charles H. Frazier (1870 –1 936)**[1077]— American neurosurgeon, on surgical division of the sensory root of the 5th cranial (trigeminal) nerve behind the Gasserian ganglion **[Johann Gasser (1723 - 65)**. 18th century] to relief or lessening of the pain in trigeminal neuralgia or tic douloureux, prevented recurrence and reduced the mortality rate (Spiller-Frazier operation, 1901). In addition, Charles Frazier contributed to the section of the anterolateral columns of the spinal cord (chordotomy) for relief of pain, introduced the illuminated surgical retractor and the upright position to decrease bleeding and controlling hemorrhage during intracranial operations.

George W. Crile (1864 – 1943 — American surgeon, excised cancer of the head and neck with a cervical lymphadenectomy (radical neck dissection, Crile operation, 1906), and later (1921) founded the Cleveland Clinic (CC), Cleveland, OH.

Alexander Tietze (1864 – 1927) — German surgeon, described idiopathic painful non-suppurative swelling of one or more costal cartilages, especially of the 2nd rib (costal chondritis or Tietze syndrome, 1921). The anterior chest pain in this condition may mimic that of coronary artery disease (CAD).

Founder of orthopedics and traumatology in Ukraine and Russia was **Carl F. Wegner (12.12.1864, city of Kamianets'-Podils'kyy, Khmelnytska Obl.,**

[1075] Pringle JH. Two cases of vein grafting for the maintenance of a direct arterial circulation. Lancet 1913;1:1795-6.

[1076] Spiller WG, Frazier ChH. The division of the sensory root of the trigeminus for relief of tic douloures; an experimental, pathological, and clinical study, with a preliminary report of one surgically successful case. Univ Penn Med Bull 1901;14:342-52. / Philadelphia Med Phys J 1901;8:1939-49.

[1077] Frazier CH, Allen AR. Surgery of the spine and spinal cord. D Appleton & Co, New York & London 1918.

Ukraine – 12.1940, city of Bern, Switzerland)[1078] — Ukrainian orthopedist and traumatologist, who established July 08, 1907, Medical-Mechanical Institute in city of Kharkiv, Ukraine, under his leadership until 1926 p. He was the first to apply a continuous skeletal traction to treat fracture of the femoral bone and metal-synthesis for the treatment of bone fractures in Ukraine (1910), laid the foundation for the functional treatment of bone fractures (1926).

An emotional excessive sweating (hyperhidrosis) is an autosomal dominant infringement of the exocrine (eccrine) sweat glands with an increased sweating of the palms, soles, axillary pits, sometimes of the face. In order to free patient from unilateral increased hyperhidrosis, **Anastas Kotzareff (1866, Ohrid, Macedonia – 1931, Paris, France)**[1079] — Macedonian and French surgeon, was first to perform in 1920 via the left thoracotomy removal of the lower half of the nervous sympathetic cervicothoracic (stellate) ganglion which is formed by the fusion of the inferior cervical (C_7) and the upper thoracic ganglion (Th_1) and removal of the 2nd and the 3rd thoracic sympathetic ganglia (Th_{2-3}) – a partial upper dorsal sympathectomy with success.

The name of **Paul Sudeck (1866 – 1945)** — German surgeon, is associated with sympathetic maintained pain syndrome (SMPS) which is a burning pain outside a peripheral nerve distribution out of proportion to the injury, most commonly occurring after sprain of carpal or tarsal bones with atrophy, exaggerated by the emotion, usually in older people (Sudeck atrophy)[1080]; reflex sympathetic dystrophy (RSD), painful disorder in which there is overactivity of the sympathetic nervous system with trophic changes in the skin; post-traumatic osteoporosis (Sudeck syndrome); location in the rectosigmoid junction where the last sigmoid arterial branch of the inferior mesenteric artery (IMA) usually form an anastomosis with a branch of the superior rectal artery, a small and often only singular anastomosis, forming a site of watershed rectal wall, susceptible for ischemic colitis, and has a particular revelance for the colorectal surgery (Sudeck point). He also introduced a metal face mask for application of ether or chloroform anesthesia which nowadays has only historical meaning.

[1078] Kamianets'-Podils'ky (Foot Note 927), Khmelnytska Obl., Ukraine. — City of Berne (name originate from first hunted animal – bear; site of Celts' fortifications from 2nd century; founded 1191), Switzerland. — KliuchevskyVV. Skeletal traction. St Petersburg 1991. — Korzh AA. Flagman of Ukrainian orthopedy and traumatology of M.I. Sytenko Institute – 100 years. International Med J 2007;1;132. — Абліцов ВГ. «Галактика Україна». Українська діаспора: видатні постаті. В-во «Кит», Київ 2007.

[1079] City of Orchid (known from the 5th-4th century BC), Macedonia. — Kotzareff A. Resection partielle du trone sympathique cervical droit pour hyperhidrose unilaterale. Rev Med Suisse Rom 1920;40:111-3.

[1080] Sudeck PHM. Über acute entzundlische Knochenatrophie. Arch Klin Chir Berlin 1900;62:147.

Ivan I. Grekov (1867 – 1934)[1081] — Russian surgeon, developed methods of resecting the sigmoid colon affected by twisting and necrosis (Grekov-1, 1910, and Grekov-2, 1925).

Operation «Grekov-I» consist of invagination of the twisted necrotic sigmoid colon with preservation of the viable efferent part with the expectation on its subsequent rejection and departure via the rectum: the section of sigmoid colon, subjected to removal is separated from its mesentery, evaginated behind the boundary of dilated anus, resected, and the ends of the bowel are sewed using rounded suturing to the skin. The outcomes of this operation were not satisfactory.

Operation «Grekov-II» is two-staged removal of diseased sigmoid colon, consisting of lying on bypass anastomosis between the afferent and efferent loop of the bowel in the boundary of healthy tissues, bringing it up to the outside through the accessory incision of necrotic bowel loop, excision of which is carried out several days later, then sewing the stump using a triple-layer dull suture: stumps are retracted to the inside of the abdominal cavity in 3-4 weeks.

In 1914 he attended 40-year-old men who had dislocated his knee joint and tore his anterior cruciate ligament (ACL)[1082], which he repaired by utilizing a free strip of iliotibial band placed through a drill hole in the femur and connected it to the ACL remnant with good result.

In 1908, **Erich Lexner (1867 – 1937)** — German aesthetic and plastic surgeon who successfully transplanted a total knee joint from one person to another.

Pancreatoduodenectomy for the carcinoma of pancreas was designed and performed by **Walther Kausch (1867 – 1928)** — German surgeon, in 1909 and by **Allen O. Whipple (1881 – 1963)** — American surgeon, in 1935, and therefore should be called Kausch operation or Kausch-Whipple operation. Allen Whipple also depicted a Whipple's triad for insulinoma and other insulin producing tumors which denotes a spontaneous hypoglycemia with a level of glucose in the blood 50 mg/100 ml, central nervous system (CNS) or vasomotor system symptoms and relief by peroral or intravenous (IV) administration of glucose.

For the treatment of hypertrophic pyloric stenosis **Conrad Ramstedt (1867 – 1963)** — German surgeon, introduced a longitudinal incision through the serosa and sphincter muscles at the pylorus to relax a stenotic or closed by spasm sphincter (Ramstedt operation or pyloroplasty, 1911-12).

In 1907, **Hermann Schloffer (1868 – 1937)** — Austrian surgeon, performed the first transsphenoidal resection of a pituitary adenoma. Unfortunately, the patient died several weeks later from a residual tumor. Nine years later (1916), Hermann Schloffer's

---

[1081] Grekov II. Sellected Works. Publ. «Medgiz», St Petersburg 1952.

[1082] [19th century. Surgery. – **Wilhelm Weber (1804 – 91) / Eduard Weber (1806 – 71)**].

student, **Paul Kaznelson (1898 – 1959)** — Czech scientist, who described red cell aplasia, realized that Werlhlof disease **(Paul Werlhof (1699 – 1767**: 18th century), or chronic idiopathic thrombocytopenic purpura (ITP) is a condition in which there is an excessive destruction of the blood platelets by the spleen. Facing a female patient suffering from chronic ITP, he advised his mentor to remove the spleen, which he did. This was the first splenectomy for ITP. After the operation women enjoyed a dramatic improvement of the well-being.

In 1912 **Julius M. Wieting (1868 – 1922)**[1083] — German surgeon, achieved successful surgical repair of the congenital chest wall deformity – thoracoabdominal ectopia cordis by primary closure of the diaphragm and abdominal fascia, but he ignored the visible left ventricular (LV) diverticulum of the heart.

Maude E. Abbott (1869 – 1940)[1084] — Canadian physician, identified a small anomalous artery, an evanescent remnant of the 5th left branchial arch, arising from the posterior wall of the distal aortic arch or the proximal descending aorta (DTA) – Abbott artery (1928), which must be ligated and divided during flap aortoplasty using the subclavian artery (SCA) for coarctation of the aorta (COA).

The first morbidity, and mortality (M&M) conferences and the evaluation of surgeon's competency and publishing surgical outcome were initiated in 1914 by **Ernest A. Codman (1869 – 1940)**[1085] — American surgeon. Subsequently, it led in 1987-88 to the development of clinical quality and safety outcomes in patient care. His later contributions to surgery include description of a benign chondromatous giant-cell tumor of the upper epiphyseal cartilage of the humeral bone, affect also other long bones, e.g., femur and tibia (Codman tumor, 1931); radiological description of an incomplete triangular shadow between growing bone tumor (osteogenic sarcoma, osteosarcoma) and normal bone (Codman triangle); description in the absence of rotator cuff function or when there is a rupture of supraspinatus tendon of the scapula when the arm can be passively abducted without pain, but when support of the arm is removed and deltoid muscle contracts suddenly, the pain

[1083] Wieting J. Eine operative behandelte Herzmissbildung. Dtsch Z Chir 1912;114:293-5.
[1084] **William F. Hamilton (1893 - 1963)** — American physician. / Hamilton WF, Abbott ME. Coarctation of the aorta of the adult type: Part I. Hamilton WF, Abbott ME. Complete obliteration of the descending arch at insertion of the ductus in a boy of fourteen, bicuspid aortic valve, impending rupture of the aorta, cerebral death. Part II. Abbott ME. A statistical study and historical retrospect of 200 recorded cases, with autopsy, of stenosis or obliteration of the descending arch in subjects above the age of two years. Am Heart J 1928;3:381-421. / Ferencz C. Maude E. Abbott – physician and scientist of international fame. Can J Cardiol 2000;16(7);889-892. / Tadeishi A, Kawada M. Abbotts' artery in coarctation of the aorta. Ann Thorac Surg 2010 Oct;90(4):1367.
[1085] Codman EA. A Study in Hospital Efficiency: as Demonstrated by the First Five Years of a Private Hospital. Private publisher, Boston, MA 1916.

cause the patient to hunch the shoulder joint and lower the arm (Codman's sign, 1934); proposed a form of pendulum-like movement of the upper extremity exercises with the aim of regaining and maintaining range of motion after fracture (Codman pendulum exercise).

The co-founders of neurosurgery were **Harvey Cushing (1869 – 1939)**[1086] — American neurosurgeon, **Mykola M. Burdenko [22.V(03.VI.1876, town of Kam'ianka, Penzens'ka Obl., Russia – 11.XI.1946, city of Moscow, Russia]**[1087] — Russian neurosurgeon of Ukrainian origin, **Walter E. Dandy (1886 – 1946)** — American neurosurgeon, **Wilder G. Penfield (1891 – 1976)** — Canadian neurosurgeon, **Arne Torkildsen (1899 – 1968)** — Norwegian neurosurgeon, and **A. Earl Walker (1907 – 1995)** — American neurosurgeon.

In 1901 Harvey Cushing denoted a rise in the systemic blood pressure (BP) as a result of an increase in the intracranial cerebral spinal fluid (CSF) pressure (Cushing phenomenon), the former to a point slightly above the pressure in the brain against the conical trunk (Cushing law), introduced a temporal decompression of the brain (1905).

In 1912 Harvey Cushing presented a 23-year-old woman whose symptoms were painful obesity, amenorrhea, abnormal hair growth (hypertrichosis), underdeveloped secondary sexual characteristics, hydrocephalus, and cerebral tension, considering it to be an endocrinological syndrome caused by a malfunction of the pituitary gland which he termed a «polyglandular syndrome». However, she survived more than 40 years after symptom presentation without any treatment, and since after her death an autopsy was refused, no cause of disease could be confirmed.

During WW I Harvey Cushing by applying an x-ray diagnosis, antiseptic, and removal of deeply located foreign bodies in craniocerebral trauma, decreased the mortality rate by 50% (1918).

In 1918 Walter Dandy introduced air into the cerebral ventriculi for diagnostic purposes (ventriculopunction and ventriculography (reported in 1921).

Then, in 1921 Walter Dandy, in 1942 Earl Walker described congenital hydrocephalus due to obstruction of the CSF outflow through the median opening of **Francois Magendie (1783 – 1855**: 19th century. Anatomy, histology, and embryology**)** and the lateral openings of **Herbert** von **Luschka (1820 – 75**: 19th century. Anatomy, histology,

---

[1086] Cushing H. The Pituitary Body and its Disorders. Clinical States Produced by Disorders of the Hypophysis Cerebra. JB Lippincott, Philadelphia 1912. — Cushing H. The Life of Sir William Osler. Clarendon Press, Oxford 1925. — Cushing H. The basophil adenomas of the pituitary body and their clinical manifestations (pituitary basophilism). Bull Johns Hopkins Hosp 1932;50:137-95. — Gazhenko AI, Gurkalova IP, Zukow W, Kwasnik Z, Mroczkowska M. Pathology. Medical Student's Library. Radom Medical University in Radom and Odesa State University in Odesa. Radom 2009. P 271-82.

[1087] Town of Kam'ianka (founded on the beginning of 18th century), Penzens'ka Obl. (area 43,200 км²), Russsia. — Burdenko NN. N.N. Burdenko. Collection of Works. Vol. 1-7. Acad Sc Publ House, Moscow 1951. — Ablitsov VH. «Galaxy Ukraine». The Ukrainian diaspora: prominent persons. Publ. «Kyt», Kyiv 2007.

and embryology) of the fourth ventricle of the brain into the cisterna magna and into the cerebro-pontine angle cistern, respectively (Dandy-Walker syndrome)[1088].

In 1929, Harvey Cushing proposed the use of electrocautery and surgical clips in neurosurgery and described the chiasmatic compression due to hypophyseal tumors.

In 1936, Mykola Burdenko and **Borys M. Klosovsky (1898 – 1976)** — Russian morphologist, physiologist and neurosurgeon, conceived operation of transection of the extrapyramidal conductive pathways of the oblongated brain, i.e., myelencephalon or bulbous (in Latin – myelencephalon, medulla oblongata, seu bulbus) and the upper spinal brain (medullary bulbotomy or tractotomy) in patients after inflammation of the brain with paralysis agitans, that is post encephalic parkinsonism **[James Parkinson (1755 – 1824)**. 19th century. Neurology], characterized by hypokinesis (restricted scope and speed of motions), tremor and stiffness (or rigidity) of muscles. Other indications for tractotomy were double disturbances of motion concordance (athetosis,), hyperkinesis – twisting or shortening (shaking spasm) of muscles, partial uninterrupted epilepsy of Kozhevnikov **[Oleksiy Kozhevnikov (1836 – 1902)**. 19th century. Neurology], and medullary tractotomy for alleviation (lessening) or removal of intolerant pain.

Mykola Burdenko also proposed test for disturbance of the blood supply in overgrown (obliterative) (endarteritis obliterans) inflammation of arteries (endarteritis), i.e., endarteritis obliterans of the lower, rarely the upper extremities, usually in heavy cigarette smoking men because allergy to nicotine; developed technique for anastomosis of the injured brachial nerve plexus; ensured immediate amputation of extremity in anaerobic infection; debrided stump of the nerve after removal of extremity using 5% solution of formalin or 96% spirt.

In 1937, Walter Dandy was the first to successfully operate on a patient with an aneurysm of the right internal carotid artery (ICA) that had caused the cranial III (oculomotor) nerve paresis by being able to apply a silver clip to the base of the aneurysm, which he then electro coagulated with diathermia.

Wilder Penfield improved the surgical treatment of epilepsy, depicted the speech area in the posterior portion of the left hemisphere and the memory area in the temporal lobe of the brain, and described a syndrome of paradoxical hypertension associated with tumors of the hypothalamus.

In 1939 Arne Torkildsen introduced the palliative procedure for the treatment of congenital hydrocephalus, that is an occlusion the CSF flow through the **Jacobus Sylvian (1478 – 1555**: Renaissance) aqueduct of the mesencephalon which connect the third ventricle in the forebrain (diencephalon) to the four ventricle in region

---

[1088] Walker AE. Lissencephaly. Arch Neurol Psych (Chicago) 1942;48:13-29.

of mesencephalon and hindbrain (metencephalon), located dorsally to the pons and ventrally to the cerebellum. The procedure consists of passing from the lateral ventricle through an occipital burr-drill opening and then down under the skin into cisterna magna to bypass (shunt) the aqueduct obstruction (ventriculocisternostomy or Torkildsen operation)[1089].

In 1922-28 **Sergei P. Fedorov (1869 – 1936)** — Russian surgeon proposed an intraoperative radiation of malignant tumors.

The treatment of scoliosis and tuberculosis of the vertebral column was revolutionized by **Russel A. Hibbs (1869 – 1932)**[1090] — American orthopedic surgeon, who in 1911 devised the spinal fusion (Hipps) operation.

On 1900-03 **Vasyl D. Dobromyslov (27.XII.1869 – 08.I.1870, Moscow Obl., Russia – 06.IX.1918, Kyiv, Ukraine)**[1091]— Russian Ukrainian surgeon, began work up an intra- / through- / trans- / via pleural and extra-pleural approaches to the posterior mediastinum and the esophagus using an artificial lung ventilation (ALV) on cadavers and dogs. Together with **Erast G. Salishchev (1851 – 1901)** — Russian surgeon and ordinary professor, performed the first three (3) operations of exploratory thoracotomy in patients affected by carcinoma of the thoracic section of the esophagus, in the Hospital Department of Clinical Surgery and Desmurgy with Instruction in Joint Dislocations and Bone Fractures of the Medical Faculty at the First Siberian University (established in 1878), city of Tomsk (founded 1604), Western Siberia, Russia. Unfortunately, all three operations resulted in death of patients: two of them died in the operating table, one – week later from purulent pleuritis (results were reported in 1907).

Fritz Pregl (1869 – 1930) — Austrian physician and medical chemist, developed methods of quantitative organic microanalysis (1910-17) and introduced iodine **[Bernard Courtois (1877 – 1838). 19th century. Biochemistry]** in the form of 5% alcohol solution or tincture of iodine as an antiseptic (Pregl's solution; 1921). Tincture of iodine is used for topical application for the prevention and treatment of wound infection, skin disinfection before and after surgery. However, iodine may cause poisoning. An inhalation of iodine vapor may induce damage to the mucosa of the respiratory tract, attacks of cough, chest pain and headaches. Swallowing of iodine

[1089] Torkildsen A. New palliative operation in cases of inoperable occlusion of the sylvian aqueduct. Actta Psychiatr Neurol Scand 1939;82:117-23.

[1090] Goodwin GM. Russel A. Hibbs. Pioneer in Orthopedic Surgery, 1869-1932. Oxford Univ Press, London 1935.

[1091] Dobromyslow VD. A case of excision the piece from the thoracic portion of the esophagus via the trans-pleural approach (preliminary report). Vrach 1900;21(28):846-9. — Dobromyslow VD. On the question of resection of the thoracic portion of the esophagus through an intrapleural route (an experimental investigation). Russk khir arch 1903;19(4):590-608. — Doromyslow W. Bemerkungen zum Sauerbruch'schen Aufsaz «Chirurgie des Brusttails der Speiserohre». Russische Medicinische Rundschau 1907;12:715-7. — Ablitsov VH. «Galaxy Ukraine». The Ukrainian diaspora: prominent persons. Publ. «Kyt», Kyiv 2007.

may induce burn of the esophagus, stomach, the mucosa may change color into a dark-brown color.

In contrast to the established in 1898-1902 principle of conservative treatment of penetrated abdominal wounds (PAW) by a «watch and wait», **Vira H. Gedroits / Vera Gedroitz [07(19).04.1870, village Slobodyshche, Dyatkovsky Region, Bryansk Obl., Russia** – died from uterine cancer March **03.1932, city of Kyiv, Ukraine)**[1092] — Ukrainian Russian surgeon of Lithuanian origin, and Lithuanian Princess, Privat-Dozent (1921-23), Professor (1923-32) and Chief (1929-30) of the Department of Surgical Faculty Clinic of the Kyiv Medical Institute, Ukraine, laid down during Russian-Japanese War in 1904-05 the need for the closeness of the hospital to the battle-field and established the principle of an early (emergency) surgical treatment of PAW by exploratory laparotomy (within 3 hours or less from the time of injury), which became the modern doctrine of military medicine.

Edville G. Abbott (1871 – 1938) — American orthopedic surgeon and orthotist[1093], who proposed the treatment of a lateral curvature of the spine (scoliosis) by the opposite lateral pulling of the vertebral column using wide bandages and pads until the full over-correction of deformity with subsequent placing of a plastic (celluloid) jacket to produce pressure, counterpressure and fixation of the vertebrae in the normal position (Abbott method, 1913).

Archbishop **Valentyn F. Vojno-Yasenetsky (9.III.1871, town of Kerch, Crimea, Ukraine – 11.VII.1961, city of Simferopol, Crimea, Ukraine)**[1094] — Ukrainian surgeon, introduced an early and radical treatment f purulent inflammations soft tissues and bones (osteomyelitis). B 1946 p. he was awarded the State Premium 1st degree of 130 karbovanets which he donated for the care of children suffered from consequences of WWII.

[1092] Village Slobodyshche (first mentioned 17th century), Diatkivsky Region, Bryansk Obl., Russia. — Gedroits VI. Report of the Mobile Advance Noble Detachment. Lecture for the Society of Bryansk Physicians July 27, 1905. Publ. «Printing House of S.V. Yakovlev», Moscow 1905. — Getroits VI. Speaking about surgery to nurses and physicians. St Petersburg 1914. — Bennett JD. Princess Vera Gedroits: military surgeon, poet, and author. Br Med J 1992:305:1532-1534. — Moldavanov G. Princess Vera Gedroits: scalpel and pen. Russkaya mysl (Paris) 12 Nov 1998;4245:1. — Ablitsov VH. «Galaxy Ukraine». The Ukrainian diaspora: prominent persons. Publ. «Kyt», Kyiv 2007. — Avramchuk N, Suchomozsky M. The (un)known Ukrainians who changed the world. Publ. «Samit-book», Kyiv 2019. — Ludi EK, Zalles MV, Antenzana LFV, et al. International female surgeon pioneers: paving the way to generations to come. J Am Coll Surg 2020 Aug;231(2):294-8.

[1093] Orthotics (in Greek – ortho = «align» or «to straighten») is a medical specialty concerned with the design, manufacture, and application of orthoses, that is an externally applied device used to modify structural and functional deformities of the skeletal and neuromuscular system.

[1094] Town of Kerch (founded 530 BC), Crimea, Ukraine. — Vojno-Yasenetsky VF. Outline of Purulent Surgery. Moscow – St Petersburg 1934. — Vojno-Yasenetsky VF. Late Resections of Infected Firearm Joint Wounds. Moscow 1944. — Ablitsov VH. «Galaxy Ukraine». The Ukrainian diaspora: prominent persons. Publ. «Kyt», Kyiv 2007.

Mykhailo M. Diterichs [10(22).XI.1871, city of Odesa, Ukraine, or **city of St Petersburg, Russia – 12.I.1941, city of Moscow, Russia]**[1095] — Ukrainian surgeon, constructed a splint for immobilization of fractures (Diderichs transport splint, 1932).

In 1907 **Peter A. Hertsen (1871 – 1947)**[1096] — Russian surgeon, constructed an antero-sternal esophagus using a small bowel.

The idea of **Yakiv O. Halpern [02(15).I.1876, city of Vilnius, Lithuania – 22.XII.1941, city of Dnipro, Ukraine]**[1097] — Ukrainian surgeon, arosed during his experiments on dogs in 1911-12, was to substitute cancerous esophagus with a greater curvature of the stomach in people.

On March 13, 1913, **Franz J.A. Thorek (1861 – 1938)** — American surgeon of German origin, consulted 67-year-old woman with squamous cell carcinoma of the esophagus below the transverse thoracic aorta (TA) / aortic arch, resecting it through the left 7th intercostal space thoracotomy (ICST) and reconstruction using a rubber tube as an external bypass between the cervical esophagostomy and gastrostomy with 13 years survival without recurrence. In 1921 he performed another resection of a cancer of the intrathoracic esophagus via an extra-pleural approach. His patient died a little over a year later of recurrent disease.

In 1913 **C.A. Wilhelm Denk [?12.VI.1854, Germany** – lost at sea when steamship «City of Athens» was sunk of near the State of Delaware coast (area 5,130 km²), Atlantic Ocean, **USA 01.V.1918]**[1098] — German surgeon, removed the esophagus for carcinoma using the trans-hiatal approach (trans-hiatal esophagectomy).

Erwin Payr (1871 – 1946) — Austrian German surgeon, the first surgeon to use ozone to eradicate and control bacteria (1935); introduced the use of absorbable magnesium sutures in vascular and nerve surgery, the first workable method of vascular anastomosis; used elderberry steams for capillary drainage of brain abscesses; believed that constipation is due to kinking of an membranous adhesions (Payr's membrane) between the transverse and descending colon (splenic-flexure syndrome or Payr's disease); designed the pyloric clamp used in the digestive tract surgery, called Payr's pyloric clamp; described a sign indicating thrombophlebitis in which pain occurs during application of the pressure to the sole of the foot (Payr's sign).

[1095] Diderichs MM. Military-Field Surgery of the Military Region. «Gosmedizdat», 1932; 2nd Ed. «Medgiz», Moscow – St Petersburg 1933; 3rd Ed. «Medgiz», Moscow – St Petersburg 1938. — Ablitsov VH. «Galaxy Ukraine». The Ukrainian diaspora: prominent persons. Publ. «Kyt», Kyiv 2007.

[1096] Lerut T. Esophageal surgery at the end of the millennium. J Thorac Cardiovasc Surg 1998;116:1-20.

[1097] City of Vilnius, a.k.a., Vilno (settled since 4,000 BC, founded 13th century), Lithuania. — City of Dnipro (Foot Note 250), Ukraine. — Lerut T. Esophageal surgery at the end of the millennium. J Thorac Cardiovasc Surg 1998;116:1-20. — Ablitsov VH. «Galaxy Ukraine». The Ukrainian diaspora: prominent persons. Publ. «Kyt», Kyiv 2007.

[1098] Denk W. Zur Radikaloperation des Oesophaguskarcinoms (vorläufige Mittelung). Zentralbl Chir 1913;40:1065-8.

Faculty of surgical stomatology and maxilla-facial surgery, initially under the name of «Faculty of operative teeth treatment» of the National Medical University (NMU) of O.O. Bohomolets **[Oleksandr Bohomolets (1889 – 1946):** 19th century. Pathology], city of Kyiv, Ukraine, was founded by in 1920 by. **M. Jo. Shapiro (1871 – 1942)**[1099] — Ukrainian practiced dentist of Jewish descent, led by him until 1931, with whom is connected the beginning of development of this specialty in Ukraine.

The triage system for sorting the sick and the wounded was applied by **Vladimir A. Oppel (1872 – 1932)** — Russian surgeon of German origin, during World War (WW)-1.

Victor N. Shevkunenko (1872 – 1952) — Russian anatomist and military surgeon, authored «A Course of Operative Surgery with Anatomo-Topographic Data» (Vol. 1-3. 1927).

Sidney Yankauer (1872 – 1932) — American otorhinolaryngologist (ORL) of Jewish German descent, improved clinical thermometer used up to this day to safely measure the body temperature (1899), and also designed the most commonly used suction instrument with a bullous firm head with large opening at its tip and surrounded also by large openings, which allows an effective suction of secretions from the oral cavity and pharynx to prevent aspiration into tracheobronchial airways, without damaging of tissues in contact, and to clear operative sites, its suctioning volume is counted as blood loss at surgery (Yankauer suction tip, 1907).

In 1907 **Jacob M. Blumberg (1873 – 1955)**[1100] —German surgeon and gynecologist of Jewish descent, described a sign in which when the right lower quadrant abdominal wall is compressed slowly with fingers and then rapidly released, the presence of rebound tenderness and defense of muscles indicate acute appendicitis (Blumberg symptom).

Alexis Carrel (1873 – 1944)[1101] — French American surgeon, improved and popularized the «end-to-end» vascular suture anastomosis, especially the vein grafts (VG's) to the arteries, a leak-proof technique without constricting the lumen or producing thrombosis, techniques of great importance in organ transplantation; with **Clyde G. Guthrie (1880 – 1931)** — American surgeon, transplanted a second heart into the neck of a dog – heterotopic heart transplantation (HT, 1905)[1102].

[1099] Ablitsov VH. «Galaxy Ukraine». The Ukrainian diaspora: prominent persons. Publ. «Kyt», Kyiv 2007.

[1100] Blumberg JM. Ein neues diagostisches Symptom bei Appendicitis. Munchener medizinische Wochenschrift 1907;54:117-8. — Kontsevyy OO. Acute appendicitis. In, LYa Kovalchuk, VF Sayenko, HV Knyshov. Klinichna khirurhiya. Vol. 1-2. Publ. «Ukrmedknyha», Ternopil 2000. Vol. 2. P. 125-363.

[1101] Carrel A. La technique operatoire des anastomoses vasculaires et la transplantation des visceres. Lyon Medecine 1902; 98:859-64. — Carrel A. The surgery of blood vessel. Johns Hopkins Hosp Bull 1907;18:18-28. — Demers P, Robbins RC, Doyle R, et al. Twenty years of combined heart-lung transplantation at Stanford University J Heart Lung Transplant 2003;21:S77.

[1102] A heterotopic transplantation is the transfer of living organs from one part of the body to another or from one individual to another.

With the development of vascular suturing technique and heterotopic HT in dog, and description in 1907 of the first heart and lung transplantation (HLT) in a feline, he predicted the eventual benefit of organ transplantation.

The reason for the HLT is irreversible dysfunction of both organs – the heart and lungs.

The first most frequent reason for HT (44% of cases) is congenital heart disease (CHD), i.e., ventricular septal defect (VSD), atrial septal defect (ASD), less commonly patent ductus arteriosus (PDA), or multiple congenital anomalies with reversal of the blood shunt into cyanotic right-to-left blood shunt and irreversible end-stage cardiomyopathy (CM), secondary pulmonary hypertension – Eisenmenger's complex or syndrome [**Victor Eisenmenger (1864 – 1932)**. 19th century. Internal medicine].

The second most common reason r for the HLT (23% of cases) is primary pulmonary hypertension (PPH) without CHD, associated with irreversible right ventricular (RV) failure – cor pulmonale.

The remainder reasons for HLT are cystic fibrosis (12% of cases), complex CHD (12% of cases), alpga$_1$-antitripsin deficiency (2% of cases), CM with pulmonary hypertension (2% of cases), single lung transplantation (1% of cases), congenital bronchiectasis (<1% of cases), pulmonary lymphangioleiomyomatosis (<1% of cases) and other disease (3% of cases).

Through the middle longitudinal sternotomy incision (MLSI), the donor inferior and superior vena cava (IVC, SVC) and the ascending aorta (AA) are divided, and the heart-lung block is dissected free from the esophagus and posterior hilar attachment of both lungs, and after the trachea is stapled and divided as high as possible, the entire heart-lung block is removed from the chest cavity. Also, via the MLSI, the recipient left, and right pneumonectomies are performed by dividing the respective inferior pulmonary ligament, pulmonary artery, and vein (PA, PV), and main-steam bronchus (MSB). Then the native cardiectomy of the recipient is performed at the left atrial (LA) cuff level around the four PV orifices, division of the IVC, SVC, the AA, and the trachea is opened one cartilaginous ring above the carina. The donor heart-lung block is lowered into the chest cavity of the recipient, placing hila anterior to each respective phrenic nerve pedicle, followed by anastomosis of the trachea one cartilaginous ring above the carina, of the IVC, SVC and AA.

To maintain donor organ allocation, the «domino-donor» HLT was adopted since 1989, where the recipient of the HLT serve as a lung donor for a second recipient with septic lung disease[1103].

[1103] Baumgartner WA, Trail TA, Cameron DE, et al. Unique aspect of heart and lung transplantation exhibited in the «domino-donor» operation. JAMA 1989;261:3121-5.

Later, Alexis Carrel developed a technique for cultivating and preservation of tissue in vitro for both organ and limb transplantation (1911-12); has demonstrated the revitalizing effect of young cells on culture of old and degenerated cells (1912); advanced experimental organ transplantation; and with **Henry D. Dakin (1880 – 1952**: 20th century. Biochemistry**)** developed the method (Carrel-Dakin) of wound treatment (1916-18) that consist with thorough exposure of the wound, removal of foreign material and devitalized tissue, cleaning and irrigating with diluted sodium hypochlorite (NaClO) or Carrrel-Dakin solution.

Alexis Carrel with **Charles A. Lindbergh (1902 – 74)** — American pilot, constructed a pump oxygenator (Lindbergh pump, 1935) for the perfusion of living organs outside the body.

Mykola O. Bohoraz [01(13).II.1874, city of Tahanrih, Rostov Obl., Russia – 1952, city of Moscow Москва, Росія][1104] — Ukrainian Russian surgeon, born in Jewish Ukrainian family, town of Bar (settled since 6,500 BC, founded 1401; during conducted by Russian Communist Fascist of Holodomor-Genocide against the Ukrainian Nation, perished 15 persons), Vinnytsia Obl. (area 265,130 km²), Ukraine, grafted the superior mesenteric vein to the inferior vena cava (IVC) for the treatment of cirrhosis of the liver with portal hypertension (Bohoraz operation, 1913) and replaced almost completely the removed femoral artery with gunshot wound (GSV, 1935).

One-stage removal of the pulmonary lobe (lobectomy) was carried out in 1927 by **Harold Brunn (1874 – 1951)** — American surgeon, without complication.

In May of 1901 **Yevhen H. Cherniakhivsky / Evhen H. Cherniachivsky (1874, village of Mazepyntsi, Bilotserkivsky Region, Kyiv Obl., Ukraine** – executed by Russian Communist Fascists NKVS **1938, city of Kyiv, Ukraine)**[1105] — Ukrainian surgeon, performed successful emergency operation on the heart suturing a life-threatening wound (first in Kyiv). Later (1914), he experimentally transplanted the kidney into the animal's inguinal region with the vascular anastomoses to the femoral artery and vein, which secreted urine for more than 48 hours. During WW-I (1914-18), beginning since 1916 performed transplantations of the bone tissue with complete healing.

[1104] City of Tahanrih / Taganrog (settled since 7th century BC, founded 1698), up to and in 1920-24 – Ukraine, since 1925 Rostov Obl. (area 100,967 km²), Russia. — Ablitsov VH. «Galaxy Ukraine». The Ukrainian diaspora: prominent persons. Publ. «Kyt», Kyiv 2007.

[1105] Village of Mazepyntsi (Foot Note 596), Bilotserkivsky Region, Kyiv Obl., Ukraine. — Cherniakhivsky EH. A case of suturing wound of the heart. Publ. «Company of Fast Printing» A.A. Levenson», Moscow 1905. — Cherniakhivsky Ye.H. The contemporary state of surgery of the blood vessels. The speech pronounced on the V Solemn Meeting of the Kyiv Surgical Society 18 Nov. 1913. — Cherniakhivsky EH. Contemporary State of Surgery of the Blood Vessels. Publ. «G.Y. Gouberman», Kyiv 1914. — Mirsky M. Ukrainian medical scientists – victims of Stalin's repressions. UHMJ «Ahapit» 1995;2:45-54. — Hanitkevych Ya. Yevhen Cherniachivsky – founder of the first Ukrainian school of surgery. J Ukr Med Assoc North Am 1996 Spring;43,2(139):112-7. — Ablitsov VH. «Galaxy Ukraine». The Ukrainian diaspora: prominent persons. Publ. «Kyt», Kyiv 2007.

In 1907 **Pierre Duval (1874 – 1941)**[1106] — French surgeon, described the sternotomy-laparotomy incision by dividing the sternum transversely opposite the 3rd costal cartilage, below medially the gladiolus and xiphoid process longitudinally and along the median line in the upper abdomen.

Allen B. Kanavel (1874, Sedwik, KS – died in a car accident **May 27, 1938, Pasadena, CA)** — American surgeon, described the triangle (Kanavel) in the middle of the palm, the site of a common flexor tendon sheath of fingers and the four (Kanavel) signs of its inflammation (tendovaginitis) – the affected finger is held in slight flexion and there is pain on passive extension, fusiform swelling and tenderness over the affected tendon (1912).

In 1926-27 **Antonio C.A.F. Egas Moniz (1874 – 1955)** — Portuguese neurosurgeon, developed cerebral angiography and produced prefrontal leucotomy (lobotomy) as a therapy for certain psychoses.

To correct fractures of the zygomatic bone and arch **Thomas E. Carmody (1875 – ?)** — American oral surgeon and **Oskar Batson (1894 – 1979**: 20th century. Anatomy, histology, and embryology**)** devised surgical reduction through an intraoral incision above the maxillary molar teeth (Carmory-Batson operation).

A continuous U-shaped suture used for inverting the gastric or intestinal wall in performing an anastomosis was introduced by **Frank G. Connell (1875 – 1968)** — American surgeon, where stitches are stacked in parallel and 4 mm from the edge of the wound and should be laid through all layers of the intestine (Connell suture). The suture goes through the wall from the serosa to the mucosa, then back from the mucosa to the serosa on the same side. The stitch then crosses the incision to the serosa on the other side and then repeats itself.

Later (1922), **Hans** von **Haberer (1875 – 1958)**[1107] — Austrian surgeon and **John Finney (1863 – 1942**; above)[1108] proposed a distal gastrectomy with a blind closure of the duodenal stump and an «end-to side» anastomosis of a distally transected stomach to a distally closed «Roux-en-Y» jejunum (von Heberer-Finney gastrojejunostomy, 1922, 1923). In addition, Hans von Haberer devised a technique for radical gastric resection after previous gastroenterostomy (1931)[1109].

Founders of mechanical surgical closure of the wounds and suturing of organs were using metallic clips, staples pincers (tweezers, forceps) were **Gaston M. Michel (1875-1937)** — French surgeon, with help of suturing devises (stapling) – **Maurice**

[1106] Duval P. Le incision median thoraco-laparatomie. Bull Med Soc Chir Paris 1907;33:15.

[1107] Von Haberer H. Terminolaterale gastroduodenostomie by der Resection Methode nach Bilroth I. Zentralbl f Chir Leipzig 1922;49:134.

[1108] Finley JMT. A new method of gastroduodenostomy, end-to-side. South Surg Transactions 1923;36:576.

[1109] Best RR. von Haberer's technic for radical stomach resection following previous gastroenterostomy. Ann Surg 1931Aug;94(2):233-41.

Jeannel (Feb. 01,1850, city of Bordeaus, France – Sep. 13, 1918, city of Paris, France)[1110] — French surgeon (1904), **Mykola Stachowsky (village Stetkivtsi, Zhytomyrsky Region, Volynsky Obl., Ukraine – 07.XII.1948, city of Praha, Czechia)**[1111] — Ukrainian surgeon (1907) and **Hümér Hültl (1868 – 1940)**[1112] — Hungarian surgeon (1908).

Metallic clips with the application of forceps were used by Gaston Michel for the approximation of the edges of the wound and to avoid bleeding from small vessels, i.e., hemostasis (Michel clips). The wire frame of Michel's skin (wound) clips holds 25 metal clips sutures; each suture has two sharp prongs to close the skin / wound (Michels' tweezers or staples).

Maurice Jeannel published 125 scientific papers and books concerned with diagnostic medical instruments, general, orthopedic, and urological surgery, introduced in 1904 suturing devises (staplers) for the use in general surgery.

Mykola Stachowsky treated bone wounds pouring in sterile warm Vaseline (1904); described in 1907 p. themethod and instrument for suturing tissue with the П-like metallic staples; served as the first head of the diplomatic mission of the Ukrainian National Republic (UNR, 1917-21) in Great Britain (29.I. – 29.IX. 1919).

Hümér Hültl constructed in 1908 mechanical stapler applicator for distal resection of the stomach (distal gastrectomy).

Sign of acute appendicitis observed by **Paul Rosenstein (1875 – 1964)** — German physician of Jewish descent, is an increased of tenderness in the right lower abdominal quadrant when the patient moves from supine position to a recumbent posture on the left side (Rosenstein sign).

In 1910 **Samuel Robinson (1875 – 1947)**[1113] — American thoracic surgeon, re-introduced the trocar drainage of chronic thoracic empyema, described in 15 AD by **Aulus Celsus (c.25 – c.50 AD**; Ancient medicine), adding to it a suctioning vacuum air pump. Later, in 1915 he established more permanent drainage of chronic

---

[1110] City of Bordeaux (region settled around 300 BC by a Celtic tribe, the Buturiged, founded 3rd centuty AD), France. — Jeannel M. Arse nal du diagnostic medical: reserches sur les thermometres, les balances, les instruments d'exploration des organs respiratores, de'appareil cardiovasculare, du systeme nerveux, les speculums uteri et les laryngoscopes. Balliere 1873. — Jeannel M. Des ligatures et sutures metalligues perdues. Arch prov chir (Paris) 1904;13:385.

[1111] Village Stetkivtsi (according to retelling founded in 14th century, first documented in writing 1601), Zhytomyrsky Region, Volyn Obl., Ukraine. — Stachowskii H. Metallic suture and ligature. Vrachebnaia gazeta 1907;33:907. — Rozhin I. Dr Mykola Stachowsky. Ukr Med Assoc North Am 1965 Jan;12,1(36):32-44. — Ablitsov VH. «Galaxy Ukraine». The Ukrainian diaspora: prominent persons. Publ. «Kyt», Kyiv 2007.

[1112] Hültl H. II Kongress der Ungarischen Gesellschaft für Chirurgie, Mai 1908. Pester Med Chir Presse 1909;45:108-10. — Haddad FS. Suturing methods and materials with special emphasis on the jaws of giant ants (an old-new surgical instrument).J Med Liban 2010; 58 (1): 53-6.

[1113] Robinson S. Acute thoracic empyema. Avoidance of chronic empyema. Rib trephining for suction drainage. Boston Med Surg J 1910;163:561-70.

tuberculous and non-tuberculous thoracic empyema by the creation of an open window thoracostomy[1114]. He made a U-shaped skin incision on the chest wall over the empyema thoracis. The cavity was entered by removing the overlying ribs and then debrided. Then, the skin edges were approximated to the empyema cavity.

In 1916 **J.H. Kenyon**[1115] —American surgeon, described a simpler drainage of the pleural empyema which consist of inserting the distal extremity of the draining tube into a liquid column, contained inside a flask, whose cap has two openings: one for the passage of the draining tube and one for ventilation (air vent), known as a water-seal syphon drainage system, without or with suction.

In 1911 **Ernst Sauerbruch (1875 – 1951**: 20th century. Anesthesiology: above**)** performed **thymectomy** in two patients with myasthenia gravis (MG), after which **E.D. Schumaker** — American surgeon, and **P. Roth**[1116] — American surgeon, noted improvement in their condition. Ernst Sauerbruch also excised in 1937 pseudoaneurysm of the right ventricle (RV).

Advances in ear surgery were made possible through the work of **Robert Barany (1876 – 1936)** — Austrian otologist of Jewish descent, in 1906-08 on the function and diseases of the vestibular apparatus of the ear, including the introduction of the caloric (Barany) test.

In 1906 **Jose Goyannes Capdevila (1876 – 1964)**[1117] — Spanish surgeon, reported on the use of the popliteal vein in situ to restore the continuity of the popliteal artery (PA) after excision of an aneurysm. To follow in the footsteps of Jose Goyannes Capdevila and **James Pringle (1863 – 1941**: above**)**, in 1913 **Ernst H. Jaeger** advocated the use of the great saphenous vein (GSV) grafts after the excision of peripheral arterial aneurysms.

Achalasia (cardiospasm) of the esophagus is the T-cell mediated degeneration and fibrous replacement of the esophageal myenteric (Auerbach)[1118] plexus of unknown etiology, characterized by a failure of relaxation of the lower esophageal sphincter and aperistalsis of the body of the esophagus[1119]. Patients complain of progressive dysphagia, regurgitation, and weight loss. Palliative treatment is directed towards

---

[1114] Robinson S. The treatment of chronic non-tuberculous empyema. Collect Pap Mayo Clin Mayo Fund 1915;7:618-44. — Robinson S. The treatment of chronic non-tuberculous empyema. Surg Gynecol Obst 1916;22:557-71.

[1115] Kenyon JH. Traumatic hemothorax: siphon drainage. Ann Surg 1916;64:728-9.

[1116] Schumaker ED, Roth P. Thymectomie bei einem Fall von Morbus basedowi mit Myasthenia. Mittel Grenzgebieten Med Chir 1912;25:746.

[1117] Goyanes J. Nuevos trabajos de cirugia vascular, substitucion plastica de las arterias por la vena o arterioplastia venosa aplicada como nuevo metodo del tratamiento de los aneurismas. El Siglo Medico 1906;53:546-61.

[1118] **Leopold Auerbach (1828 - 97)** — German anatomist and neuropathologist (19th century. Anatomy, histology and embryology).

[1119] Williams VA, Peters JH. Achalasia of the esophagus: a surgical disease. J Am Coll Surg 2009;208(1):151-62.

relaxation of the lower esophageal sphincter and improving esophageal emptying using calcium channel blockers and long-acting nitrates, and if ineffective its staged pneumatic balloon dilatation. When there still is no improvement, then the operative treatment, introduced in 1913, by **Ernst Heller (1877 – 1964)**[1120] — German surgeon, is indicated, that is a longitudinal esophagocardiomyotomy by the standard or minimally invasive abdominal or thoracic approach.

In 1923 and 1947 **Carlos S. Williamson**[1121] — American surgeon, described the underlying process of rejection and mechanism in organ failure after kidney and blood vessels transplantation in animals – autogenous transplantation, where the donor and recipient of the organ is this same individual, and allogenous (homogenous) transplantation, where the donor and the recipient of the organ are disparate individuals of this same species, although usually genetically and immunologically differs one from the other.

In 1934 **John Homans (1877 – 1954)**[1122] — American surgeon, recognized the difference between a painful but benign course of superficial phlebitis of the lower extremity, a phlebitis of the femoral vein (FV), sometimes after childbirth or acute disease with fever, swelling of the foot without redness (phlegmasia alba dolens), an acute painful lightening thrombosis of the deep vein of the leg, with a reactive arterial spasm, marked swelling of a lower limb with a strong cyanosis, purple areas and petechiae (phlegmasia cerulea dolens), and an insidious painless deep venous thrombosis (DVT) that can cause deadly pulmonary embolism (PE). Subsequently, he popularized the ligation of the great saphenous vein (GSV) at the saphenofemoral junction for the treatment of varicose veins of the lower extremity, and advocated ligation of the sub-sartorial vein to stop migration of clots causing PE. Then, he described the sign of DVT of the calf in which passive dorsiflexion of the foot with the knee straight is manifested by pain or tenderness in the calf and the back of the knee (Homans sign, 1944). However, Homans sign is positive in only about 50% of the patients. He also reported on the first instance of DVT due to prolonged sitting (1954).

[1120] Steichen FM, Ravitch MM. Ernst Heller, M.D., 1877-1964. NY State J Med 1965 Oct 1;65(19):2500-2. — Hanna AN, Datta J, Ginzberg S, et al. Laparascopic Heller myotomy vs per oral endoscopic myotomy: patient-reported outcomes at a single institution. J Am Coll Surg Apr 2018;228(4):465-73.

[1121] Williamson CS. Some observations of the length of survival and function of homogenous kidney transplants: preliminary report. J Urol 1923 Oct 1;10:275-8. — Williamson CS, Mann FC. Functional survival of autogenous and homogenous transplants of blood vessels. An experimental study. Arch Surg 1947;54(5):529-40.

[1122] Homans J. Thrombosis of the deep veins of the lower extremities, causing pulmonary embolism. N Engl J Med 1934;211:993-7. — Homans J. Circulatory Diseases of the Extremities. MacMillan Co, New York 1939. — Homans J. Diseases of the veins. N Engl J Med 1944;231:51-60. — Homans J. Thrombosis of the deep leg vein due to prolonged sitting. N Ing J Med 1954;250(4):148-9. — Touloukian RJ. Surgical mentorship of John Homans by Harvey Cushing: the untold story. J Am Coll Surg June 2019;228(6):819-30.

The compound made by **Charles Stent (1807 – 85**: 19th century. Surgery) for dental prostheses was used in 1917 by **Johannes (Ian) F.S. Esser (1877 – 1946)** — Dutch surgeon, for the facial reconstruction of the soldiers wounded during WW I, and by doing so he initiated the acceptance of the term «stent», instead of carcass, or stretcher, which later extended it into devices for dilatation of a narrowed tubular part of the body, such as ureter, esophagus, common bile duct (CBD), arteries, veins, aorta, and trachea.

The trans-hiatal esophagectomy was re-applied in 1931 by **George G. Turner (1877 – 1951)**[1123] — English surgeon, for removal in a 58-year-old miner of the cancerous thoracic esophagus with 19 months survival.

The contribution of **Nadiya Dobrovols'ka-Zavads'ka [(1878 – 1954)**. 20th century. Anatomy, histology, and embryology][1124] to vascular surgery from 1912 was to anastomose vessels of a different diameter, after transection its edges at the angle.

The first dissection lobectomy for carcinoma of the lung was performed by **Hugh M. Davies (1879 – 1965)**[1125] — Welsh surgeon, in 1912, the patient died 8 days postoperatively with empyema thoracic.

In 1933 **T. Oshawa**[1126] — Japanese surgeon, reported on resection with immediate reconstruction of the cancerous thoracic esophagus in 19 patients via the left 5th - 6th intercostal space thoracotomy (ICST) with an operative mortality (OM) about 50%.

The two-stage removal and replacement of the cancerous middle esophagus was reported by **Ivor Lewis (1895 – 1982)**[1127] — English surgeon. The operation is performed through the right 5th - 6th ICST for extirpation of the esophagus and an intrathoracic anastomosis of its distal part with the proximal part of the stomach, and the superior median laparotomy for mobilization of the stomach, pyloroplasty and jejunostomy (the Ivor-Lewis esophagectomy, operation 1946).

K. McKeown[1128] — American surgeon, described three-staged removal and replacement of the cancerous esophagus: (1) the right 5th - 6th ICST for resection of the thoracic esophagus, (2) superior median laparotomy for mobilization of the stomach, pyloroplasty and jejunostomy; and (3) the left cervical incision for anastomosis between remaining cervical esophagus and the pulled up stomach, when necessary

---

[1123] Turner G. Excision of the thoracic aesophagus for carcinoma with construction of an extrathoracic gullet. Lancet 1933;33:1315-6.

[1124] Dobrovolskaya NA. The management of injuries of vessels and traumatic aneurysms. Russk vrach 1916;49:1154; ibidem 50:1187; ibidem 51:1210; ibidem 52:1225.

[1125] Davies HM. Recent advances in the surgery of the lung and pleura. Br J Surg 1913-1914;1;228-58. — H,M. Davies Surgery of Lung and Pleura. Paul B. Hober, New York 1920.

[1126] Oshawa T. The surgery of the esophagus. Arch Jpn Chir 1933;10:604-95.

[1127] Lewis I. The surgical treatment of carcinoma of the oesophagus with special reference to a new operation for growths of the middle third. Br J Surg 1946;34:18-31.

[1128] McKeown K. Total three-stage oesophagectomy for cancer of the esophagus. Br J Surg 1976;63:259.

the left or right colon, and in the extreme situation the segment of the pedicled jejunum (modification of the McKeown three-incisional technique, 1962).

Thoracoscopy / video-assisted thoracic surgery (VATS) is a medical procedure for internal examination, targeting biopsy and/or resection of a diseased lesion in the thoracic (pleural) cavity, under sedation with local anesthesia or under general anesthesia.

Thoracoscopy / VATS was carried out in 1910 by **Hans Ch. Jacobaeus (1879 – 1937)**[1129] — Swedish internist-pulmonologist, under local anesthesia using cystoscope for diagnosis of conditions of the pleural cavity in patient with tuberculosis. Later he constructed optical device, named thoracoscope, with the aid of which he from the beginning performed inspection of the pleural cavity, and in 1910-13 performed 89 thoracoscopies. In 1913 he modernized thoracoscope by joining it to a galvanic cautery, began to proceed with thoracoscopy dividing and burning pleural tuberculous adhesions to allow therapeutic pneumothorax in collapse therapy. In 1923 he first carried out targeting biopsy of the pleura in the patient with mesothelioma, reported performing 120 thoracoscopies in patient s with pulmonary tuberculosis.

The pupil **Fredrich Trendelenburg (1844 – 1924**: above), **Martin Kirschner (Oct. 28, 1879, Wrocław, Poland – Aug. 30, 1942, Heidelberg, Germany)** — German surgeon, introduced in 1909 sterilized, sharpened, smooth stainless-steel pins to hold fractured bone or bone fragments together (pin fixation) or to provide an anchor for skeletal traction (Kirshner's wire or K-wires), and performed in 1924 a successful pulmonary embolectomy (PE) with a long-term survival of the patient.

Classification of maxillary bone of fractures of the face was developed in 1901 by **Rene Le Fort (1879 – 1951)**[1130] — French surgeon, consisting of:

Le Fort I fracture – horizontal of the alveolar process of the maxilla immediately above the teeth and palate, the most common midfacial fracture resulting from a force of injury directed on the maxillary alveolar rim in a downward direction, characterized by malocclusion of the teeth, described by **Alphonse Guerin (1816 – 95**: 19th century. Surgery**)** and named after him (Guerin fracture, 1866);

Le Fort II fracture – pyramidal trans-nasal fracture;

Le Fort III fracture – transverse transorbital fracture or craniofacial separation.

[1129] Jacobaeus HC. Über die Möglichkeit die Zystoscopie bei Untersuchung seröser Höhlungen anzuwenden. Münch Med Wochenschr 1910;57:2090-2. — Jacobaeus HC. The cauterization of adhesions in artificial pneumothorax treatment of pulmonary tuberculosis under thoracoscopic control. Proc R Soc Med 1923;16:45-62. — Katlic MR. Five hundred seventy-six cases of videoassisted thoracic surgery using local anesthesia and sedation: lessons learned. J Am Col Surg 2018 Jan;226(1):58-63.

[1130] Le Fort R. Etude experimental sur les fractures de la machoire superieure. Rec Chir Paris 1901;23:208-27.

A Le Fort fracture of the skull is a classic trans-facial fracture of the midface, involving the maxillary bone and surrounding structures in horizontal, pyramidal, or transverse directions. The hallmark is traumatic separation of the pterygoid bone from the maxillary bone between the pterygoid plates, a horseshoe shaped bone protuberance which extends from the inferior margin of the maxilla and its sinuses.

René Leriche (1879 – 1955) — French surgeon, performed the first lumbar sympathectomy, and resected thrombosed bifurcation of the abdominal aorta (AA) – Leriche syndrome (1940).

The psoas muscles symptom other than described by **Vasyl Obraztsov (1849 – 1920**: 19th century. Internal medicine**)** is the one proposed in 1921 by **Zachary Cope (1881 – 1974)** — English surgeon, in which the pain in the right iliac area is increased by straightening out the femur when the patient is lying on left side, suggesting acute appendicitis (Cope's symptom)[1131].

The surgeon-in-chief (since 1927) of the Boston Children Hospital (founded 1869), Boston, MA, **William E. Ladd (1880 – 1967)** — American pediatric surgeon, established the first pediatric surgical training program in the USA.

Named after him the Ladd's syndrome (in French – Bride de Ladd) is a congenital condition resulting from abnormal rotation of the intestine (malrotation) in which peritoneal bands, or the Ladd's bands, i.e., a stalk of peritoneal tissue, which normally attaches the cecum to the right lateral abdominal wall, when displaced cause obstruction of the duodenum (1932, 1936)[1132]. This condition is manifested by severe, usually bilious, continuous vomiting after birth or after first feeding, jaundice (30% of cases), distention of epigastrium and meconium may be excreted, or a partial obstruction with intermittent vomiting at variable time, days to years after birth

The surgical procedure performed to alleviate intestinal malrotation involves division of the peritoneal bands of Ladd, widening of the intestines mesentery, removal of the appendix (appendectomy) and conventional placing of the caecum and colon (the Ladd's procedure).

William Ladd with his associate **Robert E. Gross (1904 – 88)** — American pediatric general and thoracic, reported repair of the congenital diaphragmatic hernia (CDH) in a neonate on the second day of life with survival (1940)[1133], published pediatric surgical textbook – «Abdominal Surgery of Infants and Childhood» (1941)[1134], and described a newborn jaundice (icterus neonatorum), affecting both sexes, present from birth or

---

[1131] Augustin G. Acute Abdomen During Pregnancy. Springer, New York – London 2014.

[1132] Ladd WE. Congenital obstruction of the duodenum in children. N Engl J Med 1932;206:277-83. — Ladd WE. Surgical diseases of the alimentary in infants. N Engl J Med 1936;215(16):705-8.

[1133] Ladd WE, Gross RE. Congenital diaphragmatic hernia. NEJM 1940;233:917-25.

[1134] Ladd WE, Gross RE. Abdominal Surgery of Infants and Childhood. WS Saunder Co, Philadelphia 1941.

with onset at 2-3 weeks of life, manifested by fetor hepaticus, progressive jaundice and hepatosplenomegaly (Ladd-Gross syndrome, 1941).

Surgical treatment of recurrent dislocation of the shoulder joint was introduced in 1923 by **Victorio Putti (1880 – 1940)** — Italian surgeon, and in 1925 p. Sir **Harry Platt (1886 – 1986**) — English surgeon, by tightening the anterior capsule and subscapularis tendon with subsequent accepted loss of external rotation to increase stability of the shoulder (Putti-Platt capsulorrhaphy).

In 1935 **Leo Eloesser (1881 – 1975)**[1135] — American thoracic surgeon, re-described a U-shaped method of an open drainage for tuberculous and non-tuberculous empyema thoracis with a skin flap over a resected rib, first created in 1915 by **Samuel Robinson (1875 – 1947**: above), named the Eloesser flap.

Enrique Finochietto (1881 – 1948) — Argentinian surgeon, designed a thoracic rib/sternal retractor (Finochietto's spreader, 1936), the Finochietto forceps, the Finochietto scissors and the Finochietto pin and stirrup for the skeletal traction. The last was originally devised in the form of U-shaped steel strip, driven through the posterior appendage of the fractured heel bone and fixed to the transverse ledge, from which comes a traction.

Between 1922-33 **Olexandr H. Puchkivsky (June 07, 1881, village of Krasne, Chernihiv Region, Chernihiv Obl., Ukraine** – abducted and shot to death by Russian Communist Fascists NKVS agents **Dec. 14, 1937, city of Kyiv (?), Ukraine**, buried in Lukianivs'ke Cemetery in Kyiv)[1136] — Ukrainian surgeon-otorhinolaryngologist

---

[1135] Eloesser L. Of operation for tuberculous empyema. Surg Gynecol Obstet 1935;60:1096-7. — Eloesser L. Of operation for tuberculous empyema. Ann Thorac Surg 1969 Oct;8(4):355-7.

[1136] Village Krasne (founded 1630; due to conduction by Russian Communist Fascist of Holodomor-Genocide 1932-33 against the Ukrainian Nation not less than 26 people, including 5 children, have died from man-made hunge,), Chernihiv Region, Chernihiv Obl. (area 31,865 km²), Ukraine. — Puchkivsky O. Olimpiy ta Ahapit – pershi ukraïns'ki likari. Zbir med sekciï Ukr nauk t-va u Kyivi 1922;5:2. — Puchkivs'kyi O. Rola ukraïnciv v rozvytku medycyny v Rosii. Zapysky doslidnykh katedry pry VUAN. 1923. — Puchkivsky O. Three founders of Russian medicine (Nestor Ambodyk-Maksymovych, Petro Zahorsky, and Danylo Vellansky). Sci J Ukr Studies «Ukraine», AUASc, St Publ «Ukraine», Kyiv 1924. Book 4, P. 27-34. — Puchkivs'kyi O. Pershi medyko-sanitarni ta antropolohichni narysy Ukraïny. J Ukr Med Assoc North Am 1926:1. — Puchkivsky OM. Diseases of the Ear, Nose and Throat. Handbook for Physicians and Students. Publ. «Derzhvydav» of Ukraine, Kyiv 1926. — Puchkivsky OM. Scleroma. Publ. «Derzhmedvydav» of Ukraine, Kyiv 1930. — Puchkivsky OM. Diseases of the Organs of Hearing and the Upper Respiratory Airways in Children and Adolescens. Publ. «Derzhmedvydav» of Ukraine, Kyiv 1931. — Puchkivsky OM. Otosclerosis. Publ. «Derzhmedvydav» of Ukraine, Kyiv 1933. — Puchkivsky AM. Base date for development of otorhinolaryngology in former Russia and USSR. Handbook about Diseases of Ear, Throat and Nose. Vol. 1. Publ. «Gosmedizdat» of Ukraine, Kyiv 1936. — Puchkivsky O. Three founders of medicine. J Ukr Med Assoc North Am 1961 July.-Oct;8,(22-23):40-8. — Mirsky M. Ukrainian medical scientists – victims of Stalin's repressions. UHMJ «Ahapit» 1995;2:45-54. — Iskiv BH. Oleksandr Puchkivsky – founder of contemporary higher education in Ukraine. J Ukr Med Assoc North Am 2000 Spring; 45,1(144):53-63. — Ablitsov VH. «Galaxy Ukraine». The Ukrainian diaspora: prominent persons. Publ. «Kyt», Kyiv 2007. — O.M. Puchkiv'sky. Prominent Ukrainian otorhinolaryngologist. Zdorov'ia Ukrainy 07.2010;13-14(242-243):66.

(ORL), who researched the development of medicine in Ukraine from the ancient times through the 18th century and its advantageous influence on the beginning of medicine in Russia, introduced numerous methods of the conservative and surgical treatment of the ear, nose, throat and larynx (ENT) diseases in adult and children. Particularly, adapted galvanic acoustic treatment in ozaena or rhinitis chronica atrophica foetalis (1906) and three chlor-vinegar acid in dry perforations of the drum ear membrane (1906); made classical description and exposition for treatment of solid chancre of the nasal cavity (1907), syphilis of the larynx (1909), tuberculosis of nose and larynx, oozing blood or bleeding (hemorrhagic) inflammation of the larynx (laryngitis), adenoid (glandular) overgrowth (hypertrophy) in adult; elaborated inquiry inflammation of middle ear in suckling children and its connection with gastrointestinal (GI) disorders (1908); performed plastic operations on the ear, nose, trauma to the hard palate, restauration of the nose using prostheses made from elephant bone; closed perforation od drum membrane (1908); developed methodic for surgical treatment of injured hearing small bones (ossicles) and cystoid (cholesteroma) of the middle ear (1909); described surgical procedure remove overgrowth (hypertrophy) of nasal concha nasalis / conchae nasales; described impairment of hearing organ in acute and chronic inflammation of kidneys (1913); designed surgical approach for hypertrophic nasal conch (oyster, sea-shell) (1916); described diseases of the middle ear in the course of spotted typhus (1921); developed operative approach for entry into forehead (frontal) sinuses (1924); performed extirpation of the larynx (1924), endonasal procedure for correction of the tear deferent channel (1925) and endonasal drainage (opening) lacrimal gland pouch (1925); used fibrinolysin in dry cataracts of the middle ear; introduced the bone-plastic operation for frontal sinuses; studied includesscleroma (hardened) upper respiratory airway (1925-30). First used fibrinolysin for the treatment of narrowing of the upper respiratory gullet. The main work: «Diseases of Ear, Nose and Throat» (Kyiv 1926), «Scleroma» (Kyiv 1930), «Diseases of organs of Hearing and the Upper Respiratory Tract in Children and Adolescents» (Kyiv 1932), «Otosclerosis» (Kyiv 1933).

Major-general of the medical service **Andronyk A. Chayka (17.05.1881, village Ruchky, Hadiach Region, Poltava Obl., Ukraine – 19.07.1968, city of Kyiv, Ukraine)**[1137] — Ukrainian urologist, who developed original techniques of nephrotomy (1914) and restoration of narrowing ureter canal (1922), founded and led the faculty of urology (1929-61) of O.O. Bohomolets National Medical University, Kyiv, Ukraine.

[1137] Village Ruchky (founded 1646; suffered from conducted by Russian Communist Fascist regime of Holodomor-Genocide 1923-33 and1946-47 against the Ukrainian Nation), Hadiach Region, Poltava Obl., Ukraine. — Chayka AA. Technique of nephrostomy. Experimental investigation. St Petersburg 1914. — Ablitsov VH. «Galaxy Ukraine». The Ukrainian diaspora: prominent persons. Publ. «Kyt», Kyiv 2007.

The surgical technique was markedly improved with conceptualizing an electrosurgery in 1926 by **William T. Bovie (1882 – 1958)**[1138] — American scientist and inventor, who invented a modern medical device, known as the Bovie electrosurgical generator and dissector for application of a high-frequency (radiofrequency) alternating polarity electrical current at temperature of 60^0C, to biological tissue to cut, coagulate, desiccate or fulgurate. The device made precise cuts with limited blood loss during surgical procedures in the operating room (OR) in hospital, or in out-patient facilities. It was first employed by William Bovie and **Harvey Cushing (1869 – 1939**; above) in neurosurgical procedures in 1926.

Described by **Hamilton Drummond (1882** – killed in a car accident **1925)**[1139] — Irish/English surgeon, arteria marginalis coli, extends along the mesentery of the colon from the ilio-colic angle to the rectum, connects branches of the superior mesenteric artery (SMA) – ileocecal artery, right colic artery and middle colic artery and the branch of the inferior mesenteric artery (IMA) – the left colic artery (marginal artery of Drummond). The arterial anastomoses at junction of the SMA and the IMA near the splenic flexure (the splenic watershead), are often small or absent (52%), hence the marginal artery of Drummond in this point, to which **J.D. Griffiths**[1140] drew attention (Griffith's point), is prone to ischemia and necrosis.

Cell therapy was re-discovered by **Paul Niehans (1882 – 1971)** — Swiss surgeon, when in 1931 during thyroidectomy the patient's parathyroid gland (PTG) was damaged and developed clonic convulsions, and he decided to use trocar prepared parathyroid cells obtained from a calf and after injecting it the patient recovered (1931). He also has implanted in 1937 cerebral cells, principally of the hypothalamus and of the hypophysis, and since 1948 amplified the live or fresh cell therapy by cells from liver, pancreas, kidneys, heart, duodenum, thymus, and spleen.

In 1910-11 **Volodymyr M. Shamov [22.V.(03.VI).1882, town of Menzelinsk, Tatarstan, Russia – 30.III.1962, city of St Petersburg, Russia]**[1141] — Tatar Ukrainian military surgeon, applied electrocoagulation to destroy malignant tumor; in 1926 created retrosternal esophagus using denervated intestinal loop (Shamov operation), in 1928 successfully transfused fibrinolysed cadaver blood in dog, thus proving that a transplantation of cadaver organs is possible (city of Kharkiv, Ukraine).

[1138] Bovie WT, Cushing H. Electrosurgery as an aid to the removal of intracranial tumors with a preliminary note on a new surgical-current generator. Surg Gynecol Obst 1928;47:751-84.

[1139] Drummond H. Some points relating to surgical anatomy of arterial supply of large intestine. Proc Roy Soc Med 1913;7:185-93.

[1140] Griffiths JD. Surgical anatomy of blood supply of distal colon. Ann Roy Coll Surg, England 1956;19:241-56. / Meyers MA. Griffiths' point: critical anastomosis at the splenic flexure. Significance of in ischemia of the colon. Am J Roentgenol 1976 Jan;126(1):77-94.

[1141] Town of Menzelinsk (founded 1584-86), Tatarstan (area 68,000 km²), Russia. — Ablitsov VH. «Galaxy Ukraine». The Ukrainian diaspora: prominent persons. Publ. «Kyt», Kyiv 2007.

A key person in the development of hand surgery was **Sterling Bunnell (1882 – 1957)** — American surgeon, known for introduction of the «figure of eight suture» to restore transected tendons (Bunnell suture, 1944).

By the end of the 19th century and the beginning of the 20th century into the WW I (1914-18) the incidence of an acute empyema thoracis had increased to enormous dimensions, because the latter brought the huge number military trauma to the thorax and coincided with the influenza epidemic due to Streptococcal pneumonia.

That calamity led to the creation of the Empyema Commission under leadership of **Ewarts A. Graham (1883 – 1957)**[1142,1143,1144,1145] — American general and thoracic surgeon, who by February 1918 found that the empyema thoracic was still largely treated by rib resection and open drainage with death of the soldiers frequently occurring within 30 min. of the procedure, being attributed to the open pneumothorax and mediastinal instability with operative mortality (OM) averaging 30%.

Evarts Graham provided the Empyema Commission the principles of management of thoracic empyema which included (1) careful avoidance of open pneumothorax during the acute stage, (2) prevention of chronicity by rapid sterilization and obliteration of the space, (3) careful attention to the patient's nutritional status. He recommended closed rather than open drainage to treat an early thoracic empyema's, and advised about the ideal timing of an operation, writing «when the exudate has become frank pus instead of being merely serofibrinous…there is less danger of creating an open pneumothorax». Subsequently, OM decreased to 5%-10%.

Expanding further the work of Evarts Graham, the American Thoracic Society (ATS, established in 1905) divided in 1962 the formation of an empyema thoracis which usually progress over 3- to 6-weeks period into the acute exudative phase (Stage I), the acute fibro-purulent phase (stage II), and the chronic organization phase (Stage III).

The stage I and II should be treated with antibiotics, nutritional support and repeated thoracocenteses, if pleural effusion recurs by closed tube thoracostomy. When the lung cannot be re-expanded, there will be the need for video-assisted thoracic surgery (VATS) or open thoracotomy, debridement, decortication of the lung and closed tube thoracostomy. Stage III should be treated, if the patient's condition allows, by VATS or thoracotomy, empyemectomy and decortication of the lung,

[1142] Graham EA, Bell RD. Open pneumothorax: its relation to the treatment of acute empyema. Am J Med Sc 1918;158(6):939.

[1143] Dragstedt LR. Evarts Ambrose Graham (1883-1957). A Bibliographical Memoir. Nat Acad Sc, Washington, DC 1975:221-50.

[1144] Olch PD. Ewarts A. Graham in World War I: The Empyema Commission and service in the American Expeditionary Forces. J Hist Med Allied Sc 1989 Oct;44(4):430-46.

[1145] Mueller CB. Evarts Graham: Superstar among stars. ACS Bull 1991;3:7-11.

otherwise by a segmental rib resection and open thoracic window, space sterilization or space-filling procedures, such as muscle transposition or thoracoplasty.

Ewarts Graham[1146] and **Warren H. Cole (1898 – 1990)** — American general surgeon, initiated in 1923 cholecystography for imaging of the gallbladder by x-rays and a contrast media, an important discovery for the diagnosis of diseases of this organ.

Later, in 1933 were performed successful pneumonectomies with identification of the vessels at the hilum pulmonis: the right – by Ewarts Graham[1147] with thoracoplasty to decrease the large residual pleural cavity, left – by **Willam F. Rienhoff, Jr (1896 – 1981)**[1148] — American thoracic surgeon, and right again – by **Richard M. Overholt (1901 – 90)** — American thoracic surgeon, with subsequent thoracoplasty.

The first surgical correction of a congenital lung anomaly performed by William Rienhoff also in 1933, that being excision of a cyst in a 3-year-old boy.

Since the beginning of the 1930s Richard Overholt have warned of the high risk of lung cancer from smoking cigarettes.

The pancreatic gland (PG) beta islet cell (IC) of Langerhans [19th century. Physiology. – **Paul Langerhans (1847 – 88)**] grafting (transplantation) into another person is a method of surgical treatment of insulin-dependent diabetes mellitus (DM), i.e., juvenile-onset DM type I. The condition usually has an onset prior to the age of 25 years, where the essential abnormality is related to absolute insulin deficiency, the onset is abrupt, symptoms include abnormally great thirst (polydipsia), excessive appetite or eating (polyphagia), excessive urination (polyuria), weight loss, little to none insulin is present and beta IC antibodies present at onset in the blood, human lymphocyte antigen (HLA) is positive, excessive accumulation in the blood of ketone and ketone bodies with increased acidity of the blood (ketoacidosis), eventual development of vascular and neural pathological changes eventually.

DM type I is difficult to regulate, a standard treatment included diet and insulin.

But if DM is not controlled by the standard treatment, then the choice for further treatment is either surgical intervention in the form of organ (PG) transplantation, or transplantation of the PG beta IC of Langerhans.

[1146] Graham EA, Cole WH, Copher GH. Cholecystography: an experimental and clinical study. JAMA 1925;84:14-6. — Graham EA, Cole WH, Copher GH. The roentgenological visualization of the gallbladder by use of intravenous injection of sodium tetrabromophenolphtalein. Radiology 1925;4:83-8. — Graham EA, Cole WH, Copher GH. Cholecystography: the use of sodium tetraiodophenolphtalein. JAMA 1925;84:1175-7. — Graham EA, Cole WH, More S, Copher GH. Cholecystography: the oral administration of sodium tetraiodophenolphtalein JAMA 1925;85:953-5.

[1147] Graham FA, Singer JJ. Successful removal of the entire lung for carcinoma of the bronchus. JAMA 1933;101:1371-4.

[1148] Rienhoff WFJ, Reichert FL, Heuer GJ. Pneumonectomy: preliminary report of operative technique in 2 successive cases. Bull John Hopkins Hosp 1933;53:390-3. — Rienhoff WFJ, Reichert FL, Heuer GJ. Compensatory changes in the remaining lung following total pneumonectomy. Ibid 1935;57:373.

Organ (PG) transplantation, despite its refined surgical technique and immunotherapy is associated with severe trauma and high operative risk.

The alternative to organ transplantation is replacement of the lost function by the PG to produce insulin is by allo- or xero-transplantation of the cultured beta IC harvested from dead human embryos or fetuses and some animals.

Once the patient is transplanted the IC begin to produce insulin, actively regulating the level of glucose in the blood.

The development of the IC grafting was initiated by **F. Charles Pybus (1883 – 1975)**[1149] — English surgeon, in July 1924 when he attempted the allogenic grafting of fragments of human cadaveric PG tissue into two patients to treat DM. In one of them there was a mild reduction in glucose excretion. But eventually, both patients died after the operation.

Brother of **Oleksandr Heimanovych (1882 – 1958**: 19th century. Neurology), **Zakhar Jo. Heimanovych (1884, city of Kharkiv, Ukraine – 03.11.1948)**[1150] — Ukrainian neurosurgeon, founded Psychoneurological Institute in Kharkiv (1922), led its neurological division (1931- 48), generalized the observation of surgical treatment of vertebral-spinal trauma in monography «Military traumatic injuries of the vertebral column and spinal cord» (1934).

The one-stage surgical approach for drainage of lung abscess was proposed in 1925 by **Harold Neuhof (1884** – died of injury sustained in an automobile accident **1964)** — American thoracic surgeon. It consists of a precise radiological localization of an abscess, after which through a skin incision, a 5-7,5 cm segment of one rib was removed and the underlying lung carefully exposed. The adhesions were delicately traversed over a small area down to adherent lung. The abscess was aspirated with short aspiration needle and unroofed. Under full visualization the shell of the lung over the abscess was excised within the limits of pleural adhesions. All suppurated and clotted blood was then removed by suction, and the operation terminated by gauze tamponade of the cavity to avoid collapse and subsequent shutting off of any part of the abscess. The chest wound remained unclosed.

The parathyroidectomy for hyperfunction or tumors of the parathyroid glands (PTG) was recommended in 1924 by **Artenii V. Rusakov (1885 – 1957)** — Russian surgeon, and first performed a year later (1925) by **Felix Mandl (1892 – 1957)**

[1149] Pybus FC. Notes on suprarenal and pancreatic grafting. Lancet 1924;ii:550-1.

[1150] Heimanovych ZJ. Neurosurgery: History and experience of the Ukrainian surgeons. Khirurhiia Ukraïns'koï Radians'koï Sotsialistychnoï Respubliky, Kyiv 1938. — Heimanovych ZJ. Military traumatological wounds of the vertebral column and spinal cord. Tyumen, 1943. — Heimanovych ZJ. Firearm of peripheral nerves and their treatment. Omsk 1943. — Heimanovych ZJ. Radical operations in brain abscesses, complicated by meningitis, and the importance of sulfonamide treatment. Traumatic wounds of the central and peripheral nervous system. Kyiv - Kharkiv 1946. — Ablitsov VH. «Galaxy Ukraine». The Ukrainian diaspora: prominent persons. Publ. «Kyt», Kyiv 2007.

— Czech surgeon, on a 35-year-old street conductor suffering from osteitis fibrosa cystica due to adenoma with recovery.

Mykailo I. Sytenko (12.XI.1885, village Ryabushky, Lebedynsky Region, Sumy Obl., Ukraine – 13.Ш.1940, city of Kharkiv, Ukraine)[1151] — Ukrainian orthopedic surgeon, applied the mastlike splint for immobilization of the fractured clavicle (Sytenko splint), proposed an operation to restore the cruciate ligament of the knee joint using a strip of the thigh's fascia lata (Sytenko operation).

Max Lebsche (1886 – 1957) — German surgeon, designed a surgical device for a medial longitudinal sternotomy incision (MLSI) in the form of the knife or chisel with the paw at the end which, during cutting, skates on the posterior surface of the sternum thus preventing injury to deeper lying organs of the mediastinum (Lebsche knife). To the present time, Lebschke knife should be available as a back-up if needed in emergency situations if electric saw is not available or in difficult MLSI.

The subsequent hemorrhoidectomies are considered modification of **Frederic Salmon (1796 – 1868**: 19th century. Surgery**)** technique, among them open hemorrhoidectomy introduced in 1937 by **Edward T.C. Milligan (1886 – 1972)** — Australian surgeon and **C.N. Morgan**[1152] — English surgeons, in which after divulsion (dilatation) of the external and internal anal sphincters, the skin incision is made at the mucocutaneous junction, the hemorrhoidal plexuses, usually located at the numerical 3, 7 and 11 of the face of a clock in the supine patient, are dissected from outside inside and sewn, its vascular pedicle ligated high at the base with re-absorbable suture and removed, the skin and mucosal defect is not closed being left open and the anal canal is packed with lubricated gauze tampon (Milligan-Morgan open hemorrhoidectomy). It is a variant, introduced in 1959 by **James A. Ferguson** and **Richard J. Heaton**[1153] — American surgeons, essentially like the previous technique, with the exception that muco-cutaneous defect is closed (Ferguson closed hemorrhoidectomy).

In 1938 **Marius N. Smith-Peterson (1886 – 1953)** — Norwegian born American physician and orthopedic surgeon, designed a nail that could be inserted to hold together fractured neck of the hip (Smith-Peterson nail).

Thoracic outlet syndrome (TOS)[1154] is a condition in which there is compression of the brachial plexus (C_5-T_1), the subclavian artery (SCA) and the subclavian vein

---

[1151] Village Ryabushky (founded 1765), Lebedynsky Region, Sumy Obl., Ukraine. — Sytenko NI. Selected Works (Eds AA Korzha, EA Pankova). Kyiv 1991. — Ablitsov VH. «Galaxy Ukraine». The Ukrainian diaspora: prominent persons. Publ. «Kyt», Kyiv 2007.

[1152] Milligan ETC, Morgan CN. Surgical anatomy of the anal canal and operative treatment of haemorrhoids. Lancet 1937;2:1119-24 — Korchynsky IYo. Hemorrhoids. In, LYa Kovalchuk, VF Sayenko, HV Knyshov. Klinichna khirurhiya. Vol. 1-2. Publ. «Ukrmedknyha», Ternopil 2000. Vol. 2. P. 300-4.

[1153] Ferguson JA, Heaton JR. Closed hemorrhoidectomy. Dis Colon Rectum 1959;2(2):176-9.

[1154] Koniukh M. Thoracic outlet syndrome. J Ukr Med Assoc North Am 1981 Summer; 28,3(102):113-23.

(SCV) in the passageway from the lower neck through the cervicoaxillary canal to the upper extremity.

The cervicoaxillary canal is divided by the 1st rib into the proximal section composed by the scalene triangle and the costoclavicular space, and the distal section composed of the axilla. The proximal section is bounded superiorly by the clavicle, inferiorly by the 1st rib, anteromedially by the costoclavicular ligament and posterolateral by the scalene medius muscle and the long thoracic nerve. The scalenus anticus muscle which inserts on the scalene tubercle of the 1st rib, divides the costoclavicular space into the anteromedial compartment containing the SCV, and the posterolateral compartment containing the SCA and the brachial plexus. The latter compartment, which is bounded by the scalenus anticus muscle anteriorly, the scalene medial muscle posteriorly and the 1st rib inferiorly, is called the scalene triangle. Its dimensions are 1.2 cm at the base, 7.1 cm in height anteriorly and 6.7 cm in height posteriorly.

The superior angle of the scalene triangle may cause impingement on the upper brachial plexus (C_5 and C_6) and produce the upper type of the scalene anticus syndrome, and if the base of the scalene triangle is rise, compression of the SCA and the lower brachial plexus (C_7-T_1) occur causing the lower type of the scalene anticus syndrome.

The most common symptoms are neurological, i.e., the somatic ulnar nerve type indicating the C_5 and C_6) component involvement, of pain and paresthesia (burning, numbness, pricking or tickling) in the neck and shoulder, along the median aspect of the arm, the forearm and the hand, including the 5th finger and the median aspect of the 4th finger, in approximately 95% of the patients, motor weakness and occasionally atrophy of the hypothenar and interosseous muscles in 10%. The somatic median nerve type indicate involvement of the C_7 and C_8 component produces symptoms in the index finger and sometimes the middle finger. Entrapment of the C_5-T_1 component by cervical rib produce symptoms of various degree in distribution of these nerve. In some patients, the pain is atypical, involving the anterior chest wall or parascapular area, mimicking angina pectoris.

The dorsal and thoracic preganglionic sympathetic nerves derived from the spinal cord do not accompany the corresponding somatic nerves. The cervical nerves of C_1-C_4 are fused into a superior cervical ganglion, the cervical nerves of C_5 and C_6 into the middle cervical ganglion and the cervical sympathetic nerves of nerves of C_7 and C_8 into the inferior cervical ganglion, which combines with the thoracic sympathetic nerve of T_1 to the larger cervico-thoracic (stellate) ganglion. The preganglionic sympathetic outflow from the spinal cord to the arm are fibers from the sympathetic nerves of T_{2-9}, mostly from the sympathetic nerve of T_2. Only in about 10% of cases, the sympathetic nerve of T_1 preganglionic fibers also supply the upper extremity. For the removal of preganglionic fibers to the upper extremity in most patients, resection of the lower third of the stellate ganglion and paravertebral sympathetic nerves of T_2 and T_3 with interconnecting chain is sufficient, because the post-ganglionic fibers from those segments join and follow the nerves of the brachial plexus.

The inconstant sympathetic branches from the 2nd or 3rd, or both intercostal nerve to the 1st intercostal nerve, hence branchial plexus without passing through to the stellate ganglion were identified by **Albert Kuntz (1879 – 1957**: 20th century. Anatomy, histology and embryology**)**, and named the nerve of Kuntz.

Compression the SCA produce coldness, weakness and easy fatigability of the arm and hand, and usually diffuse pain. The secondary type of Raynaud's phenomenon, described in 1862 by **Maurice Raynaud (1834 – 81**: 19th century. Internal medicine**)**, is noted in 7.5% of patients with TOS.

Compression of the anteromedial compartment containing the SCV causes venous obstruction or occlusion, manifested by swelling, pain and bluish discoloration of the arm, is recognized as Paget-Schraeder syndrome, named in honor of Sir **James Paget (1814 – 99**: 19th century. Pathology**)** and von **Leopold Schraeder (1837 – 1908**: 19th century. Pathology**)**, or as effort thrombosis.

Physical examination should include detection of radial pulses and the three classic maneuvers:

(1) Scalene or Adson test, described in 1951 by **Alfred W. Adson (1887 – 1951)**[1155] — American physician, military officer and neurosurgeon, who excised a herniated cervical disk (1922), designed dissecting (Adson) forceps using by holding fine tissue and retractor (Beckman-Adson) using holding open surgical incisions. This maneuver tightens the anterior and middle scalene muscles and thus decreases the interspace and magnifies any preexisting compression of the SCA and brachial plexus. By rotating the patient's head to the ipsilateral side with extended neck during holding a deep inspiration the radial pulse in the arm is further decreased or lost.

(2) In the costoclavicular test (military position) the shoulders are drawn downward and backward narrowing the SCA and pressing on the brachial plexus by approximation of the clavicle to the 1st rib, manifested by decreased or lost radial pulse.

(3) Hyperabduction test. When the arm is hyperabducted to 180°, the SCA and brachial plexus are pulled around the pectoralis minor muscle tendon, the coracoid process of scapula, and the head of humerus. The radial pulse is decreases or disappears.

[1155] Adson AW. Cervical ribs: symptoms, differential diagnosis for sections of scalenus anticus muscle. J Int Coll Surg 1951;16:546-59.

Diagnosis of TOS should be confirmed by x-rays of the chest and cervical spine, electromyogram / electomyography (EMG) and ulnar nerve conduction velocity (UNCV), in cases with atypical manifestations by cervical myelography, peripheral angiography, cardiac catheterization – selective coronary angiography (CC-SCA), or phlebography.

EMG examination of each upper extremity and determination of the UNCV is done with EMG, coaxial cable with three needles or surface electrodes are used to record muscle potentials, which appears on the fluorescent screen.

The UNCV is determined by the Krusen-Caldwell[1156] technique in the supraclavicular fossa at the Erb's point[1157] over the trunk of the brachial plexus, above the elbow, below the elbow, and at the wrist. The latency period to stimulation in meters per second (msec.) from those four points to the recording electrode is obtained to the recording from the EMG digital recorder or calculated from the tracing on the screen. The distance in millimeters (mm) between two adjacent sites of stimulation is measured with steel tape. The velocities in meters per second (m/sec.) are calculated by subtracting the distal latency from the proximal latency and dividing the distance between two points of stimulation by the latency difference according to the following formula:

Velovity (m/sec.) = distance between adjacent
stimulation points (mm): difference in latency
between adjacent stimulation points (m/sec.).

The normal values for the UNCVs are 72 m/sec or greater across the outlet, 55 m/sec or greater around the elbow, and 59 m/sec or greater in the forearm. The wrist delay is 2.5 - 3.5 m/sec.

The clinical picture of TOS correlates with the UNCV across the thoracic outlet. Any value less than 70 m/sec indicates neurovascular compression. The severity is graded according to decrease of velocity across the thoracic outlet: the compression is called slight when the velocity is 66-69 m/sec., mild when the velocity is 60-65 m/sec., moderate when the velocity is 55-59 m/sec., and severe when the velocity is 54 m/sec. or less.

Most patients with TOS who have UNCVs of more than 60 m/sec. improve with conservative treatment, mainly by physical therapy. Most of symptomatic patients with the UNCV below that level, despite physical therapy, may need segmental resection of the 1st rib and correction of other bony anomalies.

[1156] **Edward M. Krusen (? - 2003)** — American physician - specializing in physical medicine and rehabilitation. / **James W. Cadwell** — American physician - specializing in physical medicine and rehabilitation. / Cadwell JW, Crane CR, Krusen EM. Nerve conduction studies: an aid in the diagnosis of the thoracic outlet syndrome. South Med J 1971 Feb;64(2):210-2. / Urshel HC, Razzuk MA, Krusen EM, Caldwell JW. The technique of measuring conduction velocity for thoracic outlet syndrome. Development in Surgery. Vol. 1 (Pain in Shoulder and Arm). Springer Netherland 1979;1:165-72.

[1157] **Wilhelm Erb (1840 - 1921)** — German neurologist. (19th century. - Neurology).

In 7.5% of the younger individuals a cervical rib is manifested by the TOS. Depending on the clinical status and the UNCV studies, patients with the TOS should be initially treated with physical therapy (PT), but patients suffering from symptomatic TOS should undergo a segmental removal of the 1st rib / without / with cervical sympathectomy.

In 1927 Alfred Adson and in 1935 **Howard C. Naffziger (1884 – 1961)** — American neurosurgeon, demonstrated that complete relief of TOS symptoms could be obtained by severing the scalene anticus muscle, regardless the presence or absence of a cervical rib. The latter also introduced the orbital decompression (Naffziger) operation for severe malignant exophthalmos.

Sympathectomy is an irreversible procedure during which by removal part of the autonomic nervous system, that is one or more sympathetic ganglions, the signals from the thalamus of the brain to the periphery are disabled. First dorsal sympathectomies were performed by **William Alexander (1844 – 1919**: 19th century. Surgery) in 1889 and 1898, and by **Anastas Kotzareff (1866 – 1931**: 20th century. Surgery: above) in 1920, and the first VATS dorsal sympathectomy for excessive sweating (hyperhidrosis) of the palms («sweaty palms») and axilla was reported by **M. Kux**[1158] in 1978. Presently, dorsal sympathectomy is performed through the axillary open incision or videoassited thoracic surgery (VATS).

The indications for dorsal sympathectomy are hyperhidrosis, typically, palmar, axillary, or plantar, commencing usually in the early childhood, Raynaud disease and phenomenon, both described by **Maurice Raynaud (1834 – 81**: 19th century. Internal medicine), complicated TOS, reflex sympathetic dystrophy (RSD) and SMPS, both described by **Paul Sudeck (1860 – 1945**: 20th century. Surgery), and vascular insufficiency of the upper extremity.

The preganglionic sympathetic fibers responsible for sudoriferous (sudorific, sudomotor) or sweating activity, of the hands are likely to arise from spinal segments of the sympathetic nerves of T_3-T_6, converging on the sympathetic nerves of T_2 and T_3 where postganglionic fibers ascend to reach the hand through connections with the stellate ganglion to the brachial plexus. Probably the sympathetic ganglion T_2 is the most common pathway responsible for palmar sweating. To be sure of success, in the surgical treatment of palmar hyperhidrosis by thoracic sympathectomy, the sympathetic chain should be resected from the immediately below the stellate ganglion to the level of just above the T_3 sympathetic ganglion. The nerve of Kuntz also should be divided, if identified, to prevent the possibility of the gray rami connecting directly from the stellate ganglion to the brachial plexus, thus avoiding the recurrence of symptoms.

[1158] Kux M. Thoracic endoscopic sympathectomy in palpar and axillary hyperhidrosis. Arch Surg 1978;113: 264-6.

Some selected cases of severe TOS may have to treated surgically by the VATS performing dorsal sympathectomy of the sympathetic ganglia T_1-T_3 with lower stellate ganglionectomy and segmental 1th rib resection.

Complications of dorsal sympathectomy include Horne's syndrome, that is inadvertent removal of the sympathetic ganglia C_7 and C_8 fibers, a part of the upper stellate ganglion, resulting in ipsilateral enophthalmos, ptosis, flushing and loss of sweating (anhidrosis) of the face, described **Francois du Petit (1664 - 1741**: 18th century**)** and **Johann Horner (1831 – 86**: 19th century. Diseases of the eye**)**, post-sympathetic neuralgia and recurrent symptoms.

Dorsal sympathectomy will change many bodily function including sweating glands, decrease of the heart rate (HR), cardiac stroke volume (SV), thyroid, baroreflex, lung volume, pupil dilatation (mydriasis), skin temperature and the other aspects of the autonomous nervous system, like the essential fight-or-flight response.

Viktor O. Pavlenko (1887, town of Nosivka, Nizhyn Region, Chernihiv Obl., Ukraine – 1937, city of St Petersburg, Russia)[1159] — Ukrainian surgeon, proposed in 1928-29 the manner for depriving sensitivity (anesthesia) of the celiac or solar anglion / plexus (in Latin – plexus coeliacus).

After departure of **M. Shapiro (1871 – 1942**: above**)**, **Solomon Weinsblat (14.I.1888, town of Malyn, Korostensky Region, Zhytomyr Obl., Ukraine – 10.VI.1965, city of Kyiv, Ukraine)**[1160] — Ukrainian stomatologist-surgeon of Jewish descent, became in 1931-53 the head by participation of Stomatology Faculty, O.O Bogomolets NMU, at the same time (simultaneously) lecturer of the Kyiv Institute of Stomatology and Faculty of Surgical Stomatology at the National Medical Academy of Post-Graduation Education (NMAPE, founded 1918). Under leadership of Solomon Weinsblat was developed in scientific school of stomatology surgeons and maxillary-facial surgeons. During surgery in oral cavity, he applied local anesthesia. In 1925 was published his monography «Conductive Anesthesia of Teeth and Maxilla», the first dedicated to local anesthesia in stomatology. At his clinics in 1935 was established laboratory for teeth technology with the aim to produce splints, complex prostheses, and apparatuses. Later, he published the second monography – «Purulent Osteomyelitis of Jaw-Bones» (1938).

[1159] Town of Nosivka (first mention 1147; suffered from Russian Bolsheviks and White Russian Guards in 1918-19 – «on streets, and chausees / paved roads or high-ways lied shot, tortured to death, deformed to extreme human bodies», and from organized by Russian Communist Fascist regime of Holodomor-Genocide 1932-33 against the Ukrainian Nation), Nizhyn Region, Chernihiv Obl., Ukraine. — Ablitsov VH. «Galaxy Ukraine». The Ukrainian diaspora: prominent persons. Publ. «Kyt», Kyiv 2007.

[1160] Town of Malyn (remains of Old-Rus town in ruins from 8th-9th century, first mention 891; Russian Communist Fascist regime of repressions and conduction of Holodomor-Genocide 1932-33 against the Ukrainian Nation destroyed part of population of towns and region), Korostensky Region, Zhytomyr Obl., Ukraine). — Ablitsov VH. «Galaxy Ukraine». The Ukrainian diaspora: prominent persons. Publ. «Kyt», Kyiv 2007.

Congenital pericardial cysts («spring water cyst») and diverticula are very uncommon lesions within the anterior mediastinum, they derive from the pericardial celom and represent different stages of a common embryogenesis. A pericardial cyst was first removed in 1931 by **Otto C. Pickhardt (1887 – 1972)** — German American surgeon, via the 8th left intercostal space thoracotomy (ICST) in a 53-year-old womam suffering from a sharp pain over the precordium, and a pericardial diverticulum measuring 6 cm in diameter located in the right cardio-phrenic angle was resected in 1943 by **Richard Sweet** via the right anterior ICST.

In 1923, **Elliot Cutler (1888 – 1947)**[1161] — American surgeon, carried out in a girl 12-year-old suffering from dyspnea and hemoptysis at rest, caused by mitral stenosis (MS) resulting from rheumatic heart disease (RHD), a closed («blind») mitral commissurotomy (MC), using a valvulotomy, through the apex of the left ventricle (LV) on a beating heart. She survived another 4½ years, dying from pneumonia. Subsequent patients fared poorly.

Two years later (1925), Sir **Henri S. Souttar (1875 – 1964)**[1162] — British surgeon, performed in a 19-year-old woman, also suffering from RHD, a «blind» digital mitral commissurotomy (MC) via the left atrial (LA) appendage. After the operation, the dyspnea and hemoptysis resolved, and she survived another 5 years. The second woman died later from cerebral embolism.

At that time the operative mortality (OM) for MC was 90% and it was abandoned in 1928.

Base on the initial experience from 1900-03 of **Vasyl Dobromyslov (1869 – 1918**: above**), Andrei G. Savinykh [6(18).XI.1888, former village Mershchyny, Orlov Region, Kirov Obl., Russia – 26.II.1963, town of Tomsk, Western Siberia, Russia]**[1163] — Russian surgeon, worked up the surgical transabdominal and extra-pleural through esophageal hiatus (opening) of the diaphragm approach to the posterior mediastinum, aiming to safety remove carcinoma of the gastric cardia and the distal portion of the esophagus, and simultaneously carrying out gastroesophageal plastic anastomosis (trans-hiatal gastrectomy with esophagoplasty) in the Surgical Division at the Hospital of the 2nd Tomsk University (established 1896), since 1992 – the Siberian State Medical University, Tomsk, Western Siberia, Russia.

[1161] Cutler E, Levine SA. Cardiotomy and valvulotomy for mitral stenosis. Experimental observations and clinical notes concerning an operated case with recovery. Boston Med J 1923;188(26):1023-7. — Cutler EC, Beck CS. The present status of the surgical procedures in chronic valvular disease of the heart. Arch Surg 1929;18:403-16.

[1162] Souttar H. The Surgical treatment of mitral stenosis. Br Med J 1925;2(3379):603-6.

[1163] Former village Mershchyny, Orlov Region (founded 1719), Kirov Obl. (area 120,800 km^2), Russia. — Savinykh AG. The radical treatment of carcinoma of the cardia and lower section of the esophagus. Work 24th All-UnionConvention of Surgeons, Kharkiv 25-31 October 1938. «Medgiz», Moscow – St Petersburg 1939. P. 516-19. — Zadorozhyby AA, Saks FF. Surgeon A.G. Savinykh (1888-1963). Publ. «Krasnoye znamiya», Tomsk3 1996.

The operation consists with an anterior sagittal incision (length 4-6 cm, maximal up to 10 cm) into the esophageal hiatus, including the muscular portion of the diaphragm, the passage into the central tendon of the diaphragm (CTD), i.e., speculum helmontii **[Jan Helmont (1580 – 1644)**: 17th century)]**, oversewing blood vessels and transverse cut through the diaphragmatic crura of the vertebral column, without fixation of the esophageal anastomosis to the diaphragm (Savinykh method, 1944).

Initially, the procedure, that is the proximal gastrectomy with resection of gastric cardia, and the lower esophagus for cancer using Savinykh method was performed in 14 patients, with success in five (5) patients.

As of 1943, Andrei Savinykh modified his procedure esophago-gastric anastomosis, replacing removed esophagus using the small intestine for the interposition anastomosis.

Subsequently, unfortunate results decreased to one percent (1%).

Theodor Hryntschak (15.07.1889, city of Vienna, Austria – 28.06.1952, city of Vienna, Austria)[1164] — Ukrainian urologist, introduced the suprapubic trans-urinary bladder prostatectomy with a primary closure of the prostatic bed and urinary bladder (Hryntschak prostatectomy, 1946).

Olexandr V. Melnykov [30.VI(12.VII).1889, town of Mezen', Arkhangel Obl., Russia – 27.V.1958, city of St Petersburg, Russia][1165] — Russian Ukrainian surgeon,

[1164] Hryntschak Th. Wiens Anteil an der Erfludung des Kystoskopes. Wien med Wochenschrift 1950;100(27-28):482-4. —Hryntschak Th. Über die Erfolge der suprapubischen Prostatektomie nach der elgenen Methode. Z Urol 1951;44(2):153-7. — Hryntschak N. Die suprapubische Prostatectomy. Maudrich Verlag, Wien 1951. — Hryntschak T. Suprapubic transvesical prostatectomy with primary closure of the bladder; improved technic and latest results. J Int Coll Surg 1951;15(3):366-8. — Meuser H. Professor Dr. Theodor Hryntschak. Z Urol 1953;46(1):1-5. — Deuticke, Nachruf für Prof. Hryntschak. Sitzung der Österreichischen Gesellschaft für Urologie am 15. Okt. 1952. Z Urol 1953;46(7):478-9. — Hryntchak T. Suprapubical Prostatectomy with Primary Closure of the Bladder by an Original Method. Charles C. Thomas, Springfiels, IL 1955. — Hornykevych M. Viden' (Wien). «Encyclopedia of Ukraine». Ed. – V. Kubijovyc, Z Kuzela, M. Hlobenko. Publ. NTSh – «Molode Zhytttia», München – New York 1955;II(1):267-9. — Hornykevych V, Zhukovsky A. Vienna. In, V Kubijovyc, DH Struk. Encyclopedia of Ukraine. Univ Toronto Press Inc., Toronto - Buffalo - London 1993;5:598-600. — Ablitsov VH. «Galaxy Ukraine». The Ukrainian diaspora: prominent persons. Publ. «Kyt», Kyiv 2007.

[1165] Town of Mezen' (first mentioned 1545), Arkhangelsk Obl. (area 589,913 km²), Russia. — Melnikov AV. Sinus Costo-diaphragmaticus. Surgical Anatomy. Theses of St Petersburg 1920. — Melnikov AV. The surgical approaches to the organs of subdiaphragmatic spaces through the lower chest. Nov Khir Arch (Kyiv) 1921;1(1). — Melnikoff A. Die chirurgische Anatomie der Sinus costo-diaphragmaticus. Arch klin Chir 1923;123:133. — Melnikoff A. Ueber extrapleural-extraperitoneale Zugänge zur den Nebennieren durch das Diaphragma. Zentralbl Chir (Leipzig) 1923;1:336-40. — Melnikoff A. Die chirurgischen Zugänge durch den unteren Rand des Brustkorbes zu den Organen des subdiaphragmalen Raumes. Deut Ztschr Chir 1923;182:83-151. — Melnikov AV. The Clinics of Pre-neoplastic Diseases of the Stomach. Moscow 1954. — Konvolinka CW, Olearchyk A. Subphrenic Abscess. Curr Probl Surg (Ed MM Rawitch). Year Book Med Publ, Chicago, Jan. 1972:1-52. —Olearchyk AS, Konvolinka CV. Subphrenic abscesses. J Ukr Med Assoc North Am 1972; 19,3 (66): 3-56. — Shevchuk MH. Peritonitis. In, LYa Kovalchuk, VF Sayenko, HV Knyshov. Klinichna khirurhiya. Vol. 1-2. Publ. «Ukrmedknyha», Ternopil 2000. Vol. 2. P. 283-96. — Andreychyn MA, Ivakhiv OL, Kim OM. Amebiasis. Ibid. P. 412-24. — Ablitsov VH. «Galaxy Ukraine». The Ukrainian diaspora: prominent persons. Publ. «Kyt», Kyiv 2007.

developed the anatomical basis for extrapleural approaches to subphrenic spaces and abscesses (1920-23), as well as the concept of precancerous lesions of the stomach (1950-54).

Norman Bethune (1890 – 1939) — Canadian surgeon of Scottish descent, who was an early proponent of universal health care; developed the Bethune rib shears (1928-36); proposed performing under local anesthesia a selective talc pleurodesis[1166] of one lobe of the lung, which would then remain inflated when resecting the other diseased lobe, without the need for positive pressure ventilation (1935)[1167]; developed mobile blood transfusion service (1936-37).

In 1920-25 working on dogs **Serhiy S. Briukhonenko (1890 – 1960)**[1168] — Ukrainian Russian physiologist and surgeon, **Sergei I. Chechulin (1894 – 1937)** — Russian physiologist and pathologist, and **Voldymyr D. Yankovsky**[1169] — Ukrainian physiologist, developed extracorporeal circulation and constructed an experimental heart-lung-machine (HLM) – Briukhonenko's autoejector (licensured in 1928-42). Specifically, Serhii Bryukhonenko and Volodymyr Yankovsky developed oxygenator (device for aeration of the blood, named «artificial lungs») for the HLM in 1937 p. On oxygenator a claim was presented for certification, and in 1942, and inventive author rights were reserved. The device was used by both to resuscitate dogs after cardiac arrest, and by **Nicolai N. Terebinsky (1880 – 1923)** — Russian surgeon, to perform open heart surgery on dogs (1930). Later (1930) Nicolai Terebinsky applied a cross-clamp on the ascending aorta (AA) to achieve a bloodless operative field.

[1166] Pleurodesis is a medical procedure in which the pleural space is artificially obliterated, usually by a talk, to create adhesions between two pleurae, the parietal pleura covering the interior wall of the thorax and the visceral pleura of the lung or its lobe.

[1167] Bethune N. Pleural poundage. A new technique for the deliberate production of pleural adhesions as a preliminary to lobectomy. J Thorac Surg 1935;4:251-61. — Katlic MR. Five hundred seventy-six cases of video-assisted thoracic surgery using local anesthesia and sedation: lessons learned. J Am Col Surg 2018 Jan;226(1):58-63.

[1168] Town of Michurinsk (up to 1932 – town of Kozlov, founded 1635), Tambov Obl. (area 34,464 km^2), Russia. — Brukhonenko SS, Terebinsky S. Experience avec la tete isole du chien. I: Techniques et conditions des experiences. J Physiol Pathol Genet 1929;27:31. — Konstantinov IE, Alexi-Meskishvili VV. Sergei S. Brukhonenko: the development of the first heart-lung machine for total body perfusion. Ann Thorac Surg 2000;69(3):962-6. — Ablitsov VH. «Galaxy Ukraine». The Ukrainian diaspora: prominent persons. Publ. «Kyt», Kyiv 2007.

[1169] Lanovenko IL, Yankovsky VD. Influence of previous adaptation to high altitude on the cardiovascular system of dogs during resuscitation after prolonged clinical «death» evoked by blood loss. Resuscitation 1973 Sep;2(3):207-20. — Lanovenko II, Yankovsky VD, Lyavinetz AS. Artificial circulation for resuscitation after prolonged clinical death caused by hypoxia. Resuscitation 1976;5(2):75-84. — Gerya YF, Yankovsky VD. Use of artificial circulation in resuscitation of drowned dogs. Resuscitation 1976-1977;5(3):145-52. — Ablitsov VH. «Galaxy Ukraine». The Ukrainian diaspora: prominent persons. Publ. «Kyt», Kyiv 2007.

In 1937 the surgical treatment of otosclerosis gained impetus by **Julius Lempert (1890 – 1968)** — American otorhinolaryngologist (ORL) performing the one-stage inaugural fenestration operation.

Founder of neurosurgery in Ukraine, **Ivan M. Ishchenko [26.VI (04.VII).1891, village Pustovarivka, Skryvsky Region, Kyiv Obl., Ukraine – 22.XI.1975, city of Kyiv, Ukraine]**[1170] — Ukrainian surgeon, major-general of medical service, in his work «Phenomenon of immunity during homotransplantation of tissues and organs» (1935) was first to turn attention in an experiment on peculiarities during acceptance the renewed grafting of tissue flaps. He proved an idea of interrelation between the condition of graft and the activism of the connective tissue, which changed under the influence of the anti-reticular cytotoxic serum (ACS) of Bohomolets [20th century. Pathology. – **Oleksandr Bohomolets (1881 – 1946)**]. In the experimental and clinical conditions were showed high activity of the ACS in the treatment of fractures, generalized and local infection, complex treatment of malignant neoplasms. He improved the treatment of cranial and brain trauma (1940), surgical approach in firearm при injuries of the peripheral nerves (1945), drainage in acute diseases of abdominal cavity (1950-60), and surgery of biliary ducts (1960-66).

The first successful surgical removal of the pituitary gland (hypophysectomy) was carried out in 1951 by **Herbert Olivecrona (1891 – 1980)** — Swedish neurosurgeon.

In 1929 **Frederic E.B. Foley (1891 – 1966)**[1171] — American urologist, first described the use of a self-retaining balloon catheter at the end for inflation with air or fluid, used initially to control a bleeding from the prostatic bed after cystoscopic prostatectomy for adenoma or carcinoma, in 1935 he demonstrated (published in 1937) an indwelling inflatable balloon catheter towards the tip of the catheter which could be inflated inside the urinary bladder to retain the it without external taping or strapping for continuing urinary drainage (Foley catheter).

After successful transfusion of cadaveric fibrinolyzed blood transfusion in a dog in 1928 by **Volodymyr Shamov (1882 – 1962**: above), **Sergei S. Yudin (1891 – 1954)**[1172] — Russian surgeon, repeated in 1930 transfusion of fibrinolyzed cadaveric blood in a patient, thus proving the feasibility of clinical transplantation of human organs; organized the first blood bank at **Mykola Sklifosovsky (1836 - 95**; 19th

---

[1170] Village Pustovarivka (settled since time of Scythes, founded in the first half of 18th century), Skvyrsky Region, Kyiv Obl., Ukraine. — Ablitsov VH. «Galaxy Ukraine». The Ukrainian diaspora: prominent persons. Publ. «Kyt», Kyiv 2007.

[1171] Foley FEB. Cystoscopic prostatectomy: a new procedure: preliminary report. J Urol 1929;21:289-306. — Foley FEB. A self-retaining bag catheter. J Urol 1937;38:140-3.

[1172] Judin S. Über die primäre Resection bei perforienten Magen- und Diodenalgeschwüren. Acta Clin Chir 1930;161:517-39. — Judine S. La transfusion du sang de cadavre a'l'homme. Masson et Cie, Paris 1933. — Judin S. Partial gastrectomy in acute perforated peptic ulcer. Surg Gynec Obstet 1937;64:63-8. —Alexi-Meskishvili V, Konstantinov IE. Sergei S. Yudin: an untold story. Surgery 2006;139(1):115-22.

century. – Surgery) Institute (1930), city of Moscow, Russia; proposed a primary partial gastrectomy for perforated or bleeding peptic ulcer (1930); and performed reconstruction of the esophagus with a small intestine.

Havrylo P. Kovtunovych [25.III(06.IV).1892, village Kyrylivka, Sosnytsky Region, Chernihiv Obl., Ukraine. – 25.V.1961, city of Lviv, Ukraine][1173] — Ukrainian surgeon, described technique of a total excision of parotid gland tumors without injury to the cranial VII (facial) nerve (1953, 1958); modified the surgical correction of prolapsed rectum by pulling and attachment of the rectosigmoid junction into the anterior wall of the abdominal wall using a strip of the aponeurosis of the external oblique abdominal muscle (colonopexy or sigmoidopexy according to Kovtunovych).

В 1933 **Mykailo S. Kolomiychenko (12.02.1892, town of Shpola, Cherkasy Obl., Ukraine – 30.05.1972, city of Kyiv)**[1174] — Ukrainian surgeon, proposed an extrapleural approach to the heart and performed pericardiectomy for adhesive constructive calcified pericarditis, and by 1958 performed 20 such operations.

Microsurgical techniques with the aid of a binocular microscope were initiated in 1921 by **Carl-Olof S. Nylén (1892 – 1978)** — Swedish otologist, who operated on a patient with chronic otitis media complicated by labyrinth fistula, and in 1922-23 by **Gunnar Holmgren (1875 – 1954)** — Swedish otolaryngologist, as an adjunct to surgery in otosclerosis.

The work of **Lester R. Dragstedt (1893 – 1975)** — American surgeon, in 1943-52 on truncal vagotomy led to its acceptance for the treatment of peptic ulcer.

Operation for the sub-sphincteric anal / rectal fistulae (Ancient medicine. *Egypt*), proposed in 1932 by **William B. Gabriel (1893 – 1975)**[1175] — English proctologist, consist of dissection of fistulae in the form of triangle from the internal opening, usually at the anal crypts (apex of the triangle) to the external opening at the skin (base of the triangle) with removal of the skin, removal the entry of infection and drainage (Gabriel operation).

[1173] Village of Kyrylivka, Sosnytsky Region (area 916 km²; populated since Neolith Epoch 5th-4th millennium BC and the epoch of Bronze 2nd millennium BC; during conduction by Russian Communist Fascist NKVS of Holodomor-Genocide 1932-33 against the Ukrainian nation many inhabitants suffered from man-made hunger), Chernihiv Obl., Ukraine. — Ablitsov VH. «Galaxy Ukraine». The Ukrainian diaspora: prominent persons. Publ. «Kyt», Kyiv 2007.

[1174] Town of Shpola (founded 1594; as result of conducted by Russian Communist Fascist NKVD of Holodomor-Genocide 1932-33 and political repressions 1937-41 against the Ukrainian Nation, majority families perished), Cherkasy Obl., Ukraine. — Ablitsov VH. «Galaxy Ukraine». The Ukrainian diaspora: prominent persons. Publ. «Kyt», Kyiv 2007.

[1175] Gabriel BG. Results of experimental and histological investigation into seventy-five cases of rectal fistulae. Proc R Soc Med 1921;14:56-61. — Gabriel WB. Fistula-in-ano. In, WB Gabriel. The Principles and Practice of Rectal Surgery. H K Klemis & Co, London 1945. P. 160-97. — Zakharash MP. Fistulae of the rectum (chronic paraproctitis). In, LYa Kovalchuk, VF Sayenko, HV Knyshov. Klinichna khirurhiya. Vol. 1-2. Publ. «Ukrmedknyha», Ternopil 2000. Vol. 2. P. 318-22.

In trans-sphincteric fistulae in case of fistula passing through the internal parts of sphincter during removal of the fistula canal the sphincter fibers are injured. Thus, it is necessary to reconstruct dissected region of the sphincter by placing knotted stitches or П-like catgut suture.

The surgical approach for the correction of idiopathic scoliosis with the use of steel springs was conceived in 1956 by **Adam Gruca (1893 – 1983)** — Polish orthopedist.

Four years later (1959-60) **Paul R. Harrington (1911 – 80)** — American orthopedic surgeon, invented a system of metal hooks and rods (Harrington) for the correction of idiopathic scoliosis, used until late 1990s.

In 1930-32 **Albert S. Hyman (1893 – 1972)** — American physiologist and cardiologist, together with his brother Charles, constructed an electro-mechanical instrument powered by a spring-wound hand cranked motor, for controlled repetitive electrostimulation of the heart, and named his device «the artificial cardiac pacemaker». It was one of the earliest electrical artificial pacemakers (PM), tested in animals and at least in one patient.

An extensive review of wartime gunshot wounds was presented by **Peter A. Kuprianov (1893 – 1963)** — Russian surgeon and **Ivan S. Kolesnikov (1901 – 85)** — Russian surgeon, in «Atlas of Gunshot Wounds» (Vol. 1-10. 1948).

A rare complication, occurring in approximately 0.1% of all patients with gallstones, described in 1948 by **Pablo L. Mirizzi (1893 – 1964)**[1176] — Argentinian abdominal and thoracic surgeon, consists of impaction of gallstone in the cystic duct or neck of the gallbladder, or in the Hartmann's pouch **[Henri Hartmann (1860 – 1952)**. Above]**[1177] between those two, causing compression of the common bile duct (CBD), resulting in obstruction in the flow of the bile with jaundice, or cholecystocholedochal fistula (Mirizzi's syndrome). Depending on the stage of the Mirizzi's syndrome, the surgical treatment may require a simple cholecystectomy, subtotal cholecystectomy, or bilioenteric anastomosis, including Roux-en-Y hepaticojejunostomy **[Cesar Roux (1857 – 1934). Above]**.

The concept of developing a dermatome, a device to cut uniform split-thickness skin graft приладу was conceived in 1930 by **Earl C. Padgett (1893 – 1946)**[1178] — American plastic and reconstructive surgeon. Aided by **George J. Hood (1877 – 1965)**[1179] — American mechanical engineer, they developed a drum-like

---

[1176] Mirizzi PL. Syndrome del conducto hepatico. J Int de chir 1948;8:731-77.

[1177] Hartmann H. Quelque points de l'anatomie et dela chirurgie des voies biliaires. Bull soc anat Paris 1891;5(5):480-00. — Hartmann H. Chirurgie des voies biliaires. Masson et cie, Paris 1923.

[1178] Ameer F, Singh AK, Kumar S. Evolution of instruments for harvesting of skin grafts. Indian J Plast Surg 2013 Jan Apr;46(1):28-35.

[1179] Hood GJ. Geometry of Drawing. Descriptive Geometry by the Direct Method. 3rd Ed. Mc Graw – Hill Book Co, Inc., New York & London 1946. — Litton C, Hood GJ. A new giant dermatome. Plastic Reconstructive Surg 1954 Mar;13(3):240-5.

(semi-cylindrical) calibrated apparatus, based on the adhesion and traction principle, a reliable instrument that rolls over the skin while a blade moves across the surface and cuts accurately a uniform sheet of a predictable split-thickness skin graft in a 10.2 cm x 20.4 cm size (Padgett-Hood drum dermatome, 1938-39). Also, he was the first to transplant skin between identical twins, with success.

The physiological basis and signs of an acute cardiac compression (tamponade) were depicted by **Claude S. Beck (1894 – 1971)** — American cardiac surgeon, who denoted a low arterial blood pressure (BP), distended the neck jugular veins and distant muffled heart sounds (Beck's triad), and also jugular venous distention with inspiration – Kussmaul sign [Adolph Kussmaul (1822-1902). 19th century. Internal medicine]. However, Beck's triad and Kussmaul's sign are present in only 10% of patients with pericardial tamponade.

To improve the blood supply to the ischemic heart he implanted part of the pectoralis major muscle into the pericardium (cardiopericardiopexy or Beck I operation, 1935) and created a vein graft (VG) bypass between the aorta to the coronary sinus (CS) for retrograde blood perfusion to the myocardium (Beck II operation, the late 1940s). Both those operations provided only borderline blood supply to the heart and were replaced in the early 1960's by the modern revascularization technique.

In 1947 he performed the first successful direct current (DC) cardiac defibrillation to reverse ventricular fibrillation (VF) in human. After repairing congenital heart disease (CHD) in 14-year-old boy and closing the chest, the patient developed cardiac arrest. The chest was reopened for direct manual cardiac massage for about 45 min., before proceeding to DC defibrillation with one paddle applied to the anterior and the other to the posterior surface of the heart, restoring normal sinus rhythm (NSR) with full recovery.

In plastic and reconstructive surgery body defect is closed by use of the «Indian method» of plastic surgery with the pedicled flap of the neighboring skin (Ancient medicine. India), and the «Italian method» of plastic surgery, where the body defect is closed by the distant pedicled flap of the skin or muscle by **Gaspar Taliacozzi (1545 – 99**: Renaissance).

Reconstruction of the oval bodily defects is performed with a neighboring skin, in which by an additional incision one can convert it into triangular form, or by using the transfer method V-Y type to elongate separate sections without tension of the skin.

Reconstruction using the matching triangles allows to increase a distance between the two points of the segment, length of which depends on the size of the angle, under which the meeting flaps are cut. The increment of the length is determined by calculating the sinuses of those angles, most commonly used flaps are cut at the angle of 60°. The mathematical base of this method was worked out by **Alexander A. Limberg (1894 – 1974)**[1180] — Russian stomatologist, maxillofacial and plastic surgeon of Estonian origin, by designing the skin flap for closure of a 60° rhomboid defect of the skin, which is an equilateral paralogue with angles of 60° and 120° (Limberg flap, 1963). This flap outline is created by extending a short diagonal in either direction in the rhomboid by distance equal in own length, and then incising third side at the angle of 60° in extended short diagonal; incised third side must equal in length a distal equal in own length. All sides of the defect and all sides of the flap are equal length, while the opposing angle of each are either 60° or 120°.

The other form of tissue transfer is the bridge-like and the «Italian» method of plastic surgery and the overturn, caterpillar, bi-layered and stalked flaps[1181].

Bridge-like plastic surgery with two stalks is used in creation of long and narrow flaps, both ends of which connected with the adjacent tissue for better blood supply.

The «Italian» method of plastic surgery of the distant body regions is used in cases absence of the tissue area for the formation of flap around the body defect. Then, it is necessary to choose such body areas where there is the possibility to compare the site of the donor flap with the site of the recipient of the flap.

The overturn is formed in the axis of the extremity, torso or proximally to the wound defect, stepping back from it for 3-5 cm. It is necessary to cut out and immobilize a rectangular cutaneous-subcutaneous flap with the correlativity of the length and the width of 1:1,5, 1:2. The flap is inverted around the frontal plain for 180° to the wound surface outside. The distal end at all breadth is put into the edge of wound defect and sewn from underneath in such a manner, that in course of 2-3 cm of its length it is lying well upon the wound. The secondary wound defect at ounce or over several days should be covered by the free skin graft (1st stage of the plastic procedure). In the interval of 2-3 weeks after the flap stalk become rooted into the edge of the wound defect the nourished stalk is cut out. The flap is returned, put to lie on the wound and sewn underneath (2nd stage of the plastic procedure). In the case of large defects the «turning over» is carried out in the 2nd - 3rd stage of the plastic procedure. This creates an opportunity to cover deep wound defect with surface area from 50 to 150-200 cm².

Caterpillar flap is cut out with the nutrient stalk close to the body defect. The flap is folded, the edges are sewn up, the end lied up to the border of wound surface of

---

[1180] Limberg FF. The Planing of the Site of Plastic Operations on the Surface of the Body. St Petersburg 1963. — Limberg AA. Design of local flaps. In, T. Gibson (ed.), Modern Trends in Plastic Surgery. 2nd Ed. Butterworths, London 1966. — Borges AF. Choosing the correct Limberg flap. Reconstructive Surgery 1971 Oct:342-5. — Bihuniak VV. Plastic surgery. In, LYa Kovalchuk, VF Sayenko, HV Knyshov. Klinichna khirurhiya. Vol. 1-2. Publ. «Ukrmedknyha», Ternopil 2000. Vol. 2. P. 359-90.

[1181] Bihuniak VV. Plastic surgery. In, LYa Kovalchuk, VF Sayenko, HV Knyshov. Klinichna khirurhiya. Vol. 1-2. Publ. «Ukrmedknyha», Ternopil 2000. Vol. 2. P. 359-90.

the secondary defect. Left over secondary defect is covered by sewing the brim of the wound or by the free skin graft. During the 2nd stage of plastic surgery, by 2-3 weeks, the flap is again unfolded to cover the defect.

Bi-layered rectangular flap directed towards the proximally located stalk is folded, and the brims of the flap are sutured. The secondary defect at the site of the flap take off is covered by free skin.

Pedicled (tubular) flap of Filatov-Gillis (1914, 1917)[1182] is formed with the skin supported by two stalks, the verge of which are sewn to construct a cylinder. The proportion of the length of the skin flap to the width is 2:1-3:1. During creation of the stalk the skin and only partially the subcutaneous tissue are used to make sure that sewing the verges of wound does not cause tension. The average length of the stalk on the abdominal wall reaches 12-15 mm, width should reach 9-10 cm. The most valuable virtue of the tubular stalk graft is the fact, that it can be prepared in any region of the body and transferred to the site of the defect without significant disturbances in the blood supply during migration.

Dermal regeneration template, i.e., the «Integra», consisted of bovine collagen, chondroitin-G-sulfate and a silastic membrane from cadaveric or living donors, and glycerol preserved allograft (GPA) date back to World War II (1939-45).

D.V. Gordon Murray (1894 – 1976)[1183] — Canadian cardiac surgeon, who described in 1947 the systolic paradoxical expansion of and acutely infarcted myocardium and correlated it with diminished arterial blood pressure (BP) and cardiac output and (CO); developed the first successful artificial kidney with a coil design (1945-46) and the second generation of an artificial kidney with a flat-plate model (1952-53); replaced arteriosclerotic (AS) disease and narrowed or obstructed the left anterior descending artery (LADA) with segments of the internal thoracic artery (ITA), axillary and carotid arteries with success (1953); replaced an aneurysm of the descending thoracic aorta (DTA) with an aortic homograft (1956).

Decorated with the orders participant of World War (WW) I and II, **Serhiy A. Verkhratsky (20.XI.1894, village of Verkhratka, Kryzhopils'kyy Region,**

[1182] **Volodymyr P. Filatov [15(27).II.1875, village Mykhailovka, Lambirsky Region, Mordovia** (area 26,200 km²)**, Russia – 30.X.1956, city of Odesa, Ukraine**: 20th century. Ophtalmology] — Ukrainian ophthalmologist and plastic surgeon, who independently of **Harold D. Gillis (1882 – 1960**: 20th century. Ophtalmology) — New Zealander and English otolaryngologist and plastic surgeon, developed the pedicled tube flap grafting, or the skin plasty using tubular migrating stalk (Filatov-Gillis flap, 1914, 1917). — Bihuniak VV. Plastic surgery. In, LYa Kovalchuk, VF Sayenko, HV Knyshov. Klinichna khirurhiya. Vol. 1-2. Publ. «Ukrmedknyha», Ternopil 2000. Vol. 2. P. 359-90. — Ablitsov VH. «Galaxy Ukraine». The Ukrainian diaspora: prominent persons. Publ. «Kyt», Kyiv 2007. — Hanitkevych YaV. The contribution of the Ukrainian physicians to the world medicine. Ukr Med Chasopys 2009 VII/VIII;4(72):110-15.

[1183] Murray G. Pathophysiology of the course of death from coronary thrombosis. Ann Surg 1947;126:523-35. — Murray G, Delorme E, Thomas N. Developmental of artificial kidney. Arch Surg 1947;55:502-22. — Murray G, Delorme E, Thomas N. Artificial kidney. JAMA 1948;137:1596-9.

Vinnyts'ka Obl., Ukraine – 23.II.1988, city of Ivano-Frankivsk, Ukraine)[1184] — Ukrainian surgeon and historian of medicine, who during WW II in Ukraine (1941-45) hold the post of the main physician-surgeon of the Evacuation Hospital, in which he organized one of the first Department of Neurosurgery for the treatment of injured soldiers and civilians.

In 1952 **Lyon H. Appleby (1895 – 1970)**[1185] — Canadian general surgeon, presented 13 patients with locally advanced gastric carcinoma who underwent an block coeliac artery resection (CAR) with the stomach, distal pancreas and spleen with ligation of the common hepatic artery to the bifurcation of the proper hepatic and gastroduodenal artery. There was no operative mortality (OM).

Further development of stereotactic apparatus and trials in the stereotactic surgery has been made in 1947 by **Ernest A. Spiegel (1895 – 1985)** — Austrian American neurologist and **Henry T. Wycis (1911 – 1972**) — American neurosurgeon, and in the stereotaxic radiosurgery (SRS) in 1949-51 by **Lars Leskell (1907 – 86)** — Swedish neurosurgeon and inventor of the SRS.

Working together, Ernest Spiegel and Henry Wycis devised a brain atlas that would be the basis for their «stereoencephalon», i.e., stereotactic apparatus, a modification of the Horsley-Clarke stereotactic animal experimental apparatus from 1908 **[Victor Horslely (1857 – 1916)**. 19th century. Surgery]**[1186]. The device was fixed to a patient head by means of a plaster cast, which were removable, allowing separate imaging (contrast radiography, ventriculography and pneumoencelography) and surgery.

Stereotactic surgery or stereotaxy is a minimally invasive surgery which uses two- or three-dimentional coordinate system to locate small pathological target in the brain and to perform on them biopsy, lesion, injection of lytic substance, ablation, stimulation, and also implantation of radioactive chips using the SRS.

During the stereotaxic surgery using «stereoencephaloton», an intraoperative radiography technique allowed the use of internal markings for each specific patient.

The first stereotactic surgery using «stereoencephaloton » was performed in 1947 by Henry Wycis on a patient with Huntington's chorea or disease (HD) **[George Huntington (1850 – 1916)**. 19th century. Neurology] by lysis with injection of alcohol into globus pallidus (pale globus) of the corpus striatum and thalamus, with successful

---

[1184] Village of Verkhratka, Kryzhopils'kyy Region, Vinnyts'ka Obl., Ukraine. —Verkhratsky SA. History of Medicine. Publ. «Zdorov'ia», Київ 1964. / Publ. «Zdorov'ia», Київ 2011.

[1185] Harrison J, Pucci MJ, Cowan S, et al. A brief overview of the life and work of Lyon Henry Appleby (1895-1970). Am Surg 2016 Nov;82(12):1151-4. — Appleby LH. The coeliac axis in the expansion of the operation for gastric carcinoma. Cancer 1953;6:704-7.

[1186] Spiegel EA, Wycis HT, Marks M, LeeAJ. Stereotaxic apparatus for operations on the human brain. Science 1947 Oct 10;106(2754):349-50.

result[1187]. Ultimately, the «stereoencephalon» was used for stereotactic surgery in variety of neurological and psychiatric disorders such as intractable pain, epilepsy, Parkinson's chorea or disease (PD) **[James Parkinson (1755 – 18240)**. 19th century. Neurology]**, electrocoagulation of the Gasserian ganglion **[Johann Gasser (1723 – 65)**. 18th century] in trigeminal neuralgia and mental illnesses.

In 1947 Lars Leksell developed a stereotactic apparatus for external cerebral marking with a frame employed three polar (cylindrical or galactocentric) coordinates (angle, depth and anterior-posterior location) system, an «arc-quadrant» device for the internal cerebral markings for the treatment of intracranial lesions[1188].

Polar coordinate system is a three dimensional, in which each point on a plane is determined by a distance from a fixed point and an angle from a fixed direction. It is based on the concept of angle and radius, used by ancient scientists of the 1st millennium BC, and by **Hipparchys (c. 190 – c. 120 BC)** — Greek astronomer, in establishing stellar positions.

In 1949 he defined the SRS as «a single high dose fraction of ionizing radiation, stereotactically directed to an intracranial region of interest»[1189], and in the SRS, the word «stereotactic» implies a three dimension coordinate system that enables accurate correlation of a virtual target seen in patient's diagnostic images with the actual target position in the patient.

The initial stereotactic tools he designed and used were probes and electrodes to treat small targets in the brain that were not amenable for a conventional surgery.

Later (1958), **Kurt Liden (1915 – 87)** — Swedish physicist radiophysicist, and **Borje Larsson** — Swedish physicist, introduced stereotactic high-energy proton beams[1190], which produced lesions in the depth of the brain[1191].

They achieved a new non-invasive method of destroying discrete pathological lesions in the brain while minimizing the effect on the surrounding healthy tissues.

In 1967 Lars Leksell, **Ladislau Steiner (1920 – 2013)** — Romanian, Swedish and American neurosurgeon, and Borje Larsson, in cooperation with **Robert W. Rand (1923 – 2013)** — American neurosurgeon, developed the Gamma Knife, a.k.a., Leksell Gamma Knife, consisted of several Cobalt-60 radioactive sources placed in a kind of helmet with central channels for irradiation with gamma rays[1192]. This prototype was designed to produce slit-like radiation lesion for functional neurosurgical procedures

---

[1187] Spiegel EA, Wycis HT, Freed H. Stereoencephalotomy: talatomy and related procedures. JAMA 1952;148(6):446-51.

[1188] Leksell L. A stereotaxic apparatus for intracerebral surgery. Acta Chir Scand 1949;99:229.

[1189] Leksell L. The stereotactic method and radiosurgery of the brain. Acta Chir Scand 1951 Dec 13;102(4):316-9.

[1190] Larsson B. The high-energy proton beams as a neurosurgical tool. Nature 1958;182(4644)1222-3.

[1191] Leksell L, Larsson B, Andersson B, et al. Lesions in the depth of the brain produced by a beam of high energy protons. Acta Radiol 1960 Oct;54:251-64.

[1192] Leksell L. A stereotaxic radiosurgery. J Neurol Neurosurg Psychiatry 1983 Sep;46(9):797-803.

to treat pain, motion, or behavioral disorders, not responding to conventional treatment. Subsequently, a second device was constructed containing 179 Cobalt-60 radioactive sources. The second Gamma Knife unit was designed to produce spherical lesions to treat primary benign and primary and metastatic malignant brain tumors, including craniopharyngioma or Rathke's pouch tumor **[Martin Rathke (1793 – 1860)**. 19th century. Anatomy and embryology]**, intracranial arteriovenous malformations (AVM), as well as tumors of the skull base.

In 1929 **Yuriy (Yurii) Yu. Voronyy or Voronoy (09.VIII.1895, village Zhuravka, Varvynsky Region, Chernihiv Obl., Ukraine** – died from heart attack **13.V.1961, city of Kyiv, Ukraine)**[1193] — Ukrainian surgeon, realized that kidney allograft rejection was an immunologic event and described complement-fixation antibodies appearing after experimental transplant operations (1929). He carried out the first human kidney allograft in 1933, to a 26-year-old woman dying from acute renal failure (RF) following sublimate (mercuric chloride) poisoning. He transplanted a kidney of a man who died from brain injury to the recipient femoral artery and the vein, the ureter was brought into the skin crevice of the thigh and sutured into it, under local anesthesia. Although initially the grafted kidney worked, she died 49 hours later from rejection. Yuriy Voronyy performed six unsuccessful trials at renal transplantation in man between

---

[1193] Village Zhuravka (ancient settlement dated by archeologists as of 12,000 - 10,000 BC, first mentioned by chronicler in 1616; due to organized by Russian Communist Fascist regime of Holodomor Genocide 1932-33 against the Ukrainian nation, at least 234 villagers have died), Varvynsky Region, Chernihiv Obl., Ukraine). — Voronyy YuYu. Ukr med Arch. 1929;4:64. — Woronoy U. Die Immunität bei Organtransplantation – II, Mitteilung. Über spezifische komplementbindende Antikörper bei freier Nierentransplantation mittelst Gefäsnacht. Arch f Klin Chir 1929;171:386-97. — Voronoy YuYu. On the question of specific complement fixing antibodies in the free transplantation of kidneys with application of a vascular suture. Ukr Med Arch 1930;6:33-46 —Voronoy U. Sobre el bloqueo del aparato reticuloendotelial del hombre en algunas formas de intoxicacio por el sublimado y sobre la transplantacion del rinon cadaverico como metodo de tratamiento de la anuria consecutiva a aquella intoxicacion. El Siglo Medico 1936;97:296-8. — Voronoy YuYu Transplantation of a conserved cadaveric kidney as a method of biostimulation in severe nephrites. Vrach delo 1950;9:813-6. — Hamilton DNH, Reid WA. The pioneer and the first human kidney allograft Yu.Yu. Voronoy. Surg Gynec Obst 1984;159:289-94. — Mirsky M. Yu.Yu. Voronoy – a pioneer of clinical kidney transplantation. UHMJ «Ahapit» 2004;14-15:10-6. — Humar A, Dunn DL. Transplantation. In, FC Brunicardi (Ed.). Schwartz's Principles of Surgery. 8th Ed. McGraw-Hill Med Publ Div, New York - Toronto 2005. — Matevossian E, Kern H, Hüser H, et al. Surgeon Yurii Voronoy (1894-1961) – a pioneer in the history of clinical transplantation: in Memoriam in the 75th anniversary of the first human kidney transplantation. Transplant International (Wiley) 2009 Dec;22(12):1132-9. — Dubenko D. Surgeon who was the first to transplant kidney in human being: Ukrainizer and participant at the Battle of Kruty. BBC News Ukraine, 29 VII. 2020 [The Battle of Kruty took place in January 29(30).I.1918, near village of Kruty railway station, today village of Pamiatne (Remembrance), Borzna Region, Chernihiv Obl., 130 km northeast of Kyiv, Ukraine, between the 400 students, cadets, and soldiers of the 1st Ukrainian Military School who were defending the capital of Ukraine against invasion of about the 4,000 soldiers mixed with thugs of Russian Communist Fascist horde]. — Ablitsov VH. «Galaxy Ukraine». The Ukrainian diaspora: prominent persons. Publ. «Kyt», Kyiv 2007. — Hanitkevych YaV. The contribution of the Ukrainian physicians to the world medicine. Ukr Med Chasopys 2009 VII/VIII;4(72):110-15.

1933-49. For his kidney transplants he introduced in 1933 the double needle armed vascular suture technique.

After placing the first steps towards mechanical suturing in surgery by **Gaston M. Michel (1875 – 1937)**, et al. (above), the development of modern staplers to improve the technique of surgical suturing began in 1945, when **Vasilii F. Gudov**[1194] — Russian engineer, constructed a vessel suturing circular apparatus (VCA), which was first used beginning in 1949, by **Yuriy Voronyy (1895 – 1961**: above) for vascular anastomoses. Thereafter, the serial production VCA was arranged. By 1983, more than 40 models of surgical staplers were constructed for vascular, thoracic, and gastrointestinal (GI), which quickly spread through all the World.

The most common parathyroid gland (PTG) pathology is hyperplasia or adenoma, rarely carcinoma [19th century. Surgery. – Sir **Richard Owen (1804 – 92)**]. Adjustive localization studies for PTG pathology include computer tomography (CT) or magnetic resonance imaging (MRI) scans (92% sensitive) and technetium (Tc 99m)-sestamibi scans (85% sensitive)

Primary hyperparathyroidism is generally managed by surgical exposure and removal of diseased PTG via the collar incision in the neck. The anterior mediastinal ectopic PTG adenomas are now exposed and excised by an open or endoscopic transcervical approach, robot-assisted operation, rarely by middle longitudinal sternal incision (MLSI), or by angiographic ablation. The middle and posterior mediastinal ectopic PTG are identified and resected via the video-assisted thoracic surgery (VATS) with the endoscope inserted through the 7th intercostal space (ICS) just anterior to the superior iliac spine and working port created in the 5th ICS at the anterior axillary line (AAL), robotic-assisted operation, or angiographic ablation. During surgery it is important to visualize, monitor and protect the phrenic and recurrent laryngeal nerves.

In 1932 **Edward D. Churchill (1895 – 1972)**[1195] — American surgeon, performed the first removal of the mediastinal PTG adenoma – mediastinal parathyroidectomy via the MLSI, then with **Oliver Cope (1902 – 94)** — American surgeon, removed in another patient, mediastinal PTG tumor with primary hyperparathyroidism, also via the MLSI.

[1194] Gudov VF. The methodic art of placing vascular sutures by mechanical means. Khrurgiia 1950,12:58-9. — Gudov VF. Technique of mechanical application of vascular suture. Khirurgia 1950;12:58-9. — Gudov VF. A New Method of Connecting Blood Vessels. Publ. Medgiz, Moscow 1950.

[1195] Churchill ED, Cope O. The surgical treatment of hyperthyroidism: based on 30 cases confirmed by operation. Ann Surg 1936 July 104(1):9-35. — Churchill ED, Belsey R. Segmental pneumonectomy in bronchiectasis: the lingual segment of the left upper lobe. Ann Surg 19939;109:481-99.

Since 1933 Oliver Cope began to carry out lobectomies, and in 1939 with **Ronald H.C. Belsey (1910 – 2007)**[1196] — English thoracic surgeon, performed the first segmented lung resections in six patients to extricate bronchiectasis and tuberculosis, thus conserving the healthy tissue and physiologic function of the lungs in those afflicted patients.

At the time when the mortality from lobectomy was about 50% due to clumsy techniques and septic complications, Edward Churchill showed how safely and selectively remove diseased parts of the lung.

During the WW 2 (1942-45) Edward Churchill developed the use of delayed primary closure and early debridement of contaminated wounds and improved the air evacuation process for treating wounded soldiers.

The screw-fastener to prevent unscrewing during vibration, designed by **William Dzus (05.I.1895, village Chernykhivtsi, Zbarazh Region, Ternopil Obl., Ukraine – 19. VI.1964, census designated place West-Islip in town of Islip, Suffolk County, N.Y., USA)**[1197] — Ukrainian inventor, named Dzus screw-fastener (1932), found wide application in the industry (1932) and for the fixation of reduced fractured bones (Dzus bone fastening device, 1945).

Temple Fay (1895 – 1963) — American neurosurgeon, lowered temperature of the patient with a malignant tumor to 30°C for several days in the hope of slowing its growth. Later he applied local hypothermia in neurosurgery to treat injuries and abscesses of the brain (1938).

For patients suffering from chronic recurrent obstruction of the small bowel, **Thomas B. Noble, Jr. (1895 – 1958)**[1198] — American surgeon, proposed to perform its horizontal (transverse) sutured folding (transverse intestinal plication, or Noble's operation, 1937, 1939).

For the palliative surgical treatment of cyanotic congestive heart disease (CHD) due to stenosis of the main pulmonary artery (MPA), specifically tetralogy of Fallot (TOF), and for stenosis of the tricuspid valve (TV), **Willis J. Potts (1895 – 1968)** — American

---

[1196] Churchill ED, Belsey R. Segmental pneumonectomy in bronchiectasis: the lingual segment of the left upper lobe. Ann Surg 1939;109:481-99.

[1197] Village Chernykhivtsi (first mentioned in chronicle 1340), Zbarazh Region, Ternopil Obl., Ukraine. — Census designated place West-Islip in town of Islip (area bought the Indians 1683), Long Island (area 3,564 km²), Suffolk County (6,160 area km²), N.Y., USA. — Bern RL. An American in the Making. The Biography of William Dzus, Inventor. William Barton Marsh Co., New York 1961. — Ablitsov VH. «Galaxy Ukraine». The Ukrainian diaspora: prominent persons. Publ. «Kyt», Kyiv 2007.

[1198] Noble TB, Jr. Plication of small intestine as prophylaxis against adhesions. Am J Surg Jan 1, 1937;35(1):41-4. — Noble TB, Jr. Plication of small intestine: second report. Am J Surg Sep 1939;445(3):574-80. — Noble TB. Evaluation of plication in the treatment of peritonitis and its aftermath of intestinal adhaesions. J Internat Coll Surg Mar 1959;31(3):286-93.

pediatric surgeon, proposed a «side-to-side» anastomosis of the descending thoracic aorta (DTA) to the pulmonary artery (PA) – Potts operation (1946).

Willis Potts performed the first neonatal lobectomy (1949), designed a straight atraumatic vascular clamp (Potts) for holding blood vessels.

In 1954, he faced a 5-year-old child suffering from the PA sling, where the left PA originated anomalously from the right PA, coursed cranially to the right main stem bronchus (RMSB) and posteriorly on the way to the left lung, compressing the RMSB and trachea. The first left PA sling repair via the right intercostal space thoracotomy (ICST) included division of the origin of the left PA and translocating it anteriorly to the tracheobronchial tree and re-anastomosing it to the distal main PA opposite to the origin of the right PA. The patient survived nearly 25-year, at which time the left PA was found at the autopsy to be occluded.

In 1931 **Rudolph Nissen (1896 – 1981)** — German, Turkish and American surgeon of Jewish origin, performed a successful one-stage ligation of the entire pulmonary hilum.

However, **Philip R. Allison (1907 – 74)** — English thoracic surgeon, was the first to perform a radical intrapericardial pneumonectomy (1941), and denoted that the right pulmonary artery (PA) passes to the right behind the ascending aorta (AA) and forms the superior border of the transverse sinus, then passes posterior to the superior vena cava (SVC) and forms the superior border of the postcaval (Allison) recess, whereas the right superior pulmonary vein (RSPV) forms the inferior border of the recess.

Operations for hiatal hernia with gastroesophageal reflux disease (GERD) were initiated by **Phillip Allison** (above) through median superior laparatomy, where a dilated esophageal hiatus is approximated by suturing the right and the left crus of the diaphragm in patients with a sliding esophageal hernia with GERD (Allison sliding hiatal hernia repair, 1951).

The surgical approach by **Ronald Belsey (1910 – 2007**: above**)** was through thoracotomy with the anterior partial fundoplication of 270^0 (Belsey-Mark IV operation, 1952).

Rudolph Nissen (above) operated via a median superior laparatomy or thoracotomy with total of 360^0) fundoplication (Nissen fundoplication, 1956).

Joseph M. Bennett[1199] approach was through a median superior laparatomy with posterior gastropexy to the median arcuate ligament as a primary, or ancillary

---

[1199] Bennett JM. Fixation of the posterior gastric wall in esophageal hiatus herniorrhaphy. Am Surg 1965 Aug;31:493-5.
— Bennett JM. Letter to the Editors. Ann Surg 2001;234(3):45.

technique (1961-64), which was popularized by **Lucius D. Hill (1921 – 2001)**[1200] — American surgeon, in 1966s – 1970s.

The recent tendency is to perform Belsey or Nissen fundoplication in the form of **Vincent M. Dor (1932 –)**[1201] — French thoracic and cardiac surgeon, anterior partial fundoplication, or **Andre Toupet (1915 – 2004)**[1202] — French surgeon, by posterior partial fundoplication, to decrease postoperative dysphagia.

In patients with severe acid GERD with a short esophagus **John L. Collis (1911 – 2003)**[1203] — English cardiothoracic surgeon, proposed a technique for lengthening a «short» esophagus by a full-thickness incision of the gastric cardia parallel to the lesser curvature, usually with a staple line to create a distal tubular neo-esophagus using the upper part of the stomach (Collis gastroplasty,1957).

Specifically, the Collis gastroplasty should be included into a repair by a fundoplication of the most frequent paraesophageal diaphragmatic hernia with the shortened esophagus to avoid postoperative recurrence and devastating gastropericardial fistula.

Paraesophageal diaphragmatic hernias are divided into: type I – a sliding hiatal hernia involving the circumferential weakening of the phrenicoesophageal ligament with symmetric displacement of the proximal stomach into the thoracic cavity, the most common type (95%) and frequently associated with shortened esophagus; type II – a rolling hernia involves weakening of the phrenicoesophageal ligament with the cardia of the stomach remaining within the abdomen and the lower esophageal sphincter remains at the level of the diaphragm, but a portion of the gastric fundus rolls into the chest either anterior or lateral to the esophagus producing its extrinsic compression; type III – a mixed hiatal hernia comprise both a sliding and a rolling components of hernia in which the lower esophageal sphincter has been displaced up into the thorax because of a shortened esophagus and then a portion of the gastric fundus rolls through the enlarged hiatal hernia, causing also extrinsic compression of the distal esophagus; type IV – a hernia with a large defect in the phrenicoesophageal membrane allowing the greater curvature of the stomach to dislocate within the

---

[1200] Hill LD, Tobias J. Morgan EH. Newer concepts of the pathophysiology of hiatal hernia and esophagitis. Am J Surg 1966; 111: 70-9. — Hill LD. An effective operation for hiatal hernia: an eight-year appraisal. Ann Surg 1967;166: 681-82.

[1201] Dor J, Humbert P, Dor V, Figarella J. L'interet de la technique modifiee la prevention du reflux apres cardiomyotomie extra muquesuse de Heller. Men Acad Chir 1962;88:881-3.

[1202] Toupet A. Technique d'oesophago-gastroplastic avec phren.-gastropexie applique'e dan la cure radicale des hernies hiateles et cumme complement de l'operatin d'Heller dan les cardiospames. Men Acad Chir 1963;89:384. — Katkhouda N, Khalil MR, Manhas S, et al. Andrew Toupet: surgeon technician par excellence. Ann Surg 2002 Apr;235(4):591-9.

[1203] Collis JL. An operation for hiatus hernia with short aesophagus. Thorax 1957;12(3):181-3. — Collis Ch. John Leigh Collis. BMJ 2003 Apr 5;326(7392):767.

hernia sac into the chest producing gastric twisting (volvulus) with gastric outlet obstruction, the so called «upside-down» stomach, as well as upwards towards the thorax displacement of the colon, spleen and small intestines.

Semen A. Holdin / Kholdin (26.12.1896, city of Odesa, Ukraine – 1975, St Petersburg, Russia)[1204] — Ukrainian Russian surgeon-oncologist of Jewish descent, after graduating in 1919 from Odesa University, in 1923-26 hold position of scientific researcher at the Odesa Clinical Institute, since 1926 worked at the Scientific Institute of Oncology in St Petersburg, Russia. He proposed a closed method of electrosurgical resection and anastomosis of the gastrointestinal (GI) tract, a radical extended mastectomy with excision of parasternal lymphatic nodes through the extrapleural approach for eradication of the breast carcinoma (Kholdin mastectomy, 1965), and electrosurgical removal of infiltrative forms of breast carcinoma.

Initially, in 1945 **Arthur H. Blakemore (1897 – 1970)** — American vascular surgeon, reported successful splenorenal shunt performed for portal hypertension using vitalium tube, a non-suture technique, and an «end-to-end» anastomosis.

In 1952 **Samuel Rosen (1897 – 1981)** — American otolaryngologist, refined a stapes mobilization operation for deafness.

Other notable developments of the 1950's were external hearing aids for deafness.

Operations for the extrasphincteric anal / rectal fistulae (Ancient medicine. *Egypt*), proposed in 1958-68 **Olexandr T. Ryzhykh (1897 – 1967)**[1205] — Russian proctologist of Jewish origin, named as operation Ryzhykh -1 and operation Ryzhykh-2, and in 1964 **N.M**. **Blinnichev**[1206] — Russian proctologist, named as Blinnichev operation.

Operation Ryzhykh-1 is used when the internal opening of fistula is in the posterior crypt. The fistulous tract in the perineal area is dissected up to the wall of rectum and cut off at the base. The stump of the fistulous tract is scraped out, processed with iodine solution, sewed back into debrided fistula tract and and covered by the surrounding tissue. The operation is terminated by calibrated sphincterotomy of the internal fibers.

[1204] Holdin SA. Malignant tumors of the rectum. St Petersburg 1955. — Holdin SA. Extended radical operations for mammary cancer. In, JM Ariel (Ed.). Progr in Clin Cancer. New York – London 1965;10:439-49. — Holdin SA, Dymarsky LYu. Extended operations for breast cancer. St Petersburg 1975. — Ablitsov VH. «Galaxy Ukraine». The Ukrainian diaspora: prominent persons. Publ. «Kyt», Kyiv 2007.

[1205] Ryzhykh *AN*. An Atlas of Operations on the Rectal and Large Bowels. 1st Ed. Publ. Meduchposobye, Moscow 1960; 2nd Ed. Publ. Meuchposobye, Moscow 1969. P. 76-104. — Zakharash MP. Fistulae of the colon chronic paraproctitis). In, LYa Kovalchuk, VF Sayenko, HV Knyshov. Klinichna khirurhiya. Vol. 1-2. Publ. «Ukrmedknyha», Ternopil 2000. Vol. 2. P. 318-22.

[1206] Blinnichev NM. Acute and chronic periproctitis. PhD degree thesis. Samara 1972. — Ryzhykh AN. An Atlas of Operations on the Rectal and Large Bowels. 1st Ed. Publ. Meduchposobye, Moscow 1960; 2nd Ed. Publ. Meuchposobye, Moscow 1969. P. 76-104. — Zakharash MP. Fistulae of the rectum (chronic paraproctitis). In, LYa Kovalchuk, VF Sayenko, HV Knyshov. Klichna khirurhiya. Vol. 1-2. Publ. «Ukrmedknyha», Ternopil 2000. Vol. 2. P. 318-22.

Operation Ryzhykh-2 is used when the internal opening of fistula is in the anterior or lateral crypt. The fistulous tract in the perineal wound is dissected as in operation Ryzhykh-1. Then, over the the internal opening, a serous flap 1-1.5 cm wide and up to 4 cm long is separated off. The internal opening of the fistula is sewed with blunt catgut, tied only after removal of the rectal mirror. Above a few stitches are placed between separated off and the partially dissected excess of the mucous membrane and distal edge of the wound. The calibrated posterior sphincterotomy is performed.

The Blinnichev operation, based on the Arthur Elting operation (1912; above), presents a segmental proctoplasty with lateral inferior translocation of the mucosal and submucosal flap of the rectum.

Brother, **Mykhailo Kolomiychenko (1892-1972**: above), of **Oleksiy S. Kolomiychenko [18(30).III.1898, town of Shpola, Cherkasy Obl., Ukraine – 17.IX.1974, city of Kyiv, Ukraine]**[1207] — Ukrainian surgeon-otorhinolaryngologist (ORL), in the 1937s - 1940s improved the treatment of infants with the ear inflammation due to toxic disturbances of digestion (dyspepsia), introduced into practice antro- and tympano-punction, removed foreign bodies from the throat, trachea and bronchi with direct laryngoscopy or bronchoscopy without tracheotomy in children. Insofar, suffering since childhood from partial deafness, he studied an extensive growth in the bones of the middle ear which interferes with transmission of sound (otosclerosis), and improved operations for deafness. Together with with the locksmith-gauger developed and manufactured the hearing aid for deaf people (1964). He established and headed Institute of the Laryngology in Kyiv which bears his name (1960-74).

Pioneer of the scientific biochemical-physiological processes of the whole blood and blood-substitutes, tied up to surgical treatment of patients, post-traumatic injuries with significant blood loss and shock, **Borys Yu. Andriyevs'kyy (July 28,1898, village of Bezsali, Lokhvytsky Region, Poltava Obl., Ukraine – Oct. 20, 1962, city of Cleveland, OH, USA)**[1208] — Ukrainian physician-surgeon, who graduated from Medical Faculty of the Kharkiv Medical Institute (KhMI) in 1924, where after the successful defense of his dissertation entitled «*Experimental and*

[1207] Town of Shpola (Foot Note 1175), Cherkasy Obl., Ukraine. — Ablitsov VH. «Galaxy Ukraine». The Ukrainian diaspora: prominent persons. Publ. «Kyt», Kyiv 2007.

[1208] Village Bezsaly (nearby are preserved remnants of an ancient town and tumulus since 10th - 11th BC, first chronicle mention 1693), Lokhvytsky Region, Poltava Obl., Ukraine — City of Cleveland (founded 1796), Ohio (OH, area 116,096 km²), USA). — Andriyevsky BYu. Diagnostics and treatment of femoral hernia. Kharkiv med archiv 1926. — Andriyevsky BYu. Postoperative Psychoses. Kharkiv 1936. — Andriyevsky BYu. Treatment of duodenal fistula after gastric resection. Works IPEPh, Kharkiv 1938. — Andriyevsky BYu. Experimental and Clinical Data About Action of Some Salt-solutions During Sudden Hemorrhage. KhMI 1939. — Andriyevsky BYu. Improvement of the extrapleural pneumothorax. Works of the All-Ukrainian Tuberculous Inst. 1940. No. 11. — Movchan J. Scientific papers of Professor Dr Borys Andriyevsky. J Ukr Med Assoc North Am 1963 Oct;10,4(31):10-17.

Clinical Data About Action of Some Salt-solutions During Sudden Hemorrhage» for the degree of Doctor of Medical Sciences (D.M.S., 1938), he was appointed Assistant / Dozenth, Department of Surgery, All-Ukrainian Institute of Postgraduate Education for Physicians (IPEPh, 1926-35) in city of Kharkiv, Ukraine, Professor Department of Surgery, Dnipro IPEPh (1938-41), and Director, Department of Surgery, Dnipro Medical Institute (DMI, 1942-43).

He developed and introduced into medical practice intravenous (IV) infusion of blood-substitute salt-solutions – sodium chloride (0,7% NaCl) solution, particularly synthesized by him complex of salt-solutions, named «Ukrainian infusyn» («Ukrinfusyn»), containing according to the chemical formula of 0.7% NaCl, potassium chloride (KCl), calcium chloride ($CaCl_2$), magnesium chloride ($MgCl_2$), sodium bicarbonate ($NaHCO_3$), glucose ($C_6H_{12}O_6$), distilled water (H_2O) and carbon dioxide (CO_2), plasma, citrated blood – blood treated with sodium citrate ($Na_3C_6H_5O_7$) or citric acid ($C_6H_8O_7$) to prevent its coagulation, in lieu (instead of) the whole blood, used during extensive surgeries, significant hemorrhages, civil or military traumatic injuries of the body, superimposed by shock (1930-38).

In his first experiments Borys Andriyevsky (1930) on 26 dogs after massive bloodletting with hemorrhagic shock, with subsequent IV infusion of blood-substitute salt-solutions (0,7% NaCl, «Ukrinfusyn») perceived the improvement of the respiratory mechanism and increase in the arterial blood pressure (ABP), concluding that some salt-solution can successfully be used in cases of very massive and unsafe for life blood loss.

In experiments from 1933 he observed changes of quantity of natrium (Na) in the blood serum during hemorrhage and after IV injection of salt-solutions and citrated blood in dogs, noting that the latter (citrated blood) was the most effective in relation as to the increase of Na level in the blood serum. Simultaneously, he observed fluctuation of albumin and globulin level in the blood serum during those studies.

In his 1934's research, he observed the influence of salt-solutions on the level of serum kalium / potassium (K) and calcium (Ca) in exsanguinated dogs.

Researching the influence of salt-solutions alkaline reserves of the blood in 37 dogs (1934), he proved that with the use of citrated blood alkaline reserves of the blood were replenished more quickly than with the infusion salt-solutions. Specially slowly blood alkaline reserves were replenished when solution of 0,7% NaCl was used (1935).

Clinical use of salt-solutions and «Ukrinfusyn» in 36 patients for the treatment of an operative or traumatic shock was accompanying by deepening of breathing, increasing of the ABP and the blood alkaline reserve, also by positive influence on fluctuation of some serum electrolytes (1934).

After significant bloodletting in 29 dogs, instead of whole blood, an IV infusion of citrated blood, «Ukrinfusyn» and solution of 0,7% NaCl, the latter was the least effective for the blood replenishment (1935).

Infused IV citrated blood has given the best outcomes as far as the blood regeneration in exsanguinated dogs is concerned (1935).

He proved that IV injection of blood substitutes causes short-lasting (45-50 days) morphologic changes in parenchymal organs (1936).

Leaning upon his laboratory and clinical data, he has come into conclusion that in regard to relative hemostatic effect of plasma, it can be used in patients as successfully as the whole blood. One of many advances of using plasma is the possibility to quickly start its IV infusion, without need for serological blood testing for the compatibility between the donor and the recipient.

Oven H. Wangensteen (1898 – 1981) — American surgeon, introduced continuous drainage by suction through a gastric or duodenal tube for treatment of an early mechanical or paralytic bowel obstruction (Wagensreen suction).

On design and suggestion by **Helen Taussig (1898 – 1986**: 20th century. Internal medicine), **Alfred Blalock (1899 – 1964)**[1209] — American pediatric cardiac surgeon, performed, with the help of **Vivien Thomas (1910 – 85)** — African American surgical technician, on a 4,5 kg cyanotic infant with a tetralogy of Fallot (TOF), the anastomosis of the right subclavian artery (SCA) to the right pulmonary artery (PA) – the Blalock-Taussig (BT) shunt (1944).

Many of surgical treatments of congenital heart disease (CHD) and congenital lung disease in children were initiated by **Clarence Crafoord (1899 – 1984**: 19th century. Anesthesia), such as resection of the atretic segment of a preductal aortic coarctation (AC) with an «end-to-end» anastomosis of the remaining ends using the proximal and distal aortic cross-clamping (1944)[1210], by **Robert Gross (1904 – 88) / William Ladd (1880 – 1967**: above), such as ligation of a patent ductus arteriosus (PDA, 1938), division and ligation of a right remnant of the double aortic arch (DAA, 1945)[1211], the first pediatric lobectomy in a child with congenital lobar emphysema (1945)[1212], pneumonectomy in a three-week old newborn with cystic disease of the

---

[1209] Blalock A, Taussig HB. The surgical treatment of malformation of the heart in which there is pulmonary stenosis or pulmonary atresia. JAMA 1945;128(3):189-202. — Faniyi L. Vivien Thomas: the unnamed father of atrial septostomy. J Am Coll Sur 2021 Dec;233(6):811.

[1210] Kvitting JPE, Olin ChJ. Clarence Crafoord: a giant in cardiothoracic surgery, the first to repair aortic coarctation. Ann Thorac Surg 2009 Jan;87(1):342-6.

[1211] Gross RE. Surgical relief for tracheal obstruction from a vascular ring. N Engl J Med 1945 Nov 15;233:586-90.

[1212] Gross RE, Lewis JEJ. Defect of the anterior mediastinum: successful repair. Surg Gynecol Obstet 1945;80:549.

lung (1946)[1213], a repair of the congenital diaphragmatic hernia (CDH) before the first 24 hours of life with good outcome (1946)[1214] and ligation of an aortopulmonary (AP) window (1948)[1215], and by **Carl C. Fischer** — American pediatric surgeon, who in an infant with congenital cystic disease of the right upper and lower lobes (RUL, RML) of the lung performed bilobectomy of the RUL and RML with recovery (1942)[1216].

The first classification for peripheral arterial (occlusive) disease (PAD) based on clinical symptoms was reported in 1954 by **Rene Fontaine (1899 – 1979)** — French surgeon, and named Fotaine's Classification[1217]:

stage 1   –   patient is asymptomatic;

stage 2 a   –   intermittent claudication appears after 200 meters of walking;

stage 2 b   –   intermittent claudication appears after less than 200 meters of walking;

stage 3   –   pain at rest;

stage 4   –   ischemic ulcer or gangrene which could be dry or humid.

Patients with the Fontaine stage 2b and higher, and noninvasive studies demonstrating evidence of significant PAD should undergo angiography to fully delineate the anatomy with the possibility of a noninvasive percutaneous (P/C) or surgical vascularization.

In 1954 Clarence Crafoord removed in a 40-year-old woman with atypical symptoms of mitral stenosis (MS), an intra-atrial myxoma, who was alive and well with 38 years of follow-up[1218].

Introduction by **Austin T. Moore (1899 – 63)** — American orthopedic surgeon, of the first metallic prosthetic hip replacement (Austin Moore prosthesis) prosthesis, in 1940 an original model, and in 1952 the current model, for surgical treatment of hip fractures, and its use decreased mortality and improved the quality of patient's life.

Edgar V.N. Allen (1900 – 61) — American physician, introduced in 1929 the test to denote the patency of the radial and ulnar arteries by temporarily compressing

---

[1213] Gross RE. Congenital cyst lung: successful pneumonectomy in a three-week-old baby. Ann Surg 1946 Feb;123(2):229:37.

[1214] Gross RE. Congenital hernia of the diaphragm. Am J Dis Child 1946;71:579-92.

[1215] Gross RE. Surgical closure of an aortic septal defect. Circulation. 1952;5:858-63. — Todurov BM. Surgical correction of aorticpulmomary window. J Ukr Med Assoc North Am 1994 Winter; 41,1(132):41-4.

[1216] Fisher CC, Tropea FJr, Barley CP. Congenital pulmonary cysts: report of an infant treated by lobectomy with recovery. J Pediatr 1943;23(2):219-23.

[1217] Fontaine R, Kim M, Kieny R. Die Behandlung der peripheren Durchblutungsstörungen. Helv Chir Acta 1954;21(5-6):499-533.

[1218] Chitwood WRJr. Clarence Crafoord and the first successful resection of a cardiac myxoma. Ann Thorac Surg 1992 Nov;54(5):997-8.

one of them and evaluation of the contralateral side for coloring of the skin, pulse by palpation or ultrasound (US) or by arterial oximetry (SaO_2) of the 1st or 2nd fingers (Allen test). It should be performed as a rule before a puncture of the radial artery for the arterial blood gases (ABG) sample, an insertion of the arterial line for a continuous monitoring of the arterial blood pressure (BP) and harvesting of the radial artery for coronary artery bypass (CAB) grafting.

Cushman D. Haagensen (1900 – 90)[1219] — American surgeon, established clinical and radiological criteria for operability of breast carcinoma, set indications and contraindications for radical mastectomy.

In 1945 **Nicolai P Sinitsyn (1900 – 75)** — Russian experimental cardiac surgeon, performed experimental heart transplantation (HT) in frogs without cardiopulmonary bypass (CPB) with long-term survival.

Mykola Ye. Dudko [06(19).12.1901, village of Kopani, Orkhivsky Region, Zaporizhian Obl., Ukraine – 21.04.1978, city of Kyiv, Ukraine)[1220] — Ukrainian surgeon, proposed to treat anaerobic infection with intravenous (IV) infusion of anti-gangrenous serum in large doses (1941).

In 1941 **Charles B. Huggins (1901 – 97)** — Canadian American urologist and **Clarence V. Hodges (1914 – 2001)** — American urologist, introduced hormonal treatment (orchiectomy and estrogen) for advanced cancer of the prostate. They monitored the prostate size and therapeutic efficacy by measuring serum prostatic acid phosphatase level, concluding that androgenic activity in the body influences prostate cancer, at least with respect to serum prostatic acid phosphatase.

One of the founders of modern orthopedics and traumatology was Sir **Reginald Watson-Jones (1902 – 72)** — English orthopedic surgeon, who authored a classical treatise «Fractures and Joint Injuries» (1940).

In 1950 **Norman R. Barrett (1903 – 79)** — Australian thoracic surgeon, described that prolonged gastroesophageal reflux disease (GERD) may lead in 5-16% of patients to acid reflux esophagitis, with subsequent abnormal change (metaplasia) in the cell of the lower portion of the esophagus, then to the replacement of normal multi-layered epithelium of the esophagus with one-layered columnal epithelium with

---

[1219] Haagensen CD. Carcinoma of the breast (II. criteria for operability). Ann Surg 1943;118:859-70,1032-51.

[1220] Village Kopani (Foot Note 968), Orkhivsky Region, Zaporizhian Obl., Ukraine. — Dudko MYe. (assoc.). Removal of foreign body from the duodenum with the aid of duodenal sound. Ukr Med Arch (Kharkiv) 1930; 5, 3. — Dodro NYe. Treatment of Gaseous Gangrene. Publ. «Derzhmedvydav» of Ukraine, Kyiv 1946. — Dudko MYe. Perforated ulcers of the stomach and duodenum. «Derzhmedvydav» of Ukraine, Kyiv 1951. — Olearczyk A. Professor M.O. Dudko – the university surgeon from Ukraine in Poland. Nasze Słowo (Warszawa) 7 Dec 1958;3,49 (121):2. — Dudko NYe New suturing method of diverticulum of the thoracic portion of esophagus. Khirugiia 1959;12. — Dudko NYe. Remote outcomes of operations for the left ventricular aneurysm. Grudnaia khirurgiia 1961; 2. — Ablitsov VH. «Galaxy Ukraine». The Ukrainian diaspora: prominent persons. Publ. «Kyt», Kyiv 2007.

goblet cells of the stomach, with/without peptic ulcer and stricture of the esophagus (columnal-cell lined or Barrett's esophagus). It can also damage the submucosa and the muscularis propria resulting in a short esophagus and development of high-grade dysplasia or intramucosal invasive adenocarcinoma. Risk factors include advanced age, smoking and obesity.

The mainstay of therapy for patients with Barrett's esophagus and esophagitis with short segment of involvement (<3 cm) is control of GERD by the acid suppressing drugs such as proton-pump inhibitors (derivatives of benzimidazole or imidazopyridine) where complete resolution of 30-50% of patients is expected at endoscopy; for those with Barrett's esophagus with metaplasia – laparoscopic antireflux procedure and long-term endoscopic surveillance; for those with Barrett's esophagus with high-grade dysplasia and long segment of involvement (>3 cm) where the incidence of invasive adenocarcinoma is high (40%) – esophagectomy, or submucosal resection in old and comorbid patients.

Russell C. Brock (1903 – 1980) — English chest and heart surgeon, who diagnosed a partial atelectasis of the right middle lobe with chronic pneumonitis (Brock syndrome); introduced an angioscopy through the aorta to visualize the interventricular septum (IVS) of the heart before and after its closure and to visualize the aortic valve (AV) after commissurotomy (1946); introduced closed valvotomy of the stenosed main PA (MPA) valve in combination with infundibulectomy of the stenosed right ventricular (RV) muscle for the surgical treatment of tetralogy of Fallot (TOF) – the Brock operation (1947-48); carried successful operations for mitral stenosis (MS) resulting from rheumatic heart disease (RHD, 1948)[1221].

Richard M. Burke (1903 – 87)[1222]— American military lung physician, diagnosed in 28-year-old United States (U.S.) veteran, complaining since 1928 of severe shortness of breath and presented with advanced cachexia, a giant bullous emphysema, from which he died in about 1935.

Giant bullous emphysema, referred as vanishing lung syndrome, is characterized by multiple bullae of the apical portions of one or both lungs, clinical resonance by the chest percussion, radiological disappearance of lung marking, absence of concomitant emphysema, relatively normal vital capacity (VC) with severe impairment of pulmonary function. Since about that time, it was treated by external drainage of the giant bulla in attempt to eliminate the space occupying lesion by collapse rather

---

[1221] Backer C, Brock RC, Campbell M. Valvulotomy for mitral stenosis: report of six successful cases. Brit Med J 1950 Jun 3;1(4665):1283-93.

[1222] Burke RM. Vanishing lung: a case report of bullous emphysema. Radiology 1937 Mar;28(3):367-71. — Cooper JD. The history of surgical procedures for emphysema. Ann Thorac Surg 1997;63:312-9. — Olearchyk AS. Diffuse bullous emphysema of the lung: conservative resection with a local application of a biological glue. J Card Surg 2004;19(6):542-3.

than excision of the bullae. Subsequently, this surgical procedure was replaced by bullectomy whereby small portion of the blowing up lung, called bulla (>1 cm in diameter) were removed allowing the remaining normal lung tissue to expand and function more effectively during respiration.

Coronary artery bypass (CAB) grafting[1223] became an effective surgical treatment of coronary artery disease (CAD) through the work of **Arthur M. Vineberg (1903 – 88)** — Canadian thoracic surgeon of Jewish origin, on the left internal thoracic artery (ITA) implants under the epicardium of the left ventricle (LV, 1946-50), of **Vladimir P. Demikhov (1916 – 98)** — Russian experimental thoracic and cardiac surgeon, on the «end-to-end» left ITA grafts to the left anterior descending (LAD) artery using **Erwin Payr (1871 – 1946)**[1224] — Austrian German surgeon, ring on beating heart in dogs (1953), of **Robert H. Getz (1910 – 2002)** — German, South-African American thoracic and cardiac surgeon, who performed the right ITA graft to the right coronary artery (RCA) with aid of the Payr ring (1960), on a beating heart, i.e., off-pump (OP) – OP-CAB, and **Vasilii (Vasyl) I. Kolesov (1904 – 92)** — Russian cardiac surgeon, performing the left and right ITA suture grafts to the LAD artery, the branches of the circumflex artery (CFA) and the RCA, OP-CAB or on cardiopulmonary bypass (CPB) (1964-68), **David C. Sabiston, Jr. (1924 – 2009)** — American cardiac surgeon, **H. Edward Garrett, Sr (1926 – 96)** — American cardiac surgeon, and **Rene H. Favaloro (1923 – suicide 2000)** — Argentinian cardiac surgeon, who carried out anastomoses between the ascending aorta (AA) and coronary arteries using the great saphenous vein (GSV), respectively, in 1962, 1964 and 1967-71.

The resistance to arteriosclerosis (AS) of the ITA and its high long-term patency, as opposed to the GSV, is attributable to: (1) absent or very thin vasa vasorum; (2) a dense internal elastic lamina with no fenestrations and thin medial layer with few smooth muscle cells; (3) secretion by endothelial cells of derived relaxing factors, prostacyclin and nitric oxide (NO), that cause vasodilatation and antagonize endogenous endothelium and calcitonin which are vasoconstrictors; (4) smooth muscle cells that exhibit little proliferation in response to platelet derived growth factor; (5) the less

---

[1223] Olearchyk AS. Coronary revascularization: past, present, and future. J Ukr Med Assoc North Am 1988;35,1 (117):3-34, 30-34. — Olearchyk AS. Vasilii I. Kolesov. A pioneer of coronary revascularization by internal mammary-coronary artery grafting. J Thorac Cardiovasc Surg 1988;96(1):13-8. — Olearchyk AS, Olearchyk RM. Reminiscences of Vasilii I. Kolesov. Ann Thorac Surg 1999; 67:273-6. — Haller JD, Olearchyk AS. Cardiology's 10 greatest discoveries. Tex Heart Inst J 2002, 29 (4):342-4. — Olearchyk AS. A Surgeon's Universe. Vol. 1-4. 4th Ed. Outskirts Press, Denver, Co 2016.

[1224] Payr E. Beträge zur Technik der Blutgefäss- und Nervennacht nebst Mitteilungen über die Verwendung eines resorbierbaren Metalles in der Chirurgie. Arch Klein Chir 1900;62:67-93. — Payr E. Zur Frage der zirkularen Vereinigung von Blutgefässe mit resorbierbaren Prothesen. Arch Klein Chir 1904;72:32-54. — The Payr's vascular prosthesis (in Greek – «prosthesis» means «addition, application, or attachment») is an absorbable extraluminal magnesium ring design for the vascular «intima-to-intima» oppositional anastomosis.

atherogenic lipid and glycosaminoglycan composition; (6) adaptation to the need for increased blood flow over time.

In 1947 **Frank Ph. Coleman**[1225] — American surgeon, reported a 40-year-old female suffering for 11 months from primary carcinoma of the right lung with invasion of the ribs, treated on February 2, 1940, by pneumonectomy and simultaneous en-block resection of the chest wall with long-term survival.

Subsequently, after en-block resection of the lung and the chest wall for malignancy with circumferential margin of 2 cm or larger margin of 4 cm on the rib resection, the defects, except those covered by scapula, were reconstructed by using the fascia lata femoris or autogenous rib grafts, cutaneous flaps of the latissimus dorsi, pectoralis major, rectus abdominis, serratus anterior, trapezius and deltoid muscles. When muscle flaps are not available or cannot be used, prosthetic materials is utilized including polypropylene mesh (PM) with or without methyl-methacrylate (MM) sandwich, polytetrafluoroethylene (PTFE) or an acryl mesh[1226].

In 1953 **John H. Gibbon (1903 – 73)**[1227] — American cardiac surgeon, closed an atrial septal defect (ASD) of the foramen ovale type in an 18-year-old woman under a direct vision, using the heart-lung machine (HLM) with a screen (modified film) oxygenator, which he himself had constructed, for cardiopulmonary bypass (CPB). Significant contribution to avoid an air embolism during open cardiotomy with the use of the HLM was made by **Bernard J. Miller (1918 – 2007)**[1228] — American Jewish surgeon, by describing in 1952 the need to introduce vent through the apex of the heart into the left ventricle (LV), and by **Frank F. Allbritten, Jr (1914 – 2005)**[1229] — American cardiac surgeon, using alternative route, i.e., by introducing the vent through the left atrium (LA) into the LV.

A year later (1954) the John Gibbon's HLM was refined by **John W. Kirklin (1917 – 2004)** — American cardiac surgeon, by improving the original pumping and oxygenator system, was brought into routine use in heart surgery, including congenital heart disease (CHD) – repair of a partial atrio-ventricular canal (AVC) canal and tetralogy of Fallot (TOF), etc. with consistent success.

[1225] Coleman FP. Primary carcinoma of the lung, with invasion of the ribs: pneumonectomy and simultaneous block resection of chest wall. Ann Surg Aug;126(2):156-68.

[1226] Le Roux BT, Shama PM. Resection of tumors of the chest wall. Curr Probl Surg 1982;20;348-86.

[1227] Gibbon JH. Application of a mechanical heart and lung apparatus to cardiac surgery. Minn Med 1954;37(3):171-85.

[1228] Miller BJ, Gibbon JH Jr, Greco VF, Cohen CH, Allbritten FF Jr. The use of a vent for the left ventricle as a means of avoiding air embolism to the systemic circulation during open cardiotomy with the maintenance of the cardiorespiratory function of animals by a pump oxygenator. Surg Forum 1953;4:29-33.

[1229] Hedlund KD. A tribute to Frank F. Allbritten Jr. Origin of the left ventricular vent during the early years of open-heart surgery with the Gibbon heart-lung machine. Texas Heart Inst J 2001;26(4):292-4.

Oleksandr I. Arutiunov (21.XII.1903 / 05.I.1904, city of Yerevan, Armenia – 05.VI.1975, city of Moscow, Russia)[1230]—Ukrainian Russian neurosurgeon of Armenian descent, made contribution in research of intracranial pressure brain edema, to set in a line functions, clinics and surgical treatment of tumors, vascular diseases, trauma to the brain. Particularly, performed the first stereotactic operation in patient with subcortical hyperkinesis (1961), introduced into neurosurgical practice method of the sub-frontal approach to sella turcica[1231] of sphenoidal bone (os sphenoidale) of the skull, made complex operations on the basal ganglia divisions of the brain – removal of meningeal-pharyngioma, cranio-pharyngioma, tumors of the lower addition of the brain (hypophysis).

In 1955 **Robert M. Zollinger (1903 – 92)** — American surgeon, and **Edwin H. Ellison (1918** – killed in a plain crush **1970)** — American surgeon, described ulcerogenic non-beta islet cell tumors of the pancreas (Zollinger-Ellison syndrome) and recommended a total gastrectomy as a treatment of choice. / Snylyk O. Zollinger-Ellison syndrome. J Ukr Med Assoc North Am 1963 Oct;10,4(31):2-9.

The zealous investigation of chronic obstructive pulmonary disease (COPD), i.e., pulmonary emphysema, was carried out by **Otto S. Brantigan (1904 – 81)**[1232] — American anatomist and surgeon.

Pulmonary emphysema is condition caused by destruction of lung parenchyma with decreasing mass of functioning lung tissue and thus decreased the amount of gas exchange. The lungs lose elastic recoil and expands in volume. Subsequently, this leads to the hyperextended chest with flattening of the diaphragm, widened intercostal spaces (ICS) and horizontal displacement of ribs. These changes led to increased work of breathing and debilitating dyspnea. The expansion of the lung may occur in heterogenous (non-uniform) or homogenous (uniform) matter. In the first instance, the most affected lung tissue can expand to crowd the relatively spared lung tissue and impair ventilation of the functioning lung. In addition, there is obstruction of the small airways caused by a combination of reversible or irreversible loss of elastic recoil by adjacent lung parenchyma.

The most common causes of COPD are idiopathic emphysema and alpha$_1$-antitrypsin deficiency (63%), cigarette smoking and environmental pollution.

[1230] City of Yerevan (founded 7th century BC), Armenia (area 29,800 km^2). — Ablitsov VH. «Galaxy Ukraine». The Ukrainian diaspora: prominent persons. Publ. «Kyt», Kyiv 2007.

[1231] Sella turcica is created within the body of the sphenoidal bone (os sphenoidale) of the cranium in the form like of concavity like the saddle, site for location of the of hypophysis..

[1232] Brantigan OC, Mueller E, Kress MB. A surgical approach to pulmonary emphysema. Am Rev Resp Dis 1959;89:194-202. / Olearchyk AS. Bullous emphysema of the lung: conservative resection, supported by the local use of the biological glue. Klinichna khirurhiya 2003;10:58-9. — Olearchyk AS. Diffuse bullous emphysema of the lung: conservative resection with a local application of a biological glue. J Card Surg 2004;19(6):542-3.

Medical treatment of pulmonary emphysema consists of cessation of cigarette smoking and avoidance of the other environmental pollutants; taking expectorants, e.g., Guaifenesin (glyceryl guaiacolater); bronchodilators for relief of symptoms or its worsening, such as Ventolin or Albuterol sulfate (Pro-air) inhalation aerosol 90 mcg per actuation, to inhale 2 puffs by mouth every 4-6 hours, and for a long-term maintenance Anoro Ellipta inhaler, one strip contains umeclidinium 62.5 mcg per blister, and the other strip contains vilanterol 25 mcg per blister, a long-acting beta$_2$-adrenergic agonist, one puff inhalation by mouth once daily; and corticosteroids, e.g., prednisone 0.5-1 mg /kg/ day for 4-12 weeks, or in case of exacerbation in an average adult prednisone 40 mg p.o. q.d. x 2 days, 30 mg p.o. q.d. x 2 days, 20 mg p.o. q.d. x 2 days, and 10 mg p.o. q.d. x 2 days.

When pulmonary function test (PFT) shows decrease of forced expiratory volume$_1$ (FEV$_1$) to below 1 liter or approximately 15% of predicted normal values and progressive elevation of carbon dioxide (CO_2), supplemental oxygen is required in excess 4 liter/minute.

With the aim of the treatment of worsening pulmonary emphysema, despite maximal medical therapy, Otto Brantigan introduced in 1957 the lung volume reduction surgery (LVRS), which is a conceptual extension of bullectomy **[Richard Burke (1903 – 87)**: above].

Indication for LVRS is emphysema with destruction and hyperinflation are marked restriction of daily living activities, failure of maximal medical treatment to correct symptoms, and marked impairment of PFT, specifically of FEV$_1$ <35% of predicted normal values. Discriminating condition favoring LVRS are marked thoracic distention, non-uniform disease, with the obvious areas in apical targets, usually upper part of the right and left upper lobe (RUL, LUL), FEV$_1$ >20% predicted and age between 60 and 70 years.

In LVRS non-uniformly destroyed lung by bullae and functionless lung is resected via bilateral sequential 4th or 5th intercostal space thoracotomy (ICST), or median longitudinal sternotomy incision (MLSI), and the remaining (20-30%) lung are spared from emphysematous destruction.

After the description of the proximal esophageal atresia (EA) by **William Durston (1670**: 17th century**)** and of the proximal EA with distal tracheoesophageal fistula (TEF) by **Thomas Gibson (1697**: 17th century**)**, **Robert Gross (1904 – 88**: above**)** proposed in 1953 the classification of congenital EA with TEF, the most common anomaly of the trachea occurring in 2,4 per 10,000 births: type A, EA without TEF (8%); type B, EA with proximal TEF (1%); type C, EA with distal TEF (87%), type D, EA with both proximal and distal TEF (1%); and type E, the absence of EA with H-like TEF (4%). Associated abnormalities are vertebral (V), anal (A), cardiac (C), tracheal (T), esophageal (E), renal R), limb (L) and duodenal atresia (D) (VACTERL+D) in 20%, including cardiovascular in 35%. The first successful primary surgical repair was of EA/TEF type C, performed in 1941 by **Cameron Haight (1900 – 70)** — American pediatric thoracic surgeon, through the right 4th extrapleural route with division of fistula, closure of the fistulous opening in the posterior (membranous) wall of the trachea and an «end-to-end» esophageal anastomosis using a single row of simple suture.

In 1948 **Jean Kunlin (1904 – 91)**[1233] — French surgeon, used a segment of the reversed great saphenous vein (GSV) for a femoro-popliteal bypass of an obstructed superficial femoral vein (SFA).

In 1937 **Isadore M. Tarlov (1905 – 77)** — American neurosurgeon, identified perineural (perineurial) or sacral nerve root cysts, usually at the sacral (S_2-S_3) level or in multiple locations, abnormal sacs filled with cerebrospinal fluid (CSF) that can cause a progressively painful radiculopathy and paresthesia.

Ivan T. Shevchenko (22.III.1905, village Novyy Starodub, Petrivsky Region, Kirovhrad Obl., Ukraine – 07.VII.1993, city of Kyiv, Ukraine)[1234] — Ukrainian surgeon, advocated a preoperative radiation for malignant tumors of the lung, stomach, bowel and soft tissue (1962) and intraoperative radio- and radium therapy of such tumors after resection (1964).

The operation for termination of bleeding from esophageal varices due to liver cirrhosis and portal hypertension proposed by **Norman C. Tanner (1906 – 82)**[1235] — English surgeon, consist of disconnection of all external porto-azygos venous and ligamentous connections of the distal esophagus, cardia and proximal stomach and of a transverse incision of the stomach in the direction below the cardia (porto-azygos disconnection, or Tanner operation, c. 1958).

It was **Alexander A. Vishnevsky (1906 – 75)**[1236] — Russian surgeon of Ukrainian descent, who in 1953 conducted under local anesthesia the first heart operation through the left antero-lateral 4th intercostal space thoracotomy (ICST), i.e., mitral commissurotomy (MC) for mitral valve (MV) stenosis.

[1233] Kunlin J. Le traitement de l'arterite obliterante par la greffe veineuse. Arch mal du Coeur 1949;42:317-71.

[1234] Village Novyy Starodub (known since 1754-59; suffered from conducted by Russian Communist Fascist regime of Holodomor-Genocide 1932-33 i 1946-47 against the Ukrainian Nation), Petrivsky Region, Kirovhrad Obl., Ukraine. — Shevchenko IT. Coming to a Head Questions of Chemotherapy of Malignant Tumors. Novyy khir archiv 1962;2:3-10. — Grenda J, Kiesz W, Olearchyk A. Nasieniak jądra z przerzutami do opłucnej. Pol Przegl Chir 1963;35(2),11a:1265-9. — Shevchenko IT. New method of surgical treatment of cancer of the lung with administration of roentgen-radium therapy during operation. Dis Chest 1964;45(1):36-9. — Shevchenko IT. Malignant Tumors and Diseases which Predeceased Them. Publ. «Derzhmedvydav» Ukraïny, Kyiv 1965. — Shevchenko IT. Malignant Tumors and Diseases which Predeceased Them. 2nd Ed. Publ. «Derzhmedvydav» Ukraïny, Kyiv 1973. — Ablitsov VH. «Galaxy Ukraine». The Ukrainian diaspora: prominent persons. Publ. «Kyt», Kyiv 2007.

[1235] Tanner NC. The late results of porto-azygos disconnection in the treatment of bleeding from aesophageal varices. Ann R Coll Engl 1961 Mar;28:153-74.

[1236] Ablitsov VH. «Galaxy Ukraine». The Ukrainian diaspora: prominent persons. Publ. «Kyt», Kyiv 2007.

A burn is an injury to skin, subcutaneous (SC) tissue, muscle and bone, or other tissue caused by heat, cold, electricity, chemicals, friction, or radiation. The severity of burn is classified into the four following degrees:

1st degree – superficial affecting epidermis.
2nd degree – superficial partial-thickness extends into superficial (papillary) dermis;
 – deep partial-thickness extends into deep (reticular) dermis.
3rd degree – full-thickness extends through entire dermis.
4th – extends through entire skin, SC tissue, muscle and bone.

The burn percentage of the skin in adults can be estimated from published in 1951 by **Alexander B. Wallace (1906 – 74)** — Scottish plastic surgeon, the «Wallace rule of nines» the total body surface area (TBSA) burn: 9% TBSA each arm, 9% TBSA each leg, 18% TBSA for the front of the torso, 18% TBSA for the back of the torso, 9% TBSA for the head and 1% TBSA for the perineum.

The severity of the skin burn is classified as minor in adults if <10% TBSA is burned, in children or old if <5% TBSA is burned; as moderate in adults if 10-20% TBSA is burned, in children or old if 5-10% TBSA is burned, and as major in adults if >20% TBSA is burned, in children or old if >10% TBSA is burned.

A year later (1952) the Evans formula[1237] was introduced for burn resuscitation based on the body weight in kilograms (kg) and the TBSA burn, followed by the Brooke and Baxter formulas[1238].

Evans formula:

1st 24 hr: colloid [Dextran 70/normal saline solution (NSS)] 1 ml/kg/% TBSA burn and crystalloid (NSS) 1 ml/kg/% TBSA burn, half given during the first 8 hr and half over the rest 24 hr, plus 24 hr maintenance water (H_2O) 2000 ml. Maximal calculated TBSA burn 50%.

2st 24 hr: colloid (Dextran 70/NSS) 0,5 ml/kg/% TBSA burn and crystalloid (NSS) 0,5 ml/kg/% TBSA burn over 24 hr, plus maintenance H_2O 2000 ml per 24 hr. Maximal calculated TBSA burn 50%.

Brooke formula:

1st 24 hr: colloid (plasma or its substitute) 0,5 ml/kg/% TBSA burn and crystalloid [lactated Ringer solution (LRS)[1239] 1,5 ml/kg/% TBSA burn, half given during the

---

[1237] Hutchar N, Haynes BM Jr. The Evans formula revisited. J Trauma 1972 Jun;12(6):453-8.
[1238] Ibid.
[1239] Ibid.

first 8 hr and half over the rest 24 hr, plus maintenance H_2O 2000 ml 24 hr. Maximal calculated TBSA burn 50%.

2st 24 hr: colloid (plasma or its substitute) 0,25 ml/kg/% TBSA burn and crystalloid (LRS) 0,75 ml/kg/% TBSA burn over 24 hr, plus maintenance H_2O 2000 per 24 hr. Maximal calculated TBSA burn 50%.

Baxter[1240] or Parkland formula:

1st 24 hr: crystalloid (LRS) 4 ml /kg/% TBSA burn, half given during the first 8 hr and half over the rest 24 hr.

2st 24 hr: that amount of H_2O necessary to maintain normal level of sodium (Na) in the blood, plus plasma 250-500 ml. No maximal calculated TBSA burn 50%.

J. Maxwell Chamberlain (1906 – May 24, 1968, killed in a car accident by decapitation while dozed momentarily on the US IS-95 No. superhighway driving from New York, NY to his family in MA, when his car crossed the grass divider and went under a truck) — American thoracic surgeon, introduced a limited left anterior parasternal mediastinotomy through the 5th horizontal incision just lateral to the sternomanubrial joint over the 2nd costal cartilage with partial subperichondrial resection or via the 2nd extrapleural ICST with/without segmental resection of the costal cartilage for biopsy of the subaortic (level 5) and para-aortic (level 6) lymphatic nodes, out of reach by suprasternal mediastinoscopy, to rule out metastases from carcinoma of the lung (Chamberlain procedure) and a segmental resection in tuberculosis of the lung.

In 1958 **Rune Elmquist (1906 – 96)** — Swedish physician, engineer and inventor, implanted, designed by himself a permanent artificial cardiac pacemaker (PM) with an electrode placed in the right ventricle (RV) and with a non-rechargeable mercury-zinc multicell battery.

One hundred after in 1857 **Ludwig** von **Buhl (1816 – 80**: 19th century. Pathology**)** discovered a direct communication between the left ventricle (LV) and the right atrium (RA) through the membranous septal defect, in 1958 **Frank Gerbode (1907 – 84)**[1241] — American surgeon, reported on a more common indirect communication between the LV and the RA through a peri-membranous ventricular septal defect (VSD) into the right ventricle (RV) and then through a defect in the tricuspid valve (TV) into the RA (Gerbobe AV defect). Specifically, the defect is in the upper membranous ventricular VSD between the left ventricular outflow tract (LVOT) and the RA, which

---

[1240] **Charles R. Baxter (1029 - 2005)** — American physician.

[1241] Gerbode F, Hultgren H, Melrose D, Osborn J. Syndrome of left ventricular-right atrial shunt. Ann Surg 1958;148:433-46.

is possible because the tricuspid valve (TV) is apical in relation to the mitral valve (MV). Acquired causes of Gerbode defect may occur after cardiac valve replacement, blunt trauma, infective endocarditis, myocardial infarction (MI) and postsurgical repair of VSD. Shunting across the Gerbode defect occurs mainly in systole, because LV systolic pressure is much higher than RA pressure; in diastole, LV pressure is only slightly higher.

Endarterectomy (thromboendarterectomy; EA) procedure was first performed by **João C. dos Santos (1907 – 75)** — Portuguese surgeon, in 1946, when he removed the complex obstructive atherosclerotic (AS) plaque superimposed by thrombosis from a superficial femoral artery (SFA).

Classification (class) of liver cirrhosis was developed by **Charles G. Child (1908 – 91)**[1242] and **Jeremish G. Turcotte (1934 –)** — American surgeons, and named after them class A, B and C:

	Class A	Class B	Class C
Total bilirubin (mg/dl)	< 2	2-3	> 3
Serum albumin (mg/dl)	> 3.5	2.8 – 3.5	> 2.8
Prothrombin time in international normalizing ratio (PT-INR)	< 1.7	1.7 – 2.3	> 2.3
Ascites	none	mild	moderate - severe
Hepatic encephalopathy	none	suppressed with drugs	refractory
Hepatic hydrothorax[1243]	none	present	present

As a medical student, **Michael E. DeBakey**, born **Michel Dabaghi (1908 – 2008)**[1244] — Lebanese American thoracic and cardiovascular surgeon, invented in 1940 the roller pump for the use in a direct «donor-to-patient» blood transfusion, which became the driving component of the heart-lung machine (HLM) to provide the patient with continuous blood flow during cardiopulmonary bypass (CPB) for open heart operation.

[1242] Child CG, Turcotte JG. Surgery and Portal hypertension. In, CG Child. The Liver and Portal Hypertension. Saunders, Philadelphia 1964. P. 50-64.

[1243] Huang PM, Chang YL, Yang CY, et al. The morphology of diaphragmatic defect in hepatic hydrothorax. Thoracoscopic findings. J Thorac Cardiovasc Surg 2005;130:141-5. — Zenda T, Miyamoto S, Murata S, et al. Detection of diaphragmatic defect as the cause of severe hepatic hydrothorax with magnetic resonance imaging. Am J Gastroenterol 1998;93;2288-9. — Temes RT, Davis MS, Follis FM, et al Video thoracoscopic treatment of hepatic hydrothorax. Ann Thorac Surg 1997;64:1468-9. — Assouad J, Barthes FP, Shaker W, et al. Recurrent pleural effusion complicated by liver cirrhosis. Ann Thorac Surg 2003;75(3):986-9.

[1244] Thompson T. Hearts. McCall Publ. Co, New York 1971.

Michael DeBakey with his mentor **Alton Ochsner (1894 – 1981)**[1245] — American thoracic and vascular surgeon, postulated a strong link between smoking and carcinoma of the lung (1939); performed a successful carotid endarterectomy (CEA) for atherosclerotic (AS) occlusion (1953).

Michael DeBakey with **Denton A. Cooley (1920 – 2016)** — American thoracic and cardiovascular surgeon, performed the first synthetic tubular graft (STG) replacement of dissecting aneurysm of the aortic arch (1955)[1246]; and Denton Cooley with Michal DeBakey reported the first homograft replacement of the entire ascending aorta (AA) under CPB (1956)[1247].

He performed a surgical transthoracic repair of the supra-aortic arterial trunks, i.e., innominate artery (IA), or brachiocephalic trunk (BCT), left common carotid artery (CCA) and both subclavian arteries (SCA) by endarterectomy (EA, 1956)[1248].

Michael DeBakey replaced an aortic arch aneurysm with homograft using antegrade cerebral perfusion (1957).

To counteract narrowing of the artery caused by carotid endarterectomy (CEA) he performed the first patch-graft angioplasty (1958); implanted into the chest of a patient in cardiogenic shock after an aortic valve replacement (AVR), the first left ventricular assist device (VAD) with a bypass inflow-outflow between the left atrium (LA) and the AA for the left ventricular failure (LV), so called «booster pump» (1963).

He developed the classification of aortic dissections (DeBakey, 1965), in which:

Type I   –   the primary intimal tear located in the AA with its involvement of the AA and beyond the origin of the innominate artery (IA) into the aortic arch and often beyond it distally (60%), being the most lethal form of dissection of the AA;

Type II   –   the primary intimal tear is in the AA and is confined to it (10-15%); and

Type III   –   the primary intimal tear located in the proximal descending aorta (DTA), distally to the left subclavian artery (SCA), rarely extends in proximally (retrogradely) into the aortic arch, but will extend distally into the DTA, abdominal aorta (AA) and beyond aortic bifurcation (20-30%).

[1245] Nather K. Der prae-oder retroperitoneale Weg zum subphrenischen Abscess als typische Operation. Arch Klin Chir 1922;122:24-99. — Nather C, Ochsner EMA. Retroperitoneal operation for subphrenic abscess. Surg Gynec Obstet 1923;37:665-73. — Ochsner A, DeBakey M. Subphrenic abscess: collective review and analysis of 3608 collected and personal cases. Internat Abst Surg 1938;66:426-38.

[1246] DeBakey ME, Cooley DA, Creech Jr D. Surgical consideration of dissecting aneurysm of the aorta. Am Surg 1955;142:586-612.

[1247] Cooley DA, DeBakey ME. Resection of entire ascending aorta in fusiform aneurysm of the aorta with cardiac bypass. JAMA 1956;11:858-9.

[1248] DeBakey ME, Morris Jr GC, Jordan GL, et al. Segmental thrombo-obliterative disease of the branches of the aortic arch: successful surgical treatment. JAMA 1958;166:998-1003.

Michael DeBakey implanted successfully into the chest of a patient in post-cardiotomy shock the first left VAD as a «bridge» to recovery (BTR; 1966) with duration of support lasting 10 days; developed VAD with the continuous-axial-flow pump between the LV and the AA (2000).

He himself developed aortic dissection type II and survived an open repair at the age 97 in 2006.

The approach to treat post-pneumonectomy empyema is that developed by **O. Theron Claggett (1908 – 90)** — American general, thoracic and cardiovascular surgeon, and **Joseph E. Geraci**[1249] where the 1st stage involved creation of an open window thoracotomy by resecting part of ribs to create an opening into the thorax to allow drainage and antiseptic irrigation of the space, and the 2nd stage, when pleural space is clean and sterile, the space is filled with 0,25% neomycine solution and closed surgically.

Brothers **Howard P. House (1908 – 2003)** — American otologist and **William F. House (1923 – 2012)** — American otologist, significantly contributed to the development of the ear surgery. The first founded in 1946 the House Ear Institute / the House Ear Research Inst. in Los Angeles[1250], CA, where he developed the cochlear implant and the auditory brain stem implant for the impaired hearing (1961). The second perfected the fenestration procedure (1940s) and stapedectomy surgery (1950-80) and implanted a prototype for the cholera implant device (1972). The device provides a sense of hearing for deaf and hard-of-hearing patients when hearing aids do not work.

The development of refined vision, that is a flexible fiberoptic endoscopy (FOE), began with the introduction of zoom lenses (1834, 1905, 1959), a coherent fiber-optics by **Heinrich Lamm (?1908 – ?1974)** — German medical student (1930), and by **Harold H. Hopkins (1918 – 94)** — English physicist, whose wave theory of aberration (1950) and design of the rod-lens flexible endoscope (1959) in which a bundle of the flexible glass fibers allows to coherently transmit an image, «opened the door» to modern «key-hole» surgery.

Frank H. Mayfield (1908 – 91) — American neurosurgeon, invented the spring and clip applier to close the blood supply to a brain aneurysm (1952), and Mayfield headrest and skull clamp (1967).

In 1962-66 **Boris V. Petrovsky (1908 – 2004)** — Russian surgeon, used a pedicled diaphragmatic flap for the reinforcement and revascularization of the post-infarction left ventricle (LV) aneurysm, and for plastic reconstruction after its resection, and

---

[1249] Claggett OT, Geraci JE. A procedure for the management of post-pneumonectomy empyema. J Thorac Cardiovasc Surg 1963;45;141-5.

[1250] City of Los Angeles (settled 1781, area 1,215 km²), California (CA), USA.

patch a perforated distal esophagus in the cardiac triangle in the syndrome described by **Herman Boerhaave (1668 – 1738**: 18th century**).**

Blunt trauma to the chest, either unilateral or bilateral with the fracture of four or more ribs at two sides, promoting enough instability as to cause a flail chest with paradoxical motion of the chest cage may lead to an acute respiratory distress syndrome (ARDS), require analgetic, internal fixation by intubation and ventilation or osteosynthesis of ribs, or both. The costal osteosynthesis was introduced by **Martin Kirshner (1979 – 1942**: above) using K-wires for a simple wiring or intra-medullary wire brace and by **Robert Judet (1909 – 80)**[1251] — French orthopedic surgeon, using Judet rib staples (1973). Also, periosteal plates are used for this purpose.

With the aim of treating Hirschprung disease [19th century. Children disease. – **Harald Hisrschprung (1830 – 1916)]**, **Orvar Swenson (1909 – 2012)**[1252] — Swedish American surgeon, in 1948-50 proposed to perform resection of the involved by disease sigmoid colon and rectum with preservation of the anal sphincter, by «pulling through» the created space in the pelvis the healthy distant descending colon and anastomosing it to the anus, or removal of congenital aganglionic megacolon and rectum, also with preservation of the anal sphincter, and anastomosing the ileum of the small bowel with the anus (ileoanal anastomosis) – the abdominoperineal (AP) Swenson rescue operation. In desperate cases it may be necessary to perform intestinal transplantation.

On June 10, 1948 **Charles P Bailey (1910 – 93)** — American cardiac surgeon, performed the closed mitral commissurotomy (MC) for rheumatic mitral stenosis (MS) through a left atrial (LA) appendage under direct digital guidance and positioning at the posteromedial, anterolateral, or both commissures with a backward cutting knife (punch) which was then drawn backward for a 2,5 cm, widely cutting a fibrosed commissure not closer than 3 mm from the MV annulus[1253]; closed tricuspid valve (TV) commissurotomy (1952)[1254]; lineal left ventricular (LV) aneurysmectomy for post myocardial infarction (MI) LV aneurysm, 1955; coronary endarterectomy (CE) of the circumflex artery (CFA) - lateral branch in a 51-year-old and 52-year-old patients, via

---

[1251] Judet R. Costal osteosynthesis. Rev Chir Orthop Reparatrice Appar Mot 1973;59:Suppl 1:334-5.

[1252] Swenson O, Bill AH, Jr. Resection of rectum and rectosigmoid with preservation of the sphincter for begin spastic lesions producing mega-colon; an experimental study. Surgery 1948 Aug;24(2):212-20. — Swenson O, Newhauser EB, Pickett LK. New concept of etiology and treatment of congenital megacolon (Hirschprung disease). Pediatrics 1949 Aug;4(2):301-9. — Swenson O. A new surgical treatment for Hirschprung disease. Surgery 1950 Aug;28(2):371-83.

[1253] Bailey CP. The surgical treatment of mitral stenosis (mitral commissurotomy). Dis Chest 1949 Apr;15(4):377-97.

[1254] Bailey CP. Tricuspid stenosis. In, CP Bailey (Ed.). Surgery of the Heart. Lea & Fibiger, Philadelphia 1955. P. 846-61.

the left intercostal space thoracotomy (ICST, 1956)[1255], all on a beating heart without cardiopulmonary bypass (CPB). Of his 5 patients whom he performed MC 4 died on the operating table, or early following the procedure.

A closed mitral commissurotomy (MC) was renewed in 1948, by **Dwight E. Harken (1910 – 93)** — American cardiac surgeon, who later, in 1960, implanted a stainless steel caged-ball cardiac valve in the aortic position with long-term survival.

In 1936-38 **Frederic E. Mohs (1910 – 2002)** — American general surgeon, developed the Mohs micrographic surgery (MMS) technique for removal skin malignancies by discovering that applying a combination of zinc chloride and bloodroot (Sanquinaria Canadensis)[1256] paste to malignant skin tissue allowed it to be removed surgically with very narrow surgical skin margin and maximal preservation of healthy tissue for microscopic examination.

The MMS technique was modified in the 1970s by **Perry Robins (1932 –)** — American dermatologist, by using fresh-tissue frozen histology. The fresh skin specimen is removed under local anesthesia, then mounted on a cryostat («cryo» meaning cold and stat meaning stable)[1257] and then frozen sections are examined.

The cure rate after the MMS for primary basal cell carcinoma is 97-99,8%, for squamous cell carcinoma 95-96%, for melanoma-in-situ 77-98%, and for malignant melanoma it is 52%.

The most common cause of diaphragmatic elevation is congenital eventration of the diaphragm and phrenic nerve paralysis.

Congenital eventration of the diaphragm occurs as result of impaired migration of the fetal myotome, affecting in mild causes only the central tendon of the diaphragm (CTD), whereas in severe causes the entire muscular portion of the diaphragm. The other cause or associate condition is an ipsilateral pulmonary hypoplasia.

Phrenic nerve paralysis in child could be caused by congenital absence of the phrenic nerve, traction injury to the phrenic nerve at the base of the neck after delivery with the use of obstetrical forceps, viral palsy, or iatrogenic injury after pediatric thoracic surgery.

The diaphragmatic elevation is poorly tolerated by neonates (in Latin – «neonatus», meaning newborn) and infants (in Latin – «infans», meaning unable to speak or speechless), because flaccid diaphragm itself compresses the lower lobe of the lung on the ipsilateral side causing atelectasis. A large displacement of the abdominal content into the negative-pressure pleural cavity causes mediastinal shift with compression of the contralateral lung, causing additional atelectasis. Those events

---

[1255] Bailey CP, May A. Lemmon WM. Survival after coronary endarterectomy in man. JAMA 1957;164:641-6.

[1256] Bloodroot (Sanquinaria Canadensis) is a perennial, herbaceous plant native to eastern North America.

[1257] Cryostat (cryo meaning cold and stat meaning stable) is a device used to maintain low cryogenic temperature of samples or devices mounted within the cryostat.

lead respiratory failure necessitating tracheal intubation and initiation of mechanical respiration to reinflate atelectatic lungs with an air enriched by oxygen (O_2).

In adult diaphragmatic elevation are usually due to phrenic nerve involvement by spinal cord disease, cervical trauma, infiltration by tumors or lymph nodes (LN) or iatrogenic injury.

In symptomatic newborn and infant surgical correction is necessary by imbrication of the CTD or a subcostal radial plication of the muscular part of the diaphragm.

In the CTD imbrication of the diaphragmatic elevation a several lineal rows of nonabsorbable sutures enforced by pledged are placed through the weak area of the diaphragm, after which the sutures are tightened and the weakened tissues plicated with the creation of a tense surgical surface.

A subcostal radial plication of the diaphragmatic elevation through the upper abdominal quadrant incision was introduced in 1949 by **David State**[1258] — American pediatric surgeon. Later (1976), **John E. Foker (1937 –)** — American thoracic, cardiac and vascular surgeon, performed subcostal radial plication via posterior thoracotomy, consisting of placing the radial non-absorbable single or pledged mattress sutures circumferentially along the muscular portion of the diaphragm in a radial manner, pulling and displacing the edge of the CTD-diaphragmatic muscle interphase towards the lateral chest wall and sewing it as unbroken band from the xiphoid process to the vertebral body, avoiding the mediastinal pleura. The sutures can anchor the endothoracic fascia inside the ribs or also pass around the ribs. If a row of stitches has been placed without achieving desired tautness of the weakened diaphragm, a second row of stitches can be placed for further tightness of the diaphragmatic muscle.

Mark M. Ravitch (1910 – 89) — American thoracic surgeon, introduced in 1949 the classic open repair (Ravitch) of pectus excavatum which entails subperichondral resection of the bent (concave) rib cartilages, a proximal anterior transverse sternal wedge osteotomy and a temporary retrosternal support with a metal bar. Then, in 1952, he introduced the classic repair (Ravitch) of the upper chondromanubrial pectus carinatum by resecting multiple costal cartilages and a double sternal osteotomy.

After original reports by **Niels Steensen (1638 – 86**: 17th century) from 1671 and **J.A. Wilson** (18th century) from 1798, **J. W. Major**[1259] — American thoracic surgeon, in 1953 and **James R. Cantrell** — American thoracic surgeon, in 1958-59, anew characterized and precisely described the congenital chest wall deformity – thoracoabdominal ectopia cordis, named the pentalogy of Cantrell, or thoraco-abdominal syndrome,

---

[1258] State D. The surgical correction of congenital eventration of the diaphragm in infancy. Surgery 1949; 25:461-8.

[1259] Major JW. Thoracoabdominal ectopia cordis; report of a case successfully treated by surgery. J Thorac Surg 1953 Sep;26(3):309-17.

consisting of a cleft in the inferior part of the sternum, defect of the midline of the abdomen with omphalocele, defective pericardium and diaphragm with communication between the pericardial sac and peritoneal cavity, ectopia cordis and cardiac anomalies – ventricular septal defect (VSD), or less often atrial septal defect (ASD), tetralogy of Fallot (TOF), or left ventricular (LV) diverticulum[1260].

James Cantrell and **H.C. Guild** — American thoracic surgeon, classified a congenital tracheal stenosis, one of the two most common congenital tracheal anomalies in infancy and children, into: (1) general hypoplasia; (2) funnel-like stenosis; and (3) segmental stenosis with a tracheal right upper lobe (RUL) bronchus, and simultaneously reported an early successful tracheal resection and re-anastomosis (TRR) in a 7-year-old child who had stenotic tracheal RUL bronchus which was excised and the carina was brought up and anastomosed to the main trachea (1964)[1261].

Tracheal stenosis is most often associated with pulmonary artery (PA) sling, tracheal RUL bronchus and in 25% with congenital heart disease (CHD).

The right posterior (interatrial, inferior or standard) incision for the exposure of the left atrium (LA) by right posterior atriotomy was proposed by **David J. Waterston (1910 – 85)** — English cardiac surgeon and **Tyge Sondergaard (1914 – 90)** — Norwegian cardiac surgeon, behind the deep fold between the systemic veins – between the superior and inferior vena cava (SVC, IVC)) and the right superior and inferior pulmonary veins and the left superior and infection pulmonary veins (RSPV, RIPV, LSPV, LIPV) at the base of the interatrial groove, i.e., the concavity on the external surface of the heart between the right atrium (RA) and the LA (Waterston's or Sondergaard's groove).

In 1952 **Paul M. Zoll (1911 – 99)**[1262] — American cardiologist of Jewish descent, paced two patients with complete heart block (CHB) with life threatening bradycardia (slow heart rate) by means of electrical stimulation through external electrodes which were applied to the anterior chest wall – the external percutaneous (PC) artificial electronic fixed rate cardiac pacemaker (PM).

A tourniquet used in thoracic and cardiovascular surgery during cannulation of the great or peripheral blood vessels and cardiac chambers for cardiopulmonary bypass (CPB), to control the bleeding or shunting of blood vessels was designed by **William R. Rumel (1911 – 77)**[1263] — American surgeon. After making a a purse-string

---

[1260] Cantrell JR, Haller JA, Ravitch MM. A syndrome of congenital defect involving the abdominal wall, sternum, diaphragm, pericardium, and heart. Surg Gynecol Obstet 1959 Nov;107(5):602-14.

[1261] Cantrell JR, Guild HC. Congenital stenosis of the trachea. Am J Surg 1964;108:297-305.

[1262] Zoll PM. Resuscitation of the heart in ventricular standstill by external electric stimulation. N Engl J Med 1952;247(20:768-71.

[1263] Welling DR, Rich NM, Burris DG, Boffard KD, DeVries WC. Who Was William Ray Rumel? World J Surg 2008Sep;32(9):2122-5.

suture on the surface of the cardiac chamber or a blood vessel or by surrounding a vessel with umbilical tape, or ligature, both double backed ends are brought through a lumen of short red rubber or plastic catheter, and the exposed loop are used to temporarily occlude the chamber of the heart or the lumen of the blood vessel (Rumel tourniquet).

Mindful of toxicity of iodine [**Bernard Courtois (1777 – 1838)**. 19th century. Biochemistry; **Fritz Pregl (1869 – 1930**: above], **Herman A. Shelanski (?1912 – c. 1954)**[1264] — American toxicologist of Jewish Lithuanian descent, developed in 1952-56 it analogue – 5%-10% alcohol solution which present complex of iodine with polyvinylpyrrolidone in the concentrate of iodine in the complex 0,1-1%, under the name of povidone- iodine (PVP-I) solution (in Latin – Povidonum-iodum, a.k.a., iodopovidone, contain 1% available iodine in water alone, forming mostly triiodides, which oxidizes water to release ions that directly act on bacterial or viral membrane proteins to exhibit microbiocidal effect. He discovered that the complex was less toxic in mice than tincture of iodine. The PVP-I solution is used for topical application to prevent and treat wound infection and skin disinfection before and after surgery. A buffered PVP-I solution of 2,5% concentration can be used for prevention on neonatal conjunctivitis, especially if caused by Neisseria gonorrhea, or Clamydia trachomatis. Also, it found application in pleurodesis for fusion of the pleura because of incessant pleural effusion.

The trademark for PVP-I is a betadine, a.k.a., povidone-iodine, which consist of 10% PVP-I, a standard topical antiinfective and antiseptic agent used to destroy microbes. In comparison to PVP-I, it is less likely to sensitize or sting the affected area. It is also soluble in water and will not permanently stain clothes. However, it may adversely affect normal cell activity and tissue repair and the potentially negative wound healing remains controversial.

In the absence of any method of accurately visualizing the mitral valve (MV) acquired insufficiency (MI), **Robert P. Glower (1913 – 61)**[1265], **Julio C. Davila (1922 – 2016)** and **Robert G. Trout (1922 – 2013)** — American cardiac surgeons, described the concept of annular size reduction by circumferential suture of the MV anulus in1955, which they used in 25 patients, claiming year later (1956) symptomatic benefit in half.

[1264] Cantor A, Schelanski HA. Furher Consideration of Germinal Capacity Testing. Pamphlet 1951. — Shelanski HA. US Patent 2,739,922 A (application / 1956 (grant). — Shelanski HA, Shelanski MV. PVP – iodine: history, toxicity, and therapeutic uses. J Int Coll Surg 1956;25(6):727-34.

[1265] Davila JC, Glover RP, Trout RG, et al Circumferential suture of the mitral ring: a method for the surgical correction of mitral insufficiency. J Thorac Surg 1955;30(3):531-60; discussion, 560-3.

Geoffrey H. Wooler (1911 – 2010)[1266] — English thoracic cardiac surgeon, performed (1957) and introduced (1962) mitral valve (MV) annuloplasty for MV insufficiency by placing two double-armed U-fashioned sutures enforced on both sides by pledged, approximating the anterolateral and posteromedial commissures to the posterior MV anulus at the outer side of posterior (P_1 and P_3) segments of the MV leaflet (Wooler mitral annuloplasty).

Repair of mitral insufficient (MI) according to technique of **E.B. Kay**[1267] — American thoracic cardiac surgeon, include the P_2 and P_3 segments of the MV leaflets annuloplasty/ plication (1955-58), and **George E. Reed**[1268] — American thoracic cardiac surgeon, which consist of combining the Kay's P_2 and P_3 segments MV leaflets annuloplasty/ plication with its own modification of placing the horizontal mattress sutures enforced by pledged in the mitral annulus without involvement of the middle scallops of the septal and mural leaflets (1965).

Separately, **Jerome H. Kay (1921 – 2015)**[1269] — American thoracic cardiac surgeon, reported (1965) on the posterior (Kay) tricuspid valve (TV) annuloplasty for TV insufficiency, at the level of the base of the posterior leaflet, using two double armed U-fashioned sutures enforced on both sides by pledged, doubling the posterior anulus over the leaflet with or without it excision. This technique was adopted from the MV annuloplasty by **Geoffrey Wooler (1911 – 2010**: above**)** and **E.B. Kay (**above**)**.

Introduced in 1957 by **H.T. Nichols**[1270] — American thoracic cardiac surgeon, popularized in 1991, by **Ottavio Alfieri**[1271] — Italian thoracic cardiac surgeon, the repair of the mitral valve (MV) insufficiency by placing one to three bow-like stitches, or clips, approximating edges of regurgitating jet of the anterior (A) and posterior (P) leaflets of the MV in the middle of the valve (e.g., segment A_2-P_2) or para-commissural area (e.g., segment A_1-P_1, or segment A_3-P_3), creates a double orifice repair (the «bow-tie» or the Alfieri stitch). Two years later, H. Nichols reported on 93 patients, treated by his technique with 27% mortality. The Nichols-Alfieri technique is suitable for the use in conventional, mini-invasive cardiac surgery (MICS) or percutaneous (PC) repair of the MV regurgitation in high-risk patients.

[1266] Wooler GH, Nixon PG, Grimshaw VA, et al. Experiences with the repair of the mitral valve in mitral incompetency. Thorax 1962;17:49-57.

[1267] Kay EB, Cross FS. Surgical treatment of mitral insufficiency. Surgery 1955;37(5):697-706. — Kay EB, Nogueira C, Head LR, et al. Surgical treatment of mitral insufficiency. J Thorac Surg 1958 Nov;36(5):677-90.

[1268] Reed GE, Tice DA, Clauss RH, et al. Asymetric exaggerated mitral annuloplasty repair of mitral insufficiency with hemodynamic predictability. J Thorac Cardiovasc Surg 1965;49:752-61.

[1269] Kay JH, Maselli-Campagna G, Tsuji KK. Surgical treatment of tricuspid insufficeincy. Ann Surg 1965;162:53-8.

[1270] Nichols HT. Mitral insufficiency: treatment by polar cross-fusion of the mitral annulus fibrosus. J Thorac Surg 1957;33(1):102-22.

[1271] Alfieri O, Moisano F, De Bonis M; et al. The double-orifice technique in mitral valve repair: a simple solution for complex problems. J Thorac Cardiovasc Surg 2001;122:674-81.

The staging system TNM = T (primary tumor or neoplasm), N [regional lymphatic nodes (LN)] and M (distant metastasis) of the malignant solid tumors was developed in 1943-52 by **Pierre Denoix (1912 – 90)** — French surgeon, according to the size and spread of the primary tumor, involvement of the regional LN and the presence of distant metastases. It was designed with the aim to provide the uniform classification of malignant neoplasm, and on that base to establish the most probable prognosis, to choose the most effective treatment for each stage of neoplasm, to assure a proper follow-up of the recovery or progression of malignant disease, evacuate modalities used to treat carcinoma of the lung and esophagus, malignant pleural mesothelioma (MPM) and other malignant tumors. It was accepted by Union internationale contre le cancer (UICC, founded in 1933), Geneva, Switzerland.

According to the International TNM Staging System (7th Ed., 2002)[1272] a Non-Small Cell Lung Cancer (NSCLC) is classified into:

Primary tumor (T)

TX   –   primary tumor cannot be assessed, or tumor proven by the presence of malignant cells in sputum or bronchial washings but not visualized imaging or bronchoscopy.

T0   –   no evidence of primary tumor.

Tis   –   carcinoma in situ.

T1   –   tumor ≤ 3 cm in the greatest dimension, surrounded by lung or visceral pleura, without bronchoscopic evidence of invasion more proximal than the lobal bronchus, i.e., not in the main steam bronchus (MSB).

T1a   –   tumor ≤ 2 cm greatest dimension.

T1b   –   tumor > 2 cm but ≤ 3 cm in greatest dimension.

T2   –   tumor > 3 cm but ≤ 7 cm, or tumor with any of the following features (T2 tumors with these features are classified T2a if ≤ 5 cm), involves MSD ≥ 2 cm distal to the carina, invades visceral pleura, associated with atelectasis or obstructive pneumonitis that extends to the hilar region but does not involve the entire lung.

T2a   –   tumor > 3 cm but ≤ 5 cm in greatest diameter.

T2b   –   tumor > 5 cm but ≤ 7 cm in greatest diameter.

[1272] Moroz ГС. Carcinoma of the lung. Ed. - LYa Kovalchuk, VF Sayenko, HV Knyshov. Clinical Surgery. Vol. 1-2. Publ. «Ukrmedknyha», Ternopil 2000. Vol. 1. P. 74-85. — Park BJ, Rusch VW Lung cancer workup and staging. In, FW Sellke, PJ del Nido, SJ Swanson (Eds). Sabiston & Spencer Surgery of the Chest. Vol. I-II. 8th Ed. ElsevierSaunders, Philadelphia 2010. P. 241-52. — Anraku M, Keshavjee S. Lung cancer: surgical treatment. Ibid. P. 253-77.

T3   –   tumor > 7 cm or one that directly invades any of the following: chest wall [including superior sulcus tumor (SST)], diaphragm, phrenic nerve, mediastinal pleura, parietal pericardium; or tumor in the MSB < 2 cm distal to the carina but without involvement of the carina, associated with atelectasis or obstructive pneumonitis of the entire lung or separate («satellite») tumor nodule(s) in the same lobe.

T4   –   tumor of any size that invades any of the following: mediastinum, heart, great vessels, trachea, recurrent laryngeal nerve, esophagus, vertebral body; SST with invasion of the vertebral body and spinal canal, or surrounding subclavian vessels, upper branches of the branchial plexus at the level of the cervical (C_8) or above; cytologically malignant pleural (or pericardial) effusion, ipsilateral separate / multiple node(s) inside the same lung or other lobe.

Regional lymph nodes (LN)

NX   –   regional LNs cannot be assessed.
N0   –   no regional LNs metastasis.
N1   –   metastasis in ipsilateral peribronchial and/or ipsilateral hilar LNs and intrapulmonary nodes, including involvement by direct extension.
N2   –   metastasis in ipsilateral mediastinal and/or subcarinal LNs.
N3   –   metastasis in contralateral mediastinal, contralateral hilar, ipsilateral, or contralateral scalene, or supraclavicular LN(s).

Distant metastases (M)

MX   –   distant metastasis cannot be assessed.
M0   –   no distant metastasis.
M1   –   distant metastasis:
M1a   –   separate tumor, or nodule in the contralateral lobe of identical histology to the primary tumor; tumor with pleural nodules or malignant pleural (or pericardial) effusion.
M2b   –   distant metastasis.

TNM Stage			Treatment
IA: T1a - T1b,	N0,	M0	Surgical + adjuvant (postoperative) platinum based combined chemotherapy (paclitaxel, carboplastin)[1273].
IB: T2a	N0	M0	Ibid.
IIA: T2b	N1	M0	Surgical + adjuvant simultaneous radiochemotherapy.
IIB: T2 - T4	N1	M0	Ibid. In SST – neoadjuvant (preoperative) simultaneous radiochemotherapy

+ *en block* resection.

IIIA: T1 - T4	N1- N2	M0	Ibid.
IIIB: T1 - T4,	N2-N3	M0	Simultaneous chemotherapy.
IV: T1 - T4,	N0-N3	M1, M1a, M1b	Chemotherapy and palliative treatment.

Subsequently, open and thoracoscopic segmentectomy was increasingly applied to early pulmonary cancers (<1 cm in size) without involvement of LNs ($T_1N_oM_o$) to patients with advanced age and marginal cardiopulmonary function.

In 1955 **Charles H. Sheldon**[1274], used the term percutaneous tracheostomy (PCT) and described the method as an alternative to the standard incisional route. Subsequently, in 1985 **Pasquale Ciaglia (1912 – 2000)**[1275] — American physician introduced percutaneous dilatational tracheostomy (PDT). The only advantage of PCT over the standard surgical tracheostomy is lessened cost.

In 1949 **John L. Madden (1912 – 99)**[1276] — American surgeon, reported two patients with atrial fibrillation (AF) and rheumatic mitral valve (MV) disease in whom he performed left atrial appendage (LAA) exclusion (excision) to prevent blood clot formation and peripheral arterial embolization. Currently, LAA exclusion is used to treat patients with AF at high risk for embolic stroke (10-20%) and limited options during cardiac surgery utilizing suture / stapling technique, or various percutaneous devices.

Clinical feature and radical surgical treatment of the superior sulcus tumor (SST), a non-small cell carcinoma of the apex of the upper lobe of the lung, accounting for less than 5% of all bronchial cancers, are influenced by their location.

[1273] Include, regardless of the size of tumor, segmentectomy, lobectomy, pneumonectomy, or radical pneumonectomy with a dissection of the mediastinal LNs (mediastinal lymphadenectomy).

[1274] Sheldon CH, Pudenz RH, Freshwater DB, et al. A new method for tracheostomy. J Neurosurg 1955;12:428-31.

[1275] Ciaglia P, Firsching R, Syniec C. Elective percutaneous dilational tracheostomy; preliminary report. Chest 1985;87:715-9.

[1276] Madden JL. Resection of the left auricular appendix – a prophylaxis for recurrent arterial emboli. JAMA 1949;140:769-72.

In 1978 **B.T Finucane** and **Hilton L. Kupshik (1950 –)**[1277] — American anesthesiologist, introduced the concept of using initially a central venous catheter, then a flexible introducer (stiletto, stylet) for replacing or exchanging an orotracheal tube during difficult «blind» intubation. Subsequently, that stiletto was given an eponym the Eschmann stylet (Eschmann Equipment, established 1830; West Sussex, UK) or the Gum elastic bougie.

Currently, most tracheal injuries are caused by a flexible stylet (Eschmann) «blind» oral intubation. Usually, diagnosis is delayed about 2 hours and injuries involves the membranous wall of the trachea or tracheobronchial junction, and the average laceration size is about 3 cm. Cervical injuries to the trachea are most common (50%), followed by intrathoracic trachea and bronchi. Subcutaneous emphysema and/or minor hemoptysis presents the usual clinical picture. A short, partial thickness membranous tracheal wall laceration can be treated conservatively with antibiotics and repositioning of the inflated endotracheal (ET) tube cuff distal to the tear. Stretching the area of injury and with the endotracheal tube cuff and high inspiratory ventilator pressure must be avoided. Concomitant esophageal injury require close observation and monitoring are essential, as surgical repair may be necessary. Sex and size have not been identified as determinants of mortality in these iatrogenic injuries. Clinical examination and endoscopic follow-up are essential.

In contrast, traumatic rupture of the tracheobronchial tree is a life-threatening situation. In blunt trauma the associated injuries may limit diagnosis and treatment options for major airway injury, which must be temporized by the tube repositioning and minimizing pulmonary barotrauma. The nature of the airway injury, the severity of subcutaneous emphysema, pneumothorax and/or pneumomediastinum, and concerns about developing mediastinitis, all contribute to risk assessment and correct decision making. Most major airway injuries from blunt chest trauma occur within 3 cm of the carina, and tears and avulsions can be extensive. The most frequent is longitudinal complete tear of the trachea. Repair requires adequate exposure, usually through right posterolateral thoracotomy, and provision for flexible options in airway control during the operation. Requirements are a minimal dissection, primary suture repair and the use of viable muscle flaps for coverage. The early extubation is desirable, but not always possible in patients with multiple injuries.

Will C. Sealy (1912 – 2001) — American cardiac surgeon, designed surgical approaches for the ablation of the conductive system of the heart (CSH) responsible for the Wolff-Parkinson-White (WPW) syndrome using the endocardial dissection of

---

[1277] Finucane BT, Kupshik HL. A flexible stilette for replacing damaged tracheal tubes. Can Anaest Soc J 1978;25(2):153-4.

the right free-wall (RV), anterior septal, posterior septal and left free-wall (LV) under cardiopulmonary bypass (CPB, 1968).

To owe **Oleksandr Mamolat [(1910 – 91)**. 19ᵗʰ century. Internal medicine], **Mykola (Nicolai) M. Amosov [06(19).XII.1913, village Olkhove, Cherepovetsky Region, Volohods'ka Obl., Russia – 12.XII.2002, city of Kyiv, Ukraine]**[1278] — Ukrainian chest (thoracic) and cardiac surgeon, became the pioneer of thoracic and cardiac surgery in Ukraine and one of the pioneers in the World, who in 1952-68 became the chief of chest surgery at the Institute of Tuberculosis and Chest Surgery (ITCS), now National Institute of Phthysiatry and Pulmonology (NIPhP) of F.H. Yanovsky, introduced mechanical suturing using staplers into thoracic surgery (1957), constructed medical computers (1960), designed and implanted cuffed caged-ball cardiac valves in the mitral position (1965). In 1983-88 established headed the Institute of Cardio-vascular Surgery (ICVS), since 2003 – National ICVS of M.M. Amosov.

The first practical artificial knitted vascular grafts were developed by **Walter M. Golaski (1913 – 96)** — Polish American engineer and inventor. It was a micro-knit polyethylene-telephtalate (PET, PETE), or Dacron arterial synthetic tubular graft (STG) – Golasky vascular prostheses (?1962).

An example of the fusion of general and cardiac surgeon was **William P. Longmire (1913 – 2003)**[1279] — American surgeon, who introduced a partial hepatectomy with intrahepatic cholangiojejunostomy for the treatment of biliary obstruction due to benign and malignant biliary obstruction (Longmire procedure, 1945) and coronary endarterectomy (CE) for the treatment of angina pectoris due to localized occlusive arteriosclerosis (AS) of coronary arteries on beating heart without cardiopulmonary bypass (CPB), i.e., off pump (OP) (1957-58). The first patient was a 52-year-old man on whom he performed on Dec. 20, 1957, through the right intercostal space thoracotomy

---

[1278] Village Olkhove [re-settled somewhere else 1937 or 1939, since 1941-47 immersed in the bottom of the Rybinsky Reservoir, constructed with water resistant structures, blocking the riverbed of Volga and Sheksna Rivers (height 102 m, dimensions 172 x 56 km, area 4,580 km², average depth 5,6 m, maximal 30 m, volume 25.4 km³)], Cherepovetsky Region, Volohods'ka Obl. (area 144,527 km²), Russia. — Amosov NM. Surgical Treatment Purulent Diseases of the Lungs. Publ. «Derzhmedvydav» of the Ukrainian Radianska Socialistic Republik, Kyiv 1956. — Amosov NM. Pneumonectomy and Lung Resection in Tuberculosis. Publ. «Medgiz», Moscow 1957. — Amosov NM. The Outlines of Thoracic Surgery. «Derzhmedvydav» Ukrainy, Kyiv 1958 — Olearchyk A. Successes of Ukrainian physicians in surgicak treatment of tuberculosis. Nasza Kultura (Warszawa) 2.1959; 2(10):14. — Amosov NM, Berezovsky KK. Pulmonary resection with mechanical suture J Thorac Cardiovasc Surg 1961;41:325-35. — Amosov NM, Sidarenko LN. The first experience in surgical treatment of mitral insufficiency. The Surgery of the Heart and Blood Vessel. «Derzhmedvydav» Ukrainy, Kyiv 1965. — Amosov NM. The Open Heart. Simon & Shuster, New York 1966. — Ablitsov VH. «Galaxy Ukraine». The Ukrainian diaspora: prominent persons. Publ. «Kyt», Kyiv 2007. — Hanitkevych YaV. The contribution of the Ukrainian physicians to the world medicine. Ukr Med Chasopys 2009 VII/VIII;4(72):110-15.

[1279] Longmire Jr WP, Cannon JA, Kattus AA. Direct-vision coronary endarterectomy for angina pectoris. N Engl J Med 1958 Nov 20;259(21):993-9.

(ICST) endarterectomy (EA) of the right coronary artery (RCA) and through the left ICST EA of the left anterior descending (LAD) artery. The patient was doing well 9 months after the operation. In total, he performed similar operations in 5 patients with one operative death (OD).

The objective aspects of metabolism in surgical patients were defined by **Francis D. Moore (1913 – 2001)** — American surgeon, in «Metabolic Response to Surgery» (1952) and in «Metabolic Care of the Surgical Patients» (1969). In 1969, he gave a detailed description of an acute respiratory distress syndrome (ARDS) in adult.

The new technique of repair of an iatrogenic[1280] perforation of the lower third of the thoracic esophagus was introduced in 1974 by **Krishna V.S. Rao**[1281] — Indian American surgeon, **Mahmood Mir** — Iranian American surgeon and **Charles L. Cogbill (1913 – 95)** — American military surgeon. It consists of closing the edges the esophageal perforation with the use of a pedicled flap of the diaphragm.

A surface hypothermic circulatory arrest (HCA) for cardiac surgery in children was introduced by **Wilfred G. Bigelow (1914 – 2005)** — Canadian cardiac surgeon, in 1950, who confirmed the initial observation of **Volodymyr Walther (1817 – 89**: 19th century. Physiology) about the decreased need of the body for oxygen during hypothermia and thus began its use in thoracic cardiovascular surgery (TCVS), and two years later (1952) by **Floyd J. Lewis (1916 – 93)**[1282] — American cardiac surgeon and **Mansur Taufic (1915 – 2005)** — American surgeon, for closure of an atrial septal defect (ASD) in a 6-year-old girl using a surface HCA and inflow occlusion of the venous blood into the right atrium (RA).

Beginning in 1949 **John A. Hopps (1919 – 98)** — Canadian electrical engineer, working with **Wilfred Bigelow** (above) and **John Callagan (1923 – 2004)** — Canadian cardiac surgeon, developed by the 1951, a prototype of combined external implantable pacemaker (PM) with a cardiac defibrillator. The device resembled a small table radio (30 cm in length), used vacuum tubes to generate pulses, and was powered by a 60-Hz household electric current, connected by developed by him transvenous catheter with electrodes, which could be passed through the external jugular vein (RJV) for the contact with the endocardium of the inferior wall of the

---

[1280] Term iatrogenia / iatrogenic (in Greek – «iatros» meaning healer and «genesis» meaning origin, or «brought forth by the healer») was introduced in 1925 by **Oswald Bumke (1877 - 1950)** — German neurologist and psychiatrist. — Bumke O. Der Artz als Unsache seelischer Strörungen. Deutsche Medicinische Wochenschrift 1925;51(1):3. — It denotes the worsening of physical or emotional condition of the patient provoked by healthcare professionals, specifically by physician, pharmaceutist, nurse, dentist, psychologist, psychiatrist, medical laboratory scientist, or therapeutant.

[1281] Rao KVS, Mir M, Cogbill CL. Management of perforations of the thoracic esophagus. A new technic using a pedicle of diaphragm. Am J Surg 1974 May;127(5);609-12.

[1282] Lewis FJ, Taufic M. Closure of atrial septal defect with the aid of hypothermia: experimental accomplishments and the report of the one successful case. Surgery 1953;33:52-9.

right ventricle (RV), without the need to open the chest cavity. However, the device was large and unhandy, tested on a dog, sometimes causing burns of the patient's body[1283].

William G. Cagan (1914 – 2001)[1284] — American thoracic surgeon, proposed four criteria for radiation-induced sarcoma (RIS) to be considered radiation induced: (1) the second neoplasm must arise in the irradiated field and be proven histologically; (2) the latent period of at least several years should elapse between radiation exposure and development of the second neoplasm; (3) there must histological and radiological evidence of the pre-existing condition, and its non-malignancy in addition to microscopic proof of a tumor; and (4) the second tumor must be histologically different from the first tumor.

In breast cancer patients the incidence of RIS following mastectomy and chest wall radiation is approximately 0.2% in 10 years, or 1.6% in 17 years.

William W.L. Glenn (1914 – 2003) — American cardiac surgeon, described the one-sided «end-to-end» superior vena cava (SVC) to the distal right pulmonary artery (PA) shunt which was created for the palliative treatment of transposition of the great arteries (TGA) with stenosis of the main pulmonary artery (MPA) – Glenn shunt (1954).

In 1951 **Charles Dubost (1914 – 91)** — French surgeon, resected abdominal aortic aneurysm (AAA) and interposed synthetic tubular grafts (STG).

In this same year (1951) **Raul Carrea** made attempt to do a carotid endarterectomy (CEA). the first successful repair of thoracoabdominal aortic aneurysm (TAAA) performed in 1955 by **Samuel N. Etheredge (1914 – 2011)** — American vascular surgeon, performed in 1955 the first successful repair of thoracoabdominal aortic aneurysm (TAAA).

The first successful intracardiac repair of a persistent truncus arteriosus (PTA) was performed in 1962 by the group lead by **Herbert Sloan (Oct. 10, 1914, Clarsburg, WW, USA – May 17, 2013, Chelsea, MI, USA)**[1285] — American cardiothoracic surgeon, using a valveless polytetrafluoroethylene (PTFE) conduit for the replacement of the main pulmonary artery (MPA).

Great vessel of the thoracic cavity accounts for a significant amount of blunt or penetrating trauma which are associated with a high mortality. Of those the most lethal is a high-speed blunt collision decelerating horizontal chest trauma with tear of

---

[1283] Hopps J. The development of the pacemaker. Pacing Clin Electrophysiol 1981;4:106-8. — Bigelow WG. The pacemaker story: a cold heart spin-off. Can Med Assoc J 1984 Oct 15;131(8):943-55. — Bains P, Chatur S, Ignaszewski M, et al. John Hopps and the pacemaker: a history and detailed overview of devices, indications, and complications. British Columbia Med J (BCMJ) 2017 Jan Feb;59(1):22-8.

[1284] Cahan WG, Woodard HQ, Higinbotham NL, et al. Sarcoma arising in irradiated bone: report of eleven cases, 1948. Cancer 1998;82:8-34.

[1285] Berhenadt DM, Kirsh MM, Stern A, et al. The surgical therapy for pulmonary artery-right ventricular discontinuity. Ann Thor Surg 1974;18:122-37.

the proximal median descending thoracic aorta (DTA) at the ligamentum arteriosum. Fortunately, usually the tear is an extra-pleural, and hemorrhage may be temporarily contained by the surrounding pleura.

Since in the most cases there is no external sign of chest trauma, the rapid diagnosis requires a high index of suspicion and prompt confirmation by the classic signs on the plain postero-anterior chest radiography, as described by **Marvin M. Kirsh** and Herbert Sloan which is accurate in 95% of cases[1286].

Unfortunately, intrapleural hemorrhage is usually lethal at the accident scene, an extra-pleural hemorrhage could any time break through surrounding pleura into the pleural cavity. Overall mortality for traumatic rupture of the DTA is about 40%.

Those classic plain chest x-ray signs are a widened mediastinum of more than 10 cm, loss of aortic knob contour, shift of the trachea and endotracheal tube (ET) to the right and nosogastric tube (NGT) to the left, elevation of the left main stem bronchus (LMSB) and depression of the right main stem bronchus (RMSB), apical capping, first rib fracture, acute left-sided hemothorax, and a retrocardiac density. Spiral computed tomography (CT) scanning with angiography has 96.2% sensitivity and 99.2% specificity for blunt traumatic aortic abnormalities.

The treatment of traumatic rupture of the DTA consist of emergency the left 4th intercostal space thoracotomy (ICT), repair of a tear with interposition of a short synthetic tubular graft (STG), occasionally with cardiopulmonary bypass (CPB), or left atrio-femoral artery shunt[1287]. Patients suffering concomitant head injury with loss of consciousness or with fractured pelvic bones should undergo repair by an intraluminal aortic graft stenting.

Continuing the leadership in the faculty of surgical stomatology at the Kyiv Institute for Physicians Educational Perfection, **Yurii Jo. Bernadsky (1915 – 2006)** — Ukrainian stomatologist and maxillary-facial surgeon, published in Ukrainian instructional textbook – «Surgical Stomatology» (Kyiv 1975).

After nine years of research **Zachary B. Friedenberg (1915 – 2011)** and **Carl N. Brighton (1931 – 2019)** — American orthopedic surgeons, and **William J. Redka**

[1286] Mattox KL. Approaches to trauma involving the major vessels of the thorax. Surg Clin North Am 1989 Feb;69(1):77-91. — Wintermark M, Wicky S, Schnyder P. Imaging of acute traumatic injuries of the thoracic aorta. Eur Radiol 2002;12:431-42.

[1287] Olearchyk AS. Complex cases in cardiac surgery. Angiology 1993 ;44,7:S 1-42. — Olearchyk AS. Traumatic tear of the proximal descending thoracic aorta. In, AS Olearchyk. A Surgeon's Universe. Publ. «Ms», Lviv 2003. P. 200-1. — Olearchyk AS . Traumatic tear of the proximal descending thoracic aorta. In, AS Olearchyk. A Surgeon's Universe. 2nd Ed. Publ. «Ms», Lviv 2006. P. 304-5. — Olearchyk AS. Traumatic tear of the proximal descending thoracic aorta (DTA). In, AS Olearchyk. A Surgeon's Universe. Vol. 3. 3rd Ed. Book Publ. «AuthorHouse», Bloomington, IN 2011. P. 948. — Olearchyk AS. Traumatic tear of the proximal descending thoracic aorta (DTA). In, AS Olearchyk. A Surgeon's Universe. Vol. 2. 4th Ed. Publ. «Outskirts Press», Denver, CO 2016. P. 680-1.

(05.05.1932, Ukraine – 08.14.2021, Williamstown, N.J., U.S.A.)[1288] — Ukrainian American inventor, pioneered in 1971 healing of the mal-union[1289] bone fracture by direct electric current stimulation from 5 to 20 milliampere.

A 51-year-old woman fractured her ankle in a fall of 1969, treated by immobilization with a cast (alabaster – calcium sulphate ($CaSO_4$), gypsum or plaster of Paris] for 13 weeks complicated by non-healing due mal-union. In February 1971 a small wire cathode was inserted into non-healed fracture under local anesthesia, an aluminum (aluminum) grid anode was taped to the skin, a cast was applied, the wires from cathode and anode were brought through cast to a small power source of batteries, resistors and transistors to the cast. Then 10 microamperes of the direct electrical current were applied for nine weeks, caused the bone pieces to grow together. She used crutches for two weeks and thereafter was walking unaided.

Carl Brighton also contributed to the understanding of endochondrial (in Greek – «endochondral» for «within cartilage») ossification during fetal (fetal) development of the mammalian skeletal system by which the bone tissue is created, the other being intra-membranous ossification. The endochondrial ossification of the cartilage is present during the rudimentary formation of long bone takes place[1290], during the growth of the long bone length[1291], and the natural healing of bone fractures[1292].

On Apr. 20, 1951, **Dan Gavriliu (1915 – 2012)** — Romanian surgeon, performed the first total resection and replacement of damaged cancerous esophagus in the human using a gastric tube at least 5 cm wide in diameter, based on the greater curvature of the stomach and supplied with the blood by the gastroepiploic artery.

In 1948 **C. Rollins Hanlon (1915 – 2011)** — American cardiac surgeon and **Alfred Blalock (1899 – 1964**; above) designed an atrial septectomy for palliation of the transposition of the great arteries (TGA).

[1288] Friedenberg ZB, Harlow MC, Brighton CT. Healing of nonunion of the medial malleolus by means of direct current. J Trauma 1971 Oct;11(10):883-5. — Brighton C, Friedenberg Z, Redke W. (inventors). Constant current power pack for bone healing and method of use. US Patent 3,884,2841A, 1971-10-29 – 2019-11-22. USA as represented by the Secretary of the Navy, Washington, DC Oct. 22, 1974. — Carl T. Brighton, Jonathan Black, William Redka (inventors). US Patent and Trademark No. 4549547. Filled July 27, 1982, Date of Patent Oct 29,1985. — Brighton CT, Pollack SR. Electromagnetics in Medicine and Biology. 1st Ed. San Francisco Pr, San Francisco 1991. — Town of Williamstown (established 1737), NJ, USA. — Ablitsov VH. «Galaxy Ukraine». The Ukrainian diaspora: prominent persons. Publ. «Kyt», Kyiv 2007.

[1289] Mal-union bone fracture (in French – «mal» for disease, disorder, evil).

[1290] Netter FH. Musculoskeletal System: anatomy, physiology, and metabolic disorders. Ciba-Geigy Corp., Summit, NJ 1987.

[1291] Brighton CT, Sugioka Y, Hunt RN. Cytoplasmic structures of epiphyseal plate chondrocytes; quantitative evaluation usig electron micrographs of rat costochondral functions with specific reference to the fate of hypetrophic cells. J Bone Joint Surg 1973;55-A:771-84.

[1292] Brighton CT, Hunt RM. Histochemical localization of calcium in the fracture callus with potassium pyroantimonate: possible role of chondrocyte mitochondrial calcium in callus formation. J Bone Joint Surg 1986;68-A:703-15.

Hryhoriy H. Hovorenko [21.I(3.II).1915, village Pishchana, Talnivsky Region, Cherkasy Obl., Ukraine – 11.X.1986, city of Kyiv, Ukraine][1293] — surgeon, who improved surgical treatment of pulmonary diseases with accompanying by diabetes mellitus (DM) and amyloidosis, worked out methods of cavernotomy, cavernoplasty, resection of lung after collapse therapy[1294], and surgical treatment of disseminated form of tuberculosis, and other disorders of the lung.

In 1957 **Ake Sennig (1915 – 2000)** — Swedish cardiac surgeon, developed an atrial switch operation for correction of the transposition of the great arteries (TGA) which consists of the creation using an interatrial septum (IAS) and the wall of the right atrium (RA) of two interatrial channels for crossing the oxygenated pulmonary veins blood return to flow through the mitral valve (MV) to the right ventricle (RV) and the ascending aorta (AA) for the systemic arterial circulation and the caval nonoxygenated venous blood return via the tricuspid valve (TV) to the left ventricle (LV) and the main pulmonary artery (MPA) for the pulmonary blood oxygenation (Sennig operation). Year later (1958) **Rune Elmqvist (1906 – 96**: above), constructed electrical pacemaker (PM), electrodes of which were attached by Ake Senning through a thoracotomy to the myocardium of the heart. The device failed after three hours, but a second device was then implanted which lasted two days.

In 1963 the Sennig operation was modified by **William T. Mustard (1914 – 87)** — Canadian cardiac surgeon, who created an intra-atrial baffle using the pericardium or synthetic patch to direct systemic and pulmonary blood flow into the right ventricle (RV) and left ventricle (LV), respectively (Mustard operation).

Peter T. Smylski (1915, city of Dauphin, Manitoba, Canada – 01.10.2002, city of Toronto, ON, Canada)[1295] — Ukrainian Canadian oral cavity surgeon, introduced the panoramic radiography into the practice of oral surgery (1967) and a method of correcting malformations of the mandible (1967-77).

In 1937 **Vladimir Demikhov (1916 – 96**: above) constructed and applied in dogs the first model of the total artificial heart (TAH), in 1937-55 – biventricular support (BVS) which became a prototype of all present models of a TAH. He carried out a heterotopic heart transplantation (HT), of the pulmonary lobe and of heart-lung

---

[1293] Village Pishchana (founded before 1914; suffered from conducted by Russian Communist Fascists regime of Holodomor-Genocide of the Ukrainian people in 1932-33 and 1946-47), Talnivsky Region, Cherkasy Obl., Ukraine. — Ablitsov VH. «Galaxy Ukraine». The Ukrainian diaspora: prominent persons. Publ. «Kyt», Kyiv 2007.

[1294] Olearchyk AS. Left-sided extrapleural thoracoplasty for pulmonary tuberculosis. J Ukr Med Assoc North Am 2001;46,1(146): 20-25.

[1295] City of Dauphin; founded 1898), Province of Manitoba (MB, area 647,797 km²), Canada. — Smylsky P. Panoramic radiography. J Ukr Assoc North Am 1968 Jul;15,3(50):21-6. — Smylsky P. Surgery for mandibular malformations. J Ukr Med Assoc North Am 1969 Jan-Apr;16,1-2(52-53):26-37. — Smylsky P. Facial asymmetry and oral surgery. J Ukr Med Assoc North Am. 1980 Summer;27,3(98):141-6. — Ablitsov VH. «Galaxy Ukraine». The Ukrainian diaspora: prominent persons. Publ. «Kyt», Kyiv 2007.

transplantation in dogs without cardiopulmonary bypass (CPB, 1946), HT, the heart-lung and of lung transplantation using staplers (1946-55), transplantation of an isolated lung (from 1947) and orthotopic HT in dogs without CPB (since 1951). Proved that in principle an orthotopic HT in the warm-blooded animal is possible with survival of several hours.

Professor dr med. **Józef Grenda (18.II.1916, village Czarkowy, Busko Region, Swiętokrzyskie Voidoship, Poland** – died from cerebral stroke with unilateral hemiplegia **06.I.1990, city of Warsaw, Poland**; Foot Notes between 1032 & 1043)[1296] — Polish general surgeon, **Wadym Kiesz (1923 – 2010)** — Polish internist and **Andrzej Olearczyk** (Foot Notes 1, between 87 & 88, 1032, 1043) — Ukrainian Polish anesthesiologist and general surgeon, combined in a 26-year-old man with an undescended right testicle, a large retroperitoneal seminoma with metastases to the mediastinum and bilateral pleural effusion, a preoperative 2-month chemotherapy, then through the median laparotomy removal of the retroperitoneal seminoma with lymphadenectomy (Oct. 06, 1961), and postoperative radiation to the mediastinum (end Oct. - Dec. 1961), with long-term survival[1297].

Still before the emergence of an open-heart surgery in 1951, **Charles A. Hufnagel (1916 – 89)**[1298] — American cardiac surgeon, designed the first mechanical heart valve that bears a close resemblance to a bottle stopper. The valve consisted of a pea-size ball synthetic (methylmetacrylate) inside a cage, that looked like a chambered metal tube – a 3,5 cm long and a 2,5 cm wide. The free-moving synthetic ball in the cage was to dislodge during the left ventricular systole (LV) systole allowing the blood to flow forwards, then during left ventricle diastole to fall back closing the cage, thus preventing regurgitation of the blood into the LV. This valve, named Hufnagel valve, became the prototype of all subsequent mechanical ball-caged heart valves. They consist of silk ring, with the attached cage, formed by curved excavations where the freely floating ball allows the passage of the blood, or closing a hole during diastole, to prevent regurgitation.

He implanted this valve, Sep. 1952, in a 30-year-old woman, suffering from aortic valve (AV) regurgitation due to rheumatic heart disease (RHD), in the proximal descending thoracic aorta (DTA), that is in non-anatomical position. Nevertheless,

---

[1296] Village Czarkowy (founded before 1250), Busko Region, Swiętokrzyskie Voivodeship (area 11,710.50 km²), Poland. — Shevchenko IT. Coming to a Head Questions of Chemotherapy of Malignant Tumors. Novyy khir archiv 1962;2:3-10. — Grenda J, Kiesz W, Olearchyk A. Nasieniak jądra z przerzutami do opłucnej. Pol Przegl Chir 1963;35(2),11a:1265-9.

[1297] Grenda J, Kiesz W, Olearchyk A. Nasieniak jądra z przerzutami do opłucnej. Pol Przegl Chir 1963;35(2),11a:1265-9.

[1298] Hufnagel CA, Harvey WP, Rabil PJ, et al. Surgical correction of aortic insufficiency. Surgery 1954;35:673-83.

this was the first mechanical heart valve replacement in a human, and since then a cardiac valve replacement became a generally accepted surgical procedure.

Charles Hufnagel also was the first to diagnose chronic pulmonary thromboembolic disease (TED) and treat it surgically by pulmonary thromboendarterectomy (PTEA, 1962)[1299]. The etiology of TED is incomplete resolution of an acute pulmonary embolectomy (PE), and in some cases previous splenectomy, permanent intravenous (IV) catheters, ventriculo-atrial shunt for treatment of hydrocephalus, chronic inflammatory conditions, sickle cell anemia, hereditary sphereocytosis, and described by **Maurice Klippel (1858 – 1942)** — French internist, and **Paul Trenaunay (1875 – ?)** — French physician, congenital capillary or cavernous hemangiomatous malformation of the aortic valve (AV) and/or lymphatic system, characterized by unilateral hypertrophy of the bone with the surrounding soft tissues, large cutaneous hemangiomas, permanent port-wine stain naevus and skin ulcerations (Klippel-Trenaunay syndrome, 1900).

Olexandr (Oleh) Z. Nechiporenko (Apr. 21, 1916, town of Malyn, Zhytomyr Obl., Ukraine – Nov. 05, 1980, town of Grodno, Belarus)[1300] — Ukrainian urologist, first to introduce resection of the urinary bladder (cystectomy) for cancer, removal of lymphatic glands (lymphadenectomy) in malignant ovarian tumors; described a test to measure the quantity of red, white, and cylindrical blood cells in 1 ml of urine (Nechiporenko's test, 1969).

Alike in the sign of **Paul Rosenstein (1875 – 1964:** above), **Mykola B. Sitkovsky (02. IX.1916, village Nemyrivka, Korostens'ky Region, Zhytomyrs'ka Obl., Ukraine – X.2003, city of Kyiv, Ukraine)**[1301] — Ukrainian pediatric surgeon-proctologist, paid attention that after turning the patient from the back to the left side, the pain in the right iliac region increases due to dislocation of the caecum and the appendix towards the left and pulling the mesentery of the later (Sitkovsky symptom).

A successful operation to correct a patient's carotid artery atherosclerotic (AS) blockage was performed in 1954 by **Felix Eastcott (1917 – 2009)** — Canadian

---

[1299] Jamieson SW, Kapelanski DP, Sakakibara N, et al. Pulmonary endarterectomy: operative experience and lessons learned in 1500 cases. Ann Thorac Surg 2003;76:1457-64. — Stuart W. Jamieson, MD interviewed by Joel Dunning, MD. Giants of Cardiothoracic Surgery. CTSNet Feb. 12, 2012.

[1300] Town of Malyn (old town in ruin from 8th-9th century, first mentioned in 891; repressions and conduction by Russian Communist Fascist regime of Holodomor-Genocide 1932-33 against the Ukrainian nation destroyed part of residents of the town and the region), Zhytomyr Obl., Ukraine. — Town of Grodno (founded 1128), Belarus. — Nechiporenko AZ. Leukocyte and erythrocyte counts in 1 ml of urine. Lab delo 1969;2:121. — Ablitsov VH. «Galaxy Ukraine». The Ukrainian diaspora: prominent persons. Publ. «Kyt», Kyiv 2007.

[1301] Village Nemyrivka (known since 945, founded 1783), Korostensky Region, Zhytomyr Obl., Ukraine. — Kontsevyy OO. Acute appendicitis. In, LYa Kovalchuk, VF Sayenko, HV Knyshov. Clinical Surgery. Vol. 1-2. Publ. «Ukrmedknyha», Ternopil, 2000. Vol. 2. P. 125-363. — Ablitsov VH. «Galaxy Ukraine». The Ukrainian diaspora: prominent persons. Publ. «Kyt», Kyiv 2007.

Australian vascular surgeon, by excision of the diseased part of the artery and then re-suturing the healthy ends together.

William Y. Inouye (1917 – 2004)[1302] — Native Indian American surgeon, devised a simplified artificial dialyzer and ultrafilter of a pressure cooker (1951).

In 1971 **Bernard Fisher (1918 – 2019)**[1303] — American surgeon, began a randomized clinical trial comparing total mastectomy and lumpectomy with or without radiation in the treatment of breast cancer. Then, in 1975, he reported the first data to show that postoperative chemotherapy could change the natural history of patients with primary operable breast cancer. His work led to subsequent research efforts that eventually overturned the paradigm of **William Halsted (1852 – 1922**: 19th century. Surgery**)** from 1882, of radical mastectomy as the standard treatment for breast cancer. Finally, in 1989 he established that manifested by a discrete mass or lump in the early stage of breast cancer could be more effectively treated by local removal or lumpectomy, in combination with radiation therapy, chemotherapy, and / or hormonal therapy, than by radical mastectomy.

Lung transplantation, i.e., full replacement of the diseases lung of one man (recipient) by the healthy lung taken from the other man (donor), that is lung allotransplantation, was undertaken for the first time by **James D. Hardy (1918 – 2003)**[1304] — American thoracic surgeon.

A 58-year-old man was admitted to the hospital with recurrent pneumonia unresponsive to antibiotics, and he also had emphysema both lungs, squamous cell carcinoma of the left lung with pleural metastases, and kidney disease. On June 11, 1963, he underwent through the left anterior lateral 4th intercostal space thoracotomy (ICST) an allotransplantation of the left lung. Died on the 18th postoperative day of renal failure (RF).

In general, the full or partial lung transplantation is performed for cystic fibrosis (CF) or mucoviscidosis, inherited disorder resulting in diffuse bronchiectasis destruction of both lungs; chronic obstructive pulmonary disease (COPD), notably idiopathic pulmonary emphysema and alpha$_1$-antitrypsin deficiency; septic lung disease; pulmonary fibrosis and restrictive lung disease, including idiopathic pulmonary fibrosis (IPF), sarcoidosis with pulmonary hypertension and obliterative bronchiolitis (OB); pulmonary hypertension secondary to congenital heart disease (CHD), or pulmonary artery (PA) thromboembolic disease; and primary pulmonary hypertension (PPH).

The reasons for lung transplantation in infants (<1 year) is CHD (32%), primary pulmonary hypertension (PPH), pulmonary vascular disease and pulmonary alveolar

---

[1302] Inouye WY, Engelberg J. A simplified artificial dialyser and ultrafilter. Surg Forum 1953 Oct;4:438-42.

[1303] Fisher B, Redmond, Poisson R, et al. Eight-year results of a randomized clinical triad comparing total mastectomy and lumpectomy with or without irradiation in the treatment of breast cancer. N Engl J Med 1989;320:822-8.

[1304] Hardy JD, Webb WR, Dalton Jr ML, et al. Lung homotransplantation in man. JAMA 1963;86(12):1065-74.

proteinosis, a disease in which an abnormal accumulation of pulmonary surfactants[1305] within the alveoli interfere with the lung's ability to exchange oxygen (O_2) from the air and carbon dioxide (CO_2) from the blood and congenital diaphragmatic hernia (CDH); in children between 1 and 10 years of life – CF (54%), PPH, CHD, pulmonary re-transplantation for OB; and in teenagers 11-17 years of life – CF (70%).

Indication for lung transplantation in COPD is destruction and hyperinflation, marked restriction of daily living activities, failure of maximal medical treatment to correct symptoms, and marked impairment of pulmonary function test (PFT), specifically of forced expiratory volume in 1 second (FEV_1) <35% of predicted normal values. Of the PFT an exercise test, i.e., maximum oxygen consumption with an exercise VO_2 max[1306] with a loading >15 ml/kg/min. is considered satisfactory, VO_2 max <10 ml/kg/min. – unsatisfactory.

Discriminating condition favoring lung transplantation are diffuse, homogeneous disease without target areas, hypercapnia with pulmonary artery (PA) CO_2 ($PaCO_2$) pressure >55 mmHg, pulmonary hypertension, age younger than 60 years, and alpha$_1$-antitrypsin deficiency.

Unilateral lung, single lobe or split-lung transplantation is done through the anterior lateral 4th ICST, in patients with significant cardiomegaly via the posterior median left 4th ICST; bilateral isolated sequential lung transplantation is through the bilateral anterior lateral 4th ICST, in patients with significant cardiomegaly via the posterior median left 4th ICST and anterior median right 4th ICST, or transverse bilateral 4th ICS sterno-thoracotomy without or with division of internal thoracic arteries (ITA), i.e., clam shell incision; en block double-lung transplantation through the transverse bilateral 4th ICS sterno-thoracotomy without or with division of the ITA (clam shell incision); and unilateral or sequential lung transplantation concomitant cardiac surgery via a median longitudinal sternotomy incision (MLSI).

In 1994 **Krisa P.** van **Meurs (1954 –)**[1307] — American neonatal pediatrician, reported on the first lobal lung transplantation for the treatment of CDH in a newborn child with success.

[1305] Pulmonary surfactants are a surface active compounds inside the microscopic air sacs (alveoli), built with membranous (phospholipids and proteins) and liquid (diluted in the water phospholipids) phases, which keep the surface stretch (tension) in the alveoli close to zero, that prevents them from sticking together at expiration, dissolve in themselves CO_2 and O_2 thus relieving its passing through the walls of the alveoli and capillaries, take part in the creation of elastic recoil of the lung, have bactericidal action.

[1306] Walsh GL, Morice RC, Putman JB, et al. Resection of lung cancer is justified in high-risk patients selected by exercise oxygen consumption. Ann Thorac Surg 1994;58:704–711. — Ferguson MK. Reeder LB, Mick R. Optimizing selection of patients for major lung resection. J Thorac Cardiovasc Surg 1995;109:275–283). — Sekela MB. Practical Thoracic Surgery. Publ «Logos». Lviv 2003. — Reilly JJ Jr. Preoperative Assessment of Patients Undergoing Thoracic Surgery. In, FW Sellke, PJ del Nido, SJ Swanson (Eds). Sabiston & Spencer Surgery of the Chest. Vol. I-II. 8th Ed. ElsevierSaunders, Philadelphia 2010. P. 39-45.

[1307] Van Meurs KP, Rhine WD, Benitz WE, et al. Lobal lung transplantation as a treatment for congenital diaphragmatic hernia. J Pediatr Surg 1994;29:1557-60.

The feasibility of an intra-aortic balloon assist (IABA) device placed in descending thoracic aorta (DTA) to augment coronary perfusion with balloon inflation during diastole was demonstrated by **Spiridon D. Moulopoulos**[1308] — Greek physician and by **Adrian Kantrowitz (1918 – 2008)**[1309] — American cardiac surgeons, with a technical support by **Arthur R. Kantrowitz (1913 – 2008)** — American scientist and engineer, in three patients, all were in cardiogenic shock, but improved during balloon pumping, one survived to leave the hospital.

The device consists of a double-lumen catheter (7 to 9,5 F) with a 25-50 ml balloon at the distal end. The internal lumen monitors arterial blood pressure (BP) and the outer lumen allows the transfer of helium gas into the balloon. It uses a contra-pulsation to increase diastolic blood flow to coronary arteries and to mechanically reduce the afterload on the heart during systole in patients with cardiogenic shock, mechanical complications of myocardial infarction (MI) – acute mitral regurgitation (MR) and ventricular septal defect (VSD), or in high-risk patients before coronary artery bypass (CAB) grafting.

The first successful blood cross-circulation was performed on March 26, 1954, by **C. Walton Lillehai (1918 – 99)**[1310] — American cardiac surgeon, whereby the father was used as a living pump and oxygenator, being connected via the artery and the vein up nearby to his 13-month-old son during a repair of a ventricular septal defect (VSD). Unfortunately, child died 11 days later of suspected pneumonia. Despite this, he repaired another atrial septal defect (ASD) in an infant with the use of the cross-circulation technique this same year. In the next year (until July 19, 1955), they continued to use the cross-circulation technique for a total of 43 open-heart operations in infants, including the first repair of a complete atrioventricular canal (AVC) and tetralogy of Fallot (TOF). Operative mortality (OM) was 17 (38.8%) children[1311]. During the 53-year follow-up 20 (44.4%) adults are alive and well.

The original description of complete AVC repair by Walton Lillehei involved attaching of the aortic valve (AV) leaflets to the interventricular septum (IVS), and the ASD was closed primarily. In its modified single-patch technique, several pledged horizontal mattresses sutured are brought through the crest of IVS, the AV valves, the autologous pericardial patch, and a narrow strip of Dacron, in that order.

[1308] Moulopoulus SD, Topaz S, Kolff WJ. Diastolic balloon pumping in the aorta: mechanical assistance to the failing heart. Am Heart J 1962;63(5):669-73.

[1309] Kantrowitz A, Tjonneland S, Freed PS, et al. Initial clinical experience with intra-aortic balloon pumping in cardiogenic shock. JAMA 1968;203:135-40.

[1310] Olearchyk AS. A Surgeon's Universe. Vol. I-IV. 4th ed. Publ. «Outskirts Press», Denver, CO 2016. Vol. I. P. 553, 569-70.

[1311] Moller JH, Shumway SJ, Gott VL. The first open-heart repairs using extracorporeal circulation by cross-circulation: a 53 year follow-up. Ann Thor Surg 2009;88(3):1044-6.

Subsequently, Walton Lillehei reintroduced **F.H. Pratt** technique **(1898**. 19th century. Physiology**)** for a retrograde coronary sinus (CS) myocardial protection using a crystalloid cardioplegia during replacement of calcific stenotic aortic valve (AV, 1955)[1312]. The development of a retrograde cardioplegia via the CS grew of concern that the existed technique of myocardial protection may not be effective unless the entire heart is perfused.

Introduced a bubble oxygenator for the heart-lung machine (HLM), reported on mitral valve (MV) annuloplasty for its insufficiency, by placing heavy silk sutures in the dilated area of the MV anulus (1957).

In 1965-68 at the laboratory of Walton Lillehai, University of Minnesota (founded 1851), city of Minneapolis (settled by Europeans and Americans beginning 1680), State of Minnesota (MN; 225,163 km²), USA, **Bhagavant R. Kalke (Nov. 24, 1927, village Kasheli, Ratnagiri District, State of Maharashtra, India – July 13, 2016, city of Mumbai, State of Maharashtra, India)**[1313] — Indian cardiac surgeon, developed contemporary models of an artificial mechanical bi-leaflets cardiac (heart) valves. His design was created on basis of the bi-leaflets irrigation (watering) dams, built with of reeds and woods in India. With the high water tide (in German – Der Flut), the flood-gates or the flood-valves were closing, with the low water tide (in German – Die Ebbe), the flood valves were opening for letting the water flow away down to the irrigation canals.

The valves and rings of the Lillehai-Kalke (1965) and Kalke-Lillehai (1968) cardiac valves were made of titanium. The first bi-leaflet artificial mechanical Kalke-Lillehai heart valve was clinically implanted in 1968, however unfortunately the patient died two days later from unknown reason.

The Kalke-Lillehai bi-leaflet cardiac valve was improved in 1976-77 by **Demetre M. Nicoloff (Aug. 31, 1933, town of Lorein, State of Ohio, USA – Aug. 05, 2003, city of Minneapolis, MN)**[1314] — Macedonian American cardiac surgeon, with participation of **Cris Possis (? - 1993)** — American bioengineer (Possis Medical,

---

[1312] Lillehei CW, DeWall RA, Gott VL, Varco RL. Direct vision correction of calcific aortic stenosis by means of pump-oxygenator and retrograde coronary sinus perfusion. Dis Chest 1956;30:123-32

[1313] Village of Kasheli, District of Ratnagiri (area 7,152 km²), State of Maharashtra (area 307,731 km²), India. — Saxena P, Konstantinov IE. Bhagavant Kalke and his pioneering work on the bi-leaflet heart valve prosthesis. Ann Thorac Surg 2009;88(1):344-7. — Olearchyk AS. A Surgeon's Universe. Vol. I-IV. 4th Ed. Publ. «Outskirts Press», Denver, CO 2016. Vol. I. P. 569-70. — City of Mumbai (first known permanent settlements 600 BC), State of Maharashtra, India.

[1314] Town of Lorain (founded 1807), State of Орайо (OH; area 116,096 km²), USA. — Olearchyk AS. A Surgeon's Universe. Vol. I-IV. 4th Ed. Publ. «Outskirts Press», Denver, CO 2016. Vol. I. P. 569-70. — City of Mumbai (first known permanent settlements 600 BC), State of Maharashtra, India.

Inc., Minneapolis, MN), under the name of St Jude valve (SJV), St Jude Medical, Inc. (established 1976), St Paul, MN, USA[1315].

On March 10, 1977, at the Minnesota University Hospital, Minneapolis, MN, Demetre Nicoloff performed in 67-year-old woman with atherosclerotic (AS) AV the first successful aortic valve replacement (AVR) using SJV. The patient survived 11 years after the operation, at which time she died from renal failure (RF).

The leaflets and ring of the SJV are made of graphite coated with pyrolytic carbon, and for that reason thus much harder, that the blood probably less adheres to its surface. However, all mechanical heart valves are sensitive foreign bodies, and after its implantation it is necessary continuously daily per oral use of anticoagulation drug, namely warfarin natrium (coumadin), to thin the blood. On the account of SJV hardness needed doses of anticoagulation drugs to thin the blood are relatively low, and this decreases the incidence of oozing or bleeding. Moreover, the SJV function more quietly than other heart valves.

Because of the marvelous hemodynamics, a long-term durability and relatively low incidence of thrombotic complications the KCI remains the best and the most common used by cardiac surgeons, artificial mechanical heart valve in the world. As of 2016 it was implanted in more than 1,3 million patients.

The foundations of the arch electric welding of the metal were made in 1802-03 by **Vasyl Petrov (1761 – 1834**: 19th century. Physics and radiology**)**, and in 1882 by **Mykola Benardos (1842 – 1905**: 19th century. Physics and radiology**)**.

However, the design of the arch electric welding of the soft tissue, i.e., the method of anastomosis of soft tissues during surgical procedures with the help of the high frequency electric current, in essence the bipolar electrocoagulation, was push forward by **Borys Ye. Paton (27.XI.1918, city of Kyiv, Ukraine – 19.VIII.2010, Ibid)** — Ukrainian scientist in the field of electric welding, metalurgy and technology of metals, from 1953 director of the E.O. Paton Electric Welding Institute of the National Academy of Science of Ukraine (NASU)[1316], worked off with the participation of its numerous employees – developers (engineers and surgeons) in experiments on animals in 1992-97 pp., successfully in clinics on patients in y 2001-15, including

---

[1315] St **Judas (Jude**) or **Thaddeus Lebbaeus (1st century AD**) — one of the apostles of Jesus Christ (Prehistoric Medicine), dedicated at the end of the 18th century, first in France and Germany, as the patron of desperate or hopeless accidents, cases or situations. — City of St Paul (founded 1849 on the site of the previous Indian settlement and embankment), MN, США.

[1316] **Yevhen O. Paton (Mar. 4, 1870, Nice, France – Aug. 12, 1953, Kyiv, Ukraine)** — Ukrainian scientist in the field of electric welding, father of Borys Paton, the founder and first director of the Electric Welding Institute of the NASU (1934-53). — City of Nice (the first known hominid settlements from 400.000 years ago; archeological site of the earliest use of fire, construction of homes and flint finding in around the year of 230.000; founded 350 BC), France. — Ablitsov VH. «Galaxy Ukraine». The Ukrainian diaspora: prominent persons. Publ. «Kyt», Kyiv 2007.

digestive tract anastomosis[1317]. The electric welding system includes a power unit comprising a power source (high frequency coagulator) with an adaptive automatic control and special software, bipolar welding tools (forceps, clamp, and laparoscopes) connected to a power source, and special assembly devices. The control is based on feedbacks. The tissue layers being joined are brought into contact over their surface layers by means of the welding tool. The surgeon clamps the tissue to be welded by the electrodes of the tool and switches on the welding current source (safe temperature in area of welding is 60-70^{0}C). Upon completion of the process, i.e. thermal denaturation of albumin molecules, control program power is turned off. Clamped tissue is the released and process repeated until complete wound closure. The device PATONMED$^{(R)}$ – EKVZ -300 (2019) has been tested in multiple experimental trials and on more than 2000 patients in the clinics and hospitals of Kyiv, Ukraine. The advantage of the device is in the formation of an attractive, smooth, thin, sutureless and neatly welded digestive tract anastomoses. It is a fumeless and odorless technology, causing no burns to surrounding tissues, shortens the average operative time, reduces blood loss, causes no organ deformation or stenosis. The electric welding of soft tissues showed promising outcomes in the construction of the safe sutureless digestive tract anastomosis, including a difficult low colorectal anastomosis.

In 1972 **Olexandr O. Shalimov (20.01.1918, village Vvedenka, Lypetsk Region, Lypetsk Obl., Russia – 28.02.2006, city of Kyiv, Ukraine)**[1318] — Ukrainian surgeon of Russian descent, headed created on basis of the 6th Kyiv Hospital, the Institute of Clinical and Experimental Surgery (ICES), since 2000 – National Institute of Surgery and Transplantology (NIST) of O.O. Shalimov, since 2006 incorporated into the National Academy of Medical Sciences (NAMS) of Ukraine. Under guidance of Oleksandr Shalimov, his son **Serhiy O. Shalimov (14.08.1943, village Nerchynsk Region, Nerchynsky Zavod Region, Chytynsk Obl., Russia –)**[1319] — Ukrainian

---

[1317] Paton BE. Electric welding of soft tissue in surgery. Paton Welding J 2004;9:6-10. — Ho Y-H, Ashour AT. Technique for colorectal anastomosis. World J Gastroenterol 2010Apr 7;16(13)1610-21. — Paton BE. «We developed the new technology of high frequency electric welding of the life tissues ». Den' (day.kyiv.ua) 21 Nov. 2011. — Ablitsov VH. «Galaxy Ukraine». The Ukrainian diaspora: prominent persons. Publ. «Kyt», Kyiv 2007.

[1318] Village Vvedenka, Lypetsk Region, Lypetsk Obl. (area 25,100 km^2), Russia. — Shalimov AA. Diseases of the Pancreatic Gland and its Surgical Treatment. Publ. «Meditsina», Moscow 1970. — Shalimov AA, Khokhola VP. About Surgeons and Surgery. Publ. «Zdorov'ia», Kyiv 1980. — Shalimov AA, Sayenko VF. Surgery of Digestive Tract. Publ. «Zdorov'ia», Kyiv 1987. — Shalimov AA, Shalimov AA, Nychytaylo ME, Radzykovsky AA. Surgery of the Pancreatic Gland. Publ. «Tavryd», Symferopol 1997. — Shalimov AA, Radzykovsky AA. Atlas Operations on Digestive Organs. Vol. 1. Publ. «Naukova Dumka», Kyiv 2003. — Shalimov AA, Radzykovsky AA. Atlas Operations on Digestive Organs. Vol. 2. Publ. « Naukova Dumka», Kyiv 2005. — Ablitsov VH. «Galaxy Ukraine». The Ukrainian diaspora: prominent persons. Publ. «Kyt», Kyiv 2007.

[1319] Village of Nerchynsk (founded 1926), Nerchynsk Zavod (Factory) Region, Chytynsk Obl. (area 431,500 km^2), Russia. — Ablitsov VH. «Galaxy Ukraine». The Ukrainian diaspora: prominent persons. Publ. «Kyt», Kyiv 2007.

surgeon, performed in 04 January 1982 transplantation of pancreatic gland (PG) in a 38-year-old man suffering from severe (brittle) insulin-dependent diabetes (IDD) mellitus. Unfortunately, by 2nd week postoperatively graft necrotized and was removed. Patient survived, and subsequently was treated with insulin. Since 1991 Serhiy Shalimov heads the Institute of Oncology (founded 1960) AMS of Ukraine.

One of the pioneers of thoracic and cardiac surgery was **Konstantyn K. Berezovsky (19.IV.1919, city of Kharkiv, Ukraine – 14.XI.1986, city of Kyiv, Ukraine)**[1320] — Ukrainian chest and cardiac surgeon, who together with **Mykola Amosov (1913 – 2002**; above**)** performed the first mechanical resections of the lung for tuberculosis and neoplasms using ushyvach korenia leheniv / sewer the root of the lung (UKL)-600 (1958), first in Ukraine implanted in 1963 Pacemaker (PM) / electrocardiostimulator (ECS) / artificial guide of the cardiac rhythm (AGCR), with Mykola Amosov conducted the first repairs of mitral valve (MV, 1970) and acquired heart diseases (AHD, 1973) under cardiopulmonary bypass (CPB).

On December 19, 1954, **Joseph E. Murray (1919 – 2012)** — American plastic surgeon, performed the successful renal transplantation on identical 23-year-old twins. He was assisted in this operation by **John H. Harrison (1909 – 84)** — American urological surgeon who performed harvesting of the kidney from the healthy patient-donor, subjected to an operation which was not for his own health benefit. The team performing this first successful kidney transplant was led by **John P. Merrill (1917 – died in a boating accident Apr. 14, 1984)** — American physician-nephrologist, leader of the team who developed an artificial kidney for the use in the treatment of acute and chronic renal failure.

In 1950 **Jaques Oudot (1919 – 53)**[1321] — French pharmacist, physician, surgeon, and alpinist, performed successful resection of an occluded bifurcation of the aorta by inserting a homograft.

Superior sulcus tumor (SST) located anterior to the anterior scalene muscle may invade the platysma and sternocleidomastoid (SCM) muscle, the external and anterior jugular veins, the inferior belly of the omohyoid muscle, the subclavian vein (SCV) and the internal jugular vein (IJV) including major branches, and the scalene fat pad.

[1320] Amosov NM, Berezovsky KK. Experience of 100 resections of the lung with the use of UKL-600. Experimental Surgery 1958; (6). — Olearchyk A. Successes of Ukrainian physicians in surgical treatment of tuberculosis. Nasza Kultura 2.1959; 2(10):14. — Amosov NM, Berezovsky KK. Pulmonary resection with mechanical suture. J Thorac Cardiovasc Surg 1961;41:325-35. — Amosov NM, Berezovsky KK. Surgical treatment of mitral diseases in the light of long-term results. Kardiologiia. 1970;10(3). — Amosov NM, Berezovsky KK. Some outcomes of 5000 operations for congenital heart disease. Klinichna hirurhiya 1973;(11). — Amosov NM, Berezovsky KK. Bi-momentary implantation of the endocardial electro-stimulator in cardiac blockade. Grudnaya khirurgiia 1982;(4). — Ablitsov VH. «Galaxy Ukraine». The Ukrainian diaspora: prominent persons. Publ. «Kyt», Kyiv 2007.

[1321] Oudot J. La greffe vasculaire dans les thromboses du carre-four aortique. Press Méd 1951;59:235-6.

SST located between the anterior and middle scalene muscle may invade the anterior scalene muscle and the phrenic nerve lying on its anterior surface, the subclavian artery (SCA) including primary branches, except the posterior scapular artery, the trunks of the brachial plexus, and the middle scalene muscle.

SST located posteriorly to the middle scalene muscle, usually spreads along the costovertebral groove, may invade the nerve roots of thoracic $(T)_1$, the posterior aspect of the SCA and vertebral artery (VA), the paravertebral sympathetic chain, the inferior cervical (stellate) ganglion, the prevertebral muscles, transverse processes, and bodies of the vertebra, the intraspinal foramen, the dura mater and spinal cord.

The surgical treatment of the SST was initiated in 1952 by **William M. Chardack (1915 – 2006**) — American thoracic-cardiac surgeon and pioneer in the development of the implantable cardiac pacemakers, and **John D. MacCallum**[1322] who performed upper lobectomy and chest wall excision followed radiotherapy with success. **Robert R. Shaw**[1323] treated the SST by the preoperative radiotherapy of 30-45 Gy in 4 weeks to the primary tumor, mediastinum, and supraclavicular area, followed by resection through the postero-lateral approach, i.e., the upper parascapular incision extending into the posterior thoracotomy. **Phillippe G. Dartevelle**[1324] — French thoracic surgeon, introduced the anterior transcervical-thoracic approach for a radical resection of the anterior and middle SST, and the combined anterior transcervical-thoracic and posterior approaches for a radical resection of the posterior SST.

On April 7, 1970, William Chardack (above), **Wilson Greatbatch (1919 – 2011)** — American engineer and inventor and **Andrew A. Gage (1922 – 2007)** — American physician, successfully implanted a permanent electrical completely internal (self-contained) pacemaker (PM) into 77-year-old man with complete heart block (CHB). PM was powered by non-rechargeable mercury-zinc oxide batteries for its energy source, driving a two transitory, convert (transform)-coupled blocking oscillator circuit. All were encapsulated in an epoxy resin in a small, flat plastic disc implanted into the body and connected to wires coupled to electrodes, sewn directly to the surface of the heart, emitting electric impulses to trigger heart's contractions. The patient died 30 months later of unrelated causes. In total the PM benefited 16 patients.

In 1953 **Henry T. Bahnson (1920 – 2003)** — American thoracic and cardiovascular surgeon and Himalayan Mountain climber, excised tangentially a saccular aneurysm of the thoracic aorta.

[1322] Chardack WM, MacCallum JD. Pancoast syndrome due to bronchogenic carcinoma: successful surgical removal and postoperative radiation. J Thorac Surg 1953;25:402-12.

[1323] Shaw RR, Paulson DL, Kee Jr JL. Treatment of the superior sulcus tumor by irradiation followed by resection. Ann Surg 1961;154(1):29-40.

[1324] Dartevelle PG, Capellier AR, Macchiarini P, et al. Anterior transcervical-thoracic approach for radical resection of lung tumors invading the thoracic inlet. J Thorac Cardiovasc Surg 1993;105:1025-34.

Resuscitative endovascular balloon occlusion of the aorta (REBOA) was introduced in 1954 by Lieutenant US Army **Carl W Hughes (?1923 – ?2007)**[1325] — American surgeon, for controlling subdiaphragmatic hemorrhage, most often after blunt torso injury, to improve cerebral and cardiac perfusion while simultaneously decreasing distal hemorrhage while improving hemodynamic stability.

To treat surgically diseased aortic root and the AA, particularly of the Marfan syndrome (19th century. Medicine), **Hugh H. Bentall (1920 – 2012)** and **A. DeBono**[1326] — English cardiac surgeons, introduced the classic operation of «inclusion» reconstruction (repair) by the insertion of composite valved synthetic tubular graft (STG) with reimplantation of coronary arteries by the «exclusion» or «button» technique and wrapping it with the remaining wall of an aortic aneurysm (Bentall operation, 1968). One attractive feature of the «button» technique is that all suture lines can be examined for ongoing hemorrhage at the completion of the repair. But the most common long-term complication from this type of composite root replacement is pseudoaneurysm formation around a coronary artery anastomosis. The wrapping of repair with the remaining wall of an aortic aneurysm has been abandoned.

By the definition development of the «sacral bedsore, decubitus or pressure ulcer means neglect and abuse of the patient», is epidemical and badly treated. This devastating condition was classified into the Braden Scale (Score or Stage)[1327] into:

Scale I – manifested by intact skin with a non-blanched redness;

Scale II – the sacral deep tissue pressure ulcer manifested partial thickness loss of dermis (skin) with a shallow open ulcer with red-pink wound bed without slough;

Scale III – a full thickness skin and subcutaneous tissue involvement down to the fascia of muscles covering the sacral bone; and

Scale IV – an infected sacral bedsore with full thickness tissue loss, including the fascia, muscles and tendons and invasion of the sacral bone, presence of slough and eschar.

[1325] Hughes CW. Use of an intra-aortic balloon catheter tamponade for controlling intra-abdominal hemorrhage in man. Surgery1954 July;36(1):65-8. — Darrabie MD, Croft ChA, Brakenridge SC, et al. Resuscitative endovascular balloon occlusion of the aorta: implementation and preliminary results at an academic level I trauma center. J Am Coll Surg 2018 Jul 227(1):127-34.

[1326] Bental H, de Bono A. Technique for complete replacement of the ascending aorta. Thorax 1968;23(4):338-9.

[1327] Developed in 1984-87 by **Barbara J. Braden (1920 - 2004)** and **Nancy Bergstrom** — American nurse-scientists. / Bergstram N, Braden BJ. Predictive validity of Braden scale among Black and White subjects. Nurs Res 2002 Nov-Dec;51(6):398-403.

There is always the possibility of developing a malignancy, i.e., squamous cell carcinoma within a bedsore – Marjolin's ulcer **[Jean N. Marjolin (1780 – 1850).** 19th century. Surgery].

In 1859 Florence Nachtingale **[Florence Nachtingale (1820 – 1910)**. 19th century. Health Welfare] said *«If he has a bedsore, it's generally not the fault the disease, but of the nursing».*

Prevention and treatment of bed sores (pressure ulcers) should include:

(1) Adequate високо протеїнове і калоричне з додатком вітаміну C, oral, total parenteral (intravenous) nutrition (TPN), total per-gastrostomy nutrition (PGN) or total enteric (via jejunostomy) nutrition (TEN).

(2) Bed sore preventing bed with flat smooth soft surface without folds, roughness (unevenness) or bundles, with innerspring, foam, or floated air mattresses; bedclothes with the base sheet, draws (drawing) sheet or the Comfort Glide® Sheet (Medline Industries, Inc., Northfield, IL) and Extrasorbs Air Permeable Drypad Underpads (ibid) to prevent a moisture associated skin damage (MASD) by timely disposing from a bed linen on emesis, urine and stool, leaking negligently connected intravenous (IV), TPN, PGN, or TEN lines / tube connections. Alternatively, one could use a sawdust bed, half filled with a soft, clean, preferably sterilized saw dust, from surface of which it is easy to remove excrements and replace sawdust.

(3) Changing pressure over the pertaining part of the body, that is the torso and other parts of the body, by turning patient every two (2) hours from the backwards (supine) position, to right, then left side (lateral) lying position, the program introduced in 1940s by Sir **Donald P. Guttmann (1899 – 1980)** — English German neurologist of Jewish descent, for prevention and treatment of paralytic (in Greek – «falling ill» or «weakening») stroke of both lower extremities (paraplegia).

(4) Patients should be moved carefully out of bed (OOB) with the help of two persons without / with the patient's mechanical lift **[R.R. Straton / Bernard Lown (1921 – 2021).** 20th century. Internal medicine] into a soft armchair three (3) times a day, or into a wheelchair. The bed and room should be cleaned every day, patient should have a sponge bath at bedside daily and a shower at least twice weekly.

(5) Systematic, thoughtful, and thorough фізіотерапія (T) should be available in all patients. The only contraindication for PT is a dead patient.

(6) Wound dressing should be done once a day under sterile precautions using fatty acid or silicone based creams, Hartmann® Argentum (Ag) / silver

atraumatic antimicrobial sterile reticular hydrophobic polyether, saturated with neutral triglyceride fatty acid ointment, well draped and penetrated by secretions (5 cm x 5 cm, 7.5 cm x 10 cm or 20 cm x 30 cm) bandages[1328].

The scale I-II bedsore is usually dressed by Desenex and Foam-Mepilex;

The scale II-IV bedsore is treated with collagenase Clostridium histolyticum (Santyl) ointment for enzymatic debridement (250 units/gram), when infected – the Dakin's solution **[Henry Dakin (1880 – 1952)**. 20th century. Biochemistry. / Surgery]**, or Derma Ginate Ag calcium (Ca^{++}) alginate wound dressing with antibacterial silver, both applied to the ulcer crater only, surrounding skin should be rubbed with Vitamin A&D ointment or protectant Z-guard paste using sterile covered by bordered gauze or sterile combine ABD pad secured to the surrounding healthy skin by pressure-sensitive 3 M Medipore H soft cloth surgical tape.

Beside of this, healing of the pressure ulcers can be accelerated by using recombinant human basic growth factor (rhbFGF) or transferrin.

Surgical treatment is indicated in septic patients for scale III-IV sacral bed sore, consist of wound debridement and an early gluteus maximus musculo-cutaneous rotational flap coverage.

Successful repair of post-infarction ventricular septal defect (VSD) was accomplished in 1956 by **Denton Cooley (1920 – 2016**: above**)** in a patient who survived several weeks after perforation of the interventricular septum (IVS) without cardiopulmonary bypass (CPB). Then, in 1958 he performed open lineal resection and closure of the post-infarction left ventricular (LV) aneurysm CPB.

He was the first to implant in 1969 into a 47-year-old man after the LV aneurysmectomy, complicated by an acute cardiac failure lasting 64 hours, a pneumatic total artificial heart (TAH), designed by **Domingo S. Liotta (1924 –)** — Argentinian experimental cardiac surgeon, as a «bridge» to transplantation (BTT). The TAH functioned for 32 hours, at which time the patient died of pneumonia.

Proposed in 1989 an endoaneurysmal repair (EAR) of the anterior LV aneurysms after myocardial infarction (MI) consist of the longitudinal incision of the aneurysm, covering of defect with using a synthetic circular patch, and wrapping it by the remaining wall of the aneurysm.

Charles G. Drake (1920 – 88) — Canadian neurosurgeon, reported in 1961, the first successful treatment of basilar artery aneurysm in the brain.

[1328] In 1818 **K. Ludwig F.** von **Hartmann (1766 - 1852)** — German industrialist, founded the Paul Hartmann Bleanching, Dyeing and Dressing of the wounds Company, which later under direction of his son **Paul Hartmann (1812 - 65)** has been and is being making a dressing material that brought fundamental breach in the wound care, especially since 1999 when the Hartmann® Ag bandages were introduced.

Henry J. Heimlich (1920 – 2016) — American thoracic surgeon of Jewish origin, invented the controversial chest drainage flutter valve which allows air and blood to drain from the chest cavity in order to allow a collapsed lung to re-expand while the patient is mobilized (Heimlich valve, 1962); and created the first aid life-saving maneuver for victims of suffocation to remove a foreign body, usually a bolus of foot from the throat or the trachea, and in saving drowning unconscious, and non-breathing victims to clear the water from the lungs, by a sudden application of five bimanual compressions (thrusts) the diaphragm on the upper abdomen by rescuers behind an upright or behind lying on a side patient (abdominal trust or Heimlich maneuver, 1974). This compresses the lung causing flow of air that carries any object or water to expel from the airway or lungs and out of the mouth.

Yurii (Rostyslav) T. Komorovsky (05.I.1920, village Svatoborice, Hodonint district, South Moravian krai, Czechia – 01.V. 2006, city of Ternopil, Ukraine)[1329]— Ukrainian surgeon, who proposed modification of primary and reconstructive gastroduodenal plasty, method of retrosternal plasty of the esophagus from combined injuries by with poisonous substances of the esophagus and the stomach, method of dissection of the large bowel band (tenotomy) with the aim of its lengthening for the esophageal plasty, modified through the mesentery creation of small intestinal folds (intestinal plication); designed modes for the risk evaluation from disease and operation in adults and old patients.

Reporting on the intravascular balloon occlusion of intracranial aneurysms in 120 patients, **Andriy P. Romodanov (11.XI.1920, town of Lubny, Poltavska Obl., Ukraine – 05.VII.1993, city of Kyiv, Ukraine)**[1330] and **Victor I. Shcheglov (1939 –)**[1331] — Ukrainian neurosurgeons, reported 73% rate of preservation of the parents vessels. Regarding the limits of balloon embolization, they stated that the endovascular operation was contraindicated in small aneurysms, in aneurysms with a wide neck, and in the acute phase after subarachnoid hemorrhage (SAH), in the presence of vasospasm, because of the 22% mortality rate among such patients.

[1329] Village of Svatoborice (founded 1349), Hodonint district, South Moravian krai (area 7,187.8 km^2), Czechia. — Komorovsky YuT. Surgical anatomy of the cervical division of thoracic duct. Dissertation for K.M.S., Lviv 1948. — Komorovsky YuT. Gastric resection using small intestinal plasty. Dissertation for D.M.S., Lviv 1964. — Ablitsov VH. «Galaxy Ukraine». The Ukrainian diaspora: prominent persons. Publ. «Kyt», Kyiv 2007.

[1330] Town of Lubny (Foot Note 200), Poltavska Obl., Ukraine. — Romadonov AP, Shcheglov VI. Intravascular occlusion of saccular aneurysms of the cerebral arteries by means of a detachable balloon catheter. In, H. Krayenbühl (Ed). Advances and Technical Standards in Neurosurgery. Vol. 9. Springer, Berlin 1982. P. 25-48. — Romadonov AP. Musiychuk MM, Tsymbaliuk VI. Neurosurgery. Publ. «Spalakh», Kyiv 1998. — Guglielmi G. History of the genesis of detachable coils. A review. J Neurosurgery 2009 July;111(1):1-8. — Tsymbaliuk IV. Academician Andriy Petrovych Romadonov – developer of the Ukrainian neurosurgery. Ukr Neurosurg J 2013;3:70-8. — Ablitsov VH. «Galaxy Ukraine». The Ukrainian diaspora: prominent persons. Publ. «Kyt», Kyiv 2007.

[1331] Ablitsov VH. «Galaxy Ukraine». The Ukrainian diaspora: prominent persons. Publ. «Kyt», Kyiv 2007.

Donald P. Shiley (1920 – 2010) — American biomedical engineer and inventor, designed number of medical devices, such as cardiac valves, tracheostomy, and endotracheal tubes for respiration.

In Nov. 1969 **Jay L. Ankeney (June 07, 1921, Cleveland, OH – Dec. 24, 2014, Naples, FL)**[1332] — American cardiac surgeon, performed coronary artery bypass (CAB) grafting without heart-lung machine (HLM), with a beating, locally stabilized heart. He also designed a retractor adapted for a midsternotomy incision used in cardiac operations (Ankeney retractor).

In 1960 **Paul J. Dzul (14.VIII.1921, village Mylne, Zborivsky Region, Ternopil Obl., Ukraine – 02.XI.2015, city of Detroit, MI, USA)**[1333] — Ukrainian American surgeon otorhinolaryngologist (ORL), initiated microsurgery of the middle ear, in 1967 proposed intracavitary treatment of maxillary cancer.

Development of a noninvasive behind (outside) the body (extracorporeal) shock-wave (striking-wave) lithotripsy (BSWL) have an interesting beginning, tied up to **Claude Dornier (1884 – 1969)** — German designer and producer of airplanes (aircrafts) who in 1911 founded Dornier Aircraft Factory (Dornier Flugzeugwerke), later its division – Dornier Medical Technical Laboratory, town of Friedrichshafen (established 1814), Germany. On its base he created in 1955 Dornier Medical Technical Company.

Independently, the credit for the developmental beginning noninvasive BSWL belongs to the pupil of **Andronyk Chayka (1881 – 1968**: above**), Yuriy H. Yedynyy**

[1332] Ankeney JL. Off-pump bypass surgery: the early experience, 1969-1985. Tex Heart Inst J 2004;31(3):210-3.

[1333] Village of Mylne (founded 1717), Zboriv Region, Ternopil Obl., Ukraine. — City of Detroit (founded 1701), MI, USA. — Dzul P. Endocarditis lenta. Diagnostic und Behandlung. Die Dissertation über Doctor der Medizin, Insbruk 1948. — Dzul PJ. Surgical treatment of deafness. J Ukr Med Assoc North Am 1962 Jan;9,1(24):8-13. — Dzul PJ. Treatment of otosclerosis in the past and now. J Ukr Med Assoc North Am 1962 Apr;9,2(25):2-5. — Dzul PJ. Menier's disease. J Ukr Med Assoc North Am 1964 Jan;11,2(33):10-2. — Dzul PJ. Carcinoma of the larynx. J Ukr Med Assoc North Am 1965 Apr;12,2(37):15-20,33. — Dzul PJ. Carcinoma of the Highmor's cavity. J Ukr Med Assoc North Am 1966 Oct;13,3-4(42-43):14-8,78. — Dzul PJ. Vasomotor rhinitis. J Ukr Med Assoc North Am 1967 Oct;14,4(47):16-22. — Benitez JT, Dzul PJ. Assessment of vestibular function by electronystegmography J Ukr Med Assoc North Am 1968 Jan;15,1(48):3-21. — Dzul PJ. Deafness in children. J Ukr Med Assoc North Am 1968 Jul; 15,3(50):3-10. — Dzul PJ. Tracheostomy. J Ukr Med Assoc North Am 1969 Jul-Aug;16,3-4(54-55):25-37. — Dzul PJ. Facial nerve paralysis. J Ukr Med Assoc North Am 1971 Jan-Apr;18,1-2(60-61):12-27. — Dzul PJ. Reconstruction of the sound conductive system of the middle ear. J Ukr Med Assoc North Am 1973 Jul;20,3(70):3-12. — Dzul PJ. Conductive deafness in children. J Ukr Med Assoc North Am 1974 Jul; 21(3): 25-40. — Dzul PJ. Diagnosis and managementof allergic rhinitis. J Ukr Med Assoc North Am 1983 Fall;30,1(108):134-146. — Dzul PJ. Rehabilitation of the paralyzed face. J Ukr Med Assoc North Am 1985 Winter;32,1(112):12-22. — Dzul PJ (Ed.). Contribution to the History of Ukrainian Medicine. Vol. II. Publ. Ukr Med Assoc North Am, Chicago IL 1988. — Dzul PJ. Personal communication 1988. — Dzul PJ. Hearing loss and methods of rehabilitation. J Ukr Med Assoc North Am 1989 Winter;36,1(119):3-16. — Truchly G, Dzul PJ. Arytenoid dislocation: literature review and case report. J Ukr Med Assoc North Am 2005 Summer; 50,1(153):27-32. — Ablitsov VH. «Galaxy Ukraine». The Ukrainian diaspora: prominent persons. Publ. «Kyt», Kyiv 2007.

a.k.a., **Yuriy G. Ediny (01.05.1921, u.t.v. Lysianka, Lysianka Region, Cherkasy Obl., Ukraine – 30.05.1991, city of Kyiv, Ukraine)**[1334] — Ukrainian physician-urologist, who first in the World designed and constructed in 1958-62 the «URAT-1» devise for noninvasive ultrasound (US) and electrohydraulic (striking-wave) crumbling (dispersion, pulverization) of the urinary bladder stones.

Yuriy Yedynyy organized in 1965 and headed until 1968 the Institute of Urology at the National Academy of Sciences (NAMS) of Ukraine (founded 1993), city of Kyiv, Ukraine.

Apparatus «УРАТ-1» became the prototype for all contemporary devices for extracorporeal lithotripsy, including the extracorporeal lithotripsy by shockwaves (ELSW, SW, HM-3 lithotripsic, Germany, 1983) for noninvasive treatment of kidney (ureteral) and hepatobiliary stones.

Afterwards, in 1972 physicists from the Dornier Medical Technical Company were able to prove that with the use of t shockwaves one can disperse into a small particle (pulverize) kidney stone.

Thereafter, in 07 February 1980 **Christian G. Chaussy (1945 –)**[1335] — German urologist, for the first time performed successfully the noninvasive BSWL of kidneys stones using the Dornier Medical Technical pulverize, Division of Urology at the Regensburg University (founded 1962), town of Regensburg (first settlements since Stone Age, founded in early 6th century), das Land Bayern (area 80.76 km²), Germany.

On average pulverization of kidneys stone require to use 800 – 2000 shock-walves.

The first successful noninvasive BSWL of the gallbladder stone was performed by **Ch. Ell**[1336]— German surgeon, using modified Dornier lithotripsic on 30.I.1985 in the Grossgadern Clinic of the Grossgaden Medical Center (opened 1999), town-district Hadern (first mentioned 11th century), Ludwig-Maximilians-Universität (founded 1472), city of München (first mentioned 1158), Bayern, Germany.

Treatment of bone fractures was improved on account of designing the distraction-compression apparatus for osteosynthesis, i.e., repositioning, and fixing bone fragments in 1950 by **Gavrilii (Gavriil) A. Ilizarov (Elizarov) (15.06.1921,**

[1334] U.t.v. Lysianka (first chronicle mention 1593), Lysiansky Region, Cherkasy Obl., Ukraine. — Yedynyy YuH. Congenital Incontinency of Urine in Women. Kyiv 1966. — Ablitsov VH. «Galaxy Ukraine». The Ukrainian diaspora: prominent persons. Publ. «Kyt», Kyiv 2007.

[1335] Chaussy C, Brendel W, Schniedt E. Extracorporally induced destruction of kidney stones by shock walves. Lancet 1980 Dec 13;2(8207):1265-8. — Chaussy C. Berührungsfreie Nierensteizertrümmerung durch extrakorporal erzeugte, focussierte Stosswellen. Karger-Verlag, Basel 1980. — Mialkovsky V. Pulverization of kidney stones using shockwalves. J Ukr Med Assoc North Am 1987 winter;34,1(115):16-23.

[1336] Ell Ch, Hochbberger J, Muller D, et al. Erste erfolgreiche endoskopisch-retrograd laser lithotripsie von Gallengansssteinchen am Menschen. Dtsch Med Wschr 1986;111:1217. — Ell Ch, Wax Hochberger J et al Lithotripsy of common bile duct stones. Gut Jun;1988;29(6):746-51. — Mialkowskyj W. Extracorporal shockwalve lithotripsy of gallstones. J Ukr Med Assoc North Am 1989 Winter;34,1(119):16-23.

village Bilovezha, Haynivsky County, Pidlashshia krai, Ukraine, Belorus, Poland – 24.07.1992, town of Kurhan, West Siberia, Russia)[1337] — Ukrainian Russian orthopedist of the Mountain Jews descent[1338], who studied in 1939-44 in the 1st Medical Faculty at Crimea State Medical University (CSMU), city of Simferopol, Crimea, Ukraine. During World War II (WW II) 1941-44, he was deported with all Crimean Tatars to town of Kyzylorda (founded 1820), Kazakhstan. After graduating from the CSMU as general physician, in 1948 became Director, since 1955 – Clinical Director of Surgical Department at the Kurhan Obl. Hospital and surgeon of the Aircraft Emergency Care, where he treated WW II veterans, suffering from open bone fractures, infected, purulent not healing, and infected complicated military wounds. In 1961 he established and headed the Russian Scientific Center «Restorative Traumatology and Orthopedics» (RSCRTO; new complex opened in 1971), until 1991.

He attained by working and popularized discovered in 1905 by **Alessandro Codivilla (1861 – 1912)** — Italian surgeon, justified by reasoning in 1926 by **Karl Wegner (1864 – 1940**: above**)** functional method of treatment fractured bones – the method distraction osteogenesis.

Research by Havrylo Ilizarov shoved that gentle and cautious cutting off (snipping) of the bone without disunion of the surrounding periosteum, slight separation of its half or parts, progressive stretching out in the vertical plane with speed 0,5-1 mm/day to maximal distance 12,5 cm, and then vigorous installation (mounting) allows the bone to regrowth, fill up the crevice, consistently lengthen and fully heal.

Practical use of the compression-distraction osteogenesis was inspired through observation of the horse harness with horse collar. With this example, Havrylo Ilizarov created in 1950 the orthopedic compressive-distractive device for throughout the bone osteosynthesis. It consists of circlets / rings with implanted holders, interconnected by parallel telescopic visible contracted (tightened) altered screw. To the holders are fastened X-like crossed straight steel wires (hooks) for reposition/fixation of an open, non-healed, infected, non-united, and other complex broken

---

[1337] Village Bilovezha (in Polish – Białowieża; existed since Neolithe Era, i.e., 3,000 BC, restricted right for hunting from 14th century, established 1699), located in South section of Bilovezhs'ka wilderness in domain of (river bed) of Rivulets of the West Buh, Niman and Pryp'iat' Rivers (general area 1,871 km², of which 1,771 km² belongs to Belarus, 100 km² belongs to Poland), Hainivsky county, Pisliashshia krai (Foot Note 693), UNR (1918-19) / Berestia Obl., Belarus. — Town of Kurhan (founded 1661), Kurhan Obl. (area 71,000 km²), Siberia, Russia. — Mirsky M. Ukrainian medical scientists – victims of Stalin's repressions. UHMJ «Ahapit» 1995;2:45-54. — Ablitsov VH. «Galaxy Ukraine». The Ukrainian diaspora: prominent persons. Publ. «Kyt», Kyiv 2007.

[1338] Mountain (North-Caucasus) Jews, Taty or Uguru settling since 4th-5th century East Caucasus. — Caucasus Mountains (highest peak Elbrus, 5,6 42 m), situated between Black Sea to Caspian Sea, divide Europe from Asia. Uguru live in Azerbaijan (area 86,600 km²) and Dagestan (area 50,300 km²), number (2004) - 150,000-270,000 person.

bones, its fragments, lengthening and restoration view of bones and arthrodesis of joints. Ilizarov device permits an early wearing of patient own weight and recovery.

Beginning 1960, **Julius H. Jacobson II (1921 – 2022)** — American surgeon, adapted the dissecting microscope to demonstrate that it was practical to reconstruct or to anastomose arteries as small as 1 mm in diameter with consistent patency.

The correctness of **William Harvey's (1578 – 1657**: 17th century**)** report that the right atrium (RA) «was the first to live, and the last to die» is supported by cardiac operations with the use of the cardioplegic solution to arrest the heart in the diastole, since during the antegrade infusion of cardioplegic solution to the ostia of coronary arteries or the retrograde infusion into the coronary sinus (CS), or both, the contractions of the RA ceases the last, and after absorption of the cardioplegic solution by the myocardium within 15-20 min., and contractions of the RA are first to return, indicating the need for a repeated infusion of the cardioplegic solution, for further safe continuation of the procedure. It is possible that the reason for this appearance is the location of the sino-atrial (SA) node, just at the junction of the superior vena cava (SVC) with the RA appendage in the terminal groove of the heart, that means in the closest vicinity to the RA.

To treat a morbid obesity, **Edward E. Mason (1920 – 2020)** — American bariatric surgeon, introduced in 1966, Roux-En-Y gastric bypass, where the stomach is reduced to a small pouch, which hold about one ounce (30 ml) of foot and the first two feet (61 cm) of the small bowel is bypassed.

In 1955 **Denis G. Melrose (1921 – 2007)**[1339] — South-African English clinical physiologist, introduced concept of «elective (cardioplegic) cardiac arrest» by rapid infusion into an aortic arch, after cross-clamping of the ascending aorta (AA) of a 2,5% solution of potassium citrate diluted in a tepid blood (Melrose solution). However, the Melrose method caused a severe necrosis of the myocardium.

Despite of this, experiments with cardioplegia has continued using mainly an intracellular and extracellular solutions. Therefore, cardioplegia was re-introduced into the clinic in 1964 by **Hans J. Bretschneider (1922 – 93)**[1340] — German physiologist and experimental cardiac surgeon, using the low-sodium, free of calcium and procaine-containing normothermic and cooled solutions.

[1339] Melrose DG, Dreyer B, Bentall HH, Baker JBE. Elective cardiac arrest. Lancet 1955;2:21-2. / Shiroishi MS. Myocardial protection: the rebirth of potassium-based cardioplegia. Tex Heart Inst J 1999;26(1):71-86.

[1340] Bretschneider HJ. Uberlebenzeit und Wiederbelebungzeit des Herzens bei Normo-und Hypothermie. Veh Disch Ges Kreislaufforsch 1964;30:11-34. / Shiroishi MS. Myocardial protection: the rebirth of potassium-based cardioplegia. Tex Heart Inst J 1999;26(1):71-86.

Then, **U. Kirsch**[1341] achieved an ischemic cardiac arrest using magnesium-aspartate-procaine solution (1972).

Finally, **David J. Hearse**[1342] — English physiologist and internist, recommended the use for cardioplegia based on extracellular solutions and outlined the basic components for the main formulation of St Thomas Hospital crystalloid cardioplegic solution No. 1 (KCl 20 mole/L; 1976), modified by him and **Mark V. Braimbridge**[1343] into St Thomas Hospital crystalloid cardioplegia No. 2 (КСl 16 моль/л; 1981). Since that time the extracellular cardioplegic solutions became standard and save method of cardiac arrest and myocardial protection during cardiac operations.

First synthetic tubular graft (STG) «Vinyon-N» was developed and used by **Arthur B. Voorhees, Jr. (1921 – 92)**[1344] — American vascular surgeon, **Alfred Jaretzky III (1919 – 2014)** — American vascular surgeon and **Arthur Blakemore (1897 – 1970:** above), initially on dogs (1950) to replace a segment of normal ascending aorta (AA, 1950), then in a human to replace abdominal aortic aneurysm, (AAA, 1953).

Accepted surgical treatment of congenital heart disease (CHD) with an upper partial anomalous pulmonary venous return (PAPVR) – sinus venosus type, in which the right superior pulmonary vein (RSPV) enters the superior vena cava (SVC), is characterized by a left-to-right shunt and if significant by cyanosis, was described in 1984 by **Herbert E. Warden (1921 – 2002)** — American cardiac surgeon. It consists of the lateral vertical incision of the right atrium (RA), transection of the SVC craniad to the entry of the RSPV, insertion of an intraatrial patch baffle to direct an atriocaval junction via a preexisting or created atrial septal defect (ASD) into the left atrium (LA), reimplantation of the transected SVC into the RA appendage and closure of the RA, thus committing the atriocaval junction exclusively to pulmonary venous blood flow (Warden procedure).

[1341] Kirsch U, Rodewald G, Kalmar P. Induced ischemic arrest. J. Thoracic Cardiovasc Surg 1972;63:121-30. / Shiroishi MS. Myocardial protection: the rebirth of potassium-based cardioplegia. Tex Heart Inst J 1999;26(1):71-86.

[1342] Hearse DJ, Stewart DA, Braimbridge MV. Cellular protection during myocardial ischemia: the development and characterization of a procedure for the induction of reversible ischemic arrest. Circulation 1976;54:193-202. / Ledingham SJV, Braimbridge MV, Hearse DJ. The St. Thomas Hospital cardioplegic solution. A comparison of the efficacy of two formulations. J Thorac Cardiovasc Surg 1987;93:240-6. / Shiroishi MS. Myocardial protection: the rebirth of potassium-based cardioplegia. Tex Heart Inst J 1999;26(1):71-86.

[1343] Braimbridge MV, Chayen J, Bitensky L, et al. Cold cardioplegia or continuing coronary perfusion? Report on preliminary clinical experience as assessed cytochemically. J Thorac Cardiovasc Surg 1977;74:900-6. / Ledingham SJV, Braimbridge MV, Hearse DJ. The St. Thomas Hospital cardioplegic solution. A comparison of the efficacy of two formulations. J Thorac Cardiovasc Surg 1987;93:240-6. / Shiroishi MS. Myocardial protection: the rebirth of potassium-based cardioplegia. Tex Heart Inst J 1999;26(1):71-86.

[1344] Smith RB III. Arthur B. Voorhees, Jr: pioneer vascular surgeon. J Vasc Surg 1993 Sep;18(3):341-8.

Warden procedure should be considered only in patient in whom the RSPV enters the SVC, otherwise it showed in 55% of them the loss of the normal sinus rhythm (NSR) which was replaced by a low atrial or junctional rhythm that did not increase normally with exercise[1345]. There is an increased risk of a new-onset atrial fibrillation (AF) after repair of ASD/PAPVR in older patients[1346]. Also, it is contraindicated in the presence of a persistent left SVC because of alternative venous blood flow and a smaller right SVC.

CHD with a low PAPVR in which the RSPV may enter the inferior portion of the RA, the inferior vena cava (IVC) – RA junction with stenosis (10% to 20%), or more commonly the IVC below the right hemidiaphragm, creating the appearance of the right-sided heart border suggestive of a Turkish sword, and therefore it is called scimitar syndrome. It is also characterized by a left-to-right shunt and cyanosis if significant, often associated with hypoplasia of the right lung, dextrocardia, systemic pulmonary arterial (PA) blood supply to the right lower lobe (RLL) of the lung, bronchial abnormalities, ASD, coarctation of the aorta (COA) and the left-sided SVC[1347].

An infantile form of scimitar syndrome can present with severe congestive heart failure (CHF), failure to thrive, tachypnea, elevated pulmonary artery pressure (PAP), e.g., >40% of the arterial systemic pressure, with a ratio of pulmonary blood flow (Qp) to systemic arterial blood flow (Qs), or Qp/Qs of up to 2:1, causing cyanosis, significant morbidity, and mortality[1348]. The clinical picture of scimitar syndrome in children and adults is usually mild.

Surgical repair of scimitar syndrome as designed by **Pablo Zubiate (1926 –)** — Peruvian American cardiac surgeon, and **Jerome Kay (1921 – 2015**: above) is to create unobstructive blood flow from the anomalous pulmonary vein (PV) to the LA, i.e., to place a long baffle in the lumen of the IVC to channel the PAPVD to the RA and then through an ASD to the LA[1349]. Another surgical approach is to divide the anomalous PV, reimplant it into the RA, and then create a baffle to divert

[1345] Stewart RD, Bailliard F, Kelle AM, et al. Evolving surgical strategy for sinus venosus atrial septal defect: effect on sinus node function and late venous obstruction. Ann Thorac Surg 2007;84:1651-5,

[1346] Cohen JS, Patton DJ, Giuffre EM. The crochetage pattern in electrocardiograms of pediatric atrial septal defect patients. Can J Cardiol 2000;16:1241-7.

[1347] Najm HK, Williams WG, Coles JG, et al. Scimitar syndrome: twenty years' experience and results of repair. J Thorac Cardiovasc Surg 1996 Nov;112(5):1161-8. — Huddleston CB, Mendeloff EN. Scimitar syndrome Adv Card Surg 1999;11:161-78. — Brown JW, Ruzmetov M, Minnich DJ, et al. Surgical management of scitimar syndrome: an alternative approach. J Thorac Cardiovasc Surg 2003 Feb;125(2):238-45.

[1348] Huddleston CB, Exil V, Canter CE, et al. Scitimar syndrome presenting in infancy. Ann Thorac Surg 1999 Jan;67(1):154-9.

[1349] Zubiate P, Kay JH. Surgical correction of anomalous pulmonary venous connection. Ann Surg 1962 Aug;156(2):234-50.

blood flow to the LA[1350]. In the third surgical approach the IVC is transected under deep hypothermic circulatory arrest (HCA) and divided transversely into posterior (anomalous PV) and anterior (IVC and hepatic vein) compartments. The posterior compartment of the IVC is sutured to the left atriotomy and the ASD is closed with pericardial patch. The anterior compartment of the IVC is anastomosed to the SVC[1351]. When the anomalous PV passed through posterior mediastinum and the right lung is hypoplastic the right pneumonectomy is justified[1352].

In a total anomalous pulmonary venous return (TAPVR) all four PV of the lung, right superior and inferior PV's (RSPV, RIPV), left superior and inferior PV's (LSPV, LIPV), instead draining into the LA, then into the left ventricle (LV) and systemic arteries, it enters the systemic venous system, creating a large left-to-right shunt, characterized by cyanosis and CHF. TAPVR occurs in 1% of all CHD, and is classified as supracardiac (45% to 55%), cardiac (15% to 20%), infracardiac (15% to 25%) and mixed (5%) types.

In the supracardiac type the PV's drain through a pulmonary venous confluence or collector behind the LA to a connecting short vertical vein to the innominate vein IV), left-or right-sided SVC. In the cardiac type, the PV's drain to a confluence, and this into coronary sinus (CS) or less frequently to the RA. In the infracardiac type the PV's drain into the pulmonary venous confluence which tend to be oriented more vertically, creating a Y-shaped confluence, then via a descending vertical vein into the portal vein of the liver, to the ductus venosus or directly to the IVC, rarely into the thoracic lymphatic duct. In the mixed type the PV drainage is bilateral and asymmetrical to multiple venous systems at the different level with different connections.

The necessary condition for saving the lives of a child is the presence of an open foramen ovale in the interatrial septum (IAS) or its defect.

The goal of the surgical correction of the TAPVD is to create unobstructed egress of an oxygenated blood from anomalous PV to the LA[1353].

In supracardiac and mixed types: (1) a longitudinal incision is created in the pulmonary venous confluence to match a corresponding incision in the posterior LA, extended out to the LA appendage and creation of an anastomosis between the LA and the pulmonary venous confluence; (2) a creation of an anastomosis through a generous atriotomy extending transversely across the RA and then across the IAS;

[1350] Schumaker Jr HB, Judd D. Partial anomalous pulmonary vein return with reference to drainage into the inferior vena cava and to an intact atrial septum. J Cardiovasc Surg (Torino) 1964 Jul-Aug;5:271-8.

[1351] Calhoun RF, Mee RB. A novel operative approach to scimitar syndrome. Ann Thorac Surg 2003 Jul;76(1):301-3.

[1352] Huddleston CB, Exil V, Canter CE, et al. Scitimar syndrome presenting in infancy. Ann Thorac Surg 1999 Jan;67(1):154-9.

[1353] Hancock Friesen CL, Zurakowski D, Thiagarajan RR, et al. Total anomalous pulmonary venous drainage connection: an anlysis of current management strategies in a single institution. Ann Thor Surg 2005 Feb;79(2):596-606.

and (3) construction of a PV-LA anastomosis between the ascending aorta (AA) and the SVC.

In the intracardiac type with a pulmonary venous confluence entering the CS, the IAS between the foramen ovale and the CS is incised, the superior aspect of the CS is unroofed until the atrial wall and CS becomes a common chamber without any separating ridge and glutaraldehyde treated or fresh autologous pericardial patch is used to reconstruct the IAS, leaving the PV oxygenated blood to flow freely through the unroofed CS into the LA. In newborn or infants with a small pulmonary venous confluence a «suture-less» technique is used, where a longitudinal incision is carried out in the posterior pericardium and the underlying pulmonary venous confluence, then a corresponding incision is made in the overlying portion of the LA and the LA edge is sutured to the pericardium circumferentially around the pericardial incision, avoiding direct suturing to the PV's, creating a «neo-atrium».

In the infracardiac type the incision into the LA is more vertical or Y-shaped to maximize the size of the nearly created LA.

On December 03, 1967 **Christiaan Barnard (Nov. 08, 1922, town of Beauford West, Western Cape Province, Union of South Africa** – died from a severe asthma attack while on holiday in **Sep. 02, 2001, ancient town of Paphos, Cyprus)**[1354] — South-African cardiac surgeon, carried out the first successful orthotopic cardiac (heart) transplantation (OHT / CT, HT) using the atrial, biatrial, or standard technique, in which the donor was a woman who died from head injury to the recipient who was a 54-year-old man, suffering from diabetes mellitus (DM) and incurable heart disease. Unfortunately, he died on the 18th day postoperatively from pneumonia.

First successful clinical OHT using bi-atrial technique, performed by Christiaan Barnard was based on the successful laboratory technique using cardiopulmonary bypass (CPB) and hypothermia, developed by **Richard R. Lower (1929 – 2008)**[1355] and **Norman E. Shumway (1923 – 2006)** — American cardiac surgeons, in the canine with survival up to 3-weeks after the operation. They achieved the long-term allograft (donor heart) survival in dogs by using immunosuppressive drugs – an antiproliferative agent (azathioprine) which inhibit lymphocyte replication through its suppressive action on the de novo and salvage pathways for purine biosynthesis, and corticosteroids after HT to combat acute rejection.

[1354] Town of Beauford West (established 1818), Western Cape Province (area 129,449 km²), Union of South Africa (area 1,221,037 km²). — Cyprus (area 9,251 km²). — Barnard CN. A human cardiac transplant: an interim report of a successful operation performed at Groote Schuur Hospital, Cape Town. S Afr Med J 1967 Dec 30;14(48):271-4.

[1355] Lower RR, Shumway NE. Studies on orthotopic homotransplantations of the canine heart. Surg Forum1960;11:18-20. — Lower RR, Stofer RC, Shumway NE. Homovital transplantation of the heart. J Thorac Cardiovasc Surg 1961;41:196-204. — Lower RR, Dong E, Shumway NE. Long-term survival of cardiac homografts Surgery 1965;58:110-9.

In 1968 Christiaan Barnard introduced an important modification in the atrial technique of an OHT, where the incision in the donor hearts right atrium (RA) was extended from the opening in the inferior vena cava (IVC) into the base of the RA appendage, rather than into the superior vena cava (SVC), avoiding the region of the sinoatrial node SAN)[1356].

Advances in immunosuppression, including the use of anti-thymocyte globulin, offered prophylaxis and treatment for acute rejection after HT[1357].

In 1973 **Philip K. Caves (1940 – 78)**[1358] — Irish Scottish cardiac surgeon, who introduced percutaneous (PC) endomyocardial biopsy (EMB), which allowed early diagnosis and treatment of acute rejection after HT. A typical surveillance PC EMB protocol includes weekly biopsies for the first weeks after HT, biweekly biopsies for the next month, and then monthly biopsies through the six months after HT. If the patient is free from rejection at 6 months, the frequency of biopsies is reduced to every 3 months.

The introduction in 1978-80 by Sir **Roy Y. Calne (1930 –)** — English transplant surgeon, of a calcineurin inhibitor (CNI)[1359] – cyclosporine (dosage according to level, with a 12-hour trough; beginning dosage, 25 mg b.i.d. / twice a day; aim for whole blood level of 300-350 ng/ml) was particularly important in the advancement of HT. He performed in 1987 the first combined liver, heart and lung transplantation, in 1994 the first successful combined stomach, intestine, pancreas, liver and kidney cluster transplant.

In 1964 **Harry J. Buncke (1922 – 2008)** — American microscopic plastic surgeon, reported a rabbit ear replantation during which he successfully anastomosed 1 mm in diameter blood vessels. This was followed by the first human microsurgical transplantation of the second toe to the thumb, performed in Feb. 1966, by **Dong-yue Yang** and **Yu-dong Gu**[1360] — Chinese microscopic orthopedic surgeon, in the city of Shanghai[1361], China.

Artificial skin («Integra») co-invented in 1981 by **John. F. Burke (1922 – 2011)** — American physician and medical researcher, together with **Ioannis V. Yannas (1935 –)**

[1356] Barnard CN. What we have learned about heart transplants. J Thorac Cardiovsc Surg 1968;56:457-68.

[1357] Griepp RB, Stinson EB, Dong Jr E, et al. Use of antilymphocyte globulin in human transplantation. Circulation 1972;45:1147-53.

[1358] Caves PK, Stinson FB, Graham AF, et al. Percutaneous transvenous endomyocardial biopsy. JAMA 1973 Jul 16;225(3):288-91.

[1359] Calcineurin is a calcium- and calmodulin-dependent phosphatase that is required for early T-cell activation and interleukin-2 (IL-2) formation — Calne RY, White DJ, Rolles K, et al. Prolonged survival of pig orthotopic heart grafts treated with cyclosporin A. Lancet 1978 Jun 3;1(8075):1183-5.

[1360] Yang DY, Gu YD. Thumb reconstruction utilizing second toe transplantation by microscopic anastomosis: report of 78 cases. Chin Med J 1979;92(5):295-309.

[1361] City of Shanghai (settled in the 5th-7th centuries, upgraded in status from a village to a market town in 1074), China.

— Greek American polymer engineer and scientist, has two layers of polymer: (1) the top synthetic layer was made of thin silicone sheet that protected the patient from dehydration and infection, the two most common causes of death in severely burned patient; and (2) the bottom layer was organic, a kind of scaffolding made from the molecular material in cow tendons and sharks cartilage which acted as a seed bed for healthy skin cells taken from other parts of the patient body. Healthy skin cells grew into the scaffolding in about a month. The cow and shark cells were absorbed by the body. After silicone layer was peeled off with tweezers, the area that has been severely burned healed as «like a sunburn».

E. Stanley Crawford (1922 – 92) — American cardiovascular surgeon, performed in 1955 the first successful excision of the left atrial (LA) myxoma under cardiopulmonary bypass (CPB), advanced a surgical transthoracic repair of the supra-aortic arterial trunks from endarterectomy (EA) **[Michael DeBakey (1908 – 2008)**. 1956: above] to the extra-anatomic synthetic tubular grafts (STG) bypass which decreased mortality from 22% to 5% (1969)[1362], in 1979 excised infected right atrial (RA) myxoma, and classified the thoracoabdominal aneurysms (Crawford Classification, 1986)[1363]:

Type I – aneurysms extended from the proximal descending aorta (DTA), i.e., from the left subclavian artery (SCA) to the suprarenal abdominal aorta (AA);

Type II – aneurysms extending from proximal DTA above the thoracic vertebra $(T)_6$ to below origin of renal arteries;

Type III – aneurysms extending from the distal DTA below T_6 to the aortoiliac bifurcation;

Type IV – aneurysms limited to the AA below the diaphragm.

Manucher J. Javid (1922 –) — Iranian American neurosurgeon, constructed a shunt (Javid shunt, 1954) to protect the brain during carotid endarterectomy (CEA) and also introduced the intravenous (IV) infusion of urea to induce osmotic diuresis, reduce intracranial (1954-56) and intraocular (1957) pressure. Optimal IV infusion of a single dose of urea is 0.5 to 1.5 mg per kilogram (kg) body weight.

By applying the methodology for suturing of the blood vessels of **Alexis Carrel (1873 – 1944**: above**)** with the inclusion anastomosis (Carrel pedicle), **Kenneth**

[1362] Crawford ES, DeBakey ME, Morris Jr GC, et al. Surgical treatment of the innominate, common carotid, and subclavian arteries: a 10-year experience. Surgery 1969;65:17-31.

[1363] Crawford ES, Crawford JL, Safi HJ, et al. The thoracoabdominal aneurysms: preoperative and intraoperative factors determining immediate and long-term results of operations in 605 patients. J Vasc Surg 1986;3:389-404.

Appell[1364] — American surgeon, performed in 1965 the «end-to-side» radial-cephalic arteriovenous fistula (RC-AVF) between the saphenous vein bifurcation and the radial artery (the branch patch technique) at the wrist of patient with an end-stage renal failure (RF) undergoing chronic hemodialysis (HD).

Transaxillary approach for the first rib resection to relieve the thoracic outlet syndrome (TOS) was introduced by **David B. Roos**[1365].

Dmytro A. Makar (23.I.1922, village Berehy, Sambirsky Region, Lviv Obl., Ukraine – 08.XII.2004, city of Lviv, Ukraine)[1366] — Ukrainian general surgeon, thoroughly studied epy pathogenesis and treatment of hyperfunction of the thyroid gland, especially thyrotoxic crises which greatly permitted improvement of treatment outcomes of this dangerous complication of toxic goiter; improved surgical treatment all other nosology's (in Greek – «nosos» means disease, and «-logia» means study) of the thyroid gland, biliary tract, pancreatic gland, inflammatory purulent diseases of the abdominal cavity; introduced original methods of treatment for an acute pancreatitis, acute complicated cholecystitis, acute intestinal obstruction, purulent peritonitis, based on the use of lymphogenic, endo-choledochal, electrophoretic and intra-sounding-lead methods of regional influence.

Predictable surgical treatment of idiopathic hypertrophic cardiomyopathy (IHSS) / hypertrophic obstructive cardiomyopathy (HOCM) / hypertrophic cardiomyopathy (HCM) / **Robert Teare (1911 – 79**: 20th century. Internal medicine**)** was introduced in 1961 by **Andrew G. Morrow (1922 – suicide by shooting himself 1982)**[1367] — American cardiac surgeon. It consists of complete transaortic septal myectomy of the left ventricle (LV) by incision just below and leftwards to the nadir of the right coronary sinus (RCS), and away from the atrioventricular node (AVN) situated below the RCS and noncoronary sinus (NCS) commissure, over to the anterior leaflet of the mitral valve (MV), then deepened and lengthened towards the apex of the heart. He came to learn that he himself suffered from HCM. Extended myectomy was introduced in 1994 by **Bruno J. Messmer** — German cardiac surgeon.

About the possibility to use the main pulmonary artery (MPA) valve for the aortic valve replacement (AVR) was first reported in 1961 by **Richard Lower (1929 – 2008:**

[1364] Brescia MJ, Cimino JE, Appell K, et al. Chronic hemodialysis using venipuncture and a surgically created arteriovenous fistula. N Engl J Med 1966;275:1089-92.

[1365] Roos DB. Transaxillary approach for first rib resection to relieve thoracic outlet syndrome. Ann Surg 1966 Mar;163(3):354-8.

[1366] Village Berehy (known since 1490), Sambirsky Region, Lviv Obl., Ukraine. — Ablitsov VH. «Galaxy Ukraine». The Ukrainian diaspora: prominent persons. Publ. «Kyt», Kyiv 2007.

[1367] Morrow AG, Brockenbrough EC. Surgical treatment of idiopathic hypertrophic subaortic stenosis. Ann Surg 1961 Aug;154(2):181-9. — Morrow AG. Hypertrophic subaortic stenosis. Operative methods utilized to relieve left ventricular outflow obstruction. J Thorac Cardiovasc Surg 1978;76;423-30. — Messmer BJ. Extended myectomy for hypertrophic obstructive cardiomyopathy. Ann Thorac Surg 1994;58:575-7.

above)[1368], when he translocated the MPA valve into the root of the aorta during the experiments on dogs.

Donald H. Ross (1922 – 2014) — Scottish South-African cardiac surgeon, replaced diseased aortic valve (AV) with the pulmonary valve autograft of the MPA by subcoronary suturing into the aortic ring and reimplanted the ostia of coronary arteries, the own pulmonary valve with the cryoprecipitated allograft / homograft (graft of tissue between individuals of the same species but of disparate genotype) of the MPA (Ross I operation, 1967), and replaced diseased mitral valve (MV) with pulmonary vein autograft of the MPA, and the own pulmonary vein with the cryoprecipitated homograft of the MPA (Ross II operation, 1967).

In 1972 Donald Ross[1369] successfully treated an aortic prosthetic valve endocarditis (PVE) by total surgical debridement of all infected tissues, reconstruction of the aortic root using an allograft, re-implantation of coronary arteries and minimal use of foreign material in infected areas.

Presently, the left cardiac sympathetic denervation (LCSD) for the long QT syndrome (LQTS) and catecholamine polymorphic ventricular tachycardia (CPVT)[1370] is performed using video-assisted thoracic surgery (VATS) with the stab incision in the middle axillary line (MAL) at the 3rd (camera), the 4th (grasper or lung retractor) and the 5th (the electrocautery hook dissector) intercostal space (ICS), transecting the left sympathetic chain medially the heads of the ribs at the level of thoracic $(T)_1$, sparing the superior aspect of the stellate ganglion and then at the level of T_2-T_5 ICS, as well as transecting the associated nerves of Kuntz (20th century. Anatomy, histology, and embryology – **Albert Kuntz (1879 – 1957)** between those levels[1371].

Congenital anomalies of the aortic valve (AV) and the aortic root include AV stenosis, subvalvular aortic stenosis (membranous, long-segment or tunnel-like stenosis and muscular stenosis) and supravalvular aortic stenosis, occurring in about 3-6% of children affected by congenital heart disease (CHD). In a tunnel-like stenosis, the length of which is more than one third of the aortic diameter and the AV is satisfactory, relief of the left ventricular outflow tract (LVOT) obstruction in older children is achieved by the modified Konno operation[1372], in the presence

---

[1368] Lower RR, Stoffer RC, Shumway NE: Autotransplantation of the pulmonic valve into the aorta. J Thorac Cardiovasc Surg 1960; 39(5):680-7.

[1369] Ross DN. Allograft root replacement for prosthetic endocarditis. J Cardiovasc Surg 1990;5:68-73.

[1370] Moss AJ, McDonald J. Unilateral cervicothoracic sympathetic ganglionectomy for the treatment of long QT interval syndrome. N Engl J Med 1971;285:903-4.

[1371] Li J, Wang LX, Wang J. Video-assisted thoracoscopic sympathectomy for congenital long QT syndromes. Pacing Clin Electrophysiol 2003;26:870-3.

[1372] Caldarone CA. Left ventricular outfow tract obstruction: the role of the modified Konno procedure. Simin Thorac Cardiovasc Surg Pediatr Card Surg Annul 2003;6:98-107.

of inadequate AV by the Konno-Rastan operation or aortoventriculoplasty[1373]. A modified Konno operation consist of transverse right ventricular (RV) infundibulum and a standard supra-annular aortic incisions, an incision into interventricular septum (IVS) below and leftward of the nadir of the right coronary leaflets or below the commissure of the right and left coronary leaflets towards an apex of the heart creating an artificial ventricular septal defect (VSD), debridement from the left side of the AV and septum, placing an augmentation synthetic patch into an artificial VSD, thus augmenting the LVOT. In the Konno-Rastan operation a vertical aortotomy incision is extended into the right ventricle (RV) infundibulum, a narrowed anterior AV annulus is excised leftward of the nadir of the right coronary leaflets with the extension into the IVS as in a modified Konno operation, aortic root and IVS are debrided, an appropriately larger sized mechanical AV is sutured to the posterior aortic annulus, a diamond-size synthetic or fixed bovine patch that matches at its midline the anterior aortic annulus, inferiorly artificially created VSD and superiorly vertical aortotomy, is sutured in its midline to the anterior ring of a mechanical AV, caudal to close artificially creating VSD and cephalad to close aortotomy incision, thus augmenting the AV size and the LVOT. For complex LVOT obstruction in a small child the Ross-Konno operation[1374] is preferable where replaced diseased AV and artificially created VSD are replaced using main pulmonary artery (MPA) pulmonary autografts with a triangular tongue of its RV wall, and the ostia of coronary arteries are reimplanted, the main pulmonary artery (MPA) – with the cryoprecipitated pulmonary homograft.

Paul Winchell (1922 – 2005) — American ventriloquist and inventor with premedical education of Jewish origin, and **Robert K Jarvik (1946 –)** — American physician-researcher, constructed the total artificial heart (TAH) «Jarvik-7» (1976-82) which **William C. DeVries (1943 –)** — American cardiac surgeon, implanted on December 02, 1982, as a permanent heart substitute into a man suffering from chronic end-stage congestive heart disease (CHF).

Three-compartment division of the mediastinum as proposed in 1983 by **Thomas W. Schields (1922 – 2010)**[1375] — American thoracic surgeon, consist of an anterior

---

[1373] Konno S, Imal Y, Iida Y, et al. A new method for prosthetic valve replacement in congenital aortic stenosis associated with hypoplasia of the aortic valve ring. J Thorac Cardiovasc Surg 1975 Nov;70(5):909-17. / Rastan H, Koncz J. Aortoventriculoplasty: a new technique for the treatment of left ventricular tract obstruction. J Thorac Cardiovasc Surg 1976;71:920-7. / **Haushang (Huschang) Rastan (1936 -)** — Iranian German cardiac surgeon. / **Josef Koncz (1916 - 88)** — German heart surgeon. / Tjindra C. Josef Koncz (1916 - 1988): Sein Leben und sein Werk. Medizinische Hochschule Hannover, Hannover 2004.

[1374] Ross DB, Trusler GA, Coles JG, et al. Small aortic root in childhood: surgical options. Ann Thorac Surg 1994;58:1617-24.

[1375] Shields TW, ed. General Thoracic Surgery. 2nd ed. Lea & Febiger, Philadelphia, PA 1983. — LoCicero III J. Thomas W. Shields, 1922-2010. J Thorac Cardiovasc Surg 2011 May;141(5):1332.

compartment, middle or visceral compartment, and posterior compartment or paravertebral sulcus. All three compartments are bounded superiorly by the thoracic inlet, laterally by the pleural space, and inferiorly by the diaphragm.

Anterior compartment is bounded anteriorly by the sternum and posteriorly by the great vessels and pericardium and contains the thymic gland (thymus), internal thoracic artery and vein (ITA, ITV), lymph nodes (LN), adipose and areolar tissue, and may contain pathologic structures such as retrosternal goiter, benign or malignant thymoma, thymic hyperplasia, thymolipoma and thymic cysts, lymphomas (Hodgkin's disease, lymphocytic, lymphocytic/histiocytic, histiocytic or undifferentiated lymphoma), germ cell tumors – seminomas account for 40% and non-seminomas account for 60% (teratoma, embryonal cell carcinoma, choriocarcinoma, yolk sac tumors and teratocarcinoma), and ectopic parathyroid gland (PTG).

Middle compartment is bounded posteriorly by the ventral surface of the thoracic space; it occupies the entire thoracic inlet and contains the majority of mediastinal structures, namely, the pericardium, heart and great vessels, descending thoracic aorta (DTA), trachea and proximal main-steam bronchi (MSB), cranial X (vagal) and phrenic nerves, esophagus, thoracic duct, proximal azygos vein, pre-tracheal LNs (level 2, 4 and 7), pleuro-pericardial LNs, adipose and connective tissue, may contain pathologic disorders such as pericardial effusion, diseased coronary arteries and cardiac valve, aneurysm of the heart's left ventricle (LV), ascending aorta (AA), transverse aorta (TA) or aortic arch and descending thoracic aorta (DTA), benign or malignant disease of the trachea, bronchi and esophagus, pericardial, bronchogenic or enteric cysts, and lymphomas.

The posterior compartment consists of paravertebral spaces along the thoracic vertebral groove that contain the sympathetic chain, proximal intercostal nerve, artery and vein, thoracic spinal ganglia, distal azygos vein, posterior paraesophageal and intercostal LNs, adipose and areolar tissue, may contain pathologic structures such as intercostal nerve tumors – neurofibroma (multiple neurofibronatosis or von Recklinghausen's disease – 19th century. Pathology), neurilemoma (Schwammoma – 19th century. Anatomy, histology and embryology), neurofibrosarcoma and neurosarcoma, sympathetic ganglionic tumors – ganglioma, ganglioneuroma, ganglioneuroblastoma and neuroblastoma (in children) and paraganglia cell tumors – paraganglioma (pheochromocytoma); esophageal disorders – neoplasms, esophageal cysts, diverticula, hiatal hernias, megaesophagus, and esophageal varices; bronchogenic, gastroenteric and neuroenteric cysts; extramedullary hematopoiesis, a giant LN hyperplasia (Castleman's disease – 20th century. Pathology); downwards spreading infections from the head and neck between the visceral and alar fascia termed the danger space, and between the prevertebral fascia and the vertebral bodies.

Recurrent dislocation of the shoulder joint treated by Putti-Platt **[Victorio Putti (1880 – 1940)** and **Harry Platt (1886 – 1961)]** capsulorrhaphy (above) was modified in 1962 by **George Truchly (27.04, 1922, city of Uzhhorod, Ukraine – 09.IX.2015, city Melbourne, FL, USA)**[1376] — Ukrainian American orthopedic surgeon, by suturing of the lateral stump and subscapular muscle as an one layer, of the capsule and subscapular tendon as one layer to the periosteal tissues along the scapular neck and overlapping it with the median portion of the capsule.

Hermes C. Grillo (1923, Boston, MA – died in a car accident **Oct. 14, 2006**, near **Ravenna, Italy)**[1377] — American thoracic surgeon, founder of the modern tracheal surgery, demonstrated in 1964 that trachea, having approximately 18-22 incomplete cartilaginous rings (4 mm wide and 1 mm thick), being on the average 11 cm long, as well as the cervical esophagus, has a segmental arterial blood supply from the branches of the inferior thyroid artery, subclavian artery (SCA), supreme intercostal arteries, internal thoracic artery (ITA), brachiocervical trunk (BCT), right bronchial artery and the left upper bronchial artery from the descending thoracic artery (DTA) at the level of the T_6-T_7. The left lower bronchial artery originates from the DTA at the level of T_8-T_9, supply the blood to the thoracic esophagus. Those small arterial branches enter the lateral tracheal wall, creating circumferential horizontal anastomotic arches between cartilaginous rings, but vertically only for distance of 1-2 cm. So that safe surgical mobilization of the trachea is possible only with anterior and posterior dissection, circumferentially for a distance not more than 1-2 cm. Usually, in cases of a short congenital segmental tracheal stenosis, less than 30% or 6-8 complete tracheal rings, a safe tracheal resection and reconstruction (TRR) via a midline longitudinal sternotomy incision (MLSI), are possible. Among selected cases, up to 50% of the human trachea could be partially resected and

---

[1376] City of Uzhhorod (founded in 19th century), Ukraine. — Town of Melbourn (c. 1867), FL, USA. — Truchly G. On problems Regarding the tendon transplantation of the leg. J Ukr Med Assoc North Am 1958 трав;5,(10):1-6. — Truchly G. Neuropathies of the upper extremity due to binding of the nerve. J Ukr Med Assoc North Am 1960 Jan;7(16):1-6. — Truchly G. Why the Hutzuls wear brace (cheres). J Ukr Med Assoc North Am 1962 Oct.;9,4(27):16-9. — Truchly G, Thompson WAL. Simplified Putti-Platt procedure. JAMA 1962;179(11):859-62. — Truchly G. Modification of Putti operation for habitual dislocation of shoulder: Results in 102 consecutive operations. J Ukr Med Assoc North Am 1968 Apr;15,2(49):9-15. — Truchly G. Use of the Kiel bone in the anterior interbody fusion of the cervical vertebrae. J Ukr Med Assoc North Am 1972 Jan-Apr;19,1-2(64-65):9- 12. — Truchly G. Reconstruction of the winging scapula deformity. J Ukr Med Assoc North Am 1980 Winter 96(1):11-5. — Truchly G, Dzul P. Arytenoid dislocation: literature review and case report. J Ukr Med Assoc North Am 2005 Summer; 50,1(153):27-32. — Ablitsov VH. «Galaxy Ukraine». The Ukrainian diaspora: prominent persons. Publ. «Kyt», Kyiv 2007.

[1377] Grillo HC, Dignan EF, Miura T. Extensive resection and reconstruction of mediastinal trachea without prosthesis or graft: an anatomical study in man. J Thorac Cardiovasc Surg 1964 Nov;48:741-9. — Grillo HC. Circumferential resection and reconstruction of the mediastinal and cervical trachea. Ann Surg 1965; 162:374-88. — Grillo HC. Surgical approaches to trachea. Surg Gynecol Obstet 1969;129:347-52.

reconstructed via a median sternotomy, if detrimental tension is eliminated by performing the laryngeal, suprahyoid or pulmonary hilar release of the trachea. During the laryngeal release of **Herbert H. Dedo (1933 –)**[1378] — American ear nose and throat physician and surgeon and **Noel H. Fishman** — American pediatric thoracic and cardiac surgeon, the cranial insertions of the thyrohyoid muscle to the hyoid bone are resected, the thyrohyoid membrane is divided, and the two (major and minor) horns of the thyroid (thyroidal) cartilage are cut. Difficulties in swallowing may persist for a long time but patients usually overcome them. Using the suprahyoid release of **William M. Montgomery (Aug. 20, 1923, Proctor, VT – Nov. 07, 2003, Boston, MA)**[1379] — American thoracic surgeon, the cranial muscle-insertions at the central hyoid bone are divided, the hyoid bone is transected just medially to its lesser cornua on both sides leaving the small and large horns, and the stylohyoid tendons are divided. He also designed T-tube for stenting trachea after complicated resections (Montgomery T-tube). First on the right, then the left, after division of the inferior pulmonary ligament, the pulmonary hilar release of **Joseph R. Newton, Jr (1969 –)**[1380] — American cardiothoracic surgeon, is performed using the U-shaped incision into pericardium below the inferior pulmonary vein (PV) 360° and around the hilum.

Trachea normally dilates slightly during inspiration and narrows during expiration.

Tracheomalacia (tracheal softening; in Greek – «malacia» means softening), one of the two most common congenital tracheal anomalies in infancy and children, is a condition characterized by flaccidity of the tracheal support by C-shaped cartilage which exaggerate physiological motion of the trachea, causing tracheal narrowing, particularly expiration, leading to tracheal collapse during or cough. The circumference of the trachea is typically normal during inspiration. The usual symptoms of tracheomalacia are stridor, dyspnea, laryngeal crow apnea.

This same condition may extend into the bronchi (tracheobronchomalacia), affect the larynx (laryngomalacia). Tracheobronchomalacia is a weakened cartilage which collapses on expiration, resulting from a postintubation injury or chronic obstructive pulmonary disease (COPD). When the patient attempts a deep breath, expiration or cough, the membranous wall of the trachea, bronchi or both approximates to the anterior, softened cartilaginous wall, causing its nearly total obstruction.

[1378] Dedo HH, Fishman N.H. Laryngeal release and sleeve resection for tracheal stenosis. Ann Otol Rhinol Laryngol 1969; 78:285-96.

[1379] Montgomery W. Suprahyoid release for tracheal anastomosis. Arch Otolaryngol 1974; 99:225-60.

[1380] Newton JR, Grillo HC, Mathisen DJ. Main bronchial sleeve resection with pulmonary conservation. Ann Thorac Surg 1991;52:1271-80.

Tracheomalacia is classified into:

Type I – congenital tracheomalacia, sometimes associated with tracheoesophageal fistula (TEF);

Type II – caused by an extrinsic compression due to vascular ring or pulmonary artery (PA) sling;

Type III – acquired due to chronic infections, prolonged intubation, or polychondritis.

Severe congenital tracheomalacia is treated by tracheostomy or mechanical ventilation of the lungs with continuous positive airway pressure (CPAP) or bi-level positive airway pressure (BiPAP), the use of sheets of polypropylene mesh to plicate the membranous wall of the trachea[1381], bronchoscopic placement of balloon-expandable metallic wire stents[1382], tracheal external stabilization with «onlay» fixation of the flaccid pars membrane, and free suspension of the malacic cartilaginous portion of the trachea within an oversized polytetrafluoroethylene (PTFE) prosthesis[1383], fixation of the trachea, innominate artery (IA) or aortic arch to the posterior surface of the sternum (tracheopexy, arteriopexy or aortopexy), or tracheobronchoplasty.

Compressive tracheomalacia is treated by the release of compressing structure, i.e., repair of a vascular ring, PA sling, and tracheopexy, arteriopexy or aortopexy.

Surgical treatment of congenital tracheal stenosis usually necessitates an approach through a median longitudinal sternotomy incision (MLSI) and cardiopulmonary bypass (CPB), include a cartilage (1982)[1384] or pericardial (1982)[1385] replacement tracheoplasty, slide tracheoplasty (1989)[1386], a homograft tracheoplasty (1996)[1387],

[1381] . Reiner WG, Newbby JP, Kelb DC. Long-term results of tracheal support surgery for emphysema. Dis Chest 1968;53:765-7.

[1382] Filler RM, Forte V, Fraga JC, et al. The use of expandable metallic airway stents for tracheobronchial obstruction in children. J Pediatr Surg 1995;30:1050-6. — Furman RH, Backer CL, Dunham ME, et al. The use of balloon-expandable metallic stents in the treatment of pediatric tracheomalacia and bronchomalacia. Arch Otolaryngol Head Neck Surg 1999;125:203-7.

[1383] Hagl S, Jacob H, Sebening C, et al. External stabilization of long-segment tracheobronchomalacia guided by intraoperative bronchoscopy. Ann Thorac Surg 1997;64:1412-21.

[1384] Kimura R, Mukohara N, Tsugawa C, et al. Tracheoplasty for congenital stenosis of the entire trachea. J Pediatr Surg 1982; 17:869-71.

[1385] Idris F, Del Leon SY, Ilbawi MN, et al. Tracheoplasty with pericardial patch for extensive tracheal stenosis in infants and children. J Thorac Cardiovasc Surg 1984;88:527-36.

[1386] Tsang V, Murray A, Gillbe C, et al. Slide tracheoplasty for congenital funnel-shaped tracheal stenosis. Ann Thorac Surg 1989;48:632-5.

[1387] Elliot MJ, Haw MP, Jacobs JP, et al Tracheal reconstruction in children using cadaveric homograft trachea. Eur J Cardiothorac Surg 1996;10:707-12.

autograph tracheoplasty where an excised piece of narrowed trachea is used as an anterior patch (1998)[1388], or a composite graft tracheoplasty (2001)[1389].

Slide tracheoplasty is indicated for a long segmental narrowing of the trachea with a complete tracheal ring. The operation is performed either through a cervical collar, or through a MLSI with/without CPB, only the midportion of the tracheal stenosis is transected and two halves of the trachea are then opened longitudinally: the upper trachea is opened posteriorly, and the lower trachea is opened anteriorly. Corners of the transected trachea are trimmed so that the leading edge will fit into the V-portion of the upper component. The two openings are then «slid» together and approximated with interrupted non-absorbable sutures, or with continuous absorbable suture.

Pectoralis major muscle whose main blood supply comes from pectoral branches of the thoracoacromial artery, a branch of the 2nd portion of the SCA, was introduced by **Maurice J. Jurkiewicz (Sep. 23, 1923, Claremont, NH – May 29, 2011, Atlanta, GA)**[1390] — American plastic surgeon, in 1977, as an island flap muscle for the coverage of anterior chest wall defects, mostly deep sternal wound infections.

George J. Magovern (1923 – 2013) — American thoracic and cardiac surgeon, together with **Harry W. Cromie (1928 –)** — American engineer and inventor, constructed the mechanical ball-caged suture-less aortic valve (AV) with a synthetic ring armed with small metallic teeth which during the activation of the clamping instrument attaches the valve to the aortic ring (Magovern-Cromie valve, 1961-63); with **Gerald E. McGinnis (1935 –)** — American engineer, constructed endotracheal and tracheal tubes with high-volume and low-pressure cuff (1971-72); introduced the clinical use of the small centrifugal bypass between the left atrium (LA) and the ascending aorta (AA) for the support of the failing left ventricle (LV) and between the right atrium (RA) and the main pulmonary artery (MPA) for the support of the failing right ventricle (RV) without systemic heparinization with a blood flow of 1,2–1,6 liters/min. (1980-82).

In order to arrest and control the temporarily massive hemorrhage from the lower esophageal and the proximal gastric varices by its tamponade, usually as a result of the liver cirrhosis, in 1930 **K. Westphal**[1391] — German surgeon, designed

[1388] Backer CL, Mavroudis C, Durham ME, et al. Repair of congenital tracheal stenosis with a free tracheal autograft. J Thorac Cardiovasc Surg 1998;115:869-74.

[1389] Backer CL, Mavroudis C, Gerber ME, et al Tracheal surgery in children: an 18-year review of four techniques. Eur J Cardiothorac Surg2001;19:777-84.

[1390] Brown RG, Fleming WH, Jurkiewicz MJ. An island flap of the pectoralis major muscle. Br J Plast Surg 1977 Apr;30(2):161-5.

[1391] Westphal K. Ueber eine Kompressionbehandlung der Blutungen aus Oesophagusvarizen. Dtsch med Wschr 1930;56(27):1135-6.

an esophageal sound, and in 1950 **Robert W. Sengstaken, Sr (1923 – 78)**[1392] — American neurosurgeon and **Arthur Blakemore (1897 – 1970**: above), designed a triple lumen tube which incorporates double balloon (esophageal and gastric) to control variceal bleeding and a third gastric cannel for suctioning, irrigation or instillation of medication to the stomach. The temporary success in arresting a hemorrhage is more than 90% of patients.

Major advance in the surgical treatment of congenital heart disease (CHD) in infants was made in 1973 by **Brian G. Barratt-Boyes [13 Jan. 1924, city of Wellington, New Zealand** – died following complications of a final third cardiac operation **08 Mar. 2006, Cleveland Clinic (CC), city of Cleveland, OH, USA]**[1393] — New Zealander cardiac surgeon, and in 1974 by **Aldo R. Castaneda (1930 – 2021)**[1394] — Italian Guatemalan American pediatric cardiac surgeon, with the use of deep hypothermic circulatory arrest (DHCA) at core temperature 15-18⁰C, relying primarily on surface cooling, with cardiopulmonary bypass (CPB) usually kept under 20 to 30 minutes.

Introduced by **Robert Dobbie (1924 – 2004)** and **Jim Hoffmeister** — American surgeons, in the middle of 1970s a nasogastric (NG) or the Dobbhoff feeding tube which is a small lumen (diameter, 4 mm) tube with a weight at the intended to pull it by gravity during insertion via an anterior nostril of the nose, past the throat and esophagus into the stomach, to roughly 5 cm below the xiphoid process, or further into the duodenum, for continuous delivery of a total enteral nutrition (TEN) for patients temporarily not able to swallow for 4 weeks or longer, if tolerated. The correct position of the Dobbhoff feeding tube could be traced by fluoroscopy, its position confirmed by the pH aspirate of 4 or below and by the plain chest and abdomen radiography. Minor complications include nosebleeds, sinusitis of the paranasal sinuses, i.e., maxillary, frontal sphenoid and ethmoid, and sore throat. Significant complications are erosion of the nose where the tube is anchored, its displacement onto the esophagus with tracheobronchial aspiration, esophageal perforation, damage to the surgical anastomosis, collapsed lung or intracranial placement of the tube.

[1392] Sangstaken RW, Blackemore AH. Balloon tamponade for control of hemorrhage from esophageal varices. Ann Surg 1950;13(5):781-9.

[1393] City of Wellington (settled 1839, capital since 1865), New Zealand (area 268,021 km²). — Cleveland Clinic (CC, established 1921), city of Cleveland (founded 1796), OH, USA. — Barratt-Boyes B. Complete correction of cardiovascular malformations in the first two years of life using profound hypothermia. In, BG Barratt-Boyes, JM Nutze, EA Harris (Eds). Heart Disease in Infancy. Churchill Livingstone, Edinburgh 1973. Page 35.

[1394] Guatemala (area 108,889 km²). — Castaneda AR, Lamberti J, Sade RM, et al. Open-heart surgery during the first three months of life. J Thorac Cardiovasc Surg 1974;68:719-31.

Member of the Chest Federation of the United States (CFUS) **Basil (Basilius) Zaricznyj (Aug. 08, 1924, Ukraine – Aug. 08, 2007, city of Springfield, IL, USA)**[1395] — Ukrainian American orthopedist and sport orthopedic surgeon, was instrumental in the development of surgical reconstruction of the shoulder, the elbow and knee joints, as well as management of sport injuries in school-aged children adolescents.

Oleksiy O. Korzh (23.IV.1924, village Obolon', Semenivs'ky Region, Poltava Obl., Ukraine – 01.XI.2010, city of Kharkiv, Ukraine)[1396] — Ukrainian physician, orthopedic and trauma surgeon, was one of the first who motivated and introduced into practice fast prosthesis fitting, immediately after amputation of an extremity developed series of methods of surgical reconstruction of the chest wall in severe form of scoliosis, of the hip joint, of congenital dislocation of the femur, of the in-grafting the head of the femur bone, and unification of the transplanted donor bone with the recipient bone.

[1395] Village of Shpykolosy (founded 1415), Zolochivsky Region, Lviv Obl., Ukraine. — City of Springfield (founded 1821), IL, USA. — Zaricznyj B. Late reconstruction of the ligaments following acromio-clavicular separation. J Bone Joint Surg Am 1976 Sep;58:792-5. — Zaricznyj B. Late reconstruction of the ligaments following acromio-clavicular separation. J Ukr Med Assoc North Am 1977 Apr;24,2(85):3-11. — Zaricznyj B. Avulsion of the tibial eminence: treatment and pinning. J Bone Joint Surg Am 1977 Dec;59(8):1111-4. — Zaricznyj B. Transfer of the distal end of the ulna for congenital radial hemimelia. J Ukr Med Assoc North Am 1977 July;24,3(86):14-29. — Zarichnyj B. Torn fractures of the tibial eminence. J Ukr Med Assoc North Am 1979 Spring;26,2(93):85-91. — Zaricznyj B, Shattuck LJ, Mast TA, at al. Sport-related injuries in school-aged children. J Ukr Med Asoc North Am 1980 Summer;27,3(98):131-40. — Zaricznyj B, Shattuck LJ, Mast TA. Sport-related injuries in school-aged children. Am J Sport Med 1980 Sep-Oct;8(5):318-24. — Zaricznyj B. Management of acromioclavicular joint injuries. J Ukr Med Assoc North Am 1982 Winter;29,1(103):3-13. — Zaricznyj B. Simple method of tightening the the spint cast. Clinical Orthopaedics and Related Research. 1982 Jun;166(Section II):141-2. — Zaricznyj B. Reconstruction of the anterior cruciate ligament using free tendon graft. Am J Sport Med 1983 May-Jun;11(3):167-76. — Zaricznyj B. Reconstruction of the anterior cruciate ligament using free tendon graft. J Ukr Med Assoc North Am 1984 Winter;31,1(109):3-21— Zaricznyj B. Surgical reconstruction for chronic scapula-clavicular instability. J Ukr Med Assoc North Am 1985 summer; 32,2(113):67-76. — Zaricznyj B. Reconstruction of the anterior cruciate ligament of the knee using a double tendon graft. Clin Orthop Res 1987 Jul;(220):162-75. — Zaricznyj B. Reconstruction of the anterior cruciate ligament of the knee using a double tendon graft. J Ukr Med Assoc North Am 1990 fall;37,2(122):3-17. — Zaricznyj B. Throwing injuries of the shoulder. J Ukr Med Assoc North Am 1993 Summer-Autumn; 40,3(131):145-166. — Zaricznyj B. Correction of habitual limb. J Ukr Med Assoc North Am 1996 Spring;43,2(139):98-100. — Zaricznyj B. Carpal tunel syndrome. J Ukr Med Assoc 1966 Autumn;43,4(141):230-40. — Zaricznyj B. Transverse divergent dislocation of the elbow. J Ukr Med Assoc North Am 2000 Autumn;45,2(149):3-10. — Ablitsov VH. «Galaxy Ukraine». The Ukrainian diaspora: prominent persons. Publ. «Kyt», Kyiv 2007.

[1396] Village Obolon' (founded 1740s; suffered from conducted by Russian Communist Fascist regime and NKVS Holodomor-Genocide against the Ukrainian Nation), Semenivs"ky Region, Poltava Obl., Ukraine. — Ablitsov VH. «Galaxy Ukraine». The Ukrainian diaspora: prominent persons. Publ. «Kyt», Kyiv 2007.

The islet-cell (IC) grafting initiated by **Charles Pybus (1883 – 1975**: above) was continued by **Paul E. Lacy (1924 – 2005)**[1397] — American anatomist and researcher in the IC transplantation, in collaboration with **Walter F. Ballinger II (May 16, 1925, city of Philadelphia, PA, USA** – died from pneumonia **May 29, 2011, city of St Louis, MI, USA)**[1398] — American general surgeon, **Mykhailo (Michael) Pavlovsky (15.XI.1930, town of Berestechko, Volyn Obl., Ukraine – 18.I.2013, city of Lviv, Ukraine)**[1399] — Ukrainian general and transplant surgeon, **DavidW Scharp**[1400] — American general and transplant surgeon, and **Andrea G. Tzakis (1950 –)**[1401] — American general and transplant surgeon of Greek descend.

Throughout the 1960s and early 1970s Paul Lacy in collaboration with Walter Ballinger II were working on the development of technique of beta IC transplantation in animals as a treatment for DM.

An international recognition received the work of Mykhailo Pavlovsky in the PG beta IC cell allotransplantation for DM, type I, particularly using the cultured IC.

In 1983 he organized Division of Endocrinology for 60 beds at Department of Surgery, Lviv Medical Institute (LMI), now Danylo Halytsky National Lviv Medical University (NLMU), city of Lviv, Ukraine, where each year about 600 patients suffering from endocrinologic disorders were treated, among them 11% patient affected by DM type I.

In 1985-95 Mykhailo Pavlovsky with colleagues performed 97 allotransplantation in patients with DM type I, using 5 – 14 days old cultures of the PG IC from dead human

---

[1397] Lacy PE. Electron microscopy of the normal islets of Langerhans. Structure in the dog, rabbit, guinea pig and rat. Diabetes 1957;6:498-507. — Lacy PE, Davies J. Preliminary studies on the demonstration of insulin in the islets by fluorescens antibody technic. Diabetes 1957;6:354-7. — Lacy PE, Kistianovsky M. Method for the isolating intact islets of Langerhans from the rat pancreas. Diabetes 1967 Jan;16(1):35-9.

[1398] City of St Louis (founded 1764), MI, USA. — Ballinger WF II, Lacy PE. Transplantation of intact pancreatic islets in rats. Surgery 1972;72:175-86. — Ballinger WF II, Lacy PE, Scharp DW et al. Isografts and allografts of pancreatic islets in rats. Brit J Surg 1973;60:313.

[1399] Town of Berestechko (founded 1445), Volyn Obl., Ukraine. — Pavlovsky MP, Boyko NI. Treatment of insulin dependent diabetes mellitus by the method of allotransplantation of islet cells of the pancreatic gland. Current Problems in Clinical Surgery. Lviv 1988. P. 83-4. — Pavlovsky MP. Surgery of the endocrine glands. J Ukr Med Assoc North Am 1990 Autumn; 37;2(122):18-25. — Pavlovsky MP, Shevchuk KO. Surgical correction of genetically determined aberration of sexual differentiation. J Ukr Med Assoc North Am 1993 Winter;40,1(129):8-14. — Pavlovsky MP, Boyko NI. Allotransplantation ofthe cultured pancreatic islands cells in surgical treatment of patients with diabetes mellitus. J Ukr Med Assoc North Am 1993 Spring;40,2(130):90-4. — Masiak BM, Pavlovsky MP, Lozynskyi YS et al. Diagnosis and treatment of pararectal teratomas cysts. J Ukr Med Assoc North Am 1993 Summer-Autumn;40,3(131:171-3. — Ablitsov VH. «Galaxy Ukraine». The Ukrainian diaspora: prominent persons. Publ. «Kyt», Kyiv 2007.

[1400] Scharp DW, Lacy PE, Santiago JV et al. Insulin independency after islet transplantation into type I diabetic patient. Diabetes 1990;39:515-8.

[1401] Tzakis AG, Ricordi C, Alejandro R, Zeng Y, Fung JJ, Todo S, Demetris AJ, Mintz DH, Starzl TE. Pancreatic islet transplantation after upper abdominal exenteration and liver replacement. Lancet 1990 Aug 18;336(8712):402-5.

fetuses of 16 – 22 weeks old, secured from the late spontaneous abortions or from artificial (therapeutic) abortions performed with medical indications. The suspension of fragments received from the PG of 4 - 8 fetuses, were injected using syringes into the rectus abdominal muscle of 76 patients, into the liver via the portal vein or its branches during intraabdominal operations in 21 patients. The interval outcomes of the PG IC transplantation studied in 95 patients showed a clear antidiabetic effect in 89 recipients, which hold out from 3 до 20 months. The 24-hours insulin dose decreased by 25% – 100%. In comparison, with intramuscular grafting, an intraportal transplantation route gave better clinical outcomes with greater decrease of necessary amount of exogenous insulin needed. Repeated transplantations of cultured beta IC were performed in approximately 3% of patients.

Subsequently, in 1989 successful beta IC grafting was performed by David Scharp into DM type I patient.

In 1990 Andrea Tzakis reported nine patients who became diabetic after upper abdominal exenteration and liver replacement were given pancreatic IC graft obtained from the liver donor (8 patients), third party donor (1 patient) and both (4 patient). Two patients died from infection 48 and 109 days after the operation, and one of malignant tumor recurrence 178 days postoperatively, five patients were insulin-free or on insulin only during night-time total parenteral nutrition (TPN). The longest survivor required neither TPN noninsulin administration.

The development of an implantable cardioverter-defibrillator (ICD) was pioneered in 1970-1980 by **Michel Mirkowski**, born **Mordechai Frydman (Oct. 24, 1924, Warsaw, Poland** – died from multiple myeloma **March 26, 1990, Baltimore, MD, USA)**[1402] — Israeli American physician, to provide an immediate, automated cardioversion or defibrillation to patients who were victims of cardiac arrest due to ventricular fibrillation (VF) or ventricular tachycardia (VT).

Distribution of lymphatic nodes (LN) in the lung cancer, defined according to the level (station) in the chest cavity is the basis of the staging system for a primary bronchogenic carcinoma, as proposed by **Clifton F. Mountain (1924 – 2007)**[1403]— American thoracic surgeon and **Carolym M. Dresler (1955 –)** — American cardiothoracic surgical oncologist. The lymphatic drainage channels run from subpleural vessels along the bronchi and pulmonary arteries towards the hilum of

---

[1402] Mirowski M, Mower MM, Staewen WE, et al. Standby automatic defibrillator. An approach to prevention of sudden coronary death. Arch Intern Med 1970 Jul;126(1):158-61. — Mirkowski M., Reid PR, Mower MM, et al. Termination of malignant ventricular arrhythmias with an implanted automatic defibrillator in human being. N Engl J Med 1980 Aug 7;303(6):322-4. — Mirowski M. The automatic implantable cardioverted-defibrillator: an overview. J Am Cardiol 1985 Aug;6(2):461-6.

[1403] Mountain CF, Dresler CM. Regional lymph node classification for lung cancer staging. Chest 1997;111(6):1718-23. / Mountain CF. Staging classification of lung cancer. A Critical evaluation. Clinics in Chest Medicine 2002 Mar 23(1):103-21.

the lung. This flow is interrupted by multiply LN along the way, mostly situated at the forking of bronchi. Each LN station has a number assigned to it as presented below:

level 1   –   highest mediastinal, pretracheal, suprasternal notch or lower cervical supraclavicular (delphian)[1404] – the prophet of mortal end of cancer;

level 2R   –   upper paratracheal (right); level 2L – upper paratracheal (left);

level 3A   –   anterior pre-vascular – anterior to the innominate artery or anterior to the superior vena vava (SVC);

level 3P   –   posterior retrotracheal;

level 4R   –   right lower paratracheal;

level 4L   –   left lower paratracheal;

level 5   –   subaortic (under the aortic arch);

level 6   –   para-aortic (ascending, transverse or proximal descending aorta, or phrenic nerve;

level 7   –   subcarinal;

level 8   –   paraesophageal (below the carina);

level 9   –   pulmonary ligament;

level 10   –   hilar;

level 11   –   interlobar;

level 12   –   lobar;

level 13   –   segmental;

level 14   –   subsegmental.

The LN levels 1 through 9 are considered mediastinal. Spread of malignant tumors to the unilateral LN's is classified as N_2, to the contralateral LN's as N_3. Anomalies of ipsilateral, centrally directed drainage pattern of cancer cells to the hilum by the alternative route, via the subaortic, para-aortic, and the anterior mediastinal channels occurs in about 33% patients, the most commonly from the left lung. Thus, up to 50% of left lower lobe (LLL) tumors and 35% of the left upper lobe (LUL) tumors have positive contralateral (N_3) LN, while contralateral metastasis is found in only 42% of the right lower lobe (RLL) tumors and 18% of the right upper lobe (RUL) tumors when N_2 LN are negative for tumor.

In March 1956, **Russel M. Nelson (1924 –)** — American pediatric cardiac surgeon, performed in a 4-year-old girl a successful total correction of tetralogy of Fallot (TOF), i.e., a longitudinal incision through the narrowed right ventricular outflow

---

[1404] Delphi is the ancient sanctuary in the upper central Greece between 510-323 BC, the seat of the oracle that was consulted on important decisions thorough the ancient world and was considered the center of the ancient classical world.

tract (RVOT), excision of obstructing muscle bundle in the RVOT, a patch closure of the ventricular septal defect (VSD) and a patch reconstruction of the stenosed RVOT.

Mikhail I. Perelman (Dec. 20, 1924, Minsk, Belarus – died from pulmonary thromboembolism **Mar. 30, 2013, Russia)**[1405] — Belarusian Russian thoracic surgeon of Jewish descent, developed the transternal approaches to the right main stem bronchus (RMSB) and the left main stem bronchus (LMSB) for stapling to eliminate a bronchopleural fistula after pneumonectomy (1961) and the video-assisted thoracic surgery (VATS) precision electrocautery excision technique to core out the deep-seated lung nodule and multiple lung metastases (metastectomy) from the surrounding lung parenchyma with adequate margins where placement of staplers is precluded by the thickness of the lung tissue and where preservation of lung tissue is important (Perelman procedure, 1986).

Osteoarthritis (in Greek – «osteo», means bone, «arthro» – joint, «-it» or «-itis» – inflammation) is a chronic hypertrophic arthritis, degenerative arthritis, or degenerative disease of joints, represents low grade inflammatory disease with degeneration of joints, most commonly of the hip and knee joint, usually in older persons, causing wear of the cartilage which cover and acts as roller (pillow) inside the joint, disappearance of the synovial fluid which lubricate joints. Loss of the bone surface protection from rubbing off a cartilage causes pain during loading on instep or lifting, whining during walking or standing, especially in the morning and after immobility. Decreasing patient's activity leads to atrophy of local muscles, ligaments become sluggish and weak. Osteoathritis is a major cause of incapability to work or invalidity.

At the initial stage of disease, the conservative treatment using pain killers, nonsteroid anti-inflammatory drugs or corticosteroids, physical therapy (PT) will alleviate symptoms.

But in advanced stage conservative treatment doesn't work, will require surgical treatment, that is replacement of the pelvic-hip bone (in Latin – os coxae) joint, or knee joint by artificial joints, made of metal, plastic, or its alloy components. During replacement of the pelvic-hip joint it is necessary to replace two components of this joint – the acetabulum, i.e., large chalice-like acetabular cavity of the hip bone and the head of the femoral bones. The acetabular cavity is composed with three hip bones – ischium, ilium, and pubis. During replacement of the knee joint, a.k.a., arthroplasty of the hip, partial or total, consist with replacement of diseased or injured weight-bearing surface of the femoral and tibial bones of the knee joint, formed with the aim of prolongation of the knee motion for approximately 25 years.

[1405] Cooper JD, Perelman M, Todd TR, et al. Precision cautery excision of pulmonary lesions. Ann Thorac Surg 1986 Jan;41(1):51-3.

In 1959 **Konstiantyn M. Syvash** a.k.a., **Konstantin M. Sivash (1924, town of Konotop, Sumy Obl., Ukraine – 1989, city of Moscow, Russia)**[1406] — Ukrainian Russian orthopedic surgeon, performed the total two component acetabulo-femoral (hip) joint replacement with a screw fixated to the acetabulum (the Syvash's hip endoprosthesis), and in 1978 the total knee joint replacement (the Syvash knee endoprosthesis). Later (1962), Sir **John Charnley (1911 – 82)** — English orthopedist, carried out a total two component hip joint replacement made of metal and plastic.

A pioneer in surgery for portal hypertension (PH), **W. Dean Warren (1924 – 89)**[1407] — American surgeon, developed the distal spleno-renal (Warren) shunt, a surgical procedure in which the distal splenic vein, a part of the portal venous system, is attached (anastomosed) to the left renal vein, a part of the systemic venous system, used to treat PH and its main complication – bleeding esophageal varices.

Independently, **Christian Cabrol (1925 – 2017)**[1408] — French cardiac surgeon, and **N.G. DeVega**[1409] described a partial encircling suture to narrow the annulus of dilated tricuspid valve (TV). The two arms of double suture enforced by a pledged are anchored near the anteroseptal commissure, then run along the base of the anterior and posterior leaflets of the TV, and anchored near the posteroseptal commissure, where they are enforced by a second pledged, pulled up and tied (Canrol-DeVega semicircular suture commissurotomy, 1972).

The sub-commissural annuloplasty for the correction of the three-leaflet's aortic valve (AV) insufficiency from the prolapse and excessive tissue of one or more leaflets, introduced by Christian Cabrol consist of using a pledged interrupted mattress sutures placed lateral to the valve annulus and deep into the aortic wall to prevent the tearing out, closing the top 0.5-1.5 mm of one to three commissures to a measured annular diameter of 19-21 mm.

[1406] Town of Konotop (site of temporary abodes since Neolithic Age, founded 1635; during organized by Russian Communist Fascists regime Holodomor-Genocide 1932-33 against the Ukrainian Nation died at least 2,471 peoples, and during political repressions in 1936-38 suffered tenths of Konotopians, and residents from surrounding villages), Sumy Obl., Ukraine. — Sivash VH. The Sivash Odyssey. Yekaterinburg 2005. — Sivash II. The Sivash Family from Tarasivka (Heshetylivka) Polohiv Zaporizhian. Kyiv 2006. — Sivash KM. New Method of Knee Joint Resection. Moscow 1960. — Sivash KM. Tuberculosis of the Bone and Joints: Prophylaxis and Treatment. Moscow 1961. — Sivash KM. Alloplastic of Hip Joint. Moscow 1967. — Sivash KM. The Sivash Hip Joint System: Description. Moscow 1974. — Sivash KM. New Technique of Osteosynthesis. Moscow 1979. — Ablitsov VH. «Galaxy Ukraine». The Ukrainian diaspora: prominent persons. Publ. «Kyt», Kyiv 2007.

[1407] Warren WD, Henderson JM, Millikan WJ, et al. Distal splenorenal shunt versus endoscopic sclerotherapy for long-term management of variceal bleeding. Preliminary report of a prospective, randomized trial. Ann Surg 1986;203(5):454-62. — Polk HC Jr, Levi D, Hutson DG. W Dean Warren, MD: iron hand and principles of steel. JACS Apr 2019;228(4);708-14.

[1408] Cabrol C. Valvular annuloplasty. A new method. Nouv Presse Med 1972 May 13;1(20):1366.

[1409] De Vega NG. La anuloplasta selective, regulable y permanente: una tecnica original para el tratamiento de la insufficiencia tricuspida. Rev Esp Cardiol 1972 Nov-Dec;25(6):555-7.

To drain blood from a «dead space» and to prevent a hematoma formation between a composite synthetic tubular graft (STG) of the aortic root and the ascending aorta (AA) – Bentall operation (above) and wrapped wall of an aortic aneurysm, Christian Cabrol[1410] proposed performing a shunt between the latter and the right atrium (RA) appendage.

He introduced an «end-to-end» anastomosis of the ostia of the left main coronary artery (LMCA) and right coronary artery (RCA) to both ends of an 8-10 mm Dacron STG with its subsequent «side-to-side» anastomosis behind to a composed STG that replaced diseased AA (Cabrol's technique, 1981). The Cabrol technique can lead to development of neointimal hyperplasia, and some patients may require reoperation, and coronary artery bypass (CAB) grafting to treat the resulting ischemia.

In 1966 **Shigeto Ikeda (1925 – 2001)**[1411] — Japanese physician, created a flexible fiberoptic bronchoscope (FOB), and subsequent improvements under his guidance included the development of video-bronchoscopy. This allows the entry and direct magnified visualization not only of the right and the left main stem bronchus (RMSB, LMSB), but also of the secondary (lobal) and the tertiary (segmental) bronchi.

FOB became an instrumental procedure for definitive diagnosis of infections, malignancy, and tracheobronchial trauma. It is achieved using routine washing of bronchi with 10 ml sterile normal saline solution (NSS), or bronchoalveolar lavage (BAL) with a sterile NSS in increments of 50 ml, aiming for a total 150-20 ml in patients in acute pneumonitis for acquiring samples to culture for microbial specification and antibiotic sensitivities; inspection of a distant airway pathology; direct endobronchial forceps or brush biopsy of tracheobronchial tumors or fine-needle aspiration (FNA) of the mediastinal lymph node (LN) sample by ultrasonic (US) guidance in preoperative staging and assessment of a primary lung malignancy for nodules less than 10 mm in size as delineated by positron emission tomography (PET) which is accurate in 80%; using an endobronchial ultrasound (EBUS) small probe fitted through a FOB port to enhance visualization of the soft tissue adjacent to the bronchial lumen to sample pulmonary nodules or LN; navigation bronchoscopy by coupling high-resolution computer tomography (CT) with electromagnetic guidance of bronchoscope delivered biopsy instruments to sample more peripheral lung lesions.

Overall sensitivity of conventional white-light bronchoscopy for detection rate of preinvasive carcinoma, i.e., carcinoma in situ is only 40%, which could be increased to 80% detection rate by employing fluorescence bronchoscopy, developed in

[1410] Cabrol C, Pavie A, Mesnildrey P, et al. Long-term results with total replacement of the ascending aorta and reimplantation of the coronary arteries. J Thorac Cardiovasc Surg 1986;91:17-25.

[1411] Ikeda S, Tabayashi K, Sunakura M, et al. Diagnosis using a fiberscope – the respiratory. Organs. Naika 1969 Aug;24(2):284-91.

1991 by **Jaclyn Hung**[1412] — Canadian physicist. The latter relies on the difference in autofluorescence spectra between normal and malignant airway epithelia. Using Helium-Cadmium (He-Cd) light amplification by stimulated emission of radiation, i.e., laser 422 nm (20th century. Radiology), normal epithelium fluorescence, green, whereas malignant tissue fluorescens brown to red, which help to increase the detection rate of carcinoma in situ.

In 1969 **Hiroshi Shinya (1935 –)** — Japanese American general surgeon, developed a flexible fiberoptic colonoscope for the examination of the large bowel, removal of colonic polyps.

Introduction of rigid and fiberoptic tracheobronchial and esophagogastroduodenoscopy led to advances and innovations not only in the endoscopic diagnosis, but also in the endoscopic therapy of the airways and the upper gastrointestinal (GI) tract diseases. Among advances and innovations in the endoscopic curative and palliative therapy of the tracheobronchial and upper GI diseases are laser ablation of endoluminal tumors, photodynamic therapy (PDT), radiofrequency (RF) ablation, brachytherapy, and endoscopic modalities for benign or premalignant esophageal disease.

Most commonly used in thoracic surgery is the solid body the neodynium:yttrium-aluminum-garnet (Nd:YAG) laser. The laser energy can be delivered through a small caliber fiberoptic conduit with the flexible bronchoscope or esophagoscopy to allow application of either low power for coagulation or high power for vaporization of diseased tissue, with the maximal depth of penetration of about 4 mm. The Nd:YAG laser is used to restore the passage through obstructed by malignant tumors airways in patients with subsequent symptomatic relief from cough, dyspnea, hemoptysis and obstructive pneumonia and accelerated weaning from mechanical ventilation. Also, malignant tumors of the esophagus, including obstructive, are amenable for lasering, usually in weekly intervals to achieve complete recanalization.

The basic tenant of PDT is the accumulation within target cells, such as tumor cells and their interstitial, of a photosensitizing substance, e.g., purified hematoporphyrin derivative (porfimer sodium) or tetra hydroxyphenyl chlorin (m-THPC) intravenously (IV) 1.5-2.0 mg/kg, with the optimal wavelengths of light absorption of 630 nm and 652 nm, respectively. After 1-4 days the Nd:YAG laser light delivered by FOB cylindrical diffuse fibers, 2,5 and 5 cm directly to cancer cells harboring the photosensitizer triggers cell destruction. A depth of penetration for porfimer sodium is approximately 5 mm. During a second and a third endoscopy performed in 48-hour intervals, necrotic debris are removed by irrigation and suction. With curative intent PDT is approved

---

[1412] Hung J, Lam S, LeRiche JC, et al. Autofluoroscence of normal and malignant bronchial tissue. Laser Surg Med 1991;11(2):99-105.

for the treatment of a small proximal endobronchial tumors, a high-grade dysplasia associated with normal stratified squamous epithelium lining of the esophagus by a single columnar epithelium with goblet cells of the stomach or Barrett's esophagus **[Norman Barrett (1903 – 79): above]**, and to treat early-stage esophageal tumors (T1-T2) who refuse or are not candidates for surgical resection. With palliative intent PDF is approved for advanced lung cancers with bronchial obstruction and advanced malignant esophageal tumors with bleeding or dysphagia. However, the FDT is complicated by post-procedural stricture of the airways and esophagus in 20% of patients.

Components RF ablation include a sizing balloon 10 to 34 mm, a 3-cm balloon-based bipolar RF electrode, and a dedicated RF generator. The RF mucosal ablation is achieved by using a power of 300 W to deliver 10-12 joules/cm^2 for the maximum ablation depth of 1 mm, which is superficial to the muscularis mucosae of the esophagus. The coagulum is removed from the esophageal surface by irrigation and suction. RF ablation is approved for the treatment of Barrett's esophagus with high-grade dysplasia.

High-dose-rate brachytherapy is performed by placement a blind-ended 6 French after-loading catheter through the working channel of bronchoscope of interstitial or intracavitary radioactive seeds (iridium Ir 192) followed by the delivery of high doses of radiation to the airway during a short period with minimal effect to the surrounding tissue. The depth of penetrations by brady-therapy is 0.5 to 2 cm, therefore this therapy is more suitable for tumors with a predominantly extraluminal component.

Deployment of expandable metal stents constructed with cobalt alloys, stainless steel, or a nickel-titanium alloy (nitinol) and silicone-based stents made of silicone rubber by rigid bronchoscope or esophagoscope for the treatment of benign or malignant airway and esophageal strictures. Concurrent interventions include before stenting of the tracheobronchial strictures are balloon dilatation, Nd:YAG laser, PDT, rigid bronchoscopy with debridement, and brachytherapy.

Both rigid and flexible endoscopes are also used for transoral endoscopic stapling of the pharyngoesophageal diverticulum of the esophagus or Zenker's diverticulum **[Friedrich von Zenker (1825 – 98)**. 19th century. Pathology] and for endoscopic plication of the lower gastroesophageal (GE) junction for gastroesophageal reflux disease (GERD).

Zenker's diverticulum is traditionally treated surgically by cricopharyngeal myotomy alone small (<2 cm), diverticulectomy with myotomy and diverticulopexy with myotomy.

But in 1993 **J.M. Collard**[1413] first published a small series of patients treated the transoral stapling of the common septum between the diverticulum and esophagus in those able to fully extend the neck without restriction, low Mallampati intubation score (1-2) / **[Seshhagiri Mallapanti (1941 –)**. 20th century. Anesthesiology], edentulous mouth and adequate diverticulum size (>2 cm). All patients should undergo flexible esophagoscopy for any associated pathological process. A guide wire may be left in place during the procedure to mark the location of the true esophageal lumen. The rigid endoscope is inserted in the patient mouth and a 5 mm straight or 30-degree thoracoscope connected to a video camera to improve visualization through the rigid endoscope. The blades of a scope are opened gently to show both the true esophageal and the lumen of diverticulum. If needed, the diverticulum is cleaned of any residual content with the help of suction and NSS washing. A traction suture is placed in the middle of the ridge of the septum to separate the true lumen of the esophagus and the diverticulum. An endo gastrointestinal apparatus (GIA)-30 is inserted through the rigid endoscope with the jaws across the septum in the midline whilst the traction suture provides contertraction on the septum, which then is stapled by GIA-30 creating of esophago-diverticulostomy.

Endoscopic fundoplication of the GE junction for the treatment of GERD is performed, to restore the angle of His **[Wilhelm His, Sr (1831 – 1904)**. 19th century. Anatomy, histology and embryology] and improve anti-reflux mechanism by creating an omega-shaped valve 3 to 5 cm in length and 200 to 300 degree in circumference, with use of multiple, full thickness endoscopic fasteners (Endophyx)[1414]; plication of a fold mucosa just below the squamous-columnar junction by using an overtube and two endoscopes, one for the suturing and the other for suction and visualization, to lace three sutures below the GE junction in a circumferential, lineal or helical pattern (Endoclinch)[1415], and a creation of a full-thickness plication of the GE junction by delivery of the suture-based plicator implant[1416].

Recently (2017), **Marcia I. Canto** — American gastroenterologist, reported of a portable, hand-held, battery-powered nitrous oxide (N_2O) balloon catheter dilatation system for the cryoendoscopic procedures with the pre-treatment of the esophageal mucosa with the Lugol solution **[Jean Lugol (1786 – 1851)**. 19th century. Microbiology and immunology] chromoendoscopy enhancement of premalignant lesion, aimed

[1413] Collard JM, Otte JB, Kestens PJ. Endoscopic stapling technique of esophagodiverticulostomy for Zenker's diverticulum Ann Thrac Surg 1993;56:573-6.

[1414] Rothstein RI. Endoscopic therapy of of gastroesophageal reflux disease: outcomes of the randomized-controlled trials done today. J Clin Gastroenterol 2008;42:594-602.

[1415] Heider M, Iqbal A, Fillipi CJ. Endoluminal gastroplasty: a new treatment for gastroesophageal reflux disease. Thorax Surg Clin 2005;15:385-94.

[1416] Pleskow D, Rothstein R, Lo S, et al. Endoscopic full-thickness plication for the treatment of GERD: 12-month follow-up follow-up for the North American open-label trial. Gastrointest Endosc 2005;61:643-9.

to eradicate early esophageal squamous cell neoplasia, including high-grade lesions. The device contains liquid N_2O, turns to gas when released into a balloon catheter in endoscope. The balloon is inflated enabling to see through it the neoplastic mucosa pre-treated with the Lugol solution, thus allowing precise cryoablation of the target lesions across the balloon wall. The gas is automatically evacuated without the need for decompression or suctioning. A single procedure produced cure in 8 of 10 patients during check- 3 months later. Two patients developed stricture which was resolved by balloon dilatation after 2-5 repeated procedures. All patients were disease free at median follow up of 10.7 months.

Frank W. Jobe (1925 – 2014) — American orthopedic surgeon, performed in 1975, a pioneering ulnar collateral ligament reconstruction, using tendon-to-ligament grafting.

Peter H. Lord (1925 – 2017)[1417] — English surgeon, introduced the surgical procedure consisting of excision and eversion (plication) of the tunica vaginalis to treat a small hydrocele testis Bold>[19th century. Surgery – **Ernst Bergmann (1835 – 1907), Mathie Jabolay (1860 – 1913), Karl Wilkenmann (1862 – 1925)**: Lord's hydrocelectomy, 1964] and re-introduced the anal dilatation or stretching for hemorrhoids [19th century. Surgery – **Joseph Recamier (1774 – 1852)**; Lord's procedure, 1970].

Hnat M. Matiashyn (11.XI.1925, village Andriivka, Slov'ianskyy Region, Donets'ka Obl., Ukraine – 30.X.1979, city of Kyiv, Ukraine)[1418] — Ukrainian general surgeon, Surgeon-in-Chief of Ukraine (1964-79).

To correct mitral regurgitation (MR) due to ruptured chordae tendinea **Dwight C. McGoon (1925 – 99)**[1419] — American cardiac surgeon, performed a partial resection and plication of the posterior leaflets of the mitral valve (MV, 1960), repaired supravalvular aortic stenosis using a rhomboid patch placed across the sino-tubular junction that extended from the base of the noncoronary sinus of Valsalva to the ascending aorta (AA, 1961). During the repair of truncus arteriosus (TA) performed

---

[1417] Lord PH. A bloodless operation for the radical cure of idiopathic hydrocele. Br J Surg 1964 Dec;51:914-6. — Lord PH. A bloodless operation for spermatocele or cyst of the epididymis. Br J Sur 1970 Sep;57(9):641-4. — Lord PH. Hemorrhoidectomy versus manual dilatation of the anus. Lancet 1972 Nov 11;2(7785):1021.

[1418] Village of Andriivka (founded 1850), Slov'iansky Region, Donetsk Obl., Ukraine. — Matiashyn IM, Gluzman AM. Reference Handbook of Surgical Operations. Publ. «Zdorov'ia», Kyiv 1979. — Ablitsov VH. «Galaxy Ukraine». The Ukrainian diaspora: prominent persons. Publ. «Kyt», Kyiv 2007.

[1419] Ellis FEJr, Frye RL, McGoon DC. Result of reconstructive operations for mitral insufficiency due to ruptured chordae tendineae. Surgery 1966;51(1):165-72. — McGoon DC, Mankin HT, Vlad P, et al. The surgical treatment of supravalvular aortic stenosis. J Thorac Cardiovasc Surg 1961;41:125-33. / **Antonio M. Valsalva (1666 - 1723]** — Italian anatomist. — McGoon DC, Rastelli GC, Ongley PA. An operation for the correction of truncus arteriosus. JAMA 1968;205:69-73.

the first valved conduit repair of the main pulmonary artery (MPA), using an aortic allograft (1967)[1420].

Between November 1963 and February 1964 **Keith Reemtsma (1925 – 2000)** — American transplant surgeon, performed xenotransplantation or cross-species transplantation of the kidneys, i.e., between two different species, from chimpanzee to series of six human recipients who suffered end-stage kidney disease (ESKD), using the «end block» technique, where two kidneys with their accompanying blood vessel, including the abdominal aorta (AA) and the inferior vena cava (IVC), were anastomosed to the recipient common iliac artery (CIA) and common iliac vein (CIV).

The chimpanzee kidneys were chosen because of their size and the ABO blood group are similar in humans. However, an unknown at that time was physiological function of the transplanted chimpanzee kidneys, because it appears to produce an excessive amount of urine, thus causing electrolyte deprivation, especially of potassium (K^{++}) which may cause a sudden cardiac death (SCD).

Postoperative immunosuppression included actinomycin C, corticosteroids, and x-ray irradiation.

Of the six operated six patients, five survived from one week to two months.

A 23-year-old woman, teacher, suffered from chronic glomerulonephritis or Bright disease [19th century. Internal medicine – **Richard Bright (1789 – 1858)**] and severe uremia (in Greek – «oupov» meaning urine and «uïque» meaning blood), with exacerbated since November 1963. She underwent on January 13, 1964, cross-species transplantation using the chimpanzees kidneys. Postoperatively antirejection treatment included azathioprine and prednisone, but she experienced excessive urination, frequently more than 20 liters (L) per day. She returned to work, but died nine months after the operation from SCD, apparently related to a low blood level of K^{++}. Autopsy revealed normal appearance of transplanted kidneys without evidence of rejection.

He also introduced an intra-aortic balloon assist (IABA) device as a bridge to the heart transplantation (HT).

In high-risk patients with fragile bone, or the off-center midline longitudinal sternotomy incision (MLSI) without deep sternal wound infection, one may consider the technique promoted by **Francis Robicsek (1925 – 2010)**[1421]— Hungarian American thoracic and cardiovascular surgeon, which involves weaving a vertical wire up and down lateral to the sternum to be included by cerclage wires of the standard closure. However, using this technique after bilateral harvesting of the

---

[1420] McGoon DC, Rastelli GC, Ongley PA. An operation for the correction of truncus arteriosus. JAMA 1968;205:69-73.

[1421] Robicsek F, Daugherty HK, Cook JW. The prevention and treatment of sternum separation following open heart surgery. J Thorac Cardiovasc Surg 1977;73:267-8.

internal thoracic artery (ITA), mastectomy and irradiation of the chest wall for cancer there is a high risk for development of ischemia and devitalization of the sternum.

The first women to perform the open heart operations with the use of under cardiopulmonary bypass (CPB) was **Lena M. Sydorenko**, a.k.a., **Lena N. Sidorenko (06.III.1925, town of Yevparoria, Crimea, Ukraine – 2006, borough of Zelenograd, Moscow Obl., Russia)**[1422] — Ukrainian thoracic and cardiac surgeon, who performed closure of an atrial septal defect (ASD) and ventricular septal defect (VSD) under CPB, and carried out the first open heart operation in the barocamera **[Ite Boerema (1902 – 80) / Dmytro Panchenko (1906 – 95**: 20th century. Internal Medicine)**]**, first with **Mykola Amosov (1913 – 2002**: above**)** carried out mitral valve replacement (MVR) using artificial ball-valve with fully veneered saddle (1965). In 1962-78 led Department of Cardiovascular Surgery in the Institute of Tuberculosis and Thoracic Surgery (ITTS), city of Kyiv, Ukraine. She authored 195 scientific publications.

Correction of a benign stricture of the lower esophagus proposed by **Alan P. Thal (1925 – 2014)** — South-African American surgeon, consist of opening the narrowed area longitudinally and the adjacent external gastric wall is patch sutured over the defect in the esophagus (Thal procedure or fundoplication).

Pioneer in the field of oral implantology, **Leonard Linkow (1926 – 2017)** — American dentist, who in 1952 placed his first dental implant with a titanium, or a titanium alloy. If the later has more than 85% titanium content, it will form a titanium biocompatible titanium oxide surface layer or veneer that encloses the other metals preventing them from contracting the bone.

In the early 1950's **Albert Starr (1926 –)** — American cardiovascular surgeon, started his career in chosen specialty, and in this same time initiated collaboration with **M. Lowell Edwards (1898 – 1982)** — American engineer and inventor, on the design of a prosthetic mitral valve (MV). By 1960 the case of the caged-ball valve was cast in a piece stainless steel, or Stellate 21. Its surface was silicone coated. The knitted Teflon cloth fixation ring was attached by Teflon spreader ring and brained Teflon thread. The ball was made of Silastic. In that way arise a caged ball (Starr-Edwards) valve for the mitral position. The first mitral valve replacement (MVR) with

---

[1422] Town of Yevpatoria (Foot Note 529), Crimea, Ukraine. — Borough of Zelenohorod (founded 1958 as the first satellite town in the North-East, 37 km, from the center of city of Moscow, now one of 12 administrative district of Moscow, Moscow Obl., Russia). — Sidorenko LN. Pleuro-pneumonectomy in the Treatment of Tuberculous Pleural Empyema. Dissetation for the title of Candidate on Medical Science (CMS), National Medical Academy of Postgraduate Education name for P.L. Shupyk, Kyiv, Ukraine 1960. — Sidorenko LN. Operations under Artificial Circulation during Repair of Intraatrial Septal Defects. Dissertation for the title of Doctor of Medical Sciences (DMS), ITTS, Київ, Ukraine, Україна 1962. — Amosov NM, Sidorenko LN. The First Experience in Surgical Treatment of Mitral Insufficiency. Surgery of the Heart and Vessels. Publ. «Derzhmedvydav» of Ukraine, Kyiv, 1965. — Ablitsov VH. «Galaxy Ukraine». The Ukrainian diaspora: prominent persons. Publ. «Kyt», Kyiv 2007

Starr-Edwards's valve was performed on August 25, 1960. In the 1961 report[1423] of 8 patients who underwent MVR with Starr-Edwards valve two patient died, two had serious complication and 4 were doing well. By 1965 the operative mortality (OM) decreased from 50% to 5%.

In 1963 **Thomas E. Starzl (1926 – 2017)** — American surgeon, performed the first successful human-to-human orthotopic liver transplantation.

Though myocardial necrosis occurred after the use of **Denis Melrose (1921 – 2007**: above) solution for cardioplegic cardiac arrest, the studies on cardioplegia were going on, using mostly an intracellular or extracellular solution. And so cardioplegia was re-introduced in 1964 by **Hans Bretschneider (1922 – 93**: above) in the form of a low salt, free of potassium and procaine containing normothermic or cold solutions.

In 1983 **Hans G. Borst (Oct. 17, 1927, München, Germany –)** — German cardiac surgeon, reported replacing the aortic arch aneurysm with TSG and leaving a 10-15 cm tube graft lying free in the proximal aneurysmal DTA for its future replacement (the elephant trunk technique).

Median arcuate ligament syndrome, a.k.a. celiac artery (CA) compression syndrome, described by **Pekka T. Harjola (1927 –)**[1424] — Finish physician in 1963, and by **Samuel A. Marable (1928 – 70)**[1425] — American surgeon, in 1965, is a rare disorder characterized by postprandial abdominal pain, nausea, and subsequent weight loss. It is caused by isolated stenosis of the CA, from the extrinsic compression by a thick fibrous band of the diaphragmatic crura called the median arcuate ligament. The diagnosis is made by exclusion, since as many as 10% to 60% healthy individuals can have stenosis of the CA without symptoms, confirmed by dynamic duplex vascular ultrasound (US), computer tomography (CT) angiography, magnetic resonance (MR) angiography, digital subtraction angiography (DSA) or formal arterial angiography. Treatment recquire decompression of the CA obstruction by arterial bypass or transection of the median arcuate ligament fibers and ganglionic tissue overlying the CA via an open laparotomy, laparoscopy with or without robotic assistance[1426]. Endovascular treatment including angioplasty and/or stenting alone has been met with poor results.

In the blunt trauma to the chest as a result of a horizontal deceleration, i.e., automobile accident, where an acute strain is placed mainly by the steering-wheel

---

[1423] Starr A, Edwards ML. Mitral replacement: clinical experience with a ball-valve prosthesis. Ann Surg 1961;154:726-40.

[1424] Harjola PT. A rare obstruction of the celiac artery. Report of a case. Ann Chir Gynaecol Fenn 1963;52:547-50.

[1425] Dunbar JD, Molnar W, Beman FF, Marable SA. Compression of the celiac trunk and abdominal angina. Am J Roentgenol Rad Ther Nucl Med 1965;95:731-44.

[1426] El-Hayek KH, Titus J, Bui A, et al. Laparoscopic median arcuate ligament release: are we improving symptoms? J Am Coll Surg 2013;216:272-9.

either on the precordial area causing the «commotio cordis» (in Latin – «agitation» of the heart), on the aortic isthmus, a slight constriction of the descending thoracic aorta (DTA) immediately distal to the left subclavian artery (SCA) at the point of attachment of the ligamentum arteriosum, or the pulmonary veins (PV). The «commotio cordis» from the steering-wheel or during the sport school hours or competition activity, is a lethal disruption of the heart rhythm due to blow directly over the heart, at a critical time during cardiac cycle heart beast with the ascending phase of the T wave, causing sudden cardiac death (SCD) from ventricular fibrillation (VF) with fatality rate of about 63% with prompt cardiopulmonary resuscitation (CPR), or over 80% without. The vertical deceleration injury to the aortic isthmus result in a retro-pleural tear or rupture of the DTA in the proximal portion of the DTA which can be confined by the aortic adventitia creating false aneurysm, or hemorrhage confined by parietal pleura creating retro-pleural hematoma, or of free intrapleural hemorrhage. The victims who survive the initial impact of this kind of blunt trauma, a hemorrhage is temporarily contained by the aortic adventitia or the parietal pleura, and they arrive to the emergency room alive in shock alive. Those with free intrapleural hemorrhage die instantly.

Blunt trauma to the chest as a result of a vertical deceleration, i.e., free fall from heights, helicopter crush, can acutely strain the ascending aorta (AA) at the junction with the innominate artery (IA), as the heart is being displaced caudally and to the left pleural cavity from its natural position in the mediastinum, causing torsion (two-way twisting) of the AA with / without rupture, or azygos vein laceration at the junction of the superior vena cava (SVC).

However, the type of deceleration trauma affecting a victim may change depending on his/hers position at moment of collision.

Techniques used in repair of blunt horizontal deceleration injury to the proximal DTA include: clamp and sew repair; pump bypass between the left atrium (LA) with cannulation site for the outflow through the LA appendage or the left-sided PV, and for the inflow via the common femoral or iliac artery; designed by **Vincent L. Gott (1927 – 2020)** — American cardiac surgeon, shunt tube for a heart lung machine (HLM) in which the heparin is ionically bound to benzalkonium chloride substrate polymer to decrease thrombogenicity, which showed to be consistent only in decreasing inflammation related to compliment activation during cardiopulmonary bypass (CPB) – Gott shunt, between the AA and the distal DTA; repair under cardiopulmonary bypass (CPB). The selected victims, i.e., unconscious or with other associated injuries such as fractured pelvis, should be treated by the non-invasive thoracic endovascular aortic repair (TEVAR).

Nina Starr-Braunwald (1928 – 92)[1427] — American cardiac surgeon, designed an artificial polyurethane cardiac valve reinforced with Dacron which had extension that could be sutured to the left ventricular (LV) wall to act as a replacement for the chordae tendineae (neo-chordae).

First valve she implanted on March 10, 1960, into a 16-year-old girl, suffering from severe mitral insufficiency (MI), in the mitral position. Despite successful outcomes in dogs, the valve failed to function, and the patient died within three days.

Next day (March 11, 1960), she implanted a second valve into a 44-year-old woman, also suffered from severe MI, in the mitral position. The patient had an uneventful recovery and did clinically well. Unfortunately, died suddenly four months following surgery, presumably from cardiac arrhythmia.

She also designed in conjunction with Cutter Laboratories (Berkeley, CA)[1428] the two types open-caged titanium prosthetic cardiac valves with cloth covered struts using Dacron fabric and polypropylene mesh that had silastic or silicone poppets for the aortic position (Braunwald-Cutter valve), used in 1968-79. There were significant number cases of fabric wear, poppet abrasions and poppet escape.

In 1970 **Fedir A. Serbynenko (May 24, 1928, village of Dmytrovske, SK, Russia – Mar. 01, 2002, Moscow, Russia)**[1429] — Ukrainian Russian neurosurgeon, who designed the remote (via peripheral artery) technique of a double balloon-catheter, one of them being a detachable microballoon, for embolization of the intracranial aneurysms (1970), since 1973 – also cerebral aneurysms. By using this technique, he successfully performed the first reconstruction of the cavernous

---

[1427] Braunwald NS, Cooper TC, Morrow AG. Complete replacement of the mitral valve. J Thorac Cardiovasc Surg 1960;40:1-11. — Braunwald NS. It will work. The first successful mitral valve replacement. Ann Thorac Surg 1989;48(3)S1-S3.

[1428] Cutter Laboratories, Berkeley, CA, founded in 1897 by **Edward A. Cutter (1840 - 1914)** — American entrepreneur. It was purchased in 1978 by Bayer AG (founded 1863), Leverkusen, Germany. Cutter Laboratories bought (1974), incorporated 1978 into German International Pharmaceutical and Life Sciences «Bayer Corporation» / «Bayer Aktien Gesellschaft (AG)» (founded 1863), Stadt Leverkuzen (known since XII century), Land Nordrhein Westfalen (area 34,084.13 km²), Germany. In «Bayer Corporation» first was produced diacetylmorphine – heroin (1898); from 1899 began production for wide use aspirin [19th century. Biochemistry– **Charles Gerard (1816 - 56)**]; since 1935 its Director **Gerhard Domagk (1895 - 1964:** 19th century. Microbiology and immunology] who in in this same year (1935) discovered the first antibiotic sulfonamide (Prontosil) bacterial Streptococcal infections, which he used to treat her own daughter, thus saved her from amputation of her hand .

[1429] Village Dmytriyevske (founded 1846), Chervonogvardiysky Region, SK (Foot Note 334), Ukraine (1918-22) / Russia. — Serbinenko FA. Balloon occlusion of cavernous portion of the carotid artery as a method of treating carotid cavernous fistula. Zh Vopr Neyrokhir 1971 Aug;6:3-9. — Serbinenko FA. Balloon catheterization and occlusion of major cerebral vessels. J Neurosurg 1974 Aug;41(2):125-45 — Teitelbaum GP, Larsen DW, Zelman V, Lysachev A, Likhtenman LB. A tribute to Dr. Feodor A. Serbinenko, founder of endovascular surgery. Neurosurgery 2000;46(2):462-9; discussion 469-70. — Guglielmi G. History of the genesis of detachable coils. A review. J Neurosurgery 2009 July;111(1):1-8. — Ablitsov VH. «Galaxy Ukraine». The Ukrainian diaspora: prominent persons. Publ. «Kyt», Kyiv 2007.

section of the internal carotid artery (ICA) at the carotid-cavernous junction. He is correctly considered the founder of endovascular neurosurgery who changed forever the practice of neurovascular surgery.

Per-Ingvar Branemark (1929 – 2014) — Swedish orthopedic surgeon, who since 1978 investigated the biological fusion of bone to a foreign material (osteointegration) and reinvigorated the field of implantology by developing and testing a type of dental implant utilizing pure titanium screws which he termed «fixtures».

Operations for the correction of cyanotic congenital heart disease (CHD) are based on the concept that the right atrium (RA) is able replace the right ventricle (RV) to maintain pulmonary circulation and the RV able to support the systemic circulation was developed by **Francois M. Fontan (1929 – 2018)** — French pediatric cardiac surgeon. In half-Fontan procedure the superior vena cava (SVC) is connected to the right pulmonary artery (PA) in the «side-to-side» fashion using homologic patches to bypass the RA (1968). The alternate way is to place a composite (valved) conduit between the RA appendage and the main pulmonary artery (MPA, 1971). In the full-Fontan's operation an internal-RA lateral synthetic tunnel-like baffle or an external lateral composite (valved) synthetic tubular graft (STG) is established between the inferior vena cava (IVC) and the right PA bypassing the RA for the in series pulmonary and systemic blood flow with the single RV (1990).

Adib D. Jatene (1929 – 2014) — Brazilian cardiac surgeon, performed the anatomical correction of D-transposition of the great arteries (D-TGA) by «switching» the great arteries (Jatene «switch» operation, 1976); introduced the method of reconstruction of the anterior post-myocardial infarction (post-MI) left ventricle (LV) aneurysm by means of a vertical septoplasty using the «U»-like sutures and by closure of the neck of the aneurysm by the endoventricular purse-string suture (1984-85).

Dmytro Ye. Babliak (Sept. 27, 1930, city of Kamyanets-Podilsky, Khmelnytsky Obl., Ukraine – March 03, 2008, city of Lviv, Ukraine)[1430] — Ukrainian thoracic and cardio-vascular surgeon, perform embolectomy of the main PA through a small incision on the affected side of its main branch (1979).

Most widely used clinical classification for thymoma [19th century. Internal medicine – **Samuel Wilks (1824 – 1911)** / 20th century. Pathology – **Mirella Marino** and **Hans Muller Hermelink (1947 –)**. Anesthesiology / Surgery – **Ernst Sauerbruch (1875 – 1951)**] based on the anatomic extend of disease at the time of surgery is that presented in 1981 by **Akira Masaoka (1930 – 2014)**[1431] — Japanese thoracic surgeon:

[1430] City of Kamyanets-Podilsky (Foot Note 927), Khmelnytsky Obl., Ukraine. — Ablitsov VH. «Galaxy Ukraine». The Ukrainian diaspora: prominent persons. Publ. «Kyt», Kyiv 2007.

[1431] Masaoka A, Monden Y, Nakahara K, et al. Follow-up study of thymomas with special reference to their clinical stages. Cancer 1981;48:2485-92. — Monden Y. Akira Masaoka (1930-2014): a great surgeon and a special musician. Gen Thorac Cardiovasc Surg 2015;63:307-8.

Stage I   –   Macroscopically completely encapsulated and microscopically without capsular invasion;

Stage II A   –   Microscopic invasion through the capsule into surrounding fatty tissues;

Stage II B   –   Macroscopic invasion into a capsule;

Stage III   –   Macroscopic invasion of neighboring organs (pericardium, great vessels, lung);

Stage IV A   –   Pleural or pericardial dissemination;

Stage IV B   –   Lymphagenous or hematogenous metastases to distant (extra-thoracic) sites.

In 1969 **Kazi Mobin-Uddin (1930 – 99)**[1432] — Pakistani American Thoracic and Cardiovascular Surgeon, conceived the first inferior vena cava filter (IVCF) or the Mobin-Uddin umbrella, for a non-invasive surgical placement through a venotomy into the internal jugular vein (IJV) under local anesthesia with the advancement it downwards into the inferior vena cava (IVC) below the entry of renal veins, in patients with deep venous thrombosis (DVT), to interrupt a thrombus passage and prevent embolization to the pulmonary artery (PA).

In 1972, **Gordon K. Danielson (1931 – 2017)**[1433] —American adult and pediatric cardiac surgeon, developed repair of the TV by annuloplasty of its posterior leaflet, construction of a monocusp valve based on its anterior leaflet coaptating it into the interventricular septum (IVS) by bringing the papillary muscles attached to the free wall of the right ventricle (RV) to the IVS, anchoring it with a pledged suture, primary partial or complete with fenestration closure of the ASD and plication of the free wall of atrialized RV.

Two years later (1974) Gordon Danielson described treatment of the aortic native valve endocarditis (NVE) with the aortic root abscess by excision of infected tissue with abscesses, an aortic root replacement by a valved synthetic tubular graft (STG) and coronary artery bypass (CAB) using the great saphenous vein grafts (GSV).

[1432] Mobin-Uddin K, McLeon R, Bolooki H, et al. Caval interruption for prevention of pulmonary embolism: long-term results of a new method. Arch Surg 1969 Dec; 99(6):711-5. — Mobin-Uddin K, Collard GM, Bolooki H, et al. Transvenous caval interruption with umbrella filter. NEJM 1972 Jan13;268(3):55-8. — Wojtowycz M, McDermott J, Crumpy A. Inferior vena cava filters. J Ukr Med Assoc North Am 1991 Winter;38,1(123):3-12.

[1433] Danielson GR, Maloney JD, Devloo REA. Surgical repair of Epstein's anomaly. Mayo Clin Proc 1979;54:185-92. — Danielson GK, Titus JL, DuShane JWL.Successful treatment of aortic valve endocarditis and aortic root abscesses by insertion of prosthetic valve in the ascending aorta and placement of bypass grafts to coronary arteries. J Thoracic Cardiovascular surg 1974;67:443-9.

Term ventricular arrhythmias refers to ventricular tachycardia (VT), ventricular fibrillation (VF) and «Torsades de pointes» / «twisting of the points» (in French – «torsade de pointes»).

In VT the heart rate (HR), i.e., the left ventricular (LV) contraction rate is accelerated to 120 beats per minute or higher, and is associated with lightheadedness, palpitation, chest pain, which may lead to VF and sudden cardiac death (SCD). Causes of VT are congenital heart disease (CHD), coronary artery disease (CAD), myocardial infarction (MI), aortic stenosis (AS), cardiomyopathy (CM), electrolyte imbalance and drug interaction. VT is classified into non-sustained lasting less that 30-second and sustained that last more than 30 second.

Torsades de pointes described in 1966 by **Francois Dessertenne**[1434] — French physician, is a polymorphic VT characterized on the electrocardiography by a prolongation of QT interval, which can increase the risk of developing of an abnormal HR and leads to SCD. His patient was an 80-year-old female with intermittent atrioventricular (AV) block with syncopal episodes due to VT with two variable opposing foci – bradycardia (in Greek «brady» – slow, «cardia» – heart), i.e., slowing of the heart beats.

First direct surgical cure of VT was reported in 1975 by **John J. Gallagher**[1435] — American cardiac surgeon, on a 54-year-old man who presented with refractory VT after two MI's which led to development of LV aneurysm. The site of earliest epicardial activity to the margin of an aneurysmal scar was delineated by epicardial mapping. Resection of an aneurysmal scar abolished VT.

In 1978 **Gerard M. Guiraudon (1931 –)**[1436] — French cardiac surgeon, introduced the encircling endocardial ventriculotomy, consisting of near transmural incision into the scar from the endocardial to the epicardial surface of the LV, to isolate the entire border zone from normal myocardium. The operation was successful in eradicating VT, but caused significant LV dysfunction.

Next year (1979) **Alden H. Harken (1942 –)**[1437] — American cardiac surgeon, reported on treatment 12 patients with malignant VT by endocardial resection

---

[1434] Dessertenne F. La tachycardie ventriculaire a deux foyers opposes. Arch mal coeur vaiss 1966 Feb;59(2):263-72. — Oksuz F, Sensay B, Sahan E, et al. The classic «R-on-T» phenomenon. Indian Heart J Jul-Aug 2015;(64(4):392-4.

[1435] Gallagher JJ, Oldham HN, Wallace AG, et al. Ventricular aneurysm with ventricular tachycardia. Report of a case with epicardial mapping and successful resection. Am J Cardiol 1975;35(5):696-700.

[1436] Guiraudon G, Fontaine G, Frank R, et al. Encircling endocardial ventriculotomy: a new surgical treatment for life-threatening ventricular tachycardias resistant to medical treatment following myocardial infarction. Ann Thorac Surg 1978;26(5):438-44.

[1437] Harken AH, Josephson ME, Horowitz LN. Surgical endocardial resection for the treatment of malignant ventricular tachycardia. Ann Surg 1979;190(4):456-60.

directed by intraoperative mapping, with one operative death and no recurrent VT in the remaining patients at one-year follow-up.

Then, in 1982 **J.M. Moran**[1438] — American cardiac surgeon, modified the endocardial resection with an extended resection of all visible scar, which abviated the need for intraoperative mapping.

Those operations resulted on operative mortality (OM) of 10-20%, success rate in more than 70% of patients, and 5-year survival about 60%.

In 1989 **Jose P. Da Silva**[1439] — American pediatric cardiac surgeon, introduced TV repair using the cone reconstruction which consist of an incision at 12 o'clock and clockwise dissection off its anterior and posterior leaflets to the apex of the RV, incision into and dissection of the proximal edge of the septal leaflet off the endocardium medially to the anteroseptal commissure, the posterior leaflet is cut and rotated clockwise to meet the proximal edge of the septal leaflet, and these two leaflets are approximated, the posterior anulus is plicated to the size of the neo-TV which is now anchored to the true anulus with interrupted sutures. Consequently, a thinned, transparent, atrialized RV is plicated, redundant right atrium (RA) is excised and ASD is closed. In older children and adults, an annuloplasty ring is placed to reinforce the neo-TV annular reconstruction.

On June 16, 1958 **Seymour Furman (1931 – 2006)**[1440] — American cardiothoracic surgeon and cardiologist, inserted the first internal temporary fixed-rate pacemaker (PM) connected to the external PM generator, by threading plastic-covered electric wires connected to the electrode, through the cephalic vein (CV) at the forearm near the elbow, into the apex of the right ventricle (RV) to treat advanced heart block in the 76-year old patient with the Gerbec-Morgagni-Stokes-Adams seizures **[Marko Gerbec (1658 – 1718)**. 18th century. **/ Giovanni Morgagni (1682 – 1771)**. 18th century**)** / **Robert Adams (1791 – 1875)**. 19th century. Internal medicine. / **William Stokes (1804 – 78)**. 19th century. Internal medicine.**]** or sick sinus syndrome (SSS), for 96 hours. After this procedure patient was walking.

First permanent transvenous PM were implanted almost simultaneously in 1962 by **H. Lagergren**[1441] and **Lennart Johansson** — Swedish cardiac surgeons, and by **Victor Parsonnet (1924 –)**[1442] — American cardiac surgeon.

[1438] Moran JM, Keohe RF, Loeb JM, et al. Extended endocardial resection for the treatment of malignant ventricular tachycardia and ventricular fibrillation. Ann Thorac Surg 1982;34(5):538-52.

[1439] Da Silva JP, Baumgratz FJ, Fonseca L, et al. The cone reconstruction of the tricuspid valve in Epstein's anomaly. The operation: early and midterm results. J Thorac Cardiovasc Surg 2007;133:215-23.

[1440] Furman S, Schwedel JB. Intracardiac pacemaker for Stokes-Adams seizures. N Engl J Med 1959;261:943-8.

[1441] Lagengren H, Johansson L. Intravenous stimulation for complete heart block. Acta chir scand 1963;125:562-6.

[1442] Parsonnet V, Zucker RI, Asa MM. Preliminary investigation of the development of a permanent implantable pacemaker utilizing an intracardiac dipolar electrode. Clin Res 1962;10;391.

However, the initial PM transvenous electrode's tip was semioval, smooth and occasionally dislodged with subsequent failure to pace. To prevent this complication, **William Chardack (1915 – 2006**: above**)** in 1961-64 developed the first helical coil electrodes, made from corrosion-resistent platinum-iridium ally for stable attachment of the electrode to the innermost serous membrane of the the right ventricle RV and right atrium (RA) of the heart.

Lubomyr I. Kuzmak (02.VIII.1931, village Balyhorod, Lis'ko county, Lemko region, Ukraine / Subcarpathian Voivodeship, Poland – died from brain ischemia due to anesthesiology misshape during decompression laminectomy and vertebral column arthrodesis for fractured vertebrae from fallen metallic closed in hospital Operative Room / OR **12.X.2006, town of Irvington, New Jersey, USA)**[1443] — Ukrainian American surgeon, developed stoma adjustable silicone gastric banding for morbid obesity (1983-86).

Transmyocardial laser revascularization (TLR) of the heart muscle (myocardium) is a surgical procedure which provides direct perfusion of ischemic myocardium for patients with (1) diffuse, severe coronary artery disease (CAD) with refractory angina pectoris, not amenable to standard medical treatment, percutaneous coronary intervention (PCI), coronary artery bypass grafting (CABG) and those who underwent CABG with incompete revascularization (25%), confirmed by selective coronary angiography (SCA); (2) those with with evidence of reversible myocardial ischemia by perfusion scanning, and (3) those with the left ventricular (LV) ejection fraction greater than 25%.

TMR of the myocardium is performed using the carbon dioxide (CO_2) laser to drill transmural channels onto a beating heart through a longitudinal median longitudinal sternotomy incision (MLSI), standard or minimally invasive video-assisted intercostal space thoracotomy (*ICST*), robotically, or percutaneously (PC).

Concept of TLR of the myocardium came after description by **Joseph Wearn (1893 – 1984**; 20th century. Physiology**)** of the trans-endocardial myocardial sinusoid

---

[1443] Village Balyhorod (first mention about town inhabitants 1615, founded 1634), Lis'ko county, Lemko region (area approximately. 8,800 km²), Ukraine / Subcarpathian Voivodeship (area 17,846 км²), Poland. — Town of Irvington (founded 1834), State of New Jersey (N.J., area 22,591.38 km²), USA. — Kuzmak LI. Intrahepatic cholangiojejunostomy as a palliative procedure in obstructive jaundice due to cancer. J Ukr Med Assoc North Am 1976 Apr;23,2(81):13-17. — Kuzmak LI. Peritoneo-venous shunt in the treatment of intractable ascites. Reversal by peritoneo-venous shunt of Levin. J Ukr Med Assoc North Am 1978;25,4(91):197-204. J Ukr Med Assoc North Am 1978;25,4(91):197-204. — Kuzmak LI. New surgical resection method of the low anterior resection of the colon. Ukr Med Assoc North Am 1980 spring;27,2(97):97-104. — Kuzmak LI. Stoma adjustable silicone gastric banding. Prob Gen Surg 1992;9:298-317. — Kuzmak LI. Stoma adjustable silicone gastric banding. Surgical Rounds 1999;(1):19-28. — Ablitsov VH. «Galaxy Ukraine». The Ukrainian diaspora: prominent persons. Publ. «Kyt», Kyiv 2007.

allowing blood flow directly from the LV into the myocardium in reptilians, in lieu of coronary arteries.

Subsequently, in 1982-83 **Mahmood Mirhoseini (1931 –)**[1444] — Iranian American cardiac surgeon, performed TLR of the myocardium in conjunction with CABG using the CO_2 laser, initially with a peak output of 80 Watts (W) which required a significant amount of time to complete a transmural channel, so that the heart had to cooled and arrested. Increasing the output of laser to 800 W allowed TMR of the myocardium to be performed on a beating heart.

CO_2 laser energy to create 1 mm in diameter channels is delivered via hollow tubes with a 20-30 Joule (J) pulse to traverse the myocardium and duration of the signal pulse of 40 msec, synchronized to occurs on the R wave of the electrocardiogram (ECG) to avoid arrhythmias. The transmural channel is reflected by mirrors to reach epicardial surface and can be confirmed by the transesophageal echocardiography (TEE). The vaporization of blood by the CO_2 laser as the laser bean enters the LV creates a distinct acoustic effect detected by the TEE.

After TLR of the myocardium a significant symptomatic improvement was seen in 25% to 76% patients, and improvement in the myocardial perfusion in previously ischemic areas at 12 months was seen in 20%. The probable mechanism for the clinical efficacy of TLR of the myocardium is the stimulation of angiogenesis, i.e., a physiological process of the formation of new vessels from pre-existing, denervation of the sympathetic nerve fibers and only after the use of the CO_2 laser an occasional channel patency.

The Passy-Muir valve (PMV)° 15 mm in internal diameter, 23 mm in outside diameter [Passy and Passy, Inc., Irvine, CA] for patients with tracheostomy and ventilator-dependent to allow them to swallow and speak, invented by **David Muir** — American quadriplegic patient[1445] and **Victor Passy (1931 -)**[1446] — American ORL, is a simple device which attaches easily to a standard 15 mm tracheostomy tube hub.

Classification for peripheral arterial (occlusive) disease (PAD) based on clinical symptoms and hemodynamic data was suggested in 1986 by **Robert B.**

[1444] Mirhoseini M, Muckerheide M, Cayton MM. Transventricular revascularization by laser. Lasers Surg Med 1982;2(2):187-98. — Mirhoseini M, Fisher JC, Cayton MM. Myocardial revascularizationtion by laser: a clinical report. Lasers Surg Med 1983;2;3(3):241-5.

[1445] Tetraplegia (in Greek – «tetra» for four, «plege» for paralysis or stroke), or quadriplegia (in Latin – «quattuor» for four, in Greek – «plege» for paralysis or stroke) is a major deficiency or paralysis of four limbs with motor and / or sensory deficit due to damage of upper cervical spine high in the level of C_{1-7}, or of the brain.

[1446] Passy V, Baydur A, Prentice W, et al. Passy Muir® tracheostomy speaking valve on ventilator-dependent patients. The Laryngoscope 1993 Jan;103(6):653-9.

Rutherford (1931 – 2013) — American vascular surgeon, and named Rutherforda Classification[1447]:

Grade	Cathegory	Clinical Description	Objective Criteria
0	0	Asymptomatic	Normal treadmill test
0	1	Mild claudication	Ankle pressure after exercise <50 mmHg, but >25 mmHg less than brachial
I	2	Moderate claudication	More moderate symptoms
	3	Severe claudication	Does not complete treadmill test Ankle pressure after exercise <50 mmHg
II	4	Ischemic rest pain	Resting ankle pressure <60 mmHg Decrease pulse volume recording
	5	Minor tissue lost Nonhealing ulcer	Resting pressure <40 mmHg Pulse volume recording moderately decreased
III	6	Major tissue loss	As noted in cathegory 5 Loss above the metatarsal, limb no longer salvageable

Patients with the Rutherford category 2 scale and higher, and noninvasive studies demonstrating evidence of significant PAD should undergo angiography to fully delineate the anatomy with the possibility of a noninvasive percutaneous (P/C) or surgical vascularization.

Walter P. Bobechko (Aug. 22,1932, city of Toronto, ON, Canada – Jan. 06, 2007, city of Dallas, TX, USA)[1448] — Ukrainian Canadian orthopedist, developed the Toronto brace for children with Legg-Calve-Perthes disease (19th century. Surgery)

[1447] Rutherford RB, Flanigan DP, Gupta SK, et al. Suggested standards for reports dealing with lower extremity ischemia. Vasc Surg 1986;4(1):80-94.

[1448] City of Toronto (settled 1750, established 1793), ON, Canada. — City of Dallas, TX (established 1841), TX, USA. — Bobechko VP, Hirsch C. Auto-immune response to nucleus pulposus in rabbit. J Ukr Med Assoc North Am 1968 Jan;15,19(48):35-41. — Bobechko VP, Hirsch C. Intervertebral discs. J Ukr Med Assoc North Am 1968 Jan;15,1(48):42-6. — Bobechko WP, McLaurin CA, Motloch WM. Toronto orthosis for Legg-Perthes' disease. Art Limbs 1968;12:36-41. — Bobechko WP. Spinal pacemakers and scoliosis. J Bone Joint Surg 1973 Feb;55:232-3. — Bobechko WP. Idiopathic scoliosis in childhood. J Ukr Med Assoc North Am 1973 Oct;20,4(71):3-14. — Bobechko WP. The Toronto Brace for Legg-Perthes Disease. Clin Orthoped 1974 Jul-Aug;102:115-7. — Ablitsov VH. «Galaxy Ukraine». The Ukrainian diaspora: prominent persons. Publ. «Kyt», Kyiv 2007.

in 1968, which allowed movement at the knees, improved mobility, and allowed temporarily remove the brace for physical therapy (PT).

Based on the thought that idiopathic scoliosis was the result of muscle imbalance he demonstrated that producing a compensatory imbalance in the opposite direction by implanting the muscle pacemaker (stimulator) could result in improvement (1972-73).

To overcome the Harrington Rod system **[Paul Harrington (1911 – 80)**: above] to dislodge from the vertebral arch lamina of the spinal column vertebrae, he and his son developed a double-hook system which not only avoided that complication but allowed the broader distribution of distraction forces.

After introduction at the end of the 1980s the extracorporeal shock valve therapy (SVT), i.e., treatment of some diseases using acoustic valves of the low frequency and high intensity, it led t to its use for the treatment of idiopathic scoliosis in children. Valves quickly pass-through soft tissues (muscles) and fluid (blood and lymph), the resistance of those tissues is low, and they do not harm them. Reaching the bone and cartilaginous tissue the SVT dissolve pathologic formations (excrescences, sheaves), stones, decreases scoliosis.

Finally, he introduced thoracoscopy for a minimally invasive anterior spinal surgery, and using it was the first to remove infected thoracic vertebra (1991).

Hryhorii B. Bodnar (22.IV.1932, village Iskra, Velykonovosilkovs'kyy Region, Donets'ka Obl., Ukraine – 27.I.2014, city of Donets'k, Ukraine)[1449] — Ukrainian surgeon, described the clutch-like esophago-intestinal anastomoses (1970-80), applied the combined intraarterial and endolymphatic polychemotherapy for the treatment of malignant tumors (1990-2000).

Vincent M. Dor (1932 –) — French thoracic and cardiac surgeon, introduced the method of reconstruction of the anterior pos-infarction left ventricular (LV) aneurysm by the exclusion of the nonfunctional part of the LV using the endoventricular circular patch-plasty (EVCPP) with/without the synthetic patch (Dor procedure, 1984-85).

In 1959 **Arthur D. Boyd (1932 – Jan. 22, 2020, city of Lexington, MA, USA)**[1450] — American thoracic and cardiac surgeon, reported about determination of cardiac output (CO) with the Fick method [19th century. Physiology – **Adolf Fick (1829 – 1906)**] and mixed venous blood oxygen (O_2) saturation (SvO_2) when catheters were placed in the left atrium (LA) and pulmonary artery (PA) in 34 children suffering

---

[1449] Village of Iskra (founded 1782), Velykonovosilkovsky Region, Donetska Obl., Ukraine. — City of Donetsk (founded 1869), Ukraine. — Ablitsov VH. «Galaxy Ukraine». The Ukrainian diaspora: prominent persons. Publ. «Kyt», Kyiv 2007.

[1450] City of Lexington (settled in 1641), MA, USA. — Boyd AD, Tremblay RE, Spencer FC, Bahnson HT. Estimation of cardiac output soon after intracardiac surgery with cardiopulmonary bypass. Ann Surg 1959 Oct;150(4):613-26.

from congenital heart disease (CHD) after open intracardiac surgical procedure on cardiopulmonary bypass (CPB). Post-operatively children who could not increase CO and continued in the state of relative low tissue anoxia and metabolic acidosis usually died. Likewise, in 1960 **Jaroslaw Barwinsky (Oct. 15, 1926, village Tustoholovy, Zboriv Region, Ternopil Obl., Ukraine – Aug. 28, 2016, city of Winnipeg, Canada)**[1451] — Ukrainian Canadian thoracic and cardiac surgeon, with associates, also found that patients after serious operations who were not able to increase CO and continued in the state of relative tissue anoxia and metabolic acidosis usually died.

Alain Carpentier (1933 –) — French cardiac surgeon, designed the biological (tissue) stentless (swine) cardiac valve preserved by the glutaraldehyde to increase its durability for the aortic valve replacement (AVR, 1965); proposed the «French resection correction» of mitral insufficiency / regurgitation (MI, MR; 1969)[1452]; introduced radial artery grafting in coronary artery bypass (CAB), although initially unsuccessful (1973); made tissue cardiac valves from swine cardiac valves or bovine pericardium preserved also by the glutaraldehyde attached to the elgiloy alloy with a Teflon sewing ring (Carpentier-Edwards valve, 1975); proposed physiological triad consisting of dysfunction, lesion and etiology for the functional classification of mitral insufficiency (MI) type I-III (1983)[1453]:

Type I
Normal leaflets motion
    Anular dilatation
    Ischemic cardiomyopathy (ICM) due to basal myocardial infarction (MI)
    Leaflet perforation
    Dilated cardiomyopathy (DCM)
    Endocarditis

Type II
Increased leaflet motion (leaflet prolapse)
    Elongation or rupture of chordae
    Degenerative leaflet disease

[1451] Village of Tustoholovy (first mentioned in 1532, founded in 1684), Zboriv Region, Ternopil Obl., Ukraine — City of Winnipeg (French traders built the first fortress on the site in 1738), MB (Foot Note 1296), Canada. — Cloves GHA Jr, Del Guercio LR, Barwinsky J. The cardiac output in response to surgical trauma. A comparison between patients who survived and those who died. Arch Surg 1960;81(2):212-22. — Ablitsov VH. «Galaxy Ukraine». The Ukrainian diaspora: prominent persons. Publ. «Kyt», Kyiv 2007.

[1452] Carpentier A. La valvuloplastie reconstitutive. Une nouvelle technique de valvuloplastie mitrale. Presse med 1969;77:251-4.

[1453] Carpentier A. Cardiac valve surgery – the «French correction». J Thorac Cardiovasc Surg 1983;86(3):323-7. — Filsoufi F, Chikwe J, Adams DH. Acquired disease of the mitral valve.

Fibroelastic deficiency

Barlow's disease[1454]
Marfan syndrome[1455]
    Endocarditis
    Rheumatic heart disease (RHD)
    Trauma.
    Elongation or rupture of papillary muscle
    ICM.

Type IIIa
Restricted leaflet motion (systole and diastole)
    Leaflet thickening or retraction
    Chordal thickening, retraction, or fusion
    Commissural fusion
    RHD
    Carcinoid disease
    Endomyocardial fibrosis

Type IIIb.
Restricted leaflet motion (systole)
    Papillary muscle displacement (one from the other, posterior and to the cardiac apex) – ICM
    Leaflet tethering – DCM.

To treat complex and extensive native valve endocarditis (NVE) of the posterior leaflet of the MV extending into its anulus with peri-annular abscess, he recommended that after excision of all infected tissue and abscess the defect to be reconstructed by placing figure-eight sutures at the atrial and ventricular edges of the defect to close the atrioventricular groove (Carpentier technique, 1996)[1456].

In 1996 Alain Carpentier performed the mini-invasive cardiac surgery (MICS) – the first videoscopic repair (valvuloplasty) of the mitral valve (MV) through a right mini-thoracotomy[1457], in 1998 – the first computer assisted robotic (rob.) repair on 52-year-

---

[1454] **John B. Barlow (1924 - 2008)** — South-African cardiologist (20th century. Internal medicine).
[1455] **Antonine B.J. Marfan (1858 - 1942)** — French pediatrician [19th century. Children diseases].
[1456] Carpentier A, Pellerin M, Fuzellier JF, et al. Extensive calcifications of the mitral valve anulus: pathology and surgical management. J Thorac Cardiovasc Surg 1996;111(4):718-29.
[1457] Carpentier A, Loulmet D, LeBret E, et al. Chirurgie a coeur ouvert par videochirurgie et mini-thoracotomie-primer cas (valvuloplastie mitrale) opere avec succes. Comptes Rendus de l'Academie des Sciences: Sciences de la Vie 1996;319(3):219-23.

old woman presenting an aneurysm and large defect of the interatrial septum (IAS), using an early prototype of the da Vinci articulated intracardiac «micro-wrist» rob. device, without complication and with discharge from the hospital on the 8th postoperative day[1458].

Ihor V. Komisarenko (15.XII.1933, city of Kharkiv, Ukraine – 23.VIII.2012, city of Kyiv, Ukraine)[1459] — Ukrainian surgeon-endocrinologist, introduced combined methods to treat tumors of the cortex of the suprarenal glands with the inhibitors of steroidogenesis, embolization of tumor and metastases.

Congenital heart disease (CHD) termed atrioventricular canal (AVC) defect refers to abnormal development of the endocardial cushions, resulting in defective atrioventricular (AV) septum and valve, affecting approximately 3% children CHD, and half of those with Down syndrome **[John Down (1828 – 96)**. 19th century. Internal Medicine], in which is seen here in 20% to 25%.

AVC defects are divided into:

(1) Partial AVC defect consisting of a large ostium primum atrial septal defect (ASD) and cleft between the left superior and inferior bridging leaflets, without ventricular septal defect (VSD), with two distinct AV valve orifices, corresponding to the mitral and tricuspid valve (MV, TV) leaflets tissues joining the left superior and left inferior leaflets together at the crest of the interventricular septum (IVS), eliminating interventricular communication. It occurs in 5% to 10% of all ASDs;

(2) Complete AVC defect consisting of ostium primum ASD, a nonrestrictive VSD and one common AV valve orifice with left and right components.

The coronary sinus (CS) ostium and AV node are displaced inferiorly in AVC defects into the so-called nodal triangle, bound by the inferior extend of the right AV valve anulus, the CS orifice, and the inferior edge of the interatrial septum (IAS)[1460].

In 1966 **Gian K. Rastelli (1933 – 70)**[1461] — Italian American pediatric cardiac surgeon, developed three types classification of a complete AVC defect.

[1458] Carpentier A, Loulmet D, Aupecle B, et al. Computer assited open-heart surgery: first case operated on with success. CR Acad Sci III 1998 May;321(5):437-442. — Carpentier A, Loulmet D, Aupecle B, et al. Computer assited open-heart surgery. Lancet 1999;353(9150):379-80. — Robot is a reprogrammed machine with complicated multiple functions with almost human characteristics.

[1459] Ablitsov VH. «Galaxy Ukraine». The Ukrainian diaspora: prominent persons. Publ. «Kyt», Kyiv 2007.

[1460] Lev M. The architecture of the conduction system in congenital heart disease. 1. Common atrioventricular orifice. Arch Pathol 1958 Feb;65(2):174-91.

[1461] Rastelli G, Kirklin JW, Titus JL. Anatomic observations on complete form of persistent common atrioventricular canal with special reference to atrioventricular valves. Mayo Clin Proc 1966;41:296-308.

In the Rastelli type A defect the superior bridging leaflet is divided into two leaflets at the crest of IVS, corresponding to the right superior leaflet (RSL) and left superior leaflet (LSL), which is the most common type (approximately 55%).

In the Rastelli type B defect the LSL bridges across the IVS and attaches to the papillary muscle of the right ventricle (RV).

In the Rastelli type C defect there is no marked bridging of the superior bridging leaflet, making it unattached to the underlying IVS and free floating.

Rastelli classification facilitated corrective surgery of complete AVC defect.

First repair of the ostium primum partial AVC defect was performed in 1957 by **John Kirklin (1917 – 99**: above)[1462] using cardiopulmonary bypass (CPB), and the first complete AVC defect was performed also in 1957 by **Walton Lillehei (1918 – 99**: above) using cross-circulation.

Presently, a complete AVC defect repair include classic single-patch technique covering both atrial and ventricular defects with partition of the superior bridging leaflet[1463], and double-patch technique using a separate atrial and ventricular patches[1464], in symptomatic infants regardless of age, and elective repair within the first 2 to 4 months of age in asymptomatic infants.

In the presence of the complex D-transposition of the great vessels (D-TGA) – VSD and main pulmonary artery (MPA) stenosis, Gian Rastelli proposed to perform a tunnel-like synthetic baffle inside the left ventricle (LV) between the base at the VSD to direct the arterial blood flow from the RV through VSD into the annulus of the right-sided aorta, and to connect the RV with the left-sided MPA using the valved conduit (Rastelli procedure or operation, 1969).

In 1971 **M. Judah Folkman (1933 – 2009)** — American surgeon and medical scientist, had postulated that tumos are angiogenesis-dependent, i.e., require a blood supply to grow. Five years later (1975), he and others reported the discovery of angiogenesis in cartilage tissue, leading to the eventual development of antiangiogenetic therapies against cancer.

In 1980 **Jean Pierre Bex (1942 –)**[1465] — French pediatric cardiovascular surgeon, strengthened the concept of aortic translocation for the surgical management of transposition of the great arteries (TGA), popularized in 1984 by **Hisashi Nikaidoh**

[1462] Kirklin JW, Daugherty GW, Burchell HB, et al. Repair of the partial form of persistent common atrio-ventricular canal: ventricular communication. Ann Surg 1955;142:858.

[1463] Rastelli GC, Ongley PA, McGoon DC. Surgical repair of complete atrioventricular canal with anterior common leaflet undivided and unattached to ventricular septum. Mayo Clin Proc 1969;44:335-41.

[1464] Carpentier A. Surgical anatomy, and management of the mitral component of the atrioventricular canal defects. In: R.H. Anderson and EA Shinebourne (Eds). Pediatric Cardiology. Churchill Livinngstone, London 1979. P. 466-90.

[1465] Bex JP, Lecompte Y, Baillot F, et al. Anatomical correction of transposition of the great arteries. Ann Thorac Surg 1980;29:86-8.

(1934 –)[1466] — Japanese American pediatric cardiac surgeon, for the surgical approach to the TGA with ventricular septal defect (VSD) and pulmonary artery (PA) stenosis, which consist of harvesting of the aortic root from the right ventricle (RV), relief of the left ventricular outlet tract (LVOT) obstruction by dividing the outlet septum, excising the PA valve, reconstructing the LVOT with the translocated aortic root and the VSD pericardial patch, reconstructing the right ventricular outflow tract (RVOT) with pericardial patch, and transfer of the coronary artery ostia (Nikaidoh procedure). A double-outlet right ventricle (DORV) with sub-pulmonary VSD (Tausing-Bing heart; 20th century. Internal medicine) is being repaired by the Jatene «switch» of the great arteries operation, the Rastelli procedure (above) with the Damus-Kaye-Stancel (DKS) procedure (below), the LeCompte's reparation l'etage ventriculare (REV)[1467] and the Nikaidoh operation[1468].

[1466] Nikaidoh H. Aortic translocation and biventricular outflow tract reconstruction. A new surgical repair for transposition of the great arteries associated with ventricular septal defect and pulmonary stenosis. J Thorac Cardiovasc Surg 1984;88:365-72.

[1467] Bex JP, Lecompte Y, Baillot F, et al. Anatomical correction of transposition of the great arteries. Ann Thorac Surg 1980;29:86-8. / **Yves LeCompte** — French pediatric cardiac surgeon.

[1468] Bacha EA. Ventricular septal defect and double-outlet right ventricle. In, FW Sellke, PJ del Nido, SJ Swanson (Eds). Sabiston & Spencer Surgery of the Chest. Vol. I-II. 8th Ed. ElsevierSaunders, Philadelphia 2010. P. 1849-63.

Mykola (Nikolay) L. Volodos (15.V.1934, village Kokoshchytsi, Slonim Region, Hrodensk, Belarus – 03.IV.2016, city of Kharkiv, Ukraine)[1469] — Ukrainian cardiovascular surgeon, the founder of intravascular surgery (1985 - 93), was one of the first to remotely insert the intraluminal (intravascular) self-expanding and self-fixing synthetic tubular graft / prosthesis (ISS-STG) for the treatment of atherosclerotic (AS) stenosis of the left external iliac artery (EIA, 1985), the first to remotely place distally the ISS-STG / thoracic endovascular aortic repair (TEVAR) of the post-traumatic pseudoaneurysm of the descending thoracic aorta (DTA, 1987) and the first to apply the remote ISS-STG / TEVAR for the treatment of bilateral aorto-iliac occlusive disease (AIOD, 1993).

[1469] Village Kokoshchytsi, Slonim Region, Hrodensk Obl. (area 25,127 km²), Bilorus. — Volodos NL, Shekhanin VE, Karpovich, et al. A self-fixing synthetic blod vessel endoprosthesis. Vest Khir im II Grekov 1986;137:123-5.— Volodos N L, Karpovich IP, Shekhanin VE, Troian VI, Yakovenko LF. A case of distal transfemoral endoprosthesis of the thoracic artery with self-fixing synthetic prosthesis in traumatic aneurysm. Grud Khir 1988;84-6. — Volodos N L, Karpovich IP, Shekhanin VE, Ternyuk NE, Yakovenko LF, et al. Self-fixing synthetic prosthesis of distant and intraoperational endoprosthetics aorta and iliac arteries. 2 Radiologie interventionelle en pathologie cardio-vasculaire Congress International, Toulouse-France 28 Fevrier, 1-er et 2 Mars 1990. Livre des resumes, P. 67. — Volodos ML. Modern methods of the remote endovascular grafting of vessels. Program and Abstracts of the 3rd Scientific Congress of the World Federation of the Ukrainian Medical Associations, Kyiv 06-09.VIII.1990 – Lviv 13-15.VIII.1990. In, AS Olearchyk. A Surgeon's Universe. Vol. I-IV. 4th Ed.Publ «OutskirtsPress», Denver, CO 2016. Vol.III. P. 152-3. — Volodos NL, Karpovich IP, Troyan VI, et al. Clinical experience of the use of self-fixing synthetic prostheses for remote endoprosthesis of the thoracic and the abdominal aorta and iliac arteries through the femoral artery and as intraoperative endoprosthesis for aorta reconstruction. Vasa Suppl 1991;33:93-5. — Volodos NL, Karpovich IP, Troyan VI, et al. Transfemoral endovascular grafting of the aortoiliac segment with the bifurcated self-affixing synthetic endoprosthesis (BSSEP). J Interventional Cardiology 1994;7(1):88. — Volodos NL. New system for intraluminal grafting of the aorta. The technique to use two approaches. Angiology and Vascular Surgery 1998;4(3-4): 103-9. — Volodos NL, Karpovich IP, Troyan VI, et al. Endovascular stented grafts for thoracic, abdominal aortic and iliac artery disease: Clinical Experience in Ukraine from 1985. Seminars in Interventional Radiology 1998;15(1).89-95. — Volodos NL, Karpovich IP, Troyan VI. et al. Treatment of thoracic aortic aneurysms complicated with an aorto-bronchial fistula and massive pulmonary bleeding. EACTS/ESTS Joint Meeting, Lisbon, Portugal September 16-19, 2001.2 P. — Cradio FJ. Nikolay Volodos and the origin of endovascular grafts. J Endovasc Ther 2012;19:568-9. — Diethrich EB. Behind the iron curtain. J Endovasc Ther 2013;20(Suppl I):1-2. — Volodos NL. The first steps in endovascular aortic repair: how it all began. J Endovasc Ther 2013;20(Suppl I):3-23. — Ivancev K. Following in the footsteps. J Endovasc Ther 2013;20(Suppl I)1-23. — Biography of Dr. Nikolay Volodos. J Endovasc Ther 2013;20(Suppl I):1-24. — Volodos NL. The 30th Anniversary of the first clinical application of endovascular stent-grafting. Eur J Vasc Endovasc Surg 2015;49:495-7. — Olearchyk AS. Mykola (Nikolay) L Volodos. In, AS. Olearchyk. A Surgeons Universe. Vol. I-IV. 4th Ed. Publ. «OutskirtsPress», Denver, CO 2016. Vol. II. P. 586-90. — Olearchyk AS. Professor Mykola L. Volodos, Medininae doctor (MD), Phylosophiae doctor (PhD) – pioneer of the remote endovascular grafting (EVG). In, AS. Olearchyk. A Surgeons Universe. Vol. I-IV. 4th Ed. Publ. «OutskirtsPress», Denver, CO 2016. Vol. III. P. 298-99. — GloviczkyP. ESVS Volodos Lecture: Innovations and the Hippocratic Oath. Eur J Vasc Endovasc Surg May 01,2018;65(5):605-13. — Ablitsov VH. «Galaxy Ukraine». The Ukrainian diaspora: prominent persons. Publ. «Kyt», Kyiv 2007.

Lazar J. Greenfield (1934 –)[1470] — American vascular surgeon, introduced a new umbrella-like inferior vena cava filter (IVCF, Greenfield filter, 1973) that consist of six guards made of stainless steel with small hooks on each end, for deep venous thrombosis (DVT) when anticoagulation is contraindicated, to prevent pulmonary emboli (PE). After it is positioned under fluoroscopic guidance into the inferior vena cava (IVC) below the origin of the renal vein at the level of lumbar vertebra (L)$_3$, it is opened and hooks fasten it into its wall. The filter catches emboli and prevents them from traveling with the venous blood into the pulmonary artery (PA).

In 1961 **Thomas J. Fogarty (1934 –)** — American thoracic and cardiovascular surgeon, invented a balloon catheter for the removal of thrombi and emboli from arteries and thrombi from veins (Fogarty balloon embolectomy catheter).

At the Amosov National Institute of Cardiovascular Surgery (NICVS; founded 1955) of the National Academy of Medical Science of Ukraine (NAMSU; created 1993), city of Kyiv, Ukraine, between 1955 – 2017, more than 200,000 operations were performed for congenital heart diseases (CHD), acquired valvular heart diseases and coronary artery disease (CAD) with operative mortality (OM) 0.7% (in the USA – 2.2%).

The favorable conditions for teaching young surgeons to perform cardiac operations under cardiopulmonary bypass (CPB) has continued after retirement of **Mykola Amosov (1913 – 2002**: above**)**, by **Hennadiy (Gennady) V. Knyshov (06.VIII.1934, town of Debaltseve, Horlivka Region, Donetsk Obl., Ukraine – 01.XI.2015, Kyiv, Ukraine)**[1471] — Ukrainian cardiovascular surgeon of Russian descent, Director (1988-2015) of the Amosov NICVS, who performed in 1972 the first in Ukraine coronary artery bypass (CAB) bypass surgery using vein grafts (VG).

He together with colleagues to prevent development of an aneurysms in the site of coarctation of the aorta (CA) recommended to preferentially perform replacement using synthetic tubular graft (STG) through the left 4th intercostal space thoracotomy (ICST) with distal blood perfusion by temporary bypass shunt, and that infected aneurysms could be primarily bypassed via the right anterior ICST for creating permanent bypass using STG between ascending aorta (AA) and descending thoracic aorta (DTA) / 1996.

Since 2001 at the Amosov NICVS during cardiac operations the artificial blood with cold cardioplegic myocardial protection is used, without a donor blood.

At the Amosov NICVS, patients with inflammation of the inner layer of the heart (endocardium), i.e., infective endocarditis, the treatment using general hyperthermia

---

[1470] Greenfield LJ, Peyton R, Crude S, et al. Greenfield vena cava filter experience: late results in 156 patients. Arch Surg 1981;116:1451-6. — Olearchyk AS. Insertion of the inferior vena cava filter followed by iliofemoral venous thrombectomy for ischemic venous thrombosis. J Vasc Surg 1987;5:645-7.

[1471] Town of Debaltseve (founded 1873), Horlivka Region, Donetsk Obl., Ukraine. — Ablitsov VH. «Galaxy Ukraine». The Ukrainian diaspora: prominent persons. Publ. «Kyt», Kyiv 2007.

[20th century. Anesthesia. – **Vitaly Maksymenko (1952 –)**] was evaluated and introduced, since it was proven that artificially created higher temperature stimulate the immune system in a man, and as result recurrences of infective endocarditis occurred five times less as in other medical centers, achieving the lowest OM of 1.9%, as compared to OM of 12-28% elsewhere (2005).

It was not earlier than in 1978 when **J. Solorzano**[1472] re-introduced the clinical use of the retrograde cardioplegia via coronary sinus (CS) for the myocardial protection during cardiac operations under cardiopulmonary bypass (CPB).

The research of **Jonathan E. Rhoads (1907 – 2002)** — American surgeon, focused on problems of nutrition among hospital patients since the 1930s, leading to the invention with **Stanley J. Dudrick (1935 – 2020)** — American surgeon of Polish descent, of an intravenous hyperalimentation (IVH) or a total parenteral hyperalimentation / nutrition (TPH)[1473]. The technique, initially successfully nourished Beagly puppies, subsequently, since 1968 newborn and infants with gastrointestinal (GI) malignancies, involves puncture and insertion of a long plastic catheter through the subclavian vein (SCV) into the superior vena cava (SVC), enabling feeding of patients who could tolerate oral or standard intravenous (IV) nutrition, a solution containing hypertonic glucose, proteins, especially essential amino acids, electrolytes, vitamins and minerals without inflammation of the vein.

Aortic valve (AV)-sparing operations for the correction of diseased dilatation of the aortic root (annuloaortic ectasia) and an aneurysmal ascending aorta (AA) include: (1) remodeling of the aortic root and (2) valve-sparing aortic root repair or replacement (VSARR).

Sir **Magdi H Yacoub (1935 –)** — Egyptian English cardiac surgeon, developed the AV-sparing aortic root repair in which the diseased (aneurysmal) noncoronary aortic sinus (NCS) of Valsalva[1474] and the AA were replaced with the polyethylene telephthalate (PETE, Dacron) synthetic tubular graft (STG) whereas the intact AV itself is left alone («uni-Yacoub» procedure, 1979), and remodeling of the aortic root with reimplantation of the ostia of the two coronary arteries into the corresponding neo-sinuses of Valsalva («three Yacoub» procedure, 1993).

In the other words, if only the NCS of Valsalva and the AA are diseased, then they are replaced with the PETE STG, while the intact AV, ostia of the left anterior descending (LAD) artery and the right coronary artery (RCA) are left alone («uni-Yacoub» procedure, 1979).

[1472] Solorzano J, Taitelbaum MG, Chiu RC. Retrograde coronary sinus perfusion for myocardial protection during cardiopulmonary bypass. Ann Thorac Surg 1978;25(3):210-8.

[1473] Dudrick SJ, Wilmore DW, Vars HM, Rhoads JE. Long-term parenteral nutrition with growth, development, and positive nitrogen balance. Surgery 1968 Jul;64(1):134-42.

[1474] **Antonio M. Valsalva (1666 - 1723)** — Italian anatomist.

However, when all three coronary sinuses of Valsalva – the right and left coronary sinuses (RCS, LCS), NCS and the AA are involved, then they are replaced with Dacron STG, while the intact AV is preserved, the ostia of the LAD artery and the RCA each with a button of the aortic wall are implanted into the neo-LCS and the neo-RCS, respectively (remodeling of the aortic root or the «three-Yacoub» procedure, 1993). In this case, the aortic root is dissected circumferentially to the down to the level of the aortic annulus in the form of teeth or festoon, and the three coronary sinuses are excised, leaving about 5 mm of tissue attached to the aortic anulus and in the aortic wall around the coronary artery orifices. The three commissures are temporarily gently suspended upwards and approximated until the three cusps coopt to evaluate its sufficiency and determine the diameter of the aortic ring, necessary for choosing the appropriate Dacron STG. The proximal end of this graft is tailored to create neo-aortic sinuses, the widthsof each are based on the distance between the commissures of each cusp when they are pulled upward to determine the diameter of the graft, whereas the heights of the neo-aortic sinuses should be approximately equal to its width. The three commissures are secured on the outside of the graft immediately above the neo-artic sinuses, the remnants of aortic wall and aortic anulus are sutured to the neo-aortic sinuses with a continuous 4-0 polypropylene suture, and the coronary arteries are reimplanted into their respective sinuses. If aortic cusp is prolapsing, its free margin is shortened, or reinforced with a double layer of a 6-0 expanded polytetrafluoroethylene (PTFE) suture.

«Three-Yacoub» procedure is relatively simple, shortens the time of myocardial ischemia, and from the physiological perspective works perfectly, since it does not place the aortic ring inside the graft, and thus preserves its motion. Nevertheless, in children, adolescent and young adult affected by genetic degenerative disease of the aorta, such as bicuspid aortic valve (BAV), may developed aneurysm within a narrow remnant of the aortic wall, unprotected from the outside by the graft.

Yevhenii (Evgenii) V. Kolesov (15.06.1935, city of Perm, Russia –)[1475] — Russian and Ukrainian cardiac surgeon, who modified the vessel-suturing circular apparatus (VCA)-4, used by **Vasilii Kolesov (1904 - 92**: above) for coronary artery bypass (CAB) during construction of the anastomosis with the internal thoracic artery graft (ITA) graft to the left anterior descending LAD) artery with the small internal diameter of 1,3 mm (1967); introduced the VCA with prolonged prong (1967) and the VCA with vacuum suction (1970) for the construction of anastomoses between the ITA and coronary arteries with an internal diameter of less than 1,3 mm; continued to use after Vasilii Kolesov (1967) staplers in CAB grafting (1985); introduced endoscopic

---

[1475] City of Pem (Foot Note 240), Russia. — Ablitsov VH. «Galaxy Ukraine». The Ukrainian diaspora: prominent persons. Publ. «Kyt», Kyiv 2007.

implantation by screwing the spiral pacemaker electrodes into the epicardium (1986); developed the technique for glue-made anastomosis to coronary arteries with the internal diameter of 1,3 mm or less (2003).

Main directions of the scientific activity of **Oleksiy V. Liulko (16.06.1935, city of Kamians'ke, Dnipropetrovska Obl., Ukraine – 12.X.2013, city of Dnipro, Ukraine)**[1476] — Ukrainian urologist, has been centered upon diagnosis and treatment of benign prostatic hyperplasia (BPH), kidney and / or ureteral stone disease, a.k.a., nephrolithiasis or urolithiasis, nonspecific inflammatory diseases of nephro-urinary and reproductive systems, and plastic, reconstructive, pediatric, and oncologic urology, as well as science about normal or pathologic conditions, diagnosis, and treatment of diseases of masculine gender (sex), andrology (in Greek – «andrikos» meaning man, «logos» means science). He made two discoveries: «Phenomenon of creation the pericapsular lymphatic capillaries on the kidney of mammals» (Diploma No. A-135 of 11.I.1999) and «Property of kidney stones as biological objective to display elastic (resilient) - stickiness (viscosity) - plastic deformations» (Diploma No. A-146 of 17.VIII.1999).

After moving on November 19, 1966, to the USA, **Andrew S. Olearchyk (**Foot Notes 1, between 87 & 88, 1032, 1043, 1297**)** — Ukrainian American general, thoracic, and cardiovascular surgeon, recognized that common habits of patients with stomach cancer were alcoholism (26,7%) and smoking (26%), described a combined treatment of advanced gastric carcinoma by resection and chemotherapy (1975)[1477]; proved that in patients with the left ventricular (LV) post-myocardial infarction (MI) aneurysm who had coronary artery bypass (CAB) grafting using vein grafts (VG), survival was better than in those without it, demonstrated the improved cardiac hemodynamics after resection of those aneurysm and the best 10-year survival achieved when LV aneurysmectomy was combined with CAB grafts to the Left anterior descending (LAD) artery or its diagonal branch (DB) due to the preservation of the high anterolateral wall and interventricular (IVS) blood supply (1979-80)[1478]; indicated that, in difficult anatomic situations when a great saphenous vein (GSV) of poor quality in the elderly, small coronary arteries in women, or diffusely diseased

---

[1476] City of Kamians'ke (during times of Kyiv Rus in the 5th-13th century through the region passed important trade routes, first settlers were Ukrainian Zaporizhian Cossacks in the 1st half of the 17th century), Dnipropetrovska Obl., Ukraine. — City of Dnipro (Foot Note 249), Ukraine. — Liulko OV, Vozianov OF. Urology. 4th ed. All-Ukrainian Specialized Publ. «Medytsyna», Kyiv 2011. — Ablitsov VH. «Galaxy Ukraine». The Ukrainian diaspora: prominent persons. Publ. «Kyt», Kyiv 2007.

[1477] Olearchyk AS. Cancer of the Stomach. J Ukr Med Assoc North Am 1977;24,4 (87):3-52. — Olearchyk AS. Gastric carcinoma: a critical review of 243 cases. Am J Gastroenterol 1978;70 (I):25-45.

[1478] Olearchyk AS, Lemole GM, Spagna PM. Left ventricular aneurysm: ten year experience in surgical treatment of 244 cases. Improved clinical status, hemodynamics and long-term longevity. J Thorac Cardiovasc Surg 1984;88:544-53.

but important coronary arteries, like the LAD artery, better graft patency and clinical success obtained with the internal thoracic artery (ITA) grafts, and in the LAD artery hypoperfusion syndrome during the ITA grafting performed an additional VG to the proximal LAD artery (1980-82)[1479]; performed in a 73-year-old man with acute ischemic venous thrombosis of the left lower extremity an emergency protected left iliofemoral thrombectomy, that is combined insertion of the inferior vena cava filter (IVCF) through the internal jugular vein (IJV) and the left iliofemoral thrombectomy under fluoroscopic guidance to prevent pulmonary artery (PA) embolization (1986)[1480]; performed (1987) in an elderly woman CAB graft using the VG to the right coronary artery (RCA), aortic valve replacement (AVR), endarterectomy (EA) of the aneurysmal / atheromatous (AM) ascending / transverse aorta (ATA) under cardiopulmonary bypass (CPB) and hypothermic circulatory arrest (HCA), and external synthetic grafting (ESG)[1481]; applied a right femoral and retroperitoneal ilio-aortic approach to remove a retained intra-aortic balloon assist (IABA) device (1989)[1482]; proved that in coronary artery disease (CAD) complicated by the post- infarction LV aneurysm and cardiomyopathy (CM) in the presence of the «hibernating» myocardium, CAB using VG to the «ghost» LAD artery, a LV endoaneurysmorrhaphy (EAR) with a rounded synthetic patch reconstruction and external wrapping by the aneurysmal wall containing some viable muscle tissues and revascularized, improves contractions and the LV systolic augmentation (1993-94)[1483]; described double clamping of the ascending aorta (AA) at coronary artery bypass (CAB) surgery (1994)[1484]; performed in patients with a significant CAD and calcified AM of the ATA, CAD grafting using ITAs with a «no-touch» technique to the aorta, an arterial cannulation of the left common femoral artery (CFA), CPB withventing of the LV via the right superior pulmonary vein (RSPV) on a beating, warm and decompressed heart during bradycardia induced

[1479] Olearchyk AS, Magovern GJ. Grafting of the internal thoracic artery. J Ukr Med Assoc North Am 1986; 33,1(114):9-24. — OlearchykAS, Magovern GJ. Internal mammary artery graftings: clinical results, patency rates and long-term survival in 833 patients. J Thorac Cardiovasc Surg 1986;92: 1082-7.

[1480] Olearchyk AS. Insertion of the inferior vena cava filter followed by iliofemoral venous thrombectomy for ischemic venous thrombosis. J Vasc Surg 1987;5:645-7.

[1481] Olearchyk AS. Endarterectomy and external prosthetic grafting of the ascending and transverse aorta under hypothermic circulatory arrest. Texas Heart Inst J 1989;16 (2):76-80. — Olearchyk AS. Endarterectomy and external prosthetic grafting of the ascending and transverse aorta under hypothermic circulatory arrest. J Ukr Med Assoc North Am 1989; 36,2 (120):83-6.

[1482] Olearchyk AS. Retained intra-aortic balloon assist device. J Thorac Cardiovasc Surg 1992; 103:1231-2.

[1483] Olearchyk AS, Shariff HM: Left ventricular endo-aneurysmorrhaphy. Grud serd-sosud khir 1994;5:10-13. — Olearchyk AS, Shariff HM: Left ventricular endoaneurysmorrhaphy. A case report. Vasc Surg 1995;29:245-50. — Bockeria LA, Gorodkov AJ, Dorofeev AV, Alshibaya MD and the Restore Group. Left ventricular geometry reconstruction in ischemic cardiomyopathy patients with predominantly hypokinetic left ventricle. Eur J Cardiothorac Surg. 2006 Apr;29 (Suppl 1):S251-8.

[1484] Olearchyk AS. Double clamping of the ascending aorta during coronary artery bypass surgery. Vasc Surg 1994;28:305-10.

by a short acting beta-blocker, or during asystole induced by short acting beta-blocker and calcium channel-blockers (1994)[1485]; in a 60-year-old woman identified, dissected, and ligated proximally and distally congenital bilateral fistulas between the RCA and left main coronary artery (LMCA) to the main pulmonary artery (MPA) on a beating heart (1996)[1486]; in 82-yearold woman performed (1995) an endo-aneurysmal repair of a congenital aneurysm of the right coronary (aortic) sinus (RCS)[1487]; in a 73-year-old woman who had AVR repaired in 1997 a pseudoaneurysm of the AA, after cannulating the CFA and common femoral vein (CFV) with a long cannula, on CPB with mild hypothermia on a beating heart[1488]; proposed in patients with a stenosed congenital bicuspid aortic valve (BAV) and an aneurysm of the AA to perform after AVR, a modified vertical reduction aortoplasty (VRA) of the proximal and middle AA and ESG of the distal AA (2002-03)[1489]; diagnosed in a 70-year-old man a clinical picture of a mimicking of the right subclavian steal syndrome (SSS) due to narrowing of the proximal right vertebral artery (VA) and an occlusion of the distal right subclavian artery (SCA), and treated him with success by a synthetic patch angioplasty of the proximal right VA and a bypass between the right common carotid artery (CCA) and the axillary artery (carotid-axillary bypass) using a STG (2003-04)[1490].

One of the pioneers in cranial reconstruction surgery, **Linton A. Whitaker (1936 –)**[1491] — American plastic surgeon, developed techniques for reconstruction of congenital deformities, primarily in children, and for aesthetic application in adults.

Chief (from 1971) of Division of Surgery for Congenital Heart Diseases (CHD) in Children of the Older Age, now the M.M. Amosov National Institute of Cardiovascular Surgery (NICVS) of the National Academy of Medical Science (NAMS) of Ukraine, **Mykhaylo F. Zin'kovsky (10.XI.1937, town of Koziatyn, Chmilnytsky Region,**

[1485] Olearchyk AS. Calcified ascending aorta and coronary artery disease. Ann Thorac Surg 1995; 59;1013-5. — Olearchyk AS. Bilateral internal thoracic artery grafting for severe coronary artery disease and calcified atherosclerosis of the ascending and transverse aorta. Klinichna khirurhiya (Kyiv) 1995;7-8 (640):41-3.*

[1486] Olearchyk AS, Runk DM, Alavi M, Grosso MA. Congenital bilateral coronary-to-pulmonary artery fistulas. Ann Thorac Surg 1997;64:233-5. — Alavi M, Olearchyk AS. Congenital pulmonary-to-coronary artery fistulas: 22 months after repair. Ann Thorac Surg 1999;67:598-9.

[1487] Olearchyk AS, Grosso MA, Danielewski GL. Endo-aneurysmal repair of a congenital right coronary sinus aneurysm. J Card Surg 1997;12:81-5.

[1488] Olearchyk AS. Repair of a pseudoaneurysm of the ascending aorta after aortic valve replacement. J Card Surg 1998;13:143-5.

[1489] Olearchyk AS. Congenital bicuspid aortic valve disease with an aneurysm of the ascending aorta in an adult: vertical reduction aortoplasty with distal external synthetic wrapping. J Card Surg 2004;19(2):144-8. — Olearchyk AS. Operative management of atherosclerotic aortic aneurysm in a patient with bicuspid aortic valve disease. J Card Surg 2004;19(4):354-5.

[1490] Olearchyk AS. Mimicking of the subclavian «steal» syndrome. J Ukr Med Assoc North Am 2005;50,2(154):48

[1491] Whitaker L. The center in human appearance; a multispecialty concept. Facial Plastic Surgery 2003 Feb;19(1):3-6.

Vinnytsia Obl., Ukraine –)[1492] — Ukrainian pediatric cardiac surgeon, have an experience on the world scale in surgical treatment of the tetralogy of Fallot [17th century. – **Nilsen Stenson (1838-86)** & 19th century. – **Étienne Fallot (1850 – 1911)**] with excellent outcomes during long-term follow-up.

Scientific and professional activity of **Anatolii V. Makarov (02.I.1937, town Inhulka, Bashtansky, Region, Mylolaïv Obl., Ukraine –)**[1493] — Ukrainian chest surgeon, concentrates on trauma to the chest, surgery to the esophagus, airways and lung, reconstructive surgery of the abdominal cavity.

In 1970 **William S. Pierce (1937 –)** — American chemical engineer and cardiothoracic surgeon, initiated development of the first pneumatic pulsatile heart assist device, that has been used in nearly 4,000 patients for right-, left- and bi-ventricular support. Subsequently, he and his group developed an implantable, permanent left ventricular assist device (LVAD) powered by a wearable, external rechargeable nickel-cadmium (NiCd) battery (invented in1899), that represented the first successful use of a transcutaneous energy transfer (TET) system, eliminating percutaneous (p/c) electric wire, and the first clinical use of a compliance sac, eliminating the need for a percutaneous (p/c) vent.

Oleksandr F. Vozianov (02.X.1938, city of Melitopol, Melitopol Region, Zaporozhian Obl., Ukraine – 07.V.2018, city of Kyiv, Ukraine)[1494] — Ukrainian urologist, developed the original operative approaches for the treatment of benign glandular tumor (adenoma) of the prostatic gland (prostate) and proposed the use of special surgical instruments, introduced in Ukraine radical removal of prostatic

---

[1492] Town of Koziatyn, (first mentioned 1731), Chmilnytsky Region, Vinnytsia Obl., Ukraine. — Zin'kovskyMF. Congenital heart defects. In, LYa. Kovalchuk, VF Sayenko, HV Knyshov. Klinichna khiruhiya. Vol. 1-2. Publ. «Ukrmedknyha», Ternopil 2000. Vol. 1. P. 184-225. — Zin'kovsky MF. Congenital Heart Defects. Publ. «Knyha Plus», Kyiv 2010. — Olearchyk AS. A Surgeon's Universe. Vol. 1-4. 4th Ed. Publ. «OutskirtsPress», Denver, CO 2016. Vol. 3. P. 268-9. — Ablitsov VH. «Galaxy Ukraine». The Ukrainian diaspora: prominent persons. Publ. «Kyt», Kyiv 2007.

[1493] C. Inhulka (remains of settlements from Bronze Age 3,000-2,000 BC and Scythes time 6th-2nd century BC, founded 1802), Bashtansky Region, Mykolaïv Obl. Ukraine. — Makarov AV. Clinics, Diagnostics and Surgical Treatment of Tracheal and Bronchial Adenomas. Dissertation for degree of CMS. Kyiv 1974. — Makarov AV, Het'man VG, Avilova OM. Thoracoscopy in Emergency Chest Surgery. Publ. «Zdorov'ia», Kyiv 1986. — Makarov AV. Emergencies in Pulmonology. Kyiv 1988. — Makarov AV. Chemical Burns of the Esophagus in Children and its Sequelae. Publ. «Vyshcha shkola», Kyiv 2002. — Makarov AV. Diseases of the Esophagus in Children. Kyiv 2004. — Makarov AV. Examination of Respiratory Organs in Children. Kyiv 2005. — Makarov AV. Thoracic Trauma. Kyiv 2006. — Makarov AV. Deformations of the Chest in Children. Kyiv 2006. — Ablitsov VH. «Galaxy Ukraine». The Ukrainian diaspora: prominent persons. Publ. «Kyt», Kyiv 2007.

[1494] City of Melitopol (known since 4th century BC, 1627, founded 1784; due to organized by Russian Communist Fascist of Holodomor-Genocide 1932-33 against the Ukrainian Nation died at least 3,743 residents), Melitopol Region, Zaporizhia Obl., Ukraine. — Vozianov OF. Atlas – Textbook of Urology. Vol. 1-3. 2001.— Liulko OV, Vozianov OF. Urology. 3rd Ed. All-Ukrainian Specialized Publ. «Medytsyna», Kyiv 2011. — Ablitsov VH. «Galaxy Ukraine». The Ukrainian diaspora: prominent persons. Publ. «Kyt», Kyiv 2007.

cancer, elaborated the classification of premalignant stages of malignant tumors of the prostate and urinary bladder, as well as the algorithm for diagnosing adenoma and cancer of the prostate with inculcation of the immunological monitoring. He established first laboratory for a thermal diagnosis and first division of the extracorporeal crushing or pulverization of kidney stones in Ukraine.

Negative pressure wound therapy (NPWT)[1495] is a therapeutic technique of using a suction dressing to remove excess of exudation and promote healing in acute or chronic wounds, soft tissue damage in open bone fractures, second- and third-degree burns, and diabetic patients with bedsores (pressure ulcers).

The therapy is based on the placing into the wound or tissue defect a sealed dressing, i.e., polyvinyl foam, to which drainage tubes are inserted. Then polyvinyl foam and adjacent skin are covered with transparent polyurethrane dressing impermeable to bacteria. The drainage tubes are connected to a suction vacuum pump with the controlled continuous or intermittent sub-atmospheric pressure, varying between -125 mmHg and -75 mmHg, depending on the material used and patient tolerance.

One form of NPWT was developed by **Katherine F. Jeter (1938 –)** — American nurse and **Mark E. Chariker (1955 –)**[1496] —American plastic and reconstructive surgeon, actually by the former, on Oct. 31, 1985, when look after an adult man with an open, deep and purulent wound in the mid-abdominal line, and successfully treated it by the use of a Jackson-Pratt[1497] drain wrapped in a gauze lightly soaked in the physiological saline (0.9% NaCl) solution, with transparent dressing placed over the top, and tubing, connected to negative (-60 mm Hg) continuous wall suction, to reduce exudate, increase blood flow and granulation tissue in the wound bed (the Chariker-Jeter technique).

The second form of NPWT was developed in 2001 by **Louis C. Argenta (1943 –)** — American general, plastic and reconstructive surgeon and **Michael Morykwas** — American bioengineer, in which they applied a system of sub-atmospheric pressure delivery to the wound site utilizing a sealed polyurethrane foam dressing attached

---

[1495] Freischmann W, Streeker W, Bombelli M, Kinz L. Vacuum sealing as a treatment of soft tissue damage in open fractures. Unfallchirurg 1993 Sep;96(9):448-92. — Cipolla J, Baillie DR, Steinberg SM, et al Negative pressure therapy. Unusual and innovative applications OPUS 12 Scientist 2008;2(3):15-28.

[1496] Chariker ME, Jeter KF, Tintle TE, Bottsford JE. Effective management of incisional and cutaneous fistulae with closed suction wound drainage. Contemp Surg 1989;34:59-63.

[1497] **Frederick E. Jackson** and **Richard A. Pratt, III** — American neurosurgeons, who designed a closed-suction medical device used as a post-operative drain for collecting bodily fluid from surgical sites. It consists of internal drain connected to a grenade-shaped collection bulb through plastic tubing, named the Jackson-Pratt drain, a.k.a. JP drain. — Jackson FE, Pratt RA. Technical report: a silicone rubber suction drains for drainage of subdural hematomas. Surgery 1971 Oct;40(1):578-9. — Jackson FE, Pratt, III RA. Silicone rubber «brain drain». Zeitschrift für Neurology Mar 1972;201(1):92-4.

by a tube to a vacuum pump. The aim of this technique was to increase blood flow to the wound area, removal of bacteria, and increase rates of granulation tissue formation. Michael Morykwas designed a suction device, known as a device for the vacuum-assisted closure (VAC) of wounds.

Both form of the NPWT result in an efficient cleansing of the wound or tissue defect, removal of microorganisms, marked increase of blood supply and granulation tissue, thus hasting the healing process.

Danylo Ya. Kryvchenia (16.III.1938, village Verkhy, Zhabynkivsky Region, Berestia Obl., Belarus –)[1498] — Ukrainian pediatric surgeon, who proposed the method of bowel resection without entering its lumen (1982); performed a tracheoplasty and aortoplasty for tracheomalacia and compressive stenosis of the trachea (1987-89); the first who for the relief of respiratory distress syndrome (RDS) in aplasia of the lung proposed and performed translocation of the diaphragm (1995).

First successful lung transplant was carried out in 1983 by **Joel D. Cooper (1939 –)** — Canadian American thoracic surgeon, who also three years later (1986) performed the first successful double lung transplantation.

Extracorporeal membrane oxygenation (ECMO), a.k.a., extracorporeal life support (ECLS) provides short-term necessary gas exchange to support life in patients who are in severe cardiac decompensation or respiratory failure (RF), or both. An outside the body ECMO circuit with membranous oxygenator re-oxygenate hypoxic venous blood, removes an excess of carbon dioxide (CO_2) from hypercarbia venous blood and infused the corrected venous blood into the artery in patients with cardiac failure or into the vein in patients with RF.

Preconditions for the development of ECMO technique were laid by **Serhii Bryukhonenko (1890 – 1960**: above) and **Volodymyr Yankovsky** (above) who developed a bubble oxygenator for their heart-lung machine (HLM) in 1937 p. **Willem Kolff (1911 – 2009**: 20th century. Internal medicine) who noted in 1943 that blood became oxygenated as it passed through the cellophane chamber of their artificial kidney. This concept was applied by **John Gibbon (1903 – 73**: above) who used in 1953 artificial oxygenator and perfusion to support blood circulation in an 18-year-old woman during the operation for congenital heart disease (CHD), namely closure of an atrial septal defect (ASD), **Walton Lillehei (1918 – 99**: above) who

---

[1498] Village Verkhy, Zhabynkivsky Region, Berestia / Brest Obl. (area 32,200 km²), Belarus. — Kryvchenia DYu, at al. Congenital Diseases in Children. Kyiv 2001. — Kryvchenia DYu, et al. Aplasia of the lung: symptomatology and indications for surgical treatment. Ukraïns'kyy pediatrychnyy zhurnal 2003;2. — Kryvchenia DYu, at al. Surgery in the Children Age. Handbook. 2nd Ed., Kyiv 2009. — Kryvchenia DYu, at al. Morphologic criteria of an early diagnosis of sepsis and prognostic risk for developmemt of sepsis in children with acute purulent destructive pneumonitis. Khirurhiia dytiachoho viku. 2011;2. — Ablitsov VH. «Galaxy Ukraine». The Ukrainian diaspora: prominent persons. Publ. «Kyt», Kyiv 2007.

developed cross circulation technique by using slightly anesthetized adult volunteer as live HLM during repair of complicated CHD, **William Rashkind (1922 – 86**: 20th century. Pediatrics)[1499] who evaluated a disposable plastic, low volume, pumpless oxygenator as a temporary lung substitute, and **W. Dorson, Jr**[1500] — American pediatric cardiac surgeon who used a bubble oxygenator as support in a neonate dying of RF, and reported the use of membrane oxygenator for cardiopulmonary bypass (CPB).

Since the 1960s **Robert H. Barlett (1939 –)**[1501] — American surgeon and medical researcher, through the laboratory research helped to develop ECMO technology for newborn and infants with potentially reversible heart and lung dysfunction.

Indication for the initiation of ECMO is potentially reversible acute cardiac and respiratory failure unresponsive to conventional management that include the following criteria: (1) hypoxemic RF with a ratio of arterial oxygen tension to fraction of inspired oxygen (Pa O_2/FiO_2 of less than <100 mmHg) despite optimization of the ventilator setting, including the fraction of inspired oxygen (FiO_2), positive end-expiratory pressure (PEEP) and inspiratory to expiratory (I:E) ratio; (2) hypercapnic RF with arterial pH <7.20; (3) refractory cardiogenic shock; (4) cardiac arrest; (5) failure to wean from CPB after cardiac surgery; (6) as a bridge to heart transplantation (HT) or ventricular assist device; (7) as a bridge to lung transplantation.

Three most common form of ECMO are the peripheral veno-arterial (VA), the central VA and the peripheral veno-venous (VV).

In the peripheral ECMO VA modality the long outflow venous cannula is placed usually inside the lumen of the right common femoral vein (CFV) with its tip maintained near the junction of the inferior vena cava (IVC) and the right atrium (RA) for drainage of the venous blood, externally connected to the transparent plastic tube for drainage outside the body through ECMO circuit with membranous oxygenator and returned via the other transparent plastic tube to the arterial outflow cannula inserted into the lumen of the common femoral artery (CFA) with the tip maintained in the external iliac artery (EIA).

The central VA ECMO may be used if CPB has already been established with cannula in the RA and ascending aorta (AA). If this being the case, CPB machine is replaced by ECMO circuit with membranous oxygenator.

[1499] Rashkind WJ, Freeman A, Klein D, et al. Evaluation of a disposable plastic, low volume, pumpless oxygenator as a lung substitute. J Pediatr 1965;66:94-102.

[1500] Dorson W Jr, Baker E, Cohen ML, et al A perfusion system for infants. Trans Am Soc Artif Intern Organs 1969;15:155-60.

[1501] Barlett RH, Gazzaniga AB, Toomasian J, et al. Extracorporeal membrane oxygenation (ECMO) in neonatal respiratory failure. 100 cases. Ann Surg 1986; Sep;204(3):236-45.

In the peripheral VV ECMO the inflow cannula is usually placed in the right CFV for extraction of the venous blood and after passing through ECMO with membranous oxygenator it is returned by using an inflow catheter into the right internal jugular vein (IJV). Alternatively, a dual-lumen catheter is inserted into the right IJV, for the simultaneous draining of the venous blood from the superior vena cava (SVC) and IVC and returning it via the RA.

Use of ECMO circuit require systemic anticoagulation with intravenous (IV) heparin.

First newborn chosen for ECMO were the premature newborn. Its use was complicated by newborn or infantile respiratory distress syndrome from hyaline membrane disease, an intracranial hemorrhage due to hypoxia, hypothermia, acidosis, systemic heparinization and alterations of cerebral circulation, all resulted in a prohibitively high mortality.

Long-term ECMO was first successfully used in 1972 in an adult with severe post-traumatic RF.

In 1977 **John C. German (1939 –)**[1502] — American pediatric surgeon, reported the first child with congenital diaphragmatic hernia (CDH) to survive after being placed on ECMO, and in 1978 Robert Barlett reported the first neonatal girl who suffered lung damage from meconium aspiration syndrome treated with ECMO. She spent 3 days on ECMO, survived and recovered

Pre-term infants are at exceptionally high risk for intraventricular hemorrhage (IVH) if ECMO is administered at a gestational age less than 32 weeks.

Most common complication of ECMO are bleeding in 30%-40% of patients which can be life-threatening and heparin induced thrombocytopenia (HIT). When CFV and CFA are used for AV ECMO there is retrograde blood flow in the descending thoracic aorta (DTA) and stasis may occur if left ventricular (LV) output is not maintained, causing thrombosis. Other complications are subarachnoid hemorrhage (SAH), ischemic infarctions, hypoxic-ischemic encephalopathy, unexplained coma and brain death.

Average survival rate with ECMO at present is 83% for newborn, when previously a predicted survival was 20% without ECMO.

General Brigadier of the Medical Department of the US Army (since 1991) **Rostyslav (Rostik, Russ) Zajtchuk (Dec. 31, 1937, village Mokre,** Sianik county,

---

[1502] German JC, Gazzaniga AB, Amlie R, et al. Management of pulmonary insufficiency in diaphragmatic hernia using extracorporeal circulation with a membrane oxygenator (ECMO). J Pediatr Surg 1977;12:905-12.

Lemko region, Ukraine / Sub-Carpathian Voivodeship, Poland –)[1503] — Ukrainian American thoracic and cardiovascular surgeon, proposed the use of flap created from the intercostal muscle (ICM) for the primary repair of esophageal atresia (EA) with tracheal-esophageal fistula (TEF, 1975). He led the development of a sophisticated telecommunication infrastructure and virtual reality to speed diagnostics, laboratory analyses, and consulting expertise worldwide (1997).

Stereolithography or a three-dimensional (3-D) printing is a solid imaging process in which material is joined or solidified under computer control to create a 3-D object with material, such as liquid molecules or powder grains, are added and fused together, was introduced in 1983-84 by **Chuck W. Hull (1939 –)** — American inventor, and in 1984 by **Alain Le Mehaute (1947 –)** —French engineer-chemist and inventor. The former defined 3-D printing as a «system for generating 3-D object by creating a cross-sectional pattern of the object to be formed».

In 1979 **Leonid L. Si(y)tar (1939, u.t.v. Yampil, Bilohirskyy Region, Khmelnytska Obl., Ukraine –)**[1504] — Ukrainian cardiovascular surgeon, performed total replacement of the aortic arch . With colleagues advised to prevent development of aneurysms at the site of coarctation of the aorta (CA), to perform its replacement using synthetic tubular graft (STG) / **(Hennadiy Knyshov (1934 – 2015**: above).

Personal 3-D printing bespoke therapy have been used beginning in the mid-1900s in planning the anatomic remodeling of the bone in the cranial, maxillary, and facial reconstruction surgery and in total replacement of the hip and other joints (1998-99)[1505].

[1503] Village Mokre (appeared before 1772), Lemko region (area 8,800 km²), Ukraine, since 1945 – Sianik povit, Sub-Carpathian Voivodship (area 17,846 km²), Poland. — Zajtchuk R, Seyfer AE, Strevey TE. Use of intercostal muscle for primary repair of esophageal atresia with tracheoesophageal fistula. Ann Thorac Surg 1975 Mar;19(3):239-41. — Heydorn LH, Zajtchuk R, Schuchmann GF. Surgical management of pectus deformities. Ann Thorac Surg 1977 May;23(5):417-20. — Edwards FH, Albus RA, Zajtchuk R, et al. A quality assurance model of operative mortality in coronary artery surgery. Ann Thorac Surg 1989 May;47(5):646-9. — Zajtchuk R (Ed), Grande CM (Ed). Anesthesia and Perioperative Care of the Combat Causalty. Publ. Combat Studies Inst., Falls Church 1995. — Zajtchuk R, Satawa RM. Medical Applications of Virtual Reality. New York 1997. — Ablitsov VH. «Galaxy Ukraine». The Ukrainian diaspora: prominent persons. Publ. «Kyt», Kyiv 2007.

[1504] U.t.v. Yampil (found 1535), Bilohirsky Region, Khmelnytska Obl., Ukraine. — Knyshov GV, Sitar LL, Glagola MD, Atamanyuk MY. Aortic aneurysms at the site of the repair of coarctation of the aorta: review of 48 patients. Ann Thorac Surg 1996;61:935-9. — Sytar LL. Aneurysms of the Thoracic Aorta (Clinics, Diagnostics, Treatment). MM Amosov Premium of the NAS of Ukraine (medicine), Kyiv 2013. — Sytar LL. Acquired diseases of the heart. In, LYa Kovalchuk, VA Sayenko, HV Knyshov. Klinichna Khirurhiya. Vol. 1-2. V-vo «Urrmedknyha», Ternopil 2000. T. 1. P. 228-42. — Ablitsov VH. «Galaxy Ukraine». The Ukrainian diaspora: prominent persons. Publ. «Kyt», Kyiv 2007.

[1505] Epplex BL, Sadove AM. Computer-generated patient models for reconstruction of cranial and facial deformities. J Craniofacial Surg 1 Nov 1998;9(6):548-56. — Erickson DM, Chance D, Schmitt S, Mathis J. An opinion survey of reported benefits from the use of stereolithographic models. J Oral Maxillofacial Surg 1 Sep 1999;57(3):1040-3.

The first 3-D printing was developed in 2003 and patented in 2006 by **Thomas Boland**[1506] — American bioengineer, utilizing the «layer-by-layer» technique to combine cells, growth factors, and biological materials to fabricate biomedical parts that would maximally imitate natural tissue characteristics, to print tissue and organs to help research drugs and pills, and to produce replacement tissue and organs.

3-D tissue printing and a bioprinting replacement was developed in 2009 for lost tissue due to arthritis and cancer.

In 2011 an 83-year-old woman underwent a successful implantation of a 3-D-printed mandible w, and in 2012 the first 3-D printing mandible made of a titanium was transplanted to a man.

In 2012 **Larry Cronin (1973 –)** — Scottish chemist proposed to use chemical ink to print medicine. Subsequently, the first 3-D printing pills were manufactured beginning 2015.

As of 2012 a 3-D printing technology has been studied for possible use in tissue engineering (in Latin – «ingenium» meaning «ability» or «inventiveness») in which organs or body parts are built in layers of living cells deposed onto a gel medium or sugar matrix and slowly built up to form 3-D structures, including vascular system, printing ear cartilage, or from living tissue the liver and kidneys.

In 2013 a 3-D printing plastic (bioabsorbable) tracheal splint was used in newborn to treat tracheobronchial malacia (in Greek – «malacia» meaning «softening»).

In 2014 a 5-year-old girl born without fully formed fingers on her left hand had a prosthetic hand made with the aid of 3-D technology based on a plaster cast o her parents, and in a boy born with a missing arm from just above the elbow, a 3-D printed myoelectric arm was uploaded.

In 2014 a 3-D printing was used to create parts to rebuild a seriously injured motorist face in a road accident, to create a titanium pelvis to implant into a cancer patient, and to create the hearing aids and dental implants.

In 2015 a hospital-based 3-D printers were added to radiology departments with the aim to improve diagnosis of a complex diseases, and for the implantation of stem cells capable of generating new tissues and organs in living humans.

Since 2016 a fused filament fabrication has been used to create microstructures with a 3-D in thermal geometry. Those structures using polylactic acid can have fully controllable porosity in the range 20%-60%. Such scaffold could serve as biochemical templets (templates) for cell culturing, or biodegradable implants for tissue engineering.

[1506] Boland T, Wilson Jr WC, Xu T. Ink-jet printing of viable cells. US patent 7051654, issued 2006-05-30.

Shafkat Anwar (1970 –)[1507] — Bangladesh[1508] American cardiologist, described in 2017 an application of imaging by 3-D printing in diagnosing complex congenital heart disease (CHD) and for planning heart and solid organ surgery.

Ira Goldsmith[1509] — Welsh cardiothoracic surgeon, consulted a 71-year-old man, because of a malignant soft tissue tumor (sarcoma) that grew from what initially was thought to be a benign cyst of the right 3rd rib in the middle clavicular line (MCL) to the size of a tennis ball involving he 2nd – 4th ribs anteriorly from the right anterior axillary line (AAL) to include right half of the sternum. After review the computer tomography (CT) of the chest performed on February 16, 2018, a wide «en block» excision of this tumor with the appropriate margin around the involved 2nd - 4th ribs to the right side of the sternum.

Individual bespoke 3-D printing titanium on lay implant to cover an extensive defect in the right anterior chest wall was designed by **Peter L. Evans** — Welsh pioneer designer of the cranio-facial and maxillofacial implants (1987) and a 3-D printing implants (2018).

Implant was sewn in to cover defect with section of the right latissimus dorsi muscle, harvested earlier from the right upper posterior chest by **Thomas Bragg** — Welsh plastic surgeon, and the wound was closed. Patient made an uneventful recovery.

This was the first case of extensive resection of the anterior chest wall «en block» with the hemi-sternum because the extend of malignant tumor with coverage of defect by a 3-D print modelled individually bespoke implant made on lay of a titanium.

In a brief note concerning a substitute for standard cotton surgical garments, **Myroslaw (Myroslau) J. Dragan (April 01, 1940, Ukraine –)**[1510] — Ukrainian American physician and pathologist, described number of substances such as talk, Lycopodium (in Greek - «lukos» meaning wolf, «podion» - meaning ground), i.e., ground (princess) pine or creeping cedar, rice, corn, starch and cotton fibers,

---

[1507] Anwar S, Singh GK, Varugheses J, et a. 3-D printing in complex congenital heart disease: across a spectrum of age, pathology, and imaging techniques. JACC: Cardiovascular Imaging Aug 2017;10(8):953-6.

[1508] Bangladesh (area 147,570 km²) — country in South Asia (independent since 1971).

[1509] Wales, Welsh or Cymru (area 20,779 km²) — country that is part of United Kingdom (UK) and the island of Great Britain.

[1510] Dragan MI, Voronka P. Incidence of Rhesus negative blood group in among Ukrainians and its anthropological significance. J Ukr Med Assoc North Am 1974 Jan;21,1(72):25-32. — Dragan MI. Idiopathic retroperitoneal fibrosis and primary sclerosing cholangitis. J Ukr Med Assoc North Am 1974 Apr;21,2(73):33-49. — Dragan MI. Silicone breast implant failure due to cellulose fibers from the disposable surgical garments. J Ukr Med Assoc North Am 1977 Jul;24,3(86): 3-13. — Dragan MJ. Wood fibers from disposable surgical gowns and drapes. JAMA 1979 May 5;241(21): 2297-8. — Dragan MJ. Post-transfusion remission in erythroleukemia pregnancy. J Ukr Med Assoc North Am 1985 Summer; 33,2(113):90-95. — Ablitsov VH. «Galaxy Ukraine». The Ukrainian diaspora: prominent persons. Publ. «Kyt», Kyiv 2007.

as well as of coniferous wood, originated from disposable surgical paper gowns, drapes and towels in surgical specimen, which introduced into soft tissue of mesothelial cavities (membrane composed of a single epithelium lining body cavities) during 18-months period in 1,000 patients evoke serious granulomatous and fibrotic reactions causing postoperative complications in 22 of them. Those complications because evident in about 6 months after operations included increased keloids, wound dehiscence, incisional hernias, chronic abscesses in the surgical field, and in two patient bouts of intestinal obstructions, due to peritoneal adhesions.

Fortunately, this report led to discontinuance of the use of a disposable surgical gowns, drapes, and towels in surgical practice, thus eliminated complications caused by its use, and the return to the previously used standard cotton ones.

Profound hypothermic circulatory arrest (HCA) for the repair of an aortic dissection type I(A) was introduce by **Randall B. Griepp (1940 –)**[1511] — American cardiac surgeon, during which the open distal anastomosis is performed with the beveled 180° synthetic tubular graft (STG) to the beveled proximal aortic arch from below the origin of the innominate artery on its greater curvature towards the distal end of its lesser curvature at the level of the left subclavian artery (SCA), using a continuous No. 4-0 polypropylene suture. Here, the longer («toe») end of the beveled STG faces the greater curvature of the beveled aortic arch and a shorter («heel») end of the beveled STG faces the lesser curvature of the beveled aortic arch (Griepp technique).

In children with a heterotaxy syndrome and congenital heart disease (CHD) – an effective single one ventricle, persistent left superior vena cava (SVC), interrupted inferior vena cava (IVC) with continuation of the systemic venous blood flow to the right SVC or azygos vein to a persistent left SVC, a definitive palliation can be accomplished by a bilateral bidirectional cavo-pulmonary anastomosis without dividing the azygos or hemiazygos veins continuation, thus routing all the systemic venous return, except hepatic and mesenteric, to the pulmonary artery (PA) – Kawashima procedure (1978)[1512].

In 1987 **Donald Nuss (1940 –)**[1513] — American thoracic surgeon, reported on a minimally invasive technique for the repair of pectus excavates by the elevation of the sternum with a convex retrosternal stainless steel bar, without resection or

[1511] Griepp RB, Stinson EB, Hollingsworth JF, et al. Prosthetic replacement of the aortic arch. J Thorac Cardiovasc Surg 1975;70:1015-63.

[1512] Kawashima Y, Kitamura S, Matsuda H, et al. Total cavopulmonary shunt operation in complex cardiac anomalies. A new operation. J. Thorac. Cardiovasc. Surg. 1984;87 (1): 74-81.

[1513] Nuss D, Kelly RJ, Croitoru D, et al. A 10-year review of a minimally invasive technique for the correction of pectus excavatus. J Pediatr Surg 1998Apr;33:545-52.

division of the costal cartilages, which is then flipped to a convex position so as to push outward on the sternum, correcting the deformity (Nuss procedure, 1987). The bar usually stays in the body for 2-5 years.

The incidence of the common bile duct (CBD) injuries during an open or laparoscopic cholecystectomy occurs in 0.15 to 0.3% of patients. The CBD injuries may occasionally take place in patients with anatomic variations of the Calot triangle **[Jean Calot (1861 – 1944)**. 19th century. Surgery]. But the most commonly they are caused by the vanishing triangle of Callot, resulting from inflammation of the gallbladder and surrounding structures, rarely from the Mirizzi syndrome **[Pablo Mirizzi (1893 – 1964**. Above], making surgical dissection very difficult and dangerous, because of distortion of normal anatomy, without possibility to obtain a critical view of safety (CVS), as outlined by **Steven M. Strasberg (1940 –)**[1514] — American general surgeon, in 2010. When the CVS cannot be achieved the surgeon should abandon further dissection in the attempt to complete cholecystectomy at the inflection point of laparoscopic cholecystectomy, and resort to «fenestrating» versus «reconstituting» subtotal cholecystectomy[1515].

Sport hernia is a condition recognized in 1980 by **Jerry Gilmore (1941 – 2019)** — English surgeon, occurs in young athletes engaged mainly in such sports as American football (58%), hockey (20%), socker (17%), basketball (12%), due to inflammation and tear of an aponeurosis of the external and internal oblique and transverse abdominal muscles, the inguinal ligament, or both, with scarring of the ilioinguinal nerve. This results from repetitive overuse of movements, e.g., sudden leg kicking, jumping, and twisting, creates shear forces across the pubic symphysis or inguinal ligament. It is characterized by pain, typically arising from the pubic symphysis, lower abdominal muscles, or inguinal region. The initial treatment is rehabilitation to strengthen the muscle of the pelvis region for 4-6 weeks, successful in majority of sportsmen. In persistent conditions a surgical reconstruction of torn aponeurosis of the abdominal muscles and release of the ilioinguinal nerve entrapment may be necessary.

[1514] Strasberg SM, Brunt LM. Rationale and use of the critical view of safety in laparoscopic cholecystectomy. J Am Coll Surg 2010;211:132-8.

[1515] Strasberg SM, Pucci MJ, Brunt LM, Deziel DJ. Subtotal cholecystectomy – «fenestrating» vs «reconstituting» subtypes and the prevention of the bile duct injury: definition of the optimal procedure in difficult operative conditions. J Am Coll Surg 2016;222:89-96.

Robert B. Dzioba (10.08.1941, city of Luts'k, Ukraine –)[1516] — Ukrainian American orthopedic surgeon, evaluated surgical outcome of standardized treatment regimen of a complete surgical excision of diseased or injured articular cartilage following by drilling of the subchondral bone plate, followed by a standardized postoperative rehabilitation protocol which included non-weight bearing for 8 weeks. Retrospective analysis revealed that the group with the best prognosis and good (95%) results consisted of small-to-medium acute partial thickness lesions of the femoral medial or lateral condyle. He also designed the original «UniPlate», an orthopedic device used in cervical spine support that streamlined surgical implantation and patient recovery.

Oleksandr S. Nykonenko (09.11.1941, village Mylovka, Ulyanov Obl., Russia –)[1517] — Ukrainian cardiovascular and transplant surgeon, was the first to perform successful heart transplantation (HT) in Ukraine, in 2011 simultaneously carried out liver, kidney and pancreas transplantation in patient with diabetes

[1516] City of Luts'k (first mentioned 1085), Ukraine. — Dzioba RB, Jackson RW. Effect of phosphate supplementation on intact and fractured femora of rats: a biochemical study. Canadian Med Assoc J 1977 Nov;117(10:73-5. — Dzioba RB, Jackson RW. Transient monoarticular synovitis of the hip joint in adult. Clin Orthop 1977;126:190-2. — Dzioba RB, Barrington TW. Short-lasting unilateral inflammation of the hip joint synovial membrane in adult. J Ukr Med Assoc North Am 1979 spring; 26,2(93):81-4. — Dzioba RB, Quinan WJ. Avascular necrosis of the shoulder glenoid. J Ukr Med Assoc North Am 1983 spring;30,2(107:81-7. — Dzioba RB. Deformations of the vertebral column in olympian gymnastics. J Ukr Med Assoc North Am 1983 fall;30,3(108:115-24. — Dzioba RB, Quilan WJ. Avascular necrosis of the glenoid. J Trauma: Injury, Infection and Critical Care 1984 May 1;24(5):448-451. — Dzioba RB, Doxey NC. Prospective investigation into the orthopedic and psychologic predictors of outcome of first lumbar surgery following industrial injury. Spine 1984 Sep;9,6:614-23. — Dzioba RB, Strokon A, Mulbry L. Diagnostic arthroscopy and longitudinal open lateral release: a safe and effective treatment for «chondromalacia of the patella». J Arthroscopy & Related Surgery. 1985 Jan 1;1,2:131-5. — Dzioba RB, Benjamin J. Spontaneous atlantoaxial fusion in psoriatic arthritis. Case report. Spine Phila PA 1976, 1985 Jan-Feb; 10(1):102-3. — Dzioba RB. Classification and treatment of acute articular cartilage lesion. J Arthroscopy & Related Surgery 1988;4:2:72-80. — Dzioba RB, Strokon A, Mulbry L. Diagnostic arthroscopy and longitudinal lateral open release in the treatment of cartilage softening. J Ukr Med Assoc North Am. 1989 fall;35,1(120):87-93. — Dzioba RB, Strokon A, Mulbry L. Diagnostic arthroscopy and longitudinal lateral release: a 4-year follow up study. J Ukr Med Assoc North Am 1990 Winter;37,1(121):3-12. — Dzioba RB, Strokon A, Mulbry L. Diagnostic arthroscopy and longitudinal open lateral release: a four-year follow-up study to determine prediction of surgical outcome. Am J Sport Medicine 1990 July 1; htts//doi.org/101177/036354659001800402. — Dzioba RB. Classification and treatment of acute articular cartilage lesions. J Ukr Me Assoc North Am 1994 winter;41,1(132):3-14. — Dzioba RB. Expanded polytetrafluoroethylene (e. PTFE) membrane cover laminotomy sites in a canine model. J Ukr Med Assoc North Am 2002 Winter;47,1(147):3-6. — Dzioba PB, Tucker TJ, Leber M, Lund P. Effect of smoking on primary instrumental lumbar fusion. J Ukr Med Assoc North Am 2002 Spring;47,2(148):24-7. — Hunter TB, Yoshino M, Dzioba RB, et al. Medical devices in the head, neck, and spine. Radiographics Jan 1 2004;24(1):257-85. Doi.org/10.1148/rg.241035185. — Ablitsov VH. «Galaxy Ukraine». The Ukrainian diaspora: prominent persons. Publ. «Kyt», Kyiv 2007.

[1517] Village Mylovka (founded 1706), Ulyanov Obl. (area 37,181 km²), Russia. — Ablitsov VH. «Galaxy Ukraine». The Ukrainian diaspora: prominent persons. Publ. «Kyt», Kyiv 2007.

mellitus (DM). He established that high serum level of beta$_2$-mictoalbumin late post-transplant predicts subsequent decline in kidney allograft function (2015), research ureteral stent in renal transplantation *«be or not to be»* (2016).

Break-through in the surgical treatment of hypoplastic left heart syndrome (HLHS) was made in 1981-98 by **William I. Norwood, Jr (1941 – 2020)**[1518] — American pediatric cardiac surgeon, who designed the three-stage surgical approach (Norwood operations).

1st stage performed through the midline longitudinal sternal incision (MLSI) include sewing of a 3,5 mm synthetic tubular graft (STG) into the innominate artery – initially as the arterial inflow for the cardiopulmonary bypass (CPB); induction of hypothermic circulatory arrest (HCA) during which the main pulmonary artery (MPA) is transected above its valve and its bifurcation is sutured, the median wall of the MPA facing the proximal ascending aorta (AA) is excised in a «U-shaped» fashion, the interatrial septum (IAS) is removed, the remnant of the patent ductus arteriosum (PDA) is dissected of the aorta, the left wall of the hypoplastic ascending and transverse aorta (ATA) is opened longitudinally from the bottom of the bordering «U-shaped» space in the median wall of the transected MPA to near the left subclavian artery (SCA), the faced walls of the proximal AA and the MPA including its valve are sutured «side-to-side» – the modified Damus-Key-Stensel (DKS: below) procedure; the distal AA and the aortic arch are anastomosed to the semicircular segment of the pulmonary homotransplant (half-MPA), thus creating the «neo-ATA»; the proximal portion of a 3,5 mm STG is detached from the arterial inflow for the CPB and anastomosed into the right pulmonary artery (PA) forming the Blalock-Taussig (BT) shunt (1981-83). Alternatively, as a first step the shunt proposed in 2003 by **Shunji Sano (1952 –)**[1519] — Japanese pediatric cardiothoracic surgeon, is performed between the single right ventricle (RV) and the MPA using the STG (Sano shunt, 2003).

2nd stage is carried out 4-6 months later, after confirmation of a proper function of the 1st stage, when through the MLSI the BT shunt is remove and the superior vena cava (SVC) is anastomosed in the «end-to-side» fashion into the right PA (semi-Fontan operation, 1989).

3rd stage is performed also via the MLSI and include the formation of the internal right atrial (RA) lateral tunnel-like synthetic or external STG between the inferior vena

---

[1518] Norwood WI, Lang P, Castaneda AR, et al. Experience with operations for hypoplastic heart syndrome. J Thorac Cardiovasc Surg 1981;82(4):511-9. — Norwood WI, Lang P, Hensen DD. Physiologic repair of aortic atresia – hypoplatic heart syndrome. NEJM 1983;308(1):23-6.

[1519] Sano S, Ishino R, Kawada M, et al. Right ventricle-pulmonary artery shunt in first-stage palliation of hypoplastic left heart syndrome. J Thorac Cardiovasc Surg 2003Aug;126(2)504-9; discussion 509-10.

cava (IVC) and the right PA (complete-Fontan operation, 1998). Thus, the Norwood operation creates the single ventricle heart and is a destination operation for surgical treatment of the HLHS.

In 1972 **Valeriy F. Sayenko (11.05.1941, u.t.v. Terny, Nedryhaylivsky Region, Sumy Obl., Ukraine –)**[1520] — Ukrainian scientific surgeon and transplantologist, Head Division of Gastrointestinal (GI) Surgery, now the National Institute of Surgery and Transplantation (NIST) of the National Academy of Medical Science (NAMS) of Ukraine, in 1989 became Director of the NIST NAMS of Ukraine. Under his leadership were develop new approaches for the diagnosis of hypersecretion state of post-surgical stomach, formed surgical procedures for peptic ulcer of different genesis (origin) and surgical correction of dumping syndrome, diversified operations for ulcers of the pylorus of the stomach and duodenum, led the intraoperative control for quality of the proximal selective vagotomy, substantiated the possibility of presentation of the pylorus in patients with closed stenosis of the duodenum, improved the treatment of perforated ulcers of the stomach and duodenum, developed concept for the treatment of external fistulae of the GI tract, tumors of the pancreas, proposed new forms of operations for the correction of chronic intestinal obstruction, developed and pathogenically substantiated the treatment of sepsis, introduced into surgery rational antibiotic therapy, formed principles of rehabilitation patients with GI diseases.

Valeriy Sayenko developed and actively introduces into practice minimal invasive surgery for the treatment of digestive tract and blood vessels, develop and improve scientific endeavor in healingdiseases of the stomach, intestines, liver, biliary tract, pancreas, portal hypertension, cardiovascular system and organ transplantation. Specifically, he performed the splitting transplantation of the liver (2001).

James L. Cox (1942 –) — American cardiac surgeon, developed methods of the surgical treatment of atrial fibrillation (AF) and atrial flutter by the I-IV Cox-Maze procedures (1985-95).

First endovascular repair of abdominal aortic aneurysm (AAA) was performed in 1991 by **Juan C. Parodi (1942 –)** — Argentinian vascular surgeon.

First open fetal surgery was performed in 1981 by **Michael R. Harrison (1943 –)** — American pediatric surgeon, to correct a dangerously blocked urinary tract in a fetus.

[1520] U.t.v. Terny (founded 1643; suffered from carried out by Russian Communist Fascist regime Holodomor-Genocide 1932-33 against the Ukrainian Nation), Nedryhaylivsky Region, Sumy Obl., Ukraine. — Kovalchuk LYa, Sayenko VF, Knyshov HV. Klinichna khiruhiya. Vol. 1-2. Publ. «Ukrmedknyha», Ternopil 2000. — Ablitsov VH. «Galaxy Ukraine». The Ukrainian diaspora: prominent persons. Publ. «Kyt», Kyiv 2007.

While researching scoliotic deformities of the backbone or vertebral column, **Valentyn V. Serdyuk (July 09, 1942, city of Mykolaïv, Ukraine –)**[1521] — Ukrainian orthopedic and trauma surgeon, found the interrelation between the development of the adult person diseases and scoliosis of the vertebral column in childhood (2005), and that this affection is related to an instability of all portion of the backbone due to asymmetric function of the brain hemisphere (2008). On this basis he developed an effective conservative treatment of scoliotic deformities. He combined surgical and magnetic-acoustic treatment[1522] of complicated infected traumatic wounds of the major joints and extremities, as well as trophic diabetic ulcers, with success

In 1975 **Paul M. Damus (1943 –)**, **M.P. Kaye** and **H.C. Stansel, Jr**[1523] — American pediatric cardiac surgeons, independently described a procedure, named the Damus-Kaye-Stensel (DKS; above) procedure, designed for the surgical treatment of subaortic stenosis or restrictive bulboventricular foramen, entailing the creation of an «end-to side» anastomosis between the end of the proximal main pulmonary artery (MPA) and the side of the ascending aorta (AA) with patch reconstruction of the additional left ventricular outflow tract (LVOT) or reconstruction of the bivalvular systemic outflow tract, performed along with the Fontan procedure (above), particularly for double outlet left ventricle (DOLV). Accordingly, the better strategy for the newborn with tricuspid valve (TV) atresia with transposition of the great vessels (TGA, 30%), restricted left ventricular (LV)-to-aortic blood flow due to ventricular septal defect (VSD) and patent foramen ovale (PFO) with right-to-left shunting, could be a single LV strategy using a DKS procedure, i.e., anastomosis of the proximal end of the transected supravalvular MPA to the AA and shunt to establish pulmonary blood flow.

Randomization of the clinical trial comparing total mastectomy and lumpectomy with or without irradiation in the treatment of carcinoma, carried out in 1972-89, by **Bernard Fisher (1918 – 2019**: above), showed that the best approach to the

---

[1521] City of Mykolaïv (Foot Note 198), Mykolaïv Obl., Ukraine. — Serdiuk VV, et al. Restoration Surgery of Destructive Forms of Bone and Joint Tuberculosis and Osteomyelitis and its Consequences. Kyiv 2002. — Serdiuk VV, et al. Magnetic Therapy. Past, Present and Future (Informative Handbook). Kyiv 2004. — Serdiuk VV, et al. Traumatology and Orthopedics (Teaching Handbook for). 2004. — Serdiuk VV, et al. Contemporary Magnetic Therapy (New Technology and Equipment). Perm 2005. — Serdiuk VV, et al. Traumatology and Orthopedics» (Guidance to the Practical Studies for Students). Odesa, 2006. — Serdyuk V. Scoliosis and Spinal Pain Syndrome. New Understanding of their Origin and Ways of Successful Treatment. Byword Books, Dehli 2014. — Serdiuk VV. Asymmetry of the Body. Scoliosis. Spinal Pain Syndrome (3rd Ed., expanded and re-done). Odesa, 2020. — Ablitsov VH. «Galaxy Ukraine». The Ukrainian diaspora: prominent persons. Publ. «Kyt», Kyiv 2007.

[1522] Magnetic-acoustic treatment is the stroke-valve therapy, i.e., an influence of the acoustic waves of low frequency and high intensity which induce the process of tissue regeneration.

[1523] Damus PS. Correspondence. Ann Thorac Surg 1975;20(6):724-5. — Kaye MP. Anatomic correction of transposition of great arteries. Mayo Clin Proc 1975;50(11):638-40. — Stansel Jr HC. A new operation for D-loop transposition of the great arteries. Ann Thorac Surg 1975;19:565-7.

successful surgical treatment of breast cancer is selective, breast-conserving one, depending on clinical stage of disease.

Selective and breast-conserving approach to the treatment of breast cancer, championed by **Jean Y. Petit**[1524] — French breast and plastic surgeon, include radical mastectomy, modified radical mastectomy, radical sectorial resection or quadrantectomy, lumpectomy and nipple-areola preserving central mastectomy with or without preoperative induction neoadjuvant radiation with polychemotherapy, postoperative (adjuvant) radiation with polychemotherapy, intraoperative radiotherapy, and conservative treatment only with polychemotherapy or poly-chemo-hormone-therapy. He also championed an integration of plastic surgery into breast-conserving surgery for cancer to decrease the radicality of breast tumor resection and improve its reconstruction.

At the early pre-clinical stages and nodular tumor up to 5 cm in the maximal dimension and in the external location of the tumor without invasion of the axillary lymph nodes (LN), the stage I according to the TNM classification (1969), are acceptable for the breast-preserving operations such as radical sectorial resection or quadrantectomy with post-operative (adjuvant) radiotherapy of retained breast tissue.

In the medial and central locations of tumor up to 5 cm in the maximal dimension without involvement of the LNs and also tumors up to 5 cm or more in the maximal dimension with movable axillary LNs, the stage II according to tumor, node and metastases (TNM) classification (1969), the initial treatment consist with neoadjuvant radiotherapy and polychemotherapy, followed by radical modified mastectomy without removal of the pectoralis major muscle, but with the removal of the pectoralis minor muscle and the involved axillary LNs with post-operative adjuvant radiation-therapy and polychemotherapy.

In tumors of any dimension with direct invasion of the chest wall, with the exception of the chest muscles and the skin, with or without involvement of the axillary or the internal thoracic artery (ITA) LNs, the stage III according to the TNM

[1524] Petit JY, Lasser P, Travagzi JP, et al. Conservative treatment of cancer of the breast. J Chir (Paris) 1987 Feb;124(2):132-5. — Petit JY, Rietjens M, Garusis C, et al. Integration of plasti surgery in the course of breast-conserving surgery for cancer to improve cosmetic results and radicality of tumor excision. Recent Result Cancer Res 1998;152:202-11. — Dryzhak VI. Cancer of the breast. In, LYa Kovalchuk, VF Sayenko, HV Knyshov. Clinical Surgery. Vol. 1-2. Publ. «Ukrmedknyha», Ternopil 2000. Vol. 2. P. 348-57. — Petit JY, Voronesi V, Orecchia R, et al. Nipple-sparing mastectomy in association with intraoperative radiotherapy (ELIOT): a new type of mastectomy for breast cancer. Breast Cancer Res Treat 2006;96(1):47-51. — Voronesi U, Stafyla V, Petit, et al. Conservative mastectomy: extending the idea of breast conservation. Lancet Oncol 2012 Jul;13(7):e 311-7. — Jatoi I, Kaufman M, Petit JY. Atlas of Breast Surgery. 2nd ed. Springer Publ Co, New York 2014.

classification (1969), the neo-adjuvant radio-chemotherapy, radical or modified radical mastectomy and adjuvant radio-chemotherapy are indicated.

In neglected cases and in extensive cancers of the breast, stage IV according to the TNM classification (1969), conservative therapy, such radiation therapy and chemo-hormono-therapy are indicated.

In 1993 **John L. Gibbs**[1525] — English cardiologist, proposed the replacement of the 1st surgical stage for the treatment of hypoplastic left heart syndrome (HLHS) by the palliative «hybrid» procedure, comprising a stenting of the patent ductus arteriosus (PDA) and the pulmonary artery (PA) banding. The subsequent «comprehensive» 2nd surgical stage, 4-6 months later, include transection of the main PA, removal of the PA bands, amalgamation of the proximal ascending aorta (AA) and the main PA with inclusion of its valve [the Damus-Kaye-Stansel (DKS) anastomosis], augmentation of the distal AA and the aortic arch, removal if the IAS and the half-Fontan operation. During the 3rd surgical stage, the full-Fontan operation is performed.

Primary biventricular repair of an interrupted aortic arch (IAA) with ventricular septal defect (VSD) consists of the left ventricular outflow tract (LVOT) reconstruction by the Damus-Kaye-Stansel (DKS) procedure and intracardiac rerouting through the VSD, along with concomitant reconstruction of the right ventricular outflow tract (RVOT) and an aortic arch – Yasui procedure (1987)[1526].

Since in 1992 **Anatoli V. Malukhin (March 02, 1943, city of Kharkiv, Ukraine –)** — Ukrainian neurosurgeon, was researching and trying to define the correlation between the blood chemical indices of human body and parameters of blood formula. Together with mathematicians from Kharkiv Physico-Technological Institute of Low Temperature (created 1960), of the Ukrainian National Institute «Kharkiv Aviation Institute» («KhAI», founded 1930) of the National Academy of Science (NAS) of Ukraine (founded 1918), PAO «Aviacontrol» the program was conducted to work out the prepared outcomes for clinical analysis of the blood and designate biochemical, hemodynamic, and metabolic signs of human vitality. Collaboration since 2002 with engineers from the Kharkiv Scientific Research Company «Biopromin», Ltd, under leadership of **Anatolii A. Pulavsky**, gave push after 2006 for the development of noninvasive diagnostic method, with time worked out the instrument for measuring temperature at biologically active points with appropriate software for the emendation of received information and studying the health of person without acquisition of the blood sample, and after successful clinical trial conducted between

[1525] Gibbs JL, Wren C, Watterson KG, et al. Stenting of the arterial duct combined with banding of the pulmonary arteries and atrial septostomy: a new approach to palliation for hypoplastic left heart syndrome. Br Heart J 1993;69:551-5.

[1526] Yasui H, Kado H, Nakano E, et al. Primary repair of interrupted aortic arch and severe aortic stenosis in neonates. J Thorac Cardiovasc Surg. 1987;93:539-45.

2007 and 2012, made it possible to register new medical instrument named the «Automatic Non-invasive Express Screen Analyzer» (ANESA) or bloodless blood analysis in Ukraine, then also in Europe, Russia and China.

To treat complex and extensive native valve endocarditis (NVE) of the posterior and the anterior leaflet of the mitral valve (MV) extending into its anulus with peri-annular abscess formation, **Tirone E. David (Nov. 20, 1944, Ribejrao-Klaro, State of Parana, Brazil** –) — Brazilian Canadian cardiac surgeon, recommended that after excision of all infected tissue and abscess the defect to be reconstructed with the use of autologous or bovine pericardium (David technique, 1987)[1527]. A semicircular patch is used to cover the defect, with one side secured to the left ventricular (LV) myocardium and the other side sewn to the left atrium (LA), or a circular patch can be used when the entire circumference of the anulus must be reconstructed.

In 1992 he proposed the second type aortic valve (AV)-preserving operation, a variant of reimplantation of the AV or valve-sparing aortic root replacement (VSARR)[1528], where a proximal Dacron synthetic tubular graft (STG), larger by one size than usual is sutured outside to the aortic ring, the three commissural posts of the AV are then resuspended high within the Dacron STG, remnants of aortic sinuses of Valsalva are reimplanted within the Dacron STG, the ostia of the coronary arteries are reimplanted and a distal Dacron STG is anastomosed to the ascending aorta (AA) – an in-graft reimplantation technique.

Initially, as in the «three-Yacoub» procedure, the aortic root is dissected circumferentially down to the level of the aortic annulus in the festoon fashion and the three coronary sinuses are excised, leaving about 5 mm of tissue attached to the aortic anulus and buttons around the coronary artery orifices. Multiply 3-0 or 4-0 polyester mattress sutures are passed from the inside to the outside of the left ventricular outflow tract (LVOT) through a single horizontal plane corresponding to the lowest portion of the aortic anulus along its fibrous components and following the scalloped shape of the anulus along its muscular component. When the fibrous component of the LVOT is thin, poytetrafluoroethylene (PTFE, Teflon) pledges should be used in those sutures. Those sutures are not hemostatic, and it is sufficient to use only three mattress sutures in the commissural line for the initial stabilization of the aortic ring. The three commissures are temporarily gently suspended upwards and approximated until the three cusps coopt to evaluate its sufficiency and determine the diameter of the sinotubular junction, necessary for choosing the appropriate

---

[1527] David TE, Feindel CM. Reconstruction of the mitral annulus. Circulation 1987;76:III 102-7.

[1528] David TE, Feindel CM, Bos J. Repair of the aortic valve in patients with aortic insufficiency and aortic root aneurysm. J Thorac Cardiovasc Surg 1995;109;345-52. — David TE. Surgery of the aortic root and ascending aorta. In, FW Sellke, PJ del Nido, SJ Swanson (Eds). Sabiston & Spencer Surgery of the Chest. Vol. I-II. 8th Ed. ElsevierSaunders, Philadelphia 2010. P. 1021-39.

Dacron STG, usually 4-6 cm greater than previous or one size greater than usual (30-34 mm). In the proximal end of the chosen Dacron STG a small triangular incision is made and the remaining portion is plicated in two or three places to reduce its diameter by 3 to 4 mm. After the insertion of the AV inside the Dacron STG, previously passed through the LVOT sutures are now passed from the inside to the outside of the tailored end of the graft and tied. If reduction of the aortic anulus is desirable, it is Done by placing wider sutures beneath the commissures of the noncoronary cusp than in the Dacron STG. All three AV commissures are simultaneously suspended by 4-0 polypropylene stay suture inside the graft to the level of full adjustment of the AV leaflets, tied, and then those same sutures are used to sew in the continuous uninterrupted fashion the remnants of the coronary sinuses of Valsalva and the aortic anulus to the internal surface of the Dacron STG. Those sutures had to be precise and fully hemostatic. The coronary artery buttons are implanted into the opening made in the conforming coronary neo-sinuses. The spaces between commissures are plicated to create a slight bulge in the neo-aortic sinuses and to reduce the diameter of the graft to that of desirable sino-tubular junction. Residual cusps prolapse are plicated or its edges reinforced. The operation is completed by anastomosing the distal end Dacron STG to the distal AA.

Postoperative complications and long-term outcomes of the VSARR are similar to the «uni-Yacoub» and «three-Yacoub» procedures.

Beside of this, he described repair of post-infarction ventricular septal defect (VSD) by endocardial patch with the infarct exclusion and wide resurfacing of the LV side of the infarction (1995)[1529].

Percutaneous (PC) endoscopic gastrostomy (PEG) is a mini-invasive surgical procedure during which a gastrostomy tube is passed under the guidance and assistance of gastroscope through the skin and anterior abdominal wall into the patient stomach and then secured tightly to the skin, as an alternative to open surgical gastrostomy. It provides a temporary or permanent total enteral nutrition (TEN) in patients in whom oral intake is inadequate or impossible due to disturbances in swallowing and recurrent aspiration pneumonitis, those who do not tolerate the Dabbhoff feeding tube, those suffering from paralysis agitans, stroke sequel, brain tumor, after surgery or radiotherapy or both for head and neck cancer.

First PEG was performed on June 12, 1979 by **Michael W.L. Gauderer (1945 –)**[1530] — American pediatric surgeon, **Jeffrey Ponsky (1947 –)** — American gastroenterologist

---

[1529] David TE, Dale L, Sun Z. Postinfarction ventricular septal rupture with infarct exclusion. J Thorac Cardiovasc Surg 1995;110:1315.

[1530] Gauderer MW, Ponsky JL, Izant RJ. Gastrostomy without laparotomy: a percutaneous endoscopic technique. J Pediatr Surg 1980;15(6):872-5. — Gauderer WM. Percutaneous endoscopic gastrostomy – 20 years later: a historical perspective. J Pediatr Surg 2001;36(1):217-9.

and endoscopist, and **James Bekeny** — American surgical resident, on a half-month-old child with inadequate oral intake.

Complications of PEG include inflammation of the skin and subcutaneous tissue (cellulitis) or structural lypodystrophic changes of the skin and subcuticular layer leading to disruption of microcirculation and drainage of the lymph around the gastrostomy site, hemorrhage, gastric ulcer, perforation of the transverse colon with peritonitis or gastrocolic fistula, puncture of the left lobe of the liver and buried bumper syndrome in which the gastric part of the tube migrates into the stomach wall.

Incidence of native valve endocarditis (NVE) in developed countries ranges about 1.7 to 7.0 patients per 100.000 persons per year. In community acquired endocarditis the more prevalent microorganism is Streptococcus viridans group, whereas in nosocomial acquired endocarditis it is more virulent Staphylococcus aureus.

The terminology of endocarditis, clinical symptoms, and diagnostic signs of the NVE were described by **Jean Bouillard [1796 – 1881). 19th century. Internal medicine], Moritz Roth (1839 – 1914). 19th century. Pathology], Moritz Litten (1845 – 1907).** 19th century. Internal medicine], **William Osler (1849 – 1919).** 19th century. Internal medicine] and **Theodore Janeway (1872 – 1917).** 19th century. Internal medicine].

Diagnostic workup in patients suspected of developing NVE may reveals positive blood cultures, the echocardiography (echoCG) – transthoracic echoCG (TTE) or esophageal echoCG (EEK) showing valvular vegetations, perivalvular leaks, annular or perianular valvular abscesses or fistulae. Computer tomography (CT) or magnetic resonance imaging (MRI) of the head and abdomen is necessary to confirm the presence of distant emboli.

Final diagnosis of NVE is based on combination of clinical findings, expressed the most commonly by the Duke's system[1531], the major criteria being positive blood culture and positive echoCG, and minor criteria being predisposing factors and clinical findings.

First and the most important treatment of NVE is aggressive antibiotic therapy monitored by blood cultures and echoCG at 2 weeks intervals.

Surgical intervention is indicated in patients developing worsening of symptoms and continuing sepsis despite adequate antibiotics treatment, those infected with Staphylococcus aureus, in congestive heart failure (CHF) NYHA class IV, in cardiogenic shock, advanced age, preoperative renal failure (RF), paravalvular leak and extension of disease.

[1531] First described in 1994 by researchers from Duke University (established 1838) and Duke University Hospital (founded 1925-30), both named in honor of **James B. Duke (1856 - 1925)** — American tobacco and electric power industrialist, city of Durham (chronicled 1701, settled mid-1700s), North Carolina (NC: area 139,390 km²), USA.

In aortic NVE the involvement of the aortic annulus requires radical surgical debridement of all infected areas and aortic root abscess, leaving a rim of healthy tissue. This should be followed by reconstruction of an anulus with autogenous or bovine pericardial patches, and larger defect with Dacron patch, and aortic valve replacement (AVR). If more than 50% of the aortic anulus has been destroyed, or there is extensive ventriculo-aortic discontinuity, then an aortic root should be replaced with valved synthetic tubular graft (STG), allograft or homograft with attached anterior mitral valve (MV) leaflet which should be implanted to exclude an abscess cavity by sewing the proximal anastomosis with attached anterior MV leaflet to the inferior border of the abscess cavity, and coronary arteries reimplanted.

In 1974 **Gordon Danielson (1931 – : above)** described the treatment of the aortic NVE with the aortic root abscesses by excision of infected tissue and abscesses and an aortic root replacement by a valved STG and coronary artery bypass (CAB) using the great saphenous vein grafts (GSV). However, in 1996 **Fabien Koskas**[1532]— French vascular surgeon, reported that in experiments in dogs the vascular homograph was significantly more resistant to infection than the STG.

Most frequent site of involvement in mitral NVE is the posterior (P_2) leaflet's segment, usually treated by quadrangular resection with sliding repair and remodeling ring annuloplasty as advocated by **Alain Carpentier (1933 – : above)**. Complex and extensive infections that extend into an MV anulus will require resection of substantial amounts of tissue, creating a sizable defect at the atrioventricular groove needing reconstruction of the MV annulus before mitral valve replacement (MVR). The MV anulus could be reconstructed using the Carpentier technique by placing figure-eight sutures at the atrial and ventricular edges of the defect to close the atrioventricular groove, or the David technique **[Tirone David (1944 –), above]** using autologous or bovine pericardium to reconstruct the MV anulus.

Prosthetic valve endocarditis (PVE) is an uncommon but serious complication after mitral or aortic valve replacement (MVR, AVR) or both (MAVR), and a consequence of spontaneous or induced bacteremia, occurring is in approximately 0.3 to 1% per patient -year. In an early PVE within the first 12 months after valve replacement, the infection is commonly caused by Staphylococcus aureus or its species, in the late PVE beyond 12 month it is commonly caused by Streptococcus viridians or its species. The clinical presentation of PVE, diagnostic signs and diagnostic workup are basically similar to those needed in patients with NVE (above).

Medical treatment with antibiotics should last at least 6 to 8 weeks.

[1532] Koskas F, Goeau-Brissonniere O, Nocolas MH, et al. Arteries from human beings are less infectible by Staphylococcus aureus that polytetrafluoroethylene in an aortic dog model. J Vasc Surg 1996;23:472-6.

Indication for surgery include heart failure; incontrollable, fungus or Gram-negative infection; prosthetic valve dysfunction; heart block; and recurrent embolism.

Infected prosthetic valve along with infected, necrotic tissue and abscess cavity should be radically surgically debrided and excised.

Mitral PVE require excision of an abscess in the posterior MV annulus or in in the region of aorto-mitral continuity, or both. The defect should be reconstructed by Carpentier or Davis technique.

Aortic PVE was successfully treated in 1972 by **Donald Ross (1922 – 2014**: above**)** using a complete surgical debridement of infected tissues, reconstruction of the aortic root with a homograft, re-implantation of coronary arteries and minimal use of foreign material in the infected area.

In repair of the aorto-mitral discontinuity, the left fibrous trigone may be reconstructed with a composite patch technique reported in 1975 by **Haushang Rastan (1936** – : above**)**[1533].

With the aim to repair mitral valve (MV) degenerative, myxomatous, mixed degenerative and myxomatous diseases, post inferior myocardial infarction (MI) with ischemic cardiomyopathy (CM), rheumatic heart disease (RHD) with sub-valvular calcifications mainly around the posteromedial commissure of the MV, and other combined diseases complicated by mitral regurgitation (MR), **Gerald M. Lawrie (1945 –)**[1534] — Australian American cardiac surgeon, developed the «American physiological non-resection repair» technique of MR (2009-16). This technique corrects all anatomical components of the «MV apparatus», outlined by **Joseph Perloff (1924 – 2014).** 20th century. Internal Medicine].

«American physiological non-resection repair» technique of mitral regurgitation (MR) uses passive inflation of the left ventricle (LV) and ascending aorta (AA) through the MV orifice by pumping up normal saline solution (NSS) given at 4 L/minute to simulate the late diastolic phenomenon of «diastolic mitral locking», defined as sustained diastolic close of the MV after left atrial (LA) systole. With continuous inflation, early isovolumic systole is simulated. Under this condition, in the normal cardiac cycle, the LV is maximally dilated in the long and short axes, the aortic root is dilated and rotated more horizontally with more acute aortic-mitral angle (normally 118^0), and the AA is dilated, elongated, and displaced upwards, outward, and horizontally from the heart. This simulates normal aortic-mitral coupling.

Forward displacement of the posterior mitral annulus during late diastole and early systole is simulated by traction forwards on posterior annular stay sutures.

[1533] Rastan A, Atai M, Hadi H, et al. Enlargement of mitral valvular ring: new technique for double valve replacement in children and adults with small mitral annulus. J Thorac Cardiovasc Surg 1981;81:106-11.

[1534] Lawrie M, Zaghbi W, Little S, et al. One hundred percent reparability of degenerative mitral regurgitation: intermediate-term results of a dynamic engineered approach. Ann Thorac Surg Feb 2016;101(2):576-84.

Through the aortic-mitral continuity, the anterior mitral annulus and leaflet are also displaced upwards and posteriorly from the left ventricular outflow tract (LVOT) towards the LA, a posterior leaflet and annulus. This enhances apposition of the mitral leaflets while moving the anterior leaflet up and out of the LV and away from the LV septum. So that this enlarges the LVOT during systole.

Because the anterior mitral leaflet forms the posterior wall of the LVOT, this posterosuperior position of the anterior leaflet away from the septum in systole minimizes the occurrence of significant systolic anterior motion (SAM), caused mainly by reduced aortic-mitral coupling. This reproducible point at the beginning of isovolumic systole is used to adjust the lengths of the artificial chordae tendinea and size the fully flexible annuloplasty ring in the three-dimensions (3-D) imaging to provide accurate positioning of the zone of leaflets coaptation premarked with dots on the LA side of the leaflets.

Technique with addition of the Kay's posteromedial commissural annuloplasty/ plication of the posterior (P_2 and P_3) segments MV leaflets (**E.B. Kay**, 1955: above) was performed for the correction of post-inferior MI in patients who developed ischemic cardiomyopathy (CM) and MR, such as type I and IIIb according to the physio-pathologic triad of MR by **Alain Carpentier (1933 – : above)**, i.e., annular dilatation or distortion, subvalvular changes (papillary muscle posterior and apical displacement – «tetheing»), LV dilatation with increased sphericity and wall motion abnormalities, apical tenting, delayed closure («loitering»), «pseudo-prolapse» and scallop mal-cooptation of the leaflets.

«American physiological non-resection repair» technique was also applied in RHD affecting MV, particularly with calcifications at the commissural edges with occasional extension posteriorly into the annulus and subvalvular apparatus, when decalcification was possible without perforation through the posterior wall of the heart.

During 10 years follow up post «American physiological non-resection repair» the reparability and freedom from recurrent significant MR in degenerative, myxomatous, mixed degenerative and myxomatous disease and in post-inferior MI ischemic CM were 100% with the overall operative mortality (OM) of 1,6% (in post inferior MI ischemic CM – 0.2%), in RHD with subvalvular calcifications at the posteromedial commissure – 56% and other combined procedures – 74.73%, with OM for later two groups of 2,3%.

Robotic surgery, a computer-assisted surgery and robotically assisted surgery, are technological creations that use rob. system to aid surgical procedures. It was developed to overcome the limitations of a minimally invasive surgery, including minimal invasive cardiac surgery (MICS) and to enhance the capabilities of surgeons in performing open heart surgery. Robotically assisted surgery debut was in 1985, when **Yik S. Kwoh (1946, Honk - Kong, China –)** — Chinese American physician, bioengineer and inventor, developed a robot PUMA 560, named «Ole», using computer soft-ware interface, assisted in placing a needle and performing a brain biopsy under computed tomography (CT) guidance, in an area that would have been difficult to reach by conventional means. This was followed by the first laparoscopic cholecystectomy, performed by **O.D. Lukichev**[1535] — Russian surgeon in 1983, **Eric Mühe (1938 – 2005)** — German surgeon (Sep. 1985), and **Phillipe Mouret (1938 – 2008)** —French surgeon (Mar. 1987), and also by **Alain Carpentier (1933 – : above)** the first robotic mitral valve (MV) repair (1998).

In 1956 **Charles Bailey (1910 – 93: above)**, and in 1957-58 **William Longmire (1913 – 2003: above)** introduced off pump (OP) coronary endarterectomy (CE), in 1960 **Robert Getz (1910 – 2002: above)**, and in 1964 by **Vassily Kolesov (1904 – 92: above)** introduced OP-coronary artery bypass (OP-CAB).

One of the pioneers of minimally invasive cardiac surgery (MICS), **W. Randolph Chitwood, Jr (1946 –)**[1536] — American cardiac surgeon, who in 1988 developed a facile retrograde cardioplegic catheter for cannulation through the ascending aorta (AA), designed a transthoracic flexible clamp for cross-clamping of the ascending aorta (AA) for MICS through a stab incision in the right 3rd intercostal space (ICS) at the middle axillary line (MAL) (Chitwood clamp, 1997) and performed the first total robotic mitral valve (MV) repair by trapezoid resection of its posterior leaflets and ring annuloplasty in the USA (2000).

Thereafter, since 1978 **Frederico J. Benetti (1947 –)**[1537] — Argentinian cardiac surgeon, has championed the resurgence, and further development of beating heart surgery, i.e., OP-CAB, initially through a longitudinal mid-sternotomy incision (LMSI).

Between May 1978 and March 1990, 700 patients underwent OP-CAB surgery with the operative morbidity of 4%, operative mortality (OM) of 1%, and the probability of survival at seven years of 90%.

One of the most important advances in an exposure of the heart for OP-CAB was made by **Ricardo C. Lima** — Brazilian cardiac surgeon, with introduction of the deep lateral pericardial retraction (Lima) sutures, which when placed under tension, create a ridge of pericardium that supports the base of the lateral left ventricle (LV) adjacent to the atrio-ventricular (AV) groove and allows the heart to be rotated rightwards to

---

[1535] Lukichev OD, Filimonov MI, Zybin IM. A method of laparoscopic cholecystostomy. Khirurgiia (Moskva) 1983;8:125-7.

[1536] Chitwood Jr WR, Elbeery JE, Chapman WH, et al. Video-assisted minimally invasive mitral valve surgery: the «micro-mitral» operation. J Thorac Cardiovasc Surg 1997;113:413-4. / Chitwood Jr WR, Nifong LW, Elbeery JE, et al. Robotic mitral valve repair: trapezoid resection and and prosthetic annuloplasty with the da Vinci surgical system. J Thorac Cardiovasc Surg 2000; 120:1171-2.

[1537] Benetti F, Naselli G, Wood M, Geffner L. Direct myocardial revascularization without extracorporeal circulation in 700 patients. Chest 1991 Aug;100(2):312-6.

assume an «apex position»[1538]. In this position the apex of the heart points towards the ceiling and protrudes through the LMSI, usually above the plane of the sternal retractor, allowing exposure of the posterior-lateral and inferior aspects of the LV before the coronary artery compression-type, suction-type or capture-type stabilizer is applied. One or two Lima sutures of 2-0 silk are placed in the pericardium posterior to the left phrenic nerve, immediately anterior to one or both, the left superior pulmonary vein and the left inferior pulmonary vein (LSPV, LIPV): one deep into the oblique sinus behind the left atrium (LA) and one to the left and posterior to the inferior vena cava (IVC).

Alternative to the placement of the Lima sutures is to use the apical suction device which allows a rapid position the heart in an «apex position».

In 1995 Friderico Benetti invented the video assisted minimally invasive direct coronary artery bypass (MIDCAB) procedure through a short (4-7-cm) left 5th antero-lateral intercostal space thoracotomy (ICST), which consist of using the left internal thoracic artery (ITA) graft to bypass the left anterior descending (LAD) artery[1539]. This approach he further refined to a xiphoid incision, being a simpler, less painful approach than a mini-left anterior ICST.

Next year (1996) he performed a mini-invasive mitral valve replacement (MIMVR), in 1997 – an aortic valve (AV) surgery via the 2nd or 3rd ICST using 3-dimentional (3-D) technology to assist the operation, and invented the combined xiphoid-lower sternotomy incision approach for performing an ambulatory MIDCAB using 3-D technology[1540], and in 2005 surgically implanted the embryonal cells into the myocardium of the patient with heart failure (HF).

Direct harvest of the ITA via the LMSI is demanding and often challenging because the inferior exposure in relation to the plane of the LMSI, often need vigorous chest wall retraction resulting in significant postoperative pain, and occasionally harvesting of an inadequate length of the ITA to perform bypass of target coronary arteries. Harvesting of the ITA which require to work under transected hemi-sternum may cause in a cardiac surgeon severe neck strain, even the prolapse of cervical disk with a subsequent invalidity.

Therefore, in lieu of a direct harvesting of the ITA for CAB grafting, **Patrick Nataf**[1541] — French cardiac surgeon, has been used regularly since 1995 the thoracosopic video-assisted harvesting of the ITA graft before making short left

---

[1538] Lima R. Revascularizacion a o da arteria circumflexa sem auxilo da CEC. XII encontro dos discipulos do Dr. F.J. Zebrini. Sessao de videos. Curitiba: Parana Outubro 1995,6b.

[1539] Benetti FJ, Ballester G, Sani G, et al. Video assisted coronary bypass surgery. J. Card Surg 1995;10:620-5.

[1540] Benetti FJ. Minimally invasive coronary surgery (the xiphoid approach). Eur J Cardiothorcic Surg 1999;16(Suppl 2):S10-11.

[1541] Nataf P, Lima L, Regan M, et al Thoracoscopic internal mammary artery harvesting: technical considerations. Ann Thorac Surg 1997;63(Suppl):104-6.

5th antero-lateral ICST, subxiphoid incision, combined xiphoid incision and lower sternotomy or LMSI for OP-CAB, MIDCAB or a standard CAB grafting procedure. In 1998 **Francis G. (Duhaylongsod) Duhay (1961 –)**[1542] — American cardiac surgeon, reviewed a multicenter experience with the thoracoscopic harvest of the ITA grafts with approving conclusions.

First robotic-assisted CAB grafting, or totally endoscopic CAB (TECAB) procedure was carried out in 1998 by **Didier F. Loulmet (1961 –)**[1543]— French American cardiac surgeon, with the da Vinci system (Intuitive Surgical, Sunnyvale, CA) on CPB and OP-CAB. Initially, the TECAB involved only the left ITG graft to the LAD artery, with subsequently increase to multivessel CAB grafting.

Through a small (1,5 mm) port incision in the 5th left intercostal space (ICS) on the anterior axillary line (AAL) a camera is introduced under the left lung collapse. Carbon dioxide (CO_2) is inflated at the target pressure of 10 mmHg. The instruments parts are then inserted through the small ports in the 3rd and 7th ICS on the middle clavicular line (MCL) under thoracoscopic vision. The entire length of the ITA graft from the subclavian artery (SCA) to the 6th ICS is harvested using a computer (the da Vinci) tele-manipulator, the LAD artery is identified and marked with a clip onto the adjacent epicardial fat.

Common femoral artery (CFA) and femoral vein (FV) are exposed in the left groin. Through the left FV the right atrium (RA) is cannulated for CPB with a 25F or a 27F venous return cannula. Then, the left CFA is cannulated with a 21F arterial cannula, a remote CPB is started, and the ascending aortic (AA) occlusion balloon is inflated for infusion of cardioplegia. Thereafter, the target LAD artery or other coronary arteries are exposed and incised longitudinally with Lancet endoscopic knife to create an opening for the anastomosis. The left ITA is sutured robotically to the LAD artery target incisional opening with a 7-0 synthetic suture, or with the aid of an automated stapler to create the distal anastomosis between the ITA or vein grafts (VG) and coronary arteries (created 1995, Cardica, Inc., Redwood City, CA)[1544].

On Apr. 01, 1992, the first video-assisted thoracic surgery (VATS) left upper lobectomy (LUL) was performed by **William S. Walker**[1545] — Scottish thoracic

[1542] Duhaylongsod FG, Mayfield WR, Wolf RK. Thoracoscopic harvest of the internal thoracic artery: a multicenter experience in 218 cases. Ann Thorac Surg 1998;66:1012-7.

[1543] Laulmet D, Carpentier A, d'Attellis N, et al. Endoscopic coronary artery bypass with the aid of robotic assisted instruments. J Thorac Cardiovasc Surg 1999;118(1):4-10.

[1544] Eckstein FS, Bonilla LF, Englberger L, et al. First clinical results with a new mechanical connector for distal coronary artery anastomoses in CABG. Circulation 2002;106(12 suppl 1):I1-4. — Gills IS, Izatani H. Internal thoracic artery to left anterior descending anastomosis performed on the beating heart by endoscopic robot assistance using a new distal connector. Can J Surg 2006 Feb;49(1):37-40.

[1545] Walker WS, Carnochan FM, Pugh GC. Thoracoscopic pulmonary lobectomy J Thorac Cardiovasc Surg 1993 Dec;106(3):1111-7. — William S. Walker, FRCS, interviewed by Joel Dunning, MD. Giants of Cardiothoracic Surgery. CTSNet Mar 30, 2012.

surgeon. The entry into the pleural cavity for the VATS lobectomy is secured by placing three ports. Its placement depends on whether lobectomy is performed using the traditional anterior dissection, i.e., first the arterial, followed by venous and bronchial, or the posterior (Walker) dissection, that is, first the bronchial, then the arterial and the venous.

For the anterior dissection, the first lateral port access for the thoracoscope lighting is placed through a 1 cm incision in the 7th or 8th intercostal space (ICS) in the middle axillary line (MAL). The second anterior access is a 4-6 cm utility incision for dissection and specimen retrieval for the pathological examination, placed in the 5th or 6th ICS between the middle clavicular line (MCL) and the anterior axillary line (AAL), just inferior to the breast. The third posterior port access to provide retraction of the lung is placed through a 1 cm port incision in the MAL lateral to the utility incision or posteriorly.

For the posterior dissection, the first initial 1 cm anterior temporary port access incision for lighting and inspection, with subsequent extension into a 4-5 cm utility incision, is placed anteriorly at the 6th-7th ICS between the MCL and the AAL in front of the anterior border of the latissimus dorsi muscle. The second 1 cm port incision is placed through the triangle of auscultation (trigonum auscultation) is bounded by the lateral border of the trapezius muscle, the superior border of the latissimus dorsi muscle and by the medial border of the scapula, the lower angle of which overlies the 7th rib posteriorly, and then the thoracoscope lighting is moved into this site. The third working port is placed in the MAL lateral to the utility incision.

In 1992 **R. J. Lewis**[1546] reported on VATS in 40 patients (20 men and 20 woman) whose ages ranged from 27 to 82 years, including a wedge resection for metastatic carcinoma (8 patients) and a lobectomy for primary carcinoma (3 patients), without operative mortality (OM).

Subsequently, in 1993 VATS lobectomies were reported by **Thomas J. Kirby** — American thoracic surgeon and **Thomas W. Rise**[1547] — American thoracic surgeon.

Leonid Ya. Kovalchuk (15.III.1947, village of Ternavka, Iz'iaslavsky Region, Khmelnytsky Obl., Ukraine – 01.X.2014, city of Ternopil, Ukraine)[1548] —

[1546] Lewis RJ, Caccavale RJ, Sisler GE, MacKenzie JW. Video-assisted thoracic surgical resection of malignant lung tumors. J Thorac Cardiovasc Surg 1992;104;1679-85.

[1547] Kirby TJ, Rice TW. Thoracoscopic lobectomy Ann Thorac Surg 1993 Sep;56(3):784-6.

[1548] Village of Ternavka (founded 1581), Iz'iaslavsky Region, Khmelnytsky Obl., Ukraine. — Kovalchuk LYa, Polishchuk VM, Velihorotskyy MM, Berehovyy OV. Choice of Methods for Surgical Treatment of Gastroduodenal Ulcers. Publ. «Vertex», Rivne - Ternopil 1997. — Kovalchuk LYa, Polishchuk VM, Nechytailo MYu. Laparoscopic Surgery of Biliary Tract. Publ. «Vertex», Ternopil - Rivne 1997. — Kovalchuk LYa, Polishchuk VM, Tsymbaliuk VI, et al. Atlas of Surgical Operations and Procedures. Publ. «Vertex», Ternopil 1997. — Kovalchuk LYa, Sayenko VF, Knyshov HV. Klinichna Khiruhiya. Vol. 1-2. Publ. «Ukrmedknyha», Ternopil 2000. — Kovalchuk LYa (Ed.). Surgery: Handbook. Publ. I.Ya. Horbachevsky Ternopil Nat Med Univ, Ternopil 2010. — Ablitsov VH. «Galaxy Ukraine». The Ukrainian diaspora: prominent persons. Publ. «Kyt», Kyiv 2007.

Ukrainian surgeon, who with the aim of refining of the surgical management of gastric ulcer disease located in the subcardial section and in the body of the stomach, but under the condition of absence of duodenal stasis, ulcer penetration into the lesser sac, injury to the nerve of **André Latarget (1877 – 1947**: 20th century. Anatomy, histology, and embryology) and malignancy, proposed the targeted resection of the ischemic gastric segment in conjunction with a selective proximal vagotomy (SPV).

Under the leadership of **Mykola O. Korzh (Mykola O. Korzh (12.VIII.1947, city of Kharkiv, Ukraine –)**[1549] — Ukrainian orthopedic surgeon, and **Volodymyr A. Filippenko**[1550] — Ukrainian orthopedic surgeon, were developed in 2006 the sapphire-sapphire endoprosthesis for a total hip replacement was developed.

Vitali I. Tsymbaliuk (26.I.1947, village Symoniv, Hoshchansky Region, Rivnens'ka Obl., Ukraine –)[1551] — Ukrainian neurosurgeon, performed neurotransplantation with healthy embryonic brain tissue in patients with organic injuries to the nervous system (1995-2009).

In January 1991 **Guido Guglielmi (1948 –)**[1552] — Italian neurosurgeon, reported the application of a detachable intravascular coil, constructed from a platinum coil soldered to a stain-less steel delivery wire, when combined with a controllable micro-guide wire, multiple coil could be inserted to fully pack brain aneurysm, and in

---

[1549] Ablitsov VH. «Galaxy Ukraine». The Ukrainian diaspora: prominent persons. Publ. «Kyt», Kyiv 2007.

[1550] Ablitsov VH. «Galaxy Ukraine». The Ukrainian diaspora: prominent persons. Publ. «Kyt», Kyiv 2007.

[1551] Village Symoniv (archeological memorables from Bronze and Early Iron Era, founded 1545), Hoshchansky Region, Rivne Obl., Ukraine –) — Tsymbaliuk VI. Neurosurgical Treatment of Spasticity in Patients with Extra-piramidal Pathology. PhD Dissertation, Kyiv 1985. — Tsymbaliuk VI. Hypophysis. In, LYa Kovalchuk, VF Sayenko, HV Knyshov. Clinical Surgery. Vol. 1-2. Publ. «Ukrmedknyha», Ternopil 2000. Vol. 1. P. 483-94. — Tsymbaliuk VI, Chebotariova LL, Yaminsky YuYa. Transplantation of the embryonal tissue as a method of the functional regeneration of the spinal cord after trauma in an experiment. Ukr Neurosurg J 2002;1:69-76. — Tsymbaliuk VI. Neurosurgery. Publ. «Medytsyna», Kyiv 2008. — Tsymbaliuk VI. Neurosurgery. Kyiv 2010. — Tsymbaliuk VI. Neurosurgery.Publ. «Nova Knyha», Vinnytsia 2011. — Salkov M, Tsymbaliuk V, Dzyak L, et al. The new concept of pathogenesis of impaired circulation in traumatic cervical spinal cord injury and its impact on disease severity: case of series of four patients. Eur Spine J 2016;25(Suppl 1):11-8. — Tsymbaliuk VI, Lazorynets VV, Lurin IA. Contemporary Clinical Organization Aspects of Surgical Treatment of Patients with Diseases and Trauma to the Heart and Magistral Vessels. Premium MM Amosov (medicine) of NAS of Ukraine. Publ. IYa Horbachevsky TNMU, Ternopil 2019. — Ablitsov VH. «Galaxy Ukraine». The Ukrainian diaspora: prominent persons. Publ. «Kyt», Kyiv 2007.

[1552] Guglielmi G, Guerrisi G, Guidetti B. L'elettrotrombosi intravasale nelle malformazioni vascolari sperimentalmente provocate. Dento, Carella A (curatore). Procedimento di III Congresso di Associazione Italiana di Neuroradiologia 1983:139-46. — Guglielmi G, Vinuela F, Sepetka MS, et al. Electrothrombosis of saccular aneurysms via endovascular approach. Part 1: Electrochemical basis, technique, and experimental results. J Neurosurg 1991 July;75:1-7. — Guglielmi G, Vinuela F, Dion J. Electrothrombosis of saccular aneurysms via endovascular approach. Part 2: Preliminary clinical results. J Neurosurg 1991 July;75:8-14. — Guglielmi G. History of the genesis of detachable coils. J Neurosurg 2009 Mar 13;11(1):1-8.

combination with electrolysis and electro-thrombosis used for the treatment of non-surgical intracranial aneurysms, named Guglielmi detachable coil (GDC).

Following **Yurii Bernadsky (1915 – 2006**: above), from 1995 the professor's chair of the Surgical Stomatology at the NMAPE heads **Vladyslav O. Malanchuk (14.VI 1949, city of Kherson, Kherson Obl., Ukraine –)**[1553] — Ukrainian stomatologist surgeon and maxilla-facial surgeon, who since 1986 erforms transplantations of the metatarsophalangeal (MTP) joint (in Latin – articulationes metatarsophalangeolis) in the form of free or vascularized graft for the treatment of the temporo-mandibular joint (TMJ) ankylosis.

Steven R. Gundry (1950 –)[1554] — American pediatric cardiac surgeon, who in 1988 developed a facile retrograde cardioplegic catheter for cannulation through the right atrium (RA) of the coronary sinus (CS) to expedite and improve myocardial protection during cardiac operations using CPB.

When in 1994 a 9-year-old Indian girl Sandeep Kaur's face and scalp were pulled off by her hair being caught by a thresher, **Abraham G. Thomas (1950 –)** — Indian plastic and reconstructive surgeon, conducted the first full-face replant operation.

Ihor (Igor) I. Huk (22.I.1952, city of Gdańsk, Poland –)[1555] — Ukrainian Austrian vascular and transplant surgeon at the Wien University, and coworkers proved than Quereitin treatment mollified ischemia-reperfusion injury to skeletal muscle by scavenging destructive superoxide and enhancing the cytoprotective nitric oxide / nitrogen monoxide (NO) concentration (1996); that L-arginine treatment decreases superoxide generation by constitutive nitrous synthetase (cNOS) while increasing NO accumulation, leading to protection from construction and reduction of edema after reperfusion (1997); that the release of NO by aceto-choline is only partially inhibited by inhibitions of NOS when used of high concentrations, and NO rather than another factor accounts fully for endothelium-dependent responses of the

---

[1553] City of Kherson (settled in Age of Copper and Bronze, founded 1778), Kherson Obl. (Foot Note 315), Ukraine. — Malanchuk VO, Surgical Stomatology and Maxillo-Facial Surgery. Vol. 1-2. Publ. «Holos», Kyiv 2011. — Ablitsov VH. «Galaxy Ukraine». The Ukrainian diaspora: prominent persons. Publ. «Kyt», Kyiv 2007.

[1554] Gundry SR, Sequera A, Razzuk AM, et al. Facile retrograde cardioplegia: transatrial cannulation of the coronary sinus. Ann Thorac Surg 1990;882-7.

[1555] City of Gdańsk (founded 997), Poland. — Huk I, Bronkovych J, Manobash V, et al. Bioflavonoid quercitin scavangers superoxide and increase nitric oxide concentration in ischemia-reperfusion injury: an experimental study. Br J Surgery doi.org/10.1046/j.1368-2168.1998.00787.x. — Huk I, Nanobashvili J, Neumayer Ch, et al. L-arginine treatment alters the kinetics of nitric oxide and superoxide release and reduces ischemia reperfusion injury in skeletal muscle. Circulation 1997 Jul 15;96:667-75. — Cohen RA, Plane F, Najibi S, Huk I, et al. Nitric oxide is the mediator of both endothelium-depending relaxation and hyperpolarization of the rabbit carotid artery. Proc Natl Acad Sci 1997Apr 15;94(8);4193-8. — Huk I., Bronkovych J, Nanobashvili J, et al. Prostoglandin E_1 reduces ischemia/reperfusion injury by normalizing nitric oxide and hyperoxide release. Schock 2000 Aug;14(2):234-42. Doi:10.1097/000 24382-2000 14020-00025. — Ablitsov VH. «Galaxy Ukraine». The Ukrainian diaspora: prominent persons. Publ. «Kyt», Kyiv 2007.

rabbit carotid artery (1997); that prostaglandin E_1 reduces ischemia-reperfusion injury by normalizing NO and hyperoxide release (2000).

Scientists of the National Aviation Institute or «Aviauniversita» (founded 1933) in Kyiv, Ukrainian National Institute «KhAI» in Kharkiv, and **Volodymyr V. Skyba / Skiba (Nov. 28, 1952, city of Rivne, Ukraine –)**[1556] — Ukrainian hepatobiliary surgeon, have constructed in 2014 «Liquid-Jet» Scalpel, which uses very high pressure, allowing to perform minimally invasive surgical procedure in the liver, stomach, removal of malignant tumors, and diseased non-muscle tissue, with virtually no damage to the vascular blood vessels. The «Liquid-Jet» end effector for manipulating tissue at the surgical site is within the body of a patient, the nozzle for ejecting a liquid jet, water, or other appropriate liquids is outside the body about at least one joint of the apparatus.

Operative management of neonates and infants with congenital heart disease (CHD), who are in congestive heart failure (CHF) due to right-to-left shunt through the foramen ovale atrial septal defect (ASD) into the left atrium (LA) and deeply cyanotic, i.e., the Ebstein anomaly **[Wilhelm Ebstein (1836 – 1912)**. 19th century. Internal medicine], include consideration for three pathways.

First pathway is biventricular of **Cris J. Knott-Craig (1953 –)**[1557] — South-African American pediatric cardiothoracic surgeon and poet, which include the tricuspid valve (TV) repair and closure of the ASD.

Second pathway is a single ventricular of **Vaugh A. Starnes (1952 –)**[1558] — American pediatric cardiac surgeon, i.e., RV exclusion which include a patch closure of the TV orifice with or without fenestration, enlarging the interatrial connection and placing systemic-pulmonary shunt, and in future the eventual Fontan procedure. In its **Sunji Sano (1953 –)**[1559] — Japanese pediatric cardiothoracic surgeon, modification during the first stage the RV free wall is resected and closed (plicated) primarily or with synthetic patch. In the second stage, 3-6 month later a bidirectional cavo-pulmonary shunt is performed.

Third pathway is heart transplantation (HT).

[1556] City of Rivne (Foot Note 978), Rivnens'ka Obl., Ukraine. — Ablitsov VH. «Galaxy Ukraine». The Ukrainian diaspora: prominent persons. Publ. «Kyt», Kyiv 2007.

[1557] Knott-Craig CJ, Overhold ED, Ward KE, et al. Repair of Epstein's anomaly in the symptomatic neonate: an evolution of technique with 7-year follow-up. Ann Thorac Surg 2002;73:1786-93. / Knott-Craig CJ, Goldberg SP, Overhold ED, et al. Repair of neonates and young infants with Epstein's anomaly and related disorders. Ann Thorac Surg 2007;84:587-93.

[1558] Starnes VA, Pitlick PT, Bernstein D, et al. Epstein's anomaly appearing in the neonate. A new surgical approach. J Thorac Cardiovasc Surg 1991;101:1082-7.

[1559] Sano S, Ishino R, Kawada M, et al. Total right ventricular exclusion procedure: an operation for isolated congestive right ventricular failure. J Thorac Cardiovasc Surg 2002;123:640-7.

Renal-cell carcinoma with distant metastases is characterized by unfavorable prediction, outcome, and a limited choice of the method of treatment. The systemic immunotherapy applied before (neoadjuvant) or after (adjuvant) radical removal of an affected kidney (radical nephrectomy) was studied by **Ihor S. Sawczuk (1953 –)**[1560] — Ukrainian American urologist, as a mode for improving results of the treatment. The median survival using immunotherapy before radical nephrectomy was 14 months, after radical nephrectomy it was 12-22 months (median, 20 months).

Hypertrophic obstructed cardiomyopathy (HOCM) is a familial disease in which the interventricular septum of the heart enlarges (became hypertrophic) without any obvious cause, creating functional impairment of the cardiac muscle due to obstruction of the blood flow from the left ventricle (LV) into the ascending aorta (AA), often associated with insufficiency of the mitral valve (MV) / **Robert Teare (1911 – 79**: 20th century. Internal medicine**)**. Advanced disease is difficult or impossible treat medically. Surgical treatment by resection of hypertrophic portion of the septum, i.e., myectomy was introduced in 1961 by Andrew Morrow, and an extended myectomy introduced in 1994 by Bruno Messmer [**Andrew Morrow (1922 – 82) / Bruno Messer**: above].

Later (2006), **Daniel G. Swistel (Mar. 04, 1953, city of Akron, OH, USA)**[1561] — Ukrainian American cardiothoracic surgeon, in complex HOCM patients supplemented resection of hypertrophic portion of the septum, i.e., septal myectomy, or extended septal myectomy, by addition of resection, or plication of an excessive overgrown of the anterior leaflet of the MV (ALMV), and release of attachments of the MV

[1560] Petrykiv A, Sawczuk I, Kennedy W et al Atoptosis in experimental model of hydronephrosis. J Ukr Med Assoc North Am 1995 Autumn;42,3(137):174-9. — Sawczuk IS, Pollard JK. Renal cell carcinoma. Should radical nephrectomy be performed in the presence of metastatic disease? Curr Opin Urol 1999;9:377-81. — Sawczuk IS, Pollard JY. Radical nephrectomy in the presence of metastatic disease. J Ukr Med Assoc North Am 2002 Spring;47,2(118):3-10. — Sawczuk IS, Burchardt T, Shadsigh A et al. Markers of carcinoma of the urinary bladder. Current availability and future standard care. J Ukr Med Assoc North Am 2001 Spring; 46,1(146):3-12. — Ablitsov VH. «Galaxy Ukraine». The Ukrainian diaspora: prominent persons. Publ. «Kyt», Kyiv 2007.

[1561] City of Akron (founded 1825), OH, USA. — Swistel DG, DeRose JJJr, Sherrid MV. Management of Patients with Complex Hypertrophic Cardiomyopathy: Resection, Plication, Release. In, I.H. Cohn, Operative Technique in Thoracic and Cardiovascular Surgery, a Comparative Atlas. WB Sanders, Philadelphia, PA 2004. P. 261-7. — Swistel DG. Balaram SK. Resection, plication, release – the RPR procedure for obstructive hypertrophic cardiomyopathy. Anadolu Kardyol Derg 2006 Dec;6 suppl 2:31-6. — Balaram SK, Swistel DG. Long-term prognosis of hypertrophic cardiomyopathy. Abadolu Kardyol Derg 2006 Dec;6 suppl 2;37-9. — Balaram SK, Tyrie I, Sherrid MV, et al. Resection, plication, release for hypertrophic cardiomyopathy: clinical and echocardiographic follow-up. Ann Thorac Surg 2008;86:1539-45. — Swistel DG. Balaram SK. Surgical myectomy for hypertrophic cardiomyopathy in the 21th century, the evolution of the RPR repair: resection, plication and release. Prog Cardiovasc Dis 2012 May;498-502. — Swistel DG, Sherrid MV. The surgical management of obstructive hypertrophic cardiomyotomy. The RPR procedure – resection, plication, release. Ann Cardiothorac Surg 2017;6(4):423-5.

restricting its motion, namely resection, plication, release (RPR) procedure, or surgical myectomy and associated procedures.

Through a median longitudinal sternotomy incision (MLSI) and a lower transverse aortotomy the resection of hypertrophic portion of the septum is initiated medially just under the ostium of the right coronary artery (RCA), proceeding laterally close to the MV, leaving a rim about 5 mm under the aortic annulus, then resecting 5-15 mm of diseased septal muscle, and subsequently the myectomy is extended to beyond the head of the anterio-lateral papillary muscle removing 30 mm wide and 40 mm long portion of the specimen. Papillary muscle head inserting directly to the ALMV, and accessory secondary chords which tent the leaflet anteriorly can be removed, if other supported structures are present to limit subsequent MV prolapse.

Plication of an excessively long ALMV with horizontal four or five mattress sutures of 5-0 Proline under the aorto-mitral curtain can shorten by 3-6 mm, and if a residual ALMV is still present then it should be excised.

Release of antero-lateral attachments of the papillary muscle, removal or excision of accessory structures, leaflets plication, and residual leaflet resection is performed to preserve normal function of the MV, thus avoiding MV replacement (MVR).

Founder of modern cryosurgery **Mykola M. Korpan (Dec. 24, 1956, village of Huziïv, Ivano-Frankivs'k Obl., Ukraine** –)[1562] — Ukrainian Austrian surgeon, discovered in 1979-84 the phenomenon of the «lunal eclipse» occurring immediately after cryosurgical freezing for the treatment of cancer, indicating the presence of angiogenesis, cryo-necrosis and cryo-apoptosis.

Remarkable advancement in the surgical treatment of hemorrhoid, especially for grade III rectal prolapse (Ancient medicine. *Egypt*) was achieved in 1993 by **Antonio Longo** — Italian surgeon, with the introduction of a stapled hemorrhoidectomy / hemorrhoidopexy in the form of stapled trans-anal rectal resection (STARR) and procedure for prolapse and hemorrhoids (PPH).

Standard approach for the treatment of broncho-pleural fistula (BPF), an opening between the lung and the pleural cavity, is that of **Claude Deschamps (1955 –)**[1563]

[1562] Village of Huziïv (first mentioned 1515), Ivano-Frankivs'k Obl., Ukraine. — Korpan NN. Hepatic cryosurgery for liver metastases. Long-term follow up. Ann Surg 1997;225(2):193-201. — Korpan NN. (ed.). Basics of Cryosurgery. Springer, Wien – NewYork 2001. — Korpan NN. (ed.). Atlas of Cryosurgery. Springer, Wien – NewYork 2001. — Korpan NN. A history of cryosurgery: its development and future. JACS 2007;204(2):314-24. — Korpan NN. Cryosurgery: ultrastructural changes in pancreas tissue after low temperature exposure. Technol Cancer Res Treat 2007;6:59-67. — Korpan NN. Cryosurgery: early ultrastructural changes in liver tissue in vivo. J Surg Res 2009;1:54-65. — Korpan NN. Cryoscience and cryomedicine: deciphering of the effect of deep low temperatures on living matter. Low Temp Med 2009;35(3):65-74. — Ablitsov VH. Galaxy Ukraine». The Ukrainian diaspora: prominent persons. Publ. «Kyt», Kyiv 2007.

[1563] Deschamps C, Allen MS, Miller DL, et al. Management of postpneumonectomy empyema and bronchopleural fistula. Semin Thorac Cardiovasc Surg 2001;13(1):13-9.

— Canadian American cardiothoracic surgeon, i.e. trimming of long bronchial stump or debriding a poorly vascularized stump and then covering the stump with, or sewing circumferentially to the stump for closure, an extra-skeletal muscle, e.g., the serratus anterior or latissimus dorsi and establishing an adequate drainage of the pleural cavity, or reopening the interspace (ICS) thoracotomy and performing open packing of the cavity in the operating room (OR) or under conscious sedation in the patient's room for several days to weeks until the cavity is clean.

Every year in the USA an estimated 5 mln people manifest deep venous thrombosis (DVT), 650,000 of them experience pulmonary embolism (PE), and between 50,000 and 200,000 dies[1564]. The risk for pulmonary embolism (PE) bed rest (is increased in patients with cancer, on prolonged (BR) and hospitalized (around 1%). The rate of fatal PE is 2%.

PE is a blockage of the main pulmonary artery (MPA), its right and left pulmonary artery (PA) divisions, then lobal, segmental and subsegmental branches. The diagnosis should be suspected in patients experiencing dyspnea, chest pain on inspiration, palpitations, increased respiratory rate (RR) and heart rate (HR), cyanosis, hypotension which leads to sudden death. Arterial blood gases (ABG) shows hypoxia. Diagnosis is confirmed by noninvasive computerized tomographic pulmonary angiography (CTPA) with radiocontrast.

Estimated prevalence of unsuspected PE is 2,6%[1565].

Treatment includes anticoagulation or thrombolysis, if contraindicated, insertion of the inferior vena cava filter (IVCF) and surgical PE.

Indication for an emergency surgical pulmonary artery embolectomy (PAE) is: (1) massive PE characterized by sustained hypotension, pulselessness, or persistent bradycardia, despite administration of tissue plasminogen activator (tPA) – alteptase and catecholamines; (2) sub-massive PE characterized by right ventricle (RV) dysfunction or myocardial necrosis, without hypotension[1566].The patient should be rapidly transferred to the operating room (OR) despite apparent hemodynamic stability. The principal features of the present-day (1999-2004) surgical technique are: (1) avoiding of the aortic cross-clamping; cardioplegic or fibrillatory cardiac arrest; (2) performing the operation on a warm beating heart; (3) avoiding of the blind instrumentation; (4) limitation of the extraction to directly visible clot; (5) improvement of visualization of the right MPA by standing on the left side of the OR table; (6) insertion of the IVCF.

[1564] Rahimtoola A, Bergin JD. Acute pulmonary embolism: an update on diagnosis and management. Current problems in cardiology 2005 Feb 30 (2): 61-114.

[1565] Fred HL. Unsuspected pulmonary thromboemboli. A continuing clinical challenge. Tex Heart Inst J 2013;40(1):9-12.

[1566] Goldhaber SZ. Surgical pulmonary embolectomy. The resurrection of an almost discarded operation. Tex Heart Inst J 2013;40(1):5-8.

Operative mortality (OM) of surgical PAE performed in 1,300 patients between 1961-2006 was 30% and in 76 patients between 1999-2004 – 3,9%.

Main direction of scientific activity of **Olexandr V. Maloshtan (30.01.1955, city of Kharkiv, Ukraine** –)[1567] — Ukrainian surgeon, has been surgery of the pancreatic gland ducts and biliary tracts, especially the influence of laparoscopic procedures on of the latter. He proposed the technology of distant laparoscopic cholecystostomy using a double balloon catheter, and original design of a needle for closure of trocar wounds.

Designed in 2003, serial transverse enteroplasty (STEP), lengthen and taper dilated proximal bowel in neonates with intestinal atresia and marginal bowel length, increases intestinal length in children with short bowel syndrome, and ameliorates bacterial overgrowth[1568]. The STEP procedure allows the creation of uniform intestinal caliber in patients with variable degrees of bowel dilatation without the need of the intestinal anastomosis and decrease risk of vascular compromise.

With his colleagues **James T. Rutka (14.01.1956, city of Toronto, ON, Canada** –)[1569] — Canadian Ukrainian pediatric neurosurgeon, the leader in application of neurosurgical techniques to children and adolescents with craniofacial anomalies, congenital malformation of the brain, tumors of the brain, and pediatric epilepsy

---

[1567] Maloshtan AV. Peculiarities of laparoscopic cholecystectomy in patients with chronic hepatitis. Experymentalna i klinichna medytsyna 2000;1(11). — Maloshtan OV. Use of a local antibiotic therapy enforced by local hyperthermia. HLV 2002;3. — Maloshtan AV, et al. Use of minimally invasive technologies in patients with acute pancreatitis. Visnyk mors'koï medytsyny. 2003;2 . — Maloshtan AV, et al. Choledocholithiasis (Diagnosis and Operative Treatment). Kharkiv 2008. — Maloshtan AV, et al. Treatment of solitary abscesses of the liver: drainage or resection. Ukr J Surg 2015;11-2. — Ablitsov VH. «Galaxy Ukraine». The Ukrainian diaspora: prominent persons. Publ. «Kyt», Kyiv 2007.

[1568] Kim HB, Fauza D, Garza J, et al. Serial transverse enteroplasty (STEP): a novel bowel lengthening procedure. J Pediatr Surg 2003;38:425-9. — Kim HB, Lee PW, Garza J, et al. Serial transverse enteroplasty for short bowel syndrome: a case report. J Pediatr Surg 2003;38:881-5.

[1569] City of Toronto (Foot Note 1449), ON, Canada. — Posnick JC, Goldstein JA, Armstrong D, Rutka JT. Reconstruction of skull defects in children and adolescents by the use fixed cranial bone grafts: long-term results. Neurosurgery 1993;32(5):785-91, discussion 791. — Rutka JT, Otsubo H, Kitano S, et al. Utility of digital camera-derived intraoperative images in the planning of epilepsy surgery for children. Neurosurgery 1999;45 (5):1186-91. — Benifla M, Otsubo H, Ochi A, Weiss S, Donner E, Shroff M, Chuang S, Hawkins C, Drake JM, Elliott I, Smith M,Snead OC, Rutka JT. Temporal lobe surgery for intractable epilepsy in children: An analysis of outcomes in 126 children. Neurosurgery 1999;59(6):1203-13, discussion 1213-4. — Stapleton SJ, Kiriakopoulos E, Mikulis D, Drake JM, Hoffman HJ, Otsubo H, Hwang PJ, Logan W, Rutka JT. Combined utility of functional MRI, frameless stereotaxy and cortical mapping in the resection of lesions in eloquent brain in children. Pediatr Neurosurg. 2003;26 (2):68-82. — Benifla M, Sala F, Jane Jr, Ostubo H, Ochi A, Drake JM, Weiss S, Donner E, Fujimoto A, Holowka S, Widjaja E, Snead OC, Smith ML, Tamber MS, Rutka JT. Neurosurgical management of intractable Rolandic epilepsy in children; role of resection in eloquen cortex. J Neurosurg Pediatr. 2009;4(3):199-216. — Ablitsov VH. «GalaxyUkraine». The Ukrainian diaspora: prominent persons. Publ. «Kyt», Kyiv 2007. — Rutka JT. Discovering neurosurgery: new frontiers. J of Neurosurgery 2011;115(6):115-66. —Tonn JC, Readon DA, Rutka JT (eds). Oncology of CNS Tumors. 3rd Ed. Springer – Verlag, Berlin 2019.

(1993); introduced digital camera technology to assist in mapping of intraoperative seizure foci (1999); among others was the first to utilize frameless stereotactic neuro-navigation technique to resect cerebral and skull base lesion in children (2003); treated surgically children with epilepsy arising from lesions within highly eloquent and critical regions of the brain (1999, 2009); used magnetoencephalography (MEG) to identify regions of epilepsy genesis amenable to neurosurgical resection.

He made several contributions to the understanding of molecular biology of human brain tumors, established several novel human brain tumor cell lines and models to study glioma multiforme invasion and has studied methods of delivering therapeutic agents across the blood-brain barrier to treat experimental brain tumors.

The most common postoperative complication after gastric surgery is an anastomotic leak in 3-7,4% patients associated with a high operative mortality (OM). To avoid this complication after a total gastrectomy for gastric cancer with an «end-to-side» Roux-an-Y gastro-jejunal anastomosis, **Ihor B. Shchepotin (Nov. 03, 1956, city of Kyiv, Ukraine –)**[1570] — Ukrainian surgeon, proposed to wrap the anastomotic site with a distal «blind» segment of the jejunum, named the Ukrainian wrap reconstruction or the Ukrainian anastomosis (1995). That decreased anastomotic leaks to 0,8%.

Illya M. Yemets (21.II.1956, town of Vorkuta, Republic of Komi, Russia –)[1571] — Ukrainian pediatric cardiac surgeon, proved that the late (average, 9,3 years) valve replacement of the main pulmonary artery (MPA) after correction of tetralogy of Fallot [17th century – **Niels Steensen (1838 – 86)**, 19th century. Internal medicine – **Étienne Fallot (1850 – 1911)**] is possible to perform with the low operative mortality (OM, 1,1%), low risk of replaced valve failure, release of the right ventricular (RV) failure, and satisfactory long-term survival (1997); introduced transfusion of the

---

[1570] Shchepotin IB, Evans SR, Buras RR, et al. Ukrainian wrap reconstruction after total gastrectomy minimizes anastomotic failure. J Surg Oncol 1995Jan; 58(1):74-5. — Ablitsov VH. «Galaxy Ukraine». The Ukrainian diaspora: prominent persons. Publ. «Kyt», Kyiv 2007.

[1571] Town of Vorkuta (founded 1939), Republic of Komi (area 415,900 km²), Russia. — Coles JG, Yemets IM, Najm HR, et al. Experience with repair of congenital heart defects using adjunctive endovascular devices. J Thorac Cardiovasc Surg 1995;110(5):1613-9. — Yemets IM, Williams WG, Webb GD, et al. Freedom RM. Pulmonary valve replacement late after repair of tetralogy of Fallot. Ann Thorac Surg 1997;64(2):526-30. — Rosti L, Murzi B, Colli AM, et al. Pulmonary valve replacement: a role for mechanical prostheses? Ann Thorac Surg 1998;65(3):889-90. — Fedevych O, Chasovyi K, Vorobiova G, Zhovnir V, Makarenko M, Kurkevych A, Maksymenko A, Yemets I. Open cardiac surgery in the first hours of life using autologous umbilical cord blood. Eur J Cardiothorac Surg 2011 Oct;40(4):985-9. — Yemets IM. Transposition of the Great Vessels (Clinics, Diagnossis, Treatment). In, IM Yemets, NM Rudenko, HM. Vorobiova. Publ. TNMU of IYa Horbachevsky«Ukrmedknyha», Ternopil 2012. — Chasovyi K, Fedevych O, McMullan DM, Mykychak Y, Vorobiova G, Zhovnir V, Yemets I. Tissue perfusion in neonates undergoing open-heart surgery using autologous umbilical cord blood or donor blood components. Perfusion 2015 Sep;30(6):499-506. — Ablitsov VH. «Galaxy Ukraine». The Ukrainian diaspora: prominent persons. Publ. «Kyt», Kyiv 2007.

placental umbilical cord autologous blood, in lieu of a donor blood, in the first hours of life of a newborn during the correction of complex congenital heart disease (CHD, 2009).

Vasyl V. Lazorishinets (25.05.1957, village Sokyrnytsia, Khust Region, Zakarpats'ka Obl., Ukraine –)[1572] — Ukrainian cardiac surgeon, recommended to carry out in patients with Ebstein's anomaly **[Wilhelm Ebstein (1836 – 1912). 19**[th] **century. Internal disease]** the preoperative electrophysiologic (EP) studies of the conduction network of the heart (CNH), to allow in the presence of the accessory atrioventricular (AV) connections, to perform combined tricuspid valve (TV) repair and surgical ablation of those accessory pathways which has been proved to be safe and effective in improving functional results (2000).

Volodymyr V. Popov (18.05.1957, City of Kyiv, Ukraine –)[1573] — Ukrainian cardio-vascular surgeon, who during his the two last student years at the National

[1572] Village Sokyrnytsia (founded before 1770), Khust Region, Zakarpats'ka Obl. (area 12,777 km²), Ukraine. — Lazorishinets VV, Knyshov HV, Novyk VM, et al. Anesthesiological Safety and Postoperative management of the Children Age Cardiosurgical Patients. ICVS AMS of Ukraine. «Accent», Kyiv 1999. — Lazorishinets VV, Glagola MD, Stychinsky AS, et al. Surgical treatment of Wolf-Parkinson-White syndrome during plastic operations in patients with Epstein's anomaly. Eur J Cardiothorac Surg 2000;18(4):487-90. — Lazorishinets VV, et al. Methods of myocardial protection during radical correction of Epstein anomaly. Klinichna khirurhiya 2001;1. — Lazorishinets VV. Plastic of the tricuspid valve septal leaflets during operative treatment of Epstein anomaly. Meditsina today and tomorrow. 2001; 2. — Lazorishinets VV. Palliative methods of surgical treatment of Epstein anomaly. Klinichna khirurhiya 2002;1. — Lazorishinets VV. Differentiate approach toward treatment methods of hypertrophic cardiomyopathy. Shchorich. nauk. pr. Asoc. Serc.-sudyn. Khirurhiv Ukraïny. 2010;18. — Lazorishinets VV, et al. Syndrome of the prolonged interval QT and pregnancy. Ukr Cardiol J 2012; 5. — Siromakha SO, Arvanitakvi SS, Rudenko SA, Lazorishinets VV. Coronary artery insufficiency during pregnancy. Epidemiology, methods of diagnosis, and treatment. Ukrainian J Perinatology and Pediatrics 2019;3(79):32-9. — Ablitsov VH. «Galaxy Ukraine». The Ukrainian diaspora: prominent persons. Publ. «Kyt», Kyiv 2007.

[1573] Popov VV. Valve preserving operations during prosthetic replacement of the mitral valve in patients with coexisting mitral-aortic heart disease. In book: 1ˢᵗ Scientific Conference of the Association of Cardio-vascular Surgeons of Ukraine. Kyiv1993. P. 73. — Popov VV. The place of valvoplasty in treatment of combined mitral-aortic valve disease. VIII World Congress of the World Society of Cardio-Thoracic Surgeons, Houston, TX 1998. P. 54. — Popov VV. Morpho-hemodynamic indicators of the left ventricular work in rheumatic mitral-aortic failure. J Ukr Med Assoc 1998;2(II):76-7. — Sytar LL, Popov VV. Declarative Patent of Ukraine for useful model № 14907 «Method of the brain protection during correction of aneurysm of the ascending and transverse aorta». Declared - № a 2005 12650, date of petition - 27.12.2005, date of publ. - 15.06.2006, bull. № 6. — Sytar LL, Popov VV. Patent for useful model № 28211 «Method of the brain protection during correction of the transverse aortic aneurysm», № petition - u 2007 09480, date of petition - 21.08.2007, date publ. - 26.11.2007, bull. № 19. — Popov VV. Patent for useful model № 28924 «Method of myocardial protection during correction of mitral valve cardiac disease», № petition - u 2007 09780, date of petition - 31.08.2007, date of publ. - 25.12.2007, bull. №21. — Sytar LL, Popov VV., Tretiak OA. Patent for useful model № 28925 «Method of the brain protection during correction of aneurysm of the aortic arch», № petition - u 2007 09781, date of petition - 31.08.2007, date publ.-25.11.2007, bull. № 21. — Popov VV. Surgical treatment of the thoracic aorta pathology. Premium President of Ukraine, Kyiv 2018. — Ablitsov VH. «Galaxy Ukraine». The Ukrainian diaspora: prominent persons. Publ. «Kyt», Kyiv 2007.

Kyiv Medical University (NKMU) from April 1978 worked in National Institute of Cardiovascular Surgery (NICVS) of Ukraine, now of M.M. Amosov (founded 1955), on the beginning as nurse at the Division for Acquired Heart Diseases (AHD), in particular in Division of Intensive Therapy, since 1980 after graduating from NKMU, he began to work as physician-surgeon at the Division of AHD. On this position in 1988 he defended his dissertation for the candidate of Medical Science (CMS), entitled «Prosthesis of the Mitral Valve in Complicated Mitral Stenosis (calcinosis, thick fibrosis, massive thrombosis of the left atrium», since 2004 he took the responsibility for fulfilment the scientific-research work (SRW) in the surgical treatment of the ascending aortic (AA) aneurysms, surgical repair of cardiac valves, removing tumors of the heart, resection of aneurysms of the transverse aorta (TA) under deep hypothermic circulatory arrest (HCA), treatment of cardiomyopathies (CM), in 2005 he defended dissertation for the degree of Doctor of Medical Science (DMS), entitled «Surgical Correction of Combined Mitral and Aortic Heart Diseases under Cardiopulmonary Bypass».

Beginning March 2005 he became Chief of Division of AHD while directing his work towards myocardial protection; bloodless surgery; protection of the brain in patients with the evidence of neurological deficit; protection of organs in comorbid onset of multiorgan insufficiency (lung, liver, kidneys) with prophylactic pathways; complex reconstruction of the mitral Valve (MV), i.e., retaining the sub-valvular apparatus during mitral valve replacement (MVR) combined with plastic of the left atrium (LA); complex combined reconstruction of both or more cardiac valves; reconstruction of narrow aortic root during aortic valve replacement (AVR); restoration of the sino-tubular junction in post-stenotic dilatation of the AA during AVR; valve replacement in patients with severely reduced myocardial contraction.

Repair of the aortic leaflet prolapse, introduced by **Hans J. Schäfers (1957 –)**[1574] — German cardiac, thoracic and vascular surgeon, consist of shortening of its free margin by a single or multiple mattress No. 5-0 or 6-0 proline «central plicating suture», when necessary enforced by a piece of pericardium, placed into thickened free leaflet's margin adjacent to the nodulus of **Guilio Arantius (1530 – 89**: Renaissance**)**, until the leaflet effective height, defined as the vertical distance from its base to the top of the central free leaflet margin, approximates 10 mm. The length of the free margin of an aortic leaflet should be about 1.5 times the length of its base, to produce adequate coaptation.

[1574] Schäfers HJ, Aicher D, Langer F. Correction of leaflet prolapse in valve-preserving aortic replacement: pushing the limits? Ann Thorac Surg 2002;74:S1762-4. / Rankin JS, Gaca JG. Techniques of aortic valve repair. Innovations: Technology and Techniques in Cardiothoracic and Vascular Surgery 2011NovDec;6(6):348-54.

Simpler than David operation (above) is introduced by **Rugher De Paulis**[1575] — Italian cardiothoracic surgeon, an aortic valve (AV) repair combined with a three-scalloped remodeled graft for the replacement of the aneurysmal aortic root with coronary reimplantation.

In the 1997's report **T. Mayumi**[1576] — Japanese general surgeon, the first six (6) patients with nonmetastatic pancreatic distal adenocarcinoma (CAE) with the body and tail of the pancreas.

Prostate specific antigen (PSA) of the blood is a sensitive but unspecific marker for the prostate cancer detection which may result in harms including overdiagnosis and overtreatment. Therefore, the research in development of new markers by **Marko (Marek) Babjuk (16.04. 1961, city of Praha, Czechy** –)[1577] — Ukrainian Czech urologist, is of absolute value. He also developed the qualitative and quantitative detection of urinary human complement factor H-related protein and fragments of cytokeratin's 8,18 as markers for diagnosis of transitional-cell carcinoma of the urinary bladder (2002).

In sport hernias with tearing of the aponeurosis of the external oblique abdominal muscle complicated by scarring of the ilioinguinal nerve, **Ihor D. Herych (15.10.1961, town of Turka, Lviv Obl., Ukraine** – died from an acute MI **15.06.2014, city of Lviv, Ukraine)**[1578] — Ukrainian surgeon, developed the operation consisting of release from the surrounding scar tissue the aponeurosis and the ilioinguinal nerve, re-routing of the nerve outside the inguinal canal through artificially created opening in the lateral segment of the aponeurosis and placing it without tension between the aponeurosis and the subcutaneous tissue, restoration of the integrity of the aponeurosis.

[1575] De Paulis R, D Matteis GM, Nardi P, et al. One-year appraisal of a new aortic root conduit with sinuses of Valsalva. J Thorac Cardiovasc Surg 2002;123:33-9. / Rankin JS, Gaca JG. Techniques of aortic valve repair. Innovations: Technology and Techniques in Cardiothoracic and Vascular Surgery 2011NovDec;6(6):348-54.

[1576] Mayumi T, Nimura Y, Kamiya J, et al. Distal pancreatectomy with en block resection of the celiac artery for carcinoma of the body and tail of the pancreas Int J Pancreatol 1997;22:15-21.

[1577] Babjuk M, Kostirova M, Mudra K, et al. Quanlitative and quantitative detection of urinary human complement factor H-related protein (BTA stat and BTA TRAK) and fragments of cytokeratins 8,18 (UBC rapid and UBC IRMA) as markers for transitional cell carcinoma of the bladder. Eur Urol 2002 Jan;41(1):34-9. — Babjuk, Matonskov M, Finek J, Petruzelka L. Konsenzualni Doporucene Postupy u Uroonkologii. Nakladatelske udaje: Galen, Praha 2009. —Babjuk M, Burger M, Kaasinen CE, et al. Guidelines on non-muscle-invasive bladder cancer (Aa, T1n and CIS). Eur Assoc Urol 2014;2014:1-48. — Ablitsov VH. «Galaxy Ukraine». The Ukrainian diaspora: prominent persons. Publ. «Kyt», Kyiv 2007.

[1578] Town of Turka (settled since the middle of Stone Age, first written mention 10th century, founded 1280 -1320), Lviv Obl., Ukraine. — Herych ID, Stoyanovsky IV, Herych HI, Cheremys OM. The new method of surgical treatment of Gilmore hernia. Ukr J Surg 2011;2(11):103-8. — Ablitsov VH. «Galaxy Ukraine». The Ukrainian diaspora: prominent persons. Publ. «Kyt», Kyiv 2007.

Andriy Jo. Nakonechmy (28.10.1961, city of Lviv, Ukraine –)[1579] — Ukrainian pediatric urologist, et al., discovered in the Western Ukrainian population from the allelic class II loci of the main human histocompatibility system immune-genotypic retention markers of testicles, determined that the specificity loci of human lymphocyte antigen (HLA) system class II and the genotype of patients with testicular retention condition their offspring's to the pathology of the reproductive system – an undescended testicle and infertility; identified in patients with testicular retention from allelic loci of the HLA system class II and their genotypes immune-genotypic markers that are associated with an inclination/resistance to the synthesis of antispermic antibodies; proven the ability to produce antispermic antibodies in boys before the puberty and established the laws of dynamics for the synthesis of antispermic antibodies, depending on the clinical form of testicular retention, patients age and timing of surgical treatment (01.03.2010 - present).

On Mar. 01, 2012, Ukrainian surgeons[1580] informed about the introduction into clinical the use of soft tissues and parenchymal organs (liver) dissection with a stream of a cold air plasma and warm nitrous oxide (N_2O).

Fold formed by the parietal pericardium and the overlying fibrous pericardium folding back onto themselves over the left-sided pulmonary veins (PVs), which is 1 mm to 3 mm wide, and runs from the inferior edge of the left inferior pulmonary vein (LIPV) to the superior edge of the left superior pulmonary vein (LSPV), was described in 2017 by **Vipin Zamvar (1961 –)**[1581] — Indian Scottish cardiac surgeon, named the Zamvar fold. A similar fold is not present on the right side. The Zamvar fold allows for the safe placement of the deep lateral pericardial retraction (Lima) sutures during off-pump coronary artery bypass (OP-CAB) surgery, to expose the left posterolateral and inferior left ventricular (LV) walls of the heart **[Frederico Benetti (1947 –)** and **Richardo Lima**: above).

[1579] Naronechny AJo, et al. Antisperm antibodies in prepubertal boys with cryptorchidism. Archives of Andrology 2006;52. — Naronechny AJo. Problem with surgical treatment of cryptorchismus in children. Practical Medicine 2009:119-127. — Naronechny AJo, et al. Cryptorchidism and long-term consequences. Reproductive biology 2010;10(1). — Naronechny AJo, Naronechny RA. Possibilities of endoscopic correction congenital diseases urinary tract in children. Bukovynskyy Medychnyy Visnyk 2011;15,1(57):211-3. — Naronechny AJo, et al. Weak association of anti-sperm antibodies and strong association of familial cryptorchidism/infertility with HLA-DRB1 polymorphisms in prepubertal Ukrainian boys. Reproductive Biology and Endocrinology 2011;9. — Naronechny AJo, et al. Killer cell immunoglobulin-like receptor gene association with cryptorchidism. Reproductive biology 2015;15(4). — Naronechny AJo, Kuzyk AS, Naronechny PA. Surgical correction of inguinal hernia in children with the use of mini-invasive method. Khirurhia Dytiachoho Viku 2016;1(50-51):78-81. — Naronechny AJo, et al. Antisperm antibodies are not frequently induced in semen of men with testicular hyperthermia. J. of Reproductive Immunology 2018;128. — Ablitsov VH. «Galaxy Ukraine». The Ukrainian diaspora: prominent persons. Publ. «Kyt», Kyiv 2007.
[1580] Ablitsov VH. «Galaxy Ukraine». The Ukrainian diaspora: prominent persons. Publ. «Kyt», Kyiv 2007.
[1581] Zamvar V. The Zamvar pericardial fold. J Cardiothor Surg 12 Sep 2017;12:84.

The best result in conservative and surgical treatment of children with diseases of the ear, nose and throat (ENT) in the USA was achieved by **Wasyl Szeremeta (1962 –)**[1582] — Ukrainian American pediatric otorhinolaryngologist (ORL), who at the same time significantly contributed to the development of this specialty of medicine.

The use of the 3-dimensional (3-D) printed implant to replace a large defect in the chest wall and half of the sternum after resection of a malignant tumor was performed on February 16, 2018, by **Ira Goldsmith** — Welsh cardiothoracic surgeon, **Thomas Bragg** — Welsh plastic surgeon, **Peter L. Evans** — Welsh manager of the maxillofacial laboratory and reconstructive scientist and **Heater Goodrum** — Welsh biomedical 3-D technician.

The patient, a 71-year-old man developed a mass involving the right 2nd rib at the middle clavicular line (MCL) which initially thought to be benign cyst. However,

[1582] Szeremeta W, Moravati SS. Isolated hyoid bone fracture. J Ukr Med Assoc North Am 1990 Autumn;37,2(122)26-31. — Szeremeta W, Moravati SS. Isolated hyoid bone fracture: a case report and review of literature. J Trauma 1991 Feb;31(2)268-71. — Szeremeta W, Sataloff RT, Maddox HE III et al. Contralateral deafness following ear surgery. J Ukr Med Assoc North Am 1992 Summer;38, 2(127):1924. — Szeremeta W, Novelli NJ, Bellinger M. Postoperative haemorrhage after tonsillectomy. J Ukr Med Assoc North Am 1993 Winter; 40,1(129):15-21. — Szeremeta W, Monsell EM, Rock JP et al. Proliferation indices of vestibular schwannomas by Ki-67 and proliferating cell nucleal antigen. Am J Otology 1995 Jan 1;16(5):616-9. — Szeremeta W.Endoscopic paranasal sinus surgery. J Ukr Med Assoc North Am 1995 Winter;42,1(135):12-7. — Bonilla JA, Szeremeta W, Yellon RF et al. Teratoid cysts of the floor of the mouth. Int J Ped Otolaryngology 1996 Dec 5;38(1):71-5. — Szeremeta W, Novelly NJ. Benniger M. Postoperative bleeding in tonsillectomy patients. Ear, Nose and Throat J 1996 Jun;76(6):373-6. — Jellon RF, Szeremeta W, Grandis JR et al. Subglottic injury, gastric juice, corticosteroids, and peptide growth factors in a porcine model. Laryngoscope 1998 Jun 1;108(6)854-62. — Kim MK, Buckman R, Szeremeta W. Penetrating neck trauma in children: an urban hospital's experience. Otolaryngology – Head and Neck Surg 2000 Jan 1;132(4):439-43. — Szeremeta W,Parameswaran MS, Isaacson G. Adenoidectomy with Laser or incisional myringotomy for otitis media with effusion. Laryngoscope 2000 Mar;110(3Pt 1):342-5. — Kumar VV, Ganghan J, Isaacson G, Szeremeta W. Oxymetazoline is equivalent to ciprofloxacin in preventing postoperative otorrhea or tympanostomy tube obstruction. Laryngoscope 2005 Feb 1;115(2):363-5. — Basha S, Bialowas C, Ende K, Szeremeta W. Effectiveness of adenotonsillectomy in the resolution of noctutnal enuresis secondary to obstructive apnea. Laryngoscope 2005 June; 115(6):1101-3. —Seidman MD, Szeremeta W, Isaccson G. Is oxymetazoline really save for middle ear use. Laryngoscope 2005 Jul 1;115(7)1321-3. — Cohen MS, Getz AE, Isaacson G et al. Intracapsular vs. extracapsular tonsillectomy: a comparition of pain. Laryngoscope 2007 Oct 1;117(10):1855-8. — Koshkareva YA, Cohen M, Gangham JP et al. Utility of preoperative hematologic screening for pediatric adenotonsillectomy. Ear, Nose and Throat J 2012 Aug 1;91(8)346-56. — Swonke ML, Smith SA, Ohlstein JF et al. Unexplained destructive nasal lesions in a half brothers: a possible case of Munchausen syndrome. Int J Ped Otorhinolaryngology 2019 Aug;123:75-8. — Lavere PF, Ohlstein JF, Smith SP et al. Preventing unnecessary tympanoscopy tube placement in children. Int J Ped Otorhinolaryngology 2019 Jul;122:40-4. — Olstein JF, Padilla PL, Garza RK et al. Ankyloglossum Superius Syndrome compromising a a neonatal airway: considerations in congenital oral airway obstruction. Int J Ped Otorhinolaryngology 2019 Feb 1;117:167-70. — Santos-Cortes RLP, Chiong CM, Frank DN et al. Gene makes some susceptible to middle ear infections. University of Colorado Anschutz Medical Campus. Science Daily Oct 25, 2018.— Lavere P, Szeremeta W, Pine H. Pediatric Embryology (Chapter 7). In, M.R. Chaaban (Ed). Otolaryngology Boards. Otolaryngology Research Advancers. Vol. 1. Publ Nova Medicine & Health, Galveston, Tx 2019. — Ablitsov VH. «Galaxy Ukraine». The Ukrainian diaspora: prominent persons. Publ. «Kyt», Kyiv 2007.

with the time it grew to the size of a tennis ball invading the right 2nd - 4th ribs and the half of the right sternum. The extent of the tumor was confirmed by computed tomography (CT) of the chest. The operation consisted of a wide excision of the tumor including the right 2nd-4th ribs from the right anterior axillary line (AAL) in block with the right hemi-sternum. Pathological examination of the resected specimen revealed malignant soft tissue sarcoma arising from the cartilage of his 2nd rib. The margin of the resection was free of tumor. An extensive defect in the right antero-superior chest wall and the sternum was replaced based of the previous design using the chest CT by a personal custom-made (bespoke) 3-D printed titanium implant, and then covered by a section of the right latissimus dorsi muscle, harvested earlier from the upper back of the patient. He made an uneventful recovery.

Borys M. Todurov (02.I.1965, city of Kyiv, Ukraine –)[1583] — Ukrainian cardiac surgeon of Greek descent, performed a heart transplantation (HT) on 02.III. 2000, Kyiv, Ukraine, in a 36-year-old patient after myocardial infarction (MI) complicated by the left ventricular (LV) failure. Unfortunately, the patient died 10 days after the operation because of poor oxygen delivery (O_2), i.e., anoxia to the brain and renal failure (RF). Between 2000 and 2007 he performed four HT, three with successful result.

In 2008 he became director of the Kyiv City Heart Center (built 2006-08 upon his initiative and co-participation), since 2016 p. – the Heart Institute, Ministry of Health of Ukraine, in which on 12.VIII.2016, he implanted an artificial heart EURO 120 K (made in Germany), performs every year approximately 5,940 cardiac operations, 1,436 coronary artery bypass (CAB) grafting, and 2,593 percutaneous (PC) cardiac procedures, i.e., balloon dilatation of stenosed coronary arteries with/without stenting, with operative mortality (OM) less than 1%, on the level of the best cardiac clinics in Europe.

Major-general of the medical service, **Ihor A. Lurin (18.VI.1968, city of Vinnytsia, Vinnytsia Obl., Ukraine –)**[1584] — Ukrainian military surgeon-proctologist, since 2004 – Deputy Chief of the Main Military Clinical Surgical Hospital – Chief Surgeon in the Department of Defense of Ukraine, who in 2014-19 served as Chief in the Department of Health Care in the Administration of the President of Ukraine.

[1583] Ablitsov VH. «Galaxy Ukraine». The Ukrainian diaspora: prominent persons. Publ. «Kyt», Kyiv 2007.

[1584] City of Vinnytsia (Foot Note 445), Vinnytsia Obl. (Foot Note 3), Ukraine. — Lurin IA, Halushka AM, 3aporozhan CP, et al. A New Look at the Medical Security of Military Employees of the Armed Forces of Ukraine and other Military Formations (with the Experience during Conduction of Antiterrorists Operations). Publ. I.Ya Horbachevsky THMU, Ternopil 2015. — Tsymbaliuk VI, Lazorynets VV, Lurin IA. Contemporary Clinical Organization Aspects of Surgical Treatment of Patients with Diseases and Trauma of the Heart and the Main Blood Vessels. The MM Amosov Price by the NAS of Ukraine (medicine). Publ. IYa Horbachevsky THMU, Ternopil 2019. — Ablitsov VH. «Galaxy Ukraine». The Ukrainian diaspora: prominent persons. Publ. «Kyt», Kyiv 2007.

Major of the medical service, Candidate of Medical Sciences (C.M.S.) **Konstantyn P. Gerzhyk**[1585] — Ukrainian military physician, specialist in thoracic surgery, Chief of the Division of Thoracic Surgery of the Military-Medical Center of the South Region, city of Odesa, Ukraine, who dedicated his dissertation for the degree C.M.S. (2010) to decide the current problems of surgery, related to the improvement of results in diagnosis and surgical treatment wounded with combat trauma of the chest cavity in 103 pearson aged from 23 to 52 years (37±1,7), received during Russian-Ukrainian War in the East Ukraine lasting since 2014, on an account of the use video-thoracoscopic interventions.

Son of **Dmytro Ye Babliak (1930 – 2008**; above), **Aleksander (Oleksandr) D. Babliak (July 09, 1970, city of Lviv, Ukraine –)** — Ukrainian cardiac surgeon, founded the first network of private clinics «Dobrobut» («Prosperity»), including the Cardiac Surgery Center which assures contemporary standards quality of medical care in the city of Kyiv and all regions of the Kyiv Oblast' (2012), introduced method of mini-invasive multiply coronary artery bypass (CAB) grafting using the anterior minimal intercostal space thoracotomy (ICST), 6-8 cm long, without transection of the chest wall bone – total coronary revascularization via anterior thoracotomy (TCRAT, 2017).

In November 2019 **Mark J. Truty (1973 –)**[1586] — American general surgeon, reported on 178 patients suffering from nonmetastatic pancreatic distal adenocarcinoma (PDAC) who underwent between January 1995 and September 2019 pancreatectomy with an block celiac artery (axis) resection (CAR) and in more locally advanced stages resection of surrounding arteries and veins, followed by corresponding revascularization. Of them 89% patients received systemic induction neoadjuvant chemotherapy. Ninety-day mortality in the most recent 50 patients decreased to 4%. Median overall survival was 36.2 months.

Gynecology and obstetrics

In 1900 **Herman J. Pfannenstiel (1862 – 1909)** — German gynecologist, proposed a long horizontal suprabubic incision made below the line of the pubic hair and above the mons veneris, i.e., the **Hill of Venus** — goddess of love (Pfannenstiel incision), as an approach for operations on the organs of the lower lower abdominal and pelvic cavity, including the Caesarean section (C-section).

[1585] Gerzhyk KP. Video-thoracoscopic surgical intervention in combat injuries and injuries of the chest organs. The dissertation for the degree of C.M.S. Ukrainian Military Medical Academy of the Ministry of Defense of Ukraine and the NMAPE of PL Shupyk, Kyiv 2020. — Ablitsov VH. «Galaxy Ukraine». The Ukrainian diaspora: prominent persons. Publ. «Kyt», Kyiv 2007.

[1586] Truty MJ, Colglazier JJ, Mendes BC, et al. En block celiac axis resection for pancreatic cancer: classification of anatomical variants based on tumor extent. J Am Coll Surg 2020 July;231(1):8-32.

In his clinic **Hryhoriy F. Pysemsky (25.I.(07.II).1862, town of Pyriatyn, Lubensky Region, Poltava Obl., Ukraine – 20.VII.1937, city of Kyiv, Ukraine)**[1587] — Ukrainian gynecologist-obstetrician, was the first who proposed and brought into practice anesthesia during the childbirth by means of pharmacological drugs, and was the first in Ukraine to perform the donor blood transfusion. Among written by him scientific publications the best known are investigation concerning the innervation of the uterus, monography about dermoid cysts of the abdominal wall, work on operative gynecology. He was the editor of the «Ukrainian Medical Herald», opened in Kyiv the first consultation for pregnant women, created farmers childbirth buildings, created the childbirth help in village localities.

First woman-physician of Ukraine and Austro-Hungary, **Sophia D. Okunevsky-Morachevsky (12.V.1865, village Dobzhanka, Ternopilsky Region, Ternopilska Obl., Ukraine** – died after acute gangrenous perforative appendicitis with diffuse peritonitis **24.II.1926, city of Lviv, Ukraine)**[1588] — Ukrainian gynecologist and obstetrician, who in 1895 graduated from the Medical Faculty at the Zürich University (founded 1833), city of Zürich (early settlements more than 4,400 BC, founded by the Romans 15 BC), Switzerland, with the degree of Doctor of Medicine. In turn, she investigated the influence of temperature on erythrocytes (1898), initiated the treatment of cancer of the uterine cervix with radium (1904).

Among people suffering from interstitial cystitis **[Alexander Skene (1837 – 1900):** 19th century. Gynecology and Obstetrics] 5-10% will develop ulcer of the urinary bladder, first described in 1915 by **Guy L. Hunner (1869 – 1957)** — American gynecologist, since then called Hunner ulcer, a deep oozing or hemorrhagic lesion which vary in size. Hunner ulcer can only be accurately diagnosed via a cystoscopy and removed using a high-frequency electric current applied with a needle electrode (fulguration).

Hans Spemann (1869 – 1941) — German embryologist, discovered the organizer effect in embryonic development (1924).

A. Watts Makepeace and **George L. Weinstein (1873 – 1946)** — American gynecologists and obstetricians, along with **Maurice H. Friedman (1903 – 91)**[1589]

[1587] Town of Pyriatyn (founded 1155; suffered from organized by Russian Communist Fascist regime of Holodomor-Genocide 1932-33 and political repressions in 1937 against the Ukrainian Nation, during which perished at least 189 residents), Lubensky Region, Poltava Obl., Ukraine. — Olearczyk A. Ukrainian scientists. In XX anniversary from death of H.F. Pysemsky (1862-1937). Nasze Słowo (Warszawa) 7.4.1957;2, 14(34):2. — Ablitsov VH. «Galaxy Ukraine». The Ukrainian diaspora: prominent persons. Publ. «Kyt», Kyiv 2007.

[1588] Village Dobzhanka (founded 1473), Ternopilsky Region, Ternopilska Obl., Ukraine. — Okunevska-Morachevska S. Influence of temperature upon the osmotic pressure of erythrocytes. Collection of the Mathematic-Biological Section of NTSh 1898;1;1-10. — Ablitsov VH. «Galaxy Ukraine». The Ukrainian diaspora: prominent persons. Publ. «Kyt», Kyiv 2007.

[1589] Makepeace AW, Weinstein GL, Friedman MH. Effect of progestin and progesterone on ovulation in the rabbit. Experimental Biology and Medicine 1936 Nov 1,35(2):269-70.

— American physician and reproductive physiology researcher, discovered that progestin and progesterone inhibit ovulation in the rabbit, leading to its application as the birth control, contraception, or fertility control agent in women. In addition, Maurice Friedman developed in the early 1930's the rabbit test to determine the presence of hormones in pregnant women (Friedman test). Then, the Friedman test was replaced by pregnancy kit that determined the presence of hormones in pregnant women without the sacrifice of animals.

In 1928 **Selmar S. Aschheim (1878 – 1965)** — German gynecologist and obstetrician of Jewish descent and **Bernhard Zondek (1891 – 1966)** — German Israeli gynecologist and obstetrician, introduced the pregnancy (Ascheim Zondek) test.

One of the early proponents of the birth contraception was **Margaret H. Sanger (1879 – 1966)** — American nurse, educator and writer who popularized the term «birth control», opened the first Birth Control Clinic in the USA (1916), founded the Birth Control League (1921) that was elevated into the status of the Planned Parenthood of America (1942).

In 1906-49 **Konstantyn I. Platonov [18(30).X.1877, city of Kharkiv, Ukraine – 06.III.1969, city of Kharkiv, Ukraine]**[1590] — Ukrainian neurologist and psychologist, and **Illia Z. Velvovsky (21.06.1899, city of Kharkiv, Ukraine – 07.03.1981, city of Kharkiv, Ukraine)**[1591] — Ukrainian psychiatrist and psychotherapist, developed and described in details the system of psychotherapeutic preparation of pregnant women for painless childbearing (delivery).

System of the painless childbearing management, developed by Konstiantyn Platonov and Illia Velvovsky arosed interest of **Fernand Lamaze (1891 – 1957)**[1592]

[1590] Platonov KI. Verba'sl inspire and its physiological substantiation for the use in painless childbearing. In collection: Problems of Psychotherapy in Obstetrics. Publ. Road-Sanitary Division of the Southern Railway, Kharkiv, 1906. — Platonov KI. Painless childbirth according to the inspire method. Scientific-informative material of the Ukrainian Psychoneurological Scientific Institute (UPSI). Work of the Conference for Painless Childbirth. Donetsk 1936. — Platonov KI. Psychotherapy. UPSI, Kharkiv 1940. — Platonov KI. Word as a Physiological and Therapeutic Factor. 1st Ed. Publ. «Derzhvydav» of Ukraine, Kharkiv 1930; 2nd Ed. Publ. «Derzhvydav» of Ukraine, Kyiv 1957; 3rd Ed., «Medgiz», Moskva 1962. — Ablitsov VH. «Galaxy Ukraine». The Ukrainian diaspora: prominent persons. Publ. «Kyt», Kyiv 2007..

[1591] Velvovskyy IZ, Platonov KI, Ploticher VA, Shugom EA. Psychoprophylaxis of a Child-bearing Pain, Medgiz, St Petersburg 1954. — Velvovskyy IZ. Word as a Physiological and Therapeutic Factor. Kyiv 1957. — Velvovskyy IZ. System of Psychoprophylactic Painless Childbirth. Medgiz, Moscow, 1963. — Velvovskyy IZ, et al. Psychotherapy in Clinical Practice: Instructive Textbook. Publ. «Zdorov'ia», Kyiv 1984. — Ablitsov VH. «Galaxy Ukraine». The Ukrainian diaspora: prominent persons. Publ. «Kyt», Kyiv 2007.

[1592] Lamaze F, Vellay P. L'Accouchement sans douleur par la méthode psychophysique. Gazette médicale de France. 1952;59(23):1445-60. — Lamaze F. Painless Childbirth: The Lamaze Method. Publ. Henry Regnery Company, Chicago 1955. — Lamaze F Painless Childbirth. Psychoproprophylactic Method. Publ. Contemporary Books, Chicago 1970. — Lamaze F. Painless Method. The Lamaze Method. Contemporary Books, Chicago 1984. — Michaels P. Lamaze: An International History. Publ. Oxford University Press, Oxford 2014.

— French neurologist and obstetrician, member of National Committee of Physicians (Comite national des medicins) with close ties to the French Communistic Party (FCP). Attributed to this, in 1951 he was able to visit and stay for some time in Kharkiv, Ukraine, where he intensely learned psychoprophylactic techniques of women prepared for the painless child-bearing management by Konstiantyn Platonov and Illia Velvovsky. After return to France, he «exported» the Ukrainian method of psychoprophylactic of preparation of women for painless childbearing, adapted it to local conditions, included it in his practice, extended this pre-labor / pre-delivery pain management among pregnant women and family pair (couple). Then, he named this system of painless child-birth preparation and management, the «Lamaze method» or «child-bearing according to Lamaze».

The name of **Ernt Gräfenberg (1881 – 1957)**[1593] — German gynecologist of Jewish origin, is remembered for studies on the first intrauterine device (IUD) or the «Gräfenberg ring» to prevent undesired pregnancy (1929) and on female ejaculation, referred to an erotic zone in the anterior wall of the vagina along the course of the urethra, named after him the «G-spot» (1950), noted previously by **Regnier de Graaf (1641 – 93**: 17th century).

George N. Papanicolaou (1883 – 1962) — Greek cytopathologist, developed the method of screening for cancer the uterine cervix (Papanicolaou's smear, 1928).

Walter Schiller (1887 – 1960) — Austrian American pathologist, introduced a test for detection of an early squamous cell carcinoma of the uterine cervix by staining it with the iodine solution: the surface of the healthy cervix became brownish, cancerous becomes white or yellow because cancer cells do not contain glycogen thus are unable to stain with iodine (Schiller's test, 1933)

In 1934 **U. Salmon**[1594] reported benign pelvic tumor associated with pleural effusion, and in 1937 **Joe (Joseph) V. Meigs (1892 – 1963)**[1595] — American gynecologist and obstetrician, reported a benign ovarian tumor (fibroma) associated with ascites and hydrothorax (Meigs-Salmons or Meigs syndrome).

The major role in the development in 1955 of the oral contraceptives, i.e., the first birth control pills, colloqially called «the pills» were played by **John Rock (1890 – 1984)** — American gynecologist and obstetrician, and by **Gregory G. Pincus (1903 – 67)** — American biologist and researcher born into a Jewish family, by a successful use of progestins to prevent ovulation in woman, under the brand name Enovid 10 mg tablet per mouth / orum (p.o.). However, the drug was discontinued in 1988 because

---

[1593] Gräfenberg E. The role of urethra in female orgasm. Int J Sexology 1950;3(3):145-8.

[1594] Salmon U. Benign pelvic tumors associated with ascites and pleural effusion. J Mt Sinai Hosp 1934;1: 169-72.

[1595] Meigs JV, Cass JW. Fibroma of the ovary with ascites and hydrothorax: with a report of seven cases. Am J Obstet Gynecol 1937;33: 249-67.

of increased risk to form an intravascular blood clot, mainly venous thrombosis, pulmonary embolism (PE) and myocardial infarction (MI). In addition, John Rock was a pioneer in «in vivo» fertilization and sperm freezing.

As an emergency first aid in the intrauterine asphyxia of the fetus, **Anatoliy P. Nikolayev (23.I.1896, town of Tarashcha, Bilotserkivsky Region, Kyiv Obl., Ukraine – 1972, city of Kyiv, Ukraine)**[1596] — Ukrainian gynecologist and obstetrician, proposed administering oxygen by the nose to the mother, intravenous (IV) injection of cardiac stimuli and glucose (Nikolayev's triad, 1952).

Maximillian Ehrenstein (1899 – 1968) — American chemist of Jewish descent, altered the molecular structure of an endogenous steroid progesterone, and synthesized a new hormone that contributed to development of anticonception pills (1944).

Progesterone is crucial metabolic intermediary in the production of other endogenous steroids, including sex hormones and the corticosteroids, and plays important role in brain function as neurohormone, and progestogen is a sex hormone involved in the menstrual cycle, pregnancy and embryogenesis of humans and the species.

To inspect the internal genitalia, **Raoul A.Ch. Palmer (1904 – 85)**[1597] — French gynecologist and obstetrician, introduced during WW 2 a laparotomy through a transabdominal approach or transvaginal approach in the «head-down position» of **Friedrich Trendelenburg (1844 – 1924**: 19th century. Surgery] through the rectouterine pouch of Douglas **[James Douglas (1675 – 1742)**: 18th century] using the Veres needle **[Jaros Veres (1903 – 79)**. 20th century. Internal medicine] for controlled pneumoperitoneum with carbon dioxide (CO_2) insufflation.

It was **Ian Donald (1910 – 87)**[1598] — Scottish English gynecologist and obstetrician, who introduced in 1958 the use of the first contact compound 2-dimentionain (2-D) ultrasound (US) scanning / imaging machine and technique for visualization gynecological tumors and for obstetrical visualization of abnormalities in pregnancy and of the fetus.

A hospital-acquired (nosocomial) aspiration pneumonia due to inhalation of oropharyngeal or gastric content into tracheobronchial tree was observed by **Charles**

[1596] Town of Tarashcha (Foot Note 621), Bilotserkivsky Region, Kyiv Obl., Ukraine. — Nikolayev AP. Prophylactics and Therapy of Intrauterine Asphyxia of a Fetus. Publ. «Gosmedizdat», Moskva 1952. — Nikolayev AP. Weak Childbed Activity and its Treatment. Publ. «Derzhmedvydav» of Ukraine, Kyiv 1956. — Ablitsov VH. «Galaxy Ukraine». The Ukrainian diaspora: prominent persons. Publ. «Kyt», Kyiv 2007.

[1597] Palmer R. Instrumentation et technique de la coelioscope gynecologique. Gynecol Obstet (Paris) 1947;46:420-31.

[1598] Donald I, MacVicar J, Brown TG. Investigations of abdominal masses by pulsed ultrasound. Lancet 1958;1:1188-95. — Campbell S. A short history of sonography in obstetrics and gynecology. Facts Vievs Vis ObGyn 2013;5(2):213-29.

C. Hall[1599] — American gynecologist and obstetrician, who identified 15 deaths of parturient women caused by pulmonary aspiration during anesthesia, and **Curtis L. Mendelson (1913 – 2002)**[1600] — American obstetrician and cardiologist, initially in obstetrics anesthesia (Mendelson syndrome or aspiration pneumonitis). Pneumonia usually results from a chemical injury with sterile acid gastric content (pH<2.5) at the volume of approximately 0,3 ml/kg or from an infected material.

The pioneers of reproductive medicine and in the *vitro* fertilization (IVF) became Sir **Robert G. Edwarts (1925 – 2013)** — English physiologist and **Patrick C. Steptoe (1918 – 88)** — English gynecologist and obstetrician. The former developed human culture to allow the IVF an early embryo to culture on the Petri dish [**Julius R. Petri (1852 – 1921)**: 19th century. Microbiology and immunology], and the later utilized laparoscopy to recover oocyte from patients with tubal infertility, which lead to the birth of the first human after conception by the *in vivo* fertilization (IVF), i.e., after fertilized egg (oocyte) was planted in her mother uterus on July 24, 1978, of an English girl named **Loise J. Brown** (weight 2,608 kg) who is alive and well.

The non-guided («blind») amniocentesis was performed in 1956 by **Fritz F. Fuchs (1919 – 95)** — Danish American gynecologist and obstetrician and his colleague **Polv Riis (1925 – 2017)** — Danish American gastroenterologist, for sex determination of the fetus and to detect fetal genetic disorders, and the treatment of pre-mature uterine contraction during pre-term pregnancy before the amnionic fluid broke.

The discovery in 1960s of Fritz Fuchs wife, **Anna-Riitta**, nee **Olson Fuchs (1926 – 2014)** — Finnish American chemist and endocrinologist, who elucidated the role of oxytocin[1601] in the onset of labor, that ethanol (ethyl or grain alcohol) inhibited labor in rabbits, led him to use this method in 1964 when she had the premature uterine contractions in her seventh months of pregnancy, by intermittent intravenous (IV) infusion of ethyl solution. The contraction stopped and child was delivered in due term. Thereafter, he apparently successfully used this method in similar clinical settings and helped to develop this treatment in use for some time. Subsequently, the studies have showed that ethanol appeared to be not better than placebo (sugar water or saline) and inferior to beta2-adrenergic agonist for prevention of preterm labor (less than 37 weeks gestation) and no longer in used due to safety concern for the mother and her child[1602].

[1599] Hall CC. Aspiration pneumonitis. JAMA 1940;114:728-33.

[1600] Mendelson CL. The aspiration of stomach content into the lung due to obstetrics anesthesia. Am J Obstet Gynecol 1946;52:191-205.

[1601] Oxytocin is a peptide hormone and neurohormone released by the posterior (neural) hypophyseal (pituitaty) lobe of the brain' hypothalamus.

[1602] Zervoudakis I, Kraus A, Fuchs F, et al. Infants of mothers treated with ethanol for premature labor. Am J Obstet Gynecol 1980;137(6):713-8,

Afterwards, in 1959-67 **Robert L. Gadd (1921 –)** — English gynecologist and obstetrician, continued the use and improved the technique of «blind» amniocentesis.

Of three (3) patients suffering from acromegaly (Ancient medicine. *Egypt*), treated by **Vasyl Truchly (27.12.1925, city of Uzhhorod, Ukraine –)**[1603] — Ukrainian American gynecologist and obstetrician, with radiation therapy, two (2) achieved resumption of ovulary cycles, and one (1) successful pregnancy.

Myroslav M. Hreshchyshyn (30.08.1927, town of Kovel, Volynska Obl., Ukraine – 24.05.1999, city of Lviv, Ukraine)[1604] — Ukrainian American gynecologist and obstetrician, treated advanced tumors of the female reproductive system, especially ovarian tumors, by a combination of resection, radiotherapy and chemotherapy (1961), proved that in menopausal women exercise (mainly daily

---

[1603] Truchly V. Colposcopy as a diagnostic method in extrauterine pregnancy. J Ukr Med Assoc North Am 1958 Sep;5,(11):1-4. — Truchly V. The use of paracervical and pudental bloc in delivery contractions. J Ukr Med Assoc North Am 1962 Oct; 9,4(27):33-5. — Truchly V. Estrogen in excessive growth of girls. J Ukr Med Assoc North Am 1964 Oct;11,3-4(34-36):8-9. — Truchly V, Young IS, Beck P, et al. Ovulation and pregnancy in acromegaly. Am J Obstet Gynecol 1966 Sep 15;96(2):174-80. — Truchly V. Hirsutism in woman. J Ukr Med Assoc North Am 1975 Apr;22,2(77):38-43. — Truchly V. Laboratory evaluation of sterility and infertility. J Ukr Med Assoc North Am 1975 Jul-Oct;22,3-4(78-79):3-12. — Truchly V. Adolescent gynecology. J Ukr Med Assoc North Am 1981winter; 28,1(100)24-8. — Truchly V. Osteoporosis. J Ukr Med Assoc North Am 1985 Winter;32,1(112):3-11. — Truchly V. Update in obstetrics and gynecology. J Ukr MedAssoc North Am 1989 Winter; 36,2(120):123-4. — Ablitsov VH. «Galaxy Ukraine». The Ukrainian diaspora: prominent persons. Publ. «Kyt», Kyiv 2007.

[1604] Town of Kovel (archeological evidence for the presence of settlements from the Bronze-Stone Age 3,000 years BC, first written mention 1518), Volyn Obl. (area 20,143 km²), Ukraine. — Hreshchyshyn MM, Holland JF. Chemotherapy in patients with gynecologic cancer. Am J Obstet Gynecol 1962 Feb 15;83:468-89. — Hreshchyshyn MM. The general review of chemotherapy of gynecological carcinoma. J Ukr Med Assoc North Am 1962 Oct;4(27):25-6. — Hreshchyshyn MM. Ovarian cancer. J Ukr Med Assoc North Am 1967 Oct.;14,4(47):3-15. — Holland JF and Hreshchyshyn MM (Ed.). Choriocarcinoma. UICC monograph series Vol. 3. Springer, Berlin 1967. — Hreshchyshyn MM, Basic considerations in chemotherapy of gynecological cancer. Clin Obstet Gynecol 1968 Jun;11(2):334-53. — Hreshchyshyn MM. Chemotherapy of gynecological malignancies (Part I). J Ukr Med Assoc North Am 1970 Jan;17,1(56):3-14. — Hreshchyshyn MM. Chemotherapy of gynecological malignancies (Part II). J Ukr Med Assoc North Am 1970 Apr ;17,2(57):3-9. — Hreshchyshyn MM. The effect of hormones and cytotoxic drugs on human trophoblast. J Ukr Med Assoc North Am 1971 Oct.;18,4(63):22-4. — Hreshchyshyn MM. Gynecologic Oncology Group Report I and II. Gynecol Oncol 1975 Sep;3(3):251-7. — Hreshchyshyn MM, Aron BS, Boronow RC. Hydrourea or placebo combined with radiation to treat stages III B and IV cervical cancer confined to the pelvis. Int J Radiat Oncol Biol Phys 1979 Mar;5(3):317-22. — Hreshchyshyn MM. The role of adjucent therapy in Stage I ovarian cancer. Am J Obstet Gynecol 1980 Sep 15;138(2):139-45. — Hreshchyshyn MM, Hopkins A, Zylstra S, et al. Effects of natural menopause, hysterectomy, and oophorectomy on lumbar spine and femoral neck bone density. Obst Gynecol 1988 Oct;72(4):631-8. — Marchetti DL, Hreshchyshyn MM. Is ovarian cancer undertreated in older women? Drugs and Aging 1994;5:81-4. — Hreshchyshyn MM. The treatment of advanced ovarian cancer. J Ukr Med Assoc North Am 1997 Winter;42,1(135):3-6. — Dzul PJ. Professor Dr Myroslaw M. Hreshchyshyn (1927-1999). J Ukr Med Assoc North Am 2000 Winter;45,2(145):67-8. — Ablitsov VH. «Galaxy Ukraine». The Ukrainian diaspora: prominent persons. Publ. «Kyt», Kyiv 2007.

walking for 1 hour) prevents the occurrence of osteoporosis, mainly of the hip joint (1989)

Valentyn I. Hryschchenko (27.XI.1928, city of Kharkiv, Ukraine. – 01.I.2011, city of Kharkiv, Ukraine)[1605] — Ukrainian gynecologist and obstetrician, who directly participated in the use of lower temperatures (hypothermia), especially in surgery with the very low temperature (cryosurgery) in his specialty, particularly he designed cryosurgical method for the treatment of dysfunctional uterine erosions and hemorrhages, proposed the use of echocardiography for the diagnosis of fetal heart abnormalities (1977-78). By turns, in 1983 he organized in Kharkiv the human reproduction laboratory, which later was remade into the Center of Human Reproduction «Implant», since 2001 – Academician V.I. Hryshchenko Clinic of Reproductive Medicine of the Academy of Medical Sciences (AMS) of Ukraine, where in 1999 for the first time in Ukraine took place successful conception *in vitro*, i.e., cycle of the extracorporeal conception (ECC), as a result of which on March 19, 1999 the girl «Katia» was born.

Report from 1972 by **Jens Bang** and **Alan Northeved**[1606] — Danish gynecologists and obstetricians, about the guided ultrasonic (US) amniocentesis opened a new chapter in the development and application of this technique.

Amniocentesis is an invasive procedure performed under the guidance of US through a puncture of the abdominal wall, pregnant uterus and the amnionic sac into the amniotic fluid, surrounding the developing fetus, from which a small amount of amniotic fluid containing fetal tissues is sampled, when a woman is between 14 and 16 weeks of gestation.

It has been used primarily in prenatal diagnosis in the fetus of genetic aberrations or disorders by examining chromosome and deoxy ribonucleic acid (DNA).

Further advances were made in 1982 when **Giuseppe Simoni (1944 –)** — Italian biologist and geneticist, performed chorionic villous (villus) sampling (CVS) – form of prenatal diagnosis to determine chromosomal or genetic deviations / disorders in the fetus, usually with fluorescence in situ hybridization (FISH) or polymerase chain reaction (PCR) which usually takes place at 10-12 weeks of gestation, earlier that amniocentesis or percutaneous (PC) umbilical cord sampling.

In addition, the amnionic fluid can be rich source of the multipotent embryonic (embryonal) stem cells (ESC) – mesenchymal, hepatic, neural, epithelial, and endothelial.

[1605] Hryshchenko VI, Shcherbyna MO (Eds). Obstetrics and Gynecology. Vol. 1-2. Publ. «Medytsyna», Kyiv 2011. — Ablitsov VH. «Galaxy Ukraine». The Ukrainian diaspora: prominent persons. Publ. «Kyt», Kyiv 2007.
[1606] Band J, Northeved A. A new ultrasonic method for abdominal amniocentesis. Am J Obstet Gynecol 1972 Nov 1;114(5):599-601.

Pediatrics

Congenital anomaly described in 1900 by **Maurice Klippel (1858 – 1942)** — French internist, and **Paul Trénaunay (1875 – ?)** — French physician, in 1907 and 1918 by **Frederick Weber (1863 – 1962)** — English dermatologist, under the name of naevus vasculosus osteohypertrophicus, is a rare condition usually affecting one extremity, characterized by hypertrophy of the bone, and surrounding soft tissues, varicose veins, hypoplasia of the lymphatic system, large skin hemangiomas, permanent port-wine stain naevi, skin ulcerations, thrombocytopenia and thromboses (Klippel-Trénaunay- Weber syndrome).

The first description of congenital absence of the trachea in 1900 belong to **W.A. Payne**[1607] — American physician. This anomaly represents a partial or complete absence of the trachea below the level of the normal larynx. If there is no connection of trachea or bronchi with the esophagus, the newborn die. **J. Floyd**[1608] classified agenesis of the trachea into three distinct types:

Type I – the trachea originates from the esophagus rather than from larynx (10%-13% of cases.

Type II – no trachea is present and the carina originates from esophagus (50%-63% of cases).

Type III – each main stem bronchus (MSB) originates from the esophagus (22%-31% of cases).

Male incidence pre-dominates in 2 to 1 ratio. In 90% of tracheal agenesis cases, there are other congenital abnormalities – the vertebral, anal, cardiac, tracheal, esophageal, renal and limb (VACTERL) or tracheal, anal, cardiac, radial, and duodenal atresia (TACRD). Neonates present with severe respiratory distress and inability to cry. They should be intubated endoesophageally with the endotracheal tube (ETT), alongside with a nasogastric tube (NGT) for gastric decompression, or undergo urgent bronchoscopy or tracheostomy, and supported with extracorporeal membrane oxygenation (ECMO). The gastrointestinal (GI) continuity can be corrected with colonic interposition or gastric conduit, and the airway reconstructed with tracheal allografts or tissues engineered replacement.

Congenital parenchymal cysts of the lung, most commonly present in the left lower lobe, often supplied by an aberrant systemic artery, which causes type

---

[1607] Payne WA. Congenital absence of the trachea. Brooklyn Hosp J 1900;14:568-70.

[1608] Floyd J., Cambell Jr DC, Dominy DE. Agenesis of the trachea. Ann Resp Dis 1962;86:557-60.

B Niemann-Pick disease, described by **Albert Niemann (1880 – 1921)**[1609] — German pediatrician, in 1914, and by **Ludwig Pick (1868 – 1944)**[1610] — German pediatrician, 1926. It is a group of inherited metabolic diseases included in the larger family of lysosomal storage diseases, resulting from deficiency of sphingomyelin phosphodiesterase and deposition sphingomyelin in the reticuloendothelial system (RES). In general, there is ataxia, dysarthria, dysphagia, sleep disorders, dementia, and seizures, enlarged livers and spleen, coxa vara with O-like bending of the femur inside and thrombocytopenia. Type B (chronic non-neuropathic) begins in the early childhood without affection of the central nervous system (CNS) or intellectual abilities, the normal lifespan is expected.

In 1938 **Olexandr A. Kysil (Kysel) [19(31).VIII.1859, city of Kyiv, Ukraine. – 08.III.1938, city of Moscow, Russia]**[1611] — Ukrainian pediatrician, in 1944 – **T. Duckett Jones (1899 – 1954)** — American physician, and in 1951 – **Anatolii I. Nesterov (1895 – 1979)** — Russian therapeutant, established the great criteria for rheumatic disease (RD) – the great Kysil, or the great Kysil-Jones-Nesterov criteria, «five absolute signs of rheumatism», of an acute RD as an outcome of infection of the upper respiratory tract by the group A beta-hemolytic Streptococcus: (1) inflammation of the heart (carditis) with signs of inflammation of the pericardial sack (pericarditis), a new cardiac murmur, and congestive heart failure (CHF); (2) migratory inflammation of joints (migrating arthritis); (3) involvement of the central nervous system (CNS) as evident by jumping (dancing) chorea – «St Vitus dance»[1612], or Sydenham chorea[1613]; (4) involuntary face grimaces and muscle jerks accompanied by an incoordinate

---

[1609] Niemann A. Ein unbekanntes Krankheitsbild. Jahrbuch für Kinderheilkunde, NF Berlin 1914;79:1-10. Niemann A. Kompendium der Kinderheilkunde. Berlin 1920.

[1610] Pick L Morbus Gaucher und die ihm ähnlichen Kranhiten (die lipoidzellige Splenomegalie Typhus Niemenn und die diabetische Lipoidzellenhypoplasie der Mitz. Ergebniss der Innerer Medizin und Kinderheilkunde. Berlin 1926;29:519-627.

[1611] Kysel AA. Rheumatism in Children. Moscow – St Petersburg 1940. — Kysel AA. Works of the Earned Figure in Science Professor A.A. Kysel. Vol. 1-2, Moscow – St Petersburg 1940-49. — Nesterov AI. Outline of Studying About Rheumatism, and Diseases of Joints. Moscow 1951. — Nesterov AI. Question of Pathogenesis, Clinics and Treatment of Rheumatism. Moscow 1956. — Kysel AA. Selected Works. Moscow 1960. — Hetiazhenko VZ, Lapin PV. 125-years of O.A. Kysil Student Scientific Society. Life and Scientific Activity of O.A. Kysil – founder of the Student Scientific Society (1859-1938). Creation and Activities of Clinical Students of Medicine Society of Imperial St Volodymyr University (1881-1918). «Clinical Society Students-Medics Male and Female in city of Kyiv» and «Ukrainian Students Medical Society in Kyiv» (1918-1922). General Activity Characteristics of Scientific Students Societies of the Higher Medical Educational Institutions (1921-1991). O.A. Kysil Student Scientific Society, O.O. Bohomolets National Medical University. Medical Cadre – Gazette of O.O. Bohomolets NMU 10 Oct. 2006;6(2501:3-6. — Ablitsov VH. «Galaxy Ukraine». The Ukrainian diaspora: prominent persons. Publ. «Kyt», Kyiv 2007. — Hanitkevych YaV. The contribution of the Ukrainian physicians to the world medicine. Ukr Med Chasopys 2009 VII/VIII;4(72):110-15.

[1612] St **Vitus (IV ст)** — Sicilian martyr, the patron of artists and dancers, who faithfully intercede (solicit) for peoples suffering from different diseases, including neurologic and psychiatric illnesses

[1613] **Thomas Sydenham (1624-89)** — English physician.

motions of the shoulder which disappears while sleeping; (5) ring-like redness of the skin (annular erythema), and authentic formation of the subcutaneous nodularity's on the back of the wrist, on the outside of the elbow, and on the front of the knees (Heberden's knots)[1614].

Olexandr Kysil made the correct assertion that chorea is one of RD's signs in children.

He established in 1881 at St Volodymyr Kyiv University, the Student Medical Scientific Society, named O.A Kysil in his honor.

Described in 1903 by **Frederich E. Batten (1865 – 1918)** — British pediatrician and neurologist, the genetic juvenile neuronal ceroid lipofuscinosis, a fatal, inherited autosomal recessive, neurodegenerative disease, named in his honor the Batten's disease, affect children, beginning insidiously between 2 and 10 years of age. Early symptoms usually appear with gradual onset of impairment of vision, seizures, personality and behavioral changes, slow learning or regression, repetitive speech, or echolalia (in Greek – «echo» meaning «echo» or «repeat», «laelia», meaning «speech» or «talk») – unsolicited repetition of localization made by another person(s), clumsiness or stumbling. Signs include retarded growth of the head in the infantile form, poor circulation in the lower extremities, decrease body fat and muscle mass, curvature of the spine, hyperventilation, and/or breath-holding spells, teeth grinding and constipation. With passing time children suffer from mental impairment, worsening seizures, progressive loss of sight, speech, and motor skills. On average death arise about a year after the first symptoms of disease. Treatment is symptomatic and supportive. The drug cerliponase was approved for the treatment of children 3 years of age or older, suffering from the late onset of the juvenal neuronal ceroid lipofuscinosis with slow walking ability.

Luis Marquio (1867 – 1935) — Uruguayan physician, described an inborn familial autosomal recessive lysosomal storage disease, due to mucopolysaccharidosis type IV (A) due to N-acetylgalactosamine-6-sulfate deficiency, is marked by a short statue, malalignment of the vertebral column, peculiar facies, corneal clouding, deafness, aortic insufficiency (AI) and hepatomegaly, and type IV(B) due to beta-galactosidase deficiency marked by a short stature with dysostosis multiplex, vertebral malalignment, corneal clouding, pectus carinatum and genu valga with X-like twisting of the knee outside (Marquio syndrome, 1929). He also discovered congenital atrioventricular (AV) block in atrial septal defect (ASD) and noted sign (Marquio) when lying down patient affected by epidemic poliomyelitis can sit up only after passive flexion of legs.

[1614] **William Heberden (1710-1801)** — English physician.

Congenital chest wall deformities, among others, are occasionally associated with a maldevelopment, described by **Pierre Robin (1867 – 1950)** — French stomatologist, characterized by unusual smallness of the jaw combined with cleft palate, downward displacement of the tongue and absent gag (pharyngeal) reflex (Pierre Robin syndrome, 1929).

John Zahorsky (13.10.1871, village of Nalepkovo, Gelnica Okres, Kosice Region, Slovakia – 05.02.1963, town of Steelville, MO, USA)[1615] — Ukrainian Slovak American pediatrician, improved incubators for premature infants, and newborns (1904), described exanthema subitem (Zahorsky disease, 1910)[1616] and herpangina (Zahorsky syndrome, 1920)[1617] in children.

The heterogenic group of inherited autosomal recessive hemolytic anemias in people or descendants from countries around the Mediterranean Sea to the South East Asia, was described by **Thomas B. Cooley (1871 – 1945)** — American pediatrician, hematologist and hygienist, in which the common is decreased rate of alpha, beta or delta chains hemoglobin (Hb, Hgb) synthesis, while homozygotic forms shows profound anemia and death in utero, and heterozygotic forms showed abnormalities of erythrocytes from moderate to severe (Cooley anemia, 1925, 1927). The term Cooley anemia was renamed by **George H. Whipple (1878 – 1976)** — American pathologist and biochemical researcher, into thalassemia major, meaning «sea water in the blood». On the contrary, thalassemia minor is a heterozygotic form of beta thalassemia, usually asymptomatic, although synthesis of Hb A is sometime sluggish on the background of moderate anemia and splenomegaly; and sickle cell-thalassemia (disease) is any inherited anemia which include simultaneous heterozygosity of Hb S and thalassemia gene, symptoms similar to sickle cell anemia seen in people living in or descendants of West Africa, include accelerated hemolysis, increased blood viscosity, and occlusion of vessels, joint and abdominal pain, lower extremity ulcerations and periodic sickle cell crises.

Causes of non-atherosclerotic (AS) coronary artery disease (CAD), among others, are accumulation of metabolic substrates, including mucopolysaccharides, like syndrome of **Charles H. Hunter (1872 – 1955)** — Scottish Canadian physician,

---

[1615] Village of Vondrisel (in German – Wagendrüssel, in Hungarian of Mereny; founded 1295), since 1948 – village of Nalepkovo [named in honor of **Jan Nalepka (1912 - 43)** — Slovakian partisan during World War II, Brigadier General], Gelnica Okres / district (area 584 km²), Kosice Region (area 6,753.26 km²), Slovakia (area 49,035 km²). — Town of Steelville (founded 1835), Missouri (MO, area 180,560 km²), USA. — Zahorsky J. From the Hills: An Autobiography of a Pediatrician. CV Mosby Co, St Louis 1950. — Ablitsov VH. «Galaxy Ukraine». The Ukrainian diaspora: prominent persons. Publ. «Kyt», Kyiv 2007.

[1616] Zahorsky J. Roseola infantilis. Pediatrics (New York):1910;22:60-4. — Zahorsky J. Roseola infantum. JAMA 1913 Oct 18; 1446-50.

[1617] Zahorsky J. Herpetic sore throat. So Med J (Birmingham - Nashville) 1920;13:871-2. — Zahorsky J. Herpangina (a specific infectious disease). Arch Pediatr (New York) 1924:41:181-8.

and syndrome of **Gertrud Hurler (1889 – 1965)** — German pediatrician, primary oxalosis and alcaptonuria, genetic lysosomal lipid storage diseases – such as disease clinically indeterminable from Tay-Sachs disease (19th century. Eye disease. / 20th century. Internal medicine), described in 1963-68 by **Kondrad Sandhoff (1939 –)** — German biochemist and **Horst Jatzkewitz (1902 – 2002)** — German biochemist, i.e., gangliosidoses involving abnormalities in hexosaminidase A and B (Sandhoff-Jatzkewitz disease), or disease of **Sidney Farber (1903 – 73)** — Polish American pediatric pathologist of Jewish parents, i.e., lipogranulomatosis syndrome (1952).

Ernst Moro (Dec. 08, 1874, city of Ljubljana, Slovenia – April 17, 1951, die Stadt Heidelberg, der Staat Baden-Württemberg, Germany)[1618] — Austrian pediatrician, described a defensive reflex consisting of the infant's drawing of its arms across its anterior surface of the chest in embracing manner in response to stimuli produced by striking the surface on which its rest (embrace reflex, 1918).

In 1906 **Clement F.** von **Pirquet (1874 – 1929)** — Austrian scientist and pediatrician with **Bélla Schick (1877 – 1967)** — Hungarian American pediatrician, developed the concept of allergy. Then, in 1907, the former reported a classical serological reaction or test (Pirquet) for tuberculosis in which a little tuberculin is applied to a superficial abrasion of the skin of the arm. A positive reaction is seen if a red papillary eruption appears several days later at the site of the inoculation. It was soon (1908-10) replaced by **Felix Mendel (1862 – 1925)** — German physician and **Charles Mantoux (1877 – 1947)** — French physician, with an intracutaneous (Mendel-Mantoux) test in which 0,1 ml of tuberculin (5 tuberculin units) is injected intradermally in the arm. The size of induration on the 2nd or 3rd day is used to differentiate between infection with tubercle bacilli or related mycobacteria.

Of the congenital anomalies of coronary arteries, which affect from 1 in 30,000 to 1 in 300,000 live births (0.25% – 0.5%), the most common is an anomalous origin of the left coronary artery, especially the left main coronary artery (LMCA), from the pulmonary artery (PA) – an anomalous origin of the left coronary artery from the pulmonary artery (ALCAPA), that is, from the posterior right or left facing or the anterior non-facing sinuses of the main pulmonary artery (MPA) or from the right or left pulmonary artery (PA). ALCAPA usually occurs in isolation, occasionally is associated by patent ductus arteriosus (PDA), ventricular septal defect (VSD), coarctation of aorta (COA) and tetralogy of Fallot (TOF).

[1618] City of Ljubljana (Ljubljana marshes in the immediate vicinity Ljubljana were settled in 2000 BC by people living in pile dwellings, the Romans built around 50 BC a military encampment that later became a permanent settlement called Iulia Aemana, first mentioned in the first half of the 12th century), Slovenia (area 20,271 km²). — Die Stadt Heidelberg (Prehistoric medicine), der Staat Baden-Württemberg, Germany. — Moro E. The erste Trimenon. Münchener medicinische Wochenscrift 1918;65:1147-59.

The anomalous origin of the LMCA from the MPA, generally from the left posterior pulmonary sinus, occurs in more than 90% of children. In the remainder percentage the right coronary artery (RCA) or both coronary arteries originate from the MPA. ALCAPA and origin of both coronary arteries from the MPA are the most common causes of myocardial ischemia and myocardial infarction (MI) in children, manifested by the left ventricular (LV) failure shortly after birth, and are always deadly if not early recognized and treated.

Initially, ALCAPA was described in 1911 by **Aleksei Abrikosov (1875 – 1955**: 20th century. Pathology), re-described in 1933, by **Edward B. Bland (1901 – 92)**[1619] — American cardiologist, **Joseph Garland (1893 – 1973)** — American physician, and **Paul D. White (1886 – 1973)** — American cardiologist, a.k.a. Bland-Garland-White syndrome.

The pathophysiology of the hemodynamics in the anomalous origin of the LMCA from the MPA was first explained in 1964 by **Jesse H. Edwards (1911 – 2008)**[1620] — American pathologist of Jewish parents, by the fall in the pulmonary artery pressure (PAP), followed by decrease blood perfusion in the anomalous coronary artery from the MPA. Unless an adequate collateral blood flow develops from the collateral coronary artery, the anomalous coronary artery causes myocardial ischemia within its territory. In 1960s he founded Jesse E. Edwards Register of Cardiovascular Diseases.

In general, the surgical treatment of the anomalous origin of the coronary artery from the MPA consist of reimplantation of the anomalous artery to the ascending aorta (AA). In cases of ALCAPA where the most common origin of an anomalous left coronary artery is from the posterior right facing sinus of the MPA, the surgical approach consists of creation of direct reattachment of the ostia of an anomalous coronary artery in the form of the Carrel's patch **[Alexis Carrel (1873 – 1944)**: 20th century. Surgery], a technique with the «end-to-side» anastomosis to the closest AA sinus. At times, when ALCAPA originate from the posterior left facing or the anterior non-facing sinuses of the MPA or from the right or the left PA, then the coronary artery ostia in continuity with a transverse strip of the MPA is formed to lengthen it with subsequent reimplantation to the closest AA sinus. An alternative for ALCAPA originating from the posterior left facing MPA sinus, especially of the LMCA, is the creation of an intrapulmonary arterial pericardial or synthetic baffle that would direct

---

[1619] Bland EF, White PD, Garland J. Congenital anomalies of the coronary arteries: report of anusual case with cardiac hypertrophy. Am Heart J 1933;8:787-801.

[1620] Edwards JE. Editorial. The direction of blood flow in coronary arteries arising from the pulmonary trunk. Circulation 1964;29:163-6.

the blood flow from the aortic sinus to the ostium of the anomalous coronary artery (Takeuchi operation)[1621].

In 1916 i 1936 **René Lutembacher (1884 – 1968)**[1622] — French cardiologist, defined a congenital heart disease (CHD) combining mitral stenosis (MS) with dilatation of the main pulmonary artery (MPA) and atrial septal defect (ASD) with left-to-right shunt of the blood (Lutembacher syndrome). It is more common in rheumatic heart disease (RHD) than non-RHD.

Described in 1916 by **Winfried R.C. Brachmann (1888 – 1969)** — German physician, and in 1933 by **Cornelia C. de Lange (1871 – 1950)** — Dutch pediatrician, a relatively common birth defect of unknown origin comprising of microcephaly with oligophrenia, short stature, multiple malformations of the face, neck and extremities, telengiectasia, ankylosing spondylitis, cutis laxa (elastosis) and ataxia, subsequently named Brachmann-de Lange syndrome or de Lange syndrome.

In 1926 **R.T. Grant** described in a 14-month-old-girl suffering from congenital heart disease (CHD) named pulmonary stenosis (PS) with an intact interventricular septum (IVS) – PS/IVS [the 18th century – **John Hunter (1728 – 93)**] the existence of a connection in the form of sinusoids or intermuscular spaces between the coronary arteries and the right ventricular (RV) myocardium, so called RV dependent coronary circulation (RVDCC)[1623].

The first palliative surgical management of children with PS/IVS was an open trans-ventricular valvotomy and infundibulotomy with or without patch successfully performed in infants by **Milton Weinberg, Jr (1925 –)**[1624] — American thoracic and cardiac surgeon, in 1962. It was later supplemented by a systemic-to-pulmonary shunt with a 3- or 3.5-mm polytetrafluoroethylene (PTFE) tube graft to prevent life-threatening hypoxia.

The possibility of the definite procedure is decided by cardiac catheterization (CC) in infants between 6 and 12 months of life during which systemic-to-pulmonary shunt is temporarily occluded. If the arterial blood oxygen saturation (SaO$_2$) remains high and after occlusion of the atrial septal defect (ASD) or patent foramen ovale (PFO) the right atrial (RA) pressure remains below 15 mmHg and cardiac output (CO) is adequate, then two-ventricle repair is carried out by closure of the systemic-to-pulmonary shunt and ASD/PFO. Otherwise, one - and one-half ventricle repair is performed by closure of ASD/PFO, taking down systemic-to-pulmonary shunt

---

[1621] Takeuchi S, Imamura H, Katsumoto K, et al. New surgical method for repair of anomalous left coronary artery from pulmonary artery. J Thorac Cardiovasc Surg 1979;78:7-11.

[1622] Lutembacher R. De la stenose mitral avec communication interauriculaire. Arch dis mal coer 1916;9:237-60. — Lutembacher R. Stenose mitra et communication interauriculaire. Arch dis mal coer 1936;29:229-36.

[1623] Grant RT. An unusual anomaly of the coronary vessels in the malformed heart of a child. Heart 1926;13:273-83.

[1624] Weinberg Jr M, Bicoff JP, Bucheleres HG, et al. Pulmonary valvotomy and infundibulotomy in infants. J Thorac Cardiovasc Surg 1962 Oct;44:433-42.

and the creation of bidirectional superior vena cava (SVC) to pulmonary artery (PA) anastomosis or the Glenn operation [20th century. Surgery – **William Glenn (1914 – 2003)**]. However, if the RA pressure exceeds 15 mm Hg, a small (4 mm) fenestration can be left with a purse string suture and adjustable snare used to incrementally close ASD/PFO postoperatively, or by catherer closure within months later. Infants with severe RV hypoplasia or RVDCC, or both, are designated at 4- to 6-months of age for one-ventricle repair, which include the Glenn operation, and at 2- to 4-years of age for an internal lateral tunnel-like baffle or external lateral synthetic tubular graft (STG) between the inferior vena cava (IVC) and the right PA or the Fontan operation [20th century. Surgery – **Francois Fontan (1929 – 2018)**]. Children with poor left ventricular (LV) function should undergo heart transplantation (HT).

Cystic fibrosis (CF) or mucoviscidosis[1625] is the most common multigene autosomal recessive inherited disorder of the endocrine glands, caused by mutation in both copies of the gene for the cystic fibrosis transmembrane conductance regulator (CFTR) protein, affecting the lungs, pancreas, liver, kidneys, sweat glands and gastrointestinal (GI) tract with the formation of fibrosis and cysts with excessive mucous secretion, most common among the people of North European descent.

The disease is characterized by shortness of breath (dyspnea), coughing up mucous, recurrent bacterial pneumonia, paranasal sinus infections, poor growth of children, fatty stools, clubbing of fingers and toes, infertility in most males and the average life expectancy of 42 to 50 years. The cause of death in 80% of patients is recurrent pneumonia.

There is no cure for CF. However, CF is being treated with inhalations of hypertonic saline solution and salbutamol, a.k.a., albuterol (Ventolin), chest physical therapy (CPT), pancreatic enzyme replacement, addition of fat-soluble vitamins, antibiotics (azithromycin) for the lungs infections, and lung transplantation if condition worsens. Recent (since 2016), gene therapy or gene modulation shown limited success, and phage therapy is being tested.

The first description of CF dates to 1595 and the first specific diagnosis was made in 1938 by **Dorothy H. Andersen (1901 – 63)**[1626] — American pathologist.

The breakthrough in the treatment of pulmonary affections by CF occurred with development in 1968-88 by **Warren J. Warwick (c. 1925 – 2016)**[1627] — American

---

[1625] Bush A, Bilton D, Hodson M (eds). Hodson and Geddes' Cystic Fibrosis. 4rd Ed. CRC Press, London 2015.

[1626] Andersen DH. Cystic fibrosis of the pancreas and its relation to celiac disease. A clinical and pathological study. Am J Dis Child 1938 Aug; 56(2):344-99.

[1627] Warwick WJ, Hansen LG. The long-term effect of high frequency compression therapy on pulmonary complications of cystic fibrosis. Pediatr Pulmonol 1990 Dec;11(3):265-71. — Warwick WJ, Wielinski CL, Hansen LG. Comparison of expectorated sputum after manual chest physical therapy and high-frequency chest compression. Biomed Instar Technol 2004 Nov;38(6):470-5.

pediatrician and **Leland G. Hansen (1970 –)**, Public Health Master (PHM) — American scientist, and associates **David N. Cornfield (1961 –)** and **Carlos Milla (1962 –)** — American pulmonologists, the chest wall oscillation device, called high frequency chest compression (HFCC) or the Vest Airway Clearance System for effective cleaning from the lung airways, i.e., bronchi and bronchioles, an excessive accumulation of mucous secretions in children with CF.

Since 1999 the portable HFCC device patented and introduced for a self-administration by patients with CF, with the inflatable chest vest connected to a compressor to provide external high frequency chest wall oscillation delivering air impulses through tubes attached to the vest.

As an extension of the HFCC device for the treatment of patients with CF, the oscillation and lung expansion (OLE) therapy was developed using continuous high frequency oscillation and continuous positive expiration pressure (CPAP) for targeted high-risk patients prone to atelectasis and postoperative pulmonary complications (PPC) to improve outcomes[1628].

The OLE therapy is gaining use in the treatment of chronic obstructive pulmonary disease (COPD), bronchiectasis of the lungs, cerebral palsy, and muscular dystrophy, in which excessive mucous secretion or impaired clearance of the lungs can block airways[1629].

Pierre Mounier-Kuhn (1901 – ?)[1630] — French physician, characterized tracheobronchomegalia as the condition of excessive dilatation of the trachea and main stem bronchi (MSB) to the third order subdivision due to absence of elastic fibers and smooth muscles in children, that lead to mucosal herniations between tracheal rings with creation of diverticula, retention of sputum with subsequent recurrent respiratory infections, bronchiectasis, emphysema, pulmonary fibrosis, respiratory failure and death (Mounier-Kuhn syndrome, 1932). It may be associated with or should be differentiated from connective tissue disorders such as a familial excessive elasticity of the skin and joints (cutis et articuli hyperelastica)[1631], Marfan syndrome [19th century. Pediatrics. – **Antoine Marfan (1858 – 1942)**], Brachmann-de Lange syndrome or de Lange syndrome and Kenny-Caffey (above). Palliative treatment

---

[1628] Huynh TT, Liesching TN, Cereda M, et al. Efficacy of oscillation and lung expantion in reducing postoperative pulmonary complications. J Am Coll Surg 2019 Nov;229(5):458-66.

[1629] Birnkraft DJ, Pope JF, Lewarski J, et al. Persistent pulmonary consolidation treated with intrapulmonary percussive ventilation: a preliminary report. Pediatr Pulmonol 1966;21(4):246-9.

[1630] Mounier-Kuhn P. Dilatation de la trachee: constatations radiographiques et bronchoscopiques. Lyon Med 1932;150:106-9.

[1631] **Edward Ehlers (1863 - 1937)** — Danish dermatologist, and **Henri A. Danlos (1844 - 1912)** — French physician and dermatologist, who after **Job J. Meekeren (1611 – 66**:17th century**)**, re-described a familial excessive elasticity of the skin and joints (cutis et articuli hyperelastica) in 1901 and 1908, subsequently.

of tracheobronchomegaly consist of stenting and T-tubes, definite – double lung transplantation with bilateral bronchial stents.

Bronchial atresia, described by **B.H. Ramsay**[1632] in 1953, is the second most common abnormality of the airway after tracheoesophageal fistula (TEF). A lobal segmental bronchus ends blindly in the lung tissue distal to the bronchial atresia. Peripheral lung tissue expands and become emphysematous as air entering through the pores of Kohn [19th century. Internal medicine – **Hans Kohn (1866 – 1835)**]. Beyond the atretic segment but proximal to the hyperinflated by the air lung, the terminal is mucous filled. The most common location is the left upper lobe (LUL), then the left lower lobe (LLL), followed by the right upper lobe (RUL). Infants usually develop respiratory distress 4 or 5 days to several weeks after birth, manifested by wheezing, stridor, and repeated pulmonary infections. Diagnosis is established by chest radiography and computer tomography (CT). Indication for segmentectomy or lobectomy of bronchial atresia include a recurrent and serious pulmonary infection, respiratory distress, and increasing size of the translucent lung.

The major contributions to pediatrics was made by **Charles C. Chapple (1903 – 79)**[1633] — American pediatrician, by discovering of a method for early diagnosis and treatment of congenital dislocation of the hip in infants, fetal abnormalities and development defects, and also by developing in the middle 1930's an improved over earlier incubators, the incubator or isolate, that carries his name which minimized the major perils that threatened of premature and feeble newborns: their vulnerability to atmospheric changes, the lack of breathing capacity, dehydration, and the susceptibility to infection[1634].

Spondylothoracic dysplasia of **Saul W. Jarcho (1906 – 2000)** — American physician and **Paul M. Levin** is an inherited malformation with an autosomal recessive trait of an early life, manifested by multiple alternating hemivertebrae and rib deformities causing a crablike configuration, a short neck and chest, relatively long arms, fixed flexion deformity of the interphalangeal joints of the little finger (campodactyly) and syndactyly, protuberant abdomen, rarely genitourinary disturbances and death from respiratory insufficiency (RF) – Jarco-Levin syndrome (1938).

The Heath-Edwards Classification[1635] of the histopathology of the pulmonary artery (PA) hypertensive vascular disease correlates with pulmonary vascular resistance (PVR) of patients with patent ductus arteriosus (PDA), atrial septal defect

---

[1632] Ramsay BH, Bayron FX. Mucocele, bronchogenic cyst. J Thorac Surg 1953;26(1):21-30.

[1633] Cartner JA. Memoir of Charles C. Chapple 1903-1979. Trans Stud Coll Physicians Phila 1973 Sep;1(3):232-3.

[1634] Chapple Ch.C. An incubator for infants. Am J Obst and Gynec 1938;35:1062-5.

[1635] Heath D, Edwards JE: The pathology of hypertensive pulmonary vascular disease: a description of six grades of structural changes in the pulmonary arteries with special reference to congenital cardiac septal defects. Circulation 18: 533, 1958; 18: 533-43.

(ASD) and ventricular septal defect (VSD): grade 1 changes are defined as media hypertrophy; grade 2 – median hypertrophy with cellular intimal reaction; grade 3 – medial hypertrophy and intimal fibrosis; grade 4 – plexiform lesions of the muscular PA and arterioles with a plexiform network of capillary-like channels within a dilated segment; 5 – complex plexiform, angiomatous and cavernous lesions and hyalinization of intimal fibrosis; 6 – necrotizing arteritis.

Virginia Apgar (1909 – 74) — American obstetrical anesthesiologist, neonatologist and teratologist[1636], who invented in 1953 the Apgar score of the newborn child condition[1637], five criteria of which are:

	Score of 0	Score of 1	Score of 2
Skin color	blue of pale all over	blue at extremities body pink (acrocyanosis)	no cyanosis, all body and extremities pink
Pulse rate	absent	< 100 beats per minute	> 100 beats per minute
Reflex irritability, grimace	no response to stimulation	grimace on suction or aggressive stimulation	cry on stimulation
Activity	none	some flexion	flexed arm and legs that resist extension
Respiratory effort	absent	weak, irregular, gasping	strong, robust, cry

The Apgar scores 7 or above are generally normal, 4 to 6 low, and 3 or below generally regarded as critically low.

Asphyxiating thoracic dystrophy in newborn was described by **Mathis Jeune (1910 – 83)** — French pediatrician, is an inherited autosomal-recessive trait form of osteochondrodystrophy, consist of a narrowing in the transverse and sagittal axes, rigid bell-shaped thorax, short and wide ribs, abundant and irregular costal cartilages, splayed costochondral junctions barely reaching the anterior axillary line (AAL), short stubby extremities, short and wide bones, elevated and fixed clavicles, small and hypoplastic pelvis, square iliac bones, polydactyly and protuberant abdomen, a little respiratory motion causing an early death from suffocation – Jeune disease or syndrome (1954).

Guido C. Currarino (1920 –) — Italian American pediatric radiologist and **Frederik N. Silverman (1914 – 2006)** — American pediatric radiologist, described

---

[1636] Teratology is the study of physiological development, i.e., the study of human congenital abnormalities.

[1637] Apgar V. A proposal for a new method of evaluation of the newborn infant. Curr Res Anesth Analg 1953;32(4):260-77.

an association of pectus carinatum and congenital heart disease (CHD) – Currarino-Silverman syndrome (1958).

Hyperlucent lung syndrome was described in 1953 by **Paul R. Swyer-James (1921 –)** — American pediatrician and **George C.W. James (20th century)** — American radiologist, and in 1954 by **William M. Macleod (1911 – 77)** — English physician. It is a congenital unilateral agenesis or hypoplasia of the pulmonary artery (PA) with a small hilum, imitating emphysema, possible partial or segmental agenesis, accessory lungs, lobes or segments, or the outcome of chronic childhood lower respiratory tract infections, caused by bronchiolitis obliterans, that lead to pulmonary vascular changes and bronchiectasis with marked obstruction and air trapping during expiration (Swyer-James or Macleod syndrome).

S. Howard Williams and **Peter E. Campbell**[1638] published on a familial congenital generalized bronchiectasis from reduction of the ring-like cartilage distally to the first division of the peripheral bronchi, probably due to alpha$_1$-antitrypsin deficiency, resulting in a disease of the airways, characterized by bronchomalacia, ciliary dyskinesis and hyperinflation of the lungs (Williams-Campbell syndrome, 1960). The condition may be associated with pulmonary sequestration or Marfan syndrome.

In 1965 **Angello M. DiGeorge (1921 – 2009)** — Italian American physician and pediatric endocrinologist, described palate-cardio-facial or of depletion of the chromosome 22q11.2 (DiGeorge) syndrome, autosomal with an abnormal embryonal development of the 3rd and 4th pharyngeal pouches, which gives the origin to the thymic and parathyroid glands and the great vessels, characterized by the phenotypic apocrine «CATCH22» or «CATCH» (*c*ardiac defects, *a*bnormal facies, *T*-cell deficit, *c*left palate, *h*ypocalcemia), or tracheal abnormalities, complex congenital heart disease (CHD), upper extremity defects and duodenal atresia[1639].

Kenny-Caffey syndrome[1640] is a hereditary autosomal dominant and autosomal recessive disorder characterized by dwarfism, abnormalities of the head and eyes, thickening of the long tubular bones ones, medullary stenosis, hypofunction of the parathyroid gland (PTG) with the intermittent hypocalcemia, which was described in 1966 by **Frederic M. Kenny (1929 –)** — American pediatrician, and in 1967 by **John P. Caffey (1895 – 1978)** — American pediatrician.

[1638] Williams SH, Campbell PE. Generalized bronchiectasis associated with deficiency of cartilage in the bronchial tree. Arch Dis Child 1960 Apr;35(180:182-91.

[1639] Evans JA, Reggin J, Greenberg C. Tracheal agenesis and associated malformations: a comparison with Tracheoesophageal fistula and the VACTERL association. Am J Med Genet 1985;21(1):21-38.

[1640] Kenny FM, Linarelli L. Dwarfism and cortical thickening of tubular bones. Am J Dis Child 1966;111:201-7. — Caffey J. Congenital stenosis of medullary spaces in tubular bones and calvaria in two proportionate dwarfs - mother and son; coupled with transitory hypocalcemia. Am J Roentgenol 1967;100:1-11.

Similar affection to the DiGeorgi syndrome was described by **Atsuyoshi Takao (1925 – 2006)** — Japanese pediatric cardiologist, the syndrome of cardiac (conotruncal) and facial anomalies with normal T- cell and parathyroid function (Takao syndrome, 1975) and by **Robert J. Shprintzen (1946 –)** — American speech pathologist, plastic surgeon, otorhinolaryngologist (ORL) and pediatrician, velocardiofacial (Shprintzen) syndrome (1978).

Described in 1961 by **John C.P. Williams (1922** – disappeared in **1972?)** — New Zealander cardiologist, congenital disorders of newborn and infants is characterized by miniature («alfin») facies with low nasal bridge, mental retardation but strong language skill and an unusual cheerful demeanor with strangers, supravalvular aortic stenosis and transient hypercalcemia (Williams syndrome).

Robert A. Good (1922 – 2003) — American physician, discovered an immune deficiency with thymoma (1962); documented the important role of tonsils in developing an immune defense system in mammals including humans (1965); transplanted in 1968 to a 5-months-old boy with a profound immune deficiency a bone marrow from his 8-year-old sister and he has grown up to become a healthy adult.

With the aim of palliation of the newborn with transposition of the great vessels (TGV), **William J. Rashkind (1922 – 86)** and **William W. Miller**[1641] — American pediatric cardiologists, proposed the enlargement of the patent foramen ovale (PFO) or the atrial septal defect (ASD) with the Miller balloon catheter (balloon atrial septostomy, or Rashkind procedure, 1966) to increase the arterial blood O_2 by shunting of the blood through an enlarged ASD, thus decreasing hypoxemia and acidosis.

This was followed by prenatal transcatheter occlusion with a foam plug of the patent ductus arteriosus (PDA) in the fetus (1967)[1642]; the first percutaneous (PC) closure of an ASD in adult using No. 23-Fr catheter to deliver a double-umbrella device (1976)[1643]; prove for the feasibility of PC transluminal balloon / transcutaneous dilatation of restenosis of coarctation of the aorta (COA) after surgical repair on postmortem[1644] or surgically resected[1645] specimens; the first static percutaneous

---

[1641] Rashkind WJ, Miller WW. Creation of an atrial septal defect without thoracotomy. A palliative approach to complete transposition of the great arteries. JAMA. 1966 June 13;196(11):991-992. — Rashkind WJ, Miller WW. Transposition of the great arteries: results of palliation by balloon atrio-septostomy in thirty-one infants. Circulation 1968; 38: 453-62. — Boehm W, Emmel M, Sreeram N. Balloon atrial septostomy: history and technique. Images Paediatr Cardiol 2006;26:8-14.

[1642] PortsmannW, Wierny L, Warnke H, et al. Catheter closure of patent ductus arteriosus, 62 cases treated without thoracotomy. Radiol Clin North Am 1971;9;203-18.

[1643] King TD, Thompson SL, Steiner C, et al. Secundum atrial septal defect: non-operative closure during cardiac catheterization. JAMA 1976;235:2506-9.

[1644] Sos T, Sniderman KW, Retlek-Sos B, et al. Percutaneous transluminal dilatation of coarctation of the aorta post mortem. Lancet 1979;2:970-1.

[1645] Lock JE, Niemi T, Burke BA, et al Transcutaneous angioplasty of experimental aortic coarctation. Circulation 1982;66(6):1280-6.

PC balloon angioplasty (BA) of congenital pulmonary artery (PA) stenosis (1982)[1646]; transluminal aortic BA for COA in the newborn (1982)[1647]; the first successful PC balloon aortic valvuloplasty in 26 patients (1984)[1648]; the first report of PC catheter mitral commissurotomy (MC) in rheumatic mitral stenosis (MS) (1985)[1649]; the fetal cardiac intervention with an attempt in utero pacing for congenital complete heart block (CHB) causing hydrops (1986)[1650]; PC treatment of PDA with an occluding device (1987)[1651]; the earliest attempt at transcatheter closure of the ventricular septal defect (VSD) using a «double-umbrella» device in six patients who were not considered candidates for operative closure, description of technical aspects of the procedure and feasibility (1988)[1652]; the first clamshell occlusion device of an ASD in children delivered through a No. 11-Fr catheter (1990)[1653]; BA for congenital MS in nine infants and toddlers which resulted in immediate gradient relief in seven of them (1990)[1654]; BA of stenotic aortic valve (AV) in the fetus (1991)[1655], prenatal pericardiocentesis in the management of an intrapericardial teratoma. (1991)[1656]; transcatheter closure with the atrial septal occlude of secundum ASD in children and adults with 100% successful implantation, a complete closure (98%), and a low incidence (2%) of complications (2002)[1657].

[1646] Kan JS, White RI, Mitchell SE, Gardner TJ. Percutaneous balloon valvuloplasty: a new method for treating congenital pulmonary valve stenosis. N Engl J Med 1982;307:510.

[1647] Singer MI, Rowen M, Dorsey TJ. Transluminal aortic balloon angioplasty for coarctation of the aorta in the newborn Am Heart J 1982;1882;103(1)131-2.

[1648] Labadini Z, Wu JR, Walls JT. Percutaneous balloon aortic valvuloplasty: results in 23 patients. Am J Cardiol 1984;53(1):194-7.

[1649] Lock JE, Kalilullah M, Shrivastava S. Percutaneous catheter commissurotomy in rheumatic mitral stenosis. N Engl J Med 1985; 313;(24);1515-8.

[1650] Carpentier Jr RJ, Strasburger JF, Garson Jr A, et al. Fetal ventricular pacing for hydrops secondary to complete atrioventricular block. J Am Coll Cardiol 1986;8(6):1434-6.

[1651] Rashkind WJ, Mullins CE, Hellenbrand WE, et al. Nonsurgical closure of patent ductus arteriosus: clinical application of the Rashkind PDA Occluder System. Circulation 1987;75;583-92.

[1652] Lock JE, Block PC, McKay RG, et al. Transcatheter closure of ventricular septal defects. Circulation 1988 Aug;78(2):361-8.

[1653] Rome JJ, Keane JF, Perry SB, et al. Double umbrella closure of atrial septal defect: initial clinical applications. Circulation 1990;82:761-8.

[1654] Spevak PJ, Bass JL, Ben Shachar G, et al. Balloon angioplasty for congenital mitral stenosis. Am J Cardiol 1990;66(4):472-6.

[1655] Maxwell D, Allan L, Tynan MJ. Balloon dilatation of the aortic valve in the fetus: a report of two cases. Br heart J 1991;65(5):256-8.

[1656] Benatar A, Vaughan J, Nicolini U, et al. Prenatal pericardiocentesis: its role in the management of an intrapericardial teratoma. Obstet Gynecol 1992;7995):856-9:(Pt 2).

[1657] Du ZD, Hijazi ZM, Kleinman CS, et al. Comparision between transcatheter and surgical closure of secundum atrial septal defect in children and adults: results of a multicenter nonrandomized trial. J Am Col Cardiol 2002;39(11):1836-44.

Discovered in 1952 by **Maurice Goldenhar (1924 – 2001)**[1658] — Belgian American ophthalmologist and general practitioner, a rare congenital oculo-auriculo-vertebral or Goldenhar syndrome of unknown etiology, is associated with anomalous development of the first and second brachial arches during the first trimester of pregnancy. Male-to-female ratio is 3:2. It is characterized by incomplete development of the ear (without / with preauricular skin tags or hearing loss), nose, soft palate, and mandible, limbal or ocular (without / with strabismus or blindness, dermoid, and with abnormalities of the vertebral column, mainly scoliosis. Typically, patients also suffer from congenital abnormalities of internal organs, including the heart, kidneys, and lungs, characteristically on the one side of the body, or bilaterally (10%), or granuloma-cell tumors. Treatment included in utero stem cell grafting, correction of congenital anomaly, de-bulging of ocular dermoid, hearing aid and glasses.

A congenital familiar autosomal dominant disorder consisting of congenital heart disease (CHD), usually a secundum type of the atrial septal defect (ASD), in association with the upper extremity anomalies in the form of polydactyly and syndactyly, resulting in short forearms and loss of opposition to the thumb or a finger-like thumb, was described in 1960 by **Mary C. Holt (1924 – 93)** — English cardiologist and **Samuel Oram (1913 – 91)** — English cardiologist, and called the heart-limb or Holt-Oram syndrome.

In 1969 **Daniel Alagille (1925 – 2005)**[1659] — French pediatrician, specializing in the study of liver disease in children (hepatology), described arterio-hepatic dysplasia, a rare autosomal dominant inherited disorder that affects the liver causing paucity or stenosis of intrahepatic bile ducts leading to liver cirrhosis and hepatic failure; congenital heart disease (CHD), coarctation of aorta (COA) or peripheral pulmonary artery (PA) stenosis; cystic kidney disease, echogenic kidney and nephrocalcinosis; posterior ocular embryotoxon which involves a thickened and centrally displaced border ring of Schwalbe [**Gustaw Schwalbe (1844 – 1916)**. 19th century. Anatomy, histology, and embryology]; triangular face; butterfly type of vertebrae (about 50% of cases) in children (Alagille syndrome). The most effective treatment is a full liver transplantation.

Jack H. Rubinstein (1925 – 2006) — American pediatrician and **Hooshang Taybi (1919 – 2006)** — Iranian American pediatric radiologist, described congenital condition characterized by a short statue, enlarged foramen magnum of the skull, small head, downward slant of the eyes, straight or beak-like nose, arched palate,

[1658] Goldenhar M. Association malformatives de l'oreil, en particulier le syndrome dermoïde epibulbaire-appendices auriculaires-fistula congenita et ses relations avec la dysostose mandibulo-faciale. J genetique humaine (Geneve) 1952;1:243-82.

[1659] Alagille D, Odievre M, Gautier M, et al. Hepatic ductular dysplasia associated with characteristic facies, vertebral malformations, retarded physical, mental, and sexual development, and cardiac murmur. J Pediatr 1975 Jan;86(1):63-71.

anomalies of the vertebra and the sternum, large wide fingers of the wrist and toes of the foot, mental and physical retardation, keloid formation in the surgical scar, congenital heart disease (CHD) – ventricular septal defect (VSD) and the main pulmonary artery (MPA) stenosis (Rubinstein-Taybi syndrome, 1963).

In contrast to atherosclerotic coronary artery disease (CAD) which involves the epicardial named coronary arteries (diameter, 1.0-5.0 mm), non-atherosclerotic CAD affects the small unnamed coronary arteries (diameter, 0.1-1.0 mm) by vasculitis, intimal proliferation or fibrosis, abnormal accumulation of metabolic substances or extrinsic compression. The incidence of non-arteriosclerotic CAD is unknown. The clinical, angiographic, and histologic differentiation between angina pectoris or acute myocardial infarction (MI) and between atherosclerotic and nonatherosclerotic CAD is difficult.

Coronary artery vasculitis or depositions are seen in polyarteritis nodosa, systemic lupus erythematosus (SLE), granulomatous polyangiitis or Wegener's granulomatosis **[Peter McBridge (1854 – 1946)**. 19th century. Internal medicine]**, nonspecific aorto-arteritis, pulseless or Takayasu's disease **[Mikito Takayasu (1860 – 1938)**. 20th century. Ophthalmology]**, mucocutaneous lymph node syndrome or Kawasaki's disease and infection; intimal proliferation or fibrosis in ionizing radiation, homocystinuria, Friedreich's ataxia **[Nicollaus Friedreich (1825 – 88)**. 19th century. Neurology]**, primary amyloidosis (50%) and after heart transplantation (HT); accumulation of metabolic substances in mucopolysaccharidoses such as Hunters and Hurlers diseases **[Charles Hunter (1872 – 1955) – Gertud Hurler (1889 – 1965)**. 20th century. Pediatrics]**, GM - gangliosides or Jatzewitz- Dandhoff's disease and GM2 - gangliosides or Dandhoff's disease **[Horst Jatzewitz (1912 – 2002)** and **Konrad Sandoff (1939 –)**. 20th century. Biochemistry]**, primary oxalosis, alkaptonuria and sphingolipidosis or Anderson-Fabry's disease **[William Anderson (1842 – 1900)** and **Johannes Fabry (1860 – 1930)**. 19th century. Dermatology]**; extrinsic coronary artery compression in aneurysms of the sinus of Valsalva **[Antonio Valsalva (1666 – 1723)**. 18th century]**, metastases to the epicardium, aortic root abscesses, dilated main pulmonary arteries (MPA), and systolic compression of the coronary arteries by muscle bridges.

Polyarteritis nodosa is a systemic necrotizing vasculitis affecting medium and small coronary arteries in two thirds of patients between ages 30 to 60 years, in whom it may form coronary occlusion with MI or aneurysm.

SLE is a chronic multisystem connective tissue disease associated with autoantibodies and immune complexes with cardiovascular system involvement in more than 50% women between the ages of 20 and 40 years, occasionally children. It commonly causes pericarditis or myocarditis. Clinical picture of pericarditis alone or with pericardial effusions, occasionally cardiac tamponade, or constructive pericarditis, is characterized by chest pain, fever, tachycardia, decreased heart sounds, pericardial friction rub, hypotension, increased jugular venous pressure, paradoxical pulse (pulsus paradoxus)[1660] and a positive Kussmaul's sign **[Adolph Kussmaul (1822 – 1902)**: 19th century. Internal medicine].

Approximately, 18% of patients with SLE exhibits valvular and subvalvular abnormalities, predominantly of the mitral valve (MV), with noninfected vegetations, named Libman-Sacks endocarditis **[Emmanuel Libman (1872 – 1946)** and **Benjamin Sacks (1873 – 1939)**. 20th century. Internal medicine]. Of those, about 8% of patients require surgical correction of the MV.

In the newborn and pre-school children **Tomisaku Kawasaki (1925 –)**[1661] — Japanese pediatrician, observed an autoimmune disorder that include acute onset febrile sickness, sterile conjuctivitis, noso-pharyngeal erythema, non-purulent cervical lymphoid adenopathy and skin desquamation of fingers and toes. About 20% of them showed an intense systemic vasculitis of the coronary artery vasa vasorum that lead to coronary artery aneurysm, thrombosis, or stenotic scarring. MI or arrhythmia causes death 1% - 2% children. It is the most common cause of non-arteriosclerotic CAD in children, named the mucocutaneous lymph node syndrome or Kawasaki's disease (1960).

In 1972 **Philip L. Townes (1927 – 2017)** — American pediatric geneticist and **Eric R. Brocks**[1662] — American medical student, described congenital autosomal dominant syndrome consisting of multiple defects of the ears (including deafness), ano-vaginal and extremities (especially of the thumb), congenital heart disease (CHD) – atrial septal defect (ASD), ventricular septal defect (VSD) and tetralogy of Fallot (TOF) – **Etienne Fallot (1850 – 1911**: 19th century. Internal medicine), and hypoplastic kidneys and cystic ovaries (Towne-Brocks syndrome).

Multiple pterygium syndrome [19th century. Children's disease – **Oskar de Kobylinski (1856 – 1926)**] was re-described in 1963 by **Jaequeline A. Noon (1928 – 2020)**[1663] — American pediatrician and cardiologist, it should be named Kobylinski-Noon syndrome.

In the normal heart, a number of morphologic differences exist between the right and left ventricle (RV, LV), including the coarse trabeculations in the apical part of the morphologically RV, and much finer trabeculations in the apical part of

---

[1660] Paradoxic (paradoxical) pulse (pulsus paradoxus) is an abnormally, more than 10-20 mm Hg, decrease in stroke volume (SV) of the heart, systolic blood pressure (BP) and pulse wave amplitude during inspiration.

[1661] Kawasaki T. Acute febrile mucocutaneous syndrome with lymphoid involvement with specific desquamation of fingers and toes. Arerugi 1967 Mar;16(3):178-222.

[1662] Townes PL, Brocks ER. Hereditary syndrome of imperforate anus with hand, foot and ear anomalies. J Pediatr 1972;81:321-6.

[1663] Noonan JA, Ehmke DA. Associated noncardiac malformations in children with congenital heart disease. J Pediatr 1963;63:468-70.

the morphologically LV, the arrangement of the leaflets of the atrioventricular (AV) valves and their tension apparatus, their shape, the thickness of their walls, and the configuration of the RV and LV outflow tract (RVOT, LVOT), which can be altered or lacking in the congenital heart disease (CHD), when making the final diagnostic decision it is important to follow «morphological method» introduced in 1980 by **Richard** van **Praagh (1930 –)**[1664] — American pediatric pathologist and cardiologist, which states that one variable feature should not be defined as the basis of another feature that is itself variable.

The situs, i.e., location, or pattern of anatomical organization of the subarterial infundibulum (SAI) and the great arteries and the degree of development of the SAI largely determine whether the great arteries are normally, or anomaly related. Situs solitus, i.e., normal location, situs invertus, i.e., a minor image of situs solitus, or situs ambiquous, i.e., uncertain location.

Infundibulum may be normally developed, absent, atretic (occluded), severely, moderately, or mildly stenotic.

Great arteries can be normally related, or inversely normally related.

When the situs of SAI and the situs of the great arteries are discordant, i.e., different, then the great arteries are abnormally located. Equations indicating the situs of the infundibulum and the situs of the great arteries show whether the infundibuloarterial (IA) situs concondance or disconcondance is present.

Typically, discordance include transposition of the great arteries (TGA), double outflow right and left ventricle (DORV, DOLV) and anatomically corrected malposition of the great arteries.

Typically, IA concordance include tetralogy of Fallot (TOF) and truncus arteriosus (TA) with hypoplasia or atresia of the subpulmonary infundibulum. The relation between the great arteries and TA is almost normal.

The IA equations demonstrate the infundubular situs, the great arterial situs, the IA situs concordance or dis-concordance, and the degree of development of the infundibulum. The infundibular situs or the great arterial situs are the formulas for each of the abnormal types of conotruncal malformations.

In 1985-95 **Jen-Tien Wung (1940–)**[1665] — Chinese American pediatrician specializing in neonatology and perineonatology, introduced the concept for the

[1664] Van Praagh R, David I, Wright GB, Van Praagh S. Large RV plus small LV is not single RV. Circulaion 1980;61;1057-8. — Val Praagh R. What determines whether the great arteries are normally or abnormally related? Am J Cardiol 2016; Nov 1;18(9):1390-8.

[1665] Wung JT, James LS, Kilchevsky E, et al. Management of infants with severe respiratory failure and persistence of the fetal circulation without hyperventilation. Pediatrics 1985;76:488-94. —Wang JT, Sahni R, Moffit ST, et al. Congenital diaphragmatic hernia: survival treated with very delayed surgery, spontaneous respiration, and no chest tube. J Pediatr Surg 1995;30:406-9.

treatment of respiratory failure (RF), primary pulmonary hypertension (PPH) and congenital diaphragmatic hernia (CDH) in the newborn and infant by spontaneous respiration, avoiding hyperventilation and minimizing barotrauma to the lungs with a significant improvement in survival.

A rare clinical disease caused by heterozygous mutation or deletions of the 2q - q23 gene, delineated in 1998 by **David R. Mowat** — Australian pediatrician and **Meredith J. Wilson**[1666]—Australian pediatrician, is characterized by typical face, moderate-to-severe mental retardation, epilepsy and variable congenital malformations, including megacolon, genital anomalies (particularly hypospadias in males), congenital heart disease (CHD), agenesis of the corpus callous (ACC) and eye defect (Mowat-Wilson syndrome). The treatment is supportive, there is no cure at the present.

In 1991 **Harry C. Dietz (1950 –)** — American pediatric geneticist and in 2005 **Bart L. Loeys (1958 –)** — Belgian physician, described a hereditary autosomal dominant disorder caused by a new (de novo) missense (nonsynonymous) mutation in the fibrillin gene, with many features similar to the syndrome of **Antoine Marfan (1858 – 1942**: 19th century. Pediatrics), but with a peculiar ocular hypertelorism – Loeys-Dietz's syndrome (LDS).

The aim of the extra-utero intrapartum treatment (EXIT) procedure, developed in 1993-95, was initially to correct offspring who had congenital diaphragmatic hernia (CDH), and then those with airway compression, including bronchopulmonary sequestration, congenital cystic adenomatous malformation (CCAM), mouth, neck or sacrococcygeal tumors such as teratoma, and lung or pleural tumors, such as pleuropulmonary blastoma and other developmental anomalies of the fetus[1667].

The EXIT procedure is an extension of a standard classical Caesarean section (C-section) [Ancient medicine. *Chinese*], when incisions are made on the lower abdominal midline (vertical) along the linea nigra («black line») or through a horizontal suprapubic Pfannenstiel route **[Herman J. Pfannenstiel (1862 – 1909)**. 20th century. Gynecology and Obstetrics] of anesthetized mother, the preterm fetus (<37 weeks of pregnancy) or the term fetus (>37 weeks of pregnancy) is delivered through the incision in the lower anterior wall of the uterus but remains attached by its umbilical cord to the placenta, while the neonatal or pediatric surgeons are establishing an

---

[1666] Mowat DR, Croaker GDH, Cass DT, et al. Hirschprung disease, microcephaly, mental retardation, and characteristic facial features: delineation of a new syndrome and identification of a locus at chromosome 2q22-q23. J Med Genet 1998;35:617-23.

[1667] Hirose S, Farmer DL, Lee H, et al. The exutero intrapartum treatment procedure: looking back at the exit. J Ped Surg 2004 Mar;39(3):375-80. — Jancelewicz T, Harrison MR. A history of fetal surgery. Clinics in Perinatology 2009 June;36(2):227-36.

airway so the fetus can breathe, or correct other abnormalities. If the fetus is preterm, it is returned with an intact umbilical cord into the uterus, the incisions in the uterus and abdomen are closed, and the fetus is then delivered in due time. Otherwise, in a term fetus after the operation is completed, the umbilical cord is clamped and divided, and the offspring is fully delivered. The placenta is removed, the uterine and abdominal incisions are closed.

In 1989-2000 **Katherine A. High (1951 –)** — American pediatrician, successfully treated dogs with hemophilia by using the gene therapy to stimulate the factor IX, a protein involved in blood clotting. Later, she and her team duplicated the dog trial in humans with similar results, and patients were able to control their disease with lesser doses of injectable factor IX.

Neonatal stroke, like to one occurring in adult, and defined as a disturbance of the blood supply to the brain in the first 28 days of life, from an ischemic event due to a blockage of the blood vessel (80%), and hypoxia from lack of oxygen (O_2) to the brain, or both[1668].

The most common manifestation of neonatal strokes are local or generalized seizures which account for about 10% of seizures in term neonates, lethargy, hypotonia, apnea and hemiparesis. Fortunately, 60% of infants have no neurological sequelae from infarction.

Seizure-like episodes can be caused by traumatic hemorrhages during the painful uterine contractions, active phase of contractions and the birth (delivery, labor) of a newborn, anatomic abnormalities, stroke, sepsis, or metabolic diseases.

In this report on neonatal stroke, **Maria P. Hrycelak (1954, city of Chicago, IL, USA –)**[1669] — Ukrainian American pediatrician, presented a 3,010 gram, blood type O+, girl born by repeated Cesarean section at 39 week of gestation to a O+, serology negative healthy mother, without complications at the delivery. The newborn was doing well until 36 hours of age, when she was observed to have multiply twitching episodes of the left leg and arm.

The initial 16-channel electroencephalography (EEG) was abnormal with spikes noted in the right temporal lobe of the brain with concomitant seizure activity – leg twitching and lip smacking. The magnetic resonance imaging (MRI) with intravenous (IV) gadolinium contrast performed at 56 hours of age revealed a right periventricular and basal ganglion infarction.

[1668] Aden U. Neonatal stroke is not a harmless condition. Stroke 2009 Jun;40(6):1948-9.

[1669] Hrycelak MR. Vertical transmission of AIDs in pediatric patients. J Ukr Med Assoc North Am 2002 Summer; 47,3(149):5-9— Hrycelak MR. Stroke – not only in old age. J Ukr Med Assoc North Am 2007-2017 May 07;52,1(156):3-11. —Ablitsov VH. «Galaxy Ukraine». The Ukrainian diaspora: prominent persons. Publ. «Kyt», Kyiv 2007.

In the first 72 hours she was treated with IV antibiotics and Acyclovir, which were discontinued when blood cultures were negative. Also, initially she was given Lorazepam 0.1 mg/kg for seizures, and when they stopped, she was given a loading doses of Phenobarbital 20 mg/kg/24/hrs and continued 5 mg/kg/day. No further seizures occurred during hospitalization.

At the follow-up 4 weeks later, there was no seizure, and a repeated EEG was normal. On the follow-up 2 months later, she was doing well without seizures.

The primary objective in the treatment of neonatal ischemic stroke is to re-establish unobstructed blood flow to the brain.

The mainstay in the management of this disease should still be careful pharmacological treatment as in the presented case.

The use of antithrombotic agents, such as tissue plasminogen activator, may cause organ and limp impairment and bleeding. However, in cerebral venous thrombosis a heparin treatment to prevent thrombus extension and recurrence has been used. In the cases of an extremely high intracranial pressure, surgical removal of hematoma or thrombus may be indicated and curative.

Therapeutic hypothermia to the head or systemic within 6 hours of birth for 72 hours has proven beneficial in reducing death and neurological deficit at 18 months of age. But it does not completely protect the injured brain and may not improve risk of death in the most severely hypoxic / ischemic neonates and pre-term infants.

It appears that in the future, depending on the severity of neonatal stroke, the treatment should consist of pharmacological drugs, hypothermia, growth factors, neural and umbilical stem cells, or its combination.

One of the most distinguished pioneers of the EXIT procedures is **Oluyinka Olutoya (1967 –)** — Nigerian American pediatric surgeon, who performed this ground-breaking procedure on the fetus in utero that could not otherwise be operated. He was one of three Nigerian surgeons who in 2015 successfully separated conjoined twins, and in 2016 lead a team of pediatric surgeons to save a 23-week-old foetus from a life-threatening tumor, that is removing a large sacrococcygeal teratoma, growing in her tail bone, after which they returned the fetus to its mother womb to complete the full gestation period of nine months.

Presently the diagnosis of the Marfan syndrome (MFS) **[Antoine Marfan (1858 – 1941)**. 19th century. Pediatrics] is based on Ghent criteria or nosology (established 1996, revised 2010)[1670].

[1670] City and municipality of Ghent (habitable since Stone and Iron Ages, founded c. 655), Flemish Region (area 13,527 km²), Belgium (area 30,528 km²). — Loeys BL, Dietz HC, Braverman AC, et al. The revised Ghent nosology for the Marfan syndrome. J Med Genetics. London BMJ Group 2010;47;(7):476-85.

In the absence of a family history of MFS:

1. Aortic root Z Score ≥ & ectopia lenses
2. Aortic root Z Score ≥ & an fibrillin (FBN) 1 mutation
3. Aortic root Score ≥ 2 % systemic score* ≥ 7 points
4. Ectopia lenses & FBN1 mutation with known aortic pathology

In the presence of a family history of MFS (as defined above):

1. Ectopia lenses Score* > 7
2. Systemic score* > 7
3. Aortic root 2 Z-Score > 2

* Points of systemic score:
Wrist & thumb sign = 3 (wrist or thumb sign = 1)**
Pectus carinatum deformity = 2 pectus excavates or chest asymmetry = 1)
Hindfoot deformity = 2 (plain pes planus =1)
Dural ectasia = 2
Protrusion acetabula =2
Pneumothorax (an air in the pleural cavity) = 2
Reduced upper segment / lower segment ratio & increased arm / height & no severe scoliosis (in Greek (in Greek – bend) is a curve of the vertebral column (spine) in the frontal (lateral) plain =1
Scoliosis or kyphosis (in Greek – hunched, humped) is a curve of the spine in the sagittal (in Latin – «sagitta», meaning an arrow), i.e., in the sagittal plain of the spine = 1
Reduced elbow extension = 1
Facial features (3/5) = 1 – dolichocephaly (in Greek – «dolichos», meaning long), i.e., elongation of the head, enophthalmos posterior displacement of the eyeball within the orbital cavity), down slanting palpebral fissures, molar hypoplasia, retrognathia (retro-position, or retrusion of the mandible in the frontal plane)
Skin striae (stretch marks) = 1
Myopia (near-sightedness) > 3 diopters = 1
Mitral valves prolapse ¼ = 1

** The Steinberg thumb sign (described in 1945 by **Parker** and **Hare**, named in 1966 for Steinberg)[1671] is elicited by asking the patient to flex thumb as far as possible

---

[1671] Pasztor P, Huttmann A. Clinical and radiological importance of the Parker-Hare sign in Marfan's syndrome. Ann Pediatr (Paris) 1971Feb 2;18(2):162-5.

and then close the fingers over it. A positive thumb sign is where the entire distal phalanx is visible beyond the ulnar border of the hand, caused by a combination of hypermobility and elongation of the thumb. The Walker-Murdoch wrist sign[1672] is elicited by asking the patient to curl the thumb & fingers of one hand around the other wrist. A positive wrist sign is when the little finger and the thumb overlap, caused by combination of thin wrist & long fingers.

Pioneer of minimal invasive fetal surgery and therapy (MIFST), **Thomas Kohl** — German pediatric interventional cardiologist, reported on the two percutaneous (PC) ultrasound-guided ballon fetal valvuloplasty, that is a type of intrauterine surgery for severe aortic valve (AV) obstruction (2000)[1673], pioneered repair of spina bifida aperta with PC mini-invasive fetoscopy method (2010-16), co-developed fetoscopy tracheal balloon occlusion for fetal congenital diaphragmatic hernia (CDH), developed chronic intermittent maternal fetal hyperoxygenation for fetuses with hypoplastic left heart syndrome (HLHS) – Kohl procedure (2012-17).

In 2008 **John P Wright**[1674] — American criminologist, known for his work in biosocial criminology, in a prospective birth cohort found that prenatal and childhood blood level concentrations were predictive of criminal arrests in an early adulthood age.

Dermatology

John T. Bowen (1857 – 1940)[1675] — American dermatologist, described squamous-cell carcinoma in situ of the skin (an early stage or intraepidermal form), which can arise anywhere on the skin surface, and on the floor of the mouth, tongue, lips, glans, or prepuce of the penis in males or the vulva in females and may be induced by human papilloma virus (HPV), named Bowen disease. Treatment is by photodynamic therapy, cryotherapy, or local chemotherapy with 5-fluorouracil (5-FU).

[1672] Walker BA, Murdoch JL. The wrist's sign. Usual findings in the Marfan syndrome. Ann Int Med 1970 Aug;126(2):276-7).

[1673] Kohl T, Sharland G, Allan LD, et al. World experience of percutaneous ultrasound-guided balloon valvuloplasty in human fetuses with severe aortic valve obstruction. Ann J Card 2000 May 15;85(10):1230-3.

[1674] Wright JP, Dietrich KN, Ris MD, et al. Association of prenatal and childhood blood lead concentrations with criminal arrests in early adulthood. Plo S Med 2008 May 27;5(5):e 101. Doi: 10.1371/journal.pmed.0050101. — Wright JP, Dietrich KN, Ris MD, et al. Association of prenatal and childhood blood lead concentrations with criminal arrests in early adulthood. J Ukr Med Assoc North Am 2013 Dec 27;53,1(157):39-65. — Wright JP, DeLisi M. Conservative Criminology. A call to restore balance to the social science. Publ. Routledge, New York & Abingdon, OX, England 2016.

[1675] Bowen JT. Precancerous dermatoses: a study of two cases of chronic atypical proliferation. J Cutaneous Dis incl Syphilis (New York) 1912;30:241-55. — Bowen JT. Precancerous dermatoses: a sixth case of a type. Recently described. Ibid 1915;33:787-801.

After description by **Job J. Meekeren (1611 – 66**: 17th century**)** of a familial excessive elasticity of the skin and joints (cutis et articuli hyperelastica), in 1901 **Edward Ehlers (1863 – 1937)** — Danish dermatologist, and in 1908 **Henri A. Danlos (1844 – 1912)** — French physician and dermatologist, re-described this condition which may be associated with congenital heart disease (CHD) that is atrial septal defect (ASD), mitral valve (MV) prolapses, annuloaortic ectasia and aortic dissection or rupture, gastrointestinal (GI), orthopedic or ocular disorders, and also found in those patients a defect in fibronectin of the connective tissue. This disease was named Ehlers-Danlos syndrome.

Congenital dyskeratosis was described by **Ferdinand Zinsser (1865 – 1952)** — German dermatologist,**Martin F. Engman (1869 – 1953)** — American dermatologist, and **Harold N. Cole (1884 – 1968)** — American dermatologist, and named the Zinsser-Engman-Cole's syndrome.

Erich Hoffmann (1868 – 1959) — German dermatologist, with **Fritz R. Schaudinn (1871 – 1906)** —German and Lithuanian zoologist, discovered in 1905 Gram-negative bacterium Spirochaete pallida, a.k.a., Treponema pallidum, intracellular parasite, causative agent of syphilis.

Emil Meirowsky (1876 – 1960) — German American dermatologist, drew attention to the darkening of existing melanin, possibly as a result of oxidation, which occurs within seconds and ends minutes or a few hours after exposure to the long-waved ultraviolet rays and a tan, i.e., darkening of the skin color after staying under the Sun or ultraviolet rays caused by the action of melanin, existing in tissue, accelerated formation of new melanin or accumulation of melanin in the epidermis due to delay of keratinization (Meirowsky's phenomenon, 1909).

Mykhailo P. Demianovych (19.X.1879, Cossack village of Labins'ka, Kuban' Obl., now **town of Labins'k, Labins'ky Region, KK, Ukraine (1917-22) / Russia – 1957, city of Moscow, Russia)**[1676] — Ukrainian dermatologist, introduced the method of treatment of a scab of the skin with the consistent applying (rubbing) of the lesion with a 60% / 100 ml solution of natrium thiosulphate and a 6% / 100 ml solution of hydrochloric acid (HCI) – Demianovych method (1947):

Solution № 1 after Demianovych:
Rp.: Solutionis Natrii thiosulfatis 60% 100 ml
D.S.: rub into affected part of the skin.

[1676] Cossack village of Labins'ka (founded 1841), Kubans'ka Obl., since 1947 – town of Labins'k, Labins'kyy Region, Krasnodars'ky kray krai (KK: Foot Note 337). KK was artificially stricken by Russian Communistic Fascist regime with Holodomor-Genocide 1932-33 in Kuban' with the aim to decrease the population, majority at that time were ethnic Ukrainians), Ukraine (1917-20) / Russia. — Ablitsov VH. «Galaxy Ukraine». The Ukrainian diaspora: prominent persons. Publ. «Kyt», Kyiv 2007.

Solution № 2 after Demianovych:

Rp.: Solutionis Acidi hydrochlorici 6% 100 ml

D.S.: rub into affected part of the skin.

Both **Albert M. Klingman (1916 – 2010)** — American dermatologist of Jewish Ukrainian background, and **James F. Fulton (1940 – 2013)** — American dermatologist, developed the retina-A- cream to effectively treat acne and superficial wrinkles (1969-75). In addition, James Fulton recommended the early use of fat transfer for breast augmentation and surgical procedures to alleviate acne scars.

Unfortunately, in 1941-75 Albert Klingman conducted unsanctioned medical experiments on about 75 prisoners by exposure them to fungal and bacterial (staphylococci) pathogens causing s ringworm, Pityriasis or tinea versicolor herpes, athlete's foot and various other infections, effect of high doses of digoxin on human body and testing psychoactive drugs.

Felix Lewandowski (1879 – 1921) — German dermatologist, and **Wilhelm Lutz (1888 – 1958)**[1677] —German dermatologist, published in 1922 on a rare inherited genetic dermatosis, characterized by chronic infection with human papilloma virus (HPV), leading to polymorphous cutaneous lesions and high risk of developing non-melanoma skin cancer Bowen disease, named epidermodysplasia verruciform is (Lewandowski-Lutz syndrome), colloquially known as a «tree-men» or a «tree-woman» illness. The disease is usually manifested in infancy (7.5% of cases), childhood (61.5% of cases) or puberty (22% of cases) with a progressive development of hyperpigmented or hypopigmented flat wart-like papules, irregular reddish-brown plaques, seborrheic keratosis-like lesions and pityriasis versicolor-like macules on sun exposed skin of the trunk, neck, face, dorsal hands, and feet. There is no permanent cure of this disease. Prevention includes avoidance of sun exposure and photoprotection, treatment – the systemic use of the 2nd generation of retinoids (acitrecin) 0.5-1 mg / day for 6 months, topical 5-FU, interferon alpha and 5-aminolevulinic acid, cryotherapy, photodynamic therapy, and surgical excision.

Hulusi Behçet (1889 – 1948) — Turkish dermatologist, described chronic inflammation of small arteries and skin with uveitis, retinal vasculitis, and atrophy of the optic nerve, aphthous ulcerations of the serosa of the mouth, throat, and external genitalia (Behçet's syndrome, 1924-25, 1936). A variant of Behcet syndrome is a very rare clinical disorder depicted in 1959 by **John P. Hughes** and **Peter G.I. Stovin**, characterized by thrombophlebitis and multiple pulmonary artery (PA) and/ or bronchial artery aneurysms, possible causes include infections and angiodysplasia

---

[1677] Lewandowsky F, Lutz W. Ein Fall einer bisher noch nicht besschriebener Hauter-Krankurg (Epidermodysplasia verruciformis) . Archiv fur Dermatologie und Syphilis (Berlin) 1922;141:193-202.

(Hughes-Stovin syndrome, 1959)[1678]. Both syndromes can cause isolated PA (Rassmussen) aneurysm previously discovered by **Fritz V. Rasmussen (1837 – 77)** — Danish physician.

Yuriy Vasylyshyn or **Jerzy Wasyłyszyn (22.06.1934, city of Lviv, Ukraine – 07.06.2011, city of Warszawa, Poland)**[1679] — Ukrainian Polish dermatologist, proved that the **Theodor Schwann (1810 – 82**: 19th century. The anatomy, histology, and embryology) cells may transform themself into the cells of the nevus and the pigmented cells of the nevus (1969) and that congenital dyskeratoses with/without leucoplakia could undergo the malignant degeneration (1974); discovered the protective filter against burns from the ultrashort rays of the Sun (1976).

Graham R.V. Hughes (1940 –) — English dermatologist and rheumatologist, described in 1983 a common genetic autoimmune prothrombotic disease, caused by the presence of antiphospholipid antibodies in the blood, affecting both men and women from adolescence to high age, characterized by increased tendency to form thromboses in the veins and the arteries, accompanied by headaches, diplopia, memory loss, ataxia, and multiple sclerosis (MS), causing increased risk for a stroke or heart attack, in some patients after starting on estrogen-containing oral contraceptive pills or in pregnancy (antiphospholipid or Hughes syndrome).

Ophthalmology

Following the description in 1858 of association between retinitis pigmentosa and perceptive deafness by Friedrich von Gröefe [19th century. Diseases of the eyes. – **Friedrich** von **Gröefe (1828 – 70)]**, in 1914 **Charles H. Usher (1855 – 1942)** — Scottish ophthalmologist, surveyed 69 patients with this condition dividing them into the three (Usher) types according to clinical presentation: Usher type 1 – children are born profoundly deaf, begin lose sight in the first decade of life, and have difficulties in maintaining balance due to abnormalities in the vestibular function of ears; Usher type 2 – children are born hard-of-hearing and begin to lose their vision in the second decade of life; and Usher type 3 – since birth the children experience progressive loss of hearing and half of them have difficulties in maintaining balance

---

[1678] Hughes JP, Stovin PGI. Segmental pulmonary artery aneurysms with peripheral venous thrombosis. Br J Dis Chest 1959;53:19-27.

[1679] Wasyłyszyn J. Elementy nerwowe skór w znamionach i znamionach barwnikowych. Dissertation for DMS, Medical University of Warsaw, May 28, 1969. — Wasyłyszyn J, Kryst L, Langer A. Congenital dyskeratosis of Engman-Cole. Przegl Dermatol 1974;61(5):687-91. — Wasyłyszyn J. Protective item against the ultraviolet rays causing the Sun's burn. Polish Patent No. 79535; 28.06.1976. — Ablitsov VH. «Galaxy Ukraine». The Ukrainian diaspora: prominent persons. Publ. «Kyt», Kyiv 2007.

as they growth into an adulthood. Since then, the disease was named the Usher syndrome in his honor.

In 1906 **Olexandr F. Shymanovsky (27.V.1860, city of Kyiv, Ukraine – 06.V.1918, city of Kyiv, Ukraine)**[1680] — Ukrainian ophthalmologist, performed transplantation of the anterior portion of the eye.

In 1908 **Mikito Takayasu (1860 – 1938)**[1681] — Japanese ophthalmologist, discovered a peculiar «wreathlike» appearance of the blood vessels in the back of the eye (retina), a nonspecific idiopathic pulseless granulomatous aorto-arteritis of the large vessels, due to immuno-mediated destruction of the medial elastic fibers, followed by scaring of the media and internal elastic lamina, which causes compensatory intimal proliferation, fibrosis and narrowing of its branches, most commonly affecting the young or middle age women of Asian descent (female of about 8-9 times more often affected than males), termed the Takayasu disease: type I – aorto-arteritis of the ascending aorta (AA), aortic arch and the origin of the innominate artery (IA), left common carotid artery (CCA) and right subclavian artery (SCA); type II – aortitis of the abdominal aorta (AA); type III – aorto-arteritis of the AA and the origin of the IA, left CCA and right SCA; and type IV – aorto-arteritis of the main pulmonary arteries (MPA) and its branches.

Allvar Gullstrand (1862 – 1930) — Swedish ophthalmologist and optician, is noted for his study of astigmatism (1890), applied the methods of mathematics to the study of the optical images and the refraction of lights in the eye (1904 - 08)[1682], improved corrective lenses for the use after cataract removal from the eye (1908), and designed reflex-free ophthalmoscope (1911).

Between 1894-1928, **Edward K. Zirm (1863 – 1944)** — Austrian Czech ophthalmologist, carried out 7,866 eye cataract operations, exchanging in 1900 the old «coaching» of cataract by its extracapsular removal; and in 1905 performed in a 45 year-old farmer whose corneas had turned white-gray and opaque while working with slaking lime, the first successful transplantation of the cornea (keratoplasty) using cornea from 11-year-old boy with penetrated injury to both eyes and iron metal foreign bodies irretrievable lodged in his eyes which had to be extracted.

In 1904 **Eugen A.A. von Hippel (1867 – 1939)** — Prussian German ophthalmologist, described angiomatosis of retinae, and in 1926 **Arvid V. Lindau (1892 – 1958)** — Swedish pathologist, reported about hemangioblastomas of the cerebellum and

---

[1680] Shymanovsky AF. Three cases of transplantation of the anterior chamber of the eyeball. Vestnik Oftalmologii 1912. — Ablitsov VH. «Galaxy Ukraine». The Ukrainian diaspora: prominent persons. Publ. «Kyt», Kyiv 2007.

[1681] Takatayasu M. Acute with peculiar changes of the central retinal vessels. Acta Societatis Ophtalmologicae Japanicae 1908;12:554.

[1682] Gullstrand A Zur Kenntnis der Kreispunkte. Acta Mathematica 1904;29:59-100.

the spinal cord, association of which later (1936) was termed von Hippel-Lindau (VHL) disease (syndrome), or hereditary cancer syndrome. This dominantly inherited cancer resulting from a mutation of a protein encoded by the VHL tumor suppressor gene on chromosome 3p25.3. It has an incidence of one in 36,000 births. Age of clinical diagnosis varies from infancy to age of 70 years (average, 26 years). The VHL syndrome predispose to a variety of benign and malignant tumors of the brain and spinal cord, eye, liver, kidney, pancreatic and adrenal glands. The most common VHL disease related tumors are hemangioblastoma of the central nervous disease (CNS, 40%), angiomatosis of the eye (37,2%), pheochromocytoma, renal cell carcinoma, pancreatic serous cyst adenoma, endolymphatic sac tumor, bilateral papillary cystadenoma of the epididymis, or the broad ligament of the uterus. VHL tumors are caused by gene mutation, known to be angiogenic, creating blood vessels that secrete erythropoietin (EPO), as proved by **Paul Carnot (1869 – 1958**: 20th century. Internal medicine) and **Clotilde Deflandre (1871 – 1946**: 20th century. Internal medicine), a hormone produced by the kidney and known to be part of the body mechanism to react to low oxygen level in the blood (hypoxia).

Mykhailo A. Levytsky (21.11.1871, town of Horodyshche, Cherkasy Region, Cherkasy Obl., Ukraine – 07.09.1942, city of Kyiv, buried in town of Bohuslav, Kyiv Obl., **Ukraine)**[1683] — Ukrainian ophthalmologist, Director of the Department of Eye Surgery at the Kyiv Medical (KMI, 1922-41), who studied the effectiveness of surgical interventions on the eye and its appendages; composed with **D.H. Goldman** — Ukrainian ophthalmologist, the projection table for ophthalmoscopic changes in the sclera that is useful to localize a break of the sclera caused by dissection (1927); developed the technique for separation from the eye of subretinal cysticercosis or blistery tapeworm (Taenia solium); improved the reconstruction of eyebrows using free pieces of skin (1936).

Significant contribution into plastic and eye surgery was made by **Volodymyr (Vladimir) P. Filatov [15(27).II.1875, village Mykhailivka, Liambirskyy**

[1683] Town of Horodyshche (founded 1050; during conducted by Russian Communist Fascist of Holodomor-Genocide 1932-33 against the Ukrainian Nation at least 173 residents died), Cherkas'kyy Region, Cherkasy Obl., Ukraine. — Town of Bohuslav (засноване 1032; during its rule Russian Communist Fascist regime introduced repressive economic sanctions against the Ukrainian peasantry in the form of so called «black boads» (1920,1928-29 and 1931), conducted of Holodomor-Genocide 1932-33 against the Ukrainian Nation, when at least 377 residents, including 6 children have died, and at the time of political stifling (1937-38) 31 people were repressed), Obukhivs'kyy Region, Kyivs'ka Obl. Ukraina. — Levytsky MA. To the question regarding reforming the higher medical education. Kyiv 1925. — Levytsky MA, et al. Designation of distant dots on the bottom of the eye using the visual angle. Ukrainian Medical Herald 1927;1. — Levytsky MA. Regarding Injuries of the Eyes by Combatant Poisonous Substances. Kharkiv 1932. — Levytsky MA. Pathological changes on the bottom of the eye and its projection on the external surface of the sclera. Work of the Eye Clinic of the KMI. To the 40-Unniversary of the Scientific, Pedagogic and Community Activities of Professor M.L. Levytsky. Kyiv 1936. Vol. 1. — Ablitsov VH. «Galaxy Ukraine». The Ukrainian diaspora: prominent persons. Publ. «Kyt», Kyiv 2007.

Region, Mordovia, Russia – 30.X.1956, city of Odesa, Ukraine][1684] — Ukrainian ophthalmologist, eye, and plastic surgeon [20th century. Surgery – **Олександр Лімберґ (1894 – 1974)]**, who performed transplantation of the eye's cornea by the method of a total keratoplasty (1911, unsuccessful); independently of **Harold D. Gillis (1882 – 1960**: 20th century. Surgery**)** — New Zealander English otolaryngologist (ORL) and plastic surgeon, developed the pedicled tube flap grafting (Filatov-Gillis flap, 1914, published 1917); performed a partial corneal transplantation which passes through transparent cornea (1923); used for transplantation of the cornea in human, a cornea taken from death person, preserved by cooling – homo-keratoplasty (1926); introduced technigue for preservation of the cadaver cornea and its use for keratoplasty (1931); pushed forwards the idea of tissue therapy by the so-called biogenic stimuli (1933); proposed transplantation of the cornea in humans using animal corneas preserved by cooling – keratoalloplasty (1936).

On April 04, 1936, Volodymyr Filatov established the Ukrainian Institute of Eye Diseases and Tissue Therapy in Odesa, Ukraine, named after his death «the Academician O.V. Filatov» in his honor. He was good, benevolent, and religious men.

In 1916 **Jan** van der **Hoeve (1878 – 1952)** — Dutch ophthalmologist, in 1926 **Irmgard Mende** — German physician, in 1929 **Nicolaas A. Halbertsma (1889 – 1968)** — Dutch physician, in 1930 **Vincenzo Gualdi (1892 – 1976)** — Italian physician, in 1947 **David Klein (1908 – 1993)** — Swiss ophthalmologist and human geneticist, in 1951 **Petrus J. Waardenburg (1886 – 1970)**[1685] — Dutch ophthalmologist and human geneticist, and in 1981 **Krishnakumar N. Shah** — Indian physician, described a rare

---

[1684] Village of Mykhailivka, Lambirskyy Region, Mordovia (area 26,200 km²), Russia. — Grandson of **Nil F. Filatov (1847-1902)]** — Russian pediatrician, described German measles / rubella (1885) and infectious mononucleosis (1885). Later, in 1890 **A.V. Belsky**, in 1895 Nil Filatov, and in 1896 **Henry Koplik (1858 - 1927)** — American pediatrician, described little gryish-white surrounding by ruddy small wreath spots on the mucosa of cheeks across the lower 2nd molar teeth or on mucosa of lips and gums during the initial (prodromal) period of measles (Belsky-Filatov-Koplik spots), 1st - 2nd days prior to tiny spotted skin eruptions. — Filatov VP. Rounded Stalk in Ophthalmology. Medgiz, Moscow 1943. — Filatov VP. Optical Grafting and Tissue Therapy. Медгиз, Moscow 1945. — Filatov VP. My Roads in the Science. Odesa Obl. Publ., Odesa 1955. — Olearchyk Ukrainian Scientists. Filatov VP — prominent ophthalmologist. Nasze Słowo (Warszawa) 27.1.1957; 2,4(24):2. — Filatov VP. Selected Works. Vol. 1-4. RE Kavetskyy (Editor-in-Chief). Publ. Academy of Sciences of Ukraine, Kyiv 1961. — Bihunyak VV. Plastic surgery. In, LYa Kovalchuk, VF Sayenko, HV Knyshov. Clinical Surgery. Vol. 1-2. Publ. «Ukrmedknyha», Ternopil 2000. Vol. 2. P. 359-90. — Ablitsov VH. «Galaxy Ukraine». The Ukrainian diaspora: Prominent persons. Publ. «Kyt», Kyiv 2007.

[1685] Waardenburg PJ. A new syndrome combining developmental anomalies of the iris and head hair with deafness; dystopia canthi medialis et punctorum lacrimarium lateroversa, hyperplasia supercillii medialis et radicis nasi, hyperchromia iridum totalis sive partialis, albinismus circumscriptus (leucismus, poliosis) et surditas congenita (surdimutitas). Am J Hum Genet 1951 Sep;3(3):195-253. — Makarevycz BA., Makarevycz-Beiger MC. A melanocyte story: Waardenburg syndrome; Vogt-Koyanogi-Harada syndrome; vitiligo and case report. J Ukr Med Assoc North Am 2003 Spring48;1(150):5-17. // **Bohdan A. Makarevycz (1945 –)** — Ukrainian American otolaryngologist and plastic neck surgeon. / **Motryia C. Makarevycz-Beiger** — Ukrainian American optometrist. — Ablitsov VH. «Galaxy Ukraine». The Ukrainian diaspora: prominent persons. Publ. «Kyt», Kyiv 2007.

hereditary complex consisting of four types condition (Waardenburg syndrome), occurring in the prevalence of one (1) in 42,000 individuals:

Type I – Characterized by congenital sensory neural hearing loss of various degree and pigmentation deficiencies manifested by bright brilliant blue eyes, different colored eyes (complete heterochromia iridium) or multiply color in an eye (sectoral heterochromia iridium), a white forelock of hair or patches of light skin, wider gap between the inner corners of the eyes (telecanthus seu dystopia canthorum);

Type II – Ibid, but high nasal bridge (poliosis), a flat nose and a unibrow (synopyrus) – a single long eye-brown when two eye-brown meets in the middle above the bridge of the nose, smaller edges of the nostrils (alae) or a smooth philtrum («love charm»); some have a defect in the iris (coloboma), small eyes (microphthalmia), hardened bones (osteopetrosis), larger than usual head (macrocephaly), albinism (in Latin albus, meaning white) and deafness, lack of sense of smell (anosmia) due to missing the olfactory bulb of the brain;

Type III – As type I, but joint contractures of the fingers (camptodactyly), fused digits (syndactyly), winged scapula, and possible developmental delay (Klein-Waardenburg syndrome);

Type IV – As type II with aganglionic megacolon or Hirschprung disease [XIX ст. 19th century. Children disease (Pediatrics). – **Harald Hirschprung (1830 – 1916)**], named Shah-Waardenburg syndrome.

This syndrome is caused by mutation in any of several genes that affects the function of neural crest cells in the embyonic development, mostly by autosomal dominant pattern of inheritance.

The only available treatment is correction of deafness [XX ст. 20th century. Surgery. – **Paul Dzul (1921 – 2015) / Wasyl Szeremeta (1962 –)**], of Hirschprung disease by resection of aganglionic colon and rectum with the «pull through» of the descending colon and anastomosing it with the anus, resection of the entire large bowel and rectum with ileo-anal anastomosis [XX ст. Surgery. – **Orvar Swenson (1909 – 2012)**], or in desperate situations an intestinal transplantation.

In 1906 **Alfred Vogt (1879 – 1943)**[1686] — Swiss ophthalmologist, in 1914 **Yoshizo Kayanagi (1910-54)**[1687] — Japanese ophthalmologist, and in 1926 **Einosuke Harada**

[1686] Vogt A. Frühzeitiges Ergrauen der Lilien und Bemerkungen über den den sogenannten plötzlichen Eintritt dieser Veraderung. Klinische Monatsblätter für Augenheikunde, Stutgard 1906;44:228-42.
[1687] Koyanagi Y. Dysakusis Apoleci e und Poliosis bei Schwerer Uveitis nicht traumatischen Ursprungs. Klinische Monatsblätter für Augenheikunde, Stutgard 1929;82:194-211.

(1892 – 1946)[1688] — Japanese ophthalmologist, described a multisystem bilateral inflammation of presumed autoimmune system that involve melanin pigmented tissue of the uvea [the middle vascular pigmented layer of eye consisting of the iris (irydocyclitis), ciliary body and the chorioid], meninges and the brain (uveo-meningo-encephalitis), associated with relapsing meningo-encephalitis, deafness, apolecia (baldness), depigmentosis of the skin and eyes, symmetrical vitiligo and poliosis circumscripta (whitening of the eyes at the ends of the hairs forming of white forelock), named Vogt-Kayanagi-Harada disease[1689]. It is epidemic in Far East, occurs usually in adult age, manifested by malaise, fever, nausea, redness of the eyes and blurring vision, tinnitus, vertigo, and impairment of hearing (hypoacusis) with the tendency for partial recovery.

A rare autoimmune disease of the connective tissue was described by **Henrik S.C. Sjögren (1899 – 1986)**[1690] — Swedish ophthalmologist, in which foreign cells are attacking and invading in the primary form – the tears and mucous glands, in the secondary form – internal organs, such as the heart, lung and liver, and also lymph nodes. It may coexist with systemic lupus erythematosus (SLE), scleroderma, rheumatoid arthritis and various mixed diseases of the connective tissue (Sjögren disease or syndrome). Nine of ten patients with Sjögren syndrome are women, and the average age of onset is after menopause in women. The overall prevalence is estimated to be 0,2% an 0,05% of the total population. There is no known cure nor a specific treatment.

In 1935 **Alexander T. Smakula (09.09.1900, village of Dobrovody, Zbaraz'ky Region, Ternopils'ka Obl., Ukraine – 17.05.1983, town of Auburn, MA, USA)**[1691] — Ukrainian physicist, invented anti-reflective lens coating based on optical interference. He also discovered the effect of brightening of the optics, essence of which consist in the substantial improvement of optical properties of lenses by placing on them a layer of magnesium fluoride (MgF_2) in thickness of a 0.25 length of light wave (Smakula effect).

[1688] Harada E. Beitrag zur Klinischen Kenntnis von nichteitriger Chorioiditis (Chorioiditis diffusa acuta). Acta Societatis Ophtalmica Japanica 1926;30:356-61.

[1689] Makarevycz BA., Makarevycz-Beiger MC. A melanocyte story: Waardenberg syndrome; Vogt-Koyanogi-Harada syndrome; vitiligo and case report. J Ukr Med Assoc North Am 2003 Spring48;1(150):5-17.

[1690] Sjögren HSC. Zur Kenntnis der Keratoconjunctivitis sicca bei hypofunktion der Tränendrüsen. Acta Ophtalmologica, Copenhagen 1933;Suppl II:1-141.

[1691] Village of Dobrovody (founded 1463), Zbaraz'ky Region, Ternopils'ka Obl., Ukraine. — Town of Auburn (settled 1714), MA, USA. — Smakula O. Monocrystals: bringing up, preparation und the use. 1962. — Holovach Yu, Honchar Yu, Krasnytska M, et al. Physics and Physicists in NTSh in Lvovi. Zhurnal Fizychnykh Doslidzhen' 2018;22(4):4003-31. / https://doi.org/10.30970/jps.22.4003sm — Holovach Yu, Honchar Yu, Krasnytska M, et al. Physics and Physicists in NTSh in Lviv. In O. Petryk, A. Trokhimchuk (Eds). Leopolis Scientifica. Lviv 2019. — Ablitsov VH. «Galaxy Ukraine». The Ukrainian diaspora: prominent persons. Publ. «Kyt», Kyiv 2007.

In 1946 **David H. Bushmych / Bushmich (March 27, 1902, city of Kropyvnytsky, Ukraine – Oct. 24, 1995, New York, N.Y., USA)**[1692] — Ukrainian ophthalmologist of Jewish origin, invented the technique of layered cornea transplantation technique and worked out for it specialized tools.

In the 1930's **D. Ramon Castroviejo (1904 – 87)** — Spanish American eye surgeon, improved corneal transplantation technique and in the 1940th designed the needle holder (Castroviejo) for the use in eye and cardiovascular microsurgery.

Sir **N. Harold L. Ridley (1906 – 2001)**[1693] — English ophthalmologist, was the first to implant in 1949 a polymethylmethacrylene intraocular lens to correct aphakia.

The combination of congenital oculomotor apraxia, i.e., an absence or anomality of horizontal motion of the eye, caused probably by an injury to the brain, rheumatic autoimmune interstitial keratitis with tingling in the ear and deafness was described in 1945 by **David G. Cogan (1908 – 93)** — American ophthalmologist, named the Cogan's syndrome, an atypical form which is associated with granulomatous aortitis and arteritis (Logan syndrome, 1963).

Daughter of **Olexandr Puchkivsky (1881-1937**: 20th century. – Surgery**)**, **Nadiya O. Puchkivsky**, a.k.a., **Nadezhda A. Puchkovskaya (12.05.1908, city of Smolensk, Russia – 15.05.2001, city of Odesa, Ukraine)**[1694] — Ukrainian physician ophthalmologist-surgeon, physician-lieutenant military-medical service and Chief Eye Division of the 4th Ukrainian Front (UF, 1943-45); in 1956-85 directed the Academician V.P. Filatov Ukrainian Institute of Eye Diseases and Tissue Therapy **[Volodymyr Filatov (1875 – 1956)**. 20th century – Surgery], city of Odesa, Ukraine; designed and combine surgical treatment and immunotherapy for the sequelae of severe burns of eyes, performed subtotal replacement of the cornea for cataract (1949); grafted cornea in complicated cataracts; developed two stage surgery for extensive adhesions of eyelids and cornea enabling to recover the sight of «hopelessly» blind patients; performed the layered peripheral keratoplasty, i.e., peripheral transplantation of the cornea (1954); introduced biological covering of an injured cornea (Puchkivsky method) and kerato-prosthesis; took active part in creating first domestic lasers, developing

---

[1692] City of Kropyvnytsky (Foot Note 899), Ukraine. — Bushmych DH. Layered grafting of the corneal stratum with a tectonic aim. Oftalmol Zhu 1947:34-8. — Ablitsov VH. «Galaxy Ukraine». The Ukrainian diaspora: prominent persons. Publ. «Kyt», Kyiv 2007.

[1693] Ridley NHL. Intraocular acrylic lenses. Trans Ophthalmol Soc UK & Oxford Ophthalmol Congress. 1951:71:617-21.

[1694] City of Smolensk on the Dnipro River (founded 863), Russia. — Puchkivsky NO. Grafting of the cornea in complicated catharacts. 1949 i 1957. — Puchkivsky NO. Epoch and My Life. Publ. «Zdorrov'ia», Kyiv 2004. — Ablitsov VH. «Galaxy Ukraine». The Ukrainian diaspora: prominent persons. Publ. «Kyt», Kyiv 2007. — Iakymenko S. Fourty-five years of keretoprosthesis study and application at the Filatov Institute: a retrospective analysis of 1,060 cases. Int J Ophthalmol 2013;6(2):375-80.

techniques for laser operations, and introducing ultrasound (US) examination into clinical ophthalmology.

Tadeusz I. Krwawicz (Jan. 15, 1910, city of Lviv, Ukraine – Aug. 17, 1988, city of Lublin, Poland)[1695] — Ukrainian Polish ophthalmologist, who experimentally that cells of the proper layer (substantia propria) of the cornea belong to the functional mesenchymal system, on that basis of which established the mechanism of its injuries, and introduced the method of cleansing its ulcers using proteolytic enzymes (1948- 49); constructed the first cryo-sound with the use of which introduced the cryotherapy (temperature, -70°C) into ophthalmology. He was the first to describe the method of intracapsular cataract extraction by cryo-adhesion, develop probe by which cataract may be grasped and extracted (Krwawicz method, 1959-61), applied cryotherapy for the treatment of virus-related diseases of the eye, glaucoma, tumors of the eye and retinal dissections.

The syndrome of **Thomas P. Kearns (1922 – 2011)** — American ophthalmologist and **George P. Sayre (1911 - 91)** — American ophthalmologist, is a mitochondrial deoxyribonucleic acid (DNA) deletion with ragged red fibers, featuring the triad of chronic progressive ophthalmoplegia, cardiomyopathy with disorder of the conductive system of the heart (CSH), i.e., the 2nd - 3rd degree atrioventricular (AV) block and retinitis pigmentosa, with onset prior to the 20th year of life (1958).

Robert Hollenhorst (1913 – 2008) — American ophthalmologist, described emboli of cholesterol crystals in the retinal blood vessels (Hollenhorst plaques, 1961), which indicate atheromatosis (AM) and increased risk of cerebral stroke, myocardial infarction (MI), an aortic aneurysm, occlusion of retinal vessels and the other peripheral arterial complications.

Special attention devoted **William Selezinka (Sep. 30, 1923, Ukraine – Jan. 13.01,2010, city of San Diego, CA, USA)**[1696] — Ukrainian American ophthalmologic surgeon, to ocular injuries which are responsible for 50% of eye emergencies and are a common cause or possible source of visual lost.

[1695] City of Lublin (settled since 6th-7th century, first mentioned in the historical document 1224), Poland. — Grzybowski A, Kanclerz P. Tadeusz Krwawicz, MD: the inventor of cryosurgery in ophthalmology. Eur J Ophthalmology May 2019;29(3):348-56. — Ablitsov VH. «Galaxy Ukraine». The Ukrainian diaspora: prominent persons. Publ. «Kyt», Kyiv 2007.

[1696] City of San Diego (established 1769), CA, USA) — Selezinka W. Ocular injuries in athletic activities. J Ukr Med Assoc North Am 1982 Spring; 29,2(104):59-71. — Selezinka W, Sandall GS, Henderson JW. Rectus muscle union in sixth nerve paralysis – Jensen rectus muscle union. Arch Ophthalmol 1974;92(5):382-6. — Wainstock MA, Selezinka W. Localization of radiopaque intraocular foreign bodies with silicone lens: description and case histories. J Ophthalmology and Strabismus 1975 May;12(2):124-30. — Rüssmann W. Augenärtzliche Operationen. Chirurgie der änsseren Augenmuskeln. Springer-Verlag, Berlin - Heildelberg 1988. Seite 399-489. — Selezinka W. The «red eye». Ukr Med Assoc North Am 1995 Autumn;42,3(137):147-52. —Selezinka W. Ocular toxicology. J Ukr Med Assoc North Am 1996 summer;43,3(140):157-70. — Ablitsov VH. «Galaxy Ukraine». The Ukrainian diaspora: prominent persons. Publ. «Kyt», Kyiv 2007.

Sviatoslav N. Fedorov (08.VIII.1927, city of Khmelnytskyy, Ukraine – died during crash of helicopter **02.VI.2000, city of Moscow, Russia)**[1697] — Ukrainian Russian ophthalmologist, advanced correction for aphakia with artificial intraocular lenses (1967), developed methods for the surgical correction of near-sightedness (myopia) and astigmatism by radical keratotomy and introduced kerato-prosthesis and keratoplasty (1981-82).

In 1963 **Leonid A. Lynnyk**, a.k.a., **Linnik (02.X.1927, city of Kam'yans'ke, Dnipropetrovs'k Obl., Ukraine – 23.IV.2012, city of Odesa, Ukraine)**[1698] — Ukrainian ophtalmologist, designed photocoagulator of the Special Construction Bureau (SCB-1)[1699], made together with the Construction Bureau (CB) the «Precise Machine »[1700], creating simulating effect of low power laser radiation on the retina of the human eye that was put in the basis of surgical treatment of dystrophic diseases of the eye. Using this device, he successfully performed the first laser coagulation («welding») of the retina at the Academician O.P. Filatov Ukrainian Institute of Eye Diseases and Tissue Therapy in Odesa, Ukraine, thus ushering the era of «laser in ophthalmology».

Later, Leonid Lynnyk with **Petro P. Chechyn** — Ukrainian ophthalmologist, developed laser ophthalmoscope CM-2000 – medical device (ТУ У14220751.007-2000), based on the powerful infrared semiconductor GaAlAs-laser[1701], which is used at the Academician O.P. Filatov Ukrainian Institute of Eye Diseases and Tissue Therapy, to perform minimally invasive iridotomy / iridectomy for the treatment of some various types of glaucoma, and laser discision surgery, i.e., non-contact incision into the eye tissue, for the treatment of secondary membranous cataracts, and for the treatment of peripheral degeneration or dissection of the retina, macular ruptures and initial or proliferative laser discision retinopathy, retino-vasculitis, neuro-retinopathy and retinitis of different etiology.

Laser microsurgery in ophthalmology was introduced by **Mikhail M. Krasnov (1929 – 2006)** — Russian ophthalmologist, for glaucoma (1972), cataract (1975) and other eye disorders (1974-77). He also constructed ultrasonic (US) instruments for eye surgery (1974).

[1697] City of Khmelnyts'kyy (settlements since Bronze Age and Scythe times, earliest documented mention 1431; during conduction by Russian Communist Fascists regime of Holodomor-Genocide 1932-33 against the Ukrainian nation died at least 786 residents), Ukraine. — Ablitsov VH. «Galaxy Ukraine». The Ukrainian diaspora: prominent persons. Publ. «Kyt», Kyiv 2007.

[1698] City of Kam'yans'ke (Foot Note 1477), Dnipropetrovs'k Obl., Ukraine. — Ablitsov VH. «Galaxy Ukraine». The Ukrainian diaspora: prominent persons. Publ. «Kyt», Kyiv 2007.

[1699] OKB 1 — Osoblyve konstruktos'ke biuro, meaning the Special Construction Bureau (SCB) 1.

[1700] CB «Tochmash» — Construction Bureau «Tochna mashyna» («Precision machine»).

[1701] Manufacturer – Nizhyn Laboratories of Scanning Devices, LLC (founded 1999), town of Nizhyn (Foot Note 457), Chernihiv Obl., Ukraine.

Cataract phaco-emulcification was proposed in 1967 by **Charles D. Kelman (1930 – 2004)** — American ophthalmologist. He also proposed cryo-extraction of cataracts, the use of freezing for the repair of retinal detachments and designed numerous ophthalmic instruments and intraocular lenses.

In 1998 **Geritt R.J. Melles** — Netherlander eye surgeon, who performed posterior lamellar keratoplasty[1702], where only a selected portion of the cornea was transplanted, and in 2006 he developed another technique that he named after **Jean Descemet (1732 – 1810**: 18th century**)** – Descemet membrane (DM)[1703] endothelial keratoplasty (DMEK)[1704]. The DMEK is a partial-thickness cornea transplant where the host DM and endothelium are replaced with donor DM and endothelium.

Significant contribution into the gene-therapy of eye diseases was made by **Samuel G. Jacobson (1944 -)**, **Jean B. Bennett (1954 –)** and her husband **Albert M. Maguire (1960 –)** — American ophthalmologists. They were among the first investigators to use viral factors to deliver transgenes to specific cells and the first to demonstrated proof-of-principle of the ocular gene therapy, developed number of strategies of the gene therapy for mediated treatment of retinal diseases.

A study conducted by him with other collaborators on the gene transfer led to reversal of blindness in a canine model, specifically in Leber's [19th century. Ophthalmology – **Theodor** von **Leber (1840 – 1917)**] congenital amaurosis[1705] (LCA), a rare congenital blinding disease affecting infants in which a mutated gene prevents the retina from making a nutrient vital to eye health. At last, Jean Bennett and Albert Maguire achieved the first successful demonstration of gene therapy in humans, i.e., they restored much of the vision in patients suffering from LCA[1706].

Since 2004 p. **Nataliya V. Pasyechnikova (15.II.1950, u.t.v. Novomykolaïvka, Novomykolaïvka Region, Zaporizhia Obl., Ukraine –)**[1707] — Ukrainian

---

[1702] Melles GR, Eggink FA, Lander R, et al. A surgical technique for posterior lamellar keratoplasty. Cornea 1998;17(6):618-26.

[1703] **Jean Descemet (1732 - 1810**: 18th century**)** — French physician.

[1704] Melles GR, Ong TS, Ververs B, et al. Descemet membrane endothelial keratoplasty (DMEK). Cornea 2006;25(8):987-90.

[1705] Amaurosis (in Greek – «amaurosis» meaning darkening or dark), a weakness or loss of vision.

[1706] Maquire AM, Simonelli R, Pierce EA, et al. Safety and efficacy of gene transfer for Leber's congenital amaurosis. N Engl J Med 2008;358:2240-8.

[1707] U.t.v. Novomykolaïvka (until 1812 – village Kocherzhka ; founded 1790; during organized by Russian Communist Fascist regime of Holodomor-Genocide 1932-33 against the Ukrainian Nation, died at least 96 inhabitants), Novomykolaïvka Region, Zaporizhia Obl., Ukraine. — Pasyechnikova NV. Theoretical and Clinical Studies about the Effectiveness of the Laser Technology in the Treatment of Pathology of the Bottom Eye. Dissertation for the degree of Dr Med Sc., P.V. Shupyk Kyiv Medical Academy of Postgraduate Education (KMAPE), Kyiv 2003. — Ablitsov VH. «Galaxy Ukraine». The Ukrainian diaspora: prominent persons. Publ. «Kyt», Kyiv 2007.

ophthalmologist, Director of the Academician V.P. Filatov Ukrainian Institute of Eye Diseases and Tissue Therapy, city of Odesa, Ukraine, founded a new direction for the use of laser radiation, i.e., selective influence upon the chorio-retinal complex, made significant contribution to high-frequency electrical welding of biological tissues, especially of vitreoretinal structures (2012).

Minimally invasive strabismus surgery (MISS), invented by **Daniel S. Majon (1963 –)**[1708] — Swiss ophthalmologist and ophthalmic surgeon, is an operative method of correcting squinting (squinted-eyes) with very small (2 mm and 3 mm) incisions.

Neurology

The main contribution of **Volodymyr (Vladimir) M. Bekhteriev (1857** – poisoned with arsenic by the Russian Communist Fascists of the USSR regime and NKVS **1927)** — Russian neurologist, is the description of ankylosis spondylitis (Bekhteriev disease, 1891), indicating the memorizing function of the hypocampus (1900) and the theory of restrain, according to which the neural energy of the brain is directed through the pathways to active fields that are located in the vicinity of the fields with a depressed activity (1903).

Hryhoriy I. Rossolimo (05(11).XII.1860, city of Odesa, Ukraine – 29.IX.1928, city of Moscow, Russia)[1709] — Ukrainian neurologist of Greek origin, introduced a reflex of plantar flexion of toes on taping its plantar surface in lesions of the pyramidal tract of the brain and the spinal cord (Rossolimo reflex, 1902).

Scientific and clinical interest of **Mykhailo M. Lapinsky (06.VI.1862, village Smolyhivka, Ripkynsky Region, Chernihiv Obl., Ukraine – 14.VI.1947, city of**

[1708] Majon DS. Comparison of a new minimally invasive strabismus surgery technique with the usual limbal approach for rectus muscle recession and plication. Br J Ophthalmol 2007;91:76-82. — Majon D, Fine H (Eds). Minimally Invasive Ophthalmologic Surgery. Springer, Berlin 2010.

[1709] Rossolimo HI. Pathological spinal reflexes. Zhurnal Neuropathology and Psychiatry (Moscow), 1902; 2:239. — Rossolimo GI: Der Zehenreflex (ein speziell pathologischer Sehnenreflex). Neurologisches Centralblatt, Leipzig, 1908; 27:452. — Ablitsov VH. «Galaxy Ukraine». The Ukrainian diaspora: prominent persons. Publ. «Kyt», Kyiv 2007.

Buenos Aires, Argentina)[1710] — Ukrainian neurologist, graduate (1891), Doctor of Medical Sciences (DMS, 1897) and Head of Department of Neurology and Psychiatry (1903-18) at St Volodymyr University, city of Kyiv, Ukraine, were in the field of experimental as well as clinical neurology. He described some forms of nervous endings in the external layer (adventitia) of the extremities arteries, particularly one of the first proved the role of infringement «external vessels of vessels» (in Latin – vasa vasorum externa) arose the nerve pathology; proved that injuries to the peripheral nerves may stipulate the primary pathological changes in vessels at the area of innervation which he explained by the lack of transmission of stimuli through nerve fibers, a secondary vasomotor disturbances, as in neuritis and neuralgias (1897); proved that ischemic paralysis the outcome of nerve ischemia (1900); one of the first described within the external and muscle layers arteries of the posterior extremities of dogs the network of myelin and not-myelin fibers (1903); made detailed description of nervous sympathetic endings in the carotid artery and first described perivascular innervation of the brain vessels with the average diameter 15-20 mm (1900-02); showed the possibility to evoke in frogs epileptic convulsions by stimulating the cortex of the brain; initiated research for dynamic localization functions in the spinal cord (in Latin – medulla spinalis), which explain of the sudden appearance spinal-cerebral (spinal, vertebral) automaticity.

Experimentally exploring the localization of motor functions in the spinal cord, Mykhailo Lapinsky proposed a new concept of spinal motor centers, according to which the nuclei of anterior horns do not form an anatomical unit, but a functional association of synergy (1901–1902). This classic work laid the foundation for the study of localization of functions in the spinal cord. He also investigated reflex functions after spinal cord injury. The question about the changes of tendon reflexes below the

---

[1710] Village of Smolyhivka (founded 1826), Ripkynsky Region, Chernihiv Obl., Ukraine. — City of Buenos Aires (established 1636), Argentina. — Lapinsky MN. To the question about the width of capillary. Press of the St. Volodymyr University of V.I. Zavadsky, Kyiv 1895. — Lapinsky MN. To the question about structure of capillaries of cerebral cortex. Questions of neuro-psychiatric medicine Kyiv 1896;2. — Lapinsky MN. Zur Frage über den Zustand der kleinen Capillaren der Gehirnrinde bei Arteriosclerose der grossen Gefässe. Neurologisches Centralblatt 1896;15:921-925. — Lapinsky MN. About diseases of vessels by damage of primary nerve trunks. Thesis of MD. Press of SV Kulzenko, Kyiv 1897. — Lapinsky MN. Diseases of nervous system in diabetes mellitus. Neurologicheskii vestnik 1901;(1):41-113. — Lapinsky MN. About the innervations of brain vessels. Korsakoff's J Neuropathol Psychiatry 1903;3. — Lapinsky MN. Nervöse Symptome auf Grund von Gallenleiden und ihre Behandlung. München med Wschr 1925;72:1560. — Lapinsky MN. Ueber Meralgie. Deutsche Zeitschrift für Nervenheilkunde 1926;94:293-311. 1926. — Lapinsky MN. Ueber zentripetale Verbindungen der Leber mit dem Rückenmark. Deutsche Zeitschrift für Nervenheilkunde 1927;97:104-111. — Vinnychuk SM, Dudenko YeH, Macheter YeL, et al. Neurological Diseases. Publ. «Zdorov'ia», Kyiv 2001. — Ablitsov VH. «Galaxy Ukraine». The Ukrainian diaspora: prominent persons. Publ. «Kyt», Kyiv 2007. — Vinnychuk SM. Mykhailo Mykytovych Lapinsky (1862-1947). J Neurology 2011 Jun 28;258(12):2300-1.

level of spinal cord injury was controversial. He experimentally proved that the reflex function is related to the mode of injury and the chronology factor: transection of the cervical spinal cord caused exaggeration of reflexes, gross trauma of the lower spinal cord resulted in diminution of reflex responses, but reflexes returned if the lower spinal cord was carefully dissected.

In 1913-14 Mykhailo Lapinsky described the pain syndrome accompanied diseases of of small pelvis organs with irradiation of the pain below («small pelvis syndrome», or Lapinsky syndrome) which include neuralgia of the sciatic (ischiatic) nerve (in Latin – nervus ischiadicus) with irritation of the small pelvis organs, roots of the tibial nerve in the intervertebral foramen (in Latin – foramen intervertebrale) or in the boundary of the small pelvis associated with diseases of the rectum, inflammation of the prostate gland (prostatitis) and roots of the tibial and peroneal nerves (1914). Pathology of the irradiated pain from illnesses of the small pelvis, he explained by alterations in the blood supply to corresponding segments of the spinal cord causing irritation of the lower extremity nerves and the development of pain syndrome.

Beside of this, Mykhailo Lapinsky described the vegetative vascular point on the internal surface of the thigh; researched visceral-reflective changes of the nervous system; proved the existence of double innervation of muscles (somatic and vegetative), dysfunction of which may cause the development of muscular dystrophic syndromes.

Concerning the treatment with water (hydrotherapy) Mykhailo Lapinsky designed in 1913-15 bathtubs for under-water and above the water massage, proposed the bathtubs for the whirlwind, rotational and current (running) water massages.

Three different neurological syndromes were named after **J. Ramsay Hunt (1872 – 1937)**[1711] — American neurologist. They are: Ramsay Hunt syndrome type 1, or Ramsay Hunt cerebellar syndrome, is a rare form of cerebellar degeneration which involved myoclonic epilepsy, progressive ataxia, tremor and dementia; Ramsay Hunt syndrome type 2, called Hesper zoster optics, is a reactivation of herpes zoster in the geniculate ganglion, caused by a lower motor neuron lesion of the cranial 7th (facial) nerve, deafness, vertigo, pain, and also a triad of ipsilateral facial nerve paralysis, ear pain and vesicles on the face and ear or on the ear; Ramsay Hunt syndrome type 3, or Hunt's disease, or Artisan's palsy, is an occupationally induced neuropathy of the deep palmar branch of the ulnar nerve.

[1711] Hunt JR. On the herpetic inflammation of the geniculate ganglion: a new syndrome and its complications. J Nerv Ment Dis 1907;34(2):73-96.

Oleksiy V. Favorsky (1873 – 1930)[1712] — Russian neurologist, developed the method of impregnating by silver nitrate of the piece of tissue to show the nerve fibers and its endings (Favorsky method, 1930).

Олексій E. Yanyshevsky or **Aleksiy E. Janishewsky (12.IV.1873, city of Kazan, Tatarstan, Russia – 09.X.1936, city of Sophia, Bulgaria)**[1713] — Ukrainian neurologist, and **Vladimir Bekhteriev** (above) described an involuntary snatching and holding of an object conveying touching irritation at the base of fingers at the wrist as a sign of a lesion in the premotor area of the cortex in the cerebral hemisphere (Yanyshevsky-Bekhteriev snatching reflex). Beside this, Aleksiy Janishewsky described a reflex of oral automatism that is a tonic squeezing of the jaws by touching with the spatula of the lips or the hard palate indicating paralysis agitans (Yanyshevsky's reflex).

Joseph Brudzinski (1874 – 1917) — Polish neurologist, observed that in meningitis, subarachnoid hemorrhage (SAH) and encephalitis the passive flexion of the lower limb on one side causes a similar movement in the opposite limb (Brudzinski reflex, 1908) and the passive flexion of the neck results in flexion of the hip, knee, and ankle (Brudzinski reflex, 1909).

Vladimir Bekhteriev (above) and **Kurt Mendel (1874 – 1946)** — German neurologist, described a cub digital reflex where a percussion of the dorsum of the foot causes a dorsal extension (normal) or plantar flexion (abnormal) of the 2nd to 5th toes (Bekhterev-Mendel reflex).

Witold Chodzko (1875 – 1954)[1714] — Polish neurologist and psychiatrist, described a multiple reflex phenomenon in which a blow with the reflex hammer is followed by a contraction of various muscles in the arm, unilateral, or bilateral.

Constantin von **Economo (1875 – 1931)** — Austrian neurologist born in Greece, discovered epidemic lethargic encephalitis (1917) and developed his architectonics of the cerebral cortex by dividing the brain hemisphere into seven lobes and 109 areas (1925).

[1712] Ablitsov VH. «Galaxy Ukraine». The Ukrainian diaspora: prominent persons. Publ. «Kyt», Kyiv 2007.

[1713] City of Kazan (founded 1005), Tatarstan (Foot Note 1142), Russia. — City of Sophia (founded 7th millennium BC), Bolgaria (area 110,993 km²). — Janischevsky A. Un cas de maladie de Parkinson avec syndrome pseudo-bullaire et pseudo-ophtalmoplégique; quelques considérations sur la pathogénie de cette maladie. Revue Neurologique 1909;17(13):823-831. — Janischevsky A. Le réflexe de préhension dans les affections organiques de l'encéphale. Revue Neurologique 1914;22:678-681. — Janischevsky A. Symptomatology of Diseases of the Nervous System. Odesa University, Odesa 1918. — Janischewsky A. Das Greifen als Symptom von Grosshirnläsionen. Dtsch Z Nervenheik 1928;102:177. — Janischewsky A. Das dystonische Phänomen der erhobenen Arme bei der Kleinhirnläsionen. Lijecn Vjesn 1928;50:1221-4. — Janischevsky A. Rhombencephalitis as a manifestation of a particular neurotropic virus. Rev neur psychiat 1932;29:129. — Janischewsky A. Das Greifen als Symptom von Grosshirnläsionen. Dtsch Z Nervenheik 1928;102:177. — Ablitsov VH. «Galaxy Ukraine». The Ukrainian diaspora: prominent persons. Publ. «Kyt», Kyiv 2007.

[1714] Chodzko W. Sur un mouveau symptome des lesions organiques du systeme nerveux central. Signe sternal. Revue neurologique, Paris 1936;65:131.

Joseph Gerstmann (July 17, 1877, city of Lviv, Ukraine – May 23, 1969, city of New York, N.Y., USA)[1715] — Ukrainian American neurologist of Jewish descent, described cognitive impairment resulting from damage, e.g., by stroke or trauma, to a specific area of the brain, i.e., the left parietal lobe in the region of the angular gyrus of the dominant hemisphere, characterized by writing difficulties or disability (dysgraphia or agraphia), lack of understanding of rules for calculation or arithmetic (dyscalculia or acalculia), inability to distinguish right from left, and inability to identify fingers (finger agnosia), named Gerstmann syndrome (1924). He with **Ernst Sträussler (1872 – 1959)** — Czech Austrian neuropathologist and judicial psychologist, and **Ilya M. Scheinker (1902 – 54)** — Russian Latvian Austrian American neurologist of Jewish descent, depicted a very rare autosomal dominant transmissible infectious familiar spongiform encephalopathy in adults, caused by prions, a class of proteins resistant to proteases, characterized by widespread deposition of amyloid plaques. Its symptoms include difficulty in speaking (dysarthia), unsteadiness (ataxia), spasmatic muscle contractions (myoclonus), involuntary fast rhythmic movements of the eyeball (nystagmus), dementia, visual disturbances, blindness, or deafness (Gerstmann-**Sträussler-Scheinker syndrome**, 1936). There is no treatment for both those syndromes. In the first syndrome symptoms diminish over the time, in the second syndrome death occur within 1-6 years.

Samuel A.K. Wilson (1878 – 1937) — American neurologist, described in 1912 a rare inherited disorder that causes excess of copper to accumulate in the body, primarily in the brain, liver, kidneys, and the cornea of the eyes, known as hepatocellular degeneration or Wilson disease.

A tingling at the distal digit during percussion of the area of nerve division was noticed by **Jules Tinel (1879 – 1952)** — French neurologist, related to a partial injury or the beginning of regeneration of that nerve (Tinel sign, 1915).

Hryhoriy I. Markelov [26.01(07.02).1880, city of Perm, Russia – 08.04.1952, city of Odesa, Ukraine][1716] — Ukrainian neurologist, who authored the concept that the light irritants influence the function of the human organs via hypothalamus and the vegetative nervous system.

Following description in 1889 by **Eugene Bamberger (1858 – 1921**: 19th century. Neurology**)**, and detailed description in 1890 by **Pierre Marie (1853 – 1940**: 19th century. Neurology**)**, hypertrophic pulmonary, bone and joints osteoarthropathy or Bamberger-Marie disease, **Borys M. Mankivsky** or **Boris N. Mankovskii** or **Boris N. Mankowsky (11.III.1883, town of Kozelets, Chernihiv Obl., Ukraine – 24.XI.1962,**

[1715] Ablitsov VH. «Galaxy Ukraine». The Ukrainian diaspora: prominent persons. Publ. «Kyt», Kyiv 2007.

[1716] Markeloff G. Paralysis agitans. Neurology 1909: 1802. — Markelov HI (Ed.). Vegetative Nervous System. Vol. 1-4. Kyiv 1938-40. — Markelov HI. Diseases of the Vegetative System. Kyiv 1948. — Ablitsov VH. «Galaxy Ukraine». The Ukrainian diaspora: prominent persons. Publ. «Kyt», Kyiv 2007.

city of Kyiv, Ukraine)[1717] — Ukrainian neurologist, identified its familiar form, known under the name of familiar dysplastic osteopathy (Mankovsky syndrome). In 1928 Borys Mankivsky described diseases of diencephalon[1718] in association with paralysis agitans. He made the first clinical description of craniopharyngioma, a rare slow growing suprasellar[1719] tumor originating from the embryonic tissue or epithelium derived from the Rathke's pouch or cleft [19th century. Anatomy, histology and embryology. – **Martin Rathke (1793 – 1860)]**, which is precursor of the anterior pituitary gland of the brain. Craniopharyngioma occurs most commonly in children with the onset at 5 to 12 years of life and in the adult with the onset at 50 to 74 years of life. As the tumor compresses the optic chiasm it may presents with bilateral inferior quadrantanopia leading to bilateral hemianopsia.

Independently, described in 1950 by **William G. Lennox (1884 – 1960)**[1720] — American neurologist and epileptologist, a pioneer in the use of electroencephalography (EEG) for the diagnosis and treatment of epilepsy, and in 1961 and 1966 by **Henri J.P. Gastaut (1915 – 95)**[1721] — French neurologist and epileptologist, a congenital, rare, complex and severe form of epilepsy that starts in children aged 3-5 years and persists into an adulthood, is characterized by recurrent seizures and impaired mental ability, right cerebral hemiatrophy, and the EEG showing a particular pattern of brain activity called slow spike-and-walve, was named Lennox- Gastaut syndrome (LGS).

In addition to independently describing the LGS, Henri Gastant in 1957 described the hemi-convulsion- hemiplegia-epilepsy syndrome, in 1981 and 1982 the late variant of the begin childhood epilepsy with occipital paroxysms.

The LGS is one of the most severe age-dependent and resistant form of infantile epilepsy, with respect to the clinical course and treatment. The main

---

[1717] Town of Kozelets (Foot Note 956), Chernihiv Obl., Ukraine. — Mankivsky BM. Clinics of Epidemic Encephalitis. Kyiv 1924. — Mankovsky BH. Case of Organic Serous Meningitis with the Clinical Picture of Ponto-Cerebellar Tumor. Kyiv 1925. — Mankovsky BM. Multiply Disseminated Sclerosis. Kyiv 1941. — Mankivsky BM. About Nervous Pathology in Hypertensive Disease. Kyiv 1960. — Mankovsky BM. Selected Works. Publ. Acad Sci Ukraine, Kyiv 1972. — Ablitsov VH. «Galaxy Ukraine». The Ukrainian diaspora: prominent persons. Publ. «Kyt», Kyiv 2007.

[1718] Diencephalon is the central, lower part of the brain that contains the basal ganglia, thalamus, hypothalamus, pituitary gland and third ventricle.

[1719] Sella turcica (in Latin-Turkish seat) is a saddle-shaped depression in the body of the sphenoid bone of the skull.

[1720] Lennox WG, Davis JP. Clinical correlates of the fast and the slow spike-wave electroencephalogram. Pediatrics 1950;5:624-44). — Datsenko I. Lennox-Gastaut syndrome: congenital malformation of the central nervous system – right cerebral hemiatrophy. J Ukr Med Assoc North Am 2007-2012 May 7;52,1(156):12-8.

[1721] Gastaut HJP, Meyer JS. Eds. Anoxia and the Electroencephalogram. CC Thomas, Springfiel, IL 1961. — Datsenko I. Lennox-Gastaut syndrome: congenital malformation of the central nervous system – right cerebral hemiatrophy. J Ukr Med Assoc North Am 2007-2012 May 7;52,1(156):12-8.

clinical symptoms of the LGS are tonic spasms, tonic generalized attacks, atypical absentia epileptic, tonic, and myoclonic attacks. The high frequency of trauma observed in these children has been traced to falling, which increase the severity of epileptic encephalopathy. Mental retardation is a consequence of this organic brain disorder and constant epileptic activity. Treatment with adrenocorticotropic hormone (ACTH) in promising for those cases. The prognosis is poor, the mortality in childhood is 5% and persistent seizures in adulthood ranges between 80% and 90%.

In the middle of an infectious genus spongiform encephalopathy was identified by **Hans G. Creutzfeldt (1885 – 1964)**[1722] — German neurologist and neuropathologist, and **Alfons M. Jacob (1884 – 1931)**[1723] — German neurologist, in 1920 and in 1921, respectively, a disease caused by prion, a protein infectious agent, human equivalent of bovine tied to significant number of the gene mutation by the prion protein **[Daniel Gajducek (1923 – 2008)**. 19th century. Microbiology and immunology], which sporadically exist in families affected by autosomal dominant and infectious forms inherited from the parents in 85%, and of about 7.5% spontaneously, a fatal degenerative disease of the brain, usually affect people at the beginning of 60-years of age року, named the Creutzfeldt – Jacob disease (CJD). It is characterized by an early onset of memory loss, behavioral changes, poor coordination, dysfunction of vision, later by dementia, unintentional motions, blindness, weakness, and loss of consciousness. The diagnosis of the CJD is usually confirmed by the electroencephalography (EEG) which may show the generalized spiked pattern of waves from the brain cortex, by evaluation of the cerebrospinal fluid (CSF) secured by a spinal canal tap which may reveal an elevation of the protein level 14-3-3 and a more sensitive (80-100%) the real-time trembling forced transformation (RT-TFT) and by magnetic resonance imaging (MRI) of the brain which demonstrate bilateral highly enforced signals from the brain's caudal nucleus and putamen at the T2 valued images. There is no specific treatment for the CJD, opioids could be used to alleviate pain, clonazepam or natrium valproate for minimizing inattentional movements.

[1722] Creutzfeldt HG. Über eine eigenartige herdförmige Erkrankung des Zentralnervensystems. Vorläufige Mitteilung. Zeintschrift für die gesampte Neurologie und Psychiatrie 1920;57:1-18. — Belay ED, Holman RC, Schonberger LB. Creutzfeldt - Jacob Disease: surveillance and diagnosis. Clin Infec Dis 2005;41(6):834-5). — Belay ED, Holman RC, Schonberger LB. Creutzfeldt - Jacob Disease: surveillance and diagnosis. J Ukr Med North Am 2005 Winter;50,2(154):3-6.

[1723] Jacob A. Über eigenartige Erkrankungen des Zentralnervensystems mit bemerkensweten anatomischen Befunde (spastische Pseudosklerose – Encephalomyelopathie mit disseminierten Degenarationsherden). Vorläufige Mitteilung. Deutsche Zeinschrift für Nervenheilkunde 1921;70:132-46. — Belay ED, Holman RC, Schonberger LB. Creutzfeldt - Jacob Disease: surveillance and diagnosis. Clin Infec Dis 2005;41(6):834-5). — Belay ED, Holman RC, Schonberger LB. Creutzfeldt - Jacob Disease: surveillance and diagnosis. J Ukr Med North Am 2005 winter;50,2(154):3-6. — Ablitsov VH. «Galaxy Ukraine». The Ukrainian diaspora: prominent persons. Publ. «Kyt», Kyiv 2007. Village of Belzhets, Belzets or Belz; in Polish

Frederic H. Lewey or **Friedrich H. Lewy (Jan. 28, 1885, Berlin, Germany – Oct. 05, 1950, Haverford, PA, USA)** — Jewish German American neurologist, discovered degenerative changes, due to the intracytoplasmic aggregates (inclusions) of abnormal brown proteins, colored by alpha-synuclein inside the neuron of the substantia nigra in the brainstem, in the locus caeruleus, a nucleus in the pons, of the brain-steam, in the cranial X (vagal) nerve and other nerves which result in the loss of dopamine (dying or death) and transmission of nervous signals (Lewy bodies, 1912), indicate the possibility of achalasia and paralysis agitans [Ancient medicine. *Egypt* –**Ebers papyrus** since 1600 BC. 19th century. Neurology – **James Parkinson (1755 – 1824)**]. The presence of Lewy bodies in the cerebral cortex indicating the possibility of dementia with Lewy bodies (DLB).

Lucia Frey (03.XI.1889, city of Lviv, Ukraine – killed by Gestapo **1942, Ibid,** or in **Belz German Nazi extermination camp, since 1945** –Tomaszów powiat, Lublin Voyevodeship, Poland)[1724] — Ukrainian Polish neurologist of Jewish origin, described the auriculo-temporal (Frey) syndrome (1923).

F.A. Bohorad[1725] — Ukrainian neurologist of Jewish ancestry, noted spontaneous tearing in parallel with normal salivation of chewing and eating on the side of injury of the cranial VII (facial) nerve (syndrome of crocodile tears or Bohorad syndrome, 1926).

Some 0,1% of the population suffers from cluster headache, which **Byaard T. Horton (1895 – 1980)** — American neurologist, related to giant-cell arteritis, a chronic nodular inflammation of the aorta (15%) and medium-sized arteries, connected with rheumatic polymyalgia (Horton disease or syndrome, 1939).

Twelve years before **Harvey Cushing (1969 – 1939**: 20th century – Surgery**), Mykola M. Itsenko / Itzenko (Dec. 17/29, 1888, village of Sofiïvka, Kryvorih Region, Dnipropetrovsk Obl., Ukraine – Jan. 14, 1954, city of Voronezh,**

[1724] Village of Belzhets, Belzets or Belz; in Polish – Bełżec (created at the beginning of 16th century), Tomaszów powiat, Lublin Voyevodeship, Poland. — The Belz German Nazi extermination camp (operating between March 17, 1942, and June 1943) — Frey L. La syndrome du nerf auriculo-temporal. Revue Neurologique 1923;2(2):97-104. — Burton MJ, Brochwicz-Lewinski. Lucja Frey and the auriculotemporal nerve syndrome. J R Soc Med 1991;84(10):619-20. — Dunbar EM, Singer TW, Singer K, et al. Understanding gustatory sweating. What have we learned from Lucja Frey and her predecessors? Clin Auton Res 2002;12:179-84. — Ablitsov VH. «Galaxy Ukraine». The Ukrainian diaspora: prominent persons. Publ. «Kyt», Kyiv 2007.

[1725] Bohorad AF. The syndrome of «crocodile tears». Likars'ka sprava / Vrachebnoye delo (Kyiv, Ukraine) 1928;11:1328-30. — Bogorad FA, (translated by Sackersen A). The symptom of crocodile tears. The Journal of the History of Medicine and Alied Sciences 1979 Jan 01;34(1):74-9. — Ablitsov VH. «Galaxy Ukraine». The Ukrainian diaspora: prominent persons. Publ. «Kyt», Kyiv 2007.

Russia)[1726] — Ukrainian neurologist, reported in 1924 two patients who had an adenoma of the anterior portion (adenophysis) of the pituitary gland with excessive adrenocorticotropic hormone (ACTH) secretion leading to production of large amounts of cortisol and other glucocorticoids by the cortex the adrenal glands. Later, in 1946, he wrote a monography in which he described 15 women and two men suffering diseases of the anterior pituitary gland and the hypothalamus (in Greek hypo – «under», thalamus – «room, camera, cut off, thalamus») causing hypercortisolism / hypercorticism.

Subsequently, eight years later (1932) Harvey Cushing wrote on a basophilic adenoma of the anterior portion of the pituitary gland causing a condition which he termed «pituitary basophiles» causing hypercortisolism. Out of 12 patients, 67% died within a few years from the onset of disease.

Thus in 1924 Mykola Itsenko, and in 1932 Harvey Cushing described a secondary hypercorticism due primary disease of the anterior portion of the pituitary gland, caused by hyperplasia, adenoma or carcinoma, most commonly by pituitary basophilic adenoma (80-90%), with excessive secretion of ACTH and secondary excessive secretion by the adrenal gland cortex of cortisol and other glucocorticoids, and a tertiary hypercorticism due to an excess production of hypothalamus corticotropin releasing hormone (CRH) that also stimulate the synthesis of cortisol and other glucocorticoids by the adrenal gland cortex, both named Itsenko-Cushing disease.

On the other side, a primary hypercorticism resulting from hyperplasia or neoplasm of the adrenal gland cortex with an excessive production of cortisol and other glucocorticoids, is named Itsenko-Cushing syndrome.

Although etiology of Itsenko-Cushing disease and Itsenko-Cushing syndrome are various, though the pathogenesis and the clinical picture is similar.

[1726] Village of Sofiivka (founded 1793), Kryvorih Region, Dnipropetrovsk Obl., Ukraine. — City of Voronezh (originated as a settlement of the Ukraine-Rus in about 12th century, created 1585, Russia). — Itsenko NM. Tumor of the hypophysis with polyglandural symptomatic complex, in connection with the review of the issue of the central innervation of vegetative functions. Yugo-Vostochnyy vestnik zdravookhraneniya 1924;3-4:136. — Itzenko N. Pluriglanduläres Syndrom mit pathologisch-anatomischen Bild. Zeinschrift für die Gesamte Neurologie und Psychiatrie (P.f.d.Neur. u.Psych.) Berlin Dez 1926;103:63-72. — Itzenko HM. About Clinics and Pathogenesis of Cerebral Syndromes in Connection with Knowledge of the Intermediate-hypophyseal System. Voronezh 1946. — Gazhenko AI, Gurkalova IP, Zukow W, Kwasnik Z, Mroczkowska M. Pathology. Medical Student's Library. Radom Medical University in Radom and Odesa State University in Odesa. Radom 2009. P. 271-82. — Ablitsov VH. «Galaxy Ukraine». The Ukrainian diaspora: prominent persons. Publ. «Kyt», Kyiv 2007.

The scientific interest of **Anatol Dowżenko (Jul. 02, 1905,** town of **Mineralni Vody, SK, Russia – Apr. 20, 1976,** city of **Warszawa, Poland)**[1727] — Ukrainian Polish neurologist, concentrated on the cerebrospinal fluid (CSF), syphilis of the nervous system, extrapyramidal system diseases, particularly multiple sclerosis (MS) and other demyelinating processes, and in epilepsy. Specifically, he found an increased level of proteins (normal 20-40 mg% of total proteins) by electrophoresis of the CSF in certain cases of MS.

Seymour S. Kety [(1915 – 2000). 20th century. Radiology– **David Kuhl (1929 – 2017)]**[1728] introduced the chelating agent citrate to relieve the children of their lead poisoning caused by chewing on their cribs, coated in paint containing lead (1942)[1729]. Citrate would help flush the lead out of the children's body through urine. It was the first chelating agent used to help treat heavy metal intoxication. With **Carl Schmidt (1893 – 1988:** 20th century. Pharmacology)[1730], who published on his bubble flow meter technique for the quantitative determination of cerebral blood flow (CBF; 1943), Seymour Kety invented the nitrous oxide (N_2O) method for the quantitative determination of the CBF, thus making it possible to physically measure of the function of the human brain (Kety-Schmidt method, 1948)[1731]. Finally, Seymour Kety produced the most definitive evidence for the essential involvement of genetic factors in schizophrenia (1975).

Juvenile myoclonic epilepsy (JME), originally described by **Theodore Herpin (1799 – 1865:** 19th century. Neurology) was re-described 100 years later (1957)

[1727] Town of Mineralni Vody (founded 1878), Stavropols'kyy krai (SK: Foot Note 334), Ukraine (1918 - 22) / Russia. — Dowżenko A. Cranial oedema as a factor irritating brain centers for blood regulation and its impact on the morphological status of bone marrow and the peripheral blood. Doctoral thesis. University of Warsaw 1949. — Dowżenko A. Choroby układu nerwowego. PWN, Poznań 1951. — Dowżenko A, Wender M, Patelski J. Beitrag zu ungewöhnlich grossen Liquorveränderungen by der multiple sclerose. Schweizer Arch f. Neurol Psych 1958;81:144-51. — Dowżenko A, Jakimowicz W, Bromowicz J, Zebrowski S. Choroby układu nerwowego: podręcznik dla studentów i lekarzy. Wyd. 3. PZWL, Warszawa 1959. — Herman EJ. Some recollections about prof. Anatoli Dowżenko. / Scientific works and publications by Prof. A. Dowżenko. Neurol Neurochir Pol 1977 Jan-Feb;11(1):1-2 / 122-8. — Wald I. In memoriam: Prof. Anatol Dowżenko (1905-1976). Neurol Neurochir Pol 1977 Jan-Feb;11(1):3-5. — Zielinski JJ.Prof. A. Dowżenko's studies and activities in the field of epilepsy. Neurol Neurochir Pol 1977 Jan-Feb;11(1):11-4. — Kulczycki J, Barańska-Gieruszczak M. Professor Anatol Dowżenko – clinical neurologist, scientist, and educator. Neurologia i Neurochirurgia Polska. 2011 Nov 11;44(6):611-3,608-10. — Ablitsov VH. «Galaxy Ukraine». The Ukrainian diaspora: prominent persons. Publ. «Kyt», Kyiv 2007.

[1728] Ablitsov VH. «Galaxy Ukraine». The Ukrainian diaspora: prominent persons. Publ. «Kyt», Kyiv 2007.

[1729] Kety SS. The lead citrate complex ion and its role in the physiology and therapy of lead poisoning. J Biol Chem 1942:181-92. — Sokoloff L. Seymour S. Kety (August 25,1915 – May 25,2000). Biographical Memoirs. Nat Acad Sci. Nat Acad Press, Washington, DC 2003;83:60-79.

[1730] Koelle GB. Carl Frederic Schmidt (July 29, 1893 – April 4, 1988). Biographical Memoirs. Nat Acad Sci. Nat Acad Press, Washington, DC 1995;68:273-88.

[1731] Kety S, Schmidt CR. The nitrous oxide method for the quantitative determination of blood flow in the man: theory, procedure, and normal values. J Clin Invest 1948;27:476-83.

by **Wofgang D. Janz (1920 – 2010)**[1732] — German neurologist, and is, a.k.a., Janz syndrome.

JME is inherited genetic disease (17-49%), a fairly common form of idiopathic generalized epilepsy, representing 50-10% of all epilepsy cases, occurring more often in females than males. It arised secondary to defective genes causing improper function of ionic channels, gamma-aminobuttery acid (GABA) receptors and receptors of combined G-proteins. Disorder typically first manifest itself between the ages of 10 and 18 with brief episodes of involuntary muscle twitching occurring early in the morning and shortly before falling asleep. Most patients also have generalized seizures that affect the entire brain and many also have absence of seizures. Diagnosis is based on patient history. Electroencephalography (EEG) shows pattern with often generalized 4-6 Hz frequent spikes and slow wave discharge, often provoke by phobic stimulation with blinking lights and sometimes by hyperventilation. Both magnetic resonance imaging (MRI) and computed tomography (CT) scans are normal.

Treatment of epilepsy has begun with bromides in the middle 1880s, then with phenobarbital in 1912, with phenytoin since 1935, and at the present time the most effective medication is valproic acid (Deparote), first synthesized in 1881-82 and used in medical practice intravenously (IV) and per orum (p.o.) since 1962.

The basis for functional imaging (scanning) of the brain was laid by **Louis Sokoloff [(1921 – 2015).** 20th century. Radiology– **David Kuhl (1929 – 2017)]**[1733], who in 1976 with **Martin Reivich (1933 –**: 20th century. Radiology) developed technique for measuring metabolic rates through the brain using the positron emission tomography (PET). Then, Louis Sokoloff with the help of the glucose analog – 2-deoxy-D-glucose (2-DG) created of a safe and effective radioactive «tracer molecule» in brain scanning (1977). This led to the application of the PET labeled by the 18-fluorodeoxyglucose (^{18}FDG) imaging (FDG-PET) for meaningful investigation of the normal and abnormal function of the human brain. The FDG-PET scanning method ultimately proven to be valuable and widespread in detection of cancerous tissue and their metastases in the human body. Many neoplastic cells metabolize glucose only to lactate, and thus utilize more glucose than normal cells to support cellular function. Thus, FDG could identify tumor cells. FDG-PET scans are routinely used in clinical medicine for detection and staging of tumors and monitoring its response to treatment.

Arvid Carlsson (1923 –) — Swedish neuropharmacologist, who demonstrated that dopamine [19th century. Neurology – **James Parkinson (1755 – 1824)]** was a neurotransmitter in the brain (1957). Subsequently, he administered L-DOPA to

---

[1732] Watts G. Wolfgang Dietrich Janz. The Lancet Apr 1-7, 2017;389 (10076):1292.

[1733] Squire LR (Ed). The History of Neuroscience in Autobiography. Louis Sokoloff. Society for Neuroscience, Washington, DC 1996. Vol. 1. P. 454-97.

animals with drug-induced by reserpine (isolated in 1952 from Rauwolfia serpentina or Indian snakeroot) Parkinson's disease (PD) symptoms which caused reduction in the intensity of the animal symptoms.

In 1960-61 **Oleh Hornykiewicz** or **Oleh Hornykevych (Nov 17, 1926, village of Sykhiv,** from 1942 incorporated into **city of Lviv, Ukraine – May 26, 2020, city of Vienna, Austria**: 20th century. Pharmacology)[1734] — Ukrainian biochemist, pharmacologist, and neuroscientist, examined results of autopsies of patients who had died with Parkinson's disease (PD) [Ancient medicine. *Egypt –* **Ebers papyrus** since **1600 BC**) and 19th century. Neurology – **James Parkinson (1755 – 1824)**. 20th century. Neurology – **Frederick Lewy (1885 – 1950)]**, suggesting that the disease is associated with, or caused by a reduction in the level of dopamine in the basal ganglia of the brain. Since dopamine itself did not enter the brain, he tried treating 20 patients with racemic mixture of dihydroxyphenylalanine (D-DOPA) which could enter the brain and be converted there to dopamine by the action of D-DOPA decarboxylase, with positive results.

Therefore, he discovered that PD was due to dopamine deficiency in the brain, and played a key role in the development of L-DOPA as a therapy for PD.

The L-DOPA was re-introduced for the treatment of of PD by **George C. Cotzias (1918 – 77)**[1735] — Greek American scientist, by starting with small doses given orally every two hours and gradually increasing the dose to stabilize patients on large doses to cause remission.

Presently, the prevention and palliative treatment of PD consist of a regular physical exercise, physical therapy (PT) and taking the drug L-DOPA (Levodopa), currently Carbidopa / Levodopa (Sinemet) 25 mg / 100 mg per orum bis in die (b.i.d.) / twice a day as an initial dose. Presently, there is no cure for PD.

Progressive proximal spinal and bulbar muscular atrophy is a debilitating neural degenerative genetic disease, first diagnosed in 1966 by **William R. Kennedy (1927 –)**[1736] —American neurologist, that directly affects neurons, a sex-linked recessive trait, an adult-onset symptoms usually appearing between ages of 30 and 50, more

[1734] Village of Sukhiv (first mentioned in 1409), from 1942 incorporated into the city of Lviv, Ukraine. — Ehringer H, Hornykiewicz O. Distribution of noradrenaline and dopamine (3-hydroxytyramine) in the human brain and their behavior in disases of the extrapyramidal system. Klin Wschr 1960;38:1236-9. — Birkmayer W, Hornykiewicz O. The L-2,4-dioxyphenylalanine (DOPA)-effect in Parkinson-akinesia. Wien Klin Wschr 1961;73:787-8. — **Walther Birkmayer (1910 – 2016)** — Austrian neurologist. — Hornykiewicz O. Biochemical-Pharmacological basis for clinical application of L-Dopa in Parkinson's Syndrome. J Ukr Med Assoc North Am 1975 Apr;21,2(77):32-7. — Oleh Hornikiewicz. In, L.R. Squire (Ed.). The History of Neuroscience in Autobiography. Vol. 4. Elsevier Acad Press, Amsterdam – Tokyo 2004. P. 240-81. — Ablitsov VH. «Galaxy Ukraine». The Ukrainian diaspora: prominent persons. Publ. «Kyt», Kyiv 2007.
[1735] Cotzias G. L-Dopa for Parkinsonism NEJM 1968;278(11):630.
[1736] Kennedy WR, Alter M, Sung JH. Progressive proximal spinal and bulbar muscular dystrophy of late onset. Neurology 1968 July;18(7):671-80.

noticeable in men, because of having only one chromosome. It is less noticeable in women having two chromosomes, and can carry disease on one of them, while the other one help to conceal the symptoms from being apparent. Women may suffer from a milder version of the disorder and be carriers. Slowly progressive symptoms initially are manifested by shaking hands or muscle cramps, the extreme weakness of the extremities, of muscles responsible for speech and swallowing with slurring of words, dysphagia, aspiration of saliva and food to the tracheobronchial airways and recurrent pneumonia. The physical and speech therapy are indicated.

Scientific and clinical activity of **Ivan S. Zozula (12.08.1939, village of Volos'ke, Derazhenians'ky Region, Khmelynytsky Obl., Ukraine –)**[1737] — Ukrainian neurologist, is concerning with the management of an emergency medical conditions.

John O'Keefe (1939 –) — American British neuroscientist, who discovered place cells[1738] by analyzing the environmental factors influencing the firing properties of individual hippocampal nerves (1971-76) and proposing the functional importance of the hippocampus, as a cognitive map for special memory function (1978), discovered of theta phase procession when he found evidence of a distinctive variation of temporal coding of information by the timing of action potential (AP) in place cells, relative to electroencephalography (EEG) cycle, known as theta rhythm as opposed to space timing within a single cells, and demonstrated that place cells spike at different place relative to theta rhythm oscillations in the local field potential of the hippocampus (1993); predicted and discovered boundary vector cells by showing shift in the position and size of a place firing fields when the barriers defining the environment were shifted (1996); presented a model of the above phenomenon predicting the existence of boundary vector cells that would respond at a specific distance from barriers in the environment (2000), and this prediction was supported by demonstrating border cells with the predicted properties in the subiculum (in Latin – «support») which is the most inferior component of the hippocampus formation (2009), and the medial entorhinal cortex of the brain[1739] where they are called border cells.

[1737] Village of Volos'ke (founded 1431), Derazhenians'ky Region, Khmelynytsky Obl., Ukraine. — Zozula IS (Ed.). Explanatory Dictionary of Medical Emergency Conditions. VOV «Medycyna», Kyiv 2009. — Zozula IS (Ed.). Medycine of Emergency Conditions. VOV «Medycyna», Kyiv. — Ablitsov VH. «Galaxy Ukraine». The Ukrainian diaspora: prominent persons. Publ. «Kyt», Kyiv 2007.

[1738] A place cells is a type of pyramidal neurons within the hippocampus (in Greek mythology – «hippocampos» meaning «horse», from words «hippos» - «кінь» і «campos» - «sea monster») in the temporal lobe of the brain, that becomes active when an animal enters t particular place in its environments, known as place field.

[1739] The entorhinal cortex of the brain (ento = interior, rhino = nose, entorhinal = interior to the rhinal cortex) is an area of the brain in the median temporal lobe which function as a centre in a widespread network for memory and navigation.

In 1991 **Kenneth H. Fischback (1950 –)** — American neurologist, discovered the genes for fragile X-syndrome – the most common form of mental retardation, Charcot-Tooth-Marie's disorder [19th century. Neurology – **Jean Charcot (1825 – 93), Pierre Marie (1853 – 1940)** and **Howard Tooth (1856 – 1925)],** Huntington's disease [19th century. Neurology – **George Huntington (1956 – 1916)]** and Kennedy's disease **[William Kennedy (1927 –):** above], and recently introduced oligonucleotide treatment of Huntington's disease[1740].

With their mentor, John O'Keefe, **Edvard I. Moser (1962 –)** — Norwegian psychologist and neuroscientist and **May-Britt Moser (1963 –)** — Norwegian psychologist and neuroscientist, pioneered research on the brain's mechanism for representing space, and on their own they discovered types of cells crucial for determining position, that is space representation, close to the hippocampus, that is important for encoding space, and also for episodic memory. Thus, they allowed physicians to gain new knowledge into the cognitive processes and special deficits in neurological conditions such as Alzheimer's disease **[Alois Alzheimer (1864 – 1915). 20th** century. Psychiatry and psychology]. Alzheimer's disease is characterised the presence of the arteriosclerotic plaques in the vessels, neurofibrillary tangles – insoluble aggregates of hyperphosphorylated tau proteins and the Lewy bodies in the brain.

Psychiatry and psychology

In 1902 **Oleksiy Meinong,** also **Alexius Meinung Ritter** von **Handschuchsheim (17.VII.1853, city of Lviv, Ukraine – 27.XI.1920, city of Graz, Austria)**[1741]— Ukrainian born philosopher and psychologist of Austrian descent, founded the theory of objects that is the intentionality or the direction of the attention to objects as a basic feature of intelligent states, drew a difference between the two elements in every experience of the objective world: «content» that differentiates one object from another and «action» through which the experience is related to their own object.

Psychoanalysis was conceived by **Sigmund Freud (1856 – 1939)** — Czech Austrian neurologist and psychoanalyst of Jewish-Ukrainian origin, with the publication of «The Interpretation of Dreams» (1900). He formulated the concepts of unconscious infantile sexuality, repression, sublimation, and the formation of the «id», «ego» and

---

[1740] Fischback KH, Weiler NS. Oligonucleide treatment of Huntington's disease. N Engl J Med 2019;380(24):2373-4.

[1741] City of Graz (founded 1128), Austria. — Ablitsov VH. «Galaxy Ukraine». The Ukrainian diaspora: prominent persons. Publ. «Kyt», Kyiv 2007.

«super-ego», and their influence on human behavior. Sigmund Freud argued that all psychiatric and psychological processes could be explained in biological terms. In 1908 **Eugene Bleuler (1857 – 1935)**[1742] — Swiss psychiatrist proposed the name dementia praecox a schizophrenia, the main symptom of which he considered to be ambivalence (in Latin – «ambo» means «both», and «valentia» meaning «force»), that is the ambiguity in relation to the experience expressed by the fact that one object summons in a patient simultaneously two opposite feelings.

Researching schizophrenia, he proposed in 1911 the term autism / autismus (in Greek – «autos» meaning «self») which he defined as reserve itself, immersion in the world of the own experiencing, and dug up from reality, the spectrum of psychiatric diseases caused by disturbances in the development of the brain, manifested by the marked and all-sided deficiency in reciprocity with the surrounding world and social relations, restricted interests, and repetitive activities.

Julius Wagner-Jauregg (1957 – 1940) — Austrian physician, who in 1917 introduced malaria infection (inoculation) by the less aggressive parasite Plasmodium vivax in the treatment of the tertiary syphilis – gummous (15%), cardiovascular (10%) and neurological (6,5%), mainly dementia progressive paralytica, also called general paresis or «paralysis of insane». This form of pyrotherapy induced prolonged and high fever. Later malaria was treated with quinine. The treatment was successful in many patients, but because of high mortality (15%) abandoned by about 1940.

After **Emil Krepelin (1856 – 1926**: 19th century. Psychiatry), in 1907 **Alois Alzheimer (1864 – 1915)** — German pathologist and psychiatrist, defined pre-senile dementia as progressively degenerative disease of the brain of unknown etiology, characterized by diffuse atrophy throughout the cerebral cortex (Alzheimer disease).

The concept of psychobiology was developed by **Adolf Meyer (1866 – 1950)** — Swiss American psychiatrist, who saw mental illness as the interaction of developmental, social, and psychological forces (1931).

Olexandr I. Yushchenko (02.12.1869, farmstead of Vodotecha near town of Hlukhiv, Hlukhiv Region, Sumy Obl., Ukraine – 13.06.1936, city of Kharkiv, Ukraine)[1743] — Ukrainian psychiatrist, brought in psychiatry the treatment by work (Psychiatric Hospital in town of Kamianets-Podilsky, Khmelnytsky Obl., Ukraine, 1897), and biochemical screening of patients (1912).

The psychology of the individual (Adlerian) was founded by **Alfred Adler (1870 – 1937)** — Austrian American psychiatrist and psychotherapist of Jewish parents, who

---

[1742] Bleuler E. Dementia praecox oder Gruppe der Schizophrenien Handbuch. Leipzig 1911.

[1743] Farmstead of Vodotecha near town of Hlukhiv (founded 992; during organized by Russian Communist Fascist regime of Holodomor-Genocide 1932-33 against the Ukrainian Nations perished at least 768 inhabitants), Hlukhiv Region, Sumy Obl., Ukraine. — Yushchenko AI. Essence of Psychiatric Diseases and their Biological-Chemical Testing. St Petersburg 1912. — Ablitsov VH. «Galaxy Ukraine». The Ukrainian diaspora: prominent persons. Publ. «Kyt», Kyiv 2007.

developed theories concerning the motivation of human behavior. The school of analytic psychology was founded by **Carl G. Jung (1875 – 1961)** — Swiss psychiatrist and psychoanalyst of German descent, who postulated a collective unconscious of mankind.

Electroconvulsive (electroshock) therapy (ECT) or shock therapy was first applied in 1938 by **Ugo Cerletti (1877 – 1963)** — Italian neurologist, as a psychiatric treatment in which seizures are electrically induced in the patients to provide relief from mental disorders, such as major drug-resistant unipolar or bipolar depression, mania or manic syndrome (state of abnormally elevated arousal, affects and energy level, or state of heightened overall activation with enhanced affective expression together with lability of affect) and catatonia **[Karl Kahlbaum (1828 – 99)**. 19th century. Psychiatry]. Typically, ECT is administered 2-3 times per week during general anesthesia until the patient is no longer suffering from symptoms. In patients with major drug-resistant depression ECT is effective in about 50% of cases with relapse in half within 12-months.

Victor P. Protopopov (22.X.1880, village of Yurky, Kremenchuk Region, Poltavska Obl., Ukraine – 29.XI.1957, м. Київ, Україна)[1744] — Ukrainian psychiatrist, described a syndrome or triad (Protopopov) which consists of tachycardia, dilated pupils and constipation in patients with bipolar disorders (1920); introduced pathophysiology into psychiatry, biochemical testing, sleep treatment, detoxication and diet therapy.

Valentin M. Hakkebush [01(13).01.1876, town of Mohyliv-Podilsky, or **1881, town of Nemyriv, Vinnytsia Obl., Ukraine – 17.10.1931, city of Kyiv, Ukraine]**[1745] — Ukrainian psychiatrist, **Tykhon A. Heyer (1875 – 1955)** — Russian psychiatrist,

---

[1744] Protopopov VP. About the Conjunction of Motion Reaction on Sound Irritation. Doctoral Dissertation. St Petersburg 1909. Protopopov VP. Somatic syndrome, perceived during maniacal-depressive psychosis. Nauchnaya medicina 1920;7:721-49. — Protopopov VP. Methods of Reflexological Examination of Humans (1923). — Protopopov VP. Pathophysiological Particularities in Activity of the Central Nervous System in Schizophrenia (1938). — Protopopov VP. Pathophysiological Basis of Rational Treatment of Schizophrenia (1956). — Protopopov VP. Selected Works. Acad Sci Ukraine, Kyiv 1961. — Ablitsov VH. «Galaxy Ukraine». The Ukrainian diaspora: prominent persons. Publ. «Kyt», Kyiv 2007.

[1745] Town of Mohyliv-Podils'kyy (Foot Note 3). — Town of Nemyriv (Foot Note 798), Vinnytsia Obl., Ukraine. — Hakkebush VM. Decentralization of the Psychiatric Help. 1908. — Hakkebush VM, Heyer TA. About Alzheimer Disease. 1912. — Hakkebush VM, Heymanovych OJ. About the System of Ischemic Psychosis of the Old Age with Atrophic Centers in the Brain. Clinics and Pathoanatomic of Alzheimer Disease. Collection of Papers. Kharkiv 1915. — Hakkebush VM. Neurological Disease Related to the Air Contusion. 1916. — Hakkebush VM, et al. Outline of Reflexology. Kyiv 1925. — Hakkebush VM. Prognosis in Psychiatry. 1927. — Hakkebush VM. Course of Judicial Psychopathology. 1928. — Hakkebush VM, et al. Self-inspiration in the Therapy of Neuroses. Kyiv 1926. — Hakkebush VM. Environment and Children Hysteria. 1930. — Hakkebush VM, et al. Experience in Learning the Person Development with the Help of Hypnotic Restraining. Collection of Works, Kyiv Psychoneurologic Institute 1930;12(2). — Ablitsov VH. «Galaxy Ukraine». The Ukrainian diaspora: prominent persons. Publ. «Kyt», Kyiv 2007.

and **Olexandr I. Heymanovych [22.07(03.08).1882, city of Kharkiv, Ukraine – 18.04.1958, city of Kharkiv, Ukraine]**[1746] — Ukrainian neurologist, defined a senile dementia due to atherosclerosis (AS) of the cortical arteries of the brain (Hakkebush-Heyer-Heymanovych syndrome, 1912-15), which remain described in 1907 by **Alois Alzheimer (1864 – 1915**: above**)** a senile progressive degeneration of the cerebral cortex characterized by the weakness of mind of unknown etiology (Alzheimer disease).

Stepan (Stefan) M.V. Baley (04.II.1885, village of Velyki Birky, Ternopil Region, Ternopil Obl., Ukraine – 13.IX.1952, city of Warsaw, Poland)[1747] — Ukrainian Polish psychologist, created the concept of personality as the basis of the modern psychology and pedagogy (1938-46).

The work of **Karl R.** von **Frisch (1886 – 1982)** — Austrian scientist in the objective study of the individual and groups animal behavior (ethologist), in 1914-27 on the behavior and the sensory perceptions of the carnicovan honey bee (Apis mellifera carnica) had an impact on psychology, specifically, he found that: (1) the sensitivity of this bee to a «sweet» taste was only slightly stronger than in humans; (2) it has color vision comparable to that of humans with the shift away from the red to ultraviolet spectrum; (3) recognizes the desired compass direction primarily by the sun, alternatively by the polarization pattern of the blue sky, and by the earth's magnetic field; (4) recognizes of the daytime position of the sun; (5) recognizes the horizontal position based on the magnetic field of the earth and the alignment of the plane of a honeycomb under construction, and sensing the vertical alignment of the honeycomb attributed to the ability to identify what is vertical with the help of

---

[1746] Ablitsov VH. «Galaxy Ukraine». The Ukrainian diaspora: prominent persons. Publ. «Kyt», Kyiv 2007.

[1747] Village of Velyki Birky (founded 1224), Ternopil Region, Ternopil Obl., Ukraine. — Baley S. Über Urteilsgefühle. Schewtschenko Gesellschaft der Wissenschaften. Lemberg (Lviv) 1916. — Baley S. About the psychology of Shevchenko's work. Shliakhy, Lviv 1916. — Baley S. Outline of logics. Shevchenko Sci Soc, Lviv 1923. — Baley S. Fever and the consciousness. Likarskyy visnyk (Lviv) 1926;2:1-7. — Baley S. Psychologia wieku dojżewania. Książnica – Atlas, Lwów 1931. — Baley S. Psychologia wychowawcza w zarysie. Książnica – Atlas, Warszawa 1938, 1948, 1951. — Baley S. Osobowość. Nakł. Lwowska Biblioteka Pedagiczna, Lwów 1939. — Baley S. Drogi samopoznania. Warszawa 1946. — Baley S. Characteryzacja i typologia dzieci i młodzieży. Wyd. III. Inst. Wyd. «Nasza Księgarnia», Warszawa 1948. — Baley S. Wprowadzenie do psychologii współczesnej. PWN, Warszawa 1959. — Baley S. Collection of Works: in 5 vol. & 2 books. Vol. 1-5. Vol. 1 - IFLIS ЛФС «Cogito» Lviv - Odesa 2002; Vol. 2 – National University «Lviv Politechnics», Lviv 2009. — Petriuk PT, Bondarenko LI. Creative and vital pathway of Academician Stepan Volodymyrovych Baley – prominent Ukrainian and Polish psychologist, physician, psychoanalytic, philosopher and pedagogue (to the 120th-anniversary of birthday). Psychichne zdorov'ia 2005;2:86-94. — Petriuk PT, Bondarenko LI. Academician Stepan Volodymyrovych Baley: bibliographic and scientific aspects (to the 125th unniversary of birthday). J Psychiat Med Psychol. 2011; 1:83-93. — Yatyshchuk AA. Problems in the psychology of education in the creative inheritance of S. Bakey. Dissertation. Academy of Pedagogical Sciences of Ukraine. H.S Kostyuk Institute of Psychology. Kyiv 2018. C. 1-249. — Ablitsov VH. «Galaxy Ukraine». The Ukrainian diaspora: prominent persons. Publ. «Kyt», Kyiv 2007.

their head used as a pendulum together with a ring of sensory cells in the neck; (6) uses dances as a language – the «round dance» provides information about a feeding place in the vicinity of the beehive at distance of 50-100 meters, and the «waggle dance» is used to relay information about more distant foot source; (7) outside the hive, the queen bee and her daughters emits pheromones (in Greek – «phero» meaning «to bear», and «hormone» meaning «impetus») is secreted or excreted that trigger a social response in members of the same species. He advocated to use eugenics (in Greek – «eugenes» – «well born», and «genes» – «race, stock, kin») – a of beliefs and practices aiming at improving the genetic quality of a mankind, to prevent hereditary diseases from reproducing through voluntary education, a family planning and sterilization (1959).

Ambrozy Kibzey (20.XII.1888, village Myshkiv, Chortkivs'kyy Region, Ternopils'ra Obl., Ukraine – 25.IV.1954, town of Sault Ste. Marie, Chippewa County, MI, USA)[1748] — Ukrainian, Canadian American physician-psychiatrist and a Maecenas, who after studies in 1917-19 at the University of Alberta (established 1908), city of Edmonton (earliest inhabitants arrived c. 3,000 BC, established 1795), Province of Alberta (area km²), and in 1919-1922 at the Faculty of Medicine and Surgery (founded 1876 and 1843, respectively) Universite de Montreal (established 1878), city of Montreal, Province Quebec, Canada, defended the degree of Medicine Doctor (1922), after specialization in psychiatry practiced it in the USA. He authored the medical handbook «The Ukrainian Physician» (Yorktown 1945; 1954) and series of papers in Likars'kyy Visnyk / Medical Herald (published 1920-39), organ of the Ukrainian Medical Society (founded 1910) and Medical Commission (established 1898) of the Shevchenko Scientific Society (ShSS, founded 1873), city of Lviv, Ukraine.

That types of physique are related to personality or temperamental traits was proposed by **Ernst Kretschmer (1888 – 1964)** — German psychiatrist, categorized as asthenic, athletic, dysplastic and pyknic types.

Ivan O. Sokolyansky (25.03.1889, village Cossack Dyns'ka, settlement Ukraïns'kyy, Dyns'ky Region, KK, Ukraine (1917-21) / Russia – 27.11.1960, city

---

[1748] Village Myshkiv (remembrances of late Paleolithic, Trypillian, Gava-Halihradsky, Cherniakhiv and Old-Rus Cultures; founded 1495), Chortkivs'kyy Region, Ternopils'ka Obl., Ukraine. — Town of Sault Ste Marie (founded 1668), Chippewa County, MI, USA. — Kibzey A. The Ukrainian Physician. Yorktown 1945 & 1954. — Kibzey A. Studies and Science in America. Likars'kyy Visnyk (Lviv) 1930;1. —Kibzey A. Tanine in the treatment of burns. Likars'kyy Visnyk (Lviv) 1935;4. — Osinchuk R. Dr Ambrozy Kibzey: Obituary. Likars'kyy Visnyk 1954;1,1(1):50-1. — Ablitsov VH. «Galaxy Ukraine». The Ukrainian diaspora: prominent persons. Publ. «Kyt», Kyiv 2007.

of Moscow, Russia)[1749] — Ukrainian pedagogue, psychologist, specialist in branch of surdo- and tyflo-pedagogue, defectologue, who designed the reading instruments (machine) for learning for the blind and deaf-mute (1936). He in 1923 assembled group of blind and deaf-mute children, created the school-clinic for blind and deaf-mute (speechless) children (1929-34), and the Institute Defectology of Ministry for the Protection of Health (MPH) of Ukraine (1929-34), city of Kharkiv, Ukraine. His contribution to the foundation of defectology, and surdo- and tyflo-pedagogue consists of the following papers: «Of so-called reading from lips by the deaf and the mute» (1925), «Articulated schemes of the receptor and effector language of the deaf and the mute» (1926), «About methods for teaching the deaf and the mute of mouth speaking» (1930), «Blind reads any (whatever) book» (1936), «Braille[1750] screen for deaf-mutes (1941), «Of new reading way by blind», «Of reading by the blind and the blind-mute the flat-type print» (1956), «Mechanical primer (reader)», et al.

The first reference to autism from 1797 belongs to **Jean Itard (1774 – 1838**: 19th century. Internal Medicine**).** In 1911 **Eugene Bleuler (1857 – 1935**: above**)** named and identified autism, which was re-described in 1925 by **Grunya Ye. / E. Sukhareva (Ssucharewa) (**Nov. 11, 1891, Kyiv, Ukraine – Apr. 26, 1981, Moscow, Russia)[1751] — Ukrainian psychiatrist of Jewish descent, in 1943 by **Leo Kanner (June 13, 1894, village of Klekotiv, Brodivs'kyy Region, Lvivs'ka Obl., Ukraine – Apr. 3, 1981,**

[1749] Village (stanytsia) Cossack Dyns'ka (found 1794), colony Ukraïns'kyy, Dyns'ky Region (area 1,361.96 km²), Krasnodarsky krai (KK: Foot Note 337), Ukraine (1917-22) / Russia. — **Olha I. Skorokhodova [18(31).VII.1911, u.t.v. Bilozirka** (remains of mounts since Copper-Bronze Age 6th-3rd millennium BC, settled by Zaporozhian Cossacks 1754; suffered by organized by Russian Communist Fascist regime of Holodomor-Genocide 1932-33 and political repressions 1937-38, established 2 names of killed), **Khersons'kyy Region, Khersons'ka Obl.** (Foot Note 315), **Ukraine – 07.V.1982, city of Moscow, Russia]** — Ukrainian pedagogue, scientist-defectologist, educated as scientist and under the leadership of **Ivan Sokoliansky**, became the only one in the world blind and deaf scientist, who despite the absence of sight and hearing, created series of scientific works in the field of development, education, and learning of the blind-deaf children. — Skorokhodova OI. How I Perceive the Surrounding World. Publ. the Academy of Pedagogue Sciences (APSc) of Russia, Moscow 1947. — Skorokhodova OI. How I Present the Surrounding World. Publ. APSc.of Russia, Moscow 1954. — Skorokhodova OI. How I Perceive, Present and Understanding the Surrounding World. Publ. AGSc of Russia, Moscow 1956. — Ablitsov VH. «Galaxy Ukraine». The Ukrainian diaspora: prominent persons. Publ. «Kyt», Kyiv 2007.

[1750] **Louis Braille (1809** – died from tuberculosis **1852)** — French educator and inventor, blind since the 3-year-old-age on one eye as a result with stitching awl accident, who invented a system of reading and writing that could bridge in communication between sighted and blind (Braille system, 1824), as well as music imaging (Braille notation, 1929).

[1751] Ssuchaveva G E. Schizophrenic psychopathy in children age. Vopr Pedology Children Psychoneurology. Issue 2. Moskva 1925;2:157-87. — Ssuchaveva G E. Die schizoiden Psychopathien Kindesalter. Monatschr Psychiatr Neurol 1926;235-61. — Ablitsov VH. «Galaxy Ukraine». The Ukrainian diaspora: prominent persons. Publ. «Kyt», Kyiv 2007.

town of Sukesville, MD, USA)[1752] — Ukrainian American psychiatrist of Jewish descent, and in 1938-44 by **Jo. Hans F.R. Asperger (1906 – 80)**[1753] — Austrian pediatrician, as a development disorder with onset during the first two-three years of children's life, affected by inherited and environmental factors, characterized by difficulties with social interaction and communication and by restricted and repetitive behavior.

In the spectrum of autistic disorders, Hans Asperger identified group of children, before two years of age and up to adulthood, with a mild form of disease, normal or above normal intelligence, where cognitive development was normal despite social difficulties in communication and social interaction, with repetitive pattern of behavior, manners, interests or activities, rigidity in thinking, a focus on rules and routines, who had limited understanding of others feelings, and were physically clumsy (named in 1981 – Asperger syndrome).

The important work by **Mykhaylo (Michail) M. Mishchenko (24.05.1896, city of Kharkiv, Ukraine – 20.05.1974, city of Minneapolis, MN, USA)**[1754] — Ukrainian American neurophysiologist and psychiatrist, was dedicated to the significance of vision in the evolution of the hand motion (1931) and evolution of mimics (1932), cerebral changes in schizophrenia by insulin shock therapy (1937) and development of experimental and a natural sleep of the human being (1938).

Wilhelm E. Reich (24.III.1897, village of Dobrianytsi, Peremyshl Region, Lviv Obl., Ukraine – died from heart failure **03.XI.1957**, US Penitentiary, **town of Lewisburg, PA, USA)**[1755] — Ukrainian American psychoanalyst of Austrian origin, created psychoanalysis of a character, and body psychotherapy (1935-39).

[1752] Village of Klekotiv (founded before St Bohoslav Church was built 1686), Brodivs'kyy Region, Lvivs'ka Obl., Ukraine. — Town of Sukesville (founded 1825-30), State of Maryland (MD, area 32,314 km2), MD, USA. — Kanner L. Autistic disturbances of affective contact. Nerv Child 1943;2:217-50.

[1753] Asperger H. Die Autischen Psychopathen im Kindesalter. Archiv für Psychiatrie und Nervekranheiten 1944;114(1)132-5.

[1754] City of Minneapolis (founded 1849-50), State of Minnesota (MN, area 225,163 km^2), USA. — Mishchenko MM. Motor function of blind children. Psychoneurologiia detskogo vozrasta. Kharkiv 1931. — Mishchenko MM. Changes in the Higher Nervous Function in Experimental and Natural Sleep. Experymentalna Medycyna. Kharkiv 1935. — Mishchenko MM. Pathophysiology of Schizofrenia. Sbornik trudov Ukrainskoi psykhonevrologicheskoi akademii. Kharkiv 1936. — Mishchenko MM. Psychomotor of the Face of Blind Children and Adolescents / Psychomotor of the Face of Blind Children and Adolescents in the Norm and Pathology. Kharkiv 1936. — Mishchenko MM. Physiologic Basis of Pathogenesis for Neurotic and Reactive States. Augsburg 1948. — Mishchenko MM. Pathogenesis of Neurotic and Reactive States. Augsburg 1949. — Mishchenko MM. Physiologic Bases of a Character (Part 1&2). Part 1 – J Ukr Med Assoc North Am 1965 Oct.;12,4(39):2-6. / Part 2 – J Ukr Med Assoc North Am 1966 Apr.;13,1-2(40-41):11-21. — Ablitsov VH. «Galaxy Ukraine». The Ukrainian diaspora: prominent persons. Publ. «Kyt», Kyiv 2007.

[1755] Village of Dobrianytsi (in German – Dobzau), Peremyshl Region, Lviv Obl., Ukraine. — US Penitentiary (opened 1932), town of Lewisburg (founded 1785), PA, USA. — Ablitsov VH. «Galaxy Ukraine». The Ukrainian diaspora: prominent persons. Publ. «Kyt», Kyiv 2007.

Hryhoriy (Hryhorii) S. Kostiuk, a.k.a., **Grigory S. Kostyuk [11(23).IX.1899, village Mohylne, Holovanivs'kyy Region, Kirovohrads'ka Obl., Ukraine – 25.I.1982, city of Kyiv, Ukraine]**[1756] — Ukrainian psychologist, who researched psychology of education and upbringing of pupils, development of their thinking and abilities, and the formation personality.

A principle of attachment principle was formulated **by Douglas Spalding (1842 – 77**: 19th century. Psychiatry and psychology), a popularized by **Konrad Z. Lorentz (1903 – 89)** — Austrian zoologist, ecologist, and ornithologist, through which in some species, i.e., ducks and geese a bound is formed between newborn and its caregiver.

As of the attachment theory, developed by **Douglas Spalding (1841 – 77**: 19th century. Psychiatry and psychology) and Konrad Lorenz (above), **E. John Bowlby (1907 – 90)** — English psychiatrist, psychologist and psychoanalyst, believed that the tendency for primate infants to develop attachments to familiar caregiver was the result of evolutionary pressures, since attachment behavior would facilitate the infants survival in the face of dangers such as predation or the elements (1988).

Narcyz Lukianowicz (20.VI.1907, city of Chernivtsi, Ukraine – 24.VII.1985, city of Cambridge, England)[1757] — Ukrainian neurologist, psychiatrist, and a poet, defended doctorate on «Mental Illness and Crime» (Medical Faculty Lviv University, city of Lviv, Ukraine), studied interrelationship «children – parents», adolescent's crime, parahalucination phenomenon, sexual deviations.

The work of **Nikolaas Tinbergen (1907 – 88)**[1758] — Dutch biologist and ornithologist, concentrated on the organization and elucidation of individual and social genetically programmed behavior pattern in animals that contributed to

---

[1756] Village Mohylne (known since 15th century), Holovanivs'ky Region, Kirovohrads'ka Obl., Ukraine. — Kostiuk HS. About the comparing value of the visual and hearing ways of testing the successfulness of pupils. Ukrainian Herald of the Experimental Pedagogue and Reflexology 1928;2(9):99-131. — Kostiuk HS (Ed.). Psychology: Handbook for the Pedagogue's Higher Educational Establishments (HEE). 2nd Ed. Publ. Radians'ka Shkola, Kyiv 1968. — Kostiuk HS. Education and psychical development of pupils. In VI Voytko (Ed.). Psychological Science, Teacher, Pupil. Publ. Radians'ka Shkola, Kyiv 1979. — Kostiuk HS. Selected Psychological Work. Publ. Pedagogika, Mosccow 1988. — Kostiuk HS. Educational-Upbringing Process and Psychical Development of Personality. Publ. Radians'ka Shkola, Kyiv 1989. — Ball HO. Hryhoriy Kostiuk – prominent psychologist of Ukraine. Education and Management 1999;3:218-24. — Ablitsov VH. «Galaxy Ukraine». The Ukrainian diaspora: prominent persons. Publ. «Kyt», Kyiv 2007.

[1757] City of Chernivtsi (Foot Note 501), Ukraine. — City of Cambridge (inhabited since Bronze Age, founded in the 1st century AD), England. — Lukianowicz N. Autoscopic phenomena. Arch Neurol Psychiat 1958;80-199-2000. — Narcyz Lukianowicz. Bulletin of the Royal College of Psychiatrics. London 1985. – No. 10. – P. 210-211. — Ablitsov VH. «Galaxy Ukraine». The Ukrainian diaspora: prominent persons. Publ. «Kyt», Kyiv 2007. — Hanitkevych YaV. The contribution of the Ukrainian physicians to the world medicine. Ukr Med Chasopys 2009 VII/VIII;4(72):110-15.

[1758] Tinbergen N. The Study of Instinct. Clarendon Press, Oxford 1951.

the relief of human suffering in psychiatric diseases such as anguish, compulsive obsession, stereotypic behavior, and catatonic posture.

Albert Ellis (1913 – 2007) — American psychologist of Jewish origin who in 1955 developed rational emotive behavior therapy.

Bold and honest **Borys Ia. Pervomaisky (06.09.1916, u.t.v. Stara Ushytsia, Staroshyc'kyy Region, Khmelnyts'ka Obl., Ukraine – 23.09.1985, м. Київ, Ukraine)**[1759] — Ukrainian psychiatrist, spoke out against the use of psychiatry as repressive tool by Russian Communist Fascist regime, hence in 1980 was forced to abandon his professional activities.

Aaaron T. Beck (1921 –) — American psychiatrist of Jewish origin who proposed cognitive negative triad for the diagnosis of depression, i.e. negative view of the world, of the future and of oneself (Beck's triad, 1976).

Psychiatry found fruitful ground for application during and after WW 2 in treating «combat psychosis». New dimensions were added to the field with the introduction of psychotherapy, psychologically active drugs (thorazine and others), electro-convulsive therapy, psychosurgery, «community psychiatry» and «behavior psychiatry».

In 1966-83 **Erik R. Kandel (07.11.1929, м. Відень, Австрія –)**[1760] — Austrian American psychiatrist, and neuroscientist of Jewish Ukrainian parents, examined biochemical changes in neurons associated with learning and memory storage in Aplysia californica, a genus of medium-sized to extremely large slug, which linked short-term memory to functional changes in existing synapses, while long-term memory was associated with a change in the number of synaptic connections.

The near-death experience (NDE) is an experience reported by a person who nearly died or who was clinically dead and revived by the cardio-pulmonary resuscitation (CPR). Regarding the NDE, it was described in 1992 by **James Mauro**[1761] — American psychiatrist, and research in 2001 by **Pim van Lommel**

[1759] U.t.v. Stara Ushycia (founded 1144; during conduction by Russian Communist Fascist regime of Holodomor-Genocide 1932-33 against the Ukrainian Nation at least 78 inhabitant died), Staroshyc'kyy Region, Khmelnyts'ka Obl., Ukraine. — Pervomays'kyy BYa. Maniacal Phase of Maniacal-Depressive Psychosis and Maniacal Symptoms of the Second Etiology (Clinic and Pathogenesis of Maniacal State). Dissertation for the degree of DMS, St Petersburg 1959. — Pervomays'kyy BYa. About Methodic of Psychiatric Diagnosis. Luhansk 1963. — Mirsky M. Ukrainian medical scientists – victims of Stalin's repressions. UHMJ «Ahapit» 1995;2:45-54. — Ablitsov VH. «Galaxy Ukraine». The Ukrainian diaspora: prominent persons. Publ. «Kyt», Kyiv 2007.

[1760] Kandel ER. In Search of Memory: The Emergence of a New Science of Mind. WW Norton & Co, New York 2006. — Ablitsov VH. «Galaxy Ukraine». The Ukrainian diaspora: prominent persons. Publ. «Kyt», Kyiv 2007.

[1761] Mauro J. Bright lights, big mystery. Psychology Today. 1992 July 1.

(15.III.1943, town of Laren, Province North Holland, the Netherlands –)[1762] — Dutch cardiologist, who conducted large-scaled prospective study to the NDE after cardiac arrest in the Netherlands. Pim van Lommel work was criticized for ignoring the science, for not refuting neurobiological explanations, misunderstanding and misinterpretation of the dying-brain hypothesis and anoxia, misplacing confidence in electroencephalography (EEG) that a flat EEG reading is not evidence of total brain inactivity. However, presently we can only say that *«The NDE often includes an out-of-body experience»* (James Mauro 1992).

Constantly increasing longevity of population is associated with increased number of geriatric patients needing psychiatric care. Therefore, the clinical research by **Andrew H. Ripeckyj [Oct. 16, 1950,** camp of displaced person (DP), **town of Berchtesgaden, Freistaat Bayern, Germany –]**[1763] — Ukrainian American psychiatrist, and his associated became very important with the passing time. They proposed the family guide to the problems of the hemorrhagic or ischemic stroke patients (1980); by evaluating and treating psychiatrically impaired elderly in the natural setting of their home, they found that the majority of patients had improved or stabilized, and most were able to be maintained in the community (1984); suggested that the age-specific geriatric modelled units are the best effective treatment format for the elderly in need of psychiatric care, and opened one in 1980 (1988).

In regard to the near-death experience (NDE), **Robert P. Lanza (11.II.1956, Boston, MA, США –)**[1764] — American physician-scientist, brought the concept of biocentric universe into domain of biology and regenerative medicine, according to which biology is the main science for understanding other sciences (2007). With this conception, known from ancient times, exactly the life is in all only illusion (mirage),

---

[1762] Town of Laren (known since late 19th century), Province North Holland (area 4,092 km²), the Netherlands (area 41,856 km²). — Lommmel P van, Wees R van, Meyer V, et al. Near-death experience in survivors of cardiac arrest: a prospective study in the Netherlands. The Lancet Dec 2001;358(9298):2039-45. — Lommel P van. Consciousness Beyond Live: Science of the Near-Death Experience. Publ HarperOne, New York 2011.

[1763] Town of Berchtesgaden (first historical note dates to 1102), Land Bayern (in German – Freistaat Bayern), Germany. — Nikolayeva N. Ulyana Kriuchenko – «Oksana». The Chronicle (Litopys) of the Ukrainian Insurgence Army (UIA). Series «Events and Peoples». Vol. 1-50. Publ. «Litopys UIA» & Publishing Committee / Volodymyr Makar Charitable Fund «Litopys UIA», Toronto – Lviv 1973-2013. — Ripeckyj A, Lazarus LW. Family guide to the problems of the stroke patient. Geriatrics 1980 Oct;35(10):47-8. — Hicks R, Dysken MS, Dewis JM, Lesser J, Ripeckyj A, Lazarus L. The pharmacokinetics of psychotropic medication in the elderly: a review. J Clin Psychiatry 1981 Oct;42(10):374-85 — Wasson W, Ripeckyj A, Lazarus LW, et al. Home evaluating of psychiatrically impaired elderly: process and outcome. Gerontologist 1984 June;24(3):238-42. — Ripeckyj A. Psychiatric Assessment. In: LB Wilson, SP Simson, CR Baxter (Eds). Handbook of Geriatric Emergency Care. University Park Press, Baltimore 1984. P. 197-201. — Corbett L, Ripeckyj A, Miller J, et al. The acute geriatric psychiatry service: a suggested model. Psychiatr Hosp 1988 Spring; 19(2):67-72. — Ablitsov VH. «Galaxy Ukraine». The Ukrainian diaspora: prominent persons. Publ. «Kyt», Kyiv 2007.

[1764] Lanza R, Berman B. Beyond Biocentrism: Rethinking Time, Space, Consciousness, and the Illusion of Death. Publ. BenBella, Dallas, TX 2016.

everything surrounding humans, create its consciousness, and death is only a self-delusion of our consciousness. Death does not exist, a human being simply traverse into a paraller world. Really, the consciousness does not vanish, but continue exist and work though already in the other similar world.

Rehabilitation

The concept of rehabilitation, propagated by **Howard Rusk (1901 – 89)** — American physician in rehabilitation medicine, arose in 1918 out of society's compassion for the mutilated veterans of WW 1.

Forensic medicine

One of the founders of this specialty was **Mykola (Nikolai) S. Bokarius (06. III.1869, city of Odesa, Ukraine – 23.XII.1931, city of Kharkiv, Ukraine)**[1765] — Ukrainian scientist-physician of forensic medicine and criminology, who proposed methods for examining the strangulation furrow (1904), formulated the classic sperm test (1925-30).

Chronic traumatic encephalopathy (CTE), originally described by in 1928 by **Harrison S. Martland (1883 – 1954)** — American physician and forensic pathologist, as «dementia pugilistica» or «punch drunk» in boxers, is a progressive degenerative disease occurring in individuals who had multiple head concussions or injuries. Beside of boxers, CTE is seen in American football players, wrestlers, ice hockey, rugby football players, sockers and veterans. It is characterized by behavioral, mood and thinking anomalies which with time can result in depression, insomnia, dementia, and suicide.

In 2002 after **Bennet Omalu (1968 –)** — Nigerian American physician, forensic pathologist and neuropathologist, by performing autopsy on the American football player found the evidence of CTE, thus increased awareness in society about this condition.

[1765] Bokarius NS. The Cristals of Florence, its Chemical Nature and Judicial-Medical Importance. Doctoral dissertation. Kharkiv 1902. — Bokarius NS. The Importance the Strangulation Furrow at Hanging. Khrkiv 1904. — Bokarius NS. The Forensic Medicine for Medics and Jurists. Juridical Publ. of Ukraine, Kharkiv 1930. — Ablitsov VH. «Galaxy Ukraine». The Ukrainian diaspora: prominent persons. Publ. «Kyt», Kyiv 2007.

Space biology and medicine

The founder of space biology and medicine in the years of 1919-31 was **Olexandr L. Chyzhevsky (07.02.1897, town of Ciechanowiec, Wysokie Mazowiecki County, Pidliashs'ke Voyevodoship, Poland – 20.12.1964, city of Moscow, Russia)**[1766] — Ukrainian Russian biophysicist. He established the interaction between metachromasia of bacterias and increased sun activities enabling solar emissions that are hazardous to man, both on Earth and space to be forecast (Chyzhevsky effect, 1935) and described the formation of erythrocytes into «coin-column» like structure under the influence of hydrodynamic forces (Chyzhevsky phenomenon, 1951).

This specialty has emerged on account of the pioneering work of **Serhiy P. Koroliov [12/30.XII.1906, city of Zhytomyr, Ukraine** – died during abdominal-perineal resection (APR) for bleeding sarcoma of the rectum as a result of inability by the anesthesiologist to intubate the trachea for artificial lung ventilation (ALV) **14.I.1966, city of Moscow, Russia]**[1767] — Ukrainian scientist in the field of astronautics, with the launching of the earth's satellite – «suputnyk» (1957) and the manned flight into the earth's orbit (1961), and through the work of **Wernher M.M.F.** von **Braun (1912 – 77)** — German American rocket designer who landed a man on the Moon (1969). Those projects needed advanced technological backup for biology and medicine, mainly in the form of computers.

On the account of the baseless accusation, Serhiy Korolov was arrested on 27. VII.1938, cruely interrogated and imprisoned till 27.VII.1944 by Russian Communist Fascist monsters of the Narodnyy komisariat vnutrishnikh sprav (NKVS: 1934-43) / in Russian – Narodnyy kommissariat vnytrennykh del (NKVD: 1934-43), through the Holovne upravlinnia vypravno-trudovykh taboriv (НУТАБ: 1918-56) / in Russian – Glavnoye upravleniye ispravitelno-trudovykh lagerey (GULAG: 1918-56), to the State Trust of the Road and Industrial Buildings (Dalbud: 1932-54) in the drainage area of the upper Kolyma River (length 2,129 km, basin 647,000 km^2), obl.-town Madagan (established 1919), Madagan Obl. (area 462,464 km^2), Far East of Russia (6,942,900 km^2), where he stayed until February 1940; the notorious Butyrka prison (established 1784), a.k.a., Butyrka inquiring isolator No. 2, or «Butyrka», Moscow, Russia, where he stayed from 02.III through 10.IX.1940; further time served in the Central Construction Bureau (29.IX.1940 – VI.1941), and in the Special Constructor Bureau (SCB), city of

---

[1766] Town of Ciechanowiec (known since 12th century), Wysokie Mazowiecki County, Pidliashs'ke Voyevodoship (area 20,187.02 km^2), Ukraine (10th century – 1947) / Poland. — Ablitsov VH. «Galaxy Ukraine». The Ukrainian diaspora: prominent persons. Publ. «Kyt», Kyiv 2007.

[1767] City of Zhytomyr (founded 884), Ukraine. — Ablitsov VH. «Galaxy Ukraine». The Ukrainian diaspora: prominent persons. Publ. «Kyt», Kyiv 2007. — Olearchyk AS. A Surgeon's Universe. Vol. I-IV. 4th Ed. Publ. «Outskirts Press», Denver, CO 2016.

Kazan, Tatarstan (from June 1941 until fried from GULAG 27.VII.1944), but he still worked in the SCB till September 1945[1768].

During one of investigations an inquirer of NKVD hit Serhiy Korolov's face with a decanter, which led to a measured fracture of the jawbone, that in primitive conditions and absence of proper medical care in the GULAG could not been repaired by an open reduction and internal fixation (ORIF). Fractured jaw- bone grew together improperly, and as a result the jaws no longer could be opened in full.

The surgical procedure on 14.I.1966 began with the patient breathing spontaneously,with addition of oxygen (O_2) via nasal cannula and sedation, by proctoscopy, that is a direct visual examination of the rectum through the anal opening, removal of polyps, that had not stopped bleeding.

At once, under an intravenous (IV) and general anesthesia using modified Esmarch mask [19th century. Anesthesiology – **Johann** von **Esmarch (1828 – 1908]**, the abdominal cavity was opened. Direct visual inspection revealed tumor (sarcoma) of the rectum measuring the fist's size. But during dissection off tumor, he developed heart failure (HF), likely due anoxia, which increased rapidly.

Emergency oropharyngeal intubation of the trachea by anesthesiologist to assure artificial lung ventilation (ALV) and supporting of O_2 the blood was unsuccessful, because of an old fracture of the jaw not treated by ORIF, caused improper healing of bones, which restricted opening of the mouth. An increasing anoxia and HF lead to death on the operative table.

The way out from this complicated situation could have been a «blind» intubation of the trachea via nostril (naris) of the nose, and in the contemporary situations – intubation of the trachea using flexible fiberoptic bronchoscopy (FOB), invented in 1966 by Shigeto Ikeda [20th century. Surgery – **Shigeto Ikeda (1925 – 2001)]**[1769].

The first biological and medical computers for space technology were developed beginning 1948 under the leadership of **Serhiy O. Lebediev (1902 – 74**; 20th century. Physics and radiology)[1770] — Russian Ukrainian computer designer, at the Institute of Cybernetics, the Academy of Sciences (AS) of Ukraine, from 1957 under the leadership of **Mykola Amosov (1913 – 2002**: 20th century. Surgery) at the F.H. Yanovsky Institute of Tuberculosis and Chest Surgery (ITCS) in Kyiv, Ukraine.

The schematic approach for prognosis of the human condition during the space flight was developed by **Mykola Syrotinin (1896 – 1977**: 20th century. Pathology) on the account of practical recommendations during hypoxia (1939-73); by **Vasyl P. Parin (1903 – 71)** — Russian space physiologist, who noted that irritations of the

---

[1768] Olearchyk AS. A Surgeon's Universe. Vol. I-IV. 4th Ed. Publ. «Outskirts Press», Denver, CO 2016.

[1769] Olearchyk AS. Cardiac surgery in patients with ankylosing spondylitis of the cervical spine case report. Vasc Surg 1992;26:426-7.

[1770] Ablitsov VH. «Galaxy Ukraine». The Ukrainian diaspora: prominent persons. Publ. «Kyt», Kyiv 2007.

pulmonary artery (PA) baroreceptors causes bradycardia, hypotension and dilatation of the spleen (Parin's reflex, 1946), authored «Introduction into Medical Cybernetics» (1966) and «Biological Telemetry» (1971); and **Oleh H. Gazenko (Dec. 12, 1918, the village of Mykolaivka, Mineralovodsky Region, SK, Russia – Nov. 12, 2008, Moscow, Russia)** — Russian space physiologist. Both were directly involved and responsible for the construction of the medical environment of the first spacecrafts, space station, training cosmonauts and the monitoring of their physiological parameters during and after space flights.

Pavlo P. Popovych (05.10.1930, town of Uzyn, Bilotserkivs'ky Region, Kyiv Obl., Ukraine – died from stroke **30.09.2009, town of Hurfuz, Yalta Region, Crimea, Ukraine)**[1771] — became the first Ukrainian and the fourth World pilot-cosmonaut, general-major of aviation, inside the cabin of the spacecraft «Skhid-4» 12-15 August 1962, who proved in the Space the lack of gravity with a ball point pen, which after release from his hand kept itself steadily free in the spacecraft cabin air.

During the span of not long life **Jerry R. Hordinsky (Aug. 03, 1943, city of Kalusch, Ivano-Frankivsk Obl., Ukraine** – died of cancer **Oct. 20, 2000, Oklahoma City, OK, USA)**[1772] — Ukrainian American physician, who after graduating from premedical and engineering curriculum at the University of Minnesota, city of Minneapolis,

---

[1771] Town of Uzyn (founded 1590; during conduction by Russian Communist Fascist regime of Holodomor-Genocide 1932-33 against the Ukrainian nation died more than 300 persons, but by the calculation of witnesses – 1563 person). — Town of Hurfuz (first mentioned in 6th century), Yalta Region, Crimea, Ukraine. — Ablitsov VH. «Galaxy Ukraine». The Ukrainian diaspora: prominent persons. Publ. «Kyt», Kyiv 2007.

[1772] City of Kalush (first mentioned in chronicle May 27, 1437), Ivano-Frankivsk Obl., Ukraine. — Oklahoma City (founded in 1889), OK, USA. — Hordynsky JR. Medical aspects of Skylab Mission. J Ukr Med Assoc North Am 1974 Jan;21,1(72):20-24. — Hordynsky JR. Manned space flights – current international effort. 1981 Spring;28,2(101):78-84. — Hordynsky JR. Medical evaluation of Skylab Mission. J Ukr Med Assoc North A 1977 Jan; 24,1(84):23-32. — O'Donnell RD, Hordinsky JR, Madakasira S, et al. A candidate automated test screening of airmen: design and preliminary validation. Office of Aviation Medicine, Washington, DC Feb 1992. — Teague S, Hordinsky JR. Tolerance of beta blocked hypertensives drug using orthostatic and altitude stresses. Office of Aviation Medicine. Washington, DC Apr 1992. — White VL, Canfield DV, Hordinsky JR. The identification and quantitation of triamterene in blood and urine from fatal aircraft accidents. J Analytical Toxicology 1993 Dec;18(1):52-3. — Canfield DV, Flemig Jo, Hordinsky JR. Unreported medications used in incapacitating medical conditions found in civil aviation accidents. Office of Aviation Medicine, Washington, DC, and Civil Aeromedical Institute FAA, Oklahoma City, OK Aug 1994. — White VL, Canfield DV, Hordinsky JR. Elimination of quinidine in two subjects ofter injection of tonic water: an exploration study. Office of Aviation Medicine, Washington, DC, and Civil Aeromedical Institute FAA, Oklahoma City, OK Aug 1994. — Canfield DV, Flemig Jo, Hordinsky JR, et al. Drugs and alcohol found in fatal civic alcohol aviation accidents between 1989 and 1993. Office of Aviation Medicine, Washington, DC, and Civil Aeromedical Institute FAA, Oklahoma City, OK Oct 1995. — Hordinsky JR. Blending the disciplines (Part II): medicine and engineering. Minnesota Alpha '63. The Bent of Tan Beta P 2001:30-34. — Pundy P. Jaroslav Hordynsky, MD (03.VIII.1942, Kalush, Ivano-Frankivsk Obl., Ukraine – 20.X.2000, Oklahoma City, OK, USA). J Ukr Med Assoc North Am 2001 Spring; 46,1(146):68. — Ablitsov VH. «Galaxy Ukraine». The Ukrainian diaspora: prominent persons. Publ. «Kyt», Kyiv 2007.

MN, with degree of baccalaureate at Feinberg Medical School of the North Western University, city of Chicago, Il, with degree of MD (1967) and occupational medicine at the Harvard University, Boston, MA, with degree of Master in Industrial Health (1972), became Diplomate of the American Board of Prevention (and Aerospace) Medicine (established in 1948), Flight Surgeon with the US Army, Flight Surgeon with the National Aeronautics and Space Administration (NASA, founded 1958), Deputy Flight Surgeon steering the research with Deutsche Gesellschafts für Biomedizinische Technik for the Skylab 1 NASA, Chief Flight Surgeon of the first USA Orbital Station, a crewed three men Skylab 2 mission of the L.B. Johnson NASA, Houston, TX (May 1973 –February 1974), Director of the Aeromedical Research Division of the Federal Aviation Administration (FAA) Civic Aeromedical Institute in Oklahoma City, OK, the co-founder of Oklahoma Friends of Ukraine, Inc., a non-profit organization dedicated to fostering academic, business and cultural ties between the citizen of Oklahoma City and Ukraine (filled date: Dec. 09, 1993).

His achievement included co-development of automated cognitive function test to evaluate air men after head injury or brain disease; research in civil aviation inflight medical needs, surveillance of civil aviation accidents and their association with specific drug and alcohol usage, hazard of using antihypertension drugs in special civic aviation settings; securing emergency medical kits within the civil aviation sector; and optimization of aircraft cabin environmental work condition.

Indebted to developed method of sanogennic asphyxia, **Vadym Berezovsky (1932 – 2020**: 20th century. – Pathology), used preventive increase resistance of the organism to anoxia in a pre-pilotage preparation of **Leonid K. Kadeniuk (28.I.1951, village of Klishkivtsi, Khotyns'ky Region, Chernivetska Obl., Ukraine** – died from an acute MI during a traditional morning run **31.I.2018, city of Kyiv, Ukraine)**[1773] — Ukrainian 1st class pilot-tester, cosmonaut / astronaut, general-major, and its double **Yaroslav I. Pustovyy (29.XII.1970, town of Kostroma, Kostroms'ka Obl., Russia)**[1774] — Ukrainian cosmonaut / astronaut and engineer, for launching on the American repeated use spacecraft (space shuttle) «Columbia» 19.XI.1997 in the laboratory module – «Space Laboratory» mission of the Space Transportation System (STS)-87: 15 days and nights, 16 hours, 35 minutes and 1 second.

In 09 September 2006 on the Space Transport System (STS)-115 «Atlantida», or space shuttle, were launched from the John F. Kennedy Center of Space Flights, Cape Canaveral, FL, USA, the command (mission) of three (3) American and one

---

[1773] Village Klishchivtsi (settled since times of Trypillian Culture 3,000 BC, founded 1434), Khotynskyy Region, Chernivetska Obl., Ukraine. — Ablitsov VH. «Galaxy Ukraine». The Ukrainian diaspora: prominent persons. Publ. «Kyt», Kyiv 2007.

[1774] Town of Kostroma (founded 1152), Kostromska Obl., Russia. — Ablitsov VH. «Galaxy Ukraine». The Ukrainian diaspora: prominent persons. Publ. «Kyt», Kyiv 2007.

(1) Canadian astronauts (all men), and one (1) woman – Commander (Cmdr), Master of Sciences (MS) **Heidemarie M. Stefanyshyn-Piper (07.02.1963, city of St Paul, MN, USA –)**[1775] — Ukrainian American astronaut, Cmdr of the Military Naval Forces of the USA and expert of the National Aerodynamics and Space Administration (NASA), in journey (voyage) to be temporarily connected to the International Space Station (ISS). The main task of those astronauts-experts was delivery and attachment (fastening) to the ISS a 17,5-tonne weighting devise, and 3,72 meters long communication bundle (bunch), containing large set of additional solar batteries, and rotational instruments for a direct contact with the Sun's rays, without which it was not possible to complete the construction of this station. This task was performed during the first time going outside the spacecraft into an open cosmic space on 12 September 2006 by **Joseph R. «Joe» Tanner (15.01.1950, town of Denville, IL, USA –)** — expert, and Heidemarie Stefanyshyn-Piper), the second one – on 13 September 2006 by one American and one Canadian astronaut, and the third one – on 14 September 2006 – again by Joseph Tanner and Heidemarie Stefanyshyn-Piper. Together, Heidemarie Stefanyshyn-Piper stayed in the open cosmic space 12 hour and 8 minutes. After disconnection from the ISS, STS-115 «Atlantida» happily landed on 21 September 2006 on John F. Kennedy Center of Space Flights, Cape Canaveral, FL, USA.

Medical statistics

In 1972 Sir **David R Cox (1924 – 2022)** — British statistician, observed that if the proportional hazard assumption holds, or is assumed to hold, then it is possible to estimate the effect parameter(s), without any consideration of the hazard function (Cox proportional hazard model).

Health care

Public health on an international scale has been provided by the International Sanitary Bureau (now Pan American Health Organization, 1902), l'Office International d'Hygiene Publique (1907), the League of Nations Health Organization (1919-46) and the United Nations (UN)' World Health Organization (WHO, 1948 - present).

[1775] City of St Paul (settled 1,000 BC., founded as military fortress 1819), MN, USA. — Ablitsov VH. «Galaxy Ukraine». The Ukrainian diaspora: prominent persons. Publ. «Kyt», Kyiv 2007.

In 1903 **Ivan Ya. Franko (1856 – 1916)**[1776] — Ukrainian poet, writer, and publicist, warned that the type of society envisaged by **Friedrich Engels (1820 – 95)** — German philosopher and sociologist, would end up as a totalitarian state.

Walter Reed [19th century. Microbiology and immunology. – **Walter Reed (1851 – 1902)**] Army Medical Center, Washington, DC, was founded in 1909.

The National Insurance Act of 1911 in Britain induced creation of the national social health insurance for the primary patient care physicians, excluding specialists and hospital care, initially for about one-third of the population which included employed workers wage earners, but not their dependents. This system continued until the British National Health System was proposed in 1942 for the Social and Allied Forces by **William H. Beveridge (1879 – 1963)** — British economist and social reformer, and included in the Ministry of Health (established 1948). It is based on centralized budged financing and provision by the government with more ability to control costs, through tax payment, just as police and libraries. The health care is provided and financed by the government. Private hospitals and physicians collect their fees from the government which controls them. Bills for medical services were never addressed to the patients. Countries using the «William Beveridge model» or variations are UK, Spain, Denmark, and New Zealand.

The scientific foundations of socialized medicine were laid by **Alfred Grotjahn (1869 – 1931)** — German physician and social hygienist. The social insurance was and is supposed to guarantee a protection of the individual against economic hazards (unemployment, old-age or disability, in which government participates or enforces the participation of employers and affected individuals (1917). Organized social welfare (1917) and work (1920) were created to provide public or private professional services, activities or methods concerned with investigation, the treatment and material aid of the economically underprivileged and socially maladjusted.

The so called «socialistic medicine» or the Soviet Health Care System of the Russian Soviet Federative Republic (RSFSR), including Ukraine and the other former soviet republic, created in 1918 by **Mykola (Nikolai) O (A). Semashko (1874 – 1949)** — Russian physician, people commissar of health care of the RSFSR (1918-30), where health care was to be controlled by the state and would be provided to its citizen free of charge. Although initially it made some progress as compared to the primitive health care of the Tsarist Russia, thought in 1964-90 it degenerated into a disastrous pseudo-system, which amused catastrophic health conditions of that time and the following generations of its citizens.

[1776] Franko IYa. The Collection of Works in Fifty Volumes. Vol. 1-50. Publ. «Scientific Thought», Kyiv 1976-86. — Nahaylo B. Ukrainian poet boldly addresses the blank spots in Western Ukraine's history. Ukrainian Weekly Aug. 31, 1988;66(34)1-2. — Ablitsov VH. «Galaxy Ukraine». The Ukrainian diaspora: prominent persons. Publ. «Kyt», Kyiv 2007.

In the USA public health assumed a local autonomous form. The establishment of the Social Security Act (1935) created the principle, practice, and program of public provisions (social insurance or assistance) for the economic security and social welfare of the individual and his family. This program includes an old-age survivor insurance, contributions to state unemployment and an old-age assistance.

Socialized medicine, organized in 1938, provides a system of medical care for the population, regulated, and controlled by the government and paid for from founds obtained usually by assessments, philanthropy or taxation.

Introduced by **John G. Diefenbacker (1895 – 1976)** — Prime Minister (PM) of Canada (1957-63), the Hospital and Diagnostic Services Act of 1957-61 shared the cost of covering hospital services, until the Canada Health Act (CHA) of 1984 paved way to creation of the Canada National Health Insurance (NHI) model, informally called Medicare, or a single-payer insurance, a mixture of the Sickness Insurance Law of of Germany from 1883 **[Otto Bismarck (1815 – 98)**. 19th century. Health Care] and the British «William Beveridge model» of 1942-48 (above). The NHI model, guided by the CHA delivers health care through a public founded health care system which uses private-sector providers, but payment comes from a government-run insurance program every citizen pays into. These universal insurance programs do not need marketing, have no financial motive, and seeks no profit, tends to be administratively simpler and cheap. They also control costs by limiting the medical services they will pay for, by making patients wait to be treated, have considerable power to negotiate for lower prices, e.g., from pharmaceutical companies for drugs. The NHI model was adopted by Taiwan (area 36,197 km^2) and South Korea (area 100,210 km^2).

In 1965 **Lyndon B. Johnson (1908 – 73)** — the 36th President of the USA (1961-69), signed the law that guarantees access to health insurance for Americans ages 65 and older and younger people with disabilities as well as people with end stage kidney disease (ESKD) (Medicare), and the program for certain people and families with low incomes and resources (Medicaid).

The out-of-pocket model of health care applies to mostly in poor, undeveloped and disorganized countries, unable to provide any kind of mass medical care. In those countries the basic rule is that the rich get medical care, the poor stay sick or die. Those countries are mostly Africa (area 30,370.000 km^2), India (area 3,287.363 km^2), Cambodia (area 1181.035 km^2) and South America (area 17,840.000 km^2). Some poor people may have access to a village healer using home-brewed herbal remedies that may or not be affected against disease. Sometimes they can scratch together enough money to pay a physician the bill out-of-pocket at the time of treatment, or pay him with their home-made foot products, or by his child care. In some of those countries if one is sick enough, he may be admitted to the emergency ward of the public hospital.

The fragmented national health insurance for profit of the USA is maintained because of existence of separate classes of the people. It combines elements of the National Insurance Law of Germany from 1883 **[Otto Bismarck (1815 – 98)**. 19th century. Health care]**, the British «William Beveridge model» from 1942-48 (above), the NHI of Canada from 1984 (above) and the out-of-pocket model of health care (above).

The established health care system is available in about 40 well developed and industrialized countries, out of the world's 200 countries. 15% of the USA population have no health insurance.

Since January 01, 2018, Ukraine began introducing the NHI model of Canada[1777]. With new on-line E-National Health System of Ukraine, each family physician is allowed to have no more than 2,000 patients, and the clinic where the physician works will receive $15 per patient. In 2018-20 the Government of Ukraine will calculate the rates at which it will purchase secondary and specialized tertiary care services. Based on these rates, contracts with providers will be signed starting in 2020. The funding will be allocated from unified social tax contributions. Ukraine hopes to eventually introduce co-payment principle into its health care as well.

According to the World Health Report of the WHO concerning life expectancy at birth for male/female (2015) is following:

Japan	80/87
Australia	81/85
Sweden	81/84
Italy	80/85
Canada	80/84
Norway	80/84
New Zealand	80/83
France	79/85
Germany	79/83
UK	79/83
USA	77/82
Poland	74/81
India	67/70
Ukraine	66/76
Russia (both sexes)	70

[1777] Bugrly M. New Analysis: Ukraine and Russia compete in health reform. The Ukrainian Weekly Dec 3, 2017;85(49):3,13. — The Future of Medicine. Breakthroughs to Improve Your Health. Tips Specials. National Geographic, Washington, DC 2021.

The USA has the most expensive health care as a percent of the gross domestic product (GDP) – 16.1%. This could be explained by the fact that in the USA are performed the greatest number of the most complicated and expensive medical research, as well as invasive medical and surgical treatment of the patients. Despite of this, the quality of medical care in the USA since the beginning of 21th century considerable worsened.

At present the best health care is being practiced in the Scandinavia (Denmark, Norway and Sweden), Finland, UK, Ireland, Germany, Austria, Switzerland, Belgium, Luxembourg, the Netherlands, France, Italia, Canada, Australia, New Zealand and Japan (1st place), USA (2nd place) and the rest of the World (3rd place).

According to data of the State Statistics Service of Ukraine (founded 1920) the average expected life longevity at birth in Ukraine for 2019 was 72,01 years, for man 66,92 years, for women 76,96 years. In the last 30 years (since 1990) the population of Ukraine has shortened by approximately 10 million, from 51,556,500 (in 1990), to 41,732,800 million (in 2020 p.). The life longevity in Ukraine became the shortest in Europe.

REFERENCES

1. Amosov NM. Ocherki torakalnoi khirurgii. «Derzhmedvydav» Ukraïny, Kyiv 1958.
2. Amosov NM, Berezovsky KK. Pulmonary resection with mechanical suture. J Thorac Cardiovasc Surg 1961;41:325-35.
3. Amosov NM, Lissov IL, Sidarenko LN. Operacii na serdce s isskustvennym krovoobrashcheniem. «Derzhmedvydav » Ukraïny, Kyiv 1962.
4. Amosov NM Regulaciia zhiznennykh funkcii i kibernetika. «Naukova Dumka»Kyiv 1964.
5. Amosov MN. Modeliravanie myshleniya i psychiki. «Naukova Dumka», Kyiv 1965.
6. Amosov MN. Modeling of Thinking and Mind. Spartan Books, New York 1967.
7. Amosov NM. The Open Heart. Simom & Shuster, New York 1966.
8. Babych I. Ukrains'kyi biotrom. Ukraïna 1960;17(272):2.
9. Bazhan MP. Ukraïns'ka Radians'ka Encyclopedia. T. 1-17. Vyd pershe. Hol red URE, Kyiv 1959-65.
10. Bazhan MP. Ukraïns'ka Radians'ka Encyclopedia. T. 1-12. Vyd druhe. Hol red URE, Kyiv 1977-85.
11. Aseev YS. Architektura Kyivs'koi Rusi. «Budivelnyk», Kyiv 1969.
12. Bazylevych I. Teofil Yanovs'kyy (1860 -1928). J Ukr Med Assoc North Am 1955;22(4):23-8.
13. Bezredka AM. Istoria odnoi idei. Tvorchestvo Mechnikova. Kharkiv 1926.
14. Bibikov CM. Narysy starodavnoi istoriï Ukraïns'koï RSR. AN Ukraïny, Kyiv 1959.
15. Bobechko WP, McLaurin CA, Motloch WM. Toronto orthosis for Legg-Perthes' disease. Art Limbs 1968;12:36-41.
16. Bobechko W. Samovilna skolioza u ditei. J Ukr Med Assoc North Am 1973;20:4(71):3-14.
17. Bogoiavlenskii NA, Drevnerusskoe vrachevanie v XI-XII vv. «Medgiz», Moskva 1960.
18. Bodrova NV, Kraiuchin BV. Vydatnyi ukraïns'kyi vchenyi Olexandr Olexandrovych Leontovych. Fiziol zh 1959;5(6):689.

19. Bovkun S, Semen V. Verba i b'ye i likuye. Express, 14-21.04.2011;40(5693):6.

20. Bratus VD, Michnev AL, Duplenko KF, Beniuchov RY. Ocherki istorii vysshego medicinskogo obrazovaniia i nauchnych medicinskich shkov na Ukraine. «Zdorov'ia», Kyiv 1965.

21. Blummelkamp WH, Boerema I, Hoogendyk L. Treatment of Chlostridian infections with hyperbaric oxygen drenching: a report of 36 cases. Lancet 1963;1:235.

22. Burskner YS. Vykorystannia pryrodnychych i likuvalnych resursiv Ukraïny. AN Ukraïny, Kyiv 1959.

23. Chagovec VY. Izbrannye trudy. AN Ukrainy, Kyiv 1957.

24. Chykalenko Y. Spohady (1861-1907). Chastyna 1-3. Lviv 1925-26.

25. Chychenko IM. Medychna osvita v Ukraïni. J Ukr Med Assoc North Am 1971;18,4(63:47-55.

26. Chynchenko IM. Viktor Drobot'ko (iz serii vyznachnych syniv Ukraïny). J Ukr Med Assoc North Am 1974;21,3(74):76.

27. Dal MK. Hryhoriy Mychailovych Minch (1835-1896). «Derzhmevydav» Ukraïny, Kyiv 1956.

28. Dejneka VL, Maryenko FS. Korotki narysy z istoriï chirurhiï v Ukraïns'kiy RSR. «Zdorov'ia», Kyiv 1968.

29. Dzeman MI. Tradition of the triuneness treatment process in Ukrainian medicine. (Part 1). The Practitioner Physician 2015;4:63-73.

30. Dumka MS. Pro medycynu skifiv. «Derzhmedvydav» Ukraïny, Kyiv 1960.

31. Duplenko KF. Materialy do istoriï rozvytku ochorony zdorov'ia na Ukraïni. «Derzhmedvydav» Ukraïny, Kyiv 1957.

32. Duplenko KF. Narysy istoriï rozvytku ochorony zdorovia na Ukraïni. «Derzhmedvydav» Ukraïny, Kyiv 1965.

33. Farb P. He found medicine in the earth. Today's Health 1959;37:54-6.

34. Farb P. Lekarstwa w łyżce ziemi. Ameryka, Warszawa 1963;52:48.

35. Felczyński Z. 500-lat szpitalnictwa w Przemyślu 1461-1964. Przemyśl 1965.

36. Filatov VP. Optychna peresadka rohivky i tkanynna terapiia. Odesa 1948.

37. Filatov VP. Moi puti v nauke. Odesskoe obl izd, Odesa 1955.

38. Filatov VP. Izbrannoye trudy . T. 1-4. AN Ukrainy, Kyiv 1961.

39. Fisher I. Biographischer Lexicon der hervorragenden Aerzte der letzer funfzing Jahre. Band 1-11. 1933.

40. Garrison FH. A Introduction to the History of Medicine. 4th Ed, WB Saunders, Philadelphia- London 1970-76.

41. Gillispie CC. Dictionary of Scientific Biography. Vol. 1-14. Charles Scribners, Philadelphia-London 1970-76.

42. Grenda J, Olearczyk A. On the problems of the psyche of the patient and the surgeon in the treatment of surgical diseases. Zdrowie Psych 1965;6,3:48-58.

43. Grenda J, Olearczyk A. Psychotherapy in Surgery. In T. Bilikiewicz, Psychoterapy in General Medical Practice. II Ed. PZWL, Warszawa 1966: 274-85.

44. Grenda J, Kiesz W, Olearczyk A. Seminoma of the testicle with pleural metastases. Pol Przegl Chir 1963;35(11a):1265-9.

45. Grenda J, Olearczyk A. Psychotherapy in Surgery. In T. Bilikiewicz, Psychoterapy in General Medical Practice. III Ed. PZWL, Warszawa 1970. P. 288-301.

46. Grmek MD. Starodawna słowiańska medycyna. J Hist Med 1959 Jan ;14(1):18-40.

47. Hanitkevych Ya. Ukrainian Physician-Scientist of the First Half of the 20th Century. Biographic Outlines and Bibliography. Lviv 2004.

48. Hanitkevych Ya. Vnesok likariv v ukraïns'ku kulturu. J Ukr Med Assoc North Am 2004;49,1(151):56.

49. Hanitkevych Ya. Istoriia ukraïns'koï medycyny v datakh ta imenakh. NTSh-Львів, I-t ukr arch dzhereloznavstva im. MS Hrushevs'koho NAN Ukraïny, Lvivs'ke viddilennia Vse-ULT, Lviv 2004.

50. Hanitkevych Ya. The contribution of Ukrainian physicians to world medicine. Acta Inter Hist Med «Vesalius» June, 2009;15(1):12-8.

51. Hanitkevych Ya. Vnesok ukraïns'kykh likariv u svitovy medycynu. Ukr med chasopys VII/VIII 2009;4(74):110-5.

52. Hirsch A. Biographisches Lexicon der hervorragenden Aerzte aller Zeiten und Volker. Band 1-V. Wien und Leipzig 1929-35.

53. Honcharenko M. Akademik Ivan Horbachevskyi (15.V.1854 – 24.V.1942). J Ukr Med Assoc North Am 1970;17,1(56):34-44.

54. Hordynsky J. Medical Aspect of Skylab Missions. J Ukr Med Assoc North Am 1974;21,1(72):20-24.

55. Hornykevych O. Biochemichni ta farmacevtychni osnovy zastosuvannia l-dopy u likuvanni chvoroby Parkinsona. J Ukr Med Assoc North Am 1975;22,2(77):32-37.

56. Hryntschak T. Suprapubic Prostatectomy with Primary Closure of the Bladder by an Original Method. Charles C. Thomas, Sprinfield, IL 1955.

57. Hubskyi IM. Aptechna sprava v Ukraïnskiy RSR. «Zdorov'ia», Kyiv 1964.

58. Hutchin P. History of blood transfusion: a tercential look. Surgery 1968;64:685-700.

59. Ihumov SM. Narys rozvytku zemskoi medycyny v hubeniach, shcho vviyshly do skladu Ukraïns'koï RSR, Besarabiï ta v Krymu. Vyd Kyïvskoho univ, Kyiv 1940.

60. Ikonnikov VS. Biograficheskiy slovar professorov i prepodavateley Imp Univ Sv Volodymyra. Kyiv 1984.

61. Isajewicz J. Jerzy z Drohobycza. Malop Sl Hist 1963.

62. Kabanov V. Sklifosovskii. Moskva 1952.

63. Kacl K. Prof. dr. Jan Horbaczewski. Casopis Lekaru Ceskich 1954;93:22-23;578-580.

64. Kalyta VT. Danylo Zabolotnyi. «Molod'» Kyiv 1981.

65. Kasymenko OK. Istoriya Kyyeva. T. 1-2. AN Ukraïny, Kyiv 1960-61.

66. Kavec'kyy RE, Balic'kyy KP. U istokov otechestvennoi mediciny. AN Ukrainy, Kyiv 1954.

67. Kaveckyi RE, Balyckyi RP. Vklad uchenych Akademiï Nauk Ukraïns'koi RSR v rozvytok medycyny. AN Ukraïny, Kyiv 1965.

68. Kaveckii RE, Gamaleya N. Laser protiv raka. Sputnik, Moskva 1972;10:118-21.

69. Klein BI. Osnovopolozhnyky odes'koi shkoly microbiolohiv. Microb zh 1958;20(2):71.

70. Klein BI. Kyïvs'ki mikrobiolohy devianostych rokiv XIX stolittia. Microbiol zh 1958;28(1).

71. Kochan I. Role of macrophages in cellular immunity. J Ukr Med Assoc North Am 1974;21,1(72):3-13.

72. Kochan I. Rolia zaliza v rehuliuvanni protybakteriynoï imunosti. J Ukr Med Assoc North Am 1975;22,2(77):17-31.

73. Kolomiychenko M. Zhyvy liudyno. «Molod'», Kyiv 1962.

74. Konovalov VV, NV Sklifosovskii. 1836-1904. Moskva 1952.

75. Konvolinka CW, Olearczyk A. Subphrenic Abscess. In, MM Rawitch (ed), Curr Probl. Surg. Year Book Med Publ, Chicago, Jan. 1972:1-52.

76. Kostyuk HS. Narysy z istoriï psycholohiï kincia XIX i pochatku XX stolittia. «Rad Shkola», Kyiv 1969. —

77. Kovalchuk LYa, Sayenko VF, Knyshov HV. Clinical Surgery. Vol. 1-2. Publ. «Ukrmedknyha», Ternopil 2000.

78. Kovner SG. Istoria mediciny. T. 1-3. Kyiv 1878-88.

79. Kovner SG. Istoria srednevekovoi mediciny. T. 1-2. Kyiv 1893.

80. Kovpanenko HT. Plemena skifs'koho chasu na Vorskli. «Naukova Dumka», Kyiv 1967.

81. Krymov OP. MM Volkovych. Kyiv 1947.

82. Koźmiński S. Słownik lekarzy polskich. W-wo «Oryginalne autora», Warszawa 1888.

83. Kubijovyc V, Kuzela Z. Encyklopedia Ukraïnoznavstva. T. 1/1-3. NTSh, Munich - New York 1949-55.

84. Kubijovyc V, Kuzela Z. Encyklopedia Ukraïnoznavstva. T. 2/1-10. NTSh / Univ Toronto Press, Paris - New York 1955-76.

85. Kubijovyc V. Ukraine. A Concise Encyclopedia. Vol. 1-2. NTSh / Univ Toronto Press, Paris – New York 1963-71.

86. Kubijovyc V, Struk DH. Encyclopedia of Ukraine. Vol. 1-5. Univ Toronto Press, Toronto – Buffalo - London 1985-93.

87. Kukulev AV. VA Betz (1834-1894). Moskva 1958.

88. Kwaskowski A. Moi profesorowie Wydziału Lekarskiego Uniwersytetu Kijowskiego (1912-19). Arch Hist Med 1962;25:221.

89. Larichev LS, Karayev PG. Kurorty Ukrainy. Kyiv 1959.

90. Larychev LS. Rozvytok sanitarno-kurortnoï spravy v Ukraïns'kiy RSR. «Zdorov'ia», Kyiv 1969.

91. Lazarenko ZK. 300 let Lvovskogo universiteta. Lviv 1961.

92. Lohaza M. Do istoriï ULT-va u Lvovi. J Ukr Med Assoc North Am 1973;20,3(70):31-8.

93. Leshchenko PD, Kaliuzhny DN, Grando AA. Materialy k istorii gigieny i sanitarnogo dela v Ukrainskoï SSR. «Gosmedizdat Ukrainy», Kyiv 1959.

94. Lohaza M. Shche do istoriï ULT u Lvovi. J Ukr Med Assoc North Am 1974;21,2(73):71-3.

95. Lohaza M. Dva sribni iuvilei (do ictoriï Ukraïns'koho Likars'koho Tovarystva u Lvovi). J Ukr Med Assoc North Am 1976;23,1(80):39-42.

96. Lyons AS, Petrucelli RJ II. Medicine. An Illustrated History. HN Abrams Publ, New York 1987.

97. Mar'yenko F. Hamaliya. Odesa 1961.

98. Martych Y. Olexandr Bohomolets. Kyiv 1951.

99. Marchenko MI. Ukraïns'ka istoriohrafia. Kyiv 1959.

100. Mayba II, Gaida R, Kyle RA, et al. Ukrainian Physicist contributed to the discovery of x- rays. Mayo Clin Proc 1997;72:658.

101. Mechnikov II. Akademicheskoe sobranie sochinenii. T. 1-15. Moskva 1950-60.

102. Michailov SS. Karavayev. St. Petersburg 1954.

103. Michniov AL, Duplenko KF. Narysy istoriï terapiï v Ukraïns'kiy RSR. «Derzhmedvydav Ukraïny», Kyiv 1960.

104. Mishchenko M. Medical Science in Ukraine under the Soviets. Ukrainian Quarterly 1949;5:310-17.

105. Mogilevskii B. Zhizn' Mechnikova. Khark obl izd. Khakiv 1955.

106. Moskovets SM. DK Zabolotyi i suchasna mikrobiolohia ta epidemiolohia. «Naukova Dumka». Kyiv 1968.

107. Muller JE. Diagnosis of myocardial infarction: historical notes from the Soviet Union and the United States. Am J Cardiology 1977;40:269-71.

108. Myrs'kyy MB. The pioneer in clinical transplantation of kidneys Yu.Yu. Voronyy. Ahapit, Kyiv 2001;14-15:10-15.

109. Nichyk VM. Folosofs'ki osnovy prac' O.O. Bohomoltsia. AN Ukraïny, Kyiv 1958.

110. Olearchyk A. Ukrainian Scientists. Olexandr Volodymyrovych Palladin – a prominent biochemist. Nashe Słovo, Warszawa 16 Dec 1956;16(18):3.

111. Olearchyk A. Ukrainian Scientists. Olexandr Bohomolets – a prominent pathophysiologist. Nashe Słovo, Warszawa 30 Dec. 1956;1(20):4.

112. Olearchyk A. V. Ukrainian Scientists. V.P. Filatov – a prominent ophtalmologist. Nashe Słovo, Warszawa 27 Jan 1957;2,4 (24):2.

113. Olearchyk A. Ukrainian Scientists. Danylo Kyrylovych Zabolotnyi. Nashe Słovo, Warszawa 17 Feb 1957;2,7(27):3.

114. Olearchyk A. Ukrainian Scientists. On the 20th anniversary from the death of H.F. Pysems'kyi (1862-1937). Nashe Słovo, Warszawa 7 Apr 1957;2,14 (34):2.

115. Olearchyk A. Ukrainian Scientists. Ivan Horbachevs'kyi (103 years from the date of birth). Nashe Słovo, Warszawa 14 Apr 1957;2,15 (35):2.

116. Olearchyk A. Ukrainian Scientists. Volodymyr O. Betz (1834-1894). Nashe Słovo, Warszawa 8 Sep 1957;2,36(56):2.

117. Olearchyk A. The contributions of Ukrainians to the development of medicine during the 40 years' existence of the Ukrainian S.S.R. Nashe Słovo, Warszawa 2 Mar 1958;3,9(81):4.

118. Olearchyk A. A Ukrainian – professor of three universities in the 15th century. Nashe Słovo, Warszawa 11 May 1958; 3,19 (91):2.

119. Olearchyk A. On current Ukrainian scientific terminology. Nasha Kultura, Warszawa 29 Jun 1958;1,2:6-7.

120. Olearchyk A. Professor M.O. Dudko – a university surgeon from Ukraine in Poland. Nashe Słovo, Warszawa 7 Dec 1958;3,49 (121):2.

121. Olearchyk A. An interview with O.V. Palladin – a world renown Ukrainian scientist. Nasha Kultura, Warszawa 30 Dec 1958;8:1-2.

122. Olearchyk A. Successes of Ukrainian physicians in the surgical treatment of tuberculosis. Nasha Kultura, Warszawa Feb 1959;2 (10):14.

123. Olearchyk A. Mykola Fedorovych Hamaliia. Nasha Kultura, Warszawa Mar 1959;3(11):16.

124. Olearchyk A. I.I. Mechnykov, O.M. Bach, I. Horbachevs'kyi Nasha Kultura, Warszawa May 1959;5(13):16.

125. Olearchyk A. Ukrainian Scientists in Medicine. Ukrajins'kyi Kalendar, Warszawa 1961: 84-90.

126. Olearchyk A. Ukrainian biological science. Nasha Kultura, Warszawa May 1961;5 (37):11-12.

127. Olearchyk A. At the origin of medicine in Ukraine. Nasha Kultura, Warszawa Jun 1965;6 (86):13.

128. Olearchyk A. Ukrainian medicine in the 14th–17th centuries. Nasha Kultura, Warszawa Sep 1965;9(89):14-5.

129. Olearchyk A. Prominent Ukrainian physicians in the 18th century. Nasha Kultura, Warszawa Nov 1965;11(91):14-5.

130. Olearchyk A. Prominent Ukrainian physicians of the 19th and the beginning of the 20th century. Nasha Kultura, Warszawa Jan 1966;1(93):14-5; Feb 1966;2 (94):15; Mar 1966;3 (95):15; Apr 1966;4 (96):15.

131. Olearchyk A. Health service in Western Ukrainian territories prior to their reunification with the Ukrainian S.S.R. Nasha Kultura, Warszawa Jun 1966;6 (98):15; Jul 1966;7(99):15.

132. Olearchyk A. History of anesthesia in Ukraine. J Ukr Med Assoc North Am 1966;13,3-4 (42-3):19-21,24,32.

133. Olearchyk A. Medicine and health welfare in the Ukrainian S.S.R. Nasha Kultura, Warszawa Nov 1966; 11 (103):15; Dec 1966;12 (104):15; Feb 1967;2 (106):15; Mar 1967;3 (107):15.

134. Olearchyk AS, Konvolinka CW. Subphrenic Abscesses. J Ukr Med Assoc North Am 1972;19,3 (66):3-56.

135. Olearchyk AS. Ulcers of the Stomach and Duodenum in Children. J Ukr Med Assoc North Am 1974;21,4 (75):3-100.

136. Olearchyk AS. To the history of medicine and health welfare in Ukraine. Ukr Med Assoc North Am 1980;27,1 (96):37-53.

137. Olearchyk AS. To the history of medicine and health welfare in Ukraine. America, Philadelphia, PA 25 Jul. 1980;69,115:2-3; / 26 Jul. 1980;69,116:2-3.

138. Olearchyk AS. History of coronary artery bypass grafting. J Ukr Med Assoc North Am 1986;33,1 (114):3-8.

139. Olearchyk AS. Internal mammary artery grafting. J Thorac Cardiovasc Surg 1987;94:312.

140. Olearchyk AS. Coronary revascularization: past, present and future. J Ukr Med Assoc North Am 1988;35,1 (117):3-34.

141. Olearchyk AS. Vasilii I. Kolesov. A pioneer of coronary revascularization by internal mammary-coronary artery grafting. J Thorac Cardiovasc Surg 1988;96:13-8.

142. Olearchyk AS. Endarterectomy and external prosthetic grafting of the ascending and transverse aorta under hypothermic circulatory arrest. Texas Heart Inst J 989;16 (2):76-80.

143. Olearchyk AS. Revascularization of the myocardium. Grud serd-sosud khir 1990;2:68- 70.

144. Olearchyk AS. About the health. Interview. Lviv Television Station, Ukraina Aug 13, 1990.

145. Olearchyk AS, Olearchyk RM. Concise History of Medicine. J Ukr Med Assoc North Am 1991;38,3 (125):1-159.

146. Olearchyk AS, Olearchyk RM. A comment on a review. Svoboda Sep. 27. 1992;99,164:2-3.

147. Olearchyk AS, Olearchyk RM. Concise History of Medicine. Errata. J Ukr Med Assoc North Am 1993;40,3(131): 203.

148. Olearchyk AS. Medicine. In V. Kubijovyc, DH Struk (Eds), Encyclopedia of Ukraine. Univ Toronto Press. Toronto - London 1993;3:363-6.

149. Olearchyk AS. Medicine in Ukraine. America, Philadelphia, PA Dec 6, 1995;84, (93):11-12. Dec. 13, 1995;84 (95-97):13. / Jan 3, 1996;85 (1):11. / Jan. 17, 1996;85 (3):18. / Jan. 24, 1996:85 (5):12.

150. Olearchyk AS. In defence of Likars'kyi Visnyk (Lik Visn) or the Journal of the Ukrainian Medical Association of North America (JUMANA). UMANA News, Chicago, IL 1996;Summer:1,3.

151. Olearchyk AS, Olearchyk RM. Reminiscences of Vasilii I. Kolesov. Ann Thorac Surg 1999:67:273-6.

152. Olearchyk AS. Left-sided extrapleural thoracoplasty for pulmonary tuberculosis. J Ukr Med Assoc North Am 2001;46,1(146): 20-25.

153. Haller JD, Olearchyk AS. Cardiology's 10 greatest discoveries. Tex Heart Inst J 2002;29(4):342-4.

154. Olearchyk AS. A Surgeon's Universe. Publ. «Medytsyna svitu» - «Ms», Lviv 2003.

155. Olearchyk AS. A Surgeonn's Universe. 2nd Ed. Publ. «Medytsyna svitu» - «Ms», Lviv 2006.

156. Olearchyk A. To the history of medicine and health welfare in Ukraine. Ukrainian Medical Association in Lviv Aug. 1, 2009;21 (www.utl.lviv.ua/index).

157. Olearchyk AS. A Surgeon's Universe. Vol. 1-4. 3rd Ed. Book Publ. Co. «AuthorHouse», Bloomington, IN 2011.

158. Olearchyk AS. A Surgeon's Universe. Vol. 1-4. 4rd Ed. Book Publ. «Outskirts Press», Denver, CO 2016.

159. Olearchyk AS, Olearchyk RM. The Evolution of Medicine. Book Publ. «AuthorHouse», Bloomington, IN 2020.

160. Oleinik SF. K otkrytiu kapsuly Shumliansjogo. Klin med 1952;30(6).

161. Oleinik SF. Perelivanie krovi v Rossii i SSSR. «Gosmedizdat« Ukrainy, Kyiv 1956.

162. Olesnyc'kyi Y. Pershi orhanizacii ukraïns'kykh likariv na zakhidnykh ukraïns'kykh zemliakh. J Ukr Med Assoc North Am 1954;1(1):13.

163. Onackyi Y Ukraïns'ka Mala Encyklopedia. T. 1-16. UAPC v Arhentyna, Buenos Aires 1957.

164. Orlovskyi ST. Istoriia khimii. «Rad Szkola», Kyiv 1959.

165. Ostapenko TA, Bulakh SM, Kirisheva OM, Kornilova LYe. Raïsa Ivanivna Pavlenko, Bibliographical Outline. Publ Nat Sci Med Lib (NSML) of Ukraine, Kyïv 2007.

166. Pavlenko RI, . Ostapenko TA, Kharchenko NS, Bulakh SM, Nikolayeva VV. The Centuries Alive in the Old Folio-Volume. The Treasures of Medical Thought. NSML of Ukraine, Kyïv 2015.

167. Palladin AV. Ocherki po istorii biokhimii na Ukraine. Vyp. 1. AN Ukraïny, Kyiv 1954.

168. Palladin OV. Rozvytok nauky v Ukraïns'kiy RSR za 40 rokiv. AN Ukraïny, Kyiv 1957.

169. Pashkin IP. Diyalnist' akademika O.O. Kovalevs'koho na Ukraïni. V, KK Kremov (red), Narysy z istorii tekhniky i pryrodoznavstva. Vyp. 2. AN Ukraïny, Kyiv 1962:105-12.

170. Pasternak Y. Archeolohiia Ukraïny. NTSh, Toronto 1961.

171. Paton BY. Istoriia Akademiï Nauk Ukraïns'roï RSR. Kn. 1-2. Ukraïns'ka Radians'ka Encyclopedia, Kyiv1967.

172. Petrov BD, Bratus' VD, Duplenko KF. Ocherki istorii medicinskoi nauki i zdravookhraneniia na Ukraine. «Gosmedizdat» Ukrainy, Kyiv 1954.

173. Petrov VP. Entohenez slov'ian. «Naukova Dumka», Kyiv 1972.

174. Picyk N. Liulyna velykoï mriï. Kyiv 1962.

175. Picyk N. Olexandr Olexandrovych Bohomolec' «Naukova Dumka», Kyiv 1971.

176. Plushch V. Outline of the History of Ukrainian Medical Science and Medical Education. Ukr Free Acad Sci in Germany. Vol. 1-2. München 1970, 1983.

177. Plushch V, Dzul P. Contribution to the History of Ukrainian Medicine. Vol. 1-2. UMANA, New York - Munich - Chicago 1975, 1988.

178. Polians'kyi Y. Marian Panchyshyn (1882-1943). «Siohochasne i Mynule». 1983;3:121.

179. Panov VM. Ucheni vuziv Ukraïns'koï RSR. Kyiv University, Kyiv 1968.

180. Prykhod'kova YK. Do 75-richchia z dnia narodzhennia akademika AN Ukraïny HV Folborta. Fiziol zh 1960;6(2):148.

181. Puchkivs'kyi O. Olimpiy ta Ahapit – pershi ukraïns'ki likari. Zbir med sekciï Ukr nauk t-va u Kyivi 1922;5:2.

182. Puchkivs'kyi O. Rola ukraïnciv v rozvytku medycyny v Rosii. Zapysky doslidnykh katedry pry VUAN. 1923.

183. Puchkivs'kyi O. Try fundatory rosiys'koï medycyny (Petro Zahors'kyi, Nestor Maksymovych-Ambodyk i Danylo Vellans'kyi-Kavunnyk). Ukraïna 1924;4:27.

184. Puchkivs'kyi O. Pershi medyko-sanitarni ta antropolohichni narysy Ukraïny. Likarskyy visnyk 1926:1.

185. Puluj Ivan. V, Studnicka FJ, Čelakovský J. Ottuv slovník naučný. J Otto, Praha 1903;20:83-4.

186. Rakovs'kyi I. Ukraïns'ka Zahalna Encyklopediia. T. 1-3. Lviv-Kolomyia 1930-35.

187. Rafes UI. Z ukraïns'ko-pols'kykh zvi'azkiv u medycyni. Kyiv 1961.

188. Renner W. Historical Data on the Beginning of Medicine. J Ukr Med Assoc North Am 1982;29,1(103):37-39.

189. Rozhin I. Professor d-r Olexandra Smyrnova-Zamkova. J Ukr Med Assoc North Am 1959;3,1(5):32-4 .

190. Rozhin I. Do istorii medycyny na Ukraïni. J Ukr Med Assoc North Am 1956;3,2(6):19.

191. Rozhin I, Rozhin V. Volodymyr Pidvysoc'kyi. J Ukr Med Assoc North Am 1957;4,1 (17):43-6.

192. Rozhin I. Materialy do istoriï ukraïns'koi veterynarno-medychnoï nauky. J Ukr Med Assoc North Am 1963;10,3(24):2-7.

193. Rostafilski J. Nasza literatura botaniczna 16 w. oraz jej autorowie i tłumacze. Pam Akad Umiejętności, Wydz Mat Przyrod. T. 14. Kraków 1888.

194. Rukin VO. Do istoriï vitchyznianoï oftalmolohiï. «Derzhmedvydav» Ukraïny, Kyiv 1957.

195. Rukin VA. Istoriia oftalmologii na Ukraine. Kharkiv 1960.

196. Russin LA. The Sivash total hip prosthesis, its principles and a clinical implantation. US Surg Corp Film Lib, USSC-16.

197. Ruchkovskii BS. Rol otechestvennykh uchenykh v razvitii eksperimemtalnoi oncologii. AN Ukrainy, Kyiv 1953.

198. Ruchkovskii BS. Ocherki razvitiia sovetskoi eksperimentalnoi onkologii. Kyiv 1959.

199. Ryan M. The Organization of Soviet Medical Care. Professional Semimar Consultant, New York 1978.

200. Samoilovych DS. Isbrannye proizvedeniia. T. 1-2. Moskva 1949-52

201. Schmidt JE. Medical Discoveries. Who and When. Charles C. Thomas Publ, Springfield, IL 1959.

202. Semenenko MT. Akademiia Nauk Ukraïns'koï RSR. «Naukova Dumka», Kyiv 1969.

203. Senycia P. Pedahoh, likar-psycholoh, d-r Stepan Baley. J Ukr Med Assoc North Am 1976;23,2(81):41-2.

204. Shamov WN. The transfusion of stored cadaver blood. Lancet 1937;233:306-9.

205. Shapiro IY. Iz istorii vyshchego medicinskogo obrazovaniia v zapadnykh zemliakh Ukrainy. Vrach delo 1957;2:211.

206. Shapiro IY. Ocherki po istorii Lvovskogo medicinskogo instituta. Lviv derzh med i-t, Lviv 1959.

207. Shymanko OI, Melnychenko PK. Orhanizaciia farmacevtychnoï spravy. «Zdorov'ia», Kyiv 1965.

208. Shishina Y. Khirurg-izobredatel Konstantin Shivash: «ne nazyvaite menia Edisonom». Sputnik (Moskva) 1972;10:100-4.

209. Shlakhtychenko M. Profesor doctor Ivan Horbachevs'kyi. J Ukr Med Assoc North Am 1958;5(9):7-13.

210. Schumlansky A. De structura renum. Tractatus physiologico-anatomicus. 1st & 2nd Eds. Strasbourg 1782 & 1788.

211. Shpilberh HI. Dytiachi kurorty Ukraïny. «Derzhmedvydav» Ukraïny, Kyiv 1959.

212. Shumada IV. Okhorona zdorov'ia v Ukraïns'kiy RSR. «Derzhmedvydav» Ukraïny. Kyiv 1963.

213. Szumowski W. Historia medycyny. PZWL, Warszwa 1961.

214. Shupyk PL, Bratus' VD, Duplenko KF. Dosiahnennia okhorony zdorov'ia v Ukraïns'kiy RSR. «Derzhmedvydav Ukraïny» 1958.

215. Shupyk PL. Okhorona zdorov'ia na Ukraïni. «Zdorov'ia» 1966.

216. Syrotynin MM. Olexandr Olexandrovych Bohomolets. AN Ukraïny, Kyiv 1959.

217. Sirotinin NN. A.A. Bogomolets. Moskva 1967.

218. Sichyns'kyi V. Medycyna na Ukraïni v kozac'kych chasach XVII-XVII st. J Ukr Med Assoc North Am 1957;2,1(3):16-9.

219. Sichkar OA. Do istoriï Kyevo-Mohylians'koi akademiï. V, JZ Shtokalo (red), Narysy z istoriï pryrodoznavstva i techniky. Vyp 11. «Naukova Dumka», Kyiv 1970:46-70.

220. Spirov MS. Kyivs'ka anatomichna shkola. «Zdorov'ia», Kyiv 1965.

221. Steichen FM, Ravitch MM. History of Mechanical Devices and Instruments for Suturing. In, MM Rawitch (ed), Curr Probl. Surg. Year Book Med Publ, Chicago 1982;19:1-52.

222. Sverg E. Pred pul stoletim u prof. Horbaczevskoho. Cas Lek Ceskych 1954;92:2223-33.

223. Szczerbak J. Kwarantanna. Pax, Warszawa 1968.

224. Szczerbak J. Transplantacja. PIW, Warszawa 1973.

225. Szumowski W. Historia medycyny. Wyd. 3. Sanmedia, Warszawa 1994.

226. Tarasov MM. Cadaveric blood transfusion. Ann NY Acad Sci 1960;87:512-21.

227. Tataryk PJ. Vellans'kyi Danylo Mykhailovych. Ukr med arch 1931;7(1).

228. Teich M. K istorii sinteza mochevoi kisloty (ot Shele k Horbachevskomu). Trudy I-ta istorii yestestvoznaniia i tekhniki AH SSSR. 1961;35:212-44.

229. Trofymenko AP. Rozvytok doslidzhen' radioaktyvnosti ta budovy atomu na Ukraïni dorevoliuciyni roky. V, JZ Sztokalo (red.), Narysy z istoriï pryrodoznavstva i tekhniky. Vyp 16. «Naukova Dumka», Kyiv 1972.

230. Turkalo JK. First Ukrainian medical journal. J Ukr Med Assoc North Am 984;31,11(109):37-39.

231. Turkalo JF. First medical book in Ukrainian language. J Ukr Med Assoc North Am 1987;34,1(115):28-30.

232. Turkevych NM, Balyc'kyi KP. Akademik AN Ukraïns'koï RSR RY Kavets'kyi. Fiziol zh 1959;5(6):845.

233. Torsuyev NA. P.V. Nikolskii (1858-1940). «Medgiz», Moskva 1953.

234. Uden F. Primae lineae fundamentorum pathologiae et therapiae. Petropoli 1809.

235. Uden F. Akademicheskie chitaniia o khronicheskikh bolezniakh. Ch. 1-7. St Petersburg 1816-22.

236. Uden F. Pharmakopea. St. Petersburg 1818.

237. Yuriev P. Yurii z Drohobycha. Nasha Kultura, Warszawa 1968;10 (126):14-5.

238. Vasiliev KG, Segal LZ. Istoriia epidemii v Rossii. Moskva 1960.

239. Velychkivs'kyi M. Under Two Occupations. Memoirs and Documents. Publ. Shevchenko Scientific Society in the US, New York 2017.

240. Vengerov SA. Kritiko-bibliograficheskii slovar russkikh pisatelei i uchenykh. St. Petersburg 1893.

241. Verkhratskii SA. Tsekhovaia medicina na Ukraine. Vrach delo 1946;9:659.

242. Verkhratskii SA. Istoriia medycyny. V-vo «Zdorov'ia», Kyiv 1964. / V-vo «Zdorov'ia», Kyiv 2011.

243. Vorobec T, Osinchuk P, Khmilevs'kyi Y. Medychnyi fakultet Tainoho universytetu u Lvovi v rokakh 1920-25. J Ukr Med Assoc North Am 1961;8(22-23):5.

244. Voronoy U. Sobre el bloqueo del aparato reticuloendotelial del hombre en algunas formas de intoxicacio por el sublimado y sobre la transplantacion del rinon cadaverico como metodo de tratamiento de la anuria consecutiva a aquella intoxicacion. El Siglo Medico 1936;97:296-298.

245. Voroncov DS. Rozvytok elektrofiziolohiï na Ukraïni. Fizjol zh 1957;3(5):29.

246. Voroncov DS. Chahovec' VY: osnovopolozhnyk suchasnoi elektrofiziolohiï. Kyiv 1957.

247. Voroncov DS, Nikitin VN, Sierkov PM. Narysy z istoriï fiziolohiï na Ukraïni. AN Ukrainy, Kyiv 1959.

248. Waksman SA. Life with the Microbes. Simon & Shuster, New York 1954.

249. Walther AP. Beitrage zur Lehre von der thierischen Warme. Virchov Arch Path Anat 1962;25:414.

250. Walther A. On the infuence of cold on the living being. Sovremennaya med 1863;45(48):51-52.

251. Walther A. The Course of Practical and Applied Anatomy of the Human Body. Part 1 & 2. Kyiv 1870-72.

252. Westaby S. Landmarks in Cardiac Surgery. Isis Medical Media, Oxford 1997.

253. Woodruff MFA. The Transplantation of Tissues and Organs. Charles C. Thomas, Springfield, IL 1960.

254. Zabludovskii DK. Prof. N.N. Diderichs. Vest khir 1941;51(3):424.

255. Zabolotnyi DK. Vybrani tvory. «Naukova Dumka», Kyiv 1969.

256. Zaluc'kyi T. Do istoriï farmaciï na Ukraïni. J Ukr Med Assoc North Am 1975;22,2(77):44-49.

257. Zhmuds'kyi OZ. Istoriia Kyivs'koho universytetu. Kyiv 1959.

258. Zhukovskyi A. Prof. d-r Mykhailo Mishchenko. J Ukr Med Assoc North Am 1975;22,1(76):46-47.

259. Zinevych TP, Kruc SI. Antropolohichna kharakterystyka davnioho naselennia terytorii Ukraïny. «Naukova Dumka», Kyiv 1968.

260. Zwoździak W. Historia Wydziału Lekarskiego Uniwersytetu Lwowskiego. Arch Hist Med 1964;27(1-2):11. / 27(3):193. / 27(4).

ABBREVIATIONS

A	—	adenine.
A	—	Ampere **[Andre M. Ampere (1775 - 1836)** — French physicist] is the unit for the measurement of the electric current force in a circuit which is directly proportional to the difference of the electric current potential across two points of the electric conductor (gradient) and inversely proportional to the resistance of the electric conductor (1820).
A	—	angiotensin.
AA	—	ascending aorta.
AAA	—	abdominal aortic aneurysm.
AAL	—	anterior axillary line.
ABB	—	acid-base balance.
ABG	—	arterial blood gases.
ABI	—	ankle-brachial index.
ABS	—	American Board of Surgery (organized 1937).
ABTS	—	American Board of Thoracic (and Cardiac) Surgery (founded 1948).
AC	—	Ante Christum; BC.
AC	—	aortic coarctation, COA.
ACA	—	anterior cerebral artery.
ACB	—	aorto-coronary bypass, CAB, CAB grafting, CABG.
ACC	—	agenesis of the corpus callosum.
ACA	—	American College of Angiology (ruling year 1982).
ACC	—	American College of Cardiology (established 1949).
ACE	—	angiotensin converting enzyme.
ACh	—	acetylcholine.
ACS	—	American College of Surgeons (founded 1913).
ACTH	—	adrenocorticotropic hormone.
AD	—	Anno Domini.
ADH	—	antidiuretic hormone, AVP.
ADPKD	—	autosomal dominant polycystic kidney disease.
ADF	—	adenosine diphosphate.
AGP	—	autonomic ganglionic plexus.
AED	—	automatic external defibrillator.
AES	—	atomic (nuclear) electro- station.
AF	—	atrial fibrillation.

AF	—	armed forces.
AGP	—	autonomic ganglionic plexus.
AHD	—	acquired heart disease.
AI	—	aortic insufficiency, AR.
AICD	—	automatic implantable cardioverter defibrillator, ICD.
AIDS	—	autoimmune deficiency syndrome.
AIOD	—	aorto-iliac occlusive disease.
AIP	—	acute interstitial pneumonia.
a.k.a.	—	also known as; nom de guerre.
AKH	—	Allgemeines Krankenhaus (established 1686, 1693, 1697), Vienna, Austria.
AL	—	antero-lateral projection.
ALG	—	antilymphocyte globulin.
ALCAPA	—	anomalous origin of the coronary artery from the pulmonary artery (PA).
ALMV	—	anterior leaflet of the mitral valve.
ALS	—	amyotrophic lateral sclerosis.
AM	—	acute margin of the heart.
AMB	—	acute marginal branch of the right coronary artery (RCA).
AM	—	atheromatosis.
AMF	—	adenosine monophosphate.
AN	—	atrio-nodal portion of the AVN.
ANH	—	atrial natriuretic hormone, ANP.
and oth.	—	and others, et al.
ANP	—	atrial natriuretic peptide, ANH.
ant.	—	anterior.
anti-AChR	—	anti-acetylcholine receptors.
AP	—	aortopulmonary.
AP	—	action potential.
AP	—	abdominoperineal.
AP	—	aspiration pneumonia.
APC	—	atrial premature contraction.
APUD	—	amine precursor uptake and decarboxylation.
ARDS	—	acute respiratory distress syndrome.
aPTT	—	activated partial thromboplastin time.
AR	—	aortic regurgitation, AI.
ARC	—	antireticular cytotoxic (Bohomolets) serum. / **Olexandr O. Bohomolets (1881 - 1946)** — Ukrainian patophysiologist.
ARM	—	arteria radicularis magna Adamkiewiczi. / **[Adalbert Adamkiewicz (1850 - 1921)** — Polish anatomist.
ARVD	—	arythmogenic RV dysplasia.
AS	—	aortic stenosis.

AS	— arteriosclerosis, atheromatosis.
AS	— aortic sac.
AS	— assist system (device).
ASA	— acetylsalicylic acid (aspirin).
ASD	— atrial septal defect.
ASI	— aortic size index.
ASM	— anterior systolic motion (mitral valve).
ASMSBL	— apparatus (device, stapler) for suturing main the stem bronchus of the lung.
ACVS of Ukraine	— Association of Cardio-Vascular Surgeons of Ukraine (created 1994).
ATA	— ascending and transverse aorta.
ATA	— anterior tibial artery.
et al.	— at alia, and others.
ATP	— adenosine triphosphate.
ATS	— American Thoracic Society (founded 1905).
AU	— astronomical unit (1 AU = length approximately equal to the distance from the Earth to the Sun (about 150 million km).
AUASc	— All-Ukrainian Academy of Science (1918-36).
AV	— arteriovenous.
AV	— atrioventricular.
AV	— axillary vein.
AVA	— an aortic valve area, is calculated during cardiac catheterization (CC) first by measurement of the cardiac output (CO) and cardiac index (CI) is based on the Ohm law[1778], obtained by Fick method or principle[1779], less often by thermodilution or dye methods, and the application of the Gorlin equation[1780] or Hakki formula[1781].

$$CI\ (L/min) = O_2\ consumption\ (mL/min/m^2)\ /\ arterio\text{-}venous\ (A\text{-}V)\ O_2\ difference$$

The normal O_2 consumption index ranges from 110 to 150 L/min/m². This must be multiplied by the patient

$$A\text{-}V\ O_2\ difference = P_A\ (\%\ saturation) - P_a\ (\%\ saturation)\ (Hgb\ in\ g/dL)\ (1.36\ mL\ O_2/g\ Hgb),\ where$$

[1778] **Georg S. Ohm (1787 - 1854)** — German physicist.

[1779] **Adolf E. Fick (1829 - 1901)** — German physiologist and physician. / Fick A. Über die Messing das die Blutquantums in den Herzventrikela. Verhandlungen der Physikalisch-medizinische Gesellschaft zu Würzburg. 1870 Juli 9;2:XVI-XVII.

[1780] **Richard Gorlin (1926 - 97)** — American physiologist and cardiologist. / Gorlin R, Gorlin SG. Hydraulic formula for calculation of the area of the stenotic mitral valve, other cardiac valves, and central circulatory shunts. Am Heart J 1951 Jan;41(1):1-29.

[1781] **Abdul H. Hakki** — Kurdish American cardiothoracic surgeon. / Hakki A, Iskandrian A, Bemis C, et al. A simplified valve formula for the calculation of stenotic cardiac valve areas. Circulation 1981;63 (5): 1050-5.

PA is the O_2 saturation in the peripheral arterial blood, PA is the O_2 saturation in the pulmonary artery, (PA), Hgb is the hemoglobin concentration in the blood, and 1.36 is the correction factor for the ability of fully saturated Hgb is to carry O_2.

In the Gorlin equation the AVA in cm^2 = CO / [(diastolic filling period (DFP) or systolic ejection period (SEP) x heart rate (HR) / using an empiric constant (C) for the aortic valve (AV) and the tricuspid valve (TV) of 44.3, and an empiric C for the mitral valve 37.7 and a square root delta P is the pressure gradient in mmHg:

$$\text{Aortic valve area} = cm^2 = \frac{CO/(DFP \text{ or } SEP)(HR)}{C(\sqrt{\Delta P})}$$

or

AVA (cm^2) = aortic valve flow (AVF) / 44,5 the aortic valve pressure gradient (AVG) in mmHg,

$$AVA = \frac{AVF}{44.5\sqrt{AVG}}$$

or

AVA (cm^2) = CO / divided by the square root of the AVG (mmHg),

or

$$AVA = \frac{CO}{\sqrt{AVG}}$$

Since in the Gorlin equation, HR and SEP are generally similar among most patients, therefore, an alternative equation, the Hakki formula, has been suggested for estimation of the AVA (cm^2), where the CO (L/min.) is divided by the square root the peak-to-peak gradient (P) across the valve (mmHg):

$$\text{Valve area} = \frac{\text{cardiac output (liters/min)}}{\sqrt{\text{pressure gradient}}}$$

Normal AVA in the transverse diameter = 2.0 – 4.0 см². Aortic valve (AV) stenosis (AS) is considered severe when the AVA is less than 1.0 cm^2, and is critical when the AVA is less than 0.7 cm^2.

According to the principle established by **Daniel Bernoulli (1700 - 82)** — Netherlander and Swiss mathematician and physicist, the increased velocity nonviscous fluid occurs simultaneously with the fall of the pressure or decrease its potential energy (Hydrodynamica, 1738), as depicted in the following equation:

$$p_1 - p_2 = \frac{1}{2} \rho \left(\mathbf{v}_1^2 - \mathbf{v}_2^2 \right)$$

where, p – denote the pressure drop (mmHg), **p** – denote density of fluid, and **v** – denote magnitude of peak velocity in the jet in m/second.

Modified Bertuolli equation, where the delta P (pressure) meaning a gradient pressure and v – meaning the flow of the fluid, is used in the continuous color valve Doppler 2-D transthoracic or transesophageal echocardiography (TTE, TEE)[1782], for evaluation of the valve area. With the decrease of an aortic valve area, the velocity of the blood flow increases, and because of this the pressure gradient across the opening of the valve increases as result of maximal increase the pressure in the left ventricle (LV) and the systolic pressure in the pulmonary artery (PA) – dP/dT. This implies that one should also take into consideration the velocity of regurgitant flow through the tricuspid valve (TV):

$$\Delta P = 4v^2$$

or

$$v = [m/c^2] \times 4$$

AV bundle	—	atrio-ventricular bundle.
AVC	—	atrioventricular canal.
AVF	—	aortic valve flow.
aVF lead	—	augmented unipolar limb lead.
AVM	—	arterio-venous malformation.
AVN	—	atrioventricular node, or AV node.
AVP	—	arginine vasopressin, vasopressin.
AVP	—	aortoventricular plasty after Kono-Rastan. – **S. Kono** — Japanese cardiac surgeon, and **Hashang Rastan (1936 -)** — Iranian German cardiac surgeon.
AVNRT	—	atrioventricular nodal reentrant tachycardia.
AVR	—	aortic valve replacement.
AVRT	—	atrioventricular re-entrant tachycardia.
B	—	bursae, or B-cells.
σ	—	osmotic reflection coefficient.
BA	—	basilar artery.
BAL	—	bronchoalveolar lavage.
BAV	—	bicuspid aortic valve (congenital).

[1782] Von **Christian A. Doppler (1803 - 53)** — Austrian mathematician and physicist. / Doppler Ch. Über das farbige Licht der Doppelsterne und einiger anderer Gestirne des Himmels. Gelesen bei der Königl. Böhm. Gesellschaft der Wissenschaften zu Prag, in der naturwissenschaftlischen Sectionssitzung vom 25. Mai 1842. In Commission bei Dorrosch & Andre, Prag 1842. V. Folge, Bd. 2, S. 465-482.

BH	—	bundle of His is collection of specialized cardiac muscle cells that transmits the electrical current and synchronize contraction of the myocardium, discovered in 1893 by **Wilhelm His, Jr (1863 - 1934)** — Swiss German cardiologist, and named in his honor.
BC	—	Before Christ, AC.
BCG	—	balistocardiography.
BCG	—	bacillus Calmette-Guerin. / **Leon A.Ch. Calmette (1863 - 1933)** — French physician, bacteriologist and immunologist, and **Camille Guérin (1872 - 1961)** — French veterinarian, bacteriologist, and immunologist.
BCT	—	brachiocephalic trunk, IA.
BD	—	birth date or DOB.
BEE	—	basal energy expenditure.
biVAD	—	biventricular assist device.
b.i.d.	—	bis in die.
bi-PAP	—	bi-level positive airway pressure.
bln	—	billion, milliard.
BMI	—	body mass index.
BMP	—	bone morphogenic protein.
BMS	—	bare metal stent.
BMCs	—	bone marrow cells.
BMR	—	basal metabolic rate, or BEE.
BNA (1st. Ed.)	—	Basle Nomica Anatomica (1895).
BNP	—	brain natriuretic peptide.
BOOP	—	(idiopathic) bronchiolitis obliterans organizing pneumonia, or COP.
BP	—	blood pressure.
BPF	—	bronchopleural fistula.
BPS	—	bladder pain syndrome.
BR	—	bed rest.
BS	—	bronchus suturing (apparatus, device, stapler).
BSA	—	body surface area (height in centimeter: weight in kilogram x 3600 = m^2.
BT shunt	—	Blalock-Taussig (shunt). / **Alfred Blalock (1899 - 1964)** — American pediatric cardiac surgeon, and **Helen B. Taussig (1898 - 1986)** — American pediatric cardiologist.
BTR	—	«bridge» to recovery.
BTT	—	«bridge» to transplantation.
BVS	—	biventricular system.
Ca^+	—	calcium.
°C	—	centigrade scale with 100 steps, which shows the degree units between freezing (0°C) and boiling (100°C) water. / **Anders Celsius (1701 - 44)** — Swedish mathematician, physicist, and astronomer.

C   —   cervical.

          —   cono-truncus.

C   —   constant (empiric).

C   —   cytosine.

c., ca   —   circa, ca. / about, approximately, around, near.

CA   —   celiac artery.

CA   —   carbohydrate antigen

CAB   —   coronary artery bypass, or ACB, CABG.

CABG   —   coronary artery bypass grafting, or ACB, CAB.

CAC or TCA   —   citric acid cycle or tricarboxylic acid cycle (Krebs cycle) – Sir **Hans Krebs (1900 - 81)** — Jewish, German English biochemist and physician.

CAD   —   coronary artery disease.

cal.   —   calorie.

cAMP   —   cyclic adenosine monophosphate.

CAP   —   chronic anterior poliomyelitis.

C-arm   —   figure of C portable radiology film apparatus.

CAR   —   coeliac artery resection.

CAT   —   computer axial tomography.

«CATCH»   —   *C*ardiac defect, *A*bnornal facies, *T*hymic aplasia, or *T*-cell deficit, *C*left palate, *H*ypocalcemia syndrome.

CAV   —   cardiac allograft vasculopathy.

CBD   —   common bile duct.

CBF   —   cerebral blood flow.

CC or HC   —   cardiac or heart catheterization.

CCA   —   common carotid artery.

CCAM   —   congenital cystic adenomatoid malformation.

CCCH   —   Central City Clinical (Olekcandrivska) Hospital (founded 1875), Kyiv, Ukraine.

CCR   —   cardiocerebral resuscitation.

CC-SCA   —   cardiac catheterization and selective coronary angiography.

CCT   —   chest computed tomography.

CC-TA   —   cardiac catheterization and thoracic angiography.

CCU   —   coronary care unit.

Cd   —   cadmium.

cdc   —   cell division cycle.

CDC   —   Centers for Disease Control and Prevention (established 1953), Atlanta, GA, USA.

CDH   —   congenital diaphragmatic hernia.

cdk   —   cell-dependent kinase.

CDT   —   catheter-directed thrombolysis.

CE or CEA	—	coronary endarterectomy.
CEA	—	carcinoembryonic antigen.
CEA	—	carotid endarterectomy.
CEA or CE	—	coronary endarterectomy.
cerebral palsy	—	condition marked by impaired spastic or paralytic muscle coordination and/or other disabilities caused by damage to the brain before or at birth.
CF	—	cystic fibrosis.
CFA	—	circumflex artery of the heart.
CFA	—	common femoral artery.
CFTR protein	—	cystic fibrosis transmembrane conductance regulator protein.
CG AV-AA	—	composite graft replacement of the aortic valve and ascending aorta with an artificial aortic valve sewn into a synthetic tubular graft (Bentall-Bono operation). / **Hugh Bental (1920 - 2012)** and **Anthony De Bono** — English cardiac surgeons.
CGS	—	centimeter (cm) - gram (gm) - second (c) system of units.
CHA	—	Canada Health Act (adopted 1984).
ChAES	—	Chornobyl Atomic Electrical Station (functioning 1971-77), Chornobyl, KY, Ukraine.
CHB	—	complete heart block.
CHD	—	congenital heart disease.
chest x-ray	—	CXR or chest radiography.
CHF	—	congestive heart failure.
CI	—	cardiac index (normal 2.0 - 3.0 L/min./m^2).
CI	—	conical (conicity) index.
CIA	—	common iliac artery.
CIV	—	common iliac vein.
CJD	—	Crentzfeld-Jacob disease.
CJS	—	candidate of jurisprudence science.
CK	—	cytokeratin.
CML	—	chronic myeloid leukemia.
CM or CMP	—	cardiomyopathy.
CMIII	—	Cox-Maze III operation (procedure). / **James L. Cox (1942 -)** — American cardiothoracic surgeon.
CMIV	—	Cox-Maze IV operation (procedure).
CMP	—	cardiomyopathy, or CM.
CMR	—	computed magnetic resonance, or MR.
CNI	—	calcineurin inhibitor.
CNH	—	conduction system (network) of the heart.
CNI	—	calcineurin inhibitor.
CNS	—	central nervous system.

CO — cardiac output, as determined by the method of **Adolph Fick (1829 - 1901)** — German physiologist and physician, where the oxygen (O_2) consumption in ml/min (the O_2 consumption index = 110-150 ml/min/m^2) is divided by the arteriovenous (AVO_2) difference in the O_2 saturation of the blood, thus the AVO_2 difference = pulmonary artery (PA, % saturation) – pulmonary vein (PV, % saturation) (Hb) (1.36) / 10 (normal 30-50 ml/min/m^2): CO = consumption O_2 (ml/min/m^2) / AVO_2 difference or the heart rate (HR) x stroke volume (SV). Normal CO is 2.5 – 4.2 L/minute. / CO, Cl.

CO_2 — carbon dioxide.

COPD — chronic obstructive pulmonary disease.

CoS — coronary sinus.

COA — coarctation of the aorta, or aortic coarctation.

CoA — coenzyme A.

QOL — quality of life.

COP — coatomer is a protein complex that coats membrane-bound transport vesicles.

COP — cryptogenic organizing pneumonia, or BOOP.

COX — cyclooxygenase.

CPA — chirped pulses amplification.

CPAP — continuous positive airway pressure.

CPAF — coronary-pulmonary arterial fistula.

CPB — cardiopulmonary bypass.

CPR — cardiopulmonary resuscitation.

CPT — chest physical therapy

CPVT — catecholamine (catecholaminergic) polymorphic ventricular tachycardia.

CRF — chronic renal failure.

CRH — corticotropin-releasing hormone.

CRC — cardiac repair cells.

CRS — congenital rubella syndrome.

CRT — cardiac resynchronized therapy.

CRT-D — CRT-defibrillator.

CS — coronary sinus.

C-section — caesarean section.

CSF — cerebrospinal fluid.

CSH, CNH — conduction system of network of the heart.

CT — crista terminalis.

C — computed tomography.

CT or HT — cardiac or heart transplantation.

CTA — computed tomographic angiography.

CTD	—	central tendon of the diaphragm or Speculum Helmontii. / **Jan B**. van **Helmont**, or van **Helmontii (1580 - 1644)** — Belgian Dutch chemist, physiologist, and physician.
CTE	—	chronic traumatic encephalopathy
CTG	—	composite tubular graft.
CTLA-4	—	cytotoxic T-lymphocyte-associated protein-4.
CUSA device	—	cavipulse ultrasonic surgical aspirator device.
CV	—	cephalic vein.
CV	—	curriculum vitae.
CVS	—	cardiovascular surgery.
CVS	—	critical view of safety.
CVA	—	cerebrovascular accident, stroke.
CVD	—	chronic venous disease.
CVS	—	cardiovascular surgery.
CVS	—	chorionic villous (villus) sampling
CVP	—	central venous pressure.
CVVH	—	continuous veno-venous hemofiltration
CXR	—	chest x-ray, chest radiography.
D	—	dimension.
D	—	diuretics.
DAA	—	double aortic arch.
DB	—	diagonal branch of the left anterior descending artery (LADA) of the heart.
DBE	—	diffuse bullous emphysema.
DC	—	direct (electric) current.
DCM	—	digital counting machine.
DCM	—	dilated cardiomyopathy.
DVD	—	digital versatile disc.
DDAVP	—	1-deamino-8-deamino-arginine vasopressor (desmopressin) – the synthetic analog of V_2 (L) – arginine vasopressor (AVP), which stimulates the release of ultra-large multimers of vWF from endothelial cells, is associated antihemophilic (VIII) factor and improves platelets function. It is administered IV at the dosage of 0.3 µg/kg in 0.9% NaCl solution over 15 min. for the treatment of the type I von Willenbrand disease (vWD) – the most common inherited bleeding disorders (easy bruising and menorrhagia) and for the treatment of perioperative bleeding after cardiac surgery.
D-DOPA	—	dihydroxyphenylalanine.
DDT	—	dichlorodiphenyltrichloroethane.
de novo	—	from the beginning.
«Derzhmedvydav»	—	State Medical Publishing.

DES	—	drug eluting stent.
DFA	—	deep femoral artery.
DFP	—	diastolic filling period.
DFV	—	deep femoral vein.
DHE-CMR	—	delayed hyper-enhancement - computed magnetic resonance.
DHCA	—	deep hypothermic circulatory arrest.
DIC	—	disseminated intravascular coagulation.
DIJ	—	distal interphalangeal joint.
DIP	—	desquamative interstitial pneumonia.
DIPNECH	—	diffuse idiopathic pulmonary neuroendocrine cell hyperplasia.
DKS	—	Damus-Kaye-Stansel (DKS) anastomosis or procedure (1975). / **Paul M. Damus (1943 -)**, **M.P. Kaye** and **H.C. Stansel, Jr** — American pediatric cardiac surgeons,
dl	—	deciliter (1000 ml).
DL	—	double lumen.
DLB	—	dementia with Lewy bodies. / **Frederic (Friedrich) H. Lewey (Lewy) (1885 - 1950)** — Jewish German American neurologist.
DLCO	—	diffusing capacity of the lung for carbon monoxide (CO), normal – 70% or more of the predicted.
DM	—	Descemet membrane. / **Jean Descemet (1732 - 1810)** — French physician.
DMEK	—	Descemet membrane endothelial keratoplasty.
DMS	—	Doctor of Medical Sciences.
DNA	—	deoxyribonucleic acid.
DO	—	Doctor of Osteopathy.
DOB	—	date of birth, or BD.
DO_2 (ml O_2/min)	—	delivery of oxygen (ml O_2/min).
DOLV	—	double-outflow left ventricle.
DORV	—	double-outflow right ventricle.
DP	—	dorsalis pedis (artery).
DP	—	displaced person.
2,3-DPG	—	2,3-diphosphoglycerate.
Dr	—	Doctor.
Dr rer. nat.	—	Doctor rerum naturarium, or Doctor of Natural Science.
DS	—	de-signo.
DS or Vd	—	dead space of the airways.
DSA	—	digital subtraction angiography.
DT	—	destination therapy.
DTA	—	descending thoracic aorta.
DTG	—	diffuse toxic goiter (goiter).

DVD	—	digital versatile disc.
DVT	—	deep venous thrombosis.
E	—	expiration, expiratory.
EA	—	esophageal atresia.
EA	—	endarterectomy.
EACA	—	epsiloaminocapronic acid (amicar).
EAR	—	endoaneurysmal, endoanerysmorrhaphy.
EBOV	—	Ebola virus.
EBUS	—	endobronchial ultrasound.
ECA	—	external carotid artery.
ECLS	—	extracorporeal life support.
ECMO	—	extracorporeal membrane oxygenation.
ECANS	—	external cardiac autonomous nervous system.
Echocardio	—	echocardiography.
EchoCG	—	echocardiography.
ECG	—	electrocardiography
ECP	—	endothelial cell precursor.
ECPCR	—	endothelial cell protein-C receptor.
ECT	—	electroconvulsive (electroshock) therapy.
EDD	—	expected date of delivery.
EDRF	—	endothelium-derived relaxing factor.
Ed.	—	editor or redactor.
EDTA	—	ethylene diamine tetra-acetic acid.
EDV	—	end-diastolic velocity.
EDV	—	end-diastolic volume.
EE	—	extraembryonic ectoderm.
EEA	—	«end-to end anastomosis».
EEG	—	electroencephalography.
EF	—	ejection fraction.
e.g.	—	exempli gratia, for example.
EGFR	—	epidermal growth factor receptor.
EG junction	—	esophagogastric junction, a.k.a., GE junction.
EIA	—	external iliac artery.
EJV	—	external jugular vein.
ECI	—	extracorporeal insemination, ECC.
ECC	—	extracorporeal conception, ECI.
DLCO	—	diffusing capacity of the lung for carbon monoxide (CO), normal – 70% or more of the predicted.

ELISA-OD — enzyme-linked immunosorbent assay - optical density.

EMB — endomyocardial biopsy.

EMC — emergency medical care.

EMG — electromyogram (electromyography).

EMR — electromagnetic radiation.

EMR — electronic medical record.

ENT — ear, nose and throat, or ORL.

EOA — effective orifice area. For the elderly patients undergoing AVR, the patient - prosthetic valve mismatch, defined as calculated prosthetic EOA/BSA less than 0,75 cm²/m², does not impair survival, especially in small patients (BSA less than 1,7 m²). The peak gradient, estimated using continued valve Doppler US[1783] as four times the square of the peak velocity, 25 mmHg or less is an acceptable level for the most prosthesis for the AVR.

EOET — extraosseous Ewing tumor. / **James S. Ewing (1866 - 1943)** — American pathologist.

EP — electrophysiology.

epi — epiblast.

EPI — echo-plantar imaging.

EPO — erythropoietin.

EPP — extrapleural pneumonectomy.

EPR — electronic paramagnetic resonance, or PMR, or NPM.

EPSP — excitatory postsynaptic potential.

ER — endoplasmic reticulum.

ER — emergency room.

ERI — enhancement of the radiological imaging.

ER Ca++ — energy redding by calciun.

ERO — effective regurgitation orifice.

ERV — expiratory reserve volume.

ESC — embryonal (embryonic) stem cells.

ESG — external synthetic (prosthetic) grafting (reinforcement, wrapping).

ESKD — end-stage kidney disease.

ESM — extensible system monitor.

ESR — erythrocyte sedimentation rate.

ESV — end-systolic volume.

ESVI — end-systolic volume index.

ESWL — extracorporeal shock wave lithotripsy.

ET tube or ETT — endotracheal tube.

ETA — endovascular treatment of aneurysms.

[1783] Von **Christian A. Doppler (1803 - 53)** — Austrian mathematician and physicist.

et al.	—	et alii, and others.
ETB	—	Ewing's tumor of the bone. / **James S. Ewing (1866 - 1943)** — American pathologist.
etc.	—	et cetera; and so on, and so forth, and other things.
ETG	—	endothoracic grafting.
ETT or ET tube	—	endotracheal tube.
EVAR	—	endovascular aortic repair.
EVCPP	—	endoventricular circular patch plasty.
EVD	—	Ebola virus disease.
EVG	—	endovascular grafting, or ISS-STG.
EVH	—	endoscopic vein harvesting.
EVI	—	endovascular intervention.
Exam.	—	examination.
EXIT	—	extrautero intrapartum treatment.

$^{\circ}F$	—	Fahrenheit mercury-in-glass thermometer in which the temperature (T) of freezing point of water is 32°F, the T of boiling point of water is 212°F at the standard atmospheric pressure at the sea level, and T of normal T of the human body measured orally is 98.2° ± 9 °F). This thermometer was invented in 1724 by **Daniel G. Fahrenheit (1686 - 1736)** — German chemist and physicist.
F	—	factor; Φ.
F-1	—	factor-1 (fibrinogen, highly molecular plasma protein which under action of active F-2a converts
into fibrin).		
F-2	—	factor 2 (prothrombin, plasma protein which under action of F-10 converts into active F-2a).
F-3	—	factor 3 (tissue factor, or tissue thromboplastin, lipoprotein which take part in the external mechanism of blood coagulation by activating F-10).
F-4	—	factor 4 (Ca^{++}, indispensable in many phases of blood coagulation).
F-5	—	factor 5 (proaccelerin, a labile substance of plasma which take part in the internal and external mechanism of blood coagulation, catalyzing F-2 into active F-2a).
F-6	—	factor 6 (not discovered).
F-7	—	factor 7 (proconvertin, a stable F that participate in the external mechanism of blood coagulation, activated by Ca^{++} and jointly with F-3 activate F-10.
F-8	—	factor 8 (antihemophiliac factor =AHF, labile F which take part in the internal mechanism of blood coagulation, acting together with F-vWB as a co-F in activation of F-10). Deficit of F-8, bind with chromosome X recessive trait, causes a classical hemophilia A.

F-9	—	factor-9 (plasma thromboplastin component = PTC, stabile f which take part in the internal mechanism of blood coagulation, after activation at once activate F-10. Deficit of F-9 is manifested by htmophilia B).
F-10	—	factor 10 (Stuart-Prower, stable F in preservation of the internal and external coagulation mechanisms by combining them at the beginning of the common pathway).
F-11	—	factor 11 (plasma thromboplastin antecedent component = PTC, stable F, included into the internal mechanism of blood coagulation).
F-12	—	factor 12 (Hageman, stable F, begin the internal mechanism of blood coagulation by activization of F-11.
F-13	—	factor 13 (fibrin stabilizing F = FSF).
F	—	French scale, used for diameter determination of catheters, probes, drainage tubes and synthetic tubular grafts (STG), where the unit of the 1F scale equals 0,33 mm in diameter, e.g., 12F = 4 mm in diameter.
f	—	frequency.
Fab	—	fragment antigen-binding.
FACS	—	Fellow of the American College of Surgery.
FBN	—	fibrillin.
FBB	—	faculty of the biotechnology and biotechnics.
Fc	—	fragment crystallizable.
FDG	—	fluorodeoxyglucose.
FE	—	foregut endoderm.
FEV_1	—	forced expiratory volume in 1 second.
F-F	—	flatter-fibrillation.
FFP	—	fresh frozen plasma.
FGF	—	fibroblast growth factor.
5-HT	—	5-hydroxytryptamine (serotonin).
FiO_2	—	a flow of the inspired oxygen percentage (%), a fractionated oxygen concentration (%).
FISH	—	fluorescence in situ hybridization.
fl.	—	falsa lectio.
FM	—	pharmaco-mechanical.
FMD	—	fibromuscular dysplasia.
FNA	—	fine-needle aspiration.
FOB	—	fiberoptic bronchoscopy.
FOE	—	fiberoptic endoscopy.
FP	—	femoro-popliteal.
ESG	—	ultrasonography, US.
FSH	—	follicular-stimulating hormone.
FTL	—	faster-than-light.

5-FU	—	5-fluorouracil.
FV	—	femoral vein.
FVC	—	forced vital capacity.
g	—	gram.
g	—	growth.
G	—	guanin.
GABA	—	gamma aminobuttery acid.
GABB	—	gases and acid-base balance.
GC-SF	—	granulocyte colony-stimulating factor.
GDC	—	Guglielmi detachable coil. / **Guido Guglielmi (1948 -)** — Italian neurosurgeon.
Gd-DTPA	—	gadolinium – diethylenetriamine pentoacetic acid.
GE junction	—	gastroesophageal junction, a.k.a., EG junction.
GERD	—	gastroesophageal reflux disease.
GFP	—	green fluorescent protein.
GFR	—	glomerular filtration ratio.
GH	—	growth hormone or HGH.
GI	—	gastrointestinal.
GIA	—	gastrointestinal apparatus.
«GIA» stapler	—	stapler for the «Gastro-Intestinal Anastomosis».
GIST	—	gastrointestinal stromal tumors.
g-LOC	—	gravitation (or acceleration) induced loss of consciousness.
GMP	—	guanosine monophosphate.
GMT	—	general military training or ROTC.
GnRH	—	gonadotropin-releasing hormone.
G&O	—	gynecology & obstetrics.
GP	—	glycoproteins.
GPA	—	glycerol preserved allograft.
GPCR	—	G protein-coupled receptors.
G-suit	—	gravitation suit.
GSV	—	great saphenous vein.
GWAS	—	genotype-wide association.
Gy	—	gray unit (SIU, 1975) define the absorption of one Joule of the radiated energy by one kilogram of matter: $1Gy = 1J/kg = 1 m^2 \times S^{-2}$. – **Louis H. Gray (1905 - 65)** — English physicist.
Gyn & Obst	—	gynecology & obstetrics.
H	—	height.
H^+	—	hydrogen.

Hand — protein genes of the heart and neural crest derivatives.

Hb, Hgb — hemoglobin or hemoglobin.

HbO$_2$ — oxyhemoglobin.

HBV — hepatitis B virus.

HCA — hypothermic circulatory arrest.

HCl — hydrochloric acid.

HCM — hypertrophic cardiomyopathy.

HD — hemodialysis.

HD — Huntington's disease. / **George Huntington (1815 - 1916)** — American physician.

HDL — high density lipoprotein.

HDL-C — LDL-Cholesterol («good» cholesterol) is test that include lipid profile along with total cholesterol and triglycerides.

He — helium.

H&E / HE — hematoxylin and eosin stain.

Herz, Hz — **Gustaw L. Hertz (1887 - 1974)** — German physicist.

HF — heart failure.

HFOV — high-frequency oscilatory (jet) ventilation.

HFCC — high frequency chest compression.

Hg — mercury, mmHg.

Hgb — hemoglobin, or Hb.

HGH — human growth hormone.

HGPRT — hypoxantine guanine phosphoribosyl transferase.

HHT — hereditary hemorrhagic telangiectasia.

HIV — human immunodeficiency virus.

HIT — heparin-induced thrombocytopenia.

HITT — heparin-induced thrombocytopenia and thrombosis or HIT.

HIF — hypoxia inducible factor.

HLA — human leucocyte (lymphocyte) antigen.

HLHS — hypoplastic left heart syndrome (Shone syndrome). / **John D. Shone (1924 - 2002)** — English, American and Canadian pediatric cardiologist.

HLM — heart-lung machine.

HLT — heart and lung transplantation.

HM — heart mesoderm.

HM — HeartMate – device to support a falling heart.

HMG-CoA
reductase — 3-hydroxy-3-methyl-glutaryl-coenzyme A reductase.

HOCM — hypertrophic obstructive cardiomyopathy, or IHSS, or HCM.

HP — high profile.

HPV — human papillary virus.

HPV	—	hypoxic pulmonary vasoconstriction.
HR	—	heart rate.
HRT	—	hormonal replacement therapy (menopause).
H_2S	—	hydrogen sulfide.
hsCRP	—	high sensitivity C-reactive protein.
HSC	—	hematopoietic stem cells.
HT	—	heart transplantation, or CT.
HT	—	hydroxy tryptamine.
HTN	—	hypertension.
5-HTP	—	5-hydroxytryptophan.
HU	—	Hounsfield unit. / Sir **Godfrey H. Hounsfield (1919 - 2004)** — English electrical engineer.
HVE	—	hepatic vein exclusion.

I	—	inotropic.
I	—	inspiration, inspiratory.
$_{53}$I$^-$	—	iodium.
IA	—	iliac artery
IA	—	infundibuloarterial.
IA	—	innominate artery.
IAA	—	interrupted aortic arch.
IABA	—	intra-aortic balloon assist device.
IARC	—	International Agency for Research on Cancer Created (formation 1965).
IAS	—	interatrial septum.
ib	—	ibid (ibidem), idem, idera (as above, the same).
IC	—	islet cells.
IC	—	intermittent claudication.
ICA	—	internal carotid artery.
ICANS	—	internal cardiac autonomic nervous system.
ICC	—	intestinal cells of Cajal. / **Santiago R. y Cajal (1852 - 1934)** — Spanish histologist and neuroscientist.
ICD	—	implantable cardioverter-defibrillator or AICD.
ICM	—	ischemic cardiomyopathy.
ICS	—	intercostal space.
ICST	—	intercostal space thoracotomy.
ICU	—	intensive care unit.
id	—	idem, ibid, ibidem, idera (as above, the same).
ID	—	internal diameter.
I&D	—	incision and drainage.

IDCM	—	ischemic dilated cardiomyopathy.
I:E ratio	—	inspiratory/expiratory ratio.
i.e.	—	id est; in other words, that is, to say.
IF	—	iliofemoral.
IGF	—	insulin-like growth factor.
IgG2a	—	immunoglobulin G2a.
IHD	—	ischemic heart disease, or CAD.
IHSS	—	idiopathic hypertrophic subaortic stenosis (historic name), or HOCM.
IIA	—	internal iliac artery.
IIP	—	idiopathic interstitial pneumonia.
IJV	—	internal jugular vein.
IL	—	interleukin.
iLA	—	interventional lung-assist.
ILD	—	interstitial lung diseases.
IM	—	intramyocardial.
IMA	—	internal mammary (thoracic) artery.
IMA	—	inferior mesenteric artery.
IMH	—	intramural hematoma.
IMIG	—	International Mesothelioma Interest Group (created 1991).
inf.	—	inferior.
iNO	—	inhalation of nitric oxide (NO).
iNOS	—	inducible NO synthetase.
INR	—	international normalized ratio.
IP	—	implantable (internal) pneumatic.
IPPB	—	intermittent positive pressure breathing.
IPPV	—	intermittent positive pressure ventilation, or IPPB.
IPF	—	idiopathic pulmonary fibrosis.
IPG	—	impedance plethysmography.
IPSP	—	inhibitory postsynaptic potential.
iPS cell	—	induced pluripotent stem cell.
IR	—	interventional radiologist.
IRV	—	inspiratory reserve volume.
ISBN	—	International Standard Book Number.
ISC	—	International Society of Cryosurgery (founded 1972).
ISS	—	International Space Station (launched 1998).
ISS-STG	—	intraluminal (transluminal) self-expanding and self-fixing synthetic tubular graft (prosthesis).
IST	—	inappropriate sinus tachycardia.
IT	—	information technology.

ITA	—	internal thoracic (mammary) artery or IMA.
ITV	—	internal thoracic vein.
ITP	—	idiopathic thrombocytopenic purpura.
IU	—	international units.
IUD	—	intrauterine device.
IV	—	iliac vein.
IV	—	innominate vein.
IV	—	intravascular.
IV	—	intravenous (injection).
IVF	—	«in-vitro» fertilization.
IVIG	—	intravenous immune globulin.
IVH	—	intraventricular hemorrhage (brain).
IVH	—	intravenous hyperalimentation, or TPH.
IVC	—	inferior vena cava.
IVCF	—	inferior vena cava filter, Greenfield filter. / **Lazar J. Greenfield (1934 -)** — American vascular surgeon.
IVCD	—	intraventricular conduction delay.
IVS	—	interventricular / intact ventricular septum.
IVT	—	ischemic venous thrombosis.

JME	—	juvenile myoclonic epilepsy.
JNA	—	Jenaer Nomiana Anatomica (1935).
Joule, Jo	—	**James P. Joule (1818 - 89)** — English physicist. The Jo is the unit of energy, work, and heat, which equal the work performed with the force of 1 newton (N) by moving a matter for distance of 1 meter (m) in the direction of the working force, where 1Jo = 1N x 1m.
jpg	—	joint photography (expert) group.
Jr	—	junior.
J-tube	—	jejunal tube.
JUMANA, or J Ukr Med Assoc North Am	—	Journal of the Ukrainian Medical Association of North America (founded 1954).

K	—	Kelvin scale. / **William T. 1st Baron Kelvin** — English physicist. Scale K is a thermodynamic absolute temperature scale where absolute zero, the theoretical absence of all thermal energy, equals to zero kelvin (0 K). He calculated that absolute zero was equivalent to -273°C on the air thermometers of the time (1848).
K^+	—	potassium.

kcal	—	kilocalorie.
KCl	—	potassium chloride.
KCM	—	Kyiv Cave Monastery or Lavra (founded 1051), Ukraine.
K-DOPA	—	karbi-3,4-dioxyphenylalanine.
Keq	—	equilibrium constant.
$K_{f,}$	—	filtration coefficient.
кg	—	kilogram, kilo.
KhNMU	—	Kharkiv National Medical University or Universitetas medicalis nationalis Charkoviensis (founded 1805), Kharkiv, Ukraine.
KhNU	—	Vasyl N. Karazin Kharkiv National University (founded 1805), Kharkiv, Ukraine. – **Vasyl (Vasily) N. Karazin (1773 - 1842**) — Ukrainian scientist and inventor.
kilo	—	kilogram, kg, kilo.
KIU	—	kallikrein inhibition unit.
KK	—	Kransnodarsky krai, Ukraine (1917 - 22) / RF.
CMS	—	Candidate of Medical Sciences.
km	—	kilometer.
kPa	—	One Pascal (1 Pa) = $1N/m^2$ = $1N/m^2$ = $1kg/m \times s^2$ = $1J/m^2$, or 1 Pa = $1N/m^2$, or $1 kg/m/s^2$, where N = Newton, m = meter, kg = kilogram, s = second, and J = Joule (1645). In other words, according to **Blaise Pascal (1623 - 62**) — French mathematician and physicist, 1 Pa is the pressure exerted by a force of magnitude one Newton perpendicularly upon an area of one square meter.
L	—	Linne / **Carl Linnaeus (1707 - 78**) — Swedish botanist, zoologist and physician, who introduced the modern taxonomy with his work «Systema Naturae» (1735).
l.	—	liter.
L	—	lumbar.
LA	—	left atrium.
LAA	—	left atrial appendage.
LAD	—	left anterior descending (artery) or LADA.
LADA	—	left anterior descending artery.
LAM	—	lymphangioleiomyomatosis.
LAO projection	—	left anterior oblique projection.
laser	—	light amplification by stimulated emission of radiation.
lat.	—	lateral.

LBA	— Lactobacillus acidophilus, meaning «acid-loving milk-bacillus»), is a gram-negative bacterium, grows in pH below 5.0 and in optimum temperature around 37°C or 99°F. It is harmless bacteria, fermenting sugars into lactic acid, present in human and animal gastrointestinal (GI) tract, mouth, and vagina. Occasionally, immune depressed patients may harbor severe Lactobacillus endocarditis (LBE)[1784].
LBBB	— left bunfle branch block.
LCA	— Leber congenital amaurosis. / **Theodor K.G. Leber (1840 - 1917)** — German ophthalmologist.
LCS	— left coronary (aortic) sinus Valsalva. / **Antonio M. Valsalva (1666 - 1723)** — Italian anatomist.
LCSD	— left cardiac sympathetic denervation.
LCX	— left circumflex coronary artery, CFA.
LDH	— lactate dehydrogenase.
LDL	— low density lipoprotein.
LDL-C	— LDL-Cholesterol («bad» cholesterol) is test that include lipid profile along with total cholesterol and triglycerides.
L-DOPA	— Levo-3,4-dioxyphenylalanine.
LDS	— Loeys-Dietz syndrome. / **Harry C. Dietz (1950 -)** — American pediatric geneticist, and **Bart L. Loeys (1958 -)** — Belgian physician.
LED	— light emitting diode.
LGL	— Lown- Ganong-Levine syndrome. / **Bernard Lown (1921 - 2021)** — Lithuanian American cardiologist, **William F. Ganong, Jr (1924 - 2007)** — American physiologist, **Samuel A. Levine (1891 - 1966)** — American cardiologist.
LH	— luteinizing hormone.
LHC	— left heart catheterization.
LHD	— literatim humaniform doctor; doctor of humane letters, doctor of humanities.
LHF	— left hear failure.
Li	— lithium.
LIFE	— light imaging fluorescence endoscopy.
LIL	— left inferior leaflet (mitral valve).
LIMA	— left internal mammary (thoracic) artery or LITA.
LIP	— lymphocytic interstitial pneumonia.
LIPV	— left inferior pulmonary vein.
LITA	— left internal thoracic (mammary) artery or LIMA.
LLL	— left lower lobe (lung).

[1784] Olearchyk AS. Lactobacillus endocarditis necessitating combined mitral and aortic valve replacement - a case report. Vasc Surg 1993;27:219-25. — Olearchyk AS. Lactobacillus endocarditis: long-term survival after replacement of mitral and aortic valves. Cardio-Vascular Surgery (Kyiv) 1999;7:222.

LMCA	—	left main coronary artery of the heart.
LMP	—	last menstrual period.
LMSB	—	left main stem bronchus.
LMWH	—	low molecular-weight heparin.
LN	—	lymph node.
LNMU	—	Danylo R. Halytsky Lviv National Medical University (founded 1784), Lviv, Ukraine. **Danylo R. Halytsky (1202 - 64)** — King of Rus-Ukraine (1253 - 64).
LNU	—	Ivan Ya. Franko Lviv National University (pre-founded 1586, founded 1661), Lviv, Ukraine. **Ivan Ya. Franko (1856 - 1916)** — Ukrainian poet and writer.
LOC	—	loss of consciousness.
LP	—	labyrinth procedure (operation), or MP.
LP	—	low pressure.
LP	—	liver phosphatase.
LP	—	lumbar puncture.
LPN	—	licensed practical nurse.
LPS	—	lipopolysaccharide.
LQTS	—	long QT syndrome.
LSL	—	left superior leaflet (mitral valve).
LSPV	—	left superior pulmonary vein.
LTACH	—	long-term acute care hospital.
Lt. Col.	—	Lieutenant colonel.
Ltd	—	limited.
L-type	—	long-lasting type.
LUL	—	left upper lobe (lung).
LV	—	left ventricle (heart).
LV aneurysm	—	left ventricular aneurysm. / The development of the LV or aortic aneurysms is explained by the law of LaPlace **[Pierre S. de LaPlace (1749 - 1827)** — French mathematician, physicist and astronomer]**, according to which an average circular tension on their wall during systole is directly proportional to the intraluminal pressure and the curvature of the radius, and reversely proportional to the wall thickness, as outlined by the following equations: the tension of the LV or aortic wall during systole (T) = intraluminal pressure (P) x radius (r) / 2 x wall thickness (h), or

$$T = P \times r / 2 \times h, \text{ where}$$

T = tension (wall stress), P = intraluminal pressure, r = radius, and h = wall thickness, delta P = pressure difference across the wall of the LV or aorta, T_1 i T_2 = tension of the LV or aortic wall in proportional directions, r_1 i r_2 = the largest and smallest principal curved radius of the LV or aortic wall.

The highest intraluminal pressure during diastole on LV or aortic wall, specifically in patients with hypertension, the thinnest the wall and increased radius of the LV or aortic wall, the highest risk of a progressive dilatation, aneurysm formation, dissection, rupture of its wall and death.

LVA	—	LV aneurysmectomy.
LVAD	—	left ventricular assist device, HM.
LVDD	—	left ventricular end-diastolic internal dimension.
LVEDP	—	left ventricular end-diastolic pressure.
LVEDV	—	left ventricular end-diastolic volume.
LVEF	—	left ventricular ejection fraction.
LVOT	—	left ventricular outflow tract.
LVRS	—	lung volume reduction surgery.
LVSD	—	LV systolic dimension.
LVSV	—	left ventricular stroke volume.
M	—	marginal.
M	—	mean.
m	—	meter.
m	—	mitosis.
MA	—	medical academy.
mAb OKT3	—	murine antibody OKT3 (muromonab-CD3).
MAL	—	middle axillary line.
M & M	—	morbidity and mortality.
MAO	—	monoamine oxidase.
MAP	—	mean arterial pressure.
MAPCA	—	major (multiple) aortopulmonary collateral arteries (anastomoses).
MASD	—	moisture associated skin damage.
maser	—	microwave amplification by stimulated emission of radiation (maser).
MAVR	—	mitral and aortic valve replacement.
MB	—	marginal branch of the circumflex artery (CFA) of the heart.
MC	—	mitral commissurotomy.
MD	—	Medicinae Doctor.
mcg, μg	—	microgram is a unit of mass equal to 1/1,000,000 of a gram (1×10^{-6}), or 1/1000 of a milligram.
MCL	—	middle clavicular line.
MD	—	Medicinae Doctor.
MEN	—	multiple endocrine neoplasia.
Met	—	mesenchymal-epithelial transcription factor.
MetS	—	metabolic syndrome, syndrome X.

MFS	—	Marfan syndrome. – **Niels Steensen, (1638 - 86)** — Danish anatomist. / **Antoine B.J. Marfan (1858 - 1942)** — French pediatrician.
mg	—	milligram.
MG	—	myasthenia gravis.
mg/dL	—	milligrams per decaliter.
MGH	—	Massachusetts General Hospital (founded 1818), Boston, MA, USA.
HC	—	major histocompatibility complex.
MI	—	medical institute.
MI	—	myocardial infarction.
MI	—	mitral insufficiency or MR.
MIBI	—	methoxy-isobutylisonitrile, sestamibi.
MICS	—	minimally invasive cardiac surgery.
MICU	—	medical intensive care unit.
MIDCAB	—	minimally invasive direct coronary artery bypass.
MISS	—	minimally invasive strabismus surgery.
MIFST	—	minimal invasive fetal surgery and therapy.
MIMVR	—	minimally invasive mitral valve replacement.
min.	—	minute.
MIRPE	—	minimally invasive repair of pectus excavates.
MCH	—	melanin-concentrating hormone.
ml	—	milliliter.
mln	—	million.
MLSI	—	median longitudinal sternotomy incision.
MM	—	methylmethacrylate.
mm	—	milimeter.
μm	—	milimicron, that equals one mln (10^{-6}) of a meter.
mmHg	—	mm of mercury, torr.
mol, mole	—	name created 1893 by **F. Wilhelm Ostwald (1853 - 1932)** — German chemist.
mmol/L	—	mili mol per litre.
MMS	—	Mohs micrographic surgery. / **Frederic R. Mohs (1910 - 2002)** — American, physician and general surgeon.
MOC	—	maintenance of certification.
MP	—	musculi pectinati.
MP	—	maze procedure (operation) or LP.
MPA	—	main pulmonary artery.
MPAP	—	mean pulmonary artery pressure.
MPM	—	malignant pleural mesothelioma.
MPNST	—	malignant peripheral nerve sheath tumor.
MPS	—	mucopolysaccharidosis.

MR	—	magnetic resonance.
MR	—	mitral regurgitation, or MI.
MRA	—	magnetic resonance angiography (arteriography).
MRCP	—	member of the Royal College of Physicians (founded 1518), London, England.
MRI	—	magnetic resonance imaging, or NMR.
mRNA	—	messenger ribonucleic acid.
MRSA	—	methicillin resistant staphylococcus aureus.
MS	—	manuscript.
MS	—	Master of Sciences.
MS	—	multiple sclerosis.
MS	—	mitral stenosis.
MS	—	microsatellite.
MSB	—	main steam bronchus.
MSC	—	mesenchymal stem cells.
MSCT	—	multislide computer tomography.
msec	—	millisecond.
MTP	—	metatarsophalangeal (joint).
mv	—	microvascular (capillary) space.
mV	—	mili-Volt.
MV	—	mitral valve.
MVA	—	mitral valve area.
MVB	—	mixed venous blood.
MVO_2	—	maximal volume of O_2 consumption (ml/kg/min).
M_vO_2	—	mixed venous O_2 concentration (saturation) in the right atrium (RA).
MVR	—	mitral valve replacement.
MVV	—	maximal voluntary ventilation.
N	—	Newton = Sir **Isaaк Newton (1642 - 1727)** — English mathematician, and physician. The N is a unit of force in which $1N = 1kg \times m/s^2$, where kg = kilogram, m = mass and c = second.
N	—	nitrogen.
N	—	nodal portion of the atrioventricular node (AVN).
n	—	number, No, Nr.
Na^+	—	natrium (sodium).
NaCl	—	natrium (sodium) chloride
NADP	—	nicotinamide-adenine diphosphate.
NAMS of Ukraine	—	National Academy of Medical Sciences of Ukraine (founded 1993), Kyiv, Ukraine.

NAS of the USA	—	National Academy of Sciences of the USA (formed 1863), Washington, DC, USA.
NAS of Ukraine	—	National Academy of Sciences of Ukraine (founded 1918).
NaUKMA	—	National University «Kyiv-Mohyla Academy» (founded 1615), Kyiv, Ukraine.
NB-H	—	nodal bundle of His is a collection of specialized cardiac muscle cells in the heart that transmits the electrical current to synchronize contraction of the heart muscle, discovered in1893 by **Wilhelm Hiss, Jr (1863 - 1934)** — Swiss anatomist and cardiologist.
NDE	—	near-death experience.
Nd;YAG laser	—	neodymium:yttrium-aluminum garnet laser.
NGF	—	nerve growth factors.
NF	—	neurofibromatosis.
ng	—	nanogram – one billionth (10-9) of a gram.
NGT	—	nasogastric tube.
NHI	—	National Health Insurance.
NIH	—	National Institute of Health (founded 1887), Bethesda, MD, USA.
NIDCM	—	non-ischemic dilated cardiomyopathy.
NIP	—	nonspecific interstitial pneumonia.
NIPhP	—	National Institute of Phtysiatry and Pulmonology, Kyiv, Ukraine.
NCC	—	nuclear core complex.
NCS	—	non-coronary (aortic) sinus Valsalva.
NLM	—	National Library of Medicine (established 1836), Bethesda, MD, USA.
NLU	—	National Library of Ukraine (established 1918), Kyiv, Ukraine.
NMAPE	—	National Medical Academy of Postgraduate Education named for Platon L. Shupyk (founded 1918), Kyiv, Ukraine. **Platon L. Shupyk (1907 - 86)** — Ukrainian physician and surgeon.
NMDA	—	N-methyl-D-aspartate.
NMMU	—	National Museum of Medicine of Ukraine (founded 1973), Kyiv, Ukraine.
NMR	—	nuclear magnetic resonance, or MRI.
NMU	—	National Medical University of Олександр O. Bohomolets (founded 1840), Kyiv, Ukraine. **Олександр O. Bohomolets (1881 - 1946)** — Ukrainian pathophysiologist.
NO	—	nitric oxide, nitrogen monoxide, nitrogen oxide, nitrogen (II) oxide.
N_2O	—	nitrous oxide, laughing gas, sweet air.
No.	—	number.
NPC	—	nuclear pore complex.
NPM	—	nuclear para-magnetism.
NPO	—	nil per os.
NREM	—	non-rapid eye movement.
NSAID	—	nonsteroidal anti-inflammatory drugs.

NSC	—	national scientific center.
NSCLC	—	non-small cell lung cancer.
NSMLU	—	National Scientific Medical Library of Ukraine (founded 1930), Kyiv, Ukraine.
NSS	—	normal (physiologic) saline solution (0,9% NaCl).
NTG	—	nitroglycerin.
NTUU-«KTI»	—	National Technical University of Ukraine «Kyiv Technical Institute» or «KTI» named Ihor Sikorsky (founded 1898). **Ihor I. Sikorsky (1889 - 1983)** — Ukrainian American aviation pioneer of both helicopters and fixed-wing aircraft.
NVE	—	native valve endocarditis
NYHA	—	New York Heart Association classification (originated 1903).

O_2	—	oxygen – odorless and colorless, component part in the free state of an air (by 20% weight) and in connections in many different substances, atomic number 8, atomic mass 15,999, needed for breathing of plants, animals and people and maintenance of burning.
OB	—	obliterative bronchiolitis.
Obl.	—	Oblast / Province.
OCP	—	other cardiac procedures.
OCPP	—	other cardiac or pulmonary procedures.
OCPs	—	oral contraceptive pills.
OD syndrome	—	oropharyngeal dysphagia syndrome.
OD	—	outside diameter.
Ohm	—	**Georg S. Ohm (1787 - 1854)** — German physicist, who stated that the intensity (energy, strength) of an electric current through a conductor between two points (**I**) in units of amperes (A) is directly proportional to the difference (gradient) of the electric current potential measured across those two points of the conductor in units of volts (**V**), and inversely proportional to the resistance (**R**) in units of Ω to the conductor, as expressed in the Ohm's law (1825-27), where **I = V/R**, **V = I x R** and **R = V/I**. In this relation the R is constant, independent of electric current. The Ohm' law is used to estimate the cardiac valve area.
OHT	—	orthotopic heart transplantation.
OLE	—	oscillation and lung expansion.
OM	—	obtuse margin of the heart.
OM branch	—	obtuse marginal branch of the CFA (heart).
OMC	—	open mitral commissurotomy.
OMT	—	optimal medical treatment.
ONU	—	Odesa National University of Ilya I. Mechnykov (founded 1865), Odesa, Ukraine. / **Ilya I. Mechnykov** or **Elie I. Metchnikoff (1845 - 1916)** — Ukrainian French microbiologist.

ONMU	—	Odesa National Medical University (founded 1900), Odesa, Ukraine.
OOB	—	out of bed.
OP	—	off-pump.
OR	—	operative room.
ORIF	—	open reduction internal fixation.
ORL	—	otorhinolaryngology or ENT.
OSR	—	open surgical repair.
OSCS	—	oversulfated chondroitin sulfate.
p	—	density of the fluid (blood).
p	—	page, pg.
p	—	p value.
post.	—	posterior.
P	—	posterior leaflets of the mitral valve.
P	—	pressure.
PA	—	peroneal artery.
PA	—	pulmonary artery.
PA	—	pulmonary atresia.
$PaCO_2$	—	partial blood carbon dioxide pressure (venous between 35 and 45 mmHg).
PAD	—	peripheral arterial disease.
PAE	—	pulmonary artery embolectomy.
PAL	—	posterior axillary line.
PaO_2	—	partial blood oxygen pressure (arterial between 80 and 100 mmHg).
$Pa\,O_2/Fi\,O_2$	—	ratio of arterial oxygen tension to fraction of inspired oxygen.
P_AO_2	—	pulmonary artery O_2 concentration.
PAP	—	pulmonary artery pressure.
PAPVD	—	partial anomalous pulmonary venous drainage, or PAPVR.
PAPVR	—	partial anomalous pulmonary venous return, or PAPVD.
PAR-1	—	protease activated receptor-1.
PAS	—	periodic acid-Schiff. / **Gugo Schiff (1934 - 1916)**[1785]— German Italian chemist.
PAS	—	Per-Apnt-Sim domain.
PAW	—	penetrated abdominal wounds.
PBAV	—	percutaneous balloon aortic valvuloplasty.
PBMV	—	percutaneous balloon mitral valvuloplasty.
PBS-8	—	protein-like blood substitute – 8.
PCA	—	posterior cerebral artery.
PC	—	para-corporeal.

[1785] Schiff H. Mittheilungen aus dem Univesitäts-laboratorium in Pisa. Eine neue Reihe organisher. Ann Chem Pharm 1864;131:118-9.

PSA	—	prostatic specific antigen.
PC (p/c)	—	percutaneous.
PCI	—	percutaneous coronary intervention = percutaneous transluminal balloon angioplasty (PTA), or percutaneous transluminal coronary angioplasty (PTCA), or balloon angioplasty (BA) with stenting, with or without thrombolysis.
PCR	—	polymerase chain reaction.
PCT	—	percutaneous tracheostomy, or PDT.
PCWP	—	pulmonary capillary wedge pressure.
PDE	—	phosphodiesterase
PD	—	Parkinson disease (syndrome), or paralysis agitans. / **James Parkinson (1755 - 1824)** — English surgeon.
PD-1	—	protein's death -1.
P&D	—	pleurectomy and decortication.
PDA	—	patent ductus arteriosus, or ductus Botallo. / **Leonardo Botallo (1518/30 - 1587/1600)** — Italian anatomist and physician.
PDA	—	posterior descending artery of the heart.
PDF	—	pigment dispersing factor.
PDF	—	portable document format.
PDGF	—	platelets derived growth factor.
PDS	—	polydioxanone suture.
PDT	—	percutaneous dilatational tracheostomy.
PDT	—	photodynamic therapy.
PE	—	pulmonary embolism.
PE	—	pulmonary embolectomy.
PEEP	—	positive end-expiratory pressure.
PEG	—	per-endoscopic gastrostomy.
PET	—	positron emission tomography.
PETE	—	polyethylene telephthalate (Dacron) grafts, patented in 1941 by **John. R. Whinfield (1901 - 66)** — English chemist, are extensively used in thoracic, cardiovascular, and other surgeries to create arterial synthetic tubular grafts (STG) and patches.
PEV syndrome	—	pre-excitation of ventricle syndrome. / WPW syndrome.
PF-4	—	platelet factor-4.
PFE	—	papillary fibroelastosis of the heart.
PFO	—	patent foramen ovale of the heart.
PFT	—	pulmonary function test (testing).
PG	—	pancreatic gland.
pg	—	picogram – one trillionth (10^{-12}) of a gram.
PG	—	pressure gradient.

PG — prostaglandins.

pH — denote the concentration of hydrogen ions (H⁺) or activity of solution as compared with accepted neutral standard solution (pH 7). The quantity of pH equals to the negative logarithm of concentration H⁺, expressed in molarity. The higher than pH 7 solution is alkalotic, lower than pH 7 is acid. Normal pH of the blood is 7.35-7.45.

PH — portal hypertension.

PH — pulmonary hypertension.

PhD — Philosophiae Doctor.

PHM — Public Health Master.

PIJ — proximal interphalangeal joint.

PL — posterolateral projection (radiology).

PLVB — posterior LV branch of the RCA or CFA of the heart.

PLC — postoperative lung complications.

PM — pacemaker.

PM — polypropylene mesh.

PML — posterior middle line.

PMN — polymorphonuclear neutrophils (PMNs).

PMR — paramagnetic resonance, or EPR.

pmv — perimicrovascular (interstitial) space.

Pmv — pressure, microvascular (capillary hydrostatic.

π_{mv} — pressure, capillary colloid osmotic.

PMV — Passy-Muir valve is one-way valve to facilitate ventilation and speech in patients who have undergone tracheostomy, devised by **David Muir** — American quadriplegic patient, and **Victor Passy (1931 -)** — American otolaryngologist.

PNA — Parisiana Nomica Anatomica (1955).

PND — paroxysmal nocturnal dyspnea.

PNET — primitive neuroectodermal tumor.

PNS — peripheral nervous system.

po — per orum.

post. — posterior.

pp — particles per.

PPV — pulse pressure variation.

PPH — primary pulmonary hypertension.

PPH — procedure for prolapse and hemorrhoids, or procedure for prolapsed hemorrhoids.

Ppmv — pressure, perimicrovascular (interstitial) fluid hydrostatic.

π_{pmv} — pressure, perimicrovascular (interstitial) colloid osmotic.

PPC — peritoneo-pericardial communication (defect, hernia) of the diaphragm central tendon.

PR   — pulmonary regurgitation.

PRA   — panel-reactive antibody.

P-R   — P-R interval (duration 120 – 200 msec.).

p.r.n.   — pro renata (as necessary, as needed, as necessary).

PS   — post scriptum.

PS   — primitive streak.

PS-8   — protein substitute PS-8 – synthesized in 1955 by **Volodymyr O. Belitser (1906 - 88)** — Ukrainian biochemist.

PS/IVS   — pulmonary stenosis with intact ventricular septum.

PSV   — peak systolic velocity.

PS/IVS   — pulmonary stenosis and intact interventricular septum.

$PaCO_2$ — partial pressure of CO_2 in the arterial blood.

$PeCO_2$ — partial pressure of CO_2 in the exhaled air.

PT   — physical therapy.

PT   — posterior tibial (artery).

PT   — prothrombin time.

PTA   — persistent truncus arteriosus, TA.

PTA   — posterior tibial artery, or PT artery.

PTA   — percutaneous transluminal angioplasty.

PTBA   — percutaneous transluminal balloon angioplasty.

PTH   — parathyroid hormone.

PTG   — parathyroid gland.

PTEA   — pulmonary thromboendarterectomy.

PT INR   — prothrombin time according to INR.

PTEA   — pulmonary thromboendarterectomy.

PTLD   — post-transplant lymphoproliferative disease.

PTPBLD   — percutaneous transabdominal puncture of the cysterna chyli, or lymphatic duct.

PTCA   — percutaneous transluminal coronary angioplasty.

PTFE   — polytetrafluroroethylene (teflon, politef, Gore-Tex), discovered in 1938 by **Roy J. Plunkett (1910 - 94)** — American chemist, at DuPont Company, Wilmington, DE, USA, founded by **Enthenee du Pont (1771 - 1834)** — French American chemist and industrialist, in 1802 (defunct in 2017). PTFE is extensively used in cardiovascular surgery to create arterial synthetic tubular grafts (STG) and patches.

publ.   — publisher, publishing.

PV   — pulmonary valve.

PV   — pulmonary vein.

PVE   — prosthetic valve endocarditis.

PVC   — premature (paroxysmal) ventricular contraction.

P_VO_2 — pulmonary vein O_2 concentration.

PVP-I   — povidone-iodine (in Latin – povidonum-iodum).

PVR   — pulmonary vascular resistance (WU). **Paul H. Wood (1907 - 62)** — West Indies, New Zealander and British cardiologist.

Q   — flow.
q.d.   — quaque die.
Q_{eff} — effective blood flow.
q.h.   — quaque hora.
q.i.d.   — quarter in die.
q.h.s.   — quaque hora sleep.
QOL   — quality of life.

$$\dot{Q}P = \frac{O_2 \text{ Consumption (ml/min)}}{[PVO_2] - [PAO_2]}$$

Q_p — denotes pulmonary blood flow, PVO_2 – denotes the partial pressure of O_2 in the pulmonary vein or the left atrium (mmHg.), PAO_2 – denotes the partial pressure of O_2 of the blood in the pulmonary artery (mmHg).

$$\dot{Q}S = \frac{O_2 \text{ Consumption (ml/min)}}{[SAO_2] - [MVO_2]}$$

Q_s — denotes the systemic blood flow, SAO_2 – denotes the systemic arterial oxygen concentration in the aorta (mmHg), MVO_2 – denotes the mixed venous oxygen concentration in the right atrium (mmHg).

$$\dot{Q}P/\dot{Q}S = \frac{[SAO_2] - [MVO_2]}{[PVO_2] - [PAO_2]}$$

Q_p/Q_s — denotes the ratio of relative blood flow in the pulmonary and systemic circulations, MVO_2 – denotes the partial pressure of O_2 in the mixed blood from the the right atrium (RA) for ventricular septal defect (VSD), from the right ventricle (RV) for patent ductus arteriosum (PDA) and for an atrial septal defect (ASD) in mmHg, and for the later it is a derived number from the Flamm equation[1786], where MVO_2 = in an ASD is 3 [(superior vena cava (SVC) +1 (inferior vena cava (IVC)] / 4.

A Q_p/Q_s greater than 2.0 is considered high and indicate possible surgical or percutaneous closure, values between 1,5 and 2,0 are intermediate, and surgical or percutaneous closure may be indicated if there is low surgical risk, or if symptomatic [silent stroke in ASD or patent foramen ovale (PFO)]. Values lower than 1,0 suggest right to-left-shunt.

[1786] **Melvin D. Flamm (1934 –)** — American military and civilian cardiologist. / Flamm MD, Cohn KE, Hancock EW. Measurement of systemic cardiac output at rest and exercise in patients with atrial septal defect. Am J Cardiol 1969;23:256- 65

Q_P/Q_S — left-to-right shunt.

$$Q_{eff} = \frac{O_2 \text{ Consumption (ml/min)}}{[PV_{O_2}] - [MV_{O_2}]}$$

Q_{eff} — denotes the effective blood flow, used to estimate the bidirectional (left-to-right and right-to-left) shunt of the blood, which is reproduced accordingly to the equation $Q_p - Q_{eff}$ and $Q_s - Q_{eff}$, respectively.

Q-wave   — is an initial part of the QRS complex with a downward (negative) de-flexion, related to the initial phase of depolarization of the interventricular septum.

QRS   — QRS complex.

Q-T   — Q-T segment.

Q_f,   — flow across the capillary wall (volume/unit time).

q.v.   — quod vide (which see).

r   — radius.

R   — resistance.

RA   — radial artery.

RA   — right atrium (heart).

RALPL   — rate adaptation (of the heart) to the level physical load.

RAM   — rectus abdominis muscle.

RAO   — right anterior oblique projection.

RAP   — right atrial pressure.

RBBB   — right bundle branch block.

RBC   — red blood cells, i.e., erythrocytes.

RC   — Red Cross International (founded 1863).

RCA   — right coronary artery of the heart.

RC-AVF   — radio-cephalic arterio-venous fistula.

RCS   — right aortic (coronary) sinus (Valsalva).

RD   — rheumatic disease (fever).

RDS   — respiratory distress syndrome.

re.   — regarding.

re-   — redo-.

REBOA   — resuscitative endovascular balloon occlusion of the aorta.

redo-   — re-done.

REM   — rapid eye movement.

RES   — reticulo-endothelial system, or SMPh.

RES   — reparation a l'etage ventriculaire (LaCompte). / **Yves Le Compte** — French pediatric cardiac surgeon.

RF   — renal failure.

RF	—	respiratory failure.
RF	—	rheumatic fever.
RFA	—	radiofrequency ablation.
RHC	—	right heart catheterization.
RHD	—	rheumatic heart disease.
RHF	—	right heart failure.
RI	—	ramus intermedius of the heart.
RIA	—	radioimmune assay.
RIMA	—	right internal mammary (thoracic) artery or RITA.
RIPV	—	right inferior pulmonary vein of the heart
RIS	—	radiation induced sarcoma.
RITA	—	right internal thoracic (mammary) artery or RIMA
RLL	—	right lower lobe (lung).
RLN	—	recurrent laryngeal nerves.
RML	—	right middle lobe (lung).
RMSB	—	right main stem bronchus.
RN	—	registered nurse.
RNA	—	ribonucleic acid.
RNFA	—	RN-first assistant.
ROA	—	regurgitation orifice area (mitral valve).
Roman numbers	—	I – 1; V – 5; X – 10; L – 50; C – 100; D – 500; M – 1000.
ROTC	—	reserve officer's training corps, GMT.
rpm	—	revolution per minute.
RR	—	respiratory rate.
RSD	—	reflex sympathetic dystrophy.
RSPV	—	right superior pulmonary vein (heart).
RUL	—	right upper lobe (lung).
RT-QuIC	—	real-time Quaking induced conversion.
RV	—	renal vein.
RV	—	residual volume.
RV	—	right ventricle of the heart.
RVAD	—	right ventricular assist device.
RVSZVE vaccine	—	recombinant vesicular stomatitis Zaire virus-EBOV vaccine.
RVDCC	—	right ventricular dependent coronary circulation.
RVOT	—	right ventricular outflow tract.
RVSWI	—	right ventricular stroke work index.
S	—	sacral.
S	—	septal (anterior) leaflets of the MV.
SAH	—	subarachnoidal hemorrhage.

SAI	—	subarterial infundibulus.
SAM	—	systolic anterior motion of the MV.
SAN	—	sinoatrial node.
SaO_2	—	saturation of the arterial blood with oxygen (95%-98%).
S_AO_2	—	systemic artery O_2 concentration.
S^2AS	—	S^2 anastomotic system.
SBLA syndrome	—	sarcoma, breast leukemia and adrenocortical carcinoma syndrome.
SBT	—	spontaneous breathing trial.
SC	—	subcutaneous.
SCA	—	selective coronary angiography.
SCA	—	subclavian artery.
SCV	—	subclavian vein.
SCD	—	sudden cardiac death.
SCEP	—	spinal cord evoked potential.
SCF	—	subcapsular proliferative foci.
SCM muscle	—	sternocleidomastoid muscle.
ScO_2%	—	central venous (mixed) oxygen saturation.
SCPP	—	spinal cord perfusion pressure.
SCS	—	subcapsular sinus.
SCUBA	—	self-contained underwater breathing apparatus.
Sci.	—	science.
ScO_2%	—	central venous (mixed) oxygen saturation.
sec.	—	second.
sec.	—	selenocysteine.
SEM	—	standard error of the mean.
SEO	—	search engine optimization.
SEP	—	systolic ejection period.
Sept.	—	septum.
Sept I	—	septum primum of the heart.
Sept. II	—	septum secundum of the heart.
SESAP	—	Surgical Education and Self-Assessment Program.
SESATS	—	Self-Education and Self-Examination in Thoracic Surgery (established 1980).
Sesta-mibi	—	is a pharmacological agent used in nuclear medicine imaging known as MIBI. The radioisitope attached to the sestambi molecule is technetium (Tc)-99^m, forming 99^{m}Tc-sestamibi (or Tc99^{m}MIBI).
SEW	—	scientific-experimental work.
SFA	—	superficial femoral artery.
SFV	—	superficial femoral vein.
Shh	—	sonic hedgehog.

SHFM	—	Seattle Heart Failure Model.
SICU	—	surgical intensive care unit.
SI	—	Systeme international d'unitees, Paris 1960.
SI	—	spherical index.
SIRS	—	systemic inflammatory response syndrome.
SJV	—	St Jude (medical) valve. **St Jude the Alostle (? - 63 AD)** — patron of desperate or lost causes, born in Jewish family.
SK	—	Stavropolsky krai, Ukraine (1918 - 22) / RF.
SKSAP/NO-0l	—	scalpel-coagulator-stimulator air-plasma.
SLD	—	Loeys-Dietz syndrome or LDS.
SLE	—	systemic lupus erythematosus.
snus	—	small ventral lateral neurons (sLNus) in the fruit fly brain.
SMA	—	superior mesenteric artery.
cm	—	centimeter, centimeter; см.
SMPS	—	sympathetic maintained pain syndrome.
SMPh	—	system of macrophagal phagocytes or RES.
SNP	—	single nucleotide polymorphism.
S1	—	1st heart sound (tone). / 31.
SO_2%	—	oxygen (O_2) content (saturation) of the blood.
soc.	—	society.
SPAMM	—	spatial modulation and magnetization.
SPECT	—	single-photon emission computed tomography.
SPN	—	solitary pulmonary nodule.
SPV	—	selective proximal vagotomy.
SC	—	stem cells.
Sr	—	senior.
SRS	—	stereotactic radiosurgery.
SS	—	steal syndrome.
SSD	—	systemic scleroderma.
SSRI	—	selective re-uptake serotonin inhibitors.
SSS	—	sick sinus syndrome.
SSS	—	subclavian steal syndrome.
SST	—	superior sulcus tumor of the lung or Pancost's tumor. **Henry K. Pancoast (1875 - 1939)** — American radiologist.
STARR	—	stapled trans-anal rectal resection.
STEMI	—	ST-elevation (Q-wave) myocardial infarction (MI).
STEP	—	serial transverse enteroplasty.
S-T	—	S-T interval (segment).
STG	—	synthetic tubular graft.
STS	—	Society of Thoracic Surgeons (founded 1964).

STSG	—	split-thickness skin graft.
SUV	—	standardized uptake value.
SUBA	—	shallow under water breathing apparatus.
SUDV	—	Sudan virus.
SV	—	stroke volume.
SVC	—	superior vena cava.
SVD	—	Sudan virus disease.
SVR	—	systemic vascular resistance.
SVT	—	supraventricular tachycardia
syn.	—	synonym.
T	—	temperature.
T	—	«T» technique for anastomosing of the end of diametrically smaller synthetic tubular graft to the side of diametrically larger tubular graft.
T	—	thoracic.
T	—	thymus.
T	—	thyroid.
T	—	thymine.
т	—	tonne.
T_3	—	triiodothyronine.
T_4	—	thyroxine.
TA	—	thoracic aorta.
TA	—	transverse aorta, or aortic arch.
TA	—	truncus arteriosus.
TAA	—	thoracic and abdominal aorta, or thoracoabdominal aorta.
TAAA	—	thoracic and abdominal aortic aneurysm.
TAAD	—	thoracic aortic aneurysm or dissection.
TAPVD	—	total anomalous pulmonary venous drainage (connection, return). / TAPVR.
TAVI	—	transcatheter aortic valve implantation.
TAVR	—	transcatheter aortic valve replacement.
TAPVR	—	total anomalous pulmonary venous return (connection, drainage), TAPVD.
TAH	—	total artificial heart.
TACRD	—	*«**t**rachea, **a**nus, **c**or, **r**adial bone and **d**uodenal»* congenital anomalies.
TBSA	—	total body surface area.
TV	—	television.
TV, V_t	—	tidal volume.
TV	—	tricuspid valve.
TVR	—	tricuspid valve repair or replacement.

TWA	—	T-wave (repolarization) alternans.
T-cells	—	thymus-cells.
TCVS	—	thoracic and cardiovascular surgery.
2,3,7,8-TCDD	—	2,3,7,8-tetrachlorodibenzo-p-dioxin.
TCRAT	—	total coronary revascularization via anterior thoracotomy.
TD	—	thoracic duct.
3-D echo	—	3-dimentional echocardiography.
TDI	—	tissue Doppler imaging.
TNMA	—	Ternopil National Medical Academy of Ivan Ya. Horbachevsky (founded 1957), Ternopil, Ukraine. **Ivan Ya. Horbachevsky (1854 - 1942)** — Ukrainian biochemist.
TECAB	—	total endoscopic coronary artery bypass.
TEVAR	—	thoracic endovascular aortic repair.
TED	—	thromboembolic disease.
TEE	—	transesophageal echocardiography, TOE.
TEG	—	thromboelastogram.
TEM	—	trans-nasal endoscopic microsurgery.
TEN	—	total enteric nutrition.
TERAV	—	total endoscopic repair of the aortic valve
TERMV	—	total endoscopic repair of the mitral valve.
TERTV	—	total repair of the tricuspid valve.
TET	—	transcutaneous energy transfer.
TEF	—	tracheoesophageal fistula.
TG	—	thyroid gland.
TGA	—	transposition of the great arteries (vessels).
TGF	—	transforming (transcription) growth factor.
TGF-alpha	—	transforming growth factor-alpha.
Thalium	—	radioactive isotope: atomic mass 201, half-life – 3,05 days, decay with the capture of electrons and emission of gamma-rays (0,135, 0,167 megaVolts), used in the form of thallium chloride to evaluate work of the myocardium.
TIA	—	transient ischemic attacks.
TIPS	—	trans-jugular intrahepatic portocaval shunt.
TLC	—	total lung volume.
TLR	—	Toll-like receptor.
TM	—	thrombomodulin.
TMVR	—	transcatheter mitral valve replacement.
TMJ	—	temporo-mandibular joint.
TNF	—	tumor necrosis factor.

TNM	—	Staging System TNM, where **T** – denotes primary neoplasm (tumor), **N** – denoted nodus (node, nodule), i.e., regional lymphatic nodes (LN), and **M** – denotes distant metastasis, was conceived 1943-52 by **Pierre Denoix (1912 - 90)**[1787] — French surgeon, according to the size and extension of the primary tumor, involvement of the LN and the presence of distal metastases, with the aim of uniform staging of neoplastic disease (I-IV), selection of the best available treatment for each stage, establishing of prognosis, follow-up and survival (surviviability) or progression disease, approved by Union internacionale contre le cancer (UICC, founded 1933), Geneva, Switzeland.
THT	—	thrinitrotoluin ($C_7H_5N_3O_6$), isolated in 1915 chemical compound with the high flamable, toxic and explosive peculiarities.
TOE	—	transoesophageal echocardiography. / TEE.
TOF	—	tetralogy of Fallot. / **Niels Steensen** or **Nicolaus Steno (1638 - 85)** — Danish scientist, and **Etienne L.A. Fallot (1850 - 1911)** — French physician.
torr	—	1 torr – the unit of pressure, that approximates 3/760 of the atmosphere, i.e., 1 mmHg. / **Evangelista Torricelli (1608 - 46)** — Italian mathematician and physicist.
TOS	—	thoracic outlet syndrome.
t-PA	—	tissue plasminogen activator.
TPG	—	transpulmonary gradient.
TRH	—	thyroid – releasing hormone.
TPN	—	total parenteral nutrition, IVH.
TPT	—	tibio-peroneal trunk.

[1787] **Clifton F. Mountain (1924 - 2007)** — American thoracic surgeon, and **Carolyn M. Dressler)** — American thoracic oncological surgeon. Mountain CF, Dresler CM. Regional lymph node classification for lung cancer staging. Chest 1997;111(6):1718-23. / Mountain CF. Staging classification of lung cancer. A Critical evaluation. Clinics in Chest Medicine 2002 Mar 23(1):103-21. — Moroz HS. Cancer of the lung. Klinichna khirurhiya. Ed. - LYa Kovalchuk, VF Sayenko, HV Knyshov. Ternopil: Publ. «Ukrmedknyha», 2000. - Vol. 1. P. 74-85. — Park BJ, Rusch VW Lung cancer workup and staging. In, FW Sellke, PJ del Nido, SJ Swanson (Eds). Sabiston & Spencer Surgery of the Chest. Vol. I-II. 8th Ed. ElsevierSaunders,Philadelphia 2010. P. 241-52. — Anraku M, Keshavjee S. Lung cancer: surgical treatment. Ibid. P. 253-77. — McNamee C, Ducko CT, Sugarbacker DJ. Pleural tumors. Ibid. P. 449-72. — Lightdale CI. Endoscopic ultrasonography in the diagnosis, staging and follow-up of esophageal and gastric cancer. Endoscopy 1992 May;1(24 Suppl):297-303. — Little VR. Staging techniques for carcinoma of the esophagus. In, FW Sellke, PJ del Nido, SJ Swanson (Eds). Sabiston & Spencer Surgery of the Chest. Vol. I-II. 8th Ed. ElsevierSaunders, Philadelphia 2010. P. 577-87. — Jurkiewicz MJ. Reconstructive surgery of the cervical esophagus. J Thorac Cardiovasc Surg 1984;88:893-7. — Affleck DG, Karwande SV, Bull DA, et al. Functional outcome, and survival after pharyngo-laryngo-esophagectomy for cancer. Am J Surg Dec 2000;180(6):546-50. — Triboulet JP, Mariette C, Chevalier D, et al. Surgical management of carcinoma of the hypopharynx and cervical esophagus: analysis of 209 cases. Ann Surg Oct 2001;136(1):1164-70. — Nishimaki T, Kanda T, Nakagawa S, et al. Outcomes and prognostic factors after surgical resection of hypopharyngeal and cervical esophagus. Int Surg Jan-Mar 2002;87(1)38-44.

TRR	—	tracheal resection and re-anastomosis.
TRH	—	thyrotropin-releasing hormone.
TS	—	tinea sagittalis.
TSC	—	tuberous sclerosis complex.
TSH	—	thyroid stimulating hormone.
TT	—	tetanus toxoid.
TT	—	tracheostomy tube.
	—	T2 weighted images.
TTE	—	transthoracic echocardiography.
TTI	—	thyreotropic immunoglobulin.
TTP	—	thrombotic thrombocytopenic purpura.
TTF	—	thyroid transcription factor.
TTFL	—	transcription-translation feedback loop model.
u	—	flow of the fluid (blood).
u	—	unit.
UFH	—	ultrafractionated heparin.
UICC	—	Union internacionale contre le cancer (founded 1933). UIMJ «Ahapit» — Ukrainian Historic-Medical Journal «Ahapit» (established 1994).
UIP	—	usual idiopathic pneumonia.
UJ	—	Uniwersytet Jagielloński (founded 1364), Kraków, Polska.
UNCV	—	ulnar nerve conduction velocity.
US	—	ultrasound.
USA	—	Unites States of America.
USDDBF	—	ultrasonic Doppler flow detector of the blood flow. **Christian A.** von **Doppler (1803 - 53)** — Austrian mathematician and physicist.
USN	—	United States Navy.
u.t.v. / u.t.s.	—	urban type village / urban type settlement.
UTI	—	urinary tract infections.
VAD	—	ventricular assist device for the support of a failing heart.
VACTERL	—	*v*ertebrae, *a*nus, *c*ardiac, *t*rachea, *e*sophagus, *r*enal, *l*imbs congenital anomalies.
VACTERL +D	—	*v*ertebrae, *a*nus, *c*ardiac, *t*rachea, *e*sophagus, *r*enal, *l*imbs + *d*uodenal atresia.
VCA	—	vessel-suturing circular apparatus. / **Yuriy (Yurii) Yu. Voronyy** or **Yuriy (Yurii) Yu. Voronoy (1895 - 1961)** — Ukrainian surgeon, who performed the first human kidney transplantation (1933) and introduced circular mechanical suturing of the blood vessels (1949).
Vd	—	dead space, DS.
VM	—	Ventuti mask / **Giovanno B. Venturi (1746 - 1822)** — Italian physicist.

VS   — vital signs.

VSD   — ventricular septal defect

VVF   — vesico-vaginal fistula.

vWD   — van Willebrand disease. **Erik A.** von **Willenbrand (1870 - 1949)** — Finish
                        physician, who described vWD in 1926, named after him.

WBC   — white blood cells. i.e., leukocytes.

WPW syndrome   — **Louis Wolff (1888 - 1972)** — American cardiologist, **John Parkinson
                        (1885 - 1976)** — English cardiologist, and **Paul D. White (1886 - 1973)** —
                        American physician and cardiologist. / PEV syndrome.

X   — tenth (vagus) cranial nerve.

XL   — extra-large.

x-ray   — radiography.

XVE   — **ex**tended entrance and exit of the electric current into the ventricular
                        assist device (VAD).

Printed in the United States
by Baker & Taylor Publisher Services